THE FAMILIES AND GENERA OF VASCULAR PLANTS

Edited by K. Kubitzki

Springer
Berlin
Heidelberg
New York
Barcelona
Budapest
Hong Kong
London
Milan
Paris
Singapore
Tokyo

Volumes published in this series:

Volume I	Pteridophytes and Gymnosperms Edited by K.U. Kramer and P.S. Green (1990)
Volume II	Flowering Plants. Dicotyledons: Magnoliid, Hamamelid and Caryophyllid Families Edited by K. Kubitzki, J.G. Rohwer, and V. Bittrich (1993)
Volume III	Flowering Plants. Monocotyledons: Lilianae (except Orchidaceae) Edited by K. Kubitzki (1998)
Volume IV	Flowering Plants. Monocotyledons: Alismatanae and Commelinanae (except Gramineae) Edited by K. Kubitzki (1998)

The Families and Genera of Vascular Plants

Edited by K. Kubitzki

III *Flowering Plants · Monocotyledons*

Lilianae (except Orchidaceae)

Volume Editor:
K. Kubitzki
in Collaboration with H. Huber, P.J. Rudall, P.S. Stevens, and T. Stützel

---

With 136 Figures

Springer

Professor Dr. KLAUS KUBITZKI
Institut für Allgemeine Botanik
und Botanischer Garten
Ohnhorststraße 18
22609 Hamburg, Germany

ISBN 3-540-64060-6 Springer-Verlag Berlin Heidelberg New York

Library of Congress Cataloging-in-Publication Data.

Flowering plants, Monocotyledons : Lilianae (except Orchidaceae) /
volume editor, K. Kubitzki; in collaboration with H. Huber . . . [et al.].
p. cm. – (The families and genera of vascular plants; 3)
Includes bibliographical references and index.
ISBN 3-540-64060-6 (hardcover : alk. paper)
1. Liliales – Classification. I. Kubitzki, Klaus, 1933–
II. Huber, Herbert, Prof. Dr. rer. nat. III. Series.
QK495.A14F55 1998
584′.3 – dc21 98-13621

Cover Design: W. Eisenschink, Heddesheim
Typesetting: Best-set Typesetter Ltd., Hong Kong
SPIN 10101858 31/3137 – 5 4 3 2 1 0 – Printed on acid-free paper

Preface

When Rolf Dahlgren and I embarked on preparing this book series, Rolf took prime responsibility for monocotyledons, which had interested him for a long time. After finishing his comparative study and family classification of the monocots, he devoted much energy to the acquisition and editing of family treatments for the present series. After his untimely death, Peter Goldblatt, who had worked with him, continued to handle further incoming monocot manuscripts until, in the early 1990s, his other obligations no longer allowed him to continue. At that time, some 30 manuscripts in various states of perfection had accumulated, which seemed to form a solid basis for a speedy completion of the FGVP monocots; with the exception of the grasses and orchids which would appear in separate volumes. I felt a strong obligation to do everything to help in publishing the manuscripts that had been put into our hands. I finally decided to take charge of them personally, although during my life as a botanist I had never seriously been interested in monocots.

The beginning of my involvement coincided with the revival of interest in monocot systematics, which became manifest in the Kew Monocot Conference and the publication of its proceedings in 1995. The latter helped me greatly in finding access to modern concepts in monocot classification and evolution; this made it necessary to make considerable changes to all the existing contributions. The results from molecular systematics have not only greatly influenced phylogenetic concepts but have also led to an increase in the number of monocot families, contrary to my original aim to reduce them. It is likely that a number of genera such as those treated here as "genera anomala" may deserve family status; further progress in molecular systematics, implying the use of additional genes and a denser sampling, will certainly result in changes of our phylogenetic and taxonomic concepts.

While working on these volumes, I tried to do justice to the most recent scientific developments and, at the same time, tried not to lose sight of the legacy left to us from earlier generations. Therefore, I often added information from important earlier works, especially from the German morphological literature that is largely unknown abroad. In floral morphology, I aimed at greater precision and clarity and made in these two volumes a consistent distinction between style and stylulus (for an explanation of the latter term, see footnote on p. 382, Vol. III). Nevertheless, I am fully aware of possible imperfections and inconsistencies in the more than 100 family contributions; any attempt at reducing their number, however, would have meant an unjustifiable delay in the publication of these volumes.

I would like to express my deep gratitude to all those who supported my work in various ways. The directors and curators of the herbaria in Berlin, Geneva, Kew, and Munich always responded quickly in putting critical and rare herbarium material at my disposal. W. Barthlott, R. Brummitt, M. Chase, J. Reveal, F. Salamini, and D.W. Stevenson, among others, provided important information as to their own or others' recent work. A. Meerow, Nam Sook Lee, J. Kadereit, M. Sazima, T. Stützel, S. Verhook, Woo-Tschul Lee, and the librarians of my own institution, of the Botanic Garden and Museum at Berlin and of the Botanische Staatssammlung München helped with rare literature. C. Bayer, P.E. Berry, C.D.K. Cook, D.A. Cooke, M. Fay, P. Goldblatt, O. Huber, M. Lock, S. Mayo, A. Meerow, I. Nordal, H.-H. Poppendieck, S.S. Renner, J. Semir, M. Sazima, H.-J. Tillich, G. Wiegleb, and others read and commented on manuscripts. C. Schirarend and H.-H. Poppendieck offered important help with translations from Russian and into English, respectively. C. Bayer very patiently tried to develop my understanding of inflorescence morphology. A special word of thanks, for his unfailing help, is due to J. Bogner, who not only reviewed various manuscripts but also provided important plant material and illustrations. For the generous permission to use illustrations from Vol. 6 of his *Plant Life* (Leningrad 1982), I am deeply indebted to A.L. Takhtajan. I am also most grateful to all those authors and copyright holders who permitted the use of their illustrations. Likewise, I wish to thank H.-D. Behnke, C.D.K. Cook, E. Köhler, J. Kress, H.P. Linder, R. Mello-Silva, S. Nilsson with E. Grafström, and M. Harley for making valuable light, scanning and transmission electron microscope photographs available.

Finally, I wish to thank all contributors for generously sharing their specialist knowledge and for their patience, with which they reacted to all my suggestions and requests. Last, but not least, my thanks go to the editorial and production staff of Springer Verlag who, as always, have done splendid work.

Hamburg, May 1998 K. Kubitzki

Contents

List of Contributors

Appel, O.	Institut für Allgemeine Botanik und Herbarium, Universität Hamburg, Ohnhorststr. 18, 22609 Hamburg, Germany
Barthlott, W.	Botanisches Institut der Universität Bonn, Meckenheimer Allee 170, 53115 Bonn, Germany
Bayer, C.	Institut für Allgemeine Botanik und Herbarium, Universität Hamburg, Ohnhorststr. 18, 22609 Hamburg, Germany
Bayer, E.	Botanischer Garten, Menzinger Str. 65, 80638 München, Germany
Bogler, D.J.	Department of Botany, University of Texas at Austin, Austin/Texas 78713-7614, USA
Bos, J.J.	Department of Plant Taxonomy, Agricultural University, Postbus 8010, 6700 ED Wageningen, The Netherlands
Clifford, H.T.	Queensland Museum, P.O.B. 3300, South Brisbane Qld 4101, Australia
Conran, J.G.	Department of Botany, University of Adelaide, Adelaide SA 5005, Australia
Eriksson, R.	Institute of Systematic Botany, University of Göteborg, Carl Skottsbergs Gata 22, 413 19 Göteborg, Sweden
Goldblatt, P.	Missouri Botanical Garden, POB 299, St. Louis, Missouri 63166-0299, USA
Harling, G	Institute of Systematic Botany, University of Göteborg, Carl Skottsbergs Gata 22, 413 19 Göteborg, Sweden
Henderson, R.F.J.	Queensland Herbarium, Meiers Road, Indooroopilly, Queensland 4068, Australia
Huber, H.	Universität Kaiserslautern, FB Biologie, Postfach 3049, 67653 Kaiserslautern Germany
Huynh, K.-L.	Botanical Institute, 2007 Neuchâtel, Switzerland
Ibisch, P.L.	Botanisches Institut, Universität Bonn, Meckenheimer Allee 170, 53115 Bonn, Germany
Kao, P.-C.	Chengdu Institute of Biology, Academia Sinica, POB 416, Chengdu 610041, Sichuan, PR China
Kubitzki, K.	Institut für Allgemeine Botanik und Herbarium, Universität Hamburg, Ohnhorststr. 18, 22609 Hamburg, Germany
Maas-van de Kamer, H.	Department of Systematic Botany, University of Utrecht, Postbus 80.102, 3508 Utrecht, The Netherlands
Manning, P.J.	Botany Department, University of Cape Town, Rondebosch 7700, South Africa

Meerow, A.W.	IFAS Research and Education Center, University of Florida, 3205 Southwest College Avenue, Fort Lauderdale, Florida 33314, USA
Neinhuis, C.	Botanisches Institut, Universität Bonn, Meckenheimer Allee 170, 53115 Bonn, Germany
Nordal, I.	Biological Institute, University of Oslo, Postboks 1045, Blindern, 0316 Oslo, Norway
Nordenstam, B.	Swedish Museum of Natural History, POB 50007, 104 05 Stockholm, Sweden
Perry, P.L.	19 Rhosfryn, Bangor, Gwynedd LL57 DL, United Kingdom
Poppendieck, H.H.	Institut für Allgemeine Botanik und Herbarium, Universität Hamburg, Ohnhorststr. 18, 22609 Hamburg, Germany
Rahn, K.	Skyttehusene 49, 2640 Albertslund, Denmark
Rudall, P.J.	Royal Botanic Gardens Kew, Richmond, Surrey TW9 3AB, United Kingdom
Simpson, M.G.	Department of Biology, San Diego State University, San Diego, California 92182-0057, USA
Smith, G.F.	The National Herbarium, Private Bag X101, Pretoria 0001, South Africa
Snijman, D.	Botany Department, University of Cape Town, Rondebosch 7700, South Africa
Speta, F.	Oberösterreichisches Landesmuseum, Museumstr. 14, 4010 Linz, Austria
Tamura, M.N.	Faculty of Science, Osaka City University, 3-3-138 Sugimoto, Sumiyoshi-ku, Osaka 558, Japan
Theisen, I.	Botanisches Institut der Universität Bonn, Meckenheimer Allee 170, 53115 Bonn, Germany
Tillich, H.-J.	Institut für Systematische Botanik der Universität München, Menzinger Str. 67, 80638 München, Germany
van Wyk, B.-E.	Botany Department, Rand Afrikaans University, POB 524, Johannesburg 2000, South Africa
Verhoek, S.	Department of Biology, Lebanon Valley College, Annville, Pennsylvania 17003, USA
Yeo, P.F.	71 Grantchester Meadows, Cambridge CB3 9JL, United Kingdom
Vogel, S.	Botanisches Institut der Universität Wien, Rennweg 14, 1030 Wien, Austria
Weustenfeld, T.	Spezielle Botanik, Ruhr-Universität, Postfach 10 21 48, 44801 Bochum, Cermany
Wilder, G.J.	Dept. of Botany, Cleveland State University, 2399 Euclid Avenue, Cleveland, Ohio 44115-2403, USA

Development and Organization

H.-J. Tillich

1. Introduction

The monocotyledons are very diverse in their external morphology. In several instances the morphological comparison and/or the homologies with corresponding organs of the dicotyledons are controversial. This problem is reflected in the descriptive literature by numerous rather vague terms and the frequent use of the prefix pseudo- (e.g., pseudobulb, pseudolamina, pseudopetiole, pseudostipule, etc.).

This chapter aims to give a brief summary of the vegetative morphology of the monocotyledons and a set of unequivocal descriptive terms.

2. Monocotyledonous Organization: Possibilities and Limitations

The great majority of the monocotyledons lack a cambium. This is, according to Holttum (1955), the basic limitation for the construction of the shoot system. Since there is no cambium in the primary root, it soon becomes too small in diameter to conduct the nutrients and water needed for the growth of the shoot system. Thus it is necessary to produce shoot-borne roots early during the seedling stage and establishment growth (Tomlinson and Esler 1973). At the same time, the axis undergoes a marked primary thickening, increasing from internode to internode and leading to the typical inverted conical shape of the basal primary axis (Fig. 1). The number of leaves per node is generally restricted to 1, since the leaf base in monocotyledons usually embraces the axis for more than half the circumference and there remains no space for further leaf primordia at a given node (see Tomlinson 1970b).

The limited conductive capacity of the axis is also responsible for the restricted branching of upright aerial shoots. The development of tall trees is possible only since the number of leaves in a given individual always remains constant, as is true e.g., for palms or *Strelitzia nicolai*. Vegetative branching, if happening at all, mainly takes place from basal nodes.

Bearing these limitations in mind, it is fascinating to see what a great diversity of growth forms and special adaptations is to be found in monocotyledons. A selection is shown in Fig. 2. The most remarkable specialized forms include the epiphytic orchids and bromeliads, the submarine members of Alismatales and Najadales, the much-reduced Lemnaceae, and the mycotrophic Burmanniales and Triuridales. In some climbing members of the araceous genera *Monstera* and *Pothos*, a peculiar skototropic growth (Strong and Ray 1975, Boyce and Poulsen 1994, Boyce and Nguyen Van Dzu 1995; Fig. 2J) helps the young plant to find a host tree. Among the rattan palms, the genus *Calamus* includes species with shoots more than 180 m long, the longest known in the plant kingdom (Dransfield 1978). Despite this great diversity, it is surprising that the monocotyledons contain only a very limited number of therophytes. They are mainly found in Poales and Cyperales; elsewhere they are very scattered, e.g., in Asparagales (*Bulbinella*, *Asphodelus p.p.*), Pontederiaceae, or Commelinaceae.

3. Leaf Morphology in Monocotyledons

At a primordial developmental stage foliage leaves in angiosperms are subdivided into a proximal leaf base and a distal leaf part. The distal part (Oberblatt) was recently named hyperphyll (see Weberling 1989: 45); similarly, the proximal part (Unterblatt) can be called a hypophyll. In many monocotyledons the hypophyll tends to be much more extended and contributes a far greater part to the leaf than in dicotyledons, but there are several exceptions to this rule.

The discussion on the interpretation of the foliage leaf of monocotyledons is dominated by the so-called phyllode theory, which goes back to de Candolle (1828) and was extensively supported by Arber (1918 ff.). According to this theory, the unifacial, often more or less cylindrical or ensi-

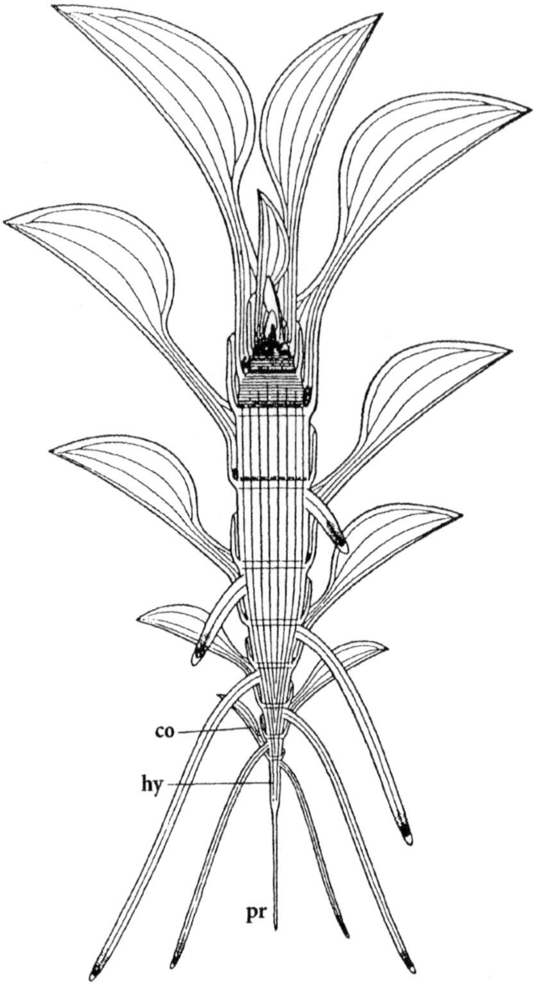

Fig. 1. Scheme of monocotyledonous organization. Meristematic regions shown in *solid black*. Note primary thickening of basal axis. *co* Cotyledon; *hy* hypocotyl; *pr* primary root. (Troll 1954)

form hyperphyll in monocotyledons represents only a petiole; the lamina is said to be totally missing (Fig. 3D,H). This view was refuted by Gaisberg (1922) and Peters (1927); more recently, the extensive and careful investigations of Kaplan (1972, 1973, 1975) have demonstrated that the unifacial hyperphyll in monocotyledonous leaves is equivalent to the complete hyperphyll of dicotyledons (Fig. 4). Thus, the term phyllode is unsuitable for monocot leaves. Recent investigations by Bharathan (1996) additionally demonstrate that the hyperphyll region may differentiate into an extended bifacial lamina and a unifacial tip. Thus it is not appropriate to consider automatically a unifacial leaf tip (Vorläuferspitze, precursory tip) as equivalent to the complete hyperphyll, and the rest of the leaf, whether or not it is differentiated into sheath, petiole, and lamina, as representing the hypophyll. In many cases, an adequate interpretation will be possible only after careful ontogenetic investigations (Bloedel and Hirsch 1979). Therefore it is preferable to use the terms petiole and lamina in a descriptive sense only to refer to a narrowed and a broadened leaf area, respectively.

In Poaceae and Commelinaceae we never find a unifacial leaf tip. On the other hand, in these families the leaves are clearly subdivided into a culm-embracing sheath and a bifacial blade, which itself may be separated from the sheath by a shorter or longer petiole (Fig. 5I). Arber (1922a, 1923a) considered the blades in both families to be phyllodes, despite their bifacial structure and a normal bundle orientation. Such an imposition of a theory onto nature brings no advantage at all. Wherever a clearly distinguishable petiole and lamina are found, such as in Alismataceae, palms, Dioscoreaceae, or bamboos, these organs should be called petiole and lamina. Terms like pseudopetiole and pseudolamina are dispensable.

Where the leaves consist of a more or less linear, parallel-veined hypophyll and a hyperphyllary tip (Fig. 5J–N), the relative contributions of the two parts to the adult foliage leaf varies considerably (see Fig. 4A,C). Leaf parts formed by the hypophyll are always bifacial, whereas the hyperphyll may be completely or in part unifacial. The extent to which each of the two leaf zones and the unfacial and bifacial regions can contribute to the construction of the adult leaf is shown in Fig. 4A–F.

The leaf base in monocotyledons only rarely bears paired appendages, which usually are termed stipules. However, true stipules, as defined by Weberling (1958, 1975), seem to be absent. Lateral appendages as found in some *Smilax* species are vaginal lobes rather than stipules. Median, often hyaline appendages of the leaf base (as in Potamogetonaceae, Poaceae, Zingiberaceae) are called ligules and can be interpreted as median vaginal lobes.

Small, few-celled appendages are regularly found in the axils of the Alismatidae, where they were discovered by Irmisch (1858a,b), who named them squamulae intravaginales. These supposedly glandular organs were studied later by Gibson (1905), Arber (1923a,b) and Tomlinson (1982). Outside the Alismatidae, similar squamules are reported from some Araceae (*Philodendron*, *Cryptocoryne*, *Lagenandra*, see Dahlgren et al. 1985).

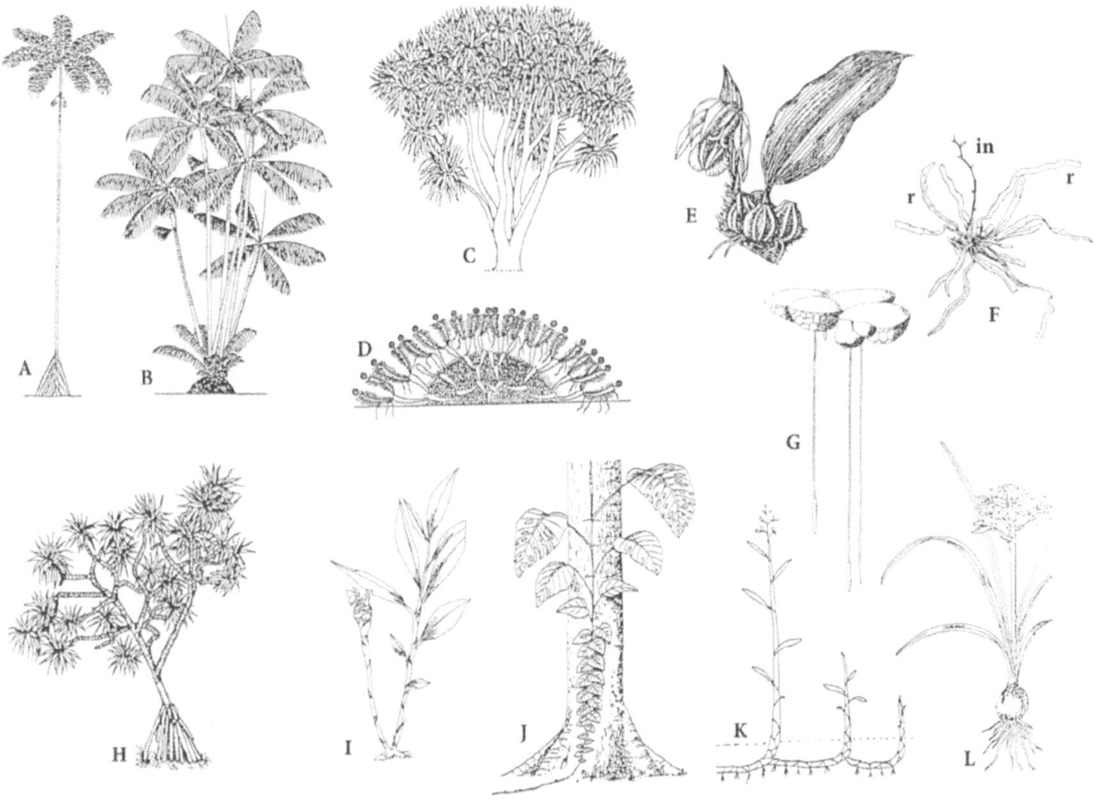

An appendage specific to some palmate and costapalmate palm leaves is the hastula, which is found at the junction between petiole and lamina, and which may be developed adaxially and abaxially as well (Fig. 5P,Q).

Palm leaves are pinnate, palmate, costapalmate, or, in *Caryota*, twice-pinnate. Developmentally, the blade is always initially undivided, but early in ontogeny the lamina exhibits a regular plication. The separation of the pleats occurs late in ontogeny. The process of blade plication and subsequent separation of the leaflets was described in detail by Kaplan et al. (1982a,b) and Dengler et al. (1982).

In Araceae the leaf shape is exceedingly diverse (see Mayo et al. 1997). It ranges from simple elliptic to highly dissected, as in *Dracontium*, *Taccarum*, or *Amorphophallus*. Fenestrate (perforated) leaves are a special peculiarity of the family; they can be found, e.g., in species of *Monstera*, *Dracontium*, and *Amydrium*. These perforations are of secondary origin, since the respective leaves initially have an entire lamina, in which certain areas undergo necrosis during early ontogenetic development (Arber 1925a, Hotta 1971). If the holes extend to the margin of the lamina, the leaf

Fig. 2A–L. Monocotyledonous growth forms. **A** *Socratea exorrhiza*, unbranched palm with prop roots. **B** *Euterpe oleracea*, basally branched palm. **C** *Cordyline kaspar*, richly branched tree with secondary thickening. **D** *Abromeitiella* sp., schematic section through a cushion. The oldest parts decaying, the younger shoots rooting in the organic matter. Inflorescences indicated as *circles*. **E** *Coelogyne salmonicolor*, epiphytic orchid with specialized storage axes. **F** *Taeniophyllum* sp., epiphytic orchid with much reduced shoots, assimilating roots (*r*), and a small inflorescence (*in*). **G** *Lemna gibba*, free-floating plant with much reduced vegetative shoot system. **H** *Pandanus* sp., richly branched shrub with strong stilt roots. **I** *Zingiber zerumbet*, vegetative leafy shoots strongly different from scale-leaved flowering shoots. **J** *Monstera dubia*, scale-leaved seedling axis growing skototropically along soil surface until reaching a trunk; when climbing upwards, producing small shingle leaves and finally foliage leaves typical of the adult plant. **K** Schematic representation of regular sympodial rhizomatous growth, each sympodial element ending in an inflorescence. This growth type is very common in monocotyledons. **L** *Scilla verna*, a bulbous member of Hyacinthaceae. Some figures modified. Not to scale. (**A**, **B** Kahn and Granville 1992; **C** Tomlinson and Fisher 1971; **D** Rauh 1990; **E** Bechtel et al. 1993; **F** Strasburger, Lehrbuch der Botanik, 32. Aufl. 1992; **G** Hegelmaier 1868; **H** Brooks 1993; **I** Wagner et al. 1990; **J** Bown 1988; **K** Holttum 1955; **L** Dahlgren et al. 1985)

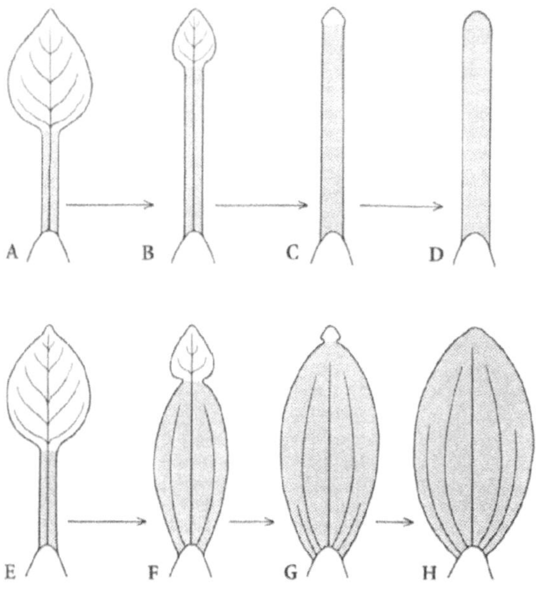

Fig. 3A–H. Illustration of the phyllode theory of Arber (1918 seq.) with a hypothetical series of monocotyledonous leaves, in which a progressive reduction of the lamina is compensated by an increase of the petiole. **A–D** Rhabdophyllodes. **E–H** Platyphyllodes. Petiole region *shaded*. (Kaplan 1975)

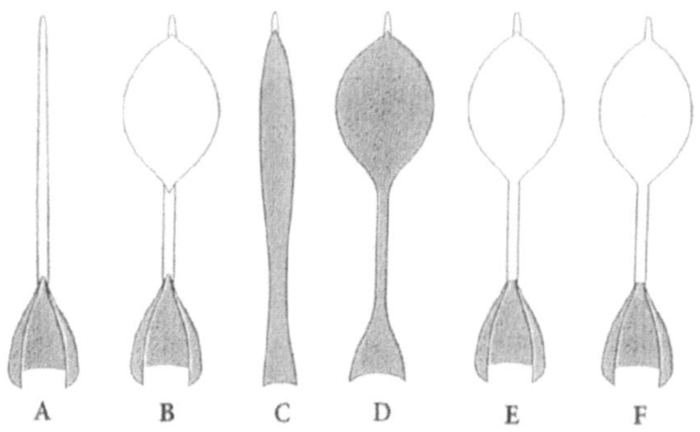

Fig. 4A–F. Modes of participation of leaf base (*shaded*) and hyperphyll (white) in the leaf construction of monocotyledons. **A** Unifacial hyperphyll (cylindrical or ensiform), not differentiated in petiole and lamina (e.g., *Juncus* spp.). **B** Hyperphyll with unifacial petiole, bifacial lamina, and unifacial tip (e.g., *Orontium aquaticum*). **C** Leaf base ribbon-like elongated, the hyperphyll a small unifacial tip (e.g., *Dracaena deremensis*). **D** Leaf base differentiated in sheathing base, petiole, and lamina, hyperphyll a small unifacial tip (e.g., *Hosta* spp.). **E** Hyperphyll differentiated in bifacial petiole, lamina, and a unifacial tip (e.g., *Dioscorea* spp.). **F** Hyperphyll completely bifacial (e.g., *Croomia japonica*). (Following Troll 1955)

becomes pinnatisect. Truly pinnate leaves typical of many dicotyledons are missing in monocotyledons, although the lamina of *Zamioculcas* resembles a truly pinnate lamina.

Leaf tendrils are developed only very sporadically in monocotyledons. They are formed by the apices of the lamina in *Gloriosa* and *Flagellaria*. In *Smilax* they are paired outgrowths of the leaf base of an obscure nature (Fig. 5F).

In the Marantaceae the junction between petiole and lamina is a swollen pulvinus or geniculum, which turns the lamina into a favorable position to light. A pulvinus is also found in some Zingiberaceae and Araceae. Curiously enough, in *Anthurium oerstedianum* and some other Araceae, the pulvinus has its position on the petiole halfway between its base and the lamina.

4. Seedling Organization in Monocotyledons

Seedlings of monocotyledons, without any exception, possess only *1* cotyledon. This cotyledon represents *1* leaf. There is no morphological or developmental indication of either syncotyly or anisocotyly (Tillich 1992). The cotyledon is subdivided, like any foliage leaf, into a leaf base (hypophyll) and an upper part (hyperphyll).

The cotyledon base embraces the seedling axis and is often tubular. The assumed ancestral form is a short sheath without or with inconspicuous appendages. In the simplest case, the hyperphyll may have only a haustorial function, i.e., it is completely hidden in the seed (Fig. 6A). Seedlings with compact cotyledons in most cases develop 1–several scalelike cataphylls as their first plumular phyllomes. From this basic model a great diversity of forms has evolved (Figs. 6,7).

A first evolutionary line leads to an elongated, green, upright cotyledonary hyperphyll capable of assimilation (Fig. 6B). The green hyperphyll is usually cylindrical, but in Trilliaceae it is distally flattened to form a small lamina. Except for some bulb or tuber-bearing species this cotyledon type is generally combined with green eophylls. The same is true for the third type, which is characterized by a well-developed coleoptile instead of an elongated hyperphyll. This organ is a secondary outgrowth from the margin of a tubular cotyledonary sheath (Fig. 6C). The elongation growth of the sheath tube is less the more strongly the coleoptile develops. At the most derived stage the sheath tube is totally reduced and the hyperphyll seems to arise directly from the base of the coleoptile, as in Poaceae, Cyperaceae, or Marantaceae (Figs. 6D, 8J). The coleoptile seems to protect the first eophylls.

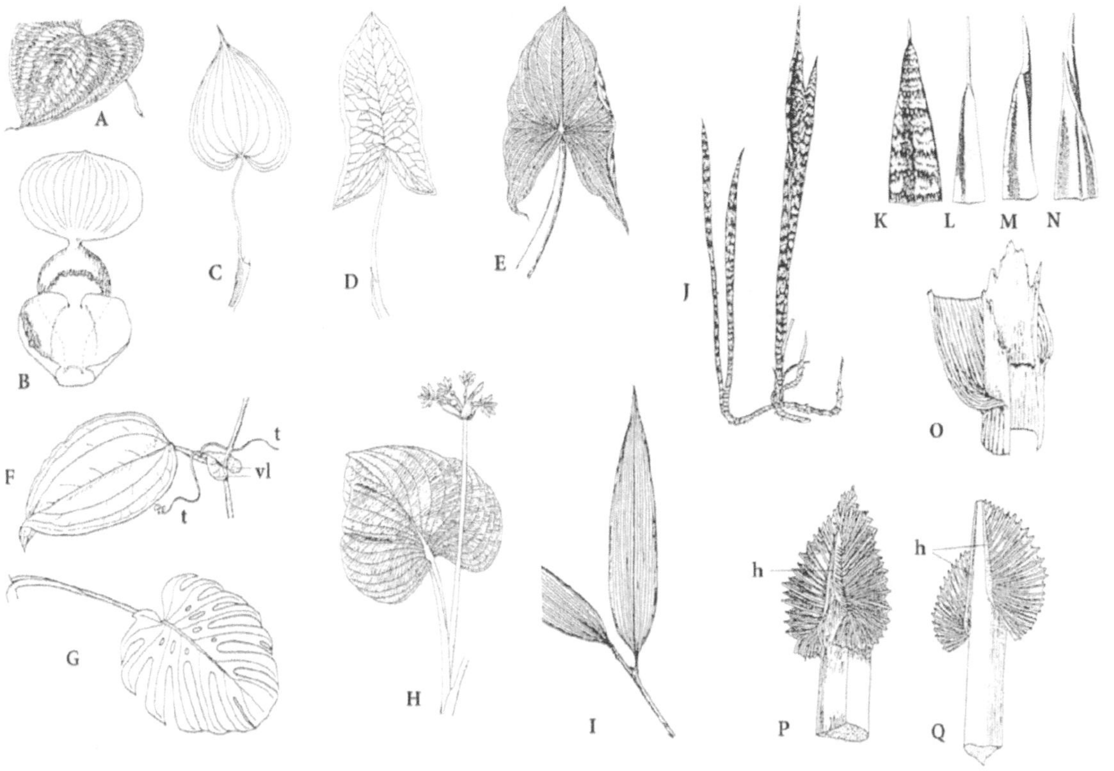

Fig. 5A–Q. Foliage leaves in monocotyledons. **A–I** Laminated leaves. **A** *Dioscorea bulbifera*. **B** *Eichhornia crassipes*. **C** *Streptolirion volubile*. **D** *Arum maculatum*. **E** *Sagittaria montevidensis*. **F** *Smilax perfoliata*. **G** *Monstera deliciosa*. **H** *Proiphys amboinensis*. **I** *Sinobambusa kunishii*. **J** *Sansevieria trifasciata* with "typical", parallel-veined monocotyledonous leaves. **K–N** Leaves with unifacial apices. **K** *Sansevieria trifasciata*. **L** *Chlorophytum comosum*. **M** *Dracaena fragrans*. **N** *Dracaena deremensis*. **O–Q** Leaves with appendages. **O** *Anthoxanthum formosanum* with well-developed ligule. **P, Q** *Sabal* sp., insertion of lamina on petiole. **P** Upper face, with prominent adaxial hastula. **Q** Lower face, with inconspicuous abaxial hastula. Some figures modified. Not to scale. *h* hastula, *t* tendril, *vl* vaginal lobes. (**A** Liu and Huang 1978; **B** after Weber 1950; **C, D** Arber 1922a, 1925a; **E** Cabrera 1968; **F** Koyama 1978; **G** Rohweder and Endress 1983; **H** Telford 1987; **I, O** Hsu 1978; **J** Dahlgren et al. 1985; **K–N** Troll 1955, **P, Q** Tomlinson 1961)

The next cotyledon type again has assimilatory capacity, but here this is achieved by a dilatation of the cotyledonary sheath to form a laminalike area, while the hyperphyll portion remains in the seed and has only a haustorial function (Fig. 6E, F). This cotyledon type is typically found in Costaceae, in some Araceae (*Philodendron*, *Colocasia*, see Tillich 1985), in Bromeliaceae-Pitcairnioideae, and in *Xyris*.

The last type to be reviewed here is typical of species, whose seeds contain no or only remnants of endosperm, so that nutrients have to be stored in the cotyledon. In these cases, the storage organ is generally the cotyledonary hyperphyll (Fig. 6G). Outside the Alismatidae the storage cotyledon is known only in several Araceae and in *Cyanastrum* (Tillich 1985, 1995b).

Some special features are exhibited by palm seedlings. The cotyledon never produces chlorophyll even if developing in permanent light. It may be of the compact type with no appendages or have a very short coleoptile (*Cocos*, *Roystonea*, *Salacca*). In other cases, the hyperphyll elongates considerably and the sheath is either tubular (*Phoenix*) or completely absent, but with a long coleoptile (*Sabal*). The first plumular leaves are aleays cataphylls (Tomlinson 1960). Martius (1823) called palm seedlings with compact cotyledons, admotive, and those with an elongated hyperphyll, remotive.

The hypocotyl in monocotyledonous seedlings is mostly inconspicuous. However, in a few cases it is well developed and may even become a small tuber (*Pinellia*, *Paris*, *Trillium*, some *Discorea*). In most families of the Alismatanae it is an important

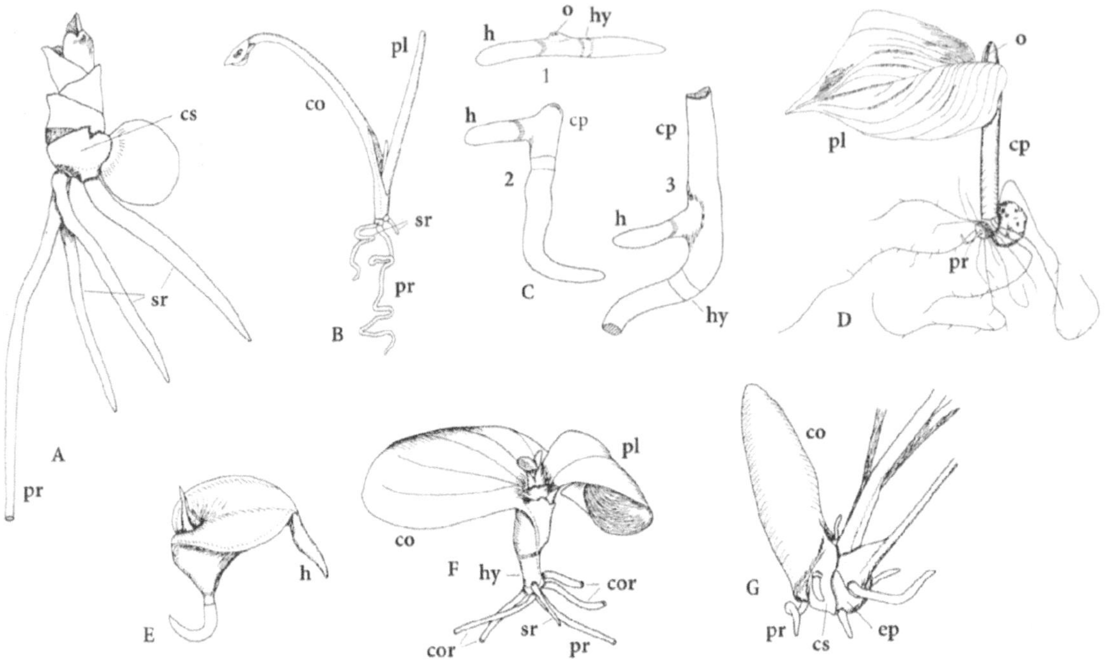

Fig. 6A–G. Cotyledon types in monocotyledons. **A** *Sansevieria grandis*, compact cotyledon. **B** *Acorus calamus*, cotyledon with elongated, assimilating hyperphyll. **C** *Bulbine semibarbata*, three stages of coleoptile development. **D** *Sarcophrynium brachystachys*, strongly developed coleoptile, the numerous collar roots growing and branching vigorously unlike the vestigial primary root. **E, F** Cotyledons with expanded, assimilating sheath. **E** *Dyckia sulphurea*. **F** *Costus megalobractea*, the cotyledon consisting of a sheathing base, a laminalike portion, and a minute unifacial haustorium (hidden in the background). **G** *Aponogeton distachyus*, storage cotyledon; tuber formation commencing with a swelling of the epicotyl. *co* Cotyledon; *cor* collar roots; *cp* coleoptile; *cs* cotyledonary sheath; *ep* epicotyl; *h* haustorium; *hy* hypocotyl; *o* opening of cotyledonary sheath or coleoptile; *pl* eophyll; *pr* primary root; *sr* shoot born root

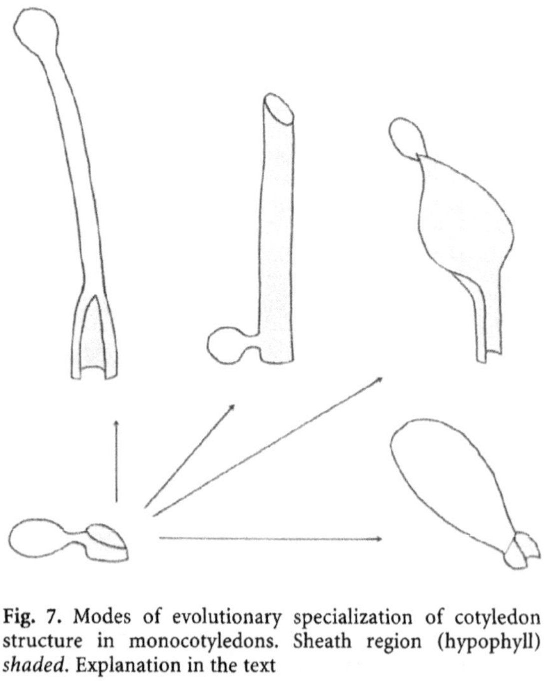

Fig. 7. Modes of evolutionary specialization of cotyledon structure in monocotyledons. Sheath region (hypophyll) *shaded*. Explanation in the text

additional storage organ of the appropriately named macropodous embryo (Fig. 8F,H).

The transition zone between the hypocotyl and the primary root is the collar, actually the lowermost part of the hypocotyl. A distinguishing feature of the collar is the development of dense rhizoids (Fig. 8F,H). Roots formed endogenously at the collar level are called collar roots (Grenzwurzeln).

The primary root is initiated exogenously, mostly exactly at the root pole of the embryo, i.e., at the attachment point of the suspensor. *Lilaea*, *Triglochin*, *Ruppia*, and *Aponogeton* are exceptions, and here the primary root is initiated at some distance from the root pole (Yamashita 1970, 1972, 1976).

Shoot born roots are always initiated endogenously. When a root develops in a young parental tissue, as is the case in seedlings, it may stimulate

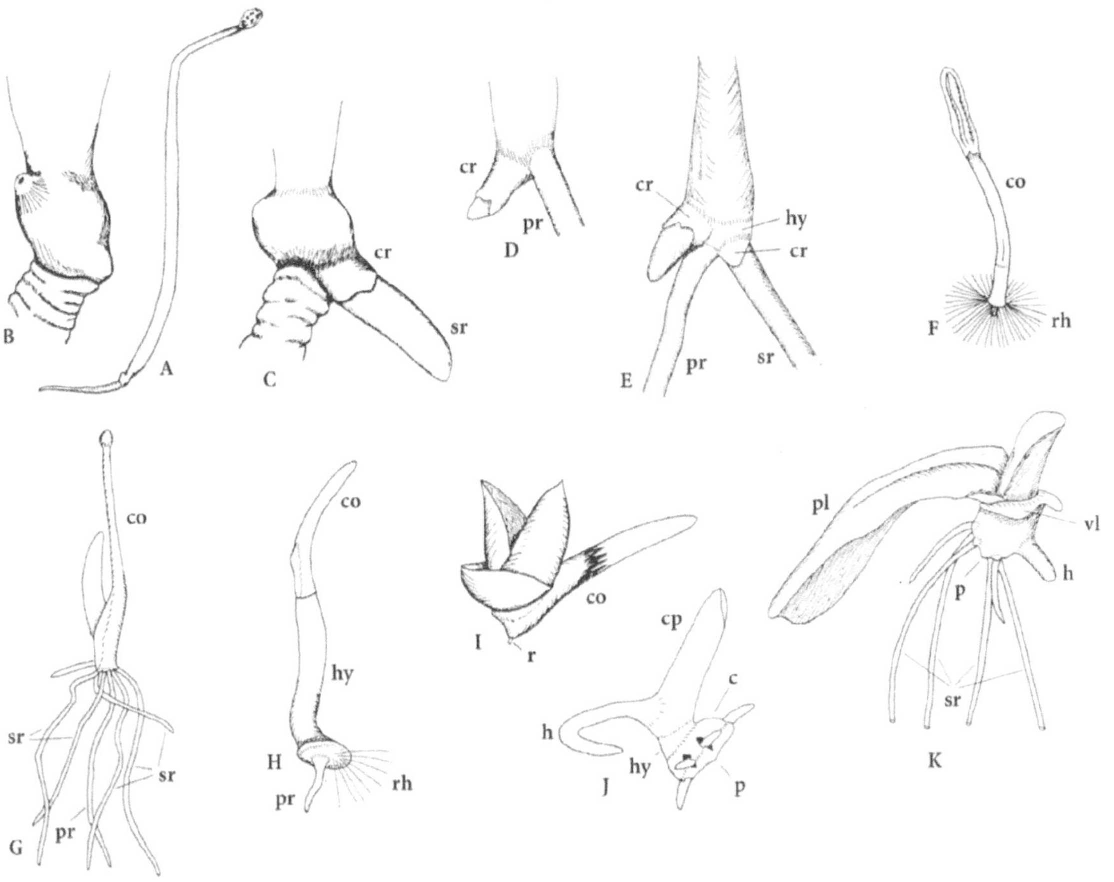

Fig. 8A–K. Collar and root formation in monocotyledonous seedlings. A–C *Galtonia candicans*. A Seedling with fully developed cotyledon, the furrowed surface of the primary root indicating root contraction. B Cotyledon base, detail of A. C Same at a somewhat later stage, the first shoot born root piercing its coleorhiza. D *Allium ascalonicum*, the first shoot born root emerging. E *Aphyllanthes monspeliensis*, emergence of shoot born roots. F *Butomus umbellatus*, collar densely covered with collar rhizoids. G *Anigozanthos manglesii*, numerous shoot born roots accompanying the weak primary root. H *Potamogeton nodosus*, the collar (only a few rhizoids indicated) covering lower surface of thick storage hypocotyl, primary root only weakly developed. I *Tillandsia butzii*, with rudimentary primary root. J *Maranta* sp., collar roots emerging from massive collar, primary root arrested. K *Pistia stratiotes*, primary root missing. *c* Collar; *co* cotyledon; *cp* coleoptile; *cr* coleorhiza; *h* haustorium; *hy* hypocotyl; *p* root pole; *pl* eophyll; *pr* primary root; *sr* shoot born root; *r* rudiment of primary root; *rh* collar rhizoids; *vl* median vaginal lobe of the cotyledonary sheath

the peripheral tissue layers of its parental organ to form a kind of bag over the growing root apex. Such a covering bag, which is later penetrated by the growing root, is called the coleorhiza (Fig. 8C,D,E).

The primary root in monocotyledons is often much reduced, although in *Strelitzia* or several Convallariaceae it grows and branches vigorously. Usually, it grows only weakly and is short-lived (Figs. 6D, 8H). In some *Tillandsia* spp. it is only a small rudiment (Fig. 8I), in Marantaceae it may fail to commence elongation growth, and the primary root is not initiated al all in Poaceae, *Zostera*, Lemnaceae, and *Pistia* (Fig. 8K).

5. Special Modes of Shoot Organization in Monocotyledons

Pseudostems. It is well-known that in Musaceae and *Veratrum* successively embracing sheaths of foliage leaves form an upright stemlike shoot – actually a special type of a leaf rosette. In the vegetative phase the axis proper is represented by the sturdy rhizome, and only with the onset of

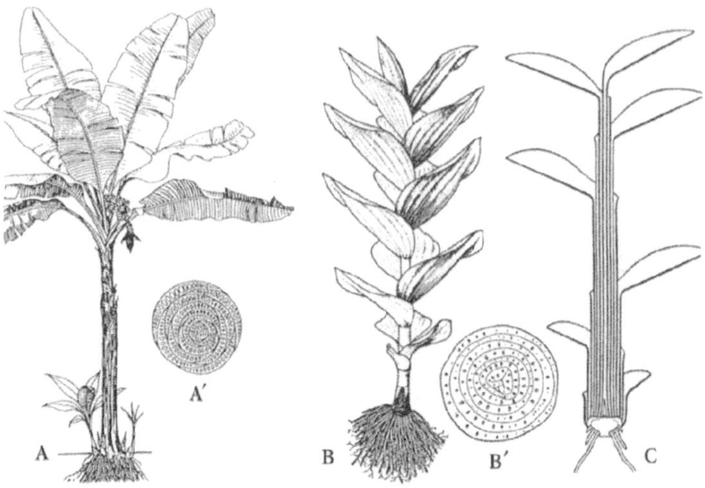

Fig. 9A–C. Pseudostems in monocotyledons. **A** *Musa* × *sapientum*, habit. **B, C** *Veratrum album*. **B** Habit. **C** Schematic longisection. **A', B'** Cross-section of the pseudostem. Not to scale. (**A, A'** Dahlgren et al. 1985; **B, B', C** Troll 1954)

flowering does the inflorescence axis push its way through the sheath complex to bring the flowers above the leaves (Fig. 9).

Special Modes of Branching. Generally, the monocotyledons bear 1 bud axillary and in a median position with respect to the subtending leaf, but there are exceptions to this rule. An interesting case was reported by Fisher (1978) and Fisher and Dransfield (1979). In *Musa* and in rattan palms (*Daemonorops* spp. and *Myrialepis scortechinii*), they found the lateral buds opposite the leaf axil (Fig. 10B,C). Since in rattans these buds are subtended by the tubular sheath of their subtending leaf, it is difficult to explain this unusual position as being caused by a pronounced concaulescence, i.e., that this bud could belong to the preceding leaf. In *Korthalsia rigida* the vegetative buds are 130° to the axil, intermediate between the normal and the leaf-opposed position.

Collateral accessory buds are scattered in monocotyledons. A well-known example is garlic (Fig. 10D,E). Here a whole group of daughter bulbs represents 1 normally positioned median axillary bulb and a number of additional collateral bulbs. Similar arrangements are found in other *Allium* species (Irmisch 1850, 1876; Fig. 10F). The numerous small corms developed at the base of a *Gladiolus* corm are also produced from collateral buds. Collateral flowers are produced in the axils of the large bracts of the *Musa* inflorescence.

Another strange mode of branching is the dichotomous forking of the apical meristem. This phenomenon was first reported by Schoute (1909) for the palm genus *Hyphaene*. It has taken more than 60 years for additional examples to be found (see Fig. 10A). Detailed analyses are now available for *Flagellaria indica* (Tomlinson 1970a, Tomlinson and Posluszny 1977), *Nypa fruticans* (Tomlinson 1971), *Chamaedorea cataractarum* (Fisher 1974) and *Strelitzia reginae* (Fisher 1976). Uhl and Dransfield (1987) listed six palm genera with dichotomizing apices. There is every indication that in all these cases the branching is initiated by a true bifurcation of the apex. This mode of branching is likely to be derived rather than ancestral in monocotyledons.

A further unusual branching pattern is found in many bamboos. The axils of the sheathing scales along the culm bear "branch complements", which often are difficult to interpret morphologically (Fig. 10H). In some cases, they appear to be a telescoped, much branched axillary shoot system which originates from a single axillary bud. In other cases, a series of accessory buds seems to be developed (McClure 1966). To understand these complex systems, detailed developmental analyses are needed.

Phyllocladia. Assimilating, axillary organs in *Asparagus* (Fig. 40), *Ruscus*, *Semele*, and *Danae* (Fig. 116) have been subject to controversial interpretations. Arber (1924a) gives an extended overview of the older literature. Several authors thought that these organs are pseudoterminal leaves, in some instances fused with inflorescences (see Velenovský 1903, Arber 1924a, Schlittler 1953,

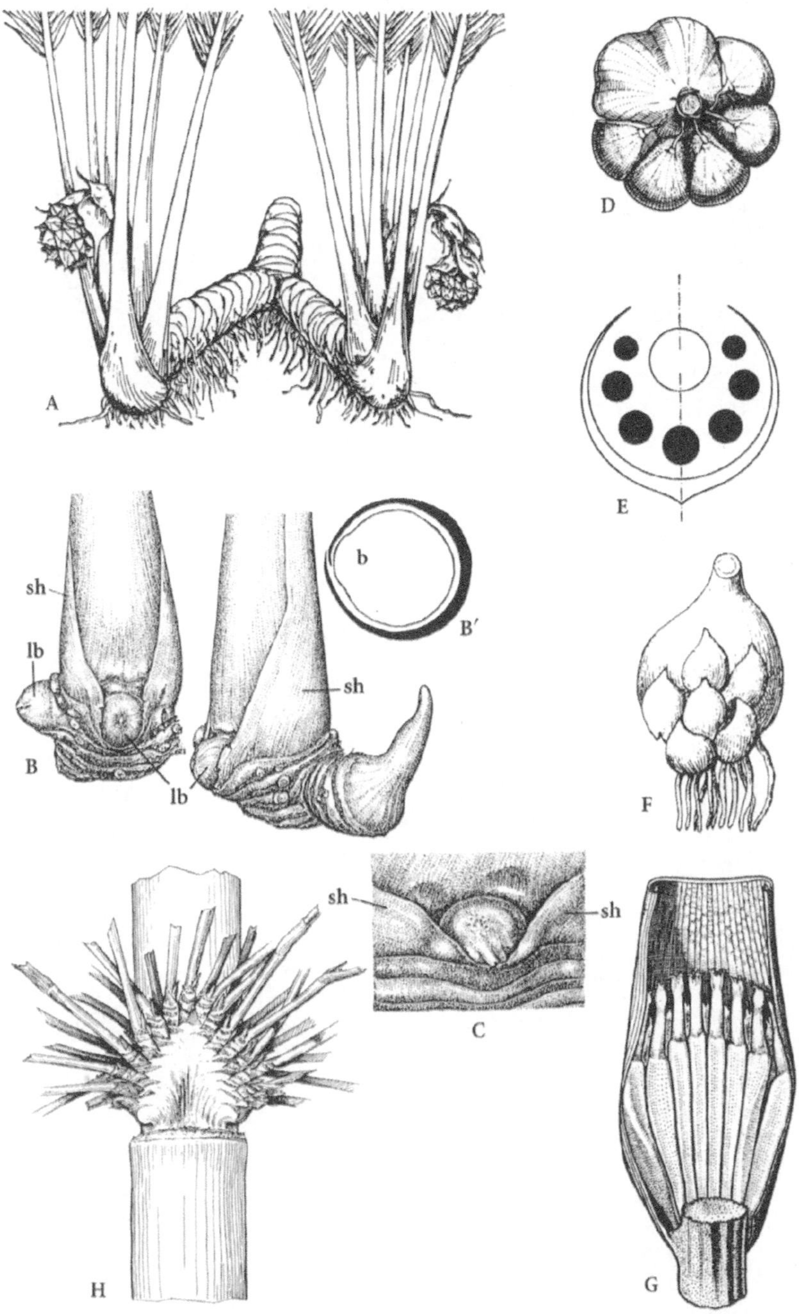

Fig. 10A–H. Special modes of branching in monocotyledons. A Dichotomous branching. **A** *Nypa fruticans*. **B, B′, C** Leaf-opposed bud initation in *Musa*. **B** *Musa coccinea*, two views of the same plant base, **B′** Diagram of bud position. **C** *Musa troglodytarum*, detail of bud. **D–H** Collateral accessory buds. **D** *Allium sativum*, bulb from above. **E** Diagram of **D**. **F** *Allium vineale*, daughter bulb with five accessory bulbs. **G** *Musa chinensis*, two series of collateral flowers in the axil of a large bract, adaxial view. **H** *Arthrostylidium harmonicum*, basal portion of a fully developed mid-culm branch complement. Some figures modified. Not to scale. *b* Bud; *sh* sheath of subtending leaf to leaf-opposed bud; *lb* leaf-opposed bud. (**A** Tomlinson 1970; **B, B′**, **C** Fisher 1978; **D, E, G** Troll 1948; **F** Irmisch 1850; **H** McClure 1966)

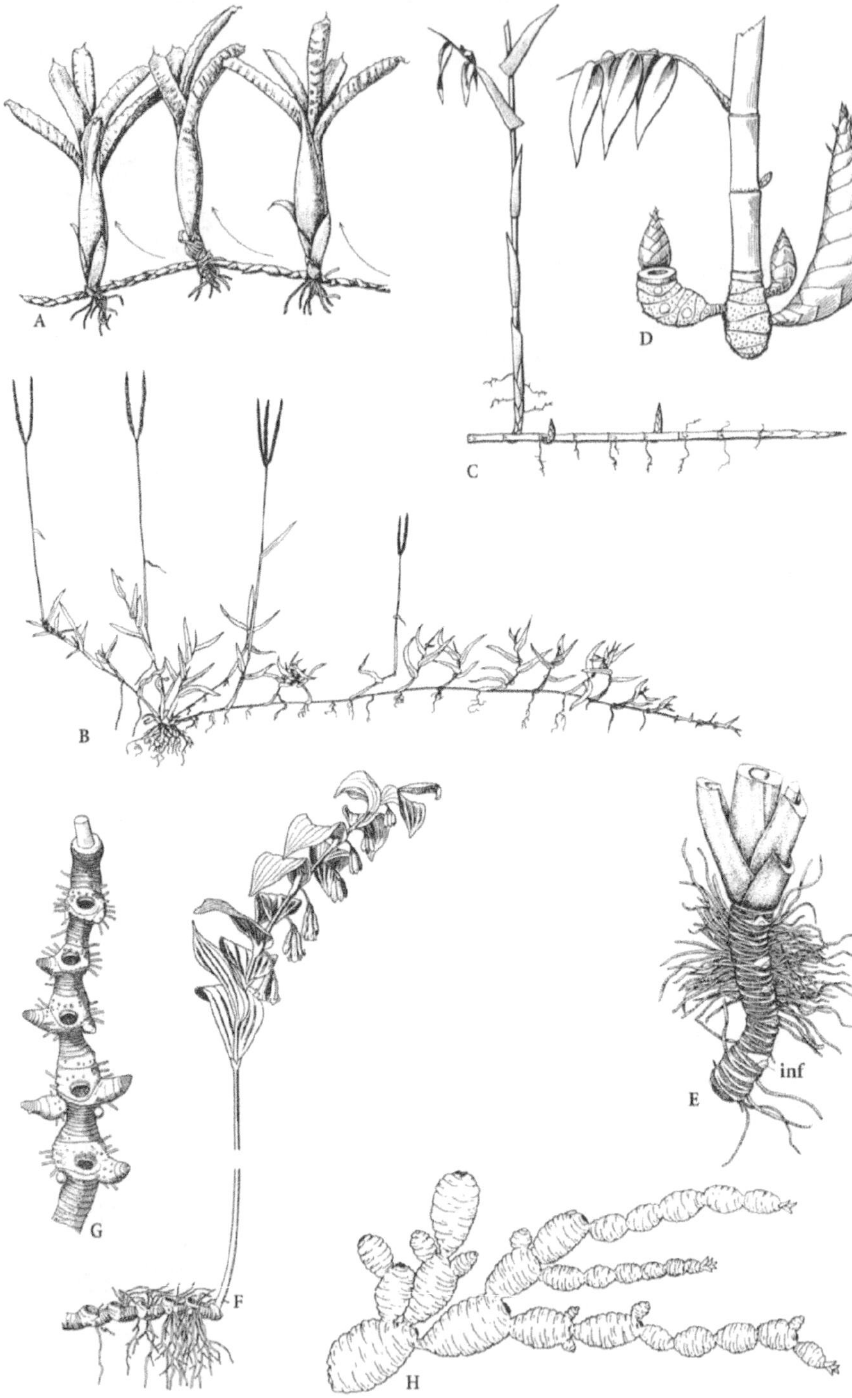

Fig. 11A–H. Rhizomes and runners in monocotyledons. **A** *Neoregelia ampullacea*, sympodially connected generations of runners. Each sympodial unit consisting of a plagiotropic, scale-leaved runner and a distal orthotropous rosette of foliage leaves. *Arrows* indicate the change in growth direction for each sympodial unit. **B** *Digitaria pertenuis.* Runner growing monopodially for a long distance. **C, D** Two principal forms of rhizomes in bamboos. **C** *Arundinaria amabilis*, monopodial, leptocaul rhizome. **D** *Bambusa beecheyana*, sympodial, pachycaul rhizome. **E** *Sagittaria lancifolia.* Monochasial rhizome, upper view, erect leaves cut off to show scars of previous terminal inflorescences (*inf*). **F, G** *Polygonatum multiflorum.* **F** Flowering plant. **G** Rhizome from above. **H** Rhizome of *Iris pseudacorus.* Depending on whether the plant comes to flower or not, annual rhizome units are added sympodially after flowering, monopodially without flowering. Some figures modified. Not to scale. (**A** Rauh 1990; **B** Gilliland 1971; **C, D** McClure 1966; **E** Tomlinson 1982; **F, G** Troll 1943; **H** Rodionenko 1961)

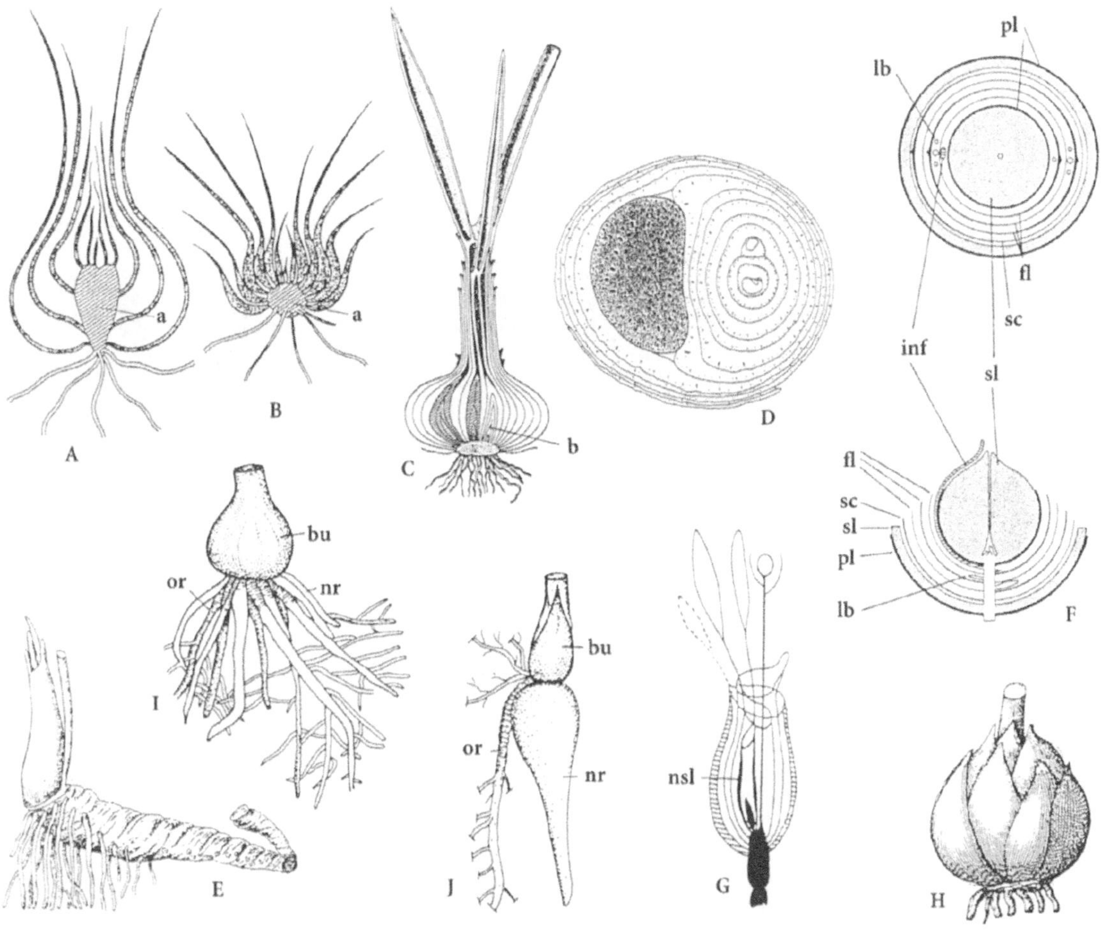

Fig. 12A–J. Bulbs in monocotyledons. **A, B** Shoot types in *Tillandsia*. **A** Pseudobulb (e.g., *T. bulbosa*). **B** Bulb proper (e.g., *T. argentea*). **C–G** Bulb structure in *Allium*. **C, D** *A. cepa*. **C** Bulb in longisection, showing the apical bud (*b*) and two lateral buds (*shaded*). **D** Flowering bulb in cross-section, inflorescence scape *shaded*. **E** *A. fallax*, rhizome bulb. Storage leaves developed at the top of a monopodial rhizome. **F** *A. neapolitanum*, schematic representation of bulb structure in longisection (*below*) and in cross-section (*above*). **G** *A. ursinum*. **H** *Lilium bulbiferum*, bulb with numerous narrow scales. **I, J** Bulbous plants with contractile roots. **I** *Eucomis punctata*, several contractile roots developed each season. **J** *Tigridia pavonia*, one contractile root of large diameter using the "tunnel effect" to pull the bulb downwards. Some figures modified. Not to scale. *a* Axis; *b* terminal bud; *bu* bulb; *fl* foliage leaf; *inf* inflorescence scape; *lb* lateral bud; *nr* new contractile root; *or* old contractile root; *pl* protective leaf; *sc* scale leaf; *sl* storage leaf; *nsl* new storage leaf. (**A, B** Rauh 1990; **C** Troll 1954; **D** Speta 1984; **E, H** Irmisch 1850; **F** Mann 1960, **G** Buxbaum 1937; **I, J** Pütz 1992)

1960). In *Asparagus* subgen. *Myrsiphyllum* Arber (1924b) interpreted these organs as being prophylls of an axillary shoot the axis of which is completely abortive, while in *Asparagus* s. str. she thought that they are axial organs. Most morphologists accepted the idea that the organs in question are axial structures, i.e., phyllocladia. This interpretation is strongly supported by developmental studies of Kaussmann (1955), Hirsch (1977), and Barykina and Gulenkova (1985).

6. Underground and Storage Organs in Monocotyledons

Runners and Rhizomes. Both are creeping shoots, and they are not always clearly distinguished. The runner is a slender axis which serves vegetative propagation. It often has elongated internodes

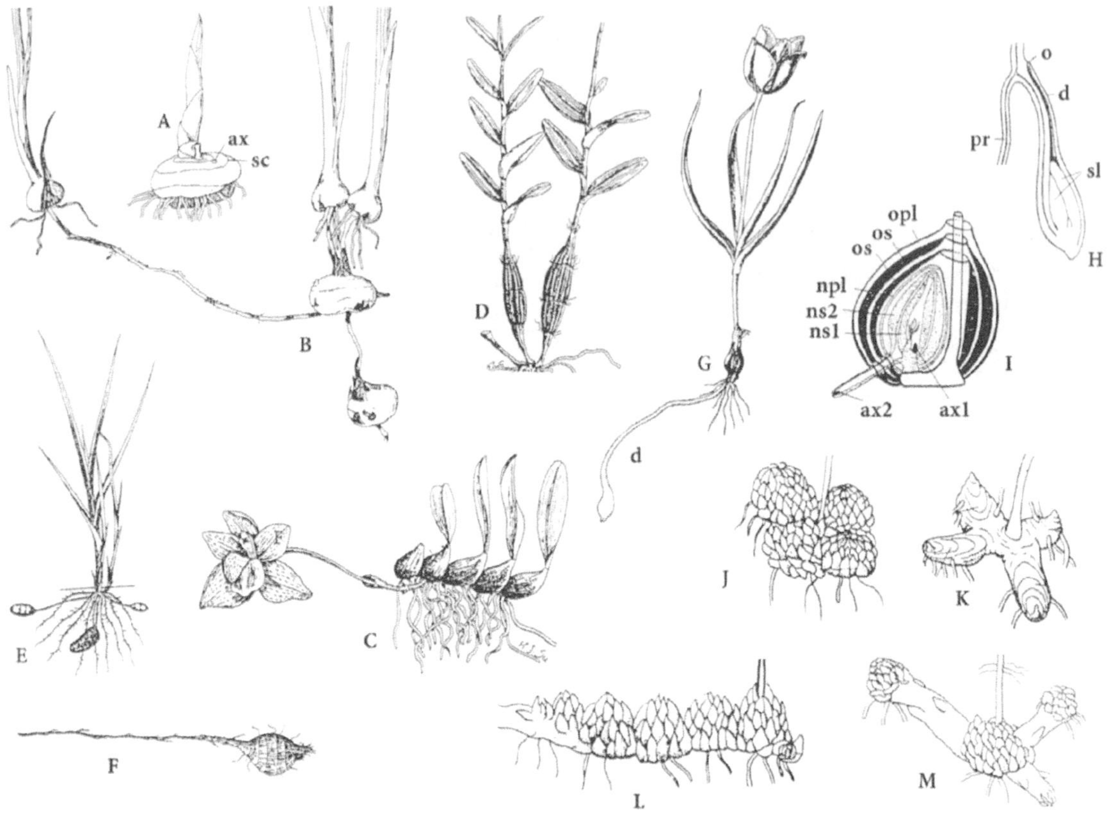

Fig. 13A–M. Storage organs and droppers in monocotyledons. **A** *Gladiolus* sp., corm with emerging shoot, remnants of last year's corm beneath. **B** *Crocosmia* × *crocosmiflora*, corms separated by runners of different length, thus being runner tubers rather than corms. **C** *Bulbophyllum transarisanense*, several generations of corms, each with one foliage leaf. **D** *Dendrobium crumenatum*, stem tubers formed by basal internodes. **E, F** *Cyperus esculentus*. **E** Plant with subterraneous runner tubers. **F** runner tuber enlarged. **G–I** Droppers of *Tulipa*. **G** *T. montana*, habit. **H** *T. gesnerana*, base of seedling plant in median longisection. **I** Schematic longisection of bulb with developing dropper, new generation stippled. **J–M** Bulb runners in *Lilium*. **J, K** *L. pardalinum*. **J** Four young bulbs around parent bulb. **K** Same after detachment of all bulb scales. **L** *L. vollmeri*, sympodial sequence of bulbs. **M** *L. canadense*, elongated runners with only a few scales. Some figures modified. Not to scale. *ax* Axillary bud; *ax1*, *ax2* axillary bud of new storage leaves; *d* dropper; *npl* new protective leaf; *ns1*, *ns2* new storage leaves; *o* opening of cotyledonary sheath; *opl* old protective leaf; *os* old storage leaf; *pr* primary root; *sc* leaf scar; *sl* storage leaf. (**A, B** Bell 1991; **C, D** Liu and Su 1978; **E** Rauh 1950; **F, G, H** Troll 1935, 1943; **I** Buxbaum 1958; **J–M** Baranova 1981)

and may grow above or below the ground, and in most cases it bears scale leaves (Fig. 11A,B). The rhizome, on the other hand, may serve vegetative propagation as well, but additionally has a storage function. The rhizome normally grows underground, but epiphytic or epilithic plants may have rhizomes that grow free on the supporting substrate. The branching of rhizomes may be monopodial or sympodial (Fig. 11), branching patterns being studied in detail by Bell and Tomlinson (1980).

Bulbs. The bulb is a very short axial body bearing 1-many leaves the bases of which store food. These leaves may be scales consisting only of the hypophyllary part or foliage leaves with a storing sheath. A bulb may be formed by only 1 of these leaf types or by a combination thereof. Additionally, protective leaves without storage function can be found. Thus, a great diversity exists in the details of bulb construction (Speta 1984; see also Fig. 12). Intermediate between a bulb and a rhizome is the rhizome bulb which is found in several *Allium* species (Fig. 12E).

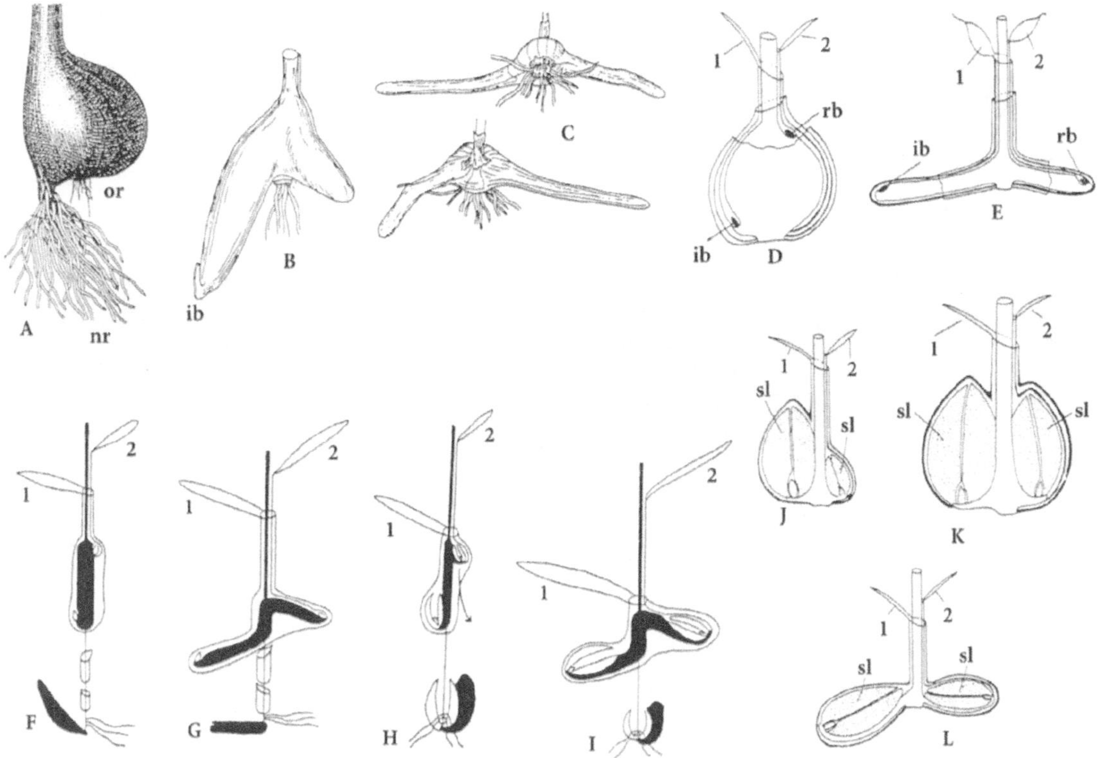

Fig. 14A–L. Corms and bulbs of Colchicaceae and *Gagea*. **A** Corm of *Colchicum autumnale*. **B** Finger corm of *Gloriosa superba*. **C** Finger corm of *Ornithoglossum glaucum* in two views. **D–L** Schematic figures to explain corm and bulb structure. **D** Corm of *Colchicum*. **E** Corm of *Gloriosa*. **F, G** Shoot structure of *Colchicum* (**F**) and *Gloriosa* (**G**), storage function restricted to the axis. **H, I** Shoot structure of *Gagea*, storage function taken over by scale leaves. **J–L** Bulb structure in *Gagea*. **J** *Gagea* subgen. *Holobolbos*. **K** *Gagea* subgen. *Didymobolbos*. **L** *Gagea* subgen. *Tribolbos*. *1* Subtending leaf of innovation bud; *2* subtending leaf of reserve bud; *ib* innovation bud; *or* roots of previous season; *nr* new roots; *rb* reserve bud; *sl* storage leaf. (**A** Troll 1935; **B–L** after Buxbaum 1936, 1937, 1958)

A very vague and misleading term is pseudobulb, often used for orchid tubers (see below). The term pseudobulb is properly used in cases when the bases of rosulate leaves are inflated and in this way imitate a bulb (Fig. 12A).

Tubers and Corms. Both terms relate to swollen, storage axes. Tubers may develop at the end of subterranean runners (runner tubers, Fig. 13E,F) or on aerial shoots (Figs. 2E, 13C,D). In typical cases a tuber is a swollen part of an otherwise unthickened axis, and may survive for several seasons. The so-called pseudobulbs of epiphytic orchids are ordinary tubers. A typical corm is a vertical subterranean shoot of only a few swollen internodes, which bears a terminal inflorescence (Fig. 13A). After flowering, the corm shrivels and decays and is replaced by 1 or a few new corms that develop in an axillary position. *Crocosmia* (Fig. 13B) shows how weak the distinction between runner, tuber, and corm is, since here the corms may be borne at the end of typical runners.

A special type of corm is found in Colchicaceae (Buxbaum 1936, 1937, 1958). The corm has only 2 buds with a swollen internode between them (Fig. 14A,D,F), as is clearly seen in *Colchicum*. From the lower of the 2 buds, the new corm later develops; the upper bud is a reserve bud. There is a strong tendency for a the corm axis to extend laterally and eventually the 2 buds are found at the end of two fingerlike outgrowths. The finger-corm is well known in *Gloriosa* and related genera (see Fig. 14B,C,E,G).

Bulbs of Liliaceae. A closely related structure is found in the subterranean shoot of *Gagea*. Here again, the shoot bears 2 buds, but 1 scale leaf per bud forms the storage organ, the axis being

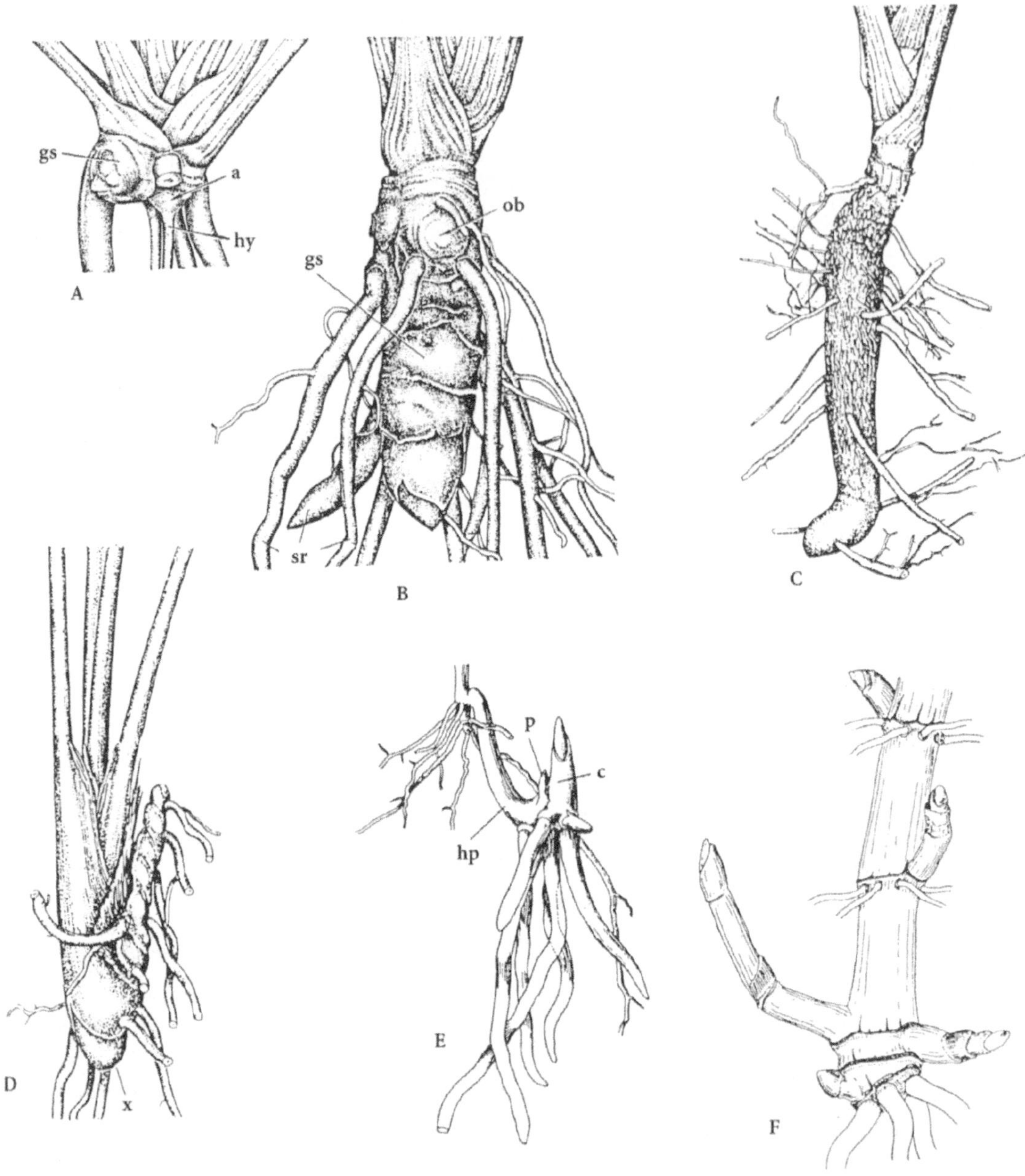

Fig. 15A–F. Geophilous shoots in monocotyledons. **A–C** *Cordyline australis*. **A** Seedling base with rhizome bud originating from seedling axis. **B** Base of an older seedling with well-developed rhizome (the parental seedling axis almost obliterated). **C** Young plant during establishment growth, with well-developed positively geotropic rhizome axially continuous with aerial stem. **D** *Rhopalostylis sapida*, young plant with positively geotropic primary axis; younger leaves breaking through a slit in the base of an older leaf at *x*. **E, F** *Tricyrtis hirta*. **E** Base of a young plant with geophilous lateral shoot, the hypopodium of the latter markedly elongated. **F** Deep buried shoot, its lowermost axillary shoots becoming runners, the upper ones growing negatively geotropic to form new aerial shoots. Not to scale. *a* Seedling axis; *c* cataphyll; *gs* geophilous shoot; *hy* hypocotyl; *hp* hypopodium; *ob* obliterated proximal seedling axis; *p* prophyll; *sr* shoot born root. (**A, B** Tomlinson and Fisher 1971; **C, D** Tomlinson and Esler 1973; **E** Troll 1943; **F** Buxbaum 1960)

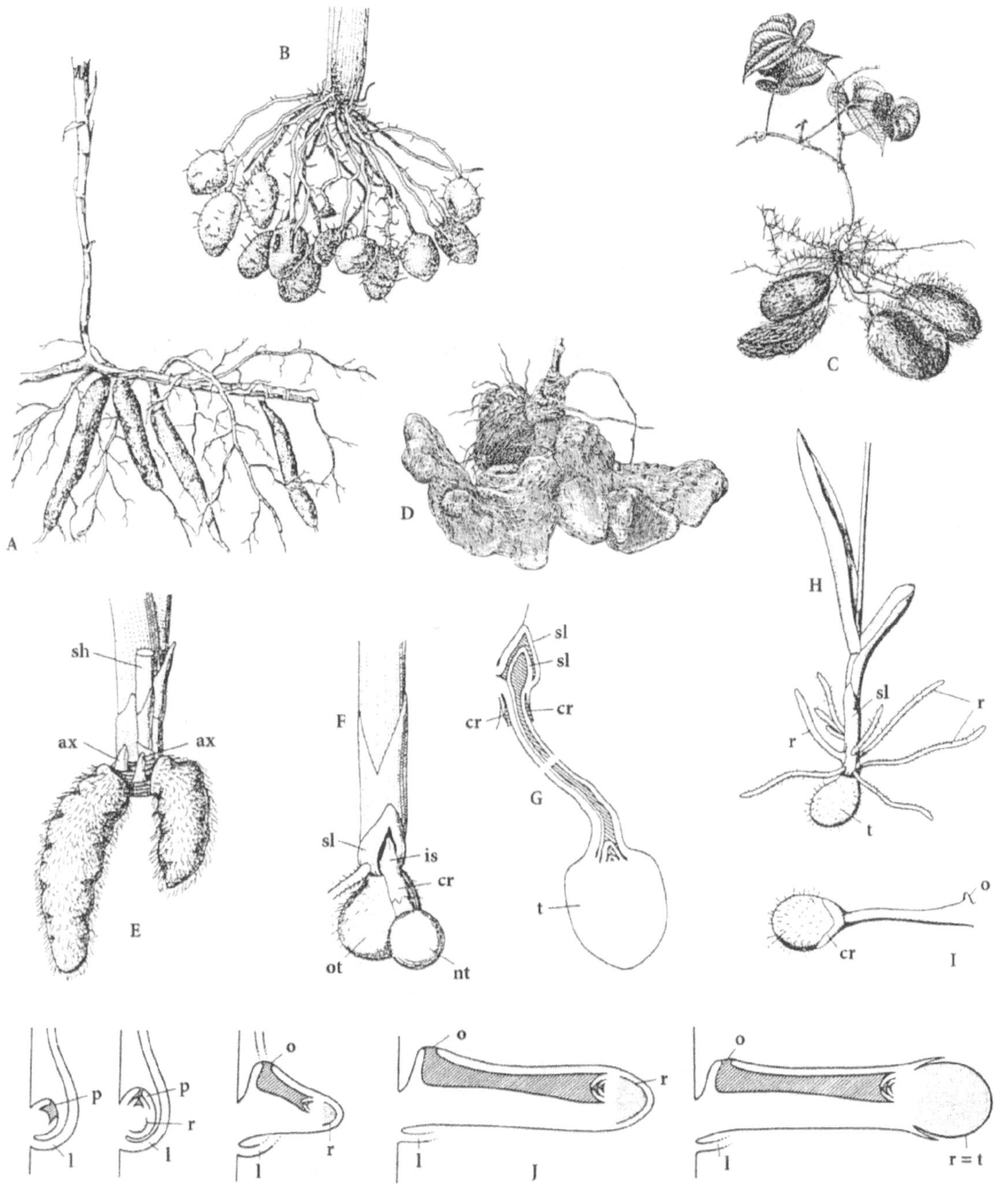

Fig. 16A–J. Root tubers in monocotyledons. **A** *Alstroemeria aurantiaca*, rhizome with swollen roots. **B** *Calathea macrosepala*, roots with distal tubers. **C, D** Tubers of *Dioscorea* (Yams). **C** *D. aculeata*. ***D*** *D. alata*. **E–J** Root tubers of orchids. **E** *Spiranthes autumnalis*, axis-born tuberous roots. **F–J** Root tubers developed at axillary buds. **F** *Ophrys insectifera*, base of a plant with old tuber (*ot*) and the developing innovation shoot (*is*) broken through its pherophyll (*sl*). The young tuber (*nt*) broken through its coleorhiza (*cr*). **G** *Serapias lingua*, innovation shoot in median longisection. After developing two scale leaves (*sl*), the axis of the axillary bud forming a long tube pushing the root tuber (*t*) into the soil. **H, I** *Herminium monorchis*. **H** Plant with slender roots (*r*) and a root tuber (*t*). **I** Runnerlike lateral elongation of the bud axis, forming with the prophyll a narrow tube. **J** Schematic explanation of the tube-forming process in *Herminium monorchis*. Some figures modified. Not to scale. *ax* Axillary bud; *cr* coleorhiza; *is* innovation shoot, l subtending leaf; *nt* new tuber; *o* opening of prophyll; *ot* old tuber; *p* prophyll; *r* root; *sh* inflorescence shoot; *sl* scale leaf; *t* tuber. (**A** Dahlgren et al. 1985; **B** Weber 1958; **C, D** Ochse 1931; **E–J** Troll 1943, 1948)

slender (Fig. 14H–L). Thus in *Gagea* generally 2 small bulbs are connected to the axis. Again, the axis tends to extend laterally, as can be seen in the double bulb of *Gagea* subgen. *Tribolbos* (Fig. 14L). In *Lilium* the bulb has another set of special features: the bulb scales are numerous and narrow, and the axis is relatively massive. In some species the bulb produces lateral outgrowths like thick, short rhizomes (Fig. 13J–M), as was first reported by Grove (1933) and studied in more detail by Baranova (1981). In *Lilium vollmeri* a sympodial sequence of bulbs may be formed, with each bulb surviving for a long period of time after flowering (Fig. 13L). This structure comes close to a sympodial rhizome. Finally, in *Tulipa* the bulb can develop the well-known dropper (Fig. 13G–I). This curious organ again is formed by an extensive lateral extension of the axis. In the course of this process an axillary bud is pushed deeper into the ground by a tubular structure. One half of the tube is formed by axis material, the other half comes from the base of the subtending leaf (Fig. 13H,I). The bud itself produces new storage scale leaves, thus forming a new bulb at the distal end of the dropper.

Geophilous Shoots. In the course of establishment growth, some monocotyledons develop shoots which grow directly downward into the soil. This phenomenon was described in detail by Tomlinson and Fisher (1971) and Tomlinson and Esler (1973) for *Cordyline australis*, *C. banksii*, and the palm *Rhopalostylis sapida*. In *Rhopalostylis*, as well as in some other palms, the primary axis grows directly downward until the apical bud is buried at a depth of about 15 cm (Fig. 15D). Thereafter the direction of growth is abruptly reversed, becoming erect. In *Cordyline* a laterally initiated axillary rhizome grows downward, thus forming an axis that is in direct continuity with the upright aerial shoot (Fig. 15A–C). The vertical, scale-bearing rhizome grows to a depth of 30–40 cm. The development of such geophilous axes is an effective means to overcome the limitations of trunk stability in large woody monocotyledons. Troll (1943) described a similar situation in the herbaceous genus *Tricyrtis* (Fig. 15E). Here an axillary branch grows downward from the seedling base with a remarkably elongated hypopodium.

Root Tubers. Tubers formed by roots may be developed in the proximal root region and then are mostly spindlelike as in *Alstroemeria* (Fig. 16A), *Asparagus*, or *Chlorophytum*. In other cases the swelling appears distally and is more or less ovoid or globose, as in several Marantaceae (Fig. 16B).

Of special interest are the root tubers of soil-rooting orchids (Fig. 16E–J). *Spiranthes* is the only known genus of the family that produces tuberous roots initiated on the main axis (Fig. 16E). In all other cases, root tubers develop from axillary buds. This may happen in two different ways. The first is exemplified by the genera *Ophrys* and *Serapias* (Fig. 16F,G). The axillary bud produces 2 scale leaves and an endogenous root, which soon breaks through its coleorhiza. Thereafter the bud axis begins to form a tube so that the apical meristem sinks to the base of the tube. The bud, together with the distal young root tuber, is pushed into the substrate by further elongation growth of the tube.

Herminium monorchis (Fig. 16H–J) is an example of the second way of forming a root tuber. Here the bud axis undergoes a strong lateral elongation, carrying the bud meristem and the root already covered by its coleorhiza away from the parental axis. The base of the prophyll participates in this process and forms a tube with the bud axis (Fig. 16I,J). The root breaks through its coleorhiza and develops a tuber only after the runner has reached its final length.

Tubers of the Dioscoreaceae. The famous *Testudinaria elephantipes* bears a large shoot tuber at ground level. Underground organs in *Dioscorea* include rhizomes and tubers of greatest diversity (Fig. 16C,D), the morphological interpretation of which is often controversial (Goebel 1905, Sharma 1974, von Teichman und Logischen et al. 1977). Although there is good evidence for the root nature of yam tubers (see Sharma 1974), a comprehensive study of the underground organs in this heterogeneous genus is urgently needed. Similarly, the axillary bulbils of aerial shoots of species of *Dioscorea* need further morphological study (Burkill 1960).

Selected Bibliography

Arber, A. 1918. The phyllode theory of the monocotyledonous leaf with special reference to anatomical evidence. Ann. Bot. 32: 465–501.

Arber, A. 1920a. Leaf-base phyllodes among Liliaceae. Bot. Gaz. 69: 337–340.

Arber, A. 1920b. On the leaf structure of certain Liliaceae, considered in relation to the phyllode theory. Ann. Bot. 34: 447–465.

Arber, A. 1921a. Leaves of the Helobiae. Bot. Gaz. 72: 31–38.

Arber, A. 1921b. The leaf structure of the Iridaceae, considered in relation to the phyllode theory. Ann. Bot. 35: 301–336.

Arber, A. 1922a. Leaves of the Farinosae. Bot. Gaz. 74: 80–94.
Arber, A. 1922b. On the nature of the "blade" in certain monocotyledonous leaves. Ann. Bot. 36: 329–351.
Arber, A. 1922c. On the leaf tips of certain monocotyledons. J. Linn. Soc. Bot. 45: 467–476.
Arber, A. 1923a. Leaves of the Gramineae. Bot. Gaz. 76: 374–388.
Arber, A. 1923b. On the "squamulae intravaginales" of the Helobiae. Ann. Bot. 37: 31–41.
Arber, A. 1924a. *Danae*, *Ruscus* and *Semele*: a morphological study. Ann. Bot. 38: 229–260.
Arber, A. 1924b. *Myrsiphyllum* and *Asparagus*: a morphological study. Ann. Bot. 38: 635–659.
Arber, A. 1925a. Monocotyledons. A morphological study. Cambridge: Cambridge University Press.
Arber, A. 1925b. On the "squamulae intravaginales" of the Alismataceae and Butomaceae. Ann. Bot. 39: 169–173.
Baranova, M.V. 1981. On the relationship of bulb structure with environment. Tr. Mosk. Ova. Ispyt. Prir. 56: 76–90. (in Russian)
Barykina, R.P., Gulenkova, M.A. 1985. Ontomorphogenesis, anatomy and the nature of leaf-like organs in *Asparagus sprengeri* (Asparagaceae). Bot. Žh. 70: 322–331. (in Russian)
Bechtel, H., Cribb, P., Launert, E. 1993. Orchideen-Atlas. Lexikon der Kultur-Orchideen. 3. Aufl. Stuttgart: E. Ulmer.
Bell, A.D. 1991. Plant form. An illustrated guide to flowering plant morphology. Oxford: University Press.
Bell, A.D., Tomlinson, P.B. 1980. Adaptive architecture in rhizomatous plants. J. Linn. Soc. Bot. 80: 125–160.
Bharathan, G. 1996. Does the monocot mode of leaf development characterize all monocots? Aliso 14: 271–279.
Bloedel, C.A., Hirsch, A.M. 1979. Developmental studies of the leaves of *Sagittaria latifolia* and their relationship to the leaf base theory of monocotyledonous leaf morphology. Can. J. Bot. 57: 420–434.
Bown, D. 1988. Aroids. Plants of the *Arum* family. London: Century Press.
Boyce, P.C., Nguyen Van Dzu 1995. *Pothos grandis* (Araceae: Pothoideae) described and validated and architectural notes on *Pothos* subgen. Pothos. Kew Bull. 50: 753–759.
Boyce, P.C., Poulsen, A.D. 1994. Notes on *Pothos insignis* (Araceae: Pothoideae). Kew Bull. 49: 523–529.
Brooks, A.K. 1993. Pandanaceae. In: Harden, G.J. (ed.) Flora of New South Wales, Vol. 4. Kensington: N.S.W. Univ. Press pp. 659–660.
Burkill, I.H. 1960. The organography and evolution of Diocoreaceae, the family of yams. J. Linn. Soc. Bot. 56: 319–412.
Buxbaum, F. 1936. Die Entwicklungslinien der Lilioideae. I. Die Wurmbaeoideae. Bot Arch. 38: 213–293.
Buxbaum, F. 1937. Die Entwicklungslinien der Lilioideae. II. Die systematische Stellung der Gattung *Gagea*. III. Die Lilioideae Engl. Bot. Arch. 38: 305–398.
Buxbaum, F. 1958. Der morphologische Typus und die systematische Stellung der Gattung *Calochortus*. Beitr. Biol. Pflanz. 34: 405–452.
Buxbaum, F. 1960. Beiträge zur Morphologie der Gattung *Tricyrtis*. Beitr. Biol. Pflanz. 35: 55–75.
Cabrera, A.L. (ed.) 1968. Flora de la Provincia de Buenos Aires, Vol. 1. Buenos Aires: INTA.
Candolle, A.-P. de 1828. Organographie végétale. Paris: Déterville.
Correll, D.C., Correll, H.B. 1982. Flora of the Bahama Archipelago. Vaduz: J. Cramer.
Dahlgren, R.M.T. et al. 1985. See general references.
Dengler, N.G., Dengler, R.F., Kaplan, D.R. 1982. The mechanism of plication inception in palm leaves: histogenetic observations on the pinnate leaf of *Chrysalidocarpus lutescens*. Can. J. Bot. 60: 2976–2998.
Dickson, A. 1885. On the occurrence of foliage leaves in *Ruscus* (*Semele*) *androgynus*; with some structural and morphological considerations. Trans. Proc. Bot. Soc. Edinb. 16: 130–149.
Dransfield, J. 1978. Growth forms of rain forest palms. In: Tomlinson, P.B., Zimmermann, M.H. (eds.) Tropical trees as living systems. Cambridge: Cambridge University Press.
Fisher, J.B. 1974. Axillary and dichotomous branching in the palm *Chamaedorea*. Am. J. Bot. 61: 1046–1056.
Fisher, J.B. 1976. Development of dichotomous branching and axillary buds in *Strelitzia* (Monocotyledoneae). Can. J. Bot. 54: 578–592.
Fisher, J.B. 1978. Leaf-opposed buds in *Musa*: their development and a comparison with allied monocotyledons. Ann. Bot. 65: 784–791.
Fisher, J.B., Dransfield, J. 1979. Development of axillary and leaf-opposed buds in rattan palms. Ann. Bot. (Lond.) 44: 57–66.
Gaisberg, E. von 1922. Zur Deutung der Monokotylenblätter als Phyllodien. Flora 115: 177–190.
Gibson, R.J.H. 1905. Axillary scales of aquatic monocotyledons. J. Linn. Soc. Bot. 37: 228–237.
Gilliland, H.B. 1971. Grasses of Malaya. In: Burkill, H.M. (ed.) A revised flora of Malaya, Vol. 3. Singapore: Government Printing Office.
Goebel, K. 1905. Morphologische und biologische Bemerkungen. 16. Die Knollen der Dioscoreen und die Wurzelträger der Selaginellen, Organe, welche zwischen Wurzeln und Sprossen stehen. Flora 95 (Ergänzungsband): 167–212.
Goebel, K. 1933. Organographie der Pflanzen. 3. Teil: Samenpflanzen. 3. Aufl. Jena: G. Fischer.
Grove, A. 1933. *Lilium michiganense*. Gard. Chron. 1933, 2: 120–121.
Hegelmaier, F. 1868. Die Lemnaceen. Leipzig: W. Engelmann.
Hirsch, A.M. 1977. A developmental study of the phylloclades of *Ruscus aculeatus* L.J. Linn. Soc. Bot. 74: 355–365.
Holttum, R.E. 1955. Growth habits of monocotyledons – variations on a theme. Phytomorphology 5: 399–411.
Hotta, M. 1971. Study of the family Araceae. General remarks. Jpn. J. Bot. 40: 269–310.
Hsu, C.-C. 1978. Gramineae (Poaceae). In: Flora of Taiwan, Vol. 5. Taipei: Epoch Publishing Co. pp. 373–783.
Irmisch, T. 1850. Zur Morphologie der monokotylischen Knollen- und Zwiebelgewächse. Berlin: G. Reimer.
Irmisch, T. 1853. Beiträge zur Biologie und Morphologie der Orchideen. Leipzig: A. Abel.
Irmisch, T. 1858a. Über einige Arten aus der natürlichen Pflanzenfamilie der Potameen. Abh. Naturwiss. Ver. Sachsen Thüringen (Halle) 2: 1–56.
Irmisch, 1858b. Über das Vorkommen von schuppen- oder haarförmigen Gebilden innerhalb der Blattscheiden bei monokotylischen Gewächsen. Bot. Zeitung (Berlin) 16: 177–179.
Irmisch, T. 1876. Ueber einige Pflanzen, bei denen in der Achsel bestimmter Blätter eine ungewöhnlich grosse Anzahl von Sprossanlagen sich bilden. Abh. Naturwiss. Ver. Bremen 5: 1–27.
Kahn, F., Granville, J.-J. de 1992. Palms in forest ecosystems of Amazonia. Berlin Heidelberg New York: Springer.
Kaplan, D.R. 1972. Comparative developmental analysis of the heteroblastic leaf series of axillary shoots of *Acorus calamus* L. (Araceae). Cellule 69: 251–290.
Kaplan, D.R. 1973. The problem of leaf morphology and evolution in the monocotyledons. Q. Rev. Biol. 48: 437–457.

Kaplan, D.R. 1975. Comparative developmental evaluation of the morphology of unifacial leaves in the monocotyledons. Bot. Jahrb. Syst. 95: 1–105.
Kaplan, D.R., Dengler, N.G., Dengler, R.F. 1982a. The mechanism of plication inception in palm leaves: problem and developmental morphology. Can. J. Bot. 60: 2939–2975.
Kaplan, D.R., Dengler, N.G., Dengler, R.F. 1982b. The mechanism of plication inception in palm leaves: histogenetic observations on the palmate leaf of *Rhapis excelsa*. Can J. Bot. 60: 2999–3016.
Kaussmann, B. 1955. Histogenetische Untersuchungen zum Flachsproßproblem. Bot. Stud. (Jena) 3: 1–136.
Knoll, F. 1948. Bau, Entwicklung und morphologische Bedeutung unifazialer Vorläuferspitzen an Monokotylenblättern. Oesterr. Bot. Z. 95: 163–193.
Koyama, T. 1978. Smilacaceae. In: Flora of Taiwan, Vol. 5: 110–137. Taipei: Epoch Publishing Co.
Liu, T.-S., Huang, T.-C. 1978. Dioscoreaceae. In: Flora of Taiwan, Vol. 5. Taipei: Epoch Publishing Co., pp. 99–109.
Liu, T.-S., Su, H.-J. 1978. Orchidaceae. In: Flora of Taiwan, Vol. 5. Taipei: Epoch Publishing Co., pp. 859–1137.
Mann, L.K. 1960. Bulb organisation in *Allium*: some species of section *Molium*. Am. J. Bot. 47: 765–771.
Martius, C.F.P. von 1823. Historia naturalis palmarum. Vol. 1: Palmas generatim tractat. Leipzig: T.O. Weigel.
Mayo, S.J., Bogner, J., Boyce, P.C. 1997. The genera of Araceae. The Trustees. Royal Botanic Gardens, Kew.
McClure, F.A. 1966. The bamboos. A fresh perspective. Cambridge (Mass.): Harvard Univ. Press.
Ochse, J.J. 1931. Vegetables of the Dutch East Indies. Buitenzorg: Archipel Drukkerij.
Peters, T. 1927. Über die Bedeutung der inversen Leitbündel für die Phyllodien-Theorie. Planta 3: 90–99.
Pütz, N. 1992. Das Verhältnis von Bewegung und Wurzelkraft bei Monokotylen. Beitr. Biol. Pflanz. 67: 173–191.
Rauh, W. 1950. Morphologie der Nutzpflanzen. 2. Aufl. Reprint 1994. Heidelberg: Quelle & Meyer.
Rauh, W. 1990. Bromelien. 3. Aufl. Stuttgart: Ulmer.
Rodionenko, G.J. 1961. Rod iris – *Iris* L. Moskau: Isdatelstwo AN SSSR. (in Russian)
Rohweder, O., Endress, P.K. 1983. Samenpflanzen. Morphologie und Systematik der Angiospermen und Gymnospermen. Stuttgart: Thieme.
Schlittler, J. 1953. Die Blütenartikulation und die Phyllokladien der Liliaceen organphylogenetisch betrachtet. Feddes Repert. 55: 154–258.
Schlittler J. 1960. Die Asparageenphyllokladien erweisen sich auch ontogenetisch als Blätter. Bot Jahrb. Syst. 79: 428–446.
Schoute, J.C. 1909. Über die Verästelung bei monokotylen Bäumen. II. Die Verästelung von *Hyphaene*. Rec. Trav. Bot. Néerl. 6: 211–232.
Sharma, O.P. 1974. Anatomy, origin and development of tuber of *Dioscorea glabra*. Phytomorphology 24: 297–305.
Speta, F. 1984. Zwiebeln – versteckte Vielfalt in einfacher Form. Linzer Biol. Beitr. 16: 3–44.
Stevenson, D.W. 1973. Phyllode theory in relation to leaf ontogeny in *Sansevieria trifasciata*. Am. J. Bot. 60: 387–395.
Strong, D.J., Ray, T.S. 1975. Host tree location bahavior of a tropical vine (*Monstera gigantea*) by skototropism. Science 190: 804–806.
Teichman und Logischen, I. von, Robbertse, P.J., Schijff, H.P. van der 1977. The subterranean intermediary organs of *Dioscorea cotinifolia* Kunth. 1. The germination, development, morphology and vegetative reproduction of the tuberous swollen and cylindrical intermediary organs. J. South Afr. Bot. 43: 41–56.
Telford, I.R.H. 1987. Proiphys. In: Flora of Australia, Vol. 45. Canberra: Austral. Governm. Publ. Service, pp. 376–379.
Tillich, H.-J. 1985. Keimlingsbau und verwandtschaftliche Beziehungen der Araceae. Gleditschia 13: 63–73.
Tillich, H.-J. 1992. Bauprinzipien und Evolutionslinien bei monocotylen Keimpflanzen. Bot. Jahrb. Syst. 114: 91–132.
Tillich, H.-J. 1995a. Seedlings and systematics in monocotyledons. In: Rudall, P.J., Cribb, P., Cutler, D.F., Humphries, C.J. (eds.) Monocotyledons: systematics and evolution. Royal Bot. Gardens Kew, pp. 303–352.
Tillich, H.-J. 1995b. Früchte, Samen und Keimpflanzen bei den Cyanastraceae Engler 1900 und einigen vermuteten Verwandten. Feddes Repert. 106: 483–493.
Tomlinson, P.B. 1960. Essays on the morphology of palms. I. Germination and the seedling. Principes 4: 56–61.
Tomlinson, P.B. 1961. Palmae. In: Metcalfe, C.R. (ed.) Anatomy of monocotyledons, Vol. 2. Oxford: Clarendon Press.
Tomlinson, P.B. 1970a. Dichotomous branching in *Flagellaria indica* (monocotyledons). J. Linn. Soc. Bot. 63, Suppl. 1: 1–14.
Tomlinson, P.B. 1970b. Monocotyledons – towards an understanding of their morphology and anatomy. Adv. Bot. Res. 3: 207–292.
Tomlinson, P.B. 1971. The shoot apex and its dichotomous branching in the *Nypa* palm. Ann. Bot. (Lond.) 35: 865–879.
Tomlinson, P.B. 1982. Helobiae (Alismatidae). In: Metcalfe, C.R. (ed.) Anatomy of monocotyledons, Vol. VII. Oxford: Clarendon Press.
Tomlinson, P.B., Esler, A.E. 1973. Establishment growth in woody monocotyledons native to New Zealand. N. Z. J. Bot. 11: 627–644.
Tomlinson P.B., Fisher, J.B. 1971. Morphological studies in *Cordyline* (Agavaceae). I. Introduction and general morphology. J. Arnold Arbor. 52: 459–478.
Tomlinson, P.B., Posluszny, U. 1977. Features of dichotomizing apices in *Flagellaria indica* (monocotyledons). Am. J. Bot. 64: 1054–1065.
Troll, W. 1935/1943. Vergleichende Morphologie der höheren Pflanzen. 1. Band. Vegetationsorgane. 1. Teil (1935), 3. Teil (1943). Berlin: Gebr. Borntraeger.
Troll, W. 1948. Allgemeine Botanik. Stuttgart: F. Encke.
Troll, W. 1954. Praktische Einführung in die Pflanzenmorphologie. 1. Teil. Jena: G. Fischer.
Troll, W. 1955. Über den morphologischen Wert der sogenannten Vorläuferspitze von Monokotylenblättern. Beitr. Biol. Pflanz. 31: 525–558.
Uhl, N.W., Dransfield, J. 1987. Genera palmarum. Lawrence (Kansas): Allen Press.
Valdes, B., Talavera, S., Fernandes-Galiano, E. (eds.) 1987. Flora Vascular de Andalucia Occidental, Vol. 3. Barcelona: Ketres Editora.
Velenovský, J. 1903. Zur Deutung der Phyllokladien der Asparageen. Beih. Bot. Centralbl. 15: 157–268.
Wagner, W.L., Herbst, D.R., Sohmer, S.H. 1990. Manual of the flowering plants of Hawai'i. Honolulu: Univ. of Hawaii Press.
Weber, H. 1950. Morphologische und anatomische Studien über *Eichhornia crassipes* (Mart.) Solms. Akad. Wiss. Lit. Mainz, Abh. Math.-Naturwiss. Kl. 1950, 6: 135–161.
Weber, H. 1958. Die Wurzelverdickungen von *Calathea macrosepala* Schum. und einigen anderen monokotylen Pflanzen. Beitr. Biol. Pflanz. 34: 177–193.

Weberling, F. 1958. Die Bedeutung blattmorphologischer Untersuchungen für die Systematik. Bot. Jahrb. Syst. 77: 458–468.
Weberling, F. 1975. Über die Beziehungen zwischen Scheidenlappen und Stipeln. Bot. Jahrb. Syst. 96: 471–491.
Weberling, F. 1989. Morphology of flowers and inflorescences. Cambridge: Cambridge University Press.
Yamashita, T. 1970. Eigenartige Wurzelanlage des Embryos bei *Lilaea subulata* Humb. & Bonpl. und *Triglochin maritimum* L. J. Fac. Sci. Univ. Tokyo, Sect. III. 10: 181–205.
Yamashita, T. 1972. Eigenartige Wurzelanlage des Embryos bei *Ruppia maritima* L. Beitr. Biol. Pflanz. 48: 157–170.
Yamashita, T. 1976. Über die Embryo- und Wurzelentwicklung bei *Aponogeton madagascariensis* (Mirbel) van Bruggen. J. Fac. Sci. Univ. Tokyo, Sect. III, 12: 37–64.

Epicuticular Wax Ultrastructure

W. Barthlott and I. Theisen

1. Introduction

Over the past decades, systematically relevant information on structure and composition of the plant cuticle based on SEM studies have been published (survey in Barthlott 1990). The cuticles of the majority of higher plants are covered with epicuticular wax secretions. They often cause a glaucous appearance. Epicuticular "waxes" occur throughout bryophytes, pteridophytes, gymnosperms, and angiosperms. They are chemically multicomponent mixtures, in which a particular compound or class(es) of compounds prevails, such as lipophilic substances like long-chain fatty acids (true waxes), as well as alkanes, ketones, exters, and cyclic compounds such as pentacyclic triterpenes, phytosterols, and flavonoids (Barthlott and Wollenweber 1981; surveys by Baker 1982; Jeffree 1986; Bianchi 1995). These substances occur usually as local projections of crystalline nature (Jeffree et al. 1975) and exhibit a high ultrastructural diversity. The genetic control of their biosynthesis has been studied by Wettstein-Knowles (1979, 1995). The dominating constituent(s) are responsible for the particular structure of the crystalloids. The correlation between micromorphology and chemistry has been proven by recrystallization experiments (Jeffree et al. 1975; e.g., Jeffree 1986; Jetter and Riederer 1994, 1995; Meusel 1997). Hence, the ultrastructure of epicuticular waxes manifests a visualized chemotaxonomy. Yet in cases of ultrastructural similarity and simultaneously uncertain relationships, it is necessary to analyze the chemical composition of the waxes in order to differentiate between ultrastructural homologies and convergencies.

A first detailed survey of monocotyledonous epicuticular waxes was published in two earlier papers (Barthlott and Frölich 1983; Frölich and Barthlott 1988). The following conclusions are based on the examination of some 400 genera and 750 species of 97 out of 102 monocotyledon families; they are interpreted in comparison with wax data of about 13000 species of angiosperms analyzed in the past two decades. The terminology used for different wax crystalloids is that of Barthlott et al. (1998).

2. Micromorphology of Monocotyledonous Waxes

The ultrastructure of monocotyledonous epicuticular waxes characterizes major groups of the monocotyledons. In particular, two wax types, more or less restricted to the monocotyledons, are of high systematic significance. The first wax type consists of parallel oriented platelets, which occasionally form striking patterns around idioblasts and stomata (Fig. 17A,B); it is called the Convallaria type, and is found in Nartheciaceae; in Liliales (Alstroemeriaceae, Colchicaceae, Luzuriagaceae, Smilacaceae); in Asparagales (Agavaceae, Anthericaceae, Aphyllanthaceae, Asparagaceae, Behniaceae, Convallariaceae, Eriospermaceae, Hemerocallidaceae, Herrereriaceae, Hyacinthaceae, Iridaceae, Johnsoniaceae, Lomandraceae, Melanthiaceae, Nolinaceae, Phormiaceae, Tecophilaeaceae); in Dioscoreales (Dioscoreaceae, Burmanniaceae). Also the Triuridaceae have parallel oriented platelets (Fig. 17A), which – consistent with molecular evidence (Chase et al. 1995) – supports their inclusion in Lilianae.

The second wax type – longitudinally aggregated rodlets, often developing thick crusts or wax chimneys around the stomata – is called Strelitzia type (Fig. 17D–F). It is recorded from many members of the Commelinanae, including all Zingiberales families, but also palms, Bromeliaceae, Eriocaulaceae, Cyperaceae, Juncaceae, Poaceae, and Typhaceae. The Convallaria type and Strelitzia type never occur in the same family and therefore support the distinction between Lilianae and Commelinanae. There is, however, a problem: in Pandanaceae and Velloziaceae, Strelitziatype wax has been found, although not consistently. Since these families are now placed in the Lilianae (Chase et al. 1995, 1995a), which is supported by data from three different plastid genes (M.W. Chase, pers. comm.), the possible rare occurrence

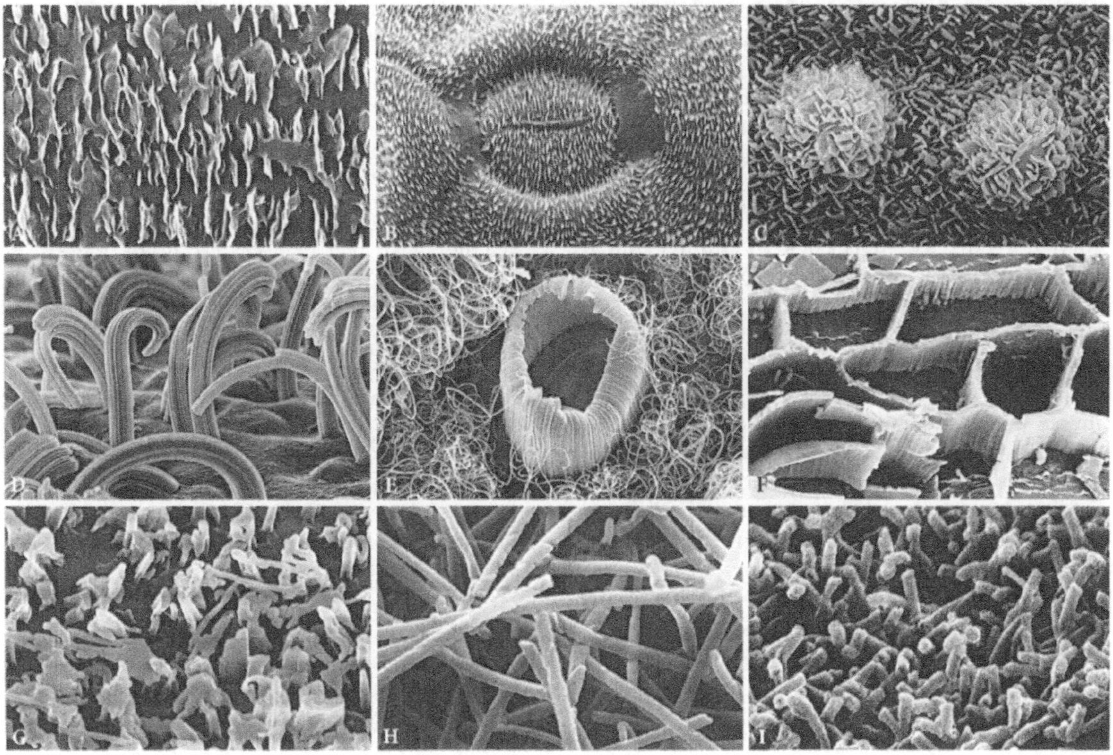

Fig. 17A–I. SEM micrographs of epicuticular wax types of monocotyledons. **A, B** Parallel platelets (Convallaria type). **A** *Sciaphila purpurea* (Triuridaceae), ×5750. **B** *Convallaria majalis* (Convallariaceae), ×1438. **C** Nonentire platelets: *Lysichiton camtschatcensis* (Araceae), ×5500. **D, F** Longitudinally aggregated rodlets (Strelitzia type). **D** *Phenakospermum guianense* (Strelitziaceae), ×950. **E** *Heliconia collinsiana* (Heliconiaceae), wax chimney around stoma, ×900. **F** *Pseudophoenix lediniana* (Palmae), massive wax projections above anticlinal cell walls, ×450. **G** Transversally ridged rodlets, *Fritillaria pallidiflora* (Liliaceae), ×4800. **H–I** Tubules. **H** Diketone tubules: *Secale cereale* (Poaceae), ×9500. **I** Nonacosan-10-ol-tubules (*Berberis* type): *Tulipa sylvestris* (Liliaceae), ×11750

of Strelitzia type wax in Lilianae needs to be explored further.

Two different kinds of wax tubules occur among the monocotyledons. Only with high-resolution SEM the differences in micromorphology or even the hollow structure can be recognized; formerly, tubules had often been misinterpreted as rodlets. On the one hand, there are often clustered, short tubules of the Berberis type (Fig. 17I) with a secondary alcohol (nonacosan-10-ol) as the main component (Jeffree 1986; Barthlott et al., subm.). This is the wax type characteristic of the Ranunculidae (Hennig et al. 1994; Barthlott and Theisen 1995) but it is also common in the other "lower eudicots" (sensu Soltis et al. 1997) (Canellaceae, Cercidiphyllaceae, Daphniphyllaceae, Eupteleaceae, Hamamelidaceae, Nelumbonaceae, Proteaceae, Trochodendraceae, Winteraceae) and characterizes almost all gymnosperms (Wilhelmi and Barthlott 1997). Within the monocotyledons this type has been found only in members of the Alliaceae, Corsiaceae, Hostaceae, and Liliaceae. The second tubule type, diketone tubules (Fig. 17H) (main component: β-diketones and hydroxy-β-diketones) (Jeffree 1986; Wettstein-Knowles 1995) is found solely within the Poaceae, but has also a sporadic occurrence in some dicotyledonous families such as Myrtaceae, Caryophyllaceae, Buxaceae, and Asteraceae.

Within the monocotyledons, transversally ridged rodlets (Fig. 17G) are confined to a few taxa within the Lilianae (Liliaceae, Luzuriagaceae, Calochortaceae, Smilacaceae, Philesiaceae and Aloaceae, Amaryllidaceae, Asphodelaceae, Dracaenaceae, Eriospermaceae, Lomandraceae, Xanthorrhoeaceae), Alismatanae (Araceae, Butomaceae), and Commelinanae (Haemodoraceae).

Although this wax type is widespread but sparsely distributed among angiosperms, it well characterizes the Magnoliidae s.str. (Aristolochiales, Magnoliales, Laurales) and distinguishes them from the Ranunculiflorae (see above).

Nonentire, irregular wax platelets (Fig. 17C) are widespread among angiosperms and therefore of limited systematic importance. Within the monocotyledons they can occur also in families that are characterized by other wax types such as the Convallaria or Strelitzia types.

Thin wax films forming continuous layers are present in all taxa and not always detectable by SEM. They are characteristic of almost all submerse taxa, especially of the Alismatales. Under wax micromorphological aspects the Alismatales and *Acorus* – according to molecular data representing the most basal branch of monocotyledons – are indistinguishable from the Piperales and Nymphaeales, which belong to the most basal angiosperms (Soltis et al. 1997). Therefore wax data do not reject an "Acoranean origin" of monocots (Chase et al. 1993; Duvall et al. 1993).

Selected Bibliography

Baker, E.A. 1982. Chemistry and morphology of plant epicuticular waxes. In: Cutler, D.F., Alvin, K.L., Price, C.E. (eds.) The plant cuticle. London, New York: Academic Press, pp. 139–165.

Barthlott, W. 1990. Scanning electron microscopy of the epidermal surface in plants. In: Cutler, D.F., Alvin, K.L., Price, C.E. (eds.) The plant cuticle. London: Academic Press, pp. 69–94.

Barthlott, W., Frölich, D. 1983. Mikromorphologie und Orientierungsmuster epicuticularer Wachs-Kristalloide: ein neues systematisches Merkmal bei Monokotylen. Plant Syst. Evol. 142: 171–185.

Barthlott, W., Theisen, I. 1995. Epicuticular wax ultrastructure and classification of *Ranunculiflorae*. In: Jensen, U., Kadereit, J. (eds.) Systematics and evolution of the Ranunculiflorae. Plant Syst. Evol., Suppl. 9: 39–45.

Barthlott, W., Wollenweber, E. 1981. Zur Feinstruktur, Chemie und taxonomischen Signifikanz epicuticularer Wachse und ähnlicher Sekrete. Trop. Subtrop. Pflanzenwelt 32: 35–97.

Barthlott, W., Neinhuis, C., Cutler, D., Ditsch, F., Meusel, I., Theisen, I., Wilhelmi, H. 1998. Classification and terminology of plant epicuticular waxes. Bot. J. Linn. Soc. 126: 237–260.

Barthlott, W., Theisen, I., Ditsch, F., Meusel, I., Neinhuis, C. Ultrastructure and chemistry of epicuticular wax crystals: systematic significance for the Classification of vascular plants. (submitted)

Bianchi, G. 1995. Plant waxes. In: Hamilton, R.J. (ed.) Waxes: chemistry, molecular biology and functions. Dundee, Scotland: The Oily Press, pp. 177–222.

Chase, M.W. et al. 1993, 1995, 1995a See general references.

Duvall, M.R. et al. 1993. See general references.

Frölich, D., Barthlott, W. 1988. See general references.

Hennig, S., Barthlott, W., Meusel, I., Theisen, I. 1994. Mikromorphologie der Epicuticularwachse und die Systematik der Magnoliidae, Ranunculidae und Hamamelididae. Trop. Subtrop. Pflanzenwelt 90: 1–60.

Jeffree, C.E. 1986. The cuticle, epicuticular waxes and trichomes of plants, with references to their structure, function and evolution. In: Juniper, B.E., Southwood, R. (eds.) Insects and plant surfaces. London: Edward Arnold, pp. 23–64.

Jeffree, C.E., Baker, E.A., Holloway, P.J. 1975. Ultra-structure and recrystallization of plant epicuticular waxes. New Phytol. 75: 539–549.

Jetter, R., Riederer, M. 1994. Epicuticular crystals of nonacosan-10-ol: In-vitro reconstitution and factors influencing crystal habits. Planta 195: 257–270.

Jetter, R., Riederer, M. 1995. In-vitro reconstitution of epicuticular wax crystals: formation of tubular aggregates by long chain secondary alkanediols. Bot. Acta 108: 111–120.

Meusel, I. 1997. Rekristallisationstudien ausgewählter epicuticularer Wachse. PhD Thesis, Rhein. Friedrich-Wilhelms-Universität Bonn.

Soltis, D.E. et al. 1997. See general references.

Wettstein-Knowles, P. von 1979. Genetics and biosynthesis of plant epicuticular waxes. In: Appelqvist, L.A., Liljenberg, L. (eds.) Advances in the biochemistry and physiology of plant lipids. Amsterdam: Elsevier, pp. 1–26.

Wettstein-Knowles, P. von 1995. Biosynthesis and genetics of waxes. In: Hamilton, R.J. (ed.) Waxes: chemistry, molecular biology and functions. Dundee, Scotland: The Oily Press, pp. 91–130.

Wilhelmi, H., Barthlott, W. 1997. Mikromorphologie der Epicuticularwachse und die Systematik der Gymnospermen. Trop. Subtrop. Pflanzenwelt 96: 1–49.

Systematics and Evolution

K. KUBITZKI, P.J. RUDALL, and M.C. CHASE

1. A Brief History of Monocot Classification

John Ray in his *Historia Plantarum* (1686–1704) was the first botanist to recognize cotyledon number as a useful means of subdividing flowering plants (Bancroft 1914). Although Linnaeus in his *Philosophia Botanica* (1751) did not explicitly mention this distinction between monocotyledons and dicotyledons, it was taken up by all later botanists. In most angiosperm classifications from Jussieu (1789) to Engler and Prantl (1887–89), monocotyledons were arranged in a position intermediate between nonangiosperms and dicotyledons, indicating a lower level of organization for monocotyledons than dicotyledons. However, with the spread of phylogenetic thinking, a ranalean origin for the monocotyledons was suggested (e.g., Bessey 1893), and Wettstein's (1901–1907) textbook was the first to acknowledge a derived position for monocotyledons by treating them after the dicotyledons.

When Engler (1892) justified his new classification of monocotyledons, family concepts in it had largely been consolidated, with the notable exception of Liliaceae. Indeed, in the second edition of *Die natürlichen Pflanzenfamilien* (Engler 1925 →), Krause's (1930) account of the Liliaceae comprised no less than 233 genera, divided into 12 subfamilies and numerous tribes, while at about the same time, Hutchinson (1934) proposed a new scheme of monocot classification in which he dismantled the Liliaceae concept. He created smaller and therefore more homogeneous families, based upon a stronger reliance on inflorescence and other morphological characters and paying less attention to ovary position which until then had been considered highly conservative.

Comparative studies in fields such as morphology, anatomy, embryology, palynology, phytochemistry, etc., such as Buxbaum's (1937) painstaking morphological analysis of Liliaceae s. l., paved the way towards a further reorganization of Krause's (1930) catch-all Liliaceae. This was achieved by Huber (1969), who, on the basis of seed-coat characters and an integration of other evidence, proposed a radically novel taxonomic scheme for the Liliiflorae (= Liliaceae sensu Krause and related families). Several new orders and numerous new families replaced the Liliaceae, and justice was done to the numerous – mostly austral – isolated lineages hidden in the traditional Liliaceae. These concepts were incorporated in classifications such as that of Takhtajan (1980), underpinned factually by Dahlgren and Clifford (1982) and conceptually by Dahlgren and Rasmussen (1983), and eventually resulted in the influential classification of Dahlgren et al. (1985). Since then, the principal aspects of this Huber-Dahlgren classification have been adopted by leading angiosperm systematists such as Takhtajan (1987, 1997), Thorne (1992), and others. The position of Cronquist (1981, 1988), who retained an all-embracing family Liliaceae in the sense of Krause (1930), became untenable. However, problems remained (and still remain) in monocot classification, since in many cases consistent gross morphological characters that effectively defined groups above the species and genus levels were sparse or even completely absent, especially in Lilianae, in which many families have simple unmodified flowers with 6 tepals and 6 stamens.

In the decade after Dahlgren, the study of monocotyledons has received new impetus from molecular and combined morphological/molecular studies, as shown in the volumes documenting the Kew monocot symposium of 1993 (Rudall et al. 1995) and subsequent papers (e.g., Rudall and Chase 1996, Fay and Chase 1996, Chase et al. 1996, Rudall et al. 1997). Some micromorphological characters, such as pollen and ovule structure and development, show close congruence with the topologies produced by analysis of molecular data from *rbc*L, leading to new insights into relationships and family circumscriptions.

2. Origin of Monocotyledons

Since the time of Hallier (1905) and Bessey (1915), the origin of monocotyledons from a ranalean ancestor has been widely accepted (for a differing view see Burger 1981). On the basis of our increasing knowledge of the ranalean dicotyledons and their characters, Huber (1977) and others stressed the idea that the primary division in angiosperms is the one that separates the ranalean dicots plus the monocots from the remaining dicots ("eudicots"), a concept in part supported by molecular studies (e.g., Chase et al. 1993), although Ranunculales themselves are eudicots. The distinction between dicotyledons and monocotyledons is sometimes blurred, as Huber (1977) and Dahlgren et al. (1985) emphasized, by "dicot characters" appearing in the monocotyledons, and "monocot characters" occurring in dicotyledons. Dicot characters include an allegedly eustelic stem organization (Dioscoreaceae, Trichopodaceae), leaves with broad, net-veined blades (several groups, e.g., Dioscoreaceae and Convallariaceae), seed coats with a crystalliferous testa (Dioscoreaceae), embryos with terminal plumule (Dioscoreaceae, Taccaceae and others), seedlings with robust, long-lived and branched primary roots (*Agave*, *Cordyline*, palms, *Strelitzia*, *Yucca*; Tillich 1995), cellular endosperm formation (*Acorus*: Grayum 1987, 1991), and successive microsporogenesis (some Asparagales, Arecaceae, and others: Rudall et al. 1997).

Character states that are common among monocotyledons but have a restricted representation in dicotyledons include trimerous flowers (e.g., *Cabomba, Hedyosmum, Lactoris*, and *Saruma*: Endress 1987, 1990, 1994.), adaxial prophylls (Aristolochiaceae, some Annonaceae, Lactoridaceae), atactostelic stem organization (Nymphaeaceae), Helobial endosperm formation (*Cabomba*), P2-subtype sieve-element plastids with cuneate protein crystalloids (Aristolochiaceae: Behnke 1976, 1981, 1995), and sulcate pollen. Among dicotyledons, these characters are largely restricted to magnolialean families, particularly Piperales, Nymphaeaceae, and Aristolochiaceae.

The scattered distribution of dicotlike characters in monocotyledons such as *Dioscorea* and *Tacca* led various authors, including Suessenguth (1921), to suggest a polyphyletic origin of monocotyledons from the magnoliids. However, analyses of both morphological and molecular data refute this theory. There remain some highly consistent synapomorphies for the monocotyledons, particularly P2-subtype sieve-element plastids (Behnke 1976, 1981, 1995) (Fig. 18), cotyledon number, and absence of vascular cambium.

The monophyly of the monocotyledons is well supported by several analyses, both morphological (Loconte and Stevenson 1991) and molecular, employing sequence data from *rbc*L (Chase et al. 1993, Qiu et al. 1993), rRNA (Hamby and Zimmer 1992, Soltis et al. 1997), and plastid DNA restriction sites (Davis 1995). These often indicate a group of primarily herbaceous taxa ("paleoherbs") including families such as Aristolochiaceae, Piperaceae, Chloranthaceae, and Nymphaeaceae as the closest relatives of the monocotyledons. However, all these groups are polymorphic with respect to plesiomorphic angiosperm characters, and so far no *single* extant dicotyledonous family has been identified as the sister to the monocotyledons; indeed, their sister group may be a large clade containing many families, perhaps all magnoliids except for Nymphaeaceae and Winteraceae. Chase et al. (1995) and others have argued for inclusion of Magnoliales and Laurales along with paleoherbs for outgroup comparison. In many most parsimonious trees, either paleoherbs or Magnoliales and/or Laurales are the sister group to the monocotyledons. Since Magnoliales/Laurales do not include herbaceous groups, the question whether paleoherbs (including monocotyledons) were primarily herbaceous or diverged at an early stage from woody Magnoliales is not yet resolved, although there is little doubt that the immediate ancestor of the monocotyledons was herbaceous.

It is futile to search among extant dicotyledons for a monocot ancestor, because monocotyledons evolved from early angiosperms. Attempts to estimate the time of divergence by applying a molecular clock have yielded widely divergent results between a calculated age of $319 + 33$ million years ago (Ma) (Martin et al. 1989) and 200 Ma (Wolfe et al. 1989). Confidence in these estimates may be lessened, however, by the considerable differences in substitution rate between various plant groups (although most studies factor out this effect by performing relative rate tests and eliminating those groups that appear to be significantly faster or slower than average). Between grasses and palms, for instance, this amounts to a fivefold difference, a value consistent with the differences in generation time (Gaut et al. 1992). Palynological evidence from the lower Cretaceous of 140 Ma (Hauterivian of Israel, Brenner 1996) indicate the latest date for the dicot/monocot split, which, of course, may extend back further.

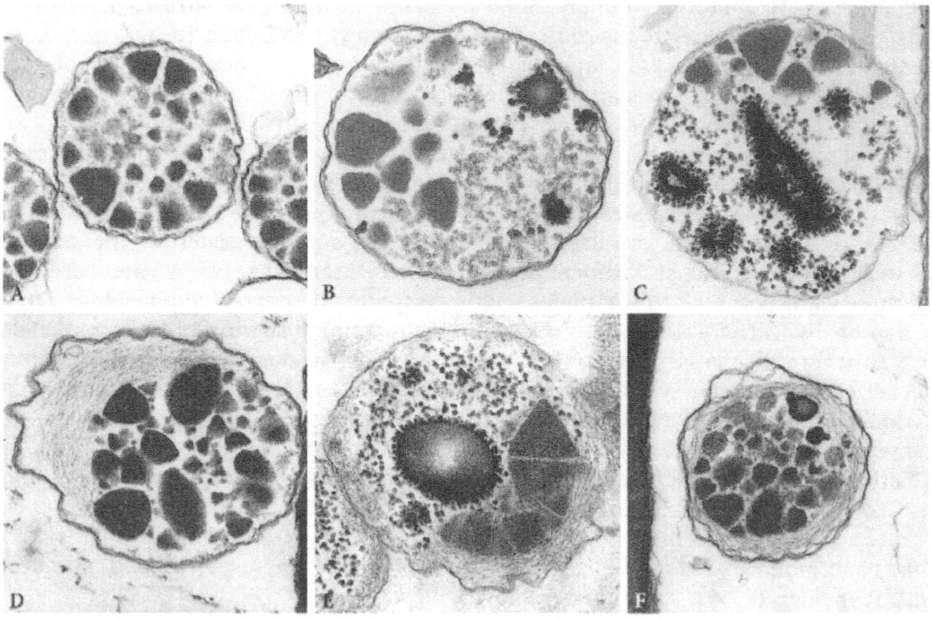

Fig. 18A–F. Sieve-element plastids of monocotyledons invariably are characterized by cuneate protein crystals (P2c type, Behnke 1981). This type is restricted to monocots (exception: *Asarum*, Aristolochiaceae). Although the pure P2c type is the most frequent form, additionally, starch grains (P2cs, in Dioscoreaceae, Araceae, Zingiberales, and palms) and/or peripheral protein filaments (P2cfs, in Musaceae and *Monstera*, P2cf, in Lomandraceae, Asphodelaceae) may be present. **A** *Acorus calamus*, P2c. **B** *Gymnostachys anceps*, P2cs. **C** *Caladium bicolor*, P2cs. **D** *Freycinetia cumingiana*, P2cf. **E** *Musa sumatrana, P2cfs.* **F** *Monstera deliciosa*, P2cfs. TEM ×2000. (Original H.-D. Behnke)

3. Ancestral Characters and Basal Taxa

The lack of a convincing dicot sister taxon to the monocotyledons makes outgroup comparison difficult in morphological analyses and presents problems in rooting and identification of "basal" monocotyledons. Although many authors previously considered Dioscoreales to be the basal monocot taxa on the basis of their dicotlike net-veined leaves and other characters (see below), it has been demonstrated that reticulate venation has evolved several times in monocotyledons, probably in response to shady woodland understory habitats, for example in the orders Asparagales (e.g., Convallariaceae and Vaniloideae of Orchidaceae), Liliales (e.g., Trilliaceae, Smilacaceae, and related taxa) and Dioscoreales (e.g., Dioscoreaceae). Several recent molecular analyses (Duvall et al. 1993, Nadot et al. 1995, Chase et al. 1995, 1995a, Davis et al. 1998) have placed *Acorus* as sister to the rest of the monocotyledons, and this view (the Acoranean hypothesis) is now widely held. However, there are few, if any, morphological characters supporting a basal position for either *Acorus* or, indeed, for that of any other monocotyledons – but, similarly, there is little to conflict with this position, since *Acorus* is an isolated taxon with many uniquely derived character states (Grayum 1987, Rudall and Furness 1997). Other taxa that were strongly supported in the molecular and combined analyses as "primitive" monocotyledons include aquatic and moisture-loving groups: Araceae, Alismatales, and some of the genera here included in Nartheciaceae (*Harperocallis, Tofieldia,* and *Pleea*). Hutchinson (1934) also considered Alismatales as basal monocotyledons.

The hypothesis of an aquatic origin for monocotyledons was advocated by Jeffrey (1917), who considered a shift into an aquatic habit a feasible explanation for the loss of cambial activity, as in the aquatic dicotyledon Nymphaeaceae. In defence of this theory, Cronquist (1988), largely based upon observations by Cheadle (1942, and numerous later publications summarized by Wagner 1977), argued that the loss of the vascular cambium eliminated all vessels with the exception of those developed in the primary tissue and left only vestigial vessels in the roots. In many monocotyledons, vessels are restricted to the roots. Vessel elements in roots of species that lack vessels elsewhere in the plant body are usually less specialized than those of species in which vessels are also widespread in stems and/or leaves. The conclusion was that vessels spread from roots into stems, leaves, and inflorescence axes, a progression paralleled by a change from vessels first with end walls with numerous long scalariform perforations to those with short scalariform plates with few bars, and finally with simple perforations. No species is known in which vessel perforations are more specialized in the shoot system than in the roots. Carlquist and Schneider (1997) found that vessel element morphology does not conflict with a basal position for *Acorus* in monocotyledons but could also be compatible with Cheadle's (1942) hypothesis of an independent origin of vessels in dicotyledons and monocotyledons.

Les and Schneider (1995) summarized these ideas, but considered a connection between Nymphaeaceae- and Alismatales-like forms improbable. However, the aquatic origin theory has gained new ground, since *Acorus* and other aquatics and helophytes such as Arales and Alismatales are now widely considered to be the basal monocotyledons. This is supported by the presence of mucilage-filled ovaries in *Acorus* and most Araceae (Rudall et al. 1998), since, as Endress (1990, 1995) demonstrated, mucilage-filled ovaries (i.e., mucilage-mediated pollen tube transmission) are found in magnoliids that occur in aquatic or moist habitats, including Nymphaeaceae, Cabombaceae (Endress, 1990, 1994), and Chloranthaceae (Endress 1987).

Huber (1969) emphasized the large number of dicotlike characters in Dioscoreales, such as the crystalliferous inner epidermis of the testa, lateral cotyledon, and terminal plumule on the embryo (also in Taccaceae and other families), circular arrangement of vascular bundles (also in other taxa), frequently opposite leaves with reticulate venation and differentiation into pedicel and blade, and elongate connective tips of *Stenomeris* and *Trichopus* (at that time included by Huber in Dioscoreaceae). However, now it appears that most of these characters are homoplasious within monocotyledons. Although Huber (1969) noted several "advanced" characters in common with other Lilianae (endosperm of reserve cellulose, steroidal saponins, and chelidonic acid), Dahlgren et al. (1985) tentatively placed Dioscoreales and consequently Lilianae as the basal group of monocotyledons.

The position of *Acorus* mirrors that of *Ceratophyllum*, which in analyses based on the plastid genes has been found to occupy a sister position to all remaining angiosperms (Les et al. 1991, Chase et al. 1993, Duvall et al. 1993). However, rRNA sequence analyses by Hamby and Zimmer (1992) and Soltis et al. (1997) placed these two genera in nearly reversed positions: *Ceratophyllum* as sister to the monocots, and *Acorus* branching off at the base of the angiosperms between Nymphaeaceae-Piperales and the remaining angiosperms. This vacillating position may be due to the long isolation of these relictual genera, which in the gene trees lie on long branches that are reputedly prone to be "attracted" into basal positions (Sytsma and Baum 1996). Thus, both *Ceratophyllum* and *Acorus* may represent ancient, largely extinct clades, but only further work can show whether they are the sister of all angiosperms and monocotyledons, respectively.

Important aspects of the vegetative organization of the monocot plant body are the presence of a primary thickening meristem near the apex (Rudall 1991), sympodial habit (Holttum 1955), and the absence of a bidirectional vascular cambium. Despite this, monocotyledons show a remarkable diversity in habit. Tomlinson (1990, 1995) found it hard to relate the monocot vascular system to that of any modern dicotyledon. For similar reasons, Burger (1981) argued for an evolutionary primacy of monocotyledons, a concept that has no support in the light of present knowledge (see above). Unifacial equitant leaves, found in basal monocotyledons such as *Acorus, Gymnostachys, Harperocallis, Narthecium, Pleea, Tofieldia*, and many others, such as Iridaceae, *Xeronema*, Xyridaceae, and Anarthriaceae, are a uniquely monocotyledonous feature. This character thus cannot be assessed as plesiomorphic by outgroup comparison. Since many taxa with this trait are near the base of the monocotyledons or sister to all other major monocotyledonous clades, the Acoranean hypothesis (Duvall et al. 1993) gains momentum (Chase et al. 1995). This hypothesis also implies that other monocotyledons with dicotlike features (e.g., *Dioscorea, Tacca*) must have acquired them independently, and not directly inherited them from the magnoliid paleoherbs. Having "freed" the issue of monocot orgins from outgroup comparisons, one can imagine monocotyledons to be derived from any type of magnoliid ancestor, which is compatible with many of the results from the DNA sequence studies that place them as sister to a large magnoliid clade composed of both woody and herbaceous taxa.

The quest for the primitive monocot flower has been an issue of continuous debate. As early as 1892, Engler directed attention to the frequent occurrence of multistaminate and multicarpellate flowers in numerous monocotyledons such as Pandanaceae, Alismataceae, Hydrocharitaceae, Gramineae-Bambusoideae, and palms, which, according to him, pointed to a descent of monocotyledons from polyandric/polycarpic dicots. However, Engler[1] rejected the similarity of the polymerous flowers in Alismataceae and Ranunculaceae as suggestive of a relationship between them because to him the anatomical gap between the two groups was unsurmountable. Nevertheless, this similarity has been used by many, among others Hutchinson (1934: 9), to suggest a close relationship between monocotyledons and Ranunculaceae, and specifically between Butomaceae and Ranunculaceae-Helleboroideae, and between Alismataceae and Ranunculaceae-Ranunculoideae. Developmental studies by Singh and Sattler (1972; see also Sattler and Singh 1978) dismissed this concept by demonstrating that floral organs in polyandric and polycarpic Alismatales are never spirally arranged, but always follow a (basically) trimerous pattern. A noteworthy trait of alismatalean flower development is the appearance of three "paired" stamen groups opposite the outer tepals, followed by an inner alternating trimerous stamen whorl. Since this A 6 + 3 pattern is found in several ranalean taxa but in monocotyledons is restricted to Alismatales, it was viewed by Erbar and Leins (1994) as linking these groups. However, the multistaminate state is taxonomically widespread among monocotyledons, and may be related to special demands of the pollination system. Nevertheless, in view of the systematic significance of *Acorus* it is difficult to think of the basal monocot flower other than in terms of a conventional trimerous and pentacyclic flower. Endress (1995) comprehensively reviewed floral structures in monocotyledons.

The evolution of monocotyledons from dicotlike progenitors implies the possession in the earliest monocots of "true" brachyblastic flowers of limited growth. The basic homology of flowers between monocots and dicots is also supported by findings on homoeotic gene action in maize (Mena et al. 1996). Here, two genes with overlapping activities, *zag 1* and *zum 2*, control the differentiation of stamens + carpels ("region C"). These genes exhibit a high degree of sequence identity with the *ag* gene that functions in a comparable way in dicotyledons.

[1] In an annotation to the work of Buchenau (1903: 66).

4. Major Clades of the Monocotyledons

A new higher-level classification of monocotyledons is not the primary aim of this book. This outline is therefore presented as the current state of our understanding of broader monocotyledon relationships (Fig. 19). Although the composition and relationships of some groups are well resolved and clear, some remain uncertain (e.g., Typhales; see below) and will inevitably change as data are added and new analyses undertaken.

a) Acoranae

The work of Tillich (1985) and particularly Grayum (1987) have brought into focus the isolation of *Acorus* from Araceae, leading to the resurrection of Acoraceae as a separate family. Its position in the molecular tree has revived interest in its allegedly dicotyledonous characters, almost all of which occur in other monocotyledons (Rudall and Furness 1997). Two of these characters, anther wall formation and cellular endosperm, both require further review. Several characters indicate that *Acorus* is a monocotyledon, but provide little further evidence of relationships; these are P2-subtype sieve-element plastids (Behnke 1995), seedling structure (Tillich 1995), and unifacial leaves. The perisperm of *Acorus* is of dermal derivation from the entire nucellar epidermis and not homologous with the perisperm in other angiosperms (Rudall 1997, Rudall and Furness 1997).

b) Nartheciaceae

Although here circumscribed as a single family (see Tamura, this Vol. p. 389/90 for an explanation of his concept of a primitive core group), Nartheciaceae are polyphyletic in analyses of existing morphological and molecular data (e.g., Chase et al. 1995, 1995a), and probably include three separate assemblages, with one group (including *Tofieldia* and *Pleea*) sister to Alismatanae, and another (including *Aletris*,

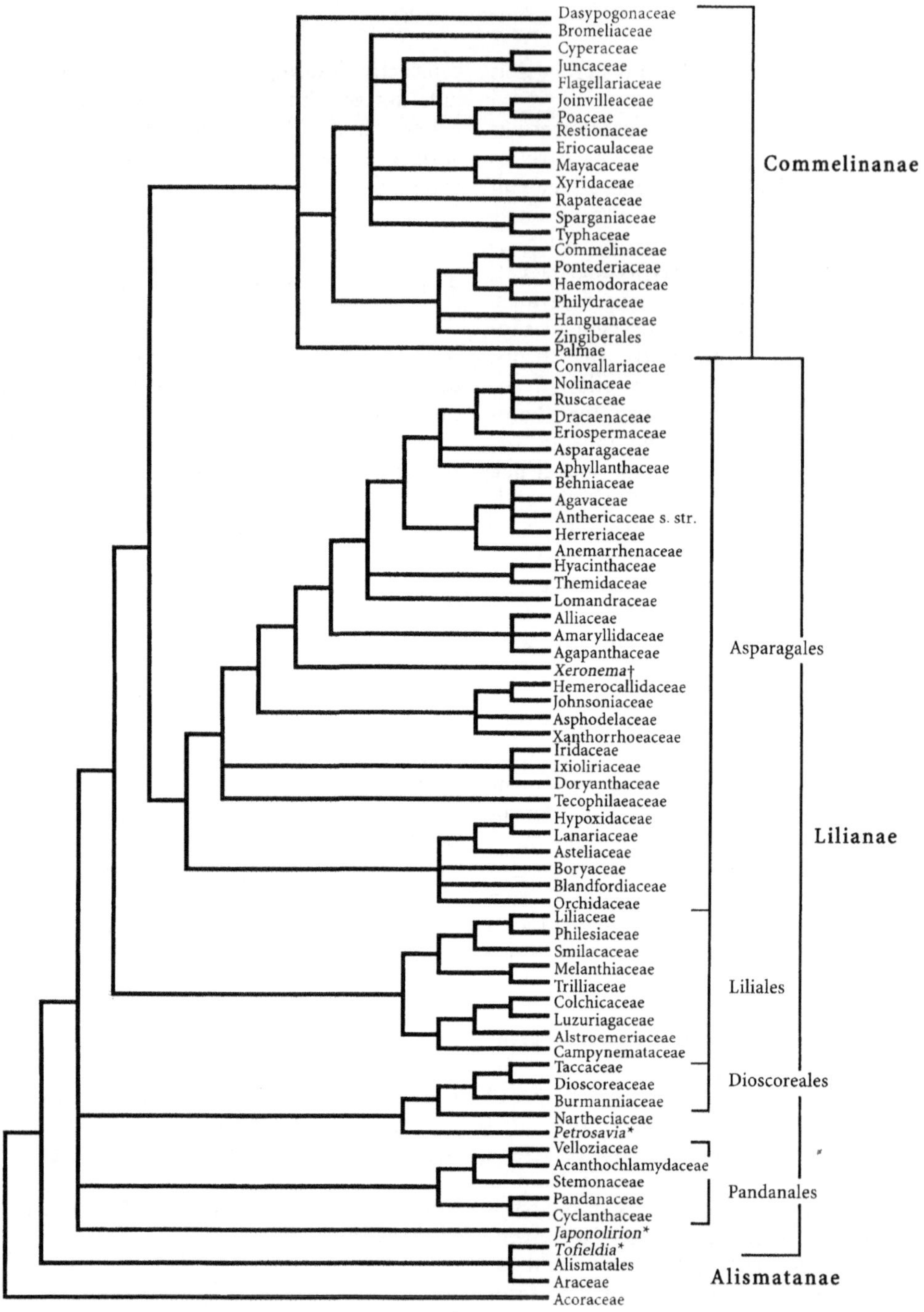

Fig. 19. Consensus phylogenetic tree of the monocotyledons summarizing DNA evidence for relationships. This is based largely on the widely sampled *rbc*L plastid locus (Chase et al. 1993; Duvall et al. 1993; Chase et al. 1995), but also includes results from plastid *atp*B (Chase, Hahn, and Sullivan, unpubl.), plastid trnL-F for Liliales, Asparagales, and Commelinales (Chase, Fay, Hung et al., unpubl.), and nuclear ribosomal 18S (Soltis et al. 1997), as well as studies using all of these data simultaneously. Acoraceae are shown as sister to the rest in accord with all plastid gene studies, whereas the 18S tree places it among the dicots; in combined analyses with plastid loci, Acoraceae resumes the position shown here. Groups that are shown as resolved (no polytomies) are well-supported patterns (as estimated with bootstrapping), whereas unsupported or variable patterns of relationships are shown as polytomies. For example, some studies have weakly supported Bromeliaceae as sister to Poales, others as sister to Juncales, and yet others as sister to Rapateaceae, so we have simply shown the interrelationships of these clades and Bromeliaceae as unresolved. *These taxa are included in Nartheciaceae in this Volume; *Pleea* and *Harperocallis* also cluster near *Tofieldia*, and with *Narthecium* fall *Aletris* incl. *Metanarthecium*, and *Lophiola*. †Although distantly related to Hemerocallidaceae, in this treatment *Xeronema* is included there as a *genus anomalum*

Narthecium and *Metanarthecium*) close to Dioscoreaceae/Taccaceae; *Petrosavia* also falls near this clade. This is a very diverse group, although members are linked by very small chromosome size (Tamura 1995, and this Vol. p. 386). *Japonolirion* is isolated, being sister to all monocotyledons except Alismatanae and Acoranae. More work is needed to clarify the systematics of this taxonomically significant group.

c) Alismatanae (Alismatales and Arales)

Alismatales represent a coherent group of water plants. Following the recognition of the close relationship between Hydrocharitaceae and *Najas* (Shaffer-Fehre 1991), the maintenance of an order Najadales separate from Alismatales has become untenable, and here only two orders, Alismatales and Arales, are recognized. For a detailed discussion of the characters and taxonomic problems in Alismatales, see Les and Haynes (1995). Mayo et al. (1995) discussed the characters supporting a sister group relationship between Alismatales and Araceae, including a periplasmodial tapetum, uninucleate tapetal cells, and embryos that both store reserves and are chlorophyllous (the latter character may be derived independently in each group).

Mayo et al. (1995) reviewed the arguments for inclusion of Lemnaceae within Araceae, and more recently Stockey et al. (1997), in an analysis including the fossil *Limnobiophyllum*, argued for inclusion of *Pistia* in Lemnaceae. Araceae are the only monocots of which fossils (other than pollen) older than Cenomanian are known.

d) Lilianae

Recent DNA sequence analyses (e.g., Chase et al. 1995) in combination with nonmolecular studies (e.g., Tamura 1995, Chase et al. 1996, Fay and Chase 1996, Rudall and Chase 1996, Rudall et al. 1997) have improved our understanding of family boundaries and affinities within Lilianae. A monophyletic Lilianae, comprising three orders (Asparagales, Liliales, and Dioscoreales), was retrieved by a combined analysis of morphological and molecular data (Chase et al. 1995a), although in analyses of molecular data (plastid *rbc*L and *atp*B) it is paraphyletic. The circumscription of Asparagales and Liliales broadly corresponds with Dahlgren et al.'s (1985) classification but with some recircumscription. For example, some of Dahlgren et al.'s (1985) Liliales, notably Iridaceae and Orchidaceae, are now regarded as asparagoid (e.g., Chase et al. 1995). This is based partly on molecular data, but is also supported by some morphological characters, such as the presence of phytomelaniferous seeds in Iridaceae, as Dahlgren et al. (1985) and Goldblatt (1995) acknowledged. The systematics of Dioscoreales and their possible allies requires further review (see below).

Liliales (incl. Melanthiaceae sensu stricto). This represents a coherent order characterized by the presence of perigonal nectaries (or the near absence of septal nectaries). As circumscribed in Chase et al. (1995, 1995a), Liliales include some taxa from Dahlgren et al.'s (1985) Dioscoreales (Smilacaceae, Trilliaceae, and Petermanniaceae) and Melanthiales (Campynemataceae and some Melanthiaceae). Genera that were formerly in Asparagales include the "net-veined" *Clintonia* group, Philesiaceae, and Luzuriagaceae (but not *Behnia* or *Eustrephus*, both of which are asparagoid).

Asparagales. The most important synapomorphy for this order, first used by Huber (1969) as a unifying character, are the characteristic black seeds caused by phytomelan incrustation of the seed coat in most of the capsular and some berry-fruited taxa. The taxa now included in this order, many of them Australasian, were relatively little known at the time of Krause (1930), who placed them in his broad Liliaceae. It was mainly the work of Schlittler (1945) that paved the way for an understanding of their intriguing morphology (including thyrses with bostryces as partial inflorescences). Schlittler (1953) also brought into focus the interesting phenomenon of articulation of flowers on the pedicels, which among monocots is concentrated in Asparagales. Articulation is sometimes accompanied by an elongated basal constriction of the flower, the pericladium, which appears superficially like the continuation of the pedicel (Velenovský 1904). Functionally, articulation effects the abscission of supernumerary flowers or young fruits.

Both molecular and morphological data indicate that Asparagales comprise a paraphyletic "lower" asparagoid grade including the early-branching families (including Orchidaceae, Iridaceae, and Tecophilaeaceae), and a "higher" asparagoid clade (including Agavaceae, Alliaceae, and Convallariaceae). Simultaneous microsporogenesis and inferior ovaries are characteristic of lower asparagoids, whereas higher asparagoids have successive microsporogenesis (Rudall et al. 1997) and generally superior ovaries. Some families have been recircumscribed in the light of new micromorphological and molecular data. For example, the family Anthericaceae was previously poorly defined and in some cases difficult or impossible to distinguish from some Asphodelaceae, except on the basis of some poorly understood and often largely ignored micromorphological characters such as microsporogenesis (Chase et al. 1996), which nevertheless allow the two to be distinguished.

Dioscoreales and Pandanales. Relationships within this "problem" group require more study. The order Dioscoreales has been variously circumscribed by different authors. Huber (1969) originally included some net-veined genera of Asparagales and Liliales, but later (1991) adopted a more restricted circumscription, including only Dioscoreaceae (incl. *Stenomeris*) and Trichopodaceae. The combined morphological/molecular analysis of Chase et al. (1995a) identified a monophyletic clade with two related groups, one comprising Dioscoreaceae, Taccaceae, and Burmanniaceae, the other Pandanaceae, Cyclanthaceae, Velloziaceae (incl. *Acanthochlamys*), and Stemonaceae. Pandanaceae and Cyclanthaceae have long been considered highly isolated families that for lack of a better alternative have been asociated with the palms (Commelinanae). Velloziaceae have also always been problematic in their affiliation. These families are clearly not commelinoid, since they lack the commelinoid features, e.g., cell-wall ferulates and epidermal silica bodies (although Strelitzia type surface waxes occur sporadically), but more data are required for a better understanding of their relationships.

e) Commelinanae

There is considerable nonmolecular evidence in support of a monophyletic Commelinanae, reviewed by Rudall and Chase (1996). Relatively consistent synapomorphies for this clade include the presence of ferulic acid in cell walls (Harris and Hartley 1980), presence of silica bodies or particles in leaves, stomata formed by nonoblique cell divisions, the presence of long and short cells in the root epidermis and epicuticular wax morphology of the Strelitzia type (summarized by Frölich and Barthlott 1988). Relationships within Commelinanae are less clear.

Arecales Palms. These are one of the most distinct groups of monocotyledons. Previously, they were considered linked with the Pandanaceae and Cyclanthaceae, with which they share some superficial similarities, especially in habit. Current evidence suggests that palms are among the first-branching groups of commelinoids.

Dasypogonales. The only family of this order, Dasypogonaceae, comprises four highly divergent genera, which were previously included in Xanthorrhoeaceae. The presence of silica and cell-wall ferulates indicated a commelinoid affinity for these four genera, and analysis of molecular sequence data from *rbc*L supported this (Rudall 1994, Rudall and Chase 1996). Their precise affinities remain uncertain, although molecular data (Chase et al. 1995) suggest that Dasypogonaceae are among the first-branching families of the commelinoid clade, close to palms. The arborescent forms (*Kingia* and *Dasypogon*) lack a secondary thickening meristem and accomplish their increased girth by means of an extensive primary thickening meristem, as in palms.

Hanguanaceae and Zingiberales. *Hanguana* has been associated with a wide range of monocot taxa, including Flagellariaceae and Asteliaceae. However, it now seems certain that *Hanguana* belongs in the commelinoid clade (sensu Chase et al. 1995; i.e., Commelinanae), perhaps close to Zingiberales or Commelinales, which it resembles in habit and other characters (Stevenson and Loconte 1995, Rudall et al., submitted). Zingiberales themselves are a coherent, widely recognized order including several distinct families.

Bromeliales, Commelinales, and Xyridales. The classification of this complex of families is notoriously difficult, as is evident when comparing the arrangements of Dahlgren et al. (1985), Thorne (1992), and Takhtajan (1997). Although a close relationship between families such as Xyridaceae, Eriocaulaceae, and Mayacaceae is obvious, the positions of others, such as Rapateaceae, are controversial (Linder and Kellogg 1995). Molecular analyses bring Rapateaceae and Bromeliaceae into a sister position, whereas Rapateaceae have much anatomical and embryological similarity with Commelinaceae and Xyridaceae, and to a lesser degree with Eriocaulaceae and Mayacaceae (see Stevenson on Rapateaceae, Vol. IV: 420). Bromeliales may contain only Bromeliaceae (or perhaps also Rapateaceae); Xyridales consist of Eriocaulaceae, Mayacaceae, Xyridaceae, and perhaps Rapateaceae; Commelinales contain Commelinaceae, Pontederiaceae, Haemodoraceae, and Philydraceae. Typhaceae (incl. Sparganiaceae) possibly also belong here (but see below).

Juncales, Typhales, and Poales. There is morphological and molecular evidence that Juncales, Typhales, and Poales, which have a common pollination syndrome, together form a major clade within the commelinoids. The relationships of Typhales were controversial until Dahlgren et al. (1985) convincingly argued for placement in the commelinoid clade. This is supported by the available molecular data (*rbc*L, *rps*4, *atp*B, 18S rDNA), which invariably place Typhaceae among the commelinoids, sometimes in proximity to Poales and Juncales, at other times nearer to Bromeliaceae or Commelinales. The cladistic analysis by Linder and Kellogg (1995) makes a sister group position of Typhaceae to Poales likely.

The close relationship between the three families of Juncales, Juncaceae, Thurniaceae, and Cyperaceae was recognized as early as 1959 by Takhtajan. Morphological and molecular analyses (Linder and Kellogg 1995, Simpson 1995) have supported them as a robust grouping. The diffuse centromeres are a particularly notable trait of these families, although, as indicated by Linder and Kellogg, they are probably derived within the group: there are indications that both holocentric and monocentric chromosmes coexist in the Cyperaceae (see Vol. IV: 149), but the records for the localized type may be wrong (Greilhuber 1995).

In the 1980s, various workers supported the coherence of the order Poales using anatomical, embryological, and palynological data. Sequence data mainly from plastid genes were evaluated together with more conventional information by Kellogg and Linder (1995) and Linder and Kellogg (1995), and the following phylogeny is suggested: *Flagellaria* with its relatively showy flowers is sister to the remainder of Poales; Restionaceae and *Ecdeiocolea* are a monophylum (*Anarthria* probably to be included), and *Joinvillea* is sister to the grasses. Centrolepidaceae, a family often considered as doubtful in its affiliation to the Poales, has now been recognized as closely related to, or embedded in, Restionaceae.

Hydatellaceae are still of unknown relationship, but on account of their operculate seeds probably belong in Commelinanae.

Outline of Classification of Monocotyledons

Acoranae
- Acoraceae

Not Ranked
- Nartheciaceae

Alismatanae

Arales
- Araceae
- Lemnaceae

Alismatales
- Butomaceae
- Alismataceae
- Limnocharitaceae
- Hydrocharitaceae
- Najadaceae
- Aponogetonaceae
- Scheuchzeriaceae
- Juncaginaceae
- Potamogetonaceae
- Ruppiaceae
- Posidoniaceae
- Zosteraceae
- Zannichelliaceae
- Cymodoceaceae

Lilianae

Liliales
- Campynemataceae
- Luzuriagaceae
- Alstroemeriaceae
- Colchicaceae
- Melanthiaceae
- Trilliaceae
- Liliaceae
- Calochortaceae
- Petermanniaceae
- Smilacaceae
- Philesiaceae

Asparagales
- Lanariaceae
- Orchidacee
- Hypoxidaceae
- Blandfordiaceae
- Asteliaceae
- Boryaceae
- Tecophilaeaceae
- Ixioliriaceae
- Doryanthaceae
- Iridaceae
- Xanthorrhoeaceae
- Asphodelaceae
- Johnsoniaceae
- Hemerocallidaceae
- Asparagaceae
- Behniaceae
- Agavaceae
- Funkiaceae
- Herreriaceae
- Anthericaceae
- Anemarrhenaceae
- Aphyllanthaceae
- Lomandraceae
- Alliaceae
- Agapanthaceae
- Amaryllidaceae
- Themidaceae
- Hyacinthaceae
- Eriospermaceae
- Convallariaceae
- Dracaenaceae
- Nolinaceae
- Ruscaceae

Triuridales
- Triuridaceae

Dioscoreales
- Dioscoreaceae
- Trichopodaceae
- Taccaceae
- Burmanniaceae
- Corsiaceae

Pandanales
- Pandanaceae
- Cyclanthaceae
- Velloziaceae
- Acanthochlamydaceae
- Stemonaceae
- Pentastemonaceae

Commelinanae

Principes
- Palmae

Dasypogonales
- Dasypogonaceae

Bromeliales
- Bromeliaceae
- ? Rapateaceae (see also Xyridales)

Commelinales
- Commelinaceae
- Pontederiaceae
- Philydraceae
- Haemodoraceae

Xyridales
- Mayacaceae
- Xyridaceae
- Eriocaulaceae
- ? Rapateaceae

Zingiberales
- Musaceae
- Strelitziaceae
- Lowiaceae
- Heliconiaceae
- Costaceae
- Zingiberaceae
- Cannaceae
- Marantaceae

Possibly Related to Zingiberales or Commelinales
- Hanguanaceae

Typhales
- Typhaceae

Juncales
- Juncaceae
- Thurniaceae
- Cyperaceae

Poales
- Flagellariaceae
- Restionaceae
- Ecdeiocoleaceae
- Anarthriaceae
- Centrolepidaceae
- Joinvilleaceae
- Poaceae

Position uncertain
- Hydatellaceae

Selected Bibliography

Bancroft, N. 1914. A review of literature concerning the evolution of monocotyledons. New Phytol. 13: 285–308.

Behnke, H.-D. 1976. The distribution of characters within an angiosperm system. Bot. Not. 129: 287–295.

Behnke, H.-D. 1981. See general references.

Behnke, H.-D. 1995. P-type sieve-element plastids and the systematics of the Arales (sensu Cronquist 1988) with S-type plastids in *Pistia*. Plant Syst. Evol. 195: 87–119.

Bessey, C.E. 1893. Evolution and classification. Bot. Gaz. 18: 329–333.

Bessey, C.E. 1915. The phylogenetic taxonomy of flowering plants. Ann. Mo. Bot. Gard. 2: 109–164.

Brenner, G.J. 1996. Evidence for the earliest stage of angiosperm pollen evolution: a paleoequatorial section from Israel, pp. 91–115. In: Taylor, D.W., Hickey, L.J. (eds.) Flowering plant origin, evolution and phylogeny. New York: Chapman and Hall.

Buchenau, F. 1903. Alismataceae. In: Engler, A. (ed.) Das Pflanzenreich IV. 15. Leipzig: W. Engelmann.

Burger, W.C. 1981. Heresy revised: the monocot theory of angiosperm origin. Evol. Theory (Chicago) 5: 189–225.

Buxbaum, F. 1937. Die Entwicklungslinien der Lilioideae. Bot. Arch. 38: 213–293, 305–398.

Carlquist, S., Schneider, E.S. 1997. Origins and nature of vessles in monocotyledons. I. *Acorus* (Acoraceae). Int. J. Plant Sci. 158: 51–56.

Chase, M.W. et al. 1993. See general references.

Chase, M.W. et al. 1995. See general references.

Chase, M.W. et al. 1995a. See general references.

Chase, M.W. et al. 1996. See general references.

Cheadle, V.I. 1942. The occurrence and types of vessels in the various organs of the plant in the Monocotyledoneae. Am. J. Bot. 29: 441–450.

Cronquist, A. 1981, 1988. See general references.

Dahlgren, R.M.T. et al. 1985. See general references.

Dahlgren, R.M.T., Clifford, H.T. 1982. See general references.

Dahlgren, R.M.T., Rasmussen, N. 1983. See general references.

Davis, J.I. 1995. A phylogenetic structure for the monocotyledons, as inferred from chloroplast DNA restriction site variation, and a comparison of measures of clade support. Syst. Bot. 20: 503–527.

Davis, J.I., Simmons, M.P., Stevenson, D.W., Wendel, J.F. 1998. Data decisiveness and data quality in phylogenetic analysis: an example from the monocots using using mitochondrial *atp*A sequences. Syst. Biol. (in press).

Duvall, M.R. et al. 1993. See general references.

Endress, P.K. 1987. The Chloranthaceae: reproductive structures and phylogenetic position. Bot. Jahrb. Syst. 109: 153–226.

Endress, P.K. 1990. Evolution of reproductive structures and functions in primitive angiosperms (Magnoliidae). Mem. N. Y. Bot. Gard. 55: 5–34.

Endress, P.K. 1994. Floral structure and evolution of primitive angiosperms: recent advances. Plant Syst. Evol. 192: 79–97.

Endress, P.K. 1995. Major evolutionary traits of monocot flowers. In: Rudall, P.J., Cribb, P.J., Cutler, D.F., Humphries, C.J. (eds.) Monocotyledons: sytematics and evolution, pp. 43–79. Royal Botanic Gardens Kew.

Engler, A. 1892. Die systematische Anordnung der monocotyledoneen Angiospermen. Abh. Preuss. Akad. Wiss. Berlin 1892. 55pp.

Engler, A. (ed.) 1925→ Die natürlichen Pflanzenfamilien, 2nd edn. Leipzig: W. Engelmann and Berlin: Duncker & Humblot.

Engler, A., Prantl, K. (eds.) 1887–1889. Die natürlichen Pflanzenfamilien. II. Teil (Gymnospermae, Monocotyledoneae). Leizig: W. Engelmann.

Erbar, C., Leins, P. 1994. Flowers in Magnoliidae and the origin of flowers in other subclasses of the angiosperms. I. The relationships between flowers of Magnoliidae and Alismatidae. Plant Syst. Evol. Suppl. 8: 193–208.

Fay, M.F., Chase, M.W. 1996. Resurrection of Themidaceae for the Brodiaea alliance, and recircumscription of Alliaceae, Amaryllidaceae and Agapanthoideae. Taxon 45: 441–451.

French, J.C., Chung, M.G., Hur, Y.K. 1995. Chloroplast DNA phylogeny of the Ariflorae. In: Rudall, P.J., Cribb, P.J., Cutler, D.F., Humphries, C.J. (eds.) Monocotyledons: sytematics and evolution, pp. 255–275. Royal Botanic Gardens, Kew.

Frölich, D., Barthlott, W. 1988. See general references.

Gaut, B.S., Muse, S.V., Clark, W.D., Clegg, M.T. 1992. Relative rates of nucleotide substitution at the *rbc*L locus of monocotyledonous plants. J. Mol. Evol. 35: 292–303.

Goldblatt, P. 1995. The status of R. Dahlgren's orders Liliales and Melanthiales. In: Rudall, P.J., Cribb, P.J., Cutler, D.F., Humphries, C.J. (eds.) Monocotyledons: systematics and evolution, pp. 181–200. Royal Botanic Gardens, Kew.

Grayum, M.H. 1987. A summary of evidence and arguments supporting the removal of *Acorus* from the Araceae. Taxon 36: 723–729.

Grayum, M.H. 1991. Systematic embryology of the Araceae. Bot. Rev. 57: 167–203.

Greilhuber, J. 1995. Chromosomes of the monocotyledons (general aspects). In: Rudall, P.J., Cribb, P.J., Cutler, D.F., Humphries, C.J. (eds.) Monocotyledons: systematics and evolution, pp. 379–414. Royal Botanic Gardens, Kew.

Hallier, H. 1905. Ein zweiter Entwurf des natürlichen (phylogenetischen) Systems der Blütenpflanzen. Vorläufige Mitteilung. Ber. Dtsch. Bot. Ges. 23: 85–91.

Hamby, K.R., Zimmer, E.A. 1992. Ribosomal RNA as a phylogenetic tool in plant systematics, pp. 50–91. In: Soltis, P.S., Soltis, D.E., Doyle, J.J. (eds.) Molecular systematics of plants. New York: Chapman and Hall.

Harris, P.J., Hartley, R.D. 1980. See general references.

Holttum, R.E. 1955. Growth habits of monocotyledons – variations on a theme. Phytomorphology 5: 399–413.

Huber, H. 1969, 1977, 1991. See general references.

Hutchinson, J. 1934. The families of flowering plants, Vol. 2. London: Macmillan.

Jeffrey, E.C. 1917. The anatomy of woody plants. Chicago: University of Chicago Press.

Jussieu, A.L. de 1789. Genera Plantarum secundum Ordines Naturales disposita. Paris: Hérissant & Barrois.

Kellogg, E.A., Linder, H.P. 1995. Phylogeny of Poales. In: Rudall, P.J., Cribb, P.J., Cutler, D.F., Humphries, C.J. (eds.) Monocotyledons: systematics and evolution, pp. 511–542. Royal Botanic Gardens, Kew.

Krause, K. 1930. See general references.

Les, D.H., Haynes, R.R. 1995. Systematics of subclass Alismatidae: a synthesis of approaches. In: Rudall, P.J., Cribb, P.J., Cutler, D.F., Humphries, C.J. (eds.) Monocotyledons: systematics and evolution, pp. 353–377. Royal Botanic Gardens, Kew.

Les, D.H., Schneider, E.L. 1995. The Nymphaeales, Alismatidae, and the theory of an aquatic monocotyledon origin, pp. 23–42. In: Rudall, P.J., Cribb, P.J., Cutler, D.F., Humphries, C.J. (eds.) Monocotyledons: systematics and evolution. Royal Botanic Gardens, Kew.

Les, D.H., Garvin, D.K., Wimpee, C.F. 1991. Molecular evolutionary history of ancient aquatic angiosperms. Proc. Natl. Acad. Sci. USA 88: 10119–10122.

Linder, H.P., Kellogg, E.A. 1995. Phylogenetic patterns in the commelinid clade. In: Rudall, P.J., Cribb, P.J., Cutler, D.F., Humphries, C.J. (eds.) Monocotyledons: systematics and evolution, pp. 473–496. Royal Botanic Gardens, Kew.

Linnaeus, C. 1751. Philosophia botanica. Stockholm: G. Kiesewetter.

Loconte, H., Stevenson, D.W. 1991. Cladistics of the Magnoliidae. Cladistics 7: 267–296.

Martin, W., Gierl, A., Saedler, H. 1989. Molecular evidence for pre-Cretaceous angiosperm origins. Nature 339: 46–48.

Mayo, S.J., Bogner, J., Boyce, P. 1995. The Arales. In: Rudall, P.J., Cribb, P.J., Cutler, D.F., Humphries, C.J. (eds.) Monocotyledons: systematics and evolution, pp. 277–286. Royal Botanic Gardens, Kew.

Mena, M., Ambrose, B.A., Meeley, R.B., Briggs, S.P., Yanofsky, M.F., Schmidt, R.J. 1996. Diversification of C-function activity in maize flower development. Science 274: 1537–1540.

Nadot, S., Bittar, G., Carter, L., Lacroix, R., Lejeune, B. 1995. A phylogenetic analysis of monocotyledons based on the chloroplast gene *rps*4, using parsimony and a new numerical phenetics method. Mol. Phylogenet. Evol. 4: 257–282.

Qiu, Y.L., Chase, M.W., Les, D.H., Parks, C.R. 1993. Molecular phylogenetics of the Magnoliidae: cladistic analysis of nucleotide sequences of the plastid gene *rbc*L. Ann. Mo. Bot. Gard. 80: 587–606.

Ray, J. 1686–1704. Historia Plantarum. London: H. Faithorne.

Rudall, P. 1991. Lateral meristems and stem thickening growth in monocotyledons. Bot. Rev. 57: 150–163.

Rudall, P.J. 1994. The ovule and embryo sac in Xanthorroeaceae sensu lato. Flora 189: 335–351.

Rudall, P.J. 1997. See general references.

Rudall, P.J., Caddick, L.R. 1994. Investigation of the presence of phenolic compounds in monocotyledonous cell walls, using UV fluorescence microscopy. Ann. Bot. 74: 483–491.

Rudall, P., Chase, M.W. 1996. Systematics of Xanthorrhoeaceae sensu lato: evidence for polyphyly. Telopea 6: 629–647.

Rudall, P.J., Furness, C.A. 1997. Systematics of *Acorus*: ovule and anther. Int. J. Plant Sci. 158: 640–651.

Rudall, P.J., Cribb, P.J., Cutler, D.F., Humphries, C.J. (eds.) 1995. Monocotyledons: systematics and evolution. 2 vols. Royal Botanic Gardens, Kew.

Rudall, P.J. et al. 1997. See general references.

Rudall, P.J., Prychid, C.J., Jones, C. 1998. Intra-ovarian trichomes in monocotyledons. In: Owens, S.J., Rudall, P.J., (eds.) Reproductive biology. Royal Botanic Gardens, Kew.

Rudall, P.J., Stevenson, D.W., Linder, H.P. Submitted. Structure and systematics of *Hanguana*, a monocotyledon of uncertain affinity. Aust. Syst. Bot.

Sattler, R., Singh, V. 1978. Floral organogenesis of *Echinodorus amazonicus* Rataj and floral construction of the Alismatales. Bot. J. Linn. Soc. 77: 141–156.

Schlittler, J. 1945. Untersuchungen über den Bau der Blütenstände im Bereich des Anthericumtypus (Asphodelinae-Anthericinae-Dianellinae). Ber. Schweiz. Bot. Ges. 55: 200–239.

Schlittler, J. 1953. Blütenartikulation und Phyllokladien der Liliaceae organphylogenetisch betrachtet. Feddes Rep. 55: 154–258.

Shaffer-Fehre, M. 1991. The position of *Najas* within the subclass Alismatidae (Monocotyledones) in the light of new evidence from seed coat structures in the Hydrocharitaceae. Bot. J. Linn. Soc. 107: 189–209.

Simpson, D. 1995. Relationships within Cyperales. In: Rudall, P.J., Cribb, P.J., Cutler, D.F., Humphries, C.J. (eds.) Monocotyledons: systematics and evolution, pp. 497–509. Royal Botanic Gardens, Kew.

Singh, V., Sattler R. 1972. Floral development of *Alisma triviale*. Can. J. Bot. 50: 619–627.

Soltis, D.E. et al. 1987. See general references.

Stevenson, D.W., Loconte, H. 1995. Cladistic analysis of monocotyledons. In: Rudall, P.J., Cribb, P.J., Cutler, D.F., Humphries, C.J. (eds.) Monocotyledons: systematics and evolution, pp. 543–578. Royal Botanic Gardens, Kew.

Stockey, R.A., Hoffman, G.L., Rothwell, G.W. 1997. The fossil monocot *Limnobiophyllum scutatum*: resolving the phylogeny of Lemnaceae. Am. J. Bot. 84: 355–368.

Suessenguth, K. 1921. Beiträge zur Frage des systematischen Anschlusses der Monokotylen. Beih. Bot. Centralbl. 38, Abt. 2: 1–79.

Sytsma, K.J., Baum, D.A. 1996. Molecular phylogenies and the diversification of the angiosperms, pp. 314–340. In: Taylor, D.W., Hickey, L.J. (eds.) Flowering plant origin, evolution and phylogeny. New York: Chapman and Hall.

Takhtajan, A. 1959. Die Evolution der Angiospermen. Jena: G. Fischer.

Takhtajan, A. 1980, 1987, 1997. See general references.

Thorne, R.F. 1992. Classification and geography of the flowering plants. Bot. Rev. 58: 225–348.

Tillich, H.-J. 1985. Keimlingsbau und verwandtschaftliche Beziehungen der Araceae. Gleditschia 13: 63–73.

Tillich, H.-J. 1995. Seedlings and systematics in monocotyledons. In: Rudalll, P.J., Cribb, P.J., Cutler, D.F., Humphries, C.J. (eds.) Monocotyledons: systematics and evolution, pp. 303–352. Royal Botanic Gardens, Kew.

Tomlinson, P.B. 1990. The structural biology of palms. Oxford: Clarendon Press.

Tomlinson, P.B. 1995. Non-homology of vascular organisation in monocotyledons and dicotyledons, pp. 589–622. In: Rudall, P.J., Cribb, P.J., Cutler, D.F., Humphries, C.J. (eds.) Monocotyledons: systematics and evolution. Royal Botanic Gardens, Kew.

Velenovský, J. 1904. Die gegliederten Blüten. Beih. Bot. Centralbl. 16: 289–300.

Wagner, P. 1977. Vessel types of the monocots: a survey. Bot. Not. 130: 383–402.

Wettstein, R.R. von 1901–1907. Handbuch der systematischen Botanik. Leipzig and Wien: F. Deuticke.

Wolfe, K.H., Gouy, M., Yang, Y.-W., Sharp, P.M., Li, W.-H. 1989. Date of monocot-dicot divergence estimated from chloroplast DNA sequence data. Proc. Natl. Acad. Sci. U. S. A. 86: 6201–6205.

Floral Biology

S. Vogel

1. Introduction

Having ranked for a long time as a taxonomic equivalent of the dicots, the monocots are now considered as representing a clade nested within the basal dicotyledonous angiosperms. Nevertheless, the monocotyledonous clade stands out from all the remaining angiosperm lineages by both its uniformity on the one hand and its high degree of diversity on the other: shoot organization, leaf structure, and the conformation of floral organs are much more homogeneous than within most dicotyledons. In contrast, diversification into a wealth of vegetative life styles, as well as pollination systems, even if based on a relatively stereotypic fundamental construction mode, appear to repeat, and equate, the diversification of the rest of the angiosperms together. This reflects the enormous evolutionary potential and success of the group which, in terms of species number, accounts for almost one fourth of the flowering plants.

Within the monocots, almost all kinds of reproductive strategies and floral contrivances found in the dicots reappear: the trend toward wind pollination, the adaptive radiation into an array of corolla shapes and zoophilous floral styles, types of floral reward, types of sex distribution, gametophytic and sporophytic incompatibility systems, etc. Also a great many of the more sophisticated floral specializations recur among the monocots just a few examples may be given:

Pseudanthia: Compositae/Eriocaulaceae; kettle traps: Aristolochiaceae/Araceae; asymmetric flowers: Phaseoleae/Marantaceae; pollinaria: Asclepiadaceae/Orchidaceae; pollen threads: Oenotheraceae/Strelitziaceae; enantiostyly: *Cassia*/*Cyanella*; secondary pollen presentation: *Campanula*/*Canna*; explosive pollen application: *Medicago*/Marantaceae; sensitive movements aiding in pollen release: *Stylidium*/Marantaceae; lever mechanism of stamens: *Salvia*/*Roscoea*; heterodistyly: *Primula*/*Nivenia*; heterotristyly: *Oxalis*/*Eichhornia*; gynostemium: Stylidiaceae/ Orchidaceae; stigmatic glue attaching pollen on vector: Apocynaceae/Orchidaceae; flickering appendages as attractants: *Ceropegia*/*Cirrhopetalum* spp.; flowers advertently pollinated by ovipositing phytophages: *Ficus*/*Yucca*.

In contrast, the number of anthecological adaptations completely or almost confined to monocots is comparatively low. With few exceptions, evolution of hydrogamy is restricted to this group; pollination by sexual deceit does not seem to occur elsewhere, and floral deception in general is much more widespread here than elsewhere. The luxuriance of flower modeling, color shades, and perfumes, as well as the degree of structural intrafloral synorganization and variety of mechanisms evolved in the orchids are unsurpassed, with hardly any equivalent among dicots. The only known case of subterranean zoophilous pollination occurs in this family (*Rhizanthella*, Dixon 1985).

Finally, we call to mind that, owing to the frequent occurrence of bulbs and corms, monocots are disproportionally more highly represented among geophytic spring and desert flowers and also among epiphytes than dicots. The wind pollination that characterizes several plant formations is due to the predominance of monocotyledoneous taxa in them.

As a basis to the following paragraphs, Table 1 surveys the distribution of essential floral features relevant to pollination in the monocots in the taxonomical classification adopted in this Volume. Emphasis is paid to properties that deviate from the regular (actinomorphic), choripetalous, hermaphrodite, and homogamous flowers that are taken as standard (and are not otherwise indicated). Data are based on records from the literature and the author's own observations (partially unpubl.). Lack of particular data signifies absence of adequate information but does not exclude their possible existence; likewise, indication of one type of plurimodal characters such as nectaries, dichogamy, etc., does not necessarily exclude the incidence of alternatives.

Since Knuth's handbook (1899, 1904), no detailed survey on pollination covering the monocots as a whole has been published. Vogel (1981) presented a short synopsis. The recent account on

Table 1. Distribution of essential floral features relevant to pollination in monocots. Classification as in this Volume

Nartheciales																																				
Nartheciaceae														n					s				w			z	α								ι	
Lilianae																																				
Liliales																																				
Melanthiaceae				d	e		g		i				m♀	n		p							w			z				δ		ζ			ι	
Colchicaceae				d	e	f	g						♀m♀	n		p	q									z				δ	ε					
Alstroemeriaceae		b				f										p										z					ε		η			
Luzuriagaceae	a															p							w			z	α									
Smilacaceae				d	e		g		i					n		p			s				w			z				δ						
Philesiaceae						f	g									p																	η			
Trilliaceae							g		i					n		p				t										δ						
Liliaceae	a					f	g						♀m♀			p								x[1]		z	α			δ	ε	ζ	η			
Calochortaceae							g									p								x[1]		z							η			
Campynemataceae													m♀			p?																				
Triuridales																																				
Triuridaceae				d	e			h					♀m♀			p														δ?					ι	
Asparagales																																				
Orchidaceae	a	b	c	d	e	f	g	h								p		r								z	α	β	γ	δ	ε	ζ	η		ι	ϰ
Hypoxidaceae							g						m♀													z	α			δ?						
Blandfordiaceae	a																																η			
Asteliaceae				d			g																	x		z								θ		
Boryaceae	a													n?																						
Tecophilaeaceae	a	b									k															z	α									
Ixioliriaceae							g																			z										
Doryanthaceae														n																			η			
Iridaceae	a	b					g			j			m♀	n	o	p								x[1]		z	α	β		δ	ε	ζ	η		ι	
Xanthorrhoeaceae														n												z							η?			
Asphodelaceae	a	b				f	g						♀m♀	n					s							z						ζ	η		ι	
Johnsoniaceae																										z	α									
Hemerocallidaceae	a	b					g						m♀	n			q									z	α				ε	?	η		ι	
Asparagaceae				d	e									n									w			z										
Behniaceae				d										n												z										
Agavaceae	a	b				f	g						m♀	n																		ζ	η	θ		
Funkiaceae	a	b				f																				z					ε?	ζ				
Herreriaceae														n									w										η			
Anthericaceae	a	b				f					k			n												z	α					ζ			ι	
Aphyllanthaceae	a						g																			z										
Lomandraceae				d		f								n?					s							z	α									
Alliaceae	a					f	g						♀m♀	n			q?	r					w			z						ζ				
Agapanthaceae	a	b												n												z					ε		η			
Amaryllidaceae	a	b				f	g			j			m♀	n												z	α				ε	ζ	η	θ		
Themidaceae	a													n			q									z						ζ	η			
Hyacinthaceae	a					f	g							n									w			z	α			δ		ζ	η			
Eriospermaceae														n									w			z										
Convallariaceae	a						g						♀m	n									w			z	α			δ		ζ				
Dracaenaceae	a												♀m	n												z						ζ				
Nolinaceae				d										n						t?						z										
Ruscaceae					e	f																	w													
Dioscoreales																																				
Trichopodaceae	a																		s											δ						
Dioscoreaceae	a			d	e	f	g							n[2]					s				w							δ					ι?	
Taccaceae	a						g							n?					s											δ						
Burmanniaceae	a	b				f							m♀	n				r								z?				δ						ϰ
Corsiaceae		b		d	e																									δ					ι	
Pandanales																																				
Stemonaceae	a				e														s											δ						
Pentastemonaceae													m♀						s											δ?						
Velloziaceae	a						g		i					n												z	α					ζ	η	θ	ι	
Cyclanthaceae					e	f			i				♀m				q							x	y											ϰ
Pandanaceae				d					i										s	t					y	z	α						η	θ		
Acorales																																				
Acoraceae													♀m						s				w													
Alismatanae																																				
Arales																																				
Araceae				d	e	f	g	h	i			l	♀m	n[3]			q		s				w	x	y	z			γ	δ					ι	ϰ
Lemnaceae						f							♀m						s		u															
Alismatales																																				
Butomaceae							g						m♀	n[4]									w													
Alismataceae				d	e		g		i					n[4]			q						w				α									
Limnocharitaceae						f	g?		i					n[4]									w													
Hydrocharitaceae				d	e		g		i			l				p	q	r	s	t	u	v	w				α									
Najadaceae				d	e														s			v														
Aponogetonaceae		b			e		g		i					n[5]									w													
Scheuchzeriaceae							g						m♀						s	t																
Juncaginaceae					e		g												s	t																
Potamogetonaceae							g						♀m						s	t		v														

Table 1. (Continued)

Ruppiaceae													m♀						s		u														
Posidoniaceae					e														s			v													
Zosteraceae				d	e								♀m						s		u	v													
Zannichelliaceae					e														s		u														
Cymodoceaceae				d															s			v													
Commelinanae																																			
Palmae				d	e	f	g		i				♀m♀	n						t			w	x	y	z								θ	ι
Dasypogonaceae	a					f								n										x		z?	α?				ε?			θ?	
Hanguanaceae				d													q		s				w												
Commelinales																																			
Commelinaceae		b			e	f	g	h										r								z	α								ι
Mayacaceae																										z	α								
Pontederiaceae	a	b					g			j	k			n				r								z	α								
Philydraceae	a	b					g	h			k															z	α								
Haemodoraceae	a	b				f	g	h			k			n												z	α		γ?				η		
Bromeliaceae	a	b				f	g						m♀	n						t?						z					ε	ζ	η	θ	
Rapateaceae																										z	α						η		
Xyridaceae	a	b										l		n[5]												z	α								ι
Eriocaulaceae					e		g	h					m	n[5]		p			s				w				α								
?Hydatellaceae					e														s	t															
Zingiberales																																			
Musaceae	a	b			e	f								n																			η	θ	
Strelitziaceae	a	b				f								n																			η	θ[6]	
Lowiaceae	a	b				f								n												z?									
Heliconiaceae	a	b				f								n																			η	θ	
Costaceae	a	b				f		h						n												z					ε		η		
Zingiberaceae	a	b				f		h						n[7]												z					ε	ζ	η		
Cannaceae			c			f		h				l		n																		ζ	η	θ	
Marantaceae			c			f		h				l		n				r								z					ε	ζ	η		
Typhales																																			
Typhaceae					e		g												s	t															
Juncales																																			
Juncaceae				d	e		g						♀m					r	s	t						z	α								
Cyperaceae				d	e		g		i				♀m						s	t						z	α								
Poales																																			
Flagellariaceae					e		g												s	t			w?												
Restionaceae				d	e														s	t															
Centrolepidaceae					e														s	t															
Joinvilleaceae																			s																
Poaceae					e		g		i				m♀					r	s	t						z	α								

No adequate information available: Acoraceae; Petermanniaceae; Lanariaceae; Anemarrhenaceae; Acanthochlamydaceae; Anarthriaceae; Thurniaceae; Ecdeiocoleaceae;

[1] "Poppy guild" (Sect. 3.7).
[2] Peristylar epithelial nectary.
[3] Stigmatic nectary.
[4] Lateral base of carpels.
[5] Back of carpels.
[6] Including lemur pollination.
[7] Emergences probably derived from septal nectaries.

Explanation of letters.

a syntepaly; b zygomorphy; c asymmetry (zoophilous flowers only), d dioecy; e unisexuality of flowers (monoecy or unknown sex distribution); f wet stigma; g dry stigma; h oligandry (<3, zoophilous flowers only); i polyandry; j heterostyly; k enantiostyly; l secondary pollen presentation; m dichogamy (sensu lato; ♀m protogyny; m♀ protandry; ♀m♀ both modes recorded); n septal nectaries (incl. other kinds of carpellary nectaries); o trichomatous perigonal nectaries; p epithelial perigonal nectaries; q androecial nectaries (epithelial); r cleistogamy; s occurrence of inconspicuous flowers (aphananthy); t anemophily; u ephydrogamy; v hyphydrogamy; w allophily (entomophilous generalists); x cantharophily; y food tissue; z melittophily; α pollen flowers; β oil flowers; γ perfume flowers (male Euglossine syndrome); δ sapromyiophily; ε psychophily; ζ sphingophily, phalenophily; η ornithophily; θ chiropterophily; τ floral deception (partial or total); ϰ osmophores; ? = doubtful or controversial records.

monocotyledonous flowers by Endress (1995) comprises much anthecological information. Daumann (1970) monographed the septal nectaries, Biedinger and Barthlott (1993) described UV-absorbing nectar guides in members of 72 families of this group. The few contributions at family level emphasizing pollination and/or particular floral organs involved include Hydrocharitaceae (Cook 1982), Orchidaceae (van der Pijl and Dodson 1960), Commelinaceae (Faden 1992), Haemodoraceae (Simpson 1990), Heliconiaceae (Kress 1984), Marantaceae (Kunze 1984), palms (Henderson 1986; Silberbauer-Gottsberger 1990), and Pandanaceae (Cox 1990). Stigma types of the Bromeliaceae were studied by Schill et al. (1988) and Brown and Gilmartin (1989), septal nectaries of this family by Boehme (1988), and those of the Haemodoraceae by Simpson (1993). Schill and Pfeiffer (1977) screened types of orchid pollinia by SEM. The majority of data, however, are found in accounts at generic and species level, or are included in taxonomic work, many papers being quoted in the particular family introductions of this and the following volumes.

2. The Perianth

The monocot perigone, main site of advertisement in the zoophilous members, consists with few exceptions of two alternating trimerous tepal whorls. In the opinion of Leinfellner (1963), it is, at least in the Liliaceae, staminodial in origin and thus would represent true petals. The majority of morphological criteria, however, argue for an origin from bracts of vaginal character; this hypothesis is advocated here. For the great similarity of certain tepals of the inner whorl with true petals (spatulate shape, fanlike innervation), another explanation must be found (Weber 1980). Thus, the monocot perigone differs from the typically heterochlamydeous perianth - which consists of phyllomes of both bracteal (calyx) and staminodial nature (petals) - in being composed of two homoeologous, oligomerous leaf whorls. While the perigone - when present - has maintained (or reacquired) its bracteose appearance (aphananthy) in the anemophilous families, in most zoophilous flowers both tepal whorls are corolline, comparable to the calycinal perianth like that of *Caltha*. On the other hand, in a series of monocot families, known in the past as Calycinae, only the inner whorl is petaloid, whereas the outer one is inconspicuous and serves the protective role of a calyx, thus paralleling heterochlamydeous dicots: Trilliaceae, *Philesia*, *Thysanotus*, Bromeliaceae, Commelinaceae, Mayacaceae, Rapateaceae, Philydraceae, Xyridaceae, and zoophilous Alismatales. Regarding the appearance and internal structure of the corolline tepals, there is a considerable variety. This indicates that their evolutionary history was quite diverse. A more primitive type as represented in the Liliales, orchids, and most Asparagales still reflects in nervature and longevity its origin from vaginal bracts. These tepals display a slow, gradual, unprotected growth, ±broad insertion, a reticulate vasculature, entire margins, a rather firm consistency, and long persistence. Flowers of certain orchids, unless pollinated, may persist several weeks. In contrast, the flowers of other groups, especially the corolline inner whorl in many Alismatales, Bromeliales, and Commelinales, but also the bicyclic perigones of Iridaceae, Haemodoraceae, and Pontederiaceae tend to have a fanlike innervation, narrow insertion, a denticulate or even fimbriate rim, a rapid extension of primordial tissue shortly before anthesis, and a delicate consistency. They are fugacious, sometimes live for a few hours only, and on wilting, are often deliquescent, probably due to pectin predominating in their cell walls. The investment in biomass is low and enables plants to expose new flowers daily. Fugacious flowers often mature under the protection of slime or watery secretions produced by enveloping spathes (Commelinaceae, Heliconiaceae, Bromeliaceae, *Costus*), or collectively covered by mucilage as in flower heads of Rapateaceae, *Guzmania*, and *Elleanthus*. Transitional forms suggest that in such fugacious flowers a gradual replacement of the ancestral bracteose leaf by neotenic marginal outgrowths of the primordium has occurred during evolution. Cell proliferation ceases prematurely, probably a process of ontogenetic abbreviation of the ancestral pattern of development (Ihlenfeldt 1971). A similar delicacy of the corolla is shown by the short-lived perianths of Zingiberaceae, Marantaceae, and Cannaceae. In these families, both tepal whorls are bracteose, less conspicuous, and mainly protective in function, the corolla being replaced by staminodial petaloid semaphylls, even by leafy lateral appendages of the sole fertile stamen. These phyllomes, considering their staminodial origin, are comparable to the true petals of dicotyledons.

Some of the more persistent perigones undergo thermonastic opening and closing movements (*Tulipa*, *Thelymitra*).

3. Zoophilic Pollination

a) Visual Attraction

About two thirds of monocot species are adapted to pollination by animals, mostly insects. As in dicots, optical signaling by colorful semaphylls, one of the main secondary attractants, is usually a function of a single or both tepal whorls, but sometimes partly or exclusively provided by exserted filaments (*Haemanthus*, *Xeronema*), staminodes (Cyclanthaceae), or stylodia (*Iris*). These showy flowers will here be termed phaneranthous. Aggregates of phaneranthous florets form conspicuous pseudanthic capitula in *Aponogeton ranunculiflorus* (Dahlgren et al. 1985), some Zingiberaceae [*Etlingera* (*Achasma*)], entomophilous Cyperaceae (e.g., *Dichromena*), umbels in *Cirrhopetalum*, or geminate pseudanthia in *Thalia*. However, several groups of monocots bearing small, inconspicuous, and mostly greenish tepals (aphananthous flowers) are animal-pollinated as well (p. 45). They rely on chemical attraction alone, or their visual advertisement is taken over by adjacent, conspicuous extrafloral organs, mainly colored bracts such as the spathes of most members of Araceae, many palms, and the Cyclanthaceae. Also in phaneranthous flowers or inflorescences, visual advertisement may be reinforced by showy bracts, as in Taccaceae, in genera of the Amaryllidaceae (*Haemanthus* spp.), Commelinaceae (*Cochliostema*, *Coleotrype*), orchids, Zingiberales, and Bromeliaceae, or by sterile flowers as in *Muscari* and *Oncidium heteranthum*. Petaloid involucral bracts form part of pseudanthia in Eriocaulaceae (*Paepalanthus* subg. *Xeractis*, *Syngonanthus*), *Etlingera* (*Phaeomeria*), and *Androcymbium*.

Contrasting nectar guides in form of single spots or patches of fine dots are mainly restricted to Liliales (including Orchidaceae), Iridaceae, Haemodoraceae, Pontederiaceae, and some Zingiberales (Dahlgren and Clifford 1982). Distinct UV-absorbing patterns, invisible to man, have been detected in species of *Hemerocallis*, *Eichhornia*, *Wachendorfia*, and *Dendrobium*, among many others (Biedinger and Barthlott 1993).

The pigments involved in flower colors, both chymochromic (vacuole soluble) and plasmochromic (plastidal), are generally similar to those in dicots. Anthocyanins are responsible for most blue, red, and purple colors, and certain anthocyaninidin moieties and types of glycosylation have been found only in monocots. For example, a complex compound consisting of six delphinidin and six flavone molecules bound to two atoms of magnesium causes the bright azure tint of *Commelina coelestis*. Three pigments consisting of rutinosides and diglucosides bound in various positions to cyanidin are responsible for the scarlet coloration of bird-pollinated Bromeliaceae, whereas pelargonidin, the anthocyanidin which is commonly responsible for this hue, is infrequent in this family. Cyanidin-3-rutinoside also contributes to the maroon coloration of araceous spathes and appendices (Harborne and Williams 1995). The red pigment haemocorin (9-phenyl phenalenon derivative) present in all vegetative parts of Haemodoraceae is also involved in flower pigmentation in this family (Simpson 1990). Carotene plasmochromes are the coloring matter in many orange-red bird-pollinated corollas, and the orange stylodia of *Crocus* are unique in containing crocetin, a glycosylated carotenoid dissolved in the cell sap.

b) Olfactory Attraction

Production of floral scents as a means of secondary attraction is widespread in monocots, the fragrances of *Convallaria*, *Hyacinthus*, and *Polyanthes* being familiar as among the most powerful. While fragrance is normally produced on the entire surface of the perigone, it may be localized at particular sites, serving as local scent cues, e.g. on the paracorolla of *Narcissus* and the labellum of *Platanthera*. A unique case of olfactory (and visual) floral dimorphism is exhibited by *Dimorphorchis lowii*, whose spikes bear basally 1–2 long-lived yellow flowers producing scent, and distally a number of purple nonscented flowers, both kinds of flowers apparently being fertile (Winkler 1906; pers observ.).

Anatomically and morphologically distinguished scent glands, (osmophores) are found in the staminate zone of the spadix or its sterile appendix in many myiophilous and cantharophilous aroids, where the volatilization of foul or pungent benzoloid odors is often reinforced by thermogenesis. In *Cryptocoryne* and *Lysichiton* the spathe is the odoriferous part. Club- or tail-shaped osmophores are frequent in the pleurothallidine, dendrobiine, and chloraeine alliances of orchids and probably in Burmanniaceae (Vogel 1990). Osmophores which exude liquid odorous substances are characteristic of four subtribes of epidendroid orchids and some aroids, where the fragrance – a blend containing up to 60 different compounds – functions as a secondary as well as a primary attractant. These aromas are gathered and stored as a reward by euglossine bees (p. 43).

c) Corolla Shapes

Apart from signaling functions, the perianth serves as a landing platform, a guide leading animals to feeding positions appropriate for pollination, and if the nectar is concealed (euphilic flowers), to restrict the access to specialized, tongued visitors. Nectariferous flowers and flowers merely pretending presence of nectar (many orchids, p. 43) have three-dimensional, cup-shaped, campanulate, or tubular perianths (most Liliales including orchids, many Asparagales, the Bromeliaceae, Velloziaceae, and Zingiberales). The molding of the corolla, however, is generally less elaborate as compared to the personate, gullet, and spurred corollas of sympetalous dicots. This may be due to morphogenetical limitations of the tepal whorls, related to their origin from bracts.

The orchid labellum, which often alone substitutes for a gullet, labiate, or spurred flower, is an exception by its greater flexibility. This could be related to a possible complex nature and morphogenetic reorganization of the organ, if it is supposed to be a fusion product of a tepal with adjacent staminodes (Endlicher 1836/40, discussed by Endress 1994). The labellum is formed by the posterior, median tepalum of the inner whorl, and not the anterior median tepalum of the outer whorl, perhaps because of the limited evolutionary potential of the latter. In order to bring the median endotepalum to its functional anterior position, the flower undergoes resupination. In a few orchids only, such as species of *Disa* and relatives with a reduced labellum, it is the posterior exotepalum that forms a spur. Here, since an erect position of the gynostemium would hamper access to the spur's entrance, the column is truncate and the anther tilted backwards (Vogel 1959; Dressler 1992).

The perigone as a whole is rarely integrated into elaborate guiding sculptures, and congenital fusion (syntepaly, or, better, symperianthy; Guédès 1979) is uncommon. Although the perigone is often fused at the base, the dilated apical part is only rarely fused. This may be due to morphogenetic problems in bridging two whorls of phyllomes, or even bridging the distance between the three members of the same whorl – contrary to the situation in the pentamerous, single-whorled dicot corolla. Advanced novel acquisitions such as the paracorolla of *Narcissus*, on the other hand, may form perfect "bell flowers" (as in *N. bulbocodium*). Simple syntepalous tubular corollas with a reduced limb occur in *Muscari*, *Polygonatum*, bird-pollinated Liliaceae, Agavaceae, Blandfordiaceae, Amaryllidaceae, and Iridaceae.

Like symperianthy, floral monosymmetry generally remains at an evolutionary level lower than in dicots, again with the exception of the orchids. In the Iridaceae (*Gladiolus*, *Antholyza*, *Babiana*), zygomorphy, including that of guide marks, is known to be determined by gravity (geomorphosis). Development of regular corollas results when such plants are kept on a clinostat. The arcuate bending of stamens and the style, frequent in Liliales and Asparagales, and facilitating sternotribic or nototribic pollination, is also a geomorphosis. In *Anigozanthos*, unilabiate zygomorphy is produced by a secondary ventral splitting of a congenitally united corolla tube. Zygomorphy, in most cases innate, is a family character of the Orchidaceae, Corsiaceae, Philydraceae, Pontederiaceae, Musaceae, Heliconiaceae, Strelitziaceae, and Lowiaceae. The radial flowers of *Iris* and *Moraea* consist of triads of zygomorphic units. Each unit is made up of a tepal of the outer whorl and a petaloid stylodium, and is functionally equivalent to a zygomorphic nototribic flower. In *Cypella* and relatives, the lips formed by the inner whorl.

The phaneranthous perigone (which itself is reduced and sepaloid) is replaced by petaloid staminodia and their fusion products in four zingiberalean families. Especially in the ginger family and Costaceae, perfect gullet and personate flowers have evolved, paralleling in shape those of orchids. The complete loss of intrafloral symmetry in the flowers of the Marantaceae and Cannaceae is utilized in building a highly advanced pollination apparatus. Individual flowers are integrated into a higher level of symmetry, whose plane is displaced outside the flower, governing a partial inflorescence, as in the pollination unit of *Thalia*, which consists of a pair of mirror-image asymmetric flowers.

d) Pollination Mechanisms

Many modes of pollen exposure known from the dicots recur in the monocots. The usually 6 or 3 stamens may be enclosed in the corolla (*Muscari*, *Polygonatum*, *Galanthus*) or, at the other extreme, be long exserted and often possessing versatile anthers, especially in brush-type flowers of collectively pollinated inflorescences as in *Agave*, *Xeronema*, *Haemanthus*, *Massonia*, *Xanthorrhoea*, *Encholirium*, *Dasypogon*, etc. Zygomorphic arrangements of upcurved or downcurved stamens are associated with sternotribic (e.g., *Hippeastrum*, *Hemerocallis*) or nototribic (e.g., *Gladiolus*) pollination. Pollen production may extend over a couple of days as stamens successively rise into functional positions and dehisce (e.g., in *Alstroemeria*). In many Iridaceae and in *Veratrum* with regular flowers the stamens converge, forming a central group with extrorse anthers. Extrorse anthers also occur in nototribic meranthia (*Iris*) and are characteristic of the Alismatales, Araceae, and Triuridaceae. Introrse anthers occur throughout the Asparagales (except Ruscaceae), Cyanastraceae, Zingiberaceae, and Orchidaceae. In other families, both anther types occur, in accordance with the intended positioning of visitors. In the basically diplostemonous androecium reduction of stamen number, enhancing pollen economy and concomitant precision of pollen deposition is common. Flowers with 3 stamens are common; reduction to two stamens is found in the Cypripediaceae, while a single fertile stamen is left in Orchidaceae, Philydraceae and Zingiberaceae. In Marantaceae and Cannaceae only a half anther is fertile. Because a fail-safe attachment of pollen upon the visitor's body is crucial in these monantherous flowers, sticky secretions of the stigma or its derivatives (rostellum) in orchids, Marantaceae, Cannaceae, and *Etlingera*, or mucilage produced by the connective (*Hedychium*, *Alpinia*) are instrumental in the pollination mechanism, serving to attach pollen masses firmly on the visitor (Vogel 1984).

Marantaceae and Cannaceae are precociously protandrous, the pollen being deposited on the style's surface in the bud. Functionally, the flowers are homogamous. The style of the Marantaceae is sensitive (Kunze 1984) and snaps down forcibly when touched; in the course of this single stroke, it (1) scrapes off from the visitor, with a sharp edge, any pollen it carries, (2) loads the stigma with this pollen, (3) smears, by means of a stylar gland, an adhesive upon the visitor, and (4) deposits fresh pollen thereupon. The process is irreversible, and the first and only efficient visitor delivers the pollen it carries and picks up new pollen (Kennedy 1978; Kunze 1984; Vogel 1984; Classen-Bockhoff 1991). Resinous secretions of the spatha or pistils mediate application of loose pollen accumulated at the bottom of the spathe in some Araceae (*Monstera*, Ramirez and Gomez 1978; *Philodendron*, Gottsberger and Amaral 1984). Secondary pollen deposition is also recorded in some Alismatales (Yeo 1993).

Where secondary polyandry occurs in zoophilous monocots, it is, with the exception of the Alismatales, usually linked with the syndrome of pollen flowers (p. 42).

The pollen grains of monocots, rather monotonously sulcate or uniporate, generally have less varied exine sculptures than are found in dicots. The exine is comparatively smooth or granulate, in some zoophilous taxa (but also in *Lemna*) it is spinulose; in most Zingiberales, for example, it is very thin or completely lacking. The zoophilous pollen is usually sticky because of tryphine; in *Cryptocoryne* it is embedded in a sticky fluid and extruded from the anther in creamy masses. Many other Araceae shed dry pollen, which then collects at the bottom of the spathe, and dry pollen is characteristic of the common pollen flower syndrome (p. 42). In most orchids, the anther's pollen content is in compact packets (pollinia), and in many tribes it is transferred in the form of pollinaria by attached, sticky rostellar products (retinacula, viscidia).

Insect-trapping is usually combined with sapromyiophily or cantharophily and deception. Homogamous flowers of Taccaceae, some Burmanniaceae, *Trichopus*, *Pentastemona*, Convallariaceae-Aspidistreae and -Ophiopogoneae, and some orchids (*Pterostylis*), provide a pollination chamber and are presumed to attract small insects which enter and stay inside more or less deliberately for a short time (semitraps), but detailed observations are lacking. In the strictly dichogamous, protogynous inflorescences of Araceae, all levels of insect-trapping mechanisms are represented, from long-lived semitraps (pollination chambers with insects that come in and go out almost without hindrance), to perfect kettle traps (visitors captive; various exactly timed schedules for retention and release; pollen export and potential import by the same individuals). In the pistillate kettle traps of dioecious *Arisaema* and female phase spathes of *Pinellia* visitors are retained until they die. Slippery surfaces and semitraps are found in bee-pollinated *Cypripedium* and *Stanhopea*.

The predominantly papillate stigmas of monocots are either of the Dry or the Wet type (Heslop-Harrison and Shivanna 1977, see p. 46), in the latter case being covered with a watery or oily secretion. As far as is known, wet or moist stigmas are characteristic of Alstroemeriaceae, Philesiaceae, Asteliaceae, Orchidaceae, Lomandraceae, and Cyclanthaceae. Groups characterized by dry stigmas include the zoophilous Alismatales, the Calochortaceae, Convallariaceae, Iridaceae, Eriocaulaceae, Philydraceae, Pontederiaceae, Velloziaceae, and most of the anemophilous monocots. Both types occur within many other families including Liliaceae, Araceae, and palms (Table 1). Among hydrogamous taxa, there are stigmas of the Dry type as well as stigmas provided with a waterproof adhesive (*Amphibolis*, Pettitt et al. 1980).

Alismatales with apocarpous polymerous gynoecia have as many styles and stigmas as carpels; most syncarpous pistils possess a single style. While in a great many zoophilous taxa the stigma is small, usually topping a slender style and emergent beyond the stamens and thus promoting herkogamy, it is subsessile and often large in the aroids and Cyclanthaceae. It is extremely enlarged and umbrellalike in *Tupistra*, where it occludes the flower entrance, leaving only narrow openings for the access and exit of tiny pollinators. The stigma may be entire or trilobed, while in various Iridaceae it is divided into three stylodia which are repeatedly bifurcate or brightly colored, thus playing a part in visual attraction. In the large, petaloid stylodia forming the upper lips of the herkogamous meranthia of *Iris* the stigma is confined to a movable flap which faces entering bees and receives cross-pollen. As the bee withdraws, the flap is forced upward, hiding its receptive surface, and so selfing is precluded. The stylodia of the Bromeliaceae are flabellate with a dilated papillose, receptive distal rim. The rim, however, is contracted to a small area by plicate or helical foldings (Brown and Gilmartin 1989; Schill et al. 1988).

Dichogamy is widespread and sometimes family-specific. Zoophilous families known to be protogynous include Convallariaceae, Dracaenaceae, Araceae, and Cyclanthaceae, while more or less proterandrous flowers are typical of Burmanniaceae, Amaryllidaceae, most Iridaceae, Hemerocallidaceae, Bromeliaceae, Melanthiaceae, and Butomaceae. Other families are predominantly homogamous, or (Alliaceae, Asphodelaceae, Liliaceae, Palmae) exhibit both kinds of dichogamy.

e) Rewarding

Nectar Flowers. The vast majority of zoophilous monocots have nectar flowers, and only nectar-producing groups have strongly adapted to specialized pollinators and undergone adaptive radiation in floral styles (p. 44). The nectar secreted by monocots is like that of dicots and as to the types of sugar solubles it exhibits no particular trends. Many orchid flowers are nectarless but, by shamming presence of nectar, also belong to this ecological class.

Following the comprehensive study of Daumann (1970), carpellary septal nectaries are both restricted to, and widespread in, the monocots, but are by no means universal in this clade (see also Schmid 1988). Their alleged occurrence in dicots (in *Buxus*, *Cneorum*) has not been confirmed (Daumann 1974), but papillate nectar epithelia occur on the carpel flanks of *Caltha* (Smets and Cresens 1988). There are no nectarial disks in monocots. The notion that the evolution of septal nectaries is a synapomorphy of the monocots is still under debate. Septal nectaries occur in at least one family of each order (Table 1), except for the Triuridales, Arales, Orchidaceae, and the anemophilous orders Juncales, Poales, and Cyperales. In their most primitive condition they are freely exposed at the carpel flanks of apocarpous gynoecia, as in entomophilous Alismatales. They are usually concealed in syncarpous ovaries, lining longitudinal clefts inside the septa and drained via three external outlets that may be situated at the top, in the middle, or at the base of the ovary. The clefts are usually separate, but may also merge centrally, forming a three-winged cleft (*Tofieldia*, *Burmannia*, Bromeliaceae, Palmae). In epigynous flowers the septal nectar is discharged distally by peristylar pores at the bottom of the corolla (Amaryllidaceae, Agavaceae, Iridaceae-Ixioideae, Haemodoraceae, Bromeliaceae, Velloziaceae, Cannaceae, Costaceae). In *Wachendorfia* and *Costus*, only two septal clefts are developed and nectariferous. In vertebrate-pollinated epigynous flowers with copious nectar secretion, the secretory surface is often displaced to sterile distal or proximal parts of the ovary and here enormously enlarged by labyrinthic ramification, reminiscent of kidneys. The peristylar nectariferous protrusions of the Zingiberaceae are possibly derived from septal glands.

Anatomically, septal nectaries are epithelial, with secretion oozing from their entire – sometimes (Alliaceae, Themidaceae) papillate – epidermis, which is usually underlain by several layers of storage tissue. Remarkably, the mesenchymatous gland type of nectary, in which nectar discharges via stomatal pores, seems to be missing in monocot flowers, while predominant in the disks of dicots.

The nonseptal nectaries found in monocots are usually epithelial and located on the tepals (perigonal nectaries). As far as known, they are restricted to nectariferous Liliaceae, Orchidaceae and Triuridaceae, Calochortaceae, Alstroemeriaceae, Luzuriagaceae, Melanthiaceae, and *Eriocaulon*; septal nectaries are absent there. In the Iridaceae, perigonal nectaries characterize the subfamily Iridoideae, while in the Ixioideae septal nectaries are developed (Goldblatt 1990). The nectariferous taxa of tribe Mariceae (*Trimezia* spp., *Neomarica*) produce nectar in perigonal gland fields, the gland fields being unicellular hairs. This type seems to be unique in monocots. There are also androecial nectaries (some Hydrocharitaceae and Alismataceae; Colchicaceae, Hemerocallidaceae, Hanguanaceae, and Cyclanthaceae), and rarely nectaries seated on the back of carpels (*Paepalanthus*; Aponogetonaceae?). Finally, in some Araceae the stigma (*Arum*, spp. of *Anthurium*) or staminodes (*Spathicarpa*) exude nectar. All these glands are of the epithelial type.

While in allophilic flowers, which are usually rotate and shallow, the nectar is freely exposed, it is concealed and stored in the depth of the corolla in flowers adapted to specialized (eutropous), tongued or beaked animals. In most cases, the container is a perigone tube. The tube may be extremely elongated and slender in moth-pollinated taxa (*Crinum*, *Hymenocallis*, *Lapeirousia*, Velloziaceae). In *Lilium*, *Strumaria*, and Themidaceae the tube is often subdivided into separate nectar pipes by prominent adnate filament bases that form septa. These compel visitors to probe them consecutively. In *Milla* three canals run down within an elongate receptacle to a distance of 10 cm below the superior ovary, where nectar is secreted. Long tubes, however, are not always nectar pipes exploited by long tongues. Those of *Crocus* merely conduct the fluid by capillarity from subterranean septal nectaries up to the mouth, where it is taken by bees. In *Gethyllis*, *Colchicum*, and some *Iris* spp., also with underground gynoecia, the long perigone tubes have no other function than to expose the limb above the ground, while those of hydrogamous *Elodea* and other Hydrocharitaceae raise the limb above the water surface, substituting a pedicel. Also in *Weldenia* (Commelinaceae) the tube replaces a pedicel. Flowers with perigonal nectaries often present the reward excentrically enough to induce insects to crawl around the (mostly nototribic) sexual organs and to exploit the sources one by one (revolver flowers; *Scoliopus*, *Tricyrtis*, *Veratrum*, *Androcymbium*). Each nectariferous inner tepal of *Neomarica* and species of *Trimezia* has a flexible joint, forming a balance that tilts, when burdened with a probing bee, towards the central sexual column. Nectariferous perigonal furrows — two in *Alstroemeria*, three in *Bomarea* and Philesiaceae, and six in *Lilium* and *Gloriosa* species — are also designed to be exploited separately. The pendent flowers occurring in the latter two genera are visited by hovering lepidopterans that, while circling the perigone, must introduce their proboscis separately into each furrow. *Tricyrtis*, *Herreriopsis*, and *Dichelostemma* store nectar in three pouches. Nectar-containing spurs are common in orchids but extremely rare outside this family; *Disporum calcaratum* with three, and *Gladiolus* (*Kentrosiphon*) *saccatus* with a single spur are examples. Scale-like nectar covers ("Saftdecken") on the perigone are found in *Vriesea* and *Barbacenia*, fimbriate fringes bordering perigonal nectaries occur in *Fritillaria* and *Calochortus*.

Flowers with Other Rewards. Nectarless flowers occur in the abiotically pollinated monocots, but are also widely scattered among the entirely or partially zoophilous orders. In all of these orders there are at least some taxa which lack floral nectaries (Table 1). Save for a few exceptions mentioned above, nectarlessness is widespread in the Arales, while Taccaceae, Tecophilaeaceae, Philydraceae, Commelinaceae, Xyridaceae, Mayaceaceae, Pandanaceae, Cypripediaceae, and Apostasiaceae are characterized by this feature. In the Dioscoreales, Commelinales, and Bromeliales, nectariferous families form the minority; most remaining zoophilous families have at least some taxa with nectarless flowers. While in several orders, such as Liliales and Asparagales, there may have been secondary loss of nectaries, their possible basic absence in groups like Commelinaceae, Dioscoreales, and Pandanales is still a matter of discussion.

Pollen Flowers. These represent the most important zoophilous syndrome involving absence of nectar. They are exclusively melittophilous and offer pollen as the reward to female bees that gather it as a provision for their larvae. They generally have flat or bowl-shaped corollas and freely exposed, conspicuous, usually yellow anthers. Usually being oligandrous, the monocots are restricted in producing a sufficient amount of surplus (fodder) pollen. A few genera have evolved secondary polyandry: *Gethyllis*, *Vellozia*, bee- and beetle-pollinated palms, Cyclanthaceae, and certain entomophilous grasses. However, most pollen flowers have enlarged the volumen of their six, three, or (Philydraceae) even the single anther. Flowers of this type (*Solanum* type, Vogel 1978; Faegri 1986) are predominantly actinomorphic, pendent, and their enlarged poricidal anthers are connivent, forming a central cone around a protruding style; they produce copious loose pollen that is discharged by vibration (buzzing). The Tecophilaeaceae, various Lomandraceae *Echeandia*, *Sowerbaea*, *Calectasia*, *Apostasia*, and

Galanthus belong here. In other pollen flowers, the androecium is more or less spreading, sometimes with poricidal anthers and dry pollen (*Dianella*, *Thysanotus*, Rapateaceae), sometimes with longicidal anthers and ± sticky pollen (*Narthecium*, *Chlorophytum*, *Arthropodium*, *Stypandra*, *Libertia*, Hypoxydaceae, Xyridaceae, Commelinaceae pp.). Stamen dimorphism of pollen flowers with conspicuous fodder anthers and fertilization anthers (often with cryptically colored pollen) (heteranthery), occurs in the monosymmetric flowers of *Cyanella*, *Monochoria*, and several genera of the Commelinaceae. Certain oligandrous pollen flowers pretend more copious reward than is really available by shamming pollen or additional anthers (partial deception). Yellow-colored pollen dummies are produced by sterile anthers (*Commelina*, *Murdannia*), anther or filament appendages (e.g., *Dianella*, *Arthropodium*, *Dichopogon*) or hair tufts associated with true anthers (*Coleotrype* and other Commelinaceae, *Bulbine*, *Narthecium*, *Tricoryne*, *Xyris*).

Oil Flowers. The flowers of a small number of nectarless monocots secrete a fatty oil, which is collected by specialized anthophorid bees for brood rearing and probably for lining the walls of their underground brood cells. The taxa involved are all neotropical. The liquid is produced by special glands (elaiophores); these are sometimes arranged in pairs, as the oil is always scraped by the legs. The elaiophores consist of glandular hairs on the tepals or filament columns in species of *Sisyrinchium*, in various genera of the iridaceous tribe Mariceae, and in at least four genera of subtribe Ornithocephalinae (Orchidaceae). In species of *Maxillaria*, *Oncidium*, and in *Sigmatostalix*, the lipid is produced by warty epithelia of the labellum (Vogel 1974).

Food Tissues. Starchy tissue forms another edible reward; in several beetle-pollinated aroids either the interior flank of the spathe, spent male flowers (and pollen), or staminodes are regularly gnawed by pollinating coleopterans; some of them even breed in the decaying inflorescences. Edible flower parts of Cyclanthaceae and palms are eaten or used for oviposition mainly by weevils (Curculionidae) which transfer the pollen. The fleshy, colored bracts next to the floral spikes of *Freycinetia* (Pandanaceae) are not only a visual attractant, but also provide food for vertebrates: predominantly birds in ornithophilous, and mainly flying foxes in chiropterophilous species.

Housing. The presentation of sleeping holes for bees as a reward in flowers of *Serapias* appears to be unique in flowering plants (Dafni et al. 1981; Paulus and Gack 1994). The inflorescences of beetle-pollinated palms, Cyclanthaceae, and aroids also provide shelter for their visitors, a feature that may play a functional role.

Perfume Flowers. This melittophilous, entirely neotropical floral syndrome is characterized by the production of powerful spicy volatiles which are exuded in small droplets on the glandular surface of flower parts. It is only male bees of the family Euglossini that are attracted and pollinate these flowers. They collect the aromas which consist of specific mixtures of terpenoids, benzoids, hydrocarbons, etc., by brushing the glands with their front legs. The males use the exudate as a pheromone in their mating display in a way which is still incompletely explored (Vogel 1966; Dressler 1992; Williams and Whitten 1983). The orchid tribes *Catasetinae*, *Stanhopeinae*, and *Lycastinae*, altogether with ca.625 species in 55 genera, produce such liquid perfumes as an attractant and reward, in some genera combined with most sophisticated sliding and catapult mechanisms. In monocots, the perfume syndrome is also known from *Spathiphyllum* and spp. of *Anthurium*. In addition, scent collecting has been observed in *Xiphidium* (Haemodoraceae).

f) Floral Deception

Totally deceptive flowers that are devoid of any real reward are disproportionally common. They are found in up to half of orchid species; this family includes 34% of all monocot species. The presence of both nutritive and non-nutritive rewards may be simulated (Vogel 1993). Among the former, false nectar flowers prevail; bees or birds visit them. The empty floral spurs of *Orchis* and *Dactylorhiza* are classical examples, already recognized by Sprengel (1793). Deception of bee females by false pollen occurs in species of *Maxillaria* and *Polystachya*. Most sapromyiophilous monocots (see Table 1) simulate nutritive rewards or even brood sites; both color and smell help elicit instinctive attempts at oviposition in dipterans (some flowers secrete small amounts of nectar which, however, is not the motive for visitation). Plants with such flowers are found in several tribes of Araceae, many orchids (Pleurothallidinae, Bulbophyllinae), Stemonaceae, Taccaceae, Trichopodaceae, *Ferraria*, *Scoliopus*, *Eucomis*, and

Pentastemona. Larvae hatching from any eggs deposited will probably die due to the inappropriate food stuff they encounter. Species of *Arisaema* and some orchids imitate mushrooms and are pollinated by mycetophilids (Vogel 1978). Sexual deceit of bees and flies is confined to monocots. Such mimics operate by imitating female mates in shape, coloration, and scent. This has been intensively studied in *Ophrys* (Borg-Karlson 1990; Paulus and Gack 1994), and also occurs in a number of mainly Australian terrestrial orchids (Dafni and Bernhardt 1990).

g) Floral Styles and Adaptive Radiation

Annonaceae, Aristolochiaceae, and Piperaceae, believed to be close to the phyletic roots of monocots, are too specialized anthecologically to suggest a direct link with syndromes found in primitive monocots. However, a strong tendency towards myiophily and cantharophily exhibited by the Dioscoreales, Aspidistreae, Araceae, and Cyclanthaceae may still reflect a trait prevalent in ages when flies and beetles were the main agents in the floral world. In Araceae, an advancement has since taken place, paralleling trends in the Annonaceae, by incorporating large-sized modern dynastid coleopterans as pollinators (*Philodendron* and allies, Gottsberger and Silberbauer-Gottsberger 1991). Participation in the beetle-pollinated "poppy guild" (Dafni et al. 1990) by members of *Aristea* (Goldblatt and Manning 1996) and *Tulipa* and *Calochortus* is probably also a relatively recent evolutionary event. Certain complicated myiophilous contrivances of the Burmanniaceae, Taccaceae, and *Aspidistra* appear to be derived as well. Pollination strategies are rather uniform in the above-mentioned families, and also in anemophilous groups. However, an intensive adaptive radiation into different highly adapted floral styles has taken place within families and genera of the Liliales, Asparagales, Zingiberales, and Velloziaceae. Here the possession of septal or perigonal nectaries has given rise to flowers adapted to the more or less complete spectrum of guilds of tongued, anthophilous insects (myiophily, melittophily, psychophily, sphingophily, phalenophily), and vertebrates (ornithophily and chiropterophily), repeating the functional floral styles of euphilic dicots. Particular adaptations reflect a common history of interaction with endemic vectors. Neotropical Marantaceae (*Calathea, Ischnosiphon, Monotagma*) are adapted to long-tongued euglossine bees, species of *Lapeyrousia* and *Disa* from the Cape are associated with long-tongued nemestrinid and pangonid dipterans; paleotropical *Antholyza, Brunsvigia, Strelitzia, Burbidgea, Xeronema*, but also the neotropical *Puya* provide special perches to passerine birds; or geoflory in *Etlingera* (*Achasma*) is connected with pollination by birds hopping on the ground. Musaceae, Strelitziaceae, and Cannaceae are almost exclusively characterized by bird, bat or lemur pollination, suggesting a long association with vertebrate pollinators. The hummingbird-adapted flowers so common in bromeliads and in *Heliconia* must have evolved in more recent, Post-Gondwanian times.

4. Aphananthy and Abiotic Pollination

a) Anemophily

About 17000 species, about one third of the monocots, display abiotic pollination by wind or water. Habitual anemophily is characteristic of the Cyperales, Juncales, Poales, Typhales, and *Pandanus*. It occurs in genera of palms and the Alismatales. It has also been reported in taxa scattered over other orders, but most of these cases are weakly supported: *Navia* (Bromeliaceae), *Paris* (Trilliaceae), *Acorus*. The occurrence of anemophilous taxa within predominantly animal-pollinated genera is much rarer than in dicots.

Wind-pollinated monocots and dicots share a characteristic syndrome of features that lack pollinator rewards, having small, usually unisexual flowers. In pistillate flowers the ovules may be reduced to one, and the pollen/ovule ratio is high. Flowers are often numerous in heads, spikes or loose panicles of spikelets. The anthers are movable, the pollen is nonsticky and loose; the stigma is large, often feathery or tufted (Juncaginaceae, Cyperaceae, Poaceae, Restionaceae, *Joinvillea*; Endress 1995). The perigone is bract-like, often glumaceous, always inconspicuous, usually greenish in color, reduced or completely missing. Wind-pollinated monocots tend to occur in large, homogeneous stands in open areas, forming grass savannas, steppes, swamp vegetation, palm and bamboo forests, etc., where chances of wind-blown pollen reaching conspecific individuals are favorable.

b) Hydrogamy

Adaptation to water as a pollen vector is nearly confined to monocots. It characterizes six alismatalean families, and also occurs in various Hydrocharitaceae (Cook 1982) and the Hydatellaceae. The syndrome shares many traits with anemophily and is probably derived from it. Hydrogamous monocots account for the bulk of submerged aquatic plants; Cymodoceaceae, Zosteraceae, and Posidoniaceae are the only marine hydrophilous angiosperms. A number of hydrophiles, including *Vallisneria*, *Elodea* spp., *Ruppia*, some species of the predominantly anemophilous genus *Potamogeton*, and several seagrasses, are ephydrogamous (Cook 1982; Cox and Knox 1989). The pollen, often collectively dispersed in rafts or abscised staminate flowers, drifts on the water surface towards the pistillate flowers, being driven by water currents or wind. Lemnaceae probably also belong here, their fronds meet as they drift together; "haptogamy"). In the remaining hydrophiles, pollen transport is underwater (hyphydrogamy). Their flowers remain submerged, and their exine-less, wettable, sometimes filiform (up to 5 mm long) pollen is floating, suspended in the water and is caught by the forked stigmas (Pettitt et al. 1981).

c) Aphananthy and the Problem of Secondary Zoophily

As already noted above, aphananthy is not restricted to abiotic pollination syndromes. It is also associated with zoophily, as in many Dioscoreaceae, Smilacaceae, Convallariaceae-Ophiopogoneae, palms, Cyclanthaceae, most Araceae, and the Eriocaulaceae. Here, the inconspicuous flowers are attractive to insects because of their scent and (or) nectar or are (in Araceae, Eriocaulaceae, *Freycinetia*) arranged in pseudanthia and associated with showy bracts and spathes.

On the other hand, a number of species in generally wind-pollinated lineages show features of an incomplete zoophily (allophily), including scent and white or yellow color of anthers, glumes, or bracts. These species are often tropical forest dwellers and are visited by insects searching for pollen; they include Poaceae (*Pariana*, Soderstrom and Calderon 1971), Cyperaceae (species of *Cyperus*, *Carex*, *Rhynchospora* sect. *Dichromena*, *Mapania*, *Hypolytrum*, *Ascolepis*, *Ficinia*, and *Chrysithrix*), Juncaceae (*Luzula* spp.), and Pandanceae with entomophilous species of *Pandanus* and *Sararanga* (Cox 1990).

Zoophily of aphananthous monocot taxa is usually interpreted as a reversal from abiotically pollinated ancestors and assumed to have evolved in response to shady habitats where air currents are weak. When considering the monocots as monophyletic, this hypothesis would imply a secondary reaquisition of the basically zoophilous condition of the whole clade. This is certainly true in particular cases, but a doubtful explanation for the occurrence of aphananthy in monocots and angiosperms in general. The very early advent of this trait, documented by Cretaceous fossils (Herendeen and Crane 1995) and including members of Araceae, Poaceae, Restionaceae, Pandanaceae, Typhales, and Palmae, argues against the view that the prototype of the monocot flower was phaneranthous and zoophilous. The initial design of monocot flowers may have been "undecided", ecologically indifferent, possessing an inconspicuous, bracteose perigone and still lacking nectaries, being open for adaptation in either direction. The palms, where the switch between anemophily and an imperfect entomophily apparently has occurred more than once (Silberbauer-Gottsberger 1990), parallel this situation. In several lineages a tendency toward floral reduction and inability to acquire nectaries led to anemophily or, as in the aroids, to the evolution of zoophily with other means of attraction.

5. Breeding Systems

Monoecy, dioecy, and polygamy are scattered over many orders, but are most frequent among the anemophilous and hydrogamous families. Some small families are entirely dioecious: Asparagaceae, Behniaceae, Stemonaceae, Hanguanaceae, and Cymodoceaceae. Dioecy is frequent (75% of species) in the hydrogamous Alismatales (Cox and Knox 1989); wind-pollinated dioecious genera occur in the Cyperales and Juncales. Among zoophilous taxa, there are dioecious genera or species in Dioscoreaceae, Smilacaceae, Colchicaceae, Corsiaceae (as androdioecy), Triuridaceae, Asteliaceae, Melanthiaceae, Ruscaceae, Nolinaceae, Bromeliaceae, and Palmae. As a rule, sex expression is genetically fixed, but in *Arisaema*, *Catasetum*, and *Cycnoches* it is labile and dependent merely on the vigor of the individual. Monoecy occurs throughout the Typhales, Musaceae, Centrolepidaceae, Pandanaceae, Cyclanthaceae, Aponogetonaceae, and Hydatellaceae; it is predominant in Alismatales, Palmae, and Eriocaulaceae, frequent in Commelinaceae (as andromonecy) and Cyperaceae, and occurs sporadically in the Triuridaceae, Melanthiaceae, and Juncaceae.

Knowledge of the distribution and genetics of self-incompatibility is still too scanty to allow wide generalizations. Self-sterility is rare or absent in Araceae, Palmae, Alismatales, Bromeliaceae, Heliconiaceae, Marantaceae, and Juncaceae; on the other hand, dichogamy (p. 41), herkogamy, or dicliny are common there. About 5% of orchid species are estimated to be self-sterile (East 1940). As is well known, interspecific fertilization barriers are weak in orchids, in part possibly due to the lack of a controlling endosperm. Intergeneric and even intertribal (usually sterile) crosses are possible but occur only rarely in nature, owing to the narrow, often species-bound pollinator specificity of this family (cf. van der Pijl and Dodson 1960). The Commelinaceae and grasses are largely self-incompatible, except for many cereals (Connor 1979; Richards 1986). Even monoecious seagrasses are probably self-sterile (Pettitt et al. 1981). Within many families tested so far, both self-incompatible and self-fertile (including habitually autogamous) taxa occur (cf. Fryxell 1957; Charlesworth 1985).

Self-incompatible monocots share with dicots, as already mentioned, both gametophytic and sporophytic incompatibility. The former is usually linked with the presence of a moist stigma (cf. Table 1) and binucleate pollen (self-pollen tube growth halted in the style), the latter is usually associated with dry stigmas and trinucleate pollen (self-pollen inhibition at the stigma surface; Heslop-Harrison and Shivanna 1977). The latter mode occurs in underwater-pollinated Hydrocharitaceae (Pettitt 1980). While in these systems the genetic determination is in most cases monogenic, (with many alleles), Poaceae deviate in possessing two-gene-, poly-allelic gametophytic incompatibility associated with a dry stigma and trinucleate pollen (De Nettancourt 1977; Richards 1986). Heterostyly (see Table 1) is rare and partly atypical in monocots. Genetics and compatibility patterns in *Nivenia* are discussed by Goldblatt & Bernhardt (1990), in the Pontederiaceae by Barrett (1993), and in *Narcissus* by Barrett et al. (1993).

The occurrence of cleistogamous flowers is known from members of Poaceae (with 70 genera being involved, Connor 1979), Alliaceae (*Allium spp.*), Hydrocharitaceae (*Ottelia*), Pontederiaceae, Commelinaceae, Burmanniaceae, Juncaceae, Marantaceae (*Calathea* spp.), and Orchidaceae. Agamospermy has been described, among others, in *Musa*, species of *Maianthemum*, *Pandanus*, some grasses and orchids.

Selected Bibliography

Barrett, S.C.H. 1993. The evolution of heterostyly. Berlin: Springer.

Barrett, S.C.H., Lloyd, D.G., Arroyo, J. 1993. Stylar polymorphisms and the evolution of heterostyly in *Narcissus*, pp. 339–376, in: Lloyd D.G., Barrett, S.C.H. (eds.) Floral biology. Studies on floral evolution in animal-pollinated plants. New York: Chapman & Hall.

Biedinger, N., Barthlott, W. 1993. Untersuchungen zur Ultraviolettreflexion von Angiospermenblüten I: Monocotyledoneae. Trop. Subtrop. Pflanzenwelt 86. F. Steiner, Stuttgart.

Boehme, S. 1988. Bromeliaceenstudien. III. Vergleichende Untersuchung zu Bau, Lage und systematischer Verwertbarkeit der Septalnektarien von Bromeliaceen. Trop. Subtrop. Pflanzenwelt (Akad. Wiss. Lit. Mainz) 62. 154 pp.

Borg-Karlson, A.K. 1990. Chemical and ethological studies of pollination in the genus *Ophrys* (Orchidaceae). Phytochemistry 29: 1359–1387.

Brown, G.K., Gilmartin A.J. 1989. Stigma types in Bromeliaceae – a systematic survey. Syst. Bot. 14: 110–132.

Charlesworth, D. 1985. Distribution of dioecy and self-incompatibility in angiosperms. In: Greenwood, P.H., Slatkin, M. (eds.) Evolution: essays in honour of John Maynard Smith. Cambridge Univ. Press, Cambridge, pp. 237–268.

Classen-Bockhoff, R. 1991. Untersuchungen zur Konstruktion des Bestäubungsapparates von *Thalia geniculata* (Marantaceae). Bot. Acta 194: 183–193.

Connor, H.E. 1979. Breeding systems in the grasses: a survey. NZJ. Bot. 17: 547–574.

Cook, C.D.K. 1982. Pollination mechanisms in the Hydrocharitaceae. In: Symoens, J.J., Hooper, S.S., Compère, P. (eds.) Studies on aquatic vascular plants. Roy. Bot. Soc. Brussels, pp. 1–15.

Cox, P.A. 1990. Pollination and the evolution of breeding systems in Pandanaceae. Ann. Mo. Bot. Gard. 77: 816–840.

Cox, P.A., Knox, R.B. 1989. Two-dimensional pollination in hydrophilous plants: convergent evolution in the genera *Halodule* (Cymodoceaceae), *Halophila* (Hydrocharitaceae), *Ruppia* (Ruppiaceae), and *Lepilaena* (Zanichelliaceae). Am. J. Bot. 76: 164–175.

Dafni, A., Bernhardt, P. 1990. Pollination of terrestrial orchids of southern Australia and the Mediterranean region. In: Hecht, M.K., Wallace, B., Macyntire, R.J. (eds.) Evol. Biol. 24: 193–252.

Dafni, A., Ivri, Y., Brantjes, N.B.M. 1981. Pollination of *Serapias vomeracea* Briq. (Orchidaceae) by imitation of holes for sleeping solitary male bees (Hymenoptera). Acta Bot. Neerl. 30: 60–73.

Dafni, A., Bernhardt, P., Shmida, A., Ivry, Y., Greenbaum, S., O'Toole, Ch., Losito, L. 1990. Red bowl-shaped flowers: convergence for beetle pollination in the Mediterranean region. Isr. J. Bot. 39: 81–92.

Dahlgren, R.M.T., Clifford, H.T. 1982. The monocotyledons, a comparative study. London: Academic Press.
Dahlgren, R.M.T. et al. 1985. See general references.
Daumann, E. 1970. Das Blütennektarium der Monokotyledonen unter besonderer Berücksichtigung seiner systematischen und phylogenetischen Bedeutung. Fedde's Repert. 80: 463–590.
Daumann, E. 1974. Zur Frage nach dem Vorkommen eines Septalnektariums bei Dicotyledonen, zugleich ein Beitrag zur Blütenmorphologie und Bestäubungsökologie von *Buxus* L. und *Cneorum* L. Preslia 46: 97–109.
De Nettancourt, D. 1977. Incompatibility in angiosperms. Berlin Heidelberg New York: Springer.
Dixon, K.W. 1985. The underground orchids of Australia – an appraisal. Orchidarian 8: 75–79.
Dressler, R. 1992. Biology of the orchid bees (Euglossini). Ann. Rev. Ecol. Syst. 13: 373–394.
East, E.M. 1940. The distribution of self-sterility in the flowering plants. Proc. Am. Philos. Soc. 82: 449–518.
Endlicher, S. (1836–1840). Genera plantarum, Vol. 5. Beck, Vienna.
Endress, P. 1994. Diversity and evolutionary biology of tropical flowers. Cambridge: Cambridge Univ. Press.
Endress, P. 1995. Major evolutionary traits of monocot flowers. In: Rudall, P.J., Cribb, P.J., Cutler, D.F., Humphries, L.J. (eds.) Monocotyledons: systematics and evolution. Royal Botanic Garden, Kew, pp. 43–79.
Faden, R.B. 1992. Floral attraction and floral hairs in the Commelinaceae. Ann. Mo. Bot. Gard. 79: 46–52.
Faegri, K. 1986. The solanoid flower. In: Alexander I.J., Gregory N.M. (eds.) Bot. Soc. Edinburgh 150th Anniversary Supple, Transactions, Edinburgh, pp. 51–59.
Fryxell, P.H. 1957. Mode of reproduction of higher plants. Bot. Rev. 23: 135–233.
Goldblatt, J.C. 1990. Phylogeny and classification of Iridaceae. Ann. Mo. Bot. Gard. 77: 607–627.
Goldblatt, P., Bernhardt, P. 1990. Pollination biology of *Nivenia* (Iridaceae) and the presence of heterostylous self-compatibility. Isr. J. Bot. 39: 93–111.
Goldblatt, P., Manning, J.C. 1996. Aristeas and beetle pollination. Veld Flora 82: 17–19.
Gottsberger, G., Amaral, A. 1984. Pollination strategies in Brazilian *Philodendron* species. Ber. Dtsch. Bot. Ges. 97: 391–410.
Gottsberger, G., Silberbauer-Gottsberger, I. 1991. Olfactory and visual attraction of *Erioscelis emarginata* (Cyclocephalini, Dynastidae) to the inflorescences of *Philodendron selloum* (Aracee). Biotropica 23: 23–28.
Guédès, M. 1979. Morphology of seed plants. J. Cramer, Vaduz.
Harborne, J.B. 1995. Small molecules and monocot classification. In: Rudall, P.J., Cribb, P.J., Cutler, D.F., Humphries, L.J. (eds.) Monocotyledons: systematics and evolution. Royal Botanic Garden, Kew, pp. 201–215.
Henderson, A. 1986. A review of pollination studies in the Palmae. Bot. Rev. 52: 221–259.
Herendeen, P.S., Crane, P.R. 1995. The fossil history of the monocotyledons. In: Rudall, P.J., Cribb, P.J., Cutler, D.F., Humphries, L.J. (eds.) Monocotyledons: systematics and evolution. Royal Botanic Garden, Kew, pp. 1–23.
Heslop-Harrison, Y., Shivanna K.R. 1977. The receptive surface of the angiosperm stigma. Ann. Bot. 41: 1233–1258.
Ihlenfeldt, H.-D. 1971. über ontogenetische Abbreviationen und Zeitkorrelationsänderungen und ihre Bedeutung für Morphologie und Systematik. Ber. Dtsch. Bot. Ges. 84: 91–107.
Kennedy, H. 1978. Systematics and pollination in the "closed-flowered" species of *Calathea* (Marantaceae). Univ. Calif. Publ. Bot. 7: 1–90.
Knuth, P. (1899, 1904). Handbuch der Blütenbiologie II,2 and III,1. Leipzig: W. Engelmann.
Kress, W.J. 1984. Pollination and reproduction biology of *Heliconia*. In: D'Arcy, W.G., Correa, A. (eds.) The botany and natural history of Panama. Monogr. Syst. Bot. Missouri Bot. Gard. 10: 267–271.
Kunze, H. 1984. Vergleichende Studien an Cannaceen- und Marantaceenblüten. Flora 175: 301–318.
Leinfellner, W. 1963. Das Perigon der Liliaceen ist staminaler Herkunft. Österr. Bot. Z. 110: 448–467.
Paulus, H., Gack, C. 1994. Signalfälschung als Bestäubungsstrategie in der mediterranen Orchideengattung *Ophrys* – Probleme der Artbildung und der Artabgrenzung. Proc. Int. Symp. European Orchids, Utrecht/Haarlem.
Pettitt, J.M. 1980. Reproduction in seagrasses: nature of the pollen and receptive surface of the stigma in the Hydrocharitaceae. Ann. Bot. 45: 257–271.
Pettitt, J., McConchie, L., Ducker, S., Knox, R. 1980. Unique adaptations for submarine pollination in seagrasses. Nature 286: 487–489.
Pettitt, H.M., Ducker, S.C., Knox, R.B. 1981. Submarine pollination. Sci. Am. 244(3): 92–101.
Pijl, L., van der, Dodson, C.H. 1960. Orchid flowers, their pollination and evolution. Fairchild Tropical Garden and University of Miami Press, Coral Gables.
Ramirez, B.W., Gomez, P.L.D. 1978. Production of nectar and gums by flowers of *Monstera deliciosa* (Araceae) and of some species of *Clusia* (Guttiferae) collected by New World *Trigona* bees. Brenesia 14/15: 407–412.
Richards, A.J. 1986. Plant breeding systems. London: George Allen & Unwin.
Schill, R., Dannenbaum, C., Jentzsch, E.-M. 1988. Untersuchungen an Bromeliennarben. Beitr. Biol. Pflanz. 63: 221–252.
Schill, R., Pfeifer, W. 1977. Untersuchungen and Orchideenpollinien unter besonderer Berücksichtigung ihrer Feinskulptur. Pollen Spores 19: 5–118.
Schmid, R. 1988. Reproductive versus extra-reproductive nectaries – historical perspective and terminological recommendations. Bot. Rev. 54: 230–239.
Silberbauer-Gottsberger, I. 1990. Pollination and evolution in palms. Phyton (Austria) 30: 213–233.
Simpson, M.G. 1990. Phylogeny and classification of the Haemodoraceae. Ann. Mo. Bot. Gard. 77: 722–784.
Simpson, M.G. 1993. Septal nectary anatomy and phylogeny of the Haemodoraceae. Syst. Bot. 18: 593–613.
Smets, E.F., Cresens, E.M. 1988. Types of floral nectaries and the concepts "character" and "character state" – a reconsideration. Acta Bot. Neerl. 37: 121–128.
Soderstrom, T.R., Calderon, C.E. 1971. Insect pollination in tropical rain forest grasses. Biotropica 3: 1–16.
Sprengel, C.K. 1793. Das entdeckte Geheimnis der Natur im Bau und in der Befruchtung der Blumen. Berlin: Vieweg.
Vogel, S. 1959. Organographie der kapländischen Ophrydeen, mit Bemerkungen zum Koaptations-Problem I, II.-Abh. Akad. Wiss. Lit. (Mainz) Math.-Naturwiss Kl. 6 and 7. F. Steiner, Wiesbaden.
Vogel, S. 1966. Parfümsammelnde Bienen als Bestäuber von Orchideen und *Gloxinia*. Österr. Bot. Z. 113: 302–361.
Vogel, S. 1974. Ölblumen und ölsammelnde Bienen. Trop. Subtrop. Pflanzenwelt 7: 283–547.
Vogel, S. 1978. Evolutionary shifts from reward to deception in pollen flowers. In: Richards, A.J. (ed.) The pollination of flowers by insects, London: Academic Press pp. 89–96.

Vogel, S. 1981. Bestäubungskonzepte der Monokotylen und ihr Ausdruck im System. Ber. Dtsch. Bot. Ges. 94: 663–675.

Vogel, S. 1984. Blütensekrete als akzessorischer Pollenkitt. Abstr. 1211 in Mitt. Tagung Dtsch. Bot. Ges. Vienna 9–14 Sept. 1984.

Vogel, S. 1990. The role of scent glands in pollination. On the structure and function of osmophores. Smithsonian Inst. Libraries and the Nat. Sci. Foundation Washington. Amerind, New Delhi.

Vogel, S. 1993. Betrug bei Pflanzen: Die Täuschblumen. Abh. Akad. Wiss. Lit. Mainz Math.-naturwiss Kl. Jg. 1993, 1. Stuttgart: F. Steiner.

Weber, A. 1980. Die Homologie des Perigons der Zingiberaceen. Ein Beitrag zur Morphologie und Phylogenie des Monokotylen-Perigons. Plant Syst. Evol. 133: 149–179.

Williams, N.H., Whitten, W.M. 1983. Orchid floral fragrances and male euglossine bees: methods and advances in the last sesquidecade. Biol. Bull. 164: 355–395.

Winkler, H. 1906. über den Blütendimorphismus von *Renanthera lowii*. Ann. Jard. Bot. Buitenzorg 20: 1–12.

Yeo, P.F. 1993. Secondary pollen presentation. Form, function and evolution. Plant. Syst. Evol. Suppl. 6. Vienna New York: Springer.

Conspectus of Families Treated in this Volume

K. KUBITZKI

A. Nartheciaceae

Rhizomatous; leaves equitant or flat; raphides usually lacking; flowers hypogynous; stamens introrse; nectar secretion, where present, from carpels; endosperm formation Helobial; chromosomes very small (0.7–1.7 μm long). Only one family. 10/72, temperate regions of Northern Hemisphere and northern S America **Nartheciaceae**

B. Lilianae

Herbaceous or woody; raphides often present; silica bodies lacking (excl. part of orchids); stomata anomocytic (excl. *Aloe*, orchids pro parte, *Ripogonum*); vessels present in the roots and less frequently in stems and leaves; perianth mostly petaloid (scarious or differentiated in calyx and corolla); 3 carpels fertile, placentation mostly marginal; endosperm development Helobial or Nuclear; endosperm without or – before maturity – with little starch (with abundant starch in Velloziaceae and Acanthochlamydaceae); epicuticular wax often in parallel platelets

1. Liliales

Geophytes or rosulate perennials (shrubs); flowers hypogynous and epigynous; nectar secretion from tepals or filament bases (nectaries also septal in Trilliaceae, Smilacaceae); anthers introrse (mostly extrorse in Colchicaceae); endosperm formation Nuclear (Helobial in part of Trilliaceae). Chromosomes >1.7 μm long, usually much larger

1 Flowers hypogynous or rarely half-epigynous
  2 Rhizomatous herbs, shrubs or climbers; fruit baccate; [leaf venation reticulate]
    3 Herbaceous; pedicels articulated; anther dehiscence by slits or pores; [pistil with a style; ovules bitegmic-tenuinucellate]. 2/6, southern S America, Australia, and New Zealand **Luzuriagaceae**
    3 Woody; pedicels not articulated; anther dehiscence by slits
      4 Climbing or erect shrubs; stamens bi- or tetrasporangiate; carpels with 3 free styluli; ovary 3-locular; nectaries perigonal and septal. 3/320, tropical and warm-temperate, widely distributed **Smilacaceae**
      4 Erect or twining shrubs; stamens tetrasporangiate; carpels forming a single style; ovary 1-locular; nectaries perigonal. 2/2, S Chile **Philesiaceae**
  2 Rhizomatous, bulbous, or cormous perennials; fruit mostly capsular
    5 Raphides present; parietal cell present
      6 Leaves not whorled, not petiolate; venation parallel; inflorescence many-flowered; nectaries perigonal; embryo sac monosporic or bisporic; fruit a capsule; seeds often compressed or winged or both; Veratrum-type alkaloids present. 12/97, temperate regions of Northern Hemisphere **Melanthiaceae**
      6 Leaves (pseudo-)whorled, often petiolate; venation reticulate; flowers solitary; nectaries perigonal and septal; embryo sac bisporic; fruit a berry or capsule; seeds not compressed or winged but enclosed in aril or with sarcotesta. 4/70, cold to warm temperate regions of Northern Hemisphere **Trilliaceae**
    5 Raphides wanting; parietal cell absent
      7 Plants from tunicated corms or rhizomes, often with crystal sand and colchicine alkaloids; stamens dorsifixed; anthers mostly extrorse; embryo sac Polygonum type; [carpels ending in free styluli]. 19/215, tropical and temperate regions of all continents except S America **Colchicaceae**
      7 Plants from bulbs or rhizomes; crystal sand and colchicine alkaloids wanting; stamens often pseudobasifixed, i.e., anther base tubular, concealing the filament tip; anthers introrse
        8 Inner and outer tepals usually similar; carpels united into a distinct style; capsules loculicidal or septicidal; embryo sac variable, mostly tetrasporic. 11/560, temperate regions of Northern Hemisphere **Liliaceae**
        8 Inner and outer tepals quite distinct; carpels ending in free styluli, rarely united into a short style; embryo sac Polygonum type. 5/100, temperate regions of Northern Hemisphere **Calochortaceae**
1 Flowers epigynous
  9 Woody, spinescent, climbing by leaf-opposed, sterile, tendrillar inflorescences. 1/1, Australia **Petermanniaceae**
  9 Non-climbing herbs
    10 Roots tuberous; floral bracts pseudowhorled; tepals brightly colored and patterned, caducous after anthesis; anthers pseudobasifixed, introrse; no parietal cell formed. 5/160, tropical to temperate C and S America **Alstroemeriaceae**
    10 Roots not tuberous; floral bracts not pseudowhorled; tepals greenish, persistent; anthers not pseudobasifixed, at least the outer extrorse; a parietal cell formed. 2/4, islands of western Pacific **Campynemataceae**

2. Asparagales

Herbaceous perennials from a rhizome, bulb, corm, or tuber, often with secondary thickening, (woody, arboreous); flowers hypogynous or epigynous, often articulated from pedicel and with or without a pericladium; septal nectaries usually present; fruit capsular, loculicidal, (septicidal; baccate; a winged or unwinged nut; a schizocarp); seeds phytomelan incrusted or not; steroid saponins consistently in "higher" Asparagales present; 32 families, many of them from Southern Hemisphere

1 Microsporogenesis simultaneous, (successive: Hypoxidaceae, Xanthorrhoeaceae; unknown: Blandfordiaceae, Boryaceae); steroid saponins scattered **("Lower" Asparagales)**
  2 Flower epigynous, not articulated from pedicel
    3 Stamens and style fused into a gynostemium; capsules septicidal and loculicidal at the same time; [perennial herbs, often epiphytic, frequently with hypogeal or epigeal storage organs]; flowers zygomorphic; ovules bitegmic, tenuinucellate; endosperm development Nuclear; seeds minute, mycotrophic]. ca. 900/>20 000, cosmopolitan **Orchidaceae**
    3 Stamens and style not fused; capsules not septicidal and loculicidal at the same time
      4 Stamens 3(2) + 0, [extrorse or latrorse; herbs or shrubs from rhizomes, bulbs, or corms, (saprophytic: *Geosiris*); leaves mostly distichous, equitant; flowers petaloid; nectaries septal (perigonal: *Isophysis*, *Geosiris*); (ovary superior: *Isophysis*); endosperm development Nuclear (Helobial: *Isophysis*, *Geosiris*); seeds brown]. 76/1750, cosmopolitan, but mostly S African **Iridaceae**
      4 Stamens 3 + 3
        5 Seeds brown; [giant rosulate perennials from vertical subterranean stem; anthers pseudobasifixed]. 1/2, Australia **Doryanthaceae**
        5 Seeds black
          6 Capsules 1-seeded; tepals connate, feathery-hairy outside; septal nectaries present; [rhizomatous herbs; anthers dorsifixed, introrse; raphides lacking]. 1/1, S Africa **Lanariaceae**
          6 Capsules many-seeded; tepals distinct; septal nectaries lacking
            7 Leaves distichous; flowers on leafy peduncles; carpels fused into a style. 1/3, N Africa, SW and inner Asia **Ixioliriaceae**
            7 Leaves tristichous; inflorescence scapose; carpels ending in free styluli or a short style. 9/100, widely distributed but mostly in Southern Hemisphere **Hypoxidaceae**
  2 Flowers hypogynous; flower articulation variable
    8 Pollen invariably trichotomosulcate (sulcate in *Hemerocallis*); [leaves distichous; flowers usually articulated; seeds black; ovules bitegmic, tenuinucellate]
      9 Rhizomatous or stilt-rooted; leaves setaceous; anthers dorsifixed, introrse-longicidal; fruit a capsule, nut or schizocarp; seeds mostly arillate. 8/50, Australia, New Guinea, Madagascar, S Africa **Johnsoniaceae**
      9 Rhizomatous or caespitose; leaves ovate to linear; flowers hypogynous or perigynous; anthers basifixed or dorsifixed, dehiscing latrorsely by pores or slits; fruit a berry or capsule; seed exarillate. 13/50, cosmopolitan **Hemerocallidaceae**
    8 Pollen sulcate, only exceptionally trichotomosulcate
      10 Rootstock a corm; [flowers often zygomorphic, not articulated; stamens sometimes partly staminodial, but never an entire whorl; stamens opening by apical pores or slits; flowers hypogynous to perigynous; seeds pale, brown or black, with endosperm (not in *Cyanastrum*, this with chalazosperm)]. 8/23, Chile, California, Africa, Madagascar **Tecophilaeaceae**
      10 Rootstock a rhizome, sometimes upright, but never a corm
        11 Seeds brown, clothed with colorless hairs often appearing white; raphides lacking; ovary on gynophore; [rhizomatous, leaves distichous; flowers articulated from pedicel with long pericladium; capsules septicidal]. 1/4, E Australia **Blandfordiaceae**
        11 Seeds black, glabrous; raphides present; gynophore absent
          12 Style short or lacking; [rosette plants from short rhizomes; flowers not articulated]. 4/35, Southern Hemisphere **Asteliaceae**
          12 Style distinct
            13 Anthers basifixed; plants lacking anthrones; caespitose or shrubby herbs, often dwarf; [flowers not articulated from pedicel, in involucrate scapose spikes or racemes]. 2/12, Australia **Boryaceae**
            13 Anthers dorsifixed; plants containing anthrones, either succulent or tall, woody
              14 Succulent herbs or trees; flowers often articulated from pedicel, with or without pericladium; seeds arillate; embryo in longitudinal axis of seed. 15/780, widely distributed, but mainly S African [in America introduced] **Asphodelaceae**
              14 Resinous perennials with thick woody stems with terminal leaf rosettes; flowers not articulated; seeds not arillate; embryo in transversal axis of seed; [perianth persistent and hardened at maturity]. 1/30, Australia **Xanthorrhoeaceae**

1 Microsporogenesis successive; steroidal saponins common (except apparently in Amaryllidaceae) **("Higher" Asparagales)**
  15 Seeds black (black or pale in Lomandraceae)
    16 One ovule/carpel; pollen spiraperturate; [plants rushlike, caespitose, with drying cataphylls and assimilating scapes; flowers not articulated]. 1/1, western Mediterranean **Aphyllanthaceae**
    16 Two or more ovules/carpel; pollen not spiraperturate
      17 Stamens 0 + 3; [rhizomatous perennials with spike-like panicle; flowers articulated; anthers basifixed; capsule in each locule with 1–2 carinate spindle-shaped seeds]. 1/1, E Asia **Anemarrhenaceae**
      17 Stamens 3 + 3
        18 Inflorescence pseudoumbellate, scapose; [flowers not articulated (except Themidaceae)]
          19 Flowers epigynous; [plants bulbous, (rhizomatous); leaves usually distichously arranged; flowers often with a corona; Amaryllidaceae alkaloids universally present]. 59/850, widely distributed, mainly S Africa, Mediterranean, and S America **Amaryllidaceae**

19 Flowers hypogynous, rarely perigynous
20 Rhizomatous; leaves distichous; flowers lacking a corona, [articulated]. 1/9, S Africa **Agapanthaceae**
20 Underground organ a bulb or corm; leaves spiral; corona present
21 Underground organ a bulb (rhizome bulb: *Allium victorialis*); alliaceous odor present; flowers not articulated. 13/600, widely distributed, mostly S America **Alliaceae**
21 Underground organ a corm; alliaceous odor lacking; flowers articulated. 12/60, western N and Central America **Themidaceae**
18 Inflorescence a raceme, spike, panicle, or thyrse, scapose or not
22 Plants with reduced leaves and flat or needle-like phylloclades; [flowers articulated; fruit a berry, rarely a nut]. 1/170–300, Old World **Asparagaceae**
22 Leaves well developed, phylloclades absent
23 Rootstock a bulb (rhizome: *Schoenolirion*); [inflorescence scapose, a spike or raceme or panicle; flowers not articulated (articulated: Chlorogaloideae)]. 45/900, widely distributed, centered in S Africa and from the Mediterranean to SW Asia **Hyacinthaceae**
23 Rootstock never bulbous
24 Shrubby or arborescent perennials; [flowers not articulated; anthers dehiscing by slits or pores; fruit a loculicidal capsule or berry; seeds pale or black; proximal nucellar region with enlarged dermal layer and a central conducting passage of elongated cells]. 14/180, widely distributed **Lomandraceae**
24 Rhizomatous
25 Capsule septicidal; perennials with branching stems or stemless rosette plants; flowers not articulated; anthers basifixed. 2/9, S America, Madagascar **Herreriaceae**
25 Capsule loculicidal; anthers dorsifixed (rarely basifixed in Anthericaceae)
26 Leaves often petiolate; flowers not articulated, white, blue, or lavender. 1/23–50, E Asia **Hostaceae**
26 Leaves not petiolate; flowers white, yellow, or greenish, never blue
27 Sympodial perennials, often with thickened roots; flowers often articulated with a pericladium; carpels fused into a style. 9/200, cosmopolitan **Anthericaceae**
27 Small to gigantic rosulate plants; flowers not articulated; carpels ending in free styluli or fused into a style. 8/300, N, Central, and northern S America **Agavaceae**
15 Seeds brown or pale (in Lomandraceae also black); [steroidal saponins common]
28 Tuberous perennials; flowers not articulated; seeds lanate, with oil-rich perisperm. 1/100, southern Africa **Eriospermaceae**
28 Rhizomatous or caulescent; flowers articulated
29 Leaves on adult plants reduced and replaced by phylloclades; [filaments fused into a tube; anthers extrorse; ovules 2 per carpel; fruit a berry]. 3/9, Macaronesia to Caucasus **Ruscaceae**
29 Leaves well developed
30 Leaf venation reticulate; tepals united for 2/3 of their length; [shrubby perennials with distichous phyllotaxy; ovules few per carpel; fruit a berry]. 1/2, S Africa **Behniaceae**
30 Leaf venation parallel; tepals largely free
31 Ovule 1 per carpel; [rhizomatous herbs, lianas, or trees; nectar secretion from septal nectaries and pedicels and inflorescence axes; fruit a berry]. 1/100, Old World, Central America, Cuba, Hawaii, widely introduced elsewhere **Dracaenaceae**
31 Ovules 2 or more per carpel
32 Rhizomatous; leaves distichous, opposite or verticillate; stigma often expanded; fruit a capsule or berry. 17/130, mainly in Northern Hemisphere **Convallariaceae**
32 Rosette-forming perennials, often treelike and the stem swollen at the base; fruit a loculicidally dehiscent capsule or a 3-winged or unwinged nut. 4/50, N and northern Central America **Nolinaceae**

3. Triuridales

Achlorophyllous, mycotrophic herbs; raphides present; stomata anomocytic; leaves scaly; flowers mostly unisexual; stamens 2–8; gynoecium of 10–numerous free carpels with (sub)basal to (sub)apical styluli and each 1 basal ovule; endosperm formation nuclear; endosperm copious, containing oil, protein and hemicellulose; epicuticular wax of parallel platelets. Only one family. 9/45, Old and New World tropics **Triuridaceae**

4. Dioscoreales

Rhizomatous or tuberous plants or small saprophytic herbs; stomata anomocytic; flowers epigynous, sapromyiophilous, (some *Burmannia* probably sphingophilous); stamens often with apically projecting connective; microsporogenesis simultaneous or successive. The expanded circumscription of this order suggested by molecular data; convincing synapomorphies still unknown

1 Seeds albuminous; leaves petiolate, often partite, with reticulate venation; endosperm development Nuclear

2 Stem vascular bundles scattered; leaves all basal; ptyxis plicate; seeds longitudinally ridged; [aril and fleshy raphe in *T. involucrata*]. Only one family. 1/ca. 11, pantropical **Taccaceae**

2 Stem vascular bundels in 2 dissimilar cycles or one of two alternating types; leaves dispersed; ptyxis conduplicate; seeds not ridged

3 Forming often vast tubers from hypocotyl and lower internodia; tannin cells mostly present; biforinelike idioblasts wanting; petiole with a pulvinus at each end; fruit mostly a capsule, rarely a berry or a 1-winged samara; seeds not ruminate; a parietal cell cut off. 20/600, widely distributed **Dioscoreaceae**

3 Tubers and tannin cells lacking; biforinelike crystal idioblasts with bundles of pseudoraphides present; petioles without pulvinus; fruit indehiscent, 1-seeded; seeds deeply ruminate; no parietal cell cut off. 2/2, Madagascar, tropical Asia **Trichopodaceae**

1 Dust seeds with endosperm vanishing before maturation; leaves not petiolate, undivided, with parallel venation; endosperm development Helobial; [autotrophic or holomycotrophic saprophytes]

4 Inflorescences cymose; stamens 6 or 3; anthers introrse. 13/130, pantropical-subtropical **Burmanniaceae**

4 Flowers solitary; stamens 6, anthers extrorse. 2/28, Australasia, S America **Corsiaceae**

5. Pandanales

Herbs, trees, or climbers; raphides mostly present, silica bodies lacking; vessels in roots and mostly also in stems and leaves, with scalariform perforations; stomata tetracytic, rarely anomocytic; flowers epigynous to hypogynous; gynoecium syncarpous; microsporogenesis successive; endosperm development Nuclear or Helobial. Structural traits holding together these families still to be explored

1 Flowers perianthless, not distinctly organized in whorls, [unisexual or with vestiges of the other sex; ovary position not discernible; woody plants, erect and on stilt roots, or climbers; raphides and cubic calcium crystals present, silica lacking; ptyxis nonplicate; endosperm development Nuclear; endosperm oily or starchy; epicuticular wax scales unoriented (Strelitzia type compound rodlets only once recorded)]. Only one family. 3/800–900, paleotropical **Pandanaceae**

1 Flowers with perianth and distinct whorls

2 Leaves petiolate; flowers 2-, 4-, or 5-merous; endosperm without starch or with little starch

3 Flowers 5-merous; [hypanthium and ovary forming a swollen, disklike structure; juicy herbs; seeds with sarcotesta and arillode]. 1/2, Sumatra **Pentastemonaceae**

3 Flowers 2- or 4-merous

4 Flowers unisexual, not articulated; perianth inconspicuous or lacking; stamens mostly numerous; endosperm oily, rarely (*Dicranopygium*) starchy; stomates tetracytic; [herbaceous or woody plants with specialized roots]. 12/230, neotropical **Cyclanthaceae**

4 Flowers usually hermaphroditic, articulated from pedicel with pericladium; perianth distinct; stamens 4; endosperm oily and starchy; stomates anomocytic; [seed with elaiosome]. 3/35, Australasia, N America **Stemonaceae**

2 Leaves not petiolate; flowers 3-merous; endosperm with copious starch surrounded by a thin aleurone layer

5 Shrubby or arborescent, the thin stem surrounded by a mantle of adventitious roots piercing the withered leaf bases; leaves multiveined; stamens 6(–48); septal nectaries present; [endosperm formation Helobial with small chalazal chamber; raphides rarely present]. 8/200, S America, Africa, Madagascar **Velloziaceae**

5 Dwarf caespitose perennial; stems lacking mantle of adventitious roots; leaves acerose, 2-veined; stamens 6; septal nectaries lacking. 1/1, SW China **Acanthochlamydaceae**

General References

Morphology and Anatomy

Behnke, H.-D. 1981. Siebelement-Plastiden, Phloem-Protein und Evolution der Blütenpflanzen: II. Monokotyledonen. Ber. Deutsch. Bot. Ges. 94: 647–662.
Czaja, A. Th. 1978. Structure of starch grains and the classification of vascular plant families. Taxon 27: 463–470.
Daumann, E. 1970. Das Blütennektarium der Monokotyledonen unter besonderer Berücksichtigung seiner systematischen und phylogenetischen Bedeutung. Feddes Rep. 80: 463–590.
Frölich, D., Barthlott, W. 1988. Mikromorphologie der epicuticularen Wachse und das System der Monokotylen. Trop. Subtrop. Pflanzenwelt 63. 135 pp. Akad. Wiss. Lit. Mainz.
Gerenday, A., French, J.C. 1988. Endothecial thickenings in anthers of porate monocotyledons. Am. J. Bot. 75: 22–25.
Heel, W.A. van 1988. On the development of some gynoecia with septal nectaries. Blumea 33: 477–504.
Heslop-Harrison, Y., Shivanna, K.R. 1977. The receptive surface of the angiosperm stigma. Ann. Bot. II, 41: 1233–1258.
Holttum, R.E. 1955. Growth habits of monocotyledons – variations on a theme. Phytomorphology 5: 399–413.
Rudall, P.J. 1995. New records of secondary thickening in monocotyledons. IAWA J. 16: 261–268.
Takhtajan, A.L. (ed.) 1985. Anatomia seminum comparativa. T.I. Liliopsida seu monocotyledones. Leningrad: Nauka. (in Russian)
Tomlinson, P.B. 1969. Anatomy of the monocotyledons. III. Commelinales-Zingiberales. Oxford: Clarendon Press.
Tomlinson, P.B. 1982. Anatomy of the monocotyledons. VII. Helobiae (Alismatidae) (including the seagrasses). Oxford: Clarendon Press.
Velenovský, J. 1904. Die gegliederten Blüten. Beih. Bot. Centralbl. 16: 289–300.
Wagner, P. 1977. Vessel types of the monocotyledons. Bot. Not. 130: 383–402.

Embroylogy and Palynology

Batygina, T.B., Yakovlev, M.S. (eds.) 1990. Comparative embryology of flowering plants: monocotyledons. Leningrad: Nauka. (in Russian)
Davis, G.L. 1966. Systematic embryology of the angiosperms. New York: Wiley.
Erdtman, G. 1952. Pollen morphology and plant taxonomy. Angiosperms. Stockholm: Almqvist & Wiksell.
Johri, B.M., Ambegaokar, K.B., Srivastava, P.S. 1992. Comparative embryology of angiosperms 2 vols. Berlin, Heidelberg, New York: Springer.
Punt, W., Blackmore, S., Nilsson, S., Le Thomas, A. 1994. Glossary of pollen and spore terminology. Utrecht: LPP Foundation.
Rudall, P. 1994. The ovule and embryo sac in Xanthorrhoeaceae sensu lato. Flora 189: 335–351.
Rudall, P.J. 1997. The nucellus and chalaza in monocotyledons: structure and systematics. Bot. Rev. 63: 140–181.
Rudall, P.J., Furness, C.A., Chase, M.W., Fay, M.F. 1997. Microsporogenesis and pollen sulcus type in Asparagales (Lilianae). Can. J. Bot. 75: 408–430.
Takhtajan, A. (ed.) 1985. Anatomia seminum comparativa. Tomus I. Liliopsida seu Monocotyledones. Leningrad: Nauka. (in Russian)
Zavada, M. 1983. Comparative morphology of monocot pollen and evolutionary trends of apertures and wall structures. Bot. Rev. 49: 331–379.

Karyology

Fedorov, Al. A. (ed.) 1969. Chromosome numbers of flowering plants. Leningrad: Nauka (in Russian).
Sen, S. 1975. Cytotaxonomy of Liliales. Feddes Rep. 86: 255–305.
Tamura, M.N. 1995. A karyological review of the orders Asparagales and Liliales (Monocotyledonae). Feddes Rep. 106: 83–111.

Phytochemistry

Bate-Smith, E.C. 1968. The phenolic constitutents of plants and their taxonomic significance II. Monocotyledons. J. Linn. Soc. Bot. 60: 325–356.
Gibbs, R.D. 1974. Chemotaxonomy of flowering plants. 4 vols. Montreal: McGill-Queen's University Press.
Harris, P.J., Hartley, R.D. 1980. Phenolic constitutents of the cell walls of monocotyledons. Biochem. Syst. Ecol. 8: 153–160.
Hegnauer, R. 1963. Chemotaxonomie der Pflanzen, vol. 2. Stuttgart: Birkhaeuser.
Hegnauer, R. 1986. Chemotaxonomie der Pflanzen, vol. 6. Stuttgart: Birkhaeuser.
Rudall, P.J., Caddick, L.R. 1994. Investigation of the presence of phenolic compounds in monocotyledonous cell walls, using UV fluorescence microscopy. Ann. Bot. 74: 483–491.

Systematics and Evolution

Baker, J.G. 1875. Revision of the genera and species of Asparagaceae. J. Linn. Soc. Bot. 14: 508–632.
Baker, J.G. 1879. A synopsis of Colchicaceae and the aberrant tribes of Liliaceae. J. Linn. Soc. Bot. 17: 405–510.
Bentham, G., Hooker, J.D. 1883. Genera Plantarum 3, 2. London: L. Reeve, Williams and Norgate.
Brummitt, R.K. 1992. Vascular plant families and genera. Royal Botanic Gardens, Kew.
Chase, M.W., Rudall, P.J., Conran, J.G. 1996. New circumscriptions of asparagoid lilies: genera formerly included in Anthericaceae. Kew Bull. 51: 667–680.
Cronquist, A. 1981. An integrated system of classification of flowering plants. New York: Columbia University Press.

Dahlgren, R., Rasmussen, F.N. 1983. Monocotyledon evolution. Characters and phylogenetic estimation. Evol. Biol. 16: 255–395.
Dahlgren, R.M.T., Clifford, H.T. 1982. The monocotyledons: a comparative study. London: Academic Press.
Dahlgren, R.M.T., Clifford, H.T., Yeo, P.F. 1985. The families of the monocotyledons. Berlin Heidelberg New York: Springer.
Goldberg, A. 1989. Classification, evolution, and phylogeny of the families of Monocotyledones. Smithson. Contrib. Bot. 71.
Huber, H. 1969. Die Samenmerkmale und Verwandtschaftsverhältnisse der Liliifloren. Mitt. Bot. Staatssamml. München 8: 219–538.
Huber, H. 1977. The treatment of the monocotyledons in an evolutionary system of classification. Plant Syst. Evol. Suppl. 1: 285–298.
Huber, H. 1991. Angiospermen. Stuttgart: G. Fischer.
Krause, K. 1930. Liliaceae, pp. 227–386. In: Engler, A. (ed.) Die natürlichen Pflanzenfamilien, 2nd edn., vol. 15a. Leipzig: W. Engelmann.
Rudall, P.J., Cribb, P.J., Cutler, D.F., Humphries, C.J. (eds.) Monocotyledons: systematics and evolution. 2 vols. Royal Botanic Gardens, Kew.
Takhtajan, A. 1980. Outline of the classification of the flowering plants (Magnoliophyta). Bot. Rev. 46: 225–359.
Takkhtajan, A.L. (ed.) 1982. Plant life, vol. 6. Angiosperms, monocotyledons. Moscow: Prosweshenie. (in Russian)
Takhtajan, A. 1987. Systema Magnoliophytorum. Leningrad: Nauka. (in Russian)
Takhtajan, A. 1997. Diversity and classification of flowering plants. New York: Columbia University Press.
Thorne, R.F. 1992. Classification and geography of the flowering plants. Bot. Rev. 58: 225–348.

Molecular Systematics

Chase, M.W., Soltis, D.E., Olmstead, R.G., Morgan, D., Les, D.H., Mishler, B.D., Duvall, M.R., Price, R.A., Hills, H.G., Qiu, Y.-L., Kron, K.A., Rettig, J.H., Conti, E., Palmer, J.D., Manhart, J.R., Sytsma, K.J., Michaels, H.J., Kress, W.J., Karol, K.G., Clark, W.D., Hedrén, M., Gaut, B.S., Jansen, R.K., Kim, K.-J., Wimpee, C.F., Smith, J.F., Furnier, G.R., Strauss, S.H., Xiang, Q.-Y., Plunkett, G.M., Soltis, P.S., Swensen, S.M., Williams, S.E., Gadek, P.A., Quinn, C.J., Eguiarte, L.E., Golenberg, E., Learn jr., G.H., Graham, S.W., Barrett, S.C.H., Dayanandan, S., Albert, V.A. 1993. Phylogenetics of seed plants: an analysis of nucleotide sequences from the plastid gene *rbc*L. Ann. Mo. Bot. Gard. 80: 528–580.
Chase, M.W., Duvall, M.R., Hills, H.G., Conran, J.G., Eguiarte, L.E., Hartwell, J., Fay, M.F., Caddick, L.R., Cameron, K.M., Hoot, S. 1995. Molecular systematics of Lilianae. In: Rudall, P.J., Cribb, P.J., Cutler, D.F., Humphries, C.J. (eds.) Monocotyledons: systematics and evolution. Royal Botanic Gardens, Kew, pp. 109–137.
Chase, M.W., Stevenson, D.W., Wilkin, P., Rudall, P.J. 1995a. Monocot systematics: a combined analysis. In: Rudall, P.J., Cribb, P.J., Cutler, D.F., Humphries, C.J. (eds.) Monocotyledons: systematics and evolution. Royal Botanic Gardens, Kew, pp. 685–730.
Clark, D.W., Gaut, B.S., Duvall, M.R., Clegg, M.T. 1993. Phylogenetic relationships of the Bromeliiflorae-Commeliniflorae-Zingiberiflorae complex in monocots based on *rbc*L sequence comparisons. Ann. Mo. Bot. Gard. 80: 987–998.
Davis, J.I. 1995. A phylogenetic structure for the monocotyledons, as inferred from chloroplast DNA restriction site variation, and a comparison of measures of clade support. Syst. Bot. 20: 503–527.
Duvall, M.R., Learn, G.H., Jr., Eguiarte, L.E., Clegg, M.T. 1993. Phylogenetic analysis of *rbc*L sequences identifies *Acorus calamus* as the primal extant monocotyledon. Proc. Natl. Acad. Sci. USA 90: 4641–4644.
Duvall, M.R., Clegg, M.T., Chase, M.W., Clark, W.D., Kress, W.J. et al. 1993. Phylogenetic hypotheses for the monocotyledons constructed from *rbc*L sequence data. Ann. Mo. Bot. Gard. 80: 607–619.
Les, D.H., Garvin, D.K., Wimpee, C.F. 1993. Phylogenetic studies in the monocot subclass Alismatidae: evidence for a reappraisal of the aquatic order Najadales. Mol. Phylogen. Evol. 2: 304–314.
Les, D.H., Cleland, M.A., Waycott, M. 1997. Phylogenetic studies in Alismatidae, II: Evolution of marine angiosperms (seagrasses) and hydrophyly. Syst. Bot. 22: 443–463.
Nadot, S., Bittar, G., Carter, L. Lacroix, R., Lejeune, B. 1995. Mol. Phylogen. Evol. 4: 257–282.
Soltis, D.E., Soltis, P.S., Nickrent, D.L., Johnson, L.A., Hahn, W.J., Hoot, S.B., Sweere, J.A., Kuzoff, R.K., Kron, K.A., Chase, M.W., Swensen, S.M., Zimmer, E.A., Chaw, S.-M., Gillespie, L.J., Kress, W.J., Systsma, K.J. 1997. Angiosperm phylogeny inferred from 18S ribosomal DNA sequences. Ann. Mo. Bot. Gard. 84: 1–49.

Acanthochlamydaceae

P.-C. Kao and K. Kubitzki

Acanthochlamydaceae Kao P.-C., Fl. Sichuan. 9: 483 (1989).

Dwarf caespitose perennial *herb*; rhizome short; roots dense, thin and long, fibriform; *leaves* acerose, dorsiventral, ventrally subsemiorbicular and 2-canaliculate, dorsally flattened and 1-canaliculate, sheathed at the base. *Inflorescence* a compound capitulum on a scape arising from the rhizome, at the base usually surrounded by 3 leaflike aristate bracts, the peduncle bearing 5–8 few-flowered capitula, the flowers subtended by aristate bractlets. *Flowers* hermaphroditic, actinomorphic, epigynous, shortly pedicellate; perianth corollinic, pink, tubular; perianth lobes 3 + 3, the inner lobes slightly smaller than the outer; stamens 3 + 3, borne upon the corolla lobes; filaments short; anthers oblong, bisporangiate, dorsifixed, introrse, dehiscing by longitudinal slits; ovary inferior, syncarpous, in lower part trilocular with axile placentation, in upper part unilocular with parietal placentation; ovules numerous, anatropous, bitegmic; style elongate, with trilobate stigma; *fruit* capsular, trigonous, shortly rostrate; seeds oblong, brown; endosperm starch-containing; embryo large, central.

Only one sp., *Acanthochlamys bracteata* P.C. Kao, subalpine xerophytic valley zone of W Sichuan and SE Tibet, SW China.

Vegetative Structures. The vascular cylinder of the root is a tetrarch, rarely triarch, actinostele lacking pith; this low number of xylem rays is unusual in monocotyledons. The root epidermis is formed by elongate cells which bear the root hairs. The cortex is composed of many cell layers and is divided into exodermis, cortex and endodermis. The cells of the pericycle are 1-rowed and densely arranged; the lateral roots originate from them.

The aerial parts are covered by elongate, thick-walled epidermal cells and bear sunken stomata which are paracytic. Massive sclerenchyma underlies the epidermis. In the leaves, the mesophyll lacks differentiation into palisade and spongy tissue. The midrib of the leaf is composed of 2 vascular bundles, which are arranged "back to back" with their xylem portions. The scape has on one side a longitudinal groove and hence in transversal section is indistinctly heart-shaped. Its anatomical structure is very peculiar insofar that it has a stele with a central tetrach vascular cylinder similar to that of the roots, which in the upper portion of the scape disintegrates into 5 or 6 collateral bundles. Outside the stelar tissue the inner part of the cortex is sclerified and underneath the longitudinal groove of the scape the cortex is transversed by 2 obliquely oriented vascular bundles (a leaf trace which supplies an involucral bract), which resemble those forming the leaf midrib but are fused with their xylem portions. The structure of the scape is similar to that of a leaf ensheathing a rhizome. Vessel elements with simple perforations and helical thickenings are present in the root, scape and leaf. Raphides and tannin cells are lacking completely (Kao 1989).

Embryology. The anther is bisporangiate (Fig. 20), and the anther wall is of the monocotyledonous type. The tapetum is glandular. Microsporogenesis is of the Successive type; the pollen tetrads are isobilateral. The mature pollen grains are bi- or trinucleate. The ovule is anatropous, bitegmic and tenuinucellate. The development of the embryo sac is of the Polygonum type or Allium type (Li et al. 1992). Fertilisation is porogamous and premitotic and endosperm development is Nuclear (Li and Gao 1993).

Pollen Morphology. The pollen of *Acanthochlamys* is sulcate, spheroidal and (under SEM) finely verrucate to reticulate (Gao 1987).

Karyology. Kao et al. (1993) reported the chromosome number $2n = 38$.

Pollination. The anthesis of *Acanthochlamys* is very short. Pollination seems to be mediated by small bees.

Seed. The seeds (Fig. 22) are brown, shining, ellipsoid, 0.75 × 0.5 mm. The epidermis of the outer integument is collapsed so that the seeds have a grooved, microreticulate surface. The other layer(s) belonging to the outer integument – certainly not more than 2 – is/are collapsed. The inner integument is represented by one well-developed and another collapsed layers that contain a red-brown content, presumably condensed tannins. Mechanical strength is provided mainly by the thickened periclinal walls of the cell layer immediately beneath the seed coat. This layer con-

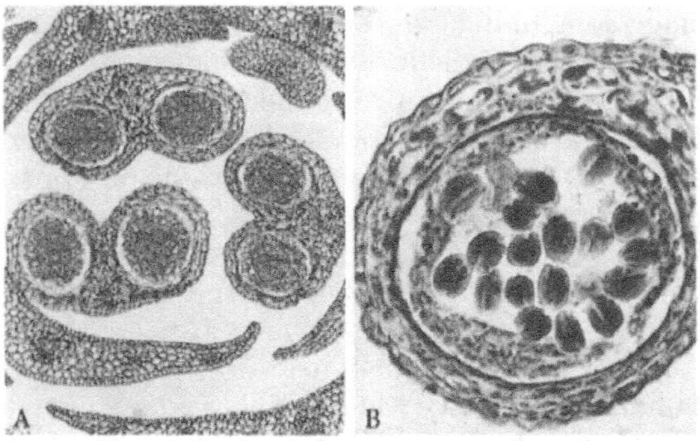

Fig. 20A,B. Acanthochlamydaceae. *Acanthochlamys bracteata*. **A** Transversal section of flower bud, showing stamens with 1-sporangiate thecae. **B** Part of stamen in transversal section; pollen mother cells after first meiotic division

tains aleuron and is free of starch. The rest of the endosperm contains starch in the form of composed starch grains. The embryo lies in the middle of the endosperm and reaches 3/4 of the length of the endosperm (Kao 1989; H. Huber, pers. comm.).

PHYTOCHEMISTRY. According to a preliminary study (B.C. Gao, unpubl.) alkaloids are lacking, but steroid saponins may be present. Fluorescent ferulate is absent from the cell walls (Rudall 1994).

AFFINITIES. Originally described by Kao (1980) as a member of the Amaryllidaceae, *Acanthochlamys* was elevated to subfamily rank in Amaryllidaceae by Chen (1981). After studying the anatomy, taxonomy, palynology, embryology, karyology, phytochemistry and ecology of *Acanthochlamys* for nearly 10 years, the profound differences separating the genus from the Amaryllidaceae led Kao (1989) to elevate it to the rank of family. A close relationship to the Velloziaceae had been proposed by Gao (1987), but subsequently, Kao (1989) compared *Acanthochlamys* with *Tofieldia* and suggested for *Acanthochlamys* an intermediate position between *Tofieldia* and Amaryllidaceae. Later, and mainly based on the agreement in chromosome base number, a close relationship of *Acanthochlamys* to Convallariaceae (*Aspidistra*

Fig. 21A–C. Acanthochlamydaceae. *Acanthochlamys bracteata*. **A** Seed, transversal section. **B** Seed, longitudinal section. **C** As A, enlarged. *a* Endosperm; *b* embryo; *c* seed coat; *d* aleurone layer; *e* cell filled with aleurone; *f* cell full of starch grains

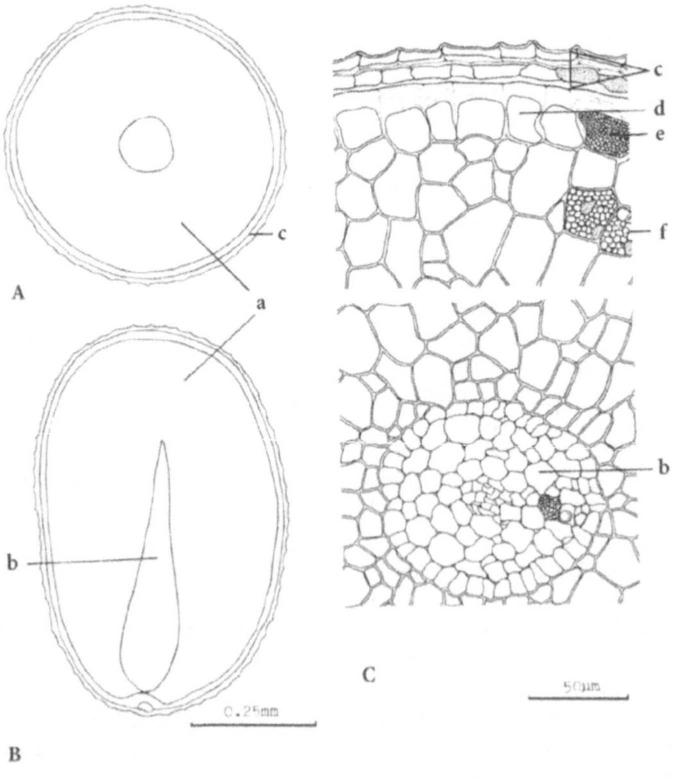

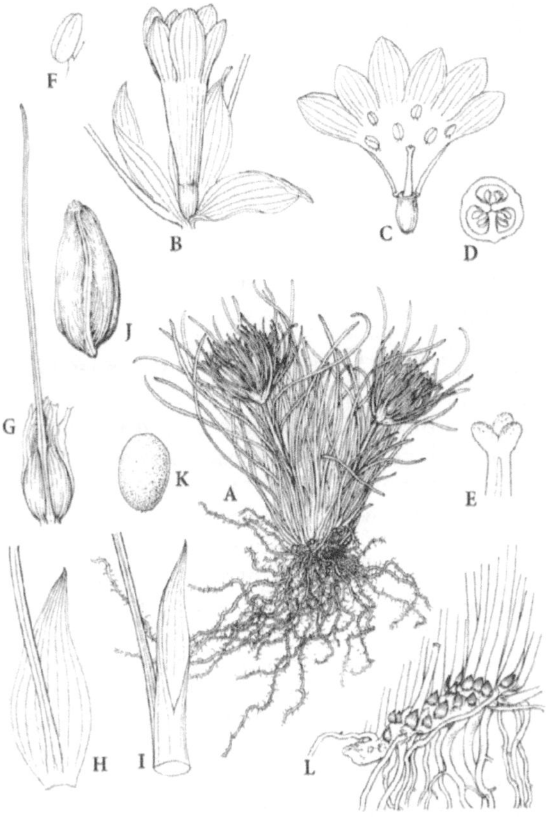

Fig. 22A–L. Acanthochlamydaceae. *Acanthochlamys bracteata.* **A** Habit. **B** Flower. **C** Flower, perianth opened. **D** Transsection of ovary. **E** Stigma. **F** Anther. **G** Flower bract. **H, I** Leaf sheath. **J** Fruit. **K** Seed. **L** Rhizome

and *Tupistra*) was considered (Kao et al. 1993); however, this suggestion is not supported by morphological evidence and is no longer upheld. In the *rbc*L analysis by Chase et al. (1995) *Acanthochlamys* appeared in a strongly supported branch together with the Velloziaceae.

Indeed, a possible relationship should be considered with all those families that share with *Acanthochlamys* the possession of epigyny and starch accumulation in the endosperm, viz. Bromeliaceae, Haemodoraceae and Velloziaceae. It is significant that these three families agree with *Acanthochlamys* in having starch grains composed of relatively few elements in the endosperm. Among these families, the Bromeliaceae differ too much from *Acanthochlamys* in their perianth differentiated into calyx and corolla and their epidermal cells containing large silica bodies. The remaining two families agree with *Acanthochlamys* in the collapsed testa and 2-layered tegmen, the Velloziaceae also in the bisporangiate anthers, and additionally the Haemodoraceae in the thickened outer periclinal wall of the aleuron layer. Also the Xyridaceae are similar in several respects, although they are truly "enantioblastic" and have hypogynous flowers.

Among all families compared here with *Acanthochlamys*, the absence of cell wall-bound ferulate is shared only with the Velloziaceae. Thus the morphological data point in the same direction as the molecular analysis, although the relationship between *Acanthochlamys* and the Velloziaceae may not be very close.

Distribution and Habitats. *Acanthochlamys bracteata* is restricted to the Hengduan Mountains at the SE margin of the Kang-Zang Plateau of SW China (Xiangcheng, Daocheng, Daofu of W Sichuan to Zhag'yab of Tibet), where it occurs in the subalpine aciculignose shrub-meadow region at an altitude of 2700–3500 m.

Only one genus:

Acanthochlamys **P.C. Kao** Figs. 20–22

Acanthochlamys P.C. Kao, Acta Phytotax. Chengdu Inst. Biol. Acad. Sin. 1: 1 (1980).

Only one sp., description and distribution as for family.

Selected Bibliography

Chase, M.W. et al. 1995. See general references.

Chen, S.-C. 1981. Acanthochlamydoideae – a new subfamily of Amaryllidaceae. Acta Phytotaxon. Sin. 19: 323–329. (In Chinese with Engl. summ.)

Gao, B.-C. 1987. The sociological characteristics and pollen morphology of *Acanthochlamys*. Acta Bot. Yunnan. 9: 401–405. (In Chinese with Engl. summ.)

Gao, B.-C., Li, P. 1993. Studies on the morphology and embryology of *Acanthochlamys bracteata* I. Morphological and anatomic studies on vegetative organs. J. Sichuan Univ. (Science ed.) 30: 534–537. (In Chinese with Engl. summ.)

Kao, P.-C. 1980. A new genus of Amaryllidaceae from China. Acta phytotax. Chengdu Inst. Biol. Acad. Sin. 1: 1–3, pl. 1–11. (In Chinese and Latin)

Kao, P.-C. 1989. Acanthochlamydaceae – a new monocotyledonous family. Fl. Sichuan (Kao, P.C., Tan, Z.-M., eds.) 9: 483–507, pl. I–IX. (In Chinese and Engl.)

Kao, P.-C., Tang, Y., Guo, W.-H. 1993. A cytological study of Acanthochlamys bracteata P.C. Kao (Acanthochlamydaceae). Acta Phytotaxon. Sin. 31: 42–44. (In Chinese with Engl. summ.)

Li, P., Gao, B.-C. 1993. Studies on morphology of *Acanthochlamys bracteata*. III. The investigation on double fertilization, embryogenesis and endosperm development of *Acanthochlamys bracteata*. J. Sichuan Univ. (Science ed.) 30: 260–263, Figs. 1–18. (In Chinese with Engl. summ.)
Li, P., Gao, B.-C., Chen, F., Luo, H.X. 1992. Studies on morphology and embryology of *Acanthochlamys bracteata*. II. The anther and ovule development. Bull. Bot. Res. 12: 389–395. (In Chinese with Engl. summary)
Rudall, P. 1994. See general references.

Agapanthaceae

K. Kubitzki

Agapanthaceae Voigt, Gesch. Pflanzenwelt 2: 440 (1850), Agapantheae.

Perennials from a tuberous, erect rhizome; roots fleshy, provided with a multiple velamen. Leaves rosulate, distichously arranged, linear, slightly contracted between the sheath and linear, flat blade, slightly fleshy or leathery, margins smooth, hyaline or ridged. *Inflorescence* a pseudo-umbel subtended by 2 deciduous involucral spathal bracts; scape stout, terete or slightly compressed; pedicels erect, spreading or cernuous, subtended by persistent, threadlike bracts. *Flowers* trimerous, hermaphrodite, hypogynous, zygomorphic; perigone tubular or infundibuliform, tepals 6, subequal, basally ± fused, blue or white, the outer face with a median ridge, the inner grooved along the same line; outer segments narrower than inner and often slightly hooded, thickened, and uncinate or bearded at the apex, somewhat fleshy; stamens 6, inserted on the tube; filaments declinate, of unequal length; anthers dorsifixed, introrse with longitudinal dehiscence; ovary superior, sessile, ovoid or oblong, trilocular; inner septal nectaries present; ovules numerous, bitegmic; style slender, hollow, declinate, stigma small, of the dry type. *Fruit* a loculicidal capsule with many flat, black, winged seeds. 2n = 30, rarely 32.

Only one genus of 9 spp. endemic to South Africa from the Cape to the Limpopo River.

Anatomy. The roots have a multiple velamen (Goebel 1933). Laticifers have been reported by Hegnauer (1963). Vessels are present only in the roots, mostly with scalariform perforations. Raphides are present.

Karyology. Various *Agapanthus* species are reported as having 2n = 30 (very rarely 32) (Satô 1942; Leighton 1965).

Pollen Morphology. Pollen of *Agapanthus* is sulcate-reticulate (Schulze 1980).

Embryology. Endothecial thickenings in the stamens consist of incomplete loops. Tapetum cells

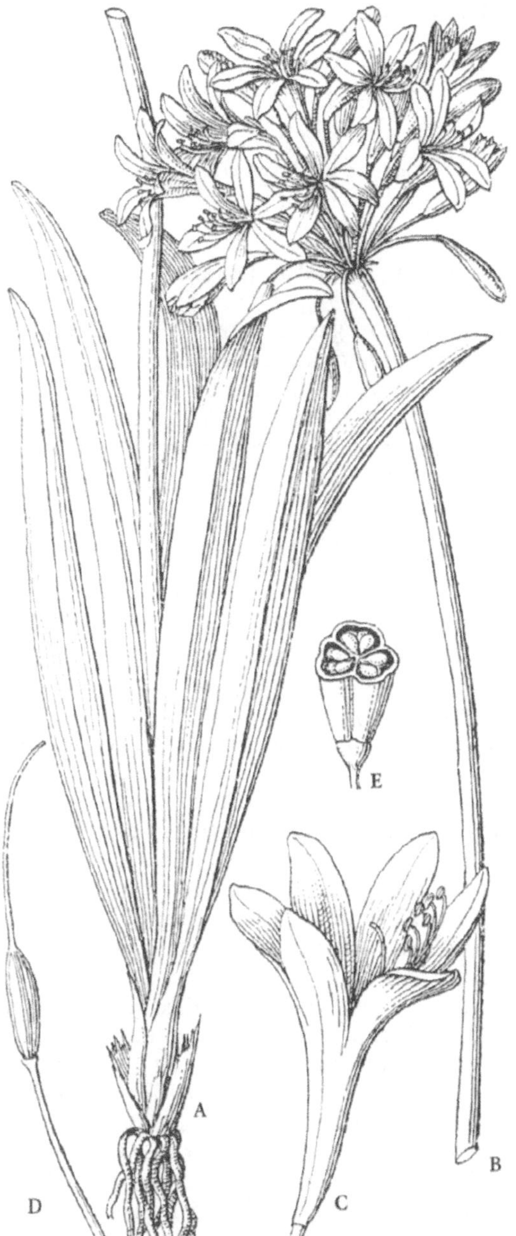

Fig. 23A–E. Agapanthaceae. *Agapanthus umbellatus*. **A** Habit. **B** Scape with pseudoumbel. **C** Flower. **D** Young fruit. **E** Capsule. (Krause 1930)

are predominantly binucleate. Microsporogenesis is successive (Stenar 1933; Rudall et al. 1997).

SEEDS. The seeds of *Agapanthus* are angulate, elongated, half-campylotropous, and winged opposite the micropyle; the phytomelan crust is thin. The endosperm stores aleuron, oil and reserve cellulose. The embryo is well developed and reaches 4/5 of the length of the endosperm (Huber 1969).

PHYTOCHEMISTRY. Various steroid saponins have been isolated from the genus (Hegnauer 1963, 1986).

AFFINITIES. The older concepts of a family Agapanthaceae as proposed by Voigt (1850) and Lotsy (1911) were based on superficial similarities and need not be discussed here. In more recent classifications, *Agapanthus* and *Tulbaghia* form a subfamily, Agapanthoideae, of Alliaceae (Huber 1969; Dahlgren et al. 1985). However, these two genera differ considerably and Dahlgren et al. (1985) have argued that they may form a heterogeneous assemblage. This is certainly true because, for instance, raphides are lacking in *Tulbaghia* (as in most other Alliaceae) and are present in *Agapanthus*; *Tulbaghia* has a corona, which is lacking in *Agapanthus*; *Tulbaghia* has a solid style, *Agapanthus* a hollow style (Fay and Chase 1996); the chromosome base number of *Tulbaghia* is $x = 6$, that of *Agapanthus* 15 and 16; the embryo sac of *Tulbaghia* develops according to the Scilla type, that of *Agapanthus* follows the Polygonum type; a parietal cell is lacking in *Tulbaghia*, but present in *Agapanthus*; *Tulbaghia* has alliaceous chemistry, *Agapanthus* accumulates steroid saponins. In short, *Tulbaghia* definitely is alliaceous, *Agapanthus* not.

The molecular (*rbc*L) studies by Chase et al. (1995) and Fay and Chase (1996) have provided support for the inclusion of *Tulbaghia* in Alliaceae and have suggested *Agapanthus* as a sister to Amaryllidaceae. Fay and Chase (1996) proposed including *Agapanthus* as a monotypic subfamily in Amaryllidaceae in order to avoid the proliferation of families. However, characters such as the superior ovary of *Agapanthus* in contrast to the inferior ovary of all Amaryllidaceae and the steroidal saponin chemistry of *Agapanthus* in opposition to the specialized amaryllidaceous alkaloids of the Amaryllidaceae do not recommend this inclusion. The elevation of *Agapanthus* to family rank seems best to reflect present knowledge and is in accord with the narrow family concept practiced in this book. It is clear that this new family is closely related to Amaryllidaceae and Alliaceae.

DISTRIBUTION AND HABITATS. The genus *Agapanthus* is distributed from the Cape Peninsula to the mountain ranges south of the Limpopo River.

Its range of altitude extends from sea level to 2300 m. Species growing in the southern part of the range of the genus in the Cape Province, where rain falls in the winter or throughout the year, are evergreen. Further to the north in the summer rain region of Natal and Transvaal, species lose all their leaves in the autumn and after lying dormant for 1–2 months produce an entirely new set of leaves in the spring (Leighton 1965).

Only one genus:

Agapanthus **L'Hér.** Fig. 23

Agapanthus L'Hér., Sert. Angl.: 17 (1788); Leighton, J.S. Afr. Bot. Suppl. Vol. 4: 1–50 (1965), rev.

Description as of family.
Nine spp., South Africa.

Selected Bibliography

Chase, M.W. et al. 1995. See general references.
Dahlgren et al. 1985. See general references.
Fay, M.F., Chase, M.W. 1996. Resurrection of *Themidaceae* for the *Brodiaea* alliance, and recircumscription of *Alliaceae*, *Amaryllidaceae* and *Agapanthoideae*. Taxon. 45: 441–451.
Goebel, K. 1933. Organographie der Pflanzen. 3. Teil: Samenpflanzen. Jena: G. Fischer.
Hegnauer, R. 1963, 1986. See general references.
Huber, H. 1969. See general references.
Johri, M.M. 1966. The style, stigma, and pollen tube. III. Some taxa of the Amaryllidaceae. Phytomorphology 16: 142–157.
Krause, K. 1930. See general references.
Leighton, F.M. 1965. The genus *Agapanthus* L'Héritier. J. South Afr. Bot., Suppl. 4: 50.
Lotsy, P.J. 1911. Vorträge über botanische Stammesgeschichte, Vol. 3, 1. Jena: G. Fischer.
Rudall, P.J., Furness, C.A., Chase, M.W., Fay, M.F. 1997. Microsporogenesis and pollen sulcus type in Asparagales (Lilianae). Can. J. Bot. 75: 408–430.
Satô, D. 1942. Karyotype alteration and phylogeny in Liliaceae and allied families. Jpn. J. Bot. 12: 57–161.
Schulze, W. 1980. Beiträge zur Taxonomie der Liliifloren V. Alliaceae. Wiss. Z. Friedrich Schiller-Univ. Jena, Math. Naturwiss. Reihe 29: 595–606.
Stenar, H. 1933. Zur Embryologie der *Agapanthus*-Gruppe. Bot. Not. 1933: 520–530.
Tamura, M.N. 1985. See general references.
Voigt, F.S. 1850. Geschichte des Pflanzenreiches 2 Vols. Jena: F. Mauke.

Agavaceae

S. Verhoek

Agavaceae Endl., Enchiridion: 105 (1841), nom. cons.

Plants small to gigantic, perennial, monocarpic or polycarpic, acaulescent or arborescent, sometimes caespitose. Roots fibrous or fleshy; rhizomes spreading or thick and upright. *Leaves* rosulate, spiral, annual or long-lived, linear, lanceolate, elliptic or ovate, fibrous, rigid or flexible, the texture thin, thickened and hard, or succulent, pale to dark green, often glaucous, maculate in some species; the apex with a short or long soft or pungent point; margins entire, denticulate to coarsely toothed or filiferous. Flowering stems terminal or axillary, sometimes huge, bracteate. Inflorescence a panicle, raceme, or spike. *Flowers* perfect, often protandrous, hypogynous, or epigynous, actinomorphic to zygomorphic, never provided with a pericladium or articulate with the pedicel. Tepals 6, petaline, semisucculent, waxy, green, greenish yellow or white, sometimes tinged with maroon, or red to coral, free or united into a long or short tube, the lobes erect, spreading or revolute. Stamens 6, with long or short filaments, included or long exserted; filaments thin or variously thickened, smooth, papillate or short-hairy, the long ones bent in bud; anthers bithecate, dorsifixed, versatile, linear, oblong, or sagittate, introrsely dehiscent with longitudinal slits. Gynoecium of 3 united carpels, ovary superior or inferior, trilocular, ovoid to cylindrical, fleshy, with septal nectaries; ovules numerous, in 2 axile rows in each locule, anatropous; style single or styluli free (*Yucca*), long or short, included or exserted, thickened in *Furcraea*, *Beschorneria*, and *Yucca*; stigma single and capitate or 3-lobed or (*Yucca*) stigmas 3 on separate branches. *Fruit* a loculicidally dehiscent, erect capsule, or (Yucceae) sometimes septicidally dehiscent, or indehiscent and dry or fleshy. Seeds mostly flat, sometimes bordered with a narrow wing, cuneate, black. Embryo cylindrical, straight or slightly curved, oriented along the long axis of the seed.

The family as presently circumscribed is entirely New World temperate and tropical, with 8 genera and ca. 300 species.

VEGETATIVE MORPHOLOGY. The large members of the family such as *Furcraea* and some *Agave* can be massive rosettes up to 4 m broad with xeromorphic leaves persistent for 8–15 years. Some species of *Yucca* attain treelike form. Genera with primarily smaller members such as *Manfreda* and *Polianthes* have softer leaves and occupy dry to mesic habitats. In *Agave*, horizontal anchoring roots spread to a considerable distance. Certain species in most genera produce horizontal rhizomes, and in *Manfreda*, *Polianthes*, and *Prochnyanthes* (Verhoek-Williams 1975) and in *Yucca elata* (J. Webber 1953) there is an upright globose to cylindrical tuberlike rhizome.

Stems vary from very short to treelike and may be branched. Branching in *Yucca* is sympodial. There is secondary thickening growth in stems of at least *Agave*, *Furcraea*, and *Yucca*. The leaves are arranged in spirals at the top of the stems, giving rise to the rosette form in short species or to a palmlike tree habit. *Yucca* approaches a 2/5 phyllotaxy (Arnott 1962).

Leaves throughout the family may be deeply channeled. The leaves in *Manfreda*, *Polianthes*, and *Prochnyanthes* are thin to only slightly succulent, but in *Yucca*, and especially *Agave* and *Furcraea*, they are thickened over their whole width and thickest at the center. Leaf tips are diagnostic. Those of *Agave*, *Hesperaloe*, and *Yucca* are a hard pungent spine, those of *Furcraea* a hard button or short point, and those of *Beschorneria*, *Manfreda*, *Polianthes*, and *Prochnyanthes* are soft points. Leaf margins may be entire or with small translucent teeth. Additionally, in *Furcraea* and *Agave* the margins may be provided with large corneous teeth which may be diagnostic for species. Filiferous margins occur in *Agave*, *Hesperaloe*, and *Yucca*.

VEGETATIVE ANATOMY. The roots contain vessels with simple and scalariform perforation plates (Wagner 1977). Vessels are absent in stems and leaves. The fleshy roots in the family have a large cortical parenchyma. A multilayered velamen is reported in *Agave* (v. Guttenberg 1968) and a thick endodermis in *Yucca* (Arnott 1962). The primary root in *Agave* is tetrarch (Boyd 1932). Primary and adventitious roots in *Yucca* have 4 to 28 protoxylem poles, depending upon the species.

Leaves in both *Yucca* and *Agave* contain bundles of acicular crystals of calcium oxalate. In *Yucca brevifolia* these crystals are 4-sided raphides, while in *Agave americana* there are 6-sided raphides with laminated sheaths, and styloids covered with suberinic sheaths (Wattendorf 1976). Cells walls of the cells containing styloids are also suberized. Sieve-tube plastids have cuneate protein crystals only (Behnke 1981).

Based upon stomatal complexity and leaf epidermal characteristics, the genera fall into two broad ecologically adapted groups (Alvarez 1990). These groups mainly follow generic lines, but particular species may align with the other group. Those genera with persistent leaves adapted to dry conditions (*Agave*, *Beschorneria*, *Furcraea*, and *Yucca*) have a thick cuticle, isodiametric epidermal cells and projections from the subsidiary cells which make complex chambers around the stomata (Blunden and Binns 1970; Blunden et al. 1973; Alvarez 1990). *Prochnyanthes* also falls in this group. In the "herbaceous" genera *Manfreda* and *Polianthes*, the epidermal cells are tabular and the stomata are without complex stomatal chambers. Papillate epidermal cells are arranged over veins in *Prochnyanthes*, *Hesperaloe*, and *Yucca* section *Hesperoyucca*. Groups of nonuniform papillate cells form ridges in *Yucca*, *Furcraea*, and *Beschorneria*. Rows of papillae are also present in *Manfreda*. Pubescence is not common in the family, but where it occurs, trichomes are simple. These characters, along with those of stomata, may be diagnostic at the species level and at the section level in *Agave* (Gentry and Sauck 1978).

Stomata are largely anomocytic in *Polianthes tuberosa* (Shah and Gopal 1970) but paracytic, tricytic, and tetracytic types also occur. Stomata are paracytic in *Manfreda*, *Polianthes*, and *Prochnyanthes* and tetracytic in *Agave*, *Beschorneria*, *Furcraea*, and many *Yucca* (Blunden et al. 1973; Gentry and Sauck 1978) and also anomocytic in *Yucca* (Blunden and Binns 1970). The stomata are sunken below the epidermal cells because the subsidiary cells are dilated above and below the guard cells. The suprastomatal chamber is guarded and sometimes compartmentalized by 6 lips in *Yucca* and 4–15 lips in *Agave*, *Beschorneria*, and *Furcraea*. *Hosta*, previously included in the family, has very different and specialized epidermal and stomatal characters. *Leucocrinum* and *Hesperocallis*, also thought to be allied to the family, are anomocytic and less specialized.

In the leaves, palisade tissue may or may not be differentiated and the mesophyll may be largely composed of groups of water-storage cells (Blunden et al. 1973).

Secondary growth in stems of *Yucca* and *Agave* results in alternating dense and parenchymatous rings (Arber 1925). In *Yucca whipplei* the primary and secondary thickening meristems are longitudinally continuous and occur functionally and histologically as phases in the activity of the same meristem (Diggle and DeMason 1983). The bark in arborescent yuccas is thick.

In the leaves, vascular and fiber bundles occur in arcs in regular or irregular rows adjacent to the epidermis or palisade layers and through the center of the leaf. Bundles of heavily lignified mechanical fibers occur mainly at the periphery of the leaf. Fibers may occur at both phloem and xylem poles of vascular bundles. The presence and degree of development in these fiber caps and the arrangement of fiber bundles is diagnostic for *Agave*, *Beschorneria*, and *Furcraea* (Blunden et al. 1973). A bundle sheath of longitudinally elongated parenchyma cells is present in *Agave* and *Yucca* but absent in *Beschorneria* and *Furcraea*. Gross fiber characters in *Agave* have been analyzed for the fiber industry (Lock 1969). Individual leaf fiber cells in species of *Hesperaloe* and *Yucca* are longer (>2 mm) and narrower (<20 μm) than those of *Agave sisalana* (McLaughlin and Schuck 1991). In species with filiferous leaf margins the margin is set off from the main part of the leaf by layers of lignified cells and abscission cells.

Inflorescence Structure. Inflorescences in the family vary from paniculate to racemose or spicate. Except in some species of *Yucca*, the flowering portion is commonly borne on a tall, bracteate peduncle (Fig. 24). Among those with a spicate or racemose inflorescence, flowers in *Manfreda* are solitary at the nodes, in *Polianthes* and *Prochnyanthes* the flowers are paired, and in *Agave* subgen. *Littaea* the flowers are commonly clustered at each node.

The panicles of *Agave* subgen. *Agave* (Fig. 25) bear the flowers clustered in smaller umbellate panicles at the ends of secondary and tertiary branches (Catalano 1931) and in some species may reach heights of 8–12 m. In *Beschorneria*, *Furcraea*, *Hesperaloe*, and *Yucca* the inflorescence is a raceme or panicle, often within the same species. The panicles are evenly floriferous with individual flowers or clusters of up to 10 flowers spread at regular intervals along the branches. Normally paniculate agaves and yuccas may become spicate by contraction of lateral branches (McKelvey 1947; Gentry 1982).

In *Polianthes* and *Prochnyanthes* each pair of flowers occurs in the axil of a single bract and each flower is subtended by a bract. In *Manfreda* there is 1, occasionally at the lower nodes 2, flowers in the axil of each bract, each pedicel having 2 lateral bracteoles (Verhoek-Williams 1975).

Fig. 24A–K. Agavaceae. **A–F** *Hesperaloe parviflora*. **A** Habit. **B** Leaf. **C** Flower. **D, E** Stamen. **F** Pistil. **G, H** *Manfreda longibracteata*. **G** Flower. **H** Bulbous rhizome. **I, J** *Prochnyanthes mexicana*. **I** Habit. **J** Opened flower. **K** *Polianthes durangensis*, flower. (A–F, I–K Takhtajan 1982)

In *Agave* and *Furcraea* the inflorescence occasionally produces plantlets.

Flower Structure. *Yucca* and *Hesperaloe* have superior ovaries and *Agave*, *Manfreda*, *Polianthes*, *Prochnyanthes*, *Furcraea*, and *Beschorneria* have inferior ovaries. Intermediates between these two conditions occur. In *Hesperaloe nocturna* the basal part of the ovary is surrounded by

Fig. 25A–C. Agavaceae. *Agave cerulata* subsp. *cerulata*. **A** Partial inflorescence. **B** Flower, longitudinal section. **C** Opening fruit. (Takhtajan 1982)

receptacular tissue. The tip of the ovary in *Agave striata*, *Manfreda guttata*, and several other species protrudes into the floral tube (Verhoek-Williams 1975).

The tepals are united into a tube in *Agave*, *Manfreda*, *Polianthes*, and *Prochnyanthes*, are free or nearly so in *Yucca* and *Furcraea* and some *Hesperaloe*, or so closely appressed as to approximate a tube in *Beschorneria* and some *Hesperaloe*. Flowers with short tubes or none are actinomorphic. Long-tubed flowers of *Polianthes* and *Manfreda*, and flowers of *Agave* section Parviflorae are frequently zygomorphic: the tubes are curved, the mouth of the flower is overarching at the top, and the stamens (and later the style) are held along the top of the tube.

The stamens are inserted either at the tepal base, at characteristic levels in the tube, or sometimes at different levels (*Manfreda potosina* and *Agave* sect. *Ditepalae*). Filaments in *Agave*, *Polianthes*, and *Manfreda* are usually long and slender, included or exserted and then initially bent in bud, but in some species of *Polianthes* and *Manfreda* the filaments are short and inserted at the mouth of the tube. The filaments in *Furcraea* are swollen in the basal portion and abruptly narrowed distally. In *Yucca* the filaments are sometimes puberulent, and the tips frequently swollen and turned outward. Anthers are oblong to linear except in *Yucca* and *Hesperaloe*, where they are sagittate to hastate.

In the ovaries, septal nectaries have been reported in *Agave*, *Beschorneria*, *Furcraea*, *Polianthes*, and *Yucca*. In *Yucca*, *Furcraea*, and *Beschorneria* the nectaries are augmented with external secretory grooves along the ovarian septa (Arber 1925). Nectar amounts in *Yucca* are small, copious in *Agave* (Schaffer and Schaffer 1977) and *Manfreda scabra* (Eguiarte and Búrquez 1987).

The style is elongate except in some species of *Yucca*. In *Beschorneria*, and more so in *Furcraea*, it is swollen into 3 basal ridges and abruptly narrowed distally. At anthesis the long style of *Agave*, *Polianthes*, *Manfreda*, and *Prochnyanthes* is not yet fully extended and may be bent downwards out of the circle of stamens. The broad style of *Yucca* is often short and not clearly demarcated from the ovary.

Stigmas in the family are 3 or there is a single 3-lobed or capitately 3-angled stigma. In *Polianthes* and a few species of *Manfreda* the stigma lobes reflex at maturity. Most *Yucca* species have 3 stigmas but *Yucca* sect. *Hesperoyucca* has a capitate 3-lobed stigma. The stigmas in *Agave*, *Furcraea*, and *Yucca* are of the Dry type with unicellular papillae and those of *Beschorneria* and some species of *Agave* are of the Wet type with low to medium papillae (Heslop-Harrison and Shivanna 1977). *Polianthes* and *Manfreda* have stigmas that are moist when receptive.

Floral Anatomy. Floral anatomy has been described in detail for *Agave lechuguilla* (Grove 1941) and *Polianthes tuberosa* (Joshi and Pantulu 1941). The floral parts arise acropetally. The stamens arise in 2 cycles of 3 but mature simultaneously. Vascular anatomy of the flower is simple. Bundles are present at the base of the ovary and continue with minimum divergence into the separate floral parts.

Embryology. In *Agave* the middle layer of the anther wall has 10–12 layers and the endothecium is multilayered with fibrous thickenings (Davis 1966). There is a 2-layered endothecium in *Polianthes*. In the family, tapetal cells are 2–4-nucleate. Microsporogenesis is successive and the pollen grains are 2-celled at anthesis.

The ovules are anatropous, bitegmic, and crassinucellate. In *Yucca* the nucellus divides and becomes a perisperm (Arnott 1962). The hypodermal archesporial cell cuts off a parietal cell, which divides to form 2 or 4 cells (Wolf 1940; Joshi and Pantalu 1941) or a single parietal layer (Grove 1941). The megaspore tetrad is linear (Mogensen 1970) or T-shaped. The chalazal megaspore divides to produce a normal Polygonum type embryo sac in *Yucca*, *Agave*, *Manfreda*, and *Polianthes*. In *Furcraea* the micropylar megaspore is the functional one. The polar nuclei fuse before fertilization. The mature embryo sac is narrowed toward the chalaza and the narrowed portion is interpreted as having a haustorial function (Wolf 1940). Endosperm formation is Helobial in *Yucca*, *Agave*, and *Beschorneria*, and nuclear in *Furcraea* (Joshi and Pantulu 1941).

Pollen Morphology. Pollen studies of 15 genera historically included in the family show 7 types based upon aperture and exine ornamentation, and help to support a narrow circumscription of the family (Ojeda and Ludlow-Wiechers 1995; see also Alvarez and Köhler 1987). Semitectate, primarily sulcate grains occur in *Agave*, *Beschorneria*, *Dasylirion*, *Hesperaloe*, *Manfreda* (Fig. 26A), and *Polianthes*. Exine ornamentation in *Furcraea* is tectate perforate to semitectate. *Yucca* has tectate perforate, sulcate grains. Tetrads occur simultaneously with monads in *Beschorneria* and *Furcraea* (Fig. 26B). Bisulcate grains have been reported in *Polianthes* and *Prochnyanthes* (Erdtman 1969). The pollen grains are prolate or subprolate and the sulcus is nearly as long as the axis of the grain.

Karyology. The base number of the family is $x = 30$, with a karyotype of 5 long and 25 short chromosomes (Granick 1944; Gomez-Pompa et al. 1971). This complement is a major distinguishing characteristic of the family as presently circumscribed. Only *Hosta* (Funkiaceae) and *Hesperocallis* have similar bimodal karyotypes. Half of the species of *Agave* counted are polyploid, most of these in subgen. *Agave*. Tetraploids are also reported for *Polianthes tuberosa*.

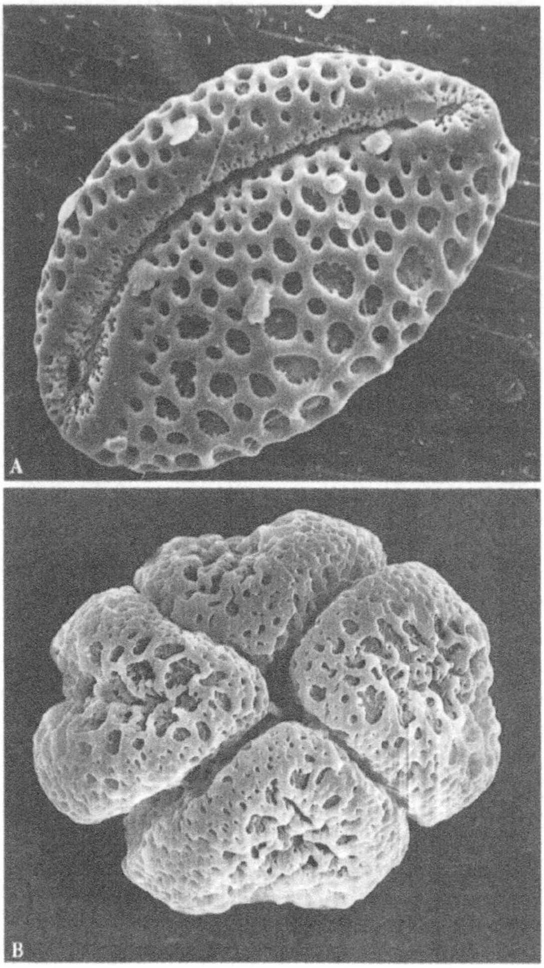

Fig. 26A,B. Agavaceae. **A** *Manfreda variegata*, distal face of pollen grain showing sulcus with narrow and compact margin, SEM ×1000. **B** *Furcraea macrophylla*, pollen tetrad. SEM ×900. (Photo Alvarez and Koehler)

Pollination and Reproductive Systems. Floral traits of color, odor, shape, nectar production, and timing of anthesis attract specific moth, hummingbird, and bat pollinators. *Yucca* spp. are pollinated by females of the yucca moth *Tegiticula* which collect pollen and pack it into stigmas of ovaries in which they oviposit (Powell and Mackie 1966; Dodd and Linhart 1994). Two species are pollinated by moth species unique to them, *Y. whipplei* by *T. maculata* (Riley) and *Y. brevifolia* by *T. paradoxa* (Riley). All other *Yucca* spp. are pollinated by *T.* (=*Pronuba*) *yuccasella* (Riley) (McKelvey 1947). There is either early abortion of infested fruits or later seed predation by the yucca moth larvae (Richter and Weis 1995). When moth populations are low, the fly *Pseudocalliope* may be an important alternate pollinator (Dodd and Linhart 1994).

Flowering and pollination have been investigated in *Agave* by Schaffer and Schaffer (1977) and Howell (1979); in *Hesperaloe* by Starr (1995); and in *Manfreda*, *Polianthes*, and *Prochnyanthes* by Verhoek-Williams (1975), Verhoek (1978a), and Eguiarte and Búrquez (1987). Most species of *Agave* and *Polianthes* are protandrous by several days. Anther dehiscence and nectar production occur at particular times of the day. Both nocturnal and diurnal flowering occur in *Agave*. *Prochnyanthes* and *Manfreda* are nocturnal. *Hesperaloe* and *Polianthes* have whitish, fragrant, nocturnal flowers or diurnal and crepuscular red flowers. Nocturnal odors in white-flowered *Polianthes* and some greenish *Manfreda* are floral and fruity; other greenish *Manfreda* flowers produce alliaceous odors.

Some species of *Agave*, mainly of subgen. *Agave*, are adapted to pollination by bats, other species by bumblebees and carpenter bees. Hummingbird and hawkmoth pollination is recorded in *Manfreda*, *Prochnyanthes*, *Polianthes*, and *Hesperaloe*. Bat pollination is also recorded for *Manfreda scabra* and *Hesperaloe*.

Protandry in the family may encourage cross-pollination but the amount may be low. Flowers bloom in succession at a node or up the spike so that male and female phase flowers are present in the inflorescence at the same time. Self-compatibility has been demonstrated in many of the genera.

Self-compatibility has been demonstrated in *Manfreda* and *Polianthes* (Verhoek-Williams 1975), in *Yucca* (Dodd and Linhart 1994; Richter and Weis 1995), and reported for *Agave* (Gentry 1982). However, seed set is lower by about 20% in self-pollinated *Manfreda scabra* (Eguiarte and Búrquez 1987). Howell and Roth (1981) have shown in *A. palmeri* that self-pollination does not usually result in fruit set.

Artificial interspecific crosses have been made in *Hesperaloe* (Starr 1995), *Polianthes* (Bundrant 1985), and *Manfreda* and between *Manfreda* and *Polianthes* (Verhoek-Williams 1976). Hybrids are postulated in *Agave* (Gentry 1982) and *Yucca* (J. Webber 1953).

Fruit and Seed. Most genera have loculicidal capsules except in several sections of *Yucca*, where the fruits may be indehiscent, and then they are dry and spongy in sect. *Clistocarpa*, and pendent, fleshy ("baccate"), and sweet with a hard endocarp in sect. *Sarcocarpa*. Capsule dehiscence in *Yucca* sect. *Chaenocarpa* is septicidal and also apically loculicidal.

The seeds are usually numerous in a capsule. They are 2–17 mm long, flat, black, and asymmetrically cuneate and sometimes narrowly winged. In the outer cell layer of the outer integument there is always a brittle phytomelan crust present; in Yucceae it is 10–40 μm thick, in the Agaveae ca. 10 μm. The epidermal cells of *Y. whipplei* seed coats contain raphides. The inner integument collapses and loses its cellular structure (Huber 1969).

Storage tissue in the seed of *Yucca* is reported as perisperm (Arnott 1962), and as endosperm in the other genera. Seeds contain lipids and proteins, and some seeds have large amounts of saponins.

The embryo is linear, cylindrical, straight, or slightly curved, extending generally along the greatest length of the seed from the chalazal end. Seeds of species of *Yucca* with baccate fruits are ruminate.

Seedlings are similar in at least 5 of the genera (Boyd 1932; Arnott 1962), with the cotyledon photosynthetic.

Dispersal and Reproductive Biology. Members of the Agavaceae are efficient vegetatively propagating plants. All genera produce new plants by lateral budding from the parent plant, either directly or from the tips of spreading rhizomes. In *Furcraea* and some species of *Agave*, multiple plantlets are produced in the inflorescence. As a result, a whole population may consist of only 1 or several clones. Commercial agaves, such as *A. sisalana* and *A. tequilana*, are propagated by vegetative offshoots.

Seed dispersal in the capsular species is primarily by wind. The flat seeds are released by waving of the flowering stem and are then collected by ground-dwelling animals. Indehiscent baccate fruits of *Yucca* sect. *Sarcocarpa* drop to the ground upon ripening. The fleshy fruits of *Y. aloifolia* are eaten by mockingbirds (H. Webber 1895). The shed dry fruits of *Y. brevifolia* are considered to be wind-dispersed "tumble-fruits" (Trelease 1893).

Phytochemistry. Steroidal saponins occur throughout the family, sometimes in high amounts. About 30 sapogenins have been identified, among which hecogenin, tigogenin, and sarsapogenin abound. Sapogenin concentration is often very high in the seed (Wall et al. 1957; Blunden et al. 1978).

Yucca glauca seeds contain linoleic acid. Cyanogenic compounds are reported in *Yucca* and chelidonic acid in *Agave* and *Yucca* (Ramstad 1953). The presence of flavonoids as kaempferol xyloside is reported from *Polianthes tuberosa* (El-Moghazy et al. 1980) and from *Agave americana* as kaempferol-3-glucoside and kaempferol-3-rutinoside (Subramanian and Nair 1970). Seed protein electrophoresis is useful in the identification of sections and species in *Yucca* (Smith and Smith 1970). Wall et al. (1957) did not find alkaloids or tannins in *Agave* and *Yucca*.

Subdivision. Genera of three tribes remain in the family following the narrower circumscription (Ojeda and Ludlow-Wiechers 1995): the Yucceae with the hypogynous genera (*Yucca*, *Hesperaloe*), the Agaveae with the epigynous, evergreen-leaved and fibrous-rooted genera (*Agave*, *Furcraea*, *Beschorneria*), and the Poliantheae composed of the epigynous genera with soft leaves and a storage rhizome (*Manfreda*, *Polianthes*, and *Prochnyanthes*). Athough the distinction between the hard xerophytic Agaveae and the softer, mesic Poliantheae is ecologically useful, molecular data do not support the latter division (Eguiarte 1995). The separation between the hypogynous and the epigynous genera is confirmed by recent molecular and cladistic analysis (Eguiarte 1995; Hernandez 1995).

Affinities. The family Agavaceae was resurrected by Hutchinson (1934) to include six tribes of xerophytic genera of mainly arborescent habit previously included in the Liliaceae and Amaryllidaceae. In the current restricted concept of the Agavaceae, only New World genera with bimodal karyotypes are included. Huber (1969), as well as Dahlgren et al. (1985), segregate Old World genera previously included (e.g., *Phormium*, *Dracaena*, *Cordyline*, *Doryanthes*) and the New World Nolineae as several other families. All of these families are included in the Asparagales, a segregate from the Liliales. In addition to the karyotype, anatomical, palynological, embryological and fruit characters, and geographical distribution support the new circumscription. The Agavaceae are composed of specialized genera adapted to arid habitats and advanced pollinators. It has been proposed that the more mesic genera may be the result of dedifferentiation from xerophytic types (Alvarez 1990).

Molecular studies of the chloroplast gene *rbc*L indicate that the Agavaceae sensu stricto are a clade within the Asparagales (Eguiarte 1995). The *rbc*L studies, as well as analysis of chloroplast DNA restriction sites (Bogler and Simpson 1995) and of internal transcribed spacers of nuclear ribosomal DNA (Bogler 1995) strongly support a clade for Agavaceae sensu stricto, with capsular fruits and phytomelan in the seed, separate from the Nolinaceae, which lack phytomelan and are grouped with genera with berries or dry berrylike fruits. All of these molecular studies, as well as cladistic analyses of morphological characters, indicate *Hosta* (Funkiaceae) as a sister group to Agavaceae, perhaps closest to *Yucca* (Hernandez 1995). Within the Agavaceae there are two well-supported clades: *Yucca* and *Hesperaloe* with ovaries superior, and *Agave*, *Manfreda*, *Polianthes*, *Prochnyanthes*, *Furcraea*, and *Beschorneria* with ovaries inferior. *Furcraea* and *Beschorneria* are closely paired in all cladistic studies and form a sister group to *Agave* and the herbaceous genera.

Distribution, Habitats, and Ecology. The Agavaceae are centered in SW USA and Mexico, but range from central USA to Panama, the Caribbean Islands, and northern montane S America (García-Mendoza and Galván 1995). *Yucca* has the widest range, followed by *Agave* and *Manfreda*. All genera except *Furcraea* occur primarily north of Guatemala. *Furcraea* is the main representative in S America. Several genera are naturalized in the Old World.

Ecologically, the Agavaceae range from xeric to mesic conditions from sea level to 2500 m. They are characteristic in desert and chaparral communities with *Larrea*, *Prosopis*, *Fouquieria*, and cacti. In mesophytic habitats they occur in open, well-drained sites such as cliffs and ravine slopes, usually on limestone-derived soils. Species of *Polianthes*, *Prochnyanthes*, and some *Agave* and *Manfreda* occur in montane pine-oak forest. Species of *Agave* and *Yucca* may be epiphytic in wet forests.

The Agavaceae are well adapted to dry environments and show high efficiency of water use (Nobel 1988). Crassulacean acid metabolism has been reported for all *Agave* spp. studied and some species of *Yucca*. Only some species are cold-tolerant.

Vesicular arbuscular mycorrhiza have been found in association with *Yucca valida* (Rose 1959). Various generalized and specific viral and fungal pathogens have been identified on *Agave*, *Yucca*, and *Furcraea* (Wellman 1977).

Several insects are specific to the Agavaceae. Among Lepidoptera, one family, Megathymidae, are exclusive feeders upon Agavaceae, with larvae of particular species feeding in stems or leaves of one or several preferred plant species (Freeman 1969). Larvae of single species in other moth genera also feed preferentially on seeds of Agavaceae (Powell and Mackie 1966).

Paleobotany. Tidwell and Parker (1990) described an arborescent monocotyledon, *Protoyucca shadishii*, from the Middle Miocene of Nevada. Although the authors did not assign this fossil to a family, they treated it as being closely related to *Yucca*.

Economic Importance. Members of the Agavaceae have been extensively used by Aztec and other aboriginal New World civilizations (Castetter et al. 1938; Bell and Castetter 1941; J. Webber 1953; Trueblood 1973; Verhoek 1978b; Gentry 1982). They provide fiber from leaf and peduncle, food from flowers and baked stem and leaf bases, beverages, soap, various medicines, ornamentals, and animal fodder. *Agave* spp. are cultivated for cordage fibers (*A. fourcroydes*, henequen; *A. sisalana*, sisal), for the making of distilled alcoholic beverages (primarily *A. tequilana*), and as a source of corticosteroid precursor hecogenin (*A. sisalana*) (Cruz et al. 1985). The long narrow fiber cells of *Hesperaloe* and *Yucca* can be used as pulped fibers for specialty papers (McLaughlin and Schuck 1991). *Yucca* has been used for fiber and soap, and provides a wetting agent in fertilizers and frothing agent in drinks and fire extinguishers. *Polianthes tuberosa* is cultivated commercially as a cut flower and in perfumery. Members of most genera are horticultural subjects.

Key to the Genera

1. Ovary superior 2
- Ovary inferior 3
2. Flowers globose to broadly campanulate; filaments pubescent or papillose, clavate, or stout at middle **1. *Yucca***
- Flowers narrowly tubular to rotate-campanulate; filaments glabrous, subulate-filiform **2. *Hesperaloe***
3. Leaves tough, long-lived; marginal teeth often corneous; roots arising from stem base 4
- Leaves soft; marginal teeth soft; roots arising from an upright rhizome ("rootstock") 6
4. Leaf apex a long, hard spine; flowers erect or horizontal; filaments exserted **6. *Agave***
- Leaf apex short or soft; flowers pendulous; filaments included 5
5. Floral bracts small; perianth campanulate; filaments swollen below, narrowed above **7. *Furcraea***
- Floral bracts large; perianth funnel-shaped; filaments slender **8. *Beschorneria***
6. Flowers single at nodes, greenish or brownish (rarely white or pink); stamens and style exserted; stigma trigonous or rarely 3-lobed **3. *Manfreda***
- Flowers usually paired at nodes, white to reddish; stamens and style included; stigma with 3 reflexed lobes 7
7. Leaves chartaceous; perianth tube narrow below, campanulate above, abruptly bent near middle **4. *Prochnyanthes***
- Leaves herbaceous; perianth narrow, straight or slightly curved **5. *Polianthes***

Genera of Agavaceae

1. *Yucca* L.

Yucca L., Sp. Pl.: 319 (1753); McKelvey, Yuccas of the SW United States 1, 2 (1938, 1947), rev., Webber, J.M. 1953. Yuccas of the Southwestern United States. Agric. Monogr. U.S.D.A. 17: 1–97.
Hesperoyucca (Engelm.) Baker (1892).
Clistoyucca (Engelm.) Trelease (1902).
Samuela Trelease (1902).

Caespitose or arborescent, stems usually branched; leaves pliant to rigid, margins entire, filiferous or serrulate, spine-tipped; inflorescence a raceme or panicle; flowers white or greenish, globose to campanulate; tepals free or basally united; filaments fleshy, clavate, or slightly swollen, pubescent or papillose; anthers small; ovary superior; stigmas 3 or 1 and capitate; fruit loculicidally or septicidally dehiscent or indehiscent. About 40 spp., mainly in arid and semiarid regions from central USA to Mexico and the Caribbean (*Y. aloifolia* L.; *Y. elephantipes* Regel in Central America possibly introduced.

Four sections distinguished by McKelvey: sect. *Sarcocarpa*, western and southern N America, with indehiscent, fleshy, pendent fruits; sect. *Clistocarpa*, SE and SW USA, with indehiscent, dry fruits; sect. *Chaenocarpa*, mainly north-central and eastern USA with septicidal (sometimes also loculicidal) capsule; section *Hesperoyucca*, containing only *Yucca whipplei* Torrey from California-Baja California, with loculicidal capsules and capitate stigma, is sometimes raised to the generic level. Major morphological and floristic works have retained the species within *Yucca*, but chloroplast DNA restriction site and *rbc*L data place it outside of *Yucca* and more closely related to *Hesperaloe* (Clary and Simpson 1995).

2. *Hesperaloe* Engelm. Fig. 24A–F

Hesperaloe Engelm. in S. Watson, Botany U.S. Geol. Explor 40th Parallel: 497 (1871); Starr, G., Madroño 44: 282–296 (1998), rev.

Acaulescent; caespitose; leaves linear, to 2 m, apex frayed or a hard spine; margins filiferous; inflorescence to 4 m tall, racemose or paniculate with 3–8 branches; flowers fascicled, tubular, campanulate, or rotate-campanulate, greenish white, maroon-streaked, or red to salmon, or rarely yellow; filaments basally inserted, filiform, included; anthers sagittate; ovary superior; stigma only slightly expanded; capsule septicidal. Five spp. and one ssp., Central Texas and N Mexico on both sides of the Sierra Madre Occidental, limestone prairies, mesquite groves, and rocky slopes.

3. *Manfreda* J.H. Salisb. Fig. 24G,H

Manfreda J.H. Salisb., Gen. Pl. Fragm.: 78 (1866); Verhoek-Williams, Study of the tribe Polianthеae (including *Manfreda*) (PhD thesis Cornell Univ. 1975), rev.

Small to medium-sized; rhizome erect with fleshy and fibrous roots; leaves annual, thin or succulent, apex soft to firm, margins entire, papillate, or dentate; inflorescence a spike or raceme; flowers solitary at nodes, erect or horizontal, green, rarely white or pinkish; tube funnel-shaped; filaments straight or bent in bud, exserted; ovary inferior; stigma trigonous or with 3 reflexed lobes; capsule loculicidal. About 26 spp., Texas and northern Mexico to Honduras and El Salvador, in moist to dry sites.

4. *Prochnyanthes* S. Watson Fig. 24I,J

Prochnyanthes S. Watson, Proc. Am. Acad. Arts 22: 457 (1887); Verhoek-Williams, Study of the tribe Polianthеae (including *Manfreda*) (PhD thesis Cornell Univ. 1975), rev.

Acaulescent, with a fleshy erect rhizome; leaves chartaceous, with a distinct midrib, linear, lanceolate to oblanceolate, narrowed at base, apical point soft, margins papillate to denticulate; inflorescence a lax raceme; flowers paired at nodes; floral tube bent at middle, narrow below, campanulate above, white tinged green or reddish; filaments inserted in tube, subulate; anthers oblong; ovary inferior; stigma with 3 reflexed lobes; capsule globose. Only one sp., *P. mexicana* (Zucc.) Rose, W Mexico, dry, rocky slopes, grasslands, or moist ravines.

5. *Polianthes* L. Fig. 24K

Polianthes L., Sp. Pl.: 316 (1753).
Bravoa Llav. & Lex.
Pseudobravoa Rose

Small to medium-sized; rhizome short, cylindrical, with fleshy roots; leaves 2–15, thin or slightly succulent, linear to lanceolate or elliptic, with a short, soft point; margin entire or minutely papillate; inflorescence a spike or raceme; flowers usually paired at nodes, red, pink, or white; tube narrowly funnel-shaped, straight to curved; filaments included; anthers oblong; stigma with 3 spreading or reflexed lobes; capsule loculicidal. About 14 spp., W and S Mexico, in well-drained, moist sites.

6. *Agave* L. Figs. 25, 26A

Agave L., Sp. Pl.: 323 (1753); Gentry, Agaves Cont. N America (1982), rev.

Small to gigantic; acaulescent or with short trunk, perennial, poly- or monocarpic; leaves thick and fibrous, linear to lanceolate or ovate, with long pungent apex, margins minutely to coarsely toothed, filiferous or entire; inflorescence a spike, raceme, or panicle, to 12 m tall; flowers in pairs or umbellate clusters, erect, usually yellow or greenish; perianth tubular or campanulate; stamens exserted, filaments subulate, anthers oblong; ovary inferior; stigma 3-lobed; capsule loculicidal. About 200 spp., SW USA to W Panama, the Caribbean and Venezuela, mainly arid and semiarid regions. Two subgenera: subgen. *Littaea*, with spikes or racemes, and subgen. *Agave* with panicles.

7. *Furcraea* Vent. Fig. 26B

Furcraea Vent., Bull. Soc. Philom. 1: 65 (1793); J.R. Drumm., Ann. Rep. Mo. Bot. Gard. 18: 25–75 (1907), rev. Orthographic variants: *Furcroea*, *Furcroya*, *Fourcroya*, and *Fourcroea*.

Plants massive, acaulescent or with trunk to 6 m high; leaves stiff or flexible, fibrous, lanceolate, apex a short firm point, margin entire, denticulate or coarsely toothed; panicle lax, to 13 m tall with long side branches, producing plantlets; flowers in clusters of 2–5, pendulous, campanulate; tepals mostly free, spreading, greenish or white; filaments included, expanded below middle; ovary inferior; style dilated below the middle; stigma capitate or shortly 3-lobed; capsules loculicidal. Perhaps 21 spp., N South America to Peru, extending to east Central Mexico, the Antilles, and E Brazil. Two subgenera: subgen. *Furcraea*, with firm-textured, marginally spined leaves, and subgen. *Roezlia*, with flexible leaves and minutely denticulate margins. In great need of revision and field work.

8. *Beschorneria* Kunth

Beschorneria Kunth, Enum. Pl. 5: 844 (1850). Garcia Mendoza, A. Monographía del género *Beschorneria* (Master's thesis, UNAM 1987), rev.

Acaulescent; leaves tough, linear-lanceolate, narrowed at base, apex a long, soft point, margins entire or minutely toothed; inflorescence a raceme or few-branched panicle; flowering stem and bracts red, pink, or yellow; bracts long; flowers 2–5 together in remote fascicles, pendulous; tepals lanceolate, free but connivent into a tube, greenish, yellowish, or red, papillose or puberulous outside; filaments as long as tepals, subulate; anthers oblong; ovary inferior; style as long as or exceeding the stamens, stigma obscurely 3-lobed; capsule loculicidal. Seven spp., Mexico, dry rocky woodland.

Selected Bibliography

Alvarez, A. 1990. El complejo estomático en la familia Agavaceae II. Epidermis adulta. Feddes Repert. 101: 113–134.

Alvarez, A., Köhler, E. 1987. Morfología del polen de las Agavaceae y algunos géneros afines. Grana 26: 25–46.

Arber, A. 1925. See general references.

Arnott, H.J. 1962. The seed, germination and seedling of *Yucca*. Univ. Calif. Publ. Bot. 35: 1–144.

Behnke, H.D. 1981. See general references.

Bell, W.H., Castetter, E.F. 1941. Ethnobiological studies in the American southwest. VIII. The utilization of yucca, sotol and beargrass by the aborigines of the American Southwest. Univ. N.M. Bull. #372, Biol. Ser. 5: 3–74.

Blunden, G., Binns, W.W. 1970. The leaf anatomy of *Yucca glauca* Nutt. Bot. J. Linn. Soc. 63: 133–141.

Blunden, G., Yi Yi, Jewers, K. 1973. The comparative leaf anatomy of *Agave*, *Beschorneria*, *Doryanthes* and *Furcraea* species (Agavaceae: Agaveae). J. Linn. Soc. Lond. Bot. 66: 157–179.

Blunden, G., Yi Yi, Jewers, K. 1978. Steroidal sapogenins from leaves of Agaveae species. Phytochemistry 17: 1923–1925.

Bogler, D.J., 1995. Systematics of *Dasylirion*: taxonomy and molecular phylogeny. Bol. Soc. Bot. Méx. 56: 69–767.

Bogler, D.J. Simpson, B.B. 1995. Chloroplast DNA study of the Agavaceae. Syst. Bot. 20: 191–205.

Boyd, L. 1932. Monocotylous seedlings: morphological studies in the post-seminal development of the embryo. Trans. Proc. Bot. Soc. Edinb. 31: 1–224.

Bundrant, L.H. 1985. *Polianthes tuberosa* and its hybrids. Herbertia 41: 55–60.

Castetter, E.F., Bell, W.H., Grove, A.R. 1938. Ethnobiological studies in the American southwest. VI. The early utilization and the distribution of *Agave* in the American Southwest. Univ. N.M. Bull. Biol. Ser. 5: 1–92.

Catalano, G. 1931. Sulla morfologia delle infiorescenze di *Agave*. R. Inst. Bot. Palermo, Lavori 2: 99–107.

Clary, K.H., Simpson, B.B. 1995. Systematics and character evolution in the genus *Yucca* (Agavaceae): evidence from morphology and molecular analyses. Bol. Soc. Bot. Méx. 56: 77–88.

Cruz, C., Castillo, L., Robert, M., Ondarza, R.N. (eds.) 1985. Biología y aprovechamiento integral del henequén y otros agaves. Centro de Investigación Científica de Yucatán, A.C.

Davis, G.L. 1966. See general references.

Diggle, P.K., DeMason, D.A. 1983. The relationship between the primary thickening meristem and the secondary thickening meristem in *Yucca whipplei* Torr. I. Histology of the mature stem. II. Ontogenetic relationship within the vegetative stem. Am. J. Bot. 70: 1195–1204, 1205–1216.

Dodd, R.J., Linhart, Y.B. 1994. Reproductive consequences of interactions between *Yucca glauca* (Agavaceae) and *Tegiticula yuccasella* (Lepidoptera) in Colorado. Am. J. Bot. 81: 815–825.

Eguiarte, L.E. 1995. Hutchinson (Agavales) vs. Huber y Dahlgren (Asparagales): análisis moleculares sobre la filogenia y evolución de la familia Agavaceae sensu Hutchinson dentro las monocotiledóneas. Bol. Soc. Bot. Méx. 56: 45–56.

Eguiarte, L., Búrquez, A. 1987. Reproductive ecology of *Manfreda brachystachya*, an iteroparous species of Agavaceae. Southwest Nat. 32: 169–178.

El-Moghazy, A.M., Ali, A.A., Ross, S.A., El-Shanaway, M.A. 1980. Phytochemical studies on *Polianthes tuberose* L. Fitoterapia 51: 179–181.

Erdtman, G. 1969. Handbook of palynology. New York: Hafner.

Freeman, H.A. 1969. Systematic review of the Megathymidae. J. Lepid. Soc. 23, Suppl. 1.

García-Mendoza, A., Galván, R. 1995. Riquezas de las familias Agavaceae y Nolinaceae en México. Bol. Soc. Bot. Méx. 56: 7–24.

Gentry, H.S. 1982. Agaves of Continental North America. Tucson: University of Arizona Press.

Gentry, H.S., Sauck, J.R. 1978. The stomatal complex in *Agave*: groups Deserticolae, Campaniflorae, Umbelliflorae. Proc. Calif. Acad. Sci. 41: 371–387.

Gómez-Pompa, A., Villalobos-Pietrini, R., Chimal, A. 1971. Studies in the Agavaceae. I. Chromosome morphology and number of seven species. Madroño 21: 208–221.

Granick, E.B. 1944. A karyosystematic study of the genus *Agave*. Am. J. Bot. 31: 283–298.

Grove, A.R. 1941. Morphological study of *Agave lechuguilla*. Bot. Gaz. 103: 354–365.

Guttenberg, H. von, 1968. Der primäre Bau der Angiospermen wurzel. Handb. Pflanzenanat. 2nd edn. 8: 134–137, 318–333.

Hernández, L. 1995. Análisis cladístico de la familia Agavaceae. Bol. Soc. Bot. Méx. 56: 57–68.

Heslop-Harrison, Y., Shivanna, K.R. 1977. The receptive surface of the angiosperm stigma. Ann. Bot. (Lond.) 41: 1233–1258.

Howell, D.J. 1979. Flock foraging in nectar-feeding bats: advantages to the bats and to the host plants. Am. Nat. 114: 23–50.

Howell, D.J., Roth, S. 1981. Sexual reproduction in agaves: the benefits of bats; the cost of semelparous advertising. Ecology 62: 1–7.

Huber, H. 1969. See general references.

Hutchinson, J. 1934. The families of flowering plants, vol. 2. London: Macmillan.

Joshi, A.C., Pantulu, J.V. 1941. A morphological and cytological study of *Polianthes tuberosa* Linn. J. Indian Bot. Soc. 20: 37–71.

Lock, G.W. 1969. Sisal, 2nd edn. London: Longmans, Green.

McKelvey, S.D. 1938, 1947. Yuccas of the Southwestern United States, parts 1 & 2. Jamaica Plain, Massachusetts: Arnold Arboretum of Harvard University.

McLaughlin, S.P., Schuck, S.M. 1991. Fiber properties of several species of Agavaceae from the Southwestern United States and Northern Mexico. Econ. Bot. 45: 480–486.

Mogensen, H.L. 1970. Megagametophyte development in *Agave parryi*. Phytomorphology 20: 16–22.

Nobel, P.S. 1988. Environmental biology of agaves and cacti. Cambridge: Cambridge University Press.

Ojeda, L., Ludlow-Wiechers, B. 1995. Palinología de Agavacee, una contribución biosistemática. Bol. Soc. Bot. Méx. 56: 25–43.

Powell, J.A., Mackie, R.A. 1966. Biological interrelationships of moths and *Yucca whipplei*. Univ. Calif. Publ. Entomol. 42: 1–46.
Ramstad, E. 1953. Über das Vorkommen und die Verbreitung von Chelidonsäure in einigen Pflanzenfamilien. Pharm. Acta. Helv. 28: 45–57.
Richter, K.S., Weis, A.E. 1995. Differential abortion in the yucca. Nature 376: 557–558.
Rose, S.L. 1959. Vesicular-arbuscular endomycorrizal associations of some desert plants of Baja-California Mexico. Can. J. Bot. 59: 1056–1061.
Schaffer, W.M., Schaffer, M.V. 1977. The reproductive biology of the Agavaceae: I. Pollen and nectar production in four Arizona agaves. Southwest. Nat. 22: 157–168.
Shah, G.L., Gopal, B.V. 1970. Structure and development of stomata in the vegetative and floral organs of some Amaryllidaceae. Ann. Bot. (II) 34: 737–749.
Smith, C.M., Smith, G.A. 1970. An electrophoretic comparison of six species of *Yucca* and of *Hesperaloe*. Bot. Gaz. 131: 201–205.
Starr, G. 1995. *Hesperaloe*: aloes of the west. Desert Plants 11: 3–8.
Subramanian, S.S., Nair, A.G.R. 1970. Chlorogenin and kaempferol glycosides from the flowers of *Agave americana*. Phytochemistry 9: 2582.
Takhtajan, A.L. 1982. See general references.
Tidwell, W.D., Parker, L.R. 1990. *Protoyucca shadishii* gen. et spec. nov., an arborescent monocotyledon with secondary growth from the Middle Miocene of northwestern Nevada, USA. Rev. Paleobot. Palynol. 62: 79–95.
Trelease, W. 1893. Further studies of yuccas and their pollination. Annu. Rep. Mo. Bot. Gard. 4: 181–226.
Trueblood, E.W.E. 1973. Omixochitl – the tuberose (*Polianthes tuberosa*). Econ. Bot. 27: 157–173.
Verhoek, S. 1978a. Floral biology of *Manfreda virginica* (L.) Rose (Agavaceae) Bot. Soc. Am., Misc. Ser., Publ. 156: 82.
Verhoek, S. 1978b. Huaco and amole: a survey of the uses of *Manfreda* and *Prochnyanthes*. Econ. Bot. 32: 124–130.
Verhoek-Williams, S.E. 1975. A study of the tribe Poliantheae (including *Manfreda*) and revisions of *Manfreda* and *Prochnyanthes* (Agavaceae). Ph. D. Thesis. Ithaca: Cornell Univ., NY.
Verhoek-Williams, S. 1976. Hybridization in *Manfreda* and *Polianthes*. Abstr. of Papers, Bot. Soc. Am., Am. J. Bot., Suppl 63.
Wagner, P. 1977. Vessel types of monocotyledons: a survey. Bot. Not. 130: 383–402.
Wall, M.E., Fenske, C.S., Kenney, H.E., Willaman, J.J., Correll, D.S., Schubert, B.G., Gentry, H.S. 1957. Steroidal sapogenins XLIII. J. Am. Pharm. Assoc. 46: 653–684.
Wattendorf, J. 1976. A third type of raphide crystal in the plant kingdom; six-sided raphides with laminated sheaths in *Agave americana* L. Planta 130: 303–311.
Webber, H.J. 1895. Studies in the dissemination and leaf reflexion of *Yucca aloifolia* and other species. Annu. Rep. Mo. Bot. Gard. 6: 91–112.
Webber, J.M. 1953. Yuccas of the Southwestern United States. Agric. Monogr. U.S.D.A. 17: 1–97.
Wellman, F.L. 1977. Dictionary of tropical American crops and their diseases. Metuchan, NJ: Scarecrow Press.
Wolf, F.T. 1940. Macrosporogenesis and the development of the embryo sac in *Yucca aloifolia*. Bull. Torrey Bot. Club. 67: 755–761.

Alliaceae

K. Rahn

Alliaceae J. Agardh, Theoria Syst. Pl.: 32 (1858), nom. cons.
Liliaceae subfam. Allioideae Engl. (1887).
Amaryllidaceae tribus Allieae Hutchinson (1934).
Milulaceae Traub, Plant Life 28: 129 (1972).

Acaulescent or short-stemmed biennial or perennial *geophytes*; alliaceous odour often present; stem usually swollen and often forming a bulb, tuberous rhizome or rarely a corm, enveloped by sheathing, dry leaves or leaf bases. *Leaves* linear, filiform, lanceolate or rarely ovate, flat, angular, terete, or fistular, often fleshy, forming closed sheaths below, and sometimes forming a pseudostem around the scape; veins parallel. Scape from apex of each shoot or bulb, sometimes also lateral, terete, flat or angular, sometimes fistular, leafless except at apex. *Inflorescence* usually umbel-like, formed of 1 or more contracted helicoid cymes, rarely reduced to a single flower, in one species a spike (*Milula*). Inflorescence subtended by 2 or less frequently 1 or several, membranous, sometimes ± united spathe bracts, enveloping the young inflorescence, its branches or individual pedicels sometimes also subtended by smaller membranous bracts. Pedicels not articulate. *Flowers* hermaphrodite, usually actinomorphic, zygomorphic in *Solaria*, *Miersia*, and *Gilliesia*. Tepals 3+3 (rarely 3 or 3+2), usually similar and petaloid, united at base, almost free, or forming a tube; tepal lobes erect, spreading or recurved. Scales or appendices between tepals and stamens, of different origin, often forming a corona. Functional stamens usually 3+3, rarely 3 or 2 and then the missing ones often transformed to staminodes. Filaments inserted on the tepals, free from each other or united, often with lateral, dorsal or apical appendices. Anthers versatile, basifixed, introrse, opening with longitudinal slits. Ovary superior or in *Allium siculum* and *A. tripedale* almost semiinferior; tricarpellary, trilocular; provided with septal nectaries; 2-several ovules in each locule. Ovules anatropous or campylotropous, bitegmic. Style solitary, erect, at apex of ovary or in *Allium* and *Milula* ± gynobasic. Stigma capitate or trilobate with a Dry or sometimes Wet (*Leucocoryne*) surface. *Fruit* a loculicidal capsule with few–numerous seeds. Seeds either rather small, ovoid or ellipsolidal to subglobose (rounded in transection), or larger, angular, semiovoid to semiglobose (triangular in transection), or in *Tulbaghia* flat. Testa with a crust of phytomelan. Endosperm with fatty oils and aleuron, but no starch. Embryo short and straight, or long and curved. Vegetative buds produced instead of flowers in some species of *Allium*.

Comprising 13 genera and about 600 species; most genera in S America, particularly Chile, the genus *Allium* with 260–700 species in the northern hemisphere.

VEGETATIVE MORPHOLOGY. The leafy part of the stem is short in all species. Each shoot ends in an inflorescence, and shoot structure is sympodial. Most species (except *Tulbaghia* and some *Allium*) have a true bulb (Fig. 27); the stem is short, often depressed conical and covered by one or several swollen sheathing leaves or leaf bases, externally with dry leaf bases. The new main bulb appears from the axil of the uppermost leaf; increase bulbs originate in the axils of lower leaves in the old bulb. The scape is often apparently lateral, because the green leaves spring from a lateral shoot or bulb that is to flower the next year. In *Tulbaghia* the stem or rhizome is more or less swollen and irregularly shaped, tuberous, covered by dry, fibrous leaves; the closed sheath formed at the base is very short. In the rhizome-bearing species of *Allium* the rhizome is covered by thin and fibrous leaf bases only and may be the only storage organ. The leaf bases often form long, closed sheaths around the younger leaves and the scape (Fig. 27, 28). The increase bulb in *Allium* (Mann 1960) produces a single contractile root which pulls the increase bulb deeper into the soil away from the mother bulb (Fig. 12F). Leaves are distichous, at least in *Milula* and some species of *Allium* (e.g. *A. ampeloprasum*), or else apparently spirally arranged.

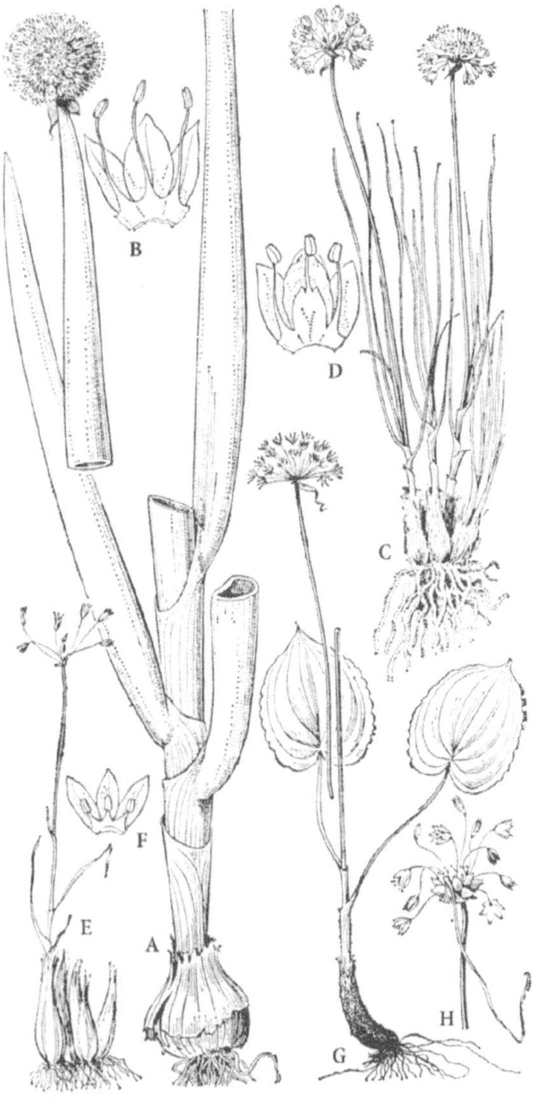

Fig. 27A–H. Alliaceae. **A, B** *Allium altaicum*. **A** Habit. **B** Tepals and stamens. **C, D** *A. polyrhizum*. **C** Habit. **D** Tepals and stamens, the stamens of inner whorl conspicuously appendaged. **E, F** *A. setifolium*. **E** Habit. **F** Tepals and stamens. **G** *A. funkiifolium*, habit. **H** *A. oleraceum*, inflorescence with flowers and bulbils. (Takhtajan 1982)

VEGETATIVE ANATOMY. Lacticifers have been found in all Alliaceae screened by Sterling and Huang (1972) (*Allium, Nothoscordum, Tristagma* and *Tulbaghia*). Raphides of calcium oxalate are recorded as being absent in *Allium, Milula* and *Tulbaghia*. The stomata are anomocytic. Epicuticular waxes are arranged in non-orientated platelets (several *Allium*) or in threadlike, reticulate projections (*Ipheion uniflorum* and *Nothoscordum bivalve*); neither the Convallaria type nor the Strelitzia type was found (Frölich and Barthlott 1988). The veins in cylindric leaves, and often also in flat leaves in *Allium*, are arranged in a ring with the phloem closest to the epidermis. The cylindric leaf is probably primitive in *Allium* (unlike, e.g. in *Juncus*).

INFLORESCENCE STRUCTURE. The scape is apical in the shoot (bulb). In some species lateral inflorescences also appear. The inflorescence in

Fig. 28A–I. Alliaceae. A, B *Allium schubertii* . A Habit. B Opened perianth with stamens. C, D *A. verticillatum.* C Habit. D Fruit. E–G *A. paradoxum.* E Habit. F Dehiscing capsule. G Seed with elaiosome. H, I *A. chamaemoly.* H Habit. I Capsule. (Takhtajan 1982)

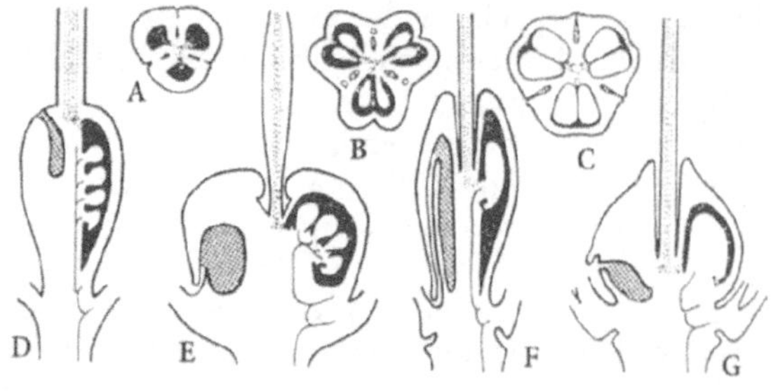

Fig. 29A–G. Alliaceae. Pistil organisation in Alliaceae. **A–C** Transversal sections at the level of the compitum. **D–G** Longitudinal sections through a plane of symmetry. Pollen transmission tissue *dotted*; septal nectaries *cross-hatched*. Note differences in style insertion, opening of nectaries, and shape of placentae. **A, D** *Nothoscordum borbonicum*. **E** *Allium nigrum*. **B, F** *A. asperum*. **C, G** *A. ampeloprasum*. (Di Fulvio 1973)

Nothoscordum, Allium and probably also in the other genera is formed of one or more contracted helicoid cymes, which are spikelike in *Milula* and umbel-like in all other genera, but sometimes reduced to 1 or 2 flowers. The inflorescence is subtended by usually 2, less frequently 1 (*Milula*, some *Allium*) or more (*Nothoscordum fictile* and *N. sessile*) membranous bracts, which envelop the young inflorescence. The spathe bracts sometimes sheath at base, and the 2 sheaths are sometimes more or less united. The short branches in the inflorescence or the individual pedicels are sometimes also subtended by smaller membranous bracts (most species of *Tulbaghia*, many *Allium* and a few *Nothoscordum*). In the inflorescence of some species of *Allium* the flowers may partly be replaced by bulbils.

Flower Structure. Septal nectaries are probably present in all species (recorded in nine *Allium* and two *Nothoscordum* species by Daumann 1970). According to Di Fulvio (1973), the opening of the nectary is close to the style in *Nothoscordum* and close to the base of the ovary in *Allium*. A very short gynophore may be found in some *Nothoscordum* species (Fig. 29).

Acording to Daumann (1970), a cuticula is present in the septal nectaries of *Nothoscordum*, but not in those of *Allium*. A nectarium parenchyma is present, and vascularisation absent in the species studied.

Embryology. In *Nothoscordum*, embryo-sac development follows the Polygonum type, while in *Allium* and *Leucocoryne* the megaspore mother cell undergoes only the first meiotic division and the chalazal dyad cell develops into an bisporic Allium-type embryo sac (Davis 1966).

In the facultatively parthenogenetic *Allium ramosum* the synergids and antipodal nuclei are supposed to fuse in pairs and to give rise to supernumerary embryos: apogamy. Adventitious embryony from nucellar tissues is recorded several times from *Allium*, *Nothoscordum borbonicum* and probably other related species (Schnarf 1931; Guaglianone 1972). In contrast to *Agapanthus*, a parietal cell is not cut off from the archesporial cell in *Allium*, *Leucocoryne*, *Miersia*, *Nothoscordum* and *Tulbaghia* (Wunderlich 1959). The endosperm is Nuclear in *Allium* and Helobial in *Nothoscordum* and *Tulbaghia*. The anther tapetum is secretory; the tapetum cells uninuclear in 4 species of *Allium*, and 2–4-nuclear in *Tulbaghia* as in most other genera in Asparagales (Wunderlich 1954). Microsporogenesis is successive, and the pollen 2-celled when dispersed.

Pollen Morphology. Schulze (1980), who studied the pollen of eight genera (*Tulbaghia*, *Allium*, *Nothoscordum*, *Ipheion*, *Tristagma*, *Solaria*, *Miersia* and *Gilliesia*) found it invariably sulcate and reticulate. The mean length of the pollen is between 30 and 45 µm, and the sulcus does not reach the proximal side of the pollen grain.

Karyology. Most species of *Tulbaghia* have $2n = 12$ (Vosa 1975). *Milula* and most species of *Allium* have $2n = 16$ (but also $2n = 32$, 14, 28 and 18 are found in *Allium*). In *Nothoscordum* (Nuñez et al. 1974; Crosa 1975; Ravenna 1978) five species have $2n = 10$, two $2n = 16$, two $2n = 18$, two $2n = 19$, and one $2n = 26$. In this genus acrocentric chromsomes are about 1/2 the length of the metacentric chromosomes. The sum of the number of acrocentric and twice the number of metacentric chromosomes are multiples of 8. Alternatively, the chromosomes can be considered combinations of sets with 4 metacentric and sets with 3 metacentric + 2 acrocentric (Crosa 1981). Four species are diploid, six tetraploid and one hexaploid when a set is considered "the basic number". Two *Ipheion* species were studied by Crosa (1975); they had chromosomes similar to, but smaller than *Nothoscordum*, $2n = 20$ in *I. sessile* (4 metacentric and 16 acrocentric) and $2n = 12$ in *I. uniflorum* (2 metacentric and 10 acrocentric). *Tristagma*, the other genus with which *Nothoscordum* has been associated, was studied by Crosa (1981). Three species have $2n = 8$ (6 meta-, 2 acrocentric); one $2n = 16$ (12 meta- and 4 acrocentric); one unidentified species $2n = 24$ (18 meta- and 6 acrocentric). The chromosomes are of a size similar to those of *Nothoscordum*. In *Leucocoryne narcissoides* Ravenna (1978) counted $2n = 24$, and $2n = 12$ or 18 has been recorded in other species of the same genus. In *Miersia* chilensis $2n = 12$, 20 and 21 were recorded. The chromosome numbers in the other genera are unknown.

Pollination. All or most species are pollinated by insects. Beetles, flies, bees and butterflies are recorded as pollinators of *Allium*. Flowers of most species of *Allium* are protandrous, a few are protogynous. Insects are attracted by the usually showy mass of flowers and by the nectar, which is produced in septal nectaries, and in *Allium siculum* and *A. tripedale* in 3 glands at the base of the inner filaments. Some species of *Allium* (e.g. *A. rotundum*, *A. sphaerocephalum*) have cleistogamous flowers and autogamy.

Fruit and Seed. The fruit is a loculicidal capsule, with 2–several seeds in each of the 3 locules. Seeds are flat and elongate with a straight embryo in *Tulbaghia*. Seeds in the other genera are angular and more or less isodiametric, 1.7–4 mm long. *Allium* often has seeds with a long embryo, which is curved at 120–180°, or the embryo is shorter and then curved at less than 90°. In the other genera the seeds have a short and thick embryo. An eleiosome may be present (*Allium triquetrum*). The epidermis of the testa has a crust of phytomelan 10–28 µm thick in *Allium* and *Milula*, thinner in the other genera. The epidermis cells are isodiametic or slightly elongate. The inner layers of the testa are compressed or collapsed, as is the tegmen. Endosperm without starch, but with aleuron and fat.

Dispersal. Seeds of some *Allium* species are dispersed by ants (e.g. *A. ursinum* and *A. triquetrum*).

Reproductive Biology. Vegetative propagation is largely restricted to the formation of increase bulbs. In some species of *Allium*, such as *A. vineale*, small inflorescence propagules (bulbils) are produced at the base of the pedicels or from transformed gynoecia. Apospory has been recorded for *Allium* and in *Nothoscordum borbonicum* and is probably also present in other related species (Guaglianone 1972).

Phytochemistry (Hegnauer 1963, 1986). Characteristic for Alliaceae is the absence of alkaloids, which are regularly found in Amaryllidaceae. Saponins are recorded from *Allium*, *Nothoscordum*, *Leucocoryne*, *Tristagma* and *Gilliesia*, and not recorded as being absent in any species of Alliaceae. The scales of onions are particularly rich in flavonoids (quercetin glycosides) and therefore are used for dyeing Easter eggs, textiles, etc. Starch has not been found in bulbs of *Allium* and is replaced by several other carbohydrates, including galactane, raffinose and the inulinelike fructane. Starch is found in all bulbs or rhizomes studied from the other genera (*Nothoscordum*, *Ipheion*, *Tulbaghia*). Calcium oxalate is not present as raphides in *Allium*, *Milula* or *Tulbaghia*, but as solitary crystals or other conglomerations in *Allium* (Jaccard and Frey 1928) and *Nothoscordum* (Guaglianone 1972: 193). The characteristic alliaceous odour is present in *Milula*; in some or most species of *Tulbaghia*, *Allium*, *Leucocoryne*, *Ipheion* and *Tristagma*; in one *Gilliesia*; but noted neither in any species of *Nothoscordum*, *Schickendantziella*, *Speea*, *Trichlora*, *Solaria*, nor *Miersia*. This odour is released from wounded or decaying tissue only, when alliin, an S-substituted cysteine derivative, under the influence of the enzyme alliinase, is degraded to the water-soluble and strong-smelling allicin, pyruvic acid and ammonia. Alliin has four homologues, allylalliin, propylalliin, methylalliin and ethylalliin, of which methylalliin is widely distributed among plants (e.g. in all Brassicaceae and within various other families). *Ipheion uniflorum* contains methyl-, propyl-, and ethylalliin. The lachrymatory factor from *Allium cepa* is propanthial-S-oxide derived from propenylsulhuric acid and is liberated from onion cells on crushing; it is almost absent in *Allium ampeloprasum* "Porrum" and *A. sativum*. All these sulphur compounds are potent feeding deterrents to animals.

Subdivision and Relationships Within Alliaceae. Krause (1930) divided his Liliaceae subfam. Alloideae into four tribes: Agapantheae with *Agapanthus* and *Tulbaghia*; Miluleae with *Milula*; Gilliesieae with *Trichlora*, *Speea*, *Erinna*, *Solaria*, *Miersia*, *Gethyum*, *Gilliesia* and *Ancruma*, i.e. our genera nos. 9 and 11–14. The tribe Allieae contained *Allium*, *Nothoscordum*, *Tristagma*, *Steinmannia*, *Leucocoryne*, and all genera now in *Themidaceae*.

Traub (1963) divided his Amaryllidaceae subfam. Allioideae into four tribes: Miluleae (with *Milula*), Allieae, Gilliesieae (with *Schickendantziella*, *Speea*, *Miersia*, *Gethyum*, *Gilliesia*, *Solaria*, *Erinna*, *Trichlora* and *Ancruma*, i.e. our genera nos. 7, 8 p.p. and 9–13); and Agapantheae (with *Agapanthus*). Allieae were divided into three subtribes: Alliinae (with *Allium*, *Nectaroscordum*, *Caloscordum*, *Steinmannia*, *Nothoscordum*, *Leucocoryne*, *Tulbaghia*, *Stemmatium* and *Tristagma*, i.e. our genera nos. 1–6 and 8 p.p.), Brodiaeinae (here in Themidaceae) and Millinae (here in Themidaceae).

After a parsimony analysis of *rbc*L sequences, Fay and Chase (1996) proposed placing *Agapanthus* in Amaryllidaceae subfam. Agapanthoideae. Alliaceae were divided into three subfamilies: Allioideae (with *Allium* and *Milula*); Tulbaghioideae (with *Tulbaghia* only); and Gilliesioideae with the rest of the genera here placed in Alliaceae.

Affinities. After the rejection of *Gagea* (Schnarf 1948), the Alliaceae (including Themidaceae and *Agapanthus*) were long considered an indisputable monophyletic taxon. The superior ovary placed it in Liliaceae, until Hutchinson (1934), mainly because of the umbellate inflorescence, placed it in Amaryllidaceae. Dahlgren et al. (1985) placed Alliaceae in Asparagales together with Hyacinthaceae, Amaryllidaceae and 28 other families. Alliaceae were separated from Amaryllidaceae by having steroidal saponins, superior ovary (like most other Asparagales), and absence of alkaloids. Based on a parsimony analysis of *rbc*L sequences, Fay and Chase (1996) found that "Alliaceae sensu Dahlgren et al. were polyphyletic, with *Agapanthus* sister to Amaryllidaceae, and Brodiaeeae (with the exception of *Petronymphe*) being more closely related to Hyacinthaceae . . . , than to the Amaryllidaceae-Alliaceae complex".

Distribution and Habitats. The majority of the Alliaceae species are found between 25° and 45° latitude N and S, *Tulbaghia* in southern Africa, *Milula* in Himalayas, *Allium* in N America, Europe, N Africa and especially the drier parts of Asia. The rest of the genera are found in S America, especially in Chile and Argentina.

The Alliaceae probably evolved in adaptation to a long dry season; but a few species are (probably secondarily) adapted to a more humid climate, e.g. *Allium ursinum* to woods in Europe.

Parasites. Cultivated *Allium* are attacked by many parasites. The larva of leek moth (*Acrolepia assectella*) eats especially the young leaves of *Allium ampeloprasum* "Porrum". The larva of the onion fly (*Delia antiqua*) destroys the bulbs or the young plants of *Allium*. The onion thrips (*Thrips tabaci*) does much harm, mainly by transferring virus. Leaves and scapes of several species are attacked by rust (*Puccinia porri*), and the young plants by a smut (*Urocystis cepulae*). *Peronospora destructor* attacks the leaves and scapes, and *Botrytis allii* the bulbs, especially of *A. cepa*.

Economic Importance. The cultivated species of *Allium* have been treated by Moore (1954/55). *Allium cepa* "Cepa", onion, is the most commonly cultivated and used as a vegetable or for flavouring food. *A. cepa* "Aggregatum", shallot (for which the name *A. ascalonicum* has been misapplied), is used in a similar way or pickled. Less common is *A. fistulosum*, Welsh onion, and *A. chinense*, rakkyo. The leaves of *A. schoenoprasum*, chive, and *A. tuberosum*, Chinese chive, are used for flavouring food, as are the increase bulbs (cloves) of *A. sativum*, garlic. *A. ampeloprasum*, "Porrum", leek, has a milder taste and is used as a vegetable. Some species of *Allium* are used as ornamentals. Less frequently, *Tulbaghia*, *Nothoscordum*, *Ipheion* and *Leucocoryne* are used for the same purpose.

Key to the Genera

1. The closed sheaths at base of leaves very short; distinct bulb structures absent; rhizome or tuber containing starch, covered by dead leaves; inflorescence umbellate; flowers actinomorphic; stamens 6, included in the perianth tube; corona fleshy, annular or of 3–6 scales in the throat; seeds many, flat; embryo straight or slightly curved, phytomelan crust rather thin. Alliaceous odour present. S Africa **1. *Tulbaghia***
- The closed sheath at base of leaves usually long; rhizome or bulb with or without starch; flowers actinomorphic or zygomorphic; stamens 2–6, free or included in the perianth tube; a corona present or absent; seeds few or many, more or less isodiametric and angular, phytomelan crust thin or relatively thick. Alliaceous odour present or absent. Northern hemisphere and S America 2
2. Style ± gynobasic, young ovary with the style in a depression at the summit; bulb without starch; flowers actinomorphic; tepals 6; 6 anthers fertile; ovules few, rarely several per locule; most species with alliaceous odour; embryo often long and curved. Northern hemisphere 3
- Style not in a depression at top of young ovary; bulbs with starch; few species with alliaceous odour; usually several ovules in each locule; tepals 3–6, 2–6 stamens fertile; embryo straight, usually short. S America 4
3. Inflorescence umbellate or flowers 1 or 2; alliaceous odour present or absent. N Africa, Europe, Asia, N America **2. *Allium***
- Inflorescence spicate of many flowers, when young covered by 1 large spathe; with alliaceous odour. Himalayas, Tibet **3. *Milula***
4. Six stamens fertile 5
- Less than 6 stamens fertile 11
5. No corona or scales between stamens and tepals; tepals united below 6
- Staminodes, scales or corona between perianth and fertile stamens 10
6. Spathe bracts free from each other or united for less than 25%; stamens free from each other or united below, in 1 or 2 series; flowers solitary or several; testa smooth to the naked eye 7
- Spathe bracts 2, united with each other at 1 margin for more than 75%, the other for more than 30%. Lower 30–50% of tepals united. Stamens free from each other, inserted in the tube in 2 series; flowers usually solitary; testa pitted. S Brazil, Uruguay, Argentina, Chile **5. *Ipheion***
7. Tepals united below for 10–65%; filaments free or slightly united below 8
- Tepals almost free; filaments united, forming a tube, enveloping the ovary 9
8. Lower 10–30% of tepals united, anthers protruding, filaments in one series, sometimes united with each other below. S America, one species cosmopolitan **4. *Nothoscordum***
- Lower 25–65% of tepals united, anthers usually included into the tepal tube; filaments free from each other, in 2 series. Chile, S Argentina **6. *Tristagma***
9. Tepals 3; stigma entire. NW Argentina **7. *Schickendantziella***
- Tepals 6; stigma slightly 3-lobed. Chile **8. *Speea***
10. Flowers actinomorphic; corona in throat; anthers in tepal tube **9. *Leucocoryne***
- Flower zygomorphic; 6 scales between staminal tube and linear tepals. Chile **12. *Miersia***
11. Three outer stamens fertile, 3 staminodes or coronar appendices; filaments free or absent 12
- Three or 2 anterior stamens fertile; the posterior are small, staminodial, or absent; filaments united 13
12. Inner and outer tepals almost equal; staminodes long, protruding from the throat; some species with alliaceous odour. Chile **9. *Leucocoryne***
- Inner tepals short, obtuse, scalelike; outer lanceolate, almost free; staminodes small. Peru **10. *Trichlora***
13. Corona absent or of 6 small violet scales; filaments forming a tube below. S Chile, S Argentina **11. *Solaria***
- Corona of irregular scales on the outside of the staminal cup; staminal tube split down one side. Chile **13. *Gilliesia***

1. *Tulbaghia* L.

Tulbaghia L., Mant. 148, 223 ("Tulbagia") (1771), orth. et nom. cons.; Vosa, Ann. Bot. (Rome) 34: 47–121 (1975), rev.; Burbidge, Notes R. Bot. Gard. Edinb. 31: 77–104 (1978), rev.

With alliaceous smell. Rhizome, corm or bulb with thick roots. Old dried or slightly fleshy leaf bases enveloping the rhizome or corm. Leaves 4–8. Flowers 3–40. Spathe bracts 2, enclosing the floral buds. Bracts subtending the individual pedicels. Perianth 7–14 mm long, forming a tube for about half its length, corona of 3–6 scales or forming a fleshy tube. Anthers in 2 whorls, sessile in the tube formed by the perianth or the corona. Style short. Stigma capitate. Seeds several in each cell. The phytomelan crust thin. Embryo in the mature seed almost straight. $x = 6$, most species diploid. Twenty two spp., S Africa.

2. *Allium* L. Figs. 12C–G, 27, 28

Allium L., Sp. Pl.: 294 (1753); Stearn, Ann. Mus. Goulandris 4: 83–198 (1978), Europ. spp.; Pastor & Valdés, Rev. genero Allium Peninsula Iberica y Islas Baleares (1983); de Wilde-Duyfjes, Belmontia 7: 1–237 (1977), African spp.
Nectaroscordum Lindl. (1836).
Caloscordum Herbert (1844).
A large number of other synonyms have been described but never generally recognised.

Most species with alliaceous odour and a tunicated bulb. Leaves basal, but their sheathing bases may form a flat or terete, sometimes fistular pseudostem. Spathe bracts 1–2, in some species also individual bracts at base of each pedicel. Inflorescence umbel-like with few to many flowers, sometimes with bulbils. Tepals 6, small, free or slightly united below, very rarely with more than 1 nerve. Stamens 6, all or rarely only 3 fertile, usually attached to the base of the tepals. Filaments often dilated at the base. Septal nectaries open in lower part of ovary, or openings absent. Ovules 1–10 in each locule, often 2. Seeds broad and triangular in transection, with a thick phytomelan crust. $2n = 16$ in most species. Estimated 260–550–690 spp., N Africa, Europe, Asia and N America, especially abundant in Central Asia (Flora USSR has 225 species).

Subdivisions controversial. *Allium siculum* Ucria and *A. tripedale* Trautv. from SE Europe and W Asia with semiinferior ovary, discoid receptacle or the pedicel expanded above, and $2n = 18$ have been referred to the genus *Nectaroscordum*. *Allium neriniflorum* (Herb.) Baker and *A. tubiflorum* Rendle from E Asia without alliaceous odour, tepals united below and forming a cup, have been placed in a separate genus *Caloscordum* or in *Nothoscordum*. Several species economically important, see above under the family.

3. *Milula* Prain

Milula Prain, Ann. R. Bot. Gard. (Calcutta) 5: 164 (1896).

With alliaceous odour. Leaves linear. One large monophyllous spathe bract enclosing the inflorescence in bud. Inflorescence a cylindric spike, flower without individual bract. Tepals 6, united in lower half. Stamens 6, free. Three outer filaments dilated below and with side teeth. Ovary with 2 ovules in each cell, style ± gynobasic, young ovary with the style in a depression at the summit. $2n = 16$. Only one sp., Himalayas and Tibet.

4. *Nothoscordum* Kunth Fig. 29A,D

Nothoscordum Kunth, Enum. 4: 457 (1843), nom. cons.
Zoellnerallium Crosa (1975).

Alliaceous odour absent. Leaves linear or filiform. Scape erect after anthesis. Spathe bracts 2 or in *N. fictile* Macbr. and *N. sessile* (R.E. Fries) Beauverd several. Inflorescence with 1–many flowers. Pedicels slender, often of unequal length. Tepals 6, united below, forming a short tube. Stamens 6, inserted in the tepal tube, sometimes united with each other below. Four to 12 ovules in each cell. $2n = 10, 16, 18, 19, 24, 26$. Twenty four to 26 spp., Argentina, Central Chile, Bolivia, S Peru, Paraguay, Uruguay and S Brazil. *N. bivalve* (L.) Britton extending to Canada, and *N. borbonicum* Kunth (= *N. gracile* auct. = *N. in-odorum* auct. = *N. fragrans* auct.) is a nearly cosmopolitan weed. *Allium siculum* Ucria and *A. tripedale* Trautv. from Asia have previously been referred to *Nothoscordum*. *N. fictile* Macbr. from Peru and *N. sessile* (R.E. Fries) Beauverd from NW Argentina deviate from all other Alliaceae in S America by having the pedicels subtended by individual bracts and by apiculate outer tepals. They may belong to an undescribed genus (Guaglianone 1972). *N. andinum* (Poepp.) Fuentes may be related; it sometimes has individual bracts (Guaglianone 1973) and was placed by Crosa (1975) into a monotypic genus *Zoellnerallium*, based on cytological and morphological characters.

5. *Ipheion* Raf.

Ipheion Raf., Fl. Tellur. 2: 12 (1837).
Beauverdia Herter (1943).

Alliaceous odour in *I. uniflorum* (Lindl.) Raf., absent in the other species. Scape curving after anthesis. Scape bracts 2, united at one side for about 1/2 their length, at the other for more. Flower solitary, rarely 2, with tepals united for 1/3–2/3. Anthers free from each other, inserted in the tepal tube at 2 levels. $2n = 12$ and 20. Three spp., Central Chile, Argentina, Uruguay, S Brazil.

Traub et al. (1955) included *Tristagma* and some *Nothoscordum* in *Ipheion*, but Traub (1963a) changed the name of this taxon to *Tristagma*. Guaglianone (1972) included some *Nothoscordum* with solitary flowers into *Ipheion* but separated *Tristagma*.

6. *Tristagma* Poeppig

Tristagma Poeppig, Fragm. Syn. Phan. 8 (1833).
Steinmannia Philippi (1884).
Garaventia Looser (1945).

Some species with alliaceous odour. Leaves usually linear and flat, rarely canaliculate, terete or filiform. Scape slender, erect after anthesis. Inflorescence with 1–8 [?] flowers. Bracts united below for less than 15% at any margin. Tepals 6, united below. Anthers 6, placed in 2 series in the tube, versatile or basifixed, filaments long to absent. Several ovules in each locule. $2n = 8$, 16 and 24. Fourteen spp., Chile and Argentina.

7. *Schickendantziella* (Speg.) Speg.

Schickendantziella (Speg.) Speg., Anal. Mus. Buenos Aires III, 2: 8 (1903).
Schickendantzia Speg. (1896) non Pax (1889).

One to 2 flowers, violet-purple, nodding. Tepals 3, caudate, almost free. Corona absent. Stamens 6, fertile, filaments united below and enveloping the ovary. Stigma thick. Many ovules in each cell. One sp., Argentina (Tucumán).

8. *Speea* Loes.

Speea Loes., Notizbl. Bot. Gard. Berlin-Dahlem 10: 63 (1927).
Geanthus (1884), non Raf. (1814).

Spathe bract 1 or 2 separate, the outer with margins united below. Flower 1–3. Tepals 6, free, caudate. Stamens 6, fertile, filaments free or united, anthers basifixed. Two spp., Chile.

9. *Leucocoryne* Lindl.

Leucocoryne Lindl., Bot. Reg.: t. 1293 (1830); Zoellner, An. Mus. Hist. Nat. Valparaiso 5: 9–83 (1972), rev.
Latace Philippi (1889).
? *Erinna* Philippi (1864).
Stemmatium Philippi (1873).
Chrysocoryne Zoellner (non Endl.) (1973).
Pabellonia Quezada & Martic. (1976).

Some species with alliaceous odour. Leaves linear. Scape slender. Spathe bracts linear-lanceolate. Inflorescence with 1–12 flowers. The 6 tepals united below and forming a tube. Stamens 3 or 6, inserted in the tepal tube; when 3, then opposite the outer tepals. Corona of 3 or 6 lobes. Seeds many in each cell. $2n = 12$, 18, 24. Fifteen to 20 spp., Chile.

L. volkmannii (Philippi) Traub has been placed in the monotypic genus *Latace*, having segments of tepals longer than tube and the 3 functional stamens exserted from the tube. Those with 6 functional stamens are named *Stemmatium* by Philippi and *Chrysocoryne* by Zoellner (non Endl.) and *Pabellonnia* by Quesada et al. Ravenna (1978) includes them in *Leucocoryne*. *Erinna* Phil. is dubious or extinct. Only one plant, the type, is known, collected near Santiago de Chile. According to the description, like *Leucocoryne*, but with smaller flowers and a short tepal tube.

10. *Trichlora* Baker

Trichlora Baker, Hook. Icon. Pl. t. 1237 (1877).

Bulb narrow, tunicated. Leaves 3–4, linear. Scape slender. Spathe bract 2, free from each other. Inflorescence with about 5 flowers. Outer tepals green, lanceolate, free; inner short, scalelike. Stamens 3 fertile, staminodes 3. Corona absent. Stigma with 3 hornlike branches. Two spp., Peru.

11. *Solaria* Philippi

Solaria Philippi, Linnaea 29: 72 (1858).
Gethyum Philippi (1873).
Ancrumia Harvey ex Baker (1877).

Leaves 1–2. Scape very short. Spathe bracts 2, linear. Inflorescence of 3–15 flowers. Tepals 6, green, united below. Corona absent or present as small scales. Functional stamens 2 or 3, staminodes 0 or 3 small, united below and with the tepal tube. Stigma undivided. Ovules 2 in each cell. $x = 7$. Five spp., S Chile and S Argentina.

12. *Miersia* Lindl.

Miersia Lindley Bot. Reg.: under t. 992 (1826).

Leaves 3–5, linear. Inflorescence with 3–7 flowers. Tepals 6, free, acuminate. Corona of 6 narrow scales outside the oblique urn-shaped staminal cup bearing 6 small anthers. Style short. Stigma small disk-like. Capsule truncate. $2n = 20$ and 21. Three spp., Chile.

13. *Gilliesia* Lindley

Gilliesia Lindley, Bot. Reg.: t. 992 (1826).

Leaves usually 2, slender towards the base. Scape slender, longer than the leaves. Spathe bracts unequal. Tepals 6, free, rarely 5. Staminal cup slit down one side, functional stamens 3, staminodes

3. Style slightly trilobed. Alliaceous odour recorded. Five spp., Chile.

Selected Bibliography

Crosa, O. 1975. Las especies unifloras del género *Nothoscordum* Kunth y el género *Ipheion* Rafinesque de la tribu Allieae (Liliaceae). Darwiniana 19: 335–344.
Crosa, O. 1981. Los chromosomas de cinco especies del género *Tristagma* (Liliaceae). Darwiniana 23: 361–366.
Dahlgren, R.M.T. et al. 1985. See general references.
Daumann, E. 1970. Das Blütennektarium der Monocotyledonen unter besonderer Berücksichtigung seiner systematischen und phylogenetischen Bedeutung. Feddes Repert. Spec. Nov. Regni Veg. 80: 463–590.
Davis, G.L. 1966. See general references.
Di Fulvio, E. 1973. Sobre el gineceo de *Allium* y *Nothoscordum*. Kurtziana 7: 241–253.
Fay, M.F., Chase, M.W. 1996. Resurrection of Themidaceae for the Brodiaea alliance, and recircumscription of Alliaceae, Amaryllidaceae and Agapanthoideae. Taxon 45: 441–451.
Frölich, D., Barthlott, W. 1988. See general references.
Guaglianone, E.R. 1972. Sinopsis de las especies de *Ipheion* Raf. y *Nothoscordum* Kunth (Liliáceas) de Entre Rios y regiones vecinas. Darwiniana 17: 159–242.
Guaglianone, E.R. 1973. *Nothoscordum andinum*, especie de Liliaceae nueva para la flora argentina. Darwiniana 18: 31–36.
Hegnauer, R. 1963, 1986. See general references.
Hutchinson, F. 1934. The families of flowering plants. Vol. 2, Monocotyledons. 2nd edn. London: MacMillan.
Jaccard, P., Frey, A. 1928. Kristallhabitus und Ausbildungsformen des Calciumoxalats als Artmerkmal. Vierteljahrschr. Naturforsch. Ges. Zür. 73, Beiheft: 126–161.
Komarov, V.L. 1968. Flora of the USSR. 4 (Russ. edn. 1935). Jerusalem: Israel Progr. Sc. Transl.
Krause, K. 1930. Liliaceae. In: Engler, A., Prantl, K. (eds.) Die natürlichen Pflanzenfamilien, 2nd edn. 15a: pp. 249–329. Leipzig: W. Engelmann.
Mann, L.K. 1960. Bulb organization in *Allium*: some species of the section Molium. Am. J. Bot. 47: 765–771.
Moore, H.E., Jr. 1954–55. The cultivated *Allium*. Baileya 2: 103–113, 117–124. 3: 137–149, 156–167.
Nuñez, O., Fraissinet, N., Rodríguez, R.H., Jones, K. 1974. Cytogenetic studies in the genus *Nothoscordum* Kunth. I The *N. inodorum* polyploid complex. Caryologia 27: 403–441.
Özhatay, N. 1978. The chromosomes of *Milula spicata* (Liliaceae) Kew Bull. 32: 453–454.
Ravenna, P. 1978. Studies in Allieae II. Plant Life 1978: 130–151.
Schnarf, K. 1931. Vergleichende Embryologie der Angiospermen. Berlin: Borntraeger.
Schnarf, K. 1948. Der Umfang der Lilioideae im natürlichen System. Österr. Bot. Z. 95: 257–269.
Schulze, W. 1980. Beiträge zur Taxonomie der Liliifloren V. Alliaceae. Wiss. Ztschr. Friedrich-Schiller-Univ. Jena, Math.-Nat. R. 29: 595–606.
Sterling, C., Huang, S. 1972. Notes on the lactifers of *Allium*, *Caloscordum*, *Nothoscordum*, *Tristagma* and *Tulbaghia*. Plant Life 28: 43–46.
Takhtajan, A.L. 1982. See general references.
Traub, H.P. 1963. *Tristagma* Poepp. Plant Life 19: 60–61
Traub, H.P., Moldenke, H.N. 1955. The genus *Ipheion*, diagnosis, key to species and synonymy. Plant Life 11: 125–130.
Vosa, C.G. 1975. The cytotaxonomy of the genus *Tulbaghia*. Ann. Bot. (Rome) 34: 47–121.
Wunderlich, R. 1954. Über das Antherentapetum mit besonderer Berücksichtigung seiner Kernzahl. Österr. Bot. Z. 101: 1–63.
Wunderlich, R. 1959. Zur Frage der Phylogenie der Endospermtypen bei den Angiospermen. Österr. Bot. Z. 106: 203–293.

Alstroemeriaceae

E. Bayer

Alstroemeriaceae Dumort., Anal. fam. pl.: 57, 58 (1829).

Erect or twining, herbaceous, mostly glabrous, rhizomatous *perennials* with storage roots or rarely (one sp.) annual herbs. Indumentum where present of 1- to 4-celled unbranched hairs. Phyllotaxis spiral. Leaves evenly dispersed on an elongated stem or crowded at its upper end, or on short stems as a rosette near to the ground. *Leaves* simple, entire, parallel- or arched-veined, thin or somewhat fleshy, generally twisted at the base and leaf blade becoming half or totally inverted (resupinate) (except *Bomarea* subgen. *Wichurea*, some *Alstroemeria* and *Schickendantzia* spp.), in the flowering stems of *Alstroemeria* often reduced to sessile bracts. *Inflorescence* terminal, 2- to many-flowered, lax or dense, bracteate, umbel-like, simple or branched, or flowers solitary. Bracts leaflike or reduced, at the base of the inflorescence a pseudowhorl of bracts. *Flowers* bisexual, actinomorphic or zygomorphic, epigynous. Tepals 3 + 3, petaloid, free, the 2 whorls similar or conspicuously different, the inner ones often basally narrowed into a canaliculate, nectariferous claw. Stamens 3 + 3, free, anthers pseudobasifixed, ± sagittate, dithecal, tetrasporangiate, latrorse, opening by 2 longitudinal slits; filaments filiform, erect or curved. Ovary inferior, turbinate, syncarpous, 3-carpellate, unilocular with parietal placentae (*Leontochir, Schickendantzia*) or trilocular with axile placentae (other genera); septal nectaries absent; style filiform, at the base triangled, apically with 3 stigmatic branches. Tepals, stamens and style caducous, except in *Leontochir*. Ovules usually many, in 2 rows, anatropous, bitegmic, tenuinucellate. *Fruit* a usually dehiscent, dry, leathery, or somewhat fleshy capsule, loculicidal, 3-valved, in *Alstroemeria* mostly explosive, rarely fruit berrylike and indehiscent. Seeds ± globose, brown, with or without a red sarcotesta.

A family comprising 5 genera and about 160 species, distributed in tropical, subtropical, and temperate S and Central America; often in dry, but also in moist habitats, from the coast to highest Andean regions.

Vegetative Morphology. All genera except one (*Taltalia*) are herbaceous perennials with a sympodial rhizome bearing triangular bracts. The rhizome can be either long, cylindrical, fleshy ± horizontal or ascending and sometimes branching, or very short, nodose, and often stoloniferous. Actively growing rhizomes usually bear long, thin, white, fibrous roots and storage roots. The starch- and water-storing roots can be long, cylindrical, white, and thickened ± equally all over their length or tuberous (Fig. 16A). Tubers can be narrowly ellipsoid to spindle-shaped and white, covered with root hairs, or ± globose or ovoid and yellowish; they rise either directly from the rhizome or are developed at the end of long fibrous roots. Usually every year after a period of dormancy, 1 or several sterile and/or fertile orthotropous shoots are produced. Aerial sterile stems are unbranched, erect, twining or procumbent, slender, sometimes somewhat fleshy or rigid, and foliate; the basal leaves are usually scaly. The leaves are alternate, sessile or constricted towards the base into a petiole, sometimes semiamplexicaulous; the leaf blades are linear, (ob-)lanceolate, (ob-)ovate, or spatulate.

Buxbaum (1951) has indicated the homology between the rhizome of *Alstroemeria* and the stoloniferous bulbs of Liliaceae.

Vegetative Anatomy. In the stems most vascular bundles are arranged inside a sclerenchymatic ring; only few bundles are found occasionally outside this ring in the cortex tissue. Bundles of different size are located in the central part of the inner parenchyma, scattered (*Bomarea*) or forming an irregular ring (*Alstroemeria*); the stems of some species show a central hole.

Vessels are present in stems and roots but lacking in the leaves. In the roots, vessels have scalariform and simple perforation plates, while in the leaves, perforation plates are exclusively scalariform.

The leaves are inversely bifacial with the palisade parenchyma located in the physiological adaxial side or equifacial without palisade parenchyma. Stomata are anomocytic; in resupinate leaves they are present on the physiological underside; in not or partly inverted leaves, stomata occur in different numbers on both sides. Calcium oxalate raphides can be found in the roots and in the leaves. When an indumentum is present, it consists of unicellular, bicellular, or multicellular trichomes with a single basal cell, which in *Alstroemeria* is often succulent and white.

Inflorescence Structure. The inflorescence is an umbel-like, condensed thyrse composed of 1–10 helicoid bracteate, seldom single-flowered cymes with ± stout axes. More rarely, solitary flowers terminate the shoot apex.

Flower Structure. Tepals of the outer whorl are nearly uniform, those of the inner whorl dissimilar in most species of *Alstroemeria* (Fig. 30) and in *Taltalia*. The outer and inner whorls differ slightly (*Leontochir*, *Schickendantzia*, some species of *Bomarea* and *Alstroemeria*) or conspicuously (*Bomarea*, *Alstroemeria*, *Taltalia*) in size, shape, color, or patterning. In all genera except *Schickendantzia* the base of all or 2 of the inner tepals is canaliculate and in *Leontochir* it is even closed to form a pocket and contains nectar from succulent trichomes. In contrast to these perigonal nectaries, records of septal nectaries seem to be erroneous (Buxbaum 1954). The stigmatic surface is of the Wet type (*Alstroemeria*).

The midveins of the tepals continue on the ovary as 6 longitudinal ribs, and the perianth, after shedding, leaves a typical circular scar around the top of the ovary. Details of floral and fruit organization of *Alstroemeria* were given by Buxbaum (1954). He was of the opinion that the ovary is fused only to the base of the tepals and not to the receptacle; therefore, he regarded the ovary as pseudoepigynous.

Fig. 30A–F. Alstroemeriaceae. *Alstroemeria aurea*. **A** Young flower. **B** Flower with dehiscing stamens. **C** Flower with receptive stigmas. **D** Young fruit. **E** Young capsule in longitudinal section. **F** Seed. (Takhtajan 1982)

Embryology. Microspore cytokinesis is successive. The ovule is anatropous, bitegmic, and tenuinucellate. The archesporial cell functions directly as the megaspore mother cell. The chalazal megaspore develops into a Polygonum type embryo sac. Endosperm formation is Nuclear (Davis 1966).

Pollen Morphology. The pollen grains are large, 53–140 × 35–75 μm, sulcate, bilateral-symmetric, ± elongated, plano-convex or more rarely crescent-shaped. In *Bomarea* the sulcus sometimes extends to the proximal side. The exine is 2–5 μm thick and differentiated into a thicker sexine and a thinner nexine. *Alstroemeria*, *Taltalia*, and *Schickendantzia* show a striate to striato-reticulate, perforate exine; the proximal pole is often a center of striation. *Bomarea* and *Leontochir* have a variably structured reticulate exine with different meshes; the size of the brochi varies between 1 and 6 μm (Erdtman 1969; Schulze 1978; Kosenko 1994).

Karyology. Chromosomes are generally large (in *Alstroemeria* 9–16 μm). The basic chromosome number is $x = 8$ in *Alstroemeria* and $x = 9$ in *Bomarea* and *Leontochir* (Bayer 1988). Polyploidy rarely seems to occur in nature, but triploid and tetraploid cultivars of *Alstroemeria* are known (Tsuchiya et al. 1987). In *Alstroemeria*, Giemsa C-banding revealed asymmetric karyotypes and considerable interspecific variation in the C-banding pattern (Buitendijk and Ramana 1996); Brazilian species show hardly any, Chilean species prominent C-bands.

Pollination and Reproductive Systems. In *Alstroemeria*, bumblebees seem to be effective pollinators, but bees and more rarely butterflies were also observed (pers. observ.). The red color and tubular shape of some species of *Bomarea* and some Brazilian species of *Alstroemeria* point to bird pollination. Only one species, *Alstroemeria caryophyllaea*, is known to be fragrant; since the intensity of the scent increases at night, moth pollination is conceivable. Self-pollination was observed (pers. observ.) in cultivated *Taltalia* and one species of *Alstroemeria* (*A. patagonica*). Due to the arrangement of stamens and style, self-pollination may occur in *Schickendantzia* and *Leontochir* as well. In *Alstroemeria* self-pollination is excluded by proterandry.

Fruit and Seed. In *Leontochir*, *Schickendantzia*, *Alstroemeria*, and *Taltalia* the fruit is a dry, ovoid, loculicidal, 3-valved capsule which is only occasionally indehiscent. In *Alstroemeria* the mature capsules open under dry and warm conditions with an audible explosion, catapulting valves and seeds away. In *Bomarea* some species have a berrylike, mealy fruit, but mostly the fruit is a leathery or somewhat fleshy globose to ovoid capsule opening with 3 valves and with the seeds staying attached to the placentae. The fruits are mostly many-seeded.

Huber (1969), who studied seeds of *Bomarea* and *Alstroemeria*, found them globose to ellipsoid and yellowish brown or brown. Cell differentiation in the outer integument is weak; the cells are colorless or contain a yellow pigment. In contrast to Colchicaceae and Iridaceae, the inner integument collapses into a thin membrane. The endosperm cells have thick pitted walls and contain aleurone and fatty oils but no starch. The cylindrical embryo is about 2/3 the length of the endosperm.

Dispersal. The presence of a red sarcotesta in *Bomarea* and the fact that the seeds remain attached to the placentae and are presented in the open capsule may point to bird dispersal; the same may be true of the orange-red berrylike fruits in *Bomarea*. The capsules of *Alstroemeria* show explosive dehiscence.

Phytochemistry. Alkaloids seem to be lacking; chelidonic acid is present. Substances with hemolyzing properties were found in the flowers and fruits of *Alstroemeria* and *Bomarea* (Hegnauer 1963) and indicate the presence of steroidal saponins. The presence of flavonol glycosides (and absence of flavones) and the presence of tupilosides links the Alstroemeriaceae with the Liliaceae s. str. (Hegnauer 1986).

Relationships Within the Family. Pollen morphology and the type of storage roots indicate a close relationship between *Alstroemeria* and *Schickendantzia* and between *Leontochir* and *Bomarea*; additionally the last two have actinomorphic flowers and the same chromosome number. *Leontochir* and *Schickendantzia* share the possession of a unilocular ovary. In spite of the great morphological similarity between some species of *Alstroemeria* and *Bomarea*, these genera are clearly distinguished by their chromosome numbers. *Taltalia* is supposed to be closely related to *Alstroemeria* and may be interpreted as annual derivation.

Nevertheless, the differences between the genera are slight, and Hunziker (1973) suggested merging all Alstroemeriaceae in a single genus.

Affinities. Structural and chemical characters mentioned above leave no doubt as to a close relationship between Alstroemeriaceae and Liliaceae/Colchicaceae, a concept that has been elaborated by Buxbaum (1954), Huber (1969), and Goldblatt (1995). The molecular (*rbc*L) data (Chase et al. 1995) are in accord with this.

Distribution and Habitats. The family is purely New World and is distributed from Central Mexico and the West Indies to Patagonia; it occurs from temperate to tropical zones, and some species are frost-tolerant. Its members are adapted to very diverse habitats such as swamps, desert areas, shady woodland, and tropical forests, and can be found from coastal regions up to high elevations of the Andes.

Economic Importance. The family is of great horticultural interest. Several species of *Alstroemeria* and some of *Bomarea* are cultivated for their ornamental value as pot plants, in greenhouses, or in mild climate outdoors. Hybrids of some *Alstroemeria* species are commercially grown as cut flowers.

The starchy storage roots of several species of *Alstroemeria* and *Bomarea* are edible. In Chile the flour extracted out of *Alstroemeria ligtu* is called el chuño (Puga Borne 1921).

Key to the Genera

1. Ovary unilocular; placentae parietal; flowers actinomorphic 2
- Ovary trilocular; placentae axile; flowers actinomorphic or zygomorphic 3
2. Flowers in a very dense umbel-like inflorescence, brilliant red; tepals, stamens and style persistent; leaves broadly ovate, resupinate **1. *Leontochir***
- Flowers solitary, yellow; tepals, stamens, and style caducous; leaves narrowly ovate, not resupinate **3. *Schickendantzia***
3. Flowers actinomorphic 4
- Flowers zygomorphic 5
4. Fruit a leathery or somewhat fleshy capsule, opening with 3 valves or fruit berrylike; seeds in the open capsule usually staying attached to the placentae; seed coat mostly fleshy, rarely dry; inflorescence pendulous or nodding **2. *Bomarea***
- Fruit dry, opening explosively and decaying; seeds are thrown out; seed coat dry; inflorescence erect **4. *Alstroemeria***
5. Perennial with rhizome and storage roots; flowers in erect, umbel-like inflorescences at the end of single stems; rarely, flowers solitary **4. *Alstroemeria***

– Annual without rhizome and storage roots; flowers solitary at the end of several stems ascending from a nodose base 5. *Taltalia*

1. *Leontochir* Phil.

Leontochir Phil., An. Univ. Chile 43: 544–546 (1873).

Herbaceous glabrous perennials with erect to procumbent, densely foliate shoots. Rhizome short; tubers yellowish, ovoid, at the end of fibrous roots. Leaves resupinate, ovate. Inflorescence very dense, umbel-like; flowers actinomorphic, bright red; outer and inner whorl of tepals differently shaped. Stamens curved, longer than the style. Ovary unilocular. Fruit an dry capsule with persisting tepals, stamens, and style. Seeds dry. $2n = 18$. Only one sp., *Leontochir ovallei* Phil., endemic to N Chile, on coastal sands.

2. *Bomarea* Mirb.

Bomarea Mirb., Hist. nat. pl. 9: 71 (1804).

Stems twining, erect or procumbent, leafy. Rhizome nodose or elongated, tubers yellowish, at the end of fibrous roots. Leaves usually resupinate, linear to broad ovate. Flowers in dense or lax pendulous or nodding inflorescences, rarely solitary, actinomorphic, white, yellow, orange, red, pink, purple, or greenish; outer whorl of tepals uniform, rarely patterned, sometimes shorter than the inner whorl; inner whorl mostly differently colored, uniform, usually patterned. Stamens usually straight. Ovary trilocular. Fruit berrylike or a leathery or somewhat fleshy capsule, very rarely with persisting tepals and stamens. Seeds with a fleshy seed coat, rarely dry. $2n = 18$. According to Baker (1888) subdivided in three subgenera, subgen. *Bomarea*, subgen. *Wichurea*, and subgen. *Sphaerine*. About 100 spp., S and Central America, in very different habitats. In need of revision.

3. *Schickendantzia* Pax

Schickendantzia Pax, Bot. Jahrb. Syst. 11: 336–337, t. VII (1889).

Herbaceous glabrous perennial with an erect stem. Rhizome nodose; storage roots narrowly ellipsoid. Leaves not twisted, narrowly ovate. Flowers solitary, erect, actinomorphic, sulfurously yellow; outer and inner whorl of tepals of slightly different shape. Stamens straight, as long as the style. Ovary unilocular. Fruit a dry capsule. Seeds dry. Only one sp., *Sch. pygmaea* (Herb.) Speg., Bolivia, Argentina, Peru, and the high Andes.

4. *Alstroemeria* L. Figs. 16A, 30.

Alstroemeria L., Pl. Alströmeria (1762); Bayer, Mitt. Bot. Staatssamml. Münch. 24: 1–362 (1987), Chilean spp.

Herbaceous, mostly glabrous perennials with erect, foliate, sterile and fertile shoots. Rhizomes long and fleshy, or short and nodose. Storage roots narrowly ellipsoid to spindle-shaped, or tuberous at the end of fibrous roots, white. Leaves mostly resupinate, more rarely half or not twisted, in fertile stems often reduced or scaly. Flowers in lax umbel-like, erect inflorescences, seldom solitary, zygomorphic, very rarely nearly actinomorphic, white, yellow, orange, red, pink, violet, or greenish; outer whorl of tepals mostly uniform, rarely patterned; inner whorl usually different, its upper 2 tepals mostly conspicuously patterned, banded with yellow and striped or spotted, the lower tepal mostly differently shaped, sometimes patterned. Stamens usually curved. Ovary trilocular. Fruit a dry capsule, opening explosively. Seeds dry. $2n = 16$. About 60 spp., W and E South America, distribution centers Chile and E Brazil, in different habitats. A revision of the Brazilian species and of the genus as a whole is lacking.

5. *Taltalia* Ehr. Bayer

Taltalia Ehr. Bayer, Sendtnera 5: 5–14 (1998).

Annual glabrous herb with several foliate stems ascending from a knotty base. Storage roots lacking. Leaves not or half-twisted, linear to lanceolate. Flowers solitary, erect, zygomorphic, white to pink; outer whorl of tepals uniform, not patterned; inner whorl different, its upper 2 tepals with a bright yellow band with brown spots, the lower tepal remarkably smaller, not patterned. Stamens straight, little longer than the style. Ovary trilocular. Fruit a dry capsule, opening mostly explosively. Seeds dry. Only one sp., *T. graminea* (Phil.) Ehr. Bayer, endemic to Northern Chile, in loma vegetation of the coastal desert.

Selected Bibliography

Aker, S., Healy, W. 1990. The phytogeography of the genus *Alstroemeria*. Herbertia 46: 76–87.
Baker, J.G. 1888. Handbook of the Amaryllideae including the Alstroemerieae and Agaveae. London: George Bell.

Bayer, E. 1987. Die Gattung *Alstroemeria* in Chile. Mitt. Bot. Staatssamml. Münch. 24: 1–362.
Bayer, E. 1988. Beitrag zur Cytologie der Alstroemeriaceae. Mitt. Bot. Staatssamml. Münch. 27: 1–6.
Buitendijk, J.H., Ramanna, M.S. 1996. Giemsa C-banded caryotypes of eight species of *Alstroemeria* L. and some of their hybrids. Ann. Bot. II, 78: 449–457.
Buxbaum, F. 1951. Die Grundachse von *Alstroemeria* und die Einheit ihres morphologischen Typus mit dem der echten Liliaceen. Phytomorphology 1: 170–184.
Buxbaum, F. 1954. Morphologie der Blüte und Frucht von *Alstroemeria* und der Anschluß der Alstroemerioideae bei den echten Liliaceae. Oesterr. Bot. Z. 101: 337–352.
Chase, M.W., et al. 1995. See general references.
Dahlgren, R.M.T., Clifford, H.T. 1982. See general references.
Davis, G.L. 1966. See general references.
Erdtman, G. 1969. Handbook of palynology. Copenhagen: Munksgaard.
Goldblatt, P. 1995. The status of R. Dahlgren's orders Liliales and Melanthiales. In: Rudall, P.J., Cribb, P.J., Cutler, D.F., Humphries, C.J. (eds.) Monocotyledons: systematics and evolution. Royal Botanic Gardens, Kew, pp. 181–200.
Hegnauer, R. 1963, 1986. See general references.
Herbert, W. 1837. Amaryllidaceae. London: J. Ridgway.
Huber, H. 1969. See general references.
Hunziker, A.T. 1973. Notas sobre Alstroemeriaceae. Kurtziana 7: 133–135.
Hunziker, J.H., Xifreda, C.C. 1990. Chromosome studies in *Bomarea* and *Alstroemeria* (Alstroemeriaceae). Darwiniana 30: 179–183.
Killip, E.P. 1936. *Bomarea*, a genus of showy Andean plants. Nat. Hort. Mag. 15: 115–129.
Kosenko, V.N. 1994. Pollen morphology of the family Alstroemeriaceae. Bot. Žh. 79: 1–8 (in Russian).
Neuendorf, M. 1977. Pardinae, a new section of *Bomarea* (Alstroemeriaceae). Bot. Not. 130/1: 55–60.
Pax, F. 1889. Beiträge zur Kenntnis der Amaryllidaceae. Bot. Jahrb. Syst. 11: 318–338.
Philippi, R.A. 1873. Descripción de las plantas nuevas incorporadas ultimamente en el herbario chileno. An. Univ. Chile 43: 479–583.
Puga Borne, F. 1921. El chuño del ligtu. Rev. Chil. Hist. Nat. 25: 313–320.
Rasmussen, H. 1983. Stomatal development in families of Liliales. Bot. Jahrb. Syst. 104: 261–287.
Schulze, W. 1978. Beiträge zur Taxonomie der Liliifloren. III. Alstroemeriaceae. Beiträge zur Phytotaxonomie, 5. Folge, Wiss. Z. Friedrich-Schiller-Univ. Jena, Math. Naturwiss. Reihe 27/1: 79–85.
Takhtajan, A.L. 1982. See general references.
Tsuchiya, T., Hang, A., Healey, W.E., Jr., Hughes, H. 1987. Chromosome studies in the genus *Alstroemeria*. I. Chromosome numbers in 10 cultivars. Bot. Gaz. 148: 519–524.
Whyte, R.O. 1929. Chromosome studies I. Relationship of the genera *Alstroemeria* and *Bomarea*. New Phytol. 28: 319–344.

Amaryllidaceae

A.W. Meerow and D.A. Snijman

Amaryllidaceae J. St.-Hil., Exp. Fam. Nat. 1: 154 (1805).

Bulbous (rarely rhizomatous), mostly geophytic, *perennials*, terrestrial, occasionally aquatic or epiphytic, rich in family-specific alkaloids. Bulbs tunicate. *Leaves* annual or persistent, sessile and linear or lorate, or petiolate and lanceolate to widely elliptic, distichous or spirally arranged; sometimes basally sheathing and forming an aerial pseudostem, usually glabrous, rarely with trichomes. *Inflorescence* scapose, pseudoumbellate (reduced helicoid cymes); scape sometimes wholly subterranean and appearing obsolete, terminated by 2 or more spathaceous, obvolute or equitant, usually marcescent bracts that enclose the flowers in bud (bracts rarely absent); inner bracteoles usually present and successively shorter and narrower. *Flowers* 1-many, perfect, frequently large and showy, sessile or pedicellate, each usually subtended by a bracteole, actinomorphic or zygomorphic, generally protandrous. Perigone crateriform, salverform, funnelform, tubular or ventricose, consisting of 3 + 3 segments (tepals) connate below into a short or long tube or rarely free to the base; inner tepals generally shorter than the outer. Outgrowth of the perigone sometimes present, forming a conspicuous corona (paraperigone), or relatively inconspicuous and consisting of a short callose rim or ring of scales or fimbriae at the throat. Stamens 3 + 3, rarely 5 or 18 or more, subequal or varying in length, inserted at the perigone throat or below, the filaments sometimes variously connate or otherwise appendaged, rarely adnate to the style; anthers usually dorsifixed, rarely centrifixed or basifixed, introrse, dehiscing longitudinally or rarely from a terminal pore. Style filiform, hollow, occasionally strumose, rarely tripartite; stigma capitate, 3-lobed or deeply trifid, usually papillate. Ovary syncarpous, tricarpellate, inferior, tri- (rarely uni-) locular, with septal nectaries; ovules axile or basal in placentation, anatropous, crassinucellate, bi-, uni- or ategmic. Mature *fruit* a loculicidally dehiscent capsule, sometimes indehiscent, rarely baccate. *Seeds* globose or subglobose and fleshy or hard, or flattened and winged, usually with a black or brown phytomelanous testa, sometimes with a caruncular elaiosome at the chalazal end; endosperm with hemicellulose and lipids, in the more derived genera rich in water and/or starch.

A widely distributed, chiefly tropical family of 59 genera and about 850 species with centers of diversity in S America and S Africa, particularly the Andean region, and in the Mediterranean.

Characters Occurring in Relatively Few Genera and Species. Bulbs are lacking in *Scadoxus*, *Clivia*, and *Cryptostephanus*. Bulb tunics and leaves produce extensible threads when torn, and the leaf margins frequently bear cartilaginous cilia in tribe Amaryllideae. A perigone nectary appears in *Galanthus* (Daumann 1970); androecial nectar glands occur in *Eucrosia* (Meerow 1987a). Stamens are adnate to the style in *Carpolyza* and *Strumaria*. Some species of *Griffinia* have only 5 stamens; *Gethyllis* may have 18 or more. Anthers dehisce by terminal pores in *Galanthus* and *Leucojum*. Pollen is bisulculate in Amaryllideae and is released as tetrads in a single species of *Stenomesson* (Meerow et al. 1986). Strumose styles occur in *Strumaria* and a few species of *Leucojum*. Phytomelan is absent from the seeds of Amaryllideae, Gethyllideae, Haemantheae (except *Crytostephanus*), *Griffinia*, *Hymenocallis*, and *Ismene*. Unitegmic, rarely ategmic (subtribe Crininae), ovules characterize the Amaryllideae. Baccate fruits characterize two African tribes, Haemantheae and Gethyllideae. Seeds germinate precociously in Calostemmateae, forming an adventitious bulbil inside the capsule.

Vegetative Morphology. All members of the Amaryllidaceae are perennials; all but three genera (*Clivia*, *Cryptostephanus*, and *Scadoxus*, Fig. 31) form tunicate bulbs. The family exhibits sympodial organization in the bulb (Müller-Doblies and Müller-Doblies 1972, 1978a, b, 1985; Müller-Doblies 1977; Aksenova and Sedova 1981; Arroyo 1984). Roots are of two types: contractile and perennial, or short-lived and fibrous. All roots are adventitious and originate from the basal plate of the bulb or from the rhizome and lower leaf axils on the few species lacking a true bulb. The leaves of most genera are sessile, linear to lorate, flat or concave, and dorsiventral; a few species of Hippeastreae have bifacial, nearly terete leaves. Petiolate leaves with well-developed laminae occur sporadically in many tribes and are usually associated with adaptation to low light habitats; some genera are polymorphic (*Crinum*, *Hippeastrum*, *Hymenocallis*, and *Stenomesson*).

Fig. 31A–G. Amaryllidaceae. *Scadoxus nutans*. **A** Flowering, rhizomatous plant. **B** Flower. **C** Perigone segment and stamen. **D** Stamen. **E** Apex of perigone segment. **F** Infructescence. **G** Berry, transverse section. (Dahlgren et al. 1985)

Phyllotaxy is generally distichous, rarely spiral. *Cybistetes* and *Ammocharis* (Amaryllideae) uniquely have a well-developed intercalary meristem in their leaves that allows rapid regrowth in response to alternating wet and dry conditions. Vernation of the leaves ranges from flat, revolute to involute.

VEGETATIVE ANATOMY. Among nine genera of the tribes Amaryllideae, Cyrtantheae, and Haemantheae examined by Arroyo and Cutler (1984), all had a 2–4-layered root velamen with the exception of *Haemanthus* and *Cyrtanthus*. The latter, along with all American tribes surveyed, had a 1-layered root rhizodermis. The root exodermis is 1-layered and consists of alternating short and long cells. The endodermis and pericycle are also 1-layered. The inner root tissue is never sclerotic (except in *Amaryllis*) and the central cylinder is usually not medullated. In all genera examined, the roots have scalariform perforation plates in their vessel elements (Cheadle 1969). Leaf surface pubescence is rare in the family; it occurs sporadically in the Amaryllideae, *Haemanthus*, *Gethyllis*, and in one species of *Pancratium* (*P. tenuifolium*, where it is limited to the base of the leaf). The hairs are mostly uniseriate and simple. Multiseriate, multicellular, and cartilaginous cilia at the leaf margins characterize *Ammocharis*, *Cybistetes*, *Boophone*, and many *Crinum* species. Clustered, medifixed hairs are unique to some *Gethyllis* species. Glaucous leaves are widespread throughout the family. The cuticle shows a wide range of sculpturing, and many genera are variable (Arroyo and Cutler 1984; Meerow 1987a, 1989). Epidermal cells are variable in shape, but usually axially elongated. Anticlinal walls tend to be straight in linear- and lorate-leafed genera, but wavy in petiolate-leafed genera with broad laminae. Papillae of varying size are found on the outer wall of epidermal cells of several genera in different tribes. Wax covers are variable and relatively unspecific; they resemble those of Asphodelaceae and Hycainthaceae, but not Convallariaceae (Frölich and Barthlott 1988). Stomata are anomocytic, usually on both abaxial and adaxial surfaces, but sometimes of lower frequency to nearly absent on the adaxial surface, especially in genera with petiolate leaves. They are usually arranged parallel to the long axis of the leaf and often in longitudinal files alternating with several rows of regular epidermal cells. Sunken stomata overlaid by a distinct stomatal crypt formed by adjacent epidermal cells have been observed in *Crossyne*, *Brunsvigia* (Arroyo and Cutler 1984), and in *Pancratium*, *Pamianthe*, and *Ismene* subg. *Elisena* (Meerow, unpubl. data). In petiolate-leafed genera with broad laminae, the stomata are often irregularly distributed. Leaves are predominantly bifacial; unifacial leaves are known only in a few Hippeastreae (*Habranthus*, *Placea*, and *Zephyranthes*). Palisade in the leaf mesophyll has been observed in the Amaryllideae, Eustephieae, Hippeastreae, Lycorideae, Pancratieae, and Narcisseae (Arroyo and Cutler 1984; Artyushenko 1989); many genera have a homogeneous chlorenchyma without distinct palisade cells. Where palisade occurs, it is found near the adaxial surface or both adaxially and abaxially. Nonvascular, helically thickened cells have been observed in the mesophyll of various Amaryllideae and one species of *Hippeastrum* (Arroyo and Cutler 1984). Lacunae are present in a few species, formed by the breakdown of parenchyma. Mucilage cells, raphides, and rhombohedral crystals are present in most species of the family (Arroyo and Cutler 1984; Dahlgren et al. 1985). Vascular bundles occur in a single arc in bifacial leaves (in a ring in unifacial leaves), with the largest bundles at the center and decreasing in size towards the margins. Vessels are absent from the leaf xylem, and the tracheids have annular and helical wall thickenings. Multiple helical thickening occurs in the tracheids of species of Amaryllideae (Arroyo and Cutler 1984; Snijman and Linder 1996). The bundle sheaths of most genera are entirely parenchymatous and surround the vascular bundle; in some Amaryllideae, they are reduced to caps of sclerenchyma at both the xylem and phloem poles or just the xylem pole. Phloem sieve tubes are narrow, thick-walled, and with transverse sieve plates (Arroyo and Cutler 1984). Sieve-tube plastids are of the PIIc type (Behnke 1981). One to two layers of parenchyma separate the xylem and phloem.

INFLORESCENCE STRUCTURE AND ANATOMY. The inflorescence of all Amaryllidaceae is a pseudoumbel formed by the reduction of a series of helicoid cymes (Müller-Doblies 1977). The umbel terminates a leafless scape that arises laterally to 2 or more leaves. The scape is variously hollow or solid (both occur in *Cyrtanthus*, *Hippeastrum* and *Leucojum*), and either terete or compressed (at times ancipitious). The flowers are enclosed by 2, rarely more, spathe bracts (Fig. 32, 33). In a number of single-flowered genera, the flower buds protrude from the spathe at the time of scape emergence from the bulb. The bracts are free to the base in most genera; in Gethyllideae, some Hippeastreae, *Paramongaia* (Stenomesseae) and Narcisseae they are fused at least basally or on one side. An enclosing spathe is absent in *Leptochiton*. The bracts of all African tribes are equitant; those of the American genera are obvolute (Arroyo 1981). The scape is nearly obsolete in a few species, remaining wholly contained

Fig. 32A–F. Amaryllidaceae. **A** *Ungernia trisphaera*, habit. **B–E** *U. sewertzowii*. **B** Stamens. **C** Pistil. **D** Infructescence. **E** Seed. **F** *Sternbergia lutea*, habit. (Takhtajan 1982)

by the bulb until the fruit is ripe (*Stenomesson humile*, *Zephyranthes pusilla*, *Gethyllis*) or is obscured by the bulb neck (*Apodolirion*, *Sternbergia*). The scape is frequently glaucous; the scapes of some *Haemanthus* and *Nerine* species bear simple trichomes. One to 2 layers of subepidermal collenchyma is found in the peduncles of all entirely American genera and *Haemanthus* and

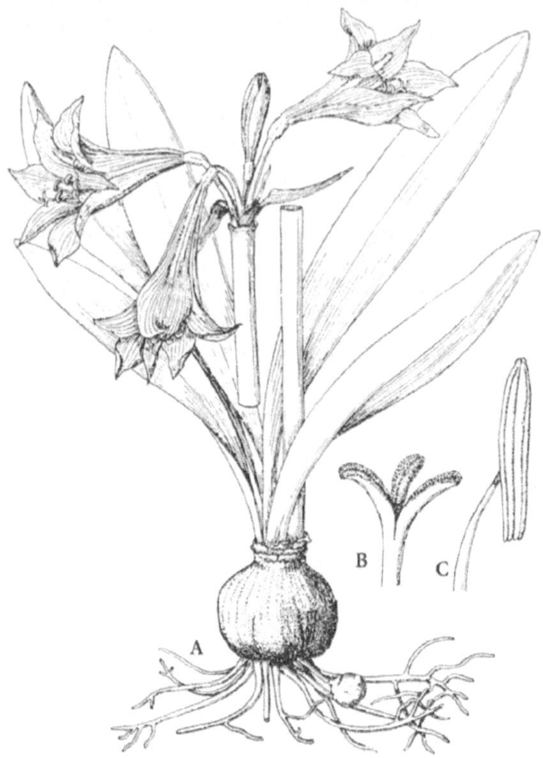

Fig. 33A–C. Amaryllidaceae. *Hippeastrum striatum*. **A** Flowering plant. **B** Style with stigmatic lobes. **C** Stamen. (Dahlgren et al. 1985)

Cyrtanthus (Arroyo and Cutler 1984). Sclerenchyma (3–10 layers) forming a sheath that separates the cortex from the central cylinder is characteristic of tribe Amaryllideae (Arroyo and Cutler 1984).

Flower Structure and Anatomy. The flowers of Amaryllidaceae are either pedicellate or sessile. The flower buds are invariably erect while contained within the spathe bracts and shortly after the bracts begin to spread apart, but at anthesis the flowers may be declinate or pendulous, either by laxness of the pedicel or curvature of the floral tube.

The perigone of all members of the family consists of 2 whorls of 3 petaloid tepals that are usually connate basally into a tube of variable length, but are occasionally free. Tepals of the outer whorl are generally longer and narrower than those of the inner series. A prominent apiculum, often papillate or hairy on the adaxial surface, is found at the apex of the outer tepals of many genera. Perigone shape varies from almost rotate, funnelform, salverform, tubular, ventricose to crateriform, and a single genus can exhibit great diversity (e.g., *Cyrtanthus*, Reid and Dyer 1984). Flowers are actinomorphic to zygomorphic. Presence of a conspicuous perigonal corona (paraperigone) with ramifying vasculature (Arber 1937; Singh 1972) occurs in *Narcissus* and one genus of Hippeastreae (*Placea*), but the presence of a putatively homologous rudimentary fringe of scales, fimbriae, or a solid callose rim is fairly widespread in Hippeastreae and also occurs in the east Asian genus *Lycoris* (Lycorideae), and *Cryptostephanus* (Haemantheae). Perigone pigmentation is diverse, though blue to near-blue coloration is known only in the genera *Griffinia*, *Lycoris*, and *Worsleya* (Meerow 1995).

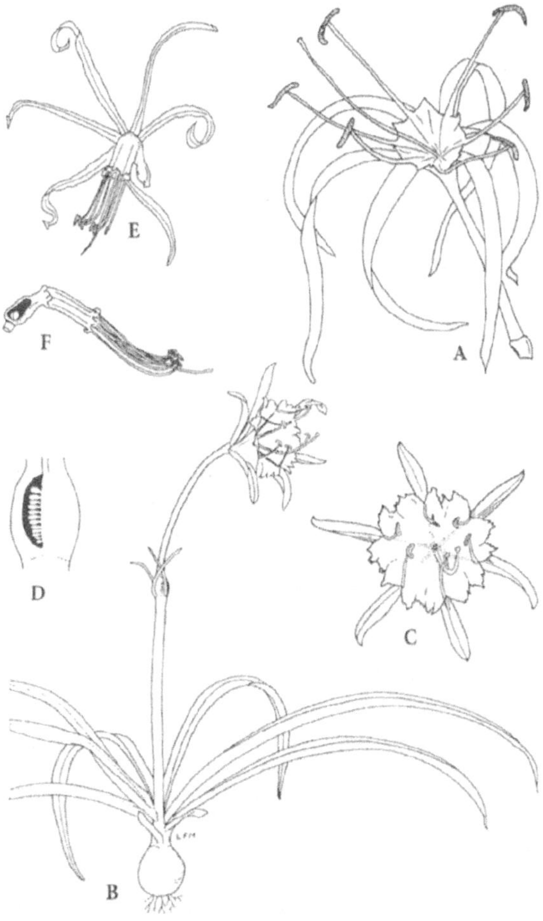

Fig. 34A–F. Amaryllidaceae, tribe Hymenocallideae. **A** *Hymenocallis pedalis*, flower. **B–D** *Leptochiton quitoensis*. **B** Flowering plant. **C** Flower. **D** Ovary, longitudinal section. **E, F** *Ismene longifolia*. **E** Flower. **F** Staminal cup and longitudinal section through ovary. (Original Meerow)

Stamens are 3 + 3 and opposite the tepals. The only variants in number occur in some *Griffinia* species with 5 stamens, and in *Gethyllis*, with 18 or more. Staminal filaments are inserted at the throat of the corolla tube or below, either free or variously connate or appendaged. Connation of the filaments, sometimes forming a conspicuous cup or paracorona, is characteristic of some Amaryllideae (subtribe Amaryllidinae), Eucharideae, Eustephieae, Gethyllideae, Hymenocallideae, some Pancratieae, Calostemmateae, Stenomesseae, and a few species of *Cyrtanthus* (Reid and Dyer 1984). In *Carpolyza* and *Strumaria*, the filaments are variously connate to the style (Snijman 1994). One species of *Hessea* has club-shaped inner filaments with an internal structural organization that matches Vogel's (1990) description of an osmophore (Snijman 1994). Androecial nectar glands at the base of each filament are found in some species of *Eucrosia* (Stenomesseae, Meerow 1987a; Fig. 35), each receiving a single vascular branch trace, but are directly connected to the septal nectaries of the ovary. One species of *Strumaria* bears false-nectar spots in a similar position (Snijman 1994). Anthers are invariably introrse, mostly dorsifixed (basifixed in Narcisseae, two genera of Galantheae, Gethyllideae), and dehiscing by longitudinal furrows, except in some Galantheae (terminal pores). Staminal filaments are typically declinate when the perigone is zygomorphic, straight to outwardly (rarely inwardly) curved when actinomorphic.

The gynoecium of Amaryllidaceae is syncarpous and tricarpellate. The style is slender in almost all genera and hollow. *Strumaria* (Amaryllideae) has strumose styles, and in some *Leucojum* (Galantheae) the style is distally strumose. In *Pyrolirion* (Hippeastreae), the style is tripartite in its distal portion. Heterostyly occurs in a single species of *Narcissus*, *N. triandrus* (Barrett et al. 1995). Stigmas are either capitate, obtusely 3-lobed or deeply trifid, and characteristically papillate. The papillae may be uni- or multi-cellular. Both Wet- and Dry-type stigmas occur in the family (Heslop-Harrison and Shivanna 1977), with the Dry type predominating.

Septal nectaries are universally found in the ovaries of all genera (Dauman 1970). Ovules of Amaryllidaceae are invariably anatropous and borne on axile placentae. Taxa with apparently basal ovules appear to represent reductions in number to a few proximally attached ovules. The

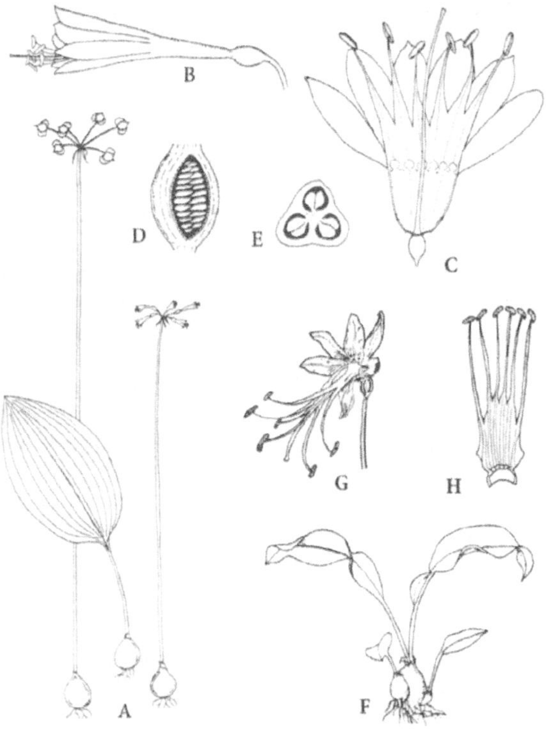

Fig. 35A–H. Amaryllidaceae. **A–E** *Eucrosia stricklandii*. **A** Habit. **B** Flower. **C** Longitudinally dissected flower. **D** Tangential section of ovary to show superposition of ovules. **E** Transversal section of ovary. **F–H** *E. bicolor*. **F** Habit. **G** Flower. **H** Dissected staminal tube. (Meerow 1987a)

ovules are globose, subglobose or compressed, and superposed.

Any one vascular strand supplies 1 tepal and its superposed stamen (Singh 1972). This strand branches into 3 bundles, the central terminating in a tepal, and the 2 laterals fusing before terminating in a staminal filament. While an individual stamen never receives more than a single trace, the tepals are often supplied by branch traces of more than 1 strand (Singh 1972). In *Haemanthus*, each tepal receives 1; in *Amaryllis*, *Eucharis*, *Hippeastrum*, and *Zephyranthes*, tepals are supplied by 3 traces. In *Pancratium*, tepal traces number 5, while in *Crinum*, the outer whorl contains 5, the inner whorl, 3 traces. Moreover, the perigonal traces differentiate at different levels in the flower (Singh 1972); the base of the ovary (*Hippeastrum*, *Pancratium*), the top of the ovary (*Eucharis*, *Zephyranthes*), or in the floral tube (*Haemanthus*, *Narcissus*). In *Amaryllis* and *Crinum*, traces supplying the outer whorl differentiate at the proximal end of the ovary, those of the inner tepals towards the distal end (Singh 1972).

FRUIT AND SEED MORPHOLOGY. Tri-locular, dehiscent capsules and seeds with phytomelanous testas (Huber 1969) are the most common fruit and seed types in the family (Dahlgren and Rasmussen 1983; Dahlgren et al. 1985; Meerow 1989). Two conspicuous character state transformations have occurred in fruit morphology: indehiscence and the evolution of a baccate fruit. Both may have occurred more than once.

Indehiscent, baccate fruits occur in two African tribes, Haemantheae and Gethyllideae, and some species of *Crinum*. The fruit of Haemantheae is a berry with 1–3 seeds, while that of Gethyllideae is a long, cylindrical structure with numerous seeds. The fruits of Calostemmateae are unilocular and also indehiscent. Capsules of *Brunsvigia* (Amaryllideae) are tardily dehiscent and require mechanical damage through tumbling to release the seeds. The dry indehiscent fruits of *Cybistetes* and *Boophone* (Amaryllideae) often rupture from the pressure of the expanding seed before the infructescence is released. Premature rupturing of the capsule may also occur in some Hymenocallideae.

Several tribes have developed fleshy seeds lacking phytomelan (Haemantheae, Amaryllideae, Hymenocallideae, Calostemmateae), though false homologies have at times been drawn between divergent types (e.g., Traub 1963). The water content of these seeds remains high and they generally lose viability if dehydrated. At least 4 distinct types of fleshy, nonphytomelanous seed occur in the family (Rendle 1901).

Seeds with phytomelan are of 2 basic types: flattened and often winged, or globose (to ovoid or ellipsoid or wedge-shaped). The flattened, winged seed is most prominent in the neotropical tribes (Hippeastreae, Stenomesseae and Eustephieae) but occurs in two Old World tribes as well, Cyrtantheae and Lycorideae (*Ungernia*), as well as in the family Agapanthaceae which resolves as sister to Amaryllidaceae in phylogenies based on chloroplast gene sequences (Fay and Chase 1996). The seeds are either obliquely winged or approximately discoid in shape with the embryo and endosperm centrally positioned. The subglobose, wedge-shaped seed typical of the Mediterranean-centered tribes Narcisseae, Galantheae, and (in part) Pancratieae resembles the seed of Hyacinthaceae. Genera of the neotropical, rainforest-dwelling Euchariedeae have seeds with phytomelanous testas, copious oily endosperm, and fairly high moisture content, suggesting, at least morphologically, an intermediate character state between the dry and fleshy type of seed. The difference between this type and the dry type may be highly plastic, however, as one species of *Hippeastrum* (*H. reticulatum*), also a rainforest understory species, has a similar type of seed, while the rest of the genus has dry, flattened seeds. The turgid, globose seeds of *Lycoris* (Lycorideae) and *Griffinia* (Hippeastreae) are similar, though the latter lacks phytomelan in the testa. Seeds are appendaged at the chalazal end with a fleshy elaiosome in some species of Pancreatieae, Narcisseae, and Galantheae (Werker and Fahn 1975; Dahlgren et al. 1985).

The fleshy seeds of Hymenocallideae are composed of the thick, chlorenchymous outer integument with a well-developed vascular system and a starch-storing embryo (Whitehead and Brown 1940). Phytomelan is absent except in *Leptochiton*. The fleshy seed of Calostemmateae is actually an adventitious bulbil, formed by the precocious germination of the seed inside the capsule (Rendle 1901). The large fleshy seeds of Amaryllideae subtribe Crinineae consist primarily of endosperm, with a thin, corky outer layer. The seeds of many Amaryllideae are chlorophyllous, though in *Amaryllis*, chlorophyll is restricted to the embryo (Markötter 1936). No detailed investigation of the succulent, nonphytomelanous Haemenatheae has been reported, but the seeds of some *Clivia* species have been noted to germinate precociously (D.A. Snijman, pers. observ.). Stomata occur on the testa of some Amaryllideae (subtribe Amaryllidinae) and *Eucrosia* (Stenomesseae). The fleshy seeds of Amaryllideae can be viviparous.

EMBRYOLOGY. In the Amaryllidaceae, the tapetum is of the secretory type, becoming 2-celled as the pollen spore mother cell begins meiosis (Davis 1966; Howell and Prakash 1990). Successive cytokinesis occurs in the pollen mother cells, so that isobilateral and decussate microspore tetrads are formed during meiotic division.

Embryo-sac formation of most Amaryllidaceae is of the Polygonum type, but in *Crinum*, and possibly *Pancratium*, Allium-type formation is reported (Davis 1966). Howell and Prakash (1990), however, characterized *Crinum flaccidum* as having the Polygonum-type embryo sac, and suggest that reports of Allium type in Amaryllidaceae may be erroneous. Ovules are crassinucellate, with a primary archesporial cell in most genera. However, in *Crinum*, *Eucharis*, and *Narcissus*, the archesporial cell functions directly as the megaspore mother cell (Davis 1966). Embryo development is either of the Asterad or Onagrad type (Davis 1966). Endosperm formation is Nuclear or Helobial (Davis 1966). Bitegmic ovules are characteristic of the family except in Amaryllideae, where unitegmic and, reportedly, ategmic ovules (Prillieux 1858; Schlimbach 1924; Tomita 1931; Markötter 1936) occur. The ategmic state is, however, controversial (Snijman and Linder 1996). The seedlings of Amaryllidaceae are distinguished by the presence of a bifacial cotyledon (Tillich 1995).

Apomixis has been documented in *Zephyranthes* and *Habranthus* (Flory 1939; Brown 1951), and has been reported for *Crinum* (Dutt 1962; Wahlstrom and Laane 1979). Polyembryony has been noted in *Hymenocallis* (Traub 1966) and *Crinum* (Dutt 1962; Wahlstrom and Laane 1979).

POLLEN MORPHOLOGY. Large, boat-shaped-elliptic, sulcate pollen grains with reticulate exine morphology, and semitectate-columellate wall structure are the most common type of pollen found in the Amaryllidaceae (Erdtman 1952; Walker and Doyle 1975; Meerow et al. 1986; Meerow and Dehgan 1988; Fig. 36C). Pollen with tectate-perforate to fossulate exines occurs in

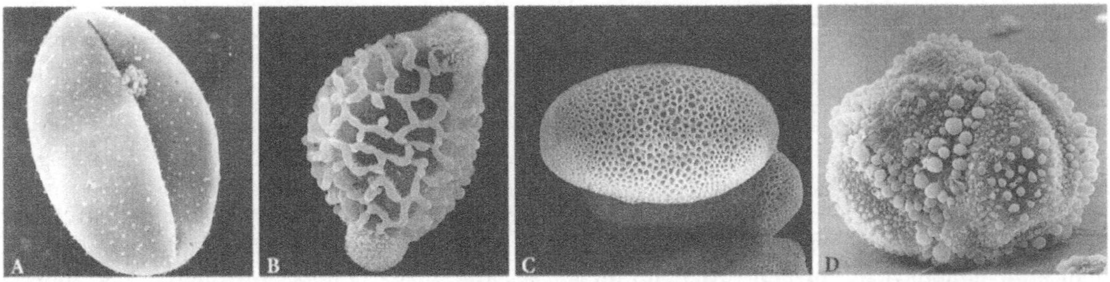

Fig. 36A–D. Amaryllidaceae. **A** *Crinum moorei*, bisulculate pollen grain with spinulose exine, characteristic of tribe Amaryllideae, ×900. **B** *Hymenocallis latifolia*, auriculate pollen with coarse reticulum, ×550. **C** *Stenomesson coccineum*, sulcate pollen with reticulate exine, ×600. **D** *Stenomesson* sp., tetrad, ×750. (Original Meerow)

some species of *Eucrosia* (Meerow 1987a). Intra- and intergeneric differences occur in pollen grain size and coarseness of the reticulum (Meerow and Dehgan 1985, 1988; Meerow 1987a, 1989). All members of the tribe Amaryllideae, endemic to sub-Saharan Africa with the exception of *Crinum*, have bisulculate pollen and spinulose exine sculpturing (Fig. 36A), with intectate-columellate wall structure (Erdtman 1952; Nordal et al. 1977; Dahlgren and Clifford 1982; Snijman 1992). A single species of *Stenomesson* releases pollen in tetrads with intectate, gemmate exines (Meerow et al. 1989, Fig. 36D). Very large, auriculate pollen occurs in *Hymenocallis*, *Leptochiton*, and *Ismene* subg. *Ismene* (Hymenocallideae, Meerow and Dehgan 1985, Fig. 36B).

KARYOLOGY. The most common base number occurring in the Amaryllidaceae is $x = 11$ (Goldblatt 1976; Flory 1977; Meerow 1987b), and $2n = 22$ characterizes many unrelated genera (Flory 1977; Meerow 1984). Two major situations are observed in amaryllidaceous karyotype evolution (Meerow 1984). Certain genera exhibit great karyotypic stability, with low frequency of polyploidy, e.g., *Crinum* (Jones and Smith 1967; Raina 1978), *Hippeastrum* (Naranjo and Andrada 1975), and *Eucharis* (Meerow 1987b). Similar chromosome morphology among the species of such genera is characteristic. Their polyploids tend to be autoploid in origin. At the other extreme, a genus may exhibit great variation in both chromosome number and morphology, e.g., *Hymenocallis* (Flory and Schmidhauser 1957; Flory 1976; Lakshmi 1978), *Lycoris* (Inariyama 1931, 1933, 1937, 1953; Bose and Flory 1963; Kurita 1986, 1987a,b,c, 1989), and *Narcissus* (Fernandes 1942, 1967, 1968). In such genera, both allopolyploidy and Robertsonian changes have been implicated as important factors in speciation. An unusual situation was reported by Ising (1970) for *Cyrtanthus*, an isolated genus of east and southern Africa with perhaps the greatest variation in floral morphology in the family. In *Cyrtanthus*, most of which are self-compatible, a great deal of internal structural change in the karyotypes has occurred while preserving the basic chromosome number of $2n = 16$.

Andean-centered genera in the tribes Eucharideae, Eustephieae, Hymenocallideae, and Stenomesseae are characterized by a somatic chromosome number of $2n = 46$ or presumptive derivations thereof (Di Fulvio 1973; Flory 1977; Williams 1981; Meerow 1984, 1987a,b). This may indicate a monophyletic polyploid origin for these tribes from an ancestor with $2n = 22$ via chromosome fragmentation or duplication and subsequent doubling (Satô 1938, Lakshmi 1978), or perhaps even reticulation between an $x = 11$ lineage and one with $x = 12$, followed by doubling. One subgroup of the neotropical tribe Hippeastreae (*Zephyranthes* and *Habranthus*) has $2n = 24$ chromosomes (Flory 1968), as do the African *Cryptostephanus* (Haemantheae, Gouws 1949) and the Mediterranean genus *Galanthus* (Galantheae). A southern Andean group of Hippeastreae (*Rhodophiala*, *Phycella*, *Traubia*) is characterized by $2n = 16$ or 18 (Flory 1968; Grau and Bayer 1991). Two baccate-fruited African genera, *Apodolirion* and *Gethyllis* have $2n = 12$ chromosomes (Wilsenach 1965; Vosa 1986). Calostemmateae are characterized by $2n = 20$ (Zaman and Chakraborty 1974). The Andean group appears to represent the only widespread polyploid lineage within the family. Assuming a base number of $x = 11$, changes in base chromosome number have been primarily downward in the major lineages. The largest somatic chromosome numbers reported for the family were 118–178 by Flory (1967) for *Sprekelia formosissima* (Hippeastreae), 138 by Meerow (1989) for *Eucharis caucana* (Eucharideae), and 104–110 by Flory (1976) for *Ismene narcissiflora* (Hymenocallideae).

REPRODUCTIVE BIOLOGY. Very little is known about the pollination biology and breeding systems of Amaryllidaceae. White, long-tubed, crateriform flowers with conspicuous false coronas formed by the basal connation of the staminal filaments occur in at least one genus each of the Pancratieae, Eucharideae, Hymenocallideae, and Stenomesseae, and appear to be adapted for sphingid moth pollination (Morton 1965; Bauml 1979; Grant 1983). *Crinum flaccidum* is also reportedly pollinated by sphingid moths (Howell and Prakash 1990). Hummingbird visitation and pollination has been observed for *Hippeastrum* (Meerow pers. observ.), and may also be important in Andean genera such as *Stenomesson* and *Phaedranassa*, the flowers of which fit an ornithophilous syndrome (Faegri and Van der Pijl 1979). *Sprekelia* (Hippeastreae) is also thought to be hummingbrid-adapted (Dahlgren et al. 1985). Euglossine bee visitation was reported for a species of *Eucharis* (Vogel 1963). Dafni and Werker (1982) reported pollination of *Sternbergia clusiana* by *Apis mellifera* and various Syrphidae. Johnson and Bond (1994) reported that *Nerine sarniensis*, *Brunsvigia marginata*, and several species of *Cyrtanthus* with large red flowers

(*Cyrtanthus elatus*, *C. guthrieae*, *C. montanus*) are pollinated exclusively by one species of butterfly, *Aeropetes tulbaghia*. Some *Brunsvigia* species (*Brunsvigia orientalis*, *B. josephinae*, and *B. litoralis*) are adapted for sunbird pollination (D.A. Snijman, pers. observ.). Bat pollination has been recently observed in Brazil on *Hippeastrum calyptratum* (J. Dutilh, pers. com.).

Most species in the family are self-incompatible with protandrous flowers, and therefore obligately xenogamous, but it is not uncommon for one to several species in a genus to be self-compatible (e.g., *Eucharis castelnaeana*, Meerow 1989; *Hippeastrum papilio*, *H. reticulatum*, Bell 1977). Meerow (1989) hypothesized that some *Eucharis* species may be partially self-compatible, wherein the incompatibility system breaks down in the last flowers to reach anthesis on a şcape if no previous flowers have been pollinated. *Strumaria* also indicates the ability to self-pollinate (Snijman 1992). *Cyrtanthus* is the only genus known to be entirely self-compatible (Ising 1970), though self-compatibility has been observed in a number of *Zephyranthes* and *Habranthus* species (Raina and Khoshoo 1972; Ghosh and Shivanna 1984; Broyles and Wyatt 1991), and is most likely more common in these two genera (and probably other genera as well) than has been reported. Raina and Khoshoo (1972) correlated self-compatibility in *Zephyranthes* with style length shorter than that of the stamens, an observation not consistent with *Z. atamasco* as reported by Broyles and Wyatt (1991).

DISPERSAL. Dispersal systems in Amaryllidaceae have been poorly documented, with the exception of the Amaryllideae and Galantheae. Wind dispersal through tumbling has evolved several times in Amaryllideae in semiarid southwestern Africa (Snijman and Linder 1996). Characters shared by these anemogeochorous taxa (*Boophone*, *Cybistetes*, *Crossyne*, *Brunsvigia*, *Hessea*, and *Strumaria*) are low stature and light-weight fruiting heads with rigid, widely radiating pedicels. The capsules remain attached to the pedicels and are often enlarged. In all the genera, other than *Boophone*, the dry scape forms part of the dispersal unit and its length relative to the pedicels determines the infructescence's tumbling, skidding or lofting ability. The derived lax, pendulous pedicels in a few species of *Strumaria* allow seeds to be scattered by wind without the infructescence breaking loose. *Crinum* has many species adapted to water dispersal (Arroyo and Cutler 1984; Howell and Prakash 1990), either along seasonal inland water systems or by offshore currents. Seeds of *Crinum* have been known to remain viable and afloat in salt water for more than 2 years (Koshimizu 1930). Biotic dispersal in Amaryllideae is unknown.

The flattened, winged type of seed characteristic of *Cyrtanthus*, Eustephieae, Hippeastreae, Stenomesseae, and *Ungernia* is adapted for wind dispersal. The berry and baccate fruits of Haemantheae and Gethyllideae are most likely attractants for animal dispersal agents, but no details are found in the literature. Seeds with an elaiosome (Galantheae, Narcisseae, Pancratieae) may also be animal-dispersed. The fleshy, orange capsules contrasting with the large, shiny black seeds of *Eucharis* subg. *Eucharis* were suggested to be mimetic by Meerow (1989), and the internally red capsule of *Hippeastrum reticulatum* (also with large, globose, shiny black seeds) may also function similarly.

PHYTOCHEMISTRY. Steroidal saponins and allyl sulfides are lacking throughout the family, but Amaryllidaceae are very well marked by the consistent presence of a large family of alkaloidal compounds, most not known to occur in any other family of plants. These alkaloids (e.g., lycorin, crinin, belladin, hemanthanin) number over 100 (Wildman 1968; Gibbs 1974) and are considered to be biogenetically related. An overproduction of tyrosine, the primary precursor component of the amaryllidaceous alkaloids, is thought to account for their biosynthesis (Hegnauer 1963). At least one of these, pancratiostatin, is highly promising as an anticancer agent (Pettit et al. 1993, 1995). Chelidonic acid is also commonly found in members of the family. The bulbs and often the foliar leaves are rich in polysaccharides; specialized mucilage cells are frequently present (Arroyo and Cutler 1984). Organic acids and soluble nitrogenous compounds are also contained in the bulbs (Dahlgren et al. 1985).

PHYLOGENETIC RELATIONSHIPS AND SUBFAMILIAL CLASSIFICATION. The four most recent infrafamilial classifications of Amaryllidaceae are those of Traub (1963), Dahlgren et al. (1985), Meerow (1995), and Müller-Doblies and Müller-Doblies (1996). Traub's scheme included Alliaceae, Hemerocallidaceae, and Ixioliriaceae as subfamilies. Within his subfamily Amarylloideae, he erected two informal taxa, "infrafamilies" Amarylloidinae and Pancratioidinae, both of which were polyphyletic. Dahlegren et al. (1985) dispensed with any subfamilial classification above the level of tribe and treated as Amaryllidaceae only those genera in Traub's Amarylloideae. Meerow (1995) resurrected Eustephieae from Dahlgren et al.'s (1985) submergence in Stenomesseae, and suggested that two new tribes may need to be recognized (Calostemmateae and Hymenocallideae). Müller-Doblies and Müller-Doblies (1996) recognized 10 tribes and 19 subtribes. Molecular data, which are accumulating rapidly, from both chloroplast and nuclear genes (Fay et al. 1995), hold great promise for a final resolution for the phylogeny of the family.

Despite a lack of consensus on generic and specific limits within Amaryllidaceae, cladistic analysis has only rarely been applied to such problems in the family, such as by Nordal and Duncan (1984) for *Haemanthus* and *Scadoxus*, two closely related, baccate-fruited African genera; Meerow (1987a, 1989) for *Eucrosia* and *Eucharis* and *Caliphruria* respectively; and Snijman (1994) and Snijman and Linder (1996) for various taxa of the tribe Amaryllideae. Developing character state matrices for the entire family is difficult due to homoplasy for many conspicuous characters within this highly canalized group (Meerow 1989, 1995), and this has hampered family-wide phylogenetic analyses. For example, petiolate leaves have evolved within independent lineages at least six times (Meerow 1987a, 1989). Likewise, staminal connation, once thought to be a highly significance character phylogentically (e.g., Traub 1963), occurs in so many disparate lineages as to be of little value.

Amaryllidaceae is inarguably monophyletic in all analyses of the chloroplast gene *rbc*L involving a larger subset of Asparagales (Fay and Chase 1996), and *Agapanthus* is sister to the family. The clades of Amaryllidaceae resolved by *rbc*L show remarkable geographic congruence (Fay et al. 1995). Tribe Amaryllideae (entirely sub-Saharan African except for a single genus, *Crinum*), the most robust monophyletic group in the family on the basis of numerous morphological autapomorphies, is the first clade resolved within the "traditional" Amaryllidaceae; it is sister to the rest of the family. The Australasian (*Proiphys* and *Calostemma*) and berry-fruited African taxa (Haemantheae) are sister groups whose ancestral node forms an unresolved trichotomy with the ancestor of the Neotrophical/Laurasian clade and *Cyrtanthus*. The placement of *Proiphys* and *Calostemma* in Eucharideae by Traub (1963) and Dahlgren et al. (1985) make that tribe polyphyletic (Meerow 1989, 1995). The position of *Cyrtanthus* as a clade distinct from the rest of Haemantheae supports Meerow's (1995) assertion that this genus is a unique, and probably basal, African lineage, with no immediate relationship to the baccate-fruited Haemantheae. The largest clade within the family represents two monophyletic subgroups (1) the Laurasian elements of the family, encompassing the Lycorideae, Pancratieae, Narcisseae, Galantheae; and (2) the Neotropical tribes. The relationships within the Laurasian clade are not yet clear; further sampling of the tribes is necessary.

Two important and very consistent Neotropical subclades resolve in the *rbc*L phylogenies. Eustephieae is sister to all the rest of the Neotropical genera. The Hippeastreae also resolve as monophyletic and represent the second most basal group within the neotropical clade. Analysis of ITS sequences (A.W. Meerow and Guy, unpubl. data) place Hippeastreae and Eustephieae as sister taxa, but still basal to the rest of the neotropical genera. With the exception of a robust clade representing Hymenocallideae, the northern Andean Amaryllidaceae (Eucharideae, Stenomesseae) are very poorly resolved by *rbc*L sequences alone. This lack of resolution is the result of the low levels of base substitution among the Andean amaryllids, among which occurs no sequence divergence rate greater than 1% (A.W. Meerow and Guy, unpubl. data). This is most likely due to the relative youth of these two lineages, the evolutionary history of which is inextricably linked to the rise of the Andean orocline to its present limits (Meerow 1985, 1987a, 1989). Preliminary ITS sequence data further support recognition of Hymenocallideae, and we are hopeful of a more resolute understanding of Eucharideae and Stenomesseae in the near future using this nuclear gene.

The affinities of the family are, with other "higher" Asparagales, most close to Agapanthaceae, which, together with Amaryllidaceae, is sister to Alliaceae, followed by Themidaceae and Hyacinthaceae (Fay and Chase 1996).

Distribution and Habitats. The Amaryllidaceae is chiefly a tropical group, with notable centers of diversity in South America (28 genera), Africa (19 genera), followed by the Mediterranean (8 genera). Only a single genus, *Crinum* (Amaryllideae), with seeds apparently well adapted for dispersal over water, is represented in both the Old and New Worlds. That the family's phylogenetic relationships so closely follow geographic distribution, with much regional endemism, adds credence to a Gondwana origin for the family at a time when the continents were much closer together (Raven and Axelrod 1974). Molecular evidence places the most ancient (though not necessarily primitive) lineages within the family in Africa (Fay et al. 1995), which would suggest an eastern Gondwana birthplace. Climatic changes in southern Africa (Goldblatt 1978) and geologic changes in South America (Meerow 1987a,c, 1989) have been important factors in the radiation of the family within its two main centers of diversity.

Despite the tropical bias in distribution, the family has adapted readily to high montane tropical climates in the Andes and extratropical temperate ecosystems in Europe and Asia. The tropical genera are chiefly adapted for seasonally dry habitats, in some cases, truly xeric environments where the bulbs may remain dormant for a period longer than they are in active growth (e.g., *Leptochiton, Paramongaia*). Some *Crinum* species form enormous populations on the floor of ephemeral shallow lake beds in sub-Saharan Africa. At the other extreme, species have colonized the understory of rainforests (Euchardeae, *Griffinia*) or an aquatic habitat (a number of *Hymenocallis, Hippeastrum angustifolium*, several *Crinum* spp.). Extremes of ecological adaptation often occur in the same genus. Fire ecology is important in the biology of some of the African genera: *Amaryllis, Haemanthus*, and *Hessea* in the South African fynbos region, and *Cyrtanthus* in grassland and savanna. This may be true of other savanna species elsewhere in the world.

Ethno- and Economic Botany. The Amaryllidaceae have little economic importance except as ornamentals. *Narcissus, Leucojum*, and *Galanthus* are among the most important temperate-zone spring-flowering bulbs in commerce. *Clivia, Crinum, Cyrtanthus, Hippeastrum, Lycoris, Nerine*, and *Zephyranthes* are also produced for ornamental use; *Hippeastrum*, in particular, has been extensively hybridized. The mashed bulbs of a number of tropical genera are utilized by indigenous cultures as poultices for treating sores, or boiled and steeped to prepare an emetic tea for stomach ailments (Schultes and Raffauf 1990; Duke and Vasquez 1994). In temperate Africa, similar preparations of the bulbs and leaves of Amaryllideae (*Amaryllis, Boophone, Brunsvigia*, and *Crinum*), and Haemantheae (*Clivia, Haemanthus*, and *Scadoxus*) are used for skin and digestive disorders, but large dosages are known to be very poisonous (Watt and Breyer-Brandwijk 1962). South Africans use the fragrant, baccate fruits of Gethyllideae as a flavor and perfume. The San in Botswana use *Pancratium tenuifolium* to induce visual hallucinations (Schultes 1976). In southern Africa a rare rock painting, attributed to the San, is thought to depict a *Brunsvigia* species, whose use probably induced psychoactive effects (Loubser and Zietsman 1994). Vargas (1981) reported floral motifs identifiable as Amaryllidaceae (*Ismene, Pyrolirion, Stenomesson*) appearing on ceremonial Inca drinking vessels called keros.

Key to the Genera

1. Leaves and bulb coats with extensible fibers when torn; scape solid; pollen bisulculate 2
- Leaves and bulb coats without extensible fibers when torn; scape hollow or solid; pollen monosulcate 12
2. Leaf margins smooth, softly pubescent or raised, reddened and fringed with long bristles or short, branched cilia; pedicels rarely shorter than flowers at anthesis; filaments more or less connate at the base, if free, then fused to the style base 3
- Leaf margins hyaline, more or less fringed with short, branched cilia; pedicels usually shorter than flowers at anthesis; filaments free to base 9
3. Leaves with a prominent midrib; flowers much longer than pedicels **1. *Amaryllis***
- Leaves without a midrib; flowers equaling or much shorter than pedicels 4
4. Flowers zygomorphic (sometimes only by the deflexed style) 5
- Flowers regular 7
5. Leaves narrow, usually less than 25 mm wide, and subsucculent; pedicels slender, rarely longer than flowers; tepal margins more or less undulate **2. *Nerine***
- Leaves broad, usually more than 25 mm wide, and leathery; pedicels stout, usually much longer than flowers; tepal margins rarely crisped 6
6. Leaves immaculate; margins smooth or fringed with short, branched cilia; pedicels obscurely 3-angled in cross-section; filaments tightly clustered **3. *Brunsvigia***
- Leaves speckled with red; margins fringed with long bristles; pedicels sharply triangular in cross-section; filaments more or less separate **4. *Crossyne***
7. Leaves 2(–4), glabrous or rarely minutely pilose; flowers persisting after anthesis; filaments connate into a short to long tube, free from style; anthers centrifixed to subcentrifixed **5. *Hessea***
- Leaves 2–6, glabrous or hairy; flowers withering after anthesis; filaments free or, if fused, then filament tube trilocular; at least one filament whorl adnate to style base; anthers subcentrifixed to dorsifixed 8
8. Leaves pubescent or glabrous, ovate to filiform; tepals free; at least outer filaments adnate to style base; style winged or swollen basally **6. *Strumaria***
- Leaves glabrous, filiform; tepals connate into a distinct tube; inner filaments adnate to style; style columnar throughout **7. *Carpolyza***
9. Leaves annual, distichous, closely abutting each other to form an erect fan; all leaf tips subacute to obtuse; fruiting head detaching from scape apex; fruit trigonous, prominently 3-ribbed **8. *Boophone***
- Leaves perennial, suberect or prostrate; usually tips of all but youngest leaves truncate; fruiting head not detaching from scape apex; fruit irregularly shaped, smooth or 6-ribbed 10

10. Pedicels remaining unchanged in fruit; fruiting head persisting and drooping; fruit membranous or rarely fleshy, smooth 11
- Pedicels elongating and radiating after anthesis, dry and stout in fruit; fruiting head detaching at ground level; fruit papery, 6-ribbed **11. *Cybistetes***
11. Leaves evenly spreading or in a vertical distichous pattern, suberect to recurved, often sheathing to form a false stem; flowers zygomorphic, sometimes only by deflexed style; perigone tube mostly curved, occasionally straight **9. *Crinum***
- Leaves biflabellate, arching or spreading on the ground; flowers actinomorphic; perigone tube straight **10. *Ammocharis***
12. Fruits baccate 13
- Fruits capsular 18
13. Inflorescence acaulescent, 1-flowered 14
- Inflorescence well developed above ground; several-many-flowered 15
14. Leaves few-many, glabrous or pubescent, often with conspicuous sheaths around foliage leaves; stamens uniseriate; anthers 6-many **20. *Gethyllis***
- Leaves 1-few, glabrous, sheaths not conspicuous; stamens biseriate; anthers 6 **19. *Apodolirion***
15. Plants with large, fleshy bulbs; leaves sword- to tongue-shaped, without a midrib, succulent, and often pubescent **16. *Haemanthus***
- Plants rhizomatous; leaves glabrous 16
16. Leaves with a petiolelike base and prominent midrib, thin-textured; ovules 1 or 2 **15. *Scadoxus***
- Leaves lanceolate, thick-textured, without a midrib; ovules more than 2 per locule 17
17. Scape strongly compressed but without wings; perigone without scales **13. *Clivia***
- Scape strongly compressed and more or less winged; perigone with bifid, lanecolate scales between filaments **14. *Cryptostephanus***
18. Capsule indehiscent, 1-loculed at maturity 19
- Capsule dehiscent, 3-loculed at maturity 20
19. Leaves sessile, linear, flowers yellow or purple **17 *Calostemma***
- Leaves petiolate, reniform or ovate, flowers white **18. *Proiphys***
20. Scape solid 21
- Scape hollow 55
21. Paraperigone never present; staminal filaments subulate or winged, broadly connate into a conspicuous corona or shortly connate near their base, or appendaged with teeth 22
- Paraperigone sometimes present; stamens completely free and unappendaged 43
22. Seeds fleshy with thick, chlorenchymous integuments, green or with a brown testa 23
- Seeds dry or turgid, but never fleshy, integuments thin, with a black or brown testa 25
23. Spathe bracts absent, flower bud subtended but not enclosed by several linear bracteoles; ovules 16–20 per locule **47. *Leptochiton***
- Spathe bracts present, enclosing the flower buds; ovules 2–10 per locule 24
24. Leaves tightly sheathing basally and forming a long, conspicuous, aerial pseudostem; flowers white, yellow, or green, declinate or subpendulous; campanulate, funnelform-campanulate or tubular, actinomorphic or zygomorphic; staminal cup often striped green along the filamental traces; free filaments incurved, declinate-ascending or straight and fasciculate; pollen yellow; ovules 2–3 per locule **48. *Ismene***
- Aerial pseudostem absent; flowers white, actinomorphic, erect or suberect, crateriform; staminal cup never green along the filamental trace; free filaments straight, spreading outward; pollen orange; ovules 2–10 per locule **46. *Hymenocallis***
25. Leaves sessile, linear or lorate, flat in vernation 26
- Leaves petiolate, with broad, elliptic or ovate laminae (rarely only lanceolate), involute or revolute in vernation 36
26. Seeds turgid, globose or wedge-shaped by compression, never flattened, sometimes bearing an elaiosome at the chalazal end **23. *Pancratium***
- Seeds flat, obliquely winged or discoid (winged on all sides), never with an elaiosome 27
27. Staminal cup conspicuous (>5 mm long) 28
- Staminal cup reduced to a membraneous connation at the base of filament, or, if stamens free, the filaments subulate, winged or appendaged with teeth 32
28. Staminal cup with a narrow, longitudinal aperture between each free filament **59. *Hieronymiella***
- Staminal cup inaperturate 29
29. Perigone tube shorter than the tepal segments; free filaments biseriate, alternately inserted at the base of the staminal cup and at the rim **55. *Pucara***
- Perigone tube as long or longer than the tepal segments; free filaments subequal, uniseriate 30
30. Flowers <10 cm long, green, or variously colored, but never white, tubular to funnelform-tubular, slightly constricted at the throat **49. *Stenomesson***
- Flowers >10 cm long, white or yellow, crateriform; tube not constricted at the throat 31
31. Leaves somewhat succulent, glaucous, flowers yellow; free filaments inserted below the rim of the staminal cup **50. *Paramongaia***
- Leaves thin, nonglaucous; flowers white, free filaments inserted at the rim of the staminal cup **51. *Pamianthe***
32. Stigma entire 33
- Stigma 3-lobed to trifid 34
33. Ovules few per locule **24. *Vagaria***
- Ovules numerous per locule **49. *Stenomesson***
34. Perigone tube shorter than the tepal segments; staminal filaments winged proximally, inserted at the throat, subequal, with a slender acute tooth on each side above the midpoint; stigma shortly 3-lobed **58. *Eustephia***
- Perigone tube much longer than the tepal segments; staminal filaments filiform or broadly subulate (winged), strongly biseriate, inserted below the throat; stigma trifid or 3-lobed 35
35. Seeds D-shaped, plants American **57. *Chlidanthus***
- Seeds obliquely winged, plants African **12. *Cyrtanthus***
36. Leaves involute in vernation **54. *Eucrosia***
- Leaves revolute in vernation 37
37. Stigma capitate 38
- Stigma obtusely 3-lobed 41
38. Ovules numerous per locule (>20), flattened 39
- Ovules 20 or less per locule, globose 40
39. Flowers zygomorphic; perigone tube not constricted at the throat; staminal cup, if well developed, edentate and irregularly cleft between each free filament, filaments declinate-ascending, at least apically; globose nectar gland usually present at the base of each filament **54. *Eucrosia***
- Flowers actinomorphic; perigone tube slightly constricted at the throat; staminal cup well developed, usually toothed, or stamens appearing free and inserted below the throat

with a toothlike process in between each that is fused to the tube; nectar glands absent **49. *Stenomesson***
40. Leaves persistent; flowers zygomorphic; perigone funnelform, white; stamens declinate-ascending; connate below into a toothed, cylindrical staminal cup; ovules 2–3 per locule, basal **44. *Plagiolirion***
- Leaves annual; flowers actinomorphic, perigone urceolate, yellow or orange and green with white margins at the apex; stamens straight, only shortly connate at the base; ovules 10–20 per locule, axile **45. *Urceolina***
41. Ovules numerous (>20 per locule); flattened **56. *Mathieua***
- Ovules 20 or less per locule; globose 42
42. Flowers crateriform or campanulate; sometimes fragrant; perigone tube longer than the tepal segments, curved, narrowly cylindrical below and abruptly dilated above its midpoint; staminal cup usually well developed or only shortly exserted from the throat, toothed or edentate, always spotted green below each free filament or flushed green on its interior surface; free filament usually much shorter than cup and broadly subulate **42. *Eucharis***
- Flowers funnelform; never fragrant; perigone tube as long or shorter than the tepal segments, straight, dilating gradually from base to throat; stamens only shortly connate below, entirely white, toothed or edentate; free filament usually narrowly subulate **43. *Caliphruria***
43. Leaves petiolate, with elliptic-lanceolate to broadly ovate laminae 44
- Leaves sessile (or very subpetiolate and appearing so), linear or lorate 47
44. Paraperigone absent 45
- Paraperigone of fimbriae, scales or callose ring present **31. *Hippeastrum***
45. Flowers blue, violet, or white; ovules few per locule, globose **33. *Griffinia***
- Flowers green, pink, orange, or yellow; ovules numerous per locule, flattened 46
46. Leaves thick and carnose, ovate, without a conspicuous midvein; petioles broad; flowers entirely green; tepal tube longer than tepal segments; stamens inserted well below the perigone throat **53. *Rauhia***
- Leaves not carnose, lanceolate to elliptic to lanceolate, with a conspicuous midvein; petioles narrow; flowers pink to orange, tipped green; perigone tube shorter than tepal segments; stamens inserted at the perigone throat **52. *Phaedranassa***
47. Paraperigone present 48
- Paraperigone absent 49
48. Spathe bracts free; perigone funnelform; paraperigone limited to a series of short fimbriae at the perigone throat **21. *Lycoris***
- Spathe bracts fused into a tube below the middle; perigone crateriform; paraperigone forming a conspicuous cup-shaped to funnelform corona or a short rim at the perigone throat **25. *Narcissus***
49. Anthers dehiscing by terminal pores 50
- Anthers dehiscing longitudinally 51
50. Tepals always free, inner whorl much shorter and broader than outer whorl **27. *Galanthus***
- Tepals free or shortly connate proximally, only slightly subequal **28. *Leucojum***
51. Scape sometimes retained within the bulb neck; spathe bracts fused into a tube below the middle **26. *Sternbergia***
- Scape always emergent; spathe bracts free fused or one side 52
52. Tepals free to the base; anthers sagittate, basifixed **29. *Lapiedra***
- Tepals at least shortly united at the base; anthers oblong, dorsifixed 53
53. Perigone slightly to strongly zygomorphic, funnelform; stamens inserted in 1 row near the perigone throat, declinate-ascending 54
- Perigone actinomorphic, funnelform-tubular; stamens inserted in 2 rows below the perigone throat, straight **22. *Ungernia***
54. Seeds lacking phytomelan, white or pink **33. *Griffinia***
- Seeds with a crustaceous coat of phytomelan, black **21. *Lycoris***
55. Scape 1-flowered (very rarely anomalously producing a 2nd flower); spathe bracts fused into a tube, at least below the middle 55
- Scape with 2 or more flowers; spathe bracts free, connate only briefly at the very base, or fused only on one side 59
56. Perigone actinomorphic; stamens straight or spreading outward 57
- Perigone zygomorphic; stamens declinate-ascending, at least distally 58
57. Style 3-branched; each branch with a spatulate stigma **38. *Pyrolirion***
- Style not branched, stigma trifid to capitate **35. *Zephyranthes***
58 Tepals lanceolate or falcate, three oriented dorsally and forming a pseudolabellum around the stamens and style **37. *Sprekelia***
- Tepals elliptic or ovate; not forming a pseudolabellum **36. *Habranthus***
59. Paraperigone absent 60
- Paraperigone present 64
60. Bulb prolonged into a very long aerial neck; leaves evergreen; flowers lilac- or amethyst-colored; stamens of 4 different lengths **32. *Worsleya***
- Bulb not prolonged into a very long aerial neck; leaves deciduous; flowers never lilac- or amethyst-colored; stamens biseriate or subequal 61
61. Stamens subequal; anthers dehiscing by terminal pores; style strumose distally **28. *Leucojum***
- Stamens biseriate; anthers dehiscing longitudinally; style filiform 62
62. Perigone tube usually longer than the limb; stamens attached at or below the throat; filaments sometimes dilated proximally, rarely connate at the base **12. *Cyrtanthus***
- Perigone tube much shorter than the limb; stamens inserted at the throat, filaments filiform and always free 63
63. Tepals keeled green; ovules 5–6 per locule **30. *Hannonia***
- Tepals keeled purple; ovules numerous per locule **40. *Traubia***
64. Paraperigone forming a conspicuous 3–6-lobed corona **39. *Placea***
- Paraperigone reduced to a callose rim or a ring of scales or fimbriae 65
65. Spathe bracts fused on one side **34. *Rhodophiala***
- Spathe bracts free, or shortly connate proximally 66
66. Leaves usually >2 cm wide **31. *Hippeastrum***
- Leaves less than 1.5 cm wide, sometimes channeled adaxially 67
67. Leaves flat adaxially; perigone funnelform, always zygomorphic; stigma trifid or 3-lobed **34. *Rhodophiala***
- Leaves channeled adaxially; perigone tubular to funnelform-tubular; actinomorphic to slightly zygomorphic; stigma capitate to obscurely 3-lobed **41. *Phycella***

1. Tribe Amaryllideae J. St.-Hil. (1805).

Bulb tunics producing highly extensible cotton-like elements when torn. Scape with a sclerenchymatous sheath. Pollen bisulculate; exine spinulose. Ovules unitegmic or ategmic. Seeds nondormant, water-rich; embryo green. Pantropical, mostly in sub-Saharan Africa.

a. Subtribe Amaryllidinae Pax (1887).

Stamens connate into a tube distally, with connation lost in certain *Strumaria* species and *Carpolyza*. Seed with a well-developed chlorophyllous integument and stomatose testa, except in *Amaryllis belladonna* L. Southern African.

1. *Amaryllis* L.

Amaryllis L., Sp. Pl. 1: 293 (1753); Dyer, Fl. Pl. Africa 30: t. 1200 (1955).

Bulb large. Leaves hysteranthous, with a midrib. Scape stout, persistent and erect, and with pedicels elongating in fruit. Flowers zygomorphic, without a tube; segments spreading. Stamens very shortly connate basally, declinate, unequal. Style declinate, stigma 3-lobed. Ovules ca. 8 per locule. Seeds compressed-globose, white to pink. $2n = 22$. Only one sp., *A. belladonna* L., S African winter-rainfall region, in fire-prone habitats. Naturalized in Mediterranean climates worldwide. *Amaryllis* shares few morphological specializations with other genera of Amaryllideae. The rudimentary filament tube places it in subfamily Amaryllidinae, where it appears to be a basal genus (Snijman and Linder 1996).

2. *Nerine* Herb.

Nerine Herb., Bot. Mag. 47: t. 2124 (1820); Herbert, Amaryllidaceae: 283–287 (1837); Traub, Pl. Life Suppl.: 1–32 (1967), rev.; Norris, Nerine Soc. Bull. 6: 7–31 (1974), rev.

Leaves annual or persistent, sometimes hysteranthous. Scape slender or robust, rarely minutely puberulous. Flowers zygomorphic, widely flared, with a short tube; segments attenuate with undulate margins. Stamens connate basally, sometimes with lateral appendages, declinate, unequal. Stigma obscurely trifid or 3-lobed. Ovules up to 4 per locule. Seeds ovoid, reddish green. $2n = 22$. About 23 spp., S Africa. Many horticultural hybrids are named. As in the closely related *Brunsvigia*, most *Nerine* species have pink flowers. The derived red flower color is an adaptation to a specific butterfly pollinator, *Aeropetes tulbaghia*.

3. *Brunsvigia* Heist.

Brunsvigia Heist., Beschr. neu. Geschl.: 3 (1755); Dyer, Pl. Life 6: 63–83 (1950), rev.; Dyer, Pl. Life 7: 45–64 (1951), rev.

Bulb usually large; tunics often cartilaginous. Leaves annual, often hysteranthous and prostrate. Scape breaking at ground level in fruit; pedicels long, stiff, radiating. Flowers zygomorphic, with short tube. Stamens clustered, shortly connate at base. Stigma 3-lobed. Ovules 3–10 per locule. Capsule large, transversally veined. Seeds ovoid, reddish green. $2n = 22$. About 20 spp., S Africa. *Brunsvigia* is the only genus of Amaryllideae in which several species have stout, somewhat tubular, red flowers that are adapted to bird pollination.

4. *Crossyne* Salisb.

Crossyne Salisb., Gen. pl.: 116 (1866); Müller-Doblies & Müller-Doblies, Feddes Repert. 105: 355–359 (1994), rev.
Boophone Herb. (1821), pro parte.

Bulb large, cartilagenous; neck transversally banded. Leaves hysteranthous, prostrate; margins bristled. Scape breaking at ground level in fruit; pedicels stiff, radiating, trigonal. Flowers many, zygomorphic; tube short. Stamens spreading, swollen and connate basally. Stigma minutely trifid. Ovules 2–4 per locule. Capsule transversally veined. Seeds ovoid, reddish green. $2n = 22$. Two spp., S African winter-rainfall region.

5. *Hessea* Herb.

Hessea Herb., Amaryllidaceae: 289 (1837) nom.cons.; Snijman, Contr. Bolus Herb. 16: 44–82 (1994), monogr.
Dewinterella Müll.-Doblies & Müll.-Doblies (1994).
Kamiesbergia Snijman (1991).
Namaquanula Müll.-Doblies & Müll.-Doblies (1985).

Leaves hysteranthous, usually 2. Scape slender, breaking at ground level in fruit; pedicels stiff, radiating. Flowers actinomorphic; segments free or connate into a short tube. Stamens proximally connate, sometimes with inwardly curved hooks; anthers centrifixed to subcentrifixed. Stigma trifid. Ovules up to 4 per locule. Capsule papery. Seeds ovoid, reddish green. $2n = 22$. Thirteen spp., S. Africa. Snijman (1994) recognized three subgenera, each distinguished by specializations of the filaments. The diminutive habit, few leaves, and actinomorphic flowers are characters which *Hessea* shares with its close relatives, *Strumaria* and *Carpolyza*.

6. *Strumaria* Jacq.

Strumaria Jacq., Collecteana 5: 49 (1797); Snijman, Contr. Bolus Herb. 16: 82–143 (1994), monogr.
Bokkeveldia Müll.-Doblies & Müll.-Doblies (1985).
Gemmaria Salisb. (1866).
Tedingea Müll.-Doblies & Müll.-Doblies (1985).

Deciduous herbs. Leaves 2 or more, usually hysteranthous, often pubescent. Scape often breaking at ground level in fruit; pedicels stiff or lax. Flowers actinomorphic; segments free. Stamens adnate to style base; filaments sometimes connate; anthers dorsifixed to subcentrifixed. Style swollen or winged; stigma trifid. Ovules up to 9 per locule. Seeds ovoid, reddish green. $2n = 20$, 22. Three subgenera are recognized (Snijman 1994) on characteristics of leaf arrangement, pubescence, and anther insertion. Twenty-three spp., S Africa in semiarid areas.

7. *Carpolyza* Salisb.

Carpolyza Salisb., Parad. lond.: 63 (1807); Müller-Doblies & Müller-Doblies, Bot. Jahrb. Syst. 107: 20 (1985); Snijman, Contrib. Bolus Herb. 16: 143–146 (1994).

Plant delicate. Leaves filiform, annual. Scape spirally twisted basally. Flowers actinomorphic; perigone tube funnelform, shorter than segments. Stamens decurrent to base of perigone, inner whorl adnate to style base; anthers subcentrifixed. Stigma trifid. Ovules 1–6 per locule. Seeds ovoid, green. $2n = 20$. Only one sp., *C. spiralis* (L'Hérit.) Salisb., S African winter-rainfall region.

b. Subtribe Crininae Pax (1887).

Fruit indehiscent, rostellate during development. Seeds sometimes apparently lacking an integument, cork-covered, with several layers of chlorophyll-containing cells below phellogen. Pantropical, mostly sub-Saharan Africa.

8. *Boophone* Herb.

Boophone Herb., Appendix: 18 (1821); Leighton, J.S. Afr. Bot. 13: 59–61 (1947); Müller-Doblies & Müller-Doblies, Feddes Repert. 105: 355 (1994).

Bulbs large. Leaves hysteranthous, spreading into an erect fan. Scape breaking distally in fruit; pedicels elongating and radiating after anthesis. Flowers actinomorphic; segments much longer than tube. Stamens free. Stigma more or less entire. Ovules 1–2 per locule. Fruit trigonal, 3-ribbed. Seeds subglobose. $2n = 22$. Two spp., sub-Saharan Africa. The bulbs of *Boophone disticha* (L.f.) Herb. have long been a source of medicine and poison to many African people (Watt and Breyer-Brandwijk 1962).

9. *Crinum* L.

Crinum L., Sp. Pl. 1: 291 (1753); Herbert, Amaryllidaceae: 242–275 (1837); Verdoorn, Bothalia 11: 27–52 (1973), rev. of S African spp.; Geerinck in Flore d'Afrique Centrale: 7–11 (1973), reg. rev.; Nordal in Flore du Gabon 28: 32–40 (1986), reg. rev.; Nordal in Flore du Cameroun 30: 10–24 (1987), reg. rev.; Nordal in Fl. trop. E. Afr.: 9–18 (1982), reg. rev.

Leaves often perennial, sometimes forming a pseudostem; mostly sessile (petiolate in one species). Flowers zygomorphic to actinomorphic; perigone tube long, cylindrical, equalling or longer than segments. Stamens free. Style declinate, stigma obscurely 3-lobed. Ovules ca. 12 per locule. Fruit sometimes beaked, rarely succulent. Seeds large, subglobose, pale to dark. $2n = 22$. About 65 spp., pantropical, mostly in sub-Saharan Africa. Baker's (1881) subgenera are no longer upheld because of many intermediate and anomalous species. The relationships between species of *Crinum*, *Ammocharis*, and *Cybistetes* are still poorly understood.

10. *Ammocharis* Herb.

Ammocharis Herb., Appendix: 17 (1821); Baker, Handb. Amaryll.: 96 (1888); Milne-Redh. & Schweick; J. Linn. Soc., Bot. 52: 159–190 (1939), rev.

Bulbs large. Leaves perennial (dying back partway with drought), often prostrate and biflabellate, with truncate apices. Flowers actinomorphic; perigone tube cylindrical, longer or slightly shorter than segments. Stamens free. Stigma undivided. Ovules 4–30 per locule. Fruit subglobose, bluntly beaked. Seeds subglobose, cream. $2n = 22$. Five spp., sub-Saharan Africa.

11. *Cybistetes* Milne-Redh. & Schweick.

Cybistetes Milne-Redh. & Schweick., J. Linn. Soc., Bot. 52: 191 (1939); Snijman & Williamson, Bothalia 24: 127–132 (1994).

Bulb large. Leaves perennial (dying back partway with drought), prostrate, biflabellate, with truncate apices. Pedicels lengthening and radiating after anthesis. Flowers weakly zygomorphic; perigone tube subcylindrical, shorter than segments. Stamens weakly declinate, free. Stigma undivided. Ovules 1–19 per locule. Fruit pyriform to fusiform, 6-ribbed, bluntly beaked. Seeds subglobose, cream. $2n = 22$. Only one sp., S African winter-rainfall region. A specialized wind-blown infructescence separates *Cybistetes* from *Ammocharis*. It seems likely that this syndrome of characters evolved from a common ancestor within *Ammocharis*. The current recognition of the monotypic *Cybistetes* seems doubtful (Snijman and Williamson 1994).

2. **Tribe Cyrtantheae** Salisb. (1866).

Description as for genus.

12. ***Cyrtanthus*** Aiton

Cyrtanthus Aiton, Hortus kew. 1: 414 (1789); Nordal in Polhill, Fl. trop. E. Afr.: 23–26 (1982), rev. of E African spp.; Dyer, Herbertia 6: 65–103 (1939); Reid & Dyer, Pl. Life Suppl.: 1–68 (1984), rev. of S African spp.
Vallota Herb. (1821).
Anoiganthus Baker (1878).

Leaves annual or persistent, sometimes hysteranthous. Scape usually hollow. Flowers 2-many (rarely 1), actinomorphic to weakly zygomorphic; tube usually longer than segments and dilated. Filaments biseriate, rarely united into a short corona. Stigma deeply trifid to capitate. Ovules numerous per locule. Seeds black, flat, usually winged. $2n = 16$ most common, 14, 22, 32. About 50 spp., sub-Saharan Africa, highly valued horticulturally. Species differ greatly in the color, size, shape, and position of the flowers.

3. **Tribe Haemantheae** (Pax) Hutchinson (1934).

Bulb sometimes lacking. Involucral bracts equitant, and more than 2. Fruit a globose berry with water-rich seeds lacking phytomelan in all but one genus.

13. ***Clivia*** Lindl.

Clivia Lindl., Bot. Reg. 14: t.1182 (1828); Baker, Handb. Amaryll.: 61–62 (1888); Vorster, Fl. Pl. Africa 53: 70–74 (1994), rev.

Rhizomatous herbs with firm, evergreen leaves. Scape strongly ancipitous. Flowers actinomorphic or weakly zygomorphic, narrowly to widely funnelform, spreading to pendulous; tube short. Filaments free. Stigma tricuspidate. Ovules 5–6 per locule. Berry subglobose, red. Seeds turgid, ivory-colored. $2n = 18, 22$. Four spp., S Africa, in coastal forests. *Clivia miniata* (Lindl.) Regel is a widely grown ornamental.

14. ***Cryptostephanus*** Welw. ex Baker

Cryptostephanus Welw. ex Baker, J. Bot. 16: 193 (1878); Baker in Dyer, Fl. trop. Afr. 7: 383–384 (1898).

Plants rhizomatous. Leaves persistent, sometimes twisted. Scape ancipitous, narrowly winged. Flowers white or dark-colored; tube subcylindrical to funnelform, longer than segments; paraperigone of linear scales arising between stamens; anthers subsessile. Stigma capitate. Ovules many per locule. Berry globose, red. Seeds turgid, with brown phytomelanous coat. $2n = 24$. About five spp., tropical Africa.

15. ***Scadoxus*** Raf. Fig. 31

Scadoxus Raf., Fl. tellur. 4: 19 (1838); Bjørnstad & Friis, Norw. J. Bot. 19: 187–222 (1972), rev., Norw. J. Bot. 21: 243–271 (1974), rev.; Friis & Nordal, Norw. J. Bot. 23: 63–77 (1976), rev.

Plants rhizomatous. Leaves annual, petiolate, often forming a false stem that is frequently mottled or spotted; blade thin-textured with distinct midrib. Spathe bracts often brightly colored and persistent. Flowers actinomorphic, funnelform to salverform; segments equaling or longer than tube. Stamens free. Stigma undivided. Ovules 1-few per locule. Berry ovoid to globose, red. Seeds ovoid, ivory-colored. $2n = 16$. Nine spp., sub-Saharan Africa.

16. ***Haemanthus*** L.

Haemanthus L., Sp. Pl. 1: 325 (1753); Friis & Nordal, Norw. J. Bot. 23: 63–77 (1976); Snijman, J.S. Afr. Bot. Suppl. 12: 1–139 (1984), monogr.

Buls large, fleshy. Leaves mostly 2, annual or persistent, often hysteranthous and spotted, succulent, frequently pubescent. Spathe bracts persistent and colored. Flowers actinomorphic, funnelform; segments longer than tube. Stamens free. Stigma minutely tricuspidate. Ovules 2 per locule. Berry subglobose. Seeds ovoid, wine-red to opalescent. $2n = 16$. Twenty-two spp., S Africa, especially the winter-rainfall region.

4. Tribe Calostemmateae D. & U. M-D. (1996).

Scape solid, terete; spathe bracts 2, free. Flowers pedicellate. Staminal cup present. Stigma capitate. Ovules 2–3 per locule. Fruit indehiscent, green, 1-loculed at maturity. Staminal cup present. Seed germinating precociously and forming an adventitious bulbil inside the capsule. $2n = 20$. Two genera, Australasia.

17. *Calostemma* R. Br.

Calostemma R. Br., Prodr.: 297 (1810).

Leaves annual, linear, sessile. Flowers 10–20, nodding; perigone tubular, purple or yellow. Staminal cup cylindrical. Two spp., Australia.

18. *Proiphys* Herb.

Proiphys Herb., App. Bot. Reg.: 42 (1821).
Eurycles Salisb. ex Schult. & Schult. f. (1830).

Leaves annual, petiolate, the laminae broadly reniform or ovate, plicate. Flowers white, erect or slightly nodding, 10–30; perigone funnelform-crateriform. Two or three spp., Maylasia, Philippines, Australia.

5. Tribe Gethyllideae Dumort. (1829).

Leaves narrow. Scape short, remaining inside bulb, bracts fused, single-flowered. Flowers actinomorphic. Fruit indehiscent, baccate or thin-walled, elongated, yellow to white, often spotted with red. Seeds hard. Although the baccate fruits might suggest placement in Haemantheae, their structure is considerably different than the globose berries and succulent seeds of that tribe, and we suspect an independent origin for this character in the Gethyllideae. Two genera, S Africa.

19. *Apodolirion* Baker

Apodolirion Baker, J. Bot. 16: 74 (1878); Baker, Handb. Amaryll.: 25–26 (1888); Hilliard & Burtt, Notes R. Bot. Gard. Edinburgh 32: 304–307 (1973); Müller-Doblies, Willdenowia 15: 465–471 (1986).

Leaves annual, mostly hysteranthous. Spathe fused. Flower 1, actinomorphic; tube long, cylindric, dilated near mouth, segments ascending. Stamens biseriate, free; anthers dorsifixed in outer series, almost basifixed in innner series. Stigma obscurely 3-lobed. Ovary usually subterranean, ovules numerous per locule. Fruit indehiscent, clavate, succulent, or thin-walled. Seeds globose, with a dark or translucent surface. $2n = 12$. About six spp., S Africa.

20. *Gethyllis* L.

Gethyllis L., Sp. Pl. 1: 442 (1753); Baker, Handb. Amaryll.: 23–25 (1888); Müller-Doblies, Willdenowia 15: 465–471 (1986).

Leaves hysteranthous, sometimes twisted, often pubescent, sometimes subtended by conspicuous sheathing cataphylls spotted brown. Spathe fused. Flower 1, actinomorphic; tube long, cylindric, solid. Stamens 6, 12, 18, or more, sometimes connate; anthers basifixed. Style often deflexed, stigma capitate. Ovary subterranean, ovules many per locule. Fruit indehiscent, clavate, succulent, or thin-walled. Seeds subglobose, pale or dark. $2n = 12$. About 32 spp., S Africa, in semiarid areas. The species differ considerably in the shape, color, and aroma of the fruits.

6. Tribe Lycorideae Traub (1963).

Leaves sessile, annual, often hysteranthous, linear or lorate, with well-developed palisade in the mesophyll. Scape solid; spathe bracts 2, free. Flowers pedicellate; perigone tube present; paraperigonal scales or fimbriae sometimes present. Stamens free. Capsules dehiscent. Seeds with a black or brown phytomelanous testa, flat, winged, irregularly discoid or globose and turgid. Two genera, temperate and subtropical East Asia and Central Asia to Iran and Afghanistan. Many species produce winter-persistent leaves after flowering in late summer or fall; the leaves senesce in the spring.

21. *Lycoris* Herb.

Lycoris Herb., Curtis' Bot. Mag 47: 5–6, pl. 2113 (1819); Traub and Moldenke, Amaryllidaceae: Tribe Amarylleae: 165–182 (1949).

Perigone nearly actinomorphic to strongly zygomorphic, funnelform; tube generally short to nearly obsolete, occasionally as long as limb; tepals sometimes closely undulate at the margins. Stamens in 1 row, inserted near the perigone throat; declinate-ascending. Stigma capitate, entire. Seeds few per locule, globose, turgid, the testa black. $2n = 22$ is considered ancestral; the many other diploid numbers found among the species are believed to have evolved by centric fusion. Perhaps 20 species, primarily east Asian, upper Myanmar through Central China to Korea and Japan. A number are important cultivated ornamentals.

22. *Ungernia* Bunge Fig. 31A–E

Ungernia Bunge, Bull. Soc. Nat. Mosc. 49: 273 (1875); Traub and Moldenke, Amaryllidaceae: Tribe Amarylleae: 158–164 (1949); Artjushenko, Pl. Life 26: 173–188 (1970).

Bulbs large, sometimes remaining joined by an elongated, branching rhizome. Perigone actinomorphic, funnelform-tubular; tube cylindrical or dilating gradually towards the throat; paraperigone absent. Stamens inserted in 2 rows below the perigone throat, biseriate in length. Stigma capitate (minutely 3-lobed). Seeds numerous per locule, flattened, winged, irregularly discoid, with a black or dark brown testa. $2n = 22$. About six spp., steppes and rocky mountains of the Turkestan region.

7. Tribe Pancratieae Salisb. (1866).

Leaves linear to lorate. Scape solid, spathe bracts 2, free. Flowers white, actinomorphic, long-tubed. Stamens connate proximally, sometimes forming a conspicuous corona, filaments often toothed. Seeds (sub)globose or flattened by compression, testa black.

23. *Pancratium* L.

Pancratium L., Sp. Pl. 1: 290 (1753); Baker, Handb. Amaryll.: 117–120 (1888); Bjørnstad, Norw. J. Bot.: 281–291 (1973), rev. African spp.

Leaves annual, sometimes hysteranthous, rarely finely pubescent. Spathe bracts 2, free or rarely fused into a tube. Flowers 1–10, generally fragrant; perigone crateriform, white; tube slender, cylindric below, funnelform above; tepals lanceolate, spreading. Staminal filaments united into a staminal cup, usually toothed between each free filament, free filament incurved or spreading outward. Stigma capitate to obscurely 3-lobed. Ovules numerous per locule. Seeds globose, subglobose, or wedge-shaped by compression, sometimes with a white elaisome. $2n = 22$. About 20 spp., Mediterranean, S Asia, sub-Saharan Africa. The genus is sorely in need of revisionary work.

24. *Vagaria* Herb.

Vagaria Herb., Amaryll. 226–227 (1837). Herklots, Plantsman 3: 226–229 (1981).

Leaves narrowly lorate, keeled. Scape ancipitous; spathe bracts 2, free. Flowers 6–8, not fragrant white, suberect; tube funnelform; tepals ascending. Stamens shortly connate proximally, the filaments broadly subulate and toothed on each side. Stigma capitate, entire, minute. Ovules 2–3 per locule. Seeds angled by compression. One or two spp., Mediterranean. Preliminary molecular data (M.F. Fay and A.W. Meerow, unpubl. data) suggest that this genus may belong in Galantheae, but greater breadth of sampling of that tribe and the genus *Pancratium* is needed.

8. Tribe Narcisseae Endl. (1836).

Leaves annual, sessile, linear to lorate, mesophyll with palisade. Scape solid; spathe bracts fused into a tube below the middle. Flowers often fragrant. Perigone mostly actinomorphic; tube short or long; paraperigone, when present, very conspicuous. Stigma capitate or shortly 3-lobed, ovules numerous per locule. Seeds phytomelanous, sometimes with an elaiosome. Two genera, centered in the Mediterranean region, but extending through Europe and Asia Minor.

25. *Narcissus* L.

Narcissus L., Sp. Pl.: 289 (1753); Fernandes, Bol. Soc. Brot. 25: 113–190 (1951), Royal Hort. Soc. Daffodil & Tulip Yearbook 1968: 37–66 (1968), rev. cult. spp.
Tapeinanthus Herb. (1837).

Leaves sometimes channeled adaxially to semiterete. Flowers 1-several, erect or declinate, pedicellate or sessile. Perigone white or yellow, sometimes bicolored, rarely green, actinomorphic to slightly zygomorphic (due to the stamens being declinate-ascending), crateriform; tube obconic to cylindrical; tepals spreading, ascending or reflexed, lanceolate; paraperigone funnel-shaped or cup-shaped, rarely reduced to a short callose rim. Stamens in 1–2 rows, attached at the throat or base of the tube, straight or declinate-ascending, anthers basifixed. Stigma minutely 3-lobed. Capsule dehiscent; seeds subglobose, angled by pressure, sometimes appendaged with an elaiosome, with a black testa. $2n = 14, 22, 26$ and numerous aneuploid and polyploid derivatives. Forty-60 spp., Europe to N Africa and W Asia. A number of species and a great many garden hybrids constitute one of the most important groups of spring-flowering bulbs in temperate climates. The species show a great deal of regional variation.

26. ***Sternbergia*** Walst. & Kit. Fig. 32F

Sternbergia Walst. & Kit. Pl. Rar. Hung. 2: 172, pl. 159 (1805); Feinbrun & Stearn, Bull. Res. Israel, Bot. 6D: 167–173 (1958).

Leaves lorate. Scape sometimes retained inside the bulb neck. Flowers 1–4, erect to suberect; perigone actinomorphic, yellow, funnelform; paraperigone absent. Stamens erect, inserted at the perigone throat; anthers dorsifixed, versatile. Stigma capitate or shortly 3-lobed. Capsule somewhat fleshy, barely dehiscent; seeds subglobose, with a fleshy elaiosome, the testa black. $2n = 22$. Seven or eight spp., from the Mediterranean to Eastern Europe, Asia Minor, the Caucasus, and Iran.

9. **Tribe Galantheae** Salisb. (1866).

Leaves annual, sessile, linear to lorate; mesophyll lacking palisade. Scape solid, except hollow in two species of *Leucojum*; spathe bracts 2, free or fused along 1 side and appearing as 1. Perigone actinomorphic; tepals free or very shortly united at their base, sometimes highly dimorphic. Paraperigone absent. Anthers dehiscing by apical pores or slits. Ovules numerous per locule. Capsule dehiscent. Seeds turgid, (sub)globose, usually with a black or brown phytomelanous testa, often with an elaiosome. Müller-Doblies and Müller-Doblies (1978b) treated this tribe as a subtribe of Narcisseae. Four genera, centered in the Mediterranean but extending through western and southern Europe to Asia Minor.

27. ***Galanthus*** L.

Galanthus L., Sp. Pl.: 288 (1753). Kamari, Bot. Jahrb. Syst. 103: 107–135 (1982), reg. rev.

Leaves sometimes plicate, straight, convolute or involute in vernation; enclosed by membranous sheath of cataphylls initially. Scape terete or compressed, 1-, rarely 2-flowered; spathe bracts 2, usually fused along one side and appearing monophyllous. Flowers nodding; tepals free, white, the inner whorl somewhat unguiculate, much shorter than the outer whorl and usually spotted green at the apex, base or both (spot rarely yellow or absent) and usually apically emarginate. Stamens very short; anthers basifixed, dehiscing by terminal pores. Stigma minutely capitate. Capsule somewhat fleshy. Seeds light brown, oblong, with a fleshy elaisome. $2n = 24$. About 17 spp., Europe through Asia Minor and the Caucacus. Traub (1963) recognized three subgenera on the basis of leaf vernation.

28. ***Leucojum*** L.

Leucojum L., Sp. Pl.: 289 (1753).

Leaves filiform to broadly lorate; enclosed by membranous sheath of cataphylls initially. Scape narrow and solid or broad and hollow, 1–5 flowered; spathe bracts 2, free or fused entirely on 1 side and appearing monophyllous. Flowers white, rarely tinged pink, bell-shaped or campanulate, nodding; tepals free or very shortly united proximally, often tipped green or yellow, only slightly subequal. Stamens filiform; anthers basifixed, dehiscing by terminal pores. Style filiform or strumose distally; stigma minutely capitate. Ovary sometimes with a lobed disk. Seeds subglobose with a black, brown or pale (phytomelan absent in the latter), often loosely adhering testa, with or without a caruncular elaiosome. $2n = 14, 16, 18, 22$. Ten-11 spp., Mediterranean through W Europe, N Africa to the Crimea and Armenia. Three to four subgenera have been recognized on the basis of scape morphology, style morphology, presence or absence of an appendage on the seed, and lobing of the ovary disk (Traub 1963; Crespo et al. 1996). These groups are supported by molecular data (Crespo et al. 1996).

29. ***Lapiedra*** Lag.

Lapiedra Lag., Nov. Sp. et Gen.: 14 (1816); Fernandes, Pl. Life 8: 60–68 (1952); Müller-Doblies and Müller-Doblies, Lagascalia 8: 13–23 (1978).

Leaves lorate, keeled. Scape terete, slender, 4–9 flowered; spathe bracts 2, free, linear. Flowers long-pedicellate, fragrant; perigone actinomorphic; tepals lanceolate, free, spreading, white with a green keel. Stamens free, filaments short; anthers sagittate, basifixed, dehiscing longitudinally. Stigma capitate. Seeds 1–2 per locule, turgid, subglobose, without an elaisome, with a lustrous black testa. $2n = 22$. Only one sp., *L. martinezii* Lag., Spain and Morocco. Müller-Doblies and Müller-Doblies (1978b) transferred this genus to Pancratieae, but this is not supported by molecular data (Crespo et al. 1996).

30. *Hannonia* Braun-Blanq. & Maire

Hannonia Braun-Blanq. & Maire, Bull. Soc. Hist. Nat. Afr. Nord. 22: 104–105 (1931).

Leaves linear, glaucous. Scape terete, hollow (some authors; solid according to Hannibal 1967), 2-flowered; spathe bracts 2, free, but appressed to the scape below their middle. Perigone tube much shorter than tepal segments; tepals lanceolate, white, keeled green on the abaxial surface. Stamens biseriate, free, inserted at the throat; filaments filiform; anthers dorsifixed, versatile. Stigma captitate. Ovules 5–6 per locule. One sp., *H. hesperidium* from Morocco. We are following Traub (1963) in placing this obscure monotypic genus in Galantheae pending molecular investigations.

10. Tribe Hippeastreae (Pax & Hoffmann) Hutch. (1931).

Scape usually hollow (solid in *Griffinia* and one *Hippeastrum* species). Flowers mostly funnelform, usually zygomorphic, some degree of paraperigonal development often present. Stamens free, not equal, usually declinate. Fruit a loculicidal capsule, seed primarily flattened, winged or D-shaped (globose and turgid in *Griffinia* and two species of *Hippeastrum*), phytomelanous (except in *Griffinia*). Eleven genera, entirely American, the majority in S America. The relationships among some of the genera appear so close as to warrant combination, while others remain problematic. The resolution of Hippeastreae will hopefully be clarified by molecular data. Zephyrantheae Salisb. (*Habranthus*, *Pyrolirion*, *Rhodophiala*, *Sprekelia*, *Zephyranthes*), Traubieae Moldenke (*Traubia*), and Griffineae Ravenna have been recognized as distinct tribes (Traub 1963; Ravenna 1974) or as subtribes (Müller-Doblies and Müller-Doblies 1996).

31. *Hippeastrum* Herb. Fig. 33

Hippeastrum Herb., App. Bot Reg. 31 (1821); Traub & Moldenke, Amaryllidaceae: Tribe Amarylleae: 107–145 rev. as *Amaryllis* (1949); Vargas, Herbertia 40: 112–134 (1984), rev. Peruvian spp. as *Amaryllis*.

Roots with thick velamen in the epiphytic species. Leaves usually annual, mostly hysteranthous, rarely persistent, sessile, rarely subpetiolate, generally more than 2 cm wide. Scape hollow (solid in *H. blumenavium* (C. Koch & Bouché ex Carr.) Sealy), terete to slightly compressed but never ancipitous; bracts 2, free. Flowers 2–13, usually large, funnelform, zygomorphic, declinate, tube usually short, paraperigonal fibriae or callose ridge present at throat. Stamens fasciculate, declinate-ascendent, of 4 lengths. Stigma trifid or shortly 3-lobed. Seeds dry, flattened, obliquely winged or irregularly discoid, rarely turgid and globose or subglobose, with a brown or black phytomelanous testa. $2n = 22$. Fifty-60 spp., Mexico and the West Indies to Argentina and Bolivia, the majority in eastern Brazil and the Andes of Peru and Bolivia. Several subgeneric or sectional schemes have been proposed for *Hippeastrum*, but do not seem very robust.

32. *Worsleya* Traub

Worsleya Traub, Herbertia 6: 118–119 (1939).

Roots thick, with velamen. Bulb mostly exposed, prolonged into long neck, leaves persistent, falcate, slightly fleshy, glaucous. Scape hollow; spathe bracts 2, free. Flowers 4–14, large and showy; perigone funnelform, zygomorphic, lilac- or amethyst-colored, tube short; paraperigone absent. Stigma capitate. Seeds black; D-shaped; the curved, outer edge thicker than the inner. $2n = 42$. Only one, lithophytic sp., *W. rayneri* (Hook. f.) Traub & Moldenke, known only from two to three populations in the Organ Mountains north of Rio de Janeiro, Brazil.

33. *Griffinia* Ker Gawler

Griffinia Ker Gawler, Edwards' Bot. Reg. 6: pl. 444 (1820); Traub & Moldenke, Amaryllidaceae: Tribe Amarylleae: 152–157 (1949); Ravenna, Pl. Life 25: 62–68 (1969); Pl. Life 30: 64–70 (1974).
Hyline Herb. (1841).

Leaves persistent or annual, subpetiolate to long petiolate; the laminae lorate, lanceolate, broadly elliptic or ovate; sometimes silver-spotted on the adaxial surface, thick or thin. Scape compressed, solid; spathe bracts 2, fused on one side. Flowers pedicellate, the pedicels short or long, suberect to declinate, sometimes nocturnal; perigone zygomorphic, white, blue, or violet; tube short or long; tepals narrow, one of the inner series dorsally declinate, the remaining five ascendent or lateral, somewhat unguiculate. Stamens 5 or 6, free, declinate-ascending. Stigma capitate. Ovules numerous per locule or reduced to 2. Seed globose, turgid, lacking phytomelan, the testa pink or white. $2n = 20$ (as so far known). Fifteen-20 spp., all endemic to Brazil in the understory of wet to seasonally dry forests. The genus can be very naturally divided into two subgenera, *Griffinia*, with often numerous blue or violet flowers, sometimes only 5 stamens, and few ovules per locule; and *Hyline* (Herb.) Rav., with few large, white, fragrant, long-tubed nocturnal flowers and numerous ovules per locule.

34. ***Rhodophiala*** Presl

Rhodophiala Presl, Bot. Bemerk.: 115 (1844); Traub & Moldenke, Amaryllidaceae: Tribe Amarylleae: 87–103 as *Amaryllis* (1949).
Rhodolirion R. A. Phil. (1857).

Leaves annual, linear, never more than 1 cm wide, flat. Scape hollow, spathe bracts 2, free or fused on 1 side. Flowers 1–7, funnelform, zygomorphic, on slender pedicels, the perigone tube short, paraperigonal bristles or scales sometimes present. Stamens fasciculate, declinate-ascendent, of 4 lengths. Stigma trifid to obscurely 3-lobed. Seeds obliquely winged, black. $2n = 18$. About 30 spp., the majority in Chile, extending to Bolivia, Argentina, Uruguay, and Brazil.

35. ***Zephyranthes*** Herb.

Zephyranthes Herb., App. Bot. Reg. 36 (1821); Sealy, J.R. Hortic. Soc. 62: 195–209 (1937); Flagg, unpubl. Ph..D dissertation, University of Virginia (1961).
Cooperia Herb. (1836).
Haylockia Herb. (1830).

Bulb generally small. Leaves annual or persistent, linear. Scape hollow, terete, rarely retained entirely inside the bulb, single-flowered, spathe bracts fused and tubular below the middle. Flower (sub)erect, actinomorphic, the tube short or long. Stamens usually biseriate in length. Stigma trifid or obscurely 3-lobed. Capsule papery with numerous or few, black flattened or compressed, D-shaped or wedge-shaped seeds. $2n = 24$, 48 most common, but $2n = 18$ is reported for *Z. pusilla* (Herb.) D. Dietr.; polyploid and aneuploid variation fairly frequent. About 50 spp., SE US to Argentina, including the West Indies. Several are cultivated ornamentals. Traub (1963) recognized the long-tubed, crepuscular species as a distinct subgenus *Cooperia*, which has also at times been proposed as a separate genus.

36. ***Habranthus*** Herb.

Habranthus Herb., Curtis' Bot. Mag., t. 2464 (1824); Sealy, J.R. Hortic. Soc. 62: 195–209 (1937); Uphof, Herbertia 13: 93–97 (1946).

Leaves annual or persistent, linear. Scape hollow, terete, single-flowered, spathe bracts fused and tubular below the middle. Flowers slightly zygomorphic, funnelform, the tube short; paraperigone of short scales or fimbriae sometimes present; tepals elliptic or ovate. Stamens fasciculate, declinate-ascendent, of 4 lengths. Stigma trifid or obscurely 3-lobed. Seeds obliquely winged, black. $2n = 24$, 48. Thirty-40 species, southern S America and Mexico and SW US. Essentially separated from *Zephyranthes* on the basis of perigone symmetry, though seeds of this genus appear to be always obliquely winged.

37. ***Sprekelia*** Heist.

Sprekelia Heist., Descr. Nov. Gen. Brunsvig.: 19 (1753).

Leaves annual, linear, bright green. Scape hollow, single-flowered; spathe bracts fused below. Flower extremely zygomorphic, red; tube very short; tepals narrow, one outer tepal dorsally erect and unguiculate, 2 inner tepals falcate and spreading laterally, the remaining tepals declinate and convolute proximally, forming a pseudolabellum that encloses the stamens and style. Stamens fasciculate, declinate ascending. Stigma trifid. Seeds black, D-shaped. $2n = >100$, widely variable. One or two spp., Mexico (purported to occur in S America, but undocumented). The genus is closely related to *Habranthus*, and has been hybridized with that genus as well as *Hippeastrum*. Similar perigone morphology occurs in two spp. of *Hippeastrum*, *H. angustifolium* Pax and *H. cybister* (Herb.) Benth. & Hook.

38. ***Pyrolirion*** Herb.

Pyrolirion Herb., App. Bot. Reg. 7: 37 (1821); Sealy, J.R. Hortic. Soc. 62: 195–209 (1947).

Leaves annual, linear to linear-lanceolate. Scape hollow, single-flowered, sometimes retained inside the bulb; spathe bracts fused below. Flower erect, actinomorphic; perigone funnelform, variously colored; tube cylindrical proximally, abruptly dilated at the throat; paraperigone of small scales sometimes present. Stamens subequal, inserted at or below the throat. Style 3-branched distally; stigmas spatulate. Seeds black, compressed, with a white raphe. $2n = 26$, 34, 51, 54. Five to ten species in Peru and Bolivia, at times treated as a subgenus of *Zephyranthes*.

39. *Placea* Miers ex Lindley

Placea Miers ex Lindley, Trav. Chile 2: 529 (1826); Edwards' Bot. Reg. 27, pl. 50 (1841). Traub and Moldenke, Amaryllidaceae: Tribe Amarylleae: 147–151 (1949).

Leaves annual, linear, bifacial. Scape hollow, spathe bracts 2, free. Perigone funnelform, slightly zygomorphic, the tepals very shortly connate at the base. Paraperigonal corona present, 3–6 lobed. Stamens inserted inside the corona, fasciculate, declinate-ascending, biseriate in length. Stigma capitate. Seeds flattened, D-shaped. Six spp., endemic to Chile.

40. *Traubia* Moldenke

Traubia Moldenke, Pl. Life 19: 55 (1963); Ravenna, Pl. Life 30: 30–33 (1974); Grau & Bayer, Mitt. Bot. Staatssamml. München. 30: 479–484 (1991).

Leaves linear, hysteranthous. Scape hollow, spathe bracts 2, free (rarely fused on one side). Flowers 1–5; perigone slightly zygomorphic, funnelform, tube very short; tepals lanceolate, white with purple abaxial keels and adaxial veins; paraperigone absent. Stamens biseriate; anthers dorsifixed but imperfectly versatile. Stigma capitate. Seeds flattened, round, with membranous edges, testa black. $2n = 16$. Only one sp., *T. modesta* Moldenke, Chile.

41. *Phycella* Lindley

Phycella Lindley, Edwards' Bot. Reg. 11: pl. 928 (1825); Traub & Moldenke, Amaryllidaceae: Tribe Amarylleae: 103–107 as *Amaryllis* subg. *Phycella* (1949); Ravenna, Pl. Life 28: 55–62 (1972).
Famatina Ravenna (1972).

Leaves annual, sometimes hysteranthous, linear, channeled adaxially. Scape terete, hollow; spathe bracts 2, free or slightly connate proximally. Flowers 2–7, pedicellate, declinate, or horizontal to the scape axis; perigone actinomorphic or slightly zygomorphic, tubular to funnelform-tubular, usually red, short paraperigone of scales or fimbriae usually present. Stamens fasciculate, of 4 lengths. Stigma capitate. Seeds black, membranously winged. $2n = 16$. About six spp., Chile, Argentina, and Uruguay.

12. Tribe Eucharideae (Pax) Hutch. (1934).

Leaves persistent or hysteranthous, petiolate with broad laminae. Scape solid, terete; spathe bracts 2, free. Capsule dehiscent. Seeds 1 to few per locule, globose, ellipsoid, or cylindical, sometimes angled by pressure, turgid, with copious endosperm and a lustrous (rarely dull) black, dark brown or blue, usually smooth, phytomelanous testa. Four genera, entirely American, mostly east Andean, below 2000 m, in the understory of rainforests.

42. *Eucharis* Planchon & Linden

Eucharis Planchon & Linden, Linden's Ann. Cat. Hort. 8: 3 (1852); Meerow, Ann. Mo. Bot. Gard. 76: 136–212 (1989), monogr.

Leaves persistent. Flowers white, sometimes fragrant, 5–10. Perigone actinomorphic, crateriform or campanulate, with a long, curved cylindrical tube abruptly dilating near the throat. Stamens always connate, at least proximally, prominent cylindrical staminal cup usually present, free filaments usually broadly subulate, straight. Stigma obtusely 3-lobed, the papillae unicellular. Ovules 2–20 per locule. Capsule leathery to slightly fleshy, green or bright orange; seeds ellipsoidal, with a black, blue or brown, smooth or rugose testa. $2n = 46, 68, 92, 138$. Seventeen spp. and several natural hybrids found in the understory of pre-montane and lower montane rainforest from Central America to Bolivia. The majority are endemic to the western Amazon basin and adjacent lower slopes of the eastern Andes. Meerow (1989) recognized two subgenera, *Eucharis* and *Heterocharis*, the former with 10 or fewer ovules per locule and leathery, orange capsules; the latter always with large, fragrant flowers and 16–20 ovules per locule.

43. *Caliphruria* Herb.

Caliphruria Herb., Edwards' Bot. Reg. 30 (misc. no. 83): 87 (1844); Meerow, Ann. Mo. Bot. Gard. 76: 212–218 (1989), monogr.

Leaves persistent or hysteranthous. Spathe bracts 2. Flowers white, never fragrant, 5–10. Perigone actinomorphic, funnelform, small; tube straight, dilating gradually from the base. Stamens only basally connate; filaments narrowly subulate, straight. Stigma obtusely trilobed, the papillae multicellular. Ovule (1-)2–7 per locule. Capsule greenish yellow. Seeds subglobose with a rugose testa. $2n = 46$. Four spp., three endemic to the Colombian Andes, one in Peru.

44. *Plagiliorion* Baker

Plagiliorion Baker, Gard. Chron. 2: 38 (1883); Meerow & Silverstone Sopkin, Brittonia 47: 426–431 (1995).

Leaves persistent. Spathe bracts 2. Flowers white, not fragrant, 10–40. Perigone zygomorphic, small; tube short, cylindrical, straight. Staminal cup cylindrical, free filaments declinate-ascending. Stigma capitate. Ovules 2 per locule, basal. Capsule leathery, yellowish green. Seeds subglobose, testa black, smooth. $2n = 46$. Only one sp., *P. horsmannii* Baker, endemic to the slopes of the Cauca Valley in Colombia.

45. *Urceolina* Reichb., nom. cons.

Urceolina Reichb. (nom. cons.), Consp.: 61 (1828).
Pseudourceolina Vargas (1960).

Leaves hysteranthous. Spathe bracts 2. Flowers yellow or orange, tipped green, not fragrant, 5–10. Perigone actinomorphic, urceolate; tube long, narrowly cylindrical, straight, abruptly dilating near the throat. Stamens only basally connate; filaments filiform. Stigma capitate. Ovules 10–20 per locule. Capsule greenish yellow at dehiscence. Seed cylindrical, slightly curved, testa smooth, black. Five to seven spp., understory of lower and mid-montane rain-forests of the south-central Peruvian Andes.

13. Tribe Hymenocallideae (D. & U.M-D.) Meerow (1998).[1]

Scape compressed to ancipitous. Stigma capitate. Seeds fleshy, the testa thick, spongy, with a well-developed vascular system; embryo with starch; phytomelan absent except in *Leptochiton*. Three genera, entirely American.

46. *Hymenocallis* Salisb. Figs. 34A, 36B

Hymenocallis Salisb., Trans. Hort. Soc. 1: 338 (1812); Sealy. J.R., Kew Bull. 2: 201–240 (1954), rev.; Bauml, unpubl. M.S. thesis, Cornell University (1979), reg. rev.

Leaves annual or persistent, sessile, linear to broadly lorate, rarely petiolate and with an elliptic lamina. Scape compressed or ancipitous. Flowers usually sessile, erect or suberect, crateriform, actinomorphic, white, fragrant. Stamens always connate below into a conspicuous funnelform or rotate pure white corona, the long free filaments straight but spreading outward. Pollen orange, the grains very large and auriculate. Stigma capitate. Ovules 2–10 per locule. Seed sometimes polyembryonic. $2n = 46, 44, 40$, plus numerous polyploid and aneuploid derivitives. About 50 spp., SE US to NE South America, including the West Indies, most prominently in Meso-America.

47. *Leptochiton* Sealy Fig. 34B–D

Leptochiton Sealy, Curtis' Bot. Mag. 160: t. 9491 (1937).

Leaves annual, linear, lax, fragile, keeled, appearing before the flowers. Scape ancipitous, single-flowered, the bud subtended by 3, linear-lanceolate bracts but not enclosed by spathe bracts. Flower large, sessile, suberect, fragrant. Perigone crateriform, white or yellow; the tube long, green, and cylindrical for most of its length. Staminal cup large, wide-spreading, rotate or campanulate, laciniate or coarsely lobed at the margin, striped green and/or yellow along the filamental trace; free filament short and incurved. Pollen yellow, the grains very large and auriculate. Stigma capitate. Ovules 16–20 per locule. Seeds with a brown phytomelanous testa. Two or three spp. of ephemeral geophytes inhabitating seasonally very dry lowland deciduous forest and scrub of SW Ecuador and NW Peru.

48. *Ismene* Salisb. Fig. 34E,F

Ismene Salisb., Trans. Hort. Soc. 1: 342 (1812).
Elisena Herb. (1837).
Pseudostenomesson Velarde (1949).

Scape ancipitous. Leaves annual, tightly sheathing proximally and forming a long, conspicuous pseudostem, appearing before the flowers, somewhat succulent. Perigone crateriform, more or less funnelform, or funnelform-tubular, white, yellow, or green, zygomorphic or actinomorphic. Staminal cup always present. Ovules usually 2 per locule and basal. Seed green, without phytomelan. Three subgenera, *Ismene* (perigone white or yellow, actinomorphic, staminal cup striped green along the filamental trace, free filaments incurved, shorter than the cup, pollen grain auriculate) *Elisena* (perigone white, zygomorphic, free filaments declinate-ascending and longer than the cup, pollen grain nonauriculate), and *Pseudostenomesson* (perigone green, tubular, pollen-grain non-auriculate) formerly treated as subgenera of *Hymenocallis* by Traub (1962) can be recognized on the basis of perigone morphology. $2n$ = most commonly 46, but up to 110. Ten to 15 spp., Andes of S Ecuador to Bolivia.

[1] Tribe Hymenocallideae (D. & U.M-D.) Meerow, stat. nov. Eucharideae (Pax) Hutch. subtribe Hymenocallidinae D. & U.M-D., Feddes Repert. 107: S.c. 1–9 (1996).

14. Tribe Stenomesseae Traub (1963).

Scape solid (sometimes with a narrow, distal lumen), ancipitous or terete; spathe bracts 2, usually free. Staminal cup usually present, or stamens at least basally connate in most genera. Capsule usually turbinate, brown and papery at dehiscence. Seeds numerous (few in *Eucrosia dodsonii* Meerow & Dehgan), flattened, obliquely winged, with black or brown phytomlanous testas. $2n = 46$. Eight genera endemic to the Andean region of S America, except for one sp. of *Phaedranassa* described from Costa Rica.

49. *Stenomesson* Herb.

Stenomesson Herb., App. Bot. Reg. 7: 40 (1821).

Leaves annual, usually linear or lorate and sessile, flat in vernation, rarely (sub)petiolate with an expanded lamina, in which case revolute in vernation. Flowers actinomorphic, funnelform-tubular, tubular, campanulate or ventricose, suberect to pendulous, without fragrance, variously colored but never white. Perigone tube longer than the limb segments, abruptly dilated at or near its midpoint and constricting slightly at the throat. Staminal cup usually conspicuous, exserted from the perigone throat, rarely reduced, often toothed between each free filament; filaments sometimes appearing free, in which case inserted below the throat but with a toothlike process in between each that is fused to the tube. Anthers sometimes briefly erect at anthesis and forming a pseudotube around the style. Stigma mostly capitate, sometimes shortly 3-lobed. The genus is in critical need of reexamination, and may be paraphyletic or polyphyletic. $2n = 46$. Thirty-five to 40 species ranging from the southern Colombian Andes to northern Chile, most of them endemic to montane Peru.

50. *Paramongaia* Velarde

Paramongaia Velarde, Lilloa 17: 489–491 (1949).

Leaves annual, narrowly lorate, keeled, deeply glaucous, somewhat succulent, emerging just before the flowers. Scape ancipitous, 1-, rarely 2-flowered; spathe bracts united below into a tube. Flowers large, fragrant, subsessile, held perpendicular to the scape axis or slightly cernuous; perigone crateriform-campanulate, yellow; tube long. Staminal cup conspicuous, funnelform-campanulate, coarsely 6-lobed with the free, incurved filaments inserted below the rim of the corona between the lobes. Stigma 3-lobed. Seeds with a black testa. Only one sp., *P. weberbaueri* Velarde, W Andes of Peru and Bolivia. The species has a very short season of active growth.

51. *Pamianthe* Stapf

Pamianthe [Stapf], Gard. Chron. ser. 3.93: 106 (1933); Stapf, Curtis' Bot. Mag. 156: sub t. 9315 (1933); Meerow, Brittonnia 36: 18–25 (1984). The actual first publication of the name was an anonymous reference in Gard. Chron., 5 months before the Bot. Mag. citation.

Epiphytic. Bulb mostly aerial, prolonged above into a long neck. Roots with velamen. Leaves persistent, broadly or narrowly lorate, thin, keeled. Scape ancipitous; spathe bracts 3–4. Flowers large and shortly pedicellate or small and very long-pedicellate, fragrant; perigone crateriform-campanulate; tube long. Staminal cup conspicuous, funnelform-campanulate, coarsely 6-lobed with the free, incurved filaments inserted at the rim of the cup between the lobes. Stigma obtusely 3-lobed. Capsule slightly woody. Seeds with a brown testa. $2n = 46$. Two spp., Peru, Ecuador, and Bolivia.

52. *Phaedranassa* Herb.

Phaedranassa Herb., Edwards' Bot. Reg. 31: misc. 16 (1845); Meerow, Flora of Ecuador 41: 32–42 (1990).

Leaves mostly hysteranthous, petiolate with lanceolate to elliptic laminae, sometimes deeply glaucous. Scape terete, solid but developing a narrow-diameter lumen in the distal half. Flowers actinomorphic, mostly tubular, (sub)pendulous, without fragrance, usually pink or red proximally and apically green. Tube constricted at its juncture with the ovary, much shorter than the tepals. Tepals concrescent for most of their length, proximally keeled. Stamens proximally fasciculate and closing off the perigone throat, free, of 4 different lengths. Stigma capitate. Seeds with a black or brown testa. $2n = 46$. Nine spp., six endemic to Ecuador; two spp. known only from Colombia and Costa Rica. Many are known only from the type localities, forming large but local populations along roadsides and in other disturbed areas, chiefly above 2000 m.

53. *Rauhia* Traub

Rauhia Traub, Pl. Life 13: 73–75 (1957).

Bulbs large. Leaves annual, shortly and widely petiolate, ovate to spatulate, very carnose, epidermis with micropapillae, sometimes tessellated with silver when juvenile. Scape terete, solid, but developing a narrow-diameter lumen in the distal half. Flowers green, numerous, pedicellate; perigone tubular, actinomorphic, or zygomorphic; tube longer than tepal segments. Stamens free, inserted below the throat of the tube, of 4 different lengths, straight and barely exserted from the perigone, or declinate-ascending and long-exserted. Stigma capitate. Seeds brown or black. $2n = 46$. Two to three spp., only in seasonally dry, open woods and rocky slopes of the middle Marañon and Utcumbamba river valleys in northern Peru at 1000–1500 m. The genus, notable for its large, fleshy leaves, is closely related to *Eucrosia* and *Phaedranassa*.

54. *Eucrosia* Ker Gawler

Eucrosia Ker Gawler, Edwards' Bot. Reg. 3: t. 207 (1817); Meerow, Syst. Bot. 12: 460–492 (1987), monogr.
Callipsyche Herb. (1842).

Leaves hysteranthous, long-petiolate, with ovate or elliptic (rarely lanceolate) laminae. Scape terete, solid but developing a narrow-diameter lumen in the distal half. Flowers without fragrance, pedicellate; perigone showy, zygomorphic; tube generally 1/2–1/5 the length of the tepals, rarely longer. Stamens usually long-exserted, declinate, and much exceeding the perigone in length, variously connate basally or rarely free, of 4 different lengths; globose nectar glands usually present at the base of each filament. Stigma capitate. $2n = 46$ and 68. Seven spp., restricted to western Ecuador and northwestern Peru, primarily in seasonally dry to xeric lowland habitats (*E. dodsonii* is adapted to cloud forest understory). The genus is considerably polymorphic in floral morphology, but most of the species have long, declinate stamens and androecial nectar glands at their base.

55. *Pucara* Ravenna

Pucara Ravenna, Ann. Mus. Hist. Nat. Valparaíso 5: 85–89 (1972).

Leaves lanceolate, possibly subpetiolate. Flowers numerous, white, funnel-shaped, ascending; perigone funnelform, the tube short. Stamens connate, forming a 3-lobed corona, each lobe concave, toothed distally; free filaments biseriate, 3 opposite the outer tepals inserted at the base of the corona, 3 opposite the inner tepals inserted between the teeth of each coronal lobe. Stigma obtusely 3-lobed. Only one sp., *P. leucantha* Rav., endemic to northern Peru.

56. *Mathieua* Klotzsch

Mathieua Klotzsch, Allg. Gartenzeitung 21: 337–338 (1853); Meerow, Taxon 36: 566–572 (1987).

Leaves long petiolate, lamina ovate. Flowers 2–4, ca. 4 cm long; perigone ventricose-campanulate, white, tube cylindrical proximally, abruptly dilated near the throat. Stamens shortly connate below, edentate. Stigma obtusely 3-lobed. Ovules compressed, numerous per locule. Fruit and seed unknown. Only *M. galanthoides* Klotzsch from a fragmentary type, the remains of a collection from dry forests of Piura, Peru. The genus is presumed extinct. The petiolate leaves would suggest some relationship to *Eucrosia*, *Phaedranassa*, and *Rauhia*.

15. Tribe Eustephieae (Pax) Hutch. (1934).

Bulbs frequently with a long neck. Leaves linear, annual, usually channeled, with well-developed palisade in the mesophyll. Scape solid, compressed to ancipitous; spathe bracts 2, free. Flowers pedicellate; perigone tubular to funnelform-tubular, tube short or long. Stamens biseriate in length, free or variously connate; free filaments often subulate or laterally toothed. Stigma trifid to obscurely 3-lobed. Capsule deeply 3-lobed, barrel-shaped, dehiscent. Seeds numerous per locule, flattened, D-shaped or discoid, winged, with a black testa. Three genera, south-central Andes of Peru, Argentina and Bolivia, usually at high elevation. The tribe has been traditionally allied with Stenomesseae, but preliminary molecular data strongly suggest either sister group status with Hippeastreae or a position ancestral to all other Neotropical tribes (unpubl. data).

57. *Chlidanthus* Herb.

Chlidanthus Herb., App. Bot. Reg. 46 (1821); Ravenna, Pl. Life 30: 71–73 (1974).
Castellanoa Traub (1953).
Sanmartina Traub (1951).

Leaves sometimes with scabrous margins. Flowers suberect to declinate; pedicels short or long; perigone funnelform or funnelform-tubular, actinomorphic or slightly zygomorphic, variously colored but never white, sometimes fragrant; tube much longer than the limb. Stamens never exserted from the perigone, sometimes very shortly conate at the base, strongly biseriate, the filaments subulate and sometimes toothed, fused to the perigone proximally. Stigma trifid. Seeds D-shaped, testa black. Six spp., S Peru through Bolivia to NW Argentina.

58. *Eustephia* Cav.

Eustephia Cav., Icon 3: 20, pl. 238 (1794).

Flowers declinate to pendent; perigone tubular, variously colored but usually tipped green; tube short. Stamens free or proximally connate, inserted at the perigone throat; the free filaments winged below, but always with a slender tooth on each side above the middle. Stigma shortly 3-lobed. Seeds discoid, testa black. $2n = 46$. Four spp., southern Peru to Bolivia.

59. *Hieronymiella* Pax

Hieronymiella Pax, Bot. Jahrb. 11: 327 (1890); Castellanos, Physis 11: 494 (1935); Hunziker, Kurtziana 5: 343-367 (1969), rev.
Eustephiopsis R. E. Fries (1905).
Androstephanos Fernández Casas & Lara (1983).

Flowers suberect to declinate; perigone funnelform or tubular, white, yellow, rose, or purple, sometimes fragrant; tube short or long. Stamens broadly connate into a conspicuous 6-lobed corona, but with a narrow, longitudinal aperture between each stamen; each coronal lobe bifid at the rim, with the short, linear free filament inserted between the teeth of each lobe, or with the anthers subsessile below the rim. Stigma mostly trifid, rarely obtusely 3-lobed. Seeds discoid, evenly or irregularly winged, the testa black or dark brown. $2n = 42, 54, 56, 60$. Four to six spp., mostly in Argentina, one in Bolivia, chiefly at high elevation.

Selected Bibliography

Aksenova, L.M., Sedova, E.D. 1981. Bulb structure and morphogenesis of some representatives of the family Amaryllidaceae. Ukr. J. Bot. 4: 41-45 (in Russian).

Arber, A. 1937. Studies in flower structure III. On the corona and androecium in certain Amaryllidaceae. Ann. Bot. II, 1: 293-304.

Arroyo, S.C. 1981. Systematic anatomical studies on Amaryllidaceae including morphological, ecological, cytological and phytogeographical considerations. Ph.D Thesis. Reading: University of Reading, 238 pp.

Arroyo, S.C. 1984. Contribucíon al conocimiento de los bulbos de Amaryllidaceae. Kurtziana 17: 55-70.

Arroyo, S.C., Cutler, D.F. 1984. Evolutionary and taxonomic aspects of the internal morphology in Amaryllidaceae from South America and Southern Africa. Kew Bull. 39: 467-498.

Artyushenko, Z.T. 1989. Aspects of research on Amaryllidaceae Jaume. Herbertia 45: 131-137.

Baker, J.G. 1881. A synopsis of the known species of *Crinum* L. Gard. Chron. 1881(1): 763.

Barrett, S.C.H., Lloyd, D.G., Arroyo, J. 1995. Stylar polymorphisms and the evolution of heterostyly in *Narcissus* (Amaryllidaceae). In: Lloyd, D.G., Barrett, S.C.H. (eds.) Floral biology: studies on floral evolution in animal-pollinated plants. New York: Chapman and Hall, pp. 339-376.

Bauml, J.A. 1979. A study of the genus *Hymenocallis* (Amaryllidaceae) in Mexico. Unpubl. M.S. Thesis. Ithaca: Cornell University.

Behnke, H.D. 1981. See general references.

Bell, W.D. 1977. More potentials in *Amaryllis* breeding. Plant Life 33: 65-69.

Bose, S., Flory, W.S. 1963. A study of phylogeny and of karyotype evolution in *Lycoris*. Nucleus 6: 141-156.

Brown, W.V. 1951. Apomixis in *Zephyranthes texana* Herb. Am. J. Bot. 38: 697-702.

Broyles, S.B., Wyatt, R. 1991. The breeding system of *Zephryanthes atamasco* (Amaryllidaceae). Bull. Torrey Bot. Club 118: 137-140.

Cheadle, V.I. 1969. Vessels in Amaryllidaceae and Tecophilaeaceae. Phytomorphology 19: 8-16.

Crespo, M.B.M., Lledó, D., Fay, M.F., Chase, M.W. 1996. Molecular phylogeny of *Leucojum* based on ITS sequences. Am. J. Bot. 6 (Suppl.): 149 (Abstract).

Dafni, A., Werker, E. 1982. Pollination ecology of *Sternbergia clusiana* (Ker-Gawler) Spreng. (Amaryllidaceae). New Phytol. 91: 571-577.

Dahlgren, R.M.T., Clifford, H.T. 1982. See general references.

Dahlgren, R., Rasmussen, F.N. 1983. Monocotyledon evolution: characters and phylogenetic analysis. Evol. Biol. 16: 255-395.

Dahlgren, R.M.T. et al. 1985. See general references.

Daumann, E. 1970. Das Blütennektarium der Monocotyledonen unter besonderer Berücksichtigung seiner systematischen und phylogenetischen Bedeutung. Feddes Repert 80: 463-590.

Davis, G.L. 1966. See general references.

Di Fulvio, T.E. 1973. Contribucíon al conocimiento cariológico de Amaryllidaceae. Estudio cromosómico en *Hieronymiella* y otras genéros afines. Kurtziana 7: 117-131.

Duke, J.A., Vasquez, R. 1994. Amazonian ethnobotanical dictionary. Boca Raton: CRC Press.

Dutt, B.S.M. 1962. A contribution to the life history of *Crinum defixum*. In: Plant embryology - a symposium. New Dehli: Council of Scientific and Industrial Research, pp. 37-48.

Erdtman, G. 1952. See general references.

Faegri, K., van der Pijl, L. 1979. Principles of pollination ecology, 2nd edn. Oxford: Pergamon Press.

Fay, M.F., Chase, M.W. 1996. Resurrection of Themidaceae for the *Brodiaea* alliance, and recircumscription of Alliaceae, Amaryllidaceae and Agapanthoideae. Taxon 45: 441-451.

Fay, M.F., Meerow, A.W., Hartwell, J., Chase, M.W. 1995. Molecular systematics and biogeography of the Amaryllidaceae. Am. J. Bot. 82 (Suppl.): 129 (Abstract).

Fernandes, A. 1942. Summary of work on cytology of *Narcissus* L. Herbertia 9: 126-133.

Fernandes, A. 1967. Contribution à la connaissance de la biosystématique de quleques espèces du genre *Narcissus* L. Portug. Acta Biol. 9: 1-44.

Fernandes, A. 1968. Improvements in the classification of the genus *Narcissus* L. Plant Life 24: 51-57.

Flory, W.S. 1939. Parthenogensis in Zephyrantheae. Herbertia 6: 196–202.
Flory, W.S. 1967. *Sprekelia* chromsomes. Plant Life 23: 54–57.
Flory, W.S. 1968. Chromosome diversity in species, and in hybrids, of tribe Zephyrantheae. Nucleus 11: 79–95.
Flory, W.S. 1976. Distribution, chromosome number and types of various species and taxa of *Hymenocallis*. Nucleus 19: 204–227.
Flory, W.S. 1977. Overview of chromosomal evolution in the Amaryllidaceae. Nucleus 20: 70–88.
Flory, W.S., Schmidhauser, T.F. 1957. Mitotic chromosome numbers in *Hymenocallis* with a consideration of factors possibly affecting numbers and karyotypes. Genetics 42: 369–370.
Frölich, D., Barthlott, W. 1988. Mikromorphologie der epicuticularen Wachse und das System der Monokotylen. Trop. Subtrop. Pflanzenwelt (Akad. Wiss. Lit. Mainz) 63: 1–135.
Ghosh, S., Shivanna, K.R. 1984. Structure and cytochemistry of the stigma and pollen-pistil interactions in *Zephyranthes*. Ann. Bot. II, 53: 91–105.
Gibbs, R.D. 1974. Chemotaxonomy of flowering plants, vol. I. Montréal: McGill-Queens Univ. Press.
Goldblatt, P. 1976. Chromosome cytology of *Hessea*, *Strumaria*, and *Carpolyza* (Amaryllidaceae). Ann. Mo. Bot. Gard. 63: 314–320.
Goldblatt, P. 1978. An analysis of the flora of southern Africa: its characteristics, relationships, and origins. Ann. Mo. Bot. Gard. 65: 369–436.
Gouws, J.B. 1949. Karyology of some South African Amaryllidaceae. Plant Life 5: 54–81.
Grant, V. 1983. The systematic and geographical distribution of hawkmoth flowers in the temperate North American flora. Bot. Gaz. 144: 439–449.
Grau, J., Bayer, E. 1991. Zur systematischen Stellung der Gattung *Traubia* Moldenke (Amaryllidaceae). Mitt. Bot. Staatssamml. Münch. 30: 479–484.
Hannibal, L.S. 1967. A primitive amaryllid – *Hannonia hesperidium*. Plant Life 23: 143–144.
Hegnauer, R. 1963. See general references.
Heslop-Harrison, Y., Shivanna, K.R. 1977. The receptive surface of the angiosperm stigma. Ann. Bot. II, 41: 1233–1258.
Howell, G., Prakash, N. 1990. Embryology and reproductive ecology of the Darling lily, *Crinum flaccidum* Herbert. Aust. J. Bot. 38: 433–444.
Huber, H. 1969. See general references.
Inariyama, S. 1931. Cytological studies in the genus *Lycoris*. Bot. Mag. (Tokyo) 45: 11–26.
Inariyama, S. 1933. Phylogeny of *Lycoris* from the karyological point of view. Rep. Jpn. Sci. Congr. 8.
Inariyama, S. 1937. Karyotype studies in Amaryllidaceae I. Sci. Rep. Tokyo Bunrika Daigaku, Sect. B 3(52): 95–113.
Inariyama, S. 1953. Cytological studies in *Lycoris*. Seiken Ziho 6: 5–10.
Ising, G. 1970. Evolution of karyotypes in *Cyrtanthus*. Hereditas 65: 1–28.
Johnson, S.D., Bond, W.J. 1994. Red flowers and butterfly pollination in the fynbos of South Africa. In: Arianoutsou, M., Groves, R.H. (eds.) Plant-animal interactions in mediterranean-type ecosystems. Dordrecht: Kluwer, pp. 137–148.
Jones, K., Smith, J.B. 1967. Chromosome evolution in the genus *Crinum*. Caryologia 20: 163–179.
Koshimizu, T. 1930. Carpobiological studies of *Crinum asiaticum* L. var. *japonicum* Bak. Mem. Coll. Sci. Kyoto Imp. Univ., Ser. B., Biol. 5: 183–227.
Kurita, S. 1986. Variation and evolution in the karyotype of *Lycoris*, Amaryllidaceae I. General karyomorphological characteristics of the genus. Cytologia 51: 803–815.
Kurita, S. 1987a. Variation and evolution on the karyotype of *Lycoris*, Amaryllidaceae II. Karyotype analysis of ten taxa, among which seven are native in China. Cytologia 52: 19–40.
Kurita, S. 1987b. Variation and evolution in the karyotype of *Lycoris*, Amaryllidaceae III. Intraspecific variation in the karyotype of *L. traubii* Hayward. Cytologia 52: 117–128.
Kurita, S. 1987c. Variation and evolution in the karyotype of *Lycoris*, Amaryllidaceae IV. Cytologia 52: 137–149.
Kurita, S. 1989. Variation and evolution of the karyotype of *Lycoris*, Amaryllidaceae V. Chromosomal variation in L. sanguina Maxim. Plant Species Biol. 4: 47–60.
Lakshmi, N. 1978. Cytological studies in two allopolyploid species of the genus *Hymenocallis*. Cytologia 43: 555–563.
Loubser, J., Zietsman, P.C. 1994. Rock painting of postulated *Brunsvigia* sp., (Amaryllidaceae) at Thaba Bosiu, western Lesotho. S. Afr. J. Sci. 90: 611–612.
Markötter, E.I. 1936. Die lewensgeskiedenis van sekere geslagte van die Amaryllidaceae. Ann. Univ. Stellenbosch XIV.A.2: 1–84.
Meerow, A.W. 1984. Karyotype evolution in the Amaryllidaceae. Herbertia 40: 139–154.
Meerow, A.W. 1985. The evolutionary significance of pancratioid floral morphology in the Amaryllidaceae. Am. J. Bot. 72: 962. (Abstract)
Meerow, A.W. 1987a. A monograph of *Eucrosia* (Amaryllidaceae). Syst. Bot. 12: 460–492.
Meerow, A.W. 1987b. Chromosome cytology of *Eucharis*, *Caliphruria*, and *Urceolina*. Am. J. Bot. 74: 1559–1575.
Meerow, A.W. 1987c. The identities and systematic affinities of *Mathieua* Klotzsch and *Plagiolirion* Baker (Amaryllidaceae). Taxon 36: 566–572.
Meerow, A.W. 1989. A monograph of the Amazon lilies, *Eucharis* and *Caliphruria* (Amaryllidaceae). Ann. Mo. Bot. Gard. 76: 136–220.
Meerow, A.W. 1995. Towards a phylogeny of the Amaryllidaceae. In: Rudall, P.J., Cribb, P.J., Cutler, D.F., Humphries, C.J. (eds.) Monocotyledons: systematics and evolution. Royal Botanic Gardens, Kew, pp. 169–179.
Meerow, A.W., Dehgan, B. 1985. The auriculate pollen grain of *Hymenocallis quitoensis* Herb. (Amaryllidaceae) and its systematic implications. Am. J. Bot. 72: 540–547.
Merrow, A.W., Dehgan, B. 1988. Pollen morphology of the Eucharideae (Amaryllidaceae). Am. J. Bot. 75: 1857–1870.
Meerow, A.W., Dehgan, N.B., Dehgan, B. 1986. Pollen tetrads in *Stenomesson elwesii* (Amaryllidaceae). Am. J. Bot. 73: 1642–1644.
Morton, J.K. 1965. The experimental taxonomy of the West African species of *Pancratium* L. (Amaryllidaceae). Kew Bull. 19: 337–347.
Müller-Doblies, D. 1977. Über den geometrischen Zusammenhang der monochasialen Verzweigungen am Beispiel einiger Liliifloren. Ber. Dtsch. Bot. Ges. 90: 351–362.
Müller-Doblies, D., Müller-Doblies, U. 1972. *Galanthus* ist doch sympodial gebaut. Ber. Dtsch. Bot. Ges. 84: 665–682.
Müller-Doblies, D., Müller-Doblies, U. 1978a. Bulbs and morphology: *Ungernia*. Lagascalia 8: 13–23.
Müller-Doblies, D., Müller-Doblies, U. 1978b. Studies on tribal systematics of Amaryllidaceae 1. The systematic position of *Lapiedra* Lag. Lagascalia 8: 13–23.
Müller-Doblies, D., Müller-Doblies, U. 1985. De Liliifloris notulae 2. De taxonomia subtribus Strumariinae (Amaryllidaceae). Bot. Jahrb. Syst. 107: 17–47.

Müller-Doblies, D., Müller-Doblies, U. 1996. Tribes and subtribes and some species combinations in Amaryllidaceae J. St.-Hil. emend. R. Dahlgren & al. 1985. Feddes Rep. 107: S.c. 1-9.

Naranjo, C.A., Andrada, A.B. 1975. El cariótipo fundamental en el genéro *Hippeastrum* Herb. (Amaryllidaceae). Darwiniana 19: 566-582.

Nordal, I., Duncan, T. 1984. A cladistic analysis of *Haemanthus* and *Scadoxus*. Nord. J. Bot. 4: 145-153.

Nordal, I., Rorslett, B., Laane, M.M. 1977. Species delimitation within the *Crinum ornatum* group (Amaryllidaceae) in East Africa. Norw. J. Bot. 24: 195-212.

Pax, F. 1887. Amaryllidaceae. In: Engler, A., Prantl, K. (eds.) Die natürlichen Pflanzenfamilien II, 5. Leipzig: W. Engelmann, pp. 97-124.

Pettit, G.R., Pettit III, G.R., Backhaus, R.A., Boyd, M.R., Meerow, A.W. 1993. Antineoplastic agents 256. Cell growth-inhibitory isocarbostyrils from *Hymenocallis*. J. Nat. Prod. 56: 1682-1687.

Pettit, G.R., Pettit III, G.B., Groszek, G., Backhaus, R.A., Doubek, D.L., Meerow, A.W. 1995. Antineoplastic agents 301. An investigation of the genus *Hymenocallis* Salisbury (Amaryllidaceae). J. Nat. Prod. 58: 756-759.

Prillieux, E. 1858. De la structure et du mode formation des graines bulbiformes de quelques amaryllidées. Ann. Sci. Nat. 4: 97-104.

Raina, S.N. 1978. Genetic mechanisms underlying evolution in *Crinum*. Cytologia 43: 575-580.

Raina, S.N., Khoshoo, T.N. 1972. Cytogenetics of tropical bulbous ornamentals. IX. Breeding systems in *Zephyranthes*. Euphytica 21: 317-323.

Raven, P.H., Axelrod, D.I. 1974. Angiosperm biogeography and past continental movements. Ann. Mo. Bot. Gard. 61: 539-673.

Ravenna, P. 1974. Griffiniae Rav., tribus nova. Plant Life 30: 65.

Reid, C., Dyer, R.A. 1984. A review of the Southern African species of *Cyrtanthus*. La Jolla: American Plant Life Society.

Rendle, A.B. 1901. The bulbiform seeds of certain Amaryllidaceae. J. R. Hortic. Soc. 26: 89-96.

Satô, D. 1938. Karyotype evolution and phylogeny. IV. Karyotype in Amaryllidaceae with special reference to SAT chromosomes. Cytologia 9: 203-242.

Schlimbach, H. von 1924. Beiträge zur Kenntnis der afrikanischen Flora V. Bull. Herb. Boiss. 4: 809-846.

Schultes, R.E. 1976. Hallucinogenic plants. New York: Golden Press.

Schultes, R.E., Raffauf, R.F. 1990. The healing forest. Portland: Dioscorides Press.

Singh, V. 1972. Floral morphology of the Amaryllidaceae. 1. Subfamily Amaryllidoideae. Can. J. Bot. 50: 1555-1565.

Snijman, D.A. 1992. Systematic studies in the tribe Amaryllideae (Amaryllidaceae). Ph.D. Dissertation. Cape Town: University of Cape Town.

Snijman, D.A. 1994. Systematics of *Hessea*, *Strumaria* and *Carpolyza* (Amaryllideae: Amaryllidaceae). Contrib. Bolus Herb. 16: 1-162.

Snijman, D.A., Linder, H.P. 1996. Phylogenetic relationships, seed characters, and dispersal system evolution in Amaryllideae (Amaryllidaceae). Ann. Mo. Bot. Gard. 83: 362-386.

Snijman, D.A., Williamson, G. 1994. A taxonomic re-assessment of *Ammocharis herrei* and *Cybistetes longifolia* (Amaryllidaceae). Bothalia 24: 127-132.

Tillich, H-D. 1995. Seedlings and systematics in monocotyledons. In: Rudall, P.J., Cribb, P.J., Cutler, D.F., Humphries, C.J. (eds.) Monocotyledons: systematics and evolution. Royal Botanic Gardens, Kew, pp. 303-352.

Tomita, K. 1931. Über die Entwicklung des nackten Embryos von *Crinum latifolium* L. Sci. Rep. Tohoku Imperial Univ., IV, Biol. 6: 163-169.

Traub, H.P. 1962. Key to the subgenera, alliances and species of *Hymenocallis*. Plant Life 18: 55-72.

Traub, H.P. 1963. Genera of the Amaryllidaceae. La Jolla: American Plant Life Society.

Traub, H.P. 1966. Polyembryony in *Hymenocallis mexicana*. Plant Life 22: 49.

Traub, H.P., Moldenke, H.N. 1949. Amaryllidaceae: tribe Amarylleae. La Jolla: American Plant Life Society.

Vargas, F.C. 1981. Plant motifs on Inca ceremonial vases from Peru. Bot. J. Linn. Soc. 82: 313-325.

Vogel, S. 1963. Das sexuelle Anlockungsprinzip der Catasetinen- und Stanhopeen Blüten und die wahre Funktion ihres sogenannten Futtergewebes. Oesterr. Bot. Z. 110: 308-337.

Vogel, S. 1990. The role of scent glands in pollination. New Dehli: Amerind.

Vosa, C.G. 1986. Chromosome studies in the genus *Gethyllis* (Amaryllidaceae). Caryologia 39: 251-257.

Wahlstrom, R., Laane, M.M. 1979. Chromosome analyses in African *Crinum* spp. Hereditas 91: 183-206.

Walker, J.W., Doyle, J.A. 1975. The bases of angiosperm phylogeny: palynology. Ann. Mo. Bot. Gard. 62: 664-723.

Watt, J.M., Breyer-Brandwijk, M.G. 1962. The medicinal and poisonous plants of southern and eastern Africa. Edinburgh: E. and S. Livingstone.

Werker, E., Fahn, A. 1975. Seed anatomy of *Pancratium* species from three different habitats. Bot. Gaz. 136: 396-403.

Whitehead, M.R., Brown, C.A. 1940. The seeds of the spider lily *Hymenocallis occidentalis*. Am. J. Bot. 27: 199-203.

Wildman, W.C. 1968. The Amaryllidaceae alkaloids. Alkaloids 11:307-405.

Williams, M.D. 1981. Chromosome count for *Paramongaia weberbaueri* Velarde. Plant Life 37: 83-89.

Wilsenach, R. 1965. On the caryology and phylogeny of some genera of Amaryllidaceae. Plant Life 21: 82-83.

Zaman, M.A., Chakraborty, B.N. 1974. Cytogenetics of Amaryllidaceae: I. Karyomorphology and meiotic behavior of inversion heterozygote *Eurycles sylvestris* Salisb. Bangladesh J. Bot. 3: 51-58.

Anemarrhenaceae

J.G. Conran and P.J. Rudall

Anemarrhenaceae Conran, Chase & Rudall, Kew Bull. 52: 995 (1997).

Hermaphrodite, perennial, terrestrial *herbs* with short erect stems from short thick rhizomes; roots thickened; *leaves* alternate, basal, numerous, more or less sheathing; leaf blades dorsiventral, linear, parallel-veined. *Inflorescence* a terminal, erect, pedunculate, leafy, spike-like condensed panicle; flowers clustered, hypogynous, small, greenish white to brownish purple, with brownish veins; perianth persistent, of 2 similar trimerous tepal whorls; tepals free, connivent basally into a short tube with linear, spreading limbs; stamens 3, attached to the inner tepals; filaments free, short, flat; anthers linear, basifixed, introrse, dehiscent longitudinally; ovary sessile, syncarpous, 3-locular, ovary ovoid, slightly 3-lobed; septal nectaries present; style short, filiform; stigma small, capitate; ovules anatropous, 2 in each locule, placentation axile; *fruit* an ovoid to oblong loculicidal capsule; seeds 1–2 in each locule, black, spindle-shaped, carinate; testa thick, phytomelan encrusted, endosperm fleshy; embryo cylindric.

A family with a single monotypic genus from SE Asia.

Vegetative Morphology. *Anemarrhena* is a herbaceous geophyte with a short thick rhizome, thick roots and a short, erect stem. During germination, the cotyledon is well differentiated, with a short apical haustorial region which remains embedded in the seed. The plumular leaf emerges through a pore in the sheathing base of the cotyledon at a considerable distance from the seed and alternates with the cotyledon. Young leaves are similar to adult leaves (Boyd 1932; Tillich 1995). Leaves are linear and grasslike, few-nerved and with slightly sheathing bases.

Vegetative Anatomy. Leaves are U-shaped in cross-section without a distinct midrib. There is a single row of vascular bundles, the large ones with a small amount of bundle-sheath sclerenchyma, mainly at the xylem poles, but not extending to the epidermis. Otherwise sclerenchyma is lacking. Mesophyll is thin-walled, with slight palisade development abaxially. Occasional enlarged, thin-walled, axially elongated, idioblastic cells are present in the mesophyll, sometimes containing bundles of raphide crystals. Leaves are amphistomatic with fewer stomata on the adaxial surface. Stomata are anomocytic. There are conspicuous papillae on both surfaces (Rasmussen 1986a,b) and especially at leaf margins, with the degree of papillar development influenced by the structure of the underlying vascular strands. The leaves were considered by Arber (1920) to have a leaf base with phyllode-like anatomy.

Inflorescence Morphology. The inflorescence in *Anemarrhena* is an erect, terete, pedunculate, bracteate, subspicate panicle (Fig. 37). The flowers are clustered at each node and each is subtended by a single bract.

Floral Morphology. The pedicel is articulated and the flower is provided with a distinct pericladium. The perianth is persistent and

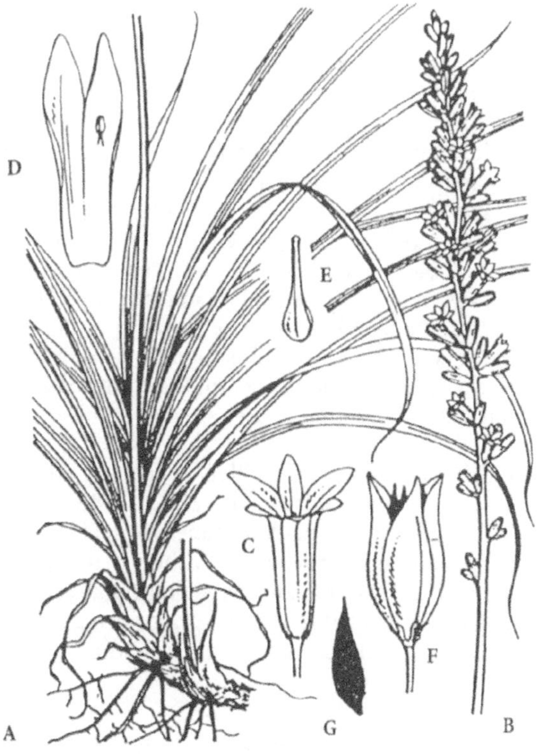

Fig. 37A–G. Anemarrhenaceae. *Anemarrhena asphodeloides.* **A** Habit. **B** Inflorescence. **C** Flower. **D** Tepals with anther attached to outer tepal whorl. **E** Pistil. **F** Dehiscing loculicidal capsule. **G** Fusiform seed. (Lee 1979)

marcescent and consists of 2 similar trimerous tepal whorls. Tepals are regular, petaloid, mostly free (but sometimes partially fused at the very base) and overlapping into a connivent campanulate tube. Tepals are more or less clawed, with spreading free lobes and 3 convergent veins. The 3 stamens are attached to the inner tepals at the connivent tube apex; the filaments are free and very short. Anthers are basifixed, with introrse, longitudinal dehiscence. The ovary is superior, sessile, syncarpous, ovoid or slightly 3-lobed, 3-locular, with axile placentation and septal nectaries opening at the style base. The style is short, filiform, thick; stigma small, capitate. There are only 2 ovules per locule, with their micropyles directed downwards. Ovules are anatropous, bitegmic and crassinucellate.

Embryology. (Data from Li and Hsu 1983; Chen et al. 1988a,b,c, 1990; Rudall et al. 1998). The tapetum is multinucleate (Chen et al. 1990). The rough endoplasmatic reticulum gives rise to pre-Ubisch bodies which are prominently aggregated, and lead to the development of the pollenkitt (Chen et al. 1988a). The tapetum is secretory and the peritapetal membrane is formed from the outer tangential wall (Chen et al. 1988b). Microsporogenesis is successive (Chen et al. 1988c).

A linear tetrad is formed, and embryo-sac development is of the Polygonum type. The small antipodals are held in a narrow constriction at the chalazal end of the embryo sac. Polar nuclei fuse prior to fertilisation. Following fertilisation, the zygote is briefly dormant, then divides once into a basal cell and apical cell. The apical cell divides transversely 2 to 3 times, forming a linear proembryo. Endosperm formation is Helobial, with the ephemeral lower chamber becoming 2- to 4-nucleate (Chen et al. 1990).

Pollen Morphology. Pollen is sulcate. Grains are shed singly, and are obovate, 38–48 μm in length, with a thin reticulate exine and sexine. The reticulum is homobrochate with short brochi, and the muri consist of simple, regularly arranged baculae (Schulze 1982). Pollen grains are 2-celled at maturity with prominent lipid bodies (Chen et al. 1988c).

Karyology. The karyotype of *A. asphodeloides* is diploid, with $2n = 22$ (Satô 1942; Li and Hsu 1983; Ma et al. 1985; Rudall et al. 1998). Rudall et al. (1998) discussed variation in reported karyotypes, and concluded that the most likely karyotype is 12m + 8sm + 2st chromosomes, although Li and Hsu (1983) recorded 18m + 4sm chromosomes, with satellites on the first pair of chromosomes.

Fruit and Seed. The fruits are loculicidal capsules enclosing 1–2 seeds per locule. The seeds are fusiform, conspiculously 3-carinate, black and phytomelan-encrusted. The outer epidermal cells are more or less isodiametric and covered with a thick coat of phytomelan, as in many other Asparagales. The inner epidermal layer is collapsed at maturity, as in many Asparagales. Endosperm cells are conspicuously pitted, with the development of large, irregular pits apparently through fusion of smaller and more or less round pits. The embryo is linear, about 3/4 of the length of the seed and strongly curved (Huber 1969).

Phytochemistry. There are numerous compounds isolated from the rhizomes including the saponins sarsasaponigen, markogenin diglycoside, timosaponins A-III and B-II, and Zhi-mu, hinokiresinol, smilageninoside and the anemarsaponins A1, A2 and B; the glycans Anemarin A, B, C, D; and the xanthone C-glycoside, mangiferin (Hegnauer 1963; Takahashi et al. 1985; Gou et al. 1991; Sato et al. 1994). There is seasonal variation in drug levels in the rhizomes, with timosaponin B-II highest in April-May, whereas the concentration of mangiferin is highest in July-August (Kizu et al. 1994).

Affinities. From analysis of *rbc*L sequence data (Chase et al. 1995) and successive microsporogenesis (Chen et al. 1988c; Rudall et al. 1997), *Anemarrhena* belongs in the "higher" Asparagales, not with Johnsonieae, as suggested by Krause (1930) and Dahlgren et al. (1985) on the basis of reduced stamen number. Chase et al. (1996) demonstrated that Anthericaceae sensu lato is polyphyletic, and placed Johnsonieae (formerly Anthericaceae) in the "lower" asparagoid family Phormiaceae sensu lato, characterised by trichotomosulcate pollen. The precise relationships of *Anemarrhena* remain unresolved. Chase et al. (1995) tentatively placed it as sister to a clade comprising *Asparagus* and *Hemiphylacus*. However, more recent *rbc*L analysis places *Anemarrhena* at a somewhat greater distance from *Asparagus* and *Hemiphylacus*, as sister to a clade including these two genera plus Agavaceae, Anthericaceae s. str. Behniaceae and Herreriaceae (M.W. Chase, pers. comm.; Rudall et al. 1998). *Hemiphylacus* and *Anemarrhena* are linked by a reduction in stamen number: 3 stamens (plus 3 staminodes) in *Hemiphylacus* and 3 only in *Anemarrhena*. Such a reduction in number is relatively rare in higher asparagoids. However, floral structure in *Anemarrhena*, with only 2 ovules per locule, their micropyles facing downwards, differs somewhat from that of both *Hemiphylacus* and *Asparagus*.

Distribution and Habitats. *Anemarrhena* is an understorey herb of moist woodlands in China and Korea.

Economic Importance. *Anemarrhena* is cultivated as an ornamental. It is also used medicinally as a febrifuge and for digestive complaints in China. The crude plant extract derived from it, known as Zhi-mu, inhibits alpha-feroprotein development in rat livers (Li et al. 1989) and there is some inhibition of ageing seen in treated mice intestines (Tang et al. 1995). Anemarsaponin B inhibits rabbit platelet aggregation *in vitro* (Dong and Han 1992), anemarans A–D show marked hypoglycaemic activity in mice (Takahashi et al. 1985), and there are cytotoxic antitumour compounds present (Park and Kim 1992).

Only one genus:

Anemarrhena Bunge Fig. 37

Anemarrhena Bunge, Mém. Sav. Étrang. Acad. Sci. St Pétersbourg 2: 140 (1831); Kitagawa, J. Jpn. Bot. 31: 303 (1956).
Terauchia Nakai, Bot. Mag. Tokyo 27: 441 (1913).

Only one sp., *A asphodeloides* Bunge, northern China and Korea. The genus *Terauchia* had been based on a monstrosity produced by a smut infection (Kitagawa 1956, see above).

Selected Bibliography

Arber, A. 1920. On the leaf structure of certain Liliaceae, considered in relation to the phyllode theory. Ann. Bot. 34: 447–465.

Boyd, L. 1932. Monocotyledonous seedlings. Trans. Proc. Bot. Soc. Edinb. 31: 1–24.

Chase, M.W. et al. 1995. See general references.

Chase, M.W., Rudall, P.J., Conran, J.G. 1996. New circumscriptions and a new family of asparagoid lilies: genera formerly included in Anthericaceae. Kew Bull. 51: 667–680.

Chen, Z.K., Wang, F.H., Zhou, F. 1988a. On the origin, development and ultrastructure of the orbicules and pollenkitt in the tapetum of *Anemarrhena asphodeloides* (Liliaceae). Grana 27: 273–282.

Chen, Z.K., Wang, F.H., Zhou, F. 1988b. The ultrastructural aspect of tapeutm and Ubisch bodies in *Anemarrhena asphodeloides*. Acta Bot. Sin. 30: 1–15 (in Chinese).

Chen, Z.K., Zhou, F., Wang, F.X., Wang, F.H. 1988c. Investigation on the development of male gametophyte in *Anemarrhena asphodeloides*. Acta Bot. Sin. 30: 569–573 (in Chinese).

Chen, Z.-K., Wang, F.-H., Li, Z.-H. 1990. Investigation on embryology of *Anemarrhena asphodeloides*. Acta Phytotaxon. Sin. 28:223–227 (in Chinese).

Dahlgren, R.M.T., Clifford, H.T. 1982. See general references.

Dahlgren, R.M.T. et al. 1985. See general references.

Dong, J.X., Han, G.Y. 1992. Studies on the active constituents of *Anemarrhena asphodeloides* Bunge. Acta Pharm. Sin. 27: 26–32.

Gou, D., Li, S., Chi, Q., Sun, W.G., Sha, Z.F. 1991. Isolation and structure determination of a new saponin of *Anemarrhena asphodeloides*. Acta Pharm. Sin. 26: 619–621 [in Chinese].

Hegnauer, R. 1963. See general references.

Huber, H. 1969. See general references.

Kizu, H., Yamamoto, M., Shimana, H., Tomimori, T. 1994. Seasonal variation of the contents of timosaponin B-II and mangiferin in the rhizome of *Anemarrhena asphodeloides* Bunge. Shoyakugaku Zasshi 47: 426–428 (in Japanese).

Krause, K. 1930. See general references.

Lee, T.B. 1979. Illustrated flora of Korea. Seoul, H.M.S.

Lee, Woo Chul 1996. Lineamenta Florae Koreae. Seoul: Academy Publ.

Li, L.C., Hsu, P.S. 1983. Karyotype analysis of Anemarrhena asphodeloides Bunge (Liliaceae). Acta Phytotaxon. Sinica 21: 445–448.

Li, P.M., Zhong, J.L., Chen, R.Q., Zhang, X.K., Ho, K.L., Chiu, J.F., Huang, D.P. 1989. Zhi-mu saponin inhibits alphafetoprotein gene expression in developing rat liver. Int. J. Biochem. 21: 15–22.

Ma, X.H., Qin, R.L., Xing, W.B. 1985. Chromosome observation of twenty species of drug plants in Xinjiang. Acta Phytotaxon. Sinica 22: 243–249.

Park, S.Y., Kim, J. 1992. Screening and isolation of the antitumor agents from medicinal plants (II). Seoul Univ. J. Pharm. Sci. 17: 1–5 (in Korean).

Rasmussen, H. 1986a. Epidermal cell differentiation during leaf development in *Anemarrhena asphodeloides*. Can. J. Bot. 64: 1277–1285.

Rasmussen, H. 1986b. Pattern formation and cell interactions in epidermal development of *Anemarrhena asphodeloides* (Liliaceae). Nord. J. Bot. 6: 467–478.

Rudall, P., Furness, C.A., Chase, M.W., Fay, M.F. 1997. Microsporogenesis and pollen sulcus type in Asparagales (Lilianae). Can. J. Bot. 75: 408–430.

Rudall, P.J., Engelman, E.M., Hanson, L., Chase, M.W. 1998. Systematics of *Hemiphylacus*, *Anemarrhena*, and *Asparagus*. Plant Syst. Evol.

Satô, D. 1942. Karyotype alternation and phylogeny in Liliaceae and allied families. Jpn. J. Bot. 12: 57–161.

Sato, S., Nagase, S., Ichinose, K. 1994. New steroidal saponins from the rhizomes of *Anemarrhena asphodeloides* Bunge (Liliaceae). Chem. Pharm. Bull. 42: 2342–2345.

Schulze, W. 1982. Beiträge zur Taxonomie der Liliifloren IX. Anthericaceae. Wiss. Z. Friedrich Schiller Univ. Jena 31: 291–307.

Takahashi, M., Konno, C., Hikino, H. 1985. Isolation and hypoglycemic activity of anemarans A, B, C and D, glycans of *Anemarrhena asphodeloides* rhizomes. Pl. Med. 1985: 100–102.

Tang, F., Nakano, A., Nakanishi, Y., Konishi, H., Kubo, M., Abe, H. 1995. Effects of Chinese medicine on morphological changes in the intestinal villi with age. Nat. Med. 49: 240–248 (in Japanese).
Tillich, H.-J. 1995. Seedlings and systematics in Monocotyledons. In: Rudall, P.J., Cribb, P.J., Cutler, D.F., Humphries, C.J. (eds.) Monocotyledons: systematics and evolution. Royal Botanic Gardens, Kew.

Anthericaceae

J.G. Conran

Anthericaceae J. Agardh, Theoria Syst. Pl.: 27 (1858).

Perennial, terrestrial *herbs* from short rhizomes, rarely with a woody stem (*Chlorophytum suffruticosum*); roots fibrous or thickened and fleshy, sometimes with distant tubers. *Leaves* spirally to distinchously arranged, basal, numerous, more or less sheathing, often surrounded by scarious leaf remnants; leaf blades dorsiventral, linear to oblong-lanceolate, parallel-veined. *Inflorescence* erect, mostly scapose, racemes or panicles, or with slender pedicels arising directly and umbel-like from a subterranean axis (*Leucocrinum* and *Anthericum angustifolium*). *Flowers* bisexual, hypogynous, pedicellate, actinomorphic or zygomorphic (*Diora*, rarely *Chlorophytum*), relatively small, white to greenish or yellow, sometimes reddish streaked or tinged, never blue or violet, solitary or several per node, with 1 to several bracts; tepals 3 + 3; both whorls more or less equal, persistent; perianth spreading to reflexed, or perigone narrowly campanulate (*Diamena*), or rarely fused basally into a long tube (*Leucocrinum, Diora*), articulated on the pedicels with pericladium or not. Stamens 6, free or shortly united basally (*Echeandia*); anthers dorsifixed and sometimes inserted into a rimmed pit or basifixed, introrse, dehiscing by longitudinal slits; thecae 2, oblong to elongate; attached to the tepal bases (*Anthericum* and *Chlorophytum*) or apex of the perigone tube (*Leucocrinum*). Gynoecium tricarpellary, syncarpous, sessile or rarely stipitate (*Chlorophytum*); ovary trilocular with 2–many anatropous or campylotropous ovules in 2 rows on axile placentas; septal nectaries present; style filiform, stigma minutely capitate; *fruit* a dry dehiscent loculicidal capsule; seeds numerous to 2 in each locule, black, rounded, angular, folded or flattened; testa thick, phytomelan-encrusted, endosperm fleshy; embryo cylindric, curved inwards; rarely with a basal tuft of hairs (*Comospermum*).

A family of worldwide distribution with 9 genera and ca. 200 species distributed mainly in Africa, Europe, Asia and the Americas, extending to N Australia.

Vegetative Morphology. The members of the Anthericaceae are mostly herbaceous and rhizomatous, with short, often vertical rhizomes or deeply buried vertical axes (*Leucocrinum* and *Anthericum angustifolium*), often sheathed with the fibrous bases of old leaves. The roots are fleshy or fibrous, often with distant tubers (Fig. 38C). In most genera, the scapes or peduncles are well developed, and are traversed by the aerial system. In *Paradisea* the peduncle lacks conspicuous nodes or bracts, other than at the apex. In *Leucocrinum* and *Anthericum angustifolium* the pedicels arise umbel-like directly from the rhizome apex.

The seedlings of the Anthericaceae are cryptocotylar sensu Boyd (1932), the cotyledon is short and haustorial, and either possesses (*Chlorophytum*), or lacks (*Anthericum*) a tubular elongation of the cotyledonary sheath caused by the presence of a closed sheath and unifacial cotyledon (Tillich 1995). There are no obvious cataphylls, the first leaf is similar to later leaves. The cotyledons have 2 vascular strands, the primary root is triarch and there is a well-developed hypodermis in at least *Chlorophytum*.

Vegetative Anatomy. Vessels with simple perforation plates occur in the roots, and vessels with scalariform or simple perforation plates are also in the aerial stems of some genera (e.g. *Anthericum*) (Wagner 1977). Sieve-tube plastids of the type PIIc are also present. A velamen is present in the roots of at least *Anthericum* and *Chlorophytum*, and there is a thickened exodermis and collenchymatous outer cortex in *Anthericum*. The roots in *Chlorophytum* have a hairy rhizodermis and, in older roots, a thickened corklike parenchyma. There is a thickened endodermis, except at the protophloem points, where there are tranfusion cells and a developed Casparian strip. The root tubers in *Chlorophytum* have a papillose or hairy rhizodermis of thin, delicate cells, with the outer wall of the exodermis thickened (von Guttenberg 1968). The endodermis is reduced and less thickened and the root parenchyma cells are greatly enlarged, and much of the central protoxylem is replaced by a parenchymatous medulla (Boyd 1932). In *Echeandia*, the rhizodermis and exodermis of the tubers are thickened and corklike (Guttenberg 1968). Calcium oxalate raphides are generally common. Silica bodies and laticifiers are lacking (Dahlgren and Clifford 1982), but there are mucilage cells and canals in at least *Echeanidia* (von Guttenberg 1968). Tannin cells are not reported. The stomata are anomocytic. Trichomes are uncommon. Epicuticular wax crystalloids are rodlike, and oriented more or less parallel towards the stomates, corresponding to the *Convallaria* type (Frölich and Barthlott 1988).

Inflorescence Morphology. The inflorescences are terminal pedunculate or scapose racemes or pedunculate, racemose, paniculate cymes (Fig. 38). The majority of genera have reduced inflorescence branches at each node. Schlittler (1945) observed that there were both true racemes (*Paradisea*), and thyrses with bostrychoid partial inflorescences (most genera). The presence of a single floral bract at each node was considered by Nordal and Thulin (1993) to be one of the definitive features of *Anthericum* s. str. from *Chlorophytum*. Schlittler (1945) regarded the single flower with a multibracteate node as representing extreme reduction from the cymose condition. In *Leucocrinum* and *Chlorophytum angustifolium*, the flowers are borne on the rhizome apex (Fig. 38F) and bear slender pedicels from multibracteate umbellike inflorescences which arise from below the soil surface.

Floral Morphology. In the Anthericaceae, the flowers are hermaphrodite and hypogynous with a perianth of 6 tepals which are free, or fused basally into a perianth tube with 6 lobes of various lengths. The 1–7-veined tepals (rarely up to ca. 12-veined in *Chlorophytum tuberosum*) are generally marcescent and persistent in all taxa. In most taxa, the pedicel is articulated, and a distinct pericladium is present (Schlittler 1943), although Nordal and Thulin (1993) differentiated the N African and European *Anthericum* species from *Chlorophytum* on the former apparently lacking articulated pedicels. Webb (1980) recorded the pedicels in the European species of the former as being articulated, although in *A. baeticum*, the articulation is very near the base of the pedicel, and difficult to see. In some *Chlorophytum* species, the articulation is apical or apparently lacking (I. Nordal, pers. comm.).

The stamens are hypogynous, inserted on the base of the tepals, or in *Leucocrinum* are inserted near the throat of the perianth tube. The filaments are free or fused (*Echeandia*) (Fig. 38E), glabrous or scabrid and introrse. In *Echeandia* and *Hagenbachia* the filaments are attached to the anther connective in a pit. The anthers can be either basifixed and introrse (*Anthericum, Chlorophytum, Diora* and *Echeandia*), or dorsifixed and usually versatile (*Diamena, Hagenbachia* and *Paradisea*).

Fig. 38A–F. Anthericaceae. **A, B** *Chlorophytum krookianum*. **A** Habit. **B** Inflorescence. **C, D** *C. tuberosum*. **C** Habit. **D** Flower. **E** *Echeandia macrocarpa, buzz-pollinated flower*. **F** *Leucocrinum montanum*, habit. (Takhtajan 1982)

The gynoecium is syncarpous and 3-locular. The style is filiform with a minute capitate or 3-lobed stigma. The stigmatic surface is Dry in all genera examined (Heslop-Harrison and Shivanna 1977; Ravenna 1987) and generally papillate with unicellular papillae. Septal nectaries are present in all genera examined. The placentas are axile and there are 2-numerous ovules per locule. The ovules are anatropous, campylotropous or rarely hemianatropous (e.g. *Anthericum ramosum*, Cruden 1987), although the latter condition is disputed by Ravenna (1987).

Embryology. The anthers of the Anthericaceae have spiral-type endothecial thickenings (Dahlgren and Clifford 1982). Microsporogenesis is successive throughout, the tapetum is secretory and the tapetal cells are predominantly binucleate. The pollen grains are 2-nucleate when shed.

The ovules are crassinucellate. The archesporial cell cuts off a parietal cell but in *Leucocrinum* the nucellar cap is formed from periclinal divisions of the apical nucellar epidermis (Cave 1948). Embryo sac formation is of the Polygonum type and the endosperm formation is Helobial. There are embryo-sac haustoria in several genera, which are developed to a greater or lesser degree (Schnarf 1931). The synergids are hooked with a filiform apparatus in *Chlorophytum*. The polar nuclei fuse prior to fertilisation, and the antipodal cells are ephemeral in a number of genera (Davis 1966).

Pollen Morphology. The pollen morphology of the Anthericaceae has been studied by Schulze (1982) and Díaz Lifante et al. (1990). The pollen grains are sulcate. The sexine is either thicker or thinner than the nexine. Exine sculpturing is reticulate and homobrochate or heterobrochate, more or less smooth (e.g. *Chlorophytum*, *Echeandia*) or with raised sculpturing (e.g. *Anthericum*, *Paradisea*). The exine of *Leucocrinum* has discontinuous reticulate ornamentation (Chung and Jones 1987).

Karyology. The chromosome numbers for the family are based on $x = 7, 8, 11, 13, 15$ (most genera) and 20 (*Comospermum*). There are diploid, triploid, tetraploid, pentaploid, hexaploid, octaploid, decaploid and dodecaploid taxa within the family, mainly in *Anthericum*, *Echeandia* and *Chlorophytum*. The karyotype of *Chlorophytum* consists of metacentric and submetacentric chromosomes (Nordal et al. 1990; Tamura 1995), whereas that of *Echeandia* has a mixture of metacentric, submetacentric and subtelocentric chromosomes, and chromosomes with satellites. There are heteromorphic bivalents, bridges, rings and chains in *Echeandia* which may reflect non-reciprocal translocations, which were thought to have been important in the speciation of three Mexican taxa (Palomino and Martinez 1994). Similarly, satellites and secondary constrictions are known for several African *Chlorophytum* species (Nordal et al. 1990).

The sometimes recognised segregate genus *Diuranthera* was considered by Li (1995) on its cytology to be derived from within the SE Asian *Chlorophytum malayense* complex, supporting its reduction to synonymy by Marais and Reilly (1978).

POLLINATION. The flowers of the Anthericaceae are insect pollinated, with the flowers of *Echeandia* buzz-pollinated by bumblebees (Bernhardt and Montalvo 1977; Fig. 38E). *Diamena cajamarcaënsis* is self-sterile (Ravenna 1987), and *Hagenbachia* is not self-pollinating, and assumed to require a biotic pollinator (Cruden 1987). The flowers of *Leuococrinum* are fragrant (Matthews 1986), and those of *Chlorophytum malayensis* are night-flowering (Jessop 1979). Kativu (1994) found that Zimbabwan *Chlorophytum* species were self-compatible, although most taxa studied required active pollination by various species of bees.

FRUIT AND SEED. The fruit in the Anthericaceae is an elongated to ovoid or trilete, loculicidal, 3-valved, sometimes angled, or deeply 3-lobed capsule (*Chlorophytum*), dehiscing apically, with few-numerous seeds. The capsules in *Diamena* are more or less secund. In *Leucocrinum*, the capsule is subterranean and more or less sessile on the rootstock.

Seed structure in the family has been investigated by Huber (1969), Oganezova (1987), Nordal et al. (1990) and Stedje and Nordal (1994). The seeds are black, and angular (most genera), round in outline and vertically flattened, discoid and thin, or flat on one side and convex on the other (some *Chlorophytum* species). In some *Chlorophytum* and *Hagenbachia* species, the seeds are also irregularly folded. The outer layer of the seed coat contains phytomelan. The epidermal cells are tabular and flattened, convex rounded, more or less hexagonal and sometimes variously papillate. In *Anthericum*, *Chlorophytum*, and *Paradisea* the testa is multilayered at maturity and the outer cell layers are thickened along their periclinal walls, with the inner cell layers thin-walled and, in *Paradisea*, somewhat compressed. The endosperm walls in *Anthericum* and *Chlorophytum* have round to elliptic pits, with no distinct size differences. Copious fats, oil and some reserve starch are stored in the endosperm of *Anthericum*, *Paradisea* and *Echeandia*.

DISPERSAL AND REPRODUCTIVE BIOLOGY. The majority of the family have limited ballistic dispersal through dehiscence of the capsules, and in some *Chlorophytum* species the peduncle is stiff and erect to assist this (Nordal et al. 1990). The discoid seeds of *Anthericum* are probably anemochorous. In *Diora* the inflorescence develops during the dry season, after the leaves have withered, and the fruits are thought to mature at the start of the wet season in late spring. In comparison, the Andean "*Anthericum*" species hold their immature capsules over winter, resulting in higher seed predation rates (Ravenna 1987). The buried fruits of *Leucocrinum* suggest that there is an unusual dispersal mechanism, requiring further study.

There is a group of *Chlorophytum* species, including the widely cultivated *C. comosum*, where the seeds germinate on the parent plant, with the development of small plantlets at the nodes of the inflorescence (Nordal et al. 1990).

PHYTOCHEMISTRY. Steroidal saponins and chelidonic acid are common in the family and cyanogenic glycosides are known from *Chlorophytum* (Hegnauer 1963; Dahlgren and Clifford 1982). Anthraquinones are absent: a feature distinguishing the Anthericaceae from the Asphodelaceae, with which they are sometimes closely related.

AFFINITIES. The Anthericaceae have been variously closely related to, or included within, the Asphodelaceae (e.g. Dahlgren and Clifford 1982) or broadly defined to include members of the Lomandraceae, Johnsoniaceae and Boryaceae (Dahlgren et al. 1985). The Anthericaceae have been shown to differ from the Asphodelaceae by the latter's possession of generally arillate seeds, anthraquinones and simultaneous microsporogenesis (Stedje and Nordal 1994). The Hemerocallidaceae differ in generally having semiequitant (schwertförmig) leaf bases or the leaves reduced to scales, simultaneous microsporogenesis with trichotomosulcate pollen, and flowers in cymose panicles (Rudall et al. 1997). In recent morphological and *rbc*L molecular studies, the Anthericaceae s. l. have been shown to be polyphyletic

(Chase et al. 1995a; Rudall and Cutler 1995; Rudall and Chase 1996), and Rudall et al. (1997) recently fragmented the family, with the Anthericaceae s. str. sister to the Agavaceae, in the higher Asparagales.

Leucocrinum has been previously associated with the Hostaceae (Dahlgren et al. 1985), the Hemerocallidaceae (Di Fulvio and Cave 1964), and by Hernández (1995) with *Hemiphylacus*, the latter now considered as close to the Asparagaceae. Studies of the pollen morphology by Chung and Jones (1987) found that the Agavaceae s. str. were distinct from *Leucocrinum* and the Hostaceae. Other studies of their pollen (Alvarez and Köhler 1987) and leaf epidermis (Alvarez de Zayas 1990) found that *Leucocrinum* was unlike both the Agavaceae and Hostaceae.

The previous association of *Petronymphe* with the Themidaceae (= Alliaceae/Brodieae) (Moore 1951), was claimed not to be supported by recent *rbc*L studies by Fay and Chase (1996), and on the basis of this, Rudall et al. (1997) placed it tentatively within the Anthericaceae near *Leucocrinum*. However, the *Petronymphe* sequence was based on material of a misidentified *Echeandia* species, and *Petronymphe* is thus retained in the Themidaceae.

Comospermum, although traditionally placed in the Anthericaceae (mostly as *Alectorurus*), was found in recent molecular studies (Chase et al. 1995a) to be nested within the Convallariaceae, although in that family only *Gonioscypha* possesses dehiscent fruits, and no other members of the Convallariaceae have hairs associated with the seeds. Because of its close resemblance to members of the Anthericaceae, and the absence of clear morphological synapomorphies to place it in the Convallariaceae, *Comospermum* is retained here within the Anthericaceae as a *genus anomalum* pending further study.

Distribution and Habitats. The family is worldwide in distribution, but with major centres of diversity in Africa, SE Asia and Central and S America. The majority of species are heliophytes, but the family occupies habitats from semiarid savanna woodland and grassland to cloud forests, rainforest and temperate deciduous woodlands and alpine meadows. Although the family is overwhelmingly terrestrial, *Hagenbachia panamensis* from montane and cloud forests in Central America occasionally occurs as an epiphyte (Cruden 1987). Flowering in *Diora* occurs after the leaves have withered, but most taxa are leafy at anthesis.

Economic Importance. Species of several genera in the Anthericaceae, in particular *Chlorophytum comosum* (spider plant) are cultivated as ornamentals. *C. comosum* is also naturalised in some urban areas of E Australia (Henderson 1987).

The roots of *Leucocrinum* and the roots, shoots, leaves and inflorescences of various *Chlorophytum* species are eaten by local indigenous peoples (Tanaka 1976; Kunkel 1984).

Key to the Genera

1. Tepals more or less free, or very shortly united at the base; inflorescence scapose or pedunculate 2
- Tepals fused into a long perianth tube; inflorescence umbel-like, bracts at base of pedicels only **8. *Leucocrinum***
2. Flowers single at each node, supported by a single bract 3
- Flowers generally several at each node, in single, then bracts 2 or more per node 4
3. Anthers dorsifixed, versatile; perianth infundibuliform, segments clawed **6. *Paradisea***
- Anthers basifixed, introrse; perianth rotate, segments not clawed **1. *Anthericum*** (p.p. Old World taxa)
4. Anthers dorsifixed 5
- Anthers basifixed 6
5. Root tubers absent; stamen filaments free; 2 to 6 ovules per locule **7. *Hagenbachia***
- Root tubers present; stamen filaments united basally; more than 8 ovules per locule **2. *Echeandia***
6. Filaments inserted in a rimmed pit; seeds not with a basal hair tuft 7
- Filaments not inserted in a rimmed pit; seeds with a basal hair tuft **9. *Comospermum***
7. Tepals overlapping to form a narrow tube **4. *Diamena***
- Tepals spreading, patent or reflexed 8
8. Ovary pubescent; flowers conspicuously zygomorphic by separation of the lower tepal from the remainder **5. *Diora***
- Ovary glabrous; flowers actinomorphic or if zygomorphic, then not as above 9
9. Seeds turgid, angular **1. *Anthericum*** (p.p. New World taxa)
- Seeds flattened, folded or compressed **3. *Chlorophytum***

1. *Anthericum* L.

Anthericum L., Sp. Pl.: 310 (1753); Obermeyer, Bothalia 7: 669–767 (1962); Webb, Fl. Europaea 5: 18 (1980); Kativu & Nordal, Nord. J. Bot. 13: 59–65 (1993).

Herbs with short rhizomes and somewhat fleshy roots. Leaves basal, sessile, conduplicate, sheathing, linear. Flowers in pedunculate compound racemes, slender thyrses or the pedicels arising directly and umbel-like from a subterranean axis (*A. angustifolium*); pericladium developed or not; flowers single per node, subtended by a single bract; tepals free or shortly united basally, white. Anthers introrse, basifixed, filaments attached to tepal bases, smooth, inserted in a rimmed pit. Ovary small, ovoid; ovules 4–8 per locule. Capsule ovoid, seeds angular and minutely papillate. $2n$ = 16, 30, 32, 48, 60, 64. About 65 spp, predominantly from Mediterranean and southern Central Europe, the Middle East, N and E African south to Tanzania; possibly in S America.

Generic limits of *Anthericum* relative to the S American species are still uncertain and the subject of ongoing research. In particular, the generic delimitation of the Old World *Anthericum* species in Kativu and Nordal (1993) and Nordal and Thulin (1993) excludes several of the New World species as defined by Ravenna (1987, 1988). At least one (and possibly up to 15) of these S American species represents an undescribed genus with affinities to *Echeandia* (R.W. Cruden and P. Ravenna, pers. comm.).

2. ***Echeandia*** Ortega

Echeandia Ortega, Nov. Pl. Descr. Decad.: 90 et 135, t. 18 (1797); Cruden, Phytologia 59: 373–379 (1986), in Contrib. Univ. Mich. Herb. 16: 129–133 (1987), in Phytologia 74: 128–137 (1993), and in Fl. Mesoamericana 6: 27–30 (1994).

Herbs with short rhizomes and tuberous roots. Stem leafy at the base, scapose, sheathed with fibrous leaf bases. Leaves sessile, conduplicate, sheathing, linear. Flowers in racemes or thyrses; pericladium present; flowers several per node, subtended by several bracts; tepals free, white or yellow, marcescent. Anthers introrse, basifixed, filaments short, broad, united basally, smooth. Ovary small, ovoid; ovules 4–8 per locule. Capsule ovoid, 3-lobed, seeds angular. $2n$ = 16, 24, 32, 40, 48, 64, 84. About 60 spp., SW North America to Central America.

The relationships between *Echeandia* and the New World species of *Anthericum* require further study. *Echeandia* has many narrowly endemic taxa (Cruden 1994), and there is evidence of evolution within the genus through chromatid exchange and lagging chromosomes combined with clonal reproduction (Palomino and Martinez 1994). The majority of the American species traditionally assigned to *Anthericum* and *Chlorophytum* were considered by Cruden (1987) to be either *Echeandia* or *Hagenbachia*.

3. ***Chlorophytum*** Ker Gawl. Fig. 38A–D

Chlorophytum Ker Gawl., Bot. Mag.: t. 1071 (1808); Obermeyer, Bothallia 7: 669–767 (1962); Marais & Reilly, Kew Bull. 32: 653–663 (1978); Kativu & Nordal, Nord. J. Bot. 13: 59–65 (1993); Nordal & Thulin, Nord. J. Bot. 13: 257–280 (1993).

Herbs with short rhizomes and more or less fibrous, fleshy or tuberous roots. Leaves in a basal rosette, rosulate or distichous, sessile, conduplicate, sheathing, linear to linear-lanceolate or ovate, rarely pseudopetiolate. Flowers in scapose or pedunculate slender thyses or racemes; peduncles leafy or bracteate; pericladium present, sometimes indistinct; flowers usually several per node, always subtended by more than 1 bract; tepals free, white or greenish, sometimes with a reddish tinge or streak. Anthers introrse, basifixed, filaments free, smooth or scabrid, inserted a rimmed pit. Ovary small, ovoid, rarely stipitate; ovules 2-many per locule. Capsule ovoid to deeply 3-lobed, seeds flattened, folded or compressed, papillate or smooth. $2n$ = 14, 16, 28, 32, 56. About 150 spp., predominantly Africa and Asia, extending to northern Australia.

4. ***Diamena*** Ravenna

Diamena Ravenna, Opera Bot. 92: 185 (1987).

Herb with a short vertical rhizome and tuberous roots. Stem leafy at the base, scapose, sheathed with fibrous leaf bases. Leaves sesssile, conduplicate, sheathing, linear-lanceolate. Inflorescence a loose simple or few-branched raceme; pericladium present; flowers narrowly campanulate, several per node, subtended by several bracts; tepals free, closely imbricate into a narrow tube, lobes densely glandular-pilose, white. Anthers dorsifixed, versatile, filaments free, smooth, inserted in a rimmed pit. Ovary small, ovoid; ovules numerous per locule. Capsule angular ovoid to trigonal, seeds angular. Only one sp., *D. stenantha* (Ravenna) Ravenna, endemic to the upper Cerro de las Cabras near Trujillo, Peru; highly endangered and may be extinct in the wild (Ravenna 1987).

5. ***Diora*** Ravenna

Diora Ravenna, Opera Bot. 92: 189 (1987).

Herb with a vary short vertical rhizome and tuberous roots. Stem leafy at the base, scapose. Leaves sessile, conduplicate, sheathing, narrowly linear, absent at flowering. Inflorescence a loose simple or few-branched raceme; pericladium present; flowers zygomorphic, narrowly tubular with patent lobes, 2 per node, subtended by several auriculate bracts; tepals free, the lower tepal separate from the remainder, white. Anthers introrse, basifixed, filaments free, smooth, appressed to the ovary, fasciculate, inserted in a rimmed pot. Ovary small, broadly elliptic, pubescent; ovules numerous per locule. Only one sp., *D. cajamarcaënsis* (Poelln.) Ravenna, Cajamarca and Junín regions in Peru.

6. ***Paradisea*** Mazzuc.

Paradisea Mazzuc., Viagg. Bot. Alpi Guilie: 27 (1814); Webb, Fl. Europaea 5: 18 (1980).

Herb with a short rhizome and fleshy roots. Stem leafy at the base, scapose, sheathed with fibrous leaf bases. Leaves sessile, conduplicate, sheathing, linear to linear-lanceolate. Flowers in loose more or less secund scapose racemes; pericladium present or absent; flowers single per node, subtended by a single bract, patent; tepals free, clawed, white. Stamens shortly attached to the tepal bases; anthers dorsifixed, versatile, filaments smooth, curved upwards. Ovary oblong-ovoid; ovules numerous per locule. Capsule oblong, 3-lobed, seeds sharply angled. $2n = 20, 32, 48, 64$. Two spp., *P. liliastrum* (L.) Bertol. and *P. lusitanica* (Cout.) Samp., mountainous meadows, woods and marshes of S Europe.

7. ***Hagenbachia*** Nees & Mart. in Schultes

Hagenbachia Nees & Mart. in Schultes, Mantissa 1: 353 (1822); Cruden, Nord. J. Bot. 7: 255–260 (1987).

Herbs with a horizontal rhizome and thick, fleshy but non-tuberous roots. Stem leafy at the base, scapose, without sheathing fibrous leaf bases. Leaves sessile, conduplicate, sheathing, linear to linear-lanceolate or narrowly ovate or oblong. Flowers in loose racemes or panicles; pedicels articulated; flowers several per node, subtended by several bracts; tepals free, white. Anthers dorsifixed, inserted into a rimmed pit, introrse or versatile; filaments free, smooth or scabrid. Ovary small, ovoid; ovules 2–6 per locule. Capsule slightly 3-lobed, seeds angular, folded or flattened, papillate. Five spp., Central and S America.

8. ***Leucocrinum*** Nutt. ex A. Gray — Fig. 38F

Leucocrinum Nutt. ex A. Gray, Ann. Lyceum Nat. Hist. N.Y. 4: 110 (1837).

Tufted acaulescent herb with a short vertical deeply buried rhizome and fleshy roots. Leaves narrowly linear, tufted, sessile, each tuft surrounded by membranous sheathing bases. Flowers in a central umbel-like cluster arising directly from the buried shoot axis, sessile and lacking a pericladium; tepals united into a very long tube, white. Stamens inserted near the top of the floral tube; anthers introrse, basifixed, filaments smooth. Ovary subterranean, ovoid; style shortly 3-lobed; ovules numerous per locule. Capsule obovoid, 3-lobed, seeds angular. $2n = 22, 26, 28$. Only one sp., *L. montanum* Nutt. ex A. Gray, W North America.

Genus anomalum

9. ***Comospermum*** Rauschert

Comospermum Rauschert, Taxon 31: 560 (1982).
Alectorurus Makino (1908) non W.P. Schimper (1869).

Acaulescent caespitose, sympodial herbs, rhizomes thick, short; roots tuberous; leaves alternate, distichous, linear; inflorescence paniculate; flowers erect; tepals free, whitish violet to pale pink; pedicels articulated; stamens 6, filaments attached to the tepal bases; carpels 3, united, style 1, stigma capitate; ovary hypogynous, 3-locular, ovules basal, 2 per locule; fruit a loculicidal capsule; seeds angular-ovoid, with a basal hair tuft. $2n = 40$. Two spp., Japan.

Selected Bibliography

Abrams, L. 1940. Illustrated Flora of the Pacific States, vol. 1. Stanford: Stanford University Press.

Alvarez, A., Köhler, E. 1987. Morfología del polen de las Agavaceae y algunos géneros afines. Grana 26: 25–46.

Alvarez de Zayas, A. 1990. El compleja estomático en la famílìa Agavaceae: 2. Epidermis adultsa. Feddes Repert. 101: 113–134.

Bernhardt, P., Montalvo, E.A. 1977. The reproductive phenology of *Echeandia macrocarpa* Greenm. (Liliaceae) with a reexamination of the floral morphology. Bull. Torrey Bot. Club 104: 320–323.

Boyd, L. 1932. Monocotyledonous seedlings. Trans. Proc. Edinb. Bot. Soc. 31: 1–24.

Cave, M.S. 1948. Sporogenesis and embryo sac development of *Hesperocallis* and *Leucocrinum* in relation to their systematic position. Am. J. Bot. 35: 343–349.

Chase, M.W. et al. 1995. See general references.

Chung, M.G., Jones, S.B.J. 1987. Pollen morphology of *Hosta* Tratt. (Funkiaceae) and related genera. Bull. Torrey Bot. Club 116: 31–44.

Cruden, R.W. 1987. *Hagenbachia*, a misplaced genus of New World Liliaceae. Nord. J. Bot. 7: 255–260.

Cruden, R.W. 1994. 3. *Echeandia* Ortega. In: Davidse, G., Sousa, M.S., Carter, A.O. (eds.) Flora Mesoamericana, Alismataceae à Cyperaceae, vol. 6. México: Universidad Nacional Autónoma de México, Instituto de Biologia pp. 27–30.

Dahlgren, R.M.T., Clifford, H.T. 1982. See general references.

Dahlgren, R.M.T. et al. 1985. See general references.

Davis, G. 1966. See general references.
Díaz Lifante, Z., Díez, M.J., Fernández, I. 1990. Morfología polínica de las subfamilias Melanthioideae y Asphodeloideae (Liliaceae) en la Península Ibérica y su importantcía taxonónomica. Lagascalia 16: 211–225.
Di Fulvio, T.E., Cave, M.S. 1964. Embryology of *Blandfordia nobilis* Smith (Liliaceae) with special reference to its taxonomic position. Phytomorphology 14: 487–499.
Fay, M.F., Chase, M.W. 1996. Resurrection of Themidaceae for the *Brodiaea* alliance, and recircumscription of Alliaceae, Amaryllidaceae and Agapanthoideae. Taxon 45: 441–452.
Frölich, D., Barthlott, W. 1988. See general references.
Guttenberg, H.V. 1968. Der primäre Bau der Angiospermenwurzel. Berlin: Borntraeger.
Hegi, G., Dunziger, G. 1909. Illustrierte Flora von Mittel-Europa, vol. 2. München: J.F. Lehmann.
Hegnauer, R. 1963. See general references.
Henderson, R.F. 1987. 38. *Chlorophytum*. In: George, A.S. (ed.) Flora of Australia, vol. 45. Canberra: Australian Government Printing Service, pp. 349–350.
Hernández, L.S. 1995. Taxonomic study of the Mexican genus *Hemiphylacus* (Hyacinthaceae). Syst. Bot. 20: 546–554.
Heslop-Harrison, Y., Shivanna, K.R. 1977. The receptive surface of the angiosperm stigma. Ann. Bot. 41: 1233–1258.
Huber, H. 1969. See general references.
Jessop, J. 1979. Liliaceae I. Flora Malesiana I 9: 189–235.
Kativu, S. 1994. Anthericaceae in Zimbabwe: a study on cytology and reproduction. In: Seyani, J.H., Chikuni, A.C. (eds.) Proceedings of the XIIIth Plenary Meeting AETFAT. Malawi, pp. 524–534.
Kativu, S., Nordal, I. 1993. New combinations of African species in the genus *Chlorophytum* (Anthericaceae). Nord. J. Bot. 13: 59–65.
Kunkel, G. 1984. Plants for human consumption. Koenigstein: Kluwer.
Li, H. 1995. The phylogenetic place and origination of the genus *Diuranthera*. Acta Bot. Yunnan. 17: 268–276 (in Chinese).
Marais, W., Reilly, J. 1978. *Chlorophytum* and its related genera (Liliaceae). Kew Bull. 32: 653–663.
Matthews, V.A. 1986. 36. *Leucocrinum*. In: Walter, S.M., Brady, A., Brickell, C.D., Cullen, J., Green, P.S., Lewis, J., Matthews, V.A., Webb, D.A., Yeo, P.F., Alexander, J.C.M. (eds.) The European garden flora, vol. 1 Pteridophyta, Gymnospermae, Angiospermae – monocotyledons (Part I). Cambridge: Cambridge University Press.
Moore, H.E., Jr. 1951. *Petronymphe*, a new genus of Amaryllidaceae. Gentes Herb. 8: 258–260.
Nordal, I., Thulin, M. 1993. Synopsis of *Anthericum* and *Chlorophytum* (Anthericaceae) in the Horn of Africa, including the description of nine new species. Nord. J. Bot. 13: 257–280.
Nordal, I., Eriksen, T.E., Fosing, M. 1990. Studies in the generic delimitation of Anthericaceae. Mitt. Staatsinst. Allg. Bot. Hamb. 23b: 535–559.
Oganezova, G.G. 1987. Characteristics of seed and fruit anatomical structure in some representatives of subfamily Asphodeloideae (Liliaceae) in connection with their systematics and phylogeny. (in Russian) Bot. Zh. 72: 436–447.
Palomino, G., Martinez, J. 1994. Cytotypes and meiotic behaviour in Mexican populations of three species of *Echeandia* (Liliaceae). Cytologia 59: 295–304.
Ravenna, P. 1987. *Diamena* and *Diora*, two new genera of Anthericaceae from Peru. Opera Bot. 92: 185–193.
Ravenna, P. 1988. Six new species of *Anthericum* (Anthericaceae) from Bolivia and Peru. Onira Bot. Leafl. 1: 24–30.
Rudall, P., Chase, M. 1996. Systematics of the Xanthorrhoeaceae sensu lato: evidence for polyphyly. Telopea 6: 185–203.
Rudall, P., Chase, M.W., Conran, J.G. 1997. New circumscriptions and new family of asparagoid lilies: genera formerly included in Anthericaceae. Kew Bull. 51: 667–680.
Rudall, P.J., Cutler, D.F. 1995. Asparagales: a reappraisal. In: Rudall, P.J., Cribb, P.J., Cutler, D.F., Humphries, C.J. (eds.) Monocotyledons: systematics and evolution. Royal Botanic Gardens, Kew.
Schlittler, J. 1943. Die Blütenabgliederung und die Perikladien bei den Vertretern des Anthericumtypus sowie ihre Bedeutung für die Systematik. Mitt. Bot. Mus. Univ. Zürich 164: 491–507.
Schlittler, J. 1945. Untersuchungen über den Bau der Blütenstände im Bereich des Anthericumtypus (Asphodelinae-Anthericinae-Dianellinae). Mitt. Bot. Mus. Univ. Zürich 174: 200–239.
Schnarf, K. 1931. Vergleichende Embryologie der Angiospermen. Berlin: Borntraeger.
Schulze, W. 1982. Beiträge zur Taxonomie der Liliifloren IX. Anthericaceae. Wiss. Z. Friedrich Schiller Univ. Jena 31: 291–307.
Stedje, B., Nordal, I. 1994. A contribution to the discussion of the family delimitation of Anthericaceae versus Asphodelaceae. In: Seyani, J.H., Chikuni, A.C. (eds.) Proceedings of the XIIIth Plenary Meeting AETFAT. Malawi, pp. 513–524.
Takhtajan, A.L. 1982. See general references.
Tamura, M.N. 1995. A karyological review of the orders Asparagales and Liliales (Monocotyledonae). Feddes Repert. 106: 83–111.
Tanaka, T. 1976. Tanaka's cyclopedia of edible plants of the world. Tokyo: Keigaku Publ.
Tillich, H.-J. 1995. Seedlings and systematics in monocotyledons. In: Rudall, P.J., Gribb, P.J., Cutler, D.F., Humphries, C.J. (eds.) Monocotyledons: systematics and evolution. Royal Botanic Gardens, Kew.
Wagner, W.M. 1977. Vessel types of monocotyledons: a survey. Bot. Not. 130: 383–402.
Webb, D.A. 1980. Liliiflorae. 183. Liliaceae. 7. *Anthericum*. In: Tutin, T.G., Heywood, V.H., Burges, N.A., Moore, D.M., Valentine, D.H., Walter, S.M., Webb, D.A. (eds.) Flora Europaea, vol. 5: Alismataceae to Orchidaceae (Monocotyledones). Cambridge: Cambridge University Press.

Aphyllanthaceae

J.G. Conran

Aphyllanthaceae Burnett, Outl. Bot.: 421, 1153 (1835).

Caespitose, slender *perennials* from short non-bulbous stock with monocotyledonous secondary thickening. Roots fibrous. Scapes thin, wiry and unbranched. *Leaves* alternate, distichous, reduced to sheathing basal scale leaves with a short ligule-like outgrowth beyond the vestigial lamina. *Inflorescence* small, compressed, capitate, surrounded by scarious bracts, 1(–3)-flowered flowers. *Flowers* showy, hypogynous, actinomorphic, sessile; perianth marcescent, not twisted after anthesis; tepals in 2 3-merous whorls, subequal, more or less clawed, free but conniventl overlapping for about 1/2 their length into a narrowly funnel-shaped tube with spreading to reflexed lobes, blue, exceptionally white. Stamens 6 in 2 whorls; filaments inserted on the tepals at the tube mouth; anthers 2-thecate, 4-sporangiate, dorsifixed, epipeltate, introrse, dehiscing longitudinally by slits; pollen grains shed singly, spiraperturate, spinulose. Gynoecium of 3 united carpels, ovary 3-locular with septal nectaries; style apically shortly tribrachiate; ovules single per locule, anatropous. *Fruit* a loculicidal capsule. Seeds 1 to 3, ovoid, slightly flattened, smooth, black; embryo linear, about as long as the seed; endosperm copious, without starch.

A monotypic family from southern France and Morocco.

Vegetative Morphology. The plants have a congested, sympodially branched, subterranean rhizome with erect, green, stemlike scapes, sheathed basally by several colorless scale leaves. The roots are fibrous.

Germination is epigaeous, with a well-developed cotyledon (Boyd 1932). There is a small, but obvious, hypocotyl, and the cotyledonary sheath has a ligule (Tillich 1995). The rhizome bears reduced, distichously arranged scale leaves with supervolute vernation. Only 2 or 3 of the leaves in a shoot appear to subtend buds. The scale leaves are dorsiventral, 3-nerved, reddish brown and resemble the leaf sheathes of grasses, only without a lamina. In the most distal leaf, there is a vestigial lamina above a short ligule (sheath extension). The scape is the major functional photosynthetic organ and is more or less terete with slight fluting distally (Tomlinson 1965).

Vegetative Anatomy. The vegetative anatomy of the Aphyllanthaceae was investigated by Tomlinson (1965). Vessels are present in the roots only, and have predominantly simple perforations, although some scalariform vessels are also present (Wagner 1977). There are PIIc-type sieve tube plastids in the roots (Behnke 1981). The roots lack a cortex at maturity, and all tissues except the phloem are heavily lignified. The rhizome stele is clearly differentiated into inner (primary), irregularly oriented traces, and outer (secondary), seriated vascular traces. In the scapes, there is a thickened, unlignified epidermis and sunken anomocytic stomata arranged in alternating bands on the upper scape. The scape vascular tissue is primary only, and the parenchymatous medulla often collapses to form an air canal. The scale leaves lack stomata and the abaxial epidermal cells are elongated and heavily thickened and lignified. Raphide cells are common in the cortex of the rhizome, scape and mesophyll of the scale leaves. The scapes have Convallaria-type wax crystalloids (Frölich and Barthlott 1988).

Inflorescence Morphology. The tall photosynthetic scapes bear a single, terminal, condensed, capitate inflorescence of 1–2, rarely 3 flowers. Each flower is subtended by 1–2 free, plus 5 fused scarious scales, and may represent a contracted cyme (Fig. 39).

Floral Morphology. The sessile flowers have free tepals which overlap and appear to fuse for about 1/2 their length into a campanulate tube. The l-veined tepals are more or less clawed, with spreading or reflexed lobes above the connivent tube. The stamen filaments are attached to the base of the tepals. The anthers are dorsifixed and epipeltate, tetrasporangiate, and dehiscence is introrse by longitudinal slits. The gynoecium is 3-carpellary, and the ovary is ovoid to slightly 3-lobed and trilocular. The stigma is filiform with 3 short stigmatic lobes. The stigmas are Dry and papillate. Placentation is axile, with a single anatropous ovule in each locule. Septal nectaries are present.

Embryology. Anther wall formation is of the monocotyledonous type, and the well-developed endothecium has spiral thickenings. The tapetum

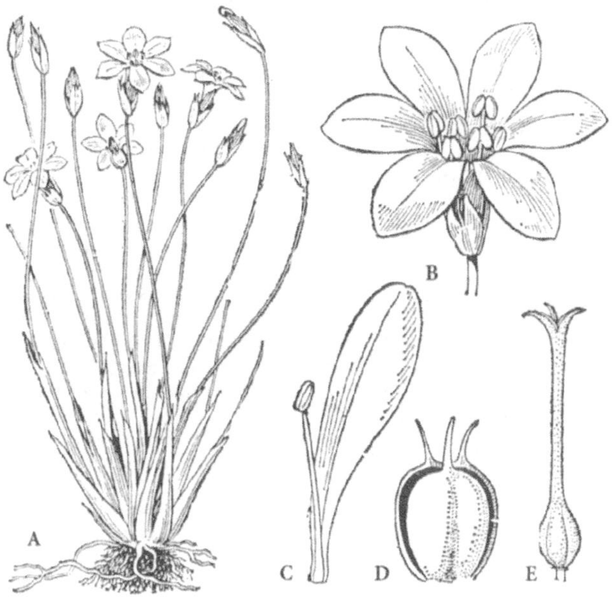

Fig. 39A–E. Aphyllanthaceae. *Aphyllanthes monspeliensis*. A Habit. B Flower. C Petal with stamen. D Capsule. E Pistil. (Takhtajan 1982)

is glandular-secretory and microsporogenesis is successive. At anthesis, the pollen grains are 2-nucleate (Chase et al. 1995b).

The ovule is anatropous or slightly campylotropous-amphitropous, bitegmic and crassinucellate. A parietal cell is cut off, and embryo-sac development is of the Polygonum type. Endosperm development is Helobial (Schnarf and Wunderlich 1939).

Pollen Morphology. The pollen grains are shed singly and are globose-ovoid spiraperturate with a non-annulate margin, echinate and 44–67 μm in diameter (Díaz Lifante et al. 1990).

Karyology. Aphyllanthes is diploid with a base chromosome number of $x = 16$ (Löve and Kjellqvist 1973).

Pollination and Reproduction. The blue flowers are bee-pollinated and exhibit strong UV absorption (Biedinger and Barthlott 1993).

Fruit and Seed. The loculicidal capsule is ovoid and slightly trilobed, with each capsule segment surmounted by a short beak derived from the style. The fruit is 3-seeded and the seeds are flattened and thin with a phytomelan coat. The outer integument has large isodiametric epidermal cells and a relatively thin phytomelan coat, and 4 to 5 layers of more or less collapsed cells beneath the epidermis. The inner integument is reduced to 2 thin-walled, more or less completely collapsed cell layers (Huber 1969; Oganezova 1987). The endosperm consists of non-starchy aleurone and lipids. The embryo is linear, straight and about as long as the seed.

Phytochemistry. The rhizomes contain the steroidal saponins genine, 25D and 25L saponogen and wax alcohols (Hegnauer 1963; Dahlgren et al. 1985). Common flavonoids and proanthocyanins are also present (Williams and Harborne 1988; Harborne and Williams 1995).

Affinities. Because of their unusual morphology, the Aphyllanthaceae have been considered to be isolated within the Lilianae. There were suggestions that they were related by this morphology to xeric members of the Eriocaulaceae (Takhtajan 1959), but this was refuted by the anatomical investigation of Tomlinson (1965). They were also considered to have some seed structure similarities with the Xanthorrhoeaceae (incl. Lomandraceae) and Anthericoid/Phormiaceae/Asphodelaceae complex (Huber 1969; Oganezova 1987). This has been further developed with observations on the similarity of the pollen to that of

Lomandra (Chanda and Ghosh 1976; Díaz Lifante et al. 1990), but this was considered by Dahlgren et al. (1985) to be the result of convergence. Similarly, Baker (1879) found common vegetative and inflorescence characters with genera now included in the Johnsoniaceae, although Huber (1969) considered some of these features to be convergent.

Molecular studies of the Lilianae using sequences of the *rbc*L gene, found that the Aphyllanthaceae were part of the higher asparagoids. They were placed in a clade with the Asparagaceae and Anemarrhenaceae (Chase et al. 1995) based on *rbc*L data only, or with the Asparagacae and Lomandraceae, using a combined molecular and morphological data set (Chase et al. 1995a). Morphologically, the family is still isolated from its molecular relatives, but there are no other morphologically and anatomically similar Lilianae. The reduction of the leaves to non-photosynthetic bracts, the unusual stomatal distibution in bands along the scape, the few-flowered multibracteate inflorescences and unusual pollen are all features which contribute to its isolation.

Distribution and Habitats. The single species *A. monspeliensis* comes from W Mediterranean Europe and Morocco, where it grows on dry and exposed rocky soils, often on limestone-dominated sites.

Only one genus:

Aphyllanthes L. Fig. 39

Aphyllanthes L., Sp. Pl.: 294 (1753).

Only one sp., *A. monspeliensis* L.

Selected Bibliography

Baker, J.G. 1879. A synopsis of the Colchicaceae and the abberant tribes of Liliaceae. J. Linn. Soc. Bot. 17: 405–510.

Behnke, H.D. 1981. See general references.

Biedinger, N., Barthlott, W. 1993. Untersuchungen zur Ultraviolett-Reflektion von Angiospermenblüten I. Monocotyledoneae. Trop. Subtrop. Pflanzenwelt (Akad. Wiss. Lit. Mainz) 86: 1–122.

Boyd, L. 1932. Monocotyledonous seedlings. Trans. Proc. Edinb. Bot. Soc. 31: 1–24.

Chanda, S., Ghosh, K. 1976. Pollen morphology and its evolutionary significance in Xanthorrhoeaceae. In: Ferguson, I.K., Muller, J. (eds.) The evolutionary significance of the exine. London, Academic Press, pp. 527–559.

Chase, M.W. et al. 1995. See general references.

Chase, M.W. et al. 1995a. See general references.

Dahlgren, R.M.T. et al. 1985. See general references.

Díaz Lifante, Z., Díez, M.J., Fernández, I. 1990. Morfología polínica de las subfamilias Melanthioideae y Asphodeloideae (Liliaceae) en la Península Ibérica y su importantcía taxonónomica. Lagascalia 16: 211–225.

Frölich, D., Barthlott, W. 1988. See general references.

Harborne, J.B., Williams, C.A. 1995. Small molecules and monocot classification. In: Rudall, P.J., Cribb, P.J., Cutler, D.F., Humphries, C.J. (eds.) Monocotyledons: systematics and evolution. Royal Botanic Gardens, Kew, pp. 201–215.

Hegnauer, R. 1963. See general references.

Huber, H. 1969. See general references.

Löve, A., Kjellqvist, E. 1973. Cytotaxonomy of Spanish plants. II. Monocotyledons. Lagascalia 3: 147–182.

Oganezova, G.G. 1987. Characteristics of seed and fruit anatomical structure in some representatives of subfamily Asphodeloideae (Liliaceae) in connection with their systematics and phylogeny. (in Russian) Bot. Zh. 72: 436–447.

Schnarf, K., Wunderlich, R. 1939. Zur vergleichenden Embryologie der Liliaceae-Asphodeloideae. Flora 133: 297–327.

Takhtajan, A. 1959. Die Evolution der Angiospermen. Jena: G. Fischer.

Takhtajan, A.L. 1982. See general references.

Tillich, H.-J. 1995. Seedlings and Systematics in Monocotyledons. In: Rudall, P.J., Cribb, P., Cutler, D.F., Humphries, C.J. (eds.) Monocotyledons: systematics and evolution. Royal Botanic Gardens, Kew.

Tomlinson, P.B. 1965. Notes on the anatomy of *Aphyllanthes* (Liliaceae) and comparison with Eriocaulaceae. J. Linn. Soc. (Bot.) 59: 163–173.

Wagner, W.M. 1977. See general references.

Williams, C.A., Harborne, J.B. 1988. Distribution and evolution of flavonoids in the monocotyledons. In: Harborne, J.B. (ed.) The flavonoids: advances in research since 1980. London, Chapman and Hall, pp. 505–524.

Asparagaceae

K. Kubitzki and P.J. Rudall

Asparagaceae Juss., Gen. Pl.: 40 (1789).

Erect or scandent, mostly glabrous herbaceous or subshrubby *perennials* with persistent, evergreen or annually withering branches growing from a compact or creeping, sympodial rhizome or rarely a tuber (*A. ovatum* and *A. undulatum*); cataphylls small or vestigial. *Roots* cylindrical or fusiform, lateral roots sometimes tuberous, covered by a multiple velamen. *Stems* erect or voluble, their branches often patent; leaves on long shoots and short shoots (if the latter present) scalelike, scariose, spurred at the base and subtending solitary or clustered phylloclades, the latter leaflike to angled or terete, rarely (*A. densiflorus*) bearing minute, reduced scale leaves; spines present or absent, formed from modified leaf spurs or branches. *Flowers* in fascicles or borne singly, bisexual, or unisexual and monoecious or dioecious, pendulous or erect, on articulated pedicels, the pericladium separated from pedicel by a swelling; tepals 6, free and spreading or basally fused and then forming a cup or tube; stamens 6, free from each other, fused to tepals, or sometimes outer stamens basally adnate, diminute and nonfunctional in pistillate flowers; filaments free, filamentous or flattened; anthers bithecate, sagittate, ± basifixed, introrse, dehiscing by slits; ovary superior, trilocular, sometimes slightly stipitate; style with 3 short stigmatic branches or a capitate or lobate stigma of the Dry or Wet type; ovules axile, 2–12 per locule, hemianatropous or almost atropous, bitegmic, crassinucellate; septal nectaries present; *fruit* a berry, rarely a nut, red, blue, or black or brown; seed coat black; endosperm containing aleurone, fat and hemicellulose but no starch; embryo slightly curved, nearly reaching the length of endosperm.

Only one genus with a controversial number of species (170–300?), throughout Africa, most of Europe, Asia and Australia.

Vegetative Morphology. Shoot architecture of *Asparagus* is highly complicated. According to Braun (1849), in the 1st year *A. officinalis* develops 3–4 shoot generations from a rhizome, followed by 5–6 generations in the 2nd and 8–10 in each subsequent year, which gradually become stronger and branch more extensively. From the 3rd to about the 15th year, the plants are fully developed, but later become weaker. Flowers are formed only from the 3rd year onward.

Flat phylloclades (Fig. 40K) are bilateral and anatomically rather leaflike, while the needlelike phylloclades (Fig. 40A,G,I) are radial throughout; both types are supplied by several or a single vascular bundle (Reinke 1898). The interpretation of the phylloclades has led to endless debates among plant morphologists (Arber 1924, 1935; Cooney-Sovetts and Sattler 1986 and literature cited therein) fostered by pecularities such as the inverted vascular bundles of bilateral symmetric phylloclades, but from their axial position it is clear that they are caulomatic. It is small wonder that in their ontogeny some traits characteristic of developing leaves are manifest.

Wenck (1935) recognised that the fascicles of phylloclades in the axil of a bract leaf are cincinni with extremely condensed axes. The flowers form part of these shoot systems as basal (first) lateral branches, and the median shoot is capable of further development.

There are 3 different kinds of spine: the indurated spur of a modified leaf, as shown in *A. albus* (Fig. 40J); indurated pungent-pointed phylloclades, as shown in *A. horridus*; and short, spinelike, sterile, real branches, as shown in *A. stipularis*.

Vegetative Anatomy. Stomata are anomocytic. Vessels with scalariform perforation plates are present in both roots and stems; vessels with simple perforation plates also occur in the roots (Cheadle and Kosakai 1971). Raphide cells are widely distributed. Hairs are mostly lacking. Cooney-Sovetts and Sattler (1986) showed that in *A. setaceus* the phylloclade is radially symmetrical throughout, with only 1 vascular bundle, whereas in *A. densiflorus* the phylloclade is dorsiventral with a bilateral portion distally, and 1 or 2 vascular bundles.

Flower Structure. *Asparagus* is a large and varied genus, including many species with bisexual flowers, and some monoecious or dioecious species (e.g. *A. officinalis*). Breitenbach (1878) observed that in the same species, individuals with pistillate flowers always have rudimentary staminodes, while individuals with staminate flowers exhibit much variation in the reduction of the pistil, ranging from fully functional to rudi-

Fig. 40A–K. Asparagaceae. **A–D** *Asparagus verticillatus*. **A** Fruiting plant. **B** Phylloclade, cross-section. **C** Flower. **D** Fruit. **E, F** *Asparagus scandens*. **E** Shoot. **F** Flower. **G, H** *Asparagus litoralis*. **G** Portion of flowering shoot, pollinated by bumble bee. **H** Flower. **I** *Asparagus schoberioides*, portion of flowering shoot. **J** *Asparagus poissonii*, shoot. **K** *Asparagus asparagoides*, flowering shoot. (Takhtajan 1982)

mentary. This was corroborated by Galli et al. (1993), who also reported a high incidence of inbreeding depression from andromonoecious plants. These observations are consistent with the hypothesis of the origin of dioecy in *Asparagus* from hermaphroditism via gynodioecy. The ovary is superior and sometimes slightly stipitate. It has septal nectaries, and 2 or more ovules per carpel. Lazarte and Palser (1979) studied the anatomy and vasculature of the flowers of *A. officinalis*. The tapetum is secretory and anther wall formation is of the monocotyledonous type (Venkataswarlu and Raju 1958; Lazarte and Palser 1979).

Embryology. Microsporogenesis is successive, and a mixture of isobilateral, T-shaped, and decussate tetrads are formed. Pollen grains are dispersed in the 2-celled state and contain starch (Venkataswarlu and Raju 1958). Ovules are crassinucellate. Embryo-sac development is according to the Polygonum type (Lazarte and Palser 1979). The mature embryo sac is asymmetrical, and this asymmetry (described by Venkateswarlu and Raju, 1958 as a diverticulum or caecum) is exaggerated during seed development. There is a persistent nucellar epidermis of enlarged cytoplasm-rich cells, as also in *Hemiphylacus* (Rudall et al. 1997a).

Pollen Morphology. Grains are sulcate (Erdtman 1952; Venkataswarlu and Raju 1958; Schulze 1982; Tamanyan 1988) and have a psilate-microperforate or slightly reticulate exine.

Karyology. A broad survey (Malcolmer and Demissew 1993) confirmed Sato's (1942) observation of $x = 10$ as the base chromosome number, with diploids, triploids, tetraploids, hexaploids and octoploids all occurring. There is no evidence for supporting a subdivision of *Asparagus* on the basis of these numbers. In the dioecious *A. officinalis* homomorphic sex chromosomes have been identified (Loptien 1979).

Pollination. Nectar secretion, which has been observed to occur even in the pistillodes of staminate flowers (in *A. acutifolius*), points to insect pollination, and although Hymenopterae and Dipterae have been observed visiting the flowers, the pollination process has not yet been documented.

Fruit and Seed. The fruit is a globose, red, blue or black berry coloured by carotenoids, enclosing 1–several seeds. Seed shape varies from globose to angular, partly depending on the number of seeds in the capsules; seed colour is deep black. Huber (1969) found the outer epidermis of the testal layer much larger than and clearly differentiated from the collapsed inner seed coat tissue. According to his description, the anticlinal walls of the epidermal cells become thicker towards the periphery, so that the melanin masses filling their lumina adopt the shape of truncated cones with hexagonal outlines, particularly well visible when the outer periclinal cell walls are detached. Malcomber and Sebsebe (1993), in contrast, emphasised the frequent presence of two well-developed cell layers, of which the outer, the epidermis, in many species is detached when the seed is removed from the berry, while it is the subepidermic cell layer that carries phytomelan. Robbins and Borthwick (1925) followed up the seed development of *A. officinalis* and their illustrations seem to support Huber's view, but there may be much variation in this large genus. The endosperm stores aleurone, lipids and reserve cellulose, but no starch. The cell walls of the endosperm are distinctly pitted.

PHYTOCHEMISTRY. Chelidonic acid and steroid saponins are recorded from various species. The rhizomes and swollen roots have been found to be particularly rich in saponins, which can be obtained in quantity from rhizomes of *A. officinalis* and *A. acutifolius* and roots of *A. thunbergianus*. Other storage compounds found in subterranean organs include inulinlike fructans and mannans (Hegnauer 1963, 1986).

AFFINITIES. Analysis of molecular data from *rbc*L (Chase et al. 1995 and pers. comm.) indicates a close relationship between *Asparagus* and *Hemiphylacus*, a Mexican genus of five species, and more weakly so with the small east Asian genus *Anemarrhena*. In the molecular analysis, *Asparagus* and *Hemiphylacus* formed a clade which is sister to a larger clade including Convallariaceae, Ruscaceae and other taxa.

The non-molecular evidence supporting a close relationship between *Asparagus* and *Hemiphylacus* is at least as strong as the morphological evidence linking them with other families. Most of the characters shared between them are plesiomorphic conditions, and therefore uninformative for systematic purposes; for example, they have crassinucellate ovules, phytomelaniferous seeds, and capsular fruits. They both have superior, trilocular ovaries with septal nectaries, as in most higher asparagoids. The shape and histology of the fertilised ovule of *Hemiphylacus* closely resemble those of *Asparagus*, with an asymmetrical embryo sac and persistent nucellar epidermis of enlarged cytoplasm-rich cells. In view of this, and the strong molecular signal linking the two genera, the latter should either be treated as separate (but closely related) monogeneric families, or *Hemiphylacus* could be included in Asparagaceae.

A relationship with *Anemarrhena* is also worth exploring, since *Hemiphylacus* and *Anemarrhena* are linked by a reduction in stamen number: 3 stamens (plus 3 staminodes) in *Hemiphylacus* and 3 only in *Anemarrhena*, a relatively unusual reduction in higher asparagoids. However, floral structure in *Anemarrhena* differs somewhat from that of *Hemiphylacus* and *Asparagus*.

Many authors have considered Asparagaceae and Ruscaceae to be closely related. Both families have berries, although these also occur in some other related taxa, such as *Aspidistra*, *Convallaria* and *Dracaena*. Their karyotypes are reputedly similar (Tamura 1995), although this aspect requires review, since the karyotypes of *Asparagus* and Ruscaceae are apparently not more closely similar than those of Ruscaeae and some Convallariaceae and Nolinaceae. Both families have phylloclades, which is a highly unusual feature. However, the phylloclades of Ruscaceae are somewhat more leaflike than those of *Asparagus*, where phylloclade organisation and vasculature is often stem-like, sometimes radially symmetrical (Cooney-Sovetts and Sattler 1986). There are many differences between *Asparagus* and Ruscaceae, including seed coat anatomy and the position of the inflorescences. Ruscaeae shares several characters with members of Convallariaceae, including lack of phytomelan in the seed coat and some aspects of floral structure, and this is currently under review.

DISTRIBUTION AND HABITATS. The family is widely distributed in the Old World. Most species are found in regions with semiarid to arid and mediterranean-type climate, and extreme xeromorphic adaptations are common. The genus is represented in South and tropical Africa, Madagascar and the Mascarenes, in Macaronesia, the Mediterranean region, particularly the eastern part, in central and E Asia and with a single species in Australia. Many species are practically leafless perennials with thick underground organs, in which they store nutrients and water. Photosynthesis is carried out by the distal green parts of the shoot.

ECONOMIC IMPORTANCE. *Asparagus officinalis*, garden asparagus, has been cultivated since ancient Greek times and today is widely spread and economically important. Useful variants can be maintained by vegetative propagation. In the Mediterranean region, all wild species are collected as vegetable, especially in the Easter week. *A. albus* (white asparagus) is a crop of N Africa. *A. setaceus* (Kunth) Jessop (=*A. plumosus*), the asparagus fern, and other species are much used by florists.

Only one genus (Malcomber and Sebsebe 1993):

Asparagus L. Fig. 40

Asparagus L., Sp. Pl.: 313 (1753); Baker, J. Linn. Soc., Bot. 14: 508–632 (1875), rev.; Malcomber & Sebsebe, Kew Bull. 48: 63–78 (1993), subgeneric classif.
Myrsiphyllum Willd. (1808); Obermeyer, Bothalia 15: 77–88 (1984), rev.
Protasparagus Obermeyer, S. Afr. J. Bot. 2: 243 (1983).

Characters as for family.

Following Malcomber and Sebsebe, only two subgenera can be maintained. Subgen. *Myrsiphyllum*: flowers hermaphrodite; filaments connivent, forming a tube around the ovary; spines absent; cladodes often flattened, leaflike. Twelve spp., S Africa, *A. asparagoides* (L.) Wight extending to S Europe; subgen. *Asparagus*: flowers unisexual or hermaphrodite; filaments free, spreading; spines present or absent; cladodes linear to filiform, ca. 160(–290?) spp.

Genus anomalum:

Hemiphylacus S. Watson[1]

Hemiphylacus S. Watson, Proc. Am. Acad. 18: 164, 165 (1882–1883), Hernandez, Syst. Bot. 20: 546–554 (1995), rev.

Perennial herbs; rhizomes vertical, with tuberous, contractile roots; leaves in a rosette, linear, dorsiventral, parallel-veined. Inflorescence a raceme or thyrse; inflorescence branches subtended by two coriaceous bracts; flowers pedicellate, bisexual, hypogynous, white to reddish; tepals 3 + 3, equal, fused to a tube for their lower half; perianth twisting after anthesis. Stamens 3, adnate to inner tepals; staminodes 3, adnate to outer tepals. Anthers dorsifixed near base, introrse; filaments inserted in a depression of the anther. Ovary superior, stipitate, syncarpous, trilocular, with septal nectaries and axile placentation; style narrow; stigma trilobate; ovules anatropous, several per locule; fruit a loculicidal capsule; seeds ellipsoid, shining black, with phytomelan in seed coat.

Five spp., Mexico, restricted to desert scrub and oak forests at the eastern and southern limits of the Chihuahuan desert and Tehuacan valley.

Flower Structure and Embryology. The style has 3 vascular traces and a single trilobed canal. The ovary has several ovules per locule, with an obturator present below each ovule, near the micropyle. Ovules are bitegmic and the inner integument forms the micropyle. The inner integument is 2 cell layers thick. The exostome borders on the hilum, but is dragged away from the hilum by tissue expansion after fertilisation. Microsporogenesis is successive (Rudall et al. 1997). Ovules are crassinucellate, with the archesporial cell forming a parietal cell. Embryo-sac development is of the Polygonum type. In the mature megagametophyte the 3 small antipodals lie in a small sinus at the distal end of the hypostase. The mature embryo sac is asymmetrical, as in *Asparagus*, being larger on the side opposite the funicle. This asymmetry becomes exaggerated during seed development. Endosperm development is Helobial. The micropylar nucellar epidermis forms a persistent uniseriate layer of enlarged cytoplasm-rich cells (as in *Asparagus*), which eventually becomes crushed in the mature seed (Rudall et al. 1998).

Pollen Morphology. Pollen is sulcate.

Karyology. *Hemiphylacus alatostylus* has an unusually high chromosome number of $2n = 112$ (Rudall et al. 1997a), indicating that this genus is a polyploid based on either $x = 14$ or $x = 7$. The chromosomes are very small.

Affinities. Hernandez (1995) and Rudall et al. (1997a) discussed the systematic relationships of *Hemiphylacus*. A relationship with the "lower" asparagoid family Asphodelaceae (Brummitt 1992) is discounted, on the basis of *rbc*L sequence data and micromorphological characters such as successive microsporogenesis (Rudall et al. 1997), which place *Hemiphylacus* amongst the "higher" Asparagales. DNA sequence analyses (e.g. Chase et al. 1995) have consistently indicated a close relationship between *Hemiphylacus* and *Asparagus*. The non-molecular evidence supporting this is sparse, but at least as strong as the morphological evidence linking them with other families. The shape and histology of the fertilised ovule of *Hemiphylacus* resemble those of *Asparagus*, with an asymmetrical embryo sac and persistent nucellar epidermis of enlarged

[1] By P.J. Rudall.

cytoplasm-rich cells. Other similarities are all plesiomorphic, and therefore uninformative: crassinucellate ovules, sulcate pollen, phytomelaniferous seeds, capsular fruits, superior, trilocular ovaries with septal nectaries.

Selected Bibliography

Arber, A. 1924. *Myrsiphyllum* and *Asparagus*. Ann. Bot. 38: 635–659.

Arber, A. 1935. The "needles" of *Asparagus* with special reference to *A. sprengeri* Regel. Ann. Bot. 49: 337–344.

Braun, A. 1849. Betrachtungen über die Erscheinung der Verjüngung in der Natur. Freiburg i. B.: H.M. Popper.

Breitenbach, W. 1878. Ueber Asparagus officinalis, eine triözische Pflanze. Bot. Zeitung 36: 163–167.

Brummitt, R.K. 1992. Vascular plant families and genera. London: Royal Botanic Gardens, Kew.

Chase, M.W. et al. 1995. See general references.

Cheadle, V.I., Kosakai, H. 1971. Vessels in Liliaceae. Phytomorphology 21: 320–333.

Cooney-Sovetts, C., Sattler, R. 1986. Phylloclade development in the Asparagaceae: an example of homoeosis. Bot. J. Linn. Soc. 94: 327–371.

Erdtman, G. 1952. See general references.

Galli, M.G., Bracale, M., Falavigna, A., Raffaldi, F., Savini, C., Vigo, A. 1993. Different kinds of male flowers in the dioecious plant *Asparagus officinalis* L. Sex. Plant Reprod. 6: 16–21.

Hegnauer, R. 1963, 1986. See general references.

Hernandez, L. 1995. Taxonomic study of the Mexican genus *Hemiphylacus* (Hyacinthaceae). Syst. Bot. 20: 546–554.

Huber, H. 1969. See general references.

Jessop, J. 1966. The genus *Asparagus* in South Africa. Bothalia 9: 31–96.

Lazarte, J.E., Palser, B.F. 1979. Morphology, vascular anatomy end embryology of pistillate and staminate flowers of Asparagus officinalis. Am. J. Bot. 66: 753–764.

Loptien, H. 1979. Identification of the sex chromosome pair in asparagus (*Asparagus officinalis* L.). Z. Pflanzenzücht. 82: 162–173.

Malcomber, S.T., Demissew, D. 1993. The status of *Protasparagus* and *Myrsiphyllum* in the *Asparagaceae*. Kew Bull. 48: 63–78.

Reinke, J. 1898. Die Assimilationsorgane der Asparageen. Jb. Wiss. Bot. 31: 207–272.

Robbins, W.W., Borthwick, H.A. 1925. Development of the seeds of *Asparagus officinalis*. Bot. Gaz. 80: 426–438.

Rudall, P.J., Furness, C.A., Chase, M.W., Fay, M.F. 1997. Microsporogenesis and pollen sulcus type in Asparagales (Lilianae). Can. J. Bot. 75: 408–430.

Rudall, P.J., Engelman, E.M., Hanson, L., Chase, M.W. 1998. Systematics of *Hemiphylacus*, *Asparagus* and *Anemarrhena*. Plant Syst. Evol.

Schulze, W. 1982. Beiträge zur Taxonomie der Liliifloren X. Asparagaceae. Wiss. Zfschr. Friedrich-Schiller-Univ. Jena, Math.-Nat. R. 31: 309–330.

Takhtajan, A.L. 1982. See general references.

Tamanyan, K.G. 1988. Pollenmorphology of Asparagus L. Flora rastitel "nost" i rastitel'nye resursy (Armenskoj SSR) 11: 96–102 (in Russian).

Tamura, M.N. 1995. See general references.

Venkataswarlu, J., Raju, C.S.K. 1958. Male and female gametophytes in four species of *Asparagus*. J. Indian Bot. Soc. 37: 290–299.

Wenck, S. 1935. Entwicklungsgeschichtliche Untersuchungen über die Assimilationsorgane von *Semele*, *Ruscus*, *Danae* und *Myrsiphyllum*. Beih. Bot. Centralbl. 53A: 1–25.

Asphodelaceae

G.F. Smith and B-E. Van Wyk

Asphodelaceae Juss., Gen. Pl.: 51 (1789).

Small to medium sized, often succulent, *herbs* or occasionally large pachycaul *trees* with leaves arranged in terminal rosettes. Stems fibrous and woody rather than succulent. Secondary thickening growth in species of *Aloe* and *Kniphofia*. Roots slightly succulent, terete, sometimes inflated and fusiform, velamen known in some genera. *Leaves* dorsiventral, lanceolate-acuminate, linear or subulate, terete, often succulent and thickly conical, spirally arranged or distichous as in some species of Alooideae, amplexicaul, margins toothed, serrate or entire, sharply pointed, parallel veins often obscure. *Inflorescence* a raceme or panicle, peduncle usually well developed and distinct from the vegetative part of the plant, leafless or bracteate. *Flowers* sometimes articulated from pedicel, petaloid, undifferentiated, regular or rarely zygomorphic (bilabiate); perigone segments fleshy or rather flimsy, tepals 3 + 3, ranging from brightly coloured (red, orange, yellow) to white, variously fused into a tube or free to the base; stamens 3 + 3, inserted below the ovary; filaments free, often the same colour as the perianth, subulate, slightly filiform-flattened, rarely hairy; anthers dorsifixed or rarely basifixed (*Eremurus*), introrse, dehiscing longitudinally; ovary superior, trilocular, tricarpellary, syncarpous; septal nectaries present; styles distinct, single; stigma minute, Dry or rarely Wet; ovules 2-many per locule, anatropous, hemianatropous, nearly orthotropous or campylotropous, crassinucellate. *Fruit* a thick-walled, loculicidally dehiscent capsule, rarely (*Lomatophyllum*) carnose and berrylike; seeds entirely enclosed by an envelope of funicular origin, usually referred to as an aril, which has the appearance of an additional integument, irregularly angled, elongate, sometimes winged, brownish grey to black; embryo straight, linear; endosperm present.

A family comprising 15 genera and ±780 species, distributed in arid and mesic regions of the temperate, subtropical and tropical zones of the Old World with the main centre of distribution in southern Africa. The genus *Hemiphylacus* is excluded since it has been shown to be more closely related to Asparagaceae (Rudall et al. 1997). Similarly, *Simethis* sits uncomfortably in the Asphodelaceae and will be included in the treatment of Phormiaceae. *Paradisea* was mistakenly assigned to the Asphodelacaeae in Brummitt (1992) and is currently referred to the Anthericaceae.

Vegetative Morphology. Species of the Asphodelaceae display a wide range of habit, from geophytes, small, highly specialised rosulate leaf succulents, acaulescent herbs, shrubs and climbers to small to large trees (Figs. 41–43). The dorsiventral leaves of most species are strongly tufted in terminal or basal rosettes. The non-succulent leaves of most species of Asphodeloideae are usually densely congested in basal rosettes and often occur in distinct ranks. In most species of Alooideae the leaves are closely compressed, whether on a leafy stem or in a basal rosette. A number of unrelated species distributed between both subfamilies have window-leaves (Cutler 1985). All the species of Alooideae have succulent leaves, whereas only some asphodeloid taxa, particularly in *Bulbine* and *Trachyandra*, have this faculty. Most of the other taxa of the subfamily Asphodeloideae have a less succulent leaf consistency, but often have fusiform, contractile roots and tend to be geophytic. Probably as a consequence of the development of leaves with a succulent consistency, cross-sections of the leaves of Alooideae species are crescent-shaped to cymbiform. In contrast, the non-succulent leaves of some Asphodeloideae are often keeled. The leaves of most Alooideae species carry distinctive, white or concolorous tubercles or creamish white spots. Such spots or protuberances are usually lacking in species of the Asphodeloideae.

Vegetative Anatomy. Both the internal and external fine structure of the leaves of many species of Asphodelaceae have received the attention of taxonomists and anatomists alike over the last few decades. Studies of leaf epidermal characters have yielded useful taxonomic information, mainly at the species level (e.g. Baijnath 1980; Brandham and Cutler 1981; Baijnath and Cutler 1993). Of special interest is the presence of thin-walled parenchymatous cells in the inner bundle sheaths of most genera (except *Kniphofia*) of this family. In the case of *Kniphofia* (and sporadically in some other genera), these cells are lignified (Beaumont et al. 1985). They have been termed aloin cells because of their association with the copious exudate which appears when the leaves of many *Aloe* species are cut (Beaumont et al. 1985).

Fig. 41A–G. Asphodelaceae. A–C *Asphodelus albus*. A Habit. B Dehisced capsule. C Seed. D, E *A. fistulosus*. D Habit. E Capsule. F, G *Asphodeline taurica*. F Habit. G Flower. (Takhtajan 1982)

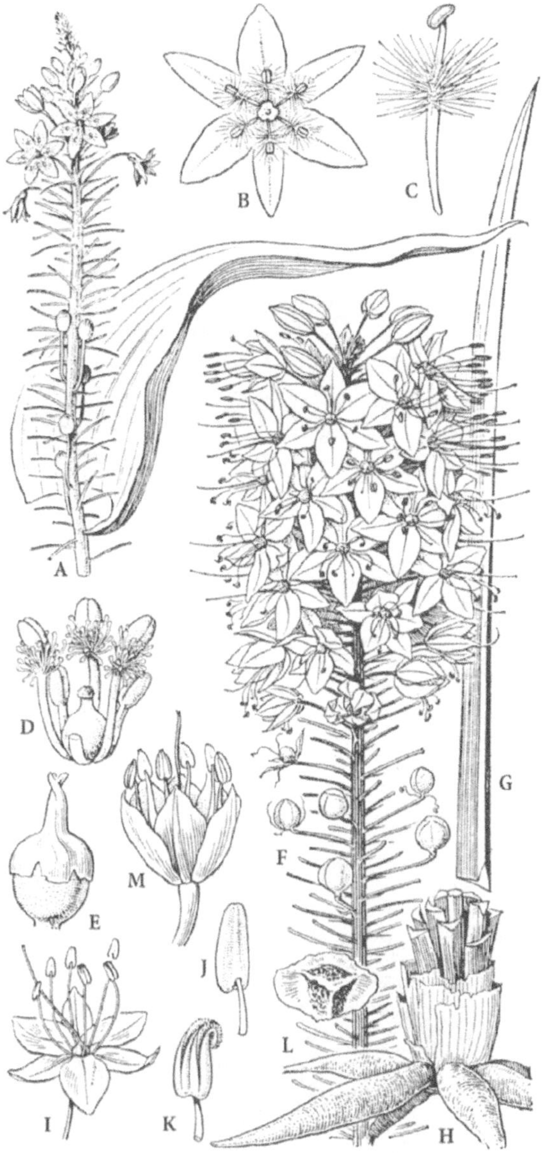

Fig. 42A–M. Asphodelaceae. A–C *Bulbine latifolia*. A Inflorescence. B Flower. C Stamen. D, E *B. semibarbata*. D Young flower, perianth and 1 stamen removed. E Fruit. F–M *Eremurus olgae*. F Inflorescence. G Leaf. H Rootstock with storage roots. I Flower. J Stamen, dorsal view. K Dehiscing stamen, ventral view. L Seed. M *E. fuscus*, flower. (Takhtajan 1982)

Summaries of the distribution of leaf anatomical characters are presented by Beaumont et al. (1985) and Smith and Van Wyk (1992). These characters show general trends in the family, but no distinct discontinuities at the generic level.

In the Asphodelaceae secondary thickening growth is not directly related to plant habit and several geophytic taxa (species of *Chortolirion*, *Bulbine*, *Kniphofia* and *Trachyandra*) also show this faculty (Coetzee and Van der Schijff 1969). In these instances it is, however, less pronounced than in the arborescent species of *Aloe*.

Flower Structure. The flowers of all the genera are bisexual, hypogynous and trimerous (3 + 3). In the subfamily Alooideae and *Kniphofia* the 6

Fig. 43A–G. Asphodelaceae. **A–C** *Kniphofia uvaria*. **A** Habit. **B** Basis of shoot with buds and stolons. **C** Flower. **D** *Aloe mendesii*, habit. **E** *A. ortolopha*, plant with secund raceme. **F** *A. richardsieae*, basis of shoot with thickened roots. **G** *A. arborescens*, habit. (Takhtajan 1982)

tepals are usually variously fused into a tube. These flowers are more or less cylindrical, often with a slight constriction above the ovary. Clavate, cylindric-campanulate or basally swollen flowers are also found in the Alooideae. Somewhat to distinctly bilabiate (zygomorphic) flowers occur in *Chortolirion* and *Haworthia*, while the flowers of *Gasteria* are variously curved. Flower colour varies greatly, from dull whitish green or powdery pink to orange and bright red or yellow. Buds and open flowers often differ in colour. The tepals tend to become coherent above when wilted, thereby forcing the nectar to form glistening droplets at the mouth of the flower. This might serve as an additional attractant for pollinators. Only in the case of the monotypic *Poellnitzia* are the flowers apically connivent and not "open" in the true sense of the word (Smith et al. 1992; Smith 1995).

In the other genera, particularly of the Asphodeloideae, the tepals are free and usually bent backwards towards the pedicel (Wendelbo 1964; Tuzlaci 1987). A tendency towards zygomorphy has been recorded in the free tepals of the genus *Bulbine* (Rowley 1967). Nocturnal closure of flowers has been recorded in, amongst others, *Bulbine*. Although the basic morphology of the flowers of species of Asphodelaceae is very simple and fairly constant, the occurrence of a number of aberrant floral morphologies has previously led to the splitting off of monotypic genera from, amongst others, *Aloe*. All these segregates are, however, currently included in *Aloe*.

Three asphodeloid genera, *Asphodelus*, *Asphodeline*, and *Bulbine* are characterised by having jointed pedicels, visible as an articulation below each flower. This character is absent from the morphologically similar *Trachyandra*.

Floral Anatomy. An investigation of the floral anatomy of five species of *Aloe*, *Haworthia* and *Gasteria* showed that both whorls of the flower are similar (Vaikos et al. 1978). In all the species studied, the tepals of the perianth are 3-traced. This study also suggested that there is a tendency in the Alooideae towards the development of an inferior ovary, based on, amongst others, adnation of the outer floral whorls with the ovary, particularly in species of *Haworthia*. In addition, adnation of the filaments to the ovary suggests a trend towards the development of a gynostemium. A detailed characterisation of the septal nectaries was given by Daumann (1970).

Embryology. Two embryological characters are of special diagnostic value in the family. Firstly, all members of the Asphodelaceae (as far as is known), have an aril-like structure which develops as an annular invagination of the funicle, and has the appearance of a 3rd integument around the ovules (Stenar 1928; Schnarf 1929). Secondly, microsporogenesis is simultaneous, which is a significant difference from the Anthericaceae, where it is successive.

One-many ovules are found per locule, and placentation is axile. Ovules are anatropous, hemianatropous, orthotropous or campylotropous (Stenar 1928). The nucellus is crassinucellate, and has parietal tissue between the megaspore mother cell and the nucellar epidermis (McNaughton and Robertson 1974).

In a study of megasporogenesis and megagametogenesis in *Aloe africana*, McNaughton and Robertson (1974) found that the ovules are campylotropous and have 3 integuments. During megasporogenesis the megaspore mother cell gives rise to either a T-shaped or a linear tetrad of megaspores. Only the chalazal megaspore remains functional to produce an embryo sac of the Polygonum type (Joshi 1937; Schnarf and Wunderlich 1939; McNaughton and Robertson 1974). Endosperm formation is of the Helobial type.

Pollen Morphology. Eleven genera are well studied palynologically (Schulze 1975; Smith 1991a), and sulcate pollen is universal in the family. In species of *Haworthia* and *Bulbinella* Schulze (1975) found sulcate pollen to co-occur with trichotomosulcate grains. The tectum varies from sparsely perforate in most genera to densely perforate; in *Eremurus*, pollen is distinctly reticulate. The ektexine is composed of a foot layer, infratectal columellae and tectum. In some genera there is an abrupt transition from a perforated sexine to a smooth surface lacking perforations in the apertural region (Smith and Tiedt 1991; Smith and Van Wyk 1992).

Karyology. The species of Alooideae have been the subject of detailed cytotaxonomic studies (summarised by Brandham 1983, Smith 1991b and Tamura 1995). Chromosome counts for all the genera are available. These studies were initiated early in the 20th century and have resulted in a considerable bibliography on the cytology of this group (for reviews see Riley and Majumdar 1979; Brandham 1983; Smith 1991b). The Alooideae are one of the most uniform groups as regards chromosome number and the markedly bimodal karyotype. All species have the same basic chromosome number ($x = 7$), with 4 long and 3 short chromosomes. In the entire subfamily the basic diploid karyotype ($2n = 14$) is only very rarely altered (Brandham 1983; Smith 1991b). In cladistic terms this character represents a synapomorphy for all the taxa of Alooideae. A large number of intra- and intergeneric hybrids have been produced in the Alooideae (Rowley 1982). This clearly testifies to the close cytogenetical relationship which exists amongst species of the subfamily. Based on overall genome size and increased bimodality, Brandham (1983) has shown that, at least in *Aloe*, there is a gradation from smaller chromosomes in species which have retained a number of plesiomorphic characters (*A. tenuior*: actinomorphic flowers; weak scandent stems; mesophytic) to larger ones in species with morphological apomorphies (*A. peckii*: stemless; extreme xerophyte).

In contrast to the Alooideae, *Kniphofia* has a basic set of 6 chromosomes ($2n = 12$). This represents a sharp discontinuity between this asphodeloid genus and the Alooideae, and suggests a distinct barrier to gene interchange (De Wet 1960). However, some of the other asphodeloid genera (*Bulbinella*, *Eremurus* and *Trachyandra*) share a basic chromosome number of $x = 7$ with the Alooideae, but $x = 14$ in the case of *Asphodelus* and *Asphodeline*. Basic chromosome numbers of 7, 12, 13 and 14 have been reported for *Bulbine* (Watson 1986), while the number for *Jodrellia* is currently unknown (Fedorov 1969).

Pollination. Flowers are borne erectly spreading to horizontal on a terete and herbaceous or massive and woody peduncle that can reach a length of up to 3 m in the genus *Aloe*. Easy access is therefore possible for any floral visitor and potential pollinating agents. An inflorescence usually carries few to many spirally arranged flowers and buds, with only a limited number open simultaneously (Smith et al. 1992). This ensures cross-pollination in that visitors are forced to visit several plants in their quest for sufficient nectar or pollen rewards. The tubular flowers of Alooideae species are usually showy and act as excellent reservoirs for the copious amounts of nectar that most species produce.

Species of Asphodelaceae are visited by numerous microfaunal species, as well as birds (Hoffman 1988; Smith et al. 1992; Ratsirarson 1995; McCabe 1995). Most species appear to be either ornithophilous or entomophilous, or utilise a combination of birds and insects to ensure effective pollination. Detailed studies, however, are lacking for most genera.

All genera of the Asphodelaceae except *Bulbine* and *Bulbinella* have septal nectaries (presence in *Trachyandra* confirmed by J.C. Manning, pers. comm.). Small amounts of nectar are produced in the Asphodeloideae and for many species pollen appears to be the main reward for insect pollinators.

Reproductive Systems. Cross-pollination is required by all groups, but selfing has been reported for a few species (Smith et al. 1992). Cleistogamy has not been recorded to date.

Fruit and Seed. The fruits are thick-walled, almost leathery, loculicidally dehiscing capsules. Only one genus, *Lomatophyllum*, has fleshy fruits which are best described as dehiscent berries. Seeds are invariably enclosed by an envelope, usually referred to as an aril, albeit inconspicuously so in most species. The aril may appear to be an additional integument. Seed colour varies from brownish grey to black due to the presence of phytomelan in the outer epidermis of the outer integument. Seed shape is fairly constant, with most species having variously angled, elongate to ovoid seeds. Distinct, white to rather obscure wings are sometimes present. For details of seed morphology and anatomy see Huber (1969).

Dispersal. The drab colour, structure and disposition of the dry capsules which turn chartaceous to woody when dry suggest wind as the primary dispersing agent (Smith et al. 1992; Van Jaarsveld 1992). Few details seem to be known in the literature.

Phytochemistry. Since the work of Hegnauer (1963) and Rheede van Oudtshoorn (1964), *Aloe* and related genera have been the subject of several chemotaxonomic and phytochemical investigations (e.g. Reynolds 1985a,b; Dagne and Yenesew 1994). The genus *Aloe* is chemically the best known and accumulates a wide variety of compounds in the leaves, including anthrones, chromones, phenylpyrones, phenolic amines and alkaloids (Reynolds 1985a,b; Nash et al. 1992). At the generic level, the most important progress has come from root metabolites (Dagne et al. 1994, 1996; Seigler and Rauwald 1994a,b; Van Wyk et al. 1995b,c), notably the presence of a wide range of tetrahydroanthracenones, many of which are as yet unidentified. The occurrence of these lipophylic anthranoid aglycones, such as chrysophanol and its dimer, asphodeline, has interesting chemotaxonomic implications (Van Wyk et al. 1995b,c). Chrysophanol is present in most genera and may be provisionally accepted as a useful chemical marker, although the presence of this anthrone in *Trachyandra* needs investigation. Asphodeline and 10,7′-bichrysophanol have been found in *Asphodelus*, *Asphodeline*, *Bulbine*, *Bulbinella* and *Kniphofia*. Knipholone and related compounds appear to be confined to the genera *Bulbine*, *Bulbinella* and *Kniphofia* (Van Wyk et al. 1995b). In contrast to the leaf compounds, the root metabolites are relatively conservative, and hence of value to determine relationships at the generic level. It is interesting that the subterranean anthranoid metabolism in *Aloe* (Seigler and Rauwald 1994a,b; Van Wyk et al. 1995c) and also in *Lomatophyllum* (Van Wyk et al. 1995c) is not similar to the above-ground metabolism as in *Gasteria* and other genera (Dagne et al. 1996) but that it is strongly differentiated. 1-Methyl-8-hydroxyanthraquinones are present in the roots of most *Aloe* species (Van Wyk et al. 1995c), while the leaves have anthrone-C-glycosides. In the leaves, stems and roots of *Gasteria* and other genera, further oxydation and hydroxylation of 1-methyl-8-hydroxy-anthraquinones is evident (Dagne et al. 1996). Further progress will no doubt come from rigorous comparisons of various other classes of compounds, such as chromones and flavonoids.

Subdivision and Relationships. Huber (1969) suggested a classification of the family in three tribes, Bulbineae, Aloineae and Asphodeleae. Van Wyk et al. (1995a) and Smith and Van Wyk (1991). recognised two subfamilies, the Asphodeloideae and Alooideae. These two groups appear to be linked by the genus *Kniphofia*, which shares characters of both groups, namely the usually tubular petaloid flowers and fusion of the perigone segments.

The succulent genera, subfamily Alooideae (Aloaceae sensu Brummitt 1992) are monophyletic, as is evidenced by several unifying characters. These include anthrone-C-glycosides in the leaves and 1-methyl-8-hydroxyanthraquinones in the roots, to which may be added the succulent leaf consistency, the widespread occurrence of secondary thickening growth and the bimodal karyotype of $2n = 14$ chromosomes (Brandham 1983). Another interesting character is the nectar sugar composition (some genera have a balanced glucose-fructose ratio, while others have much more glucose than fructose; Van Wyk et al. 1993). Despite general uniformity in karyotype, the Alooideae are unexpectedly variable in morphology and chemistry, and display unusual patterns of variation among populations and inconsistent intergradations among species.

In contrast, the subfamily Asphodeloideae (Asphodelaceae sensu Brummitt 1992) appears to be a paraphyletic assemblage, as there are no convincing morphological and chemical synapomorphies (Van Wyk et al. 1995a). This is supported by *rbc*L sequence data (De Bruijn et al. 1995). Rigorous taxonomic studies of all the genera may give a new perspective on the circumscription of the subfamilies and the relationships among the genera included therein. It may be sensible to abandon the rank of subfamily in favour of a more natural tribal classification in which only monophyletic taxa are recognised. However, we are hesitant to formally propose such a classification and provisionally recognise the following informal groups:

Group 1. Trachyandra group (*Trachyandra* only)
Group 2. Asphodelus group (*Asphodelus* and *Asphodeline*)
Group 3. Eremurus group (*Eremurus* only)
Group 4. Bulbine group (*Bulbine*, *Bulbinella* and *Jodrellia*)
Group 5. Kniphofia group (*Kniphofia* only)
Group 6. Aloe group (*Aloe*, *Astroloba*, *Chortolirion*, *Gasteria*, *Haworthia*, *Lomatophyllum*, *Poellnitzia*)

Affinities. The Asphodelaceae as defined here (Asphodeloideae sensu Dahlgren and Clifford 1982) are undoubtedly monophyletic and this family is supported by at least three convincing synapomorphies: (1) the parenchymatous inner bundle-sheath cells (in all except *Bulbinella* and *Kniphofia*); (2) microsporogenesis, which is reported to be of the simultaneous type and not the successive type as in the Anthericaceae (Dahlgren and Clifford 1982); (3) the peculiar seed appendage (aril or strophiole), which arises as an annular invagination at the distal part of the funicle (Stenar 1928; Schnarf 1929). The presence of this type of appendage in *Trachyandra* was first mentioned by Huber (1969). To this may be added the presence of chrysophanol in roots, which has been found in most genera (Van Wyk et al. 1995b,c).

Based on overall morphological similarities, the Anthericaceae appear to be the family most closely related to the Asphodelaceae, but there are no obvious synapomorphies, and molecular data indicate a more distant relationship than would be expected (De Bruijn et al. 1995).

Distribution and Habitats. Genera of the subfamily Alooideae are widespread throughout the temperate and subtropical regions of Africa, Saudi-Arabia, Madagascar and some of the Mascarene islands off the east coast of Africa. As could be expected of the essentially succulent-leaved species of this subfamily, they usually grow in arid habitats.

Aloe occurs over much of sub-Saharan Africa, ranging from the southern tip of Africa to west and northeast Africa and the Arabian Peninsula, and some of the Indian Ocean Islands off the African mainland (Socotra and Madagascar). The smaller alooid genera (*Astroloba*, *Chortolirion*, *Gasteria*, *Haworthia*, *Poellnitzia*) are more or less restricted to southern Africa. *Lomatophyllum* is a floristic component of Madagascar and a few of the Mascarene Islands off the southeast coast of Africa.

The Asphodeloideae are more widely distributed than the Alooideae and also occur in Europe, Asia, Australia and New Zealand. The distribution of *Kniphofia* coincides very closely with that of *Aloe*, but the genus favours moister habitats and occurs mainly along mountain ranges. *Bulbine* and *Bulbinella* occur in southern Africa and Australia and southern Africa and New Zealand, respectively. *Trachyandra* is more or less restricted to southern Africa. The *Bulbine* look-alike genus, *Jodrellia*, occurs in southern and central tropical Africa. The remaining ashpodeloid genera (*Asphodelus*, *Asphodeline*, *Eremurus*) are distributed mainly through the Mediteranean region, the middle and near East and central Asia.

Palaeobotany. Fossil remains of asphodelaceous species are unknown. However, it has been suggested that representatives of the family have been around since the early Cretaceous (Holland 1978; Smith and Van Wyk 1991).

Economic Importance. The genus *Aloe* is of considerable medicinal and cosmetic value and forms the basis of large industries. Two species are of particular commercial significance, namely *Aloe vera* and *Aloe ferox* (Reynolds 1994). The dried leaf sap, produced by the outer part of the leaf, is marketed in a wide range of traditional medicines, mainly as laxatives. *Aloe vera* is grown extensively in the southern part of N America and the West Indies and the product from this plant is known as Curaçao aloes. The leaf sap of *A. ferox* is harvested from natural populations (Van Wyk et al. 1995d) and large quantities are exported annually from S Africa as Cape aloes. The main purgative principle in both products is the anthrone-C-glycoside, aloin (also known as barbaloin), which occurs as two diastereoisomers at levels of between 18 and 30% of the dried product (British Pharmacopoeia 1993; United States Pharmacopeia 1995). The leaf sap of many

Aloe species is used as a first aid treatment for burns, abrasions and other wounds and is also used for the weaning of babies. Folk uses are many and varied and include, amongst others, the use of powdered dried leaves as snuff.

The central fleshy part of the leaf does not contain any bitter substances and is used to produce *Aloe* gel (Grindlay and Reynolds 1986). The gel is a complex mixture of polysaccharides, amino acids, minerals, trace elements and so-called biologically active substances, such as enzymes. It is used in a wide range of skin and hair care products and also forms the basis of health drinks and tonics.

Many species of the family are cultivated as ornamentals. The most popular genera are *Aloe*, *Asphodeline*, *Asphodelus*, *Bulbine*, *Gasteria*, *Haworthia* and *Kniphofia*. The aesthetic appeal and horticultural value of aloes, enhanced by their tolerance of dry climates, have led to their cosmopolitan cultivation. Inflorescences of *Kniphofia* (red not pokers) are sold as cut flowers.

Conservation. Many species of the subfamily Alooideae are threatened due to their popularity in amenity and commercial horticulture. Urban and regional development as well as wanton habitat destruction, for example in Madagascar, have resulted in the inclusion of numerous species on the Appendices of CITES. In southern Africa, a total of 31% of the taxa of Alooideae are threatened to some degree (Hilton-Taylor and Smith 1994). A comprehensive action plan detailing steps to be taken to curb this trend has been proposed by Van Jaarsveld and Smith (1997). The Mascarene and Madagascan alooid genus *Lomatophyllum* also contains a number of threatened species (Strahm 1989, 1993).

Data on Asphodeloideae held in the threatened plants data base at the World Conservation Monitoring Centre in Cambridge, United Kingdom, indicate that eight species of *Asphodeline*, one species of *Asphodelus* and five species of *Eremurus* are globally threatened. One species, *E. korovinii* from Tien Shan, central Asia, is doubtfully extinct. Data on the southern African Asphodeloideae have been included in a comprehensive list of southern African threatened plants (Hilton-Taylor 1996). Most of these species belong to the genera *Bulbine* and *Kniphofia*.

Key to the Genera

1. Fruit a berry — **10. *Lomatophyllum***
– Fruit a capsule — 2
2. Leaves distinctly succulent, often margined with prickly teeth — 3
– Leaves non-succulent (mesophytic) or leathery, margins lacking distinct spines — 10
3. Capsule apically acuminate, undergound parts distinctly bulbous — **15. *Chortolirion***
– Capsule apically rounded or obtuse, underground parts rhizomatous — 4
4. Filaments hairy — 5
– Filaments glabrous — 6
5. Flowers white, inflorescence shorter than the leaves — **6. *Jodrellia***
– Flowers yellow or orange (if rarely white, then inflorescence exceeding the leaves in length) — **5. *Bulbine***
6. Perianth segments apically connivent — **12. *Poellnitzia***
– Perianth segments apically spreading or redurved — 7
7. Flowers pendulous at anthesis, perianth tube curved upwards — **11. *Gasteria***
– Flowers erect, suberect or spreading at anthesis, perianth tube straight or curved downwards — 8
8. Perianth bilabiate, <15 mm long, mouth not upturned — **14. *Haworthia***
– Perianth regular (if rarely weakly bilabiate then flowers >15 mm long, mouth upturned) — 9
9. Flowers usually brightly coloured, fleshy, stamens as long as or longer than the perianth — **9. *Aloe***
– Flowers dull-coloured, flimsy, stamens included — **13. *Astroloba***
10. Leaves V-shaped in cross-section — **8. *Kniphofia***
– Leaves crescentiform in cross-section — 11
11. Pedicels articulated below the flower — 12
– Pedicels not articulated below the flower — 13
12. Stamens more or less equal — **2. *Asphodelus***
– Stamens unequal (outer 3 smaller than inner 3) — **3. *Asphodeline***
13. Anthers basifixed — **4. *Eremurus***
– Anthers dorsifixed — 14
14. Perianth caducous, racemes simple or branched, sparsely flowered, flowers white, rarely yellow or pink, tepals usually dark-keeled — **1. *Trachyandra***
– Perianth persistent, racemes simple, densely flowered, flowers yellow, orange or white, tepals usually not dark-keeled — **7. *Bulbinella***

Genera of Asphodelaceae

1. *Trachyandra* Kunth

Trachyandra Kunth, Enum. Pl. 4: 573 (1843); Obermeyer, Bothalia 7: 711–759 (1962), rev.

Herbaceous, rarely succulent, perennials with short, non-bulbous rhizome (rarely suffrutices) and fibrous, fusiform or tuberous roots. Leaves rosulate or rarely distichous, basal or cauline, variously shaped, often terete, with tubular, sheathing base. Flowers in lax or dense racemes or panicles; usually white, rarely yellow, pink or mauve. Pedicels articulated. Perianth segments free, stellate or recurved. Stamens equal, exceeding the tepals; filaments straight, retrorsely scabrid, not expanded at the base. Fruit a loculicidal capsule; seeds few or numerous. About 50 spp., southern Africa, with a few extending into tropical Africa.

2. *Asphodelus* L. Fig. 41A–E

Asphodelus L., Sp. Pl.: 309 (1753); Diaz Lifante & Valdés, Boissiera 52: 1–189 (1996), Fev. W Medit. spp. *Glyphosperma* S. Wats. (1882).

Herbaceous annuals or shortly rhizomatous perennials, with roots cylindrical or fleshy. Leaves linear, in a basal rosette, expanded at base into a membranous sheath. Flowers numerous, in a dense raceme or panicle; white or pale pink. Pedicels articulated. Perianth segments free or united only at the base, erect. Stamens free, shorter than perianth; filaments expanded at the base. Fruit a loculicidal capsule; seeds 3 or 6. About 12 spp., Mediterranean to western Asia, one introduced in Mexico.

3. *Asphodeline* Rchb. Fig. 41F,G

Asphodeline Rchb., Fl. Germ. Excurs.: 116 (1830); Tuzlaci, Candollea 42: 559–576 (1987).

Herbaceous perennial, biennial or annual, with short, non-bulbous rhizomes and cylindrical or fleshy roots. Leaves usually basal, linear, with wide scarious, sheathing base. Flowers zygomorphic, numerous, in groups or rarely solitary, in a dense raceme or panicle; white, pinkish white, yellow or yellow-orange; each bract subtending up to 5 flowers. Pedicels articulated. Perianth segments narrow, shortly connate at base, erect to recurved. Stamens decurved, unequal, outer 3 smaller than inner 3; filaments glabrous, expanded at the base. Fruit a loculicidal capsule; seeds 6. About 14 spp., Mediterranean to western Asia.

4. *Eremurus* M. Bieb. Fig. 42F–M

Eremurus M. Bieb., Cent. Pl. Ross. Merid.: ad t. 61 (1810); Wendelbo, Acta Univ. Bergensis, Ser. Math.-Nat. 5: 1–45 (1964), reg. rev.

Herbaceous, scapose, tall, perennials with short, non-bulbous rhizome and fusiform, tuberous roots. Leaves basal, linear, with scarious, sheathing base, carinate beneath. Flowers very numerous, rotate, in a long, dense raceme; white or yellow. Pedicels not articulated. Perianth segments free, slightly cup-shaped. Stamens equal, exceeding the tepals; filaments straight, glabrous, not expanded at the base; anthers basifixed. Fruit a loculicidal capsule; seeds up to 8 in each locule. About 40 spp., steppes of the high plateaus in central Asia, up to 6000 m in the Himalayas, occurring particularly in Afghanistan, Iran, western Pakistan, Iraq, Turkey and Jordania.

5. *Bulbine* Wolf Fig. 42A–E

Bulbine Wolf, Gen. Pl.: 84 (1776).

Herbaceous, succulent, miniature or medium-sized perennials, caulescent, rhizomatous or solitary, with swollen or hard roots. Leaves basal or cauline, variously shaped. Flowers numerous, stellate, in a lax or dense raceme; yellow or rarely orange or white. Pedicels not articulated. Perianth segments free, spreading or reflexed. Stamens equal, as long as or shorter than the tepals; filaments straight, densely hairy in upper half. Fruit a loculicidal capsule; seeds few to many in each locule. Species about 60, mostly in southern Africa with a few spreading into tropical Africa and Australia.

6. *Jodrellia* Baijnath

Jodrellia Baijnath, Kew Bull. 32: 571 (1978).

Herbaceous, scapose, medium-sized perennials with short, non-bulbous rhizome and terete or fusiform roots. Leaves basal, sheathing, linear. Flowers numerous, rotate, in a very short, dense raceme; white. Pedicels not articulated. Perianth segments free, slightly cup-shaped at the tips. Stamens equal, shorter than the tepals; filaments straight, numerous hairs massed at apical end. Fruit a loculicidal capsule; seeds up to 3 in each locule. Three sp., tropical and subtropical S, Central and E Africa, from Zambia and Zimbabwe northeastward to Kenya, Ethiopia and Somalia.

7. *Bulbinella* Kunth

Bulbinella Kunth, Enum. Pl. 4: 569 (1843); Perry, S. Afr. J. Bot. 53: 431–444 (1987).

Herbaceous, scapose, medium-sized perennials with unbranched, erect rhizomes and swollen or hard roots. Leaves basal, linear, surrounded by a fibrous, sheathing neck. Flowers numerous, stellate, in a tall, dense raceme; yellow, orange or white. Pedicels not articulated. Perianth segments free. Stamens equal, as long as or shorter than the tepals; filaments straight, glabrous. Fruit a loculicidal capsule; seeds 1 to 2 in each locule. About

22 spp., 16 in the winter rainfall region of S Africa and 6 in New Zealand.

8. *Kniphofia* Moench Fig. 43A–C

Kniphofia Moench, Methodus: 631 (1794); Codd, Bothalia 9: 363–513 (1968).

Herbaceous, scapose, medium-sized perennials with branched or solitary rhizomes (rarely somewhat caulescent) and fibrous or fleshy roots. Leaves basal, linear. Flowers numerous, tubular or campanulate, in tall, dense or lax racemes; white, yellow, orange, brownish or various shades of red. Pedicels not articulated. Perianth segments fused. Stamens equal, usually longer than the tepals; filaments straight, glabrous. Fruit a loculicidal capsule; seeds numerous. About 70 spp., mostly Africa, with one in Madagascar and one in southern Arabia.

9. *Aloe* L. Fig. 43D–G

Aloe L., Sp. Pl.: 319 (1753); Reynolds, The aloes of tropical Africa and Madagascar (1966); Reynolds, The aloes of South Africa, 4th edn (1982); Van Wyk & Smith, Guide to the aloes of South Africa (1996).

Herbaceous or woody succulent perennials, with roots cylindrical or rarely fusiform. Leaves rosulate or distichous, ovate to linear-lanceolate, often maculate and spiny. Flowers few to numerous, in a dense or lax raceme or panicle, tubular, more or less straight; yellow, orange, red, brown, whitish or greenish. Pedicels not articulated. Perianth segments united or rarely free. Stamens and style straight; filaments glabrous. Fruit a loculicidal capsule; seeds numerous. About 400 spp., with ± 130 in S Africa, widespread in Africa, Arabia, Socotra and Madagascar, naturalised in the Mediterranean region, India and China.

10. *Lomatophyllum* Willd.

Lomatophyllum Willd., Ges. Naturf. Freunde Berlin Mag. Neuesten Entdeck. Gesammten Naturk. 5: 166 (1811).

Herbaceous or woody succulent perennials, with roots cylindrical. Leaves rosulate, ovate to linear-lanceolate, spiny. Flowers few to numerous, in a lax raceme or panicle, tubular, more or less straight; yellow, orange or red. Pedicels not articulated. Perianth segments united. Stamens and style straight; filaments glabrous. Fruit a fleshy, semi-dehiscent berry; seeds numerous. About 14 spp., Madagascar and some of the Mascarene Islands.

11. *Gasteria* Duval

Gasteria Duval, Pl. Succ. Horto Alencon.: 6 (1809); Van Jaarsv., Gasterias of South Africa: 33 (1994); Van Jaarsv. et al., S. Afr. J. Sci.: 468 (1994), rev.

Herbaceous, succulent perennials, with roots cylindrical or rarely fusiform. Leaves rosulate or distichous, linear-lanceolate, lorate or deltoid, usually maculate and often rugulose, never spiny. Flowers few to numerous, in a dense or lax raceme or panicle, tubular, curved, with a basal swelling; red, pink, or rarely greenish yellow. Pedicels not articulated. Perianth segments united. Stamens and style curved; filaments glabrous. Fruit a loculicidal capsule; seeds numerous. Sixteen spp. and six varieties, southern Africa, mainly in the Eastern and Western Cape Provinces.

12. *Poellnitzia* Uitew.

Poellnitzia Uitew., Succulenta 22: 61 (1940); G.F. Smith, Bothalia 25: 35–36 (1995).

Herbaceous, succulent perennial, with roots cylindrical. Leaves rosulate, deltoid-acuminate, smooth. Flowers numerous, in a lax raceme, tubular, erect, more or less straight; red; connivent at tips. Pedicels not articulated. Perianth segments united. Stamens and style straight; filaments glabrous. Fruit a loculicidal capsule; seeds numerous. Only one sp., *P. rubriflora* (L. Bolus) Uitewaal in the Western Cape Province of S Africa.

13. *Astroloba* Uitew.

Astroloba Uitew., Succulenta 1947(5): 53 (1947); Groen, Succulenta 65: 19–23; 66: 51–55, 82–87, 110–113, 162–167, 171–174, 261–263 (1986–1987).

Herbaceous, succulent perennials, with roots cylindrical. Leaves rosulate, deltoid-acuminate, smooth or rugulose. Flowers numerous, in a lax raceme, tubular, more or less straight; greenish yellow or whitish. Pedicels not articulated. Perianth segments united, sometimes with inflated tissue. Stamens and style straight; filaments glabrous. Fruit a loculicidal capsule; seeds numerous. Seven spp. in the Eastern and Western Cape Provinces of S Africa.

14. *Haworthia* Duval

Haworthia Duval, Pl. Succ. Horto Alencon.: 7 (1809); Bayer, New *Haworthia* handbook (1982); Scott, Genus *Haworthia*, a taxonomic revision (1985).

Herbaceous succulent perennials, with roots cylindrical or rarely fusiform. Leaves rosulate or distichous, ovate to linear-lanceolate, sometimes retuse or truncate, often tuberculate, never spiny. Flowers few to numerous, in a lax or rarely dense raceme, tubular, zygomorphic, bilabiate, more or less straight; brownish or greenish white. Pedicels not articulated. Perianth segments united. Stamens and style straight; filaments glabrous. Fruit a loculicidal capsule; seeds numerous. About 70 spp., widely distributed in S Africa, extending into Swaziland and southern Namibia.

15. *Chortolirion* A. Berger

Chortolirion A. Berger in Engler, Pflanzenreich IV. 33: 72 (1908); G.F. Smith, Bothalia 25: 31–33 (1995).

Herbaceous succulent perennial, with roots fusiform from a bulbous base. Leaves rosulate, graminoid, minutely spiny. Flowers few, in a lax raceme, tubular, zygomorphic, bilabiate, more or less straight; brownish or greenish white. Pedicels not articulated. Perianth segments united. Stamens and style straight; filaments glabrous. Fruit a loculicidal capsule; seeds numerous. Only one sp., *Chortolirion angolense* (Baker) A. Berger, widely distrbuted in southern Africa.

Selected Bibliography

Baijnath, H. 1980. A contribution to the study of leaf anatomy of the genus *Kniphofia* Moench (Liliaceae). In: Brickell, C.D., Cutler, D.F., Gregory, M. (eds.). Petaloid monocotyledons. Horticultural and botanical research. Linn. Soc. Symp. Ser. 8: 89–103. London: Academic Press.

Baijnath, H., Cutler, D.F. 1993. A contribution to the anatomy and surface details of leaves in the genus *Bulbine* (Asphodelaceae) in Africa. S. Afr. J. Bot. 59: 109–115.

Beaumont, J., Cutler, D.F., Reynolds, T., Vaughan, J.G. 1985. The secretory tissue of aloes and their allies. Isr. J. Bot. 34: 265–282.

Brandham, P.E. 1983. Evolution in a stable chromosome system. In: Brandham, P.E., Bennett, M.D. (eds.) Kew Chromosome Conference II. London: Allen & Unwin, pp. 251–260.

Brandham P.E., Cutler, D.F. 1981. Polyploidy, chromosome interchange and leaf surface anatomy as indicators of relationships within *Haworthia* section *Coarctatae* Baker (Liliaceae-Aloineae). J. S. Afr. Bot. 47: 507–546.

British pharmacopoeia. 1993. Vol. 1. London: HMSO, p. 30.

Brummitt, R.K. 1992. Vascular plant families and genera. London: Royal Botanic Gardens, Kew.

Chase, M.W. et al. 1995. See general references.

Coetzee, J., Van der Schijff, H.P. 1969. Secondary thickening in geophytic Liliaceae. J. S. Afr. Bot. 35: 99–108.

Cutler, D.F. 1985. *Haworthia* spp. (Liliaceae) with window leaves: SEM studies of leaf surface adaptations. Parodiana 3: 203–223.

Dagne, E., Yenesew, A. 1994. Anthraquinones and the chemotaxonomy of the Asphodelaceae. Pure Appl. Chem. 66: 2395–2398.

Dagne, E., Yenesew, A., Asmellash, S., Sebsebe, D., Mavi, S. 1994. The distribution of some anthraquinones, pre-anthraquinones and isoeleutherol in the roots of some *Aloe* species. Phytochemistry 35: 401–406.

Dagne, E., Van Wyk, B-E., Mueller, M., Steglich, W. 1996. Three dihydroanthracenones from *Gasteria bicolor*. Phytochemistry 41: 795–800.

Dahlgren, R.M.T., Clifford, H.T. 1982. See general references.

Daumann, E. 1970. Das Blütennektarium der Monokotylen unter besonderer Berücksichtigung seiner systematischen und phylogenetischen Bedeutung. Feddes Repert. 80: 463–590.

De Bruijn A., Cox, A.V., Chase, M.W. 1995. Molecular systematics of Asphodelaceae (Asparagales; Lilianae). Am. J. Bot. 82(6): 124 (Abstr.).

De Wet, J.M.J. 1960. Chromosome morphology in *Kniphofia*. Bothalia 7: 295–297.

Fedorov, A.A. (ed.) 1969. See general references.

Grindlay, D., Reynolds, T. 1986. The *Aloe vera* phenomenon: a review of the properties and modern uses of the leaf parenchyma gel. J. Ethnopharm. 16: 117–151.

Hegnauer, R. 1963. See general references.

Hilton-Taylor, C. 1996. Red Data list of southern African plants. Strelitzia 4: 1–117. Pretoria: National Batanical Institute.

Hilton-Taylor, C., Smith, G.F. 1994. The conservation status of Aloaceae in southern Africa. In: Huntley, B.J. (ed.) Botanical diversity in southern Africa. Strelitzia 1: 287-303. Pretoria: National Botanical Institute.

Hoffman, M.T. 1988. The pollination ecology of *Aloe ferox* Mill. S. Afr. J. Bot. 54: 345–350.

Holland, P.G. 1978. An evolutionary biogeography of the genus *Aloe*. J. Biogeogr. 5: 213–226.

Huber, H. 1969. See general references.

Joshi, A.C. 1937. Megasporogenesis in *Aloe vera* Linn. J. Indian Bot. Soc. 15: 297–300.

Matthews, V.A. 1984. *Asphodelus* L., *Eremurus* Bieb. In: Davis, P.H. (ed.) Flora of Turkey and the East Aegean Islands 8: 85–87. Edinburgh: University Press.

Matthews, V.A., Tuzlaci, E. 1984. *Asphodeline* L. In: Davis, P.H. (ed.) Flora of Turkey and the East Aegean Islands. 8: 88–97. Edinburgh: University Press.

McCabe, S. 1995. Pollination of *Aloe polyphylla*. Cact. Succ. J. (U.S.) 67: 166–168.

McNaughton, J.E., Robertson, B.L. 1974. Megasporogenesis and megagametogenesis in *Aloe africana* Mill. J. S. Afr. Bot. 40: 75–79.

Nash, R.J., Beaumont, J., Veitch, N.C., Reynolds, T., Benner, J., Hughes, C.N.G., Dring, J.V., Bennet, R.N., Dellar, J.E. 1992. Phenylethylamine and pyperidine alkaloids in *Aloe* species. Planta Med. 58: 84–87.

Ratsirarson, J. 1995. Pollination ecology of *Aloe divaricata*, Berger (Liliaceae): an endemic plant species of southwest Madagascar. S. Afr. J. Bot. 61: 249–252.

Reynolds, T. 1985a. The compounds in *Aloë* leaf exudates: a review. Bot. J. Linn. Soc. 90: 157–177.

Reynolds, T. 1985b. Observations on the phytochemistry of the *Aloë* leaf exudate compounds. Bot. J. Linn. Soc. 90: 179–199.

Reynolds, T. 1994. A chromatographic examination of some old samples of drug aloes. Pharmazie 49: 524–529.

Rheede van Oudtshoorn, M.C.B. 1964. Chemotaxonomic investigations in Asphodeleae and Aloineae (Liliaceae). Phytochemistry 3: 383–390.

Richardson, I.B.K., Smythies, B.E. 1980. *Asphodelus*. In: Tutin, T.G. (ed.) Flora Europaea 5: 17. Cambridge: Cambridge University Press.
Riley, H.P., Majumdar, S.K. 1979. The Aloïneae: a biosystematic survey. Lexington: University Press of Kentucky.
Rowley, G.D. 1967. Zygomorphy in *Bulbine*. Natl. Cact. Succ. J. 22: 10.
Rowley, G.D. 1982. Intergeneric hybrids in succulents. Natl. Cact. Succ. J 37: 2–6, 45–49, 76–80, 119.
Rudall, P.J., Chase, W.M., Furness, C.A. 1997. Microsporogenesis and pollen sulcus type in Asparagales. Can. J. Bot. 75: 408–430.
Schnarf, K. 1929. Die Embryologie der Liliaceae und ihre systematische Bedeutung. S. B. Akad. Wiss. Wien I, 138: 69–92.
Schnarf, K., Wunderlich, R. 1939. Zur vergleichenden Embryologie der Liliaceae-Asphodeloideae. Flora, Jena 33: 297–327.
Schulze, W., 1975. Beiträge zur Taxonomie der Liliifloren I. Asphodelaceae. Wiss. Z. Friedrich- Schiller-Univ. Jena 24, 4: 403–415 (Beitr. zur Phytotaxonomie 4).
Sigler, A., Rauwald, H.W. 1994a. *Aloe* plants accumulate anthrone-type anthranoids in inflorescence and leaves, and tetrahydroanthracenes in roots. Z. naturforsch. 49c: 286–292.
Seigler, A., Rauwald, H.W. 1994b. Tetrahydroanthracenes as markers for the subterranean anthranoid metabolism in *Aloe* species. J. Plant Physiol. 143: 596–600.
Smith, G.F. 1988. A scanning electron microscopic investigation of the pollen morphology of *Chortolirion* Berger (Aloineae; Liliaceae). S. Afr. J. Sci. 84: 428–430.
Smith, G.F. 1991a. Studies on the reproductive biology and palynology of *Chortolirion* Berger (Asphodelaceae: Alooideae) in southern Africa. Taxon 40: 61–73.
Smith, G.F. 1991b. The chromosomes of *Chortolirion* and *Poellnitzia* (Asphodelaceae: Alooideae). Bothalia 21: 171–175.
Smith, G.F. 1995. FSA contributions 3: Asphodelaceae/ Aloaceae, 1028010 *Poellnitzia*. Bothalia 25: 35–36.
Smith, G.F., Tiedt, L.R. 1991. A rapid, non-destructive osmium tetroxide technique for preparing pollen for scanning electron microscopy. Taxon 40: 195–200.
Smith, G.F., Van Wyk, B-E. 1991. Generic relationships in the Alooideae (Asphodelaceae). Taxon 40: 557–581.
Smith, G.F., Van Wyk, A.E. 1992. Systematic leaf anatomy of selected genera of southern African Alooideae (Asphodelaceae). S. Afr. J. Bot. 58: 349–357, 543.
Smith, G.F., Theron, P.D., Loots, G.C. 1992. Notes on the microfaunal complement and pollination mechanisms of *Poellnitzia rubriflora* (Asphodelaceae: Alooideae): an example of mite-flower domatia association. Taxon 41: 437–450.
Stenar, H. 1928. Zur Embryologie der *Asphodeline*-Gruppe. Ein Beitrag zur systematischen Stellung der Gattungen *Bulbine* und *Paradisia*. Sven. Bot. Tidskr. 22: 145–159.
Strahm, W. 1989. Plant red data book for Rodrigues. Königstein: Koeltz Scientific Books.
Strahm, W. 1993. The conservation and restoration of the flora of Mauritius and Rodrigues. Unpubl. Ph.D. Dissertation Reading: University of Reading.
Takhtajan, A.L. 1982. See general references.
Tamura, M.N. 1995. A karyological review of the orders of the Asparagales and Liliales (Monocotyledonae). Feddes Repert. 106: 83–111.
Tuzlaci, E. 1987. Revision of the genus *Asphodeline* (Liliaceae) – a new infrageneric classifcation. Candollea 42: 559–576.
United States Pharmacopeia. 1995. 23rd edn. Rockville: United States Pharmacopeial Convention, Inc, p. 46.
Vaikos, N.P., Markandeya, S.K., Pai, R.M. 1978. The floral anatomy of the Liliaceae. The tribe Aloineae. Indian J. Bot. 1: 61–68.
Van Jaarsveld, E.J. 1992. A synopsis of the genus *Gasteria*. Aloe 29: 1–32.
Van Jaarsveld, E.J. 1994. Gasterias of South Africa. Cape Town: Fernwood Press and the National Botanical Institute.
Van Jaarsveld, E.J., Smith, G.F. 1997. Aloaceae. In: Oldfield, S. (comp.) Cactus and succulent plants – status survey and conservation action plan. IUCN/SSC Cactus and Succulent Specialist Group. Gland and Cambridge: IUCN. pp. 10–14, 131.
Van Jaarsveld, E.J., Smith, G.F. 1997. Aloaceae. In: Oldfield, S. (ed.) IUCN/SSC action plan for the conservation of cacti and succulents. Gland: IUCN. (in press).
Van Jaarsveld, E.J., Smith, G.F., Van Wyk, B-E. 1994. A cladistic analysis of *Gasteria* (Aloaceae). S. Afr. J. Sci. 90: 467–470.
Van Wyk, B-E., Whitehead, C.S., Glen, H.F., Hardy D.S., Van Jaarsveld, E.J., Smith, G.F. 1993. Nectar sugar composition in the subfamily Alooideae (Asphodelaceae). Biochem. Syst. Ecol. 21: 249–253.
Van Wyk, B-E., Viljoen, A.M., Dagne, E. 1995a. Chemotaxonomic studies in African Aloaceae and Asphodelaceae. Proceedings of the 6th NAPRECA Symposium on Natural Products, Kampala, Uganda, pp. 15–18.
Van Wyk, B-E., Yenesew, A., Dagne, E. 1995b. Chemotaxonomic significance of anthraquinones in the roots of Asphodeloideae (Asphodelaceae). Biochem. Syst. Ecol. 23: 277–281.
Van Wyk, B-E., Yenesew, A., Dagne, E. 1995c. The chemotaxonomic significance of anthraquinones and preanthraquinones in the genus *Lomatophyllum*. Biochem. Syst. Ecol. 23: 805–808.
Van Wyk, B-E., Van Rheede van Oudtshoorn, M.C.B., Smith, G.F. 1995d. Geographical variation in the major compounds of *Aloe ferox* leaf exudate. Planta Med. 61: 250–253.
Watson, E.M. 1986. Cytoevolutionary studies in the genus *Bulbine* Wolf (Liliaceae). I. The Australian perennial taxa (*B. bulbosa* s.l.). Aust. J. Bot. 34: 481–504.
Webb, D.A. 1980. *Asphodeline* Reichenb., *Eremurus* Bieb., *Simethis* Kunth. In: Tutin, T.G. (ed.), Flora Europaea 5: 17–19. Cambridge: Cambridge University Press.
Wendelbo, P. 1964. On the genus *Eremurus* (Liliaceae) in southwest Asia. Acta Univ. Bergensis, Ser. Math.-Nat. 5: 1–45.

Asteliaceae

C. Bayer, O. Appel, and P.J. Rudall

Asteliaceae Dumort., Anal. Fam. Pl.: 59, 61 (1829).

Perennial, short stemmed *herbs*, often rhizomatous, occasionally epiphytic; dioecious, gynodioecious or hermaphroditic (*Milligania*). *Leaves* spirally arranged, with open or closed sheaths and linear to lanceolate laminas, usually densely pubescent, at least when young, sometimes whitish to silvery, woolly, especially on the broadened base. *Inflorescences* terminal, composed of racemes, few-flowered in some spp., sometimes densely covered with silky-woolly, whitish to silvery hairs. *Flowers* trimerous or up to heptamerous, pedicellate without articulation to sessile, actinomorphic, pentacyclic. Tepals in 2 whorls, free or basally fused, membranous and bractlike or fleshy. Stamens 6, filaments free or basally adnate to tepals in *Milligania*. Anthers dorsifixed to basifixed, longitudinally dehiscent, introrse; female flowers with staminodes. Ovary superior, 3(–7)-carpellate, trilocular with axile placentation, or rarely (in *Astelia* section *Astelia*) unilocular with parietal placentation; mucilage-secreting intraovarian trichomes often present (*Astelia*, *Collospermum*); septal nectaries present; ovules several per locule, anatropous, funicles hairy in some species; styluli short (*Milligania*) or lacking; stigmas usually (sub-)sessile, sometimes confluent; male flowers with pistillodes, or sometimes bearing (sub-)fertile ovules. *Fruit* a berry, or capsule in *Milligania*, few- to many-seeded. Seeds more or less ovate, often angular, sometimes embedded in mucilaginous hairs; testa hard, black; endosperm abundant.

A family of four genera and about 35 spp. from Australia, New Zealand, the Pacific Islands, Mauritius, Réunion and southern S America.

Vegetative Structures. Asteliaceae comprise densely caespitose herbs with rosette-like arranged leaves (Fig. 44). The epiphytic forms in humid forests sometimes resemble neotropical Bromeliaceae. As in many Bromeliaceae, the leaves of some Asteliaceae form funnel-shaped, water-collecting cisterns (Moore and Edgar 1970). A peculiar whitish to silvery indumentum is usually present on both surfaces of at least young leaves and axes, although tending to rub off in older leaves. The trichomes are scalelike or stellate, long, multicellular and arising from a multicellular base, with marginal cells peeling off to form a matted, untidy "wool" over the surface (Fig. 44C) (McCarthy 1928; Skottsberg 1934a,b; Moore 1966; Rudall et al. 1997c). Betts (1920) assumed that the trichomes serve to absorb water since they rapidly become filled with water when moistened. In some species the leaf blades may reach more than 2 m in length. The basal portion of the leaf is usually broad and often conspicuously pubescent on the adaxial surface. A petiole is lacking, but a narrower zone with enrolled margins between the broadened base and the lamina may be found in older leaves of some species. Leaves are V-shaped in cross-section, but often

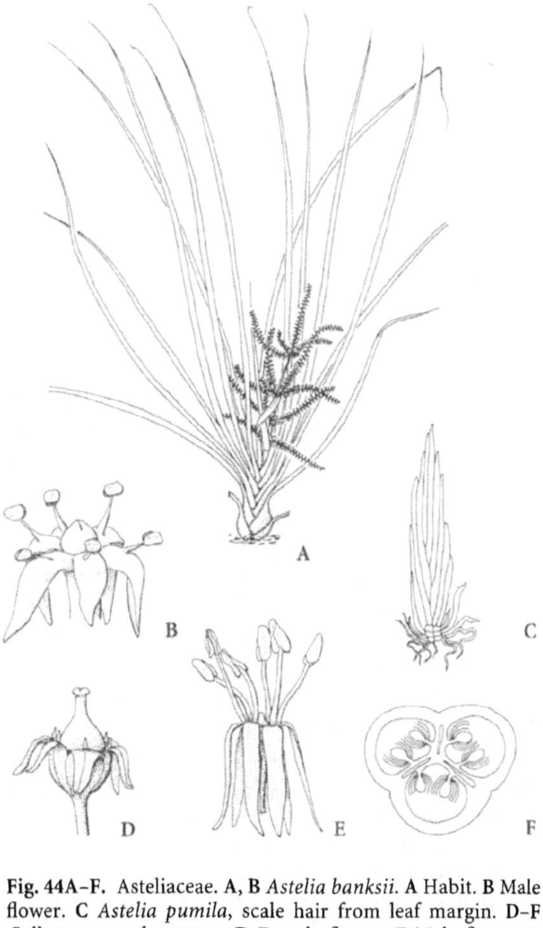

Fig. 44A–F. Asteliaceae. **A, B** *Astelia banksii*. **A** Habit. **B** Male flower. **C** *Astelia pumila*, scale hair from leaf margin. **D–F** *Collospermum hastatum*. **D** Female flower. **E** Male flower. **F** Ovary, transversal section

with 2 prominent lateral bundles, which are often larger than the midrib. There is usually a hypodermal layer of 1–2 cells present beneath the adaxial epidermis. Mucilage canals are present in leaves of some species of *Astelia* and *Collospermum*, associated with the xylem poles of some vascular bundles. Sclerenchyma is often present at the xylem poles of vascular bundles, sometimes extending round to the phloem poles. At the 2 main lateral bundles the sclerenchyma often extends to both adaxial and abaxial epidermises as girders (Rudall et al. 1997c). Stomata are paracytic, with 2 subsidiary cells, and usually present on the adaxial surface only (Stebbins and Khush 1961; Rudall et al. 1998b).

A secondary thickening meristem is lacking in *Astelia* (Rudall 1995). Vessels are restricted to the roots and have scalariform perforation plates (Cheadle and Kosakai 1971). Raphides are normally present in idioblasts in the mesophyll, and crystals of various shapes have been described for several species of *Astelia* and *Collospermum* (Rohert and Zalenski 1899; Skottsberg 1934a; Rudall et al. 1998b).

Reproductive Structures. The inflorescence consists of one terminal and several lateral racemes (or spikes), although in small species inflorescences are often highly reduced. The lower side branches of the inflorescence may be composed of several racemes. Although these additional racemes usually arise in transversal position at or near the base of the first order lateral raceme, they probably do not arise from the collateral accessory bud, since in some cases they are subtended by bracts. Similarly, the first flowers of distal, simple racemes arise in transversal positions. Prophylls are absent from the pedicels.

Asteliaceae are probably entomophilous. This is indicated by the structure and fragrance of the flowers and supported by the observation of honeybees and large flies visiting *Astelia* (Skottsberg 1934b; Moore 1966). Extensive, branching septal nectaries are present in both male and female flowers of many species of *Astelia* and *Collospermum*, though very reduced in unilocular species of *Astelia*. In *Milligania*, septal nectaries are short, limited to the upper part of the ovary (P.J. Rudall, unpubl.; J.G. Conran, pers. comm.). In *Astelia* and *Collospermum*, droplets of nectar are exuded from furrows or slits on the upper part of the ovary (Moore 1966), around the base of the short style. Heslop-Harrison and Shivanna (1977) reported dry stigmas with unicellular papillae for *Astelia*. According to Moore (1966), stigmatic papillae are much longer in *Collospermum* than in *Astelia*. In *Astelia*, although flowers are unisexual, male flowers bear ovules which may occasionally be fertile, so dioecy is not complete.

Ovaries of *Astelia* and *Collospermum* are superior (as also in *Blandfordia*), in contrast to most other "lower" Asparagales, such as Hypoxidaceae, *Doryanthes* and *Ixiolirion*, which are epigynous. In *Milligania* the ovary is semiinferior, as in the related genus *Lanaria*. In *Astelia* and *Collospermum* the gynoecium is often relatively short and broad, with a broad stigmatic region. Ovaries are generally trilocular with axile or submarginal placentation (Fig. 44F), except in *Astelia* subgenus *Astelia*, which has unilocular ovaries and parietal placentation. Many species of *Astelia* and *Collospermum* (but not *Milligania*) have intraovarian trichomes emerging from the funicles and placentae. These trichomes secrete mucilage (Rudall et al. 1998a) which fills the ovary locules in these species and provides a medium for pollen-tube growth. In *Collospermum* mucilaginous trichomes enclose the mature seeds and may facilitate seed dispersal of epiphytic species (Skottsberg 1934a).

The fruit is a berry in *Astelia* and *Collospermum*, and a capsule in *Milligania*. Berries of many species are orange to dark red, juicy and sweet, and are probably dispersed by birds.

The outer testal epidermis loses its cell structure and gives rise to a thick continuous phytomelan layer (Danilova et al. 1995). In *Astelia*, the inner integument is reduced to a compressed, but not collapsed, layer of empty, colourless cells. Endosperm cells are thin-walled and isodiametric. They contain abundant oil and aleurone, but no significant amounts of starch or reserve cellulose are found. The embryo is well developed, slender and straight or slightly bent (Huber 1969).

Embryology. Microsporogenesis is simultaneous in *Milligania*, as in most other "lower" Asparagales (Rudall et al. 1997). The ovules of *Astelia* are crassinucellate (P.J. Rudall, unpubl.), in contrast to those of the related family Hypoxidaceae.

Pollen Morphology. Pollen grains of *Astelia*, *Collospermum* and *Milligania* are usually sulcate (Erdtman 1952; Cranwell 1953; Radulescu 1973), although in *Collospermum* both sulcate and trichotomosulcate grains have been found in the same anther (Roth et al. 1987), as in occasional other "lower" asparagoid taxa with simultaneous microsporogenesis (Rudall et al. 1997). Grains are sphaeroidal to oblate or ellipsoidal and measure ca. 15 to 50 μm in diameter. In *Astelia* the exine is thin to moderately thick, but much thickened in the apertural region. Radulescu (1973) considered the exine to be foveolate. Pollen grains of *Astelia* are echinate with short conical spines; spines are shorter in grains of *Collospermum*, and indistinct or lacking in *Milligania* (Cranwell 1953).

KARYOLOGY. According to Satô (1942), the chromosomes of *Astelia* are of medium size (4–6 μm) and probable base numbers of the genus are $x = 8$, 30 and 35. These results were questioned by Wheeler (1966), who reported small chromosomes and found a base number of $x = 35$ to be most likely. Polyploid series with $2n$, $4n$ and $6n$ have been recorded. During meiosis, diploid species show regular pairing and segregation of chromosomes, whereas in the hexaploid *A. trinervia* and *A. cockaynei* ($n = 105$) multivalents were observed (Wheeler 1966).

PHYTOCHEMISTRY. Saponins have been recorded for the roots, leaves and flowers of *Astelia pumila* (Ricardi et al. 1958). Williams and Harborne (1988) reported quercetin, kaempferol and isorhamnetin. Seeds of *Astelia* and *Collospermum* contain oil; Morice (1967) reported contents of up to almost 50% (*Collospermum hastatum*). In some *Astelia* species, palmitic, oleic and linoleic acid are the main seed oils, whereas in others and two *Collospermum* species linoleic and linolenic acids are more abundant than palmitic and oleic acid. The distribution of fatty acids does not support Skottsberg's (1934a) generic delimitation, since linolenic acid is present in *Astelia* section *Desmoneuron* and *Isoneuron*, and in two species of *Collospermum*, while it is lacking in other sections of *Astelia* (Morice 1975).

RELATIONSHIPS WITHIN THE FAMILY. *Neoastelia*, *Astelia* and *Collospermum* are very closely related and their separate generic status is questionable. Intergeneric hybrids between *Astelia* and *Collospermum* can be obtained by artificial pollination (Moore and Edgar 1970). In contrast, *Milligania* is markedly different, with its lack of mucilage canals, hermaphroditic flowers, semiinferior ovaries, distinct stylar branches and loculicidal capsules. Analyses of molecular data from *rbc*L (Chase et al. 1995a; Rudall et al. 1998b) supported the affinities between *Milligania* and *Astelia*/*Collospermum* but suggested even closer relationships between *Milligania* and *Blandfordia*. However, a position for *Blandfordia* within Asteliaceae is contrary to the morphological evidence (Rudall et al. 1998b), except that *Blandfordia* and Asteliaceae both have superior ovaries (semiinferior in *Miligania*).

Skottsberg (1934a) divided the large genus *Astelia* into three subgenera, *Euastelia*, *Asteliopsis* and *Tricella*, of which the two former were further subdivided into sections. Skottsberg (1934a) considered the isolated American species *A. pumila* to be more closely related to the Hawaiian species than to those from New Zealand and Australia. He regarded the Mascarene species, *A. hemichrysa*, as a relict species, closest to *A. papuana* (New Guinea), *A. linearis* (subantarctic) and *A. alpina* (Australia).

AFFINITIES. Members of Asteliaceae, with a phytomelaniferous seed coat, clearly belong in the order Asparagales. Although they have been linked with various asparagoid taxa, including *Dracaena* (Bentham and Hooker 1883) and *Cordyline* (Dahlgren et al. 1985), analysis of molecular data from *rbc*L (Chase et al. 1995a,b; Rudall et al. 1998b) indicated that the taxa most closely related to Asteliaceae are Hypoxidaceae, *Blandfordia* and *Lanaria*, which together form a clade that is sister to the orchids, at the base of the order Asparagales. Although Hypoxidaceae and Asteliaceae differ from each other in several respects, especially gynoecium position and pollen sculpturing, there are some unusual anatomical characters linking the two families, notably branched hairs and mucilage canals (Rudall et al. 1998b).

DISTRIBUTION AND HABITATS. Apart from several species of *Astelia* on the Hawaiian archipelago, Asteliaceae is an entirely southern hemisphere family. *Milligania* is endemic to Tasmania. *Collospermum* occurs in New Zealand, Fiji and Samoa. *Neoastelia* is a monotypic segregate from *Astelia*, endemic to NE New South Wales. *Astelia* has the widest distribution, ranging from subantarctic islands to New Zealand, Australia, New Guinea, Pacific islands, Mascarenes, Chile and Hawaii. It was presumably dispersed to Hawaii by birds, as it has attractive fruits (in common with a relatively high proportion of the Hawaiian flora). Habitats range from coasts to montane or alpine vegetation on rocks, in bogs or forests. Most species grow on the ground, although some species of *Astelia* and *Collospermum* are exclusively or facultatively epiphytic on trees.

Palaeobotany. According to Mildenhall (1980), the fossil record of *Astelia* starts in the upper Eocene from New Zealand (cf. Couper 1960).

Economic Importance. The fibres of *Astelia grandis* are used in New Zealand. Fruits of *A. nervosa* are edible (Dahlgren et al. 1985).

Key to the Genera

1. Flowers penta- to heptamerous — 1. ***Neoastelia***
- Flowers trimerous — 2
2. Flowers hermaphrodite; fruit dry, dehiscent — 2. ***Milligania***
- Flowers usually unisexual; fruit fleshy, indehiscent — 3
3. Leaf sheaths sharply keeled or plants very small; perianth of male and female flowers of equal size; tepal tube fleshy or tepals free; anthers elliptic to ovate, dorsifixed, versatile — 3. ***Astelia***
- Leaf sheaths rounded, not keeled; perianth of male flowers larger than that of female flowers; tepal tube membranous and deeply cup-shaped; anthers linear, sagittate, (sub-) basifixed, immobile — 4. ***Collospermum***

Genera of Asteliaceae

1. *Neoastelia* J.B. Williams

Neoastelia J.B. Williams, Fl. Australia 45: 493 (1987).

Tufted, rhizomatous perennials with hidden leafy stems; leaves keeled, with adpressed whitish indumentum on abaxial surface; flowers hermaphroditic, pedicellate; tepals fused at base, membranous; stamens connate with tepal tube; placentation axile; ovules numerous in 2 distinct rows per locule; stylar branches elongated with decurrent stigmas; seeds numerous, small, smooth, glossy. Only one sp., *N. spectabilis* J.B. Williams, endemic to NE New South Wales.

2. *Milligania* Hook. f.

Milligania Hook. f., J. Bot. Kew Gard. Misc. 5: 296 (1853), nom. cons.; Williams, Fl. Australia 45: 169–173 (1987), rev.

Tufted rhizomatous perennials, stems short, usually hidden; leaves keeled; flowers pedicellate with basally fused tepals; filaments of stamens more or less fused with tepals; ovary superior or basally adnate to tepal tube, 3-locular; placentation axile; ovules numerous; styles or stylar branches 3; capsules globose to ovoid-oblong, splitting from the apex; seeds several, glossy. Five spp., Australia (Tasmania).

3. *Astelia* Banks & Sol. ex R. Br. Fig. 44A–C

Astelia Banks & Sol. ex R. Br., Prodr. 291 (1810), nom. cons.; Skottsb., K. Sven. Vetenskapsakad. Handl. III, 14 (2): 1–106, 24 pl. (1934), rev.

Dioecious or rarely gynodioecious, rhizomatous perennials, large to small and cushionlike, sometimes epiphytic; leaves rosulate, sheaths broad, usually keeled; lower branches of inflorescence usually with basal ramifications; flowers sessile or pedicellate; tepals basally fused (fleshy if tubular) or free, persistent; ovary 3(–4)-locular with subapical placentas, or unilocular with parietal placentation; funicles sometimes hairy; stigmas (sub-)sessile, occasionally confluent; embryo straight. $n = 35, 105, 2n = 70, 140$. About 25 spp., widely distributed on Pacific Islands (Hawaiian Islands, Marquesas, Tahiti, Society Islands, Samoa, Fiji, New Caledonia, Réunion, Mauritius, Auckland, Campbell, and Chatham Islands), in New Guinea, New Zealand, temperate SE Australia, Tasmania, the Falkland Islands, S Chile and Argentina.

4. *Collospermum* Skottsb. Fig. 44D–F

Collospermum Skottsb., K. Sven. Vetenskapsakad. Handl. III, 14 (2): 72 (1934).

Dioecious, epiphytic or rupestral perennials with short stems; leaf sheaths rounded, not keeled; inflorescence composed of simple racemes; perianth membranous, basally fused, forming a deeply cup-shaped tube with narrow, reflexed lobes; anthers persistent, sterile and flat in female flowers; ovules on the septa, few to many; style short, thick, stigmas 3, long papillose, confluent; seeds embedded in mucilaginous funicular hairs. $n = 35$. Four spp., New Zealand, Fiji, Samoa.

Selected Bibliography

Arber, A. 1925. See general references.
Badawi, A.A., Elwan, Z. 1986. A taxonomic study of Liliaceae sensu lato 1 & 2. Phytologia 60: 201–221.
Bartlott, W., Frölich, D. 1983. Micromorphologie und Orientierungsmuster epicuticularer Wachs-Kristalloide: ein neues systematisches Merkmal bei Monokotylen. Plant Syst. Evol. 142: 171–185.
Bentham, G., Hooker, J.D. 1883. Genera Plantarum, vol. 3(2). London: Reeve & Co.
Betts, M.W. 1920. Notes on the autecology of certain plants of the peridotite belt, Nelson: Part 1. Structure of some of the plants (No. 3). Trans. Proc. N. Z. Inst. 52: 276–314.
Chase, M.W. et al. 1995a. See general references.
Chase, M.W. et al. 1995b. See general references.
Chase, M.W., Rudall, P.J., Conran, J. 1996. New circumscrip-

tions and a new family of asparagoid lilies: genera formerly included in Anthericaceae. Kew Bull. 51: 667–680.
Cheadle, V.I., Kosakai, H. 1971. Vessels in Liliaceae. Phytomorphology 21: 320–333.
Couper, R.A. 1960. New Zealand Mesozoic and Cenozoic plant microfossils. N. Z. Geol. Surv. Paleontol. Bull. 32: 1–88.
Cranwell, L.M. 1953. New Zealand pollen studies. The monocotyledons. Bull. Auckl. Inst. Mus. 3: 1–91.
Dahlgren, R. et al. 1985. See general references.
Danilova, M.F., Nemirovich-Danchenko, E.N., Komar, G.A., Lodkina, M.M. 1995. The seed structure of monocotyledons. In: Rudall, P.J., Cribb, P.J., Cutler, D.F., Humphries, C.J. (eds.) Monocotyledons: systematics and evolution. Royal Botanic Gardens, Kew, pp. 461–472.
Erdtman, G. 1952. See general references.
Herriott, E.M. 1905. On the leaf structure of some plants from the Southern Islands of New Zealand. Trans. Proc. N. Z. Inst. 38: 377–422.
Heslop-Harrison, Y., Shivanna, K.R. 1977. The receptive surface of the angiosperm stigma. Ann. Bot. (Lond.) 41: 1233–1258.
Huber, H. 1969. See general references.
McCarthy, E.M. 1928. The structure and development of *Astelia nervosa* var. *sylvestris*. Trans. Proc. N. Z. Inst. 59: 343–360.
Metcalf, L.F. 1972. Some characteristic New Zeland Liliaceae. Lilies 1973: 72–93.
Mildenhall, D.C. 1980. New Zealand Late Cretaceous and Cenozoic plant biogeography: a contribution. Palaeogeogr. Palaeoclimatol. Palaeoecol. 31: 197–233.
Moore, L.B. 1966. Australasian asteliads (Liliaceae) with special reference to New Zealand species of *Astelia* subgenus *Tricella*. N. Z. J. Bot. 4: 201–240.
Moore, L.B. 1980. Hybrid asteliads (Liliaceae). N. Z. J. Bot. 18: 37–42.
Moore, L.B., Edgar, E. 1970. Flora of New Zealand, Vol. 2. Wellington: Shearer, pp. 27–40.
Morice, I.M. 1967. Seed fats of *Astelia* and *Collospermum*, family Liliaceae. J. Sci. Food Agric. 18: 343–346.
Morice, I.M. 1975. Seed fats of further species of *Astelia*. Phytochemistry 14: 1315–1318.
Radulescu, D. 1973. Recherches morpho-palynologiques sur la famille Liliacaeae. Lucr. Gradinii Bot. Bucur. (1973) 133–248.
Ricardi, M., Marticorena, C., Silva, M., Torres, F. 1958. Detección de saponinas en angiospermae chilenas. Bol. Soc. Biol. Concepción 33: 25–94.
Rohert, W., Zalenski, W. 1899. Über eine besondere Kategorie von Krystallbehältern. Bot. Centralbl. 80: 1–11, 33–50, 97–106, 145–156, 194–204, 241–251.
Roth, J.L., Walker, J.W., Walker, A.G. 1987. The distribution and systematics of trichotomosulcate pollen within the lilialean complex. Am. J. Bot. 74: 751.
Rudall, P.J. 1995. New records of secondary thickening in monocotyledons. IAWA J. 16: 261–268.
Rudall, P.J., Furness, C.A., Chase M.W., Fay, M.F. 1997. Microsporogenesis and pollen sulcus type in Asparagales (Lilianae). Can. J. Bot. 75: 408–450.
Rudall, P.J., Prychid, C., Jones, C. 1998a. Intra-ovarian trichomes in monocotyledons. In: Owens, S.J., Rudall, P.J. (eds.) *Reproductive biology*. Royal Botanic Gardens, Kew. (in press)
Rudall, P.J., Chase, M.W., Cutler, D.F., Rusby, J. 1998b. Anatomical and molecular systematics of Asteliaceae and Hypoxidaceae. Bot. J. Linn. Soc. (submitted).
Satô, D. 1942. Karyotype alteration and phylogeny in Liliaceae and allied families. Jpn. J. Bot. 12: 57–161.
Skottsberg, C. 1934a. Studies in the genus *Astelia* Banks et Solander. K. Sven. Vetenskapsakad. Handl. III, 14: 1–106, 14 pl.
Skottsberg, C. 1934b. *Astelia* and *Pipturus* of Hawaii. Bernice P. Bishop Mus. Bull. 117: 1–77, 38 plates.
Skottsberg, C. 1960. *Astelia* on Mauritius. Sven. Bot. Tidskr. 54: 477–482.
Stebbins, G.L., Khush, G.S. 1961. Variation in the organization of the stomatal complex in the leaf epidermis of monocotyledons and its bearing on their phylogeny. Am. J. Bot. 48: 51–59.
Wheeler, J.M. 1966. Cytotaxonomy of the large Asteliads (Liliaceae) of the North Island of New Zealand. N. Z. J. Bot. 4: 95–113.
Williams, C.A., Harborne, J.B. 1988. Distribution and evolution of flavonoids in the monocotyledons. In: Harborne, J.B. (ed.) The flavonoids: advances in research since 1980. London: Chapman and Hall, pp. 505–524.

Behniaceae

J.G. Conran

Behniaceae Conran, Rudall & Chase, Kew Bull. 52: 996 (1997).

Dioecious, slender *shrubby perennials* from short rhizomes. Roots fibrous. Rhizome with monocotyledonous secondary thickening. Stems thin, branching, erect or dextrorsely twisting. *Leaves* alternate, distichous, shortly petiolate, without a sheathing base; blades ovate; veins numerous, parallel, close, with many cross-veins, midrib prominenet. *Inflorescence* an axillary bostryx or flowers borne singly. *Flowers* small, hypogynous, actinomorphic, articulated on the pedicels with a short pericladium; perianth marcescent, not twisted after anthesis; tepals in 2 whorls of 3, subequal, united for about 2/3 of their length, white or pale green. Stamens 6 in 2 whorls; filaments adnate to the tepals; anthers 2-thecate, 4-sporangiate, basal-dorsifixed, introrse, dehiscing longitudinally by slits. Gynoecium of 3 united carpels, ovary 3-locular with septal nectaries; style simple, stigma 3-lobate; ovules few per locule, anatropous, in 2 rows on axile placentas. *Fruit* a berry. Seeds few to numerous, medium-sized, angular, pale brown/yellow drying blackish brown; embryo clavate, about 2/3 as long as the seed; endosperm copious, without starch.

A monotypic family from southern Africa.

Vegetative Morphology and Anatomy. The plants are short-rhizomatous and have erect, branching stems which may twine (Fig. 45). The aerial stems are wiry, tough and supple, up to 5 mm in diameter, and climb by twining dextrorsely. The main stems and lateral branches show determinate growth. The roots are fibrous; climbing roots are absent.

Germination is eligulate. The cotyledon is exposed and little differentiated. The 1st leaf alternates with the cotyledon and the primary axis initially bears sheathing scale leaves. Later leaves are entire and multicostate, ovate and bluntly acute. They have a prominent midrib, well-developed blade and tapering base. The branches bear reduced scale leaves at their lower nodes. Venation is acrodromous with proximal convergence of the primary veins. There is pronounced acropetal weakening of the primary veins, and the secondary veins are pinnate. Higher vein orders are transverse, and the areoles are elongated transversely to the leaf axis. Vernation is supervolute (mostly after Conover 1983; Conran 1985).

The flowers are axillary and borne in reduced cymes (botryces) or singly. The peduncles may be large and well developed or reduced and bear 1-several cataphylls. Inflorescence production appears to be associated only with new growth.

Vessels are present in the roots and stems and there are tracheids in the leaves. The perforation plates are simple or scalariform (Wagner 1977). The roots have a uniseriate exodermis which is unthickened. The cortex is parenchymatous. The epidermal cells of the stem are small, with slightly thickened outer walls and a thick cuticle. A ring of fibrous bundles and small collateral bundles is found in the cortex, the outer part of which may be slightly collenchymatous. Scattered bundles occur in the centre of the stem. In the rhizome there is a well-developed and heavily lignified secondary xylem and parenchymatous secondary phloem similar to that reported by Rudall (1995) in *Herreria* (Herreriaceae) rhizomes. The epidermal cells are isodiametric to shortly elongate with non-sinuous periclinal walls. Stomata are present on the abaxial leaf surface, anomocytic, and the subsidiary cells show oblique division. The mesophyll is undifferentiated. Silica bodies are lacking, but calcium oxalate raphides bundles and, more rarely, crystal druses occur in all parts of the plants. Tannin cells, mucilage cells and laticifers are absent (Conran 1985).

Flower Structure. The perianth is 3-merous, regular and petaloid, fused into a syntepal tube with reflexed free lobes. Venation of the tepals is 3-nerved and convergent. The flowers have a short pericladium. The filaments are filiform and in female flowers are attached to the tepals and reduced in size. Anther dehiscence is introrse. The anthers are basal-dorsifixed and dehisce by longitudinal slits. The gynoecium is 3-carpellary and the ovary 3-locular with axile placentas. The style has a bluntly 3-lobed, minutely papillose and Wet stigma. The flowers are of two types, the plants being apparently functionally dioecious (Fig. 45B,C). The male flowers have well developed stamens and an apparently non-functional ovary which is shorter than the perianth, lacks loculi, but has well-developed septal nectaries. In female flowers, the stamens are small and attached to the tepals for about 2/3 of their length, the ovary is

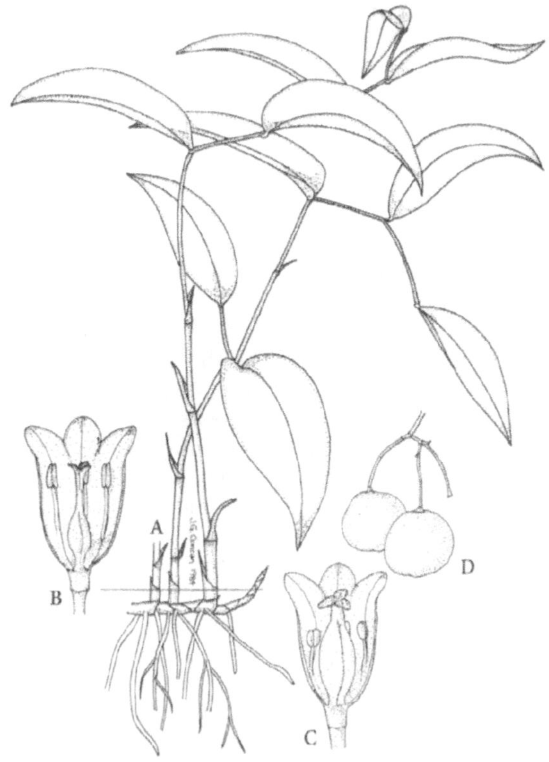

Fig. 45A–D. Behniaceae. *Behnia reticulata*. **A** Habit. **B** Functionally male flower. **C** Functionally female flower. **D** Berry. (Original Conran)

large, equalling or slightly exceeding the tepal tube.

Embryology. The well-developed endothecium has spiral thickenings, and the tapetum is binucleate. The ovule is bitegmic and tenuinucellate. Microsporogenesis is successive and the pollen grains are 2-celled at anthesis.

Pollen Morphology. The pollen grains are shed singly and are obovate and sulcate with a reticulate exine and 27–60 μm in length (Schulze 1982).

Pollination and Reproduction. The flowers are functionally dioecious and probably insect-pollinated. Flowers of both sexes have well-developed septal nectaries. The late arrested development of the anthers in female flowers after the anther is formed but apparently before microsporogenesis is similar to some of the genotypes of *Asparagus officinalis* examined by Lasarte and Palser (1979). Further study of the breeding mechanisms in the family is needed.

Fruit and Seed. The fruits are berries. The perianth is persistent. Mature fruits are pale yellow or greenish, drying black. The pericarp is thick and fleshy and encloses up to about 6 subovoid angular seeds. The seed surface is smooth in fresh seeds, becoming wrinkled on drying. The seeds are pale yellow when fresh, drying dark brown to almost black, but are not phytomelan-encrusted. The seed structure is uniform, with all cell layers thin-walled. The seeds sometimes shed the outer epidermis of the testa, which collapses during its development. The endosperm is massive and consists of cells with thick, pitted walls containing aleurones and oils but is not starchy.

Dispersal. The berries are assumed to be dispersed by animals.

Phytochemistry. Saponins are present (Conran 1985). Alkaloids are not reported.

Affinities of the Behniaceae. The family has been related by Dahlgren et al. (1985) on features such as the berry fruits, raphides, pedicel articulation, habit and anatomy to the asparagalean families Ruscaceae, Herreriaceae, and Hemerocallidaceae, but also to the Colchicaceae (part), Luzuriagaceae, Smilacaceae and Philesiaceae (Liliales) on features such as the lack of phytomelan on the seeds and reticulate-veined leaves and "woody" stems (e.g. Cronquist 1981; Dahlgren and Clifford 1982; Conran 1989; Obermeyer 1992). Conover (1983, 1991) and Conran et al. (1994) noted a close similarity between the leaf architecture of *Behnia* and *Stemona* (Stemonaceae). Schlittler (1953), although including *Behnia* within the Liliaceae: Luzuriagoideae, considered its leaf, inflorescence and floral articulation features to represent an evolutionary intermediate on a developmental line towards *Asparagus*. Nevertheless, it now appears that most of these associations were based on convergent character states. The molecular studies of Chase et al. (1995) found instead that *Behnia* occupied a basal position to a clade containing the Agavaceae, Anthericaceae and Herreriaceae. This supports Dahlgren's (1989) assertion that *Behnia* is sufficiently different to warrant family status.

Distribution and Habitats. The Behniaceae are confined to SE Africa in Zimbabwe, Swaziland

and eastern S Africa. They grow from moist, shady forest to dry vine forests.

Only one genus:

Behnia **Didr.** Fig. 45

Behnia Didr., Vid. Meddel. Kjøbenhavn: 182 (1854).

See family description for characters. A single species, *B. reticulata* (Thunb.) Didr., S and SE Africa.

Selected Bibliography

Chase, M.W. et al. 1995. See general references.
Conover, M.H. 1983. The vegetative anatomy of the reticulate-veined Liliiflorae. Telopea 2: 401–412.
Conover, M.H. 1991. Epidermal patterns of the reticulate-veined Liliiflorae and their parallel-veined allies. Bot. J. Linn. Soc. 107: 295–312.
Conran, J.G. 1985. The taxonomic affinities of the genus *Drymophila* (Liliaceae s.l.). Ph.D. Thesis. St. Lucia: University of Queensland.
Conran, J.G. 1989. Cladistic analyses of some net-veined Liliiflorae. Plant Syst. Evol. 168: 123–141.
Conran, J.G., Christophel, D.C., Scriven, L.J. 1994. *Petermanniopsis angleseaënsis*: An Australian fossil net-veined monocotyledon from Eocene Victoria. Int. J. Plant Sci. 155: 816–827.
Conran, J.G., Rudall, P.J., Chase, M.W. 1997. Two new monocotyledon families: Anemarrhenaceae and Behniaceae (Lilianae: Asparagales). Kew Bull. 52: 995–999.
Cronquist, A. 1981. See general references.
Dahlgren, G. 1989. An updated angiosperm classification. Bot. J. Linn. Soc. 100: 197–203.
Dahlgren, R., et al. 1985. See general references.
Lasarte, J.E., Palser, B.F. 1979. Morphology, vascular anatomy and embryology of pistillate and staminate flowers of *Asparagus officinalis*. Am. J. Bot. 66: 753–764.
Obermeyer, A.A. 1992. Luzuriagaceae. In: Leistner O.A. (ed.) Flora of Southern Africa, Vol. 5(3). Pretoria: National Botanical Institute, pp. 83–84.
Rudall, P. 1995. New records of secondary thickening in monocotyledons. IAWA J. 16: 261–268.
Schlittler, J. 1949. Die systematische Stellung der Gattung Petermannia F.v. Muell. und ihre phylogenetischen Beziehungen zu den Luzuriagoideae Engl. und den Dioscoreaceae Lindl. Vierteljahresschr. Naturforsch. Ges. Zür. 94: 1–28.
Schlittler, J. 1953. Blütenartikulation und Phyllokladien der Liliaceae organphylogenetisch betrachtet. Feddes Repert. 55: 154–258.
Schulze, W. 1982. Beiträge zur Taxonomie der Liliifloren VIII. Smilacaceae. Wiss. Z. Friedrich Schiller Univ. Jena 31: 285–289.
Wagner, W.M. 1977. Vessel types of monocotyledons: a survey. Bot. Not. 130: 383–402.

Blandfordiaceae

H.T. Clifford and J.G. Conran

Blandfordiaceae Dahlgren & Clifford in Dahlgren et al., Fam. Monocot.: 155 (1985).

Perennial, tufted *herbs* with fibrous roots arising from a more or less tuberous, short rhizome. *Leaves* alternate, distichous, sessile, sheathing the rhizome; blades linear; veins numerous, parallel, close, with few to many cross-veins; midrib prominent. *Inflorescence* a terminal raceme to more or less 1.5 m. *Flowers* bisexual, hypogynous, actinomorphic, pedicellate, articulated with a long, well-developed pericladium; perianth marcescent, not twisted after anthesis; pedicels with 2 very narrow bracteoles at the base, arising in the axil of a small bract; tepals 6, petaloid, fused to form a campanulate corolla-tube with broad corolla lobes, red and yellow, orange or yellow. Stamens 3 + 3; filaments attached to lower 1/3 or 1/2 of corolla tube; anthers 2-thecate, 4-sporangiate, dorsifixed, dehiscing latrorsely by slits. Gynoecium of 3 united carpels; ovary 3-locular, tapering below into a gynophore; septal nectaries present as deep exterior grooves; style single, erect, short with a single, apically 3-grooved stigma. Ovules numerous, anatropous, in 2 rows on axile placentas. *Fruit* a septicidal capsule. Seeds brown and covered with hairlike papillae. Endosperm copious, lacking starch; embryo linear, slightly curved and about 1/2 the length of the seed.

A monotypic subtropical to temperate endemic eastern Australian family with four species.

Vegetative Morphology and Anatomy. Shortly rhizomatous, tufted herbs with erect, leaves to about 1 m long and 8 mm wide, spreading basally to more or less sheath the rhizome and stem base. The roots are fibrous. The seedlings have a long, assimilating cotyledonary hyperphyll, there is no cotyledonary sheath and germination is epigeal. The leaves are linear, acute distichously imbricate and sheathing at the base, giving the appearance of a bulb. Venation is parallel. The sheath is small relative to the lamina, and a petiole is absent. The leaves are more or less 7-costate. Vernation is flat-curved to slightly concave.

The pedunculate inflorescence is an erect, terminal unbranched raceme bearing up to 20 flowers and several distant, short, erect bracts (Fig. 46).

In young roots at least, there is a uniseriate velamen which is unthickened, and an extensive cortex with unthickened cells. The vascular strands of the roots contains vessels with simple or scalariform perforations, but the stems and leaves have tracheids only. The epidermal cells of the peduncle are small with unthickened outer walls, but a very thick cuticle. The collenchymatous cortex is reduced and there is a thick ring of heavily lignified fibrous bundles and small collateral bundles in the inner cortex. Scattered bundles and lignified parenchyma occur inside the fibrous rings, but the centre of the stem is hollow. The epidermal cells of the leaves are elongate parallel to the leaf axis, (irregularly?) polygonal with more or less straight walls. Stomata are present on the abaxial leaf surface; they are anomocytic. The mesophyll is uniform and densely packed. The leaf vascular bundles are surrounded by a ring of fibres.

Calcium oxalate raphides do not appear to be present. Starch is abundant in the root cortex and sheathing leaf bases, but is absent from the laminas.

FLOWER STRUCTURE. There is a long, pedicel-like pericladium below the floral tube, and the tepals have 3 primary veins with reticulate secondary venation. The stamens are free from one another and attached to the middle or lower 1/3 of the corolla tube by their filaments. The filaments are

Fig. 46A–D. Blandfordiaceae. *Blandfordia grandiflora.* **A** Inflorescence. **B** Leaf. **C** Fruit. **D** Seed. (Takhtajan 1982)

thread-like. The anthers dehisce latrorsely by longitudinal slits and are dorsifixed, but appear to be basifixed because the connective is hollowed out in the lower part and surrounds the filament as a sheath almost to the base of the anther. The gynoecium is conspicuously stipitate on a gynophore, and there are deep nectiferous septal grooves on the outside of the ovary and gynophore. Internal septal nectaries are absent. The ovary is 3-locular.

EMBRYOLOGY. The embryology has been studied by Di Fulvio and Cave (1964). The endothecium is well developed, radially elongated and provided with fibrous thickenings. The tapetum is glandular. The pollen grains are binucleate when shed.

The ovule is bitegmic and crassinucellate with the inner integument shortly exceeding the outer. The archesporial cell cuts off a parietal cell, and a nucellar cap is formed by periclinal divisions of the epidermis. Embryo-sac formation is of the Polygonum type and endosperm formation is Helobial.

POLLEN MORPHOLOGY. The pollen grains are shed singly and are ellipsoidal and sulcate with a granular exine and 28–50 μm (Schulze 1982; Kosenko 1994).

KARYOLOGY. The basic chromosome numbers for the family appear to be $x = 17$ and $x = 27$, with diploid numbers of $2x = 34$ reported for *B. nobilis* and $2x = 54$ in *B. cunninghamii* (Di Fulvio and Cave 1964). In contrast, Smith-White (1959) observed that there were 3 diploid taxa ($2x = 34$) and 1 tetraploid taxon. The chromosomes are apparently uniform, with no marked size differences. Further work is needed.

POLLINATION. The flowers of most species are visited by honey-eating birds (Meliphagidae), and the flowers of *Blandfordia nobilis* Sm. are self-incompatible, requiring visits by pollinators to effect pollination (Pyke et al. 1988; Zimmerman and Pyke 1988a,b). Nectar volume and concentration were also found to be highly variable both between plants and within plants from one season to the next (Zimmerman and Pyke 1988a). In contrast, *B. grandiflora* R.Br. is highly self-fertile in the Blue Mountains populations, but largely self-incompatible in lowland coastal populations (Ramsey et al. 1993). Flowering in this species is also correlated with soil nutrient conditions within 5 years of a fire, most plants only flowering once or twice during this period (Johnson et al. 1994).

Fruit and Seed. The fruits of *Blandfordia* are capsules with septicidal dehiscence. The seeds are ellipsoidal and brown, and their surface is covered with short papillae.

Dispersal. The dispersal mechanisms are unknown for the family, but the erect dehiscent fruits suggest that the seeds are initially dispersed by a censer mechanism.

Phytochemistry. Chelidonic acid is present (Hegnauer 1963). The leaves of *Blandfordia* contain flavone C-glycosides (Harborne and Williams 1995), and cyanidin glycosides (Gascoigne et al. 1948).

Affinities of the Blandfordiaceae. Di Fulvio and Cave (1964) argued against the more traditional associations with the Hemerocallideae (=Hemerocallidaceae) of Krause (1930) or the Kniphofieae (Asphodelaceae) of Hutchinson (1959) on embryological grounds. Dahlgren et al. (1985) related them to the Dasypogonaceae, but observed that they also share some similarities with the Haemodoraceae. Takhtajan (1980) included *Blandfordia* within the Phormiaceae, albeit noting its apparent isolation, and Kosenko (1994) in a study of its pollen morphology removed it from that family, considering it instead to be an isolated monogeneric family. In contrast, the molecular study of Chase et al. (1995) placed *Blandfordia* as sister to the Asteliaceae s.str. in a clade with the Hypoxidaceae and Lanariaceae, and more distantly the Orchidaceae and Boryaceae.

Distribution and Habitats. *Blandfordia* is restricted to E Australia in SE Queensland, New South Wales and Tasmania. It is commonly associated with heathlands, especially seasonally waterlogged sites, and occurs mostly in coastal areas below 1000 m, although there are some isolated populations in the Blue Mountains area west of Sydney.

Economic Importance. Some species are cultivated as ornamentals. Only one genus:

Blandfordia Sm. Fig. 46

Blandfordia Sm., Exot. Bot. 1, 5. t. 4 (1804), nom. cons.

For characters see family. Four spp., eastern Australia.

Selected Bibliography

Chase, M.W. et al. 1995. See general references.
Dahlgren, R.M.T. et al. 1985. See general references.
Di Fulvio, T.E., Cave, M.S. 1964. Embryology of *Blandfordia nobilis* Smith (Liliaceae) with special reference to its taxonomic position. Phytomorphology 14: 487–499.
Gascoigne, R.M., Ritchie, E., White, D.E. 1948. A survey of anthocyanins in the Australian flora. J. R. Soc. N. S. W. 82: 44–70.
Hegnauer, R. 1963. See general references.
Hutchinson, J. 1959. The families of flowering plants, Vol. 2 Monocotyledons, 2nd edn. Oxford: Clarendon Press.
Johnson, K.A., Morrison, D.A., Goldsack, G. 1994. Post-fire flowering patterns in *Blandfordia nobilis* (Liliaceae). Austr. J. Bot. 42: 49–60.
Kosenko, V.N. 1994. Morfologiya pyl'tsy semeistv Phormiaceae, Blandfordiaceae i Doryanthaceae. Bot. Zh. 79: 1–12.
Krause, K. 1930. Liliaceae. In: Engler A., Prantl K. (eds.) Die natürlichen Pflanzenfamilien, 2nd edn, Vol. 15a. Leipzig: W. Engelmann, pp. 227–386.
Porter, C.L., Morrison, D.A., Johnson, K.A. 1992. Morphological variation within the genus *Blandfordia* (Liliaceae) in relation to its environment. Austr. Syst. Bot. 5: 373–386.
Pyke, G.H., Day, L.P., Wale, K.A. 1988. Pollination ecology of Christmas bells (*Blandfordia nobilis* Sm.): Effects of adding artificial nectar on pollen removal and seed set. Austr. J. Ecol. 13: 279–284.
Ramsey, M., Prakash, N., Cairns, S. 1993. Breeding systems of disjunct populations of Christmas bells (*Blandfordia grandiflora* R. Br., Liliaceae): Variation in self-fertility and an ovular mechanism regulating self-fertilisation. Austr. J. Bot. 41: 35–47.
Schulze, W. 1982. Beiträge zur Taxonomie der Liliifloren IX. Anthericaceae. Wiss. Z. Friedrich Schiller Univ. Jena 31: 291–307.
Smith-White, S. 1959. Cytological evolution in the Australian flora. Quant. Biol. 24: 273–289.
Takhtajan, A.L. 1980. Outline of the classification of flowering plants (Magnoliophyta). Bot. Rev. 46: 225–359.
Takhtajan, A.L. 1982. See general references.
Zimmerman, M., Pyke, G.H. 1988a. Pollination ecology of Christmas Bells (*Blandfordia nobilis*): effects of pollen quantity and source on seed set. Austr. J. Ecol. 13: 93–100.
Zimmerman, M., Pyke, G.H. 1988b. Pollination ecology of Christmas Bells (*Blandfordia nobilis*): patterns of standing crop nectar. Austr. J. Ecol. 13: 301–310.

Boryaceae

J.G. CONRAN

Boryaceae Rudall, Chase & Conran, Kew Bull. 52: 416 (1997).

Perennial, tufted, caespitose or *shrubby herbs* from short rhizomes, often dwarf; roots fibrous, wiry, sometimes stilt-like; *leaves* sessile, spirally arranged, numerous, sheathing, sometimes caducous; leaf blades dorsiventral, linear, parallel-veined. *Inflorescence* a terminal scapose involucrate spike or raceme; *flowers* not articulated, bisexual, hypogynous, showy, relatively small, white; perianth of two similar trimerous whorls, persistent, marcescent, free or united into a tube, petaloid. Stamens 6, hypogynous or attached to the tepals. Anthers basifixed, hardly longer than wide, dehiscing introrsely by longitudinal slits, sometimes glandular. Ovary syncarpous, 3-locular; septal nectaries present as deep external nectiferous slits on the ovary wall; ovules numerous per locule. Style single; stigma capitate. Ovules anatropous. *Fruit* a loculicidal capsule. Seeds few, black; testa thin, with phytomelan, ornamented; endosperm copious, without starch; embryo small, ovoid.

Two genera and about 12 species, endemic to Australia.

VEGETATIVE MORPHOLOGY. The plants are short rhizomatous, tufted-caespitose or shrubby herbs with single or branched prostrate to aerial stems (Fig. 47). The rhizomes are subterranean with scale-like leaves, or aerial with well-developed leaves, and supported by stilt roots in some *Borya* species. The roots are fibrous and strongly mycorrhizal (Keighery 1984b).

The leaves are strongly sheathing and occasionally caducous above the sheath. The leaves are linear to filiform with expanded sheath bases and in some *Borya* species, with an articulation joint between the lamina and sheath. The leaf apices are acute to pungently acicular, and in some taxa the margins are minutely serrulate. The sheath margins may be ciliate or fimbriate. In cross-section the leaf blades are triangular or rhomboid in outline.

VEGETATIVE ANATOMY. The roots in both genera have a lignified hypodermis and well-developed cortex. The endodermis is thickened so as to be almost fully occluded, and there also is a sclerenchymatous inner cortex, and the central medulla is lignified. The stems in both genera anatomically resemble xeromorphic rhizomes. In young *Borya* stems, there is an ephemeral cortex, heavily lignified endodermis, and the vascular strands are located peripherally in a single ring below the endodermis, superficially resembling dicotyledonous stem architecture, but all tissues inside the endodermis are lignified, and the medulla consists of almost fully occluded fibres. In older stems, there is monocotyledonous secondary growth similar to the type reported for *Thysanotus* (Lomandraceae) and *Herreria* (Herreriaceae) by Rudall (1995). *Alania* stems have a similar heavily thickened endodermis, but the stele is less heavily lignified, the medulla is parenchymatous, and there is no obvious secondary development.

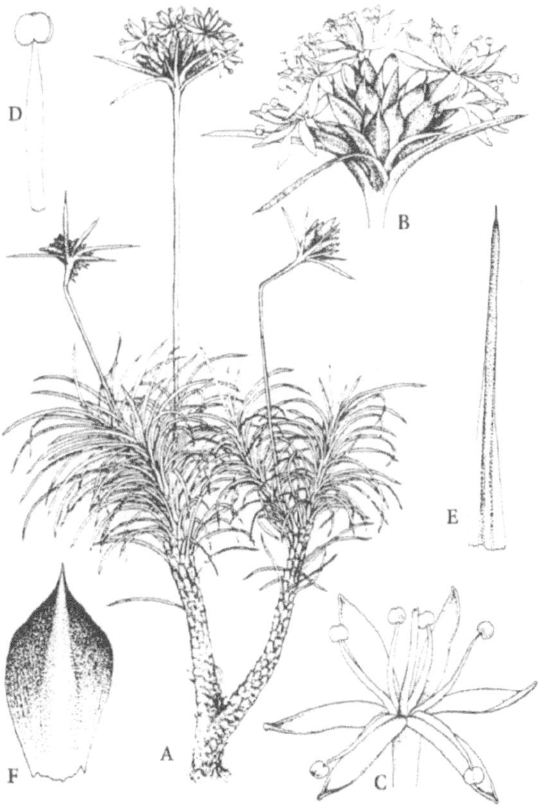

Fig. 47A–F. Boryaceae. *Borya mirabilis*. **A** Habit. **B** Inflorescence. **C** Flower. **D** Stamen. **E** Outer involucral bract. **F** Inner involucral bract. (Churchill 1984)

Anania also has a persistent parenchyma cortex, suberised epidermis and 2- to 3-seriate, slightly lignified hypodermis.

In the leaves, there are bands of heavily lignified fibres forming a single (*Alania*) to multiseriate hypodermis (*Borya*). The inner epidermal cell walls in both genera are lignified; the outer walls are thickened but unlignified, although Gaff (1981) considered that they were impregnated with cutin rather than having a clearly defined cuticle. The epidermal cells of *Borya* are shortly papillate, and the stomata are restricted to the sides and bottom of 3 to 4 deep, papilla-lined grooves in the leaf. There are 2, single files of stomata in grooves on the abaxial lamina of *Alania*, although the epidermal cells are unornamented (Thongpukdee 1989). The mesophyll in *Borya* is differentiated, with multiseriate palisade mesophyll externally. *Alania* has undifferentiated mesophyll. There is a single, heavily lignified central vascular trace surrounded by multilayered, thickened fibres in the leaves of either genus, and in both genera, there are 2 additional lateral phloem poles.

Vessels are restricted to the roots and have scalariform perforations (Wagner 1977). Raphide cells are known from both genera.

INFLORESCENCE MORPHOLOGY. The inflorescence in the Boryaceae is a terminal scapose involucrate spike or raceme. The outer involucral bracts are uniformly scarious (*Alania*), or differentiated into subulate, leaflike outer, and scalelike inner bracts (*Borya*) (Fig. 47).

FLORAL MORPHOLOGY. The flowers are hypogynous, pedicellate and subtended by a bract and several membranaceous laciniate bracteoles (*Alania*), or sessile and subtended by a bract and single membranaceous bracteole (*Borya*). The outer and inner tepal whorls are similar, free (*Alania*), or united basally into a tubular perigone which closely covers the ovary (*Borya*). The tepals are uninerved in *Borya* and 3-nerved in *Alania*. The perianth is marcescent.

The stamens are free from each other, and in *Borya* are attached to the mouth of the perigone tube. The filaments are filiform; anthers are 2-locular, 4-sporangiate and equal in number to the tepals. They are unusual in the Lilianae in being about as long as they are wide. Anther dehiscence is longitudinal and introrse in both genera.

Both genera have small, ovoid, 3-locular ovaries with numerous ovules per locule. There are external septal nectaries present as deep grooves in the ovary walls which bear nectiferous tissue. Internal septal nectaries are absent. The styles are filiform with minute capitate stigmas; the style in *Alania* is strongly geniculate.

EMBRYOLOGY. The anther wall in *Borya* is 4-layered, with an ephemeral epidermis, endothecium, flattened and short-lived middle layer, and tapetum. Microsporogenesis is simultaneous and the pollen tetrads are tetrahedral. The tapetum is binucleate and shows glandular-secretory degeneration. The endothecium has spiral thickenings. Pollen grains are 2-nucleate at anthesis.

The ovules are anatropous. In *Borya*, no parietal cell is formed, and the proximal nucellar region is 1-layered and does not undergo periclinal division. A linear tetrad is formed, and the 3 micropylar megaspores degenerate. Embryo-sac development is of the Polygonum type, and endosperm formation is Helobial. There is a small, weakly lignified hypostase. The embryo-sac nuclei are relatively small, and the antipodals are reduced and sit at the bottom of the hypostasal region. Data from Conran and Temby (in prep.).

POLLEN MORPHOLOGY. The pollen is sulcate and 23–40 μm long. The exine is reticulate, and in *Borya*, homobrochate or rarely heterobrochate, with generally complex but very variable muri. In *Alania*, the exine is reticulate, and more or less muricate with simple bacula (Schulze 1982).

KARYOLOGY. The chromosome numbers are $n = 11$ in *Alania* and $n = 14$ or 28 in *Borya*. The chromosomes are small, and metacentric to acrocentric (Keighery 1984a).

POLLINATION. The flowers are small and long-lived. Nectar is produced in both genera, and *Borya* is pollinated by flies (Keighery 1984b).

FRUIT AND SEED. The fruits are loculicidal capsules with few seeds. The seeds are small (ca. 0.5 mm long) and dull black. The seed coat is prominently papillose or striate sculptured in many *Borya* species (Keighery 1984b), and the anticlinal cell walls are polygonal in *Borya* and sinuous in *Alania*. The outer integument is only 2 cell layers thick in *Alania*, with the outer epidermal walls thickened in both genera. The inner integument is degenerate at maturity. Phytomelan is present on the testa of both genera; thin in *Alania* and thick in *Borya* (Huber 1969). The embryo in both genera is very small and ovoid (Thongpukdee 1989).

Phytochemistry. Chelidonic acid has been reported for *Borya* (Gibbs 1974).

Systematics and Affinities. The Boryaceae represent an isolated and distinct lineage in the lower Asparagales close to the orchids. They are morphologically different from most of their near relatives, with the lignified roots, secondary monocotyledonous thickening in *Borya*, strongly sheathing leaf bases, bracteate scapose inflorescences, small flowers, isomorphic anthers and numerous very small seeds tending to isolate their position. In particular, the xeromophic adaptations in both genera, but especially in *Borya*, make it difficult to associate the family readily with other Asparagales. The family has traditionally been associated with taxa in the Anthericaceae (Dahlgren et al. 1985), especially those genera placed in the Johsoniaceae (Krause 1930) and now included in the Johsoniaceae, although Keighery (1984b) placed *Borya* and *Alania* in the Liliaceae: Boryeae, separate from the remainder of the Johnsonieae, on morphological, anatomical and cytological grounds.

Molecular studies using sequences of the chloroplast gene *rbc*L by Chase et al. (1995) identified a clade, sister to the orchids, consisting of *Borya* and *Alania*, the latter mislabelled as *Herpolirion*. More recently, this clade has been proposed as the new family Boryaceae (Chase et al. 1997), which shows affinities to the orchids, Asteliaceae, Blandfordiaceae and Hypoxidaceae. The Boryaceae agree with the lower Asparagales in having simultaneous microsporogenesis without trichotomosulcate pollen, and they share similar nectaries with the Asteliaceae and Blandfordiaceae. Further research is needed.

Distribution and Habitats. The family is endemic to Australia. *Borya* is widespread, but uncommon throughout all mainland states except South Australia; *Alania* localised in mountainous areas W of Sydney. The plants are generally associated with rocky habitats. *Borya* is usually found in locally xeric environments, often on exposed granite rock surfaces. In contrast, *Alania* grows in sheltered positions on rock faces which are more or less constantly moist.

The ecophysiology of desiccation-tolerant *Borya* species has been studied in detail (Gaff 1981; Gaff and Loveys 1984; Gaff and Ziegler 1989). During dehydration, the leaves fold up over the apical shoots; the chloroplasts and mitochondria become degraded and ultrastructurally disorganised; chlorophyll is destroyed; and there are elevated levels of abscissic acid. ATP levels remain high until the moisture level in the leaves falls below 30%, but ATP activity stays low for about 16 h following rehydration. The chloroplasts and mitochondria reorganise themselves during rehydration. Although some *Borya* species are stilt plants, they appear to lack the ability of other stilt-rooted taxa such as *Laxmannia* to maintain hydration of the elevated tissues (Pate et al. 1984).

Key to the Genera

1. Tepals fused; floral tube closely enclosing the ovary; anthers attached to the apex of the floral tube; style straight **1. *Borya***
– Tepals free, spreading and exposing the ovary; anthers hypogynous; style geniculate **2. *Alania***

Genera of the Boryaceae

1. *Borya* Labill. Fig. 47

Borya, Labill., Nov. Holl. Pl. 1: 81, t. 107 (1805); Churchill, Muelleria 6: 1–8 (1984); Keighery, Flora 175: 103–108 (1984); Churchill, Fl. Australia 46: 268–279 (1987).

Short-tufted herbs, with simple or few-branched, erect or procumbent stems and sometimes stilt roots. Leaves perennial or caducous, sessile. Flowers in scapose, involucrate spikes; flowers each subtended by a bract and bracteole; tepals united basally, white, persistent. Anthers inserted at the mouth of the perigone tube, sometimes glandular. Ovary small, ovoid. Style filiform; stigma minutely capitate. Capsule ovoid-fusiform, seeds, small, ellipsoid, smooth or papillate, dull black. $2n = 28$, 56. Eleven spp., widespread throughout, but endemic to Australia.

Many species are drought-tolerant, and capable of re-greening the desiccated leaves following rehydration. Other species are drought-avoiders, with annually caducous leaves, articulating above the leaf sheath.

2. *Alania* Endl.

Alania Endl., Gen. Pl. 1: 151 (1836); Keighery, Flora 175: 103–108 (1984); Telford, in Fl. Australia 45: 279 (1987).

Short herbs, with simple or few-branched, more or less procumbent stems. Leaves perennial. Flowers in scapose involucrate racemes; flowers each subtended by a bract and several bracteoles; tepals free, membranous, white, persistent. Anthers free, hypogynous. Ovary small, ovoid. Style short, filiform, geniculate; stigma capitate. Capsule ovoid, seeds ellipsoid, smooth, black. $2n = 22$. Only one sp., *A. endlicheri* Kunth, endemic to the mountains W of Sydney.

Selected Bibliography

Chase, M.W. et al. 1995. See general references.
Chase, M.W., Rudall, P., Conran, J.G. 1997. New circumscriptions and a new family of asparagoid lilies: genera formerly included in the Anthericaceae and associated taxa. Kew Bull. 51: 667–680.
Conran, J.G., Temby, A. The embryology 2nd relationships of the Boryaceae (Lilianae: Asparagales). (in preparation)
Dahlgren, R.M.T. et al. 1985. See general references.
Forster, P.I., Thompson, E.J. 1997. Borya inopinata (Anthericaceae), a new species of resurrection plant from north Queensland. Austrobaileya 4: 597–600.
Gaff, D.F. 1981. The biology of resurrection plants. In: Pate, J., McComb, A.J. (eds.) The biology of Australian plants. Nedlands: University of Western Australia, pp. 114–146.
Gaff, D.F., Loveys, B.R. 1984. Abscissic acid content and effects during dehydration of detached leaves of desiccation-tolerant plants. J. Exp. Bot. 35(158): 1350–1358.
Gaff, D.F., Ziegler, H. 1989. ATP and ADP contents in leaves of drying and rehydrating desiccation-tolerant plants. Oecologia 78: 407–410.
Gibbs, D. 1974. See general references.
Huber, H. 1969. See general references.
Keighery, G.J. 1984a. Chromosome numbers of Australian Liliaceae. Feddes Repert. 95: 523–532.
Keighery, G.J. 1984b. The Johnsonieae (Liliaceae): biology and classification. Flora 175: 103–108.
Krause, K. 1930. See general references.
Pate, J.S., Weber, G., Dixon, K.W. 1984. Stilt plants - extraordinary growth form of the Kwongan. In: Pate, J.S., Beard, J.S. (eds.) Kwongan: plant life of the sandplain. Nedlands, Western Australia: University of Western Australia Press, pp. 101–125.
Rudall, P. 1995. New records of secondary thickening in monocotyledons. IAWA J. 16: 261–268.
Schulze, W. 1982. Beiträge zur Taxonomie der Liliifloren IX. Anthericaceae. Wiss. Z. Friedrich Schiller Univ. Jena 31: 291–307.
Thongpukdee, A. 1989. The taxonomic position of the genus *Tricoryne* R.Br. Ph.D. Thesis. St. Lucia: University of Queensland.
Wagner, W.M. 1977. See general references.

Burmanniaceae

H. Maas-van de Kamer

Burmanniaceae Bl., Enum. Pl. Javae: 27 (1827), nom. cons.
Thismiaceae J. Agardh (1858).

Small, saprophytic and colourless or rarely autotrophic and green *annuals or perennials.* Roots filiform, vermiform, branched and creeping, or sometimes coralloid. Rootstock a cylindric ± tuberous rhizome, sometimes with a clump of small tubers at each node, or a subglobose tuber. Flowering stems mostly unbranched, terete or occasionally bisulcate. *Leaves* of aerial stems alternate, sessile, entire, in saprophytic species small and scalelike, in autotrophic spp. green and small to rather large and often rosulate. *Inflorescence* a terminal 1-many-flowered cyme, usually a bifurcate cincinnus. *Flowers* perfect, sympetalous, usually actinomorphic, variously coloured, usually pedicellate; perianth basally tubular; floral tube persistent or upper part caducous or basally circumscissile, sometimes with longitudinal wings or ribs, sometimes with an ornamented annulus around the throat; tepals 6(-8) in 2 whorls, rarely 3, usually entire, rarely the inner ones connate into a mitre (*Thismia*). Stamens 6, or 3, and then opposite the inner tepals, erect or recurved to pendent, usually inserted in the apical part but sometimes below the middle of the floral tube; anthers 2-thecate, 4-sporangiate, introrse, mostly on short filaments, occasionally alternating with interstaminal lobes; thecae often separate with locules superposed and transversely dehiscent or thecae united with locules parallel and longitudinally dehiscent; connective usually dilated, with various appendages, occasionally connate into an anther tube (*Thismia*). Gynoecium of 3 ± united carpels; ovary inferior, 3-locular with axile placentation to 1-locular with parietal placentation, or with 3 free placental columns in the centre of the ovary; ovules anatropous, numerous and very small; style cylindric, usually 3-branched at apex; stigmatic branches opposite the outer tepals, or stigma capitate (*Thismia*); stigmas sometimes with appendages. Nectaries apical on the ovary or in upper part of septa. *Fruit* usually a capsule, sometimes crowned by the whole or part of the persistent perianth, rarely fleshy (*Thismia*), longitudinally or transversely dehiscent by slits or valves, or irregularly dehiscent. Seeds numerous, small (0.2–1 × 0.1–0.6 mm), fusiform to subglobose, sometimes laterally flattened; testa usually reticulate.

A family of pantropical and warm-temperate distribution with 13 genera and about 130 species.

VEGETATIVE MORPHOLOGY. The plants are either saprophytic with reduced, achlorophyllous leaves or non-saprophytic with well-developed, green leaves (*Burmannia* spp., Fig. 48A).

The subterranean parts for long periods constitute the principal portion of the plant, housing mycorrhizal fungi. The following types of underground structure are present:

1. A more or less tuberous, cylindric rhizome of a few cm in length, densely covered with exogenous roots and imbricate scales [*Apteria*, *Campylosiphon* (Fig. 48D), *Cymbocarpa*, *Dictyostega*, *Gymnosiphon* (Fig. 49A), *Hexapterella*, *Marthella* and *Miersiella*).
2. A globose to narrowly ovoid tuber bearing exogenous roots, often forming small new tubers (most neotropical *Thismia*) (Fig. 50A).
3. A clump of short, coralloid roots (*Haplothismia*, Asiatic species of *Thismia*).
4. A cylindric rhizome bearing at each node a clump of small tubers, each provided with 1 rootlet (*Afrothismia*).
5. Vermiform, branched, horizontally creeping, endogenous, cylindric roots (*Thismia*).
6. Filiform roots, rhizome wanting (autotrophic *Burmannia* spp., Fig. 48A).

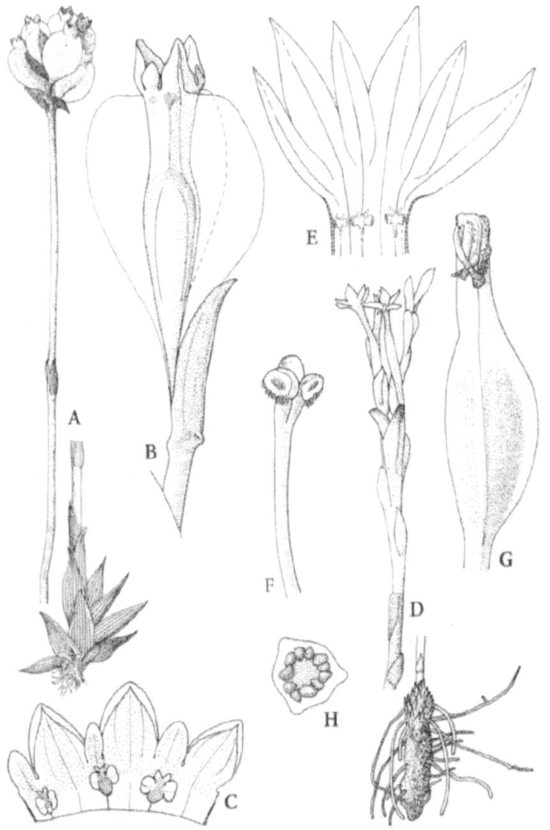

Fig. 48A–H. Burmanniaceae. **A–C** *Burmannia stuebelii*. **A** Habit. **B** Flower. **C** Opened floral tube with tepals and stamens. **D–H** *Campylosiphon purpurascens*. **D** Habit. **E** Opened floral tube with tepals and stamens. **F** Style and stigmas. **G** Fruit. **H** Fruit in transversal section. (Maas et al. 1986)

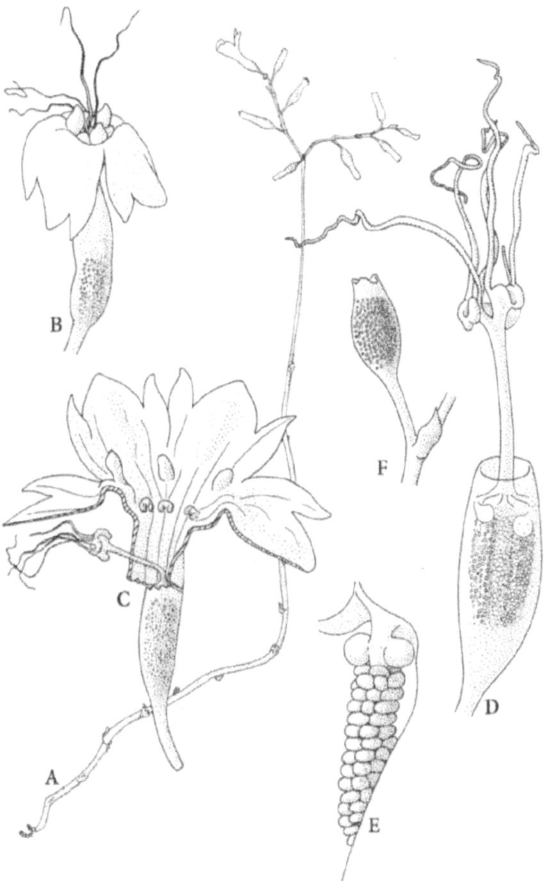

Fig. 49A–F. Burmanniaceae. *Gymnosiphon suaveolens*. **A** Habit. **B** Flower. **C** Dissected flower with outer and inner tepals, stamens, style, and appendaged stigmata. **D** Gynoecium. **E** Detail of ovary with nectar glands and ovules. **F** Fruit. (Maas et al. 1986)

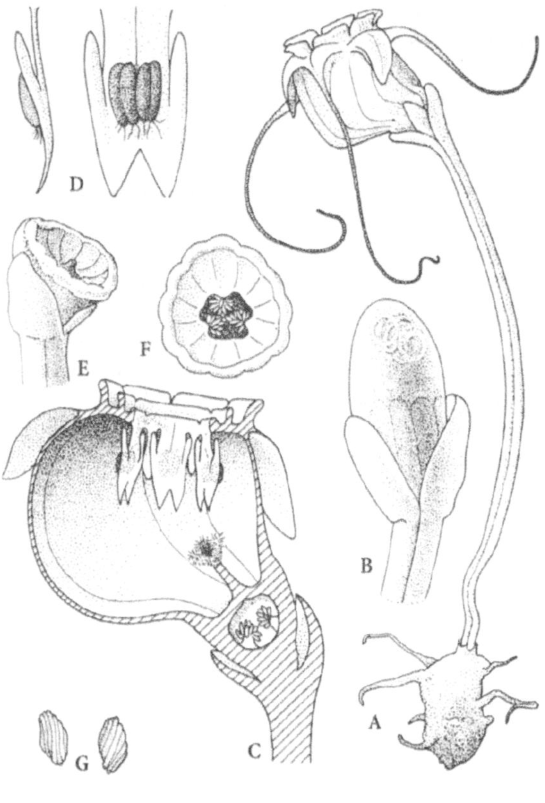

Fig. 50A–G. Burmanniaceae. *Thismia panamensis.* **A** Habit. **B** Floral bud. **C** Dissected flower, showing outer and inner tepals, annulus, stamens, interstaminal lobes, style, stigma, and ovary. **D** Stamen in side and front view. **E** Young fruit. **F** Mature fruit, seen from above. **G** Seeds. (Maas et al. 1986)

The roots lack root hairs. Those of the first two types also lack a root cap. In *Dictyostega* the margins of the rhizome scales have dense long hairs, which possibly function as root hairs. Vegetative propagation occurs in Burmannieae by branching of the rhizome and in Thismieae by the formation of new tubers or by branching and the formation of buds from roots.

The stems vary in length from a few cm (several species of *Thismia*) to ca. 0.7 m (autotrophic species of *Burmannia*). In the tuberous *Thismia* species (type 2 above) flowering stems are usually solitary. In species with a vermiform root system each flowering stem arises together with a secondary root (type 5 above). In *Thismia* (type 5) the stems are terete, in type 2 they are always longitudinally bisulcate. Both types of stem originate endogenously, the tuber or primary root forming a ca. 1 mm high sheath around the stem base.

In saprophytic Burmannieae the leaves are very small and scattered along the stem, whereas in the green-leaved species of *Burmannia*, they are often concentrated to the basal part of the stem and may be up to 50 cm long.

Vegetative Anatomy. Most Burmanniaceae show reductions and adaptions that are clearly related to saprophytism. Vessels with scalariform perforation plates may occur in roots, and in at least some chlorophyllous *Burmannia* spp. also in stem and leaves. Stomata, when present, have no subsidiary cells (anomocytic).

Sieve-tube plastids have not been found. Raphides and starch are known in *Thismia* throughout the plant.

In the roots a 1-layered epidermis and 3–12 layers of adjacent parenchyma house mycorrhizal fungi in definite layers. The well-developed endodermis has U-shaped wall thickenings and/or Casparian strips. It encircles a vaguely 3-4-arch central cylinder composed of only 1–6 vessels with spiral or annular thickenings and 3–8 peripheral phloem groups. The pericycle is almost completely reduced.

The stem consists of an epidermis almost without stomata (except in species of *Burmannia*), well-developed subepidermal tissue, a sclerenchyma sheath (lacking in *Thismia*), an endodermis similar to that of the roots, a ring of 3–20 reduced collateral vascular bundles comprising xylem elements with annular and spiral thickenings, phloem and a central pith.

The leaves possess only 1 collateral vascular bundle consisting of a few tracheids (rarely vessels) with spiral thickenings. Epicuticular wax deposits are of the Convallaria type (Barthlott and Frölich 1983).

Inflorescence Structure. In most Burmanniaceae the inflorescence is a terminal, cymose, bracteate, bifurcate cincinnus, rarely (*Apteria*) a single cincinnus. The cincinni are sometimes few-flowered or reduced to a single flower (e.g. *Thismia*). They may be lax or contracted, sometimes headlike.

Flower Structure. In the tubular to funnel-shaped flowes of *Apteria*, *Burmannia* (most spp.), *Dictyostega*, *Marthella*, *Miersiella*, *Haplothismia* and *Oxygyne* the 6 tepals are ascending, whereas in *Campylosiphon*, *Cymbocarpa*, *Gymnosiphon* and *Hexapterella* the tepals are spreading and the flowers salver-shaped. In Burmannieae the inner tepals are generally much smaller than the outer ones, although in *Campylosiphon* they are subequal; in *Marthella* they are absent. The outer tepals are usually entire, but 3-lobed in *Cymbocarpa* and *Gymnosiphon*. The inner tepals of Burmannieae are usually very small, entire and swollen or fleshy; in *Hexapterella* they are sometimes 3-lobed. In *Thismia* the outer tepals are reflexed, the inner ones being larger, sometimes connate or bearing elaborate appendages.

In the flower buds of Burmanniaceae the outer tepals completely cover the inner ones and are induplicate or valvate. They are valvate also in most species of *Thismia*, whereas the linear-triangular inner tepals of *Thismia* subgen. *Ophiomeris* sect. *Ophiomeris* are spirally included in the bud. The tepals in species with a mitre are all valvate.

The floral tube is well developed in all Burmanniaceae. It is generally tubular, sometimes campanulate or urceolate and circular, trigonous or rarely hexagonal in section. In *Burmannia* the floral tube often has 3 wings running from its apex to the ovary or often continuing downwards over the ovary (Fig. 48B). In *Thismia* (and probably also in *Afrothismia* and *Oxygyne*) the base of the floral tube has 12 swellings, whose function is uncertain but probably nectarial.

The perianth in Burmanniaceae is completely or partly caducous or persistent. In *Oxygyne* and *Thismia* it is circumscissile just above its swollen base, leaving a rim on the fruit. In *Cymbocarpa*, *Gymnosiphon* and *Hexapterella* the perianth is circumscissile just below the insertion of the stamens. This makes it difficult to identify these genera to species in postfloral stages. In *Hexapterella* even the last remnants of the perianth are ultimately shed, leaving only 6 ribs on the fruit. In *Afrothismia* both perianth and capsule wall are deciduous, leaving the placental column and seeds.

In Burmannieae there are only 3 stamens inserted opposite to and just below the inner tepals. The thecae comprise 2 ± superposed, globose microsporangia divided by a horizontal or oblique septum and dehiscing introrsely by a transverse slit. The filaments are variously developed or the anthers are sessile.

Sessile to very shortly stipitate anthers occur in *Burmannia*, *Campylosiphon*, *Cymbocarpa*, *Dictyostega*, *Gymnosiphon* and *Miersiella*. The connective is well developed and swollen, completely separating the 2 thecae; it is usually broadly obovoid and may have minute appendages; it is placed at right angles to the corolla tube or is slightly deflexed.

Filaments occur in *Apteria*, *Hexapterella* and *Marthella*. In most genera of Burmannieae the anthers lie close to each other and to the stigmas, the appendages to the connectives fitting into grooves of the stigmas, thus almost occluding the throat. The thecae often dehisce in bud, so that the stigmas are densely covered with pollen.

The 6 stamens of Thismieae (only 3 in *Oxygyne*) are inserted in the floral tube and are recurved to pendent. Their anthers then become extrorse (Fig. 50C). The thecae are united except when the connectives are connate into an anther tube. In *Haplothismia* and *Thismia* sect. *Ophiomeris* the stamens alternate with minute interstaminal lobes. The ring or pendent stamens and interstaminal lobes tends to converge towards the axis of the flower, thus forming a funnel. The stamens may also be connivent or even connate by their connectives, forming an anther tube (*Thismia* subgen. *Thismia*), in which case the connectives are ± dilated and have various appendages, hairs and glands (Groom 1895).

The ovary in the Burmannieae is 3- to 1-locular and Jonker's (1938) subtribes "Euburmannieae" (3-locular) and "Apterieae" (1-locular) are based on this character. Truly 3-locular ovaries occur only in *Burmannia*, where there are 3 axile placentas. In *Campylosiphon* and *Hexapterella* the young ovary has 3 placentas on long-intrusive septa. Later, the ovary becomes at least basally trilocular with axile placentation. In *Hexapterella* the ovules are borne only on the middle of the placentas.

In *Apteria*, *Cymbocarpa*, *Marthella*, *Miersiella*, *Gymnosiphon* and *Dictyostega* the ovary is ± 1-locular with 3 parietal placentas. In *Gymnosiphon* the placentas are intrusive on short, thick septa and in *Dictyostega* the ovary is only apically 1-locular and trilocular at the base. In Thismieae the ovary is 1-locular and the placentas are essentially parietal. In *Haplothismia* and *Thismia* subgen. *Ophiomeris* the parietal placentas extend to the length of the ovary or are restricted to the base, while in *Afrothismia*, *Oxygyne* and *Thismia* subgen. *Thismia* they are separate from the ovary wall, forming 3 free columns connected only with the bottom and top of the ovary.

The gynoecium terminates in a hollow, cylindric style divided apically into 3 stigmatic branches. In Burmannieae the style is about as long as the floral tube. The apex of each stigmatic branch is dilated and ± funnel-like and covers the thecae of 2 adjacent stamens. In *Cymbocarpa* and some species of *Gymnosiphon* the stigmatic branches bear orange, glandular, tortuous threads that extend beyond the white floral tube. In Thismieae the style is thick and much shorter than the floral tube. It is 3-branched or undivided, with 1 capitate stigma.

Nectaries are present in the complete septa of the trilocular ovaries and also in the incomplete partial septa of unilocular ovaries in *Burmannia* and *Campylosiphon*. In the unilocular ovaries of *Apteria*, *Dictyostega* and *Hexapterella* they are located in the large placentas either as perpendicular canals or as furrow on the distal side of the placenta. In *Cymbocarpa* and *Gymnosiphon* the partial septa and their cavities are broadened at both sides, as is also the case in *Burmannia densiflora*. Two types, connected by intermediates, can be recognised: sphaerical glands protruding from both sides of the partial septa have a common cavity which opens from a canal above the partial septum in the apical ovary wall, or spherical glands lacking a connection with the partial septa and each originating with its own cavity from the apical ovary wall. In intermediate forms the cavities of the 2 glands on both sides of 1 partial septum are incompletely separated (*Cymbocarpa refracta*, *Gymnosiphon divaricatus*). Finally, in *Miersiella* and *Marthella* the 6 glands are located outside the ovary, united in pairs above the placentas.

In Thismieae septal nectaries are absent, but other parts of the flower may have glandular, probably nectariferous tissue, e.g. the anther tube (Groom 1895; Pfeiffer 1918) or pits at the perianth base. Pollinators may also be attracted by osmophores on filiform appendages (Vogel 1962).

In Burmannieae the nectar glands all seem to be derived from septal nectaries (Rübsamen 1986).

For floral anatomy see Rao (1970) and Rübsamen (1986).

Embryology. Most of the genera were embryologically studied by Rübsamen (1986), in whose work the extensive older literature is cited, which mostly deals with the genera *Burmannia* and *Thismia*.

Anther wall formation follows the monocotyledonous type. In some species of *Burmannia* an endothecium is lacking and the epidermis is persistent and consists of thick-walled cells. Usually, however, the wall has a well-developed endothecium.

Microsporogensis is successive and microspore tetrads are mostly isobilateral or decussate. The mature pollen grains are mainly 3-celled and occasionally 2-celled in some species. Both conditions have been observed even within one genus, as in *Burmannia* and *Thismia*.

The ovules are anatropous, bitegmic and tenuinucellate. The 1-layered nucellus becomes crushed during the development of the embryo sac (Burmannieae) or the endosperm (Thismieae) except in the chalazal region. The archesporial cell functions directly as the megaspore mother cell. A linear or T-shaped tetrad or triad (the upper cell of the dyad often remaining undivided) is formed. The embryo sac is generally of the Polygonum type, but the Allium (Scilla) type is reported for some *Burmannia* and *Thismia*. Apomixis of the *Alchemilla* type occurs in *Thismia javanica* and of the *Antennaria* (*alpina*) type in *Burmannia coelestis*. Polyembryony has also been described. Endosperm formation is Helobial, but in *Thismia* sometimes Cellular. For further details and references see Rübsamen (1986) and Rudall and Morley (1992).

Pollen Morphology. Pollen grains are mostly shed singly but tetrads occur in *Apteria aphylla*, *Burmannia sphagnoides* and *B. aprica* (also dyads) (Guinet 1965; Chakrapani and Raj 1971). The pollen grains are sometimes sulcate (*Cymbocarpa*, *Hapolothismia*, *Hexapterella*), but usually 1- or 2-porate (*Burmannia*, *Campylosiphon*, *Dictyostega* spp., *Miersiella*, *Thismia*, Fig. 51A,C), occasionally polyforaminate (*Burmannia*, *Dictyostega*, *Thismia* spp.; Ikuse 1956; Chakrapani and Raj 1971; Huang 1972), or inaperturate (*Apteria*, *Gymno-*

Fig. 51A–C. Burmanniaceae. **A** *Thismia bicolor*, diporate pollen grain. **B** *Gymnosiphon suaveolens*, inaperturate pollen grain. **C** *Thismia luetzelburgii*, monoporate pollen grain. *Bar* 5 μm. (Photo T. Rübsamen-Weustenfeld)

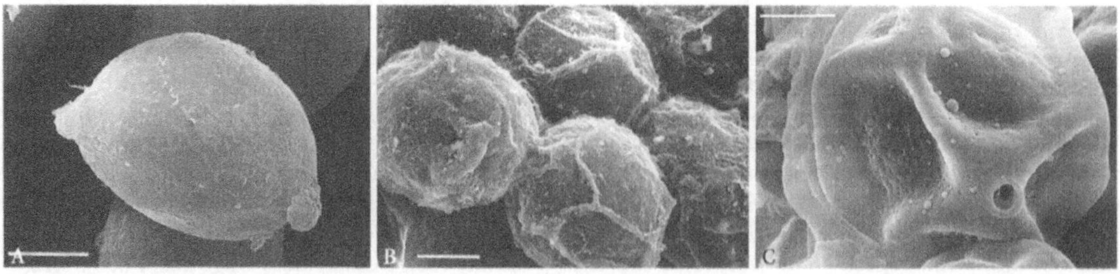

siphon, Fig. 51B). The exine is often psilate (*Burmannia*, *Campylosiphon*, *Dictyostega*, *Hexapterella*, *Thismia*), undulate or folded (*Apteria*, *Gymnosiphon*) or reticulate (*Afrothismia*) (Erdtman 1952; Chakrapani and Raj 1971). The size ranges from 13 to 54 µm. The shape is ovoid, occasionally planoconvex, biconvex, sphaeroidal or ovoid-reniform.

KARYOLOGY. Rübsamen (1986) summarised all chromosome numbers counted up to then, which range from 6 and 8 to 68 and 87–99. It is safe to assume that polyploidy is involved, while the base number(s) (6 or 8) remain uncertain.

POLLINATION. Although only self-pollination has been observed in the family, the occurrence of protandry, the flower structure and the presence of nectar-producing structures in various groups would point to cross-pollination. Indeed, Vogel (1962) supposed that several species of *Thismia* are pollinated by small Diptera attracted by scent and falling into the urceolate flowers. Vogel (1978) also suggested that *Thismia fungiformis* may be pollinated by fungus gnats tricked into laying eggs in the fungus-mimicking flower. Rübsamen (1986) twice found an egg or larva inside the nectaries of a *Gymnosiphon* flower, possibly of a visiting insect.

Self-pollination appears to be the rule in numerous members of the family and is facilitated by the close proximity of anthers and stigma. Premature dehiscence of the anthers in bud, germination of the pollen inside the thecae, and pollen tubes reaching from the anthers to the stigma and even penetrating the stigma, have been recorded for species of most genera.

According to Miers (1866), the flowers of *Thismia hyalina* are cleistogamous. *Burmannia coelestis* and *Thismia javanica* are reported by Ernst and Bernard (1910) to be apogamous.

FRUIT AND SEED. There is much variation in the dehiscence of the capsules. Most species of *Burmannia* open transversely by 1 to several slits in the membranous pericarp between the wings. Irregular dehiscence by withering of the membranous wall is known from *Burmannia* spp., *Campylosiphon*, *Gymnosiphon* spp., and *Hexapterella*. Indehiscent fruits occur in Asiatic species of *Gymnosiphon*. All other fruits open by longitudinal slits, loculicidally in *Dictyostega*, and species of *Gymnosiphon*, *Cymbocarpa* and *Miersiella*; septicidally in *Apteria*, *Gymnosiphon*, and *Burmannia tenella*. The fruit of *Thismia* is fleshy, hyaline and cup-shaped. For further details see Maas et al. (1986).

The seeds are numerous, minute and well adapted for dispersal by air or water. They are usually compact, and subglobose, ellipsoid, ovoid or fusiform (Fig. 52). The raised anticlinal walls of the epidermal cells form a network, the thin outer periclinal walls having collapsed, sometimes arching outwards. The surface of the thin outer periclinal wall is usually smooth. Sculpturing caused by cuticle wall thickenings is mostly lacking. The seed coat is formed by both integuments (*Burmannia*, *Thismia*) or mainly the outer integument (especially its outer layer) (*Gymnosiphon*, *Hexapterella*). Often the 2 layers of the inner integument are compressed and contain tannin. For further details on micromorphology of seed surface see Rübsamen in Maas et al. (1986).

The endosperm consists of relatively few cells with walls thickened by hemicellulose. They contain many single starch grains in the premature stage. Later, these are generally replaced by pro-

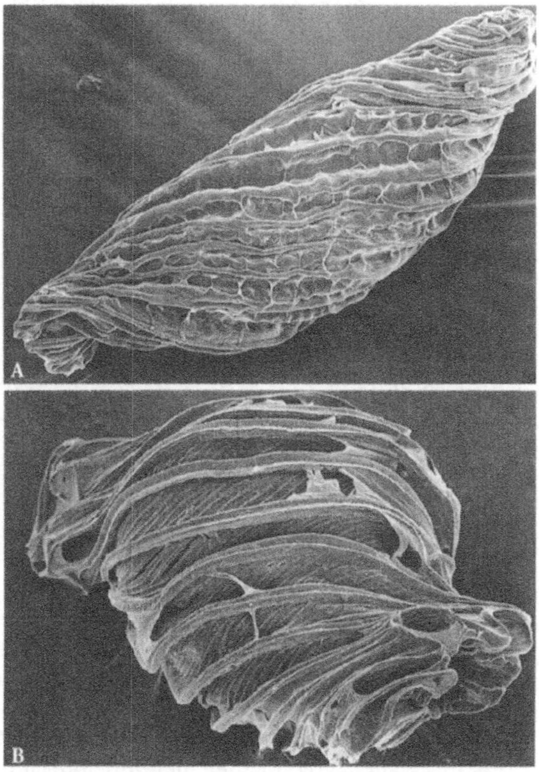

Fig. 52A,B. Burmanniaceae. SEM micrographs of seed surfaces. **A** *Burmannia juncea*, ×273. **B** *Thismia panamensis*, ×191. (Photo Rübsamen-Weustenfeld)

tein globules or crystalloids and fat. Also the small undifferentiated embryo stores starch and fat.

DISPERSAL. The minute dust seeds are produced in large quantities. In the autotrophic species of *Burmannia* they may be wind-dispersed. In most saprophytic forest-dwellers, however, they are dispersed by water (Fleischer 1929). The reticulate surface of the seeds provides mechanical resistance and air cavities. The surface is also supposed to be water-repellent. According to Stone (1980), the seeds of *Thismia* spp. may be dispersed by rain splash out of the fruit cup. Beccari (1890) supposed that the seeds of Burmanniaceae might also be dispersed by birds who have eaten earth worms that ingested them.

CLASSIFICATION. Various subdivisions have been proposed in the past; they are reviewed by Rübsamen (1986) and Maas et al. (1986). The division adopted in this treatment follows Jonker's (1938) arrangement, although the subtribes he distinguished in tribe Burmannieae are not upheld. Of the two tribes, the Thismieae seem to be more advanced on account of their pendent stamens, connate filaments, free central placentas, inner tepals connate into a mitre and bearing various appendages, zygomorphic flowers, etc. Gynoecium structure and stamen morphology has guided the arrangement of genera.

AFFINITIES. A historical review of the systematic history of the family has been given by Rübsamen (1986) and Maas et al. (1986). Bentham and Hooker (1883) associated the Burmanniaceae with the orchids in an order Microspermae. Characters such as mycotrophy, epigyny, numerous small tenuinucellate ovules, tiny seeds and the small endosperm, formerly believed to be absent, have often been used in support of such a position. However, these characters may have been developed by convergence as well. Many features, among them many embryological, are strikingly different between the two families. At present, the Burmanniaceae, together with the Corsiaceae, are considered as close allies of the Lilianae, particularly the Melanthiaceae and Campynemataceae (Dahlgren et al. 1985). In the *rbc*L analysis (Chase et al. 1995) *Burmannia* appears loosely attached to *Dioscorea*/*Tacca*.

Although the Triuridaceae show similar habitat preferences and agree in their heterotrophic life style, flower morphology and embryology do not support a closer relationship with the Burmanniaceae.

DISTRIBUTION AND HABITATS. The Burmanniaceae have a wide distribution in tropical and warm-temperate regions. The genera *Campylosiphon*, *Cymbocarpa*, *Dictyostega*, *Hexapterella*, *Marthella* and *Miersiella* occur in tropical America, *Haplothismia* in Asia and *Afrothismia* and *Oxygyne* in Africa. *Gymnosiphon*, *Burmannia* and *Thismia* are pantropical; *Thismia* subgen. *Ophiomeris* is restricted to tropical America, subgen. *Thismia*, sect. *Sarcosiphon* and sect. *Thismia* occur in tropical Asia, and sect. *Rodwaya* is extra-tropical. The latter section consists of *T. rodwayi* in Australia and New Zealand and *T. americana* from Illinois, USA, a remarkable disjunction (Thorne 1972). According to Jonker (1938), all *Burmannia* species are confined to either Asia, Africa or America, but the species on different continents are often similar and distinguishable only with difficulty.

Several species of this family are extremely rare. The species of *Haplothismia*, *Marthella* and *Oxygyne*, and several *Thismia* have been collected only once or twice. Also *Afrothismia* and most neotropical *Thismia* are known only from very few specimens.

Most species are saprophytes and grow on decaying wood or leaves in the shade of damp primary forests. Some non-saprophytic species of *Burmannia* occur in wet grassland. Most species are confined to low altitudes, but some *Apteria*, *Burmannia* and *Gymnosiphon* are occasionally found up to 3600 m in S America.

SAPROPHYTISM. The achlorophyllous Burmanniaceae live symbiotically with mycorrhizal fungi. They are mycotrophic plants, commonly called saprophytes. The green-leaved Burmanniaceae are hemisaprophytes, constituting a transitional form between autotrophy and heterotrophy. Mycotrophic representatives of several families are often found together, e.g. Burmanniaceae, Triuridaceae, Gentianaceae, Polygalaceae and Orchidaceae. This phenomenon may be related to the ecological requirements of the fungi, or to the association of several plant species with one common fungus.

Achlorophyllous Burmanniaceae are adapted to their extreme ecological niche by strong modifications: reduction of stomata, reduction of root hairs, reduction of vegetative organs (leaves), and strong development of underground parts to accommodate the fungus. The fungi are located in special layers of cells in root-derived structures, i.e. not in rhizomes. They are reported to be septate and/or non-septate, sometimes provided with supposedly reproductive vesicles. These fungi are supposed to belong to the group of Vesicular Arbuscular Mycorrhiza (VAM).

Key to the Genera

1. Stamens 6 — 2
- Stamens 3 — 4
2. Stamens inserted in the throat of the floral tube (perianth basally circumcissile, forming a rim on top of the cup-shaped fruit) — **13. *Thismia***
- Stamens inserted below the middle of the floral tube — 3
3. Underground parts a clump of short, coralloid roots; inflorescence many-flowered. S India — **12. *Haplothismia***
- Underground parts a cylindric rhizome, bearing at each node a clump of small tubers; flowers solitary. Africa — **11. *Afrothismia***
4. Stamens with filament, recurved to pendent; style much shorter than the floral tube. Africa — **10. *Oxygyne***
- Stamens with or without filament, erect; style as long as the floral tube — 5
5. Plants autotrophic, with green stems and leaves (floral tube mostly 3-winged) — **1. *Burmannia***
- Plants saprophytic, with scalelike leaves — 6
6. Upper part of perianth caducous, leaving a naked floral tube — 7
- Perianth persistent — 9
7. Outer tepals entire; anthers stipitate (flowers salver-shaped, basal part slightly winged to ribbed) — **3. *Hexapterella***
- Outer tepals 3-lobed; anthers sessile — 8
8. Capsule held at ca. 90° to the inflorescence axis, dehiscing longitudinally by 1 slit — **6. *Cymbocarpa***
- Capsule in line with the pedicel, indehiscent or withering with age, or dehiscing longitudinally by 3 or 6 valves — **7. *Gymnosiphon***
9. Inner tepals as large as the outer ones — 10
- Inner tepals smaller than the outer ones or absent — 11
10. Flowers erect, blue, salver-shaped; anthers subsessile — **2. *Campylosiphon***
- Flowers nodding, purplish, funnel-shaped to campanulate; anthers on basally decurrent filaments — **8. *Apteria***
11. Inner tepals absent; anthers on basally decurrent filaments — **9. *Marthella***
- Inner tepals present; anthers sessile — 12
12. Inflorescence umbel-like; ovary 1-locular, apically with 3 2-lobed glands — **5. *Miersiella***
- Inflorescence a bifurcate cincinnus; ovary incompletely 3-locular, without apical glands — 13
13. Flowers and fruit erect; flowers mostly 3-winged — **1. *Burmannia***
- Flowers and fruit nodding; flowers never winged — **4. *Dictyostega***

Genera of the Burmanniaceae

I. Tribe Burmannieae Miers (1847).

Stamens 3, erect, with or without filament; thecae transversely dehiscent and separate; style as long as the floral tube.

1. *Burmannia* L. Fig. 48A–C, 52A

Burmannia L., Sp. Pl.: 287 (1753).

Autotrophic or saprophytic herbs; rhizome mostly absent; roots filiform or swollen and coralloid; leaves green and often rosulate or colourless and scale-like; cincinnus bifurcate, 1-27-flowered; flowers often 2-coloured, white, yellow or blue, tubular to salver-shaped; perianth persistent; inner tepals smaller than outer ones or rarely lacking; tube wingless to broadly 3-winged; anthers sessile; ovary 3-locular; capsule dehiscing transversely or longitudinally and septicidally, or irregularly; seeds brown. About 63 spp., pantropical, Asia (30, 19 saprophytic), Africa (11, 3 saprophytic), America (20, 1 saprophytic) and Australia (2, both autotrophic). Autotrophic species in grassland, saprophytes in lowland rainforests.

2. *Campylosiphon* Benth. Fig. 48D–H

Campylosiphon Benth. in Hook., Icon. Pl. III, 14(4): 65 (1882).

Small, pale blue to white herbs; rhizome tuberous, cylindric, covered with scalelike leaves and roots; cincinnus bifurcate, 1-14-flowered; flowers pale blue to white, salver-shaped; perianth persistent; outer tepals as long as inner; floral tube narrowly 3-ribbed; style ± as long as the floral tube; anthers subsessile; ovary 1-locular above, 3-locular below; capsule dehiscing by withering of fruit wall; seeds brown. Only one sp., *C. purpurascens* Benth., tropical S America, in lowland rainforests often bordering creeks (often confused with *Voyria* of the Gentianaceae).

3. *Hexapterella* Urban

Hexapterella Urban, Symb. Ant. 3: 451 (1903).

Small herbs; rhizome cylindric, covered with scales and roots; leaves scalelike; cincinnus bifurcate, 1-8-flowered; flowers white to purple, salver-shaped to tubular; upper part of perianth caducous or persistent; inner tepals much smaller than outer ones, sometimes 3-dentate; tube slightly 6-winged to 6-ribbed; stamens slightly exceeding floral tube or sessile inside tube; ovary 1-locular above, 3-locular below; capsule crowned by perianth(-base); dehiscent by transverse slits and/or withering of fruit wall; seeds brown. Two spp., northern S America in lowland forest and open vegetation at high altitude.

4. *Dictyostega* Miers

Dictyostega Miers, Proc. Linn. Soc. Lond. 1: 61 (1840).

Small, whitish herbs; rhizome cylindric, slightly tuberous, densely covered with imbricate, ciliate scales; leaves scalelike; cincinnus bifurcate, 6-32-flowered; flowers white, nodding, tubular; perianth persistent; outer tepals as long as inner; anthers sessile; ovary 1-locular above, 3-locular below; capsule dehiscing longitudinally and loculicidally from the apex; seeds white, fusiform, with thin transparent testa. Only one sp., the variable *D. orobanchoides* (Hook.) Miers, Mexico to Bolivia and SE Brazil, in forests.

5. *Miersiella* Urban

Miersiella Urban, Symb. Ant. 3: 439 (1903).

Small herbs; rhizome cylindric, slightly tuberous, densely covered with imbricate scales and roots; leaves scalelike, almost peltate; inflorescence umbel-like, 4-10(-22)-flowered; flowers deep lilac to white, tubular; perianth persistent; inner tepals smaller than outer; anthers sessile; ovary 1-locular, apically with 3 2-lobed glands; capsule dehiscing longitudinally and loculicidally; seeds brown. Only one sp., *M. umbellata* (Miers) Urban, throughout tropical S America, in forests.

6. *Cymbocarpa* Miers

Cymbocarpa Miers, Proc. Linn. Soc. Lond. 1: 61 (1840).

Small, white herbs; rhizome cylindric, slightly tuberous, densely covered with roots and scales; leaves scalelike; cincinnus bifurcate, 1-8-flowered; flowers salver-shaped, white; upper part of perianth caducous, inner tepals much smaller than outer; anthers sessile; stigmas with filiform appendages; ovary 1-locular; capsule boat-shaped, held at ca. 90° to inflorescence axis, crowned by perianth base, dehiscing longitudinally and loculicidally by 1 slit; seeds brown. Two spp., Central America, the Greater Antilles and tropical S America, in lowland rainforests.

7. *Gymnosiphon* Blume Figs. 49, 51B

Gymnosiphon Blume, Enum. pl. Javae 1: 29 (1827).
Desmogymnosiphon Guinea (1946).

Small, whitish herbs; rhizome cylindric, slightly tuberous, densely covered by scales and roots; leaves scalelike; cincinnus bifurcate, 1-48-flowered; flowers white, occasionally partly yellow or blue, salver-shaped; upper part of perianth caducous; outer tepals 3-lobed, rarely entire, inner very small; anthers sessile; stigmas sometimes with filiform appendages; ovary 1-locular; capsule crowned by perianth base, dehiscing loculicidally from apex to base, or longitudinally and septicidally from base to apex, or irregularly, or capsule wall persistent and reticulately perforated (Asia); seeds greyish black. About 24 spp., rainforests of tropical Asia (7), tropical Africa (3) and neotropics (14).

8. *Apteria* Nuttall

Apteria Nuttall, J. Acad. Nat. Sci. Philadelphia 7(1): 64 (1834).

Small herbs; rhizome cylindric, slightly tuberous, with dense scales and roots; leaves scalelike, cincinnus 1-5-flowered; flowes purple (to white), funnel-shaped to campanulate, perianth persistent; tepals subequal; stamens with basally decurrent 2-winged filaments ending in a lunate pouch; ovary 1-locular; capsule nodding, dehiscing longitudinally and septicidally from the base; seeds brown. Only one sp., the variable *A. aphylla* (Nuttall) Barnhart ex Small, southern USA and West Indies to central S America, in wet forests.

9. *Marthella* Urban

Marthella Urban, Symb. Ant. 3: 448 (1903).

Small herbs; rhizome cylindric, slightly tuberous, densely covered with imbricate scales; leaves scalelike, few; cincinnus contracted, bifurcate, 1-18-flowered; flowers yellowish, tubular; perianth persistent; inner tepals absent; stamens alternating with tepals, filaments basally decurrent into a lunate pouch; stigmas peltate; ovary 1-locular, with 3 shortly stipitate apical, 2-lobed glands; dehiscence of capsule unknown; seeds brown. Only one sp., *M. trinitatis* (Johow) Urban, forest on Mt. Tucuche on Trinidad.

II. Tribe Thismieae Miers (1847).

Stamens 6 (3 in *Oxygyne*), recurved to pendent, with filament; thecae longitudinally dehiscent and united (separate when forming an anther tube); style much shorter than the floral tube.

10. *Oxygyne* Schlechter

Oxygyne Schlechter, Bot. Jahrb. Syst. 38: 140 (1906).
Saionia Hatus. (1976).

Small herbs; underground parts unknown; leaves scalelike; flowers solitary, funnel-shaped; outer and inner tepals equal; tube basally circumscissile, throat with annulus; stamens 3, recurved, inserted on the annulus; ovary 1-locular, with 3 free placental columns; fruit cup-shaped, with unknown dehiscence. Four species, 2 from Africa and 2 from Japan.

11. *Afrothismia* Schlechter

Afrothismia Schlechter, Bot. Jahrb. Syst. 38: 138 (1906).

Small herbs; rhizome cylindric, each node with a cluster of small tuberous bulbils, each with root; leaves scale-like; cincinnus 1-4-flowered; flowers brown, red and yellow, zygomorphic; outer and inner tepals subequal in size; tube urceolate, with an internal flange at the middle and with an annulus in the throat; stamens inserted in lower part of floral tube, incurved; anthers handing, connective connivent with the funnel-shaped stigma; ovary 1-locular, placentas 3, parietal, basally connate into a central, sterile column, apically free and fertile; all parts except placentas caducous; seeds ellipsoid. About four spp., tropical Africa; rare in mountain forests.

12. *Haplothismia* Airy Shaw

Haplothismia Airy Shaw, Kew Bull. 1952: 277 (1953).

Small herbs with a clump of short, coralloid roots; leaves scalelike; inflorescence of 3-6-flowered monochasia; flowers cream, nodding, funnel-shaped to campanulate; perianth persistent; outer tepals as long as inner; filaments adnate to floral tube, free, apical part recurved, alternating with 6 minute interstaminal lobes; ovary 1-locular; fruit apparently capsular and loculicidal. Only one sp., *H. exannulata* Airy Shaw, S India; collected once in evergreen forest.

13. *Thismia* Griffith Figs. 50, 51A,C, 52B

Thismia Griffith, Proc. Linn. Soc. London 1: 221 (1845).
Bagnisia Beccari (1878).
Geomitra Beccari (1878).
Glaziocharis Taubert (1895).
Triscyphus Taubert (1895).
Scaphiophora Schlechter (1921).
Mamorea de la Sota (1960).

Small herbs with a globose tuber-bearing small roots, or with vermiform, creeping, branched, cylindric roots, or with a clump of short, coralloid roots; leaves scalelike; few; flowers solitary or rarely 2–6, in a cincinnus, variously coloured, actinomorphic or occasionally zygomorphic, outer and inner tepals dissimilar, in 2 distinct whorls, sometimes at different levels, the inner sometimes connate into a mitre; tube urceolate to campanulate, often ± closed by an annulus, basally circumscissile; stamens inserted in the throat, pendent, occasionally alternating with interstaminal lobes; connectives dilated, often with appendages, nectaries or hairs, connate into a tube with thecae separate, or connectives free and thecae united; stigmatic branches 2-lobed or stigma capitate; ovary 1-locular with 3 parietal placentas or with 3 free placental columns; fruit fleshy, cup-shaped; dehiscing apically by withering; seeds brownish. About 32 spp., tropical America (10), tropical Asia (19), Japan (1), USA, Illinois (1), New Zealand and SE Australia (1), in forests. Two subgenera and various sections distinguished by Maas et al. (1986).

Selected Bibliography

Airy-Shaw, H.K. 1952. A new genus and species of Burmanniaceae from South India. Kew Bull. 1952: 277–279.
Barthlott, W., Frölich, D. 1983. Mikromorphologie und Orientierungsmuster epicuticularer Wachs-Kristalloide: ein neues systematisches Merkmal bei Monocotylen. Plant Syst. Evol. 142: 171–185.
Beccari, O. 1890. Le Triuridaceae della Malesia. Malesia 3: 318–344.
Bentham, G., Hooker, J.D. 1883. Genera Plantarum, vol. 3. London: Williams and Norgate, pp. 455–460.
Chakrapani, P., Raj, B. 1971. Pollen morphological studies in the Burmanniaceae. Grana Palynol. 11: 164–179.
Chase, M.W. et al. 1995. See general references.
Dahlgren, R. et al. 1985. See general references.
Erdtman, G. 1952. See general references.
Ernst, A., Bernard, C. 1910. Beiträge zur Kenntnis der Saprophyten Javas III. Ann. Jard. Bot. Buitenzorg 23: 48–61.
Fleischer, E. 1929. Zur Biologie feilspanförmiger Samen. Bot. Arch. 26: 88–132.
Groom, P. 1895. On *Thismia aseroe* (Beccari) and its mycorhiza. Ann. Bot. (Lond.) 9(34): 327–361.
Guinet, P. 1965. Remarques sur les pollens composés à paroid internes perforées. Pollen Spores 7: 13–18.
Huang, T.C. 1972. Pollen flora of Taiwan. Taipai (Taiwan).
Ikuse, M. 1956. Pollen grains of Japan. Tokyo: Hirokawa.
Jonker, F.P. 1938. A monograph of the Burmanniaceae. Meded. Bot. Mus. Herb. Rijksuniv. Utrecht 51: 1–279.
Maas, P.J.M., Maas-van de Kamer, H., van Benthem, J., Snelders, H.C.M., Rübsamen, T. 1986. Burmanniaceae, Flora Neotropica 42: 1–189.
Miers, J. 1866. On *Myostoma*, a new genus of the Burmanniaceae. Trans. Linn. Soc. Lond. 25: 461–476.
Pfeiffer, N.E. 1918. The sporangia of *Thismia americana*. Bot. Gaz. (Crawforsville) 66: 354–363.
Rao, V.S. 1970. Certain salient features in the floral anatomy of Burmannia, Gymnosiphon and Thismia. J. Indian Bot. Soc. 48: 22–29.

Rübsamen, T. 1986. Morphologische, embryologische und systematische Untersuchungen an Burmanniaceae und Corsiaceae (Mit Ausblick auf die Orchidaceae-Apostasioideae). Diss. Bot. 92: 1–310 (98 Tafeln).
Rudall, P., Morley, S. 1992. Embryosac and early post-fertilisation devolopment in *Thismia* (Burmanniaceae). Kew Bull. 47: 625–632.
Schlechter, R. 1921. Die Thismieae. Notizbl. Bot. Gart. Mus. Berlin-Dahlem 8: 31–45.
Stone, B.C. 1980. Rediscovery of *Thismia clavigera* (Becc.) F. v. M. (Burmanniaceae). Blumea 26:419–425.
Thorne, R.F. 1972. Major disjunctions in the geographic ranges of seeds plants. Q. Rev. Biol. 47: 365–411.
Vogel, S. 1962. Duftdrüsen im Dienste der Bestäubung. Akad. Wiss. Mainz, Abh. Math.-Naturwiss. Kl. 10: 601–763.
Vogel, S. 1978. Pilzmückenblumen als Pilzmimeten. I. Flora 167: 329–366; 367–398.

Calochortaceae

M.N. Tamura

Calochortaceae Dumort., Anal. Fam. Pl.: 53 (1829).
Tricyrtidaceae Takht. (1997).
Scoliopaceae Takht. (1996).

Rhizomatous or bulbiferous perennials. Stem often leafy, sometimes scapiform, usually aerial, rarely subterranean, usually erect, rarely descending, simple or branching. Foliage leaves alternate, sometimes with much reduced internodes, linear to elliptical, apically obtuse to long caudate-acuminate, subsessile or sessile to amplexicaul, rarely perfoliate, often parallel-veined, sometimes reticulate-veined, sometimes dark-spotted. Inflorescence terminal or axillary, often 1-flowered, or umbelliform, sometimes a thyrse or a thyrsoid, rarely a raceme. Flowers bisexual, hypogynous, actinomorphic, erect or nodding, often showy. Perianth often campanulate to subrotate, sometimes tubular-campanulate, infundibular or cupular, sometimes recurved. Tepals 3 + 3, free, the whorls similar or highly different, white, green, yellow, red, lavender or purple, often purple-spotted, purple-striped, or patterned with specks of contrasting color, usually nectariferous. Stamens usually 3 + 3, rarely 3 + 0 (*Scoliopus*), free, inserted at tepal base. Filaments often subulate or flattened, dilated at base, sometimes filiform. Anthers often oblong, sometimes sagittate, lanceolate or linear, dorsifixed versatile or basifixed, extrorse or latrorse. Carpels 3; ovary often 3(–1)-locular; ovules 2–many in each locule; style long to very short, apically subentire to 3-parted. Fruit a septicidal or irregularly rupturing capsule or a berry; seeds globose, oblong, irregular or flattened, sometimes with elaiosome.

A family comprising 5 genera and ca. 100 species, subtropical to arctic zones of the Northern Hemisphere, abundant in western N America.

Vegetative Morphology and Anatomy. *Calochortus* has a deep-seated, tunicate bulb with membranous or fibrous-reticulate coats (Fig. 53B). The other genera have horizontal rhizomes, varying from thin to thick and long to short. The rhizome of *Tricyrtis* is stoloniform and mostly de-

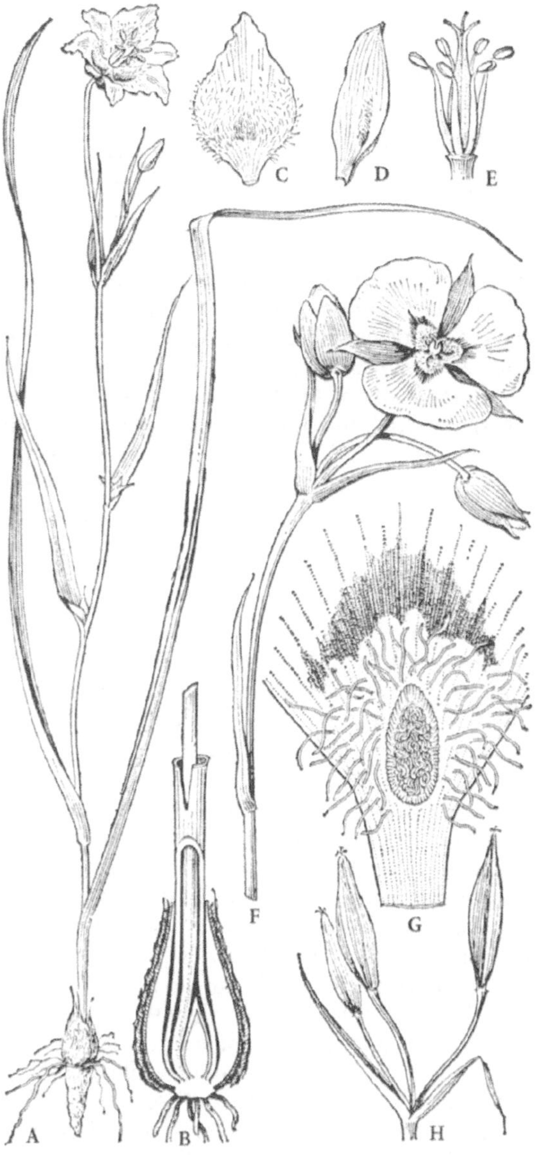

Fig. 53A–H. Calochortaceae. *Calochortus* sp. **A** Flowering plant. **B** Bulb in longisection. **C** Inner tepal. **D** Outer tepal. **E** Androecium and gynoecium. **F** Inflorescence. **G** Base of inner tepal with nectarium. **H** Infructescence. (Takhtajan 1982)

caying every year, rarely thickened and surviving for a few years (*T. macropoda*). The roots of *Scoliopus* are elongate, thick, contractile and long-lived, while its rhizome is short, thin, and hard. The roots of *Calochortus* and *Tricyrtis* have vessels only with scalariform perforation plates, which in *Tricyrtis* may have more than 40 bars (Cheadle and Kosakai 1971).

The stems of *Calochortus*, *Tricyrtis*, *Streptopus*, and *Prosartes* are mostly aerial, often erect, sometimes descending (*T.* sect. *Brachycyrtis*, *T. perfoliata*, *T. suzukii*). The vertical stem of *Scoliopus* is short and mostly subterranean and seems to be adapted to the high frequency of flooding in its environment (Utech 1992). The stem of *Scoliopus* is renewed every year. The stem of *Tricyrtis* contains vessels with scalariform perforations, while that of *Calochortus* lacks vessels and has only tracheids (Cheadle and Kosakai 1971).

Foliage leaves range from 2 (*Calochortus* spp., *Scoliopus*) to many. They are alternate, sometimes with much reduced internodes (*Scoliopus*), flat, dorsiventral, linear to elliptical, apically obtuse (*Scoliopus*) to long caudate-acuminate (*Prosartes ovalis*), subsessile or sessile to amplexicaul, rarely perfoliate (*Tricyrtis perfoliata*), often parallel-veined, sometimes reticulate-veined (*Tricyrtis*), sometimes with numerous transverse veins (*Scoliopus*), sometimes dark-spotted. Bulbils are sometimes produced in the axils of the lower leaves (*Calochortus* spp.).

In *Tricyrtis*, *Scoliopus*, and *Prosartes*, the stomata are restricted to the abaxial leaf surface and surrounded by 4 stomatal contact cells, 2 cells lateral to the guard cells and 1 at each pole (Conover 1991).

Inflorescence Structure. The inflorescence is often terminal, sometimes axillary (*Streptopus*), often 1-flowered or umbelliform, sometimes a thyrse or a thyrsoid (*Tricyrtis*), rarely a raceme (*T. ishiiana*).

The inflorescence of *Scoliopus* is umbelliform, i.e., contracted with 1–3 axillary flower fascicles. Each fascicle is composed of 2–5 pedicellate flowers separated by short internodes and interpreted as a bostryx or cincinnus (Berg 1959) or as a raceme (Utech 1992). The flowers of the upper fascicles bloom later than those of the lower. A foliage leaf is associated with each fascicle. Bracts are lacking throughout the inflorescence.

The 1–2-flowered inflorescence of *Streptopus* was interpreted by Fassett (1935) as axillary; the lower part of the peduncle is more or less fused with the stem. Goebel (1931) interpreted the inflorescence as terminal and the apparent main stem as sympodial. Here I provisionally follow the interpretation of Fassett (1935), but a more detailed examination of this strange inflorescence would be desirable.

Flower Structure. The flowers are bisexual, trimerous, hypogynous, actinomorphic, erect or nodding, usually pentacyclic, those of *Scoliopus* tetracyclic, lacking the inner stamen whorl.

The tepals are white, green, yellow, red, lavender or purple, often purple-spotted (*Tricyrtis*, *Prosartes maculata*), purple-striped (*Scoliopus*), or patterned with specks of contrasting color (*Calochortus* spp.), usually nectariferous. Sometimes the tepals of both whorls are similar, but sometimes the outer (*Tricyrtis* spp., *Scoliopus*), sometimes the inner tepals (*Calochortus*) are much larger. The difference extends also to structure. In *Calochortus* the tubular connective more or less surrounds the filament tip (pseudobasifixed anthers) so that the anthers are not capable of moving freely.

In *Scoliopus* the gynoecium is strongly triangular and is formed by limited fusion along the ventral/septal margins, which subsequently form the corners. Its dorsal regions are laterally flattened or compressed. The gynoecium is unilocular, as protruding septal wings are lacking and placentation is parietal. The ovules are loosely arranged in 2 ascending rows in each of the 3 ventral corners (Utech 1992). In *Tricyrtis*, the gynoecium is more or less unilocular with parietal placentation.

The ovary of *Calochortus* is 3-angled or 3-winged. The angles or wings are developed along the carpel midribs. The ovary is trilocular; placentation is not axile but parietal for the following reasons: (1) The placentae protrude to the center of the ovary where they join with their innermost swollen parts, thus making the ovary trilocular, although the placental fusion is quite loose and incomplete. (2) A placental vascular strand is located in the peripheral half of each placenta. (3) In each loculus the ovules are arranged in 2 rows, but the rows of a single loculus are not found side by side on the same placenta but opposite each other on 2 different placentae (Berg 1960). In *Streptopus* and *Prosartes*, the ovary is trilocular and placentation is axile.

In *Calochortus* and *Streptopus* spp. the carpels end in styluli. Elsewhere, a style of different length is developed, which ends in a subentire stigma (*Streptopus* spp., *Prosartes* spp.) or 3 stylodia (*Scoliopus*, *Tricyrtis*). In the latter genus, the stylodia are apically branched. Stigmatic surfaces are papillate, Dry (*Calochortus*, *Scoliopus*) or Wet (*Tricyrtis*) (Heslop-Harrison and Shivanna 1977). Neither septal nectaries nor crystal raphides are found.

Floral vasculature was studied in *Tricyrtis* (Sterling 1978; Takahashi 1980), *Scoliopus* (Utech 1979, 1992), and *Calochortus* (Berg 1960).

Embryology. The ovule is anatropous and bitegmic, with the micropyle formed only by the inner integument at least at anthesis. Although the embryology of *Streptopus* is not well analyzed (cf. McAllister 1914), embryological characters in *Calochortus*, *Tricyrtis*, *Scoliopus*, and *Prosartes* are similar to each other (Berg 1959, 1960, 1962; Ogura 1964; Björnstad 1970) and support the present circumscription of the Calochortaceae. The family is held together by the following characteristics: (1) Septal glands and raphides are lacking. (2) The cells of the placental epidermis are strongly papillate, i.e., a typical stigmatoid tissue is present. (3) The nucellus is of the Scoliopus type (Björnstad 1970), in which the nonepidermal part of the nucellus base is built up only of longitudinal cell rows without a radial arrangement of the lateral cells. (4) The nucellus epidermis divides periclinally. (5) The archesporial cell functions directly as the megaspore mother cell without cutting off a parietal cell. (6) A single nucleus is observed in the megaspore. (7) The embryo sac is long and narrow and of the monosporic Polygonum type with filiform apparatus. (8) The polar nuclei fuse near the egg apparatus before fertilization. (9) The endosperm is Nuclear.

On the other hand, the following embryological characteristics are variable within the Calochortaceae: (1) *Calochortus* and *Prosartes* lack a clearly defined funiculus, while *Scoliopus* has a very long funiculus. (2) *Scoliopus* has hypotropous ovules, *Calochortus* pleurotropous or slightly hypotropous ones, and *Prosartes* epitropous ones. (3) The nucellus is usually large with a long basal part, but in *Tricyrtis* it is small with a short basal part. (4) The nucellus cap of *Prosartes* shows resistance to the growing embryo sac, while the nucelli of *Calochortus* and *Scoliopus* are destroyed earlier. (5) The synergids are vacuolated in *Calochortus*, *Tricyrtis*, and *Prosartes*, while they lack vacuoles in *Scoliopus*. (6) The antipodals are ephemeral in *Calochortus*, *Scoliopus*, and *Prosartes*, yet persistent in *Tricyrtis*.

Although the Calochortaceae are often considered related to the Liliaceae, with which I agree, the embryological characteristics of the latter differ considerably (Berg 1962; Björnstad 1970). In the Liliaceae, (1) the nucellus is usually small with a short basal part, as in *Tricyrtis*. (2) Periclinal divisions in the nucellus epidermis are rare. (3) The nucellus cap shows resistance to the growing embryo sac, as in *Prosartes*. (4) The embryo sac is tetrasporic (Fritillaria type, Clintonia type, Adoxa type, or Drusa type, usually without filiform apparatus; see under Liliaceae). (5) The synergids lack vacuoles, as in *Scoliopus*. (6) The polar nuclei do not fuse prior to fertilization. This embryological evidence supports the separation of Calochortaceae from Liliaceae.

Tricyrtis and *Prosartes* are sometimes included in the Uvulariaceae (= Colchicaceae s. l.) (Dahlgren et al. 1985), but the embryological characteristics of the Colchicaceae s. l. differ from those of *Tricyrtis* and *Prosartes* in having 2 nuclei per megaspore and synergids without vacuoles (Grassmann 1884; Alden 1912; Sulbha 1954; Berg 1962; Björnstad 1970). By and large, the embryological characteristics of the Calochortaceae are similar to those of the Colchicaceae s. l.

Scoliopus is sometimes included in the Trilliaceae (Dahlgren et al. 1985), but the embryological characteristics of the Trilliaceae (Berg 1962) differ strongly from *Scoliopus*. In the Trilliaceae (1) raphides are present. (2) The cells of the placental epidermis are only slightly papillate, i.e., a typical stigmatoid tissue is lacking. (3) The archesporial cell cuts off a parietal cell. (4) The embryo sac is bisporic (Scilla type). (5) The polar nuclei do not fuse prior to fertilization. This evidence supports the inclusion of *Scoliopus* in Calochortaceae but not in Trilliaceae.

Karyology. The basic chromosome numbers of Calochortaceae are highly variable: *Prosartes*, x = 6, 8, 9, 11; *Scoliopus*, x = 7, 8; *Calochortus*, x = 6–10; *Streptopus*, x = 8, 27; *Tricyrtis*, x = 12, 13, although x = 6 of *Calochortus* (Beal and Ownbey 1943, Cave 1970) and x = 12 of *Tricyrtis* (Ishikawa 1916; Nawa 1928; Ogihara 1971; Takahashi 1991) are exceptional. x = 6, 7, or 8 might be considered as the original basic chromosome number of this family, and *Tricyrtis* as a tetraploid derivative.

The ploidy levels are as follows: *Prosartes* and *Scoliopus*, 2x; *Calochortus*, 2x, 3x, 4x; *Tricyrtis*, 2x, 4x; *Streptopus*, 2x, 3x, 4x, 6x. Chromosome size in this family is also variable: *Calochortus*, *Prosartes* and *Streptopus*, small (1.2–5 μm); *Tricyrtis*, medium (2.5–6.5 μm); *Scoliopus*, large (6–15 μm) (see Tamura 1995), although the largest chromosomes at meiotic metaphase I stages in PMC of the species complex of *Calochortus venustus*, *C. superbus*, *C. vestae*, and *C. luteus* are rather large (up to 10 μm) (Cave 1970). The original chromosome size of this family might be considered as small.

The karyotypes of this family often includes two relatively large chromosome pairs (*Calochortus* with x = 7–10, Cave 1970; *Scoliopus* with x = 7, Johansen 1932; *Tricyrtis* with x = 13, Takahashi 1991; *Prosartes* with x = 9, Tamura et al. 1992). Sometimes they include 3 relatively large chromosome pairs (*Scoliopus* with x = 8, Cave 1970; *Tricyrtis* with x = 12, Takahashi 1991; *Prosartes* with x = 8, Tamura et al. 1992) and rarely 5 relatively large chromosome pairs (*Prosartes* with x = 6, Tamura et al. 1992). The karyotype of *Streptopus* is more strongly asymmetric than that of *Prosartes*, but meiosis in these two genera resembles each other in its main features (Therman 1956), which underlines the close relationship between these genera. The karyotypes with x = 12 and 13 of *Tricyrtis* are connected with each other through Robertsonian rearrangement (Ogihara 1971; Takahashi 1991; Tamura 1995).

In contrast to Calochortaceae, the Liliaceae, though closely related, differ karyologically in having high numerical stability (x = 12) and fairly large chromosomes. However, *Scoliopus* is similar to *Medeola* (x = 7, 8–17 μm) and *Clintonia* (x = 7, 8–15 μm) (Liliaceae-Medeloideae) and *Tricyrtis* to *Tulipa* (x = 12, 3.5–6 μm) and *Gagea* (x = 12, 3–8 μm) (Liliaceae-Lilioideae-Tulipeae) in chromosome number, size, and karyotype (Tamura 1995).

The Calochortaceae are also considered to be related to the Smilacaceae. Actually, the chromosome number and karyotype of *Tricyrtis* are similar to those of *Smilax* (x = 13, 15, 16, 1.5–5 μm) (Smilacaceae), although the chromosomes of the former are slightly larger than those of the latter (Tamura 1995).

Pollen Morphology. The pollen grains of the Calochortaceae are sulcate and spheroidal. They are medium sized in *Prosartes* and *Streptopus*, but large (to 74 × 112 μm) in *Tricyrtis*. The exine is reticulate (*Prosartes*, many *Tricyrtis*, some *Streptopus*) or rugulate (some *Tricyrtis*, many *Streptopus*); the sulcus has no marginal thickening or is operculate (Shimakura 1973; Nakamura 1980; Hara 1988; Lee and Yeau 1990; K. Handa et al. unpubl.).

Pollination. At least *Calochortus*, *Tricyrtis*, *Scoliopus*, and *Streptopus* have perigonal nectaries. Since the anthers in this family are extrorse or latrorse, outbreeding seems to be promoted.

The flower of *Tricyrtis* is morphologically very peculiar. It opens usually for 2 days (observed in 8 spp. of sect. *Tricyrtis*). The primary pollinator is *Bombus diversus*; *B. honshuensis* and *Amegilla* sp. (Anthophoridae) sometimes act as additional pollinators.

Most pollen is removed on the 1st day through contact of the back of the bumblebee with the dehisced anthers, when the bee is seeking nectar in the spur. The anthers of *Tricyrtis* are so versatile that they can bring the dehisced pollen sacs into full contact with the back of the bumblebee. The stylodia, which are sometimes used by the bumblebee as a step for landing on the flower, are fully expanded and susceptible on the 2nd day. The stigmata on the stylodia are touched by the pollen-carrying back of the bumblebee and pollination is accomplished (Takahashi 1984, 1993, 1994). The UV-absorbing, orange-yellow and sometimes purple spots near the base of the tepals act as nectar guides for the bumblebee.

Tricyrtis is self-compatible and can produce seeds by self-pollination if no pollinators visit it (Takahashi 1987), but autogamy is usually avoided by pronounced protandry. Geitonogamy seems to occur rather frequently, because 1st-day and 2nd-day flowers bloom often on a single plant and are visited successively by the insects.

Tricyrtis nana (sect. *Tricyrtis*) has many features of an autogamous derivative from an allogamous stock (Takahashi 1987): its flower is homogamous and opens for only 1 day. The pendulous flowers of *T. macranthopsis* and *T. ishiiana* var. *surugensis* (sect. *Brachycyrtis*), which open for 4 or 5 days, from the point of view of pollination biology are considered derived from the upright flowers of sect. *Tricyrtis* (Takahashi 1993).

The floral morphology and geometry of *Scoliopus* are related to an outbreeding pollination system (Utech 1992): a single *Scoliopus* flower comprises 3 independent pollination units, a feature usually associated with *Iris* (Berg 1959; Utech 1992). Each pollination unit consists of a recurved stylodium, an extrorse anther, a flattened dorsal surface of the unusual triangular gynoecium, and a large nectariferous outer tepal. The pollen-ovule ratios of 600–750 confirm outbreeding (Utech 1992).

Scoliopus is considered as self-incompatible (Berg 1959; Mesler et al. 1980). Fungal gnats (*Mycetophila*, *Sciara*, *Corynoptera*) contact the anthers and stigmas while feeding and are regarded as pollinators (Mesler et al. 1980; Utech 1992).

Fruit and Seed. The capsules are orbicular to linear, 3-angled or 3-winged. They are septicidal in *Calochortus* and most species of *Tricyrtis*, and irregularly rupturing in *Tricyrtis lasiocarpa* and *Scoliopus*. The fruits of *Streptopus* and *Prosartes* are globose to subcylindric berries and are usually red to yellow, rarely blackish; they are considered to be derived from capsules.

The seeds are ellipsoidal to ovoidal, slightly incurved (*Scoliopus*, *Streptopus*, *Prosartes*) or irregular or flattened (*Calochortus*, *Tricyrtis*). In *Scoliopus*, *Streptopus*, and *Prosartes ovalis*, there are longitudinal stripes. In *Streptopus* and *P. ovalis*, the seed surface is smooth and translucent, and a straw-colored subdermal layer with dark brown spots in the micropylar and chalazal regions is visible. The seeds of *Tricyrtis* are placed like piles of coins on top of each other in the capsule in the same manner as in some liliaceous genera. The seeds of *Scoliopus* have elaiosomes.

In seed anatomy, the genera are somewhat heterogeneous (Huber 1969; Fukuhara and Shinwari 1994), although the cellular structure of both testa and tegmen at seed maturity is retained. The testa is usually 2 cells thick, but rarely 3–5 cells thick (*Prosartes ovalis*). The tegmen is invariably 2-layered. A cuticula is present between the endosperm and the tegmen (inner cuticle) and between the tegmen and the testa (middle cuticle), respectively, although in *Tricyrtis* the latter is inconspicuous.

Dispersal. The flat seeds of *Calochortus* and *Tricyrtis* point to wind dispersal. In *Scoliopus*, the elongated fruiting pedicels twist and recurve to the ground, and the capsules dehisce usually in the upper duff layers of the moist coniferous forests. The seeds are provided with eleiosomes and are dispersed by ants (Berg 1959; Utech 1992). The high density of seedlings and adults, which is often observed, may be due to this dispersal pattern.

Phytochemistry. *Tricyrtis* always possesses quercetin and often kaempferol but always lacks luteolin and apigenin (Williams 1975). Chelidonic acid seems to be absent.

Distribution and Habitats. The Calochortaceae are distributed from the subtropical to the arctic zone of the Northern Hemisphere (S Europe, E Asia, and N America); they are abundant in western N America.

Streptopus is widely distributed in the temperate to arctic zones of Eurasia and N America, where it grows preferably in coniferous forests and alpine meadows.

Prosartes is an Arcto-Tertiary element distributed in the temperate zone of eastern and western N America and NE Asia (NE China and Korea), where it grows in forests and moist shady places.

Calochortus is restricted to western N America and is basically a desert plant, growing in a wide variety of open vegetation types, sometimes entering woodland, coniferous forest, and damp to wet grassy meadows. *Calochortus macrocarpus* in

Oregon and Nevada is being seriously reduced by grazing pressure.

Scoliopus is restricted to the coastal montane region of western N America, where the two species grow in coniferous forests (Utech 1992).

Tricyrtis is distributed only in E Asia, extending from subtropical to cool-temperate zones of the Himalayas to Japan, where it is abundant in humid warm-temperate regions on the forest floor and at forest margins. *Tricyrtis* sect. *Brachycyrtis* and *T.* sect. *Flavae* are distributed in the so-called Sohayaki region (Koidzumi 1938) and are considered as relicts.

Subdivision and Relationships Within the Family. The genera united in this family are morphologically quite distinct from one another. *Calochortus* (Dahlgren et al. 1985), *Tricyrtis* (Krause 1930; Huber 1969; Dahlgren et al. 1985) and *Scoliopus* (Baker 1875; Watson 1879) have often been accommodated in monotypic tribes or monotypic families, respectively. The isolated taxonomic positions of *Calochortus* (Ownbey 1940; Berg 1960) and *Scoliopus* (Berg 1962; Utech 1992) were also supported anatomically, embryologically, and cytologically.

In the present treatment of the Calochortaceae, two tribes, Calochorteae and Tricyrtideae, are distinguished. Because of the highly distinct autapomorphies exhibited by the five genera, their interrelationships are difficult to gauge. However, the close relationship between the two baccate genera, *Streptopus* and *Prosartes*, is supported by cytological data (Therman 1956) as well as *rbc*L sequence comparisons (Shinwari et al. 1994a,b; Chase et al. 1995; Rudall et al. 1997). *Prosartes ovalis* (Ohwi) M.N. Tamura, which in the present treatment is transferred into the genus *Prosartes*, possesses some intermediate characteristics between *Prosartes* and *Streptopus*; this confirms the close relationship between *Prosartes* and *Streptopus*.

The possession of a bulb separates *Calochortus* from the other genera, which are rhizomatous; it is distinguished at tribal level. This character is also important in the subdivision of Liliaceae.

Affinities. *Calochortus* was often considered as belonging to the *Lilium* group (Krause 1930; Hutchinson 1934), *Tricyrtis* to the *Uvularia* group (Bentham and Hooker 1883; Dahlgren et al. 1985; Thorne 1992) or to the *Lilium* group (Tamura 1995), *Scoliopus* to the *Trillium* group (Watson 1879; Krause 1930; Dahlgren et al. 1985) or to the *Uvularia-Colchicum* group (Torrey 1857; Baker 1879), *Streptopus* to the *Polygonatum* group (Bentham and Hooker 1883; Krause 1930) or to the *Uvularia* group (Dahlgren et al. 1985), and *Prosartes* to the *Polygonatum* group (Krause 1930; Hutchinson 1934) or to the *Uvularia* group (Bentham and Hooker 1883; Dahlgren et al. 1985).

However, the inclusion of *Calochortus* in the *Lilium* group on morphological, embryological, and karyological grounds was questioned by Buxbaum (1937), Ownbey (1940), Cave (1941, 1953), Maheshwari (1946), Schnarf (1949), and Berg (1960), the inclusion of *Scoliopus* in the *Trillium* group by Berg (1959, 1962), Utech (1979, 1992), and Tamura (1995), and the inclusion of *Streptopus* and *Prosartes* in the *Polygonatum* group by Björnstad (1970) and Tamura (1995).

Berg (1962) and Björnstad (1970) emphasized the great embryological similarity between *Calochortus*, *Tricyrtis*, *Scoliopus*, and *Prosartes*. Tamura (1995) pointed to the karyological similarity between *Calochortus*, *Streptopus*, and *Prosartes*. Also *rbc*L sequence comparisons support the monophyly of *Calochortus*, *Tricyrtis*, *Scoliopus*, *Streptopus*, and *Prosartes* (Chase et al. 1995). On the basis of this combined embryological, karyological, and molecular evidence I propose a revised family circumscription of Calochortaceae with the latter five genera.

Berg (1959) and Utech (1992) indicated that *Scoliopus* is morphologically and anatomically similar to the *Uvularia* group (Colchicaceae). Berg (1960, 1962) and Björnstad (1970) emphasized the embryological similarity of *Calochortus*, *Tricyrtis*, *Scoliopus*, and *Prosartes* with the *Uvularia* group. Tamura (1995) suggested that *Tricyrtis* is karyologically similar to Tulipeae (Liliaceae) and to *Smilax* (Smilacaceae) and *Scoliopus* similar to Medeoloideae (Liliaceae). Chase et al. (1993, 1995), Shinwari et al. (1994a), and Rudall et al. (1997) presented *rbc*L sequence data according to which the Calochortaceae are included in the Liliaceae-Smilacaceae clade.

This evidence, taken together with the morphological similarities, supports the broader relationship of the Calochortaceae with the Liliaceae, Colchicaceae, and Smilacaceae, but not the Trilliaceae and *Polygonatum* group of the Convallariaceae.

Economic Importance. *Calochortus* is widely noted for its remarkable beauty, but it is difficult to cultivate it successfully. *Calochortus nuttallii* is the official flower of Utah, USA and was a source of food for the Indians and early pioneers

(Cronquist et al. 1977). *Tricyrtis* is frequently grown as an ornamental.

Key to the Genera

1. Underground stem a bulb; foliage leaves linear; outer 3 tepals much smaller than inner 3 (**Calochorteae**) **1. *Calochortus***
- Underground stem a rhizome; foliage leaves lanceolate to ovate or elliptical; outer 3 tepals similar to or much larger than inner 3 (**Tricyrtideae**) 2
2. Leafy stem unbranched; foliage leaves often dark-spotted, sometimes not so; anthers dorsifixed versatile; fruit a capsule 3
- Leafy stem often branched, sometimes unbranched; foliage leaves not dark-spotted; anthers basifixed; fruit a berry 4
3. Leafy stem mostly aerial; bracts present; pedicels short, straight or curved; tepals with purplish spots; nectary glands globosely pouched or shortly spurred; stamens 6; stylodia bifid; fruit usually septicidal, rarely irregularly dehiscent; seeds flattened, without elaiosomes **2. *Tricyrtis***
- Leafy stem mostly subterranean; bracts absent; pedicels long, twisted; tepals with red-brown stripes; nectary glands elevated or lined; stamens 3; stylodia unbranched; fruit irregularly dehiscent; seeds narrowly ellipsoidal, slightly curved, with elaiosomes **3. *Scoliopus***
4. Inflorescence axillary; filaments flattened at base, less than 3 mm long; ovules 6–8 in each locule **4. *Streptopus***
- Inflorescence terminal; filaments filiform, more than 6 mm long; ovules 2–6 in each locule **5. *Prosartes***

I. Tribe Calochorteae Melchior (1964).

1. *Calochortus* Pursh Fig. 53

Calochortus Pursh, Fl. Am. Sept. I: 240 (1814); Buxbaum, Beitr. Biol. Pflanzen 34: 405–455 (1958); Ownbey, Univ. Wash. Publ. Biol. 17: 765–779 (1969), rev.

Bulb tunicate. Aerial stem scapiform or leafy, simple or branched, bulbiferous in axils of the lower leaves. Foliage leaves linear, the single basal leaf large; the cauline leaves few, reduced. Inflorescence 1-flowered or umbel-like, few-flowered. Flowers large, showy, usually campanulate, white, yellow, red, lavender or purple, usually variously tinged, erect or nodding. Tepals 6, outer 3 ovate to lanceolate, green or colored similarly to inner 3, inner 3 larger, broader, cuneate to clawed, hairy in lower part, with an often densely bearded, depressed gland or spot near base. Anthers pseudobasifixed, latrorse. Ovary 3-locular; ovules numerous; stigma sessile, 3-fid. Capsules orbicular to linear, 3-angled or 3-winged, septicidal, erect or nodding; seeds numerous, irregular or flattened. $x = 7–10$. Three sections, about 65 spp., western N America, abundant in California.

II. Tribe Tricyrtideae K. Krause (1930).

2. *Tricyrtis* Wall.

Tricyrtis Wall., Tent. Fl. Napal.: 61, t. 46 (1826); Hr. Takahashi, Sci. Rep. Educ. Gifu Univ. (Nat. Sci.) 6: 583–635 (1980), rev.
Brachycyrtis Koidz. (1924).

Rhizome usually stoloniform, rarely thickened. Aerial stem often erect, sometimes descending. Foliage leaves dispersed along aerial stem. Inflorescence a thyrse or a thyrsoid, rarely a raceme. Flowers often erect, sometimes nodding. Perianth often infundibular, sometimes campanulate or horizontally patent, rarely recurved. Tepals white or yellow with purplish spots; outer ones at the base with a nectary, which is often globosely pouched or sometimes shortly spurred, similar in width or wider in comparison to inner tepals. Anthers ellipsoid, dorsifixed versatile, often extrorse, sometimes latrorse. Style trifid, stylodia apically bifid. Ovary narrowly ellipsoid, 3-locular. Ovules numerous. Fruit a linear ellipsoid, trigonous, usually septicidal, rarely irregularly rupturing capsule. Seeds flattened. $x = 13$, rarely 12. Four sections, 18 spp., Himalayas to Japan, from subtropical to cool-temperate zones, abundant in humid warm-temperate Japan.

3. *Scoliopus* Torrey

Scoliopus Torrey, Pacif. Rail. Rep. 4: 145, pl. 22 (1857); R.Y. Berg, Skr. Norske Vidensk.-Akad. Oslo I. Mat.-Naturv. Kl. 4: 1–56 (1959); F.H. Utech, Ann. Carnegie Mus. 48: 43–71 (1979); F.H. Utech, Ann. Missouri Bot. Gard. 79: 126–142 (1992).

Rhizome slender, short, horizontal. Leafy stem mostly subterranean, short, upright, unbranched. Foliage leaves 2–4, more or less sessile, often mottled with purple spots. Inflorescence contracted with several axillary fascicles of long, twisting pedicels. Outer tepals petaloid, many-nerved, striped, spreading, with nectar glands at base; inner 3 linear, 3-nerved, erect. Stamens 3, opposite and adnate at base to outer tepals; anthers dorsifixed versatile, extrorse. Carpels 3, connate except in upper part; ovary 1-locular, 3-angled; ovules 20–38; style very short, stylodia unbranched, horizontally spreading to recurved, stigmatic on upper inner side. Capsules oblong to lanceolate, 3-angled, opening irregularly; seeds 10–22, oblong, slightly curved, sulcate-striate longitudinally, with elaiosomes. $x = 7, 8$. Two spp., coastal montane regions of western N America.

4. *Streptopus* Michaux

Streptopus Michaux, Fl. Bor. Am. I: 200, t. 18 (1803); Fassett, Rhodora 37: 88–113 (1935), rev.

Rhizome creeping. Aerial stem simple or branched. Foliage leaves sessile, sometimes amplexicaul at base. Flowers axillary, 1–2, campanulate or subrotate, white, greenish pink, yellowish brown to deep purple, nodding; peduncles slender, often adnate to 1 internode, often geniculate in middle part. Tepals lanceolate, acute, recurved. Perigonal nectaries present on all tepals, but more strongly developed on base of inner tepals. Anthers sagittate, basifixed, extrorse. Ovary ovoid, 3-locular; ovules 6–8 in each locule; style columnar to absent; stigma subentire to 3-cleft. Berries globose to ovoid, usually red to yellow, rarely blackish; seeds many, oblong, often slightly curved, usually striate longitudinally. $x = 8, 27$. About 7 spp., temperate to subarctic zones of Europe, E Asia, and N America.

5. *Prosartes* D. Don

Prosartes D. Don, Proc. Linn. Soc. Lond. 1: 48 (1839).
Disporum sect. *Prosartes* (D. Don) Q. Jones, Contr. Gray Herb. 173: 1–40 (1951), rev.
Disporum sect. *Ovalia* Hara (1988).

Rhizome creeping. Aerial stem erect, sparsely branched in upper part, pubescent. Foliage leaves subsessile or rarely amplexicaul (*Prosartes ovalis*[1]). Inflorescence terminal, 1-flowered or umbel-like. Flowers subrotate to campanulate or rarely tubular, yellowish, greenish, white, sometimes purple-spotted, drooping. Tepals lanceolate to oblong, acute or long caudate-acuminate at apex, slightly gibbose or truncate at base. Anthers more or less sagittate, extrorse. Ovary ellipsoidal to obconic, 3-locular; ovules 2–6 in each locule; style long, slender; stigma subentire to 3-lobed. Berries ovoid to subcylindric, red or yellow; seeds 3–6 (–18), globose. $x = 6, 8, 9, 11$. Six spp., five in N America, one, *P. ovalis* in Korea and N China.

Prosartes ovalis is somewhat intermediate between *Prosartes* and *Streptopus*. Although *P. ovalis* agrees with all other *Prosartes* spp. in floral and fruit structure, its pollen exine ornamentation (Lee and Yeau 1990) and seed-coat anatomy (Fukuhara and Shinwari 1994) are similar to that of *Streptopus*.

[1] *Prosartes ovalis* (Ohwi) M.N. Tamura, comb. nov., basionym: *Disporum ovale* Ohwi, Bot. Mag. Tokyo 45: 385, 1931.

Selected Bibliography

Alden, I. 1912. A contribution to the life history of *Uvularia sessilifolia*. Bull. Torrey Bot. Club 39: 439–446.

Baker, J.G. 1875, 1879. See general references.

Beal, J.M., Ownbey, M. 1943. Cytological studies in relation to the classification of the genus *Calochortus*. III. Bot. Gaz. 104: 553–562.

Bentham, G., Hooker, J.D. 1883. Genera Plantarum, Vol. 3. London: Reeve; Williams & Norgate.

Berg, R.Y. 1959. Seed dispersal, morphology, and taxonomic position of *Scoliopus*, Liliaceae, Skr. Nor Vidensk. Akad. Oslo I. Mat.-Naturvidensk. Kl. 4: 1–56.

Berg, R.Y. 1960. Ovary, ovule, and endosperm of *Calochortus amabilis* with remarks on the taxonomic position of *Calochortus*. Nytt Mag. Bot. 8: 189–206.

Berg, R.Y. 1962. Contribution to the comparative embryology of the Liliaceae: *Scoliopus*, *Trillium*, *Paris*, and *Medeola*. Skr. Nor. Vidensk.-Akad. Oslo I. Mat.-Naturvidensk. Kl. Ny Ser. 4: 1–64.

Björnstad, I.N. 1970. Comparative embryology of Asparagoideae-Polygonateae, Liliaceae. Nytt Mag. Bot. 17: 169–207.

Buxbaum, F. 1937. Die Entwicklungslinien der Lilioideae. Bot. Arch. 38: 218–293, 305–398.

Cave, M.S. 1941. Megasporogenesis and embryo sac development in *Calochortus*. Am. J. Bot. 28: 390–394.

Cave, M.S. 1953. Cytology and embryology in the delimitation of genera. Chron. Bot. 14: 140–153.

Cave, M.S. 1970. Chromosomes of the California Liliaceae. Univ. Calif. Publ. Bot. 57: 1–58.

Chase, M.W. et al. 1993, 1995. See general references.

Cheadle, V.I., Kosakai, H. 1971. Vessels in Liliaceae. Phytomorphology 21: 320–333.

Conover, M.V. 1991. Epidermal patterns of the reticulate-veined Liliiflorae and their parallel-veined allies. Bot. J. Linn. Soc. 107: 295–312.

Cronquist, A., Holmgren, A.H., Holmgren, N.H., Reveal, J.L., Holmgren, P.K. 1977. Intermountain flora. Vascular plants of the intermountain west, USA, Vol. 6. New York: Columbia Univ. Press.

Dahlgren, R.M.T. et al. 1985. See general references.

Fassett, N.C. 1935. Notes from the herbarium of the University of Wisconsin–XII. A study of *Streptopus*. Rhodora 37: 88–113.

Fukuhara, T., Shinwari, Z.K. 1994. Seed-coat anatomy in Uvulariaceae (Liliales) of the northern hemisphere: systematic implications. Acta Phytotaxon. Geobot. 45: 1–14.

Goebel, K. 1931. Blütenbildung und Sprossgestaltung (2. Ergänzungsband zur Organographie der Pflanzen). Jena: G. Fischer.

Grassmann, P. 1884. Die Septaldrüsen. Ihre Verbreitung, Entstehung und Verrichtung. Flora 67: 113–136.

Hara, H. 1988. A revision of the Asiatic species of the genus *Disporum* (Liliaceae). Univ. Mus., Univ. Tokyo, Bull. 31: 163–209.

Heslop-Harrison, Y., Shivanna, K.R. 1977. See general references.

Huber, H. 1969. See general references.

Hutchinson, J. 1934. The families of flowering plants, Vol. 2. London: Macmillan.

Ishikawa, M. 1916. A list of the number of chromosomes. Bot. Mag. (Tokyo) 30: 404–448.
Johansen, D.A. 1932. The chromosomes of the Californian Liliaceae. I. Am. J. Bot. 19: 779–783.
Koidzumi, G. 1938. Florae in the eastern Kii Mountains. Acta Phytotaxon. Geobot. 7: 120–123 (in Japanese).
Krause, K. 1930. See general references.
Lee, N.S., Yeau, S.H. 1990. A palynological study on *Streptopus ovalis* (Ohwi) Wang et Y.C. Tang and the relative species (tribe Polygonateae, Liliaceae). Korean J. Plant Taxon. 20: 81–94 (in Korean with English abstract).
Maheshwari, P. 1946. The *Fritillaria* type of embryo sac. A critical review. J. Indian Bot. Soc., Iyengar Comm. Vol. pp. 101–119.
McAllister, F. 1914. The development of the embryo sac in the Convallariaceae. Bot. Gaz. 58: 137–153.
Mesler, M.R., Ackerman, J.D., Lu, K.L. 1980. The effectiveness of fungal gnats as pollinators. Am. J. Bot. 67: 564–567.
Nakamura, J. 1980. Diagnostic characters of pollen grains of Japan, parts I, II. Osaka: Osaka Museum of Natural History.
Nawa, N. 1928. Some cytological observations in *Tricyrtis*, *Sagittaria* and *Lilium*. Bot. Mag. (Tokyo) 42: 33–36.
Ogihara, R. 1971. Interspecific differentiation of basic chromosome number from $x = 13$ to $x = 12$ in natural populations of *Tricyrtis hirta*. Rep. Fac. Technol. Kanagawa Univ. 9: 26–30 (in Japanese with English summary).
Ogura, H. 1964. On the embryo sac of two species of *Tricyrtis*. Sci. Rep. Tohoku Univ. Ser. IV (Biol.) 30: 219–222.
Ownbey, M. 1940. A monograph of the genus *Calochortus*. Ann. Mo. Bot. Gard. 27: 371–560.
Rudall, P.J. et al. 1997. See general references.
Schnarf, K. 1949. Der Umfang der Lilioideae im natürlichen System. Oesterr. Bot. Z. 95: 257–269.
Shimakura, M. 1973. Palynomorphs of Japanese plants. Osaka: Osaka Museum of Natural History.
Shinwari, Z.K., Kato, H., Terauchi, R., Kawano, S. 1994a. Phylogenetic relationships among genera in the Liliaceae-Asparagoideae-Polygonatae s. l. inferred from *rbc*L gene sequence data. Plant Syst. Evol. 192: 263–277.
Shinwari, Z.K., Terauchi, R., Utech F.H., Kawano, S. 1994b. Recognition of the New World *Disporum* section *Prosartes* as *Prosartes* (Liliaceae) based on the sequence data of the *rbc*L gene. Taxon 43: 353–366.
Sterling, C. 1978. Comparative morphology of the carpel in the Liliaceae: Hewardieae, Petrosavieae, and Tricyrteae. Bot. J. Linn. Soc. 77: 95–106.
Sulbha 1954. The embryology of *Iphigenia indica* Kunth. Phytomorphology 4: 180–191.
Takahashi, Hr. 1980. A taxonomic study on the genus *Tricyrtis*. Sci. Rep. Fac. Educ., Gifu Univ. (Nat. Sci.) 6: 583–635.
Takahashi, Hr. 1984. The floral biology of *Tricyrtis latifolia* Maxim. (Liliaceae). Bot. Mag. (Tokyo) 97: 207–217.
Takahashi, Hr. 1987. A comparative floral and pollination biology of *Tricyrtis flava* Maxim., *T. nana* Yatabe and *T. ohsumiensis* Masamune (Liliaceae). Bot. Mag. (Tokyo) 100: 185–203.
Takahashi, Hr. 1991. Karyotype variations of *Tricyrtis hirta* Hook. (Liliaceae). Acta Phytotaxon. Geobot. 42: 113–124.
Takahashi, Hr. 1993. Floral biology of *Tricyrtis macranthopsis* Masamune and *T. ishiiana* (Kitagawa et T. Koyama) Ohwi et Okuyama var. *surugensis* Yamazaki (Liliaceae). Acta Phytotaxon. Geobot. 44: 141–150.
Takahashi, Hr. 1994. Floral biology of *Tricyrtis macropoda* Miq. (Liliaceae). Acta Phytotaxon. Geobot. 45: 33–40.
Tamura, M.N. 1995. A karyological review of the orders Asparagales and Liliales (Monocotyledonae). Feddes Repert. 106: 83–111.
Tamura, M.N., Utech, F.H., Kawano, S. 1992. Biosystematic studies in *Disporum* (Liliaceae-Polygonateae) IV. Karyotype analysis of some Asiatic and North American taxa with special reference to their systematic status. Plant Species Biol. 7: 103–120.
Therman, E. 1956. Cytotaxonomy of the tribe Polygonatae. Am. J. Bot. 43: 134–142.
Thorne, R.T. 1992. See general references.
Torrey, J. 1857. Report on the botany of the expedition. Pac. Railroad Rep. 4: 1–195.
Utech, F.H. 1979. Floral vascular anatomy of *Scoliopus bigelovii* Torrey (Liliaceae-Parideae = Trilliaceae) and tribal note. Ann. Carnegie Mus. 48: 43–71.
Utech, F.H. 1992. Biology of *Scoliopus* (Liliaceae) I. Phytogeography and systematics. Ann. Mo. Bot. Gard. 79: 126–142.
Watson, S. 1879. XV. Contributions to American botany. I. Revision of the North American Liliaceae. II. Descriptions of some new species of North American plants. Proc. Am. Acad. Arts Sci. 14 (N. S. 6): 213–303.
Williams, C.A. 1975. Biosystematics of the Monocotyledoneae-flavonoid patterns in leaves of the Liliaceae. Biochem. Syst. Ecol. 3: 229–244.

Campynemataceae

K. Kubitzki

Campynemataceae Dumort., Anal. Fam. Pl.: 57, 58 (1829).

Small, erect, glabrous, terrestrial *herbs* from a short rhizome or tufted crown with a single basal leaf or several basally clustered, dorsiventral, lanceolate to linear leaves, sheathed at the base, sometimes 3-dentate at the apex. *Inflorescence* 1–several, the peduncle bracteate, either poorly paniculate or seemingly umbellate, the lateral flowers subtended by bracts and their pedicels provided each with an abaxial bracteole beneath the flower; *flowers* small, trimerous, pentacyclic, epigynous; tepals greenish, persistent, enlarging after fertilisation, provided with tepal nectaries (*Campynemanthe*); stamens inserted at base of tepals; anthers tetrasporangiate, basifixed or dorsifixed, dehiscing (latro-)extrorsely by longitudinal slits. Ovary 3-locular, inferior, with free styluli; placentation axile, ovules few to many per locule; septal nectaries lacking. *Fruit* a capsule, dehiscing by decay of the lateral or dorsal walls, or indehiscent; seeds 2–numerous, storing hemicellulose and fatty oil; embryo minute.

A western Pacific bigeneric family of four species.

Vegetative Anatomy. The leaves and probably other tissues as well contain calcium oxalate raphide bundles in special idioblasts (Goldblatt et al. 1984), in contrast to Iridaceae, which have styloid calcium oxalate crystals and lack raphide bundles. *Campynema* lacks vessels in the shoot and has very "primitive" vessels in the roots (or lacks them entirely).

The stomata lack subsidiary cells (Dahlgren and Lu 1985).

Inflorescence Structure. The "pseudoumbellate" inflorescence of *Campynemanthe* (Fig. 54) is not well understood; it consists of 3–6 densely crowded lateral units each subtended by a larger basal bract, a long-pedicellate flower and several flowers subtended by smaller bracts (Dahlgren and Lu 1985).

Embryology. The tapetum is secretory, the tapetal cells being 2-nucleate in *Campynemanthe* and multinucleate in *Campynema*. Microsporogenesis is successive. The ovules are anatropous, bitegmic, and weakly crassinucellate. The parietal cell is formed from the archesporial cell and gives rise to parietal tissue. The embryo sac is of the Polygonum type, and endosperm development of the Nuclear type (Dutt 1970; Dahlgren and Lu

Fig. 54A–G. Campynemataceae. **A–E** *Campynemanthe viridiflora*. **A** Habit. **B** Leaf apex. **C** Part of inflorescence. **D** Flower, 3 petals and 3 stamens removed. **E** Petal and opposite stamen in longitudinal section to show callosity. **F, G** *C. neocaledonica*. **F** Flower, longitudinal section. **G** Dehiscing capsule. (Dahlgren and Lu 1985)

1985; Lowry et al. 1987). In this latter trait the Campynamataceae differ from Melanthiaceae and Nartheciaceae, which have Helobial endosperm development as far as is known.

POLLEN MORPHOLOGY. The pollen grains are shed as monads and are sulcate in *Campynema* (Erdtman 1952; Dutt 1970) and inaperturate (Erdtman 1952) or indistinctly sulcate (Dahlgren and Lu 1985) in *Campynemanthe.*

KARYOLOGY. The chromosome number of *Campynemanthe neocaledonica* and *C. viridiflora* is $n = 11$. This number does not fit with any member of Melanthiaceae or Nartheciaceae (Lowry et al. 1987).

POLLINATION. *Campynemanthe neocaledonica* and probably also *C. viridiflora* are protandrous. While there appears to be no overlap in the male and female phases within a single flower, flowers of both phases occur within the same inflorescence. After abscission of the anthers, the filaments recurve and twist counterclockwise. These movements are also known from *Campynema lineare* (Lowry et al. 1987).

SEED. The seeds are nearly globose to angular in *Campynemanthe* and have long persistent funicles but in *Campynema* the seeds are flattened, sometimes almost discoid, and have a spongy outer coat. The anatomical structure is only known for *Campynemanthe viridiflora* (Huber 1969). The seeds have a testa formed by the outer and the inner integuments (Huber 1969). Both the testal and tegmic layers are 2 cell layers thick, and in the former it is the outer epidermis, in the latter the inner epidermis that is large-celled and not compressed. The brown colour of the seeds is due to the red-brown phlobaphenes in the endotegmic layer.

Although Huber (1969) considered the seed structure to be very similar to Iridaceae-Iridoideae in all details, this may be due to convergence or simply plesiomorphic, but probably does not imply close relationship (Goldblatt 1986).

AFFINITIES. The members of this bigeneric family are held together by the unusual green accrescent tepals and the unusual combination of an inferior ovary and free styluli. Presence of perigonal nectaries is uncertain, but septal nectaries are lacking at least in *Campynemanthe.* Nuclear endosperm development suggests a lilialean affinity, but a parietal cell is present, unlike other Liliales sensu Dahlgren et al. (1985) except Iridaceae, in which presence of a parietal cell is plesiomorphic, although not universal.

The morphologically based cladistic analysis of Liliales (incl. Melanthiales) by Goldblatt (1995) shows Campynemataceae close to the Liliales s. str., a result supported by molecular (*rbc*L) studies (Chase et al. 1995).

KEY TO THE GENERA

1. Rosulate plant with numerous leaves; inflorescence pseudo-umbellate. New Caledonia 1. ***Campynemanthe***
– Plant usually with a single basal leaf; inflorescence a 1–few-flowered botryoid. Tasmania 2. ***Campynema***

1. *Campynemanthe* Baill. Fig. 54

Campynemanthe Baill., Bull. Men. Soc. Linn, Paris 2: 1106 (1893); Goldblatt, Bull. Mus. Nat. Hist. Nat. B, Adansonia IV, 8: 121 (1986), rev.

Herb with ascending, short rhizome; leaves dorsiventral, rosulate, many, apically 3-fid; flowering stems lateral, 1–4, unbranched, bearing reduced and progressively smaller leaves above; flowers in apical, many-flowered pseudoumbellate clusters; tepals with a pad of (secretory?) tissue in lower half; anthers basifixed; ovules 2–several per locule; styluli short, tapering; fruit capsular; seeds brown, angular. Three spp., New Caledonia, growing in mesic forested habitats.

2. *Campynema* Labill.

Campynema Labill., Novae Holland. Pl. Specim. 1: 93–94 (1805); Goldblatt, Bull. Mus. Nat. Hist. Nat. B. Adansonia IV, 8: 125 (1986).

Herb from a tufted crown; cataphylls few, basal; a single large basal leaf followed by progressively smaller leaves on flowering stem; inflorescence 1–3(–4)-flowered; tepals lacking apparent nectaries; anthers dorsifixed; styluli free but thickened and contiguous below; fruit dry, probably indehiscent; seeds numerous, discoid; outer coat spongy. Only one spp., *C. lineare* Labill., Tasmania, growing in grassy meadows and among heath and cushion plants.

Selected Bibliography

Chase, M.W. et al. 1995. See general references.
Dahlgren, R., Lu, A.-M. 1985. *Campynemanthe* (Campynemaceae): Morphology, microsporogenesis, early ovule ontogeny and relationships. Nord. J. Bot. 5: 321–330.

Dahlgren, R.M.T. et al. 1985. See general references.
Dutt, B.S.M. 1970. Comparative embryology of angiosperms: Haemodoraceae, Cyanastraceae, Amaryllidaceae, Hypoxidaceae, Velloziaceae. Bull. Ind. Nat. Sci. Acad. 41: 358–374.
Erdtman, G. 1952. See general references.
Goldblatt, P. 1986. Systematics and relationships of the bigeneric Pacific family *Campynemataceae* (Liliales). Bull. Mus. Nat. Hist. Nat. B, Adansonia IV, 8: 117–132.
Goldblatt, P. 1995. The status of R. Dahlgren's orders Liliales and Melanthiales. In: Rudall, P.J., Cribb, P.J., Cutler, D.F., Humphries, C.J. (eds.) Monocotyledons: systematics and evolution, Vol. 1. London: Royal Botanic Gardens, Kew pp. 181–200.
Goldblatt, P., Henrich, J.E., Rudall, P. 1984. Occurrence of crystals in Iridaceae and allied families and their phylogenetic significance. Ann. Mo. Bot. Gard. 71: 1013–1020.
Huber, H. 1969. See general references.
Lowry, P.P., II, Goldblatt, P., Tobe, H. 1987. Notes on the floral biology, cytology, and embryology of *Campynemanthe* (Liliales: Campynemataceae). Ann. Mo. Bot. Gard. 74: 573–576.

Colchicaceae

B. Nordenstam

Colchicaceae DC. in Lamarck & Candolle, Fl. Franç. 3: 192 (1805), nom. cons.
Uvulariaceae A. Gray ex Kunth, Enum. Pl. 4: 199 (1843), nom. cons. prop.

Perennial *herbs* with an underground corm or rhizome; stems rarely lignescent (*Kuntheria*). Roots sometimes tuberous (*Burchardia*). Stem erect, leafy, simple or branching, sometimes scandent or scapose, or reduced to a short underground portion. *Leaves* distichous, cauline, alternate, or subopposite to verticillate or almost radical, sessile and often sheathing, or shortly pseudopetiolate (*Kuntheria*, *Disporum* sp.); blades dorsiventral (bifacial), ovate or lanceolate to linear or subulate, sometimes cuspidate or cirrhose, parallel-veined with often distinct midrib, in broad leaves ± arching and rarely with additional reticulate venation. *Flowers* hypogynous, bisexual or seldom unisexual (*Wurmbea* spp.), sessile or pedicellate, actinomorphic or somewhat zygomorphic, in terminal racemes or cymes, sometimes umbellate, capitate or solitary, with or without bracts; tepals 6 (rarely 7–12), equal or somewhat unequal, connate for some length or free from the base, sometimes spotted or variegated, usually with perigonal or androecial nectaries, caducous or persistent. Stamens 6, dorsifixed (often hypopeltate), extrorse or sometimes latrorse to introrse, dehiscing by longitudinal slits, versatile or not; thecae 2, elongate. Gynoecium tricarpellary (occasionally 2 or 4 carpels), completely or partly syncarpous; ovary trilocular with few to many anatropous or campylotropous ovules on axile placentas; carpels ending in free or partly united styluli, or fused into a single style; *fruit* usually a dry or somewhat fleshy septicidal or loculicidal capsule (or both), rarely baccate (*Disporum*); seeds subglobose or ovoid to subangular, with or without a dry strophiole or fleshy aril; embryo linear, straight.

The family in the present circumscription comprises 19 genera and ca. 225 species and is distributed in temperate to tropical zones of Africa, Europe, Asia, Australia and N America.

VEGETATIVE MORPHOLOGY. The members of the family are generally perennial herbs with an underground tunicated corm (Fig. 55, 56) or a rhizome. Old tunics, and also leaf bases, may remain for several years, forming a protecting cover to renewal buds and young shoots. The corm sometimes forms stolons or droppers and typically carries 2 buds, 1 of which is axillary to the lowermost leaf (forming the tunic). The other bud is axillary to the second leaf and serves as a reserve bud. If both buds are developing into new plants, vegetative propagation takes place. When new corms are formed from apical buds on stolons, the effectiveness of such vegetative reproduction is increased. When droppers are formed, the corms are gradually positioned deeper in the ground (Nordenstam 1982).

The underground organ of *Gloriosa* has recently been termed a hypopodial tuber rather than a corm (le Roux and Robbertse 1994). According to Dahlgren and Clifford (1982: 58), the distinction between corms and tubers is that new growth takes place from apical buds in tubers, as compared to basal buds in corms. In the wurmbaeoid genera, the innovation bud and the reserve bud are principally basal, but may have an apical position on corm lobes, which are horizontally or downwards directed (e.g. *Ornithoglossum*, cf. Nordenstam 1982). In unlobed corms (e.g. *Wurmbea*, cf. Nordenstam 1986) the reserve bud may have a subapical position, but the renewal bud is basal. The underground organ of *Gloriosa* seems homologous with that of other wurmbaeoid genera (Buxbaum 1936), and I prefer to describe it as a corm.

The uvularioid genera generally have a horizontal rhizome, which may be thickened or rather thin, sometimes nodular (Fig. 57).

The stem is erect or climbing, or reduced to an underground portion (*Colchicum*, *Androcymbium* spp.).

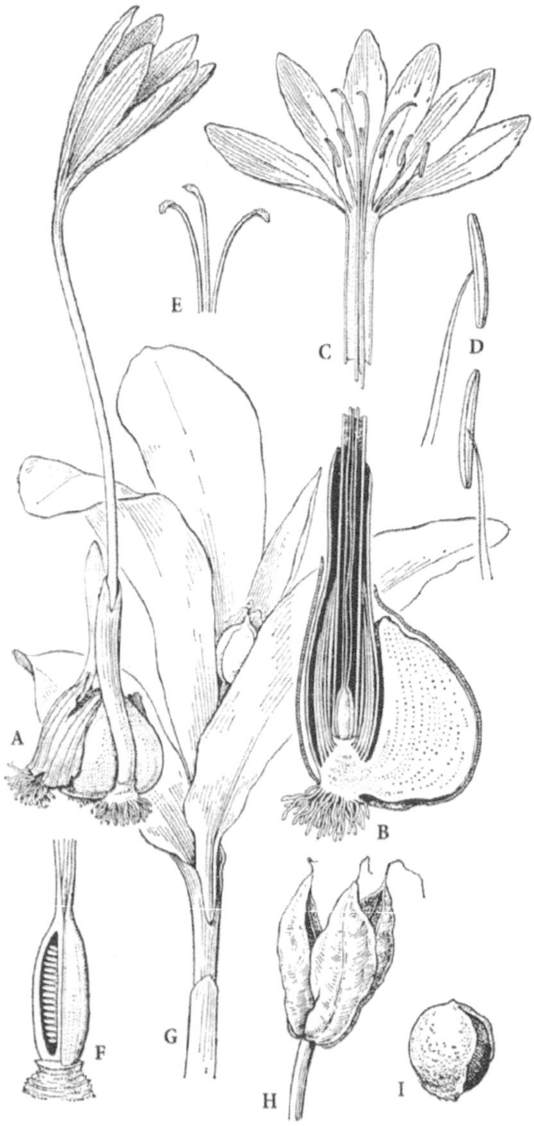

Fig. 55A–I. Colchicaceae. *Colchicum bornmuelleri*. **A** Flowering plant. **B** Longitudinal section of corm. **C** Opened flower. **D** Stamens. **E** Styluli. **F** Ovary, partly opened. **G** Fruiting plant. **H** Dehisced capsule. **I** Seed. (Takhtajan 1982)

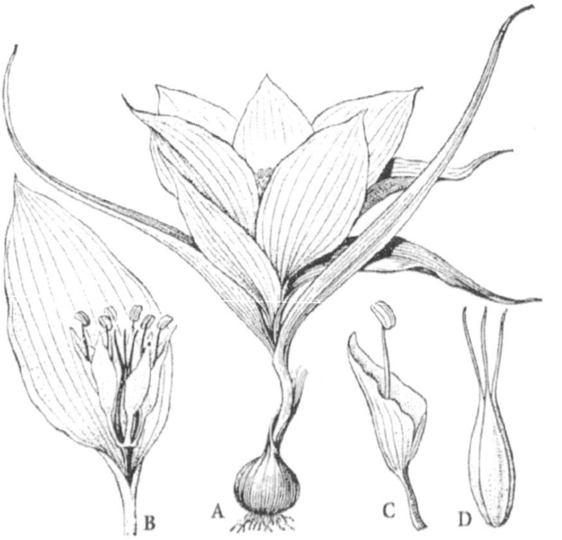

Fig. 56A–D. Colchicaceae. *Androcymbium melanthioides*. **A** Flowering plant. **B** Flower with subtending bract. **C** Tepal and stamen. **D** Pistil. (Takhtajan 1982)

Fig. 57A–I. Colchicaceae. **A–D** *Uvularia grandiflora*. **A** Flowering plant. **B** Opened flower. **C** Stamen, ventral view. **D** Pistil. **E–I** *Iphigenia oliveri*. **E** Flowering and fruiting shoot. **F** Flower. **G** Pistil, the ovary partly opened. **H** Dehiscing capsule. **I** Seed. (Takhtajan 1982)

The green foliage consists of 2-many alternate sheathing sessile leaves, which are bifacial, linear to ovate and parallel-veined (arching and convergent veining in broad leaves), flat or canaliculate, sometimes with tendrils. Broad-leaved uvularioid taxa such as *Kuntheria* and *Disporum* (Fig. 58) have a reticulate venation pattern in between the parallel nerves. Occasionally, the lower leaves may be subopposite or ternately verticillate (*Littonia*). In some genera a sheathing basal cataphyll is present, without a developed lamina, and rhizomatous genera may have several scalelike cataphylls. The leaves grade into bracts subtending flowers, or the bracts may be quite distinct from the other foliage, or absent. In some *Colchicum* spp. the leaves are hysteranthous.

VEGETATIVE ANATOMY. Vessels with scalariform perforation plates occur generally in the roots (Cheadle and Kosakai 1971). Vessels are lacking in aerial stems and rhizomes of the uvularioid genera, but occur in the stems of at least some wurmbaeoid genera (e.g., *Sandersonia*). Calcium oxalate raphides are generally lacking, at least in the wurmbaeoid genera, although crystal sand may be present (Goldblatt 1995). Silica bodies and laticifers are lacking, as far as is known (Dahlgren and Clifford 1982). The anomocytic stomata lack subsidiary cells, as generally in the Liliales. Trichomes are not widespread, but multicellular hairs occur in some genera on leaves or bracts, as well as shorter trichomes and papillae.

INFLORESCENCE AND FLOWER STRUCTURE. The flowers are axillary, forming a few- to many-flowered raceme or cyme, or sometimes are borne singly. The inflorescence may be condensed to a pseudoumbel (*Hexacyrtis*, *Burchardia*, *Kuntheria*, *Schelhammera multiflora*, some *Disporum* spp.) or a capitulum (*Androcymbium* spp.). In *Wurmbea* the flowers are sessile, forming a spikelike inflorescence, in reality a scorpioid ebracteate cyme. The perianth is actinomorphic or slightly zygomorphic with normally 6 tepals (3 + 3), which are free or basally connate to various extent, sometimes forming a long tube (*Colchicum*, *Wurmbea* spp.). The tepals may be basally clawed (unguiculate), and rarely provided with a basal spur or sac (*Hexacyrtis*, *Sandersonia*, *Disporum* spp.).

Septal nectaries are constantly lacking, but perigonal nectaries are generally present (Daumann 1970). They usually have a basal position on the tepal, and are variously shaped like spots, pockets, grooves or furrows. When the nectariferous tissue is situated on stamen filaments, as in *Colchicum* and *Androcymbium*, the term androecial nectary is used. There seems to be no absolute distinction between these two nectary types, however, since tepal and stamen form an ontogenetic unit ("tepal-stamen complex" in Endress 1995). Nectary structures may be formed by combined tepal and filament bases, such as the

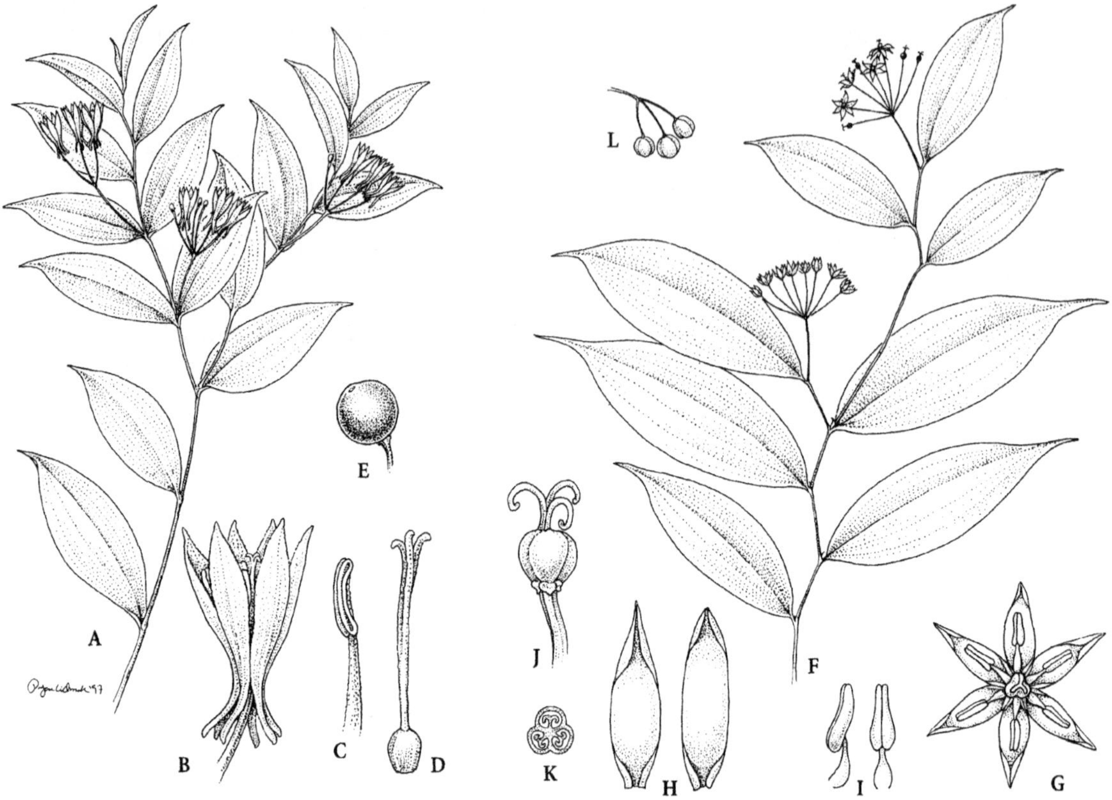

Fig. 58A–L. Colchicaceae. A–E *Disporum cantoniense*. A Habit. B Flower. C Stamen. D Gynoecium. E Fruit. F–L *Kuntheria pedunculata*. F Habit. G Flower. H Tepals. I Stamens. J Gynoecium. K Ovary, transversal section. L Fruits. (A–F original P. Lidmark; G–L after Clifford and Conran, Fl. Austral. 45. 1987)

tepal-enclosed or tepal-adnate androecial nectaries of *Colchicum* spp., or the nectariferous claws of *Onixotis*, *Baeometra* and *Ornithoglossum* spp. The close association of tepal and stamen is seen in the flower buds of some genera (e.g. *Kuntheria*, *Tripladenia*), where the young stamen is enclosed in the involute tepal (Endress 1995: Figs. 3, 4). In *Colchicum* the basal tepal-stamen complex may be very long and partly subterranean, with free or connate tepals.

The 6 stamens are free or epitepalous with terete or flattened filaments, which are sometimes swollen around the middle or basally, or provided with basal appendages. Anthers are extrorse or latrorse to introrse with longitudinal dehiscence. Although basifixed anthers have been reported in several genera, they are probably always dorsifixed, though often hypopeltate.

Normally, the flower is bisexual, but male flowers occur in a few species, and some populations of *Wurmbea dioica* may be constantly dioecious.

The gynoecium is partly or wholly syncarpous, with each carpel tapering into a stylulus, or styluli placed on top of the ovary, free or partly connate, rarely a simple erect style present (*Camptorrhiza*). The punctiform or elongate stigmas are Wet or Dry (Dahlgren and Clifford 1982: 177), a character of apparently limited taxonomic value in this family (both types are recorded in *Colchicum*).

Floral Anatomy. The following information relies heavily on the studies by Sterling (1972–77). Vascular patterns in the carpels are to some extent correlated with degree of fusion and septal invagination, but display considerable infra- and intrageneric variation. There are several to many axile ovules, which are mostly anatropous or campylotropous, or intermediate between these conditions ("weakly campylotropous"). Ovular orientation is variable (epi-, plagio- or basitropic).

Embryology. Some embryological data were summarised by Björnstad (1970). Embryo-sac formation follows the Polygonum type. Two integuments are generally present, but monotegmic ovules have been reported from some genera (*Onixotis, Neodregea, Wurmbea* spp.). The ovules are generally tenuinucellate and lack a parietal cell. The nucellar epidermis may undergo periclinal divisions (*Colchicum, Iphigenia, Gloriosa*) or not (*Uvularia*). Microsporogenesis is successive in both wurmbaeoid and uvularioid genera. The nuclear endosperm is well developed, firm and white, containing starch, or not. The embryo is usually linear and straight.

Pollen Morphology. Pollen grains are mostly sulcate, which seems to be the plesiomorphic condition, sometimes 2-sulcate (*Uvularia*; Kosenko 1988) or 2–4-sulculate or -foraminate (*Colchicum, Androcymbium*; Radulescu 1973) as an apomorphic state. (The report of spiraperturate pollen grains in *Colchicum* by Dahlgren et al. 1985 seems to be erroneous, perhaps due to confusion with *Crocus*?). The exine is crassisexinous with a usually microreticulate to eureticulate surface, homo- or sometimes heterobrochate (i.e. with alternating large and small lumina). A foveolate-spinulate sculpture has been observed in *Uvularia* and *Iphigenia* (Kosenko 1988). In *Hexacyrtis* the exine is tightly reticulate and somewhat striate (Nordenstam 1982; Kosenko 1988). The columellate sexine is usually thicker than the solid nexine layer.

Karyology. There is great deal of variation in basic chromosome numbers in the family. Published reports indicate that $x = 5, 6, 7, 8, 9, 10, 11$, and 12. For *Burchardia* $x = 24$ has been reported, which, however, may alternatively be interpreted as $x = 12$. In *Colchicum* the cytology is particularly complicated (cf. Feinbrun 1958) with a seemingly continuous aneuploid range of all basic numbers between 6 and 12, and an additional 19. Somatic numbers up to 182 have been reported in this genus. B-chromosomes have also been reported and may explain some of the irregular numbers published. Perhaps the complex karyology and high chromosome numbers in *Colchicum* and other genera with reported polyploidy (e.g. *Gloriosa*) is due to the presence of colchicine. The antimitotic effect of colchicine, i.a. inducing chromosome duplication, is well known.

It is noteworthy that *Disporum* s. str. resembles other members of the Colchicaceae in having large chromosomes, whereas the N American segregate genera *Prosartes* and *Streptopus* (now in the Calochortaceae) have small chromosomes (Therman 1956).

Pollination and Reproductive Biology. Adaptations to insect pollination are obvious and varied. The perigonal or androecial nectaries present in most genera are often combined with a variegated or spotted tepal colour pattern or with nectar guides shaped like bands or spots. Many species are distinctly scented, either pleasantly fragrant or sometimes with a dung- or sewagelike odour (*Wurmbea* spp., Nordenstam 1986). Also the petaloid (white, striate or coloured) bracts of several *Androcymbium* spp. may attract pollinating insects. Protandry has been noted in *Tripladenia* and *Schelhammera* (J.G. Conran, unpubl.), and dioecious flowers occur in some Australian species of *Wurmbea*. Most or all genera are likely to be insect-pollinated, although self-pollination may occur in some genera.

Fruit and Seed. The fruit of most genera is a dry or somewhat fleshy many-seeded capsule, with loculicidal or septicidal dehiscence, or both. Only *Disporum* has baccate fruits, with 1–many seeds. The seeds are globose or ovoid to angular or somewhat flattened, with a firm testa without phytomelan. Keel- or winglike structures (caruncles) may be present as an outgrowth of the raphe (*Ornithoglossum* spp.; Nordenstam 1982), and in some genera a strophiole or fleshy aril is developed (*Gloriosa, Littonia, Uvularia, Kuntheria, Tripladenia, Schelhammera*). Seed dispersal is not well studied, but there are no obvious adaptations to wind dispersal. The presence of caruncles, strophioles or arils on seeds of some genera and the baccate fruit of *Disporum* suggest zoochory, with ants or birds as likely vectors.

Phytochemistry. The presence of colchicine-type alkaloids with a tropolone ring is characteristic for most genera (Hegnauer 1963, 1986; Wildman and Pursey 1968; Raffauf 1970; a.o.). Many other alkaloids without a tropolone ring have also been identified and named from wurmbaeoid as well as uvularioid genera (androcymbine, floramultine, kreysigine, etc.).

Especially the colchicine type alkaloids are very poisonous, and severe damage is caused to livestock in many regions (e.g. *Ornithoglossum* spp. in Southern Africa, *Androcymbium* in the Sahara, and possibly *Gloriosa* in Central Africa; Watt and Breyer-Brandwijk 1962). They have also found wide applications in medicine and pharmacology, as well as in biological laboratories. Chelidonic acid is generally present, whereas steroidal saponins are lacking (Hegnauer 1963, 1986), probably having been lost when alkaloids developed. Some flavones have been reported, mainly commonly distributed compounds with little systematic significance. The underground organs may accumulate starch in varying degree.

Delimitation and Affinities. The concept of this family is here expanded from a previously narrow, well-defined circumscription with 13 genera (Nordenstam 1982; Dahlgren et al. 1985) to a larger and inevitably more heterogeneous concept. Most genera of the now dismantled Uvulariaceae are included as well as part of *Disporum* of the Convallariaceae. The genus *Burchardia* has been variously treated in the past. In spite of its fleshy roots and absence of alkaloids, it seems to fit into the family, perhaps best with the wurmbaeoids, as indicated by recent molecular evidence (Chase et al. 1993, 1995).

This expansion of the family calls for a new subfamilial taxonomy. Probably 2 subfamilies can be distinguished (comprising the wurmbaeoid and uvularioid genera, respectively), and a number of tribes, but further investigation is needed, including more chemical and molecular data. The wurmbaeoid genera are characterised by corms, parallel-veined sheathing leaves, dry capsules, and colchicine type alkaloids as well as other alkaloids without a tropolone ring. The uvularioid genera have rhizomes, non-sheathing leaves sometimes with reticulate venation, dry or fleshy capsules or berries, and alkaloids mainly without a tropolone ring.

The uvularioid genera may be regarded as basal in the family. Sterling (1977) has pointed to primitive features in carpellary morphology (open sutures, presence of an obturator and 2 distinct integuments), and suggested that syncarpy in *Schelhammera* may be a recent development from an apocarpous condition. Furthermore, the uvularioid genera have no vessels in stems and rhizomes, and the vessel elements of the roots have scalariform perforation plates, which is regarded as a primitive feature (Cheadle 1942; Cheadle and Kosakai 1971; Wagner 1977). On the other hand, the "hemisyncarpy" of many wurmbaeoid genera (especially *Neodregea* and *Onixotis*) has been regarded as a primitive feature (Sterling 1972). *Disporum* was previously placed in the Convallariaceae on account of its habit, white flowers and baccate fruits, but it is now obvious that these similarities should be regarded as convergences. *Disporum* s.str. (i.e. sect. *Disporum*) is better positioned in the Colchicaceae, whereas the other sections go as separate genera, *Prosartes* and *Streptopus*, into the Calochortaceae. These taxonomic dispositions have support from recent molecular studies (Shinwari et al. 1994; Chase et al. 1995). There are obvious similarities between *Disporum* and *Kuntheria* (Fig. 58), and the baccate fruit of the former and the lignescent perennial stems of the latter are to be seen as apomorphies within the same monophyletic lineage. The berries may have evolved from fleshy capsules as an adaptation to more effective bird dispersal, and the lignescent habit of *Kuntheria* may be an adaptation to a constant rainforest habitat.

The overall relationships within the Liliales are still not satisfactorily known, however. The Alstroemeriaceae have been mentioned as possibly related to the Colchicaceae (Huber 1969; Chase et al. 1995), but affinities to the Luzuragiaceae, Calochortaceae and Liliaceae s. str. should be further studied. The Campynemataceae and Iridaceae including the Geosiridaceae have also been considered, but seem more distantly related in a combined perspective (cf. e.g. Goldblatt 1986).

Distribution and Habitats. The Colchicaceae are distributed in temperate to tropical regions mainly of the Old World. The wurmbaeoid genera occur in winter rainfall areas and subtropical to tropical regions of Africa, the Mediterranean and Asia. They have a pronounced centre in southern Africa, where a dozen genera are represented. *Wurmbea* provides an interesting phytogeographical link between Africa and Australia, being equally well represented in both continents and nowhere else. The distribution of *Iphigenia* is to some extent comparable, but less widely disjunct with occurrences also in Madagascar, Socotra and India. Interestingly, its close relative *Camptorrhiza* occurs only in S Africa and India.

The uvularioid genera, on the other hand, are concentrated in forest regions of SE Asia and Australia, with a centre in Australian rainforest areas. The only genus in the New World, viz. *Uvularia*, is an example of the phytogeographical link between eastern N America and E Australasia, where the related genera occur.

Palaeobotany. A fossil corm or rhizome from Oeningen, Switzerland, was named *Gloriosites* Heer due to its resemblance to the corms of *Gloriosa*, but its affinities are impossible to ascertain, and it may not even be a monocotyledon. In a recent review of fossil monocotyledons (Herendeen and Crane 1995) no taxa referable to the Colchicaeae are mentioned.

Economic Importance. Two genera are especially well known in horticulture, viz. autumn-flowering *Colchicum* and the splendid *Gloriosa*. However, also *Androcymbium*, *Baeometra*, *Littonia*, *Sandersonia* and *Uvularia* are sometimes grown, preferably in greenhouses, but also outdoors in suitable climates. The highly toxic alkaloids of most genera cause considerable losses to livestock, but they are also used in medicine, pharmacology and laborative biology, as mentioned above. The industrial demand has been increasing in recent years, the main supply being derived from wild or cultivated species of *Colchicum*, *Iphigenia* and *Gloriosa*.

The statement by Krause (1930: 241) that corms of *Gloriosa*, *Littonia* and *Sandersonia* were used as food by native African people seems doubtful in view of the toxicity. However, *Gloriosa* and *Iphigenia* corms are locally used for medicinal purposes, but also reputedly for suicide and homicide (Watt and Breyer-Brandwijk 1962). Fatal accidents by confusion with yams have been reported from Sri Lanka. Young shoots and corms of *Uvularia* are sometimes used for nutrition in North America.

Key to the Genera

1. Fruit baccate **19. *Disporum***
- Fruit a capsule 2
2. Stems perennial, becoming woody; leaf venation reticulate with transversely elongated meshes **18. *Kuntheria***
- Stems annual, herbaceous; leaf venation, if reticulate, with meshes elongated longitudinally 3
3. Underground organ a thickened tunicated corm 4
- Rhizome, or many fleshy roots from a short corm 16
4. Acaulescent or shortly scapose herbs 5
- Caulescent herbs 6
5. Leaves radical, hysteranthous or synanthous, flowers large, with a long tube, solitary or fascicled **1. *Colchicum***
- Leaves cauline, though often congested at ground level, always synanthous, grading into bracts, which are often coloured or white; flowers stellate or campanulate in capitate or corymbose inflorescences, rarely solitary **2. *Androcymbium***
6. All flowers pedicillate in a bracteate raceme, pseudoumbel, or solitary 7
- Flowers (at least upper ones) sessile in an ebracteate raceme 13
7. Stems scandent; leaf tips with tendrils 8
- Stems erect; leaf tips without tendrils 10
8. Tepals free, strongly reflexed, 4–10 cm long **9. *Gloriosa***
- Tepals ± connate, shorter 9
9. Perianth urceolate with much connate tepals, free lobes curling outwards **10. *Sandersonia***
- Perianth campanulate with erect segments; tepals connate only basally **11. *Littonia***
10. Flowers subumbellate, long pedicellate; tepals basally spurred **7. *Hexacyrtis***
- Flowers racemose (cymose) or solitary; tepals not spurred 11
11. Perianth persistent; styluli discrete, long, spreading **8. *Ornithoglossum***
- Perianth caducous; style short, erect, simple at least basally 12
12. Style 3-branched apically **12. *Iphigenia***
- Style simple (unbranched) **13. *Camptorrhiza***
13. Tepals basally connate into a tube 1–17 mm long **6. *Wurmbea***
- Tepals free 14
14. Ovary 3-lobed with apically free carpels, each with a simple recurving stylulus; plant <10 cm high **5. *Neodregea***
- Ovary rounded-ovoid, fully syncarpous, with 3 styluli 15
15. Lower flowers pedicellate and bracteate; tepals clawed; ovary cylindric **3. *Baeometra***
- All flowers sessile; tepals not distinctly clawed; ovary obovoid **4. *Onixotis***
16. Many fleshy roots from a short vertical corm **14. *Burchardia***
- Rhizome, usually horizontal 17
17. Tepals basally with stalked glandular appendages **17. *Tripladenia***
- Tepals without stalked glandular appendages 18
18. Perianth campanulate; pedicels recurved **15. *Uvularia***
- Perianth stellate; pedicels straight **16. *Schelhammera***

1. Colchicum L. Figs. 14A,D,F, 55

Colchicum L., Sp. Pl.: 341 (1753); Stefanov, Sb. Bulgarsk Akad. Nauk Sofiya 22: 1–100 (1926, rev.); Feinbrun, Palest. J. Bot., Jerus. Ser. 6: 71–95 (1953, As. min. spp.); Persson, Fl. Iranica 170: 1–40, 14 Tab. (1992).
Bulbocodium L. (1753).
Merendera Ramond (1801).

Acaulescent herbs with tunicated corm and membranous cataphyll. Leaves radical, linear to ovate, synanthous or hysteranthous. Flowers solitary or fascicled. Perianth funnel-shaped to stellate; tepals long unguiculate, free or basally connate. Anthers introrse. Ovary subterranean with discrete styluli or simple 3-lobed style. Capsule septicidal, ellipsoid to subglobose; seeds usually appendaged. $2n$ = 12, 14, 16, 18, 20, 21, 22, 24, 36, 38, 40, 42, 44, 46, 48, 50, 52, 54, 56, 60, 76, 90, 102, 106, 108, 110, ca. 120, 138, 140, 142, 144, 146, 148, 182. About 90 spp., from the Mediterranean Region and S Central Europe to N Africa, and SW and Central Asia. A karyologically complicated genus with a high frequency of polyploidy. There has been no comprehensive monograph since Stefanov (1926). The genera *Merendera* and *Bulbocodium* are sometimes upheld, on perianth and style characters. The genus is well known in horticulture (Autumn Crocus, Naked Ladies etc.), and there are a number of garden hybrids and cultivars. The autumn-flowering *C. autumnale* is the best known and the hardiest species, tolerating temperatures around −20 °C.

2. **Androcymbium** Willd. Fig. 56

Androcymbium Willd., Ges. Naturf. Freunde Berlin Mag. Neuesten Entdeck. Gesammten Naturk. 2: 21 (1808); Krause, Notizbl. Bot. Gart. Berlin-Dahlem 7: 512–526 (1921), rev.

Stemless or shortly scapose herbs; corm with firm tunics. Leaves distichous or subrosulate, linear to ovate, grading into green or coloured bracts. Flowers in terminal capitulum or corymb, seldom solitary. Perianth stellate-campanulate; tepals free, clawed. Anthers introrse. Ovary ovoid, with 3 styluli. Capsule septicidal, subglobose-obpyriform; seeds minutely warty. $2n = 16, 18$. About 30 spp., South and North Africa and the Mediterranean (Canary Islands to the Middle East). Although non-hardy, some species are cultivated in rock gardens, otherwise in glasshouses.

3. **Baeometra** Salisb. ex Endl.

Baeometra [Salisb., Trans. Hort. Soc. 1: 330 (1812), non rite publ.]; Salisb. ex Endl., Gen. 1: 136 (1836).

Herb with tunicated corm. Stem erect, simple. Leaves distichous, sheathing. Flowers in a terminal bracteate scorpioid cyme, sometimes single. Tepals discrete, yellow and purplish inside, clawed. Anthers introrse to latrorse. Ovary cylindric with 3 short styluli. Capsule septicidal, cylindric; seeds globose, firm. $2n = 22$. Only one sp., *B. uniflora* (Jacq.) G. Lewis, confined to southwestern Cape Province of S Africa, sometimes cultivated for its striking colour combination of the perianth. Naturalised in W Australia.

4. **Onixotis** Raf.

Onixotis Raf., Fl. Tell. 2: 32 (1837) ("1836").
Dipidax Laws. ex Salisb. (1866).

Herbs with tunicated corm. Stem erect, simple. Leaves 1–2, sessile, sub-basal, linear. Flowers sessile in a racemelike ebracteate scorpioid cyme. Tepals discrete, clawed, white to pink. Anthers introrse. Ovary trisulcate with 3 filiform styluli. Capsule septicidal, ovoid; seeds subglobose, firm. $2n = 20$. Three spp., southwestern Cape Province.

5. **Neodregea** C.H. Wright

Neodregea C.H. Wright, Bull. Misc. Inform. 1909: 308 (1909); C.H. Wright in Hook. Ic. Plant. 30 t. 2931 (1911, icon.); Garside, Kew Bull. 1935: 292–298 (1935, morph.).

Small herb with tunicated corm. Stem simple. Leaves distichous, sessile, sheathing, usually 3. Flowers sessile in an ebracteate raceme (scorpioid cyme). Tepals discrete, clawed. Anthers introrse. Ovary ovoid with 3 styluli. Capsule septicidal[1], trigonous, apically 3-lobed with outwardly curved style bases; seeds globose or somewhat flattened. Only one sp., *N. glassii* C.H. Wright, southern Cape Region of S Africa.

6. **Wurmbea** Thunb.

Wurmbea Thunb., Nova Gen. Pl.: 18 (1781); Nordenstam, Notes R. Bot. Gard. Edinb. 36: 211–233 (1978, rev. Afr. spp.); Macfarlane, Brunonia 3: 145–208 (1980, rev. Austr. spp.); Nordenstam, Opera Bot. 87: 1–41 (1986, rev. Cape spp.); Macfarlane, Fl. Australia 45: 387–404 (1987, Austr. spp.); Bates, J. Adelaide Bot. Gard. 16: 33–53 (1995, S Austr. spp.).
Anguillaria R. Br. (1810), nom. cons. (over *Anguillaria* J. Gaertn. 1794).

Herbs with tunicated corm. Stem erect, simple. Leaves cauline, seesile, sheathing. Flowers in an ebracteate spikelike scorpioid cyme. Perianth actinomorphic, persistent. Tepals basally connate forming a short or long tube. Anthers introrse to latrorse. Ovary syncarpous or hemisyncarpous, apically 3-lobed. Capsule septicidal or more rarely loculicidal or both. Seeds globose, firm. $2n = 10, 14, 20, 22, 40$. About 40 spp., Africa (mainly Cape Region) and Australia, with about 20 spp. in each continent. Natural hybrids are reported from S Africa (Nordenstam 1986). Many species are scented.

7. **Hexacyrtis** K. Dinter

Hexacyrtis K. Dinter, Repert. Spec. Nov. Regni Veg. 30: 84 (1932).

Herb with tunicated corm. Stem erect, simple or branching above. Leaves distichous, sessile, sheathing. Inflorescence composed of several 2–6-flowered bracteate umbelliform cymes. Flowers purplish green outside, pedicellate, nodding; tepals discrete, basally shortly spurred. Anthers extrorse. Style simple, 3-branched. Capsule loculicidal; seeds globose, firm. $2n = 22$ (P. Goldblatt, pers. comm.). Only one sp., *H. dickiana* K. Dinter,

[1] The capsule of *Neodregea* has been described as loculicidal (Wright 1909, 1911; Krause 1930) or endocidal (Buxbaum 1936), or septicidal (Sterling 1972). The discrepancy was explained by Sterling (l.c.), who found that dehiscence takes place along the commissural lines, especially in the apical region with free carpels.

Namibia and adjacent NW Cape Province, in sandy desert with the corms deeply buried in the ground.

8. **Ornithoglossum** Salisb.

Ornithoglossum Salisb., Parad. Lond. Plate 54 (1806); Nordenstam, Opera Bot. 64: 1–51 (1982), monogr.

Herbs with tunicated corm. Stem erect, simple or branched. Leaves distichous, sessile, sheathing, 2–12. Flowers reflexed to pendulous in bracteate terminal racemes. Perianth yellow-greenish-dull brown or blackish, often bicoloured; tepals discrete, clawed, becoming patent or reflexed. Anthers extrorse. Ovary syncarpous with 3 discrete styluli. Capsule loculicidal; seeds globose, firm. $2n = 24$. Eight spp., southern and tropical Africa north to Tanzania. S African species, locally known as "slangkop", are reported to cause severe loss of livestock (Watt and Breyer-Brandwijk 1962).

9. **Gloriosa** L. Fig. 14E,G

Gloriosa L., Sp. Pl.: 305 (1753).

Perennial herb with a rounded corm. Stem erect or usually climbing. Leaves sessile, alternate-opposite or whorled, with an apical tendril. Flowers axillary, pedicellate, pendulous; tepals discrete, becoming reflexed, yellow-orange-red or bicoloured, often with undulate margins. Anthers introrse. Style filiform, bent at a right angle basally, apically 3-branched. Capsule loculicidal; seeds with a fleshy red strophiole. $2n = 14, 22, 44, 66, 77, 84, 88, 90$. Probably only one sp., although more than a dozen have been described. *G. superba* L. is native to the tropics of Africa and Asia and is a well-known greenhouse plant in various forms and colours (glory lily, flame lily, climbing lily, etc.). All parts of the plant, but especially the corm, are very poisonous (cf. pp. 179, 181).

10. **Sandersonia** Hook.

Sandersonia Hook., Bot. Mag.: t. 4716 (1853).

Perennial herb with tunicated corm. Stem erect, simple. Leaves sessile, clasping, lanceolate, usually with an apical tendril. Flowers axillary, pedicellate, pendulous. Perianth urceolate-campanulate; tepals orange-coloured, largely connate, apically curling outwards, basally shortly spurred. Anthers introrse. Style 3-branched with recurving stylodia. Capsule septicidal, ovoid-oblong; seeds globose, verrucose. $2n = 24$. Only one sp., *S. aurantiaca* Hook., native to S Africa.

11. **Littonia** Hook.

Littonia Hook., Bot. Mag.: t. 4723 (1853).

Perennial herbs with tunicated corm. Stem climbing. Leaves sessile, alternate (lower leaves opposite or sometimes 3-whorled), ovate to linear, with an apical tendril. Flowers axillary, pedicellate, nodding. Perianth campanulate, yellow or orange-coloured; tepals free or shortly connate, basally saccate. Anthers extrorse. Style simple with 3 linear stylodia. Capsule septicidal, ovoid-oblong; seeds globose, with sarcotesta. $2n = 18, 22$. About eight spp., S Africa north to Senegal and Arabia.

12. **Iphigenia** Kunth Fig. 57E–I

Iphigenia Kunth, Enum. 4: 212 (1843).

Herbs with tunicated corm. Stem erect, sometimes flexuous. Leaves sessile, linear-filiform. Flowers in a racemelike or subumbellate scorpioid cyme, rarely solitary, on recurved to reflexed pedicels; tepals discrete, linear, white-pink-brownish red. Anthers extrorse. Ovary ovoid-oblong; style short with 3 falcate stylodia. Capsule loculicidal, ovoid-cylindrical; seeds globose-oblong, papillate. $2n = 20, 22, 26, 33, 44$. About 15 spp., Africa, Madagascar, Socotra, India, Australia and New Zealand. The chromosome no. $2n = 20$ refers to the New Zealand species, *I. novae-zelandiae* (Hook. f.) Baker, which has some deviating floral features and an uncertain generic position (cf. Moore 1970: 42).

13. **Camptorrhiza** Phill.

Camptorrhiza E.P. Phillips, Flow. Pl. South Afr. 15: Plate 575 (1935). (As author of *Camptorrhiza* is often cited J. Hutchinson, Fam. Fl. Plants 2: 101 (1934). Hutchinson mentioned the generic name with two key characters, but certainly he did not intend to describe a new genus. He merely cited Phillips' new genus, of which he had advance knowledge).
Iphigeniopsis F. Buxbaum (1936).

Herbs with tunicated corm. Leaves sessile, sheathing, linear. Flowers in a racemelike scorpioid cyme on a flexuous rachis; pedicels recurved; tepals discrete, greenish or pink, soon reflexed. Filaments terete or swollen around the middle; anthers latrorse. Ovary subglobose-ovoid, with a simple erect style. Capsule loculicidal, globose-ellipsoid; seeds globose, firm. $2n = 22$. Two spp., one in S Africa, Namibia, Botswana, north to Zimbabwe and Mozambique, the other in India.

14. **Burchardia** R. Br.

Burchardia R. Br., Prodr.: 272 (1810), nom. cons.; Macfarlane, Fl. Australia 45: 405–410 (1987).

Herbs with fleshy tuberous roots. Stem leafy below, scapose above. Leaves sessile, conduplicate, sheathing, linear. Flowers in terminal involucrate umbels or rarely solitary; tepals discrete, white or pink. Anthers extrorse. Ovary 3-angled; styluli short, free or basally shortly connate. Capsule septicidal, elliptic-oblong; seeds angular. $2n = 48, 96$. Five spp., Australia. *Burchardia* was in the past often placed with the wurmbaeoid genera, or Colchicaceae s. str. (Baker 1879; Baillon 1894; Krause 1930; Hutchinson 1934), whereas Buxbaum (1925) associated it with the uvularioid genera. The genus was not mentioned in Dahlgren et al. (1985). Although lacking the colchicine-type alkaloids and differing in underground organs, *Burchardia* seems to fit into the expanded concept of the Colchicaceae. This view is also corroborated by recent molecular data (Chase et al. 1995).

15. **Uvularia** L. Fig. 57A–D

Uvularia L., Sp. Pl.: 304 (1753); Wilbur, Rhodora 65: 158–188 (1963, rev.).

Caulescent herbs with a slender rhizome. Stem erect, usually forking. Leaves sessile, sometimes perfoliate, ovate-lanceolate. Flowers axillary, pedicellate, drooping. Perianth narrowly campanulate; tepals free, white-greenish yellow. Anthers extrorse. Style simple, 3-branched above. Capsule septicidal, ovoid-elliptic, 3-angled; seeds subglobose, arillate or with a swollen strophiole. $2n = 12, 14$. Five spp., E and S North America.

16. **Schelhammera** R. Br.

Schelhammera R. Br., Prodr.: 273 (1810), nom. cons.; Cliff. and Conran, Fl. Australia 45: 413–414 (1987).
Kreysigia H.G.L. Reichenbach (1830).

Prostrate to erect herbs with scaly-nodose rhizome. Leaves distichous, sessile, ovate to lanceolate. Flowers pedicellate, solitary, axillary, or forming a pseudoumbel. Perianth white-mauve. Anthers extrorse. Style simple below, 3-branched above. Capsule loculicidal, ovoid, somewhat fleshy; seeds globose-ovoid, with or without a distinct strophiole. $2n = 36$. Two spp., *S. undulata* R. Br. and *S. multiflora* R. Br.[2], in moist open forests of eastern Australia, the latter also in New Guinea and Malesia. *Kreysigia* has sometimes been kept separate, but is here included in *Schelhammera*.

17. **Tripladenia** D. Don

Tripladenia D. Don, Proc. Linn. Soc. Lond. 1: 46 (1839); Clifford and Conran, Fl. Australia 45: 416 (1987).

Erect herb with a knotty scaly rhizome. Leaves distichous, ovate-lanceolate, with half-clasping cordate base. Flowers pedicellate, solitary or few together in axillary pedunculate cymes. Perianth pink or mauve; tepals discrete, basally with stalked nectariferous glands. Anthers extrorse. Style simple below, 3-branched above. Capsule loculicidal, pyriform-rounded, somewhat fleshy; seeds globose, strophiolate. $2n = 14$. Only one sp., *T. cunninghamii* D. Don, confined to rainforests and wet sclerophyll forests of eastern Australia. (Previously often confused with *Schelhammera multiflora*, q.v.).

18. **Kuntheria** J.G. Conran & Cliff. Fig. 58F–L

Kuntheria J.G. Conran & Cliff., Fl. Australia 45: 490, 416 (1987).

Perennial with branching lignescent stems up to 2 m from a thickened horizontal rhizome. Leaves distichous, shortly petiolate; lamina ovate-elliptic. Flowers in terminal pedunculate ebracteate umbels. Perianth pink; tepals discrete, spreading. Anthers introrse. Style short and simple below, branching into 3 longer recurving stylodia. Capsule loculicidal, rounded-oblong, somewhat fleshy; seeds with a fleshy strophiole. $2n = 14$. Only one sp., *K. pedunculata* (F. Muell.) Conran & Cliff., wet rainforests of northern Queensland, Australia.

19. **Disporum** Salisb. Fig. 58A–E

Disporum R.A. Salisbury, Trans. Hort. Soc. 1: 331 (1812), also ex D. Don, Prodr. Fl. Nepal.: 50 (1825); Hara, Bull. Univ. Mus. Univ. Tokyo 31: 163–209 (1988, rev.).

---

[2] *Schelhammera multiflora* R. Br. (syn. *Kreysigia m.*) has been repeatedly confused with *Tripladenia cunninghamii* D. Don. Thus descriptions and illustrations of "*Kreysigia*" in Hooker (1841), Baker (1879) and Krause (1930) refer solely to *Tripladenia*.

Erect herbs with rhizome. Leaves distichous, sessile or shortly petiolate, ovate to linear. Flowers single or paired, or umbellate, often pendulous. Perianth campanulate-tubular; tepals free, white-green-purple, with basal sacs or spurs. Anthers extrorse. Style 3-branched with recurved stylodia. Fruit a blue-black berry; seeds 1-many. $2n = 14$, 16, 18, 24, 28, 30, 32. About ten spp., forests of S and E Asia, Japan. In spite of its baccate fruits, *Disporum* fits well among the uvularioid genera, and *Kuntheria* may be the closest relative. It should be noted that *Disporum* is here taken in a narrow sense (i.e. sect. *Disporum*), whereas *Prosartes* and *Streptopus* are treated as genera of the Calochortaceae.

Selected Bibliography

Baillon, H. 1894. Histoire des plantes T. 12. Paris: Hachette.

Baker, J.G. 1879. A synopsis of Colchicaceae and the aberrant tribes of Liliaceae. J. Linn. Soc. Lond. Bot. 17: 405–510.

Björnstad, I. 1970. See general references.

Buxbaum, F. 1925. Vergleichende Anatomie der Melanthioideae. Feddes Repert. Beih. 29: 1–80.

Buxbaum, F. 1936. Die Entwicklungslinien der Lilioideae. I. Die Wurmbaeoideae. Bot. Arch. 38: 213–293.

Chase et al. 1993. See general references.

Chase et al. 1995. See general references.

Cheadle, V.I. 1942. The occurrence and types of vessels in the various organs of the plant in the Monocotyledoneae. Am. J. Bot. 29: 441–450.

Cheadle, V.I., Kosakai, H. 1971. Vessels in Liliaceae. Phytomorphology 21: 320–333.

Dahlgren, R., Clifford, T. 1982. See general references.

Dahlgren, R.M.T. et al. 1985. See general references.

Daumann, E. 1970. See general references.

Endress, P.K. 1995. Major evolutionary traits of monocot flowers. In: Rudall, P.J., Cribb, P.J., Cutler, D.F., Humphries, C.J. (eds.) Monocotyledons: systematics and evolution. London: Royal Botanic Gardens, Kew, pp. 43–79.

Feinbrun, N, 1958. Chromosome numbers and evolution in the genus *Colchicum*. Evolution 12: 173–188.

Goldblatt, P. 1986. Systematics and relationships of the bigeneric Pacific family Campynematacaeae (Liliales). Bull. Mus. Natl. Hist. Nat. 8, B, Adansonia 2: 117–132.

Goldblatt, P. 1995. The status of R. Dahlgren's orders Liliales and Melianthales. In: Rudall, P.J., Cribb, P.J., Cutler, D.F., Humphries, C.J. (eds.) Monocotyledons: systematics and evolution. London: Royal Botanic Gardens, Kew, pp. 181–200.

Hegnauer, R. 1963, 1986. See general references.

Herendeen, P.S., Crane, P.R. 1995. The fossil history of the monocotyledons. In: Rudall, P.J., Cribb, P.J., Cutler, D.F., Humphries, C.J. (eds.) Monocotyledons: systematics and evolution. London: Royal Botanic Gardens, Kew, pp. 1–21.

Hooker, W.J. 1841. *Kreysigia multiflora*. Curtis's Bot. Mag. 68, t. 3905.

Huber, H. 1969. See general references.

Hutchinson, J. 1934. The families of flowering plants, vol. 2. Monocotyledons. London: Mcmillan.

Kosenko, V.N. 1988. Pollen morphology in Chionographideae, Uvularieae, Tricyrtideae, Scoliopeae, Anguillarieae, Iphigenieae, Glorioseae, Colchiceae (Melanthiaceae). Bot. Zh. 73: 172–185 (In Russian).

Krause, K. 1930. See general references.

le Roux, L.G., Robbertse, P.J. 1994. Tuber ontogeny, morphology and vegetative reproduction of *Gloriosa superba* L. S. Afr. J. Bot. 60: 321–324.

Moore, L.B. 1970. Liliaceae. In: Moore, L.B., Edgar, E. (eds.) Flora of New Zealand, vol. 2. Wellington: A.R. Shearer, pp. 18–32.

Nordenstam, B. 1982. A monograph of the genus *Ornithoglossum* (Liliaceae). Opera Bot. 64: 1–51.

Nordenstam, B. 1986. The genus *Wurmbea* (Colchicaceae) in the Cape Region. Opera Bot. 87: 1–41.

Radulescu, D. 1973. Recherches morpho-palynologiques sur la famille Liliaceae. Acta Bot. Horti Bucur. 1972–73: 133–248.

Raffauf, R.F. 1970. A handbook of alkaloids and alkaloid-containing plants. New York: Wiley-Interscience.

Schulze, W. 1975. Beiträge zur Taxonomie der Liliifloren. II. Colchicaceae. Wiss. Z. Friedrich-Schiller-Univ. Jena Math.-Nat. R. 24: 417–428.

Shinwari, Z.K., Terauchi, R., Utech, F.H., Kawano, S. 1994. Recognition of the New World *Disporum* section *Prosartes* as *Prosartes* (Liliaceae) based on the sequence data of the rbcL gene. Taxon 43: 353–366.

Sterling, C. 1972. Comparative morphology of the carpel in the Liliaceae: Neodregeae. Bot. J. Linn. Soc. 65: 163–171.

Sterling, C. 1973a. Comparative morphology of the carpel in the Liliaceae: Wurmbaeae. Bot. J. Linn. Soc. 66: 75–82.

Sterling, C. 1973b. Comparative morphology of the carpel in the Liliaceae: Colchiceae (*Colchicum*). Bot. J. Linn. Soc. 66: 213–221.

Sterling, C. 1973c. Comparative morphology of the carpel in the Liliaceae: Colchiceae (*Androcymbium*). Bot. J. Linn. Soc. 67: 149–156.

Sterling, C. 1974a. Comparative morphology of the carpel in the Liliaceae: *Baeometra*, *Burchardia* and *Walleria*. Bot. J. Linn. Soc. 68: 115–125.

Sterling, C. 1974b. Comparative morphology of the carpel in the Liliaceae: Iphigenieae. Bot. J. Linn. Soc. 68: 283–290.

Sterling, C. 1975. Comparative morphology of the carpel in the Liliaceae: Glorioseae. Bot. J. Linn. Soc. 70: 341–349.

Sterling, C. 1977. Comparative morphology of the carpel in the Liliaceae: Uvularieae. Bot. J. Linn. Soc. 74: 345–354.

Takhtajan, A.L. 1982. See general references.

Therman, E. 1956. Cytotaxonomy of the tribe Polygonatae. Am. J. Bot. 43: 134–142.

Wagner, P. 1977. See general references.

Watt, J.M., Breyer-Brandwijk, M.G. 1962. The medical and poisonous plants of Southern and Eastern Africa, 2nd edn. London: E. & S. Livingstone.

Wildman, W.C., Pursey, B.A. 1968. Colchicine and related compunds. In: Manske, R.H.F. (ed.) The alkaloids, chemistry and physiology 11. London: Academic Press, pp. 407–457.

Wright, C.H. 1909. Diagnoses africanae: xxx. *Neodregea*. Bull. Misc. Inf. R. Bot. Gdns Kew 1909: 308–309.

Wright, C.H. 1911. *Neodregea glassii*, C.H. Wright. Hook. Ic. Plant. 30: t. 2931.

Convallariaceae

J.G. Conran and M.N. Tamura

Convallariaceae Horian., Prim. Lin. Syst. Nat.: 53 (1834).

Terrestrial or rarely epiphytic *herbs* with caulescent, perennial or annual aerial stems, or rosettes of leaves to 1 m tall arising from subterrancean or exposed rhizomes with fibrous or tuberous roots. *Leaves* alternate, distichous, opposite or verticillate, sessile or more or less petiolate, with or without sheathing base; blades entire, ovate to linear, sometimes sagittate, with 1-few prominent parallel main veins, and between them, parallel or transverse venation. *Inflorescences* axillary or terminal racemes or spikes, or terminal reduced dibotrya or panicles, or single or several flowers in leaf axils. *Flowers* hermaphrodite or rarely unisexual and dioecious, actinomorphic to slightly zygomorphic, hypogynous or perigynous, erect to pendulous; tepals petaloid, in 2 whorls of 3 (most genera), 4–5 or 2 dimerous whorls, both whorls of similar length, sometimes united to form a perianth tube, articulating at the base of the ovary or along the pedicel, often with septal nectaries, white, sometimes spotted with purplish red, green, yellow mauve to purple, or purplish brown. Stamens 4, 6 (most genera), 8, 10 or 12 in 2 whorls, filaments free, connate, adnate to the tepals or attached to a corona; anthers introrse, basifixed or dorsifixed, dehiscing longitudinally. Gynoecium syncarpous, ovary 2–5-locular with axile intrusive placentae; style erect, filiform or stout; stigmas capitate, 3-lobate or expanded into a shieldlike structure; ovules anatropous, campylotropous or more or less atropous borne in 2 rows. *Fruit* a blue, purple, brown, black, red or yellow berry, dry dehiscent capsule (*Gonioscypha*), drupe (*Tricalistra*) or dry, papery, indehiscent structure which ruptures during seed development (Ophiopogoneae); seeds single to many, more or less ovoid to globose, yellow-brown or berrylike, blue or black at maturity and with a fleshy sarcotesta. Embryo linear, at least 1/2 as long as seed; embedded in a copious starchless endosperm.

Seventeen genera and about 130 spp., N and Central America, Europe, India and Asia, south to about Bali and Lombok.

Vegetative Morphology and Anatomy. The plants are rhizomatous with thin, wiry or stout, thick rhizomes and have erect or sometimes pendulous unbranched stems to about 1 m, or a stemless rosette of leaves. The aerial stems are perennial or annual, herbaceous and show limited growth. In *Maianthemum paludicolum* the lower portions of the aerial stems consist of the sympodially branched rhizome and are anatomically similar to the subterranean rhizomes of the other species in the genus (LaFrankie 1986b). The roots are swollen and fleshy in some *Ophiopogon* species. The seedlings are admote and eligulate; the first leaf alternates with the cotyledon and is a cataphyll, as are several of the successive leaves (Boyd 1932; Conran 1985; Tillich 1995). The following leaves are linear and grasslike, ovate to lanceolate or sagittate, bluntly acute, sessile to petiolate, 3–7-costate, entire, with a prominent midrib in some genera. In *Polygonatum* and *Disporopsis* the apical leaf on the shoot is pseudoterminal and generally reduced in size, indicating that the shoots originally possessed indeterminate growth (Eames 1961). Venation is parallel or acrodromous with apical or proximal convergence of the primary veins, and slight acropetal weakening. The secondary veins and higher vein orders are parallel to the leaf long axis, with secondary veins sometimes directly entering the midvein. Free vein endings are absent and the areoles are elongate parallel or transverse (*Peliosanthes*) to the leaf long axis. The leaves are sheathing or non-sheathing, spirally (distichous and non-distichous), and oppositely or verticillately arranged. The petiole, if present, is small to large relative to the lamina, and canaliculate; stipulelike structures are present in some forms of *Peliosanthes* above the sheath (Larsen 1966). Vernation is flat/curved (most Ophiopogoneae), supervolute (*Theropogon* and *Aspidistra*) or conduplicate (Conran 1985).

Silica bodies are lacking. Calcium oxalate raphides occur in all genera. Tannin cells are present in at least *Polygonatum* and *Maianthemum*. The roots of *Aspidistra, Maianthemum, Ophiopogon* and *Tupistra* have a multiseriate unthickened exodermis (Mulay and Deshpande 1959), and an extensive, unthickened cortex. The rhizodermis of most genera shows differentiation into long cells and short cells (Zweigelt 1913). The endodermis is thickened. Vessel elements with scalariform or simple perforation plates and tracheids occur in the roots and rhizomes of most genera. Vessels with scalariform perforation plates sometimes occur in the aerial stems of the

Polygonateae, and the leaves of all taxa lack vessels (Wagner 1977). The epidermal cells of the rhizome are small and unthickened with a thin cuticle, and those of the aerial stem are small, with thickened outer walls and thick cuticle. The epidermal cells of the leaves are elongate parallel to the leaf axis, with straight walls in all genera. All genera examined have undifferentiated armcell mesophyll. The stomata are anomocytic and oriented parallel or transverse to the leaf axis (Conover 1991). In *Ophiopogon* and *Liriope* the leaf epidermal cells are ridged and sculptured, with the subsidary cells forming a boxlike structure around the guard cells in some species. Both genera also possess hypodermal fibres, a feature which they share with *Peliosanthes* (Cutler 1992).

INFLORESCENCE STRUCTURE. The flowers are borne singly in leaf axils or reduced axillary racemes (*Disporopsis* and *Polygonatum*, Fig. 59), or terminal or axillary spikes or racemes, occasionally in terminal racemose dibotrya (*Maianthemum* and the Ophiopogoneae) or panicles (*Maianthemum*).

FLORAL STRUCTURE. The flowers are hypogynous or perigynous, pedicellate or sessile, and are subtended by a bract which in *Campylandra*, *Gonioscypha* and *Theropogon* is longer than the flowers. The outer and inner tepal whorls are more or less similar and free or united basally into an urceolate, campanulate or tubular perigone. The tepals are either uninerved through fusion of several veins, plurinerved (*Reineckea*, *Speirantha*) or multinerved (*Aspidistra*) and have parallel venation. The perianth is caducous or marcescent and articulation is either at the base of the ovary or along the pedicel. The nectaries are septal in *Polygonatum* and *Ophiopogon* (Daumann 1970). *Theropogon* was reported as having tepal nectaries by Vaikos et al. (1989), in contrast to Utech (1979), where they are reported as septal. Nectaries are reported to be absent in *Convallaria*, which instead possesses tepaline oil glands which function as osmophores (Vogel 1990). In *Tupistra squalida*, the surface of the entire pistil appears to secrete nectar. The tepals are unspotted except in *Aspidistra*. The floral vascular anatomy of *Convallaria*, *Maianthemum* and *Theropogon* has been investigated and is summarised by Utech (1979).

The stamens are free, sometimes attached to the perianth tube, or they may be fused basally (*Ophiopogon*) or attached to a corona (*Disporopsis* and *Peliosanthes*, Fig. 60). The filaments may be filiform, broad or absent, and are twisted to one

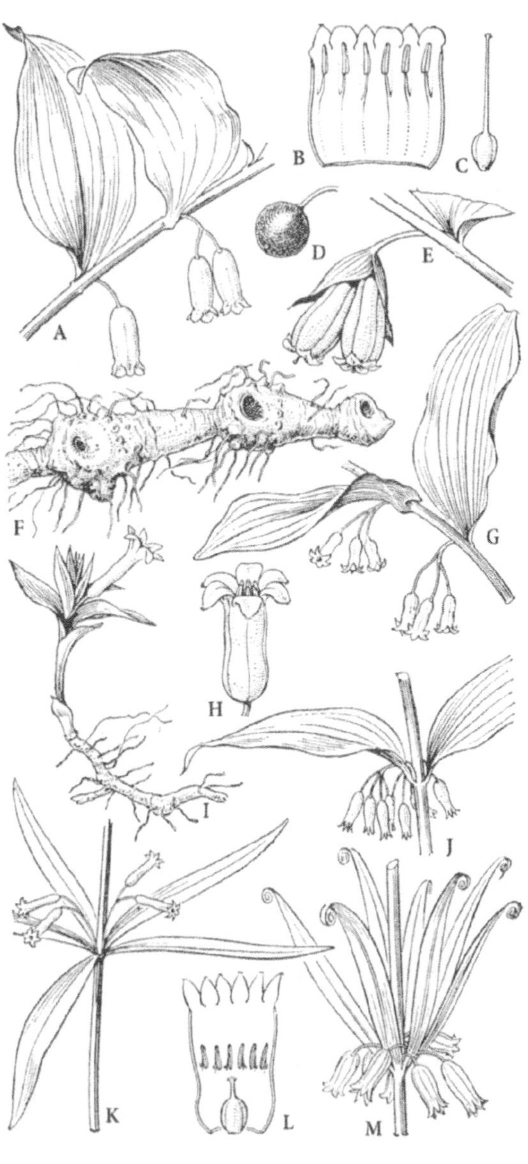

Fig. 59A–M. Convallariaceae. **A–D** *Polygonatum odoratum*. **A** Flowering branch. **B** Floral tube. **C** Pistil. **D** Fruit. **E** *P. involucratum*, partial inflorescence. **F** *P. multiflorum*, rhizome. **G, H** *P. polyanthemum*. **G** Flowering stem. **H** Flower. **I** *P. hookeri*, whole plant. **J** *P. oppositifolium*, partial inflorescence. **K, L** *P. roseum*. **K** Partial inflorescence. **L** Opened flower. **M** *P. kingianum*, partial inflorescence. (Takhtajan 1982)

side of the flower in *Liriope*. The projections and hairs on the filaments of *Polygonatum* are variable in shape an distribution and important in the taxonomy of the genus (Tamura 1991, 1993). The anthers are 2-locular, 4-sporangiate and equal in number to the tepals or double in some *Aspidistra* species (Fig. 61) which are occasionally separated

Fig. 60A–E. Convallariaceae. *Peliosanthes teta*. **A** Habit. **B** Partial inflorescences. **C** Flower, seen from above. **D** Same, longitudinal section. **E** Fruit. (Takhtajan 1982)

as a segregate genus, *Evrardiella* (Gagnepain 1934a). Anther dehiscence is longitudinal and introrse in all genera. The stigmatic surface is Dry in most genera (Heslop-Harrison and Shivanna 1977).

Pollen Morphology. The pollen grains are shed singly and are obovate and sulcate with a reticulate or perforate exine and 30–51 µm in length along the longer axis (most genera) or ovoid or orbicular and inaperturate with a reticulate exine and 40–50 µm in diameter in *Aspidistra* (Erdtman 1952; Schulze 1982).

Embryology. Anther wall formation is of the monocotyledonous type, the well-developed endothecium has spiral thickenings, and the tapetum is binucleate in most genera and glandular in at least *Ophiopogon* (Rao and Kaur 1979a). Microsporogenesis is successive in all genera examined. The ovules are anatropous (most genera), campylotropous or more or less atropous; bitegmic, crassinucellate or tenuinucellate and a parietal cell may or may not be cut off. Embryo-sac formation is commonly of the Polygonum or Allium types, but in *Maianthemum* is of the Scilla or Drusa type. Endosperm formation is of the Nuclear or Helobial type (*Ophiopogon*). In *Maianthemum stellatum*, although pollination is common and pollen tubes are formed, fertilisation does not occur. Instead, several of the mature megagametophytes degenerate, with the remainder producing adventitious embryos from the nucellus. Polyembryony occurs frequently (Ono 1928; Gorham 1953; Stenar 1953; Björnstad 1970; Rao and Kaur 1979a,b; Yang and Hong 1993).

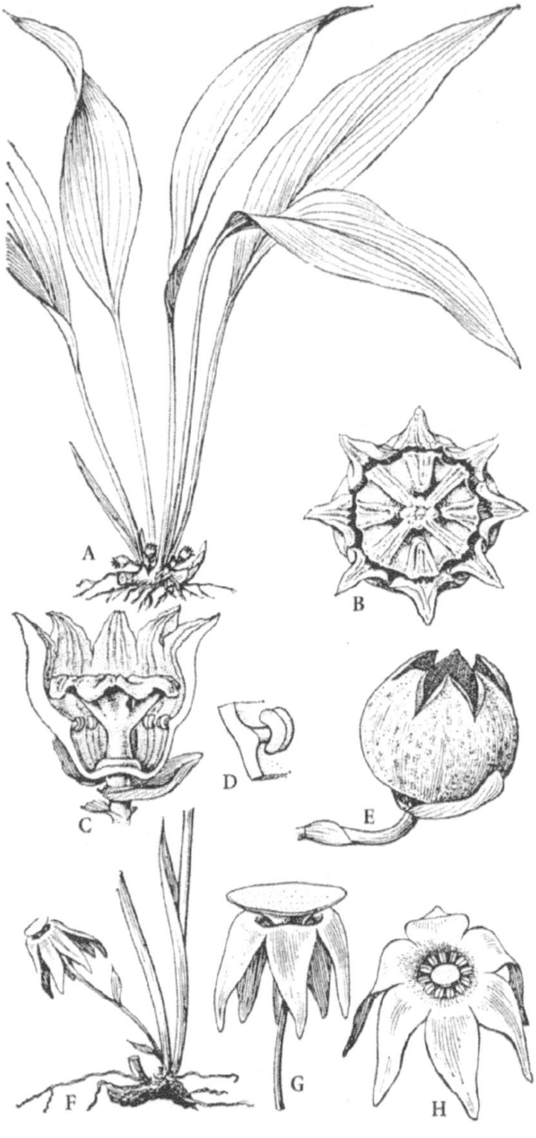

Fig. 61A–H. Convallariaceae. **A–D** *Aspidistra elatior*. **A** Habit. **B** Flower seen from above. **C** Flower, floral tube opened. **D** Stamen. **E** *A. typica*, capsule. **F–H** *A. dodecandra*. **F** Basal part of plant. **G** Female flower with large stigma. **H** Male flower. (Takhtajan 1982)

Karyology. The basic chromosome numbers for the Convallariaceae are $x = 9$–15 (*Polygonatum*), 16 (*Heteropolygonatum*), 18 (*Maianthemum*,

Liriope, *Ophiopogon*, *Peliosanthes*), 19 (*Speirantha*, *Convallaria*, *Reineckea*, *Rohdea*, *Campylandra*, *Tupistra*, *Tricalistra*, *Aspidistra*) and 20 (*Disporopsis*, *Theropogon*). Numbers are highly variable in the Polygonateae ($x = 9$–16, 18, 20), while they are constant in the Ophiopogoneae ($x = 18$), and in the Convallarieae ($x = 19$), except for *Theropogon*, whose taxonomic position is questionable (see under description). Recently the chromosome number of *Tricalistra* was determined as $2n = 38$ (J. Yamashita and M.N. Tamura, unpubl.), confirming the chromosomal stability of the Convallarieae. $x = 20$ of *Disporopsis* seems to be of tetraploid origin.

The chromosomes of the Convallariaceae are comparatively large: the largest one of a complement often exceeds 10 µm in length (Tamura 1995). Among Polygonateae, karyotypes of *Polygonatum* sect. *Polygonatum*, *Disporopsis* and *Maianthemum* are more or less symmetric and monomodal, while those of *P.* sect. *Verticillata* and *Heteropolygonatum* are asymmetric and bimodal (Kawano et al. 1967; Tamura 1990, 1993; Tamura et al. 1997). Among Ophiopogoneae, Karyotypes of the three genera are similar to one another (Tamura 1995), which supports the coherence of this tribe. Among the Convallarieae, karyotypes of *Convallaria*, *Reineckea*, *Rohdea*, *Campylandra* and *Tupistra* sect. *Nutantes* are more or less symmetric and monomodal, while those of *T.* sect. *Tupistra*, *Tricalistra* and *Aspidistra* are asymmetric and bimodal (Huang and Li 1990; Tamura 1995; Yamashita and Tamura, unpubl.). Diploid, triploid, tetraploid and hexaploid taxa are reported for *Polygonatum*, *Liriope* and *Ophiopogon* (Sato 1942; Fedorov 1969; Chang and Hsu 1974; Tamura 1995).

Pollination. *Polygonatum* is "buzz"-pollinated by bumblebees (*Bombus* species) with the introrse anthers and perigone tube creating a resonating chamber (Corbet et al. 1988). *Rohdea* and *Aspidistra lurida* are said to be snail- or slug-pollinated (Loew 1895; Richards 1986), the flowers of *R. japonica* smelling like rotting bread. *Rohdea japonica* was also reported to be ant-pollinated (Migliorato 1910), but this is disputed by van der Pijl (1955). Similarly, Kato (1995), observed that *Aspidistra elatior* in its natural habitat is pollinated by the terrestrial crustacean *Platorchestia japonica* (Amphipoda: Talitridae). The amphipod enters the flowers to feed on the pollen via narrow pores between the expanded stigma and the corolla tube, and specimens leaving the flowers were found to carry pollen on their bodies. In *Tupistra squalida*, there are transparent windows at the base of the floral tube which function as a light-orientation trap to confuse pollinators once they are inside the chamber formed by the peltate stigma and fused corolla tube. In this species, the pistil begins to secrete nectar 1 day prior to anthesis.

Jessop (1979) reports that *Ophiopogon caulescens* is self-fertile. The flowers of *Maianthemum bifolium* and *M. dilatatum* were reported as self-incompatible (Bierzychudek 1982), whereas those of *M. racemosum*, although frequently pollinated, were never fertilised, and form apomictic seeds (Gorham 1953). Piper (1989) found that both *M. stellatum* and *M. racemosum* were visited by hoverflies (Syrphidae), bees (Apidae, Bombyliidae and Halictidae) and various Lepidoptera, and were apparently self-incompatible.

Fruit and Seed. The fruits are berries in most genera. In *Gonioscypha* the fruits are described as fleshy capsules (Gagnepain 1934b), in *Tricalistra* they are drupes and in most of the Ophiopogoneae are papery and indehiscent, with the ovary wall rupturing at an early stage due to seed growth. The seeds of these Ophiopogoneae have a well-developed sarcotesta, and the seeds develop as if berries. The perianth is persistent in most genera. Mature fruits are black, brown, purple or deep blue, red or yellow (*Rohdea* and *Tupistra*). The pericarp is thick and fleshy, and encloses 1-ca. 10 seeds which are ovoid to globose. The seed surface is smooth to rugulose, and pale brownish yellow in all genera with berries examined. The seed structure is uniform, with all cell layers first thin-walled. The testal layer, which lacks phytomelan, disintegrates and separates from the tegmic layer. The tegmic layer collapses, but the original cellular structure is retained; the cells are filled with a homogeneous red-brown substance, which later is passed to the seeds (Huber 1969). The endosperm is massive, and consists of cells with thick pitted rugose walls. It contains aleurones and oils, but is non-starchy.

Dispersal. The fruits are assumed to be animal-dispersed. Birds were suggested as the likely dispersal agents for Japanese *Maianthemum* species (Kawano et al. 1968), and are known to disperse the fruits of *M. stellatum* (Piper 1986a,b). The number of flowers and size of the ramets significantly influenced fruit set in *M. racemosum*, whereas there were no significant relationships between seed set and environmental parameters in *M. stellatum* (Piper 1989), probably due to its apomictic reproductive strategy, although Piper appears to have been unaware of this aspect of its biology.

Phytochemistry. Most genera contain steroidal saponins and sapogenins, and those in *Reineckea* show possible anticancer activity (Kanmoto et al. 1994). Cardenolide glucosides occur in the fruits of *Convallaria*, *Speirantha* and *Polygonatum* (Pauli 1995), with spirostanol glycosides reported from *Peliosanthes*. The rhizomes of *Aspidistra* contain aspiridin (diosgenin 3-O-beta-lycotetraoside) and show antifungal activity against food-borne fungi (Koketsu et al. 1996). Tannins are reported for *Polygonatum* and *Maianthemum*. All genera studied contain chelidonic acid (Ramstad 1953), and most have azetidine-2-carboxylic acid (Fowden 1955, 1959; Fowden and Steward 1957; Farnsworth et al. 1971).

Classification. The Convallariaceae have been variously constituted and subdivided (Krause 1930; Schulze 1982; Dahlgren et al. 1985) with numerous tribes and even segregate families being recognised (e.g. Nakai 1936). The present treatment divides the family into three previously recognised tribes.

The Polygonateae (*Polygonatum*, *Heteropolygonatum*, *Disporopsis* and *Maianthemum*) mostly have sympodial rhizomes with erect or rarely pendulous (in epiphytic species) aerial shoots, terminal racemose, simple or compound inflorescences, or the flowers are borne singly or in condensed racemes in the leaf axils. The tepals may be free or fused, the anthers free or adnate to the tepals.

The Ophiopogoneae (*Lirope*, *Ophiopogon*, *Peliosanthes*) generally have spreading sympodial rhizomes with tufted clusters of often grasslike leaves. The inflorescences are terminal, and are reduced dibotrya or racemes. The flowers are hypogynous in *Liriope* and perigynous in *Ophiopogon* and *Peliosanthes*. The fruits are papery, rupturing early to allow the fleshy seeds to develop externally.

The Convallarieae (*Theropogon*, *Speirantha*, *Convallaria*, *Reineckea*, *Rhodea*, *Gonioscypha*, *Campylandra*, *Tupistra*, *Tricalistra*, *Aspidistra*) have monopodial shoots and racemose or spicate axillary inflorescences. The anthers are adnate to a fused perianth and the pedicels articulate along their length in *Convallaria*, and the anthers and tepals are free and basally attached and articulated in the remainder. They lack true aerial stems, but in *Convallaria* the sheathing leaf bases form a false stem, through which the apparently terminal, axillary inflorescence emerges. Although *Theropogon* is traditionally placed close to *Convallaria*, Utech (1979) found that the fruits (a major reason for their association) were anatomically quite different in origin, and that the two genera were not as closely related as previously postulated.

The genera *Aspidistra*, *Campylandra*, *Tupistra*, *Tricalistra*, and *Rohdea* are sometimes segregated as the tribe Aspidistreae, but are here retained within the Convallarieae. These genera differ by only a few characters. In particular, *Campylandra*, *Tupistra*, *Tricalistra* and *Rohdea* are separated by only a few apomorphies. Gagnepain (1934a) subdivided *Aspidistra* recognising three segregate genera, *Antherolophus*, *Colania* and *Evrardiella*, on features such as the anther number, whether the rhizome was thin or thick and whether it grew above or below the ground. Jessop (1979) considered that there were probably too many small genera in the Aspidistreae, mostly defined on very few characters and in the Iconographia Cormophytorum Sinicorum (1976) *Campylandra* and *Tricalistra* were reduced to synonymy under *Tupistra*, and the *Aspidistra* segregates under *Aspidistra*. The present treatment accepts the reduction of the *Aspidistra* segregates, but retains the other genera until more detailed studies have been carried out.

Affinities. The relationships of the Convallariaceae have been discussed by Dahlgren et al. (1985) and Conran (1985, 1987) and the family was related to other non-phytomelanic Asparagales. The lack of phytomelan in the seeds and the baccate fruits of the Convallariaceae suggests affinities with the Ruscaceae. The Ophiopogoneae were included in the Convallariaceae by Dahlgren et al. (1985), although they were recognised as a separate family by Lotsy (1911) and Nakai (1936). The Ophiopogonaceae were initially one of the associated families in a cladistic study involving the Convallariaceae and related Asparagales (Conran 1989), relating them (at family rank) to the Asparagaceae and Ruscaceae in the Asparagales, but their affinities were somewhat remote, and their inclusion in the Convallariaceae is still doubtful. In particular, the perigynous flowers and the unusual development of the fruit tends to isolate them from the remainder of the Asparagales. Bentham and Hooker (1883) related them to the Haemodoraceae, but this was disputed by Rao and Kaur (1979a) and Herr (1984) on embryological grounds. The recent *rbc*L molecular study by Chase et al. (1995) partially supports Conran's placement of the Ophiopogoneae near the Ruscaceae/Asparagaceae clade and separate from the Convallariaceae, although in the former, *Liriope* and *Peliosanthes* were somewhat distant from each other, with *Liriope* closer to *Nolina*, *Calibanus* (Nolinaceae), *Aspidistra* (Convallariaceae), *Danae* (Ruscaceae) and *Comospermum* than to *Peliosanthes* and *Polygonum* (Convallariaceae). If the *rbc*L data are to be accepted, the inclusion of this tribe within the Convallariaceae should possibly also mean the inclusion of the Nolinaceae and Ruscaceae.

Similarly, *Comospermum* was associated with the Convallariaceae rather than the Anthericaceae where it has been traditionally placed. On the other hand, a parsimony analysis of DNA sequence data of an upstream region (414 bp) of *mat*K gene of 24 spp. of *Polygonatum*, *Heteropolygonatum*, *Disporopsis*, *Maianthemum*, *Convallaria*, *Reineckea*, *Tricalistra*, *Aspidistra*, *Liriope*, *Ophiopogon*, *Peliosanthes* (Convallariaceae), *Nolina*, *Dasylirion* (Nolinaceae), and *Asparagus* (Asparagaceae as an out-group) (J. Yamashita and M.N. Tamura, unpubl.) indicates monophylies of the Convallariaceae and the Nolinaceae, respectively, which agrees not with Chase et al. (1995) but with Dahlgren et al. (1985). Also the *mat*K (a part) analysis confirms the coherence of the *Liriope* and *Peliosanthes* which share very unique characteristics of fruit development as well as similar karyotypes with $x = 18$ basic chromosome number (Tamura 1995). Sequences of *mat*K or other rapidly evolved genes also of Ruscaeae and *Comospermum* should be analyzed in order to know more exact taxonomic positions of convallariaceous, nolinaceous and ruscaceous genera as well as *Comospermum*.

Drymophila (Luzuriagaceae) and *Clintonia*, *Disporum* and *Streptopus* (Colchicaceae) were removed from the Convallariaceae by Dahlgren et al. (1985) and Conran (1987, 1989), and this was supported by karyological (Tamura 1995) and molecular evidence (Chase et al. 1995). Although *Drymophila* was retained in the Convallariaceae by Brummitt (1992), its position in the Luzuriagaceae is the subject of ongoing study, as the molecular results of Chase et al. (1995) placing it as sister to the Geitonoplesiaceae were based on misidentified material.

Distribution and Habitats. The family is confined to the Northern Hemisphere. It extends from Europe across Siberia to Asia, the Himalayas, Indochina, Malesia and America. *Convallaria*, *Polygonatum* and *Maianthemum* are widespead over most of this range. The other genera extend from the Himalayas to China, Japan, Indochina and Malesia. *Ophiopogon* was reported for Australia by Bailey (1884) but no collections exist and the plant is suspected to be possibly naturalised (Clifford 1987). The Convallariaceae generally grow as terrestrial understorey herbs in temperate moist forests. In tropical areas, they mostly grow at moderate altitudes on the floor of rainforests. *Maianthemum scilloideum* from Mexico frequently grows as an epiphyte in wet forests (Emons 1945), as do species of *Heteropolygonatum* and *Polygonatum*. The ecology of *Maianthemum* (e.g. Kawano et al. 1968, 1971), *Reineckea* (Yamamura 1984) and several North American genera (Bierzychudek 1982) have been investigated, all behaving as temperate understorey herbs. *Disporopsis* is an important component of the autumn diet of the white-necked long-tailed pheasant in China (Peng et al. 1994).

Palaeobotany. There are two records from Canada for possible Convallariaceae-like plants. These include *Maianthemophyllum* from the Palaeocene of Alberta (Bell 1949), and *Soleredera* from the Middle Eocene of British Columbia (Erwin and Stockey 1991). Although the latter was regarded as incertae sedis, an affinity with the Convallariaceae amongst other rhizomatous herbaceous lilies was postulated.

Economic Importance. The flowers and buds of *Convallaria keiskei* are eaten in Asia, and the entire plant is reportedly used in soups. *C. majalis* is poisonous; the rhizomes are used medicinally and the flowers for scent and in snuffs. The fruits of *Maianthemum racemosum* and *M. stellatum*, rhizomes of the latter and shoots of *M. japonicum* and the fruits and young shoots of *M. dilatatum* are edible. The rhizomes and young shoots of 12 *Polygonatum* species are eaten, and crushed *Polygonatum* is used medicinally to stop bruising. The flower spikes of *Tupistra nutans* are eaten as a vegetable in Sikkim. The rhizomes of *Liriope* species and the tuberous roots of *Ophiopogon* species are eaten, and the roots of *L. platyphylla* are used medicinally (Tanaka 1976; Kunkel 1984; Mabberley 1987).

A number of the Convallariaceae are used horticulturally, especially *Convallaria majalis* (lily of the valley), *Aspidistra elatior* (Aspidistra, cast-iron plant), *Liriope* species, *Maianthemum* species, *Ophiopogon japonicus* (Mondo grass),

Polygonatum species (Solomon's seal) and *Rohdea japonica*. *Maianthemum dilatatum* is an economically important weed in cranberry swamps in the United States (Patten and Wang 1994).

Key to the Genera

1. Plant with upright, ascending or descending (when epiphytic) stem. Stem not sheathing or sheathing only at base (Polygonateae) 2
- Plant usually tufted, rarely with procumbent stem (*Ophiopogon* spp., *Peliosanthes sinica*). Stem sheathing at each node 5
2. Inflorescence only terminal. Tepals usually free, rarely connate (*Maianthemum henryi*, *M. szechunicum*) **4. *Maianthemum***
- Inflorescence usually only axillary, rarely both axillary and terminal. Tepals connate 3
3. Flowers fasciculate or solitary; corona present **3. *Disporopsis***
- Flowers in a raceme, sometimes umbel-like or solitary; corona absent 4
4. Terminal inflorescence present. Tepals imbricate. Stamens outer short, inner long **2. *Heteropolygonatum***
- Terminal inflorescence absent. Tepals valvate. Stamens equal in length **1. *Polygonatum***
5. Plant sympodial. Inflorescence a reduced dibotyrum or a raceme. Fruit rupturing in early development, exposing the seeds with a sarcotesta (Ophiopogoneae) 6
- Plant monopodial. Inflorescence a spike, a raceme or 1-flowered. Fruit usually a berry, rarely a loculicidal capsule (*Gonioscypha*) or a drupe (*Tricalistra*); seeds without a sarcotesta (Convallarieae) 8
6. Flowers hypogynous, slightly zygomorphic by twisting of the stamens and standing erect of the 2 tepals; filaments long. Seeds blackish **5. *Liriope***
- Flowers perigynous, actinomorphic; filaments very short. Seeds blue 7
7. Leaves longitudinally whitish striped, transverse veins inconspicuous. Corona absent **6. *Ophiopogon***
- Leaves hollow along main longitudinal veins, transverse veins conspicuous. Corona present **7. *Peliosanthes***
8. Inflorescence a raceme. Tepals free 9
- Inflorescence often a spike, sometimes a raceme or 1-flowered. Tepals united 10
9. Leaves grasslike, lacking a petiole. Scape standing in the centre of a tuft of the leaves. Bracteoles present. Staminal filaments broad; anthers basifixed **8. *Theropogon***
- Leaves ovate-lanceolate, pseudopetiolate. Scape standing outside a tuft of the leaves. Bracteoles absent. Staminal filaments filiform; anthers dorsifixed versatile **9. *Speirantha***
10. Leaves forming a false stem. Inflorescence a raceme. Flowers nodding **10. *Convallaria***
- False stem absent. Inflorescence a spike or 1-flowered. Flowers erect 11
11. Inflorescence 1-flowered **17. *Aspidistra***
- Inflorescence a spike 12
12. Rhizome thin, stoloniform. Leaves distichous, linear to narrowly lanceolate **11. *Reineckea***
- Rhizome thick. Leaves alternate, narrowly lanceolate to ovate 13
13. Perianth lobes very short, inflexed. Stigmas 3, divergent. **12. *Rohdea***
- Perianth lobes short to long, more or less spreading to recurved. Stigma 1, trilobed, peltate to fungilliform or hippocrepiform 14
14. Floral bracts usually longer than flowers, lanceolate, rarely shorter, deltoid to ovate (*Campylandra delavayi*, *C. tui*). Anthers higher than or as high as stigma. Stigma small, trilobed 15
- Floral bracts shorter than flowers, deltoid to ovate. Anthers lower than stigma. Stigma large, fleshy, peltate to fungilliform or hippocrepiform 16
15. Filaments very short. Style long, columnar **13. *Gonioscypha***
- Filaments long to short. Style absent or short. **14. *Campylandra***
16. Style long; stigma peltate to fungilliform. Fruit a berry. **15. *Tupistra***
- Style absent; stigma hippocrepiform. Fruit a drupe. **16. *Tricalistra***

Genera of the Convallariaceae

I. Tribe Polygonateae Benth. (1883).

1. *Polygonatum* Miller Fig. 59

Polygonatum Miller, Gard. Dict. abridg. ed. 4 (1754); Ownby, Ann. Mo. Bot. Gard. 31: 373–413 (1944); Abramova, Bot. Žh. 60: 490–497 (1975); Y. C. Tang in Fl. Reipubl. Pop. Sin. 15: 52–81 (1978); Jeffrey, Kew Bull. 34: 435–471 (1980), 37: 335–339 (1982); M. N. Tamura, Bot. Jahrb. Syst. 115: 1–26 (1993).

Caulescent sympodial herbs, rhizomes thick to thin, internodes short to long, spreading; leaves lateral and pseudoterminal, distichous, opposite or verticillate, more or less sessile, ovate-elliptic to linear-lanceolate; inflorescences axillary, often racemose, sometimes umbel-like or 1-flowered; floral bracts herbaceous, membranaceous or absent; flowers hypogynous, tubular to campanulate, funnel-shaped or urceolate, usually nodding, rarely erect; pedicels articulated at the apex; tepals 6, valvate, united, lobes short, corona absent; stamens 6, attached to the tepals, filaments very short to long, often filiform, sometimes flattened, smooth, papillose or hairy, anthers basifixed to dorsifixed versatile; carpels 3, united, style 1, filiform, stigma capitate; ovary 3-locular, ovules 4–6 per locule; fruit a berry, globose, few- to 18-seeded, often dark purple, sometimes orange or red; seeds subglobose, pale brown; x = 9–15. Fifty-seven spp., temperate parts of Eurasia and N America, concentrated in the Himalayas, China to Japan.

There have been a number of attempts to subdivide the genus, although Jeffrey (1980) and Owenby (1944) found that the variation was continuous for most gross-morphological characters. Nevertheless, Tang (1978) and subsequently, Jeffrey (1982), recognised eight series gross-morphologically for the E Asian species. Tamura (1993) recognised two sections and three series for the genus based on the correlation between phyllotaxis and flower morphology and chromosomes.

2. *Heteropolygonatum* M.N. Tamura et Ogisu

Heteropolygonatum M.N. Tamura et Ogisu, Kew Bull. 52: (1997).

Caulescent sympodial herbs, rhizomes thick, internodes short, spreading; leaves lateral, distichous, more or less petiolate, lanceolate or fasciate-falcate; inflorescences axillary and terminal, often 2-flowered, sometimes 1- or 3–6-flowered, racemose or umbel-like; floral bract absent; flowers hypogynous, tubular to campanulate, nodding; tepals 6, imbricate, united for half their length or a little more, corona absent; stamens 6, the 3 outer shorter than the 3 inner, filaments filiform, adnate to tepals for most their length, smooth or verrucose, anthers basifixed; carpels 3, united, style 1, filiform, stigma capitate to trilobed; ovary 3-locular; fruit a berry, orange; $x = 16$. Two spp., China, both epiphytic.

3. *Disporopsis* Hance

Disporopsis Hance, J. Bot. 21: 278 (1883).

Caulescent sympodial herbs, rhizomes thick, internodes less short, spreading; leaves lateral and pseudoterminal, distichous, ± sessile, ovate-lanceolate; flowers axillary, fasciculate or solitary, hypogynous, campanulate to funnel-shaped, lateral to erect or nodding; floral bracts absent, pedicels articulated at the apex; tepals 6, united for 1/2 their length or a little less, white or yellowish, rarely brown-spotted inside; stamens 6, attached to the tepals by a corona at the apex of the perianth tube, anthers attached at the corona margin; carpels 3, united, style 1, stigma capitate to less trilobed; ovary 3-locular; ovules 4–6 per locule; fruit a berry, globose, 1–5-seeded, usually dark or bluish purple, rarely white; seeds pale brown; $x = 20$. Six spp., Indochina, Philippines, S China and Taiwan.

4. *Maianthemum* Wigg.

Maianthemum Wigg., Prim. Fl. Holsat.: 15 (1780). nom. cons.; LaFrankie, J. Arnold Arbor. 67: 371–439 (1986), and Taxon 35: 584–589 (1986); H. Li, Acta. Bot. Yunnanica Suppl. 3: 1–12 (1990).
Smilacina Desf. (1807); H. Hara, J. Fac. Sci. Univ. Tokyo III 14: 137–159 (1987).
Oligobotrya Baker (1886).

Caulescent sympodial herbs, rhizomes thick to thin, internodes short to long, spreading; leaves lateral, distichous, more or less petiolate, ovate to lanceolate or sagittate (sect. *Maianthemum*); inflorescence terminal, racemose, simple or conpaniculate; flowers often trimerous, sometimes dimerous (sect. *Maianthemum*), bi- to unisexual, erect, pedicels not articulated; tepals 4 or 6, free or united (sect. *Oligobotrya*), white, yellow or purplish brown, spreading, reflexed or cupuliform, rarely campanulate (sect. *Oligobotrya*); stamens 4 or 6, filaments filiform; carpels 2 or 3, united, style 1, filiform, stigma capitate to more or less trilobed; ovary 2–3-locular, ovules few; fruit 1–4-seeded, red; seeds pale brown; $x = 18$. Twenty eight spp., N Europe, Siberia, the Himalayas, China, Japan, and N and Central America.

The formerly recognised genera *Smilacina* and *Oligobotrya* have been included under *Maianthemum* by LaFrankie (1986a), with *Oligobotrya* having been previously included under *Smilacina* by Wang and Tang (1978). Recent DNA sequence data of a part of *mat*K (J. Yamashita and M.N. Tamura, unpubl.) support the inclusion of *Smilacina* in *Maianthemum*.

II. Tribe Ophiopogoneae Endl. (1836).

5. *Liriope* Lour.

Liriope Lour., Fl. Coch.: 200 (1790); L.H. Bailey, Gent. Herb. 2: 3–37 (1929).

Acaulescent caespitose, sympodial herbs, rhizomes thin, spreading; roots fibrous, often tuberiferous; leaves alternate, linear to narrowly lanceolate, sessile; inflorescence a reduced dibotyrum, more or less straight; flowers opening wide to flat, slightly zygomorphic, erect to ascending or nearly sessile, pedicels articulated; tepals 6, fused basally or free, white or violet; stamens 6, attached to the tepals basally, filaments long, curved around style, anthers obtuse, basifixed; style 1, long, stigma capitate, curved; ovary hypogynous, 3-locular, ovules axile, 2 per locule; fruit 1-seeded, rupturing in early development, exposing the seed; seed globose or ellipsoid, blackish, fleshy; $x = 18$. Eight spp., Indochina, Philippines, China to Japan.

The use of the names for the numerous commercial cultivars of some species both of *Liriope* and *Ophiopogon* were discussed by Hume (1961).

6. *Ophiopogon* Ker Gawler

Ophiopogon Ker Gawler, Bot. Mag. 27: tab. 1063 (1807), nom. cons.; Bailey, Gent. Herb. 2: 3–37 (1929); Hume, Baileya 9: 134–158 (1961); F.T. Wang et al. in Fl. Republ. Pop. Sin. 15: 130–164 (1978); Jessop, Flora Males. I. 9: 189–235 (1979).

Acaulescent or caulescent caespitose, sympodial herbs, rhizomes thick, often lignified or thin, spreading; roots fibrous often tuberiferous or lignified; aerial stem if present, unbranched, usually procumbent; leaves alternate, linear, sessile (sect. *Ophiopogon*) to oblong, petiolate (sect. *Peliosanthoides*), whitish striped on the abaxial surface; inflorescence a raceme or a reduced dibotyrum, more or less curved; flowers perigynous, campanulate to opening flat, drooping to ascending, pedicels articulated; tepals 6, fused basally or free, white or violet; stamens 6, sometimes connate, attached to the tepals basally, filaments very short, anthers long-pointed, basifixed, adpressed to the style; style 1, long, stigma capitate; ovary 3-locular, ovules basal, 2–6 per locule; fruit 1-seeded, rupturing in early development, exposing the seed; seed globose or ellipsoidal, blue, fleshy; $x = 18$. Fifty-four spp., India, the Himalayas, Indochina, Malesia, China to Japan.

7. *Peliosanthes* Andrews Fig. 60

Peliosanthes Andrews, Bot. Repos. 10: t.605 (1810); Larsen, Bot. Not. 119: 196–200 (1966); Jessop, Blumea 23: 141–159 (1976); Jessop, Flora Males. I. 9: 189–235 (1979).
Lourya Baillon (1888).
Neolourya K. Larsen (1966).

Acaulescent, rarely caulescent sympodial herbs, rhizomes short to long; roots fibrous; leaves alternate, sheathing, petiolate sometimes stipulate, sublinear to ovate-oblong, main longitudinal veins hollow, transverse veins conspicuous; inflorescence a raceme or a reduced dibotyrum (*P. teta* Andrews); flowers erect, pedicels articulated at the apex; tepals 6, fused, lobes 1/3–2/3 as long as tepals; stamens 6, filaments very short, attached to an expanded, fleshy, epipetalous corona, anthers basifixed; carpels 3, united, style 1, short to long, stigma capitate or trilobed; ovary perigynous, 3-locular, ovules basal, 1–5 per locule; fruit 1–3-seeded, rupturing in early development, exposing the seeds; seeds ellipsoidal, blue, fleshy; $x = 18$. About ten spp., India, Indochina, Malesia, S China and Taiwan.

Jessop (1976, 1979) reduced the numerous taxa in *Peliosanthes*, *Lourya* and *Neolourya* to synonymy under *Peliosanthes teta* Andrews, recognising only two, highly variable subspecies. Nevertheless, more recent accounts by Wang and Tang (1978) and Liu and Ying (1978) suggest that there are approximately ten species in *Peliosanthes*. Further study would seem to be indicated.

III. Tribe Convallarieae

8. *Theropogon* Maxim.

Theropogon Maxim., Bull. Acad. St. Pétersbourg 15: 89 (1871); Utech, Ann. Carnegie Mus. 48: 25–42 (1979).

Acaulescent monopodial herbs, rhizomes short, thick; leaves distichous, sheathing, sessile, linear; inflorescence axillary, racemose, standing in the centre of a tuft of leaves; flowers campanulate, nodding; floral bracts linear-subulate, bracteoles linear, pedicels decurved, articulated at the apex; tepals 6, free, ovate, outer tepals nectariferous; stamens 6, basally epipetalous, inserted, filaments short, broad, anthers basifixed; carpels 3, style 1, slender columnar, stigma small, capitate; ovary 3-locular, septal groove present, raphides absent, ovules 6–10 per locule; fruit a berry, many-seeded, black; $x = 20$. Only one sp., *T. pallidus* Maxim., SW China to the Himalayas.

The taxonomic position of *Theropogon* is controversial. Here it is tentatively included in the Convallarieae, but Utech (1979) stressed the dissimilarity between *Convallaria* and *Theropogon* from the floral anatomical point of view, and the chromosome number of *Theropogon* ($x = 20$) also differs from that of the remaining genera of the Convallarieae ($x = 19$).

9. *Speirantha* Baker

Speirantha Baker, J. Linn. Soc. 14: 562 (1875).

Acaulescent monopodial herbs, rhizomes thick, elongate, sometimes stologniferous; leaves rosulate, sheathing, ovate-lanceolate, pseudopetiolate; inflorescence axillary, racemose, standing outside a tuft of leaves, inside sheaths; floral bracts submembranaceous, lanceolate, bracteoles absent, pedicels articulated at the apex; flowers hypogynous, erect; tepals 6, free, lanceolate, white; stamens 6, free, filaments filiform, anthers dorsifixed, versatile; carpels 3, united, style 1, filiform, stigma small, capitate; ovary 3-locular, ovules 3–4 per locule; fruit a berry; $x = 19$. Only one sp., *S. gardenii* Baker, SE China.

10. *Convallaria* L.

Convallaria L., Sp. Pl.: 314 (1753).

Acaulescent monopodial (?) herbs, rhizomes thin, spreading; leaves usually 2, distichous, sheathing, forming an aerial false stem, ovate-elliptic; inflorescence axillary, racemose, standing outside a tuft of leaves, inside sheaths; flowers hypogynous, campanulate, nodding; floral bracts membranaceous, lanceolate, bracteoles absent, pedicels articulated near the apex; tepals 6, united, lobes short, sweet-scented; stamens 6, attached to the tepals, inserted, filaments short, filiform, anthers basifixed; carpels 3, style 1, filiform, stigma small, capitate; ovary 3-locular, septal groove absent, raphides present, ovules 4–8 per locule; fruit a berry, many seeded, red; seeds pale brown; $x = 19$. A circumboreal genus of three closely related spp., often treated as a single species, *C. majalis* L.

11. *Reineckea* Kunth

Reineckea Kunth, Abh, Akad. Wiss. Berlin 1842: 29 (1844), nom. cons.

Acaulescent monopodial (?) herbs, rhizomes thin, spreading; leaves distichous, linear to narrowly lanceolate, sessile; inflorescence axillary, spicate; flowers hypogynous, erect; floral bracts short, up to 7 mm, ovate-deltoid, membranaceous; tepals 6, united, lobes long, reflexed, pale purplish red; stamens 6, exserted, filaments long, filiform, attached apically to the floral tube; carpels 3, united, style 1, long, filiform, stigma capitate to trilobate; ovary 3-locular, ovules 2 per locule; fruit a berry, few-seeded, red; $x = 19$. Sometimes the flowers lack a pistil. Only one sp., *R. carnea* Kunth, China and Japan.

12. *Rohdea* Roth

Rohdea Roth, Nov. pl. spec.: 196 (1821).

Acaulescent monopodial herbs, rhizomes thick, spreading; leaves alternate, more or less rosulate, sessile, narrowly lanceolate, basally dilated; inflorescence axillary, spicate, fleshy; floral bracts short, up to 6 mm, ovate, membranaceous; tepals 6, fleshy, united, lobes very short, inflexed, pale yellow; stamens 6, attached to the tepals, inserted, sessile; carpels 3, united, style 1, short, stigmas 3, divergent; ovary 3-locular, ovules 2 per locule; fruit a berry, globose, single-seeded, red; $x = 19$. Two spp. 2, China and Japan, *R. japonica* Roth ex Kunth cultivated as an ornamental.

13. *Gonioscypha* Baker

Gonioscypha Baker, J. Linn. Soc. 14: 581 (1875).

Acaulescent monopodial herbs, rhizomes thick, spreading; leaves alternate, more or less rosulate, ovate-lanceolate, petiolate, basally dilated; inflorescence axillary, spicate, floral bracts large, more or less subulate; tepals 6, fleshy, united, urceolate, lobes short, spreading, undulate; stamens 6, attached to the tepals, inserted, sessile; carpels 3–4, united, style 1, filiform, stigma small, capitate; ovary 3-4-locular, ovules 2 per locule; fruit apparently a fleshy loculicidal capsule (or a berry?), purple; seeds few, fleshy; x = ? Two spp., E. Himalayas and Laos.

14. *Campylandra* Baker

Campylandra Baker, J. Linn. Soc. 14: 582 (1875).
Tupistra subgen. *Campylandra* (Baker) H. Li et J.L. Huang, Acta Bot. Yunnanica Suppl. 3: 49–61 (1990).

Acaulescent monopodial herbs, rhizomes thick, spreading; leaves alternate, often narrowly lanceolate, sessile, sometimes ovate, pseudopetiolate (*C. wattii* C.B. Clarke); inflorescence axillary, spicate; floral bracts often long, lanceolate, sometimes short, deltoid to ovate, bracteoles often absent, sometimes present (sect. *Annulatae*); tepals 6, fleshy, united, lobes short, often yellow, sometimes white, green or orange, sometimes annulated (sect. *Metatupistra*, sect. *Annulatae*); stamens 6, attached to the tepals, filaments long to short, more or less broad, decurved, anthers positioned higher than or as high as stigma; carpels 3, united, style 1, indistinct or short, stigma small, trilobed; ovary 3-locular, ovules 2–4 per locule; fruit a berry, few-seeded. $x = 19$. About 14 spp., Himalayas, Indochina and China.

15. *Tupistra* Ker-Gawler

Tupistra Ker-Gawler, Bot. Mag.: t.1655 (1814); *T.* subgen. *Tupistra* J.L. Huang and H. Li, Acta Bot. Yunnanica Suppl. 3: 49–61 (1990).

Acaulescent monopodial herbs, rhizomes thick, sometimes tuberous, spreading; leaves alternate, ovate to narrowly lanceolate, pseudopetiolate, basally dilated, midrib prominent; inflorescence axillary, spicate; floral bracts short, up to 7 mm, deltoid to oval; tepals 6 or 8, fleshy, united, lobes short, purplish, sometimes with basal transpaent "windows" in the floral tube; stamens 6 or 8, attached to the tepals, more or less sessile, anthers positioned lower than stigma; carpels 3–4, united, style long, up to 12 mm, single, stigma large, fleshy, peltate to fungilliform, up to 6 mm diam.; ovary 3–4-locular, ovules 2 per locule; fruit a berry, single-seeded; $x = 19$. Fifteen spp., E Himalayas (sect. *Nutantes*) to S China and Malesia (sect. *Tupistra*).

16. *Tricalistra* Ridley

Tricalistra Ridley, J. Fed. Mal. St. Mus. 4: 83 (1909); Jessop, Flora Males. I. 9: 189–235 (1979).

Acaulescent monopodial herbs, rhizomes thick, lignified, spreading; leaves alternate, more or less rosulate, ovate to lanceolate-oblong, pseudopetiolate, basally dilated, midrib prominent; infloresence axillary, spicate; floral bracts short, up to 6 mm, deltoid to ovate; tepals 6, fleshy, united, lobes 1/2 as long as tepals, spreading to recurved, dull yellow to purplish; stamens 6, attached at basal part of tepals, subsessile; carpels 3, united, stigma sessile, large, hippocrepiform; ovary unilocular, ovules 2–4 and sometimes 2 degenerate locules; fruit a drupe, 1-seeded, green. $x = 19$. Only one sp., *T. ochreata* Ridley, Malay Peninsula.

17. *Aspidistra* Ker Gawler Fig. 61

Aspidistra Ker Gawler, Bot. Reg.: t.628 (1823); Tao, Acta Phytotaxon. Geobot. 43: 121–123 (1992).
Antherolophus Gagnepain (1934).
Colania Gagnepain (1934).
Evrardiella Gagnepain (1934).

Acaulescent monopodial herbs, rhizomes spreading, thick to thin; leaves solitary or in fascicles or 2–4, alternate, sheathing, ovate-elliptic to linear-lanceolate, pseudopetiolate, midrib prominent; flower solitary, axillary, shortly stalked, campanulate to urceolate, erect; tepals usually 6–8, rarely from 4 to 10, fleshy, united, lobes short, purple, pale yellow or white with dark spots; stamens (4–)6–12, epipetalous, usually attached at basal part of tepals, filaments very short and broad, or stamens sessile; carpels 3–5, united, style 1, stigma large, peltate; ovary 3–5-locular, ovules 1-many per locule; fruit a berry, blobose, few-seede, blue, black or green; seeds pale brown; $x = 19$. About 11 spp., Himalayas, Indochina, S China to SW Japan.

Selected Bibliography

Bailey, F.M. 1884. Contribution to the Queensland flora. Proc. R. Soc. Queensl. 1: 8–19.

Baker, J.G. 1875. Revision of the genera and species of Asparagaceae. J. Linn. Soc. Bot. 14: 508–632.

Bell, W.A. 1949. Uppermost Cretaceous and Paleocene floras of Western Alberta. Bull. Can. Dep. Mines Res. Geol. Surv. 13: 1–231.

Bentham, G., Hooker, J.D. 1883. Liliaceae. In: Bentham, G., Hooker, J.D. (eds.) Genera Plantarum. London: L. Reeve pp. 748–836.

Bierzychudek, P. 1982. Life histories and demography of shade-tolerant temperate forest herbs. New Phytol. 90: 757–776.

Björnstad, I. 1970. Comparative embryology of Asparagoideae – Polygonateae, Liliaceae. Nytt. Mag. Bot. 17: 169–207.

Boyd, L. 1932. Monocotyledonous seedlings. Trans. Proc. Edinb. Bot. Soc. 31: 1–24.

Brummitt, R.K. 1992. Vascular plant families and genera. London: Royal Botanic Gardens Kew.

Chang, H.-J., Hsu, C.-C. 1974. A cytotaxonomical study on some Formosan Liliaceae. Taiwania 19: 58–71.

Chase, M.W. et al. 1995. See general references.

Clifford, H.T. 1987. Liliaceae (part). In: George, A.S. (ed.) Flora of Australia, Vol. 45. Canberra: AGPS pp. 148–421.

Conover, M.H. 1991. Epidermal patterns of the reticulate-veined Liliiflorae and their parallel-veined allies. Bot. J. Linn. Soc. 107: 295–312.

Conran, J.G. 1985. The taxonomic affinities of the genus *Drymophila* (Liliaceae s.l.). Ph.D. Thesis. St. Lucia: University of Queensland.

Conran, J.G. 1987. A phenetic study of the relationships of the genus *Drymophila* R.Br. within the reticulate-veined Liliiflorae. Aust. J. Bot. 35: 283–300.

Conran, J.G. 1989. Cladistic analyses of some net-veined Liliiflorae. Plant Syst. Evol. 168: 123–141.

Corbet, S.A., Chapman, H., Saville, N. 1988. Vibratory pollen collection and flower from: bumblebees on *Actinidia*, *Symphytum*, *Borago* and *Polygonatum*. Funct. Ecol. 2: 146–156.

Cutler, D.F. 1992. Vegetative anatomy of Ophiopogoneae. Bot. J. Linn. Soc. 110: 385–419.

Dahlgren, R.M.T. et al. 1985. See general references.

Daumann, E. 1970. Das Blütennektarium der Monocotyledonen unter besonderer Berücksichtigung seiner systematischen und phylogenetischen Bedeutung. Feddes Repert. 80: 463–590.

Eames, A.J. 1961. The morphology of the angiosperms. New York: McGraw Hill.

Emons, R.W. 1945. A revision of the Central American species of *Smilacina*. Ann. Mo. Bot. Gard. 32: 395–411.

Erdtman, G. 1952. See general references.

Erwin, D.M., Stockey, R.A. 1991. *Soleredera rhizomorpha* gen. et sp. nov. A permineralized monocotyledon from the Middle Eocene Princeton Chert of British Columbia, Canada. Bot. Gaz. 152: 231–247.

Farnsworth, N.R., Blomster, R.N., Quimby, M.N., Schermerhorn, J.W. (eds.) 1971. The Lynn index, Vol. 7. New York: N.R. Farnsworth.

Fedorov, A.A. (ed.) 1969. See general references.

Fowden, L. 1955. Azetidine-2-carboxylic acid: a new constituent of plants. Nature 176: 347–348.

Fowden, L. 1959. Nitrogenous compounds and nitrogen metabolism in the Liliaceae. 6. Changes in the nitrogenous composition during the growth of *Convallaria* and *Polygonatum*. Biochem. J. 71: 643–648.

Fowden, L., Steward, F.C. 1957. Nitrogenous compounds and nitrogen metabolism in the Liliaceae. 1. The occurrence of soluble nitrogen compounds. Ann. Bot. 21: 53–67.

Gagnepain, M.F. 1934a. Les aspidistrées d'Indo-Chine. Bull. Mus. Natl. Hist. Nat. Paris 2: 189–192.
Gagnepain, M.F. 1934b. Quelques liliacées nouvelles d'Indo-Chine. Bull. Soc. Bot. Fr. 81: 286–289.
Gorham, A. 1953. The question of fertilization in *Smilacina racemosa*. Phytomorphology 3: 44–50.
Herr, J.M., Jr 1984. Embryology and taxonomy. In: Johri, B.M. (ed.) Embryology of the angiosperms. Berlin, Heidelbery, New York: Springer, pp. 647–696.
Heslop-Harrison, Y., Shivanna, K.R. 1977. The receptive surface of the angiosperm stigma. Ann. Bot. 41: 1233–1258.
Hitchcock, C.L., Cronquist, A., Ownbey, M., Thompson, J.W. 1969. Vascular plants of the Pacific Northwest, Vol. 1. Washington: University of Washington Press.
Huang, J., Li, H. 1990. Study on the taxonomic system of the genus *Tupistra*. Acta Bot. Yunn., Suppl. 3: 49–61 (in Chinese with English summary).
Huber, H. 1969. See general references.
Hume, H.H. 1961. *Ophiopogon-Liriope* complex. Baileya 9: 134–158.
Iconographia Cormophytorum Sinicorum, Vol. 5. Beijing: Science Press.
Jeffrey, C. 1980. The genus *Polygonatum* (Liliaceae) in eastern Asia. Kew Bull. 34: 435–471.
Jeffrey, C. 1982. Further note on eastern Asian *Polygonatum* (Liliaceae). Kew Bull. 37: 335–339.
Jessop, J. 1976. A revision of *Peliosanthes* (Liliaceae). Blumea 23: 141–159.
Jessop, J. 1979. Liliaceae I. Fl. Males. I. 9: 189–235.
Kanmoto, T., Mimaki, Y., Sashida, Y., Koike, K., Ohmoto, T. 1994. Steroidal constituents from the inhibitory activity on CAMP phosphodiesterase. Chem. Pharmac. Bull. (Tokyo) 42: 926–931.
Kato, M. 1995. The aspidistra and the amphipod. Nature 377: 293.
Kawano, S., Ihara, M., Suzuki, M. 1968. Biosystematic studies in *Maianthemum* (Liliaceae-Poygonateae). II. Geography and ecological life history. Jpn. J. Bot. 20: 35–65.
Kawano, S., Suzuki, M., Kojima, S. 1971. Biosystematic studies in *Maianthemum* (Liliaceae-Poygonateae). V. Variation in gross morphology, karyology and ecology of North American populations of *M. dilatatum* sensu lato. Bot. Mag. Tokyo 84: 299–318.
Kawano, S., Ihara, M., Suzuki, M., Iltis, H.H. 1967. Biosystematic studies on *Maianthemum* (Liliaceae-Polygonateae). I. Somatic chromosome number and morphology. Bot. Mag. (Tokyo) 80: 345–352.
Koketsu, M., Kim, M., Yamamoto, T. 1996. Antifungal activity against food-borne fungi of Aspidistra elatior Blume. Phytochemistry 44: 301–303.
Krause, K. 1930. Liliaceae. In: Engler, A., Prantl, K. (eds.) Die natürlichen Pflanzenfamilien 2nd edn., Vol. 15a. Leipzig: W. Engelmann pp. 227–386.
Kunkel, G. 1984. Plants for human consumption. Koenigstein: Kluwer.
LaFrankie, J.V. 1986a. Transfer of species of *Smilacina* to *Maianthemum* (Liliaceae). Taxon 35: 584–589.
LaFrankie, J.V.J. 1986b. A new species of *Maianthemum* (Liliaceae) from Costa Rica with an upright and aerial rhizome. Am. J. Bot. 73: 1258–1260.
Larsen, K. 1966. Two new Liliaceae from the Khao Yai National Park. Bot. Not. 119: 196–200.
Liu, T., Ying, S. 1978. Liliaceae. In: Li, H., Liu, T., Huang, T., Koyama, T., DeVol, C.E. (eds.) Flora of Taiwan, Vol. 5. Taipei: Epoch Publishers pp. 40–84.
Loew, E. 1895. Das Leben der Blüten. Berlin: Dümmler.
Lotsy, J.P. 1911. Vorträge über botanische Stammesgeschichte Cormophyta Siphonogamia. Jena: G. Fischer.
Mabberley, D.J. 1987. The plant book. Cambridge: Cambridge University Press.
Migliorato, E. 1910. Sull'impollinazione di *Rohdea japonica* per mezzo delle formiche. Ann. Bot. 8: 241–242.
Mulay, B.N., Deshpande, B.D. 1959. Velamen in terrestrial monocots – I. Ontogeny and morphology in the Liliaceae. J. Indian Bot. Soc. 38: 383–390.
Nakai, T. 1936. Subdivision of Convallariaceae. J. Jpn. Bot. 12: 145–150.
Ono, T. 1928. Endosperm formation in the Liliaceae. Bot. Mag. Tokyo 42: 445–449.
Owenby, B.P. 1944. The liliaceous genus *Polygonatum* in North America. Ann. Mo. Bot. Gard. 31: 373–413.
Patten, K., Wang, J. 1994. Metham controls false lily-of-the-valley (*Maianthemum dilatatum*) in cranberry (*Vaccinium macrocarpon*). Weed Tech. 8: 270–276.
Pauli, G.F. 1995. The cardenolids of *Speirantha convallarioides*. Planta Medica 61: 162–166.
Peng, C., Chu, G., Guo, J. 1994. Autumn food composition of the silver pheasant and white-necked long-tailed pheasant in Nianzhu Forest Farm, Daganshan, Jiangxi Province. For. Res. 7: 574–578 (in Chinese).
Piper, J.K. 1986a. Effects of habitat and size of fruit display on removal of *Smilacina stellata* (Liliaceae) fruits. Can. J. Bot. 64: 1050–1054.
Piper, J.K. 1986b. Seasonality of fruit characters and seed removal by birds. Oikos 46: 303–310.
Piper, J.K. 1989. Light, flowering, and fruiting within patches of *Smilacina racemosa* and *Smilacina stellata* (Liliaceae). Bull. Torrey Bot. Club 116: 247–257.
Ramstad, E. 1953. Über das Vorkommen und die Verbreitung von Chelidonsäure in einigen Pflanzenfamilien. Pharm. Acta. Helv. 28: 45–55.
Rao, P.R.M., Kaur, A. 1979a. Embryology and systematic position of *Ophiopogon intermedius*. Proc. Indian Natl. Sci. Acad. B. 45: 175–187.
Rao, P.R.M., Kaur, A. 1979b. Sporogenesis and gametophtyes of *Polygonatum cirrhifolium*. Phytomorphology 29: 93–97.
Richards, A.J. 1986. Plant breeding systems. London: Allen and Unwin.
Ross-Craig, S. 1972. Drawings of British plants, Vol. 29. London: Bell and Sons.
Satô, D. 1942. Karyotype alternation and phylogeny in Liliaceae and allied families. Jpn. J. Bot. 12: 57–161.
Schulze, W. 1982. Beiträge zur Taxonomie der Liliifloren X. Asparagaceae. Wiss. Z. Friedrich Schiller Univ. Jena 31: 309–330.
Stenar, H. 1953. The embryo sac type in *Smilacina*, *Polygonatum* and *Theropogon*. Phytomorphology 3: 326–338.
Takhtajan, A.L. 1982. See general references.
Tamura, M.N. 1990. Biosystematic studies on the genus *Polygonatum* (Liliaceae) I. Karyotype analysis of species indigenous to Japan and its adjacent regions. Cytologia 55: 443–466.
Tamura, M.N. 1991. Biosystematic studies on the genus *Polygonatum* (Liliaceae) II. Morphology of staminal filaments of species indigenous to Japan and its adjacent regions. Acta Phytotaxon. Geobot. 42: 1–18.
Tamura, M.N. 1993. Biosystematic studies on the genus *Polygonatum* (Liliaceae) III. Morphology of staminal filaments and karyology of eleven Eurasian species. Bot. Jahrb. Syst. 115: 1–26.
Tamura, M.N. 1995. A karyological review of the orders

Asparagales and Liliales (Monocotyledonae). Feddes Repert. 106(1–2): 83–111.
Tamura, M.N., Ogisu, M., Xu, J.M. 1997. *Heteropolygonatum*, a new genus of the tribe Polygonateae (Convallariaceae) from West China. Kew Bull. 52: 949–956.
Tanaka, T. 1976. Tanaka's cyclopedia of edible plants of the world. Tokyo: Keigaku Publ.
Tillich, H.-J. 1995. Seedlings and systematics in monocotyledons. In: Rudall, P.J., Cribb, P., Cutler, D.F., Humphries, C.J. (eds.) Monocotyledons: systematics and evolution. London: Royal Botanic Gardens, Kew.
Utech, F.H. 1979. Floral vascular anatomy of the Himalayan *Theropogon pallidus* Maxim. (Liliaceae-Convallarieae). Ann. Carnegie Mus. 48: 25–41.
Vaikos, N.P., Markandaya, S.K., Pai, R.M. 1989. Floral anatomy of the Liliaceae: tribe Convallarieae. Proc. Indian Acad. Sci. (Plant Sci.) 99: 91–95.
van der Pijl, L. 1995. Some remarks on myrmecophytes. Phytomorphology 5: 190–200.
Vogel, S. 1990. The role of scent glands in pollination. Washington: Smithsonian Institution Libraries.
Wagner, W.M. 1977. Vessel types of monocotyledons: a survey. Bot. Not. 130: 383–402.
Wang, F.T., Tang, T. (eds.) 1978. Flora Reipublicae Popularis Sinicae Tomus 15. 280 pp. •• Beijing: Science Press. (in Chinese).
Yamamura, Y. 1984. Matter production processes of *Reineckea carnea* Kunth, an evergreen forest floor herb in the warm-temperate region of Japan. Bot. Mag. Tokyo 97: 179–191.
Yang, Q.E., Hong, D.Y. 1993. Karyotype study of the genus *Theropogon*. Acta Bot. Yunnan. 15: 263–268. (in Chinese).
Zweigelt, F. 1913. Vergleichende Anatomie der Asparagoideae, Ophiopogonoideae, Aletroideae, Luzuriagoideae und Smilacoideae nebst Bemerkungen über die Beziehungen Ophiopogonoideae und Dracaenoideae. Denkschr. Akad. Wiss. Wien 88: 397–476.

Corsiaceae

C. Neinhuis and P.L. Ibisch

Corsiaceae Becc. in Malesia 1: 238 (1878), t. 9, nom. cons.

Myco-heterotrophic, achlorophyllous perennial *herbs*. Shoot one or several emerging from a short creeping rhizome (*Corsia*) or from tuberlike roots (*Arachnitis*), unbranched, up to 40 cm high, pale red, purplish to flesh-coloured. Leaves 3–7, simple, ovate, alternate, almost distichous, amplexicaulous, 3–7-nerved, up to 5 cm long. *Flowers* solitary, terminal, bisexual, protandric. Perianth consisting of 6 tepals, 2 outer and the inner 3 filiform, lanceolate or ovate, 1- or 3-nerved; the median outer broad ovate, spathulate or acuminate, upright with a basal "callus" (*Corsia*) or reflexed with 2 rows of bullate structures (*Arachnitis*), covering the flower during anthesis. Stamens 6, with short filaments, anthers dorsifixed, extrorse. Ovary inferior, syncarpous, 3-carpellate. Placentas 3, parietal, bifid, protruding ± into the centre of the ovary on partial septa, which may be partially fused. Ovules numerous, pendulous. Styluli 3, free, or connate to form a style. Stigmas 3, free, papillose. *Fruit* capsule, with 3 valves opening down to the base (*Corsia*) or with terminal aperture (*Arachnitis*). Seeds small, dustlike, numerous, with embryo and endosperm multicellular (*Corsia*), or reduced to few cells (*Arachnitis*).

A family of 2 genera and 28 species in New Guinea, Solomon Islands, N Australia, and S America.

Vegetative Morphology and Anatomy. The anatomy of Corsiaceae is poorly investigated. Data are available only for *Arachnitis uniflora* (Reiche 1907; Colloza 1910; Solereder and Meyer 1930; Linsbauer and Ziegenspeck 1942; Minoletti 1986). *Corsia* possesses a short rhizome with small, pale white, amplexicaulous leaf sheaths, producing filiform unbranched roots that grow horizontally just below the surface (Fig. 62). The roots can extend over a considerable distance into the substrate and are quite sizeable in relation to the entire plant (van Royen 1972). The leaves are glabrous and cover the shoot almost completely. In *Arachnitis* the roots are short, about 15 mm long and 8 mm wide, with a central cylinder surrounded not by an

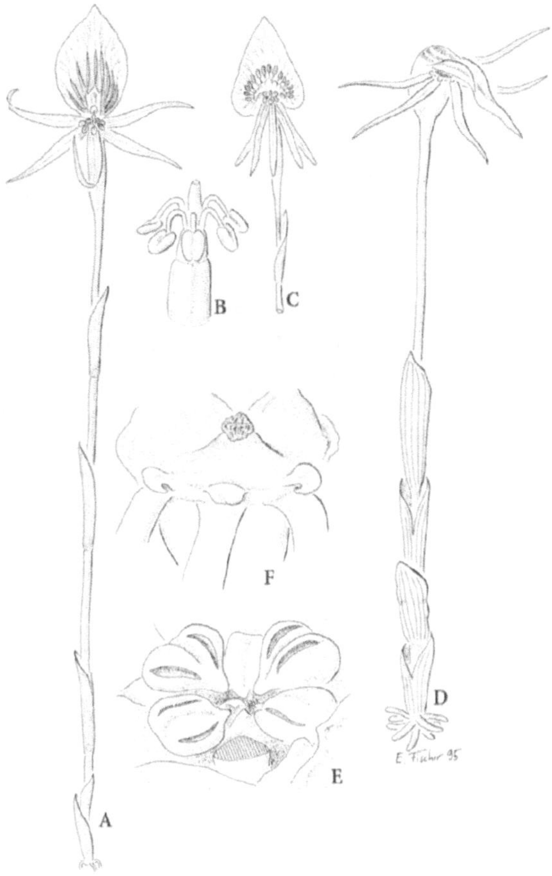

Fig. 62A–F. Corsiaceae. **A, B** *Corsia lamellata*. **A** Flowering plant. **B** Reproductive organs. **C** *Corsia cordata*, flower. **D–F** *Arachnitis uniflora*. **D** Flowering plant. **E** Flower in male stage, 1 anther removed to show stigmatic papillae not yet developed. **F** Flower in female stage, anthers dropped and stigmatic papillae well developed

endodermis but by a thick parenchymatous tissue. The leaves cover rarely more than 1/2 of the fully grown shoot, both are lacking stomata. Knowledge of the mycorrhiza of the Corsiaceae is very poor (Minoletti 1986; Leake 1994).

The collateral or concentric vascular bundles of the stem form a circle, each of them may be surrounded by a weak sheath of sclerenchyma. While the xylem elements of *Corsia* are woody, this is true for *Arachnitis* only in the ovary wall (Rübsamen 1986).

Reproductive Structures. The flowers are terminal in single-flowered inflorescences. The most common colour is reddish in different tones, sometimes intensive, yellowish or purplish. Both *Corsia* and *Arachnitis* have a well-developed zygomorphic perianth with a prominent outer median tepal. It covers the flower during the whole anthesis. In *Corsia* it varies in size and shape and represents, together with its basal "callus", the main character for species delimitation (van Royen 1972). The "callus" represents a thickened tissue of irregular shape and with various appendages and is usually glabrous, rarely papillose or hairy. At least in *Arachnitis*, the perianth does not wither until the seeds are dispersed. During the female stage of anthesis, the anthers are dropped and only short remnants of the filaments remain, which are often misinterpreted as staminodes. While *Arachnitis* has 3 free styluli with papillose stigmas, *Corsia* has a single style with free stigmas. During the male stage of anthesis, the stigmas of *Arachnitis* are not yet developed. Since the flowers are extremely protandric and show a considerable growth during anthesis (Minoletti 1986; Ibisch and Neinhuis 1996), *Arachnitis* has often been described as unisexual. A similar behaviour has been described for *Corsia* (Smith 1909).

Embryology. The embryology of the Corsiaceae is not completely known (Rübsamen 1986). The anther wall is 4-layered, the endothecium has I- or U-shaped wall thickenings. Pollen grains are 2-celled when shed. The ovules are anatropous and bitegmic with each integument 2-layered. Nucellus 2-layered. Embryo sac 8-nucleate and 7-celled. The embryo of *Corsia* is about 50-celled with a short suspensor, the endosperm multicellular with irregularly thickened cell walls composed of storage hemicellulose besides protein and oil. The embryo of *Arachnitis* is at least 3-celled and the endosperm consists of 3–5 cells with thick smooth cell walls composed of storage hemicellulose as well as starch.

Pollen Morphology. Pollen grains of both genera are sulcate with a reticulate tectum and punctate muri (Fig. 63A). The muri of *Corsia* pollen are slightly wider than those of *Arachnitis* (Erdtman 1952).

Karyology. Chromosome numbers are only known for two species (Kores et al. 1978). Both *C. cornuta* and *C. clypeata* have $2n = 18$ chromosomes.

Ultrastructure. The ultrastructure of the epicuticular wax crystals of *Corsia* revealed no clear results (Frölich and Barthlott 1988), *Arachnitis* shows small granule-like or tubular crystals similar to certain orchids and Liliaceae contrasting

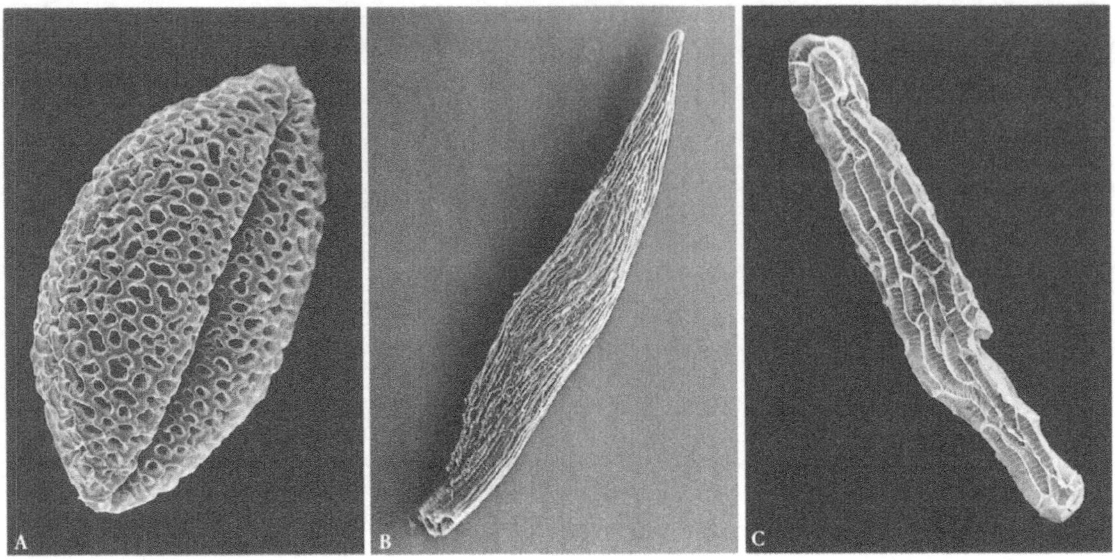

Fig. 63A–C. Corsiaceae. A *Arachnitis uniflora*, pollen grain, ×1360. B *Corsia torricellensis*, spindlelike seed, ×25. C *Arachnitis uniflora*, scobiform seed, ×65

with the Burmanniaceae, that are characterised by parallel oriented platelets (Rübsamen 1986).

Pollination. The pollination ecology of Corsiaceae has not yet been studied. Vogel (1978) predicts *Arachnitis* to be pollinated by fungus gnats. The colour and marking of the flowers and their microstructures are indeed very similar to plants that are known to be pollinated by fungus gnats.

Fruit and Seed. Both genera have capsules dehiscing with 3 lateral slits (*Corsia*) or with one terminal aperture (*Arachnitis*). *Corsia* seeds are elongated spindlelike (Feilspansamen) and about 2.5 to 3 mm long and 0.3 mm wide (Fig. 63B). The testa cells are prosenchymatic, elongated with convex periclinal cell walls. *Arachnitis* seeds are very small (Staubsamen), about 0.6–1 mm long and 0.1–0.2 mm wide (Fig. 63C). The testa cells are often quadrangular or rectangular to elongated. The membranous outer periclinal wall collapses and becomes tightly appressed to the anticlinal and inner periclinal wall tracing the helical secondary cell wall thickenings.

Dispersal. Although growing in damp shady habitats normally characterised by low windspeeds, both genera are supposed to be anemochorous. Since the tepals of *Arachnitis* do not wither until the seeds are dispersed, it seems possible that the pollinator also visits the flower in the fruiting stage and serves as a disperser as well (Ibisch and Neinhuis 1996).

Phytochemistry. *Arachnitis uniflora* roots and shoots are proven to have haemolytic effects (Ricardi 1958) and therefore are predicted to contain saponins (Hegnauer 1963). No data are available for *Corsia*.

Distribution and Habitat. *Corsia* is restricted to New Guinea (23 spp.), the Solomon Islands (2 spp.), and N Australia (1 sp.), tending to local endemism. The species grow in humus of moist beech and oak forests in montane regions from 900 to 1500 m (Cribb 1985; van Royen 1972). *Arachnitis* occurs in America mainly in the humid forests of S Chile and adjacent Argentina from sea level to 1000 m with a disjunct appearance on the non-wooded Falkland Islands (Cribb et al. 1995) as well as in semihumid and humid tropical Andean forests of Bolivia from 1900–2500 m a.s.l. (Ibisch and Neinhuis 1996). In contrast to *Corsia*, the two spp. of *Arachnitis* are widely distributed.

Affinities. The family Corsiaceae was established by Beccari (1878). However, several authors accepted only a tribe *Corsiae* within Burmanniaceae. Jonker (1938) again separated the Corsiaceae from Burmanniaceae, which was supported by the intensive study of Rübsamen (1986).

Some authors discuss the separation of *Arachnitis* from *Corsia* and a closer relationship to Ochidaceae (Ibisch and Neinhuis 1996). Usually, Corsiaceae and Burmanniaceae (incl. *Thismia*) are treated as the monophyletic order Burmanniales with affinities to Orchidales, Liliales and Melanthiales (Dahlgren et al. 1985). A recent cladistic analysis supports the close relationship between Corsiaceae and Burmanniaceae and of Burmanniales to Liliales but not to Orchidales (Rasmussen 1995).

Key to the Genera

1. Several shoots emerging from short creeping rhizome, flower with erect median sepal, capsule opening with lateral slits; New Guinea, Solomon Islands. Australia 1. ***Corsia***
- Single shoot emerging from short tuberous roots, flower with reflexed median sepal, capsule with one terminal aperture. America 2. ***Arachnitis***

1. *Corsia* Becc. Figs. 62A–C, 63B

Corsia Becc., Malesia 1: 238–254

Shoots several, emerging from short creeping rhizome; single terminal flower with enlarged median tepal upright; style 1; capsule with three lateral slits; seeds spindle-like, up to 3 mm long; embryo multicellular. About 26 spp., restricted distribution in New Guinea, Solomon Islands and Australia.

2. *Arachnitis* Phil. Figs. 62D–F, 63A,C

Arachnitis Phil., Bot. Zeitung 22: 217, nom. cons.

Single shoot emerging from short tuberlike roots; single terminal flower with enlarged median tepal reflexed; styluli 3; capsule with one terminal aperture; seeds scobiform, up to 1.2 mm; embryo 3-celled. Two spp., Patagonia, Falkland Islands and Bolivia.

Selected Bibliography

Beccari, O. 1878. Descrizione di una nuova e singolare pianta parassita - Burmanniaceae. Malesia 1: 238–254.

Colloza, A. 1910. Contributo allo studio anatomico delle Burmanniaceae. Boll. Soc. Ital. 1910: 106–115.

Cribb, P.J. 1985. The saprophytic genus *Corsia* in the Solomon Islands. Kew Mag. 2: 320–323.

Cribb, P.J., Wilken, P., Clements, M. 1995. Corsiaceae: a new family for the Falkland Islands. Kew Bull. 50: 171–172.

Dahlgren, R.M.T. et al. 1985. See general references.

Erdtman, G. 1952. See general references.

Frölich, D., Barthlott, W. 1988. Mikromorphologie der epicuticularen Wachse und das System der Monokotylen. Trop. Subtrop. Pflanzenwelt 63: 279–409.

Hegnauer, R. 1963. See general references.

Ibisch, P., Neinhuis, C., Rojas N.P. 1996. On the biology, biogeography, and taxonomy of *Arachnitis* Phil. nom. cons. (Corsiaceae) in respect to a new record from Bolivia. Willdenowia 26: 321–332.

Jonker, F.P. 1938. A monograph of the Burmanniaceae. Meded. Bot. Mus. Herb. Rijks Univ. Utrecht 51: 1–279.

Kores, P., White, D.A., Thien, L.B. 1978. Chromosomes of *Corsia* (Corsiaceae). Amer. J. Bot. 65: 584–585.

Leake, J.R. 1994. The biology of myco-heterotrophioc ("saprophytic") plants. Tansley Review No. 69. New Phytol. 127: 171–216.

Linsbauer, K., Ziegenspeck, H. 1942. Das Vorkommen von Spaltöffnungen bei heterotrophen Blütenpflanzen im Licht der Physiologie und Stammesgeschichte. Biol. Gen. 17: 511–565.

Minoletti, M.L. 1986. *Arachnitis uniflora* Phil. una curiosa monocotiledónea de la flora chilena. Bol. Soc. Biol. Concepción (Chile) 57: 7–20.

Philippi, R.A. 1864. Zwei neue Pflanzen aus Chile. Bot. Zeitung 22: 217.

Rasmussen, F.N. 1995. Relationship of Burmanniales and Orchidales. In: Rudall, P.J., Cribb, P.J., Cutler, D.F., Humphries, C.J. (eds.) Monocotyledons: systematics and evolution. London: Royal Botanic Gardens, Kew, pp. 227–241.

Reiche, K. 1907. Grundzüge der Pflanzenverbreitung in Chile. In: Engler, A., Drude, O. (eds.). Die Vegetation der Erde, vol. VIII. Leipzig: Wilhelm Engelmann.

Ricardi, M. 1958. Detección de saponinas en angiospermae chilenas. Bol. Soc. Biol. Concepción (Chile) 33: 29–94.

Royen, P. van 1972. Corsiaceae of New Guinea and surrounding areas. Webbia 27: 223–255.

Rübsamen, T. 1986. Morphologische, embryologische und systematische Untersuchungen an Burmanniaceae und Corsiaceae (Mit Ausblick auf die Orchidaceae-Apostasioideae). Diss. Bot. 92. Berlin: J. Cramer.

Smith, J. 1909. Burmanniaceae, Corsiaceae. Nova Guinea 8: 193–197.

Solereder, H., Meyer, F.J. 1930. Systematische Anatomie der Monocotylen, Heft VI. Berlin: Borntraeger.

Vogel, S. 1978. Pilzmückenblumen als Pilzmimeten. Flora 167: 329–398.

Cyclanthaceae

G. Harling, G.J. Wilder and R. Eriksson

Cyclanthaceae Poit. ex A. Rich. in Bory, Dict. Class. Hist. Nat. 5: 222 (1824), nom. cons.

Monoecious, perennial, terrestrial or sometimes epiphytic *herbs*, or herbaceous vines or lianas. Stem very short to long and slender, unbranched or branched, rhizomatous to aerial, usually ± lignified. *Leaves* spiral, spirodistichous, or orthodistichous; adult leaves always with leaf sheaths, petiolate in most species; leaf blades generally bifid, more seldom flabelliform-parted or entire, unicostate or tricostate. *Inflorescence* an axillary or terminal, unbranched, pedunculate spadix subtended and initially enveloped by 2–11 conspicuous, foliaceous or petalaceous spathes. *Flowers* unisexual, densely crowded, generally (subfam. Carludovicoideae) in spirally arranged groups (partial inflorescences), each group consisting of 1 pistillate flower surrounded by 4 staminate flowers, or (subfam. Cyclanthoideae) in alternate cycles, seldom spirals, of staminate and pistillate units. Staminate flowers in the Carludovicoideae usually with marginally situated perianth lobes on abaxial side to all around the receptacle, lobes usually in 1 whorl, rarely in 2 whorls or lacking; staminate flowers in the Cyclanthoideae as linear rows of stamens; stamens mostly numerous; anthers basifixed, bithecate and tetrasporangiate, usually longitudinally dehiscent, rarely with a special 2-phase dehiscence mechanism; filaments basally ± connate and usually swollen into so-called basal bulbs, filaments or basal bulbs rarely lacking. Pistillate flowers in the Carludovicoideae free or usually partly connate with each other; tepals 4, free or connate, usually epigynous to perigynous, more seldom almost hypogynous, of different shape, often enlarged and indurate in fruit; staminodes 4, flexuously filiform, opposite the tepals; ovary 4-carpellate, unilocular, entirely protruding from or more often ± embedded in the rachis, with 4 parietal, subapical, or apical placentae, or with one apical placenta, each placenta bearing numerous anatropous, heterotropous ovules; styles lacking or 4 free styluli, or these ± concrescent into a common style; stigmas 4, alternating with the tepals, of different shape; styluli and stigmas generally persisting and enlarging during fruit development; pistillate flowers in the Cyclanthoideae not discernible, coalescent into cycles (rarely spirals) consisting of double pistillate rows with locules confluent into a continuous ovarian chamber with numerous parietal placentae, each pistillate row having 1 row each of very much reduced perianth, staminodes, and carpels. *Fruits* in the Carludovicoideae ± fleshy, free or united to a syncarp; in the Cyclanthoideae united to ± dry, hollow rings (or spirals) filled with seeds. Seeds numerous, small to rather large, flat to terete; testa variable; endosperm copious, fat and fleshy, usually lacking starch; embryo medium-sized, straight.

A neotropical family of 12 genera and about 230 species.

Vegetative Morphology. The plants either have short, rhizomatous to aerial, herbaceous or lignified ± vertical stems (which may be nearly lacking), or form true lianas (Fig. 64, 65). The liana stems rarely exceed 10 m in height, but may reach ca. 30 m in *Thoracocarpus*. Transitions to liana stems exist in species of *Evodianthus, Asplundia, Ludovia* and *Sphaeradenia* in the form of ± creeping, relatively long stems.

Wilder (1992b) recognised 5 principal types of growth habits in the Cyclanthaceae, but a more general description is provided here. In *Carludovica, Evodianthus, Dianthoveus* and *Cyclanthus* (Fig. 66) the individual vegetative axes are sympodial, being composed of renewal shoots consisting of a basal portion that belongs to the vegetative axis and a terminal inflorescence that during expansion of the next higher-order renewal shoot is displaced to one side (Wilder 1977b, 1981a, 1988, 1989a). The remaining genera exhibit monopodial vegetative axes, which bear axillary inflorescences (Wilder 1977a, 1979, 1988). Each sympodial vegetative axis may have monomorphic renewal shoots (e.g. *Cyclanthus*) or may form discrete long and short renewal shoots (e.g. *Evodianthus*). Monopodial vegetative axes are mostly monomorphic, seldom developed as discrete long and short shoots. Each foliage leaf of a vegetative axis subtends a lateral bud. On sympodial axes this bud is normally vegetative, and is either non-renewal or expands into renewal shoots. On monopodial axes it is typically either a vegetative, non-renewal bud or an inflorescence bud.

Adventitious roots develop from the vegetative stems, and produce 2 kinds of lateral roots (Wilder 1992a). Many Cyclanthaceae, particularly terrestrial species, produce monomorphic adven-

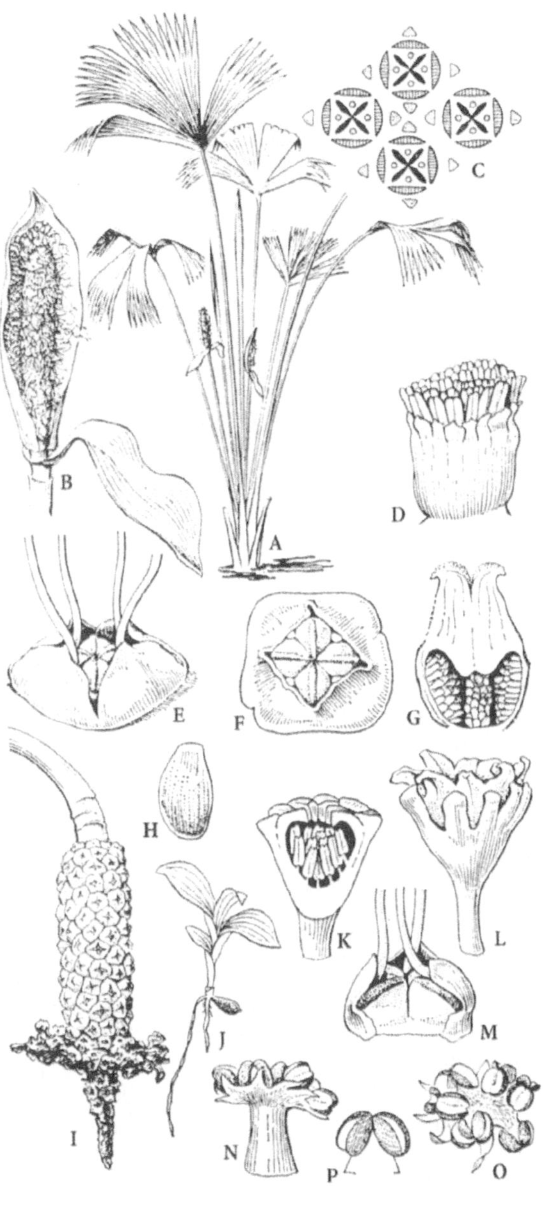

Fig. 64A–P. Cyclanthaceae. **A–J** *Carludovica palmata.* **A** Habit. **B** Flowering spadix with flexuous staminodes. **C** Diagram of flower arrangement, each pistillate flower (stigmas *black*, staminodes *open circles*, tepals *hatched*) surrounded by 4 male flowers (*stippled*). **D** Staminate flower. **E** Pistillate flower, the staminodes cut off. **F** Pistillate flower in fruiting stage. **G** Pistil, tangential section showing 3 of the 4 placentas. **H** Seed. **I** Fruiting spadix. **J** Seedling. **K–M** *Evodianthus funifer.* **K** Young staminate flower in tangential section. **L** Opened staminate flower. **M** Pistillate flower, one tepal removed. **N–P** *Dicranopygium pygmaeum.* **N** Staminate flower in side view. **O** Same, from above. **P** Stamen. (Takhtajan 1982)

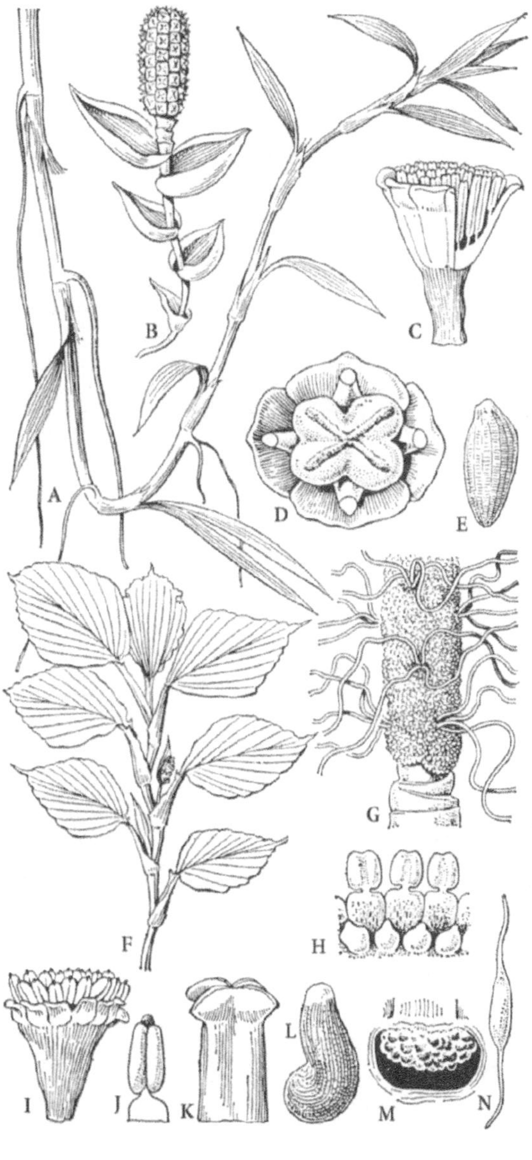

Fig. 65A–N. Cyclanthaceae. **A–E** *Thoracocarpus bissectus.* Liana stem with aerial roots, leaves entire or bifid. **B** Pedunculate inflorescence, the lower spathes not reaching the spadix. **C** Staminate flower, two tepals removed. **D** Pistillate flower in fruiting stage. **E** Seed. **F–H** *Ludovia bierhorstii.* **F** Flowering shoot. **G** Portion of spadix, note flexuous staminodes. **H** Part of staminate flower with small, glandular tepals and stamens with broadened filament bases. **I, J** *Chorigyne ensiformis.* **I** Staminate flower. **J** Stamen with secretion globule at apex of anther. **K, L** *Sphaeradenia chiriquensis.* **K** Style and stigmas. **L** Seed. **M** *S. angustifolia,* opened ovary with apical placenta. **N** *Stelestylis surinamensis,* seed. (Takhtajan 1982)

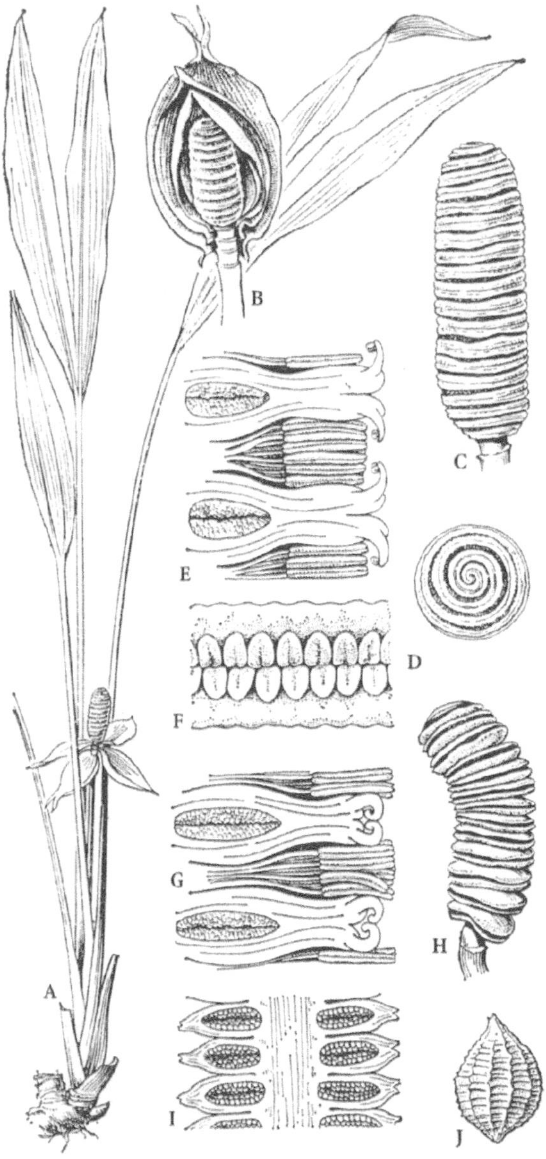

Fig. 66A–J. Cyclanthaceae. *Cyclanthus bipartitus.* **A** Flowering plant, the four spathes reflexed. **B** Spadix enveloped by spathes. **C** Flowering spadix. **D** Same, top view. **E** Spadix cut longitudinally at early anthesis (female stage) showing the alternating staminate and pistillate cycles. **F** Detail of pistillate cycle during early anthesis with staminodial lamellae, seen from the surface. **G** Spadix cut longitudinally at late anthesis (male stage), staminodial lamellae rolled up. **H** Fruiting spadix. **I** Same, longitudinal section. **J** Seed. (Takhtajan 1982)

titious roots. In contrast, certain lianas (*Evodianthus, Thoracocarpus* and some species of *Asplundia*) produce dimorphic ones, viz. anchoring (climbing) roots and absorbing (feeding) roots. Anchoring roots are short and entirely aerial, whereas the absorbing roots are long and extend downward, eventually reaching the ground and penetrating the soil.

During germination the cotyledon is enclosed in the seed or exposed, and is little differentiated. The next-formed leaf is larger, but rather similar in shape. The following leaves are generally oblanceolate or obovate, acute, sheathing at base, parallel-veined, and lack a petiole (observed in *Carludovica, Asplundia, Dicranopygium* and *Sphaeradenia*). With increasing differentiation, the leaves attain the shape of those of the adult plant, where the leaves vary in length from 10–20 cm in some species of *Dicranopygium* to more than 4–5 m in, e.g. *Carludovica palmata* and *Sphaeradenia gigantea.* In genera of the Carludovicoideae where the adult plants have distichous phyllotaxy, the first leaves produced by the juvenile plants are spirally arranged (investigated in *Sphaeradenia*; Eriksson 1995). In remaining genera of this subfamily (*Carludovica* and *Asplundia* investigated), the condition is the opposite. The adult leaves are generally petiolate, but are non-petiolate in, e.g. *Ludovia lancifolia.* The leaf sheath is of the open type, i.e. has margins that overlap. It is of variable shape, clasping in early stage, and exhibits a thickened median part and laminar marginal portions that may later partly disintegrate. The blades are usually more or less deeply cleft, due to tearing late in development. In *Carludovica* 3 incisions divide the leaf into wedge-shaped lobes, which are dentate or lobate distally, and are suggestive of some fan palms. In most other genera the leaves are bifid or bipartite, and always entire only in *Ludovia.* Tissue pads, which may be involved in tearing or expansion of the lamina, are found in all genera except *Ludovia.* Because of prolonged production of juvenile leaves, upper portions of shoots of *Evodianthus* and *Thoracocarpus* may exhibit pronounced heterophylly with both entire juvenile leaves and bifid adult leaves.

The leaves have 1 or 3 main ribs, i.e. are unicostate or tricostate. The central costa appears as a continuation of the petiole or thickened median portion of the sheath, ending close beneath the leaf tip (*Ludovia*) or in the basal end of the median sinus (remaining genera). Tricostate leaves, where each lateral costa extends into a lamina half, occur in *Carludovica, Schultesiophytum*, most species of *Asplundia*, three species of *Dicranopygium*, and sometimes also in *Evodianthus* and *Dianthoveus.* In *Cyclanthus*, where the leaf appears to be bicostate, the central costa is extremely short whereas the lateral costae are long, characteristically curved, and extend almost to the tip of the lamina segments. The lateral costae of other Cyclanthaceae terminate substantially below the leaf tip.

Leaf vernation is plicate in the Carludovicoideae, whereas *Cyclanthus* has a unique type of vernation (Wilder 1981b). Wilder (1981c) discussed correlations between particular orientations of folds and longitudinal veins on one hand, and the presence of entire vs. divided laminae on the other hand. In most Carludovicoideae poorly to well-defined ridges extend along the folds on the adaxial as well as abaxial surfaces of the adult leaves.

Hastulae and hastula-like structures are rare (Wilder 1981c). *Asplundia* sp. nov. aff. *multistaminata* exhibits a deeply cleft adaxial hastula and an abaxial hastula of two completely separate parts. In *Carludovica palmata* an adaxial hastula is sometimes present, composed of a pair of completely separate pegs arising at the base of the lamina.

The pedunculate inflorescences, whether terminal or axillary, may be borne separately or in compact clusters. The peduncles may grow particularly long in *Dianthoveus*, *Schultesiophytum*, *Stelestylis*, and species of *Dicranopygium*, *Sphaeradenia* and *Chorigyne*; in *Dicranopygium* they are slender, strong and elastic. During fruit maturation the peduncles gradually become lignified.

Peduncle sheaths develop with axillary inflorescences, but not with terminal ones, and are thus absent in *Carludovica*, *Evodianthus*, *Dianthoveus* and *Cyclanthus* (although in certain of these genera a number of bracteate leaves may intervene between members of a cluster of terminal inflorescences). In some species of *Asplundia* and *Sphaeradenia* the peduncle sheaths may be as long as the peduncles which they cover; they are often brown or greenish, but are sometimes more brightly coloured.

The spathes are 2–11 per inflorescence, generally 3–4. Their number, arrangement, shape and size sometimes characterise the genera. They may be crowded next to the base of the spadix or situated at some distance from each other. They are cymbiform with a lanceolate to ovate outline, and their colour green, whitish green, white, yellow, orange, reddish, purple or violet-brown. The lower spathes are always more pronouncedly foliaceous than the upper ones. Generally the spathes fall off during or immediately after anthesis.

Wilder (1992b) has reviewed the mirror-image symmetry of shoots, buds, leaves and inflorescences in various species of the Cyclanthaceae.

Vegetative Anatomy. Accounts of vegetative anatomy were published by Solereder and Meyer (1928), Surange (1950), Harling (1958), Wilder and Harris (1982), French et al. (1983), Tomlinson and Wilder (1984), Wilder (1984, 1985a,b,c,d, 1986a,b, 1989b, 1992a) and Wilder and Johansen (1992).

Epidermal cells of the vegetative stems are small, sometimes with differently thickened walls. Fibrous bundles, collateral fibrovascular bundles and endodermis have all been reported from the cortex. Cortical vascular bundles almost entirely represent leaf traces originating from the central cylinder, which has the vascular bundles scattered within the ground tissue. A small pith may sometimes be present. In the Carludovicoideae, but not in the Cyclanthoideae, cork is often formed subepidermally or at least from peripheral layers of the cortex; in the Cyclanthoideae there is a girdle of sclereids below the epidermis.

French et al. (1983) surveyed the three-dimensional distribution of stem vascular bundles. They found that crowded peripheral and diffuse central bundles occur within the central cylinder. The peripheral bundles are solely simple, and collateral or amphivasal. In contrast, the central bundles are always compound, each composed of discrete simple vascular bundles with separate xylem and phloem enclosed in a common sheath. Such a compound bundle is bipolar to multipolar, and results from the delayed fusion of a leaf trace and axial bundles, all these becoming associated immediately after entry of the leaf trace into the central cylinder.

Adventitious roots of the Cyclanthaceae have persistent or nonpersistent epidermis, and often wound or non-wound periderm. Cortical sclerenchyma is generally present in the endodermis, but occurs in the exodermis of relatively few species. Between endodermis and exodermis the sclerenchyma may compose a circumcortical ring, intervening tissue, and/or a circumstelar ring. The circumstelar sclerenchymatous ring is present in all genera except *Schultesiophytum* and *Cyclanthus*. In addition to parenchyma cells, the cortex sometimes contains also extracellular mucilage canals (delimited by epithelium), tubular cells and raphide sacs. Roots containing styloid sacs are restricted to *Evodianthus* and *Dianthoveus*.

The stele of adventitious roots consists of a pericycle that is 1–several cell layers thick, primary xylem, primary phloem, homogeneous or heterogeneous interfascicular tissue and sometimes pith. Raphide sacs are sometimes present. All xylem and at least some phloem develop within xylem fascicles and phloem fascicles, respectively, and always apart from each other. The following conditions may exist: (1) all xylem and phloem alternate within a peripheral ring, (2) both xylem and phloem occur also centripetal to a peripheral ring, (3) all phloem is confined to the peripheral ring, whereas both peripheral and non-peripheral xylem fascicles are found. In *Evodianthus*, *Sphaeradenia* and *Stelestylis* the sieve elements of non-peripheral phloem are dispersed among fibres. Four main types of true pith occur, depending on arrangements and kinds of parenchyma and sclerenchyma tissue.

The petiole has numerous peripheral fibre bundles, and the large vascular bundles have thick sclerenchyma sheaths. In the Cyclanthoideae the petiole has air canals arranged in the form of a V in transections.

In the interridge areas of non-costal portions of the lamina, ordinary epidermal cells are generally more homogeneous in shape (4-sided or polygonal in top view) in the adaxial epidermis than in the abaxial. Stomata are most abundant in the abaxial epidermis, which exhibits stomatal bands and 4 kinds of interstomatal bands. The stomata are typically enclosed by 2 polar and 2 non-polar subsidiary cells, and their guard cells are kidney-shaped in top view. Cuticular papillae or ridges are often found. Epidermal appendages commonly occur, and many intermediates exist between lumen papillae and elaborate multiseriate hairs. Widely scattered regions of periderm are sometimes present.

The adaxial region of the mesophyll is sometimes a palisade, whereas the more abaxially situated regions are always spongy. Between boundary layers (e.g. the abaxial and adaxial hypodermis and bundle sheaths), monomorphic ordinary parenchyma cells develop in all species except two of *Dicranopygium*, which have dimorphic cells. In all genera ordinary parenchyma cells may contain starch grains, crystals, or tannin (the latter lacking in *Cyclanthus*); in a few species these cells also have star figures composed of tannin. Extracellular mucilage cavities, enclosed by uniseriate epithelium, occur in certain species of, e.g. *Asplundia*. Parenchyma-like dead cells with strongly birefringent walls are found in all studied species of *Ludovia*, *Sphaeradenia* and *Stelestylis*, and such without birefringent walls in several species of *Asplundia*. Raphide sacs are found in all species, and styloid sacs in at least some species of all genera except *Thoracocarpus*, *Schultesiophytum* and *Ludovia*. In *Evodianthus* and *Dianthoveus* the styloid sacs are oriented in all directions, resulting in scabrous leaves as dry. Non-articulated laticifers occur solely in *Cyclanthus*, where they are known from all organs but the root. Lysigenous air spaces in the mesophyll are restricted to Cyclanthoideae, the Carludovicoideae having schizogenous ones. In all species the mesophyll contains strands of 1 or more fibres.

Interridge areas of the lamina contain longitudinal veins and commissural veins. In both subfamilies, these longitudinal veins are generally of well-defined orders, and up to a given order the number of veins is twice that of the next-lower order. In interridge areas and ridges, longitudinal veins are normally collateral, but bicollateral, amphivasal and amphicribral veins may also be present. Expansion tissue and presumed expansion tissue occur in Carludovicoideae and Cyclanthoideae, respectively.

Inflorescence Structure. The inflorescence is superficially an unbranched spadix, varying in shape from narrowly cylindrical to broadly ellipsoid, or even subspherical. It consists of a central rachis on which the flowers are located, either free from each other (*Evodianthus*, *Dianthoveus*, *Schultesiophytum* and *Chorigyne*), or more or less connate. Particularly multiflorous inflorescences occur in *Carludovica*, *Cyclanthus* and some other genera, while in *Dicranopygium* they are few-flowered.

In the Carludovicoideae the staminate and pistillate flowers are arranged in a regular chessboard mosaic (Fig. 64C), the pistillate flowers running in a flat spiral from base to apex. Each pistillate flower is surrounded by 4 staminate flowers; these 5-flowered groups seem to represent suppressed lateral branchlets, the whole complex being then a thyrse (Harling 1946, 1958). This is supported by the fact that these partial inflorescences are subtended by a minute bract in early developmental stages of some genera (Wilder 1977a).

In the Cyclanthoideae discrete flowers are not identifiable. The carpels and stamens compose pistillate and staminate units, respectively. In turn, these units constitute a regular, alternating arrangement of pistillate and staminate cycles (or spirals) on the spadix (Fig. 66). This configuration is no doubt derived. Wilder's (1981a) study of inflorescence development in *Cyclanthus* shows that each staminate cycle develops from a vertically arranged pair of meristems or paired group of meristems. This evidence favours Harling's (1958) interpretation of the evolutionary origin of this inflorescence from such found in the Carludovicoideae.

FLORAL STRUCTURE. In the Carludovicoideae the staminate flowers, which are usually shed after anthesis, consist of a pedicel of variable length, and a fleshy floral receptacle that is at least partly surrounded by the perianth and bears a number of stamens. The flowers may be symmetrical, the pedicel being then centrally attached to the receptacle and the perianth segments being equal all around, or they may be asymmetrical, the pedicel attached eccentrically to the receptacle and the tepal lobes being unequal around the periphery, sometimes wholly reduced on one side. Symmetrical flowers occur in *Carludovica*, *Dianthoveus*, *Asplundia* subgen. *Choanopsis*, *Thoracocarpus*, *Schultesiophytum* and *Ludovia*. In these genera the subequal tepal lobes are ca. 10–20 in number (rarely up to 30, as in *Ludovia*). Each perianth lobe is externally provided with a more or less well-developed glandule, except in *Dianthoveus*, where these glandules are lacking. In *Evodianthus*, which also has symmetrical flowers, the tepal lobes are long and imbricate, but in 2 whorls instead of 1, the outer being shorter than the inner, which lacks glandules. The other genera (e.g. *Dicranopygium* and *Chorigyne*) have asymmetrical flowers, where tepal lobes on the distal side of the flower in relation to the pistillate flower are shorter or even lacking. In *Dicranopygium* subgen. *Gleasonianthus* the perianth is completely reduced and is replaced by the sterilised marginal stamens.

The stamens are indefinite in number; they range from less than 10 in some species of *Dicranopygium* and *Sphaeradenia* up to ca. 100 in *Ludovia* and ca. 150 in *Asplundia multistaminata*. The filament is in most genera swollen below, possessing a basal bulb, while the distal part is usually thin (rarely the filament is absent). The basal bulbs may be fused or otherwise specialised. In *Dianthoveus* the anthers are ± sessile, lacking basal bulbs, but surrounded at base by papillae forming a cupule. The anthers are elongate (rarely spherical-cubic), and the thecae joined by a usually rather thin connective. In some species of *Dicranopygium*, the connective is broader at the base of the anther and the thecae are divergent. Dehiscence is longitudinal, or in *Dicranopygium mirabile* with a unique 2-phase mechanism, where the rupture of the inner anther wall layer is delayed (Wilder 1987). An apical secretion globule is present on the anthers of *Chorigyne, Stelestylis*, and most species of *Sphaeradenia.*

In the Cyclanthoideae, the staminate flowers are strongly reduced and transformed; they appear as cycles of numerous stamens with no visible limit between the flowers. Four rows of stamens generally occur in each cycle. The filaments of the stamens are basally fused and without basal bulbs.

The pistillate flowers in the Carludovicoideae are tricyclic and generally tetramerous (rarely 5–8-merous), with 1 perianth whorl, 1 whorl of staminodes and 1 of carpels. The staminodes are opposed to the tepals, while the carpels alternate with these. In most genera the tepals are distinct, but in species of *Asplundia* and *Dicranopygium* they are reduced to truncate fleshy structures, and in *Ludovia* they are barely visible. The tepals are generally free, but connate in some cases. In *Asplundia* the tepals are sometimes distinctly 3-cleft, in other species they bear fleshy appendages. The external side of tepals in *Sphaeradenia* may be provided with a glandule towards the apex. Rarely, as in *Ludovia*, a broad mass of tissue encircles the pistillate flower peripheral to the tepals. Wilder (1978) speculated that this mass, as well as the tepals, might be staminodial in origin. Garcin (1958), on different evidence, suggested that each tepal of the pistillate flowers actually represents a group of coalesced staminodes.

The staminodes are obtusely subulate or vermiform, and are basally adnate to the tepals. When shed they leave a scar which sometimes swells up to a "staminodial protuberance". The staminodes are generally 2–5 cm long, but vary between 0.5 and 25 cm. They are usually white, but sometimes red, orange or yellow. Rudimentary anthers may be found on the tips of the staminodes and occasionally produce normal pollen (Harling 1946). Apical secretion globules may also be present on the staminodial anthers in some species of *Sphaeradenia.*

In all members of the Carludovicoideae the gynoecium is 4-carpellate, although the ovary is unilocular. It is generally inferior or subinferior, but almost superior in, e.g. some species of *Chorigyne*. Placentation is variable in the family but constant in the genera (Harling 1958; Eriksson 1994b). The placentae are usually 4 in number and parietal, but they are subapical in *Ludovia* and *Chorigyne*, and apical in *Stelestylis*; *Sphaeradenia* has only 1 apical placenta.

There are 4 distinct styluli in most species of *Asplundia*. In the other taxa the carpel tips are connate to concrescent into a common style, except in *Evodianthus*, *Thoracocarpus*, *Schultesio-*

phytum, most species of *Dicranopygium*, some species of *Asplundia*, and *Sphaeradenia rhodocephala* where the stigmas are usually sessile or subsessile.

In the Cyclanthoideae, a ring-shaped pistillate unit bears 2 peripheral thin lists that are probably homologous to a perianth. To the inside of these lists there occur 2 larger, grooved staminodial lamellae, each bearing a row of more or less abortive anthers. Medially, the pistillate unit exhibits 2 alternating rows of carpels, which are so intimately fused beneath their stigmas that all of their ovaries form 1 coherent cavity. This bears on its inside densely crowded placentae with numerous ovules, which seem to clothe the ovarian cavity almost entirely.

EMBRYOLOGY. Anther wall formation is of the monocotyledonous type (Harling 1958). The epidermis of the microsporangia in most species of *Asplundia* and in *Evodianthus funifer* bursts apart and degenerates at a rather early stage, so that the endothecium appears as a "pseudo-exothecium", while in, e.g. *Carludovica* and *Cyclanthus*, the epidermis persists in mature microsporangia. The endothecium is well developed, radially elongated, and provided with fibrous thickenings. The tapetum is secretory and its cells become 2-, 3-, or 4- (or more-) nucleate. The tapetal cells usually degenerate during microsporogenesis, but persist in *Dicranopygium mirabile* as a distinct layer. Microsporogenesis is of the successive type. During the homotypic division the 2 spindles are generally parallel so that isobilateral tetrads arise. Rarely, as sometimes in *Carludovica drudei*, they are at right angles and the tetrads become decussate. The mature pollen grains are 2-celled.

The ovules are always anatropous (mixed epi- and apotropous), bitegmic, and weakly crassinucellate in general type (Harling 1958). In *Stelestylis* they have an unusually long funicle and a vermicular appendage in the chalaza. The outer integument is always more strongly developed and also longer than the inner, so that the micropyle consists of both exo- and endostome. The inner integument consists of 2 cell layers, which are more or less swollen at the distal end; the outer integument has 3–5 cell layers. Raphides are often present in the ovules. The young nucellus consists of the epidermis, 1 subepidermal archesporial cell, 1 layer of subepidermal somatic cells, and a variable number of somatic cells in the basal part.

The archesporial cell functions directly as the megaspore mother cell. In the Carludovicoideae there are periclinal divisions in the nucellar epidermis resulting in a nucellar cap of 2–3 cell layers, arranged in a typical, radiating way. In *Cyclanthus*, however, the nucellar epidermis does not divide periclinally, and the embryo sac lies directly under the single-layered epidermis. The embryo sac is of monosporic origin and the development follows the Polygonum type. The antipodals are invariably 3, 1-nucleate, and usually early degenerating. Endosperm formation is Helobial in the cases known (*Carludovica, Evodianthus, Dicranopygium* and *Cyclanthus*; Harling 1958).

The ripe seeds are provided with copious endosperm containing fatty oil and protein. In *Dicranopygium* only, there is abundant starch in the endosperm as well as in the embryo. The embryo is small to medium-sized, cylindrical or somewhat tapering at the apex, and generally straight (Harling 1946). It often contains crystal raphides, while these are absent from the endosperm.

POLLEN MORPHOLOGY. Pollen morphology has been surveyed by Erdtman (1954), and in more detail by Harling (1958) and R. Eriksson (1995, unpubl.). The pollen grains are dispersed singly, and are usually boat-shaped with a broadly elliptic to elongate outline, 13–42 μm long and 8–24 μm broad. They are tectate with a thick foot-layer and no endexine. The exine sculpturing is foveolate in all genera except *Evodianthus* and *Dianthoveus*, where it is psilate. The pollen grains are sulcate in *Thoracocarpus, Schultesiophytum, Ludovia, Chorigyne, Stelestylis* and most species of *Sphaeradenia*; vary from sulcate to ulcerate in *Asplundia* and *Dicranopygium*; are ulcerate to sulcoidate in *Cyclanthus* and a few species of *Sphaeradenia*; and have an end-situated pore in *Carludovica*. In *Evodianthus* and *Dianthoveus* the pollen grains are inaperturate.

KARYOLOGY. Chromosome numbers so far known in the family are $2n = 30$ for *Carludovica palmata, Asplundia* aff. *latifolia, Asplundia* sp., *Sphaeradenia hamata* and *S. stenosperma*, and $2n = 18$ for *Cyclanthus bipartitus* (Harling 1946; Eriksson 1995). Other reports are $2n = 18$ for *Carludovica palmata* and $2n = 32$ for "*Carludovica insignis*" (Janaki Ammal in Darlington and Janaki Ammal 1945).

POLLINATION AND REPRODUCTIVE SYSTEMS. The pollination biology of the Carludovicoideae has been studied in detail by Gottsberger (1990, 1991), Eriksson (1994a) and Seres and Ramírez (1995). They are pollinated by small weevils (Curculionidae), which are attracted by scent emitted by the staminodes and use the inflorescence for feeding, shelter, mating and oviposition. The inflorescence exhibits adaptations for restricting the effective pollinators to these beetles, e.g. by having pollination chambers with narrow entrance, and minimising possible damage caused by their behaviour, e.g. by promoting oviposition in the dispensable and usually caducous staminate flowers. Most species investigated have protogynous inflorescences, with short and simultaneous anthesis of the respective gender. They usually show a 2-day flowering rhythm, where the pistillate and staminate phases are distinctly separated. This condition ensures xenogamy, because usually only 1 inflorescence per genet is well developed at a time. Some species are, however, known to have overlapping pistillate and staminate antheses, and in these, geitonogamy is prevailing, promoted by visiting weevils as pollen vectors.

The Cyclanthoideae are pollinated by scarabs (Scarabaeidae), which are attracted by food-producing spathes that also serve as shelter against predators while feeding and mating (Beach 1982; Seres and Ramírez 1995). Xenogamy is ensured by a 2-day flowering period with non-overlapping pistillate and staminate antheses of the protogynous inflorescence.

Fruit and Seed. The fruits of Cyclanthaceae are always ± baccate, and, on each spadix, often united in 1 or more syncarps of various types. Only in *Evodianthus*, *Dianthoveus*, *Schultesiophytum* and *Chorigyne* are the fruits free from each other. The pericarp is usually soft, except in the upper part, on which the perianth, styles and stigmas are borne. In this part there is more sclerenchyma and often the tissue is lignified. The tepals normally undergo postfloral increase in size and become fleshy or lignified. The styles and stigmas also change their appearance during fruit development.

In several genera the individual fruits are generally circumscissilely dehiscent by an apical cap, i.e. they are fleshy pyxidia where the seeds remain exposed in the lower part or are attached to the cap (e.g. *Asplundia*, *Thoracocarpus* and *Sphaeradenia*), while in others they are more or less indehiscent (e.g. *Dianthoveus*, *Ludovia* and *Chorigyne*). In *Dicranopygium* a pressure is built up in the spadix at maturity, and the seeds explosively spurt out from irregular ruptures at the tepal bases. In *Carludovica* the fruit layer of the ripe infructescence splits in irregular flaps, which are rolled off with the bright orange to scarlet inside of the fruits turned outwards. This splitting is usually from the apical part, but sometimes from the basal.

The ripe spadices in, e.g. *Asplundia* and *Dicranopygium*, are dull brownish green or vivid green on the surface, those in, e.g. *Sphaeradenia*, are often bright yellow to orange, white, or reddish.

Cyclanthus deviates from the other genera, because each fruiting ring forms a syncarp. The hollow interior of the syncarp is filled with seeds which are embedded in mucilage. The ripe, leaning infructescence becomes bent towards the ground and each of the fruiting rings splits transversely into 2 equal halves, which loosen from each other and from the rachis. On each of the open halves the seeds are borne on the inner surface.

The cyclanthaceous seeds are extraordinarily variable. They are strongly flattened to terete, and their shape varies from subglobose to narrowly oblong, slightly crescent-shaped, or even strongly curved (*Sphaeradenia laucheana* and *S. chiriquensis*); in *Chorigyne* they are fusiform and in *Stelestylis* provided with a vermicular-subulate appendage at each end. The surface is smooth, striate, rugulose, or in *Ludovia* and *Cyclanthus* prominently ridged-reticulate. The seed coat anatomy is also very variable, but rather constant within individual genera (Harling 1958; Hammel and Wilder 1989; Eriksson 1989, 1994b). In *Dicranopygium*, *Ludovia* and *Cyclanthus* the innermost cell layer of the outer integument develops into a distinct palisade layer, while the outer layers of the same integument remain thin-walled and undifferentiated (endotestal seed coat). In other genera, the palisade layer is more or less reduced (*Carludovica*, *Evodianthus*, *Dianthoveus*, *Asplundia* and *Thoracocarpus*), or mechanical strength of the seed coat is transferred to the outer layers of the outer integument (*Schultesiophytum*, *Sphaeradenia*, *Chorigyne*, and *Steletylis*).

Dispersal. On account of the often brilliantly coloured opened spadices it has repeatedly been assumed that the Cyclanthaceae are dispersed by animals. The few actual observations on the Carludovicoideae refer to endozoochory by bats (Wilson 1971 fide Croat 1978; Hammel and Wilder 1989), monkeys (Hladik and Hladik 1969 fide Croat 1978; Terborgh fide Gentry and Dodson 1987), and birds (Olson and Blum 1968), explosive fruit dehiscence (Hammel 1986; G. Harling, unpubl.), passive flush dispersal or seeds germinating in the debris of old fallen inflorescences

(Hammel and Wilder 1989; Eriksson 1995). Certain taxa may also be dispersed by ants. Thus, *Ludovia integrifolia* has a distinct caruncle on the seed, and ants have been observed to visit the fragmenting infructescence of *Cyclanthus bipartitus* (Harling 1958). Moreover, it is likely that seeds to some extent are dispersed by water, as many species grow along watercourses. The seeds of several such species are small and light, e.g. in *Carludovica*, *Evodianthus*, *Dicranopygium* and *Cyclanthus*, and it has been shown that they are buoyant (Harling 1958).

PHYTOCHEMISTRY. Silica bodies are lacking in the family. Crystals presumed to be calcium oxalate occur in the form of scattered raphides, raphide bundles, solitary styloids or styloid bundles, other types of crystals and coarse crystal sand. Starch grains are accumulated in the embryo and endosperm in *Dicranopygium*, but starch is otherwise absent in the seed. Saponins and alkaloids have not been found in the family, and Harris and Hartley (1980) did not detect any accumulation of ferulic or diferulic acid.

SUBDIVISION AND RELATIONSHIPS WITHIN THE FAMILY. Within the Cyclanthaceae, the genus *Cyclanthus* deviates considerably in many respects, not only vegetatively and in inflorescence structures but also in several anatomical features, such as the production of latex, and in embryology, as stated above. Without doubt, this genus is characterised by many derived features. *Cyclanthus* is treated in a separate subfamily, the Cyclanthoideae, and the remaining genera form the subfamily Carludovicoideae. These subfamilies are supported as monophyletic sister taxa (Hammel and Wilder 1989; Eriksson 1994b).

Harling (1958) informally divided the Carludovicoideae into the *Asplundia* group (*Carludovica*, *Evodianthus*, *Asplundia*, *Thoracocarpus*, *Dicranopygium* and *Schultesiophytum*) and the *Sphaeradenia* group (*Ludovia*, *Sphaeradenia* and *Stelestylis*), mainly on the basis of pistillate floral structures and phyllotaxy. In this scheme the recently described *Dianthoveus* would belong to the *Asplundia* group, and *Chorigyne* to the *Sphaeradenia* group.

The *Sphaeradenia* group is certainly monophyletic, where *Ludovia* is the sister group to the other three genera, and *Sphaeradenia* most probably the sister group to *Stelestylis* and *Chorigyne* (Eriksson 1994b). The relationships of the genera in the *Asplundia* group are largely uncertain, except that *Evodianthus* and *Dianthoveus* are certainly sister taxa (see Harling 1958; Hammel and Wilder 1989; Eriksson 1994b). The probable evolution of morphological and anatomical characters in the Cyclanthaceae is discussed by Eriksson (1994b).

AFFINITIES. The Cyclanthaceae are generally placed in a separate order, Cyclanthales (or Synanthae), to express their isolated position. The affinities have been much debated, with most suggestions being in favour of the Araceae, Palmae and Pandanaceae.

The Araceae have recently been considered only distantly related to the Palmae, Cyclanthaceae and Pandanaceae (e.g. Dahlgren et al. 1985). The Palmae differ from both the Cyclanthaceae and Pandanaceae, e.g. in having exclusively simple stem vascular bundles and silica bodies in specialised cells, and in never exhibiting sympodial branching below a terminal inflorescence (Tomlinson and Wilder 1984). Furthermore, the superficially similar leaves in the Palmae and Cyclanthaceae apparently have a different ontogeny (Wilder 1976).

The agreement between the Cyclanthaceae and Pandanaceae that extends to, e.g. floral and seed structures, details of vegetative anatomy, embryology, and pollen morphology, suggests a relationship between these two families (see, e.g. Harling 1958). This is also supported by phylogenetic analyses based on *rbc*L data (Chase et al. 1993; Duvall et al. 1993).

DISTRIBUTION AND HABITATS. The distribution of the family extends from S Mexico through Central America, parts of the West Indies (mainly the Lesser Antilles), and tropical S America to central Bolivia and subtropical SE Brazil (Sta Catarina). The northern Andes and the humid lowland west of them are included in the distributional area. The Cyclanthaceae are lacking in the broad arid or semiarid corridor extending SSW from the Brazilian states Piauí and Ceará. Ten of the 12 genera have more extensive ranges, while *Dianthoveus* is restricted to southern Colombia and northern Ecuador, and *Chorigyne* to Costa Rica and Panama. Colombia is the centre of diversity for the family.

Nearly all Cyclanthaceae require wet and more or less shady habitats, and are therefore concentrated in rainforests, both tropical and montane. The few exceptions include *Carludovica palmata*, which often grows in rather dry and sunny places, sometimes in secondary habitats such as pastures, sparse scrub and on roadsides, and *Cyclanthus bipartitus*, which sometimes grows in similar situations, although its natural habitat is wet and even swampy forest. Some species of *Sphaeradenia* also occur in dryer and more open habitats than most Cyclanthaceae.

Climbing and epiphytic species of the family often grow in dense forest. This is also the case with certain terrestrial species, but many of them grow in moist glades or on riversides (most *Dicranopygium*, many terrestrial *Asplundia* and *Sphaeradenia*). Species growing in the spray of waterfalls with permanently moist air include *Asplundia luetzelburgii* and *Dicranopygium grandifolium*. A number of climbing *Asplundia* species and other taxa are found in freshwater swamps, which are also the normal habitat for *Cyclanthus bipartitus*. There are a few species which may occur in more marginal habitats, e.g. *Ludovia integrifolia* and *Sphaeradenia killipii* in mangrove, *Sphaeradenia amazonica* in white-sand areas and *Dianthoveus cremnophilus* and *Sphaeradenia gigantea* on very steep cliffs.

PALAEOBOTANY. *Ludoviopsis* from the Eocene of France (Saporta 1868), and *Cyclanthodendron* (Sahni and Surange 1944), *Viracarpon* (Sahni 1944), and *Tricoccites* (Trivedi and Verma 1978) from the Deccan Intertrappean series of India have provisionally been referred to Cyclanthaceae. A reexamination of *Cyclanthodendron* by Biradar and Bonde (1990) shows, however, that this genus (including the *Tricoccites* fruits) is closely related to the Strelitziaceae. *Ludoviopsis* probably belongs to the Palmae (see Harling 1958), while the affinities of *Viracarpon* are entirely unknown (Bande and Awasthi 1986). Additional papers pertaining to palaeobotany are listed by Wilder (1981c) and Eriksson (1994b).

Infructescences with seeds from the Eocene of Germany (Grube Messel) have a fruit organisation like *Cyclanthus* and the seeds are similar to this but not fully identical (M. Collinson, pers. comm. to J. Bogner).

ECONOMIC IMPORTANCE. *Carludovica palmata* is used in several countries of Central and S America for making the so-called Panama hats, most of the hat production and marketing nowadays taking place in Ecuador. The species is cultivated on a large scale in the hot coastal plains of Ecuador. The raw material, toquilla straw, has been exported to a minor extent and is used also for making mats and baskets.

Leaves of *Carludovica*, *Asplundia* and *Sphaeradenia* are used for thatching roofs. "Forest ropes" are sometimes made from liana stems of *Evodianthus funifer* and *Thoracocarpus bissectus*. The young inflorescences of *Carludovica palmata* and *Asplundia* spp. are sometimes eaten by natives (e.g. Hazlett 1986), and so are also fruiting spadices of, e.g. *Evodianthus funifer* (Tucuso), and the basal portion of unopened leaf buds of *Carludovica* (Bennett et al. 1992). Furthermore, the petiole of *Asplundia* sp. is chewed against snake bites (Ayala Flores 1984), an alkaline powder is produced from *Sphaeradenia* sp. for coating pellets of the hallucinogenic *Virola theiodora* (Schultes and Raffauf 1990), fish traps are made from *Carludovica* petioles (Bennett et al. 1992) and *Cyclanthus* may be used by native Indian women for perfume. Some species are also grown in greenhouses as ornamentals.

KEY TO THE GENERA

1. Leaf blades deeply bipartite, nonplicate, tricostate with lateral costae extending to segment apex; latex present; staminate and pistillate flowers in separate, alternating cycles or spirals; placentae numerous (subfam. Cyclanthoideae) **12. *Cyclanthus***
- Leaf blades bifid, entire, or flabelliform-parted, plicate, unicostate or tricostate with lateral costae ending below segment apex; latex absent; staminate and pistillate flowers in spirally arranged groups; placentae 4 or 1 (subfam. Carludovicoideae) 2
2. Phyllotaxy spiral; petiole dorsiventrally or adaxially flattened; placentation parietal 3
- Phyllotaxy distichous; petiole elliptic in cross-section; placentation subapical or apical 9
3. Leaf blades with (3–)4(–5) segments; petioles 3–5 times as long as blade; fruits in a layer irregularly splitting from rachis; pollen with end-situated pore **1. *Carludovica***
- Leaf blades usually with 2 segments, each segment sometimes secondarily split into irregular lobes, blades seldom entire; petioles less then 2 times as long as blade; fruits indehiscent or apically dehiscent, not in irregularly splitting layer; pollen monoulcerate to monosulcate, or inaperturate 4
4. Pistillate flowers and fruits connate 5
- Pistillate flowers and fruits free 7
5. Spathes clustered just below spadix; seeds terete **6. *Dicranopygium***
- Spathes dispersed along peduncle; seeds strongly flattened 6
6. Spathes 3–5(–8), diminishing in size upwards along peduncle; seed coat smooth **4. *Asplundia***
- Spathes 8–11, diminishing in size downwards along peduncle; seed coat striated **5. *Thoracocarpus***
7. Leaf blades distinctly tricostate, not scabrous when dry, black when dry; stamens with basal bulb; pollen monosulcate, foveolate **7. *Schultesiophytum***
- Leaf blades unicostate or indistinctly subtricostate, scabrous when dry, not black when dry; stamens ± lacking basal bulb; pollen inaperturate, psilate 8
8. Staminate flowers with perianth lobes in 2 whorls, lobes of outer whorl with glandules; receptacle of staminate flowers funnel-shaped, not papillate; tepals of pistillate flowers acute to shortly acuminate; seeds strongly flattened **2. *Evodianthus***

- Staminate flowers with perianth lobes in 1 whorl, lobes without glandules; receptacle of staminate flowers ± flat, papillate; tepals of pistillate flowers long-acuminate; seeds somewhat flattened **3. *Dianthoveus***
9. Leaf blades entire; anthers without secretion globule; tepals of pistillate flowers reduced to low wales **8. *Ludovia***
- Leaf blades bifid; anthers usually with apical secretion globule; tepals of pistillate flowers well developed 10
10. Pistillate flowers and fruits free; placentation subapical; fruits indehiscent **10. *Chorigyne***
- Pistillate flowers and fruits connate; placentation apical; fruits usually dehiscent by apical cap 11
11. Placenta 1; seeds broadly ellipsoid to narrowly oblong **9. *Sphaeradenia***
- Placentae 4; seeds fusiform with chalazal end long-caudate **11. *Stelestylis***

Genera of the Cyclanthaceae

I. Subfam. Carludovicoideae Harling (1958).

For characters see Key to the Genera.

1. *Carludovica* Ruiz & Pav. Fig. 64A–J

Carludovica Ruiz & Pav., Fl. Peruv. Chil. Prodr. 146 (1794); Harling, Acta Horti Bergiani 17: 40–41 (1954); Harling, Acta Horti Bergiani 18: 127–139 (1958).

Tall, terrestrial, and short-stemmed plants; branching sympodial. Leaves spiral; leaf blades flabelliform-parted, tricostate; leaf segments (3–)4(–5), wedge-shaped, apically dentate or lobate. Spathes 3–4, clustered immediately below spadix. Spadix cylindrical to ellipsoid. Staminate flowers massive, sessile; perianth lobes 15–20, rotundate, apiculate, glanduliferous; stamens numerous. Pistillate flowers and fruits partly connate; placentae 4, parietal; style single or lacking. Four spp., Guatemala to C Bolivia.

2. *Evodianthus* Oerst. Fig. 64K–M

Evodianthus Oerst., Vidensk. Meddel. Dan. Naturhist. Foren. Kbh. 194 (1857); Harling, Acta Horti Bergiani 18: 262–274 (1958).

Root-climbing lianas with slender, branched stem, or sometimes terrestrial plants with short, mostly unbranched stem; branching sympodial. Superterrestrial roots as in *Thoracocarpus*. Leaves spiral; leaf blades bifid and rather large, sometimes entire and small, unicostate or indistinctly subtricostate, scabrous when dry. Spathes ca. 3, clustered immediately below spadix. Staminate flowers symmetrical, funnel-shaped; perianth lobes in 2 alternating whorls, with inflexed apices. Pistillate flowers and fruits free; placentae 4, parietal; style(s) short or nearly lacking. One variable sp., *E. funifer* (Poit.) Lindm., Nicaragua to Peru and Brazil.

3. *Dianthoveus* Hammel & Wilder

Dianthoveus Hammel & Wilder, Ann. Mo. Bot. Gard. 76: 112–113 (1989).

Terrestrial, tall herbs with rather short stem; branching sympodial. Leaves spiral; leaf blades bifid, unicostate or indistinctly subtricostate, scabrous when dry. Spathes of 2 kinds, 3 conspicuous clustered below the spadix and up to 6 inconspicuous inserted just below or slightly above the spadix base. Spadix cylindrical. Staminate flowers symmetrical; perianth lobes narrowly triangular-acuminate, eglandular; stamens 19–32, at base surrounded by cupule-forming papillae, anthers with flattened, acute, sterile tip. Pistillate flowers and fruits free; placentae 4, parietal; styles short. Only one sp., *D. cremnophilus* Hammel & Wilder, S Colombia and N Ecuador.

4. *Asplundia* Harling

Asplundia Harling, Acta Horti Bergiani 17: 41–42 (1954), nom. cons.; Harling, Acta Horti Bergiani 18: 140–254 (1958).

Root-climbing lianas or terrestrial herbs, with short to long stem; branching monopodial. Leaves spiral; leaf blades bifid, unicostate, subtricostate or tricostate. Spathes 3–5(–8), on upper half of peduncle, not clustered. Spadix cylindrical to subspherical. Staminate flowers symmetrical or asymmetrical, sometimes funnel-shaped; pedicel distinct, central or eccentric; perianth lobes glanduliferous; stamens few to many. Pistillate flowers and fruits party connate; placentae 4, parietal; styles 4, ± free, rarely 1 style. About 100 spp., S Mexico through Central America, the Lesser Antilles, and Trinidad to central Bolivia and S Brazil. Divisible into subgen. *Choanopsis* Harling with symmetrical and subgen. *Asplundia* with asymmetrical flowers.

5. *Thoracocarpus* Harling Fig. 65A–E

Thoracocarpus Harling, Acta Horti Bergiani 18: 254–260 (1958).

Root-climbing lianas, or sometimes hemiepiphytes, with slender stem up to ca. 30 m bearing short climbing roots and ropelike feeding roots; branching monopodial. Leaves spiral; leaf blades bifid and rather large as well as entire and small, both unicostate. Peduncle short; spathes 8–11. Spadix shortly cylindrical to ellipsoid; staminate flowers large, symmetrical, funnel-shaped; perianth lobes generally 10–15, glanduliferous, with fleshy, pink apices. Pistillate flowers and fruits basally connate; placentae 4, parietal, stigmas sessile. Only one sp., *Th. bissectus* (Vell.) Harling, S Costa Rica and Trinidad to S America as far as Bolivia and Brazil.

6. ***Dicranopygium*** Harling Fig. 64N–P

Dicranopygium Harling, Acta Horti Bergiani 17: 43–44 (1954); Harling, Acta Horti Bergiani 18: 274–333 (1958).

Usually small to medium-sized terrestrial herbs, with ± short, fleshy stem; branching monopodial. Leaves spiral; leaf blades bifid, generally unicostate, seldom tricostate. Peduncle commonly rather long, slender, elastic, spathes 2–3(–5), clustered immediately below the small pauciflorous spadix. Staminate flowers generally asymmetrical, pedicellate; perianth lobes short, usually on abaxial side only, glanduliferous. Pistillate flowers and fruits connate, rather flat; tepals mostly strongly reduced; placentae 4, parietal; style(s) mostly lacking. About 50 spp., S Mexico to central Peru and Surinam. Divisible into four subgenera, *Dicranopygium*, *Uribanthus* Harling, *Glesonianthus* Harling, and *Tomlinsonianthus* Wilder, according to, e.g. form of the perianth of the staminate flower, and anther appearance and dehiscence.

7. ***Schultesiophytum*** Harling

Schultesiophytum Harling, Acta Horti Bergiani 18: 260–262 (1958); Wilder, Opera Bot. 92: 277–279 (1987).

Terrestrial plant, stem rather short; branching monopodial. Leaves spiral; leaf blades bifid, tricostate. Spathes 3–4, often clustered below spadix. Spadix cylindrical. Staminate flowers ± symmetrical, pedicellate; perianth lobes glanduliferous; stamens 15–30, anthers narrowly elongate, with flattened, acute, sterile tip. Pistillate flowers and fruits free; tepals well-developed; placentae 4, parietal; style(s) lacking or very short. Only one sp., *Sch. chorianthum* Harling, S Colombia to central E Peru.

8. ***Ludovia*** Brongn. Fig. 65F–H

Ludovia Brongn., Ann. Sci. Nat., Sér. 4, 15: 361 (1861), nom. cons.; Harling, Acta Horti Bergiani 18: 333–338 (1958).

Root-climbing vines, epiphytes, or terrestrial herbs; branching monopodial. Leaves orthodistichous; leaf blades entire, unicostate. Peduncle usually short; spathes 3–5, dispersed in upper half of peduncle. Spadix cylindrical to somewhat fusiform; staminate flowers rather large, almost symmetrical, shortly pedicellate or sessile; perianth lobes 20–30, glanduliferous. Pistillate flowers and fruits entirely connate; tepals reduced, forming indistinct wales encircled by a broad mass of tissue; placentae 4, subapical; styles lacking. Three spp., lowlands from Nicaragua to Peru, Brazil, and French Guiana.

9. ***Sphaeradenia*** Harling Fig. 65K–M

Sphaeradenia Harling, Acta Horti Bergiani 17: 1–6 (1954); Eriksson, Opera Bot. 126: 1–106 (1995).
Pseudoludovia Harling (1958).

Medium-sized to tall terrestrial or epiphytic herbs, rarely vines, with short to long stem; branching monopodial. Leaves orthodistichous; leaf blades bifid, unicostate. Spathes 2–5, usually in upper half of peduncle. Spadix narrowly cylindrical to broadly ellipsoid. Staminate flowers ± asymmetrical, with distinct pedicel; perianth lobes all around the receptacle or lacking on adaxial side; stamens few to numerous; anthers usually with apical secretion globule. Pistillate flowers and fruits connate; tepals well developed; placenta 1, apical; styles usually entirely concrescent to 1 distinct style, seldom free or lacking. Fifty spp., S Nicaragua to W Bolivia, and eastwards into Venezuela and adjacent Brazil.

10. ***Chorigyne*** R. Erikss. Fig. 65I,J

Chorigyne R. Erikss., Nord. J. Bot. 9: 31–45 (1989).

Epiphytic or terrestrial herbs with short stem. Leaves orthodistichous; leaf blades bifid, unicostate. Spathes 2–5, usually in upper half of peduncle. Spadix cylindrical to subglobose. Staminate flowers asymmetrical, distinctly pedicellate, perianth lobes usually on abaxial side only, glanduliferous; anthers with apical secretion globule. Pistillate flowers and fruits free; tepals well developed; placentae 4, subapical; styles entirely concrescent to 1 style, at least in mature fruit. Seven spp., restricted to Costa Rica and Panama.

11. ***Stelestylis*** Drude Fig. 65N

Stelestylis Drude in Mart., Fl. Bras. 3(2): 229 (1881); Harling, Acta Horti Bergiani 18: 393–399 (1958).

Short-stemmed terrestrial or epiphytic herbs; branching monopodial. Leaves orthodistichous; leaf blades bifid, unicostate. Peduncle rather long and slender; spathes 3–5, sparsely distributed in upper half of peduncle. Spadix cylindrical to broadly ellipsoid. Staminate flowers asymmetrical, with distinct, eccentrically attached pedicel; perianth lobes usually on abaxial side only, glanduliferous; stamens few to numerous; anthers usually with apical secretion globule. Pistillate flowers and fruits basally connate, with distinctly developed tepals; placentae 4, apical; 1 well-developed style present. Four spp., Venezuela, Guayana, Surinam and E Brazil.

II. Subfam. Cyclanthoideae Harling (1958).

For characters see Key to the Genera

12. *Cyclanthus* Poit. ex A. Rich. Fig. 66

Cyclanthus Poit. ex A. Rich. in Bory, Dict. Class. Hist. Nat. 5: 221 (1824); Harling, Acta Horti Bergiani 18: 399–406 (1958).

Tall, usually terrestrial herb with short stem; branching sympodial. Leaves spirodistichous or orthodistichous; leaf blades deeply bipartite, central costa forked above base into 2 lateral costae running to apex of leaf segments. Staminate flowers in each cycle (or spiral) fused to a unit; perianth lacking; stamens numerous. Pistillate flowers of each cycle (or spiral) entirely coalescent; perianth strongly reduced, visible as 2 lists, each connate with a staminodial lamella; carpels of each cycle in 2 parallel rows, connate and forming an annular chamber with numerous parietal placentae. Only one sp., *C. bipartitus* Poit. ex A. Rich., N Guatemala through Central America, the Lesser Antilles, and Trinidad to N South America as far as Peru and Brazil.

Selected Bibliography

Ayala Flores, F. 1984. Notes on some medicinal and poisonous plants of Amazonian Peru. Adv. Econ. Bot. 1: 1–8.

Bande, M.B., Awasthi, N. 1986. New thoughts on the structure and affinities of *Viracarpon hexaspermum* Sahni from the Deccan Intertrappean beds of India. Stud. Bot. Hung. 19: 13–22.

Beach, J.H. 1982. Beetle pollination of *Cyclanthus bipartitus* (Cyclanthaceae). Am. J. Bot. 69: 1074–1081.

Bennett, B.C., Alarcón, R., Cerón, C. 1992. The ethnobotany of *Carludovica palmata* Ruiz & Pavón (Cyclanthaceae) in Amazonian Ecuador. Econ. Bot. 46: 233–240.

Biradar, N.V., Bonde, S.D. 1990. The genus *Cyclanthodendron* and its affinities. Proceedings, 3IOP Conference, Melbourne, pp. 51–57.

Chase, M.V. et al. 1993. See general references.

Croat, T.B. 1978. Flora of Barro Colorado Island. Stanford: Stanford University Press.

Dahlgren, R.M.T. et al. 1985. See general references.

Darlington, C.D., Janaki Ammal, E.K. 1945. Chromosome atlas of cultivated plants. Aberdeen.

Duvall, M.R. et al. 1993. Phylogenetic hypotheses for the monocotyledons constructed from *rbc*L sequences. Ann. Mo. Bot. Gard. 80: 607–619.

Erdtman, G. 1954. Pollen morphology and plant taxonomy. Bot. Not. 65–81.

Eriksson, R. 1989. *Chorigyne*, a new genus of the Cyclanthaceae from Central America. Nord. J. Bot. 9: 31–45.

Eriksson, R. 1994a. The remarkable weevil pollination of the neotropical Carludovicoideae (Cyclanthaceae). Plant Syst. Evol. 189: 75–81.

Eriksson, R. 1994b. Phylogeny of the Cyclanthaceae. Plant Syst. Evol. 190: 31–47.

Eriksson, R. 1995. The genus *Sphaeradenia* (Cyclanthaceae). Opera Bot. 126: 1–106.

French, J.C., Clancy, K., Tomlinson, P.B. 1983. Vascular patterns in stems of the Cyclanthaceae. Am. J. Bot. 70: 1386–1400.

Garcin, H. 1958. Contribution à la connaissance de la structure florale des Cyclanthacées. Nat. Monspel. Sér. Bot. 10: 7–32.

Gentry, A.H., Dodson, C.H. 1987. Diversity and biogeography of Neotropical vascular epiphytes. Ann. Mo. Bot. Gard. 74: 205–233.

Gottsberger, G. 1990. Flowers and beetles in the South American tropics. Bot. Acta 103: 360–365.

Gottsberger, G. 1991. Pollination of some species of the Carludovicoideae, and remarks on the origin and evolution of the Cyclanthaceae. Bot. Jahrb. Syst. 113: 221–235.

Hammel, B. 1986. Cyclanthaceae. In: Wilbur, R.L. (ed.) The vascular flora of La Selva Biological Station, Costa Rica. Selbyana 9: 196–202.

Hammel, B., Wilder, G.J. 1989. *Dianthoveus*, a new genus of Cyclanthaceae. Ann Mo. Bot. Gard. 76: 112–123.

Harling, G. 1946. Studien über den Blütenbau und die Embryologie der Familie Cyclanthaceae. Sven. Bot. Tidskr. 40: 257–272.

Harling, G. 1958. Monograph of the Cyclanthaceae. Acta Horti Bergiani 18: 1–428.

Harris, P.J., Hartley, R.D. 1980. Phenolic constituents of the cell walls of monocotyledons. Biochem. Syst. Ecol. 8: 153–160.

Hazlett, D.L. 1986. Ethnobotanical observations from the Cabecar and Guaymí settlements in Central America. Econ. Bot. 40: 339–352.

Olson, S.L., Blum, K.E. 1968. Avian dispersal of plants in Panama. Ecology 49: 565–566.

Sahni, B. 1944. Takli, near Nagpur. Genus *Viracarpon* Sahni. Paleobotany in India V. Proc. Natl. Acad. Sci. India 14: 80–82.

Sahni, B., Surange, K.R. 1944. A silicified member of Cyclanthaceae from the Tertiary of Deccan. Nature 154: 114.

Saporta, G. de. 1868. Prodrome d'une flore fossile des travertins anciens de Sézanne. Mém. Soc. Géol. Fr. II, vol. 8.

Schultes, R.E., Raffauf, R.F. 1990. The healing forest. Portland, OR: Dioscorides Press.

Seres, A., Ramírez, N. 1995. Biología floral y polinización de algunas monocotiledóneas de un bosque nublado venezolano. Ann. Mo. Bot. Gard. 82: 61–81.

Solereder, H., Meyer, F.J. 1928. Systematische Anatomie der Monokotyledonen, 3. Berlin: Borntraeger.
Surange, K.R. 1950. A contribution to the morphology and anatomy of the Cyclanthaceae. Trans. Nat. Inst. Sci. India 3: 159–209.
Takhtajan, A.L. 1982. See general references.
Tomlinson, P.B., Wilder, G.J. 1984. Systematic anatomy of Cyclanthaceae (Monocotyledoneae) – an overview. Bot. Gaz. 145: 535–549.
Trivedi, B.S., Verma, C.L. 1978. *Cyclanthodendron* remains from the Deccan Intertrappean beds of Madhya Pradesh, India. Palaeobotanist 25: 529–542.
Wilder, G.J. 1976. Structure and development of leaves in *Carludovica palmata* (Cyclanthaceae) with reference to other Cyclanthaceae and Palmae. Am. J. Bot. 63: 1237–1256.
Wilder, G.J. 1977a. Structure and symmetry of species of the *Asplundia* group (Cyclanthaceae) having monopodial vegetative axes: *Schultesiophytum chorianthum*, *Dicranopygium* sp. nov., *Asplundia rigida*, and *Thoracocarpus bissectus*. Bot. Gaz. 138: 80–101.
Wilder, G.J. 1977b. Structure and symmetry of species of the *Asplundia* group (Cyclanthaceae) having sympodial vegetative axes: *Evodianthus funifer* and *Carludovica palmata*. Bot. Gaz. 138: 219–235.
Wilder, G.J. 1978. Two new species and a new subgenus of Cyclanthaceae. J. Arnold Arb. 59: 74–102.
Wilder, G.J. 1979. Structure and symmetry of species of the *Sphaeradenia* group (Cyclanthaceae): *Stelestylis stylaris*, *Sphaeradenia woodsonii*, and *Ludovia* spp. Bot. Gaz. 140: 338–355.
Wilder, G.J. 1981a. Structure and development of *Cyclanthus bipartitus* Poit. (Cyclanthaceae) with reference to other Cyclanthaceae. I. Rhizome, inflorescence, root, and symmetry. Bot. Gaz. 142: 96–114.
Wilder, G.J. 1981b. Structure and development of *Cyclanthus bipartitus* Poit. (Cyclanthaceae) with reference to other Cyclanthaceae. II. Adult leaf. Bot. Gaz. 142: 222–236.
Wilder, G.J. 1981c. Morphology of adult leaves in the Cyclanthaceae (Monocotyledoneae). Bot. Gaz. 142: 564–588.
Wilder, G.J. 1984. Anatomy of noncostal portions of lamina in the Cyclanthaceae (Monocotyledoneae). V. Tables of data. Bot. Mus. Leafl. Harvard Univ. 30: 103–133.
Wilder, G.J. 1985a. Anatomy of noncostal portions of lamina in the Cyclanthaceae (Monocotyledoneae). I. Epidermis. Bot. Gaz. 146: 82–105.
Wilder, G.J. 1985b. Anatomy of noncostal portions of lamina in the Cyclanthaceae (Monocotyledoneae). II. Regions of mesophyll, monomorphic and dimorphic ordinary parenchyma cells, mesophyll fibres, and parenchyma-like dead cells. Bot. Gaz. 146: 213–231.
Wilder, G.J. 1985c. Anatomy of noncostal portions of lamina in the Cyclanthaceae (Monocotyledoneae). III. Crystal sacs, periderm, and boundary layers of the mesophyll. Bot. Gaz. 146: 375–394.
Wilder, G.J. 1985d. Anatomy of noncostal portions of lamina in the Cyclanthaceae (Monocotyledoneae). IV. Veins of interridge areas, expansion tissue, and adaxial and abaxial ridges. Bot. Gaz. 146: 545–563.
Wilder, G.J. 1986a. Anatomy of first-order roots in the Cyclanthaceae (Monocotyledoneae). I. Epidermis, cortex, and pericycle. Can. J. Bot. 64: 2622–2644.
Wilder, G.J. 1986b. Anatomy of first-order roots in the Cyclanthaceae (Monocotyledoneae). II. Stele (excluding pericycle). Can. J. Bot. 64:2848–2864.
Wilder, G.J. 1987. Contributions to taxonomy and morphology of *Schultesiophytum chorianthum* Harl. and *Dicranopygium mirabile* Harl. (Cyclanthaceae). Opera Bot. 92: 277–291.
Wilder, G.J. 1988. Inflorescence position as a taxonomic character in the Cyclanthaceae. Bot. Gaz. 149: 110–115.
Wilder, G.J. 1989a. Morphology of *Dianthoveus cremnophilus* (Cyclanthaceae). Can. J. Bot. 67: 2450–2464.
Wilder, G.J. 1989b. Anatomy of *Dianthoveus cremnophilus* (Cyclanthaceae). Can. J. Bot. 67: 3580–3599.
Wilder, G.J. 1992a. Comparative morphology and anatomy of absorbing roots and anchoring roots in three species of Cyclanthaceae (Monocotyledoneae). Can. J. Bot. 70: 38–48.
Wilder, G.J. 1992b. Orthodistichous phyllotaxy and dorsiventral symmetry on adult shoots of *Cyclanthus bipartitus* (Cyclanthaceae, Monocotyledoneae). Can. J. Bot. 70: 1388–1400.
Wilder, G.J., Harris, D.H. 1982. Laticifers in *Cyclanthus bipartitus* Poit. (Cyclanthaceae). Bot. Gaz. 143: 44–93.
Wilder, G.J., Johansen, J.R. 1992. Comparative anatomy of absorbing roots and anchoring roots in three species of Cyclanthaceae (Monocotyledoneae). Can. J. Bot. 70: 2384–2404.

Dioscoreaceae

H. HUBER

Dioscoreaceae R. Br., Prodr.: 294 (1810), nom. cons.
Stenomeridaceae J. Agardh (1858), nom. cons.

Geophytic, rhizomatous or more commonly tuberous perennials, the tuber with or without secondary growth; aerial shoot rather short-lived, twining or sometimes erect or procumbent. *Leaves* alternate, opposite or partly whorled, with a pair of fugaceous stipules or estipulate; the petiole typically with a pulvinus at each end; the blade mostly conduplicate in bud, frequently broad and often cordate, entire, palmately lobed or digitately compound; principal veins palmate, acro-, actino- or camptodromous; secondary veins transversal, connected by reticulate veinlets. *Flowers* usually dioecious (perfect in *Stenomeris*, which doubtfully belongs here), epigynous, usually inconspicuous, pedicellate with bracteolate or ebracteolate, not articulated pedicels, or sessile, arranged in axillary panicles, racemes, or spikes; the ultimate inflorescence units monochasial but often reduced to a single flower; perianth regular, trimerous, 2-whorled, tepals mostly similar; stamens arranged in 2 whorls of 3 or the inner whorl sterile or wanting; filaments distinct or attached to the tepals, rarely connate into a staminal column; anthers tetrasporangiate, introrsely or extrorsely dehiscent, the thecae sometimes widely separated by a broad connective, the latter usually exappendiculate (*Stenomeris*, see above, excepted); ovary syncarpous, 3-locular; style trifid or 3-lobed, rarely style branches 2-lobed or bifid, or stigmas nearly sessile; placentas axile; ovules anatropous, bitegmic, when more than 1 superposed and normally 2 in each loculus but numerous in *Stenomeris*. *Fruit* 3-locular with 2 seeds in each loculus (seeds numerous in *Stenomeris*), rarely 2 loculi sterile and the fertile one 1-seeded, capsular or (*Tamus* only) baccate, a unilocular samara in *Rajania*. Seeds spherical or compressed, in the latter case usually winged, with abundant horny, starch-free endosperm, the latter extremely hard by hemicellulose in *Tamus*; embryo small with a terminal plumula and a broad lateral cotyledon; seed coat with the mechanical tissue often derived from both integuments.

A family of – conventionally treated – three (four, when *Stenomeris* included) to more than 20 genera and about 600 species, mostly in the tropics, poorly represented in Europe, Australia and Oceania.

VEGETATIVE MORPHOLOGY. In probably all species correctly assigned to Dioscoreaceae and definitely in all members of Dioscoreoideae, the seedling produces a persistent tuberous swelling (Fig. 16C,D), derived either from the epicotyl, as in *Tamus* (Bucherer 1889; Tillich 1980), or from the hypocotyl, as in *Testudinaria* and perhaps commonly in *Dioscorea* (e.g. Koch and Bruhn 1962; Trouslot et al. 1993); subsequent internodes may participate in the building of the tuber (e.g. Sharma 1976). The swelling may be radial, as in *Borderea*, *Epipetrum* or *Testudinaria*, radial in the first stage but soon one-sided in *Dioscorea* sect. *Apodostemon* and *Heterostemon* (Koch and Bruhn), or one-sided from the beginning, as in *Dioscorea* sect. *Enantiophyllum* (Trouslot et al. 1993), *D. floribunda* (Sharma 1980), *D. sinuata* and *Tamus*.

In the mature ground organs of Dioscoreaceae, De Bary (1877) recognised three categories, which still today cover practically the entire variation: (1) annually renewed tubercules as found in *Dioscorea opposita* (and the other species of sect. *Enantiophyllum*), considered modified roots by De Bary but shown to originate as lateral outgrowths of the hypocotyl by Queva (1894); (2) solitary persistent tubers growing by means of a cambium, as long known from *Tamus*; and finally (3) plagiotropic sympodial rhizomes as found in *Dioscorea quinqueloba* and *D. villosa*. In the former species, according to Queva (1894), the rhizome rises from the axillary bud of the primary leaf.

The persistent tuber is varied as to its secondary growth; this may expand over the entire surface and thus produce a discoid, ellipsoidal or spherical body as in *Epipetrum* and many American species of what traditionally is referred to as *Dioscorea*. In these instances, the entire surface of the tuber is covered with roots. Alternatively, secondary growth may be restricted to the distal surface of the tuber, which then expands upwards (*Borderea*) or upwards and horizontally, markedly so in *Dioscorea mexicana* and *Testudinaria elephantipes*. Another possibility is that the secondary growth deviates from the concentric increase of the *Testudinaria* model towards an eccentric and mainly horizontal growth as in *Dioscorea* sect. *Apodostemon* (*D. composita*,

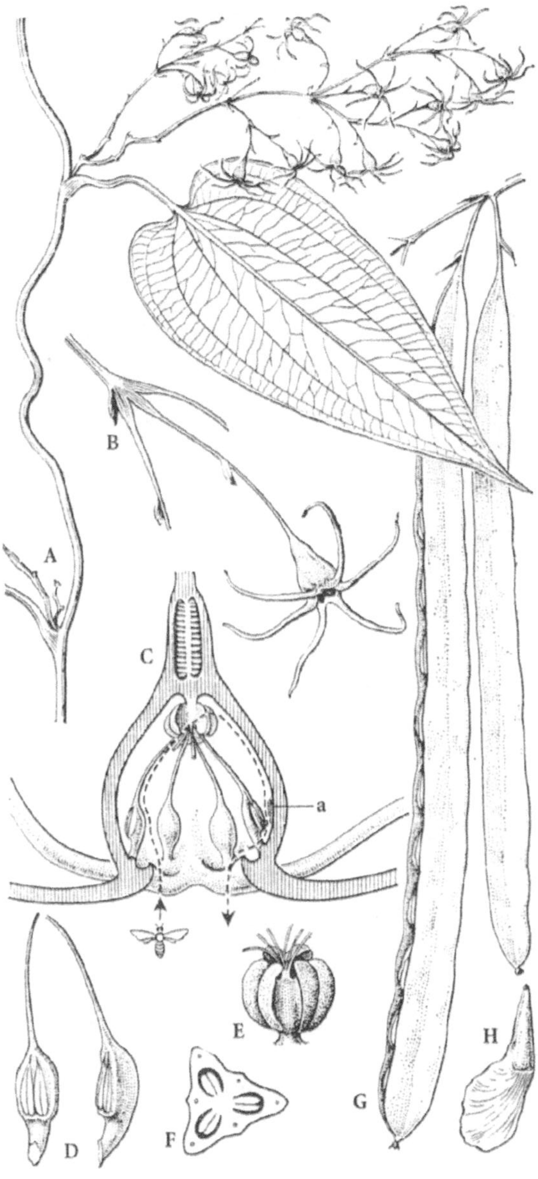

Fig. 67A–H. Dioscoreaceae. *Stenomeris dioscoreifolia.* **A** Flowering shoot. **B** Flower. **C** Flower, longitudinal section, with indication of course of pollinator. *a* Secondary pollen presentation on perianth tube. **D** Stamen in ventral and lateral view. **E** Style with stigmatic lobes. **F** Ovary, transversal section. **G** Capsule. **H** Seed. (Takhtajan 1982)

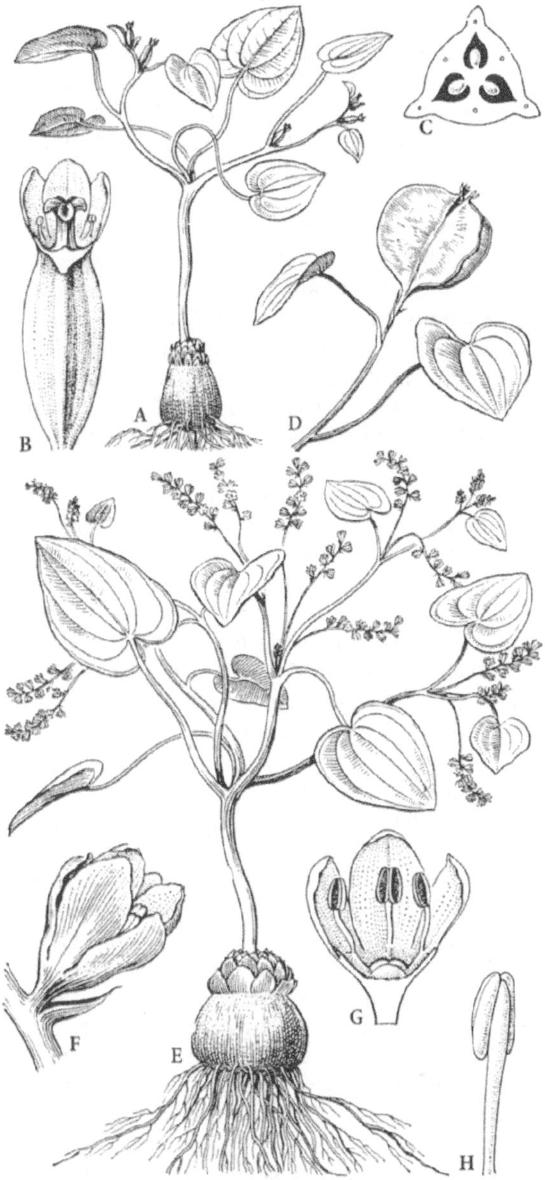

Fig. 68A–H. Dioscoreaceae. *Borderea pyrenaica.* **A** Pistillate plant. **B** Pistillate flower. **C** Ovary in transversal section. **D** Dehiscing capsule. **E** Staminate plant. **F** Staminate flower. **G** Same, longitudinal section. **H** Stamen. (Takhtajan 1982)

spiculiflora), *Brachystigma* (*D. sinuata*) and *Heterostemon* (*D. floribunda*). In this case, as well as in *Testudinaria*, root formation is confined to the under surface of the growing tuber periphery. The tuber of *Tamus* with its positively geotropic scrotiform extrusion or extrusions is basically different and bridges the gap from the persistent tuber to the annual tubercule, mostly elongate and strongly geotropic. Nevertheless, in *Tamus* the entire tuber, including its extrusions, is perennial. Further examples of this model are *Testudinaria glauca* and *Dioscorea* sect. *Perennia* (von Teichman et al. 1975).

The homology of the annually renewed extrusions is less evident, but *Tamus* is instructive. Persistent tubers are necessarily bifunctional as they give rise to the aerial shoot and serve as storage organs, obviously so in *Tamus*. A first step towards a division of functions is illustrated by *Dioscorea cayenensis* (Burkill 1960). Here the distal portion of the tuber close to the soil surface is hard and impalatable and supports the aerial shoot, whereas the proximal extrusion is converted into a fleshy annual storage organ increasingly deeply buried in the soil. A still more perfect stage in the separation of shoot-forming and storage organs is observed in *Dioscorea esculenta* and many other species with "filipendulous", i.e. long-stalked annual tubercules (Knuth 1924). Since the time of Duchartre (1854), these tubercules frequently have been mistaken for roots, understandibly, as long as the ground organs were not subjected to a comparative analysis. There seems to be no transition from species with tubercules to those with a rhizome, but Burkill (1960) draws attention to *Dioscorea perdicum*, which combines a persistent tuber with the plagiotropic orientation of the rhizomatous species. This, however, does not suggest a first step towards true rhizomes as found in sections *Macropoda*, *Stenophora* and apparently *Illigerastrum* as well, which are defined by distal increase and proximal decay in balance, and, in contrast to the persistent tuber, dispose of a far greater potential to produce aerial shoots; Burkill (1960) considers this an adaptation to severe winters. Most Asiatic dioscoreas possess annual tubercules as described above; these have been believed not to occur in the truly American sections, but the so-called tuberous or intumescent roots mentioned by Barroso et al. (1974) of *Dioscorea altissima*, *D. macrocapsa* and other species of *Dioscorea* sect. *Chondrocarpa*, *Lasiogyne*, *Sphaerantha* and *Spinosa* almost certainly belong here; another example is *D. dodecaneura*, one of the few species with edible tubercules native to America.

Thorny roots resembling those of certain palms are a striking feature in *Dioscorea* sect. *Combilium* and *Paramecocarpa*; spinescent roots also in some species of sect. *Enantiophyllum*.

All Dioscoreaceae are geophytes with the consequence of short-lived aerial shoots, and "all in all, a 12-month may be taken as the approximate limit of the life . . ." (Burkill 1960). At least in *Dioscorea* sect. *Heterostemon* the primary axis remains rudimentary, bearing merely scale leaves; the aerial shoots originate from the axils of the latter (Koch and Bruhn 1962). Stems may be ridged, winged

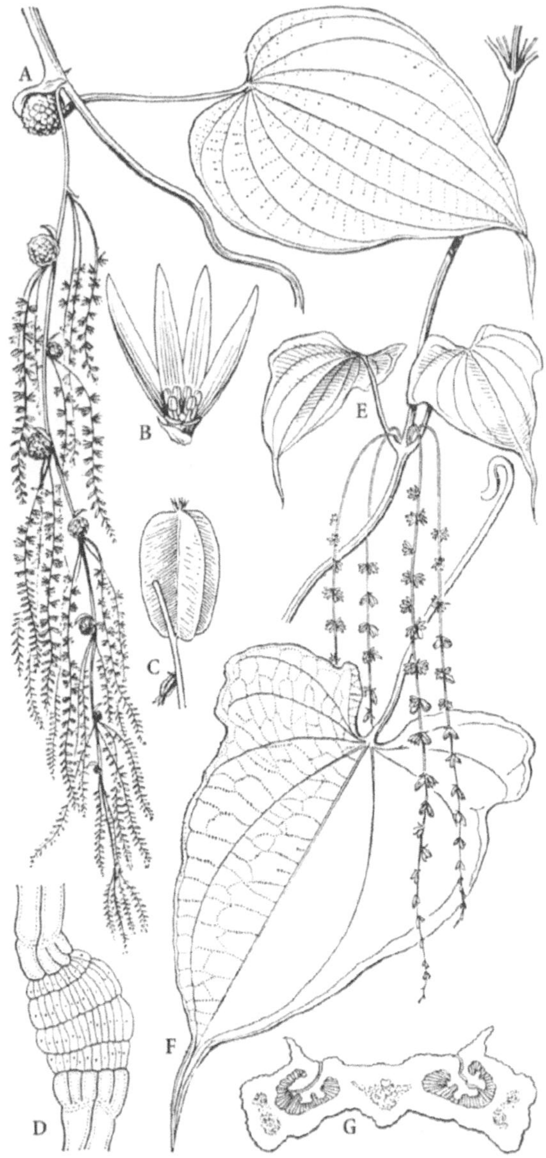

Fig. 69A–G. Dioscoreaceae. **A–D** *Dioscorea* (sect. *Opsophyton*) *bulbifera*. **A** Inflorescence with axillary tubers. **B** Staminate flower, 2 tepals removed. **C** Capsule. **D** Phloem glomerulus of a nodal ganglion, longitudinal section. **E–G** *D.* (sect. *Macroura*) *sansibarensis*. **E** Branch with inflorescence. **F** Leaf with drip-tip. **G** Cross-section of drip-tip with mucilaginous glands. (Takhtajan 1982)

or aculeate and, according to Uline (1898), in *Testudinaria elephantipes* entire branchlets can be transformed into thorns. The stem bases rarely exceed a diameter of 1 cm, with the exception of *Dioscorea basiclavaticaulis*, recently described from Brazil. A universal feature of the family is twining but as "a plant which twines cannot begin the twining until it has length enough for circumnutation" (Burkill 1960), dwarf species necessarily never climb. Uline (1897) has shown that the direction of twining is a constant feature in what he believed sections while in most other plants this is a family character. Sinistrorse twining (the term used here in accordance with Uline, Knuth, Burkill, and Matuda, but not with Barroso et al.) is the more common habit in Dioscoreaceae and the plant world throughout, but dextrorsely twining sections are several in America, few in Africa (sect. *Asterotricha, Cotinifoliae, Enantiophyllum*), and in Asia there is only one (*Enantiophyllum*).

Aerial or axillary tubers, commonly but inadequately referred to as bulbils or bulblets – they have nothing to do with bulbs – are a striking feature of several species of *Dioscorea*, particularly in sect. *Enantiophyllum* (several species), *Lasiogyne, Lasiophyton* (*D. melanophyma*), *Macroura* (*D. sansibarensis*), and *Opsophyton* (*D. bulbifera*); they are more common in sections with annually renewed tubercules than in those with a persistent tuber and are absent from the rhizomatous sections. Karsten (1847) recognised the homology of aerial tubers with axillary buds and Sharma (1978) has shown that they represent a swelling of the hypopodium.

Due to their commonly opposite leaves, *Dioscorea* sect. *Asterotricha, Enantiophyllum, Macroura* and *Xylinocapsa* are among the most puzzling monocotyledons. They are a permanent stimulus to revive the old debate on the number of cotyledons in *Dioscorea*, although Solms-Laubach (1881) and Smith (1916) have demonstrated the existence of only 1. Heterocotyly respectively the existence of 2 unequal cotyledons, postulated as early as 1835 by Dutrochet, has recently been advocated by Lawton and Lawton (1967), Sharma (1980) and Trouslot et al. (1993).

Stems and more frequently leaf blades of *Stenomeris, Tamus, Testudinaria* and at least ten sections of *Dioscorea* possess sessile secretory glands or extrafloral nectaries. These are mostly superficial, but deeply immersed in the parenchyma in some of the more derived sections of *Dioscorea*, namely *Combilium, Enantiophyllum*, and *Paramecocarpa*; in palaeontology, they allow recognition of dioscoreaceous affinity of leaf fossils (see p. 224). Delpino (1888) thought these glands to produce nectar, while Gentner (1905) suggested that they may function as hydathodes. The glandular structures abound in the forerunner tips of *Dioscorea sansibarensis*, which may reach a length of up to 7 cm and harbour nitrogen-fixing bacteria (Burkill 1960). Acarodomatia have been reported by De Wildeman (1904) from two African species of *Dioscorea* sect. *Enantiophyllum*.

The petiole is provided with 2 pulvini, 1 distal, the other 1 proximal; the latter, responsible for abscission, may be dislocated apically, leaving the stem furnished with indurescent petiole bases.

The blade is simple and entire, often cordate, but there are exceptions: 3- or more-lobed blades occur, in the Americas uncommonly and almost confined to Mexico and Chile, in the Old World mostly in Africa. Compound leaves are still rarer than the lobed ones in America (three species only and all trifoliolate), but are well represented in Africa and tropical E Asia by species with tri- or more-foliolate leaves belonging to *Dioscorea* sect. *Cardiocapsa, Illigerastrum* and *Lasiophyton*. When one takes into consideration the Dioscoreaceae-feeding diurnal moth *Episteme* (Agaristidae), centred in tropical E Asia, these leaves may well be aposematic.

VEGETATIVE ANATOMY. Mucilage idioblasts with raphide bundles are widespread; tannin cells common in *Tamus, Testudinaria* and many sections of *Dioscorea* but present neither in *Stenomeris* nor in Trichopodaceae. Vessels have scalariform perforation plates in the roots, the aerial stem, and the petiole but are usually replaced by tracheids in the leaf, rhizomes, and ground and aerial tubers. The lamina is mostly hypostomatic; anomocytic stomata prevail and the guard cells when viewed from outside often seem "sporned" by a projecting ledge, but this is a feature of Dioscoreales rather than of Dioscoreaceae, and obvious in Trichopodaceae as well. Ling et al. (1982) and Purnima and Srivastava (1985) found a considerable variation in stomata of Chinese, respectively Indian, species of *Dioscorea*, where the anomocytic construction is often accompanied by other types, the frequency of which seems to be characteristic of the sections.

Vegetative anatomy reveals impressively the degree of isolation of this order not only within the monocotyledons but within seed plants throughout. The first to note this was von Mohl (1845), who discovered the unique vascular bundles and their arrangement in the stem. They are unmistakable in three respects:

1. All species possess 2 kinds of stem bundles alternating with each other: smaller bundles (common vascular bundles), often V-shaped in transverse section, and larger bundles (cauline

vascular bundles), elliptic in transverse section and receding inwards. In thick-stemmed species additional medullary bundles may occur, but not constantly.
2. In all Dioscoreaceae, both common and cauline bundles contain at least 2 clearly and often widely separated phloem strands; their arrangement and number is partly diagnostic.
3. At the nodes, the stem bundles are interrupted by "glomeruli", in which the vessels of the successive internodes do not join with each other but are replaced by very numerous short tracheids of various size and orientation, and the sieve tubes by "phloem glomeruli", composed of thin-walled cells without sieve pores but with sieve areas not visible in the light microscope (Behnke 1965).

The following anatomical details have been found diagnostic for natural entities in Dioscoreaceae:

Tannin cells are wanting, apart from *Stenomeris*, in *Dioscorea* sect. *Dematostemon*, *Lasiogyne* and a few more (Ayensu 1972).

Hairs, although absent from more than half of the species of Dioscoreaceae (Burkill 1960), may be simple, 1- or 2-celled, as in *Rajania* and a few (mostly Old World) sections of *Dioscorea*. T-shaped hairs define sections *Combilium* (*D. esculenta*) and *Macrocarpaea* (*D. preussii*), stellate hairs are useful for recognising sect. *Asterotriche* (incl. *Syntepaleia*), and dendritic hairs occur in some species of sect. *Enantiophyllum*.

In contrast to hairs, *foliar glands* are widespread in the twining species. They are composed of an 1-celled stalk and a many-celled head arising by anticlinal and periclinal divisions. This kind of gland has been reported from numerous species of *Dioscorea* as well as from *Rajania* and *Tamus*. In *Stenomeris*, they are replaced by glands with a 2-celled head.

Larger *stem bundles* (cauline vascular bundles) with 1 or 2 major phloem strands on the inner side of the innermost pair of the large metaxylem vessels; 1 strand in *Stenomeris*, *Tamus* and most Neotropical sections of *Dioscorea*, in *Rajania* and the majority of Old World sections 2 strands in the mentioned position; *Testudinaria* is ambiguous: *T. elephantipes* following the former, *T. sylvatica* the latter type (Ayensu 1972).

The aerial stem usually presents a somewhat modified *endodermis* (endodermoid layer, differing from a true endodermis by lack of Casparian strips ab initio), but this is replaced in *Dioscorea bulbifera* (sect. *Opsophyton*) by a layer of tangentially compressed brachysclereids (Ayensu 1972). Xifreda (1991) described an "endodermoid sheath" with Casparian stripes at the inner boundary of the stem cortex of ten species of *Dioscorea* from Argentina. In the mature and sclerotic stems of two unrelated species (*D. hassleriana* and *D. multiflora*) this layer is converted into a crystalliferous sheath composed of compartimented cells that contain crystals of Ca-oxalate.

Starch grains in the storage organs appear as a promising tool for an improved classification of the Dioscoreaceae (Seidemann 1966). The starch grains are mostly simple, rather large, many-layered, and vaguely to strongly eccentric with the longer diameter up to 25 (–60) µm in most cultivated species, but compounded of isodiametric small partial grains without any layering in *Dioscorea* sect. *Combilium* (*D. esculenta*) and sect. *Lasiophyton*. Simple and rather large many-layered grains are widespread and have been reported from *Tamus* and *Testudinaria* (Czaja 1978), three Old World sections of *Dioscorea* (*Enantiophyllum*, *Macroura*, *Opsophyton*) and 14 Brazilian sections of the same genus (Seidemann 1966; Barroso et al. 1974).

Ultrastructure. Contrary to the Asparagales or Liliales, the Dioscoreales (i.e. Dioscoreaceae plus *Avetra*, *Stenomeris* and *Trichopus*) possess only non-oriented epicuticular wax crystalloids (Barthlott and Frölich 1983). Sieve-tube plastids in Dioscoreaceae s. str. contain numerous cuneate protein crystalloids of different size and starch grains; insofar they agree with most Scitamineae but not with *Trichopus*, nor Asparagales or Liliales, where starch grains are wanting (Behnke 1981).

Floral Structure. Apart from *Stenomeris*, all Dioscoreaceae are dioecious, but there are a few reports of monoecy, most of them referring to American species, one to *Tamus*, but "to date we do not have a single proved case of the two sexes being fertile on the one individual" (Burkill 1960). The number of tepals is invariably 6, they tend to fuse proximally, in the staminate more decidedly so than in the pistillate flowers but rarely a tubular or even urceolate perianth is found, as in a few Madagascan, Mexican and S American species; urceolate, however, is the perianth of *Stenomeris*, a genus quinxotic in many a respect. The fused portion of the perianth, variously referred to as torus or disc, often becomes somewhat succulent such as in *Dioscorea* sect. *Enantiophyllum* and *Sarcantha*. Staminate flowers of the African

Dioscorea hirtiflora are peculiar by their inner tepals which are apically coherent by entangled hairs, and leave only a narrow slit for entrance. The androecium of Dioscoreaceae is usually hexandrous, but in what is conventionally referred to as *Dioscorea*, the inner whorl is frequently sterile and replaced by 3 staminodia or is entirely wanting. The latter case is quite common in Central and S American but known from only one Old World species, the Japanese *Dioscorea asclepiadea*. *Dioscorea monandra* from Argentina is the only monandrous species known. The filaments are distinct or more frequently proximally attached to the perianth; quite often they are rudimentary and the anthers nearly sessile. Connate filaments occur in *Dioscorea* sect. *Higinbothamia*, *Margarethia*, and *Monadelphia*, in which a staminal column results. The anthers are mostly introrse but extrorse dehiscence occurs in *Epipetrum* and at least seven sections of *Dioscorea*, all of them American; the only Old World member of the family with extrorse anthers seems to be *Tamus* (Brenner 1914). In staminate flowers style rudiments are usually inconspicuous but strongly developed rudiments define *Dioscorea* sect. *Lasiogyne* and *Macrogynodium* and in *Epipetrum* they spectacularly exceed the stamens.

The trilocular ovary is a constant feature as long as the plants are in flower; in fruit, 2 of the loculi may abort, and the fertile loculus develops only 1 of its 2 ovules, as is known from *Rajania*, but also from a few Mexican species of *Dioscorea*; a strictly American innovation. Otherwise, ovules and seeds are 2 and superposed in the loculi, with *Stenomeris*, where they are numerous, and *Dioscorea gaumeri* as exceptions; in the latter species they range from 3 to 4 in each loculus, on which fact the genus *Higinbothamia* had been founded. Septal nectaries, in both pistillate and staminate flowers, have been described for *Dioscorea balcanica*, *D. quinqueloba* and *Tamus communis* (Dauman 1970).

Embryology. Cytokinesis of pollen mother cells takes place simultaneously and the grains are binucleate when shed. The micropyle of the bitegmic and crassinucellar ovules is formed by both integuments. Embryo-sac development corresponds to the Polygonum type and endosperm formation follows the Nuclear model.

Pollen Morphology. According to Erdtman (1952) and Su (1987), the pollen is sulcate throughout the rhizomatous sections *Macropoda* and *Stenophora* (19 species studied) as well as in sect. *Illigerastrum* and in *Stenomeris* (2 species); in the latter genus the grains are biconvex in contrast to dioecious Dioscoreaceae. Bisulculate pollen occurs in *Borderea pyrenaica*, *Tamus* (both species), in 7 old World sections of *Dioscorea* as well as in 14 New World sections. Bi- or occasionally trisulculate grains have been reported from neotropical *Dioscorea densiflora* (sect. *Oxypetalum*) and from *Rajania*. As Su (1987) has shown, pollen of sections with annual tubers (namely *Enantiophyllum*, *Lasiophyton*, *Opsophyton* and *Shannicorea*, to which *Brachyandra*, not studied by Su, must be added) is smaller, with the longest axis not exceeding 27 μm, in contrast to almost all the genera and sections with persistent tubers (*Borderea*, *Tamus*, *Testudinaria* and 14 out of the 15 tropical American sections of *Dioscorea* scrutinised) as well as the rhizomatous sections *Macropoda* and *Stenophora* with the longest pollen axis ranging from 27 to 40, rarely 50 μm. There is up to now only one example of a plant with annual tubers and large (42–μm) grains, *Dioscorea esculenta* (sect. *Combilium*) and the only tropical American Dioscoreaceae with pollen of the smaller category are *Dioscorea sinuata* (sect. *Brachystigma*) and *Rajania*. The exine is mostly reticulate; a striate exine prevails in sect. *Macropoda* (e.g. *D. villosa*), and has been reported from sect. *Brachyandra*, endemic to Madagascar, and the neotropical sections *Apodostemon*, *Periandrium* and *Cryptantha*, in the latter alongside with reticulate ornamentation (Barroso et al. 1974).

Karyology. The chromosomes in Dioscoreaceae tend to be small and numerous, in one count up to 400; this is particularly the case with the food yams, where aneuploidy is well known from *Dioscorea alata, bulbifera, cayenensis, dumetorum* and *oppositifolia*. Chromosome numbers are based on $x = 9$, 10 or 12 with the only exception of *Dioscorea mexicana* (p. 230), apparently based on $n = 8$. Numbers based on 12 are so far known to be confined to the western Eurasian species of *Borderea* and *Tamus*, whereas *Rajania* and four out of nine neotropical sections of *Dioscorea* studied have exclusively $n = 9$ and out of nine palaeotropical sections studied probably all are based on $n = 10$. The latter number is also shared with the holarctic section *Macropoda* with the sole exception of *Dioscorea quaternata*, a N American plant with $2n = 36$ or 54 (Al Shehbaz and Schubert 1989). Unfortunately, most sections of *Dioscorea* remain unknown in this respect and counts of the more isolated genera *Stenomeris* and *Testudinaria* are lacking.

POLLINATION. The flowers are usually unattractive both as to size and colour: they are green or yellowish, less frequently blackish purple as in several species of *Dioscorea* and in *Tamus edulis*, and very rarely exceed a diameter of 7 mm; no wonder that they did not evoke the curiosity of floral biologists. In this respect, the perhaps best-known member of the family is *Tamus communis*, the flowers of which, according to Brenner (1914), emit a faint fragrance of honey especially towards the evening and at least pistillate flowers produce nectar. According to the same authority, *Tamus* is visited, at least in Switzerland, mostly by *Empis* (Empididae, dance flies) and to a smaller degree by solitary bees, minute beetles, exceptionally by *Apis mellifera*. Burkill (1946), who gave a comprehensive list of visitors, found *Anaphothrips tamicola* (Thysanoptera) the most abundant active visitor of *Tamus*, concluding that "insects which lick the surfaces of nectaries, such as butterflies, bees, syrphids and most of the muscids, seem to be, on the whole, less frequent visitors than the tissue-penetrating insects above (i.e. thrips)". Sweet fragrance has also been reported from African species of *Dioscorea* and Burkill (1960) presumes "the tropical dioscoreas in a large measure seem to be pollinated at night by agents which reach the flowers guided by scent". In the Botanic Garden of Rio de Janeiro, Barroso et al. (1974) observed about a dozen species of *Dioscorea* and found only one, *D. martiana*, with perfumed flowers. The latter are visited by Meliponini, a tribe of inaculeate bees, perhaps a species of *Hypotrigona*. This genus and the related *Trigona* have also been observed visiting the flowers of *D. macrocapsa*, *pseudomacrocapsa* and *ovata*, the latter species with staminate flowers that exsude abundantly nectar.

To ensure pollination, the species studied by Barroso et al. follow alternative strategies: those possessing small flowers have a campanulate, funnel-shaped or tubular but rarely rotate perianth, pale colours, minute anthers, little and small-grained pollen but produce nectar; these flowers are arranged in many-flowered inflorescences and are all opened simultaneously for a long period. In contrast, the species with larger flowers have often a contrastingly coloured staminal column, a rotate perianth, large anthers with abundant pollen but lack nectar; within 1 inflorescence only 1–3 flowers are simultaneously opened.

In *Epipetrum*, Reiche (1908) failed to find nectar or visitors and concluded that pollination takes place only when staminate and pistillate specimens grow densely together, which is favoured by a kind of fructification that virtually precludes dispersal.

SEED STRUCTURE. A survey of 60 species (H. Huber, unpubl.), unfortunately including few African and none from Madagascar, shows that the seed coat is endotestal and exotegmic. The inner epidermis of the usually 3-layered outer integument (endotesta) is crystalliferous and in each cell a single crystal of Ca-oxalate is embedded in a moiety of reddish brown phlobaphene (Fig. 70). The invariably 2-layered inner integument consists of an outer layer with the cell walls pitted or sculptured and an inner layer, which is uniformly thin-walled.

Both a crystalliferous endotesta and an exotegmen distinguish the Dioscoreales from all other monocotyledons with few exceptions: in *Stenomeris* both features are wanting; in *Borderea* and five sections of *Dioscorea* – all American – crystals are rare or absent; in 2 instances (*D. leptostachya* and *D. sinuata*) the endotesta is nearly colourless, a unique case in Dioscoreoideae; and in *Epipetrum* and all species of

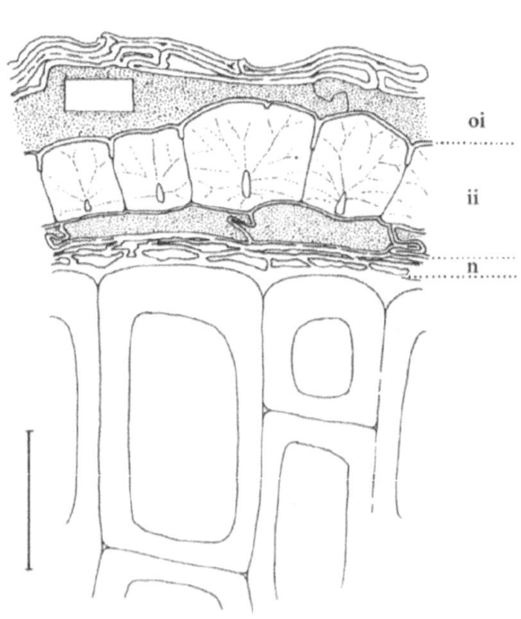

Fig. 70. Dioscoreaceae. *Tamus communis*, transversal section of seed coat and armoured endosperm. Note triple fortification by crystalliferous endotesta, sclerotic exotegmen, and stone celled endosperm, providing a unique example of overfortification in berry-fruited monocotyledons. *oi* Outer integument; *ii* inner integument; *n* nucellar tissue. (Huber 1991)

Dioscorea with proximally alate seeds, two African species of sect. *Lasiophyton* excepted, the walls of tegmic cells are delicate and unsculptured throughout. The first authors to use seed characters in generic classifications of this family were Salisbury (1824, in description of *Testudinaria*, see p. 228) and Kunth (1850). The former became aware that *Testudinaria* differs from all the Dioscoreaceae known to him by distally alate seeds, and Kunth defined a new genus, *Helmia*, with proximally alate seeds in contrast to redefined *Dioscorea* with the seeds winged all round. Apart from the above-mentioned difference in endotesta and exotegmen, in the groups with applanate seeds the course of the raphe, which may cross the seed body or not, and the pattern of epidermal cells on the seed wing were found diagnostic (H. Huber, unpubl.).

A raphe that transverses the flat side of the seed body defines *Testudinaria* and a few sections of *Dioscorea*, such as sect. *Brachystigma, Cryptantha, Higinbothamia, Lychnostemon, Macrogynodium* and *Strutantha*, in the New World, and *Dioscorea nipponica* (sect. Stenophora) and sect. *Perennia* and *Rhacodophyllum* (von Teichman et al. 1975). This feature is associated with the more conservative groups and known neither from the advanced sections such as *Enantiophyllum*, *Lasiophyton*, *Macropoda*, or *Pygmaeophyton* nor from any of the sections with proximally winged seeds.

The epidermis of the seed wings mostly consists of isodiametric cells or cells radially stretched when seen from outside, with straight anticlinal walls, but sinuously lobed anticlinals are frequent in sect. *Enantiophyllum* and in two species of sect. *Lasiophyton* they are extremely long and faintly curved.

The endosperm of Dioscoreaceae stores lipids, protein and most commonly hemicellulose in form of cell-wall thickening. The extent of hemicellulose storage is correlated with seed shape: in the turgid seeds of *Tamus* (and similarly of *Avetra* and *Trichopus*) the walls are enormously thick, not or at most indistinctly pitted, the seeds therefore extremely hard and the wall thickening occupies at least 1/2 of the cell volume, whereas in the flat-seeded genera cell walls are but moderately thickened and *Stenomeris* stands apart within the whole order with quite delicate cell walls entirely deprived of hemicellulose.

Dispersal. *Tamus*, the only berry-fruited genus in Dioscoreaceae, is obviously adapted to bird dispersal, notably to blackbirds (Brenner 1914); the red berries contain 1 to 6, mostly 4 or 5 seeds. On the other hand, *Rajania*, *Testudinaria* and nearly all species of *Dioscorea*, except a few dwarf species from Mexico and S America, are anemochorous. Depending on shape and orientation of the wing (seed wing in *Dioscorea*, *Stenomeris* and *Testudinaria*, fruit wing in *Rajania*), Burkill (1960) distinguished gliding and whirling diaspores. The former include the sections of *Dioscorea* with bialate and circularly winged seeds, the latter those with the wing proximally or distally inserted as well as *Rajana*, *Stenomeris* and *Testudinaria*. *Epipetrum* is remarkable for its antitelechory: not only is the number of flowers in pistillate inflorescences strongly reduced, mostly to 1, the long and slender positively geotropic pedicels coil spirally and deposit the capsules on the soil, but do not pierce it (Reiche 1908).

Distribution and Habitats. The Dioscoreaceae are an essentially tropical family but maximum diversity, in consonance with the pronounced storage organs, is found in seasonal (monsoon) climates just within or without the tropics. Thus the greatest concentration of species is in Central and S Mexico (about 70 species), Central and S Brazil (probably beyond 120 ssp.), Chile (about 40 ssp.), Madagascar (37 ssp.), Indochina (51 ssp.), China (48 ssp.).

Central and S American Dioscoreaceae (and among them the Chilean in particular) stand apart by an excessive abundance of species, most of them with a permanent tuber. Dwarf species with winged or unwinged seeds occur in Mexico, Brazil (Minas Gerais) and Chile; in S Brazil and Argentina also plants adapted to open grassland are found. The situation in Europe and Africa is in some way similar, but the number of species is much smaller; here again we find minute oreophytes (European *Borderea*) and characters such as unwinged seeds (*Borderea*, *Tamus*) and persistent tubers (*Borderea*, *Tamus*, *Testudinaria*, and *Dioscorea* sect. *Macroura*, *Perennia* and *Rhacodophyllum*), but also a member of the holarctic sect. *Macropoda*, *D. balcanica* (Albania and Montenegro), which is not even distantly related to *Borderea* or *Tamus*. In Africa south of the Sahara species with annually renewed tubercules are frequent and in Madagascar all truly native species follow this model. Asia (apart from the Near East) and Australia entirely lack Dioscoreaceae with persistent tubers with the possible exception of *Stenomeris*, but the ground organs of this genus are insufficiently known. Whereas in tropical Asia sections of *Dioscorea* with annual tubercules dominate, subtropical and temperate

Asia contributes the rhizomatous sections *Macropoda* and *Stenophora*, the former extending to Europe and N America, the latter strictly E Asian. There are two endemic species in Australia, both related to and best included in the widespread sect. *Enantiophyllum* of *Dioscorea*.

PALAEOBOTANY. The fossil record is poor and unreliable throughout except a leaf fragment with secretory glands described by Kräusel and Weyland (Palaeontographica Abt. B, 96: 118–120, 1954) from Tertiary deposits in the Rheinland. In their latest review, Herendeen and Crane (1995) do not know of any dioscoraceous fossil.

AFFINITIES. Adanson (1763) considered *Tamus* a member of the *Famille des Aristoloches*, whereas A.L. de Jussieu (1789) included *Dioscorea* and *Tamus* together with *Smilax* and *Ruscus* in his order Asparagi, from which R. Brown (1810) released the genera with epigynous flowers under the name Dioscoreae. The undue stress traditionally laid on floral structures and an entire neglect of both vegetative and seed characters favoured a classification that brought Dioscoreaceae close to Amaryllidaceae and Iridaceae.

A conspicuous deviation from the established view has been proposed by Huber (1969). He pays attention to the isolated position of Dioscoreaceae by recognising an order Dioscoreales which embraces the three closely related families, Dioscoreaceae, Stenomeridaceae and Trichopodaceae. This order is distantly related to the rest of the Liliiflorae (Lilianae), from which it departs by a series of characteristics not otherwise known in monocotyledons but common in dicotyledons, although hardly related to any of them, a concept in harmony with Burkill's opinion that Dioscoreaceae "are neither primitive nor ancestral to any other family" and that was closely followed by Dahlgren (1975) but not any longer by Dahlgren (1980) and later, when Stemonaceae, Taccaceae and Trilliaceae, since 1985 (Dahlgren et al. 1985) also Petermanniaceae and Smilacaceae have been indiscriminately mingled in Dioscoreales: families which have nothing to do with this order but are perfectly at home in Asparagales (Taccaceae excepted).

After the addition of the genera *Stenomeris* *Trichopus*, and *Avetra* to the Dioscoreaceae, the morphological range of this formerly well-defined and unmistakable family was so much broadened that, without the aid of anatomy, the dioscoreaceous nature of the above-mentioned genera could not be perceived any longer. Therefore Hutchinson (1934) was led to keep separate Trichopodaceae and Stenomeridaceae. Later studies have confirmed this concept, at least as to Trichopodaceae; while *Stenomeris*, poorly understood still today, remains uncertain; in view of our limited knowledge here it is reluctantly included in Dioscoreaceae.

SUBDIVISION. The major challenge in Dioscoreaceae, however, is not the definition of families but the subdivision of the dioecious branch, here referred to as subfamily Dioscoreoideae, which comprises three Linnean genera: *Dioscorea* (with capsular fruits), *Rajania* (with an 1-seeded samara), and *Tamus* (with berries). This arrangement perfectly fulfills the requirements of a tidy classification and in consequence is fiercely advocated by the more conservative students of the family, especially those primarily interested in plant resources, with Burkill (1960) as the most authoritative protagonist. Nevertheless, the essential step towards a more natural classification of dioecious Dioscoreaceae is owed to this author.

Substantial misgivings as to the Linnaean concept of *Dioscorea* arose for the first time when the differences in seed wings became apparent (Salisbury 1824; Kunth 1850) and the genera with capsular fruits but unwinged seeds (*Epipetrum*, *Borderea*) came to light. Fresh perspectives have been opened by Uline (1897), who drew attention to the direction of twining, and in the present century by Prain and Burkill (e.g. 1934) and later by Burkill (e.g. 1947), who incorporated De Bary's (1877) observations of the three different models of ground organs in dioecious Dioscoreaceae into his classification, correctly as to the facts, disputably as to phylogenetic interpretation. This must be seen in the light of Burkill's life-long engagement with the Old World species of *Dioscorea*: whereas in 1947 he considered the species with few and annually renewed tubercules "most ancestral", in 1960 he applied that distinction to the rhizomatous sections, while genera and sections with a solitary persistent tuber were thought to be advanced. In fact, not the rhizomatous species and still less those with fugaceous tubercules – here considered the most derived – represent the original ground organs of dioecious Dioscoreaceae, but those with persistent tubers. This becomes evident when the wide array of characteristics is taken into consideration which are correlated with such tubers, such as (1) twining, both sinistrorsly and dextrorsly, which is frequent in plants with persistent tubers, exclusively dextrorse twining being confined to three sections with annually renewed tubercules only and absent from the rhizomatous sections; (2) capsular, baccate and otherwise indehiscent fruits, as well as winged and unwinged seeds in genera and sections with persistent tubers; short-lived tubercules and rhizomes being constantly correlated with capsules and winged seeds; (3) steroidal sapogenins, which are widespread and perhaps ubiquitous in genera and sections with persistent tubers, present in at least all the rhizomatous Dioscoreas of China, but have

Table 2. Conspectus of genera and genus-equivalents in Dioscoreoideae, largely based on fruiting specimens and ground organs

	Ground organ a persistent tuber lacking annually renewed storage extrusions	Ground organ with one or several annually renewed storage extrusions (tubercules)	Ground organ a rhizome
Fruit a berry; seeds unwinged	2. *Tamus*		
Fruit capsular, longer than wide; seeds distally winged (i.e. towards the apex of the capsule)	3. *Testudinaria*	4. *Dioscorea* sect. *Combilium*, *Paramecocarpa* and *Cotinifoliae* 5. *Dioscorea* sect. *Shannicorea*	6. *Dioscorea* sect. *Stenophora*
Fruit capsular, longer than wide; seeds distally and proximaly winged	7. *Dioscorea* sect. *Macroura* 8. *Dioscorea* sect. *Rhacodophyllum*	9. *Dioscorea* sect. *Macrocarpaea* (possibly misplaced here)	
Fruit capsular, usually longer than wide; seeds winged all round in the plane of the body, or unwinged	10. *Dioscorea* sect. *Apodostemon* and numerous related sections, incl. *Higinbothamia*		
As above but capsule as wide as long or wider than long; seeds winged or – *Borderea* only – unwinged	11. *Dioscorea* sect. *Periandrium* and related sections 12. *Borderea*	a) twining to the left 13. *Dioscorea* sect. *Lasiogyne*. Compare also sect. *Spinosa* under nr. 24 14. *Dioscorea* sect. *Stenocorea* 15. *Dioscorea* sect. *Cardiocapsa* and *Madagascarienses* a') twining to the right 16. *Dioscorea* sect. *Asterotricha* 17. *Dioscorea* sect. *Enantiophyllum*	18. *Dioscorea* sect. *Macropoda*
Fruit capsular, longer than wide; seeds proximally (i.e. towards the base of the capsule) winged or wings wanting	a) anthers introrse b) twining to the left 19. *Dioscorea* sect. *Hypoerocarpa* 20. *Dioscorea* sect. *Perennia* b') twining to the right 21. *Dioscorea* sect. sections *Trigonobasis* and related sections a') anthers extrorse; twining to the right or not twining 22. *Dioscorea* sect. *Dematostemon* 23. *Epipetrium*	a) leaf blade entire or lobed 24. *Dioscorea* sect. *Sphaerantha* and related sections 25. *Dioscorea* sect. *Brachyandra* and related sections 26. *Dioscorea* sect. *Opsophyton* a') leaf blade ternately or digitately compound 27. *Dioscorea* sect. *Lasiophyton*	28. *Dioscorea* sect. *Illigerastrum* (possibly misplaced here)
Fruit indehiscent, a samara	29. *Rajania*		

not been found in the Chinese representatives of sections with annually renewed tubercules, i.e. sect. *Enantiophyllum*, *Lasiophyton*, *Opsophyton*, and *Shannicorea* (Pei Chien et al. 1979).

These details may suffice to show that the Linnean genera most inadequately reflect the diversity of dioecious Dioscoreaceae (Table 2). Therefore those sections of *Dioscorea* that to the present writer's mind merit generic status are keyed out and briefly characterised in the treatment that follows. For several of these infrageneric entities generic names have been in use (*Testudinaria, Boderea, Epipetrum*); these are used in our treatment. For other sections, such names exist but have never been in use in our century (*Oncus* for sect. *Combilium*, *Botryosicyos* for sect. *Lasiophyton*, etc.) and will not be taken up here either. For the majority of the sections no generic name is available. The incomplete state of knowledge of the African and New World dioscoreas does not allow, at this time, the necessary taxonomic and nomenclatural changes. Therefore, the subgeneric and sectional names will be retained throughout this treatment with the understanding that the generic status of most of them will be formalised in a future revision.

Uses. Species of *Dioscorea* provide staple food for large populations in the tropics, most notably in Upper Guinea, the so-called Yam Zone, and as early as 1890 Watt "recognised that the practice of yam cultivation was indigenous to Africa as well as to Asia and tropical America". It has been shown "that different species of *Dioscorea* were brought into cultivation independently in the three areas of the world" and Coursey (1976) presumes that domestication of *Dioscorea* started in W Africa as early as 11000 B.P., where *D. cayenensis*, *dumetorum* and *rotundata* are most important. Another centre of yam domestication is S China and adjacent tropical Asia, which contributed several valuable species, in particular *D. alata, bulbifera, esculenta, japonica, oppositifolia* and *pentaphylla*. Two of them, *D. alata* and *esculenta*, are true cultigens unknown in the wild and were early brought to Africa and, together with *D. cayenensis* and *rotundata*, later to America; no African species is grown in Asia, and *D. trifida* is the only valuable food yam of American origin, but rarely planted in the Old World. As a consequence of ancient cultivation, pistillate specimens of *D. esculenta* flower rarely and hardly ever set fruit (Prain and Burkill 1934).

Within the last decades, species of *Dioscorea* which often accumulate in their tubers or rhizomes great quantities of steroidal saponins – up to 15% of the dry weight – have attracted the attention of the pharmacentical industry for semisynthetic production of corticosteroid and sexual hormones, including oral contraceptives. Traditional civilisations made use of these saponins for stupefying fish, poisoning vertebrates, including criminal practices, preparation of arrow poison, and for washing. Besides that, *Dioscorea* is one of the exceptional genera of monocotyledonous plants used for tanning (Al Shehbaz and Schubert 1989).

Key to the Subfamilies

1. Flowers perfect. Anther connective produced into a caudate appendage. Ovules and seeds numerous in each loculus **Subfam. Stenomeridoideae**
– Dioecious. Anther connective exappendiculate or at most apiculate. Ovules and seeds (1 or) 2 (very rarely up to 4) in each loculus **Subfam. Dioscoroideae**

I. Subfam. Stenomeridoideae Dahlgren, Clifford & Yeo (1985).

Flowers perfect, pedicellate, arranged in an axillary panicle; pedicels bracteolate. Tepals fused into an urceolate tube. Fertile stamens 6, attached below the mouth of the tube, refracted, anther produced into a caudate appendage, pollen unisulcate. Ovules and seeds numerous in each loculus. Fruit a linear capsule. Seeds distally winged, the body lanceolate, faintly longitudinally ridged.

1. *Stenomeris* Planchon Fig. 67

Stenomeris Planchon, Ann. Sci. Nat. Bot. III, 18: 319 (1852); Burkill in Fl. Males. I, 4: 294 (1951).

The ground organ is "an elongated swollen stem, rich in stored starch" (Burkill 1960), otherwise poorly known. Aerial stem twining to the left. Plant unarmed, glabrous except for a minute scurf on the inflorescence. Leaves alternate, blade entire, mostly cordate. Two spp., peninsular Malaya, N Borneo and the Philippines.

Key to the Genera and Genus-Equivalent Sections of Dioscoreoideae

1. Seeds unwinged. Fruit a berry, capsule, or samara 2
– Seeds winged. Fruit capsular 8
2. Fruit a 1–6-seeded berry, red when ripe. Macaronesia to SW Asia **2. *Tamus***
– Fruit dry when ripe 3
3. Fruit 1-seeded. Aerial stem twining. American 4

- Fruit with 3 fertile locules. Aerial stem mostly not twining 5
4. Fruit unilocular when fully developed; samara strongly compressed. Seeds flattened. Antillean Is. **29. *Rajania***
- Fruit 3-locular, 2 locules empty. Seed not strongly flattened. Mexico ***Dioscorea matudae*** and sect. ***Polyneuron*** partly, see under no. 10
5. Pedicels longer than flowers and fruits. Chile **23. *Epipetrum***
- Pedicels shorter than flowers and fruits, or flowers sessile 6
6. Staminate flowers in racemosely arranged cymules. Anthers extrorse. Americas 7
- Flowers racemosely arranged, i.e. cymules reduced to a single flower. Anthers introrse. Capsule transversely striate. Europe, Africa **12. *Borderea***
7. Perianth tube longer than lobes. Fertile stamens 6 **22. *Dioscorea*** sect. ***Dematostemon*** partly
- Perianth tube shorter than lobes. Fertile stamens 3 ***Dioscorea*** sect. ***Polyneuron*** partly, see under no. 10
8. Seeds distally winged (i.e. towards the apex of the capsule only). Leaves never compound. Old World 9
- Seeds distally and proximally winged or only proximally or winged all round in the plane of the body. Throughout the range of the family 13
9. Tuber epigeal or half-epigeal, the upper surface covered by a massive cork layer; annually renewed tubercules wanting. Flowers distinctly pedicellate. S and SE Africa **3. *Testudinaria***
- Rhizomatose, or tuber subterraneous and with annually renewed tubercules. Flowers mostly sessile or short-pedicellate 10
10. Stem twining to the right. Plant glabrous, unarmed. SE Africa ***Dioscorea*** sect. ***Cotinifoliae***, see under no. 4
- Stem twining to the left. Plant glabrous or more commonly hairy 11
11. Flowers usually in racemosely arranged cymules (fascicles, glomeruli). Plant unarmed. E and SE Asia 12
- Flowers racemosely arranged, i.e. partial inflorescences reduced to a single flower. Stem base often spiny; thorny roots present or not. SE Asia to Oceania, cultivated elsewhere. **4. *Dioscorea*** sect. ***Combilium*** and ***Paramecocarpa***
12. Plant producing annually one tubercule. Leaf blade entire **5. *Dioscorea*** sect. ***Shannicorea***
- Plant rhizomatose; annually renewed tubercules wanting. Leaf blade lobed or margin undulate **6. *Dioscorea*** sect. ***Stenophora***
13. Capsule not longer than wide, circular or transversely elliptic in outline. Seeds winged all round in the plane of the body 14
- Capsule longer than wide, occasionally as wide as long 20
14. Twining to the right. Leaves often opposite. Annually renewed tubercules 1- several. Stem prickly or unarmed. Staminate flowers short-pedicellate or sessile, racemosely arranged, i.e. cymules reduced to a single flower 15
- Twining to the left. Stem unarmed 16
15. Indumentum of stellate hairs. Fertile stamens 6 or 3 and then alternating with 3 staminodia. Capsule negatively geotropic. Africa **16. *Diosorea*** sect. ***Asterotricha***
- Glabrous or with simple or dendritic, but not stellate hairs. Fertile stamens usually 6. Capsule positively geotropic. Old World tropics, extending into temperate E Asia and Australia, introduced in the Americas **17. *Dioscorea*** sect. ***Enantiophyllum***
16. Tuber entirely persistent, without annual storage extrusions. Leaves alternate, trifid. Perianth syntepalous, tubular or hypocrateriform. Fertile stamens 3. Brazil **11. *Dioscoreae*** sect. ***Periandrium***
- With fleshy annual storage organs or plant rhizomatous. Perianth deeply cleft. Fertile stamens mostly 6 17
17. Plant rhizomatous, annually renewed tubercules absent. Leaves alternate, opposite, or whorled. Europe, temperate and subtropical Asia, temperate N America **18. *Dioscorea*** sect. ***Macropoda***
- Not rhizomatous, annually renewed tubercules present. Leaves alternate 18
18. Staminate flowers mostly in few-flowered, racemosely arranged cymules. Filaments very short or anthers almost sessile. Leaf blade entire, 3-lobed, or 3-foliolate. Madagascar, Aldabra, Assumption **15. *Dioscorea*** sect. ***Cardiocapsa*** and ***Madagascarienses***
- Staminate flowers racemosely arranged, i.e. cymules reduced to a single flower. Leaf blade entire 19
19. Inflorescence axis, tepals and ovary pubescent. Pedicel ebracteolate. Filaments slender, longer than anther. S America **13. *Dioscorea*** sect. ***Lasiogyne***
- Inflorescence and flowers glabrous. Staminate flowers with a bracteole above midlength. Filaments thickened, short. SE Asia **14. *Dioscorea*** sect. ***Stenocorea***
20. Seeds 2-winged, one wing distal, the other one proximal; wings not or hardly confluent (but compare sect. *Rhacodophyllum*). Leaves alternate or opposite. Africa 21
- Seeds proximally winged or winged all round in the plane of the body. Leaves alternate 23
21. Plant pubescent to tomentose. Fertile stamens 3. Leaves alternate, blade entire **9. *Dioscorea*** sect. ***Macrocarpaea***
- Glabrous throughout. Fertile stamens mostly 6. Leaf blade entire or lobed 22
22. Leaves opposite in all mature plants except at the base of the stem; blade entire or feebly 3–5-lobed, long-acuminate **7. *Dioscorea*** sect. ***Macroura***
- Leaves alternate; blade palmately 3–7-lobed. Seeds bialate, the wings confluent, or winged all round **8. *Dioscorea*** sect. ***Rhacodophyllum***
23. Seeds winged all round in the plane of the body. America **10. *Dioscorea*** sect. ***Apodostemon***
- Seeds proximally winged, i.e. towards the base of the capsule 24
24. Stem twining to the right. America, but compare *Dioscorea inopinata* Prain & Burkill from Thailand (p. 233) 25
- Stem twining to the left 26
25. Flowers racemosely arranged, rarely paired, i.e. cymules usually reduced to a single flower. Perianth deeply cleft. Fertile stamens 6 or 3, anthers introrse. Mexico to S America **21. *Dioscorea*** sect. ***Trigonobasis*** and related sections
- Staminate flowers in racemosely arranged cymules, less commonly the latter reduced to single flower. Tepals connate into a tube. Fertile stamens 6, anthers extrorse. S America **22. *Dioscorea*** sect. ***Dematostemon***
26. Leaf blade at least partly 3- or more-foliolate 27
- Leaf blade entire, rarely 3-lobed 29
27. Filaments longer than anthers. S America ***Dioscorea*** sect. ***Stenocarpa*** and ***Trifoliatae***, see under no. 19
- Filaments shorter than anthers, at least those of the inner whorl. Old World tropics 28
28. Plant glabrous. Perianth rotate. Staminate flowers in racemosely arranged cymules, the latter reduced to a single flower towards the apex of the inflorescence. Stamens 6, of two sizes. S China, Vietnam **28. *Dioscorea*** sect. ***Illigerastrum***

- Plant pubescent or tomentose. Perianth globose. Staminate flowers racemosely arranged, i.e. cymules reduced to a single flower throughout. Fertile stamens 6 or 3 **27.** ***Dioscorea*** sect. ***Lasiophyton***
29. Tuber persistent, without annually renewed tubercules. Fertile stamens 3 or 6 30
- Ground organ with annually renewed tubercules. Fertile stamens 6 31
30. Staminate flowers in racemosely arranged cymules. Fertile stamens 3. S America **19.** ***Dioscorea*** sect. ***Hyperocarpa***
- Staminate flowers racemosely arranged, i.e. cymules reduced to a single flower. Fertile stamens 6. Plant unarmed, but often verrucose. S Africa **20.** ***Dioscorea*** sect. ***Perennia***
31. Annual extrusions of the perennial tuber spherical or pear-shaped, thick-necked, broad-based, or absent but often replaced by aerial tubers. Plant unarmed. Staminate flowers racemosely arranged, i.e. cymules reduced to a single flower **26.** ***Dioscorea*** sect. ***Opsophyton***
- Tubercules long and slender, or long-stipitate. Staminate flowers often in racemosely arranged cymules 32
32. Flowers sessile or almost so. Filaments attached to the perianth tube. S America **24.** ***Dioscorea*** sect. ***Sphaerantha*** and related sections
- Flowers sessile to long-pedicellate. Filaments not adnate to perianth tube. Madagascar **25.** ***Dioscorea*** sect. ***Brachyandra*** and related sections

II. Subfam. Dioscoreoideae Dahlgren, Clifford & Yeo (1985).

Dioecious, flowers pedicellate or sessile, the pedicels often ebracteolate. Tepals distinct or fused into a tube, fertile stamens 6 or 3, very rarely 1, erect or spreading, filaments distinct or attached to perianth tube, rarely connate; anther not appendaged, rarely apiculate; pollen mostly bisulculate, rarely trisulculate, sulcate in *Dioscorea* sect. *Illigerastrum*, *Macropoda* and *Stenophora*. Fruit a berry, samara or most commonly a capsule, the latter oblong, oblanceolate, elliptic, circular, deltoid or transversely elliptic in outline. Seeds winged or not, the body oblong to circular, spherical or compressed, not ribbed.

2. *Tamus* L.

Tamus L., Spec. Pl.: 1028 (1753).

Tuber hypogeal, persistent. Plant unarmed, glabrous, twining to the left, without axillary tubers. Leaves alternate, with a pair of fugaceous subulate stipules, blade entire or three-lobed. Flowers pedicellate, recemosely arranged or in few-flowered cymules arranged in an axillary spikelike raceme; perianth deeply divided with the base cup- or saucer-shaped. Fertile stamens 6, anthers extrorse. Fruit a 1- to 6-seeded berry, red at maturity. Seed spherical, unwinged. $2n = 48, 96$. Two spp., W Europe and the Mediterranean from Macaronesia to Iran and Iraq.

3. *Testudinaria* Salisb.

Testudinaria Salisb. in Burchell, Travels S Africa 2: 147 (1824); Rowley, Nat. Cactus Succ. J. 8: 46–50 (1953).
Dioscorea subg. *Testudinaria* (Salisb.) Uline (1898).
Dioscorea sect. *Testudinaria* (Salisb.) Burkill (1952); von Teichman et al., Boissiera 24: 222 (1975).

Tuber at or above soil surface, persistent, covered by a thick, often deeply fissured cork layer, growing horizontally, or horizontally and distally, or distally and proximally and then lobed beneath, without short-lived fleshy extrusions. Plant not prickly, glabrous, twining to the left or not twining, without axillary tubers. Leaves alternate, entire of feebly 3-lobed, bearing extrafloral nectaries of the superficial type. Flowers pedicellate, in racemosely arranged 2(–1)-flowered cymules. Fertile stamens 6, anthers introrse. Capsule obcordate, longer than wide. Seeds compressed, unevenly winged all round, strongly so towards to apex of the loculus. Three to six spp., S and SE Africa extending northwards to Simbabwe and Sambia.

4. *Dioscorea* L.[1] sect. *Combilium* Prain & Burkill

Dioscorea sect. *Combilium* Prain & Burkill, J. Asiat. Soc. Bengal. 10: 7 (1914), and in Fl. Gén. Indo-Chine 6: 701, 713–714 (1934); Burkill & H. Perr. in Fl. Madag. 44e Fam.: 7–9 (1950); Burkill in Fl. Males. I, 4: 307 (1951).
Alternative generic name: *Oncus* Lour. (1790).

Perennial tuber close to soil surface, producing annually numerous stipitate edible tubercules protected by spinescent roots, or unarmed in cultivated races. Plant abundantly hairy with T-shaped hairs, twining to the left, prickly at the base, without axillary tubers. Leaves alternate, blade entire, with immersed extrafloral nectaries. Staminate flowers short-pedicellate or sessile, almost always racemosely arranged, i.e. cymules reduced to a single flower, the spikes ascending and rigid; fused portion of perianth a shallow cup. Fertile stamens 6, anthers introrse. Pistillate spikes recurved. Capsule (very rarely produced) reflexed, about twice as

[1] Seven of the eight *Dioscorea* species named by Linnaeus 1753 belong to sect. *Enantiophyllum, Lasiophyton, Macropoda* and *Opsophyton*, one – *D. sativa* – being a mixture of five dioscoreaceous and two dicotyledonous components (Prain and Burkill 1919), and with an infallible instinct for the worst possible solution, Britton and Brown (1913) selected this one to lectotypify *Dioscorea*, hence no attempt is made to alter any of Uline's well-known section names.

long as wide. Seeds compressed, distally winged. $2n = 40$, 80, 100. Only sp. *D. esculenta* (Lour.) Burkill, native to continental SE Asia, for its edible tubercules of ancient cultivation, early introduced to other parts of tropical Asia, to Oceania and Madagascar, less frequently grown in Africa and America. Very similar is sect. *Paramecocarpa* Prain & Burkill, from Indochina eastwards to Palau Islands; more distantly related sect. *Cotinifoliae* Burkill from SE Africa with distally winged seeds and the stem twining to the right.

5. ***Dioscorea*** L. sect. ***Shannicorea*** Prain & Burkill

Dioscorea sect. *Shannicorea* Prain & Burkill, J. Asiat. Soc. Bengal. 10: 7 (1914), and in Fl. Gén. Indo-CHine 6: 702, 715–716 (1934).

Tubercules solitary, annually renewed, deeply descending in the soil. Plant unarmed, pubescent or tomentose, twining to the left, apparently without axillary tubers. Leaves alternate, entire. Staminate flowers sessile or nearly so, in racemosely arranged cymules. Fertile stamens 6, anthers introrse. Capsule about twice as long as wide. Seeds compressed, distally winged. $2n = 30$, 40, 60, 80 and more. Few spp., China and adjacent E Asia.

6. ***Dioscorea*** L. sect. ***Stenophora*** Uline

Dioscorea sect. *Stenophora* Uline in E.-P., Nat. Pflanzenfam., Nachtr. zu II, 5: 84 (1897), not sect. *Stenophora* as understood by Burkill, who amalgamates sect. *Stenophora* Uline with sect. *Macropoda* Uline.

Plants rhizomatous, unarmed, pubescent or glabrous, twining to the left, lacking axillary tubers. Leaves alternate, blade lobed or with undulate margin. Staminate flowers pedicellate or almost sessile, in racemosely arranged cymules; fused portion of perianth cupular. Fertile stamens 6, anthers introrse. Capsule little longer than wide. Seeds compressed, distally winged. $2n = 20$, 40. Several spp., temperate and subtropical E Asia, including *D. nipponica* Makino.

7. ***Dioscorea*** L. sect. ***Macroura*** Burkill

Dioscorea sect. *Macroura* Burkill, Blumea, Suppl. I: 232 (1937); Burkill & H. Perr. in Fl. Madag. 44e Fam.: 4–7 (1950).

Tuber at or near soil surface, persistent, lobed or branched with age, strongly poisonous, without a cork layer and lacking annual fleshy extrusions. Plant glabrous, twining to the left, aculeate at base or unarmed, with or without axillary tubers. Leaves opposite in mature plants except at the base of the stem, blade entire or feebly 3- to 5-lobed, long-acuminate. Staminate flowers almost sessile; tepals nearly distinct. Fertile stamens 6, anthers introrse. Capsule large, about twice as long as wide. Seeds distally and proximally winged. $2n = 40$. Few spp., native to tropical Africa.

8. ***Dioscorea*** L. sect. ***Rhacodophyllum*** Uline ex Knuth

Dioscorea sect. *Rhacodophyllum* Uline ex Knuth in Engler, Pflanzenreich IV.43: 183 (1924); von Teichman et al., Boissiera 24: 219 (1975).

Tuber perennial, without annually renewed storage extrusions. Plant unarmed, glabrous, twining to the left. Leaves alternate, blade palmately 3- to 7-lobed. Staminate flowers racemosely arranged, i.e. cymules reduced to a single flower; tepals shortly connate. Fertile stamens 6, rarely 3, anthers introrse. Capsule longer than wide. Seeds bialate, the wings confluent, or winged all round. Three spp., S and SE Africa.

9. ***Dioscorea*** L. sect. ***Macrocarpaea*** Uline

Dioscorea sect. *Macrocarpaea* Uline in E.-P., Nat. Pflanzenfam., Nachtr. zu II, 5: 86 (1897); Burkill, Bull. Jard. Bot. Brux. 15: 346 (1939).

Ground organ unknown, according to Burkill "thought to be an annually renewed tuber", apparently somewhat poisonous. Plant unarmed, woolly-hairy, twining to the left; axillary tubers seem to occur (Burkill 1939). Leaves alternate, the blade entire, long-acuminate. Staminate flowers short-pedicellate or sessile, the fused portion of perianth somewhat expanded. Fertile stamens 3, anthers introrse, staminodes present. Capsule and seeds similar to sect. *Macroura*. $2n = 40$. Few spp., tropical Africa. T-shaped hairs resembling those of sect. *Combilium* in *D. preussii* Pax, which belongs to sect. *Macrocarpaea*.

10. ***Dioscorea*** L. sect. ***Apodostemon*** Uline

Dioscorea sect. *Apodostemon* Uline in E.-P., Nat. Pflanzenfam., Nachtr. zu II, 5: 85 (1897); Barroso et al., Sellowia no. 25: 82 (1974).
Alternative generic names: *Higinbothamia* Uline (1899) and *Nanarepenta* Matuda (1961), neither suitably covering the whole assemblage.

Tuber persistent, mostly hypogeal, without annually renewed storage extrusions. Plant unarmed, glabrous, twining to the left. Leaves alternate, blade entire. Staminate flowers short-pedicellate or sessile, in racemosely arranged cymules; tepals at most shortly connate, style rudiment minute. Fertile stamens 6, filaments short or anthers almost sessile, the latter usually introrse. Capsule longer than wide. Seeds compressed, winged all round. $2n = 64$ (*D. mexicana* Guillemin). Numerous spp., Mexico to Chile.

For the characters of this diagnosis numerous deviations are known which cannot be listed here. More than 20 sections related to sect. *Apodostemon* have been distinguished, many of them founded on single characters, and delimitation from 11. sect. *Periandrium* Uline and similar sections is often arbitrary. Ground organs poorly known, but where described showing remarkable differences.

11. *Dioscorea* L. sect. *Periandrium* Uline

Dioscorea sect. *Periandrium* Uline in E.-P., Nat. Pflanzenfam., Nachtr. zu II, 5: 87 (1897); Barroso et al., Sellowia nr. 25: 82 (1974).

Tuber persistent, without annual storage appendages. Plant unarmed, glabrous, twining to the left. Leaves alternate, trifid. Staminate flowers pedicellate, somewhat succulent, racemosely arranged, i.e. partial inflorescences reduced to a single flower, perianth syntepalous, tubular or hypocrateriform. Fertile stamens 3, with or without staminodia; anthers introrse. Capsule as long as wide. Seeds compressed, winged all round. Perhaps only one sp., SE Brazil.

12. *Borderea* Miégev. Fig. 68

Borderea Miégev., Bull. Soc. Bot. Fr. 13: 374 (1866); Gaussen, Bull. Soc. Hist. Nat. Toulouse 100: 383–398 (1965).

Tuber conical, spherical or depressed with a flat base, rooted from proximal half, without fleshy extrusions. Plant unarmed, glabrous. Aerial stem short and not twining in the European species, up to 1.5 m tall and twining to the left in the African *D. gillettii* Milne-Redh. Leaves alternate, blade entire. Staminate flowers shortly pedicellate or almost sessile, racemosely arranged; pistillate flowers solitary or in up to 3-flowered racemes; perianth deeply divided. Fertile stamens 6 in the European species, 3 fertile and 3 staminodia in *D. gillettii*, anthers introrse. Capsule with transversely striate valves. Seeds slightly compressed, unwinged. $2n = 24$. Two spp., the Pyrenees, one in N Kenya and S Ethiopia.

13. *Dioscorea* L. sect. *Lasiogyne* Uline

Dioscorea sect. *Lasiogyne* Uline in E.-P. Nat. Pflanzenfam., Nachtr. zu II, 5: 87 (1897); Barroso et al., Sellowia nr. 25: 83 (1974).

Stoloniferous (Barroso et al.), with long-stipitate globose tubercules. Plant unarmed, glabrous except inflorescence, perianth and ovary, twining to the left, with or without axillary tubers. Leaves alternate, entire. Staminate flowers pedicellate, racemosely arranged; perianth deeply cleft. Fertile stamens 6, filaments long and slender, anthers introrse. Capsule manifestly wider than long. Seeds compressed, winged all round. $2n = 40$. Only one sp., S America.

14. *Dioscorea* L. sect. *Stenocorea* Prain & Burkill

Dioscorea sect. *Stenocorea* Prain & Burkill, J. Asiat. Soc. Bengal 10: 40 (1914), and in Fl. Gén. Indo-Chine 6: 701, 711–713 (1934); Burkill in Fl. Males. I, 4: 305 (1951).
Alternative generic name: *Peripetasma* Ridl. (1920).

Tubercules renewed annually, deeply descending in the soil, one or more, not edible unless carefully treated. Plant unarmed, glabrous, twining to the left; axillary tubers not known. Leaves alternate, entire. Staminate flowers pedicellate, racemosely arranged, the pedicels with a bracteole above midlength; perianth deeply cleft, the fused portion cup-shaped. Fertile stamens 6, filaments short. Capsule broader than long, remarkably large (at least in *D. sumatrana* Prain & Burkill), positively geotropic. Seeds compressed, winged all round. Few spp., Thailand, Cambodia, Malay Peninsula and Sunda Islands.

15. *Dioscorea* L. sect. *Cardiocapsa* Uline

Dioscorea sect. *Cardiocapsa* Uline in E.-P., Nat. Pflanzenfam., Nachtr. zu II, 5: 87 (1897), not *Cardiocarpa*, as Knuth writes; Burkill & H. Perr. in Fl. de Madag. 44e Fam.: 12–19 (1950).

Persistent tuber small, close to soil surface, producing annually 2–4 long-stipitate tubercules. Plant unarmed, glabrous or pubescent, twining to the left. Leaves alternate, the blade 3-foliolate, 3-lobed, or entire. Staminate flowers minute, short-pedicellate or sessile, in racemosely arranged cymules or the latter reduced to a single flower; perianth deeply divided. Fertile stamens 6, filaments very short or anther sessile. Capsule broader than long. Seeds compressed, winged all round. Four species in Madagascar, Aldabra and Assumption. Sect. *Madagascarienses* Knuth with the single species *D. arcuatinervis* Hochr., also from Madagascar, is closely related but glabrous in all parts, with an aerial stem rarely aculeate at base, leaf blades entire and 3 fertile stamens alternating with 3 staminodia. *D. pseudomacrocapsa* Barroso et al. from Brazil may key out here but is more closely related to no. 24, sect. *Sphaerantha*.

16. *Dioscorea* L. sect. *Asterotricha* Uline

Dioscorea sect. *Asterotricha* Uline, in E.-P., Nat. Pflanzenfam., Nachtr. zu II, 5: 87 (1897); Burkill, Bull. Jard. Bot. Brux. 15: 347 (1939).

Tubercules renewed annually, 1-several, not heavily poisonous. Plant unarmed, rarely stem aculeate at base; stem, leaves and inflorescences beset with stellate hairs at least when young. Aerial stem twining to the right; axillary tubers seem to occur rarely. Leaves opposite or occasionally alternate, blade simple. Staminate flowers short-pedicellate or sessile, sometimes somewhat fleshy, racemosely arranged; perianth deeply cleft. Fertile stamens 6 or 3 and then alternating with 3 staminodia, anthers introrse. Ovary densely stellate-pubescent. Capsule broader than long. Seeds compressed, winged all round. $2n = 40$ or 80. Few spp., tropical Africa including Socotra.

17. *Dioscorea* L. sect. *Enantiophyllum* Uline

Dioscorea sect. *Enantiophyllum* Uline in E.-P., Nat. Pflanzenfam., Nachtr. zu II, 5: 187 (1897); Prain & Burkill in Fl. Gén. Indo-Chine 6: 703–706, 729–745 (1934); Burkill, Bull. Jard. Bot. Brux. 15: 348 (1939); Burkill & H. Perr. in Fl. de Madag. 44ᵉ Fam.: 69–73 (1950); Burkill in Fl. Males. I, 4: 320 (1951); Nkounkou, Belg. J. Bot. 126: 45–70 (1993).

Persistent tuber woody, giving rise to one or few annually renewed, mostly edible tubercules and in some species to spinescent roots. Plant glabrous or with simple or dendritic but not stellate hairs. Stem unarmed or aculeate, twining to the right (one sp. excepted), with or without axillary tubers. Leaves opposite or the lower alternate or alternate throughout, blade simple or somewhat 3-lobed, with glandular acumen and immersed extrafloral nectaries. Staminate flowers sessile, racemosely arranged, globose, scarcely opening at anthesis; tepals distinct, floral axis fleshy and not enlarging into a cup. Fertile stamens 6 (3 in a few Asiatic species), short, anthers introrse. Capsule wider than long. Seeds compressed, winged all round. $x = 10$, mostly $2n = 40$, only in *D. cayenensis* $2n$ ca 36, 54, 140. Numerous spp., tropical Asia, fewer in Africa, two native to Australia, these with alternate leaves; including some of the most widespread and utilised food yams, namely *D. alata* L., *D. cayenensis* Lam., *D. japonica* Thunb., *D. oppositifolia* L. and *D. rotundata* Poir.

18. *Dioscorea* L. sect. *Macropoda* Uline

Dioscorea sect. *Macropoda* Uline in E.-P., Nat. Pflanzenfam., Nachtr. zu II, 5: 85 (1897).
Dioscorea sect. *Stenophora* Prain & Burkill (not of Uline) (1934).
Alternative generic name: *Merione* Salisb. (1866) unless *Dioscorea* is lectotypified by *D. villosa* L.

Plant rhizomatous, unarmed, glabrous or pubescent, twining to the left, without axillary tubers. Leaves alternate, opposite or whorled, blade entire or palmately lobed. Staminate flowers short-pedicellate or sessile, in few-flowered, racemosely arranged cymules; perianth deeply cleft, with the fused portion cup-shaped or campanulate. Fertile stamens 6, anthers introrse. Capsule reflexed, as long as wide or slightly wider than long. Seeds compressed, winged all round. $2n = 20$, 40, 60. Several spp., W Balkan Peninsula, W Caucasus, the Himalayas, temperate E Asia and temperate N America.

Burkill united sect. *Macropoda* Uline with sect. *Stenophora* Uline, but this must not be followed not only because of the difference in the seed wings but also in the raphe, which transverses the flattened seed body in *Stenophora* but not at all in *Macropoda*.

19. *Dioscorea* L. sect. *Hyperocarpa* Uline

Dioscorea sect. *Hyperocarpa* Uline in E.-P., Nat. Pflanzenfam., Nachtr. zu II, 5: 83 (1897).
Alternative generic name: *Hyperocarpa* (Uline) Barroso et al. (1974).

Tuber persistent, without fleshy annual appendages. Plant unarmed, glabrous, twining to the left; aerial tubers absent. Leaves alternate, entire. Staminate flowers sessile, in racemosely arranged glomeruli; perianth deeply divided, the fused portion flat. Fertile stamens 3, staminodia absent, anthers introrse. Capsule at least in the typical species oblanceolate, much longer than wide. Seeds compressed, winged towards the base of the capsule. One to four spp., Central and S Brazil and adjacent Argentina.

The S American sections *Trigonocarpa* Uline, *Stenocarpa* Uline and *Trifoliatae* Knuth apparently related, but their ground organs unknown.

20. *Dioscorea L.* sect. *Perennia* Archibald

Dioscorea sect. *Perennia* Archibald, J.S. Afr. Bot. 33: 9, 15 (in English) (1967); von Teichman et al., Boissiera 24: 222 (1975).

Tuber perennial, without annually renewed fleshy appendages. Plant unarmed but often verrucose, glabrous, twining to the left, without axillary tubers. Leaves alternate, blade entire. Staminate flowers sessile or almost so, racemosely arranged. Fertile stamens 6, anthers introrse. Capsule longer than wide. Seeds compressed, winged towards the base of the capsule. Three spp., Cape Province, S Africa.

21. *Dioscorea* L. sect. *Trigonobasis* Uline

Dioscorea sect. *Trigonobasis* Uline in E.-P., Nat. Pflanzenfam., Nachtr. zu II, 5: 83 (1897).

Alternative generic name: *Helmia* Kunth (1850) if lectotypified with *H. convolvulacea* (Cham. & Schlechtend.) Kunth, as herewith suggested.

Tuber persistent, without annual storage extrusions. Plants unarmed, glabrous, puberulous or pubescent, twining to the right. Leaves alternate, blade entire or three-lobed. Staminate flowers pedicellate, racemosely arranged; perianth deeply cleft. Fertile stamens 3, staminodes wanting, filaments rather long, anthers introrse. Capsule longer than wide. Seeds compressed, winged towards the base. $2n = 36$ (*D. convolvulacea* Cham. & Schlechtend.) or 48, 52 (*D. galeottiana* Kunth). About 14 spp., Mexico.

Sections *Centrostemon* Griseb., *Cycladenium* Uline, *Margarethia* Barroso and *Monadelphia* Uline replace sect. *Trigonobasis* in S America; contrary to the latter, plants often glabrous or almost so, leaf blade invariably entire, flowers – often long – pedicellate; furthermore, *Centrostemon* and *Margarethia* differing by 6 fertile stamens, *Margarethia* and *Monadelphia* by the filaments connate into a staminal column. *Dioscorea monandra* Haumann from Tucumán is unique in the family with 1 fertile stamen only, with stems twining to the left.

22. *Dioscorea* L. sect. *Dematostemon* Griseb.

Dioscorea sect. *Dematostemon* Griseb. in Mart., Fl. Brasil. 3, 1: 27 (1842); Barroso et al., Sellowia nr. 25: 82 (1974).

Tuber persistent, without annual fleshy extrusions. Plants unarmed, glabrous or pubescent, especially in the inflorescence. Stem twining to the right, rarely erect. Leaves alternate, blade entire. Staminate flowers pedicellate, sometimes long-pedicellate, rarely almost sessile; racemosely arranged in cymules, less commonly the latter reduced to a single flower; the tepals connate into a tube. Fertile stamens 6, the filaments long and slender, sometimes coherent; anthers usually extrorse. Capsule longer than broad. Seeds compressed, proximally winged, rarely exalate. About 40 spp., mostly S American, with centre of distribution in Central and S Brazil.

23. *Epipetrum* Philippi

Epipetrum Philippi, Linnaea 33: 253 (1864); Reiche, Bot. Jahrb. 42: 178–190 (1908).

Tuber persistent, without annual fleshy extrusions, rooted over the whole surface. Plants small, glabrous. Stem not twining. Leaves alternate, entire. Flowers long-pedicellate, perianth deeply divided, tepals usually recurved; staminate inflorescence 2- to 5-flowered, pistillate flowers solitary, rarely paired. Fertile stamens 6, filaments extremely short, anthers extrorse; style rudiment in staminate flowers cylindric, with 3 recurved distal branches. Fruiting pedicels mostly coiled. Capsule longer than broad. Seeds not flattened, unwinged. Three spp., Chile.

24. *Dioscorea* L. sect. *Sphaerantha* Uline

Dioscorea sect. *Sphaerantha* Uline in E.-P., Nat. Pflanzenfam., Nachtr. zu II, 5: 83 (1897); Barroso et al., Sellowia nr. 25: 81 (1974).

Tuber woody, giving rise to long-stipitate spindle-shaped tubercules. Plant glabrous or pubescent. Stem aculeate at base, twining to the left. Leaves alternate, blade entire. Staminate flowers sessile or almost, in racemosely arranged glomeruli; perianth deeply divided, with the fused portion cup-shaped. Fertile stamens 6, filaments attached to perianth tube, anthers introrse. Capsule longer than wide. Seeds compressed, winged proximally. Several spp., S America, mainly Brazil.

Closely related is sect. *Chondrocarpa* Uline from the same region, but leaves sometimes opposite or almost so and staminate flowers spicately arranged.

25. *Dioscorea L.* sect. *Brachyandra* Uline

Dioscorea sect. *Brachyandra* Uline in E.-P., Nat. Pflanzenfam., Nachtr. zu II, 5: 83 (1897); Burkill & H. Perr. in Fl. de Madag. 44e Fam.: 35–60 (1950).

Persistent ground organ finally deeply subterraneous, giving rise annually to only one elongate, mostly fusiform tubercule. Plant glabrous or pubescent, strongly viscose in *D. nako* H. Perr. Stem aculeate at base or more commonly unarmed, twining to the left, often erect in *D. hexagona* Baker. Leaves alternate, the blade entire or 3-lobed. Staminate flowers in few-flowered cymules or the latter reduced to a single flower, tepals shortly connate but perianth open, more or less rotate. Fertile stamen 6, anthers introrse. Capsule reflexed, longer than wide. Seeds compressed, proximally winged. Thirteen spp., Madagascar and Comoro Islands (Mohéli).

Four intimately akin sections, all from Madagascar, have been distinguished by Burkill and Perrier.

26. *Dioscorea* L. sect. *Opsophyton* Uline Fig. 69

Dioscorea sect. *Opsophyton* Uline in E.-P., Nat. Pflanzenfam., Nachtr. zu II, 5: 84 (1897); Prain & Burkill in Fl. Gén. Indo-Chine 6: 703, 728, 729 (1934); Burkill, Bull. Jard. Bot. Brux. 15: 346 (1939); Burkill & H. Perr. in Fl. de Madag. 44e Fam.: 24–26 (1950); Burkill in Fl. Males. I, 4: 311 (1951).

Opsophyton Uline becomes sect. *Dioscorea* when the genus is lectotypified with *D. bulbifera* L., as proposed by M.L. Green (1929).

Persistent ground organ close to soil surface, producing one annually renewed "thick-necked" fleshy tubercule, broadly attached to the primary tuber but tubercules absent and replaced by axillary tubers in varieties of *D. bulbifera* L. Plant unarmed, usually glabrous, twining to the left (but see *D. inopinata* Prain & Burkill below), frequently with warted aerial tubers in the leaf axils. Leaves alternate, blades entire; extrafloral nectaries superficial. Staminate flowers sessile or pedicellate, recemosely arranged, the inflorescence mostly facing earthwards; tepals distinct, long and narrow, floral axis not expanded. Fertile stamens 6, anthers introrse. Capsule longer than wide. Seeds compressed, winged towards the base of the capsule. $2n$ = mostly 40, 60, 80, 100, 120. Five or six spp., Old World tropics, best known *D. bulbifera* L., the tubercules poisonous in the wild plant, edible in cultivated races and introduced to America and Oceania. *D. inopinata* Prain & Burkill, twining to the right, is aberrant in this section.

27. *Dioscorea* L. sect. *Lasiophyton* Uline

Dioscorea sect. *Lasiophyton* Uline in E.-P., Nat. Pflanzenfam., Nachtr. zu II, 5: 84 (1897); Prain & Burkill in Fl. Gén. Indo-Chine 64: 702–703, 717–727 (1934); Burkill, Bull. Jard. Bot. Brux. 15: 347 (1939); Burkill & H. Perr. in Fl. Madag. 44e Fam.: 20–23 (1950); Burkill in Fl. Males. I, 4: 313 (1951); von Teichman et al., Boissiera 24: 219 (1975); Guan & Chamberlain, Edinb. J. Bot. 49: 77–87 (1992).

Alternative generic name: *Botryosicyos* Hochst. (1844).

Persistent ground organ close to soil surface, producing annually several, mostly poisonous tubercules which may be long and slender or short and roundish, sometimes long-stipitate. Plant pubescent or tomentose with simple 2-celled hairs. Stem prickly at the base in most species, twining to the left, with or without axillary tubers. Leaves alternate, the blade digitately compound, extrafloral nectaries superficial. Staminate flowers pedicellate or sessile, the pedicel carrying up the bract with the flower, these racemosely arranged; perianth globose, opening little, tepals nearly distinct or the fused portion saucer-shaped. Fertile stamens 6 or 3 and 3 staminodes, filaments very short, anthers introrse. Capsule longer than wide. Seeds compressed, winged towards the base of the capsule. $2n = 20, 40, 60, 80$ [except *D. dumetorum* (Kunth) Pax with $2n = 36, 40, 54$]. More than 20 spp., tropical Asia extending to S China and the Philippines, few in Africa. Strains of *D. pentaphylla* L. are grown in Asia and Pacific Islands. Starch grains have been studied by Seidemann (1966) in two species and are multiply compound. Uline (1897) distinguished among others the closely related sections *Lasiophyton* and *Botryosicyos*, perhaps rightly, the latter defined by the staminate flowers arranged in dense spikes, 3 stamens and – uniquely within the family – the epidermal cells of the seed wings strongly elongate and faintly curved.

28. *Dioscorea* L. sect. *Illigerastrum* Prain & Burkill

Dioscorea sect. *Illigerastrum* Prain & Burkill, Kew Bull. 1933: 241 (1933), and in Fl. Gén. Indochine 6: 701, 714–715 (1934).

Ground organ apparently rhizomatous (Chin et al. 1985; Ting et al. 1985). Plant probably unarmed, glabrous, twining to the left, axillary tubers not known. Leaves alternate, at least partly trifoliolate. Staminate flowers pedicellate, in racemosely arranged cymules towards the base of the inflorescence; perianth rotate, circular in outline. Fertile stamens 6, the inner whorl shorter than the outer. Capsule longer than wide. Seeds compressed, winged towards the base of the capsule. $2n = 20$. Two or three spp., S China and N Vietnam.

Chin et al. (1985) found pollen and chromosome number of sect. *Illigerastrum* to agree with no. 6, sect. *Stenophora* and no. 18, sect. *Macropoda*, which they do not separate, and suggest to cancel *Illigerastrum*.

29. *Rajania* L.

Rajania L., Spec. Pl.: 1032 (1753); Knuth in Engler, Pflanzenreich IV. 43: 330 (1924).

Ground organ allegedly a tuber. Plant unarmed, mostly glabrous. Leaves alternate, blade entire. Staminate flowers minute, pedicellate, rarely sessile, in racemosely arranged cymules; perianth cleft half way down or nearly to the base. Fertile stamens 6, anthers introrse. Fruit a unilocular 1-seeded samara, compressed, unilaterally winged, the wing first proximal, finally overtopping the perianth. Seed flattened, scarcely winged. About 25 spp., all Antillean. A poorly known genus, particularly as to ground organs, poisonous qualities and direction of twining.

Selected Bibliography

Adanson, M. 1763. Familles des plantes. 2 vol. Paris: Vincent.

Al-Shebaz, I.A., Schubert, B.G. 1989. The Dioscoreaceae in the Southeastern United States. J. Arnold Arbor. 70: 57–95.

Archibald, E.E.A. 1967. The genus *Dioscorea* in the Cape Province west of East London. J. S. Afr. Bot. 33: 1–46.

Ayensu, E.S. 1972. Dioscoreales. In: Metcalfe, C.R. (ed.) Anatomy of the monocotyledons, vol. 6. Oxford: Clarendon Press.

Barroso, G.M., Sucre, D., Guimarães, E.F., Carvalho, L.F. de, et al. 1974. Flora de Guanabara. Familia Dioscoreaceae. Sellowia 25: 9–256.

Barthlott, W., Frölich, D. 1983. Mikromorphologie und Orientierungsmuster epicuticularer Wachs-Kristalloide: ein neues systematisches Merkmal bei Monokotylen. Plant Syst. Evol. 142: 171–185.

Bary, A. de 1877. Vergleichende Anatomie der Vegetationsorgane der Phanerogamen und Farne. Leipzig: W. Engelmann.

Beccari, O. 1870. Nota sull embryone della Dioscoreacee. Nuov. G. Bot. Ital. 2: 149–155, pl. 4.

Behnke, H.-D. 1965. Über das Phloem der Dioscoreaceen unter besonderer Berücksichtigung ihrer Phloembecken. I. Lichtoptische Untersuchungen zur Struktur der Phloembecken und ihrer Einordnung in das Sproßleitsystem. Z. Pflanzenphysiol. 53: 97–125. II. Elektronenoptische Untersuchungen zur Feinstruktur des Phloembeckens. l.c.: 214–244.

Behnke, H.-D. 1981. Sieve-element characters. Nord. J. Bot. 1: 381–400.

Bouman, F. 1995. Seed structure and systematics in Dioscoreales. In: Rudall, P.J., Cribb, P.J., Cutler, D.F., Humphries, C.J. (eds.) Monocotyledons. Systematics and evolution vol. 1. London: Royal Botanic Gardens, Kew, pp. 139–156.

Brenner, W. 1914. Dioscoreaceae. In: Kirchner, O., Loew, E., Schroeter, C. (eds.) Lebensgeschichte der Blütenpflanzen Mitteleuropas, Bd 13. Stuttgart: Enke.

Britton, N.L., Brown, A. 1913. An illustrated flora of the northern United States, Canada and the British possessions, 2nd edn., vol. 1. New York: Charles Scribner's Sons.

Brown, R. 1810. Prodromus florae novae Hollandiae. vol. 1. London: J. Johnson & Co.

Bucherer, E. 1889. Beiträge zur Morphologie und Anatomie der Dioscoreaceen. Bibl. Bot. Bd. 3, Heft 16. 34 pp.

Burkill, I.H. 1937. The life cycle of *Tamus communis* L. J. Bot. (Lond.) 75: 1–12, 33–34, 65–74.

Burkill, I.H. 1939. Notes on the genus *Dioscorea* in the Belgian Congo. Bull. Jard. Bot. Brux. 15: 345–392.

Burkill, I.H. 1946. Flies of the family Empidiae and other insect-visitors to the flowers of *Tamus communis*. Proc. Linn. Soc. Lond. 157: 99–102.

Burkill, I.H. 1947. A plea for a description from life of the African *Dioscorea minutiflora* Engl. Proc. Linn. Soc. Lond. 159: 77–81.

Burkill, I.H. 1951. Dioscoreaceae. In: van Steenis, C.G.G.J. (ed.) Fl. Malesiana I, 4: 293–335. Djakarta: Noordhoff.

Burkill, I.H. 1960. The organography and the evolution of Dioscoreaceae, the family of the yams. J. Linn. Soc. Bot. 56: 319–412.

Burkill, I.H., Perrier de la Bâthie, H. 1950. Dioscoréacées. In: Humbert, H. (ed.) Flore de Madagascar 44e Famille.

Castillon, L. 1927. Las dioscoreáceas argentinas. Bol. Mus. Hist. Nat. Univ. Nac. Tucumán 11: 1–41.

Chin, H.-C., Chang, M-C., Ling, P.-P., Ting, C.-T., Dou, F.-P. 1985. A cytotaxonomic study on Chinese *Dioscorea* L. Acta Phytotaxon. Sin. 23: 11–18.

Coursey, D.G. 1976. The origins and domestication of yams in Africa. In: Harlan, J.R., De Wit, J.M.J., Stemler, A.B.L. (eds.) Origins of African plant domestication. The Hague, Paris: Mouton Publ., pp. 383–408.

Czaja, A.T. 1978. Stärke und Stärkespeicherung bei Gefäßpflanzen. Stuttgart: G. Fischer.

Dahlgren, R. 1975. A system of classification of the angiosperms to be used to demonstrate the distribution of characters. Bot. Not. 128: 119–147.

Dahlgren, R. 1980. A revised system of classification of the angiosperms. J. Linn. Soc. Bot. 80: 91–124.

Dahlgren, R. et al. 1985. See general references.

Daumann, E. 1970. See general references.

Delpino, F. 1888. Funzione mirmecofila, Estratto della Ser. IV. T. VII delle Memorie della Reale Accad. delle Scienze dell'Ist. Bologna 1886: 46 & sequ.

De Wildeman, E. 1904. Sur l'acarophytisme chez les monocotylédones. C. R. Acad. Paris 139 (2): 551–552.

Duchartre, P.E. 1854. Observations sur l'Igname batate. Bull. Soc. Bot. Fr. 1: 200.

Gentner, G. 1905. Über die Vorläuferspitze der Monokotylen. Flora 95: 327–383, pl. 9.

Herendeen, P.S., Crane, P.R. 1995. The fossil history of Monocotyledons. In: Rudall, P.J., Cribb, P.J., Cutler, D.F., Humphries, C.J. (eds.) Monocotyledons. Systematics and evolution, vol. 1. London: Royal Botanic Gardens, Kew, pp. 1–21.

Huber, H. 1969, 1991. See general references.

Hutchinson, J. 1934. The families of flowering plants. Vol. 2. London: Macmillan.

Karsten, H. (G.K.W.H.) 1847. Vegetationsorgane der Palmen, note 2. Berlin: F. Schneider, p. 158.

Jussieu, A.-L. de 1789. Genera plantarum. Paris: Hérissant et Barrois.
Knuth, R. 1924. Dioscoreaceae. In: Engler, A. (ed.) Das Pflanzenreich, IV.43. Leipzig: W. Engelmann.
Koch, W., Bruhn, Ch. 1962. Über die Morphologie der Speicherorgane einiger mittelamerikanischer Dioscoreaceen. Flora 152: 670–678.
Kunth, C.S. 1850. Dioscoreineae. Enumeratio Plantarum 5. Stuttgart J.G. Colla, pp. 323–456.
Lawani, S.M., Odubanjo, M.O. 1976. A bibliography of yams and the genus *Dioscorea*. Ibadan.
Lawton, J.R.S. 1967. A key of the *Dioscorea* species of Nigeria. J. W. Afr. Sci. Assoc. 12: 2–9.
Lawton, J.R.S., Lawton, J.R. 1967. The morphology of the dormant embryo and young seedling of five species of *Dioscorea* from Nigeria. Proc. Linn. Soc. Lond. 178: 153–159.
Ling, P.-P., Wu, Z.-J., Chin, H.-C. 1982. Some observations on the epidermis of Chinese *Dioscorea* and its taxonomical significance. Bull. Nanjing Bot. Gard. 1982: 9–16, pl. 1–5.
Matuda, E. 1953. Las dioscoreas de México. Anal. Inst. Biol. Méx. 24: 279–390.
McVaugh, R. 1989. Dioscoteaceae. In: Anderson, W.R. (ed.) Flora Novo-Galiciana 15: 355–388. Ann Arbor: University of Michigan Herbarium.
Milne-Redhead, E. 1975. Dioscoreaceae. In: Fl. of Tropical East Africa. London: Crown Agents.
Mohl, H. von 1845. Der Mittelstock von *Tamus Elephantipes*. In: Vermischte Schriften. Tübingen: L.F. Fries, pp. 186–194.
Orr, M.Y. 1926. On the secretory organs of Dioscoreaceae. Notes Bot. Gard. Edinb. 15: 133–146. pl. 215.
Pedrali, G. 1992. A familia Dioscoreaceae (R. Br.) Lindley no Rio Grande do Sul, Brasil. Caderno Pesquisas Ser. Bot. (Sta. Cruz do Sul) 4: 13–77.
Pei Chien, Ting, C.-T., Chin, H.-C., Su, P., Tang, S.-Y., Chang, H.-C. 1979. A preliminary systematic study of *Dioscorea* L. sect. *Stenophora* Uline. Acta Phytotaxon. Sin. 17, 3: 61–72.
Prain, D., Burkill, I.H. 1919. "*Dioscorea sativa*". Bull. Misc. Inf. Kew 1919: 339–375.
Prain, D., Burkill, I.H. 1934. Dioscoreaceae. In: Lecomte, H., Humbert, H., Gagnepain, P. (eds.) Flore Générale d'Indo-Chine 6: 698–745.
Prain, D., Burkill, I.H. 1936–1939. An account of the genus Dioscorea in the east. Part I. The species twining to the left. Part II. Species twining to the right. Ann. R. Bot. Gard. Calcutta 14, Pt. 1 & 2. Bengal Governmt. Press: Alipore. 528 pp., 150 pl.
Purnima, ••, Srivasava, A.K. 1985. Leaf epidermal studies in *Dioscorea* Linn. Acta Bot. Indica 13: 171–180.
Queva, C. 1894. Recherches sur l'anatomie de l'appareil végetatif des taccacées et des dioscorées. Thèse. Lille.
Ramírez Rodrígues, R., Téllez Valdéz, O. 1992. Las dioscóreas (Dioscoreaceae) del Estado de Morelos, México. Anal. Inst. Biol. Méx. 63: 67–99.
Reiche, K. 1908. Zur Kenntnis der Dioscoreaceen-Gattung *Epipetrum* Phil. Bot. Jahrb. Syst. 42: 178–190.
Rübsamen, T. 1986. Morphologische, embryologische und systematische Untersuchungen an Burmanniaceae und Corsiaceae (Mit Ausblick auf die Orchidaceae-Apostasioideae). Diss. Bot. 92.
Santapau, H. 1951. The genus *Dioscorea* in Bombay State. J. Bombay Nat. Hist. Soc. 49: 624–636.
Seidemann, J. 1966. Stärkeatlas. Hamburg: P. Parey.
Sharma, O.P. 1976. Tuber versus bulbil of *Dioscorea glabra* – origin, development, sprouting and morphology. Phytomorphology 26: 87–96.
Sharma, O.P. 1978. Ontogeny and morphology of bulbil and its seedling tuber in *Dioscorea melanophyma*. Phytomorphology 28: 192–197.
Sharma, O.P. 1980. Ontogeny and morphology of the tuber of *Dioscorea floribunda* Mart. et Gal. Proc. Indian Acad. Sci. (Plant Sci.) 89: 47–55.
Smith, P.M. 1916. The development of the embryo and seedling of *Dioscorea villosa*. Bull. Torrey Bot. Club 43: 545–558.
Solms-Laubach, H. Graf zu 1881. Über monocotyle Embryonen mit scheitelbürtigem Vegetationspunkt. Bot. Zeitung 36: 66–76, 81–93, pl. 4.
Sparshott, E.N. 1935. Observations on the formation, development, and structure of the tuber of *Testudinaria elephantipes*, and on the origin of the vegetative shoot. J. Linn. Soc. Bot. 49: 593–610, pl. 32.
Su, P. 1987. Pollen morphology of *Dioscorea* in China. Acta Phytotaxon. Sin. 25: 357–365, pls. 1–4 (in Chinese).
Takhtajan, A.L. 1982. See general references.
Teichman, I. von, van der Schijff, H.P., Robbertse, P.J. 1975. The genus *Dioscorea* in South Africa. Boissiera 24: 215–224.
Telford, I.R.H. 1986. Dioscoreaceae. In: George, A.S. (ed.) Fl. Australia 46: 196–202. Canberra: Australian Governm. Publish. Service.
Tillich, H.-J. 1980. Vergleichend-morphologische Untersuchungen an Keimpflanzen von Liliopsida (Monocotyledoneae). Thesis (Diss. Prom. B). Mühlhausen: Pädagogische Hochschule.
Ting, C.-T., Chang, M.-C., Ling, P.-P. 1985. Dioscoreaceae. In: Pei, C., Ting, C.-T. (eds.) Fl. Republ. Pop. Sin. 16 (1): 54–120.
Trouslot, M.F., Champagnat, M., Tort, M., Loiseau, M., Eraud, C. 1993. Developmental morphology of seedlings of *Dioscorea cayenensis-D. rotundata* complex. Phytomorphology 43: 49–57.
Uline, E.B. 1897. Dioscoreaceae. In: Engler, A., Prantl, K. (eds.) Nat. Pflanzenfam., Nachtr. zu II, 5: 80–87. Leipzig: W. Engelmann.
Uline, E.B. 1898. Eine Monographie der Dioscoreaceen. Bot. Jahrb. Syst. 25: 126–165.
Watt, G. 1890. A dictionary of the economic products of India, vol. 3. London: Allen, pp. 115–136.
Xifreda, C.C. 1991. Sobre la presencia de una "vaina cristalífera" en el tallo de algunas especies de *Dioscorea* (Dioscoreaceae). Darwiniana 30: 137–142.

Doryanthaceae

H.T. Clifford

Doryanthaceae Dahlgren & Clifford in Dahlgren, Clifford & Yeo, Fam. Monoc.: 175 (1985).

Giant, tufted, rosulate *perennials*, the older plants forming dense clumps with a short vertical subterranean stem. *Leaves* bifacial, basal, spirally arranged, linear to narrow-lanceolate, with massive starch-filled bases forming a bulb; margin of sheath thin, filmy, completely encircling the axis; leaf tip a cylindrical, obtuse, brown prolongation the surface of which is ultimately shed leaving a tuft of threadlike fibres. *Inflorescence* a sparse or densely contracted thyrse attaining a height of 5.5 m and terminating an axis which towards its base bears abundant leaves similar to, but much smaller than, those of the rosette. Large, often bright red bracts subtending the inflorescence branches and flowers. *Flowers* large, epigynous, with short pedicels each of which bears a prophyll. Tepals 6, petaloid, 5-nerved, nearly equal and up to 18 cm in length, fused at the base to form a nectar cup; the free portions spreading, eventually caducous; inner 3 tepals slightly dilated at base. Stamens 3 + 3, epitepalous, with linear or basally dilated erect filaments extending beyond the tepals; anthers to 3 cm long, tetrasporangiate, peltate with the apex of the filament enclosed in a tube formed by the connective, extrorse, bilocular, dehishing longitudinally by slits. Gynoecium syncarpous, 3-carpellate, 3-locular; septal nectaries well developed, opening around the base of the style; style simple, slender, with an obtuse, triangular stigma; ovules numerous, borne on axile placentas in 2 rows per locule. *Fruit* a loculicidal capsule. Seeds numerous and variable in size but usually provided with a lateral wing. Testa yellow to red-brown due to the presence of the breakdown products of phlobaphene (Huber 1969); endosperm composed of thin-walled cells rich in fats and aleurone. Embryo straight, with a broad obtriangular cotyledon.

A single genus with two species in E Australia.

Vegetative Morphology. The rosette leaves are inserted on a 5/13 spiral and may be up to 2.5 m long. Their laminae have entire margins and terminate in a tuft of dry threads which derive from an initial white and pithy structure that has been called a plug tip (Goebel 1905). The function of the plug tip has been much debated (Newman 1928; Troll 1939). Each leaf axil has a vegetative bud and after flowering, those of the lower leaves develop into secondary bulbs thereby forming a cluster about the parent plant. About 10 years elapse between bud initiation and anthesis (Newman 1928).

Vegetative Anatomy. The stomata are paracytic and the anticlinal walls of the epidermal cells are straight (Blunden et al. 1973). Idioblasts occur in unsuberised cells adjacent to the epidermis and contain single, double or triple large calcium oxalate crystals. Suberised cells containing sacs of raphides are lacking (Huber 1969). Scattered tannin cells are present in the mesophyll, which in the keeled regions of the proximal parts of the lamina develop lignified pitted parenchyma between the vascular bundles. These have a conspicuous parenchymatous sheath within which the phloem and xylem is embedded in fibres (Blunden et al. 1973).

Inflorescence Structure. The inflorescence is a panicle either loose or contracted into a spherical head. The flowering stem is clothed with scale leaves and may attain a height of 5.5 m (Fig. 71). The branches of the inflorescence and the individual flowers are subtended by bracts. Each pedicel bears a small bract below the flower.

Floral Structure. The perianth is very large (up to 18 cm in length), and the tepals, which are fused to the ovary and to each other for a short distance at the base, form a nectar cup. At length the free portions of the tepals become caducous. The stamens are fused to the centres of the tepals. The anthers are basifixed and stand erect on long filaments projecting beyond the spreading tepals.

Embryology. Anther wall formation is not known. The endothecium has well-developed spiral thickenings. The tapetum comprises 2 layers, an outer of uninucleate cells and an inner of multinucleate cells. Microsporogenesis is simultaneous and the mature pollen grains are 2-celled.

The ovules are anatropous and bitegmic. The megaspore mother cell developes immediately below the epidermal layer and cuts off a parietal cell, further divisions of which produce a nucellar cap. Embryo-sac formation is of the Polygonum type and early embryogenesis is unknown. En-

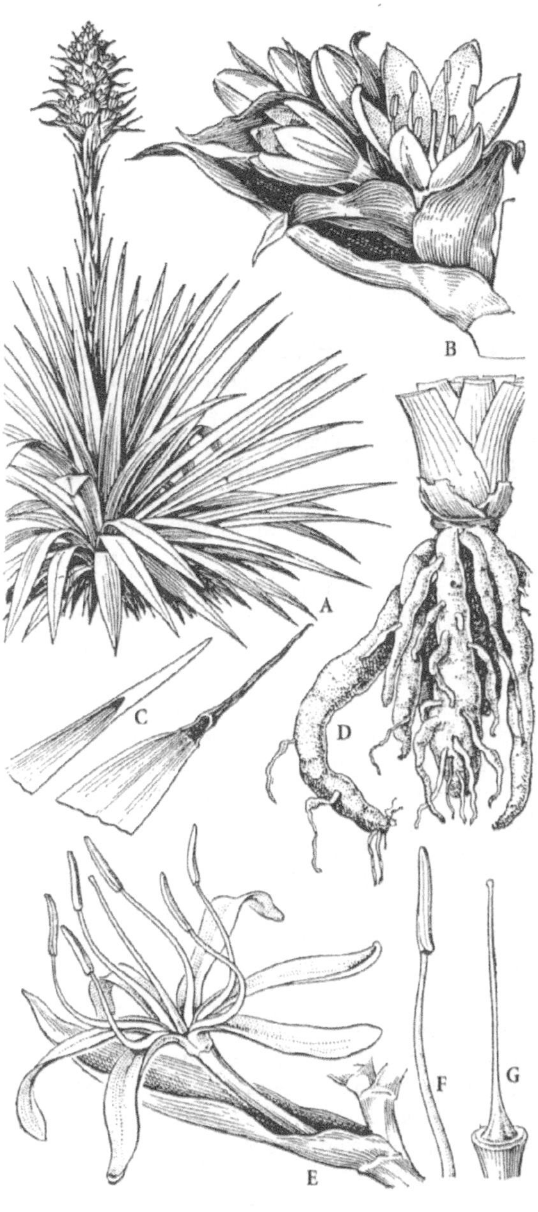

Fig. 71A–G. Doryanthaceae. **A–C** *Doryanthes palmeri*. **A** Habit. **B** Partial inflorescence. **C** Leaf apices with precursory tip. **D–G** *Doryanthes excelsa*. **D** Storage roots. **E** Flower. **F** Stamen. **G** Pistil. (Takhtajan 1982)

dosperm formation is Helobial (Newman 1928, 1929).

Pollen Morphology. The pollen grains are elliptical (45 μm × 20 μm), sulcate with a reticulate exine (Newman 1929).

Karyology. The basic chromosome number has been given as $n = 18$ (Sato 1938), $n = 22$ (Newman 1929) and $n = 24$ (Cave 1955).

Pollination. The large red flowers produce up to 10 ml nectar per day and so atract many birds which act as pollinators. Bees and flies also visit the flowers (Newman 1929).

Fruit and Seed. The fruit is woody and up to 10 cm long and 5 cm broad. The seeds are numerous and are shaken from the capsules as the infructescences sway in the wind. The more distant dispersal of the seeds depends upon the extent to which they are winged.

Phytochemistry. Unknown apart from the presence of steroid saponins in the rhizomes (Hegnauer 1963, 1972).

Affinities. Cave (1955) considered *Doryanthes* the only close relative of *Phormium* but Henderson and Clifford (1984) do not include it in Hemerocallidaceae. That the genus merited family rank was recognised by Huber (1969) and formally proposed by Dahlgren et al. (1985), who placed it in the Asparagales, where it belongs to a group of families characterised by simultaneous microsporogenesis, Helobial endosperm formation, and rhizomatous subterranean parts. Molecular data place it in close proximity to the Iridaceae (Chase et al. 1995).

Distribution and Habitats. *Doryanthes* is restricted to eastern Australia between Jervis Bay (35 °S) and the Burdekin River (20 °S), where it grows in rocky gullies beneath eucalypts in lowland localities or exposed amongst rocks on mountain tops. It rarely grows further than 100 km from the coast.

Economic Importance. *Doryanthes excelsa* is widely cultivated as an ornamental.

Only one genus:

Doryanthes Corr. Serr. Fig. 71

Doryanthes Corr. Serr., Trans. Linn. Soc. Lond. 6: 211, t. 23–24 (1802).

Description as for the family. Two spp.

Selected Bibliography

Blunden, G., Yi Yi, Jevers, K. 1973. The comparative leaf anatomy of *Agave*, *Beschorneria*, *Doryanthes*, and *Furcraea* species (Agavaceae: Agaveae). Bot. J. Linn. Soc. 66: 157–179, 9 figures.
Cave, M.S. 1955. Sporogenesis and the female gametophyte of *Phormium tenax*. Phytomorphology 5: 247–253.
Chase, M.W. et al. 1995. See general references.
Dahlgren, R.M.T. et al. 1985. See general references.
Goebel, K. 1905. Organography of plants. Pt II. Engl. Transl. (I.B. Balfour). Oxford: Clarendon Press.
Haberlandt, G. 1914. Physiological plant anatomy. Engl. Transl. (Drummond). London: Macmillan.
Hegnauer, R. 1963, 1972. See general references.
Henderson, R.F.J., Clifford, H.T. 1984. A recircumscription of the Phormiaceae Agardh. Taxon 33: 423–427.
Huber, H. 1969. See general references.
Newman, I.V. 1928. The life history of *Doryanthes excelsa*. Part 1. Some ecological and vegetative features and spore production. Proc. Linn. Soc. N.S.W. 53: 499–558, pl. 32–35.
Newman, I.V. 1929. The life history of *Doryanthes excelsa*. Part II. The gametophytes, seed production, chromosome number and general conclusion. Proc. Linn. Soc. N.S.W. 54: 411–435, pl. 17–29.
Satô, D. 1938. Karyotype alteration and phylogeny IV. Karyotypes in Amaryllidaceae with special reference to SAT-chromosomes. Cytologia 9: 203–242.
Takhtajan, A.L. 1982. See general references.
Troll, W. 1939. Vergleichende Morphologie der höheren Pflanzen. 1. Band, 2. Teil. Berlin: Borntraeger.

Dracaenaceae

J.J. Bos

Dracaenaceae Salisb., Gen. Pl.: 73 (1866), nom. cons.

Trees, shrubs, sometimes scandent, unbranched suffrutices or rhizomatous geophytes, sometimes succulent, glabrous, from less than 10 cm to over 40 m tall. Aerial stems often showing characteristic persistent leaf scars. *Leaves* alternate, distichously or spirally arranged, entire, ovate, strap-shaped or ensiform, usually amplexicaul with a sheathing base, occasionally exceeding 2 m in length, usually much shorter, sometimes variegated, the upper ones gradually to rather abruptly transformed into the inflorescence bracts. *Inflorescence* terminal, sometimes apparently axillary on leafless dwarf shoots, racemose, capitulate to paniculate; floral bracts and bracteoles often obsolescent. *Flowers* often grouped together on articulated pedicels with a longer or shorter pericladium, usually white, sometimes flushed with green or purple, fragrant, nocturnal. Perianth with a shorter or longer tube and 6 equal segments. Stamens 6, as long as the free perianth lobes and inserted at their base in the throat; filaments often inflated and spindle-shaped; anthers versatile, opening latrorsely. Ovary superior, cylindrical to bottle-shaped, with 3 locules, each containing a single ovule; septal nectaries present; the style equaling the stamens; stigma capitate to shallowly 3-lobed. *Fruit* a thin-walled berry, (depressed-) globose or ellipsoid, sometimes 2- or 3-lobed, usually bright orange when ripe, containing up to 3 large seeds lacking a testa but enclosed in an obsolescent layer of sclerotic endocarp cells.

A family of one genus of some 100 species native to the warmer regions of the Old World, odd species in Hawaii, Central America, Cuba and Macaronesia. Widely introduced and established in tropical climates elsewhere.

Vegetative Structures. Many species have bright orange-coloured roots, which occasionally may form tubers (Bos 1980). Arborescent species are capable of girth increment by secondary growth (Cheadle 1937; Zimmermann and Tomlinson 1970; Dahlgren et al. 1985). Aerial roots, when present, are invariably produced at the base of branches (Bos 1984). The leaves are frequently contracted in rosettes either at the end of aerial branches or from the apex of a subterranean rhizome. Leaf bases and leaf scars may produce a dark red exudate, which is glasslike when dry and known as dragon's blood. The leaves are strictly parallel-veined, the midrib being a concentration of primary veins. In ovate leaves the primary veins diverge from the leaf base to converge at the apex, there are never veins that diverge from the midrib. In some species the lamina may more or less abruptly narrow into a pseudopetiole in their lower half or towards the very base. There are no hairs, but few species may be partly scabrid, e.g. *Dracaena ombet* on leaf edges and *D. scabra* on the inflorescence axes (Bos 1984). Variegation in leaves is generally caused by the absence of chlorophyll in 1 or more layers of palisade parenchyma resulting in pale green and yellow to pure white colours in strands, bands, dots or rings. In very few species the underside of the leaves and particularly the inflorescence bracts may be dark purple-red, due to the presence of anthocyanins. Apparently, variegation is often caused by mutation and occasionally single individuals are sectional chimaeras, showing striking differences in the variegation on different branches (Bos et al. 1993).

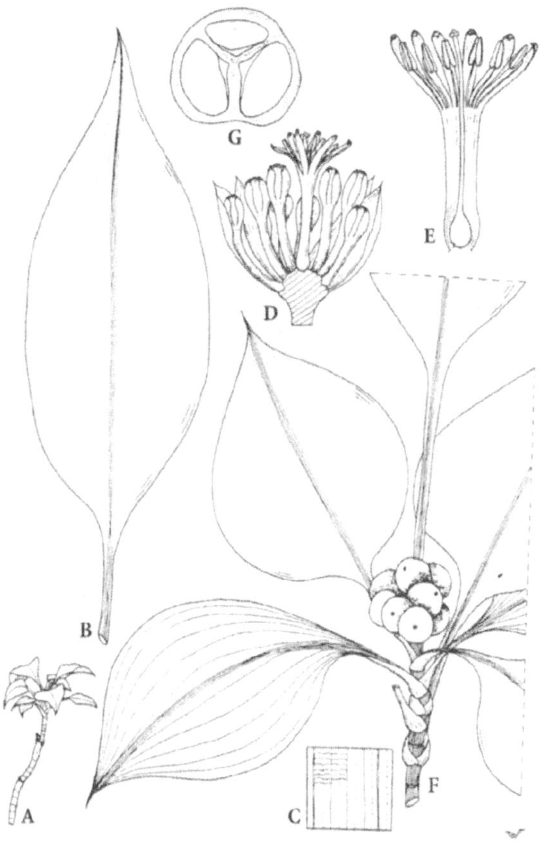

Fig. 72A–G. Dracaenaceae. *Dracaena cristula*. **A** Plant, much reduced. **B** Leaf blade. **C** Leaf venation. **D** Longitudinal section of partial inflorescence. **E** Flower laid open. **F** Infructescence. **G** Fruit, cross-section. (Bos 1984)

Reproductive Structures. The inflorescence is always strictly terminal. In cases where it appears to be axillary, the base of the peduncle will invariably prove to contain several nodes lacking appendages and thus representing a dwarf shoot with a terminal inflorescence (Bos 1984). Bracts tend to rapidly decrease in size from the basal branches in the inflorescence towards individual pedicels. Often the flowers inserted at the apex of (partial) inflorescences have no bracts at all. Bracteoles are very rare, if present, they are minute and accompany the articulation in the pedicel. Apart from the septal nectaries, inconspicuous extrafloral nectaries are usually present at the base of the pedicels on the axis of the inflorescence; they produce copious quantities of nectar. Flowering takes place during the night and under very low light intensities in daytime. After developing flower buds have reached a certain length, their growth is arrested. In the 24 h preceding anthesis, their size is abruptly doubled. At anthesis, the perianth lobes separate and recurve gracefully towards the tube, causing the stamens to assume a direction deviating about 45° from the style.

In each ovary, usually only 1 of the 3 ovules develops into a viable seed, the others being suppressed completely. The seed is covered by a flimsy scarious envelope. According to Huber (1969 and pers. comm.) this is not an obsolete testa but the innermost layer of the endocarp, containing sclerotised cells.

Anatomy, Embryology. In so far as publications on this subject are available, it is not clear whether these studies truly involved *Dracaena*. Since the middle of the 19th century, this genus has repeatedly been confused with quite unrelated genera such as *Agave*, *Yucca* and particularly with *Cordyline*. In the latter genus hundreds of cultivated varieties have received *Dracaena* names, and in living collections a correctly named *Dracaena* was the exception rather than the rule. As research material usually was obtained from such collections, and names in published research were not vouched for, it is not possible to ascertain whether the results pertain to *Dracaena*.

Pollen Morphology. The pollen grains are sulcate (Erdtman 1952).

Karyology. Reliable and vouched for counts as reported by Goldblatt (1975–1985) range between 2n = 38 – 42. According to Arends (pers. comm.), it is very difficult to establish an exact count of the very small chromosomes due to the difficulty of arranging them on one plane.

Pollination. Nocturnal flowering, strong fragrance during the late evening, and copious production of nectar on the inflorescence point to pollination by night-active animals. Support was found when *Dracaena arborea* flowered in an old conservatory with a leaky roof in Wageningen, the Netherlands. At about 22.00 h, when the fragrance of flowering was at its peak, a number of large hawkmoths (Sphingidae) were observed, hovering above the leaky panes over the flowering dragon tree. In *Dracaena* spontaneous self-pollination is quite possible and has successfully been effected artifically on several occasions.

Phytochemistry. Of all publications on this subject only those concerning *Dracaena draco*, and particularly those dealing with the composition of the red substance called dragon's blood, seem to be reliable (Hegnauer 1963, 1986).

Affinities. However variable, *Dracaena* is a homogeneous group of considerable ancestry. Because of the habit of the arborescent species, they have often been placed in relation to other so-called palm lilies, particularly in Agavaceae, as advocated by Hutchinson (1973). This must be dismissed as being due to convergence. Also the persistent insistence on relationship with *Cordyline* is erroneous; cf. e.g. the multiovuled locules and phytomelan-encrusted seeds of *Cordyline* with the single-ovuled locules and lack of phytomelan in *Dracaena*. If true affinities are sought, particularly Convallariaceae should be considered. In respect to its present classification within the Asparagales, it is interesting to note that Loefling advised Linnaeus on the similarity of the flowers of *Dracaena draco* to those of *Asparagus*, causing his master to name *D. draco* initially *Asparagus draco* (Linnaeus 1762).

Distribution and Habitats. All species are purely tropical except *D. draco*, that can survive in a mediterranean climate. This possibly also holds for *D. aletriformis*, which extends to the SE coast of the Cape Province in S Africa (Bos and Obermeyer 1992). Their natural distribution is worldwide, with the notable exception of S America. Habitats are either semidesert situations or tropical rainforests. Few species survive when their forest habitat degrades, and most forest species are vulnerable because of this. Some species occupy rather specialised habitats; they may be restricted to escarpments or occupy the first line of vegetation of forest on the seashore, and some are rheophytes in the riverbeds of rain rivers with strongly variable water levels.

Economic Importance. Mainly due to leaf variegation, a limited number of species are among the major foliage ornamentals, particularly as indoor plants in the temperate zone. Cultivars have been established for many variegation patterns (Bos et al. 1993) that can be maintained only by clonal propagation and careful elimination of spontaneous chimaeral variants. In Africa, several species of *Dracaena* have a strong reputation as fetish plants, probably gained in connection with their longevity and vitality. It is supposed that this quality led to their early introduction to S America, particularly to Brazil. They must have come with the earliest African immigrants, prior to the advent of the slave trade. Since those days, most introductions into Europe came from S America rather than from Africa. At present *Dracaena* takes second place after *Ficus* as ornamental indoor foliage plant in Europe.

Only one genus:

Dracaena L. Fig. 72

Dracaena L., Syst. nat. ed. 12. 2: 246; Mant. pl. 1: 63 (1767).
Pleomele Salisb., Prodr. Stirp. Chap. Allerton: 245 (1796).
Sansevieria Thunb., Prodr. Fl. Cap.: 65 (1794), nom. cons., syn. nov.

Description as for the family.

The distinction between the genera *Dracaena* and *Sansevieria* is purely traditional and not tenable. The succulence of the leaves of many species of *Sansevieria* is equaled and even surpassed by *D. ombet*. Some of the traditional *Dracaena* spp. such as *D. braunii* have the same architecture as *Sansevieria* including the rhizomes. Inflorescences, flowers or fruits do not provide any argument to segregate both genera (Bos 1984). There are several revisions of *Sansevieria* in Africa under way and they can be expected to result in serious reductions of the number of accepted species. In order to avoid making unnecessary new combinations in *Dracaena*, it would be preferable for the necessary combinations to be made by the present revisors.

Selected Bibliography

Baker, J.G. 1875. Revision of the species and genera of the Asparagaceae. J. Linn. Soc. Bot. 14: 508–632.

Bos, J.J. 1980. *Dracaena surculosa* Lindl. Misc. Pap. Landbouwhogesch. Wageningen 19: 71–79.

Bos, J.J. 1984. *Dracaena* in West Africa. Agric. Univ. Wag. Pap. 84-1: 1–125.

Bos, J.J., Cullen, J. 1986. *Dracaena*. In: Walters, S.M., Brady, A., Brickell, C.D., Cullen, J., Green, P.S., Lewis, J., Matthews, V.A., Webb, D.A., Yeo, P.F., Alexander, J.C.M. (eds.) European garden flora 1: 285–287. Cambridge: Cambridge Univ. Press.

Bos, J.J., Obermeyer, A.A. 1992. Dracaenaceae. In: Leistner, O.A. (ed.) Flora South. Africa 5, 3. Pretoria: Nat. Bot. Inst. pp. 1–9.

Bos, J.J., Graven P., Hetterscheid, W.L.A., Wege, J.J. van de 1993. Wild and cultivated *Dracaena fragrans*. Edinb. J. Bot. 49: 311–331.

Brown, N.E. 1915. Sansevieria: a monograph of all the known species. Bull. Misc. Inf. R. Bot. Gard. Kew 1915: 185–261.

Gave, M.S. 1955. Sporogenesis in the female gametophyte of Phormium tenax. Phytomorphology 5: 247–253.

Cheadle, V.I. 1937. Secondary growth by means of a thickening ring in certain monocotyledons. Bot. Gaz. 98: 535–555.

Dahlgren, R.M.T. et al. 1985. See general references.

Erdtman, G. 1952. See general references.

Goldblatt, P. (ed.) 1981. Index to plant chromosome numbers 1975–1978. Monogr. Syst. Bot. Mo. Bot. Gard. 5: 309–310.

Goldblatt, P. (ed.) 1985. Index to plant chromosome numbers 1982–1983. Monogr. Syst. Bot. Mo. Bot. Gard. 13: 121.

Goldblatt, P. (ed.) 1988. Index to plant chromosome numbers 1984–1985. Monogr. Syst. Bot. Mo. Bot. Gard. 23: 137.

Hegnauer, R. 1963, 1986. See general references.

Huber, H. 1969. See general references.

Hutchinson, J. 1973. The families of flowering plants. Oxford: Clarendon Press.

Linnaeus, C. 1762. Species Plantarum, 2nd edn. Holmiae (Stockholm): Impensis Direct. Laurentii Salvii.

Zimmermann, M.H., Tomlinson, P.B. 1970. The vascular system in the axis of *Dracaena fragrans* (Agavaceae). 1. Distribution and development of secondary vascular tissue. J. Arnold Arbor. 51: 478–491.

Eriospermaceae

P.L. Perry and P.J. Rudall

Eriospermaceae Endl., Ench. Bot.: 90 (1841).

Geophytic, tuberous perennial *herbs* with leaves hysteranthous or synanthous. Tuber hypocotyledonary in origin, single or multiple, sometimes with slender, rhizomelike outgrowths, flesh white, yellow, pink to maroon-red, with the growing point single or multiple, apical, basal or lateral. *Leaves* 1-several, with petiolelike leaf sheath, with age forming a fibrous or membranous neck; blade linear to ovate or orbicular, erect or prostrate, glabrous or pubescent, producing complex enations from the adaxial surface in some species. Flowering stem a scape, erect, with a sheathing cataphyll either subterranean or aerial and green. *Inflorescence* a simple raceme, straight or flexuose, few- to many-flowered. *Flowers* hermaphrodite, few to numerous, sometimes scented; bracts minute, membranous; pedicels accrescent, very short to long, erect-patent. Perianth white, pink or yellow, persistent; tepals 6, in 2 whorls, connate shortly at the base, rotate, campanulate or cyathiform, equal or dimorphic with outer spreading and inner erect. Stamens 6, adnate to base of tepals; filaments filiform to broadly lanceolate; anthers bithecate and tetrasporangiate, dorsifixed, versatile, subglobose to oval, introrse. Ovary superior, sessile, 3-locular, with basal septal nectaries; ovules axile, 3–6 per locule; style terete; stigma apical, small. *Capsule* turbinate or ovoid, emarginate, loculicidally 3-valved, valves persistent. Seeds 6–12 per capsule, oval to comma-shaped; testa fawn, densely covered with white, unicellular hairs up to 8 mm long, turning brown with age; embryo large; perisperm replacing an endosperm in the mature seed, at least in some species.

Only one genus, *Eriospermum*, with ca. 100 species confined to sub-Saharan Africa.

Vegetative Morphology. The plants are deciduous geophytes ranging in height above ground from less than 0.1 m to about 0.4 m. The underground storage and perennating organ is a hypercotyledonary tuber (Fig. 73A). A fully grown tuber may vary in size from that of a pea in *E. spirale* to 10 cm and more in length, as in *E.*

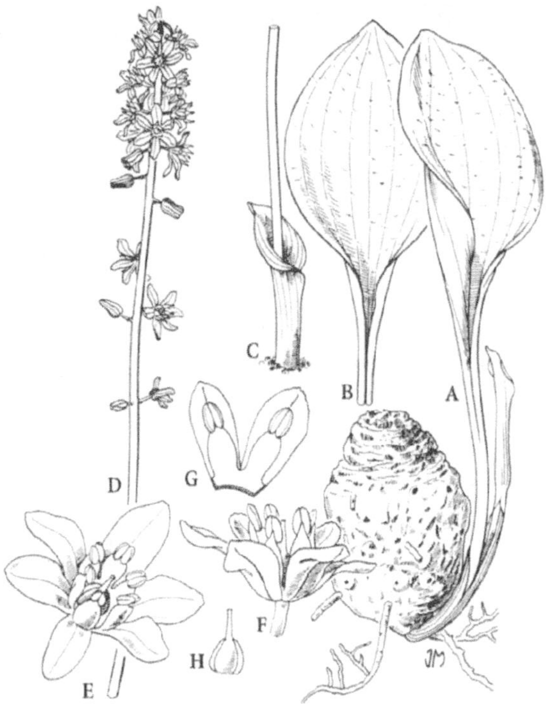

Fig. 73A–H. Eriospermaceae. *Eriospermum kiboense*. **A** Tuber and leaf. **B** Leaf, face view. **C** Peduncle with bract. **D** Inflorescence. **E, F** Flower. **G** Outer tepal (*left*) and inner tepal (*right*) with stamens. **H** Pistil. (Perry 1994)

capense and *E. porphyrium*. The shape of the tuber also shows considerable variation, although it is possible to distinguish 2 basic types: globose to subglobose with an apical growing point; and pyriform to irregularly shaped with a lateral to basal growing point. The skin of the tuber may be very thin and light-coloured or become rough and darkish. The inner flesh may be a soft translucent white, a tough bright yellow, or contain varying amounts of anthocyanin pigments giving a pink to maroon-red appearance. Many species have forms which proliferate by means of slender underground rhizomes up to about 10 cm long.

Most species, especially those from the winter rainfall area of southern Africa, produce a single leaf annually from each growing point. In these species a single inflorescence is produced after the leaves are dry. Some tropical and subtropical species produce 2 or more leaves per growing point with up to 5 inflorescences at the same time.

A petiolelike structure connects the leaf blade and tuber. This persists after the leaf has died at the end of the season and forms a fibrous or membranous sheath through which the new leaf and inflorescence emerge and are thus protected in their passage through the soil. The contemporary leaf sheath may remain entirely below ground in species with a prostrate blade, or it may extend up to 10 cm above ground, holding the blade in an erect or suberect position. The blade may be linear, ovate or orbicular to cordate, glabrous or pubescent with simple or compound hairs (Baijnath and Perry 1980).

In about ten species distinctive projections or enations are produced from the upper surface of the leaf blade. The blade is then usually considerably reduced in size. The enations appear similar to the blade regarding colour, texture and pubescence. In two species, *E. paradoxum* and *E. dregei*, the appendage consists of a single rachis of variable length with numerous featherlike branches. In the other species there is a variable number of separate branched or unbranched appendages.

Inflorescence Structure. The penduncle is a radical and leafless scape, the base of which is surrounded by a tubular sheath. The sheath may be greatly reduced and not visible above ground or considerably extended, forming a small blade sometimes almost indistinguishable from the normal foliage leaves (Fig. 73C,D).

The inflorescence is a simple raceme, either cylindrical, corymbose or occasionally secund, with pedicels up to 10 cm long. In a few species the pedicels are barely 1 mm, the inflorescence being then subspicate. The flowers number from 3 to over 100 in different species. The bracts are small, membranous, and $<$2 mm in length, or are rarely vestigial to obsolete. The pedicels are accrescent, arising singly from the axils of the bracts.

Floral Structure. The perianth is persistent with 6 tepals, connate at the base for approximately 1 mm. The 6 stamens are adnate to the base of the tepals. The dorsifixed, introrse anthers vary little throughout the family. The trilocular ovary with axile placentation is similar to that of many related families. The style is terete, about equal in length to the ovary with a small apical, papillose stigma. Septal nectaries are present in the lower part of the ovary, with openings at the ovary base.

There are several distinct flower types. In species with rotate flowers the tepals are equal to subequal, 5–6 mm long, white or greenish and spreading at anthesis. The stamens form an erect ring around the ovary and have broad, oblong, apiculate, usually white filaments. In *E. bifidum* the filaments are orange and bifid.

The campanulate or recurved flowers found in several species have equal to subequal, ligulate tepals, the upper half often becoming recurved in strong light. In such flowers the filaments are equal, filiform, erect-patent and the anthers small. The tepals may be bright yellow with filaments also yellow or white, rarely pink with white filaments. Cyathiform or cup-shaped flowers are found in a small number of species with broad tepals and white or greenish and broadly lanceolate filaments.

The flowers most typical of the genus have distinctly dimorphic tepals, the outer spreading and the inner erect, surrounding the stamens and gynoecium, with a pinched-in fimbriate apex. In these flowers the filaments vary in size, colour, shape and extent to which they adnate to the tepals.

EMBRYOLOGY. The following account is based on Lu's (1985) study of three species, *E. abyssinicum Baker* (*E. luteo-rubrum* Baker), *E. zeyheri* R.A. Dyer and *E. capense* (L.) Thunb.

The tapetum is glandular, microsporogenesis of the successive type. The tetrads are isobilateral. The ovules are anatropous, bitegmic and weakly crassinucellate with the micropyle formed by the inner integument. The archesporial cell cuts off a parietal cell, but the parietal layer is very restricted and ephemeral. The nucellar epidermis does not divide periclinally. The megaspore tetrad is linear and the chalazal megaspore develops into an embryo sac of the Polygonum type. A distinct hypostase is present. Endosperm formation is probably nuclear and embryogeny conforms to the Solanad type. The embryo enlarges successively to become conical-cylindrical and much longer and more extensive than the oil-rich tissue formed by the nucellar tissue, i.e. a perisperm, which remains around at least the chalazal 2/3 of the embryo.

Obturators are developed from the ovary wall adjacent to the micropyles of the ovules.

POLLEN MORPHOLOGY. Pollen is sulcate (Schulze 1982), with a foveolate or reticulate exine (Simpson 1985).

CHEMISTRY. Apparently unknown.

KARYOLOGY. Chromosome numbers so far known are $2n = 10, 12, 14, 20, 24, 28, 42$ (Jones and Smith 1967; Mogford and Perry 1981; Mogford et al. 1981; Johnson and Brandham 1997; c. Vosa, pers. comm.). The basic number appears to be $x = 7$. The karyotypes are sometimes bimodal. As Tamura (1995) commented, there is a remarkably wide range of chromosome number, size and karyotype recorded for this genus.

POLLINATION. Little is known of pollination in *Eriospermum*. A wide range of insect visitors have been noted, including carrions and other flies, small butterflies and small coleoptera (pers. observ.). While the bright yellow flowers of some species are fairly conspicuous, those of most species are inconspicuous. Some species have a musklike odour and most produce copious nectar.

FRUIT AND SEED. All 3 locules of the capsule may develop evenly or 1 or 2 may be suppressed.

The shape and structure of the seeds (largely from Huber 1969) is as follows. The chalazal end is rounded and ± ovoid, the micropylar end conical or shortly flask-shaped; the cross-section is circular. All epidermal hairs of the testa are produced into long unicellular air-filled trichomes. Their walls are colourless or clothed with a reddish brown cytoplasmic layer. Apart from the epidermis the testa consists of 2 cell layers, which are either strongly flattened and reddish brown (from phlobaphene pigments) or have a distinctive, air-filled lumen and then appear white. The tegmen is colourless and collapsed and generally has lost its cellular structure.

The endosperm is strongly reduced in favour of the embryo and perisperm, which encloses the embryo either completely or only for the chalazal 2/3. Even in the chalazal part the perisperm may be restricted to a few cell layers, the peripheral of which consist of cells that are usually weakly elongated in the radial direction, while the others are subisodiametric. These walls are thin or somewhat thickened and have elliptic pores. The peripheral layers of the perisperm contain aleurone and fatty oils.

The embryo is straight and massive. Starch is lacking.

DISPERSAL AND REPRODUCTIVE BIOLOGY. The hairy seeds are wind-dispersed. The capsules are often produced on long scapes held well above other vegetation. Seeds ripen towards the beginning of the rainy season and appear to lose viability after a few months.

AFFINITIES. Relationships and taxonomic history of *Eriospermum* were discussed by Dahlgren and van Wyk (1988). *Eriospermum* has been associated with various families, such as Tecophilaeaceae and Asphodelaceae, but its affinities have long been obscure, and most authors since Baker (1875) have put it into its own family, Eriospermaceae. As Dahlgren et al. (1985) mentioned, the genus is extraordinary in several aspects, such as the peculiar leaf appendages and hairy seeds with a large embryo and oily perisperm. Huber (1969) suggested an association with *Walleria* and *Cyanastrum* (Tecophilaeaceae) on the basis of the presence of a corm or tuber, and some seed characters (hairy seeds in the former and chalazosperm in the latter). However, as Lu (1985) has pointed out, the seed characters are not homologous, and there are many differences in seed structure, such as the thin seed coat in *Eriospermum*. Simpson (1985) noted that the pollen ultrastructure of *Eriospermum* is very different from that of Tecophilaeaceae. The presence of successive microsporogenesis (Lu 1985) together with an analysis of *rbc*L sequence date (Chase et al. 1995a) indicate placement amongst the higher asparagoids. Dahlgren and van Wyk (1988) tentatively suggested a relationship with Hyacinthaceae on the basis of similar scapose racemose inflorescences. Molecular data from *rbc*L (Chase et al. 1995a) weakly indicate an association with a clade comprising Convallariaceae, Nolinaceae, Ruscaceae and *Comospermum*, but there is currently little morphological evidence to support this. *Comospermum* is a poorly known Japanese genus which also has hairy seeds, although the morphology and development of these seed hairs differs from those of *Eriospermum* (P.J. Rudall, in prep.)

Distribution and Habitats. *Eriospermum* is confined to the African continent, where at least one species occurs as far north as Somalia, and several species are widespread in Central and E Africa and subtropical S Africa. A concentration of species occurs in the southwestern Cape, extending up the west side of Africa, through Namibia into Angola. Many endemics with very limited distributions occur there.

Many species prefer semiarid habitats, where they are frequently found in bare or stony patches between small shrubs. Subtropical species are more often found in grassland.

Some tropical and subtropical species are synanthous but more frequently plants are hysteranthous. In the winter rainfall area of the Cape, flowers are produced in the drier summer months and leafing occurs in the winter. In the remainder of Africa, hysteranthous species produce flowers towards the end of the dry winter and leaves shortly afterwards when rains commence.

Economic Importance. Tubers of some species of *Eriospermum* have been used as food by the early inhabitants of southern Africa and also for various medicinal remedies by different ethnic groups (Watt and Breijer-Brandwijk 1962). A few species have been cultivated ornamentally in the past (Marloth 1929) but mostly for curiosity value.

Only one genus:

Eriospermum Jacq. ex Willd.

Eriospermum Jacq. ex Willd., Sp. Pl. 2, 1: 110 (1799); Perry, Contr. Bolus Herb. 17: 1–320 (1994), monogr.

Characters as for family. Perry (1994) has divided the genus into three subgenera and a number of sections.

Selected Bibliography

Baijnath, H., Perry, P.L. 1980. Preliminary observations of leaf surface structures in *Eriospermum* Jacq. Electron Microsc. Soc. South Afr. Proc. 10: 39–40.

Chase, M.W. et al. 1995a. See general references.

Dahlgren, R.M.T., van Wyk, A.E. 1988. Structures and relationships of families endemic to or centred in Southern Africa. Monogr. Syst. Bot. Mo. Bot. Gard. 25: 1–94.

Dahlgren, R. et al. 1985. See general references.

Huber, H. 1969. See general references.

Johnson, M.A.T., Brandham, P.E. 1997. New chromosome numbers in petaloid monocotyledons and in other miscellaneous angiosperms. Kew Bull. 51: 121–138.

Jones, K., Smith, J.B. 1967. The karyotypes of twenty-five tropical species. Kew Bull. 21: 31–33.

Lu, A.M. 1985. Embryology and possible relationships of *Eriospermum*. Nord. J. Bot. 5: 229–240.

Baker, J.G. 1875. Revision of the genera and species of Asparagaceae. J. Linn. Soc., Bot. 14: 508–632.

Marloth, R. 1929. Eriospermums. S. Afr. Gardening Country Life 19: 249–261, 325–327.

Mogford, D.J., Perry, P.L. 1981. Chromosome studies in the Southern African flora 7–9. J. S. Afr. Bot. 47: 293–296.

Mogford, D.J., Hartley, D.H., Perry, P.L. 1981. Chromosome studies in the Southern African flora 22–24. J. S. Afr. Bot. 47: 751–754.

Perry, P.L. 1994. A revision of the genus *Eriospermum* (Eriospermaceae). Contr. Bolus Herb. 17. 320 pp.

Schulze, W. 1982. Beiträge zur Taxonomie der Liliifloren. IX. Anthericaceae. Wiss. Z. Friedrich-Schiller-Univ. Math.-Nat. R. 31: 291–307.

Simpson, M.G. 1985. Pollen ultrastructure of the Tecophilaeaceae. Grana 24: 77–92.

Tamura, M.N. 1995. See general references.

Watt, J.M., Breijer-Brandwijk, M.G. 1962. Medicinal and poisonous plants of southern and eastern Africa.

Hemerocallidaceae

H.T. Clifford, R.J.F. Henderson, and
J.G. Conran

Hemerocallidaceae R.Br., Prodr.: 295 (1810).
Phormiaceae J. Agardh (1858).

Perennials, caespitose or rhizomatous, terrestrial, herbaceous or shrubby; aerial stems prostrate or to about 2 m tall. *Leaves* distributed along elongated or shortened stems or only in terminal tufts, distichous with sheathing bases which are usually keeled and strongly compressed, occasionally rounded. *Inflorescence* terminal, of a solitary flower or a variously branched panicle; *flowers* actinomorphic or zygomorphic (*Phormium* and *Hemerocallis*), articulated on pedicels; perianth wholly or partially marcescent, not twisting after anthesis (except *Agrostocrinum*, *Pasithea* and *Xeronema*); tepals 6, free or shortly united into a tube, the whorls subequal or the inner slightly longer than the outer; stamens 6, free or slightly connate at the base; if free, hypogynous or inserted on tepal bases, if connate, 3 hypogynous and 3 affixed to the tepals (*Eccremis*); anther filaments ornamented or unornamented; anthers basifixed or peltate, dehiscing latrorsely by pores or slits, erect or recurved, especially after anthesis; ovary syncarpous, superior, half-inferior (*Pasithea*) and either 3- or 1-locular; style simple, terminating in a small capitate stigma or tuft of short hairs; septal nectaries except in *Phormium* and *Hemerocallis*; ovules anatropous, on axile or parietal placentas; ovules 1-numerous per placenta. *Fruit* a berry or capsule; seeds black without an elaiosome.

A family of worldwide distribution throughout the tropics, subtropics and temperate regions with 13 genera and about 50 species.

Vegetative Morphology. The plants are generally long or short rhizomatous or tufted-caespitose, and single or branched aerial stems, climbing in *Geitonoplesium*. In many taxa, the roots are wiry or thick, fleshy or tuberous. The leaves are alternate, and in most genera are distichous or spiro-distichously arranged. The leaves are annual (*Simethis* and *Hemerocallis*) or perennial (most genera), linear or strap-shaped and range from a few mm to 2 m in length, although some *Geitonoplesium* biotypes have broadly lanceolate to ovate leaves, and the leaves of *Xeronema* are unifacial, isobilateral and equitant. In cross-section, the leaf blades are flat or keeled, sometimes with recurved margins. In *Dianella* and *Phormium* the leaves are semiequitant or sword-like, in which the basal and upper parts are bifacial and sheathing, but between these parts there is a short, flattened, isobilateral region. Vernation is mostly conduplicate to flat-conduplicate.

Seedlings in the Hemerocallidaceae have a strongly developed primary root and a short, closed cotyledonary sheath; the short, unifacial cotyledon produces a tubular elongation of the cotyledonary sheath. In *Phormium* and *Rhuacophila* there is some elongation of the cotyledon, and in *Dianella*, a photosynthetic epigeous hyperphyll is developed (Thongpukdee 1989; Tillich 1995). The first leaves are either scalelike (*Phormium*) or linear.

Vegetative Anatomy. The anatomy of the family has been investigated variously by Schulze (1893), Bouvier (1915) and Schlittler (1940). The roots have a very thickened inner cortex and an endodermis with conspicuous U-thickenings. The xylem vessels have simple or scalariform perforation plates. The stomata are anemocytic and occur on either or both leaf surfaces. The vascular bundles form girders and are supported above and below by sclerenchyma which extends to the epidermal layer. A hypodermis is commonly present. Calcium oxalate in the form of cubes, raphides or crystal sand is widespread throughout the family.

Inflorescence Structure. The inflorescence is branched in all genera except *Herpolirion* (and rarely *Hemerocallis*) whose flowers are solitary, and *Xeronema* which bears its flowers in a secund raceme. The inflorescences of most genera are panicles with bostryces (Schlittler 1940, 1945). In *Phormium* the massive inflorescence is a racemose central axis with the lateral axes as secund bostryces (Moore and Edgar 1970). The inflorescences of *Simethis* and *Hemerocallis* are cymose panicles and that of *Pasithea* a racemose panicle (Fig. 74).

Floral Structure. The perianth consists of 6 more or less similar tepals which are 6–20 mm long except for *Phormium*, where they are up to 50 mm in length. The number of veins per tepal is 3–9, except for *Xeronema* (1), *Phormium* (11–15) and *Hemerocallis* (20–25). The tepals are either

Fig. 74A–G. Hemerocallidaceae. **A–E** *Phormium tenax*. **A** Flowering plant. **B** Partial inflorescence. **C** Flower, 3 tepals cut away. **D** Flower bud. **E** Sectioned capsule. **F, G** *Simethis planifolia*. **F** Flowering plant. **G** Flower. (Takhtajan 1982)

free and spreading radially, or are shortly fused at the base and then adpressed forming a zygomorphic tube (*Phormium*), or are fused to as much as 1/2 their length and then expanded to a bell-shaped corolla (*Hemerocallis*). The flowers are mostly articulated near the pedicel apex, and a pericladium is present (*Geitonoplesium*) and the petals wither and persist (base only in *Agrostocrinum*). The anther filaments are free, except in *Eccremis*, where they are shortly united at the base. The filaments are unornamented (*Phormium*, *Hemerocallis*, *Geitonoplesium*, *Xeronema* and *Eccremis*), variously swollen into a conspicuous struma near the middle (*Rhuacophila*) or immediately below the anthers (*Dianella*) (Fig. 75C), papillose-hairy more or less throughout (*Thelionema*), have a towlike tuft of hairs immediately below the anthers (*Stypandra*), or are shortly pubescent or puberulous except at the apex (*Herpolirion, Pasithea, Simethis*). In *Agrostocrinum* there are membranous anther appendages in the bud stage which are absent at anthesis. The gynoecium is sessile or more or less stipitate (*Xeronema*), tricarpellary and 1-3-locular with 2 rows of ovules per placenta. The style is solitary and terminated by a small, capitate, Dry stigma (most genera) or Wet stigma with a tuft of short hairs (*Hemerocallis*). Septal nectaries are present in most genera; there are reports of staminal nectaries in *Dianella* (Daumann 1970).

Embryology. Anther-wall formation is unknown. The endothecium is well developed with fibrous thickenings and the tapetum is glandular with binucleate cells. Microsporogenesis is simultaneous and the pollen grains are 2-celled and lack starch. The ovules are anatropous, bitegmic and tenuinucellate. The inner integument extends beyond the outer which is fused to the funicle. The archesporial cell functions directly as the megaspore. Periclinal divisions beginning in the epidermal layer produce a nucellar cap. Endosperm formation is Helobial and the embryo sac is of the Polygonum type (Schnarf and Wunderlich 1939; Cave 1955; Raju 1957).

Pollen Morphology. The pollen grains are consistently trichotomosulcate (Fig. 76) with the exception of *Hemerocallis* and *Xeronema*, and rounded triangular in polar view with a clavate-baculate exine forming a reticulum on the polar face (Couper 1960; Schulze 1982). In *Xeronema* and *Hemerocallis* the pollen is sulcate with a reticulate exine (Cranwell 1952; Erdtman 1952). Kosenko (1994) recognised five different exine

Fig. 75A–E. Hemerocallidaceae. A–D *Dianella caerulea*. A Flowering shoot. B Flower. C Stamens in front view and lateral view. D Fruit. E *D. nemorosa*, shoot. (Takhtajan 1982)

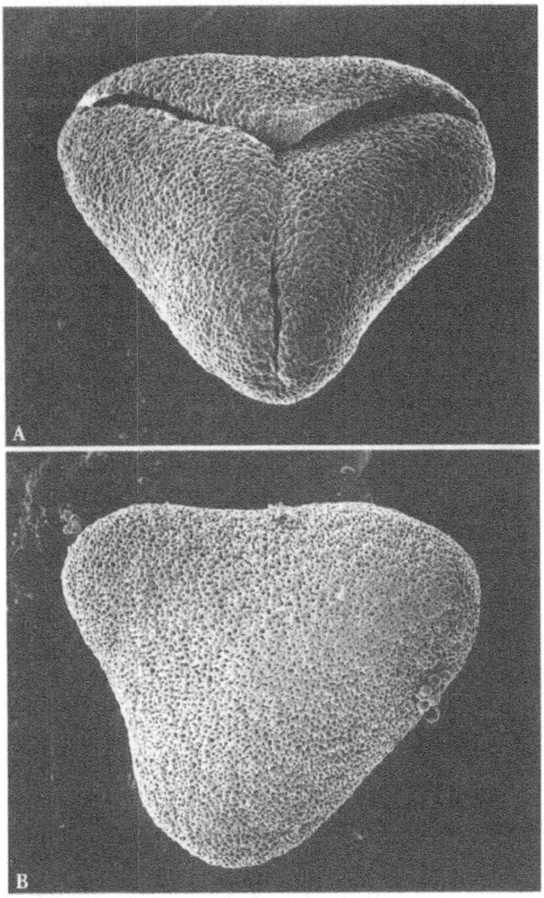

Fig. 76A,B. Hemerocallidaceae. *Stypandra glauca*. A Distal pole, B proximal pole of pollen grain. SEM ×2200. (Photo Palynological Laboratory Stockholm)

types within the family: macroreticulate (*Phormium*), reticulate (*Phormium* and *Xeronema*), microreticulate (*Stypandra, Thelionema, Eccremis, Rhuacophila* and *Dianella*), perforate (*Pasithea* and *Dianella*) and perforate-tuberculate (*Dianella* and *Agrostocrinum*).

Karyology. Somatic chromosome numbers of 16, 18, 28, 32, 48, 76, 80 and 84 have been reported for *Dianella* (Curtis 1952; Fedorov 1969; Henderson 1987); 16, 32 for *Stypandra* and *Thelionema* (Henderson 1987); 8, 16 for *Agrostocrinum* (Keighery 1984); 32 for *Phormium* (Hair and Beuzenberg 1966), 34 and 36 for *Xeronema* (Moore and Edgar 1970) and 22, 33 and 44 for *Hemerocallis* (Fedorov 1969). These numbers give base numbers of 8 or 11 for most of the family, but $n = 17$ and 18 for *Xeronema* (Tamura 1995).

Pollination. In New Zealand *Phormium* flowers are visited by honey-eaters and parakeets which are regarded as active pollinators (Godley 1969), although two species of *Hoplodactylus* gecko lizards are also common nectar feeders and possible pollinators (Eifler 1995). The flowers are protandrous hermaphrodites which exhibit partial dichogamy and herkogamy, and nectar production is greatest during the male phase of the flowers. There is preferential response to xenogamic pollen and allocation of resources to outcrossed zygotes, and many flowers function only as pollen donors (Craig and Stewart 1988, Craig 1989). In Australia, *Dianella* flowers are reported to be visited by megachilid bees (Rayment 1935). However, the nectarless flowers of *Dianella caerulea* var. *assera* in eastern Australia are buzz-pollinated by female bees primarily in the families Anthophoridae (*Exoneura* species) and Halictidae (*Lasioglossum, Nomia* species). In addition, bees less than 6 mm long (e.g. *Homalictus holochorous* and *Trigona* species) removed pollen from anthers, but did not contact the stigmas while foraging. Examination of pollen loads indicated that the bees were polylectic foragers that had visited at least one nectar-secreting taxon before foraging on *D. caerulea* (Bernhardt 1995). In the Philippines, *Rhuacophila* is visited by hoverflies (Syrphidae) and native bees (J.G. Conran unpubl. observ.)

FRUIT AND SEED. The fruit is either a berry or capsule. If a capsule, it is either dry with loculicidal dehiscence (most genera), or fleshy with a loculicidal exocarp and septicidal endocarp in *Eccremis* (Krause 1930). The seeds are ovoid and black with a shiny or a dull testa surface; the seeds in *Phormium* are winged and those in *Xeronema* are more or less spiny or papillose. The hilum is linear or punctiform. The embryo is linear-cylindrical and about $\frac{1}{3}$ or more as long as the seed. The inner integument collapses during development and the seed has a thick phytomelan coat. The endosperm is copious with very little or no hemicellulose or starch.

DISPERSAL AND REPRODUCTIVE BIOLOGY. The blue, succulent berries of *Dianella* species are eaten by several groups of birds (Rose 1973) or collected by male birds of *Ptilonorhynchus violaceus* to decorate their bowers (Gilbert 1939). The seeds may therefore be assumed to be bird dispersed, although they are also used in the nest mound decoration of the large *Myrmecia* ants (Formicidae) in southern Australia (Conole 1993). The seeds of the dry fruited species are shaken from the capsule and have no apparent adaption for dispersal, except for those of *Phormium*, which are winged.

PHYTOCHEMISTRY. Palmitic and lineolic acid are abundant in the seeds (Morice 1962; Hegnauer 1963) and naphthoquinones, phenolic glycosides and eicosanyl arachidate have been reported from the roots (Briggs et al. 1975). Chrysophanol and nepodin were isolated from the roots of *Simethis* (Cadavid et al. 1985), dianellidin, dianellidone, imbricatonol, stypandrone and stypandrol from *Dianella*, and the latter two compounds were found to be amongst the toxic principles in both *Dianella* and *Stypandra* (Colegate et al. 1987). Stypandrol is also identical to the previously incorrectly characterised toxin hemerocallin in *Hemerocallis* (Wang et al. 1989). Saponins are reported from the aerial parts of *Pasithea* (Ricardi 1958).

AFFINITIES. The Hemerocallidaceae are clearly within the Asparagales. They share simultaneous microsporogenesis with the Asphodelaceae, Doryanthaceae, Ixioliriaceae and Tecophilaeaceae (see Dahlgren and Rasmussen 1983) and Helobial endosperm formation with Asphodelaceae and Doryanthaceae. The Hemerocallidaceae, like most Asphodelaceae and Doryanthaceae, are rhizomatous. Accordingly, it is suggested the closest relatives of Hemerocallidaceae are amongst the Asphodelaceae or Doryanthaceae, a viewpoint supported by *rbc*L data (Chase et al. 1993), although their closest sister family appears to be the Johnsoniaceae (Rudall et al. 1996).

A revised circumscription of the Hemerocallidaceae in which five genera were included was proposed by Henderson and Clifford (1984) and a further sixth genus, *Thelionema* was later added (Henderson 1985). Brummitt (1992) further added *Agrostocrinum, Thelionema* and *Xeronema*, although from both *rbc*L and morphological data, *Xeronema* is an isolated lower asparagoid genus with sulcate pollen (Cranwell 1952) and with many unique or unusual features, such as waxy, ensiform leaves and profuse crimson flowers (Rudall et al. 1996). It is included here in the Hemerocallidaceae as a *genus anomalum* pending the results of ongoing investigations, and *Agrostocrinum, Herpolirion, Hemerocallis, Simethis* and *Geitonoplesium* are also included in the family. *Dianella* and *Phormium* are linked by having "semiequitant" (swordlike) leaves. Arber (1925) considered this unusual leaf type to be non-homologous with the isobilateral equitant leaves of the Iridaceae or *Xeronema*, partly because the seedling leaves and young leaves of *Dianella* and *Phormium* are entirely open and sheathing. Cave (1955, 1975) pointed out that ovule and embryological characters such as the lack of a parietal cell link *Dianella, Stypandra* and *Phormium*, and that the chromosome number is based on 8 or 16.

Cranwell (1952) noted that pollen grains of *Herpolirion* are "almost indistinguishable from those of *Dianella*", and pollen morphology generally supports much of the present family circumscription (Kosenko 1994). Although *Hemerocallis* is more generally placed in its own family (e.g. Dahlgren et al. 1985), there is serological support for its inclusion, along with *Geitonoplesium* and *Simethis* into the *Phormium* alliance (Chupov 1987), as well as from the *rbc*L chloroplast DNA data of Chase et al. (1995). In addition to *Xeronema*, the generally included S American phormioid genus *Eccremis* is retained here as a *genus anomalum* for although it fits well on morphological grounds, *rbc*L data place it within the Iridaceae, and futher work is needed to resolve its position (Rudall et al. 1996).

Distribution and Habitats. The family is widely distributed in tropical to temperate latitudes occurring throughout Eurasia, Australia, New Zealand and several Pacific Islands including Hawaii, the Andes (Venezuela-Peru-Chile) and Madagascar. The single record of *Dianella* from Zimbabwe (Wild 1953) is from a locality which suggests the species is endemic there rather than introduced. *Phormium* has become locally naturalised in parts of S Australia and SW England. Members of the family are plants of open forests, heathlands, coastal dunes, headlands and alpine meadows, from sea level to about 3500 m between latitudes 50 °N to 51 °S.

Palaeobotany. Fossil pollen of *Phormium* has been recorded from the Upper Miocene of New Zealand (Couper 1960).

Economic Importance. The leaves of *Phormium tenax* (New Zealand flax) have been widely used as a source of fibre for clothmaking in New Zealand, and more recently the species has been planted elsewhere for this purpose (Henderson and Clifford 1984), as well as a source of honeylike sap (Kunkel 1984). The berries of *Dianella* species have been used for dying kapa, a cloth made by Hawaiians (Uphof 1968) and the leaves of *D. laevis* and *D. revoluta* in Australia were used by native peoples to make string (Cribb and Cribb 1982; Gott and Conran 1991). A rat poison is also extracted from *Dianella* rhizomes (Uphof 1968), although the roots of *D. ensifolia* are used by Australian Aborigines to treat dysuria (Lassak and McCarthy 1990) and the fruits are eaten by children in SE Asia (Tanaka 1976). Several species of *Hemerocallis* are of considerable horticultural significance, as well as being important herbal drug plants in traditional Chinese medicine, with extracts of *Hemerocallis citrina* inhibiting human fibroblast proliferation (He 1994). The young leaves, tuberous roots and flowers of many *Hemerocallis* species are eaten throughout Asia (Kunkel 1984), and the young shoots of *Geitonoplesium* are eaten in Australiasia by indigenous peoples (Cribb and Cribb 1976). In Australia, *Dianella* species and *Stypandra glauca* are responsible for stock losses in sheep and goats following ingestion (Everist 1981).

Key to the Genera

1. Fruit a capsule 2
- Fruit a berry 11
2. Tepals red, orange or yellow 3
- Tepals blue, white-mauve, or partly green 5
3. Tepals narrow, forming a curved tube; flower secund 4
- Tepals spathulate, forming an open funnel; flowers not secund **2. *Hemerocallis***
4. Leaves with the basal and upper parts bifacial and sheathing, but between these parts there is a short, flattened, isobilateral region **1. *Phormium***
- Leaves unifacial, isobilateral and equitant for their whole length **13. *Xeronema***
5. Perianth twisting postanthesis 6
- Perianth not twisting postanthesis 7
6. Stamens equally long; ovary superior **3. *Agrostocrinum***
- Stamens alternately long and short; ovary half-inferior **4. *Pasithea***
7. Flowers terminal, solitary **5. *Herpolirion***
- Flowers in terminal panicles 8
8. Stamen filaments united basally, hairs or papillae absent **12. *Eccremis***
- Stamen filaments free, hairy or papillate 9
9. Stamen filaments papillate **6. *Thelionema***
- Stamen filaments hairy 10
10. Ovules 1–2 per locule; leaves basal; herbaceous geophyte **7. *Simethis***
- Ovules several per locule; leaves cauline; shrubby perennial **8. *Stypandra***
11. Stamen filament with swollen struma 12
- Stamen filament without a struma **9. *Geitonoplesium***
12. Seed with a linear hilum; struma located along the filament, distant from the anther; leaves not semiequitant **10. *Rhuacophila***
- Seed with a punctiform hilum; struma adjacent to the base of the anther; leaves semiequitant **11. *Dianella***

1. ***Phormium*** J.R. Forst. & G. Forst. Fig. 74A–E

Phormium J.R. Forst. & G. Forst., Char. Gen.:48, t.24 (1776); Baker, J. Linn. Soc. Bot. 11: 357 (1870); Moore and Edgar in Flora of New Zealand 2: 51–52 (1970).

Tufted perennial plants with short stout rhizomes; leaves to 3 m long, semiequitant and strongly compressed at the base; inflorescence to 15 m, the peduncle sheathed with a series of deciduous bracts of progressively smaller size each subtending a short lateral bostryx; corolla slightly zygomorphic, tubular and curved, tepals to 5 cm long, shortly fused at the base otherwise connate, petals slightly longer than sepals, reflexed and spreading at the tip, mostly dull red; stamens attached to tepal bases; fruit a capsule, 10–20 cm long. Two, spp., both widespread in New Zealand (N and S Isl., Stewart Isl., Chatham Isl., Auckland Isl.), one extending to Norfolk Isl.

2. *Hemerocallis* L.

Hemerocallis L., Sp. Pl.: 324 (1753); Baker, J. Linn. Soc. 11: 357 (1870); Bailey, Gentes Herb. 2: 143–156 (1930).

Tufted perennial plants with short rhizomes and radical, narrow linear leaf blades; inflorescence a cymose panicle or a solitary flower terminating a leafless scape to 1.5 m tall; corolla slightly zygomorphic, tepals yellow, orange or reddish, up to 16 cm long, spathulate, united into a narrow tube for up to ½ their length, then expanding considerably; stamens inserted centrally on the tepals at top of tube; fruit capsule (rarely formed). About 15 spp., Central Europe to China and Japan.

3. *Agrostocrinum* F. Muell.

Agrostocrinum F. Muell., Fragm. 2: 94 (1860); Henderson, Fl. Australia 45: 230–233 (1987).

Herbaceous perennials, tufted with erect or ascending leafy aerial stems to 1 m tall; inflorescence a terminal panicle; flowers erect; tepals to 2 cm long, blue, outer whorl abaxially glandular-setose, inner whorl glabrous; anther filaments glabrous, tapering distally; anthers not coiled after anthesis; fruit a globular capsule subtended by persistent bases of tepals. Only one sp., *A. scabrum* (R. Br.) Baillon, SW Australia.

4. *Pasithea* D. Don

Pasithea D. Don, Edin. New Philos. J. 8: 236 (1832); Baker, J. Linn. Soc. Bot. 15: 320–321 (1876).

Shortly rhizomatous perennial herbs to 15 cm tall; stem unbranched; leaves narrow-linear, mostly basal, bases sheathing; inflorescence a racemose panicle terminating a leafless scape, bracts short; tepals to 12 mm long, very shortly fused at base and adnate to ovary for ½ its length, spirally twisted after anthesis blue; stamens 6, filaments alternately long and short, hairy at base; stigma distinctly 3-lobed; fruit capsule. Only one sp., *P. coerulea* Ruíz & Pav., Chile.

5. *Herpolirion* Hook. f.

Herpolirion Hook. f., Fl. N.Z. 1: 258 (1853); Moore & Edgar, Fl. New Zealand 2: 51–52 (1970); Hewson, Fl. Australia 45: 242 (1987).

Herbaceous perennials, mat-forming; leaves distichous, with membranous inflated sheaths and linear longitudinally folded blades to 11 cm long; inflorescence a solitary flower, subtended by bractlike leaves; tepals to 20 mm long, blue, white or lilac; anther filaments filiform to narrowly fusiform, puberulous except at the base and the apex; fruit a globular capsule. Only one sp., *H. novae-zealandiae* J.D. Hook, restricted to alpine regions of southeastern Australia including Tasmania, and New Zealand (N and S Isl.).

6. *Thelionema* R.J.F. Hend.

Thelionema R.J.F. Hend., Austrobaileya 2: 109 (1985); Henderson, Fl. Australia 45: 228–230 (1987).

Tufted herbaceous perennials; leaves mostly radical; inflorescence a loose panicle terminating a mostly leafy peduncle to about lm tall; flowers erect; tepals blue, pale yellow-green or white, up to 12 mm long; stamen filaments papillose-barbellate for most of their length; anthers coiled post anthesis; fruit a capsule; seeds black, shiny. Three spp., Tasmania, Victoria, South Australia, New South Wales and southern Queensland.

7. *Simethis* kunth Fig. 74F.G.

Simethis Kunth, Enum. 4: 618 (1843); Webb, Fl. Europaea 5: 19 (1980).

Shortly rhizomatous perennial; leaves mostly basal, linear; inflorescence a paniculate cyme terminating a leafless scape, to 40 cm tall; tepals to 10 mm long abaxially purple, adaxially whitish; stamens 6, filaments hairy at base; fruit capsule, tardilly dehiscent. Only one sp., *Simethis planifolia* (L.) Gren. & Godron, France, Spain, Portugal, Corsica, Sardinia, Italy Morocco, Algeria and Tunisia.

8. *Stypandra* R. Br.

Stypandra R. Br., Prod. Fl. Nov. Holl.: 278 (1810); Henderson, Fl. Australia 45: 225–228 (1987).

Herbaceous to shrublike perennials to 1.5 m tall; stem unbranched or branching from the base, leafy throughout; inflorescence a loose panicle with leaflike bracts supporting lower branches; flowers pendulous; tepals darkblue, to 12 mm long; anther filaments with a tuft of dense hairs immediately below anther; anthers coiled after anthesis; fruit a capsule. Only one sp., *S. glauca* R. Br., Western Australia, Victoria, South Australia, New South Wales and southern Queensland.

9. *Geitonoplesium* Cunn. ex R. Br.

Geitonoplesium Cunn. ex R. Br., Bot. Mag. t. 3131 (1832); Conran & Clifford, Fl. Australia 46: 194 (1986); Conran, Muelleria 6: 363–369 (1987); Laferrière, Fl. Males. 12: 733–736 (1996).

Rhizomatous vines climbing to 10 m, stems branched; leaf lamina resupinate; inflorescence a cyme or cymose panicle; tepals to 11 mm long, white-mauve with a green apex; stamens 6; fruit berry, black. Only one sp., *G. cymosum* (R. Br.) Cunn., E Australia (excluding Tasmania), Papua New Guinea, Malesia, the Philippines and Pacific Islands.

10. *Rhuacophila* Blume

Rhuacophila Blume, Enum. Pl. Java: 13 (1827); Jessop, Fl. Males. 9: 209 (1979) as *Dianella*.

Herbaceous perennials to 2 m tall, stems branched or unbranched; leaves dispersed evenly along stems or clustered at the ends of branches; leaf bases not keeled. Inflorescence a panicle; bracts grading gradually in size from foliage leaves to scales; tepals to 12 mm long, blue, lilac, white or yellow; stamen filaments with a papillose, fusiform swollen struma about the middle; fruit a berry, yellow to black. Only One sp., *R. javanica* Blume, throughout Malesia, New Caledonia (including the Ile des Pins) and Fiji but absent from Australia.

11. *Dianella* Lam. ex Juss.

Dianella Lam. ex Juss., Gen. Pl.: 41 (1789); Schlittler, Mitt. Bot. Mus. Univ. Zürich 163: 1–283 (1940), ibid. 207: 1–38 (1957); Henderson, Fl. Australia 45: 194–225 (1987).

Herbaceous or shrublike perennials to 2 m tall; stems branched or unbranched; leaves semi-equitant, basal, distributed evenly or clustered on the stem at the branch ends; inflorescence a loose panicle; bracts throughout small; tepals to 10 mm long, blue to white; anther filament with swollen struma immediately below anther; fruit a berry, glossy, blue. About 20 spp., widespread in Japan, Hawaii, India, SE Asia, India, Malesia, Madagascar, Australia, New Zealand and many islands of the Pacific. In Africa known only from single record from Zimbabwe (Wild 1953).

Genera anomala

12. *Eccremis* Willd. ex Bak.

Eccremis Willd. ex Bak., J. Linn. Soc. Bot. 15: 319 (1876).

Rhizomatous or short-stemmed perennials to 1 m tall; leaves basal; leaf bases strongly keeled; inflorescence a loose corymbose panicle; bracts small; flowers articulating at the pedicel apex; tepals to 10 mm long, blue; stamen filaments united basally, 3 hypogynous and 3 affixed to the tepals, unornamented; fruit a loculicidal capsule with a septicidal endocarp. Only one sp., *E. coarctata* (Ruíz & Pav.) Bak., Andes of Ecuador Colombia and Peru.

The arrangement of the stamens, as well as the highly unusual fruit structure reported by Krause (1930) require further investigation, as does the molecular placement of this genus within the Iridaceae, despite its apparently close morphological resemblance to a capsular-fruited *Dianella*.

13. *Xeronema* Brongn. & Gris

Xeronema Brongn. & Gris, Bull. Soc. Bot. Fr. 11: 316 (1864); Baker, J. Linn. Soc. Bot. 15: 319 (1876); Moore, Pac. Sci. 11: 355–362 (1957); Moore and Edgar in Flora of New Zealand 2: 25–27 (1970).

Shrublike, stemless rhizomatous perennials to 1 m tall; leaves basal, broadly ensiform, unifacial, equitant and isobilaterally flattened, leaf bases strongly sheathing; inflorescence a dense, brushlike spike; bracts throughout small; tepals to 20 mm long, bright red; stamen filaments unornamented; fruit a dry loculicidal capsule. Two spp., one endemic to New Zealand and the other to New Caledonia.

Selected Bibliography

Arber, A. 1925. Monocotyledons, a morphological study. Cambridge: Cambridge University Press.
Bernhardt, P. 1995. The floral ecology of *Dianella caerulea* var. *assera* (Phormiaceae). Cunninghamia 4: 9–20.

Bouvier, W. 1915. Beiträge zur vergleichenden Anatomie der Asphodeloideae (Tribus: Asphodeleae und Hemerocallideae). Akad. Wiss. Wien Math.-Naturwiss. Kl. Denkschr. 91: 539–577.

Briggs, L.H., Briggs, L.R., King, A.W. 1975. New Zealand phytochemical survey 14. Constituents of *Dianella nigra* Col. N.Z. J. Bot. 18: 559–563.

Brummitt, R.K. 1992. See general references.

Cadavid, I., Gato, A., Callej, J.M. 1985. Chrysophanol and nepodin constituents of *Simethis bicolor*. Int. J. Crude Drug Res. 23: 13–15.

Cave, M.S. 1955. Sporogenesis and the female gametophyte of *Phormium tenax*. Phytomorphology 5: 247–253.

Cave, M.S. 1975. Embryological studies in *Stypandra* (Liliaceae). Phytomorphology 25: 95–99.

Chase, M.W. et al. 1993, 1995. See general references.

Chupov, V.S. 1987. Taxonomic position of the genera *Geitonoplesium* and *Simethis*. Bot. Zh. (Moskow) 72: 904–908 (in Russian).

Colegate, S.M., Dorling, P.R., Huxtable, C.R. 1987. Stypandrone: a toxic naphthalene-1,4-quinone from *Stypandra inbricata* and *Dianella revoluta*. Phytochemistry 26: 979–982.

Conole, L.E. 1993. Observations of nest mound decoration by the "bulldog" ant *Myrmecia forficata* and other *Myrmecia* species in south-west Victoria. Victorian Nat. 110: 217–218.

Couper, R.A. 1960. New Zealand Mesozoic and Cenozoic plant microfossils. Palynol. Bull. N.Z. Geol. Surv. 32: 1–87.

Craig, J.L. 1989. Seed set in *Phormium*: interactive effects of pollinator behaviour, pollen carryover and pollen source. Oecologia 81: 1–5.

Craig, J.L., Stewart, A.M. 1988. Reproductive biology of *Phormium tenax*: a honeyeater-pollinated species. N.Z. J. Bot. 26: 453–464.

Cranwell, L.M. 1952. New Zealand pollen studies: The monocotyledons. Bull. Auckl. Inst. Mus. 3: 1–91.

Cribb, A.B., Cribb, J.W. 1976. Wild food in Australia. Sydney: Fontana Collins.

Cribb, A.B., Cribb, J.W. 1982. Useful wild plants in Australia. Sydney: Fontana Collins.

Curtis, W.M. 1952. Variation in certain Tasmanian plants. New Phytol. 51: 398–414.

Dahlgren, R.M.T., Rasmussen, F.N. 1983. Monocotyledon evolution characters and phylogenetic estimation. In: Hecht, M.K., Wallace, B., Prance, G.T. (eds.) Evolutionary biology, vol. 16. New York: Plenum Publ. pp. 255–395.

Dahlgren, R.M.T. et al. 1985. See general references.

Daumann, E. 1970. Das Blütennektarium der Monocotyledonen unter besonderer Berücksichtigung seiner systematischen und phylogenetischen Bedeutung. Feddes Repert. 80: 463–590.

Eifler, D.A. 1995. Patterns of plant visitation by nectar-feeding lizards. Oecologia 101: 228–233.

Erdtman, G. 1952. See general references.

Everist, S.L. 1981. Poisonous plants in Australia. London: Angus and Robertson.

Fedorov, A.A. (ed.) 1969. See general references.

Gilbert, P.A. 1939. The bower-painting habit of the Satin Bower-Bird (*Ptilonorhynchus violaceus*). Emu 39: 18–22.

Godley, E.J. 1969. Flower biology in New Zealand. N.Z. J. Bot. 17: 441–466.

Gott, B., Conran, J.G. 1991. Victorian koorie plants. Hamilton, Victoria: the aboriginal keeping place.

Hair, J.B., Beuzenberg, E.J. 1966. Contributions to a chromosome atlas of the New Zealand flora – 7. N.Z. J. Bot. 4: 256–266.

He, C.X. 1994. Effects of extract from *Hemerocallis citrina barroni* (EHCB) and epidermal growth factor (EGF) on human dermal fibroblast proliferation. Zhonghua Pifuke Zazhi 27: 218–220, 269 (in Chinese).

Hegnauer, R. 1963. See general references.

Henderson, R.J.F. 1985. *Thelionema*, a new genus of the Phormiaceae from Australia. Austrobaileya 2: 109–111.

Henderson, R.J.F. 1987. *Dianella, Stypandra, Thelionema*. In: George, A.S. (ed.) Flora of Australia. Canberra: AGPS pp. 193–230.

Henderson, R.J.F., Clifford, H.T. 1984. A recircumscription of the Phormiaceae Agardh. Taxon 33: 423–427.

Keighery, G.J. 1984. Chromosome numbers of Australian Liliaceae. Feddes Repert. 95: 523–532.

Kosenko, V.N. 1994. Morfologiya pyl'tsy semeistv Phormiaceae, Blandfordiaceae i Doryanthaceae. Bot. Zh. 79: 1–12.

Krause, K. 1930. Liliaceae. In: Engler, A., Prantl, K. (eds.) Die natürlichen Pflanzenfamilien, 2nd edn., Vol. 15a. Leipzig: W. Engelmann pp. 227–386.

Kunkel, G. 1984. Plants for human consumption. Koenigstein: Kluwer.

Lassak, E.V., McCarthy, T. 1990. Australian medicinal plants. Melbourne: Mandarin Australia.

Moore, L.B., Edgar, E. 1970. Flora of New Zealand Vol. 2. Wellington: Government Printer.

Morice, I.M. 1962. Seed fats in New Zealand Agavaceae. J. Sci. Food Agric. 13: 666–669.

Raju, M.U.S. 1957. Some aspects of the embryology of *Dianella nemorosa*. J. Indian Bot. Soc. 36: 223–226.

Rayment, T.R. 1935. A cluster of bees. Sydney: Endeavour Press.

Ricardi, M. 1958. Detección de Saponinas en Angiospermae Chilenas. Bot. Soc. Biol. Concepción 33: 29–94.

Rose, A.B. 1973. Food of some Australian birds. Emu 73: 177–183.

Rudall, P., Chase, M.W., Conran, J.G. 1996. New circumscriptions and a new family of asparagoid lilies: genera formerly included in Anthericaceae. Kew Bull. 51: 667–680.

Schlittler, J. 1940. Monographie der Liliaceengattung *Dianella* Lam. Mitt. Bot. Mus. Univ. Zürich 163: 5–283.

Schlittler, J. 1945. Untersuchungen über den Bau der Blütenstände im Bereich des Anthericumtypus (Asphodelinae-Anthericinae-Dianellinae). Mitt. Bot. Mus. Univ. Zürich 174: 200–239.

Schnarf, K., Wunderlich, R. 1939. Zur vergleichenden Embryologie der Liliaceae-Asphodeloideae. Flora 133: 297–327.

Schulze, R. 1893. Beiträge zur vergleichenden Anatomie der Liliaceen, Haemodoraceen, Hypoxidaceen und Velloziaceen. Bot. Jahrb. 17: 295–334.

Schulze, W. 1982. Beiträge zur Taxonomie der Liliifloren IX. Anthericaceae. Wiss. Z. Friedrich Schiller Univ. Jena 31: 291–307.

Takhtajan, A.L. 1982. See general references.

Tamura, M.N. 1995. A karyological review of the orders Asparagales and Liliales (Monocotyledonae). Feddes Repert. 106(1–2): 83–111.

Tanaka, T. 1976. Tanaka's cyclopedia of edible plants of the world. Tokyo: Keigaku Publ.

Thongpukdee, A. 1989. The taxonomic position of the genus *Tricoryne* R.Br. Ph.D. Thesis St. Lucia: University of Queensland.

Tillich, H.-J. 1995. Seedlings and systematics in Monocotyledons. In: Rudall, P.J., Cribb, P., Cutler, D.F., Humphries, C.J. (eds.) Monocotyledons: systematics and evolution. London: Royal Botanic Gardens, Kew pp. 303–352.

Uphof, J.C.T. 1968. Dictionary of economic plants. Lehre: J. Cramer.
Wang, J.H., Humphreys, D.J., Stodulski, G.B.J., Middleton, D.J., Barlow, R.M., Lee, J.B. 1989. Structure and distribution of a neurotoxic principle, hemerocallin. Phytochemistry 28: 1825–1826.
Wild, H. 1953. *Dianella ensifolia*: a new generic record for continental Africa. Kew Bull. 8: 251–252.

Herreriaceae

J.G. Conran

Herreriaceae Endl., Ench. Bot.: 90 (1841).

Perennials with branching stems, sometimes twining, from tuberous rhizomes, or stemless rosette plants from an erect rhizome; roots fibrous. Aerial stems, if present, leafy, thin, spiny, branching, climbing, to about 5 m tall, or absent. *Leaves* alternate, spirally arranged, fascicled, cladode-like, ovate-lanceolate to linear with numerous parallel main veins and between them parallel secondary veins, non-sheathing. *Inflorescence* an axillary or terminal raceme or few branched panicles. *Flowers* bisexual, numerous, erect, actinomorphic, hypogynous; tepals in 2 whorls of 3, both whorls of similar length, with basal nectaries (*Herreriopsis*), red, yellow, brown or pale green, but never blue. Stamens 3 + 3; filaments free; anthers basifixed, introrse, dehiscing longitudinally. Gynoecium of 3 united carpels; ovary 3-locular with axile, intrusive placentae and septal nectaries (*Herreria*); style single, erect short to filiform; stigma capitate, trilobed; ovules anatropous, borne in 2 rows. *Fruit* a trilobed, septicidal capsule; seeds few to many, flattened and winged, black, with a phytomelan crust; endosperm copious, lacking starch; embryo small, linear and straight.

The family is restricted to S America and Madagascar and has two genera and nine spp.

Vegetative Morphology and Anatomy. The plants are tuberous-rhizomatous and have tall branching stems which climb to several metres, or in some species of *Herreria*, rosettes of leaves with little or no aerial stems and an erect rhizome (Fig. 77). The stems are tough and supple, climb by twining dextrorsely and bear prickles. The main and lateral branches appear to show determinate growth. Climbing roots are absent. The stems in *Herreria montevidensis* have secondary thickening (Rudall 1995).

The seedlings of *Herreria* have a long, assimilating cotyledonary hyperphyll, there is no cotyledonary sheath, germination is epigeal, and the first leaf is similar to successive leaves. The following leaves are linear to lanceolate, non-sheathing and spirally alternate in fascicles on short lateral

Fig. 77A–H. Herreriaceae. A–D *Herreria stellata*. A Flowering and fruiting shoot. B Flower. C Dehiscing capsule. D Seed. E–H *Herreriopsis elegans*. E Flowering shoot. F Flower. G Dehisced capsule. H Seed. (Takhtajan 1982)

branches. The leaves are sessile and the blades are entire and multicostate. The primary venation is parallel. The secondary venation is parallel and enters the midrib, with the acropetal higher vein orders transversely parallel to the leaf axis as are the areoles. There are no free vein endings. Vernation in *Herreria* is supervolute. Data from Conran (1985).

The peduncles are well developed and have reduced, or in some species of *Herreria*, large subtending bracts. The roots of *Herreria* have a uniseriate exodermis with unthickened cells and an unthickened cortex. The epidermal cells of the stem are small, with a thick cuticle and walls. A ring of fibrous bundles and small collateral bundles are found in the cortex. Scattered vascular bundles occur in the centre of the stem (Schulze 1893). Vessels with scalariform perforation plates occur in the stems of *Herreria* (Wagner 1977). The stomata in *Herreria* present on the adaxial surface of the lamina are in parallel rows. They are anomocytic. The leaves of *Herreria* have undifferentiated mesophyll. Calcium oxalate raphides are present in all parts of the plant. Tannin cells are present but mucilage canals or laticifers are absent.

Flower Structure. The flowers are trimerous and pentacyclic. The perianth is persistent and the outer and inner tepal whorls are not differentiated. The tepals are 3-veined and unspotted. *Herreria* has septal nectaries, whereas in *Herreriopsis*, saccate nectaries are present at the base of the tepals. The ovules are borne in 1–2 rows on axile placentae.

Pollen Morphology. The pollen grains in Herreria are obovate, 30–45 μm, sulcate with a reticulate exine (Schulze 1982).

Karyology. The chromosome number for Herreria is $n = 27$ with 1 large and 26 short chromosomes (Satô 1942).

Fruit and Seed. The fruits of the Herreriaceae are trilobed, septicidal capsules. The perianth is semipersistent. The capsule wall is thin, coriaceous and encloses the few to numerous flattened, winged seeds. The seed surface is rugulose, black and phytomelan encrusted at maturity. The outer epidermis of the testa collapses during its development. The endosperm is massive and consists of cells with non-pitted walls. It contains aleurone and fatty oils, but not starch.

Phytochemistry. *Herreria montevidensis* var. *bonplandii* contains phenols, flavonols, water-soluble steroids and tannins (Bandoni et al. 1976). The saponin gitogenin is reported for the roots (rhizomes?) of *H. stellata* (Farnsworth et al. 1971).

Affinities. The Herreriaceae were related to the Asparagaceae and Ruscaceae in the Asparagales (Dahlgren et al. 1985) but were also thought to show affinities with Hemerocallidaceae and Luzuriagaceae (Conran 1989). Recent molecular studies by Chase et al. (1995) confirmed the association with all of these families in the higher Asparagales except the Luzuriagaceae (Liliales).

Distribution and Habitats. The Herreriaceae occur in temperate to subtropical South America (Brazil, Uruguay, Paraguay, Argentina and Chile) (*Herreria*) and Madagascar where *Herreriopsis* is endemic. They are understorey shrubs in moist-temperate to subtropical forests. *Herreriopsis* grows along river margins.

Economic Importance. The roots (rhizomes?) of *Herreria montevidensis* var. *bonplandii*, *H. sarsaparilha* and *H. stellata* are used in folk medicine as stimulant, antipyretic and for treating venereal disease (Bandoni et al. 1976; Hoffmann 1982; Basualdo et al. 1995).

Key to the Genera

1. Outer tepal whorl unmodified, resembling the inner whorl; ovules few or single per loculus; S America **1. *Herreria***
- Outer tepal whorl with dilated saclike bases, inner tepal whorl lacking saclike bases; ovules numerous; Madagascar **2. *Herreriopsis***

1. *Herreria* Ruíz et Pav. Fig. 77A–D

Herreria Ruíz et Pav., Prodr.: 48 (1794).
Clara Kunth (1848).

Acaulescent to climbing herbs; leaves basally rosulate or clustered on short lateral shoots, ovate-lanceolate to linear-lanceolate; inflorescence racemose or paniculate; seeds few; $2n = 54$. Eight spp. in Brazil, Uruguay, Paraguay, Argentina and Chile.

A separate genus, *Clara* Kunth, has been recognised on the absence of an erect stem, but Smith (1958) and Cabrera (1968) both place the stemless species in *Herreria*.

2. *Herreriopsis* Perrier Fig. 77E–H

Herreriopsis Perrier, Bull. Soc. Bot. Fr. 81: 819 (1934 ["1935"]).

Perennial, erect herb; leaves clustered on short lateral shoots, bluntly acute, linear-lanceolate, multinerved, convergently veined; flowers few to numerous in simple racemes; outer whorl of tepals with saclike bases; seeds numerous. Only one sp., *H. elegans* Perrier, with two varieties on Madagascar.

Selected Bibliography

Bandoni, A.L., Mendiondo, M.E., Rondina, R.V.D., Coussio, J.D. 1976. Survey of Argentinian medicinal plants – folklore and phytochemical – screening. 2. Econ. Bot. 31: 161–185.
Basualdo, I., Zardini, E.M., Ortiz, M. 1995. Medicinal plants of Paraguay: underground organs, II. Econ. Bot. 49: 387–394.
Cabrera, A.L. 1968. Flora de la Provincia de Buenos Aires. Part 1. Buenos Aires: I.N.T.A.
Chase, M.W. et al. 1995. See general references.
Conran, J.G. 1985. The taxonomic affinities of the genus *Drymophila* (Liliaceae s.l.). Ph.D. Thesis. St. Lucia: University of Queensland.
Conran, J.G. 1989. Cladistic analyses of some net-veined Liliiflorae. Plant Syst. Evol. 168: 123–141.
Hoffmann, A.J. 1982. Flora Silvestre de Chile zona austral. Santiago: Fundacion Claudio Gay.
Perrier de la Bâthie, H. 1934. Deux liliacées nouvelles de Madagascar. Bull. Soc. Bot. Fr. 81: 819–823.
Rudall, P. 1995. New records of secondary thickening in monocotyledons. IAWA J. 16: 261–268.
Satô, D. 1942. Karyotype alternation and phylogeny in Liliaceae and allied families. Jpn. J. Bot. 12: 57–161.
Schulze, R. 1893. Beiträge zur vergleichenden Anatomie der Liliaceen, Haemodoraceen, Hypoxidaceen und Velloziaceen. Bot. Jahrb. 17: 295–334.
Schulze, W. 1982. Beiträge zur Taxonomie der Liliifloren X. Asparagaceae. Wiss. Z. Friedrich Schiller Univ. Jena 31: 309–330.
Smith, L.B. 1958. Notes on South American phanerogams. J. Wash. Acad. Sci. 48: 282–283.
Takhtajan, A.L. 1982. See general references.
Wagner, W.M. 1977. Vessel types of monocotyledons: a survey. Bot. Not. 130: 383–402.

Hostaceae[1]

K. Kubitzki

Hostaceae B. Mathew, Kew Bull. 43: 302 (1988).
Funkiaceae Horan. (1834), nom. illeg.

Perennial herbs from a large, clumpy, horizontal rhizome with a thick fibrous covering; roots thickened, fleshy; *leaves* spiral, basal, numerous; leaf blades dorsiventral, strongly veined, linear-lanceolate, or cordate-ovate, often petiolate. *Inflorescence* a bracteate raceme, usually subsecund, surpassing the leaves in most species, the peduncle arising from the rhizome; *flowers* hermaphroditic, hypogynous, white to blue or lavender, showy, relatively large; tepals 6, similar, united below into a tube with wide throat and spreading, funnel-shaped or campanulate limb; stamens 6, inserted either at the base of the tube or at the apex of the ovary; filaments declinate; anthers dorsifixed, introrse, dehiscent longitudinally, filament attached to connective in pit; pistil sessile, syncarpous, 3-locular, ovary oblong; septal nectaries present; style filiform; stigma small; ovules anatropous, numerous in each locule, placentation axile; *capsules* narrowly oblong to linear, pendent when mature, loculicidal; *seeds* black, flattened, winged; testa thin, papery, encrusted with phytomelan, endosperm fleshy; embryo applanate.

Only one genus of 23–25 (–50?) species from E Asia.

Vegetative Morphology. *Hosta* has large, clumpy rhizomes with prominent leaf scars, which are often clothed with a thick fibrous covering representing the remains of old leaf bases. Fleshy roots form the root system.

During germination, the embryo becomes plump and firm and the black testa continues to cover the seed, which remains below ground. The pale cotyledon remains enclosed by the seed during germination and is little differentiated. The plumular leaf emerges through a pore in the sheathing base of the cotyledon at a considerable distance from the seed. The young leaves are miniature forms of the adult leaves. Additional leaves are produced spirally increasing in size as the plant ages. The seedlings flower in the 2nd year following germination.

The leaf is sheathing at the base and in some species is narrowed into a petiole. The leaf blades vary in shape from nearly linear or oblong-lanceolate to ovate or cordate. The primary veins arise from the base with a midrib confluent between well-spaced lateral primary veins which are connected by transverse reticulate veins (Fig. 78).

Vegetative Anatomy. Xylem vessels are absent in the leaves and stems but are present in the roots. The xylem vessels are of a primitive type, having scalariform perforation plates (Cheadle and Kosakai 1971). Any cell of the piliferous layer of the roots of *Hosta* may develop root hairs. *Hosta* has PIIc-type sieve-tube plastids with cuneate (triangular) crystalloid bodies (Behnke 1981). The epidermal cells of the leaves contain rodlike crystals which are not oriented (Frölich and Barthlott 1988).

Inflorescence Morphology. The inflorescence is a terminal, usually subsecund, elongated, bracteate raceme borne on a leafless or bracteose peduncle. The raceme may be included within the foliage or borne well above the basal cluster of leaves. The bracts subtending the pedicels are highly variable in size and shape and may be diagnostic for some species. The bracts on the peduncle are subfoliose to foliose.

Floral Morphology. The flowers are bisexual and hypogynous with a perianth of 6 tepals fused into a cylindrical, campanulate, or funnel-shaped perianth with 6 lobes of variable width and length. The stamens are inserted on the tube of the perianth or in some but not all *Hosta* at the apex of the ovary. The filaments are declinate and attached to the anther connective in a pit. The anthers are dorsifixed.

The pistil is syncarpous and 3-locular. The style is filiform with a minute, capitate, or 3-lobed stigma. The stigmatic surface is of the Wet type, with the secretions released near the base of the papillae more or less simultaneously over the stigmatic surface (Heslop-Harrison and Shivanna 1977). Septal nectaries are present (Daumann 1970).

Embryology. Microsporogensis is successive throughout, the tapetum secretory and binucleate. A parietal cell and hypostase are formed, and embryo-sac development follows the Polygonum

[1] Partly based on information provided by S.B. Jones.

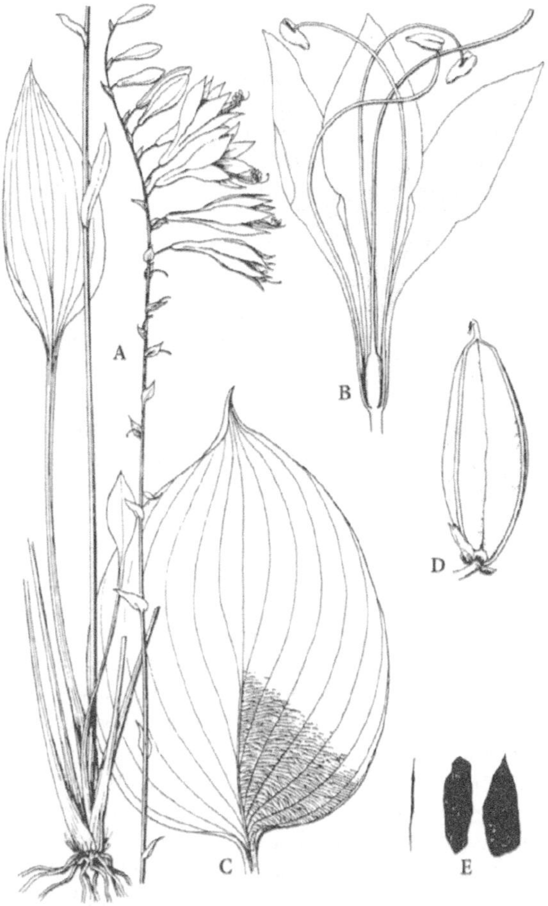

Fig. 78A–E. Hostaceae. A, B *Hosta japonica*. A Flowering plant. B Flower, longitudinal section. C–D *H. ventricosa*. C Leaf. D Capsule. E Seeds in different views. (Dahlgren et al. 1985)

type. The chalazal end of the embryo sac grows first and is relatively long and narrowed, whereas the micropylar end is rather broad. The antipodals later become mononucleate. Endosperm formation is Helobial (Schnarf and Wunderlich 1939). In all these traits *Hosta* resembles closely the Agavaceae and *Hesperocallis* but not *Leucocrinum* (Cave 1948). Adventive polyembryony has been described in the tetraploid *Hosta ventricosa*. One to 3 embryos develop from cells of the nucellus around the top of the embryo sac (Davis 1966). During germination, 1 of the embryos typically dominates and the others are retarded in development.

Pollen Morphology. The pollen grains are subspheroidal and sulcate. Exine sculpturing in most *Hosta* species is rugulate or granulate and

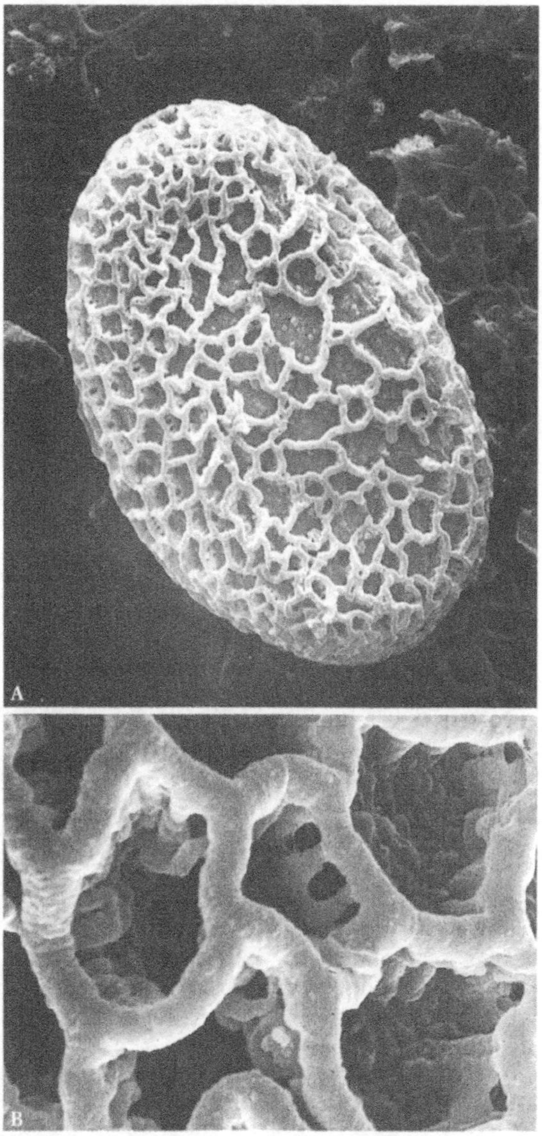

Fig. 79A,B. Hostaceae. *Hosta plantaginea*. A Pollen, proximal face, SEM ×800. B Reticulum with unibaculate muri, SEM ×5600. (Photo Alvarez and Koehler)

hemitectate. *Hosta plantaginea*, a species regarded as isolated from others in the genus *Hosta*, has reticulate exine sculpturing and tectate exine structure (Schulze 1983; Fig. 79, Chung and Jones 1989). The similarity in the unibaculate muri of *Hosta plantaginea* and *Hesperocallis* (Fig. 80) is particularly notable (Alvarez and Köhler 1987).

Karyology. Numbers reported in *Hosta* (with ca. 20–23 species) are based on $x = 30$ with $2n = 60$ for the majority of the species (Sen 1975). *Hosta*

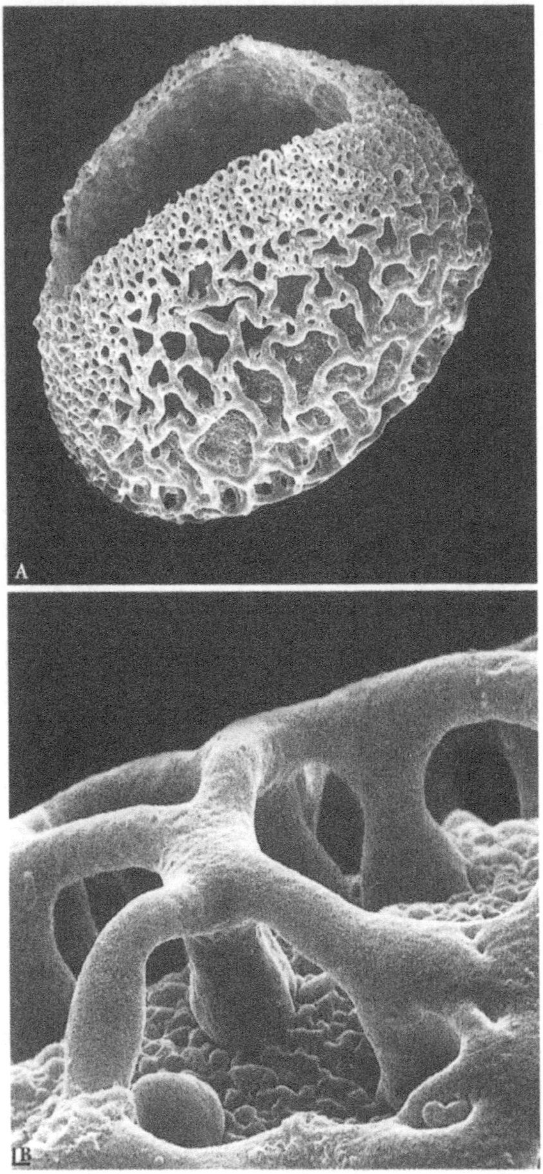

Fig. 80A,B. Hostaceae. *Hesperocallis undulata*. **A** Equatorial view of pollen showing sulcus with well developed margin, SEM ×1000. **B** Reticulum with curved and bending muri and distantly spaced bacula; lumina ornamented, SEM ×6000. (Photo Alvarez and Koehler)

clausa is triploid with $2n = 90$; its flowers do not open and it reproduces by means of vigorous rhizomes. *Hosta ventricosa* is a tetraploid with $2n = 120$ and is agamospermic. Aneuploids of $2n = 62$, 63 have been reported in several old cultivars of garden origin; these aneuploids seldom set fruit or seed and have reduced pollen fertility. The karyotype of *Hosta* is bimodal with 4 pairs of large chromosomes, 2 pairs of medium-sized chromosomes and 24 pairs of small ones. The karyotype of the monotypic *Hesperocallis* (formerly included in Hostaceae) is also heteromorphic with 5–6 large chromosomes and 18–19 small chromosomes. The large chromosomes of both genera agree in having a number of localised chiasmata and in conspicuous SAT chromosomes. The chromosomes of the monotypic *Leucocrinum* (at one time also included in the family) are relatively uniform in size, and their number has been determined as $n = 14$ (Cave 1948).

Pollination and Reproduction. *Hosta plantaginea* is the only night-blooming species and the only one with fragrant flowers. Its flowers have particularly long tubes (Schmid 1991). All other species are day-bloomers, the flowers of which begin to open the evening before the day of full anthesis. *Hosta sieboldii* and *H. sieboldiana* have homogamous, 1-day flowers pollinated by bumblebees. The abdomens touch the stigma on the extended style when they land on the flower, and autogamy is effectively prevented. However, the flowers seem to be fairly self-compatible, and geitonogamy may occur rather freqently because 2 or more flowers of an inflorescence may bloom at the same time (Takahashi et al. 1994). In the southeastern United States, cultivated *Hosta* are pollinated by various Hymenoptera, Lepidoptera and even hummingbirds. Bumblebees, in particular, are attracted to *Hosta* flowers in large numbers. Only one species of *Hosta*, *H. plantaginea*, is fragrant; several cultivars (e.g. Honeybells and Royal Standard) of *Hosta* having *H. plantaginea* as one of the parents are also fragrant.

Not only artificial hybrids between *Hosta* species are commonly made by breeders attempting to develop new cultivars, also intersectional hybrids are known to occur under natural conditions (Takahashi et al. 1994). Some of the older cultivars introduced to the West from Japan are sterile, exhibiting problems at meiosis. Many of the more recent hybrids regularly set seed. Seedlings grown from isolated plantings of "species" hostas often exhibit a great deal of phenotypic variation which appears to have a genetic basis. After maturity in the autumn, the seeds of *Hosta* can be sown in a greenhouse for rapid germination, or in an outdoor bed with germination the following spring.

Fruit and Seed. The fruit of Hostaceae is an elongated, oblong, subglobose or obovoid, loculicidal, 3-valved, angled capsule dehiscing from the apex and containing numerous seeds. The seeds are black and flattened or compressed and at the chalazal end the outer integument is protracted into a wing. The capsules become pendent at maturity and the seeds hang downwards on either side of the placenta, becoming visible as the capsule dehisces. In the seed coat, the testal portion retains its cellular structure and is encrusted by a phytomelan layer, similarly as in *Agave*. In the tegmen, cellular structure collapses completely, similarly as in *Yucca*. The mature endosperm is composed of isodiametric, thin-walled, pitted cells (Huber 1969).

Phytochemistry. Apart from common phenolics various steroid saponins have been characterised from *Hosta* species (Hegnauer 1986).

Affinities. As evidenced by the phytomelan encrusted seed coat and the successive microsporogenesis, *Hosta* is clearly placed in the "higher" Asparagales, in which Agavaceae are its closest relatives. Both groups share the same chromosome base number and a bimodal karyotype, but differ because *Hosta* has 4, *Yucca*/*Agave* 5 long chromosome pairs (Tamura 1995). The close relationship between *Hosta* and the Agavaceae has received support by the *rbc*L analysis (Rudall et al. 1997), in which *Hosta* appears embedded within the Agavaceae. However, the different seed morphology, the lack of cutinised crystal cells, the attachment of the anther filaments in a pit of the connective such as in Asphodelaceae, the bluish flowers and the distribution different from Agavaceae may justify their separate family status.

Leucocrinum and *Hemerocallis*, formerly associated with *Hosta*, are here treated in the Anthericaceae and Hemerocallidaceae, respectively.

Hesperocallis has often been considered as a close relative of *Hosta* on the basis of its raceme, applanate seeds, parietal cells, successive microsporogenesis, the probably helobial endosperm and the bimodal karyotype (see above), but differs in the underground organ, the chromosome number, the articulation of the flowers from the pedicel, and its North American distribution. Therefore, it is treated here among the genera anomala and, if better known, may be elevated to family rank, but certainly forms part of "higher" asparagoids.

Distribution and Habitats. *Hosta* is native to Japan, China and Korea, where it grows in temperate shady forests of the mountains and the coast, in bogs and also on well-drained rock outcrops, the various species differing in their moisture requirements.

Economic Importance. *Hosta* is an important herbaceous perennial for shady gardens in the temperate zone of western Europe, New Zealand, North America and eastern Asia. Primarily grown for its foliage, the flowers are also rather attractive. Hundreds of cultivars are available in the nursery trade with leaves that are variegated, golden, blue or green. They provide an array of leaf and clump sizes, form, leaf shape and texture. New cultivars often sell for $300 (US) per plant; however, the application of tissue culture propagation techniques has made the newer cultivars affordable for the average gardener. Much of the breeding work on *Hosta*, largely done by amateurs, has taken place in the United States since 1950 and many cultivars have been exported to Europe and Japan. Recently, a number of cultivars have been registered that appeared as off-types associated with commercial tissue culture propagation. *Hosta* is a major crop in perennial plant nurseries in the United States, England, Holland, Germany and New Zealand. The horticultural aspects are fully dealt with by Schmid (1991), who gives a comprehensive treatment of the taxonomy of the species and cultivars.

The leaves of *Hosta* are cooked and eaten in Korea and Japan. In Korea, the species have become uncommon in the wild due to collecting for use as food.

Only one genus:

Hosta Tratt., nom. cons.

Hosta Tratt., Arch. Gewächsk. 1: 56 (1812), nom cons., the most comprehensive taxonomic treatment by Maekawa (1940).
Funkia Sprengel (1817), nom. illeg.

Characters as for family. Twenty three–26 spp. (Maekawa 1969; Fujita 1976), but many more when a narrow species concept is applied (39 spp., Maekawa 1940, more than 40, Schmid 1991), China, Japan and Korea; 300–500 cultivars widely planted in shady gardens in E Asia, Europe, N America and New Zealand. Two (Fujita 1976) or three (Schmid 1991) subgenera, *H. plantaginea* (Lam.) Aschers. isolated (monotypic subgen. *Hosta*).

Genus anomalum:

Hesperocallis A. Gray

Hesperocallis A. Gray, Proc. Am. Acad. 7: 390 (1867).

Erect perennial with alliaceous odour from deep-seated, tunicated bulb; stem stout, straight, simple, glaucous; leaves alternate, largest at the base and upward grading into the scarious bracts subtending the flowers; leaf blades linear, folded lengthwise, strongly crisped along the white margins; inflorescence an elongated terminal, subsecund bracteate raceme; pedicels short, constricted immediately below flower; tepals 6, alike, ca. 7 cm long, united up to the middle into a tube, each tepal with a greenish blue stripe on the back, the perianth and floral bracts withering and persistent after anthesis; stamens 6, inserted in perianth throat, filaments filiform, anthers linear, dorsifixed; ovary superior, sessile, 3-celled, with numerous ovules in each locule; style filiform, persistent, stigma capitate and slightly 3-lobed; capsule subglobose, deeply 3-lobed, loculicidal, ca. 2 cm long, with style persistent; seeds numerous, dull black, compressed, horizontal in locules. $2n = 48$. Only one sp., *H. undulata* A. Gray, western N America, SE California and W Arizona, in sandy flats of creosote bush desert scrub, flowering only after the infrequent rains.

See the above discussion for the relationships of this genus.

Selected Bibliography

Aden, P. 1988. The *Hosta* book. Portland, OR: Timber Press.
Alvarez, A., Köhler, E. 1987. Morfología del polen de las Agavaceae y algunos géneros afines. Grana 26: 25–46.
Bate-Smith, E.C. 1968. See general references.
Behnke, H.-D. 1981. See general references.
Cave, M.S. 1948. Sporogenesis and embryo-sac development in *Hesperocallis* and *Leucocoryne* in relation to their systematic position. Am. J. Bot. 35: 343–349.
Cheadle, V.I., Kosakai, H. 1971. Vessels in Liliaceae. Phytomorphology 21: 320–333.
Chung, M.G., Jones, S.B., Jr. 1989. Pollen morphology of *Hosta* Tratt. and related genera. Bull. Torrey Bot. Club 116: 31–44.
Dahlgren, R.M.T., Clifford, H.T. 1982. See general references.
Dahlgren, R.M.T. et al. 1985. See general references.
Daumann, E. 1970. See general references.
Davis, G.L. 1966. See general references.
Engler, A. 1888. Liliaceae. In: Engler, A., Prantl, K. (eds.) Die natürlichen Pflanzenfamilien II, 3. Leipzig: W. Engelmann.
Frölich, D., Barthlott, W. 1988. See general references.
Fujita, N. 1976. The genus *Hosta* in Japan. Acta Phytotax. Geobot. 27: 66–96 (in Japanese with English summary).
Hegnauer, R. 1986. See general references.
Heslop-Harrison, Y., Shivanna, K.R. 1977. The receptive surface of the angiosperm stigma. Ann. Bot. (Lond.) 4: 1233–1258.
Huber, H. 1969. See general references.
Maekawa, F. 1940. The genus *Hosta*. J. Fac. Sci. Imp. Univ. Tokyo III, Bot. 5: 317–425.
Maekawa, F., Kaneko, K. 1968. Evolution of karyotype in *Hosta* (Liliaceae). J. Jpn. Bot. 43: 132–140 (in Japanese with English summary).
Rudall, P.J., Chase, M.W., Furness, C.A. 1997. Microsporogenesis and pollen sulcus type in Asparagales (Lilianae). Can. J. Bot. 75: 408–430.
Schmid, W.G. 1991. The genus *Hosta*. Portland, OR: Timber Press.
Schnarf, K., Wunderlich, R. 1939. Zur vergleichenden Embryologie der *Liliaceae-Asphodeloideae*. Flora (Jena) NF 33: 297–327.
Schulze, W. 1983. Beiträge zur Taxonomie der Liliifloren XII. Der Umfang der Agavaceen. Wiss. Z. Friedrich Schiller-Univ. Jena, Math.-Nat. R. 32, 6: 965–979 (Beitr. Phytotax. 11).
Sen, S. 1975. Cytotaxonomy of Liliales. Feddes Repert. 86: 255–305.
Takahashi, K., Niwa, S., Ukai, A. 1988. Germination characteristics of *Hosta sieboldiana* (Lodd.) Engler, *H. albomarginata* (Hook.) Ohwi and *H. longissima* Honda. Sci. Rep. Fac. Educ., Gifu Univ. (Nat. Sci.) 12: 3–10.
Takahashi, H., Goto, Y., Kanematsu, S., Niwa, S., Mori, K., Nozaki, K. 1994. Pollination biology of *Hosta sieboldiana* (Lodd.) Engler and *H. sieboldii* (Paxton) J. Ingram (Liliaceae). Plant Species Biol. 9: 23–30.
Tamura, M.N. 1995. See general references.

Hyacinthaceae

F. Speta[1]

Hyacinthaceae Batsch, Disp. gen. pl. ienens.: 48 (1786), Tab. aff. regni veget.: 136–137 (1802).

Mostly bulbous plants, rarely (*Schoenolirion* and *Chlorogalum*) rhizomatic. Bulbs subterraneous or epigeal, formed by thickened cataphylls and/or foliage leaf bases, sometimes with intercalated nonstoring cataphylls. Scales in adult bulbs basally vaginate, imbricate, or more or less connate. Foliage leaves 1-many, rarely (*Bowiea*, *Schizobasis*) lacking in the adult bulb, spirally or more rarely distichously arranged, all forming part of the bulb, filiform to elliptic or circular, very rarely (*Ledebouria petiolata*, *L. rupestris*, *Drimiopsis*) petiolate, grooved or more rarely flat or terete. Leaf blades shed after 2–8 months at a preformed area at the apex of the bulb. Raphide cells and mucilage in all tissues copiously present. Feeding roots functioning for 8–12 months, the thicker ones often branched; contractile roots often present and thicker than the feeding roots; storage roots rarely present (*Ledebouria*). Inflorescence scapose, an (open) simple raceme or rarely a spike, rarely a compound raceme, very rarely more strongly branched (*Bowiea*, *Schizobasis*), varying in length, bearing 1-hundreds of flowers. Bracts of varying shape and size, sometimes lacking. Prophylls present or lacking. Flowers not articulated from the pedicel (except Chlorogaleae), hermaphrodite, hypogynous, trimerous, actinomorphic or rarely zygomorphic; the (pseudo-) terminal flowers often sterile and sometimes of different shape and coloration. Tepals 3 + 3, free or more or less connate, white, blue, red, yellow, brownish or greenish, inner and outer usually ± equal in shape, position, and color. Stamens 3 + 3, with sometimes 1 sterile whorl, of various shape, sometimes appendaged, the filaments sometimes connate at the base, or rarely forming a paracorolla. Anthers introrse, epipeltate, dehiscing longitudinally. Ovary superior to rarely (*Bowiea*) hemi-inferior. Pistil syncarpous, tricarpellate, ± trilocular, with innerseptal nectaries (*Autonoë* excepted). Style with a simple stigma, sometimes thickened, the stigma Wet or Dry; ovules anatropous, bitegmic, crassinucellar, in each locule 2 ovules superjacent, or 1 or 2 side by side, or several to numerous in 1 or 2 rows or disordered. Capsule ± wide loculicidally opening, with the locules separating ventrally as well, dry or fleshy. Seeds spherical, ellipsoidic, drop-shaped, rounded or flattened and angular, sometimes alate, yellow, brown or black, rarely pink.

A family comprising about 67 genera and 900 species with the greatest diversity in S Africa and the Mediterranean, extending to NW Europe, Central Asia and E Asia; the small subfam. Chlorogaloideae in N America, the monotypic subfam. Oziroëideae in Andine S America.

Vegetative Morphology. Different types of blastogeny can be found in Hyacinthaceae (Boyd 1932, Tillich 1992, 1995): (1) The rather short and hypogeal cotyledon is followed by a foliage leaf; (2) the long and mostly epigeal cotyledon is followed by a cataphyll; (3) the long and mostly epigeal cotyledon is followed by a foliage leaf. These leaves have a sheathing base and are distichously arranged. After the withering of the cotyledon, the young bulb is thus formed by a cataphyll or the sheath of a foliage leaf. It takes 2–several years for the juvenile bulb to develop the first inflorescence in monopodial growth. During this time, phyllotaxis usually switches from distichy to spiral arrangement, and the sheathing leaf bases are gradually reduced. In the adult bulb growth is sympodial. The bulbs are variable as to type, number, and age of the storage leaves (1–2 or 3–4 or up to 12 years) as well as to their arrangement (Figs. 81, 82A). While their structure is usually specific for genera and species, in rare cases it may change within the same bulb from year to year (Müller-Doblies and Müller-Doblies 1996). In *Brimeura* and *Hyacinthoides*, nonstoring cataphylls may form part of the bulb (Speta 1987). Every year a single inflorescence or various inflorescences may be formed. The foliage leaves appear together with the flowers (synanthous) or before them (proteranthous). In some genera there is a resting period between the decay of the leaves and the formation of the flowers (*Prospero*, *Barnardia*, *Urginea*).

Vegetative Anatomy. Fuchsig (1911) was the first to underpin with anatomical data the separation of Hyacinthaceae ("Scilleae") from Liliaceae s. str. ("Tulipeae"). Raphide cells and mucilage are highly characteristic for the Hyacinthaceae as well as large rhexigenetic lacunae in the stem and leaf.

[1] Translation by H.-H. Poppendieck

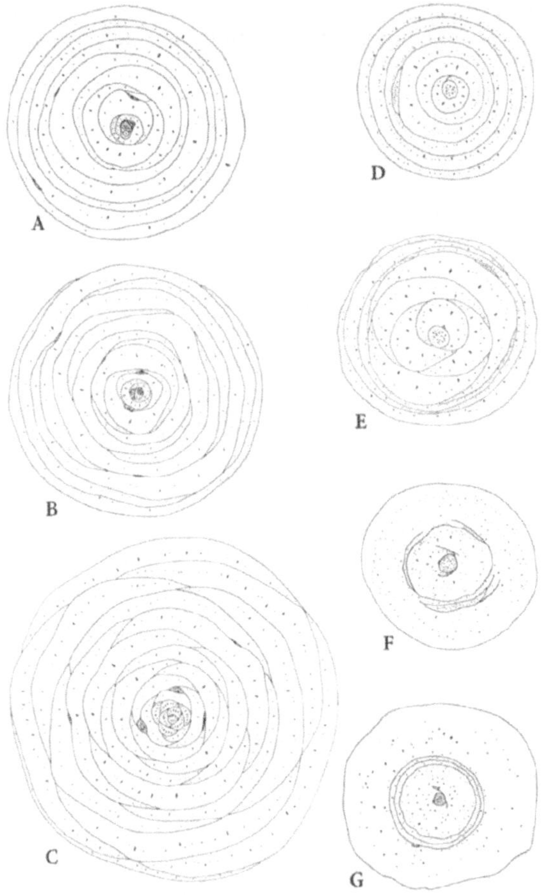

Fig. 81A–G. Hyacinthaceae. Transverse section of bulbs. **A–C** Bulbs comprising bulb leaves of 3 vegetation periods. **A** *Tractema monophyllos*. Bulb leaves vaginate; 1 scape per year. **B** *Hyacinthella millingenii*. Bulb leaves imbricate, 1 or 2 scapes per year. **C** *Prospero elisae*. Bulb leaves imbricate, 2–3 scapes per year. **D–G** Bulbs comprising leaves of 2 vegetation periods, forming annually 1 scape. **D** *Ornithogalum mysum*. Bulb leaves vaginate. **E** *O. pascheanum*. Bulb leaves imbricate. **F** *O. joschtiae*. Bulb leaves concrescent. **G** *Brimeura amethystina*. Bulb leaves vaginate and concrescent, membranous and storing. (Speta 1984)

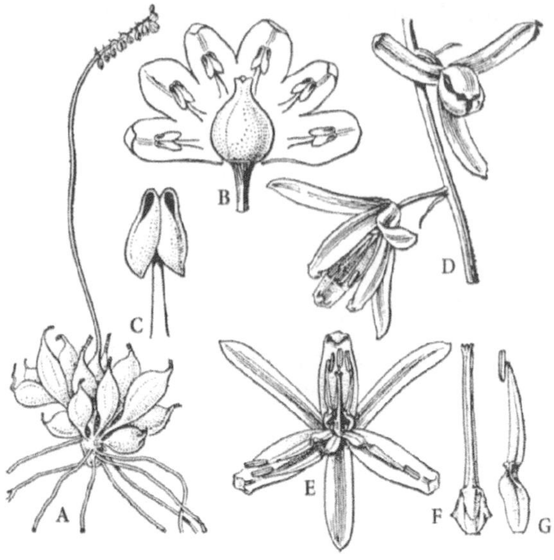

Fig. 82A–G. Hyacinthaceae. **A–C** *Rhadamanthus secundus*. **A** Flowering plant, note the stalked, loose bulb scales. **B** Dissected flower. **C** Anther, ventral view, with apical, porelike slits. **D–G** *Albuca angolensis*. **D** Two flowers, showing the connivent inner tepals. **E** Flower, inner tepals spread out. **F** Pistil. **G** Stamen. (Takhtajan 1982)

In many genera the roots lack a calyptra. Vessels in the roots are arranged in radial rows. The foliage leaves of adult plants are bifacial with the vascular bundles arranged beneath the upper leaf surface. The asssimilating tissue of the leaves consists of isodiametric or palisade cells. In some genera the scape is reinforced by mechanical tissue.

Epicuticular waxes (12 genera studied) occur in the form of fine platelets, which are mostly unordered, or more rarely are arranged parallel to each other (Convallaria type) (Frölich and Barthlott 1988).

INFLORESCENCE STRUCTURE. The flowers are usually arranged in 1-many-flowered simple racemes or more rarely spikes. A prominent scape is usually present (Fig. 83). In *Chlorogalum*, *Camassia*, and *Pseudoprospero* additional racemes branch from the axil of the lowest bracts of the primary raceme, while in *Oncostema* and *Barnadia* rudimentary racemes can be found at the base of the scape (Chouard 1931, Speta 1987). *Bowiea* and *Schizobasis* have highly branched inflorescences (Fig. 84E). The sequence of flowering is acropetalous except for *Hyacinthoides lingulata* and *H. aristidis*, where it is basipetalous. The presence, shape, and structure of bracts and prophylls is highly variable among the genera. The bracts are small and spurred in the Urgineoideae. *Eucomis* stands apart by its terminal tuft of sterile bracts.

FLOWER STRUCTURE. The flowers are only rarely articulated from the pedicel (in *Chlorogalum*, *Schizobasis*, and *Schoenolirion*, Huber 1969). In many, mostly South African, genera unfertilized flowers are shed after blooming (F. Speta, pers. observ.), but the anatomical basis for this is unknown. The outer and inner tepals are more or

Fig. 83A–S. Hyacinthaceae. **A–C** *Puschkinia scilloides*. **A** Flower. **B** Part of paracorolla with stamens. **C** Seed. **D–H** *Scilla luciliae*. **D** Flower. **E** Dissected floral tube. **F, G** Outer and inner stamen, lateral view. **H** Seed with elaiosome. **I–L** *Lachenalia verticillata*. **I** Flowering plant. **J** Flower in lateral and **K** in upper view. **L** Seed with elaiosome. **M; N** *Litanthus pusillus*. **M** Flowering plant. **N** Dissected flower. **O** *Androsiphon capense*, flower. **P–S** *Veltheimia bracteata*. **P** Flowering plant. **Q** Dissected flower. **R** Pergamentaceous fruit. **S** Seed. (Takhtajan 1982)

less equal in shape, the outer ones usually being slightly longer but considerably shorter in *Lachenalia*. They are spurred below the apex in *Dipcadi* subg. *Zuccagnia*. The tips may bear distinct hairs which develop into a robust tuft holding together in bud in *Albuca*. The tepals are frequently connate to various degrees, nearly completely so in *Muscari*. The flowers are mostly actinomorphic, rarely zygomorphic; in *Daubenya*, the lower flowers are zygomorphic and the upper (central) ones actinomorphic (Vogel 1954). Variation in flower color is striking and extends to tepals, filaments, and the ovary. Occasionally the upper sterile flowers are different in shape and color, as in *Muscari comosum*.

The stamen filaments are filiform to narrowly triangular or ribbon-shaped, and often basally connate to the perigone. In *Puschkinia* they form a paracorolla. Apical appendages (as in *Honorius*, Fig. 85B,C) are rarer than basal ones which usually are associated with the outlet of the nectar. In *Eliokarmos* the inner filaments form peculiar square, alate, fringed, corniculate, and cordate protuberances which are contiguous with the ovary (Obermeyer 1978, Müller-Doblies and Müller-Doblies 1996). In *Albuca*, the outer filaments are ribbon-shaped and provided with fertile or sterile anthers, while the inner are always fertile and form a hood over the nectar outlet. Similarly, in *Thuranthos* the flat and convex bases form a Chinese lantern with longitudinal slits (Stirton 1976). In bird-flowers such as *Polyxena* the firm and terete filaments in pollination function as a brush (Vogel 1954).

The anthers are mostly dorsifixed and introrse.

With the exception of *Bowiea*, which has a hemi-inferior ovary, the ovary is always superior and either completely syncarpous or, due to the partial fusion of the carpels, syncarpous at the base and paracarpous at the apex. The ontogeny of the gynoecium has been studied in *Ledebouria socialis* (Sattler 1973) and *Stellarioides longebracteatum* (Tilton and Horner 1983, van Heel 1988). The lateral margins of the three crescent-form primordia eventually meet in the center of the bud and fuse centrally. The septal nectaries are formed by a recess in the partial epidermal fusion of the flanks of the carpels. The degree of longitudinal growth of the outer ovary wall determines the level of the outlet of the nectaries.

The ovaries show different degrees of fusion of the stylar channels. Completely syncarpic ovaries and a style with 3 separate stylar channels are found in *Eucomis*, *Veltheimia*, *Polyxena*, *Lachenalia*, *Ledebouria*, *Barnardia*, and *Bowiea* (see Wunderlich 1937, Buchner 1948). In *Urginea*, *Drimia*, *Dipcadi*, and *Galtonia* the stylar channels are fused to form a coherent 3-lobed stylar channel at the apex but remain separate at the base.

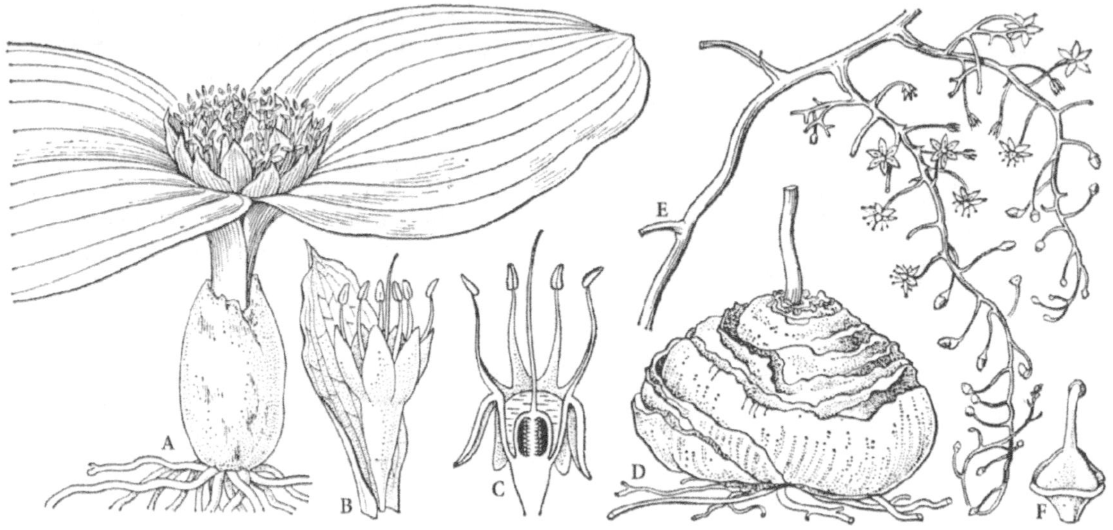

Fig. 84A–F. Hyacinthaceae. A–C *Massonia depressa*. **A** Flowering plant. **B** Flower with bract. **C** Flower, vertical section. **D–F** *Bowiea volubils*. **D** Bulb. **E** Inflorescence. **F** Pistil. (Takhtajan 1982)

The majority of the genera have a coherent 3-lobed stylar channel, which, in some of these genera (e.g., *Prospero*, *Autonoë*), continues into an apically paracarpous gynoecium.

The septal nectaries in Hyacinthaceae are of a simple type (Daumann 1970). They may rarely be connected at the base, as in *Drimia* sp. Nectar drops are rarely seen at the nectarial outlets, because the nectar is drained to the floral base in open grooves or grooves covered by simple hairs (*Oncostema*, *Camassia*) or minute tubes (Speta 1986, 1987). In *Brimeura* and *Tractema* the nectar outlet is at the middle at the ovary (Speta 1987); in *Eucomis*, an outlet may occasionally be lacking; in *Autonoë*, the septal nectary is lacking altogether. In *Dipcadi*, in addition to the septal nectaries a stylar nectary is present (F. Speta, unpubl.).

The number of ovules per locule is important for generic delimitations (Kunth 1844, Salisbury 1866). In each locule, Dipcadieae have numerous ovules arranged in 2 rows; *Muscari* and *Bellevalia* have 2 superposed ovules; *Prospero*, *Autonoë*, *Chouardia*, and *Ledebouria* have 2 ovules side by side; *Barnardia* has a single ovule.

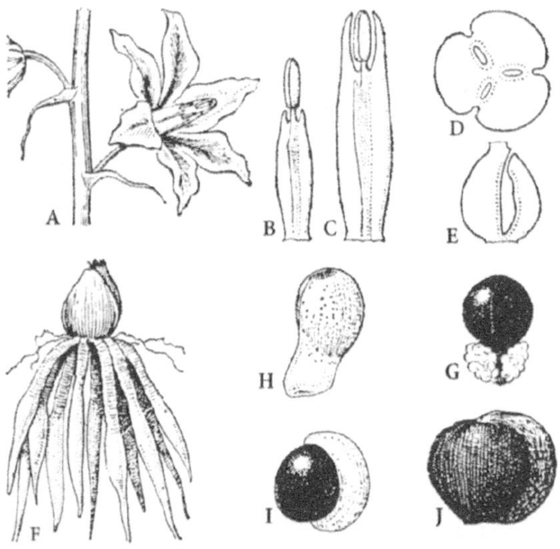

Fig. 85A–J. Hyacinthaceae. **A–E** *Honorius boucheanus*. **A** Flower. **B** Outer and **C** inner stamens with "stipellate" filaments. **D** Ovary, transversal section showing septal nectaries. **E** Ovary, longitudinal section, showing subapical outlet of septal nectary. **F, G** *Scilla bifolia*. **F** Bulb with contractible roots. **G** Seed with elaiosome. **H–J** Seeds with elaiosomes. **H** *Othocallis rosenii*. **I** *O. mischtschenkoana*. **J** *O. melaina*. (Takhtajan 1982)

Pollen Morphology. The pollen is invariably sulcate with the sulcus usually extending to the proximal side of the grain. In most species the reticulum is heterobrochate and exhibits a striking similarity with the Alliaceae. The diameter varies between 20 and 125 μm; the extremely large grains of *Pseudogaltonia*, *Dipcadi*, *Galtonia*, and some *Albuca* differ in exine structure from the rest (Radulescu 1973, Schulze 1980).

The pollen is usually tenui-exinous. In the few examples studied 2 exine types could be distinguished, a Simplisulcate type in *Muscari* spp. and

Galtonia candicans, and an Insulare type in *Puschkinia scilloides* (Halbritter and Hesse 1993). When the anthers open, the wet grains are more or less spherical but during presentation and transport lose water and by invagination of the sulcus become boat-shaped.

EMBRYOLOGY. In the mature pollen grain, the shape of the generative cell and its nucleus shows 2 different configurations: in genera such as *Muscari*, *Puschkinia*, *Barnardia*, *Prospero*, *Scilla* s. str., *Hyacinthoides*, *Hyacinthus*, *Eucomis*, *Loncomelos*, and *Honorius* an elongated nucleus with helical chromatin is embedded in a barely visible cell. A clearly visible, spindle-shaped generative cell containing an ellipsoidic nucleus was observed in *Albuca*, *Galtonia*, *Pseudogaltonia*, and *Camassia* (Wunderlich 1936, Speta 1985, 1986).

Due to the efforts of the Austrian school of embryologists (Schnarf 1929, Wunderlich 1937, Buchner 1948, Svoma 1981, Svoma and Greilhuber 1988 and 1989, Ebert 1993) the development of the female gametophyte and endosperm in Hyacinthaceae is well known. Both the embryo sac and endosperm development can vary considerably between closely related genera and even within them. A monosporic embryo sac and Helobial endosperm is considered to be primitive.

Six embryo sac types have been found so far in Hyacinthaceae: (1) the Polygonum type (normal type) in *Muscari*, *Schnarfia*, *Puschkinia*, *Fessia*, *Scilla* s. str., *Nectaroscilla*, *Autonoë*, *Prospero*, *Drimiopsis*, *Ledebouria*, *Camassia*, *Albuca*, *Dipcadi*, *Galtonia*, *Stellarioides*, *Ornithogalum* s. str., *Eliokarmos*, *Bowiea*, and *Urginea*; (2) the Hyacinthoides (=Scilla, Endymion) type in *Hyacinthoides non-scripta* and *H. hispanica*; (3) the Allium type in *Zagrosia*, *Othocallis*, and *Loncomelos*; (4) the Adoxa type in *Scilla drunensis* subsp. *buekkensis*; (5) the Drusa I type in *Othocallis rosenii*, eight *Scilla* spp., one *Prospero* sp.; and (6) the Fritillaria type in *Hyacinthoides vincentina*. Helobial endosperm is formed, e.g., in *Muscari*, *Puschkinia*, *Prospero*, *Drimiopsis* (?), *Lachenalia*, *Veltheimia*, *Eucomis*, *Galtonia*, *Bowiea*, *Urginea*, and *Ornithogalum* s. str., Nuclear endosperm in *Hyacinthus*, *Othocallis* p.p., *Scilla* s. str., *Hyacinthoides italica*, *Oncostema*, *Tractema*, *Autonoë*, *Camassia*, and *Fessia*.

KARYOLOGY. Hyacinthaceae are karyologically highly diverse, well studied, and present an excellent example of the taxonomic importance of karyology. Polyploidy reaches the levels of deca- and dodekaploidy, such as in *Ornithogalum umbellatum* aggr. and *Muscari neglectum* aggr. ($2n = 12x = 108$) or in *Prospero* ($2n = 10x = 70$). While genera like *Scilla*, *Albuca*, and *Muscari* (all $x = 9$) or *Bellevalia* ($x = 4$) have fixed basic numbers, dysploid series have been found within various genera like *Prospero* ($x = 4, 5, 6, 7$), *Barnardia* ($x = 8, 9$), *Hyacinthella* ($x = 9, 10, 11, 12$), and *Stellarioides* ($x = 2, 3, 4, 5, 6, 7, 8, 9$). In genera, tribes, or even subfamilies a basal karyotype can be identified, even if the chromosome numbers vary, as has been demonstrated already by Delone (Delaunay 1915, 1926) for *Muscari* and *Bellevalia*.

In some groups, the larger chromosomes of a karyotype are acrocentric and subtelocentric (e.g., in Dipcadieae), while in others such as *Ornithogalum* they are metacentric and submetacentric.

The subdivision of the large Linnaean genus *Hyacinthus* has been impressively confirmed by karyology. In *Scilla*, karyology was the starting point for the recent subdivision, for which later additional morphological evidence could be presented (Speta 1978, 1986, 1987). The analysis of *Ornithogalum* is still to be attempted. In *Muscari*, however, the consideration of all available evidence does not favor the splitting into four genera (Speta 1982) that had been suggested on karyological grounds by Garbari and Greuter (1970).

Chromosome banding of metaphasic chromosomes has provided important data mostly at the level of species. Genera such as *Scilla* s. str., *Othocallis* and *Muscari* have a high proportion of heterochromatin, while *Urginea* has very little. Studies in *Scilla* have led Greilhuber (1982) to postulate that heterochromatin is an additive component of the genome.

Furthermore, the structure of the interphase nucleus can be of great importance and permits species distinction in groups such as *Scilla*, which have a constant basic chromosome number ($x = 9$) and in which chromosome rearrangements are insignificant. The diversity of chromosomes is apparent also in the interphase nucleus, and uniformly euchromatic chromomeric nuclei, which are present in *Scilla* s. str., *Othocallis*, and *Ornithogalum* s. str., contrast with an irregular distribution of chromomeres such as in *Muscari*.

Genomic DNA values may vary significantly between species, as in *Scilla*, where closely related species can be identified by their DNA value even when the basic number is constant and few chromosomal rearrangements have occurred. Using this line of evidence, the merging of *Chionodoxa* with *Scilla* could be supported (Speta 1971).

Endomitotic differentiation is widespread in the Hyacinthaceae, and Lauber (1947) demonstrated that the nuclei in the fleshy capsules of *Honorius* are 8- and 16-ploid; also the seed epidermis was found to be polyploid. Later, from the root cortex of *Ornithogalum umbellatum*, *Muscari botryoides*, and *M. neglectum* (Holzner 1952) polyploid nuclei were reported. Endopolyploid nuclei were also demonstrated to occur in elaiosomes (Speta 1972), in *Lachenalia tricolor* 4 *n*, *Othocallis sibirica* 16 *n*, *Puschkinia libanotica* 128 *n*, and in *Scilla* s. str. up to 4096 *n* (!). Hasitschka-Jenschke (1962) reported from *Scilla* endomitotically enlarged antipodal nuclei, but Svoma (1981) found that only monosporic embryo sacs such as those of *Scilla* s. str. and *Schnarfia* are able to produce highly polyploid antipodals, while bisporic embryo sacs as in *Othocallis* remain haploid. Fischer and Nagl (1979) measured the degrees of polyploidy in various tissues within a single species of *Scilla*.

Protein crystals in the interphase nucleus are another important character, which up till now has been found only in subfam. Ornithogaloideae (Fig. 86; Speta 1985, and unpubl.).

POLLINATION. Pollination systems in Hyacinthaceae are both specific and highly diverse even within genera but have only occasionally been studied in detail. Pollen and nectar are offered as floral rewards, color and scent are chief floral attractions. In southern Africa various pollination types have been found (Vogel 1954): melittophilous flowers in *Massonia*, *Polyxena corymbosa*, *Lachenalia pusilla*, *L. contaminata*, *Rhadamanthus*, *Urginea*, and *Litanthus*; psychophilous flowers in *Massonia jasminiflora*, *M. heterandra* and *Dipcadi*; ornithophily in *Massonia depressa*, M. *angustifolia*, *Lachenalia tricolor*, *L. rubicunda*, *L. pendula* and *L. luteola*, *Veltheimia*, and *Daubenya*. In addition there are sapromyiophilous flowers emitting a putrid scent in *Eucomis bicolor* (Pascher 1959). Some flowers open only for a few hours, others are more persistent or perform diurnal movements of opening which, like nectar secretion and scent emission, may be triggered environmentally: flowers of *Ornithogalum umbellatum* have pollen sensitive to rain (Hansgirg 1904); they open in sunlight at 11.00 h and close in rain, shading, or in the evening. Flowers of *Chlorogalum* open for only one night from about 17.00 or 19.00 to 23.00 h and are visited by *Xylocopa tabaniformis orpifex* and *Bombus* spp., which collect pollen and nectar (Stockhouse and Wells 1978, Jernstedt 1980). Those that are not pollinated abscise after 3 or 4 days on a preformed point below the receptacle. Similarly, the nodding and strongly scented flowers of *Thuranthos macranthum* open only once between 17.30 and 18.00 h and close between 4.00 and 5.00 h., when they are visited by the noctuid *Diaphone eumela* (Stirton 1976).

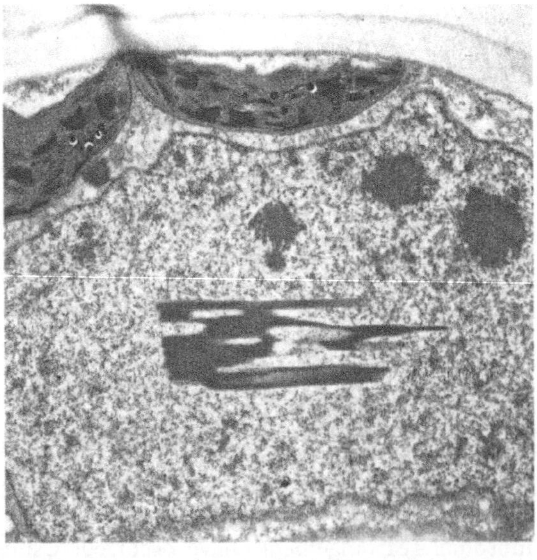

Fig. 86. Hyacinthaceae. Protein crystal in the nucleus of *Stellarioides longebracteatum*, TEM ×8000. (Photo M. Weber)

REPRODUCTION. Sexual reproduction predominates. Agamospermy (in pentaploid forms) and vegetative reproduction by bulbils in seedless strains are known from *Ornithogalum umbellatum*. Bulbils may be formed throughout the family, e.g., as daughter bulbs in polyploid strains of *Muscari neglectum* aggr. and *Ornithogalum* subg. *Ornithogalum* (sometimes even in diploid *O. undulatum*), in *Eliokarmos mater-familias*, or in *Stellarioides longebractatum*, where they are adnate to the sheathing leaf base and produced in different tiers (Lonay 1902). Bulbils are also produced on leaves outside the bulb, e.g., on subterraneous leaf blades of *Brimeura fastigiata*, or on leaf tips of various species of *Ledebouria*, or anywhere on the leaf blade of *Eliokarmos thyrsoides* and *Drimiopsis kirkii* (Lenski 1958). Under favorable light and water conditions bulbils may be produced even on detached leaves along the margin of the leaf base in *Hyacinthoides*, *Tractema*, *Brimeura*, *Hyacinthus*, *Lachenalia*, *Muscari*, *Honorius*, *Loncomelos*, *Galtonia*, etc. Bulbils produced from roots are known only from *Oncostema hughii* and *Urgineopsis puffii*.

Fruit and Seed. The loculicidal capsules are spherical, ellipsoidic, or ovate, occasionally trigonous (*Muscari, Bellevalia*) or 3-lobed (*Stellarioides, Cathissa, Thuranthos*, etc.). Their orientation is upright, even in species in which the flowers were patent or nodding (*Hyacinthoides non-scripta, Galtonia, Pseudogaltonia, Dipcadi, Albuca, Coilonox, Thurantos*, etc.). More rarely, the fruits are patent (*Muscari, Bellevalia*) or pendulous (*Honorius, Ornithogalum oligophyllum, O. alpigenum, Eucomis, Veltheimia*). The mature pericarp may be thin and pergamentaceous or succulent and mucilagenous. The valves may open completely, as in *Prospero* and *Chouardia*, so that the seeds are presented on the same level with the septa; a censerlike structure is formed if the loculicidal opening is restricted to the apex of the capsule. The narrow, papery, and alate capsules of *Veltheimia* and *Eucomis* open incompletely, separate from the inflorescence, and are dispersed by the wind.

The seeds are often globose, ovate, or guttiform. Elaiosomes are rarely formed; when present, they may be formed from the exostome (*Scilla* s. str.), from part of the raphe close to the hilum (*Schnarfia*), from the exostome and neighboring parts of the raphe (*Hyacinthus orientalis, Othocallis sibirica, Ornithogalum termessense*, and *Lachenalia* spp.), or from the complete raphe in *Othocallis mischtschenkoana, O. melaina*, etc.) (Sernander 1906, Speta 1972); *Puschkinia* has a sarcotesta.

Ornithogaloideae may have edged or flattened seeds; very small, comma-shaped, glabrous, or hairy seeds are found in *Eliokarmos*. Epidermal seed characters may be used to identify *Ornithogalum* spp. (Zahariadi 1962). Urgineoideae have shiny black seeds with ridges or wings; their subepidermal tissue disintegrates, resulting in a brittle testa loosely enclosing the endosperm.

Anatomically, the seeds were studied by Huber (1969). According to him, the outer integument is multilayered; only in some taxa with small or discoid, winged seeds is the number of cell layers reduced; in *Puschkinia*, apart from the epidermis, only 3 layers were observed. The epidermis is usually phytomelan-encrusted, although the phytomelan layer is much thinner (4–10, rarely up to 15 μm) than in the Alliaceae, for example. Phytomelan is lacking only occasionally such as in *Puschkinia* and some *Scilla* spp. The outer integument hardly provides mechanical strength; the inner integument consists of completely collapsed cells.

The endosperm stores aleuron, fat oil, and reserve cellulose; starch is usually absent. The embryo is cylindrical, straight, or very rarely slightly curved and usually attains 4/5 of the length of the endosperm.

For further details, see Bresinsky (1963) (elaiosomes), and Jessop (1975), Obermeyer (1978, 1980), Müller-Doblies and Müller-Doblies (1996) (REM studies on seed surface).

Dispersal. Atelechory and myrmekochory are frequent and may interact variously (Sernander 1906, Bresinsky 1963, Speta 1972). Seeds are laid out close to the parent plant by means of flaccid inflorescences, nodding capsules (e.g., *Honorius, Ornithogalum oligophyllum, O. alpigenum*), or short-scaped capsules which open at ground level (*Massonia, Daubenya, Polyxena, Lachenalia, Ornithogalum lanceolatum*, etc.). Seeds with elaiosomes may then be further dispersed by ants, or capsules may be actively buried in or at least on the ground through geotropic movements of the pedicel (*Ornithogalum* spp.), possibly also with subsequent ant dispersal (Sernander 1906). In species with rigid or elastic upright scapes, seed dispersal from the censerlike capsules may be induced by the wind, passing animals, or probably also raindrops.

Phytochemistry. Cardiacally active stereoids of the Hyacinthaceae comprise bufodienolids and cardenolids. In the former, a hexamerous lactone ring is attached to the sterane moiety at C17, while in the latter it is a pentamerous lactone ring. Bufodienolids so far appear to be restricted to the Urgineoideae (*Urginea, Drimia, Bowiea*, synopsis: Krenn 1994), cardenolids to the Ornithogaleae (though not occurring in all species). Cholestane glycosides have been found in *Zahariadia saundersiae* and *Eliokarmos thyrsoides* (Mimaki et al. 1992, 1996, Kubo et al. 1992). Bulbs of *Camassia cusickii* contain steroid saponins (Mimaki et al. 1991, 1992).

Flavonoids have been found in various genera (Skrzypczak 1977). Homoisoflavanones (Heller and Tamm 1981) have been found in many genera of the Hyacinthoideae and *Camassia* (Chlorogaloideae); elsewhere they are known from *Ophiopogon*. Fructanes are the major carbohydrate storage compounds of the bulbs, sometimes accompanied by starch (Hegnauer 1963, 1986). Chelidonic acid is also known from Hyacinthaceae.

Subdivision and Affinities. The way for the separation of Hyacinthaceae from the Liliaceae s.l. has been paved by anatomical (Fuchsig 1910) and embryological (Schnarf 1929, Wunderlich 1937, Buchner 1948) studies. Huber (1969) supported this separation with seed anatomical evidence and put it into a systematic perspective, which was formalized in a new classification by Dahlgren et al. (1985). At present we understand the Hyacinthaceae as part of a major branch of the Asparagales, which is held together by successive microsporogenesis, and which in the *rbc*L topology (Chase et al. 1995, Fay and Chase 1996) appears as sister to the Themidaceae in close relationship with the Agavaceae and Anthericaceae.

Characters relevant for a subdivision of the family include pistil structure, seed morphology, bracts, prophylls, karyology, phytochemistry, and molecular evidence. Flower color has proven to be of limited value; the degree of fusion of the perigone is worthless. The Hyacinthaceae are here divided into five major groups, the coherence of four of which is also supported by molecular evidence (Chase et al. 1995, Fay et Chase 1996, Speta and Pfosser, unpubl. molecular work), while one, Chlorogaloideae, according to serological (Čupov 1994, 1995, Čupov and Kutjavina 1981) and molecular (Fay et Chase 1996) evidence appears as an alien element in Hyacinthaceae that may be closer to Agavaceae:

1. Subfam. Chlorogaloideae Speta (1998). Homoisoflavanones; nucleus without protein crystals.
2. Subfam. Oziroëoideae Speta (1998). Chemical data still missing; protein cystals not yet found.
3. Subfam. Urgineoideae Speta (1998). Bufodienolides; nucleus without protein crystals.
4. Subfam. Ornithogaloideae Speta (1998). Cardenolides, or nucleus containing protein crystals (Fig. 86).
5. Subfam. Hyacinthoideae Link (1829). Homoisoflavanones; nucleus without protein crystals.

Distribution and Habitats. The family has centers of diversity in southern Africa and the Mediterranean. The Chlorogaloideae are restricted to North America; *Oziroë* is found only in Andine South America. The Hyacinthoideae have the widest distribution, extending to E Asia (*Barnardia*), India (*Ledebouria*), NW Europe (*Tractema*), and Central Asia (*Fessia*, *Bellevalia*, *Muscari*). The Ornithogaloideae extend from S Africa (*Dipcadi*) and southern France to Arabia and India, with *Loncomelos* ranging from the Mediterranean to the Ural mountains. The Urgineoideae are distributed from S Africa to the Mediterranean and further to Arabia, India, and Burma.

The Hyacinthaceae prefer open, sunny habitats with dry and hot vegetation periods. In temperate regions they appear as spring geophytes in deciduous broadleaved forests. Their vertical distribution can extend from sea level to mountain regions. Some taxa such as *Prospero* and *Loncomelos* grow in swampy habitats, where they may cooccur with *Isoëtes*.

Economic Uses. *Urginea maritima*, the sea onion, has been in medicinal use since earliest times and was mentioned as early as 1554 B.C. in the *Papyrus Ebers* of the Middle Empire of Egypt as a cure for dropsy (Speta 1980). Bufodienolids from this species and from *Thurarthos indicum* aggr. are used for the production of cardiacally active substances. Cardenolides from *Ornithogalum* have apparently not yet been utilized. There are occasional reports of an ethnomedicinal use of *Barnardia* in China, and *Ledebouria* spp. and *Albuca abyssinica* in E and S Africa. In Central Europe, *Stellarioides longebracteatum*, "false sea onion", seems to be used in the same way as the genuine sea onion despite its totally different chemistry. In S Africa, several species such as *Zahariadia saundersiae*, *Eliokarmos thyrsoides*, *Ledebouria inguinata*, *L. revoluta*, *L. ovatifolia*, *L. cooperi*, and various Urgineae are poisonous to grazing animals. The toxic scillirosid (a bufodienolid) is used for poisoning rats because it attracts these rodents, while domestic animals show little interest for it.

Hyacinthaceae are only occasionally used for human consumption. Thus in Greece the bulbs of *Muscari comosum* are eaten pickled; in France the inflorescences of *Loncomelos pyrenaicus* are eaten as a vegetable, and bushmen in Africa eat the bulbs of *Ledebouria apertiflora* and *L. revoluta*. The western species of *Camassia* once yielded a food called quamash or camas by some North American tribes of indians.

More important is the use of various species as ornamentals and cut flowers. Species of *Scilla*, *Othocallis*, *Muscari*, *Hyacinthus*, *Puschkinia*, *Chouardia*, *Hyacinthoides*, and *Camassia* are spring flowers in Northern Hemisphere parks and gardens. In southern Africa, species of *Eucomis*, *Veltheimia*, *Galtonia*, and others are cultivated as ornamentals. *Eliokarmos thyrsoides* and related species are important as cut flowers.

Key to the Genera

1. Seed coat very brittle, shiny black or brown; seeds flattened and fringed or winged; endosperm flattened or ellip-

soid; at least the lower bracts spurred; prophylls smaller than bracts, but usually present; flowers in various colors but never blue, violet, pink, yellow, or orange **(Urgineoideae)**
- Seed coat firmly adhering; seeds globose, drop or pear-shaped, sometimes edged but never winged; bracts large to minute or lacking; prophylls present or lacking; flowers in various colors 11
2. Inflorescences highly ramified 3
- Inflorescences a 1- to many-flowered raceme 4
3. Scape wiry; inflorescences up to 35 cm long; ovary superior; tepals forming a persistent cap over the capsule **7. *Schizobasis***
- Scape succulent; inflorescence longer than 35 cm, always containing sterile branches; ovary ± inferior; tepals persistent at the base of the capsule **6. *Bowiea***
4. Foliage leaves distichous **9. *Urgineopsis***
- Foliage leaves spirally arranged 5
5. Rachis long and twining; bracts often with 2 flowers **8. *Igidia***
- Rachis not twining; bracts with a single flower 6
6. Tepals connate to at least 1/3 of their length 7
- Tepals only slightly connate to nearly free 9
7. Bulb very small; 1 foliage leaf; peduncle short, thin, with 1 (−2)-flowered raceme; pedicel nodding, erect in fruit **11. *Litanthus***
- Bulbs larger; peduncle with more than 2 flowers 8
8. Anthers with apical pore or slit extending to the middle **10. *Rhadamanthus***
- Anthers with a longitudinal slit extending over the whole length **12. *Rhadamanthopsis***
9. Anthers basifixed 10
- Anthers dorsifixed 11
10. Bulb leaves concrescent, enveloped by sheathing cataphyll; the firm filaments and the style geniculate near towards stigma **14. *Tenicroa***
- Bulb leaves imbricate, not enveloped by a cataphyll; raceme lax; anthesis nocturnal; style widened towards stigma **13. *Thuranthos***
11. Tepals connate up to 1/4, free lobes completely reflexed **15. *Drimia***
- Tepals only slightly connate at base, free lobes stellate, plate-shaped or campanulate **16. *Urginea* s.l.**
12. Seeds large, 3–9 mm long and flattened, orbiculate, D-shaped, or elongate; or minute, 0.5–1 mm long and papillate, pilose or glabrous, sometimes edged; or medium-sized, 1.5–2.5 mm long and globose to ellipsoid (similar to many Hyacinthoideae), but then tepals white with a green stripe outside; bracts usually large; prophylls lacking except for *Pseudogaltonia*; flowers white, green, yellow to orange **(Ornithogaloideae)** 13
- Seeds pear-shaped to globose or ellipsoid, not distinctly edged, carinate or alate 25
13. Seeds flattened, orbiculate, D-shaped, or elongate; tepals green, brown or yellow, or whitish, with a green streak **(*Dipcadieae*)** 14
- Seeds minute, comma-shaped, glabrous or shortly pilose, rarely edged, or large, elongate, with irregular edges, or globose to ellipsoid; tepals white, with or without a green stripe, or yellow or orange **(Ornithogaleae)** 19
14. Tepals forming a distinct tube; flowers nodding 15
- Tepals ± free 17
15. Perigone tube slightly curved; bracts and prophylls present; flowers green; bulb large, with a dense reticulate tunic **20. *Pseudogaltonia***
- Perigone tube straight; prophylls lacking 16
16. Tepals connate into a tube for 1/2, tips not reflexed, white or green **22. *Galtonia***
- Tepals connate for 1/4 to 1/3 **21. *Dipcadi***
17. Inner tepals connivent over stigma, outer tepals ± spreading; fertile stamens 3 or 6, all or only the inner ones compressed in the lower 1/3; carpels ridged; style thick, stigma hairy **19. *Albuca***
- Tepals ± stellate, spreading, with green stripe on the outside; carpels not ridged; style thin 18
18. Tepals greenish, whitish or yellowish, 6–10 mm long; filaments ovate-acuminate, tapering in the upper 1/3 **17. *Stellarioides***
- Tepals yellow, (8−) 10–20 mm long, filaments subulate **18. *Coilonox***
19. Seeds minute; tepals white, yellow, or orange, without green stripe **26. Eliokarmos s.l.**
- Seeds large and edged, or globose to ellipsoid 20
20. Capsule large, fleshy, pendent; at least the inner filaments ligulate, firm, "stipellate" at apex **28. *Honorius***
- Capsule ± thin-walled 21
21. Inflorescence umbellate or comose 22
- Inflorescence an elongated raceme 24
22. Tepals cream to white, without green stripe; ovary globose to elongate, blackish green 23
- Tepals white with or without green stripe outside; foliage leaves with or without white stripe; carpels with or without 1 or 2 longitudinal ridges; seeds globose to ellipsoid **29. *Ornithogalum***
23. Unfertilized flowers dropping after anthesis; bulb leaves vaginate; seeds large, edged **23. *Zahariadia***
- Unfertilized flowers persistent; bulb leaves imbricate **24. *Melomphis***
24. Pedicels at anthesis usually much shorter than 1 cm; tepals white without green stripe; perigone usually erect campanulate **25. *Cathissa***
- Pedicels longer than 1 cm; tepals white, yellowish or greenish, usually with a green stripe outside **27. *Loncomelos***
25. Tepals multinerved, white, greenish or blue; stigma usually distinctly 3-lobed; inflorescences double racemes, sometimes reduced to simple ones; N America **(Chlorogaloideae)** 26
- Tepals 1-nerved (except *Avonsera*); stigma inconspicuous, rarely indistinctly 3-lobed 29
26. Stem a vertical, fleshy, thick rhizome extending up to 12 cm below the leaf bases, with or without prominent apical bulb; ovary sessile, globulose; locule with 2 ovules side by side; style short, stigma minute, entire or ovary slightly 3-lobed; seeds globose, testa smooth, glossy black **1. *Schoenolirion***
- Stem not extending below the bulb; bulb without rhizome; perigone 5–40 mm long, white, pink, purple or blue; ovary shortly stipitate, style with a distinctly 3-parted stigma; seeds elongated, testa rough and dull black 27
27. Locules with ca. 10 ovules; pedicels 10–50 mm long; perigone large, bluish violet, blue, or white, 7–40 mm long; tepals connivent over capsule after anthesis or withering separately at base of capsule **3. *Camassia***
- Only 2 ovules per locule, standing side by side 28
28. Inflorescence racemose; pedicels 2–3 cm long; perigone white, creamy white, or dark purple, 5–12 mm long; tepals withering separately at base; style stout, ± persistent **2. *Hastingsia***
- Inflorescence paniculate; pedicels 2–35 mm long; perigone white to pink or blue; perianth segments twisting over capsule at drying; style threadlike, deciduous **4. *Chlorogalum***

29. Flowers whitish or greenish; bracts sometimes with 2 flowers; western S America (**Oziroëoideae**) **5. *Oziroë***
- Flowers blue, violet, purple, pink, yellow, green, brown, whitish, in all shades; Old World: Europe, Africa, Asia (**Hyacinthoideae**) 30
30. Raceme with apical tuft of sterile green bracts **33. *Eucomis***
- Tuft different or absent 31
31. Tepals free or slightly connate 32
- Tepals connate to at least 1/4 53
32. Perigone ± stellate 33
- Perigone ± campanulate with a short or long pseudotube 49
33. Bracts long, ± broad at base; perigone blue 34
- Bracts short to lacking 36
34. Prophylls lacking; bulb small to medium-sized **51. *Tractema***
- Prophylls present 35
35. Prophylls very short, concealed by large bracts; bulbs large, formed by imbricate storage leaves of 3 vegetation periods **48. *Oncostema***
- Prophylls ± long; bulbs small to medium, formed by partially connate storage leaves of 1 vegetation period **49. *Hyacinthoides***
36. Bract and prophylls subulate 37
- Bract and prophylls very short, collar-shaped, irregularly shaped, or lacking 40
37. Tepals blue; ovary white; seeds vinaceous when fresh, light brown when dry **30. *Merwilla***
- Tepals pink or whitish 38
38. Locules with several ovules **31. *Schizocarphus***
- Locules 1–2-ovulate 39
39. Locules 1-ovulate **46. *Barnardia***
- Locules 2-ovulate; lowest bract of peduncle occasionally with a lateral raceme **32. *Pseudoprospero***
40. Each locule with 2 ovules side by side 41
- Each locule with 2 to several superimposed ovules 44
41. Bracts and prophylls lacking; tepals violet to pink; ovary bottle-shaped, light blue; Northern Hemisphere, autumn flowering, leaves proteranthous **54. Prospero**
- Bracts and prophylls present 42
42. Peduncle smooth, succulent, later rigid; tepals pinkish or blue; inner septal nectaries lacking; fruit succulent, orange or blackish vinaceous; cotyledon short and hypogeal **47. *Autonoë***
- Peduncle ± wiry or long and thick; fruit not succulent, thin-walled; inner septal nectaries present 43
43. Peduncle 30–80 cm long, smooth; raceme many flowered; bulb large; cotyledon epigeal **60. Nectaroscilla**
- Peduncle distinctly shorter, 20–30 cm long, wiry, sulcate; cotyledon ± long, hypogeal **61. *Chouardia***
44. Locules with 2 ovules superimposed **63. *Zagrosia***
- Locules with (2–)several superimposed ovules 45
45. Bracts very short, subulate or lacking; scape 1, terete; seeds with elaiosome formed by exostome; testa thin and smooth **66. *Scilla*** s. str
- Bracts short, collar-shaped or irregularly shaped; scapes 1-several, ± semiterete 46
46. Style 2–4 mm long; raceme dense, rachis and pedicels tinged violet 47
- Style 4.5–20 mm long; raceme lax 48
47. Seeds yellow; elaiosome formed by part of raphe bordering the hilum; testa smooth **62. *Schnarfia***
- Seeds brownish black; elaiosome lacking; testa ± distinctly papillose **58. *Pfosseria***
48. Bracts irregular, sometimes slightly "spurred"; seeds dark brown, without elaiosome; testa smooth **57. *Fessia***
- Bracts collar-shaped; seeds yellow, brown or black, with or without elaiosome; testa papillose **56. *Othocallis***
49. Tepals long, blue; bracts and prophylls long; 2-few ovules in each locule **49. *Hyacinthoides***
- Tepals short, greenish, sometimes tinged white or vinaceous, or pink; bracts short or lacking; locules with 2 ovules side by side 50
50. Tips of tepals distinctly reflexed, perigone with a short pseudotube; ovary stipitate, gugelhupf-shaped; filaments filiform, contiguous with pistil; anthers and style at least partially vinaceous; bracts and prophylls small, but usually present; foliage leaves often with vinaceous spots **35. *Ledebouria***
- Tepals not or only slightly reflexed; ovary not stipitate 51
51. Perigone whitish or greenish, nearly closed at anthesis, succulent and persistent in fruit **36. *Drimiopsis***
- Perigone pink-purple or greenish brown 52
52. Tepals 1-nerved **37. *Resnova***
- Tepals many-nerved **34. *Avonsera***
53. Perigone tube more than 2 cm long; flowers nodding **38. *Veltheimia***
- Perigone tube considerably shorter 54
54. Bracts ± large; prophylls lacking 55
- Bracts small or lacking 62
55. Filaments adnate to perigone and connate to each other to form a ± long tube 56
- Filaments adnate to perigone but free at base 59
56. Flowers green and succulent; inflorescence elongate; tube of filaments short **39. *Whiteheadia***
- Flowers not green and succulent; inflorescence usually capitate 57
57. Filaments fused to form a tube under 10 mm long **43. *Massonia***
- Filaments fused to form a tube at least 10 mm long 58
58. Filaments attached to perianth tube below the middle of the tube **40. *Amphisiphon***
- Filaments attached to perianth tube at the mouth **41. *Androsiphon***
59. Flowers zygomorphic 60
- Flowers actinomorphic; bases of foliage leaves surrounded ± distinctly by a sheathing cataphyll 61
60. Lower flowers strongly zygomorphic, red or yellow; seeds without elaiosome **42. *Daubenya***
- Lower flowers slightly zygomorphic; perigone tube rather short; outer tepals often considerably shorter than inner; seeds often with an elaiosome formed by the exostome and bordering parts of raphe **45. *Lachenalia***
61. Bracts small, inconspicuous; flowers pink or white; style 1–18 mm. S Africa **44. *Polyxena***
- Bracts long; flowers blue or whitish pink; style 3–4 mm. Western Mediterranean **50. *Brimeura***
62. Bracts minute or lacking; filaments broad, usually ligulate; seeds globose, with elaiosome formed by exostome **66. *Scilla*** s. str.
- Bracts and prophylls very short; filaments filiform, narrowly triangular 63
63. Filaments forming a paracorolla; seeds with sarcotesta **55. *Puschkinia***
- Paracorolla lacking; seeds without sarcotesta 64
64. Perigone longer than 15 mm, firm and stiff, free tips ± long and reflexed **59. *Hyacinthus***
- Perigone shorter than (10–)13 mm, usually highly connate 65
65. Capsule ± globose; each locule with 2 ovules side by side, rarely 1 or 2 additional ones superimposed; filaments in 1 row 66

- Capsule trigonous; each locule with 2 superimposed ovules 67
66. Perigone uniformly blue, erect **52. *Hyacinthella***
- Perigone two-colored, purple-violet or green-white, nodding 53. ***Alrawia***
67. Filaments in 2 rows **64. *Muscari***
- Filaments in 1 row **65. *Bellevalia***

I. Subfam. Chlorogaloideae Speta (1998).

Bulb leaves present (except for some *Schoenolirion* spp.), imbricate, short-lived. Scale and foliage leaves with little mucilage. Scape terete, sometimes with additional bracts below the rachis, inflorescence a panicle often reduced and appearing as racema. Bracts medium-sized to large, prophylls lacking. Pedicels 2–5 mm long. Perigone 4–40 mm long, free, white, cream, greenish, blue, or violet. Ovary with septal nectaries; nectar ducts tubular or covered by hairs. Seeds comparatively large. Embryo straight, nearly as long as the endosperm. Cotyledon short, hypogeal, followed by 1 foliage leaf.

For the proposed relationship of this subfamily with *Agave*, see p. 268.

1. *Schoenolirion* Torrey ex E.M. Durand

Schoenolirion Torrey ex E.M. Durand, J. Acad. Nat. Sci. Philadelphia II, 3: 103 (1855), nom. cons.; Sherman & Becking, Madroño 38: 132–135 (1991), rev.

Rhizome extending 1–12 mm below the leaf bases, with or without distinct bulb, with contractile roots lasting 1–2 vegetation periods. Leaves grasslike. Raceme simple or branched, on terete scape. Bracts short. Flowers yellow or white. Unfertilized flowers soon caducuous. Tepals ± free, (4–)4.5–6.5(–7.5) mm long, 3–7-nerved, withering and persisting separately. Stamens equal. Anthers 1–2 mm long. Ovary sessile, globose, locules with 2 ovules side by side. Style conical, short. Stigma small, entire or slightly 3-lobed. Capsule globose. Seeds black, smooth. $2n = 24, 30, 32, 49$. Three spp., N America.

2. *Hastingsia* S. Watson

Hastingsia S. Watson, Proc. Amer. Acad. Arts Sci. 14: 217, 242 (1879); Sherman & Becking, Madroño 38: 135–138 (1991), rev.

Bulb slender. Bulb leaves imbricate. Tunic fibrose. Foliage leaves grasslike, prominently keeled. Scape with hypsophylls; inflorescence a dense simple or branched raceme. Pedicels 2–3 mm long. Tepals 3-nerved, white, cream, dark purple, 5–12 mm long, reflexed, withering and persisting with the fruit. Filaments 3 + 3. Ovary stipitate, oblong, locules with 2 ovules side by side. Capsule 6–10 mm long. Seeds rugose. $2n = 52$. Four spp., N America.

3. *Camassia* Lindl.

Camassia Lindl., Bot. Reg. 18: 1486 (1832), nom.cons.; Gould, Amer. Midl. Naturalist 28: 712–742 (1942), rev.
Sitocodium Salisb. (1866).

Roots branched. Base of foliage leaves enclosed by 1–2 vaginate and 1–2 imbricate, membranaceous scale leaves. Tunic brown or black. Foliage leaves narrowly lanceolate. Inflorescence a panicle often seemingly a raceme with rudimentary branches in the axils of the large lower bracts, scape firm. Pedicels 1–5 cm long. Tepals 1.5–4 cm long, 3–9-nerved, blue, violet, or white, withering and persisting contoreted above the fruit. Filaments long. Ovary ellipsoid, yellowish green, densely pubescent. Locules up to 12-ovulate. Septal nectaries opening with downward directed duct covered by hairs. Capsule compact, subglobose to oblong. Seeds pear-shaped, smooth or rugose. $2n = 30$. Five spp., western N. America.

4. *Chlorogalum* (Lindl.) Kunth

Chlorogalum (Lindl.) Kunth, Enum. 4: 681 (1843), nom. cons.; Hoover, Madroño 5: 137–147 (1940), rev.

Roots thick, branched. Bulb leaves imbricate. Tunic fibrose. Foliage leaves few, prominently keeled. Inflorescence a slender double raceme; scape with reduced leaves. Bracts present. Anthesis diurnal or nocturnal, unfertilized flowers caducous. Tepals white to pink or blue, 3-nerved, reflexed at anthesis, withering and persisting contorted above the fruit. Filaments filiform. Ovary globose, locules with 2 ovules side by side. Capsule subglobose. $2n = 30, 34, 36, 60$. Five spp., N America.

II. Subfam. Oziroëoideae Speta (1998).

Few white and branched feeding roots and thickened contractible roots present. Bulb leaves imbricate, mucilaginous. Foliage leaves few, thick and grooved, synanthous. Scape with prominent bracts, each bearing 1–2(–3) flowers. Pedicels straight, pointing obliquely upwards. Perigone nearly free, about 6 cm long, whitish, stellate. Filaments patent. Anthers dorsifixed. Ovary conical, locules 2–6-ovulate. Style present, stigma inconspicuous. Capsule drg, dehiscence loculicidal and ventrifragous. Seeds up to 6–7 mm long, pear-shaped, black, rugose; hilum lateral at apex. Embryo straight, nearly as long as the endosperm. Cotyledon epigeal. $2n = 30, 34$. Nucleus without protein crystals. Only one genus, western S America from Peru and Bolivia to Chile and Argentina.

5. *Oziroë* Raf.

Oziroë Raf., Fl. Tellur. 3/VII: 53, 1836 (1837).
Scilla sect. *Hesperoscilla* Benth. (1883).
Fortunatia G.F. Macbride (1931).

Description as for subfamily. About 10 spp.

III. Subfam. Urgineoideae Speta (1998).

Bulb size ranging from smallest (*Litanthus*) to largest (*Urginea, Drimia*) in the family. Bulb leaves usually imbricate, lasting ca. 3 vegetation periods. Scapes usually solitary. Bracts narrow, the lowest distinctly spurred; prophylls present. Flowers white with reddish, brownish, or greenish stripe, or green, reddish, or brownish. Filaments not appendaged. Ovary oblong, locules with many ovules in 2 rows. Stigma inconspicuous. Seeds usually compressed, sometimes alate, shiny black; testa loose and fragile; endosperm compressed or ellipsoid. Nucleus without protein crystals. Bufodienolids present, hence poisonous.

The definition of genera in this subfamily is often unsatisfactory.

6. *Bowiea* Harvey ex J.D. Hooker

Bowiea Harvey ex J.D. Hooker, Bot. Mag.: t. 5619 (1867), nom. cons.

Adult bulb above ground, without foliage leaves. Roots thick, branched. Bulb Leaves greenish, vaginate. Bases of cataphylls lasting several vegetation periods. Inflorescence green, thin, twining, scape strongly branched. Flowers dioecious or monoecious. Tepals greenish, reflexed, drying and persisting. Filaments short, narrowly triangular. Ovary semiinferior, green, locules ca. 5-ovulate. Style short. Capsule papyraceous. Seeds black, oblong, compressed. Cotyledon epigeal, followed by 1 foliage leaf. $2n = 20$. Two spp., southern Africa to Namibia and E Africa.

7. *Schizobasis* Baker

Schizobasis Baker, J. Bot. 11: 105 (1873); Jessop, J. South Afr. Bot. 43: 309–312 (1977).

Bulbs formed by imbricate bulb leaves. Adult bulb without foliage leaves. Scapes 1–3, thin, straight or twining, highly branched. Bracts shortly spurred. Tepals nearly free, ± patent, 1-nerved, drying and persisting. Filaments filiform or lanceolate, anthers oblong. Ovary globose, locules 4–8-ovulate, style short. Capsule membranaceous, subglobose, trigonous. Seeds black, compressed, globose, ovoid, or edged. $2n = 20$. Nine spp., southern Africa to Angola and Zimbabwe.

8. *Igidia* Speta

Igidia Speta, Phyton (Austria) 37 (1998).

Bulb compressed-globose. Inflorescence with terete scape 25–30 cm long bearing a 25–30-flowered raceme with twining rachis. Bracts linear-lanceolate, 8–30 × 1–3 mm, equalling or exceeding the pedicels. Tepals nearly free, patent, 10 × 3.5 mm, obscurely 3–5-nerved. Filaments 6–7 mm long, conspicuously broadened (1.5 mm) below the middle. Ovary 3-lobed, locules ca. 20-ovulate. Style 2 mm long. Caspule obtusely obovoid-trilobate. Seeds compressed, alate, 11 × 6 mm. $2n = 12$. One sp., *I. volubilis* Speta, Madagascar.

9. *Urgineopsis* Compton

Urgineopsis Compton, J. Bot. 68: 107 (1930); Speta, Linzer Biol. Beitr. 12: 205–206 (1980).

Bulb leaves formed by cataphylls and bases of foliage leaves, white, vaginate and imbricate, distichous, lasting 2 vegetation periods. Tunic thin, light brown. Roots branched, with root shoots. Raceme many-flowered on a solitary scape 1.5 mm thick. Bracts short, the lowermost spurred. Prophylls lacking. Pedicels ± short. Tepals white with a vinaceous stripe outside, 5–6 mm long, stellate, ± reflexed at full anthesis. Filaments white, patent. Anthers yellow. Ovary globose, light green, locules ca. 4-ovulate, style 2.5 mm long, stigma inconspicuous. $2n = 14$. Only one sp., *U. salteri* Compton, southern Africa.

10. *Rhadamanthus* Salisb.

Rhadamanthus Salisb., Gen. Pl.: 37 (1866); Nordenstam, Bot. Not. 123: 155–182 (1970), rev.
Rhodocodon Baker (1881).

Roots fibrous. Bulb leaves sometimes only loosely imbricate or even separate and stalked, apically often with short or rarely elongated papyraceous sheaths. Leaves proteranth. Raceme solitary, few- to many-flowered, on slender scape. Bracts shorter than pedicels, lower bracts usually spurred. Pedicels erect-patent, spreading or ± curved near the apex (upwards in bud and fruit, downwards during anthesis). Perigone campanulate to subglobose, connate basally or up to above the middle. Anthers often ± connivent around the gynoecium, basifixed or medifixed, dehiscing introrsely with incomplete longitudinal slits, which remain apical and porelike or proceed down to the middle of the theca or somewhat more. Ovary sessile; ca. 5–20 ovules per locule. Style thick and short or lacking. Capsule ovoid to subglobose, ± triquetrous. Seeds compressed, discoid, black, or blackish brown. $2n = 18$. About 20 spp., southern Africa to Namibia, Madagascar.

11. ***Litanthus*** Harvey

Litanthus Harvey, London J. Bot. 3: 314 (1844); Jessop, J. South Afr. Bot. 43: 309 (1977).

Bulb small, globose, storage leaves imbricate. Inflorescence 1(–2)-flowered on a solitary terete scape. Bracts spurred. Prophylls present. Pedicels nodding, erect in fruit. Flowers nocturnal. Tepals connate for 1/2, free parts very short, white. Filaments very short, inserted at the mouth of perigone tube. Anthers ovoid, 1 mm long, opening longitudinally. Style filiform, with short 3-lobed stigma. Capsule subglobose, membranaceous. Seeds black, compressed. Cotyledon short, hypogeal, followed by 1 foliage leaf. Several spp., southern Africa.

12. ***Rhadamanthopsis*** (Obermeyer) Speta

Rhadamanthopsis (Obermeyer) Speta, Phyton (Austria) 37 (1998).
Rhadamanthus subgen. *Rhadamanthopsis* Obermeyer (1980).

Bulbs large, compact. Roots thick, branched. Raceme lax on solitary thick scape. Pedicels short, ± nodding, in fruit curved upward. Tepals basally slightly connate, campanulate, pale mauve, ca. 9 mm long. Filaments 1.5 mm, anthers dehiscing longitudinally over the whole length. Ovary ovoid; locules 8-ovulate; style short. Capsule ellipsoid, 5–7 mm long. Seeds compressed, narrowly obovoid, 4–5 mm long, black, shiny. $2n = 12, 16, 18$. •• spp., Cape Province to S Namibia.

13. ***Thuranthos*** C.H. Wright

Thuranthos C.H. Wright, Bull. Misc. Inform 1916: 233 (1916); Stirton, Bothalia 12: 161–165 (1976); Jessop, J. South Afr. Bot. 43: 272–276 (1977), sub *Drimia*; Deb & Dasgupta, Bull. Bot. Surv. India 16: 116–123 (1974), sub *Urginea*.

Bulb plate plane at base or ± protruding. Bulb leaves imbricate, pink. or white. Foliage leaves proteranthous. Flowers produced after a resting period, opening by night and closing by day. Raceme solitary, lax on terete scape. Bracts spurred, prophylls present. Pedicels 3–5 cm long, ± patent. Tepals brown or green, basally connate for 1–2 mm. Ovary ovoid, locules 12-ovulate. Style (3–)5–22 mm long, thickened towards stigma. Capsule subglobose to oblong, 2–3 cm long, deeply 3-lobed. Seeds black, compressed, 6–14 mm long. $2n = 20, 30, 40, 60$. About 10 spp., southern Africa to Morocco, India, Nepal, and Birma.

14. ***Tenicroa*** Raf.

Tenicroa Raf., Fl. Tell. 3: 52, 1836 (1837); Obermeyer, Bothalia 13: 111–114 (1980), sub *Sypharissa*.
Sypharissa Salisb. (1866).

Bulb formed by membranaceous cataphylls and concrescent bases of foliage leaves of 2 vegetation periods, often with wooden axis, roots strong. Foliage leaves synanthous, enclosed by sheathing cataphylls. Raceme few- to many-flowered on slender scape. Lower bracts navicular with a long, acute basal spur. Anthesis diurnal. Perigone stellate, white; tepals with reddish brown or green stripe. Filaments long, bent, accumbent to pistil. Anthers basifixed, opening longitudinally. Ovary oblong, locules many-ovulate. Style long, incurved. Capsule papyraceous. Seeds compressed, ovate, dark and shiny. $2n = 20$. Five spp., southern Africa from Cape Province to S Namibia.

15. ***Drimia*** N.J. Jacq. ex Willd.

Drimia N.J. Jacq. ex Willd., Sp. Pl. 2(1): 165 (1799), Jessop, J. South Afr. Bot. 43: 269–307 (1977); Stedje, Nord. J. Bot. 7: 655–666 (1987), rev.

Bulb plate ± protruding at base. Roots thick, branched. Bulb leaves imbricate, formed by cataphylls and bases of foliage leaves of ca. 3 vegetation periods. Raceme solitary. Bracts spurred, prophylls present. Tepals connate for 1/3, free part completely reflexed. Filaments ± broadened at base, inserted in the upper part of the tube, accumbent with pistil. Anthers small. Ovary ovoid, style long, locules 7–15-ovulate. Capsule ellipsoid, seeds compressed, ± edged. $2n = 18$. About ten spp., southern and tropical Africa.

16. *Urginea* Steinheil

Urginea Steinheil, Ann. Sci. Nat. II, 1: 321 (1834); Speta, Linzer Biol. Beitr. 12: 197 (1980); Jessop, J. South Afr. Bot. 43: 269–307 (1977); Stedje, Fl. Trop. E Afr.: 15–21 (1969).

Bulb leaves imbricate, formed by bases of cataphylls and foliage leaves of ca. 3 vegetation periods. Roots thick and branched. Raceme solitary, few- to many-flowered, on thin or thick scape. Bracts narrow, spurred. Prophylls present. Pedicels ± short. Tepals basally slightly connate, stellate, whitish with reddish or greenish stripe, or pink, drying and persistent. Filaments compressed, narrowly triangular or filiform. Ovary cylindric, locules few- to many-ovuled. Capsule ellipsoid, 3-lobed, cartilaginous. Seeds compressed, oblong, black, shiny. Cotyledon epigeal, followed by 1 foliage leaf. $2n = 20, 30, 40, 60$, and also 14, 18, 22, 24. About 50 spp., Mediterranean; if broadly conceived, the genus would extend to S Africa and S Iran. *Urginea*, often included in *Drimia*, is a heterogeneous, poorly understood genus in which major rearrangements are to be expected.

IV. Subfam. Ornithogaloideae Speta (1998).

Bulb leaves usually lasting only 1–2 vegetation periods. Peduncles terete. Innovation usually annually, rarely iterative. Bracts large, prophylls lacking (except for *Pseudogaltonia*). Flowers white, with or without green stripe on the tepals, or yellow, orange, greenish, or brownish. Tepals tending to be moveable at anthesis. Filaments often appendaged. Ovary globose to oblong, occasionally with ridges (*Ornithogalum* s. str. p.p., *Albuca*). Locules usually with numerous ovules, in 1–2 rows or irregularly arranged. Stylar hairs distinct. Seeds usually black, often compressed (disk- or D-shaped) and edged, ridged, or carinate, never hollow, less often small, comma- or pear-shaped, verrucose or pilose; rarely globose to ellipsoid. Nucleus with protein crystals. Cardenolids partly present. This subfamily may be further subdivided into the tribes Dipcadieae Rouy and Ornithogaleae Rouy. While the genera of the further tribe are reasonably well understood, in the large genus *Ornithogalum* systematically relevant characters, species delimitation, and possible generic limits are unsatisfactorily known.

17. *Stellarioides* Medicus

Stellarioides Medicus, Hist. & Commentat. Acad. Elect. Sci. 6 (Phys.): 369 (1790); Obermeyer, Bothalia 12: 361–363 (1978), sub *Ornithogalum*.

Bulbs at soil level. Storage leaves vaginate, greenish, formed by bases of foliage leaves, lasting for ca. 2 vegetation periods. Foliage leaves with long terete tip. Raceme solitary, many-flowered, on thick scape. Bracts long, prophylls lacking. Pedicels straight, erectly patent. Tepals yellowish, whitish, greenish, with a green stripe on the outside, stellate, free. Filaments ligulate, with a shortly acuminate tip incurved in bud. Anthers yellow. Ovary green, ovoid, locules with 9–18 ovules in 2 rows. Style ± short, stigma capitate. Capsule papyraceous. Seeds large, black, oblong, compressed. Cotyledon epigeal, followed by a foliage leaf. $2n = 4, 6, 8, 10, 12, 18, 24, 36, 54$. About ten spp., S and E Africa.

18. *Coilonox* Raf.

Coilonox Raf., Fl. Tellur. 2: 28, 1836 (1837); Obermeyer, Bothalia 12: 366–371 (1978); U. Müller-Doblies & D. Müller-Doblies, Feddes Repert. 107: 493–518 (1996), rev.
Osmyne Salisb. (1866).
Ornithogalum subgen. *Osmyne* (Salisb.) Baker (1873).

Bulb leaves imbricate. Roots branched. Solitary lax, few-flowered raceme on terete scape. Bracts long, narrowly triangular, prophylls lacking. Pedicels straight, directed upward. Flowers occasionally ± nodding. Tepals ± elliptic, whitish or yellow, outside with a broad dark or green stripe, slightly connate, ± stellate. Filaments narrowly triangular, erect, patent. Ovary oblong, green; locules with many ovules in 2 rows. Style white, long; stigma capitate, pilose. Capsule ovoid to oblong-globose, obtuse, thin, coriaceous. Seeds compressed, semidiscoid, ca. 4 mm in diameter. Cotyledon epigeal, followed by 1 foliage leaf. $2n = 18$. About 30 spp., southern Africa: Cape Province to S Namibia.

19. *Albuca* L.

Albuca L., Sp. Pl. ed. 2: 438 (1762); U. Müller-Doblies, Feddes Repert. 105: 365–368 (1994), 106: 353–370 (1995), rev.

Bulb leaves imbricate, rarely vaginate, lasting 2–4 vegetation periods. Roots white, branched. Raceme solitary, lax, on terete scape. Bracts large. Prophylls lacking. Flowers nodding, greenish or yellowish. Inner tepals pilose at tips, connivent over style, outer tepals ± patent. Filaments ligulate, all or only inner ones ± compressed in lower 1/3, all or only central ones fertile. Ovary green, oblong, with 2 ridges joining at base, locules 14–42-ovulate. Style stout, often strongly pilose, stigma 3–4-lobed, hirsute. Seeds black, compressed, D-shaped. Cotyledon epigeal, followed by 1 foliage leaf. $2n = 18, 36$. About 60 spp., Africa south of the Sahara, Arabia.

20. *Pseudogaltonia* (Kuntze) Engler

Pseudogaltonia (Kuntze) Engler in Engler & Prantl, Natürl. Pflanzenfam. II, 5: 158 (1888); Speta, Bot. Jahrb. Syst. 106: 123–134 (1985).
Lindneria Durand & Lubbers (1889).

Large bulbs formed by white, imbricate bulb leaves lasting several vegetation periods. Tunic reticulate. Solitary comose raceme on elongated scape. Pedicels incurved at anthesis, erect in fruit. Bracts long and narrow, prophylls present. Tepals connate for 2/3, tube incurved, green, free parts ± patent, whitish with green streak. Filaments broadly triangular inserted in the upper part of the tube, anthers green. Ovary oblong, locules with more than 30 ovules in 1 row. Style 3 cm long, stigma hirsute. Seeds compressed, black, endosperm ± compressed. Cotyledon epigeal, followed by 3 foliage leaves. $2n = 18$. Only one sp., *P. clavata*, Namibia.

21. *Dipcadi* Medicus

Dipcadi Medicus, Hist. Commentat. Acad. Elect. Sci. Theod. – Palat. 6: 431 (1790); Obermeyer, Bothalia 8: 117–137 (1962), rev.; Deb & Dasgupta, J. Bombay Nat. Hist. Soc. 75: 50–70 (1978).

Bulbs small. Roots thin. Bulb leaves imbricate or vaginate, lasting ca. 2 vegetation periods. Raceme lax on solitary scape. Rachis nodding in bud stage, erect at anthesis. Pedicels erect in bud stage, incurved at anthesis, erect in fruit. Flowers green or brown, occasionally ± yellow, reddish, brown, or white at the tip of the inner tepals. Tepals connate for 2/3, free lobes reflexed and occasionally apically with a ± long, rounded spur, inner tepals not spurred, occasionally forming an additional tube. Filaments flat, inserted in the perigone tube; anthers large. Ovary oblong, locules 20-ovulate. Style ± short. Capsule compact. Seeds compressed, orbicular, black. Cotyledon epigeal, followed by 1 foliage leaf. $2n = 8, 12, 18, 20, 24, 36, 64$. About 30 spp., W Mediterranean to Central Asia, Arabia, Africa, Madagascar, Sokotra, Ceylon, and India.

22. *Galtonia* Decaisne

Galtonia Decaisne, Fl. Serres Jard. Eur. 23: 32 (1880); Hilliard & Burtt, Notes R. Bot. Gard. Edinburgh 45: 95–104 (1988), rev.

Roots white, ca. 25 cm long and 2 mm in diameter, apically branched. Bulb leaves formed by cataphylls and bases of foliage leaves, white, vaginate. Tunic brown. Racemes many-flowered, 1(–2), on thick, terete scapes. Bracts large, lanceolate, prophylls lacking. Pedicels nodding, erect in fruit. Perigone white, yellowish green or light green. Tepals connate for 1/2, straight. Filaments ± biseriate, inserted at the middle of tube, anthers green. Ovary cylindric to ovoid, style ca. 10–15 mm long, locules more than 20-ovulate. Capsule cylindric or conical-oblong, edged, thick cartilagineous. Seeds black, lunate, compressed. Cotyledon epigeal, followed by 1 foliage leaf. $2n = 16$. Four spp., southern Africa, from E Cape Province to Transvaal.

23. *Zahariadia* Speta

Zahariadia Speta, Linzer Biol. Beitr. 30 (1998).

Bulb large, white, Storage leaves imbricate, lasting for 2(–3) vegetation periods. Roots thick, up to 70 cm long, whitish, branched. Foliage leaves broad, erect, not terete at apex. Subcorymbose to umbellate raceme on solitary scape. Bracts large, prophylls lacking. Pedicels straight, patent, erect. Tepals broadly ovate, creamy white, stellate. Filaments erect, short deltoid-acuminate. Anthers greenish, long. Ovary globose, blackish, locules ca. 14-ovulate. Style short, stigma small. Unfertilized flowers caducous. Capsule compact, ovoid-triangular. Seeds black, 3–4-edged, oblong, 6–8 mm long. Cotyledon short and thick, hypogeal, followed by 1 foliage leaf. $2n = 12$. One or two spp., southern Africa: E Transvaal, Swaziland to Zululand, W Cape Province.

24. *Melomphis* Raf.

Melomphis Raf., Fl. Tellur. 2: 21: 1836 (1837).

Bulb leaves formed by cataphylls and bases of foliage leaves, imbricate, white, storing for ca. 2 vegetation periods. Solitary, umbellate raceme elongating during anthesis on thick scape. Bracts large, prophylls lacking. Tepals broadly ovate, cup-shaped, white to yellowish white, without green stripe. Filaments narrowly acuminate. Ovary globose, black, locules with numerous ovules. Style short, stigma inconspicuous. Unfertilized flowers persisting. $2n = 28, 51$. More than two spp., southern Mediterranean.

25. *Cathissa* Salisb.

Cathissa Salisb., Gen. Pl.: 34 (1866); Wittmann, Stapfia 13: 83–98 (1985), rev., sub *Ornithogalum*.

Bulb leaves imbricate or vaginate, formed by cataphylls and bases of foliage leaves, lasting 2 vegetation periods. Raceme solitary, 1–20-flowered, on terete scape. Pedicels 1–10 mm long, erect in fruit. Perigone obconical, white, externally with a green stripe only at the tip. Filaments broad, acute. Ovary oblong, without ridges, dorsally arched, locules 7–15-ovulate. Style 3–5 mm long. Capsules oblong. Seeds pyramidal, black. Cotyledon epigeal, followed by a cataphyll or a foliage leaf. $2n = 32, 34, 36$. Four spp., Morocco, Portugal, Spain.

26. *Eliokarmos* Raf.

Eliokarmos Raf., Fl. Tell. 2: 24, 1836 (1837); Obermeyer, Bothalia 12: 333–357 (1978), rev.; U. Müller-Doblies & D. Müller-Doblies, Feddes Repert. 107: 378–467 (1996), rev., sub *Ornithogalum*.
Aspasia Salisb. (1866).

Bulbs small, subglobose. Roots white, thin, ± branched, up to 35 cm long. Bulb leaves vaginate, formed by cataphylls and bases of foliage leaves of ca. 1 vegetation period. Tunic brown. Foliage leaves 1-many, proteranthous, rarely synanthous. Scape solitary, terete. Bracts large, broad, prophylls lacking. Flowers cyathiform or stellate. Tepals white, yellow, orange, red, without dark median nerve. Filaments 3 or 6, variously appendaged or thickened at base. Ovary green or blackish, subglobose, locules with numerous multiseriate ovules. Unfertilized flowers drying and persistent. Capsule ellipsoid, thin-walled. Seeds very small, black, compressed or pear- or comma-shaped, papillose to echinate. Cotyledon epigeal. $2n = 12$. About 65 spp., southern Africa: Cape Province.

27. *Loncomelos* Raf.

Loncomelos Raf., Fl. Tell. 2: 24, 1836 (1837); Wittmann, Stapfia 13: 15–83 (1985).
Beryllis Salisb. (1866).
Ornithogalum subgen. *Urophyllon* (Salisb.) Baker (1873), p.p.
Ornithogalum subgen. *Spetagalum* U. Müller-Doblies & D. Müller-Doblies (1996).

Bulb leaves imbricate, formed by cataphylls and bases of foliage leaves of 2 vegetation periods. Raceme solitary, many-flowered, elongated, on long scape. Bracts medium-sized, prophylls lacking. Pedicels patent, incurved upwards, sometimes erect in fruit. Perigone stellate, white, whitish, or yellowish, usually with a green stripe. Ovary globose or ovoid, without ridges, yellow or green, locules 5–18-ovulate. Style simple. Capsule compact, thin-walled. Seeds large, edged. Cotyledon epigeal. $2n = 16, 18, 22, 24, 26, 28, 32, 36, 54$. About 20 spp., Mediterranean region, from western Europe to Ural Mts., Asia Minor, Iran.

28. *Honorius* S.F. Gray

Honorius S.F. Gray, Nat. Arr. Brit. Pl. 2: 177 (1821).

Bulb formed by imbricate bulb leaves of 2 vegetation periods. Foliage leaves linear, canaliculate, ± long and wide, with white median stripe. Scape solitary, terete, thick, succulent, rachis elongated, bracts large. Pedicels short. Flowers erect in bud, slightly nodding at anthesis. Tepals long, with broad green median stripe, basally ± connate, campanulate when open. Androecium forming a tube around the pistil. Filaments ligulate, internally with 1 longitudinal ridge, apically 2-dentate (at least the inner ones), anthers ocher. Ovary ellipsoid, dark green, with 6 longitudinal furrows especially distinct on the back of the carpels, style thick with white, 3-lobed, capitate stigma. Ovules numerous. Capsules large, pendent, succulent, green, surrounded by dried tepals. Seeds ellipsoid with large-celled epidermis. Cotyledon epigeal. $2n = 14, 28, 30, 35, 42$. Eight spp., SE Europe, W Turkey, Near East.

29. *Ornithogalum* L.

Ornithogalum L., Sp. Pl.: 306 (1753).

Roots thin, unbranched, white, rarely yellow. Bulb leaves imbricate, vaginate or concrescent, formed by bases of foliage leaves and sometimes also by cataphylls, of 1 vegetation period. Tunic brown. Foliage leaves with or without white median stripe. Racemes solitary, umbellate or capitate, on a soft, terete scape. Bracts large, prophylls lacking. Pedicels patent or erect at anthesis, patent, erect, or incurved downward in fruit. Flowers stellate. Tepals free, white, externally with green stripe. Filaments narrowly triangular, anthers creamy white. Ovary ± ovoid, with or without 2 longitudinal ridges outside. Style and stigma inconspicuous. Capsule papyraceous. Seeds globose to oblong, black to brown. Cotyledon epigeal, followed by a cataphyll, or shortly hypogeal, followed by 1 foliage leaf. $2n$ = 12, 14, 16, 18, 20, 22, 24, 26, 28, 32, 36, 45, 52, 54, 72, 90, 108. About 50 spp., from Mediterranean to W Afghanistan.

Ornithogalum in the present narrow circumscription comprises subgenera *Ornithogalum*, *Ophiogalum* Zahar., *Leptotesta* Zahar., *Oreogalum* Zahar., *Anosmium* Zahar., *Hypogaeum* Zahar., *Amphibolum* Zahar., and *Amphigalum* Zahar.

V. Subfam. Hyacinthoideae Link (1829).

Bulb leaves lasting 2–3 vegetation periods. Scapes 1-several, terete or semiterete. Bracts usually small, rarely large or lacking, occasionally with irregular spur. Prophylls present or absent. Flowers black to light blue, violet, purple, pink, white, green, or yellow, all shades of brown, occasionally 2-colored. Ovary globose to elongate, locules 1, 2, or 2-several ovules side by side, superimposed, or irregularly arranged. Stigma inconspicuous. Capsules ± globose to ellipsoid, occasionally trigonous or trialate. Seeds globose, drop-shaped or ellipsoid, not edged. Elaiosomes formed from various parts of the testa. Nucleus lacking protein crystals. Homoisoflavones present.

The Hyacinthoideae may be further subdivided into tribes Massonieae Baker (genera 30–45), which occur from Africa south of the Sahara to India, and Hyacintheae, which are Eurasian and northern African.

30. *Merwilla* Speta

Merwilla Speta, Phyton (Austria) 37 (1998).

Bulbs rather large, partly subterraneous. Bulb leaves compact, imbricate, or ca. 4 vegetation periods. Roots thick, branched. Tunic light yellow, gray, not fibrous. Foliage leaves broad. Scape terete, sometimes pilose, raceme few- to many-flowered. Bracts distinct, narrow. Prophylls very small. Pedicels upward patent. Tepals free, blue, stellate, 5–6 mm long. Unfertilized flowers caducous. Filaments narrowly triangular, connate at base, white, patent. Anthers small. Ovary subglobose, white, locules 4–20-ovulate. Style 1 mm. Capsule globose, acuminate. Seeds oblong, compressed, vinaceous when fresh, brownish when dry. Cotyledon unknown. $2n$ = ? About five spp., southern Africa.

31. *Schizocarphus* Merwe

Schizocarphus Merwe, Fl. Pl. South Afr. 23: t. 904 (1943).

Bulb plate plane with a dense ring of branched roots. Bulb leaves imbricate, compact, apically often fibrous. Tunic fibrous. Racemes many-flowered, dense, on 1–2 terete scapes. Pedicels rather short, upwards patent, often pilose. Bracts narrow, prophylls distinctly shorter. Tepals free, stellate, light blue to whitish or greenish white, ca. 4 mm long. Unfertilized flowers caducous. Filaments white, patent. Anthers short, bluish green. Ovary bluish green, very shortly stipitate, locules 2(–6)-ovulate. Capsule globose, ca. 5 mm in diameter. Seeds ovoid, black. Cotyledon epigeal. $2n$ = 28, 43, 56. About five spp., southern Africa to Tansania.

32. *Pseudoprospero* Speta

Pseudoprospero Speta, Phyton (Austria) 37 (1998).

Bulb plate slightly protruding. Bulb up to 4 cm in diameter. Storage leaves compact, whitish. Foliage leaves up to 6, lanceolate, applanate, canaliculate, 0.5 cm wide and up to 30 cm long. Scape terete, 30–40 cm long, raceme lax, occasionally branched at base. Bracts narrow, lowest 1 cm long, prophylls minute. Tepals stellate, pale pinkish mauve, ca. 5 mm long. Anthers small. Ovary globose, locules with 2 ovules side by side. Style 2 mm long. Unfertilized flowers caducous. Cotyledon hypogeal, followed by 1 foliage leaf. $2n$ = 18. Only one sp., *P. firmifolium* Speta, eastern S Africa.

33. *Eucomis* L'Hér.

Eucomis L'Hér., Sertum Angl.: 11 (1789), nom. cons.

Stem large, distincly protruding beneath the bulb. Roots thick, branched. Bulb leaves formed by cataphylls and bases of foliage leaves of 4 vegetation periods. Raceme solitary, dense, on stout scape. Bracts large, prophylls minute, only occasionally present. Tepals whitish, greenish, green with purple margin or purple, remaining succulent until fruiting stage. Filaments triangular to filiform, connate at base, forming a short tube with perigone. Anthers yellow, green, short. Ovary ovoid, locules 6–7-ovulate. Pistil ca. 5 mm long. Capsule papyraceous, ± trialate. Seeds ovoid, brown. Cotyledon short, hypogeal, followed by 1 foliage leaf. $2n$ = 20(?), 30, 60. About ten spp., southern Africa to tropical E Africa.

34. *Avonsera* Speta

Avonsera Speta, Phyton (Austria) 37 (1998).

Bulb broadly ovoid. Foliage leaves 2–4, petiolate. Raceme solitary, 15–30-flowered, on terete scape. Axils of lower bracts 2-flowered. Bracts and prophylls narrow. Pedicels straight, directed upward. Tepals connate at base, 5–7-nerved, 11 mm long. Filaments narrowly triangular, connate at base, forming a short tube with perigone. Ovary obovoid, style 3–5 mm long; locules with 2 ovules side by side. Capsule obtusely trigonous. Seeds ovoid, large. Only one sp., *A. convallarioides* Speta, Madagascar, eastern S Africa.

35. *Ledebouria* Roth

Ledebouria Roth, Nov. Pl. Spec. Indiae Or.: 194 (1821); Jessop, J. S. Afr. Bot. 38: 233–266 (1972), rev.

Stem plane at base or protruding beneath bulb, roots thick and branched. Storage leaves imbricate, white or green. Foliage leaves often with vinaceous spots. Scapes 1–14, ± compressed, succulent; racemes dense, pedicels short, flowers nodding. Bracts and prophylls small, sometimes lacking. Perigone free, campanulate, brownish green to purple. Filaments basally connate with perigone, upper part usually vinaceous, contiguous with the style. Ovary green, distinctly stipitate, above the stipe with a plate forming the base of the gugelhupf-shaped upper part of the ovary. Locules with 2 ovules side by side. Style usually at least partly vinaceous. Capsule globose to oblong. Seeds drop-shaped, black. Cotyledon short, hypogeal, followed by 1 foliage leaf. $2n = 20, 24, 26, 28, 30, 32, 34, 36, 44, 45, 46, 54, 58, 60, 66$. About 50 spp. Africa S of the Sahara, Madagascar, S Arabia (?), India.

36. *Drimiopsis* Lindl. & Paxton

Drimiopsis Lindl. & Paxton, Paxton's Fl. Gard. 2: 73 (1851); Jessop, J. South Afr. Bot. 38: 151–162 (1972); Stedje, Nordic J. Bot. 14: 45–50 (1994); Fl. Trop. East Africa: 6–10 (1996).

Stem not or only little protruding at base. Roots branched. Bulb leaves imbricate, white or greenish. Foliage leaves green, cordate, ovate or lanceolate, petiolate or not, often spotted with white. Scapes succulent, 1–2, racemes dense. Bracts and prophylls lacking. Pedicels short to very short. Flowers globose or short cylindric, white, yellowish green or green, opening only a little, unfertilized flowers caducous. Tepals short, connate for up to 2 mm, not reflexed. Filaments white, triangular, connate with the perigone at base or up to the apex of the tube. Ovary green, subglobose, sessile. Locules with 2 ovules side by side. Capsule small, subglobose. Cotyledon short, hypogeal, followed by 1 foliage leaf. $2n = 24, 60, 64, 68$. About 20 spp., Africa S of the Sahara.

37. *Resnova* Merwe

Resnova Merwe, Tydskr. Wetensk. Kuns 6: 46 (1946).

Roots comparatively thick, branched. Bulb leaves imbricate. Scape terete, solitary, with dense raceme. Bracts and prophylls lacking. Pedicels short, bent down. Tepals shortly connate at base, connivent and forming a pseudotube, tips reflexed, pink purple, bluish, pink greenish, whitish. Filaments narrowly triangular. Ovary greenish, 1,5–2 mm long, locules with 2 ovules side by side. $2n = 10$. About ten spp., southern Africa, Madagascar.

38. *Veltheimia* Gled.

Veltheimia Gled., Mem. Akad. Berlin 1769: 66 (1771).

Roots thick, branched. Bulb leaves formed by cataphylls and bases of foliage leaves of 3–4 vegetation periods. Scape solitary, elongated, with dense raceme. Bracts narrow. Prophylls filiform. Pedicels nodding, soon pendent, ± short. Perigone tube long, tips patent. Filaments inserted at middle of tube. Ovary green, elongated, locules 3–4-ovulate. Capsule pergamentaceous, 3-alate. Seeds subglobose, black. Cotyledon short, hypogeal, followed by 1 foliage leaf. $2n = 40$. Two spp., southern Africa from Cape to S Namibia.

39. *Whiteheadia* Harvey

Whiteheadia Harvey, Gen. S. Afr. Pl., ed. 2: 396 (1868); Jessop, J. South Afr. Bot. 42: 433–435 (1976).

Bulb leaves vaginate, from cataphylls and bases of foliage leaves of 2 vegetation periods. Scape solitary with dense, subspicate raceme. Bracts ovate-acuminate, 25–40 cm long; prophylls lacking. Pedicels 3–4 mm long. Tepals white or green with white margin, succulent, spreading, 3–5-nerved, connate for 4–7 mm, free part 6–7 mm long. Filaments connate for 2 mm above the perigone tube, free parts 6–7 mm long. Ovary ovoid, 4 mm long, Style 1–3 mm long. Locule many-ovulate. Capsule membranaceous, strongly trialate. Seeds small, black, drop-shaped. Only one sp., *W. bifolia* Harv., southern Africa from western Cape area to S Namibia.

40. *Amphisiphon* Barker

Amphisiphon Barker, J. S. Afr. Bot. 2: 19 (1936); Jessop, J. South Afr. Bot. 42: 432–433 (1976).

Bulb globose. Tunic dark brown. Scape terete, inflorescence capitate, 20–40-flowered, hardly exserted above the 2(3) foliage leaves. Bracts lanceolate, to 1 cm long. Pedicels 2–3 mm long. Tepals yellowish green above, white at base, up to 2 cm long, forming a tube, free tips 2 mm long. Filaments inserted at the middle of the perigone tube, connate into a tube exceeding the perigone tube for ca. 1 cm; free parts of filaments yellow, ca. 1 cm long. Anthers yellow. Ovary oblong, locules many-ovulate. Style ca. 4 cm long, yellow. Capsule membranaceous, ovoid. Seeds globose, shiny, black. $2n = 32$. Only one sp., *A. stylosa* Barker, southern Africa: Cape region.

41. *Androsiphon* Schltr. Fig. 83O

Androsiphon Schltr., Notizbl. Bot. Gart. Berlin-Dahlem 9: 147 (1924); Jessop, J. South Afr. Bot. 42: 432 (1976).

Bulb ovoid. Foliage leaves 2, ± elliptic, spotted dark green. Scape solitary, raceme capitate. Lowest bracts lanceolate, 25–30 mm long. Prophylls lacking. Pedicels 15–30 mm long. Flowers golden yellow, actinomorphic. Perigone-filament tube 6–11 mm long, exceeded by filament tube 11 mm long. Free part of tepals erect or erect-patent, 9–13 mm long. Anthers short. Ovary ellipsoid, locule 8–15-ovulate. Style 20–26 mm long. Capsule obovoid, apically subtruncate, membranaceous. Seeds globose. One sp., *A. capense* Schltr., southern Africa: Cape region, Karroo.

42. *Daubenya* Lindl.

Daubenya Lindl., Bot. Reg. 21: t. 1813 (1835); Jessop, J. South Afr. Bot. 42: 431–432 (1976).

Bulb subglobose. Tunic brown. Foliage leaves 2, ovate, lying flat on soil. Scape terete, short, subterraneous. Raceme capitate, at soil level. Lowest bracts oblong, 15–30 mm long. Prophylls lacking. Pedicels short. Flowers zygomorphic, lowest (outer) ones with 3 enlarged tepals, red, orange or yellow, showy; upper (inner) flowers with short free tips. Perigone-filament tube 2–4 cm long, filaments inserted at the margin of the tube. Anthers short, yellow. Ovary oblong, with long style, locules many-ovulate. Capsule ovoid, not winged. Seeds globose. Only one sp., *D. aurea* Lindl., southern Africa: Cape region, Karroo.

43. *Massonia* Houtt. Fig. 84A–C

Massonia Houtt., Natuurl. Hist., pt. 2, 2: 424 (1780); Jessop, J. South Afr. Bot. 42: 406–426 (1976), rev.
Neobakeria Schltr. (1924).

Roots white, ± thin. Bulb leaves vaginate. Foliage leaves usually 2, broadly oblong or elliptic. Scape solitary, with a condensed subcapitate raceme close to soil level. Inflorescence surrounded by large bracts, upper bracts much smaller. Flowers actinomorphic, pink, white, or green, rarely yellow or red, fragrant. Tepals connate at base to form a 3–20-mm-long perigone-filament tube, free part characteristically folded and reflexed. Filaments sublinear to deltoid, inserted at upper margin of the tube, free or forming a ± long tube. Ovary estipitate, oblong, obovoid, locules ca. 10–30-ovulate. Style 3–30 mm long. Capsule obovoid, trialate. Seeds black, subglobose, 1–2 mm in diameter. $2n = 18, 22, 26$. Eight spp., S Africa: Cape Province, SW Transvaal, Orange Free State, Usotho.

44. *Polyxena* Kunth

Polyxena Kunth, Enum. Pl. 4: 294 (1843). Jessop, J. South Afr. Bot. 42: 426–431 (1976), rev.
Periboea Kunth (1843).

Bulb leaves from cataphylls and bases of foliage leaves of 2 vegetation periods, bases of storage leaves enclosed by 1 sheathing membranaceous cataphyll. Scape solitary, raceme occasionally corymbose. Bracts short and broad, prophylls lacking. Flowers actinomorphic, pedicellate, usually pink or white. Tepals forming a long perigone-filament tube, free parts ± oblong. Filaments in 2 rows. Ovary ellipsoid, locules with several ovules. Style long. Capsule with numerous subglobose black seeds. Cotyledon short, hypogeal, followed by 1 foliage leaf. Two spp., S Africa: Cape Province.

45. *Lachenalia* Jacq. f. ex Murray Fig. 83I–M

Lachenalia Jacq. f. ex Murray, Syst. Veget. ed. 14: 314 (1784); Crosby, Plantsman 8: 129–166 (1986); Duncan, Ann. Kirstenbosch Bot. Gard. 17: 1–71 (1988), rev.

Bulb leaves formed by cataphylls and bases of foliage leaves of 1(–2) vegetation periods. Scape solitary with dense spike or raceme. Bracts medium-sized, base broad, prophylls lacking. Pedicels nearly lacking to distinct and nodding. Flowers zygomorphic, white, green, yellow, purple, or pink. Perigone ± connate at base, free part narrowly or broadly campanulate, often the outer tepals distinctly shorter. Capsule subglobose to ellipsoid, green, yellow, locules few- to many-ovulate. Seeds (sub-)globose, black, smooth, often with elaiosome formed by exostome and bordering part of the raphe. Cotyledon short, hypogeal, followed by 1 foliage leaf. $2n = 10, 14, 15, 16, 18, 20, 21, 22, 24, 26, 27, 28, 29, 30, 32, 40, 42, 44, 56$. About 110 spp., southern Africa, from the Cape Province to S Namibia and Transkei.

Subgen. *Brachyscypha* (Baker) Baker has concrescent bulb storage leaves. A modern subdivision is lacking, but should be promising if *Polyxena* is included.

46. *Barnardia* Lindl.

Barnardia Lindl., Bot. Reg. 12: t. 1029 (1826); Speta, Phyton (Austria) 37 (1998).

Bulb leaves formed by the bases of foliage leaves and scape bracts and hypsophylls. Scape compact, with rudimentary branches at base; raceme dense. Flowering in autumn. Bracts small and narrow, spurred, prophylls minute. Perigone stellate, very small, pink, rarely whitish. Filaments broadened at base. Anthers yellow, dark red, or blue. Ovary shortly stipitate, nectar duct covered by short unicellular hairs. Locules 1-ovulate. Capsule papyraceous, oblong, opening only at apex. Seeds elongate, dark brown. Cotyledon epigeal, followed by 1 foliage leaf. $2n = 16, 18, 26, 27, 34, 35$. Eight spp., NW Africa, Ibiza, Formentera, S China to Japan, Korea, and Wladiwostok.

47. *Autonoë* (Webb & Berth.) Speta

Autonoë (Webb & Berth.) Speta, Phyton (Austria) 37 (1998).

Stem plane at base, rarely protruding somewhat under the bulb. Roots thick and branched. Storage leaves white, imbricate, lasting for 3–12 vegetation periods. Scape solitary, smooth, terete, raceme dense. Bracts short, prophylls minute. Rachis and the short pedicels colored like perigone. Perigone star-shaped, small, pink to violet. Filaments triangular to filiform. Ovary depressed globose, without septal nectaries. Locules with 2 ovules side by side. Capsule globose, succulent, orange or black purple when ripe. Cotyledon short, hypogeal, followed by 1 foliage leaf. $2n = 28, 62$. Six spp., Canary Islands, Madeira, W Morocco.

48. *Oncostema* Raf.

Oncostema Raf., Fl. Tell. II, 1: 13 (1837); Speta, Phyton (Austria) 26: 263–272 (1987).

Bulbs large, bulb stem protruding, with thick branched roots. Bulb leaves imbricate, formed by cataphylls and bases of foliage leaves, lasting for 3 vegetation periods. Tunic brown. Scape solitary, terete, with rudimentary branches at base. Bracts long and broad, prophylls minute. Pedicels long, straight. Flowers stellate. Tepals blue, persistent after anthesis. Filaments blue, ± thickened in lower 1/3, connate at base. Anthers yellow. Pistil bottle-shaped. Ovary green, locules 4–6-ovulate. Nectariferous duct covered with short hairs. Capsule compact. Seeds subglobose, black. Cotyledon short, hypogeal, followed by 1 foliage leaf. $2n = 14, 16, 28, 32$. About ten spp., southwestern Mediterranean area to S Italy, N Africa and Near East.

49. *Hyacinthoides* Medicus

Hyacinthoides Medicus, Ann. Bot. (Usteri) 2: 9 (1791); Speta, Phyton (Austria) 26: 281–294 (1987).
Endymion Dumort. (1927).
Somera Salisb. (1866).

Stem very small, roots white, unbranched, short-lived, partly protruding through the bases of bulb leaves. Bulb leaves annual, first 1–3 imbricate, followed by 1–3 vaginate non-storing cataphylls and by the imbricate, highly connate bases of foliage leaves. Scape solitary, raceme lax. Bracts large, sharply triangular, prophylls somewhat smaller. Pedicels straight upwards or slightly incurved, upright in fruit. Perigone blue, only basally connate, stellate or campanulate. Filaments filiform, all equal or inner distinctly shorter. Anthers long. Ovary ovate to globose, locules 2–6-ovulate. Capsule papyraceous. Seeds subglobose, black. Cotyledon hypogeal, followed by 1 foliage leaf. $2n = 16, 24$. W Mediterranean area from Morocco to England.

The distinction of two subgenera, subg. *Somera* (Salisb.) Speta and subgen. *Hyacinthoides* is supported by studies on nucleotide sequences (Pfosser and Speta, unpubl. results).

50. *Brimeura* Salisb.

Brimeura Salisb., Gen. Pl.: 26 (1866); Speta, Phyton (Austria) 26: 254–263 (1987).

Roots white, unbranched. Bases of outer cataphylls or outer foliage leaves vaginate to the base, not storing. Storage leaves from bases of inner foliage leaves, vaginate only at apex, strongly connate at base, lasting 1 year. Tunic thin, brown. Scape solitary. Bracts large, prophylls lacking. Flowers erect or slightly nodding, pinkish white or blue. Perigone connate for 1/3 to 2/3. Filaments arranged in 1 or 2 rows. Ovary globose, thickest at the septae, nectar secreted at the middle. Locules with several ovules. Seeds black, globose. Cotyledon short, hypogeal, followed by 1 foliage leaf. $2n = 28, 56$. About five spp., W Mediterranean area from Corsica and Sardinia to Spain and S France.

51. ***Tractema*** Raf.

Tractema Raf., Fl. Tell. 2, III: 58 (1837).
Oncostema subgen. Tractema (Raf.) Speta, Phyton (Austria) 26: 272–281 (1987).

Bulb leaves imbricate or vaginate, white or yellow, formed by cataphylls and bases of foliage leaves of 3 vegetation periods. Scape terete, becoming rigid. Bracts large, prophylls lacking. Pedicels directed straight upward. Perigone ± free, stellate, blue. Filaments white, broadened in lower 1/3. Anthers blue. Ovary yellow, green, with downward directed nectar duct, locules ca. 2–4-ovulate. Capsule pergamentaceous. Seeds drop-shaped or globose, black. Cotyledon very short, hypogeal, followed by 1 foliage leaf. $2n = 20, 22, 24$. About six spp., W Europe from S Norway to NW Africa (Morocco).

52. ***Hyacinthella*** Schur

Hyacinthella Schur, Österr. Bot. Wochenbl. 6: 227 (1856); Feinbrun, Bull. Res. Counc. Israel 10D: 324–347 (1961), rev.; Persson & Wendelbo, Candollea 36: 513–541 (1981), 37: 157–175 (1982), rev.

Bulb leaves vaginate and imbricate, white, formed by the bases of foliage leaves of 3–6 vegetation periods. Tunic whitish brown, farinaceous inside. Scapes 1–2, ± wiry, terete. Bracts and prophylls small. Pedicels short, straight. Perigone whitish blue to violet, tubular, free lobes ± patent. Filaments filiform, uniseriate. Anthers small, dark blue. Ovary globose. Locules with 2 ovules side by side, sometimes with 3–4 ovules. Capsule small and firmly pergamentaceous. Seeds black, elongate. Cotyledon epigeal. $2n = 16, 18, 20, 22, 24$. Eighteen spp., Balkan Peninsula to W Persia.

53. ***Alrawia*** (Wendelbo) K. Persson & Wendelbo

Alrawia (Wendelbo) K. Persson & Wendelbo, Bot. Not. 132: 201 (1979); Wendelbo, Fl. Iraq 8: 111–113 (1985), Fl. Iranica 165: 139–140 (1990).

Bulb leaves imbricate, lasting ca. 3 vegetation periods. Tunic grayish outside, often violet inside. Scape solitary, terete, firm, raceme few- to many-flowered. Bracts and prophylls small. Pedicels short, nodding at anthesis, elongated and patent in fruit. Perigone tubular, connate for 1/4 to 1/2, zygomorphic, polychromatic. Filaments ± filiform, uniseriate. Each ovary locule with 2 ovules side by side, rarely with a 3rd superimposed ovule. Capsule subglobose without prominent edges. Seeds globular to broadly ovate, black. Cotyledon unknown. $2n = 12$. Two spp., N Iraq, W Iran.

54. ***Prospero*** Salisb.

Prospero Salisb., Gen. Pl.: 28 (1866).

Bulb plate plane with a ring of white, branched roots of 1 vegetation period. Tunic light to reddish brown. Bulb leaves imbricate, white, yellowish or pink, lasting for ca. 3 vegetation periods. Foliage leaves filiform or strap-shaped, appearing in autumn and withering in spring. Scapes 1–4, terete, sulcate, racemes dense, many-flowered. Pedicels patent. Bracts and prophylls lacking. Tepals free, pink to violet, rarely whitish, 4–10 mm long. Filaments filiform to ligulate, colored like perigone. Anthers short, purple. Pistil light blue, bottle-shaped. Locules with 2 ovules side by side. Capsules small, membranaceous. Seeds ± oblong, dark brown, smooth. Cotyledon epigeal. $2n = 8, 10, 12, 14, 26, 28, 42, 54, 70$, ca. 150. About 25 spp., S England, Mediterranean basin to Caucasus and N Iraq.

55. ***Puschkinia*** Adams — Fig. 83A–C

Puschkinia Adams, Nova Acta Acad. Sci. Imp. Hist. Acad. Petrop. 14: 164 (1805).

Bulb leaves vaginate at base, formed by cataphylls and the bases of storage leaves lasting for 3 vegetation periods. Scape solitary, terete, flaccid in fruit. Bracts and prophylls small. Perigone white, greenish or blue, connate for 1/2, free parts patent. Filaments short, inserted in the middle of the paracorolla. Anthers small, yellow. Ovary green, bottle-shaped. Locules with several ovules. Capsule fleshy. Seeds ellipsoid, yellow, with sarcotesta, myrmecochorous. Cotyledon epigeal.

$2n = 10$. Three to four spp., Near East to Caucasus and W Iran.

56. *Othocallis* Salisb. Fig. 85H–J

Othocallis Salisb., Gen. Pl.: 28 (1866). Mordak, Bot. Zhurn. 56: 1449–1453 (1971); Speta, Naturk. Jahrb. Stadt Linz 25: 70–77, 85–141 (1981) (sub *Scilla*, rev.).

Roots short-lived, unbranched, and white, or turnip-shaped and contractible. Tunic blackish, ± purple inside. Bulb leaves imbricate, white, from cataphylls and bases of foliage leaves of 2–3 vegetation periods. Foliage leaves synantherous. Scapes 1–8, semiterete, in fruit flaccid; racemes 1–8-flowered, lax. Bracts and prophylls short, collar-shaped. Pedicels short, often nodding. Tepals blue, only slightly connate, caducous. Filaments slender, anthers blue, long. Ovary yellow or green, style distinct, 4–10 mm long, locules with 4–29 globose ovules. Capsules remaining succulent until maturity. Seeds ± globose, sometimes with elaiosome formed by raphe or exostome and neighboring parts of raphe. Testa thick, yellow, brown, or black, epidermis papillose. Cotyledon epigeal. $2n = 12, 18, 24, 36$. About 20 spp., Ukraine to Caucasus, W Iran, N Iraque, E Turkey, Near East.

57. *Fessia* Speta

Fessia Speta, Phyton (Austria) 37 (1998); Speta, Naturk. Jahrb. Stadt Linz 25: 141–169 (1981), sub *Scilla*, rev.

Bulb leaves white, imbricate, formed by cataphylls and bases of foliage leaves of 3 vegetation periods. Tunic silvery gray or black, purple inside. Scapes 1-many, semiterete. Bracts short, prophylls even shorter, irregular, occasionally ± spurred. Pedicels straight or ± curved. Perigone ± free, blue, whitish, violet. Anthers comparatively large, ice-blue. Ovary globose to oblong, locules 2–7-ovulate. Capsule succulent. Seeds globose or drop-shaped, black. Cotyledon epigeal, only in *F. greilhuberi* short and hypogeal. $2n = 10, 18$. Ten spp., Iran to Central Asia and NW India.

58. *Pfosseria* Speta

Pfosseria Speta, Phyton (Austria) 37 (1998); Speta, Naturk. Jahrb. Stadt Linz 25: 77–85 (1981), sub *Scilla*, rev.

Bulb leaves white, imbricate, formed by cataphylls and bases of foliage leaves of 3 vegetation periods. Tunic brown, purple violet inside. Scapes 1–2(–3), semiterete, becoming flaccid after anthesis, raceme dense, 3–20-flowered. Pedicels bent straight upwards, pedicels and rachis colored like perigone. Bracts and prophylls short, slightly spurred. Tepals blue, stellate, 6–10 mm long, persistent in fruit. Anthers blue. Ovary subglobose, rugose, blue, locules 2-5-ovulate. Style 2–4 mm long. Capsule subglobose, succulent. Seeds black, ± distinctly papillose. Cotyledon epigeal. $2n = 12$. Only one spp., *P. bithynica*, eastern Bulgaria, NW Turkey.

59. *Hyacinthus* L.

Hyacinthus L., Sp. Pl.: 316 (1753); Baranova, Giazint (1965); Wendelbo, Fl. Iranica 165: 133–136 (1990).

Bulb leaves imbricate, white, formed by cataphylls and bases of foliage leaves of 2–3 vegetation periods. Tunic brown or blackish, purple inside. Scapes 1(–3), succulent, terete; raceme lax. Bracts and prophylls short. Pedicels comparatively short. Perigone blue, connate 3/5 to 2/3, free parts reflexed. Ovary green, ellipsoid. Locules 6–8-ovulate. Style short. Capsule succulent. Seeds globose, blackish brown, with or without elaiosome formed by exostome and parts of the raphe. Cotyledon epigeal. $2n = 16, 18$. Three spp., Near East to S Turkmenistan and NE Iran.

60. *Nectaroscilla* Parl.

Nectaroscilla Parl., Nuov. Gen. Sp.: 26 (1854).

Bulbs large. Bulb leaves imbricate, white. Tunic whitish or black. Scape elongated, terete, raceme many-flowered. Rachis and pedicels violet. Bracts and prophylls small. Flowers blue, stellate, tepals ± free, 6–7 mm long, drying and persisting. Anthers blue. Ovary globose, violet, locules with 2 ovules side by side, style 2 mm long. Capsule pergamentaceous. Seeds black. Cotyledon epigeal. $2n = 20$. One or two spp., E Mediterranean area.

61. *Chouardia* Speta

Chouardia Speta, Phyton (Austria) 37 (1997); Speta, Naturk. Jahrb. Stadt Linz 25: 173–181 (1981), sub *Scilla*.

Bulb leaves white, imbricate, lasting for 3 vegetation periods. Tunic brown. Scapes 1–2(–4), stiff and somewhat sulcate; raceme dense. Pedicels straight, blue. Bracts and prophylls very small. Tepals blue, short, stellate. Filaments blue, anthers dark blue. Ovary globose, blue, locules with 2 ovules side by side. Capsule small, pergamentaceous. Seeds drop-shaped, trigonous, brownish black. Cotyledon ± long, hypogeal, followed by 1 foliage leaf. $2n = 26$. Two spp., SE Central Europe: Dinaric Alps, Dalmatia.

62. *Schnarfia* Speta

Schnarfia Speta, Phyton (Austria) 37 (1998); Speta, Naturk. Jahrb. Stadt Linz 25: 67–70 (1981), sub *Scilla*.

Bulb leaves imbricate, white, formed by cataphylls and bases of foliage leaves of (2–)3 vegetation periods. Tunic brown. Racemes 1–3, the scapes semiterete, succulent, flaccid in fruit. Bracts and prophylls small. Pedicels straight. Perigone blue, nearly free at base, 6–7 mm long, stellate. Anthers purplish blue. Ovary blue, rugose, locules with 1-several ovules. Capsule slightly succulent. Seeds globose, smooth, yellow, testa thin, epidermis smooth. Elaiosome formed from part of raphe close to hilum. Cotyledon epigeal. $2n = 18, 20$. Two spp., Albania, Greece.

63. *Zagrosia* Speta

Zagrosia Speta, Phyton (Austria) 37 (1998); Speta, Naturk. Jahrb. Stadt Linz 25: 169–173 (1981), sub *Scilla*.

Bulb leaves imbricate or vaginate, white, formed by cataphylls and bases of foliage leaves of 2–3 vegetation periods. Tunic brown. Racemes 1–3 on rigid scapes. Bracts and prophylls small. Pedicels straight. Perigone small, ± free, blue. Anthers blue. Ovary globose, rugose, blue, locules with 2 superimposed ovules. Capsule pergamentaceous. Seeds drop-shaped, black. Cotyledon epigeal. $2n = 8$. Only one sp., *Zagrosia persica* (Hausskn.) Speta, Zagros Mountains (E Turkey, N Iraq to W Iran).

64. *Muscari* Mill.

Muscari Mill., Gard. Dict. Abr. ed. 4 (1754).
Leopoldia Parl. (1845).
Pseudomuscari Garbari & Greuter (1970).

Stem plane at base or slightly protruding under bulb. Roots thick and branched to thin and scarcely branched. Bulb leaves imbricate, white, rarely pink, formed by cataphylls and bases of foliage leaves of about 3 vegetation periods. Tunic brown. Racemes 1-few, dense, often many-flowered, with a ± distinct crown of sterile flowers, on terete scapes. Bracts and prophylls small. Pedicels mostly very short, nodding at anthesis, later patent. Perigone highly connate, blue, brown, yellow, green, globose to elongated, free lobes short. Filaments adnate to perigone, short, in two rows. Anthers small, globose, dark blue. Ovary green, trigonous, locules with 2 superimposed ovules. Fruits trigonous to trialate. Seeds black, globose. Cotyledon epigeal. $2n = 18, 27, 36, 45, 54, 72, 90$. About 50 spp., Mediterranean region and temperate Europe to Central Asia.

DNA sequence data suggest monophyly of *Muscari* (Pfosser and Speta, unpubl. data). It can be divided into the subgenera *Muscari*, *Leopoldia* (Parl.) Peterm., *Moscharia* (Baker) Chouard, and possibly also *Pseudomuscari* Stuart.

65. *Bellevalia* Lapeyr.

Bellevalia Lapeyr., J. Phys. Chim. Hist. Nat. Arts 67: 425 (1808); Feinbrun, Palest. J. Bot., Jerusalem Ser., 1: 42–54, 131–142, 336–409 (1938–40); Wendelbo, Fl. Iranica 165: 149–165 (1990).
Strangweja Bertol. (1835).

Bulb leaves white, formed by cataphylls and the bases of foliage leaves. Racemes 1-several with terete, smooth scapes. Tunic brown. Bracts small, spurred, prophylls minute. Pedicels straight, elongated, and patent in fruit. Flowers usually nodding, rarely erect. Perigone tubular, connate for 1/2 to 2/3, free lobes ± reflexed, colorful: greenish white, yellow, brown, purple, violet, and blue, often impure and mixed in a characteristic fashion, and often changing continuously from bud to fruit. Free parts of filaments arising from upper part of perigone tube. Anthers short, blue. Locules with 2–4 superimposed ovules. Capsule papyraceous, trigonous, winged. Seeds usually ± globose and pruinose, rarely oblong and shiny. Cotyledon epigeal. $2n = 8, 16, 24, 32$. About 50 spp., Mediterranean area to Central Asia and Afghanistan.

66. *Scilla* L. Figs. 83D–H, 85F,G

Scilla L., Sp. Pl.: 308 (1753); Speta, Naturk. Jahrb. Stadt Linz 19: 9–54 (1974), 21: 9–79 (1976), 22: 9–63 (1977), 25: 30–67 (1981), rev., Linzer Biol. Beitr. 8: 293–322 (1976).
Chionodoxa Boiss. (1844).

Bulb leaves imbricate, white, yellowish, rarely pink, of cataphylls and bases of foliage leaves of 3 vegetation periods. Tunic brown. Raceme solitary with terete scape. Bracts minute or lacking, prohylls lacking. Tepals blue, nearly free to connate for 2/5. Filaments filiform to ligulate. Anthers long, yellow or blue. Ovary bottle-shaped, blue, somewhat rugose, locules with several ovules. Capsules succulent on flaccid peduncles. Seeds globose, yellow, brown, black, with elaiosome from exostome. Testa thin, smooth. Cotyledon epigeal. $2n = 18, 27, 36, 54$. About 30 spp., from France and N Spain to Asia Minor and Caucasus.

Incompletely known and of uncertain affinity (possibly belonging to Ornithogaloideae):

67. *Neopatersonia* Schönl.

Neopatersonia Schönl., Rec. Albany Mus. 2: 251 (1912).

Perianth tube short; segments subequal, spreading. Stamens sub-1-seriate, attached to the throat of the perianth; filaments broadened and connate at the base, gradually narrowed above; anthers dorsifixed, with introrse dehiscence. Ovary sessile, subovoid, 3-furrowed, with many ovules in each chamber; style filiform; stigma of 3 very short spreading branches. Fruit membranous 3-grooved, with loculicidal dehiscence. Seeds numerous, flask-shaped, rough. Only one sp., *N. uitenhagensis* Schönl., S Africa.

Selected Bibliography

Baker, J.G. 1870. Monograph of *Scilla*: *Ledebouria* and *Drimiopsis*. Saunder's Ref. Bot. 3, Appendix: 1–18.

Baker, J.G. 1871. A revision of the genera and species of herbaceous capsular gamophyllous Liliaceae. J. Linn. Soc. (London) 11: 349–436.

Baker, J.G. 1873. Revision of the genera and species of Scilleae and Chlorogaleae. J. Linn. Soc. Bot. 13: 209–293.

Batsch, A.I.G.C. 1786. Dispositio generum plantarum ienensium secundum Linnaeum et familias naturales. Ienae: Heller.

Boyd, L. 1932. Monocotylous seedlings. Morphological studies in the post-seminal development of the embryo. Trans. Proc. Bot. Soc. Edinburgh 31: 5–224.

Bresinsky, A. 1963. Bau, Entwicklungsgeschichte und Inhaltsstoffe der Elaiosomen (Studien zur myrmekochoren Verbreitung von Samen und Früchten). Bibl. Bot. Heft 126. 54 pp.

Buchner, L. 1948. Vergleichende embryologische Studien an *Scilloideae*. Österr. Bot. Z. 95: 428–450.

Cave, M.S. 1970. Chromosomes of the California Liliaceae. Univ. Calif. Publ. Bot. 57: 1–48, t. I–VIII.

Chouard, P. 1931. Types de développement de l'appareil végétatif chez les Scillées. Ann. Sci. Nat. Bot. X, 13: 131–323.

Čupov, V.S., Kutjavina, N.G. 1981. Serologiceskie issledovanija v porjadke Liliales I &. II. Bot. Zhurn. 66: 75–81, 408–416.

Čupov, V.S. 1994. Filogenija i sistema porjadkov Liliales i Asparagales. Bot. Zhurn. 79(3): 1–12.

Čupov, V.S. 1995. Kompensirujuscij vozvrat priznakov v primitivnoe sostojanie. Bot. Zhurn. 80(4): 1–7.

Daumann, E. 1970. See general references.

Delaunay, L.N. 1915. Étude comparée caryologique de quelques espèces du genre *Muscari* Mill. Mem. Soc. Nat. Kiew 25: 33–62.

Delaunay, L.N. 1922. Vergleichende Untersuchungen einiger *Muscari* Mill. und *Bellevalia* Lapeyr. Arten. Vestn. Tiflis Bot. Sada II, 1: 24–55.

Delaunay, L.N. 1926. Phylogenetische Chromosomenverkürzung. Z. Zellf. Mikrosk. Anat. 4: 338–364.

Ebert, I. 1993. Systematische Karyologie und Embryologie von *Prospero* Salisb. und *Barnardia* Lindl. (Hyacinthaceae). Diss. Univ. Wien, 525 pp.

Fay, M.F., Chase, M.W. 1996. Resurrection of Themidaceae for the Brodiaea alliance, and recircumscription of Alliaceae, Amaryllidaceae and Agapanthoideae. Taxon 45: 441–451.

Fernandez, A., Daviña, J.R. 1991. Heterochromatin and genome size in *Fortunatia* and *Camassia* (Hyacinthaceae). Kew Bull. 46: 307–316.

Frisch, B., Nagl, W. 1979. Patterns of endopolyploidy and 2 C nuclear DNA content (Feulgen) in *Scilla bifolia* (Liliaceae). Plant Syst. Evol. 131: 261–276.

Frölich, D., Barthlott, W. 1988. See general references.

Fuchsig, H. 1911. Vergleichende Anatomie der Vegetationssorgane der Lilioideen. Sitzungsber. Kais. Akad. Wiss. Wien, Math.-Naturw. Kl., 120: 1–43, I–III.

Garbari, F., Greuter, W. 1970. On the taxonomy and typification of *Muscari* Miller (Liliaceae) and allied genera, and on the typification of generic names. Taxon 19: 329–335.

Greilhuber, J. 1982. Trends in der Chromosomenevolution von *Scilla* (Liliaceae). Stapfia 10: 11–51.

Greilhuber, J., Deumling, B., Speta, F. 1981. Evolutionary aspects of chromosome banding, heterochromatin, satellite DNA, and genome size in *Scilla* (Liliaceae). Ber. Deutsch. Bot. Ges. 94: 249–266.

Halbritter, H., Hesse, M. 1993. Sulcus morphology in some monocot families. Grana 32: 87–99.

Hansgirg, A. 1904. Pflanzenbiologische Untersuchungen nebst algologischen Schlußbemerkungen. Wien: A. Hölder.

Hasitschka-Jenschke, G. 1962. Notizen über endopolyploide Kerne im Bereich der Samenanlage von Angiospermen. Österr. Bot. Z. 109: 125–137.

Heel, W.A. van 1988. On the development of some gynoecia with septal nectaries. Blumea 33: 477–504.

Heller, W., Tamm, Ch. 1981. Homoisoflavanones and biogenetically related compounds. Progr. Chem. Org. Nat. Prod. 40: 105–152.

Holzer, K. 1952. Untersuchungen zur karyologischen Anatomie der Wurzel. Österr. Bot. Z. 99: 118–155.

Huber, H. 1969. See general references.

Hutchinson, J. 1934. The families of flowering plants. II. Monocotyledons. London: Macmillan.

Jernstedt, J.A. 1980. Anthesis and floral senescence in *Chlorogalum pomeridianum* (Liliaceae). Amer. J. Bot. 67: 824–832.

Jessop, J.P. 1975. Studies in the bulbous Liliaceae in South Africa: 5. Seed surface characters and generic groupings. J. South Afr. Bot. 41: 67–85.

Krenn, L. 1994. *Urginea*. In: Hagers Handbuch der pharmazeutischen Praxis, Hsg. F. v. Bruchhausen, 5. Aufl., Bd. 6: 1030–1050.

Kubo, S., Mimaki, Y., Terao, M., Sashida, Y., Nikaido, T., Ohmoto, T. 1992. Acylated cholestane glycosides from the bulbs of *Ornithogalum saundersiae*. Phytochemistry 31: 3969–3973.

Kubo, S., Mimaki, Y., Sashida, Y., Nikaido, T., Ohmoto, T. 1992. New cholestane bisdesmosides from the bulbs of *Ornithogalum thyrsoides*. Bull. Chem. Soc. Japan 65: 1120–1124.

Kunth, C.S. 1843. Enumeratio plantarum, vol. IV. Stuttgardiae et Tubingae: J. G. Cottae.

Kunth, C.S. 1844. Über die natürliche Gruppe der Liliaceen im weitesten Sinne des Worts. Physik. Abn. König. Akad. Wiss. Berlin 1842: 1-55.

Lauber, H. 1947. Untersuchungen über das Wachstum der Früchte einiger Angiospermen unter endomitotischer Polyploidisierung. Österr. Bot. Z. 94: 30–60.

Lenski, I. 1958. Über die Brutzwiebelbildung auf den Blättern von *Drimiopsis kirkii* Bak. Planta 50: 579–621.

Lonay, H. 1902. Recherches anatomiques sur les feuilles de l'*Ornithogalum caudatum* AIT. Mém. Soc. Roy. Sci. Liège III, 4, (9) 1–82, 5 t.

Medicus, F.K. 1791. Ueber Linnés Hyacinthen Gattung. Ann. Bot. (Usteri) 2: 5–20.

Mimaki, Y., Sashida, Y., Kawashima, K. 1992. New steroidal constituents of the bulbs of *Camassia cusickii* Wats. Chem. Pharm. Bull. 40: 148–152.

Müller-Doblies, U., Müller-Doblies, D. 1996. Revisionula incompleta Ornithogalorum Austro-Africanorum. Feddes Repert. 107: 361–548.

Obermeyer, A.A. 1978. *Ornithogalum*: a revision of the southern African species. Bothalia 12: 323–376.

Obermeyer, A.A. 1980. The genus *Sypharissa* (Liliaceae). Bothalia 13: 111–114.

Pascher, A. 1959. Zur Blütenbiologie einer aasblumigen Liliacee und zur Verbreitungsbiologie abfallender geflügelter Kapseln. Flora 148: 153–178.

Radulescu, D. 1973. Recherches morpho-palynologiques sur la famille Liliaceae. Acta Bot. Hort. Bucurestensis 1972/73: 133–248.

Rafinesque, C.S. 1837. Flora telluriana. Philadelphia: H. Probasco.

Salisbury, R.A. 1866. The genera of plants. A fragment containing part of Liriogamae. London: J. v. Voorst.

Sattler, R. 1973. Organogenesis of flowers. Toronto & Buffalo: Univ. of Toronto Press.

Savile, D.B.O. 1962. Taxonomic disposition of *Allium*. Science 196: 792.

Sernander, R. 1906. Entwurf einer Monographie der europäischen Myrmekochoren. K. Svenska Vet. Akad. Handl. 41/7: 1–410.

Schnarf, K. 1929. Die Embryologie der *Liliaceae* und ihre systematische Bedeutung. Sitzungsber. Akad. Wiss. Wien, Math.-Naturwiss. Kl. 138, Abt. I: 69–92.

Schulze, W. 1980. Beiträge zur Taxonomie der Liliifloren. VI. Der Umfang der Liliaceae. Wiss. Z. Friedrich-Schiller-Univ. Jena, Math.-Naturwiss. Reihe 29: 607–636.

Skrzypczak, L. 1977. Zwiazki flawonoidowe w chemicznej taksonomii rodziny Liliaceae. Herba Polon. 22: 336–349.

Speta, F. 1971. Beitrag zur Systematik von *Scilla* L. subgen. *Scilla* (inklusive *Chionodoxa* Boiss.). Österr. Bot. Z. 119: 6–18.

Speta, F. 1972. Entwicklungsgeschichte und Karyologie von Elaiosomen an Samen und Früchten. Naturk. Jahrb. Stadt Linz: 9–65, t. I–X.

Speta, F. 1979. Karyological investigations in *Scilla* in regard to their importance for taxonomy. Webbia 34: 419–431.

Speta, F. 1980. Karyosystematik, Kultur und Verwendung der Meerzwiebel (*Urginea* Steinh., Liliaceae s.l.). Linzer Biol. Beitr. 12: 193–238.

Speta, F. 1982. Über die Abgrenzung und Gliederung der Gattung *Muscari* und über ihre Beziehungen zu anderen Vertretern der Hyacinthaceae. Bot. Jahrb. Syst. 103: 247–291.

Speta, F. 1984. Zwiebeln - versteckte Vielfalt in einfacher Form. Linzer Biol. Beitr. 16: 3–44.

Speta, F. 1985. Nomenklatorische, morphologische und karyologische Bemerkungen über *Lindneria clavata* (Masters) Speta (Hyacianthaceae). Bot. Jahrb. Syst. 106: 123–134.

Speta, F. 1986. Über die herbstblühenden Scillen des Mittelmeerraumes. Linzer Biol. Beitr. 18: 399–416.

Speta, F. 1987. Die verwandtschaftlichen Beziehungen von *Brimeura* Salisb.: ein Vergleich mit den Gattungen *Oncostema* Rafin., *Hyacinthoides* Medic. und *Camassia* Lindl. (Hyacinthaceae). Phyton (Austria) 26: 247–310.

Speta, F. 1998. Systematische Analyse der Gattung Scilla (Hyacinthaceae). Phyton (Austria) 37: ••–••.

Stockhouse, R.E., Wells, H. 1978. Pollination ecology of *Chlorogalum pomeridianum* (DC.) Kunth (Liliaceae). Bull. South. Calif. Acad. Sci. 77: 124–129.

Svoma, E. 1981. Zur systematischem Embryologie der Gattung *Scilla* L. (Liliaceae). Stapfia 9: 124 pp.

Svoma, E., Greilhuber, J. 1988. Studies on systematic embryology in *Scilla* (Hyacinthaceae). Plant Syst. Evol. 161: 169–181.

Svoma, E., Greilhuber J. 1989. Systematic embryology of the Scilla siberica alliance (Hyacinthaceae). Nord. J. Bot. 8: 585–600.

Stirton, C.H. 1976. *Thuranthos*: notes on generic status, morphology, phenology and pollination biology. Bothalia 12: 161–165.

Tillich, H.-J. 1992. Bauprinzipien und Evolutionslinien bei monocotylen Keimpflanzen. Bot. Jahrb. Syst. 114: 91–132.

Tillich, H.-J. 1995. Seedlings and systematics in Monocotyledons. In: Rudall, P.J., Cribb, P.J., Cutler, D.F., Humphries, C.J. (eds.) Monocotyledons: systematics and evolution: 303–352. Royal Botanic Gardens, Kew.

Tilton, V.R., Horner, H.T. 1983. Carpel development, anatomy, and function in the reproductive process in *Ornithogalum caudatum* (Liliaceae). Flora 173: 1–31.

Vogel, St. 1954. Blütenbiologische Typen als Elemente der Sippengliederung dargestellt anhand der Flora Südafrikas. Bot. Stud. 1. 338 pp.

Wunderlich, R. 1936. Vergleichende Untersuchungen von Pollenkörnern einiger Liliaceen und Amaryllidaceen. Österr. Bot. Z. 85: 30–55.

Wunderlich, R. 1937. Zur vergleichenden Embryologie der Liliaceae-Scilloideae. Flora 132: 48–90.

Zahariadi, C. 1962. Caractères morphologiques, anatomiques et biologiques dans la taxonomie du genre *Ornithogalum*. Rev. Biol. 7: 5–41.

Hypoxidaceae

I. Nordal

Hypoxidaceae R.Br. in Flinders, Voy. Terra Austr. 2: 576 (1814), nom. cons.

Herbaceous plants with a ± tuberous rhizome or a corm, often carrying fibrous remains of former leaves. *Leaves* basal, 1-many, rosulate and 3-ranked, with a more or less sheathing base, sessile or with a distinct pseudopetiole (in *Curculigo*, *Hypoxidia* and *Molineria*), outermost leaves often reduced to cataphylls, lamina linear to lanceolate, pubescent with simple and compound hairs, sometimes glabrous; the lamina prominently parallel-veined, often V-shaped in cross-section, sometimes plicate (mainly in the pseudopetiolate taxa), rarely terete. *Inflorescences* usually on hairy scapes, often several per plant, arising from leaf axils, spicate to corymbose or umbel-like, sometimes reduced to solitary flowers. *Flowers* most often pedicellate, sometimes sessile, from the axils of small or larger and leafy bracts (*Curculigo*, *Molineria*), hermaphroditic (rarely unisexual in *Curculigo*), actinomorphic. Tepals 3 + 3 (rarely 2 + 2 in *Hypoxis* and *Spiloxene*), ± persistent, subequal, free (in most genera) or fused to form a tube (in *Hypoxidia*, *Pauridia*, *Rhodohypoxis*, *Saniellia*), most often yellow to orange, sometimes reddish, pinkish or white (*Rhodohypoxis* and *Pauridia*) or reddish brown (*Hypoxidia*), often green and pubescent abaxially. Stamens 3 + 3 (2 + 2), most often uniseriate, biseriate in *Rhodohypoxis* and *Hypoxidia*), or only 3, then opposite the inner tepals (*Pauridia*); filaments short to almost lacking, arising from the base of perianth segments or from the mouth of the perianth tube; anthers elongate, basifixed or (low) dorsifixed, slightly versatile or not, opening introrsely or latrorsely by longitudinal slits, in some cases with apical appendages. Ovary inferior, trilocular with axile placentation or unilocular with 3 parietal placentas (*Empodium*), transitional forms found in *Hypoxidea* and some species of *Spiloxene* with trilocular lower part merging to unilocular upper part, ovary often prolonged upwards into an epigynous beak (*Curculigo*, *Empodium*, *Saniellia*). Style 1, short with 3(6) stigmas either as free lobes or as groves along the style. Ovules often numerous, sometimes only few. *Fruit* a capsule with circumscissile or loculicidal dehiscence, or sometimes more or less fleshy and indehiscent or irregularly dehiscent, often beaked. Seeds with a thick phytomelan crust, globose to ellipsoid, with a more or less protuding micropyle, and a more or less prominent funicle, sometimes with a prominent strophiole (*Curculigo*, *Empodium*); testa most often black, sometimes brownish, glossy or dull, smooth to papillate or verrucose; embryo small; endosperm copious, non-farinaceous.

Nine genera and about 100 species (a very approximate number, as one of the genera, *Hypoxis*, is largely apomictic, accordingly with problematic species delimitation); widely distributed in tropical and subtropical regions of all continents particularly of the southern hemisphere, transgressing slightly into the northern hemisphere on all continents, but particularly in America.

Vegetative Morphology. The Hypoxidaceae are herbaceous geophytes, sprouting flowers short after the first rains. Nutrition is stored in ± tuberous rhizomes or in corms (Figs. 87, 88), the former being prevalent in the tropics and subtropics, the latter in the southern temperate regions of Africa. In the elongated to tuberous rhizomes new tissue is added gradually by the apical meristem. According to Hilliard and Burtt (1978), this "vertical stock, which is virtually a slow-growing underground trunk, is the primitive type. It is found in *Hypoxis* and *Curculigo*; in *Rhodohypoxis* and *Saniellia* there is an abbreviated vertical stock which is more fleshy and less fibrous than those of *Curculigo* and *Hypoxis*". The mainly Asiatic *Hypoxidia* and *Molineria* have elongated rhizomes similar to those of *Curculigo*. Annual corms, with mainly apical innovation, thus creating monopodial structures with new discrete tubers added apically each season, are found in *Empodium*, *Pauridia* and *Spiloxene*. These corms represent swollen stems consisting of a number of internodes. In some species (*Empodium* and partly in *Spiloxene*) both apical and basal innovation is recorded (Thompson 1976). When new corms appear from axillary buds on the base of the corm, vegetative propagation may become prevalent. [Dahlgren and Clifford (1982) wanted to restrict the term *corm* to structures with basal innovation only, i.e for exclusively sympodial structures. In Hypoxidaceae basal and apical innovation may be combined in the same plant, and the term corm thus appears appropriate also for the monopodial structures.] In the genera with corms, a considerable variety of reticulate corm

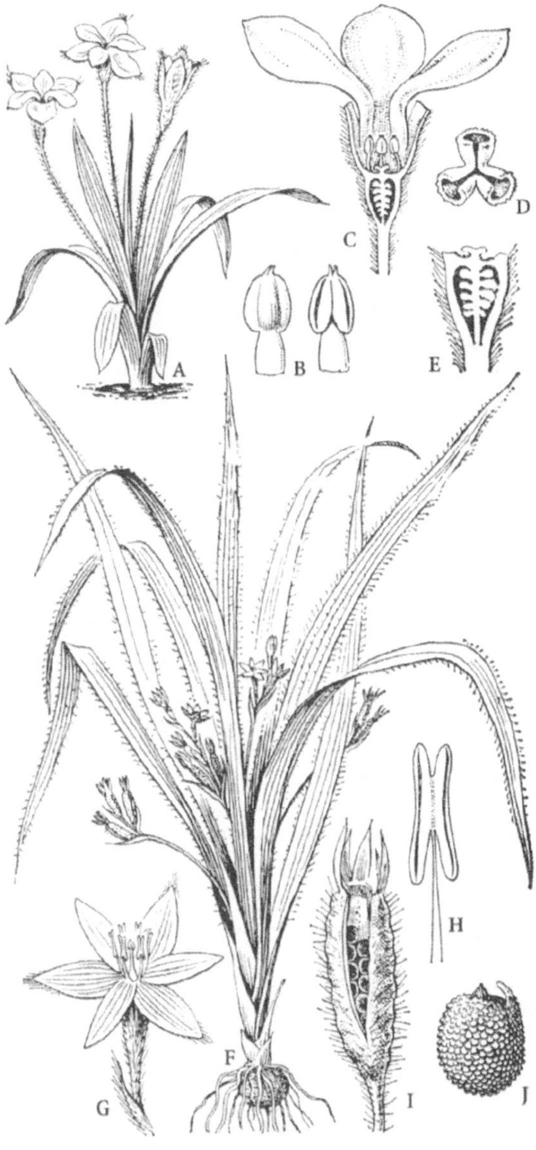

Fig. 87A–J. Hypoxidaceae. A–E *Rhodohypoxis baurii*. **A** Habit. **B** Flower, longitudinal section. **C** Stamen, dorsal and ventral view. **D** Stigma. **E** Ovary, longitudinal section. **F–J** *Hypoxis decumbens*. **F** Flowering plant. **G** Flower. **H** Stamen. **I** Dehiscing capsule. **J** Seed. (Takhtajan 1982)

tunics is displayed. The not-tunicated corms are sometimes classified as tubers (Thompson 1976). Vegetative propagation by stolons is rare but has been recorded in *Rhodohypoxis* (Hilliard and Burtt 1978). The rhizomes and corms are withheld in the ground by conspicuous contractile roots differently positioned: in *Hypoxis* situated in a narrow zone equatorially, in *Curculigo*, *Molineria* and *Hypoxidea* scattered along the rhizome, in *Spiloxene* and *Pauridia* at the base of the new corm, and in *Empodium* laterally on the corm.

Phyllotaxis is tristichous, sometimes distorted during the development to mask the 3-ranked organisation. Ptyxis is plicate to conduplicate-plicate (Cullen 1978). Leaves may be unifacial and terete or bifacial and carinate, canaliculate or plicate (Thompson 1976). The petioles found in *Curculigo*, *Hypoxidia* and *Molineria* are of the Bambusa type (cf. Dahlgren and Clifford 1982). Old leaves die back in the dry seasons and are replaced by new ones produced successively through the growing season, inside the old ± disintegrated fibrous leaf remnants. Well protected subterranean parts rich in stored nutrients and apical meristem deeply hidden among old fire-resistant leaf remnants enable some genera (e.g. *Curculigo* and *Hypoxis*) to survive heavy grassland fires.

Inflorescences. The scapes are produced continuously through the growing season from the axils of the succesively produced leaves. The inflorescences are probably indeterminate, but further morphological analyses should be undertaken. They appear as different version of principally racemose types: corymbose, when the lower pedicels are conspicuously longer than the upper, capitate (*Molineria*) to umbellate (*Spiloxene*), when the rhachis is reduced, spicate, when the pedicels are reduced, and sometimes only with solitary flowers (*Pauridia*, *Curculigo* and rarely in *Hypoxis*). In taxa of *Hypoxis* both acropetalous and basipetalous flower development have been recorded. Further, the narrowing of the rhachis from the base towards the apex is often stepwise with clusters of flowers at each node. The basic type of the inflorescence in *Hypoxis* may be transitional between a rhipidium and a bostryx. Bracts sometimes appear to fuse with the rhachis/pedicels to further obscure the basic structure of the inflorescenes (Nordal et al. 1985). One to 2 bracts per flower are usually present, only in *Empodium* no bracts develop, and the single flower is borne on a structure without differentiation of peduncle and pedicel (Thompson 1978).

Vegetative Anatomy. General for the family are seedlings with reduced open sheaths and non-assimilating cotyledons with haustorial tip; a coleoptile is present. The primary root in the seedling is ephemeral (Tillich 1995). The rhizomes/corms have a more or less defined inner core of vascular bundles, running irregularly

Fig. 88A–I. Hypoxidaceae. A–D *Molineria capitulata*. A Habit. B Flower. C Stamens, ventral and dorsal view. D Fruit. E–I *Curculigo orchioides*. E Staminate flower. F Pistillate flower. G Hermaphroditic flower. H Anthers in ventral and dorsal view. I Seed. (Takhtajan 1982)

through the ground tissue, intermingled with fibrous tissue, surrounded by a spongy storage area of parenchymatic tissue rich in starch and transversed by numerous mucilage canals, with yellowish or whitish sap, which sometimes darkens when exposed to air (Thompson 1976; Zimudzi 1994). The anatomical structure of the roots is similar to that of normal monocotyledonous root (Tomlinson 1980). The root hair development takes place from any cells, and velamen is absent. Root vessels are present, most often with scalariform perforation (Cheadle 1968).

The leaf epidermis is single-layered with most often thin-walled cells, in species with carinate and plicate leaves bulliform cells are found adaxially to the main vascular strand(s). Epicuticular wax depositions are absent (Thompson 1976; Hilliard and Burtt 1978). A typical palisade tissue is absent or weakly developed; the vascular bundles have complete or incomplete bundle sheaths and larger bundles have sclerenchyma caps, making the leaf nerves prominent. The stomata are in most cases paracytic, rarely tetracytic, they may be scattered on both leaf surfaces or confined to the intercostal zones (Thompson 1976; Dahlgren et al. 1985). Leaf vessels are absent. Mucilage canals above the vascular bundles are present in *Pauridia*, *Saniellia* and *Spiloxene*, but have not been found in the leaves of *Hypoxis*. *Empodium* is unique in having mucilage canals scattered in the mesophyll (Thompson 1976).

An indumentum is almost always present (except in *Pauridia Rhodohypoxis* and *Saniellia*). In general, a combination of multicellular simple and compound hairs is found; the latter may be bifurcate or multifurcate, sometimes on a short stem and often with very different arm lengths. In *Empodium*, appearing almost glabrous, only unicellular hairs are found (Thompson 1976).

The scape has only 1 circle of vascular bundles, vessels are lacking (Dahlgren et al. 1985). Sieve-tube plastids contain protein and are of type PIIc (Behnke 1981). Calcium oxalate raphides are generally present, silica bodies always absent.

Flower Structure. The flowers are structured according to the basic trimerous, pentacyclic monocotyledonous pattern. The tetracyclic *Pauridia* with a single inner whorl of stamens is the only exception; according to Thompson (1978) the extra 3 short lobes of the stigma found in this genus might represent reduced stamens which have fused with the gynoecium, an interpretation substantiated by an observed "freak flower", producing almost complete anthers from the short lobes of the stigma.

The perianth is undifferentiated, radial symmetrical, most often slightly heterochlamydeous, and always epigynous; perianth segments most often free, but fused to a tube in *Hypoxidia*, *Pauridia*, *Rhodohypoxis* and *Saniellia*. Nectaries appear to be lacking. Stamens, when diplostemonous, are mostly uniseriate, but biseriate in *Rhodohypoxis* and *Hypoxidia*. Filaments are subulate to filiform, often rather short, arising from the base of perianth segments or from the mouth of the perianth tube. The anthers are bithecate and tetrasporangiate. The basic morphology of the stamens varies considerably within the family and probably renders basic diagnostic characters for generic delimitation within the family. After much confusion in the application of terms like basifixed and dorsifixed (e.g. Nel 1914; Geerinck 1969), Hilliard and Burtt (1978) reclassified the terminology and classification: "In *Empodium, Molineria, Rhodohypoxis* and *Saniellia* the filament joins the anther low down on the outer face. The anthers are different on the inner and outer faces: this may be obvious, as in *Molineria* and *Rhodohypoxis*, or the difference may be slight as in *Empodium* and *Saniellia*." Further, they described sagittate anthers in the other group: "in *Curculigo, Hypoxis, Pauridia* and *Spiloxene* the anthers look the same on inner and outer faces: the filament joins the anther medianly in the sinus between its basal lobes". In the first group a cross-section of the anther is ± asymmetrical (connected to introrse dehiscence), in the second group it is symmetrical (connected to latrorse dehiscence). Later, *Hypoxidia* was described by Friedman (1984) as having basifixed/low dorsifixed anthers without sagittate base. The sagittate anthers with attachment in the sinus between the lobes are sometimes able to move in a versatile way. In some cases, the anthers are provided with apical appendages (*Empodium, Hypoxidia*).

The ovary is in most genera trilocular with axile placentation, only *Empodium* has a unilocular ovary with 3 parietal placentas. *Hypoxidia* and some species of *Spiloxene* appear to represent an intermediate stage with trilocular axile placentas in the lower part of the ovary, intergrading into unilocular parietal placentas in the upper part (Thompson 1978; Friedman 1984). The style is short and stout, the stigma often consists of 3 (6 in *Pauridia*) free lobes, sometimes of 3 vertical furrows along the upper part of the style; the stigmatic surfaces are dry. A particularly interesting feature within the family is the tendency to develop a flexible structure to separate the ovary from the perianth. This structure has been differently interpreted as a perianth tube and an ovary extension. Confusion is still apparent in Dahlgren et al. (1985) when they report the tepals to be "free or fused into a sometimes long and narrow tube above the ovary". This long and narrow structure is certainly of ovarian origin as an epigynous beak or a rostrum, as found in *Curculigo, Empodium* and *Saniellia*, occasionally also in *Molineria* and *Rhodohypoxis*, whereas a tube of perianth origin is present in *Hypoxidia, Pauridia, Rhodohypoxis* and *Saniellia*. The development of an ovary beak is connected with the reduction of the scape and the pedicels which leads to ± subterranean, protected ovaries. This system will lead to a reduced flower exposition to insects. There seems to an evolutionary tradeoff between ovary protection and flower advertisement. The flexible growth of apical ovary meristem appears to be the answer to this dilemma. As Hilliard and Burtt (1973, 1978) have recognised, the ovary beak must have arisen independently in different genera within the family.

Embryology. The endothecial walls of the anthers exhibit spiral thickenings at least in *Curculigo* and *Hypoxis* (Dahlgren and Clifford 1982). According to Dahlgren et al. (1985), "The tapetum becomes periplasmodial at an early stage, and may be classified as amoeboid, which is exceptional in Asparagales". Glandular (secretory) tapetum has, however, also been reported (cf. Davis 1966; Johri et al. 1992). Microsporogenesis is successive, regarded as a reversal of the simultaneous state common in related groups (Rudall et al. 1997). The pollen grains are dispersed as monades in the 2-celled stage.

The ovules are bitegmic and mainly anatropous, but possibly hemianatropus to campylotropous in *Pauridia* and *Empodium* (Davis 1966; Johri et al. 1992). The micropyle is formed by the 2 integuments. The archesporial cell functions directly as the megaspore mother cell; parietal tissue is thus lacking; in this respect the ovule may be classified as tenuinucellate. De Vos (1948, 1949) recorded seeds with well-developed nucellar tissue, which makes the classification "tenuinucellate" somewhat misleading. The nucellar epidermis divides periclinally to produce a nucellar cap, so that the ovule might also be classified as pseudocrassinucellate. Embryo-sac formation (Davis 1966; Dutt 1970; Johri et al. 1992) is most often monosporic (Polygonum type), only in *Empodium* is bisporic formation recorded (Allium type). Endosperm formation is most often Helobial (e.g. in *Hypoxis* and *Curculigo*), but Nuclear in *Pauridia*, where the chalazal part is somewhat haustorial. Both types have been recorded for *Spiloxene* (Davis 1966). The endosperm is composed of thin-walled, isodiametric cells, storing aleurone and fatty oils, but no starch. Embryogeny has been recorded to be of the

Onagrad, the Solanad or the Asterad type. The embryo axis is straight, and the embryo type is said to be differentiated in *Hypoxis* and undifferentiated in *Curculigo* (Johri et al. 1992). These aspects require further study.

POLLEN MORPHOLOGY. Pollen grains are mostly sulcate, but 2-sulcate grains are reported in *Pauridia* (Thompson 1979). The grains are small and biconvex; the exine is fairly thin and usually finely reticulate and columellate (Erdtman 1952), sometimes only pitted (Thompson 1979).

KARYOLOGY. Wilsenach (1967) and Wilsenach and Papenfuss (1967) reported $2n = 16, 32, 36, 72$ and 96 in South African *Hypoxis*, concluding that polyploidisation has played a role in the speciation of this genus. Naranjo (1975) reported for the same genus 2 diploids with $x = 7$ and $x = 8$, respectively, and postulated further the existence of another basic number of $x = 9$, due to observed polyploids with $2n = 36, 54$ and 72 and $x = 19$ due to observed $2n = 38$ in one species. Nordal et al. (1985) and Zimudzi (1994, 1995) supported a basic number of $x = 7$ in *Hypoxis*, in a series including di-, tetra-, hexa-, and octoploid taxa. From a putative decaploid level a more or less continuous aneuploid variation up to more than 150 somatic chromosomes has been observed. Diploids and tetraploids seem to be sexual. At higher polyploidy levels apomictic reproduction takes over (Zimudzi 1994, 1995). Darlington and Wylie (1955), supported by Fedorov (1969), reported $2n = 18$, and a basic number $x = 9$, for *Curculigo*, a result later supported by Lakshmi (1979) and Zimudzi (1994, 1995). Mitra (1966) reported also tetraploids ($2n = 36$) in the genus. Studies by Saito (e.g. 1975) have shown that *Rhodohypoxis* includes diploids (and spontaneous) triploids based on $x = 6$. No information on chromosomes has been published from the remaining six genera.

POLLINATION. Flowers are reported to be protandrous in *Hypoxis* and *Curculigo* (Zimudzi 1994). The flowers display no obvious pollination adaptations and appear to be lacking nectar. Pollen is thus the only offer to pollinators, nevertheless *Hypoxis* flowers have been observed to be visited by bees (*Apis* sp.) and butterflies (*Precis octavia*) in tropical Africa (Zimudzi 1994). In *Rhodohypoxis* the stamens and stigma are covered by the perianth constrictions, in a way that probably requires specialised insects to perform successful pollinations. The flower syndrome displayed by *Hypoxidia* (evil smell and brownish red colour) might indicate that carcass insects are involved in pollination.

FRUIT AND SEED. There are generally three different fruit types in the family: (1) capsules with circumscissile dehiscence (pyxes), where a lid carrying the persistent perianth falls off at seed maturity; (2) capsules with loculicidal (3-valved) dehiscence; or (3) fruits with more or less succulent to membranaceous fruit walls (often denoted baccate), indehiscent or with irregular dehiscence/disintegration. The last type is often produced at or below the ground and is correlated with the development of the particular ovary beak. Some genera include more than one of the types (e.g. *Hypoxis*, 1 and 2 and *Rhodohypoxis*, 1, 2 and 3). The seeds are always provided with a prominent black phytomelan crust, the pigments mainly being stored in the cell walls of the palisadelike exotesta (Danilova et al. 1995). They have a more or less protruding micropyle and a distinct raphe transgressing into a persistent funicular part, sometimes with a prominent strophiole connected to the hilum (*Curculigo, Empodium*). The phytomelan crust often makes the seeds black and glossy, but sometimes they are brownish and more dull, due to a particularly folded cuticula (some *Hypoxis* spp., cf. Nordal et al. 1985). The seeds are very variable as to testa sculpturing, from almost smooth to papillate, verrucose, mammillate or spinose.

DISPERSAL. Hilliard and Burtt (1978), with reference to American/Asiatic *Hypoxis* spp. and to S African *Pauridia, Rhodohypoxis* and *Spiloxene* argued "that when species dwell in rather damp habitats, free dispersal of the seeds may be lost; the peduncle or pedicel tends to bend downwards after flowering and the fruit is thin-walled, moulded to the seeds within and showing rather irregular dehiscence". In general, the taxa with pyxidal capsules have erect fruits and strong, erect scapes. They have all the traits characteristic of ballistic dispersal, meaning that the seeds are withheld on the mother plant until a strong push throws them away. The loculicidal opening is often connected with lax and deflexed scapes; unless animals or water flushes take hold of these seeds, they will not be dispersed. Taxa with strong testa cuticular folding in *Hypoxis* often have lax peduncles, and it should be examined whether this particular structure might be of interest to small herbivores. Particularly intriguing is seed dispersal of underground seeds in the more or less baccate fruits of *Curculigo, Empodium, Rhodohypoxis* and

Saniellia. Some of them produce seeds with a distinct strophiole, probably acting as an elaiosome. Ants or related insects might thus be involved.

PHYTOCHEMISTRY. Dahlgren et al. (1985) reported that alkaloids and steroidal saponins seem to be lacking, at least in *Hypoxis*. According to Hegnauer (1986), this information is ambiguous, stating that the phytochemistry of the family is unsatisfactorily known. According to Ramstad (1953) chelidonic acid is present in *Hypoxis*. Stevenson and Loconte (1995) postulated chelidonic acid to be absent and steroidal saponins present. The family lacks cell wall-bound ferulic acid in cell walls. Most species have mucilage canals containing pectic compounds of mucopolysaccharides (Thompson 1976).

CLASSIFICATION. The generic delimitation within the family has varied considerably through the years, and the problem probably still needs more work to be solved satisfactorily. Chase et al. (1995) analysed chloroplast DNA (*rbc*L) from *Curculigo*, *Hypoxis*, *Rhodohypoxis* and *Spiloxene* and concluded that the family is monophyletic. Due to the haplostemous stamens and nuclear endosperm, the position of *Pauridia* has been questioned. Thompson (1978, 1979), however, has convincingly shown that it belongs in the Hypoxidaceae in close proximity to *Spiloxene*. Hilliard and Burtt (1978) recognised two groups of genera based on stamen morphology: one with (low) dorsifixed anthers, different on the inner and outer faces, including *Molineria*, *Rhodohypoxis*, *Empodium* and *Saniella*; and one with sagittate "basifixed' anthers (filament attachment in the basal sinus), not different on the inner and outer faces, including *Curculigo*, *Hypoxis*, *Pauridia* and *Spiloxene*. To the eight genera distinguished by Hilliard and Burtt (1978), a ninth genus, *Hypoxidia*, has been added by Friedman (1984). This genus, according to these stamen characters, appears to belong in the first group. Even if the stamens provide important taxonomic information, these are not decisive. For example, *Hypoxis* and *Rhodohypoxis* are closely related despite the difference in stamens (Thompson 1978); this is further corroborated by the occurrence of intergeneric crosses among the two (Hilliard and Burtt 1978). Also *Curculigo* and *Molineria* appear to be closely related, despite their different stamen structure.

Geographical distribution supported by basic morphology might indicate that the genera *Curculigo*, *Hypoxidia* and *Molineria*, which are centred around the Indian Ocean, form a natural group. The widespread pantropical genus *Curculigo* might represent the basic branch within this putative clade. Another probably natural group (*Empodium*, *Hypoxis*, *Pauridia*, *Rhodohypoxis*, *Saniellia* and *Spiloxene*) has a clear centre of variation in southern Africa, with the widespread pantropical genus *Hypoxis* possibly representing a basic branch. Characters of the perianth, the stamens, the ovary and the fruit display a more or less reticulate variation pattern within this putative clade. The particularly well-protected ± baccate fruits with an ovary beak are shared among the S African summer rain taxa of the *Hypoxis* clade and the representatives of the Asia centred *Curculigo* clade. A more thorough analysis of characters is needed to reach to an unambiguous generic delimitation and to understand the phylogenetic relations between them.

AFFINITIES. For a long time the family was treated as a tribe of the family Amaryllidaceae. Hutchinson (1934) restricted the Amaryllidaceae to bulbous plants with an umbellate inflorescence, thereby excluding the tribe Hypoxideae and reinstating the family Hypoxidaceae. Hutchinson (1959) placed the family as one of the more advanced monocotyledonous families in the order Haemodorales in proximity to the Orchidaceae. Takhtajan (1969) classified the family in the order Liliales (in the wide sense) and also proposed a particular connection to the orchids. Huber (1969) studied the anatomy of the seeds of *Curculigo* and *Hypoxis* spp. and found them to be phytomelan-incrusted; he considered the Hypoxidaceae to represent a predominantly African branch of the Asparagales, with a closer relationship to the Asteliaceae. This has been accepted by subsequent authors and is now also supported by *rbc*L analysis (Rudall et al. 1997).

DISTRIBUTION AND HABITATS. The distribution of the family is in tropical and subtropical regions mainly of the southern hemisphere, in Africa, Asia, Australia and America, including N America. Two genera have a pantropical distribution, *Curculigo* and *Hypoxis*, the latter genus has even spread and speciated in temperate S, Central and N America, particularly along the Atlantic side from northern Argentina to the northern USA (Brackett 1923). The main centre of variation is, however, found in S Africa with five endemic genera (*Empodium*, *Pauridia*, *Rhodohypoxis*, *Saniellia*, *Spiloxene*) in addition to the widespread *Hypoxis*. Two of these genera are centred in the winter rainfall area of W Cape: *Pauridia* and

Spiloxene, and two in the summer rainfall area of S Africa in the Drakensberg mountains: *Rhodohypoxis* and *Saniellia*. The fifth genus, *Empodium*, has a wider distribution in S Africa and occurs in both the winter and the summer rains areas. *Hypoxidia* is a narrow endemic of the Seychelles, and *Molineria* is more widely distributed in tropical Asia (commonly cultivated in the Old World tropics). The family mainly includes typical geophytes, and as such, most representatives belong in regions with strong seasonality between dry and wet periods (savanna grasslands in the widest sense). Two genera have transgressed into tropical forests, the mainly Asiatic *Hypoxidia* and *Molineria*. The genus *Hypoxis* with its very wide distribution covers a large range of habitat types from tropical coral outcrops to alpine boggy meadows. Also the Drakenberg genera have high alpine representatives.

Key to the Genera

1. Flowers with 3 stamens, 6 stigmatic lobes; perianth segments united at the base into a tube; corm surrounded by reticulate bristly tunics; plants glabrous **6. *Pauridia***
– Flowers with 6 stamens, 3 stigmatic lobes more or less fused with the style; perianth segments united at the base into a tube or not; rhizome/corm variable; plants most often hairy 2
2. Ovary usually beaked, unilocular with 3 parietal placentas; seeds strophiolate; inflorescences 1-flowered, without bracts (no distinction peduncle/pedicels) **4. *Empodium***
– Ovary beaked or not, trilocular with axile placentas (at least in lower part); seeds not strophiolate (except *Curculigo*); inflorescence mostly multiflowered, if 1-flowered then with 1–2 bracts 3
3. Corms annual; plants glabrescent, sometimes with few simple hairs **5. *Spiloxene***
– Rhizomes perennial, plants more or less pubescent (except *Saniellia*) mostly from furcate and stellate hairs 4
4. Leaves pseudopetiolate with plicate lamina; fruits more or less baccate without distinct dehiscence; plants medium to very robust 5
– Leaves without a distinct pseudopetiole, lamina ± canaliculate, but not plicate; fruits capsules with longitudinal or circumscissile dehiscence (sometimes without distinct dehiscence in the dwarfed to small *Saniellia* and *Rhodohypoxis*) 7
5. Involucral bracts not persistent and leafy; flowers infundibuliform with a distinct tube of fused brownish red to yellowish pink perianth segments **9. *Hypoxidia***
– Involucral bracts large and leafy, covering the ovary completely; flowers more or less stellate without a tube (but often with an ovary beak); perianth segments yellow 6
6. Large robust plants with flowers and fruits in capitate inflorescences; perianth not separated from the ovary by a long beak (a short one may be present); anthers asymmetric in cross section with introrse opening; seeds with uneven contour, without an expanded hilum and strophiole **8. *Molineria***
– Small to moderately robust plants with 1-few-flowered inflorescences; perianth separated from the ovary by a long beak; anthers symmetric in cross-section with latrorse opening; seeds with a smooth and glossy surface, with an expanded hilum and a strophiole **7. *Curculigo***
7. Perianth segments free, strong yellow **1. *Hypoxis***
– Perianth segments fused in lower part to a tube, pinkish to whitish 8
8. Perianth segments different, inner ones with a constriction in lower part, thereby covering the biseriate stamens **2. *Rhodohypoxis***
– Perianth segments similar; stamens exposed **3. *Saniellia***

Genera of Hypoxidaceae

1. *Hypoxis* L. Fig. 87F–J

Hypoxis L. Syst. Nat., ed. 10, 2: 986 (1759); Nel, Bot. Jahrb. Syst. 51: 234 (1914), rev. Trop. Afr. spp; Brackett, Rhodora 69: 120, 151 (1923), rev. Amer. spp.; Nordal et al., Nord. J. Bot. 5: 15 (1985), rev. E Afr. spp.

Small to medium-sized robust plants. Rhizomes often large and ± tuberous; roots stout, arranged in an equatorial zone on the rhizome. Leaves pubescent along abaxial midrib and leaf margin, sometimes also on the lamina. Perianth segments free, ovate, patient, yellow adaxially, persistent. Stamens uniseriate, exposed, with short subulate to filiform filaments from the base of the perianth segments; anthers basifixed, more or less sagittate, versatile or not, latrorse dehiscence; theca fused or free apically. Ovary 3-locular. Capsules with circumscissile or loculicidal dehiscence. About 50 spp. (number uncertain due to polyploid apomicts), widespread in Africa, America, Asia and Australia.

2. *Rhodohypoxis* Nel Fig. 87A–E

Rhodohypoxis Nel, Bot. Jahrb. Syst. 51: 257 (1914); Milne-Redh., Bot. Mag. tab. 9412 (1935); Hilliard & Burtt, Notes R. Bot. Gard. Edinb. 36: 43 (1978).

Small herbs with a tuberous erect rhizome; roots partly thick, partly thin with a villous covering of root hairs. Leaves pubescent with compound hairs to glabrescent. Scape often short and included in the leaf sheaths; flowers 1–2, subtended by small bracts. Flowers red, pink or white. Perianth forming a cupular tube, the inner segments clawed at the base with distinct inflexed knees closing the mouth of the tube and covering the stamens, inner and outer segments spreading in upper part. Stamens biseriate, arising from the perigone tube at 2 levels; filaments short, attached to the versatile anther low down on the back and continued upwards as thick connective swelling; anther thecae with acute, usually free tips, dehiscing introrsely. Ovary trilocular with a beak separating the ovary from the perianth when scape reduced, if scape present then beak lacking. Capsule circumscissile if borne on erect peduncle; thin-walled and rupturing irregularly when remaining underground or at soil surface. Six spp., S Africa (Drakenberg), at high altitudes.

3. *Saniellia* Hilliard & Burtt

Saniellia Hilliard & Burtt, Notes R. Bot. Gard. Edinb. 36: 70 (1978).

Dwarf herbs with a vertical fleshy rhizome. Leaves glabrous, deeply channelled, somewhat fleshy. Flowers solitary, on reduced scape and very short pedicellate. Perianth raised on a relatively long ovary beak. Flowers whitish, pale yellow at the base. Perianth segments subequal, without constrictions, fused to a tube at the base, spreading above. Stamens arising near the base of the perianth tube; filaments attached to the anther low down on the back, connective prominent on the outside. Ovary trilocular with a prominent apical beak. Fruit subterranean, thin-walled, breaking irregularly. Only one sp., *S. verna* Hilliard & Burtt, S Africa, from high altitudes in the Drakensberg area.

4. *Empodium* Salisb.

Empodium Salisb., Gen. Pl.: 43 (1866); Nel, Engl. Bot. Jahrb. 51: 243, 287 (1914); Burtt, Notes R. Bot. Gard. Edinb. 32: 308 (1973); Thompson, Bothalia 12: 111 (1976) and 429 (1978).
Forbesia Nel (1914).

Small, often hysteranthous plants with annual corms, new discrete tubers added apically and sometimes also basally each season, covered with fibrous tunics; roots thin, protuding laterally on the corm. Leaves glabrescent or with unicellular short hairs, often plicate. Flowers single on a scape with no bracts and no differentiation into peduncle and pedicel. Perianth segments free, yellowish. Stamens close together with short filaments, attached to the non-versatile anthers in the sinus between the basal lobes, anthers sometimes with apical appendages. Ovary exserted or hidden, unilocular; ovules on long funicles. Fruit indehiscent, irregularly fusiform, rostrate; seeds with or without a swollen white funicle. About ten spp., S Africa from the southwestern Cape to Natal and S Transvaal.

5. *Spiloxene* Salisb.

Spiloxene Salisb., Gen. Pl.: 44 (1866); Hilliard & Burtt, Notes R. Bot. Gard. Edinb. 36: 72 (1978); Thompson, Bothalia 12: 111 (1976) and 429 (1978).
Ianthe Salisb. (1866).

Small herbs with annual corms, the old ones persisting as dentate disks at the base of the new ones. Roots covered with long root hairs. Leaves glabrous or with few simple hairs. Flowers in 1-7-flowered umbels on the peduncle. Perianth stellate, orange, yellow or white, tinged with red or green, occasionally with a dark centre; segments free, persistent. Stamens biseriate, exposed; filaments short, attached between the basal lobes of the slightly sagittate non-versatile anthers. Ovary trilocular, with many axile ovules. Capsule narrowly cylindrical, sometimes elongated into a beak, with circumscissile dehiscence or, less commonly, indehiscent. About 30 spp., southwestern Cape in the winter-rainfall region.

6. *Pauridia* Harv.

Pauridia Harv., Gen. S. Afr. Pl.: 341 (1838); Thompson, J.S.Afr. Bot. 38: 163 (1972); Hilliard & Burtt, Notes R. Bot. Gard. Edinb. 36: 72 (1978); Thompson, Bothalia 12: 111 (1976); 429 (1978) and 621 (1979).

Small plants from flat-based, annual corms with reticulate, bristly tunics. Leaves glabrous forming a tubular membranous sheath below. Scape with 1–2 flowers with 2 opposite, filiform bracts. Perianth white to pale pink, regular, persistent, united to a tube. Stamens 3, arising from the perianth-throat, opposite the inner lobes; filaments short, attached to the non-versatile anthers in the basal sinus, thecae not fused apically. Ovary trilocular; stigma 6-lobed, 3 lobes long, linear, erect, often cohering, 3 alternate, short, recurved. Capsule indehiscent, thin-walled. Two spp., southwestern Cape in the winter-rainfall region.

7. *Curculigo* Gaertn. Fig. 88E–I

Curculigo Gaertn., Fruct. 1: 63, tab. 16 (1788); Nel, Bot. Jahrb. Syst. 51: 258 (1914), rev. Afr. spp.; Brackett, Contr. Gray Herb. 69: 155 (1923), rev. Amer. spp.; Hilliard & Burtt, Notes R. Bot. Gard. Edinb. 36: 72 (1978).

Medium-sized robust plants. Rhizome elongated and vertical, often branched, with fleshy roots scattered. Leaves pseudopetiolate with sheathing leaf bases; lamina plicate, sparsely pilose. Scapes short, mostly completely enveloped by the cataphylls. Flowers subsessile, mostly single, sometimes few in a ± umbellate inflorescence supported by leafy involucral bracts; perianth segments free, patent, yellow adaxially. Stamens uniseriate from the base of the perianth segments; filaments filiform, attached in the sinuses between the anther thecae; anthers sagittate, latrorse. Ovary, trilocular, surrounded by the large bracts and old leaf remnants, often ± subterranean, separated from the perianth by a conspicuous beak. Fruit indehiscent, more or less succulent at maturity, crowned with the persistent ovary beak. Seeds with funicle expanded to form a prominent structure plugged by a whitish strophiole. About ten spp., often widespread, tropical Africa, Asia and America.

8. *Molineria* Colla Fig. 88A–D

Molineria Colla, Hort. ripul. App. 2: 331 (1826); Brackett, Contr. Gray Herb. 69: 155 (1923); Hilliard & Burtt, Notes R. Bot. Gard. Edinb. 36: 72 (1978).

Large, robust plants. Rhizome elongated, covered by sheaths of the old leaf bases. Outer leaves cataphyll-like, inner large, pseudopetiolate, with a distinctly plicate lamina, pubescent when young. Scape pubescent, carrying a large capitate inflorescence. Flowers relatively small, yellowish adaxially. Stamens uniseriate, filaments long attached to the anther low down on the outer face; anther with introrse dehiscence. Ovary most often without a beak, if present, only very short. Fruits covered by the bracts, baccate. Five to seven spp., Asiatic monsoon area, best known *M. recurvata* (Dryand.) Nel, largely cultivated in Africa, Asia to Australia.

9. *Hypoxidia* F. Friedmann

Hypoxidia F. Friedman, Bull. Mus. Nat. Hist. Paris B (Adansonia) 4: 454 (1984).

Robust, large plants from elongated rhizomes. Leaves up to 3 m long, pseudopetiolate with sheathing leaf bases; lamina plicate, pubescent when young. Inflorescence racemose with few large evil-smelling flowers; corolla reddish brown to yellowish pink, infundibuliform with a distinct tube. Stamens biseriate from the middle of the tube; anthers subsessile, not sagittate, with filament attachment probably low on the outside, with an elongated sterile part apically and latrorse opening. Ovary (incompletely) trilocular. Fruit ± baccate. Two spp., the Seychelles.

Selected Bibliography

Behnke, H.-D. 1981. Sieve element characters. Nord. J. Bot. 1: 381–400.

Brackett, A. 1923. Contributions from the Gray Herbarium of Harvard University. I. Revision of the American species of *Hypoxis*. Rhodora 69: 120–147.

Chase, M.W. et al. 1995. See general references.

Cheadle, V.I. 1968. Vessels in Haemodorales. Phytomorphology 18: 413–420.

Cullen, J. 1978. A preliminary survey of ptyxis (vernation) in the angiosperms. Notes R. Bot. Gard. Edinb. 37: 161–214.

Dahlgren, R.M.T., Clifford, H.T. 1982. See general references.

Dahlgren, R.M.T. et al. 1985. See general references.

Danilova, M.F., Nemirovich-Danchenk, E.N., Kmar, G.A., Lodkina, M.M. 1995. The seed structure of monocotyledons. In: Rudall, P.J., Cribb, P.J., Cutler, D.F., Humphries, C.J. (eds.) Monocotyledons: systematics and evolution. Royal Botanic Gardens, Kew, pp. 461–472.

Darlington, C.D., Wylie, A.P. 1955. Chromosome atlas of flowering plants, 2nd edn. London: Allen & Unwin.

Davis, G.L. 1966. See general references.

Davis, J.I. 1995. See general references.

De Vos, M.P. 1948. The development of the ovule and seed in the Hypoxideae. 1. *Ianthe* Salisb. J. S. Afr. Bot. 14: 159–169.

De Vos, M.P. 1949. The development of the ovule and seed in the Hypoxideae. 2. The genera *Pauridia* Harv. and *Forbesia* Ecklon. J. S. Afr. Bot. 15: 13–22.

Dutt, B.S.M. 1970. Hypoxidaceae. In: Symp. Comparative embryology of angiosperms. Indian Natl. Sci. Acad. Bull. 41: 368–372.

Erdtman, G. 1952. See general references.

Fedorov, A.A. (ed.) 1969. See general references.

Friedmann, F. 1984. Note sur les Hypoxidaceae des Seychelles et description du genre noveau *Hypoxidia*. Bull. Mus. Nat. Hist. Paris B (Adansonia) 4: 453–460.

Geerinck, D. 1969. Genera des Haemodoraceae et des Hypoxidaceae. Bull. Jard. Bot. Natl. Belg. 39: 47–82.

Hegnauer, R. 1986. See general references.

Hilliard, O.M., Burtt, B.L. 1973. Notes on some plants from southern Africa, chiefly from Natal: III. Notes R. Bot. Gard. Edinb. 32: 303–387.

Hilliard, O.M., Burtt, B.L. 1978. Notes on some plants from southern Africa, chiefly from Natal: VII. Notes R. Bot. Gard. Edinb. 36: 43–76.

Huber, H. 1969. See general references.

Hutchinson, J. 1934. The families of flowering plants, 1st edn., Vol. 2, Monocotyledons. London: Macmillan.

Hutchinson, J. 1959. The families of flowering plants, 2nd edn., Vol. 2, Monocotyledons. London: Macmillan.

Johri, B.M., Ambegaokar, K.B., Srivastava, P.S. 1992. Comparative embryology of angiosperms, Vol. 2. Berlin, Heidelberg, New York: Springer.

Lakshmi, N. 1979. Cytology, cytotaxonomy, interchange heterozygosity and position effect studies in *Curculigo crassifolia* Hook. f. Cytol. Genet. 14: 173–178.

Mitra, A. 1966. Karyological studies in *Curculigo orchioides* Gaertn. Sci. Cult. 32: 201–202.

Naranjo, C.A. 1975. Chromosomal studies in *Hypoxis* L. Karyotype of *H. decumbens* L. Phyton 33: 45–49.

Nel, G.C. 1914. Studien über die Amaryllidaceae-Hypoxidaea, unter besonderer Berücksichtigung der afrikanischen Arten. Bot. Jahrb. Syst. 51: 234–338.

Nordal, I., Laane, M.M., Holt, E., Staubo, I. 1985. Taxonomic studies of the genus *Hypoxis* in East Africa. Nord. J. Bot. 5: 15–30.

Oganezova, G.G. 1995. On the systematical position of the families Haemodoraceae, Hypoxidaceae and Taccaceae (in Russian). Bot. Zh. 80: 12–25.
Ramstad, E. 1953. Über das Vorkommen und die Verbreitung von Chelidonsäure in einigen Pflanzenfamilien. Pharm. Act. Helv. 28: 45–57.
Rudall, J.P. et al. 1997. See general references.
Saito, K. 1975. Studies on the occurrence of polyploidy and its contributions to the flower plant breeding. XII. On the nature and sterility in the spontaneous triploid cultivars of *Rhodohypoxis baurii* Nel. Jpn. J. Breed. 25: 355–362.
Stevenson, D.W., Loconte, H. 1995. Cladistic analysis of monocot families. In: Rudall, P.J., Cribb, P.J., Cutler, D.F., Humphries, C.J. (eds.) Monocotyledons: systematics and evolution. Royal Botanic Gardens, Kew, pp. 543–578.
Takhtajan, A. 1969. Flowering plants: origin and dispersal. New York: Columbia Univ. Press.
Takhtajan, A.L. 1982. See general references.
Thompson, M.F. 1976. Studies in the Hypoxidaceae. I. Vegetative morphology and anatomy. Bothalia 12: 111–117.
Thompson, M.F. 1978. Studies in the Hypoxidaceae. II. Floral morphology and anatomy. Bothalia 12: 429–435.
Thompson, M.F. 1979. Studies in the Hypoxidaceae. III. The genus *Pauridia*. Bothalia 12: 621–625.
Tillich, H.J. 1995. Seedlings and systematics in Monocotyledons. In: Rudall, P.J., Cribb, P.J., Cutler, D.F., Humphries, C.J. (eds.) Monocotyledons: systematics and evolution. Royal Botanic Gardens, Kew, pp. 303–352.
Tomlinson, P.B. 1980. Monocotyledonous habit in relation to morphology and anatomy. In: Brickell, C.D., Cutler, D.F., Gregory, M. (eds.) Petaloid monocotyledons. Linn. Soc. Symp. Series 8. London: Academic Press, pp. 7–29.
Wilsenach, R. 1967. Cytological observations on Hypoxis. 1. Somatic chromosomes and meiosis in some *Hypoxis* species. J. S. Afr. Bot. 33: 75–84.
Wilsenach, R., Papenfuss, J.N. 1967. Cytological observations on Hypoxis. 2. Pollen germination, pollen tube growth and haploid chromosome numbers in some *Hypoxis* species. J. S. Afr. Bot. 33: 111–116.
Zimudzi, C. 1994. Studies of the family Hypoxidaceae in South Central Africa with emphasis on variation patterns in the genus *Hypoxis*. Ph.D.Thesis. Zimbabwe: Dept. of Biological Sciences, University of Zimbabwe.
Zimudzi, C. 1995. The cytology and reproduction of the genus *Hypoxis*. In: Seyani, J.H., Chikuni, A.C. (eds.) Proc. XIIIth Plenary Meeting AETFAT, Malawi, Vol. 1: 535–543.

Iridaceae

P. Goldblatt with J.C. Manning and P. Rudall

Iridaceae Juss., Gen. Pl.: 57 (1789), nom. cons.
Geosiridaceae Jonker (1939).

Perennial evergreen or deciduous herbs, occasionally shrubs with anomalous secondary growth (*Klattia*, *Nivenia*, *Witsenia*), rarely annuals (*Sisyrinchium* spp.), or an achlorophyllous saprophyte (*Geosiris*). *Rootstock* a rhizome, corm or bulb, or, when shrubs, a woody caudex or rootstock indistinct. *Leaves* basal and cauline, sometimes the lower 2–3 without blades, sheathing the stem base and reaching shortly above the ground (thus cataphylls), mostly distichous, the bases usually imbricate, sheaths of foliage leaves open or closed, usually contemporary with the flowers, occasionally produced later, rarely already dry at flowering, occasionally the leaves of the flowering stem with reduced to entirely sheathing blades, then foliage leaves sometimes produced from separate shoots; the blades either unifacial and oriented edgewise to the stem, parallel-veined, without or with a distinct central vein, this sometimes thickened, the surface plane, ribbed, plicate or occasionally terete, very rarely pseudopetiolate, narrow below and abruptly expanded above, or bifacial and oriented with the adaxial surface facing the stem, then channelled to flat and without a median vein; margins sometimes undulate to crisped, or thickened and fibrotic or raised into wings held at right angles to the surface; leaves achlorophyllous and scalelike in *Geosiris*. Flowering stems aerial or subterranean at anthesis, then emerging in fruit, simple or branched, terete or compressed, then often angled or winged, occasionally pubescent. *Inflorescence* either composed of 1-many umbellate monochasial cymes (i.e. rhipidia) (Nivenioideae, Iridoideae) arranged in panicles or spikes or variously clustered terminally or on short lateral branches, the rhipidia paired and partly fused in Nivenioideae; rhipidia comprising pedicellate, occasionally subsessile, flowers each subtended by a single bract, enclosed, usually until anthesis, in large opposed sheathing bracts (spathes), the rhipidia occasionally solitary flowered; or a spike of sessile flowers subtended by opposed bracts (Ixioideae), sometimes reduced to a single flower (e.g., *Romulea*, *Xenoscapa*), the inner (adaxial) bract forked apically and usually smaller, sometimes deeply divided, occasionally to the base (*Babiana* spp.). *Flowers* hermaphrodite, usually large and showy, with a petaloid perianth of 2 whorls of 3 tepals each, rarely the inner whorl suppressed, actinomorphic or zygomorphic (many Ixioideae), then usually bilabiate, the posterior tepal usually largest and inclined to hooded, the lower 3 often smallest, variously coloured, often with contrasting markings, when bilabiate the lower tepals marked with nectar guides; perianth of 6 petaloid tepals in 2 equal or more or less strongly different whorls, occasionally inner whorl reduced or lacking (especially *Patersonia*), tepals free to the base (most Iridodeae) or united in a tube (Nivenioideae, Ixioideae), the tube straight or curved, cylindric or funnel- to trumpet-shaped; nectaries septal (Ixioideae, Nivenioideae), perigonal (Iridoideae), on the base of the outer tepals (most Old World spp.) or the inner tepals (New World spp.) or nectaries lacking (e.g. most *Aristea*, *Isophysis*); oil glands (elaiophores) often present on the inner tepals on New World Iridoideae. Stamens 3 (2 in the Australian *Diplarrhena*), inserted at the base of the outer tepals, or in the tube, symmetrically disposed or unilateral and arcuate, or sometimes declinate; filaments filiform, free or partly to completely united, threadlike and not supporting the anthers in some New World genera; anthers basifixed to sub-basifixed or centrifixed, occasionally sagittate or versatile, 2-thecate, 4-sporangiate, extrorse to latrorse, opening by longitudinal slits, occasionally apically dehiscent. Gynoecium of 3 united carpels. Ovary inferior (but superior in the Tasmanian *Isophysis*), 3-locular with axile placentation, rarely 1-locular with parietal placentation (*Hermodactylis*); ovules anatropous or campylotropous, many to few, especially numerous and small in *Geosiris*, in 2 rows per locule, rarely in 1 row (*Aristea*); style terminal, filiform, usually 3-branched above, or 3-lobed, rarely simple (*Zygotritonia*), the style branches either filiform to distally expanded, sometimes each divided in the upper half, stigmatic towards the apices, or the branches thickened or flattened and petaloid, the stigmas then abaxial below the apices. *Fruit* a loculicidal capsule, rarely indehiscent, firm to cartilaginous, occasionally woody, xerochastic or rarely hygrochastic. Seeds often large, globose to angular or discoid, sometimes broadly winged, usually dry, brown, occasionally blackish, rarely the seed coat fleshy or an aril present, rugulose or smooth, shiny or mat; endosperm hard, with reserves of hemicellulose, oil and protein, rarely also starch; embryo small.

A family comprising ca. 70 genera and 1750 spp., cosmopolitan in distribution, but most abundant and diversified in southern Africa. More than ½ the species belonging to only 6 genera: *Gladiolus* (255), *Iris* (ca. 225), *Moraea* (200), *Romulea* (90), *Geissorhiza* (85), *Crocus* (80), and *Sisyrinchium* (ca. 80). These as well as *Freesia* (15), *Ixia* (45), *Sparaxis* (13), and *Tigridia* (ca. 30) are well known as ornamentals.

Vegetative Morphology. *Roots.* Typical monocotyledonous adventitious roots are found in most members of the family. Many species, especially those with bulbs and corms, also have conspicuous, thick contractile roots at certain stages of their growth cycle. Distinctive fleshy roots (or root tubers) are found in *Sisyrinchium* sect. *Echthronema* and *Iris* subgen. *Nepalensis* and *Scorpiris* (*Juno*), which also has a bulb. Remarkable spiny roots that loosely surround the corm are produced in *Moraea ramosissima*.

Rootstock. Prostrate and creeping to suberect rhizomes are probably basic and are found widely in the family. The rhizome is sometimes represented by a short erect crown, e.g. *Sisyrinchium* spp. The shrubby African Cape genera *Klattia*, *Witsenia* and *Nivenia* have a woody underground caudex and the rhizome of some species of the Australasian *Patersonia* also exhibits secondary growth. The rhizome produces terminal leaves from the shoot apex, and then the same year or the next, a terminal flowering stem. The rhizome continues growth from buds lateral to the terminal shoot that produces the flowering stem. In *Trimezia* (neotropical) the rhizome is ± erect, somewhat swollen and often surrounded by distinct and sometimes slightly fleshy leaf bases.

True bulbs, comprising a few fleshy leaf bases, and dry, brown to grey or red tunics occur in the New World Tigridieae (sensu Goldblatt 1982b). Bulbs also occur in *Iris* subgen. *Xiphium*, *Scorpiris* (*Juno*) and *Hermodactyloides* (= sect. *Reticulata*). In *Hermodactylis* (Eurasian) the rootstock is a peculiar, slender annual tuberlike rhizome.

Corms of two basic types are found in Iridaceae (Fig. 89). In Ixioideae the corms have a distinct central stele and produce roots from the lower half. In Watsonieae and Pillansieae the corm is derived from a lateral bud at the base of the flowering stem (but from the apical bud when a flowering stem is not produced; de Vos 1977). In the remaining tribe, Ixieae, the corm develops from the base of the flowering stem, which enlarges

Fig. 89A–J. Iridaceae. Corm types. **A–D** Basal-rooting corms of Ixioideae. **A** *Watsonia* sp. (tunics partly removed). **B** *Pillansia* sp., with corms of past season persisting. **C** *Geissorhiza aspera*, with woody, imbricate tunics. **D** *Lapeirousia dolomitica*, with woody tunics and corm with a flat base. **E–J** Apically rooting corms of Iridoideae-Irideae showing variation in corm tunics. **E** *Moraea* sp., with tunics partly removed to show root initiation from shoots. **F** *Moraea saxicola*, finely fibrous corm tunics. **G** *Moraea margaretae*. **H** *Moraea lugubris*. **I** *Moraea lurida*. **J** *Moraea fuscomontana*

during and after flowering. In this type of corm the apical bud of the next season's corm is axillary in origin. Corms of Ixioideae may consist of 1 to several internodes. The *Watsonia*-type corm development has been likened to a sympodial system and that in the Ixieae to a monopodial system (Rudall 1995).

In African Iridoideae the corms lack a central stele and produce roots from the base of the apical shoot. In *Ferraria* the corm consists of more than 1 internode, the old corms are not resorbed annually but persist for many years, and the tunics are evanescent. It is derived from the terminal base of the flowering axis. In *Moraea* (incl. *Galaxia Homeria*, etc.) the corm is derived from an axillary bud at the base of the flowering stem (Lewis 1954; de Vos 1977; Goldblatt 1990), consists of only 1 internode and is resorbed annually. The tunics are well developed, woody to somewhat fibrous, and accumulate from year to year. The tunics are often distinctive in genera or even species and provide important taxonomic characters.

Corms of subfamily Ixioideae vary in size, shape, number of internodes and longevity (Lewis 1954). The corms of *Dierama*, *Pillansia* and *Watsonia* spp. are large, persistent (thus not completely resorbed annually) and consist of several internodes. Corm tunics in these genera consist of the thickened fibrous remains of cataphylls and lower leaf bases. More specialised are the annually replaced corms of most other Ixioideae. These tend to be smaller and have distinctive tunics, often of fine netted fibres, thickened clawlike fibres or unbroken layers of various texture. Roots are produced randomly from the lower part of the corm in most Ixioideae, but in *Geissorhiza*, *Hesperantha*, *Syringodea*, most spp. of *Romulea* and a few of *Crocus*, the roots emerge from a basal ridge or projection (de Vos 1977; Goldblatt 1982a). The corms of these genera (except *Crocus*) are further specialised in having tunics of woody, concentric or imbricate layers, and in being composed largely of 1 or few internodes. In *Lapeirousia* and some spp. of *Romulea* and *Hesperantha* the corm is bell-shaped with a flat base.

Flowering stems. Erect flowering stems bearing several to a few leaves with blades of decreasing size or entirely sheathing (thus bractlike) are predominant. The flowering stem is often branched, but in several genera simple stems are produced. In several genera the stem may be very short and subterranean (*Crocus*, *Duthieastrum*, *Syringodea*, spp. of *Hesperantha*, *Ixia*, *Lapeirousia*, *Romulea*, *Tritonia* (all Ixioideae); and spp. of *Iris*, *Moraea* and *Olsynium* (Iridoideae). Stems are generally terete, but flattened and winged in spp. of *Aristea* (Nivenioideae), *Neomarica* and *Sisyrinchium* (Iridoideae) and *Lapeirousia*, *Radinosiphon*, *Savannosiphon*, a few spp. of *Gladiolus* and *Freesia viridis* (Ixioideae).

Leaves. The basic and most common leaf type consists of a sheathing base, the sheath either open or the margins united around the stem, and a unifacial blade oriented edgewise to the stem. In genera with corms or bulbs each shoot is enclosed by 1 to 3 basal and entirely sheathing, usually membranous cataphylls that extend a short distance above ground level. Cataphylls may be ephemeral or persistent, then accumulating as a fibrous neck around the underground part of the stem. In species of *Moraea* the cataphyll forms a distinctive fibrous network around the lower part of the solitary foliage leaf.

Foliage leaves typically have a flat lamina with a series of parallel veins. A prominent central vein (pseudomidrib) is characteristic of Ixioideae except *Pillansia* and the mature leaves of species of *Dierama* and some Iridoideae (*Neomarica*, *Trimezia*). Leaf shape ranges from linear to lanceolate or occasionally ovate and the leaves are usually erect or falcate but occasionally prostrate. Plicate leaves without distinct midribs are found in several genera of Ixioideae (*Babiana*, *Savannosiphon*, ssp. of *Crocosmia*) and most genera of Iridoideae-Tigridieae.

Among the old World Iridoideae, *Moraea*, *Iris* subgen. *Scorpiris* (*Juno*) and *Xiphium* have leaves with bifacial, dorsiventral leaf blades that are interpreted as extended sheaths (Arber 1921). In *Moraea* these leaves have a short, often conspicuous unifacial apex but are otherwise channelled to flat.

Terete (centric) leaves occur in species of genera which otherwise have either unifacial leaf blades (*Hesperantha*, *Geissorhiza*, *Gladiolus*, *Bobartia*, *Olsynium*) or bifacial leaf blades (*Iris*, *Moraea*). Distinctive square leaves are characteristic of *Hermodactylis* and *Iris* subgen. *Hermodactyloides*. In *Devia*, *Romulea* and species of *Geissorhiza*, *Gladiolus* and *Tritonia* the leaf margins or midribs or both are strongly raised or thickened, giving rise to leaves of unusual shape in transection (cruciform, dumbbell). The midribs and margins may be so thickened that the leaf appears terete, but 4 ± obscure longitudinal grooves are always present, 2 on each side of the leaf. This leaftype is the ancestral type in *Romulea*. Among Ixioideae *Crocus* and *Syringodea* have channelled to flat leaf blades, thus morphologically dorsiventral. In *Crocus* the blade typically has a prominent ridged keel and a central white adaxial band.

Leaves are often pubescent in *Babiana* and a few species of *Hesperantha*, *Geissorhiza*, *Gladiolus* and *Romulea* (Ixioideae) and *Moraea* (Iridoideae) amongst others. Undulate to twisted blades occur in species of *Geissorhiza* and *Moraea* and undulate leaf margins characterize a few species of *Freesia*, *Ixia*, *Tritonia* (Ixioideae) and *Moraea*. The ± terete leaf of *M. tortilis* is coiled helically. In *Geissorhiza* sect. *Geissorhiza*, spp. of *Gladiolus* sect. *Hebea*, and *Lapeirousia* subgen. In *Lapeirousia* the leaves are conspicuously several-ribbed.

Most species have several leaves but in several lineages the leaf number is reduced and constant. The majority of species of *Moraea* have a single long-bladed foliage leaf. Many species of *Gladiolus*, *Hesperantha*, *Geissorhiza* and *Lapeirousia* have only 1 basal leaf and 1–2 more partly sheathing cauline leaves. In some African genera and a few in S. America, the leaves are produced at different times from the flowering stems, usually after flowering and on separate shoots. Such hysteranthous leaves are best known in *Gladiolus* but are also found in *Tritoniopsis* and species of *Moraea* and *Watsonia*.

Vegetative Anatomy. (in collaboration with Paula Rudall). Distinctive styloid crystals are present either in idioblasts in the mesophyll or in cells of the outer vascular bundle sheath of leaves and other organs of most species, including *Isophysis* (Isophysidoideae) (Goldblatt et al. 1984) but they are lacking in all but 1 of 20 spp. of *Sisyrinchium* examined and in the related *Olsynium*. Styloids are present in the flowering stem and rhizome of the achlorophyllous *Geosiris* but not in its scalelike leaves. Many genera contain tannin in occasional mesophyll or epidermal cells, particularly in the sheathing leaf base.

Vessels are largely restricted to the roots. Primitive-type scalariform perforation plates (usually over 25 perforations) are reported for *Aristea*, the woody *Nivenia*, *Klattia* and *Witsenia*, the Australasian *Patersonia*, and *Geosiris* (all Nivenioideae) as well as in *Isophysis* (Isophysidoideae). Numerous other genera examined (Cheadle 1963) have vessels with simple perforation plates. Vessels are found in stems and leaves in two of five spp. of *Sisyrinchium* examined.

P-type sieve-tube plastids of the monocot type with cuneate protein crystals are reported only in one species each of *Crocus*, *Gladiolus*, *Moraea* and *Orthrosanthus* (Behnke 1981). In *Nivenia corymbosa* plastids belong to a subtype with the addition of less densely packed crystals.

Roots. The roots have a clearly defined rhizodermis, cortex and central stele. The epidermis is often replaced in older roots by an exodermis, formed by lignification of outer layers of cortical cells. The cortex is sometimes further divided into inner, middle and outer regions based on cell width and size and frequency of intercellular spaces, the innermost cells being the smallest. Root contraction is accomplished by active radial growth of middle and outer cortical cells in a discrete part of the root, causing root shortening, and a collapse or buckling of the cortical tissues. The central vascular stele is bounded by an endodermis, the cells of which usually develop thickened lignified anticlinal and inner periclinal walls with age. The pericycle consists of usually 1, or occasionally up to 3, layers of thin-walled cells. Phloem strands alternate with protoxylem strands in a ring within the pericycle, and wide metaxylem vessels are arranged in a ring surrounding a central parenchymatous or lignified pith.

Stem. The outer cortical region is bounded by an epidermis or periderm, and an inner region consisting of closed vascular bundles scattered in a parenchymatous ground tissue. In many perennating organs, particularly in corms, which have an enlarged cortex, the parenchymatous ground tissue contains numerous starch grains. In rhizomes the central vascular region is sometimes bounded by a lignified endodermoid layer, although the presence of this layer is variable, and it is always absent in *Dietes*, *Neomarica* and *Trimezia* (Rudall 1984). In flowering stems there are several layers of thick-walled cells at the outside edge of the central vascular region.

Primary vascular bundles are randomly oriented, and usually larger towards the centre of the stem. In underground stems they are most commonly amphivasal almost to the stem apex, although occasionally (e.g. in rhizomes of species of *Iris*, *Trimezia* and *Neomarica* and in many flowering stems) they are mainly collateral.

Rhizomes in Iridaceae typically have a pericyclic primary thickening meristem (PTM) near the stem apex, initiating root primordia and a limited amount of stem vasculature (Rudall 1984). In species of *Orthrosanthus* the PTM is extensive and produces a pericyclic vascular plexus. A secondary thickening meristem (STM), similar to that in *Dracaena* and *Aloe*, produces radial chains of individual or linked vascular bundles in the erect stems of the shrubby South African genera *Klattia*, *Witsenia* and *Nivenia* (Scott and Brebner 1893; Adamson 1926) and the rhizomes of some species of the herbaceous *Patersonia* (Rudall 1984).

Leaf. Epicuticular wax platelets of the Convallaria type, characteristic of the Liliiflorae, are reported for a few species of Iridaceae (Barthlott and Frölich 1983). Cuticular striations are rare. The epidermis consists of a single layer of cells, which are most commonly elongated along the long axis of the leaf (longer in costal than in intercostal areas). The cell walls are either straight-sided or in many species of Ixioideae deeply undulating in surface view. In some genera of Tigridieae there are often small groups of large (bulliform) cells. Papillae are relatively frequent in Ixioideae, each epidermal cell usually with 2 or more papillae in a single row (e.g. in species of *Gladiolus*, *Ixia*), but less common elsewhere in the family, and usually 1 per epidermal cell (e.g. in species of *Iris*). Many genera have marginal hairs.

Stomata are often evenly distributed in the intercostal areas of the leaf blade. In some species, e.g. of *Crocus* and *Iris*, stomata are confined to 1 surface, and, where leaves have distinct ridges or marginal thickenings stomata, are typically restricted to the sunken or unthickened areas. As in most other Asparagales, stomata are anomocytic (agenous or hemimesogenous) except in *Dietes* and *Diplarrhena* (Rudall 1983), where subsidiary (perigene) cells are formed by oblique divisions of cells neighbouring the meristemoids. In transverse section stomata are sometimes slightly sunken, and usually with an outer cuticular ridge only, or in a few genera (e.g. *Diplarrhena*, *Pillansia*), with both inner and outer cuticular ridges. Some genera (e.g. *Libertia*, *Dietes*) have a raised rim around the stomatal aperture.

Marginal epidermal and subepidermal structure is often characteristic of genera (Rudall 1995), some (e.g. *Neomarica*) having very thick-walled marginal epidermal cells and others (e.g. *Dietes*, *Ferraria*, *Cipura*) subepidermal marginal sclerenchyma which in Ixioideae is associated with a marginal vein (e.g. *Gladiolus*, *Ixia*, *Watsonia*). Genera of Ixioideae that lack submarginal sclerenchyma usually have instead thick-walled, columnar epidermal cells at the margins. Only *Pillansia* appears to lack cellular differentiation along the margins. In *Isophysis* both thick-walled epidermal cells and subepidermal sclerenchyma are present at the leaf margins. A peculiar lignified marginal "rib" is present in species of the Australian *Patersonia*.

Mesophyll cells are also often characteristic of groups of species or genera. In many species of Ixioideae, mesophyll cells are periclinally elongated along the horizontal axis of the leaf, a condition rare elsewhere in the family. In *Sisyrinchium* sect. *Sisyrinchium* (= sect. *Bermudiana*), the outer layers of mesophyll cells are anticlinally elongated and oval in both TS and LS, whereas in species of section *Echthronema* the mesophyll cells are longitudinally elongated and lobed in LS (Rudall et al. 1986). In thick leaves of many genera the outer 2–3 layers only are chlorenchymatous and the central cells are large and colourless. In leaves of *Crocus* the characteristic white stripe along the central "keel" is caused by large colourless cells (Rudall 1995).

Vascular bundles in the sheathing leaf base are most commonly in a single row, orientated with their xylem poles towards the adaxial epidermis, although in *Dietes* (Rudall 1983) there is also an extra row of bundles immediately within the adaxial epidermis, and in some species of *Neomarica* an extra row within the abaxial epidermis. In both cases the extra bundles are almost entirely sclerenchymatous. The upper isobilateral leaf blade has vascular bundles either in 2 rows with xylem poles towards the leaf centre, or in 1 row with alternating orientation, or more rarely in 1 row with groups of 3–5 bundles alternating in orientation across the leaf (e.g. in the Tigrideae with plicate leaves). In terete leaves, such as those of species of *Bobartia* (Strid 1974) and several other genera, the vascular bundles are in a single ring around the edge of the leaf, with xylem poles towards the centre. In *Crocus* leaves, which have a central "keel" and two lateral "arms", vascular bundles are in 1 row in the arms, all similarly orientated, and extend around the projecting edge of the keel. Leaves of many genera have a single bundle near 1 leaf margin orientated at right angles to the rest, which is continuous with the median bundle of the bifacial sheathing leaf base. Arber (1921) compared the isobilateral equitant leaves of Iridaceae with *Acacia* phyllodes and noted the tendency to reduction or loss of the median bundle in Iridaceae.

Vascular bundles may have an inner sheath and an outer sheath as in many monocotyledons, the outer bundle sheath usually consisting of 1 layer of small parenchymatous cells which are sometimes indistinct from the surrounding mesophyll cells. The inner sheath, where present, is sclerenchymatous and often discontinuous, although in a few genera, such as *Libertia*, the vascular bundles may be completely enclosed in sclerenchyma. In most cases sclerenchyma is confined to the phloem pole of the vascular bundle, although in species of *Sisyrinchium* sect. *Sisyrinchium* and the small allied genera *Olsynium* and *Solenomelus*, sclerenchyma is confined to the xylem pole (Rudall et al. 1986). In some species of *Patersonia* the phloem is completely surrounded by sclerenchyma and in isolated species of *Bobartia*, *Dietes* and *Moraea* the phloem tissue is interspersed with fibres (Rudall 1983). Occasionally, sclerenchyma may be completely absent (e.g. spp. of *Sisyrinchium*). Vascular bundles may either be separated from the epidermis by 1 or more layers of mesophyll cells, or linked to it by a sclerenchyma cap or girder, this feature usually being characteristic of a species or group of species.

Inflorescence Structure. Inflorescences of 2 basic types are found in Iridaceae, spikes (or solitary sessile flowers) in Ixioideae (Fig. 90); and 1 or more, variously arranged, umbel-like monochasial cymes called rhipidia in Nivenioideae (Fig. 91) and Iridoideae (Fig. 92). The rhipidia consist of a few to several pedicellate flowers (sessile in *Galaxia*, most spp. of *Moraea* sect. *Hexaglottis* and a few of sect. *Moraea*) arising from a branch apex and enclosed (at least in the young stages) by a pair of large opposed bracts, the spathes. Each pedicel is subtended by a membranous, 2-keeled, apically bifurcate bract (a prophyll) except for the outermost, which is subtended by the inner of the 2 spathes, and is usually green and never bifurcate. The outer spathe is interpreted as the bract subtending the entire inflorescence. The spathes are obscure in most spp. of *Libertia*. In a few species of *Aristea* and *Nivenia* the flower bracts much exceed the spathes and are distinctively dry and wrinkled and sometimes fringed or lacerate. Flowers of each rhipidium open centripetally, and the pedicels are extended sequentially so that only 1–2 flowers are usually open at one time. The spathes tend to be short, leathery and partly dry in *Aristea* and *Orthrosanthus*. In some *Iris* and *Moraea* spp. the spathes often become dry and membranous, and notably late in the flowering season in *Moraea* sect. *Gynandriris*, in which the intercostal part of the spathes are transparent. The rhipidia occasionally have only a single flower in *Iris* sect. *Oncocyclus*, *Moraea cooperi*, *Moraea* sect. *Gadaxia* and in a few other species including *Isophysis*. Contractile pedicels are found in *Moraea* sect. *Acaules* in which the ovary is withdrawn into the lower part of the inflorescence after fertilization.

Rhipidia are generally single and terminal on the main and lateral branches and the latter are sometimes short or obsolete, notably in most species of *Moraea* sect. *Hexaglottis* in which the lateral rhipidia are sessile, thus spicate in arrangement. In *Neomarica* and most species of *Bobartia* and *Orthrosanthus*, the rhipidia are clustered apically or laterally and their arrangement often appears complex. In genera of Nivenioideae the rhipidia are compound and consist of 2 partly

Fig. 90A–F. Iridaceae. Synflorescence types in Ixioideae. A *Pillansia templemannii*, panicle with all flowers pedicellate. B *Lapeirousia dolomitica*, branched spike. C *Gladiolus huillensis*, simple spike. D *Lapeirousia erythrantha*, multi-branched spike (pseudopanicle), the terminal flowers sessile. E *Xenoscapa fistulosa*, flowers solitary on branches. F *Romulea setifolia*, flowers solitary on branches, main inflorescence axis underground

fused (binate) rhipidial units (Fig. 91C–E). Flower number per binate rhipidium is often 2 (*Klattia*, *Witsenia*, spp. of *Nivenia*) or even 1 (some *Nivenia*). In several species of *Nivenia*, flowering stems bear many binate rhipidia in a corymblike arrangement while in *Klattia* they are arranged in laterally compressed heads enclosed in enlarged, often colourful, subapical leaves.

In Ixioideae the inflorescence comprises sessile flowers subtended by opposed outer (abaxial) and inner (adaxial) bracts, the latter usually forked apically and 2-veined or 2-keeled. The inner bract is sometimes deeply divided (spp. of *Watsonia*), or completely so (spp. of *Babiana*). The flowers are usually aggregated in spikes and arranged spirally, especially when actinomorphic, or when zygomorphic are secund or occasionally distichous (*Watsonia*, *Micranthus*, spp. of *Gladiolus*). The spikes are erect or inflexed, except in *Dierama*, in which they are usually pendent on slender, wiry peduncles. The shape, texture and size of the bracts is an important taxonomic character in Ixioideae, and especially valuable at the generic level. Herbaceous bracts are characteristic of most species of *Geissorhiza*, *Gladiolus*, *Hesperantha* and *Melasphaerula*. In *Romulea* the firm-textured bracts usually have scarious edges, especially the inner one. Scarious, crinkled bracts distinguish *Dierama*, *Ixia* and *Sparaxis*. In *Tritoniopsis* and *Zygotritonia* the bracts are relatively short and coriaceous to dry, and the inner always larger than the outer. Distinctive, fleshy, keeled and occasionally dentate bracts are found in species of *Lapeirousia* subgen. *Lapeirousia*.

In *Pillansia* and many species of *Lapeirousia*, the inflorescence resembles a panicle rather than a spike. In *Lapeirousia* this is best interpreted as a highly branched flowering axis, the terminal branches of which are spikes. In *Pillansia*, however, the inflorescence appears to be a true

Fig. 91A–E. Iridaceae. Complex synflorescences and growth form in Nivenioideae. **A** *Nivenia corymbosa*, pseudocorymbose synflorescence, note woody stems and leaf scars. **B** *Nivenia stenosiphon*, pseudocorymbose synflorescence. **C** *Witsenia maura*, racemose panicle of paired flowers in binate rhipidia. **D, E** *Klattia flava*. **D** Capitulate synflorescence of binate rhipidia. **E** Separate rhipidium removed to show each unit with 1 flower

panicle and all the flowers are stalked. Reduced inflorescences are found in several genera. In *Romulea* and *Xenoscapa* the flowers are best interpreted as solitary on branches of the flowering stem. The spike axis may be subterranean at anthesis in *Romulea*, several species of *Lapeirousia* and *Hesperantha* and is always so in *Crocus* and *Syringodea*. In *Crocus*, *Romulea* and *Syringodea* the flowering stems are raised above ground as the capsules ripen.

Floral Morphology. The flowers of nearly all Iridaceae are perfect and have petaloid tepals, in 2 whorls of 3, and 3 stamens inserted opposite the outer tepals. An inferior, 3-carpellate ovary is present in all genera except *Isophysis* which has a superior ovary, and for this reason is segregated in subfam. Isophysidoideae.

Septal nectaries appear to be ancestral in Iridaceae. They are found in most Nivenioideae and all Ixioideae that produce nectar (Daumann 1970). The nectaries consist of slender pockets in the septal walls of the ovary and open to the exterior through narrow ducts at the base of the style opposite the inner tepals. *Isophysis* lacks nectaries, as do *Geosiris* and most species of *Aristea*. Perigonal nectaries are present in 3 of 4 tribes of Iridoideae and in one species of *Aristea*, *A. spiralis*. In the predominantly Old World Irideae they are usually located at the base of the outer tepals, sometimes also at the base of the inner tepals. In the New World Mariceae and Tigridieae, so-called nectaries are usually located on the inner tepals or sometimes also on the outer tepals (Molseed 1970). These glandular areas may consist of true sugar-secreting nectaries or oil-secreting glands, the elaiophores (Vogel 1974; Simpson and Neff 1981). Elaiophores usually consist of unicellular and clavate to cushion-shaped trichomes and are often concentrated on the middle or lower part of the inner tepals and concealed by folds or pockets of the tepal surface, or they may be fully exposed. The glandular hairs common on the filaments of species of *Sisyrinchium* may also produce oil. The tepals are free in *Isophysis*, more or less so in *Aristea* and *Geosiris* (Nivenioideae), and usually so in Iridoideae. The flowers are radially symmetric in Isophysidoideae, Nivenioideae and Iridoideae, with the exception of *Diplarrhena* (Australian) in which bilateral symmetry is developed and the flowers are secund. In *Moraea* sect. *Galaxia*, scattered species of *Bobartia* and *Moraea* sects. *Hexaglottis* and *Moraea* a hypanthium tube is present. These tubes are closed at the apex and their function is to raise the flower above the bracts

Fig. 92A–F. Iridaceae. Adaptive radiation in southern African *Gladiolus*. **A** *G. serpenticola*, short-tubed pink flowers, bee-pollinated. **B** *G. emiliae*, long-tubed brown-speckled, scented flowers, moth-pollinated. **C** *G. maculatus*, long-tubed, brown, sweetly scented flowers, moth-pollinated. **D** *G. angustus*, flowers cream with red nectar guides and elongate tube, long-tongued fly-pollinated. **E** *G. uysiae*, cryptically colored and marked, anthophorid bee-pollinated. **F** *G. priori*, red flowers with elongate, wide tube, sunbird-pollinated

and inflorescence spathes and in *Moraea* sect. *Galaxia* also above the ground. Most species of *Iris* have a true perianth tube, which may be several cm long, in which nectar is often stored.

The tepal whorls are either subequal (especially *Orthrosanthus* and *Sisyrinchium* (Fig. 93A) and its allies, but are dimorphic in *Libertia* and many other Iridoideae (e.g., *Iris, Moraea, Trimezia*) (Fig. 93B–D). The outer tepals smaller than the inner in *Libertia* (occasionally vestigial) or more often larger than the inner and usually provided with nectar guides. The smaller inner tepals are sometimes erect (many *Iris* spp., a few spp. of *Moraea*). In a few species the inner tepals may be much reduced and reflexed (*Iris* subgen. *Scorpiris*) or tricuspidate, aristate or lacking (spp. of *Moraea, Patersonia*). The tepals in Irideae and Tigridieae are often distinctly unguiculate and then the claws, especially of the inner whorl, are often hairy or variously marked and spotted. In *Iris* subgen. *Iris* the outer tepals have the nectar guide covered with long multicellular hairs (the beard).

In Ixioideae, in contrast, a true perianth tube is always present and is usually well developed. The ancestral condition, present in several genera, is an actinomorphic, radially symmetric flower. Many genera are, however, characterised by zygomorphic, bilabiate flowers with unilateral, arcuate stamens (e.g. *Babiana, Freesia, Gladiolus, Tritoniopsis, Watsonia*). In most of these genera there are also a few species with actinomorphic flowers, presumably a secondary condition. In species with zygomorphic flowers the uppermost tepal is usually enlarged and often hoodlike and the lower 3 tepals together form a platform. In extreme examples of zygomorphy, often associated with bird pollination, the tube may be slender and cylindric below and abruptly expanded, curved and broadly

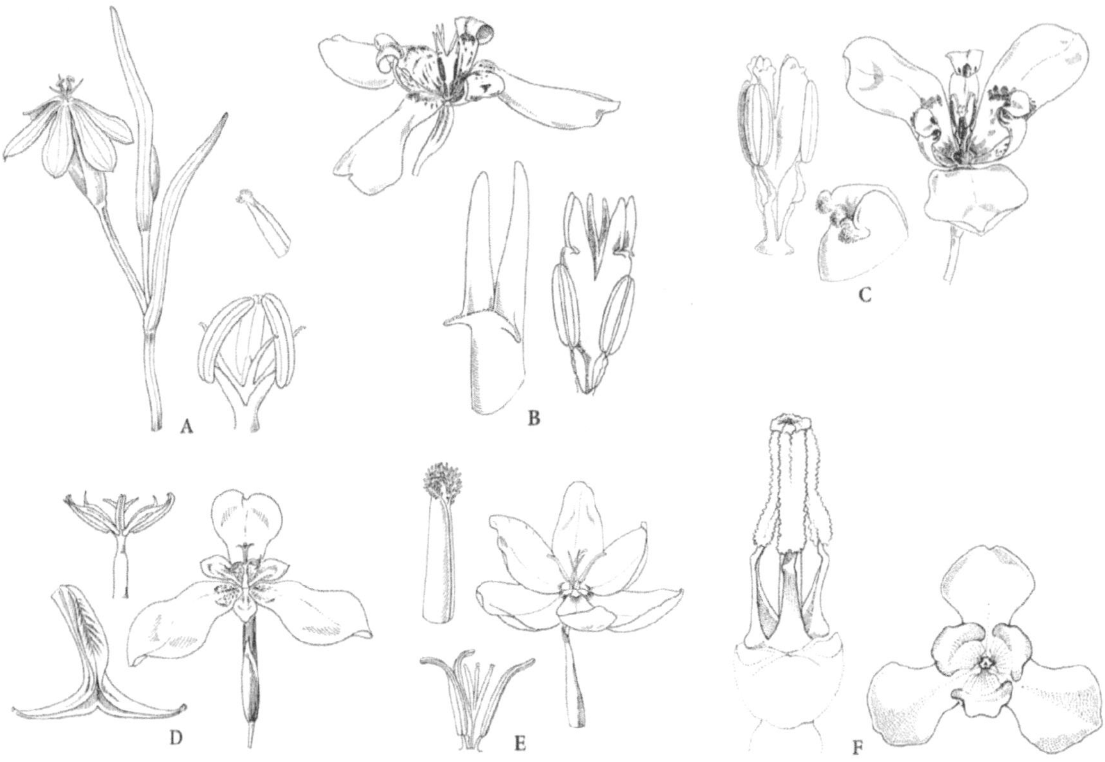

Fig. 93A–F. Iridaceae. Flowers of New World Iridaceae. **A** *Sisyrinchium convolutum*, tepal whorls undifferentiated and without claws, filaments united below with style branches extending between stamens. **B** *Trimezia steyermarkii*, tepals clawed and whorls differentiated, the inner tepals folded back at base of limb, stamens free with weak filaments, style branches thickened with transverse stigma and paired crests. **C** *Trimezia martinicensis*, similar to above but style crests reduced in size. **D** *Herbertia lahue*, filaments united, inner tepals reduced, anthers appressed to spreading style branches. **E** *Calydorea* sp., tepals not clawed, filaments free and style branches undivided and apically stigmatic. **F** *Cipura campanulata*, inner tepals erect, filaments free, style branches vestigial

cylindric above. In bird-pollinated species there is a tendency for the enlargement of the upper tepal and often an associated reduction of the lower tepals, notably in *Chasmanthe* and species of *Gladiolus*. In many genera of Ixioideae the perianth tube may be greatly enlarged, sometimes reaching 15 cm. Such long tubes are associated with pollination by long-tongued flies (Nemestrinidae, Tabanidae) or moths (especially Sphingidae).

Androecium. The 3 stamens are inserted opposite the outer tepals and either free, joined to the base of the outer tepals, or, when a perianth tube is present, the stamens are variously inserted within or at the mouth of the tube. Often the tube widens into a throat above the point of insertion. *Diplarrhena* is unique in Iridaceae in having 2 stamens, with the anterior (lower) one lacking. The filaments are united in a few species of Ixioideae, notably *Ixia* section *Ixia*, and in many genera of Iridoideae, either partly (most Sisyrinchieae, species of *Moraea*) or entirely (*Moraea* sect. *Homeria*, *Tigridia*). The filament column may be smooth and cylindric, enlarged below, and sometimes papillate to pubescent, notably in species of *Moraea* and *Sisyrinchium* subgen. *Sisyrinchium*. In Mariceae and species of Tigridieae with free stamens, the filaments are extremely thin and weak and do not support the anthers, each of which is attached to the abaxial surface of a style branch. The filaments are extremely short in *Cobana* and ± obsolete in *Sessilanthera*, both Tigridieae.

The anthers are 4-sporangiate and usually extrorse and dehisce longitudinally in the majority of species. Anthers are typically oblong to linear, basifixed or sub-basifixed and sometimes sagittate. The connectives are narrow but often mucronate above. The anthers are unusual in being pandurate with a broad connective, and latorse in several neotropical genera, e.g. *Alophia*, *Cypella* spp. In species of *Ferraria* the anther sacs are divergent and contiguous only at the apices. Apical dehiscence occurs in *Cobana*, a genus defined largely by this character. In species of *Ixia* subgen. *Dichone* the anthers are bent at right angles just above the base, and the connectives are described as ± obsolete (i.e. subdidymous).

In many Ixioideae the stamens are asymmetrically disposed and either unilateral, erect and arcuate, and then usually with parallel, contiguous anthers (e.g. spp. of *Babiana*, *Gladiolus*, *Watsonia*) or are laxly to firmly decumbent and declinate with horizontal to upright anthers (spp. of *Geissorhiza*, *Hesperantha* and *Watsonia*). The anthers dehisce inward in species with unilateral stamens.

Gynoecium. The 3-carpellary ovary is 3-locular with axile placentation in nearly all genera. *Hermodactylis* has a unilocular ovary with parietal placentation, and is distinguished from *Iris* partly on this character. The ovary is superior in *Isophysis*, but inferior in all other genera. Typically globose to oblong, the ovary is strongly trigonous in several Iridoideae and in *Melasphaerula* (Ixioideae). In *Moraea* sect. *Gynandriris* and *M. herrei* the upper part of the ovary is sterile and forms an elongated tube which raises the flower above the inflorescence spathes.

The style is single below and ± filiform, and variably divided and elaborated above. The stigma surface is of the Dry type in all Iridaceae (Heslop-Harrison and Shivanna 1977). Digitiform, unicellular papillae are found in most species examined, and this appears to be a family characteristic. Simple filiform style branches stigmatic along the upper surface are presumably ancestral and occur in *Isophysis*, most Ixioideae and *Nivenia* (Nivenioideae). In other Nivenioideae the styles are either minutely trifid apically (*Klattia*, *Witsenia*, spp. of *Aristea*), or divided into 3 broad, ciliate to fringed lobes (spp. of *Aristea*, *Geosiris*, *Patersonia*). Iridoideae have hollow, conduplicate style branches that often divide below the level of the anthers, a feature characteristic of Sisyrichieae (except *Diplarrhena* which may not belong here). In other tribes of Iridoideae the style branches are usually complex structures, either flattened tangentially (spp. of Irideae) or thickened and compressed radially, in either case bearing paired apical appendages.

Most striking is the development in Irideae of tangentailly flattened petaloid style branches, characteristic of *Iris*, for example. Here, each of the 3 broad style branches divides apically into paired appendages (crests) above a transverse stigmatic lobe. The condition is similarly developed in *Dietes*, *Hermodactylis*, *Pardanthopsis* and most species of *Moraea*. This elaborate structure is progressively reduced in several lineages of *Moraea* into short, but tangentially compressed and bilobed style branches, stigmatic at their tips. In the S African *Moraea* sect. *Hexaglottis* the style divides shortly above the base into 3, each branch immediately forked into 2 long filiform and horizontal, apically stigmatic arms. A convergent modification is found in the N American *Nemastylis*. In the S African *Moraea fistulosa* the style branches are filiform and lie alternate to the stamens, presumably a reversal to the ancestral state for the subfamily.

In New World Iridoideae, elaborate style branches are present in several genera including *Neomarica*, *Trimezia* (Fig. 93B,C) and *Cypella*, in which they are thickened rather than flattened and petaloid. Distal to the transverse stigma lobes are paired acute crests and sometimes a second pair of crestlike appendages produced from the edges of the stigma lobes. Variation in the stamen-style branch apparatus in the New World Iridoideae is considerable. The *Cypella* type where stamens have weak filaments and the anthers are loosely attached to the abaxial side of the style branch is most likely ancestral. The pattern is progressively modified in various genera. In *Tigridia* and close allies the filaments are united in a tube and each style branch is divided nearly to the base with the paired arms extending outwards on either side of the opposed anther. The style branches are simple and filiform in *Eleutherine* and several species of *Calydorea* and extend between the anthers. In species of *Cipura* the style branches are barely developed and the stigmatic lobes terminate the thickened style (Fig. 93F). The numerous, mostly small, genera of Tigridieae are defined almost exclusively by various configurations of the style branches and stamens.

The position of the style arms opposite or alternate to the anthers was regarded as important in some early classifications of Iridaceae and was used to separate the genera close to *Sisyrinchium* from Iridoideae, as Sisyrinchioideae. The character has no fundamental significance and is the result of a small twist in the style branch just above the style apex which occurs after the flower opens. During early ontogeny of the flower, the style branches are opposite the stamens. In most Sisyrinchieae the slender style arms extend between the filaments at the base of the anthers.

In Ixioideae, the level of division and orientation of the style branches is often characteristic. *Hesperantha* is distinguished from the allied *Geissorhiza* largely by its long ± spreading style branches that divide at the apex of (rarely within) the perianth tube. *Gladiolus* is distinguished by enlarged and bilobed style branch apices. Several genera including *Freesia*, *Lapeirousia*, *Romulea*, *Savannosiphon* and *Watsonia* have deeply bifurcate and recurved style branches. In *Crocus* the style branches are particularly variable, and may be short to long, simple, bifurcate or multifid, and sometimes distally flattened. In *Ixia* the style is typically well exserted from the tube but divides at the mouth in several species including those of subgen. *Dichone*, in which the style branches are conduplicate, tubular and stigmatic at the apices. In a few isolated species of *Geissorhiza*, *Hesperantha*, *Ixia*, *Romulea* and *Watsonia*, the stamens and the style branches are completely included within the perianth tube.

EMBRYOLOGY. The anther wall consists of 4 layers, an outer epidermis, a well-developed endothecium, typically with helical thickenings, an inner layer several cells thick and crushed at maturity, and a tapetum. Helical endothecial thickenings are most likely ancestral but U-shaped thickenings characterise *Isophysis*, most Nivenioideae, including *Geosiris* and Iridoideae: Sisyrinchieae (Manning and Goldblatt 1990). In *Klattia* and *Nivenia* the endothecial bands anastomose, and in *Patersonia* there is a complete base plate present. The tapetum is secretory, and generally binucleate. Microsporogenesis is successive in *Geosiris* but simultaneous in *Sisyrinchium*, *Iris* spp. and several genera of Ixioideae that have been studied (summary by Rübsamen-Weustenfeld et al. 1994). The condition in *Isophysis* and Nivenioideae except *Aristea* is unknown. The microspore tetrads are isobilateral or tetrahedral. Pollen grains are 2-celled except in rare cases.

The ovule is bitegmic, primitively anatropous, but campylotropous in Ixioideae, and usually semicrassinucellate. The micropyle may be formed by the inner or outer integument. The archesporial cell produces a parietal cell which may form 1–2 layers of parietal tissue but in *Sisyrinchium* the parietal cell degenerates without further division. In contrast, in the only 2 genera of Tigridieae known embryologically (*Gelasine* and *Eleutherine*) the archesporial cell functions directly as the megaspore mother cell and these species are thus tenuinucellate. The primary sporogenous cell forms a linear or T-shaped tetrad and the embryo-sac development is of the Polygonum type. The outer integument is usually several cells thick and the inner integument is 2 cells thick for most of its length. In *Crocus* and *Romulea* the nucellus disintegrates before fertilisation, so that the synergids extend into the micropylar canal. In Ixioideae a large refractive hypostase at the chalazal end of the embryo sac is formed from the nucellus and surrounds the antipodal cells (e.g. Rudall et al. 1984). The antipodal cells are generally large and persistent in Ixioideae and relatively smaller in Iridoideae, and there are records of the proliferation of the antipodals (Haeckel 1930). The polar nuclei often fuse at an early stage before fertilisation to form a fusion nucleus which is usually closely associated with the antipodals. The 2 synergids have a filiform apparatus and are sometimes hooked.

Endosperm formation is Helobial in *Isophysis* and *Geosiris* (Rübsamen-Weustenfeld et al. 1994) but Nuclear in other genera examined including one species of *Aristea* (Nivenioideae), and the few genera of Iridoideae and Ixioideae known (Wunderlich 1959). Embryogeny is poorly known and both the Asterad and Caryophyllad types are recorded.

POLLEN MORPHOLOGY. Pollen grains are normally shed singly, and are ellipsoid to spherical and sulcate in most genera and species. The intine is reported to be unusually thick (Heslop-Harrison 1977). The exine is tectate and primitively reticulate, presumed to be the basic type in the monocots (Zavada 1983), in Isophysidoideae and most Nivenioideae and Iridoideae (Schulze 1971). The lumina of the reticulum are usually nearly equal but sometimes markedly unequal, e.g. in *Ferraria* spp. Lumina vary considerably in size, and in some species the exine is microreticulate (lumina less than 0.5 mm in diameter). Several species of *Aristea* have areolate to rugulate exine and *Patersonia* and *Diplarrhena* have intectate exine with scattered verrucae supported by short columellae. Most Ixioideae have perforate, microscabrate exine except for *Micranthus* and one species of *Thereianthus*, which have reticulate exine grading to perforate at the aperture margins.

While monosulcate apertures are most common, several other aperture types occur. In *Aristea*, sect. *Aristea* has trisulcate (or trichotomosulcate) grains, sect. *Eucapsulares* has zonosulculate grains with the aperture partially obscured by exine fragments, and sect. *Pseudaristea* has dizonosulculate grains. *Diplarrhena* and two species of *Patersonia* examined have inaperturate grains. In most species of Ixioideae, the apertures have a 2-banded operculum (Goldblatt et al. 1991). In *Ixia* and a few species of *Freesia*, *Romulea* and *Thereianthus*, the operculum consists of a single band or is somewhat disorganised. In *Crocus* and *Syringodea* grains are inaperturate or (*Crocus*) sometimes spiraperturate; in Iridoideae *Tigridia* and its close allies, *Cobana* and *Sessilanthera*, grains are bisulcate. In a few genera of the subfamily, including species of *Cipura*, *Dietes* and *Iris*, grains are zonosulculate. In most species of *Iris* subgenera *Scorpiris* (*Juno*) and *Nepalensis*, the grains are spherical and have unconventional apertures. *Moraea* (*Galaxia*) *citrina* has ± typical reticulate exine while other two species have clavate exine. There is a tendency for the exine to become retipilate in other species of *Moraea* through loss of the tectum.

Karyology. Both chromosome number and size vary considerably in Iridaceae (Goldblatt 1971, 1982b; Kenton and Heywood 1984). Small chromosomes are probably ancestral. The number in the monotypic Isophysidoideae is unknown. In Nivenioideae *Aristea* and the woody *Nivenia*, *Klattia* and *Witsenia* have $x = 16$. Among unspecialised genera of Iridoideae, *Libertia* has $x = 19$, *Orthrosanthus* probably $x = 9$. *Olsynium* may have an ancestral base of $x = 10$ and *Sisyrinchium* probably $x = 9$, but both genera have other base numbers, the latter $x = 8$, 17, 6 and 5 and frequent polyploidy, especially in N American species. Small to moderate-sized chromosomes are characteristic of these genera. Other Iridoideae have medium to large chromosomes. *Bobartia*, *Dietes*, *Ferraria* and *Moraea* have $x = 10$, while *Iris* is very variable cytologically and its base number is uncertain. Extensive aneuploidy has occurred in *Moraea* and haploid numbers range from $n = 10$ through 5, with $n = 6$ the most common number in the genus and the base number in sections *Gynandriris*, *Hexaglottis* and *Homeria*. *Iris* has considerable polyploidy as well as aneuploidy with $x = 12$ and 10 the most common base numbers. Among the New World Iridoideae, *Neomarica* and *Trimezia* are poorly known and base numbers include $x = 10$, 9, 8 and 7. In the bulbous genera (Tigridieae) $x = 7$ is evidently ancestral and strongly conserved. Most species of *Tigridia* and its immediate allies *Cobana*, *Fosteria* and *Sessilanthera*, as well as the less closely related *Alophia* are polyploid with $n = 14$, but at least two species of *Tigridia* are diploid, $n = 7$.

Most Ixioideae have relatively small chromosomes and although the subfamily is variable cytologically, a single base number is characteristic of most genera or groups of genera. *Gladiolus* and *Radinosiphon* have $x = 15$; *Hesperantha* and *Geissorhiza*, $x = 13$; *Crocosmia*, *Freesia*, *Tritonia*, and *Xenoscapa*, $x = 11$; *Dierama*, *Ixia* and *Sparaxis*, $x = 10$; *Babiana* and *Zygotritonia*, $x = 7$. *Crocus* and *Romulea* have extensive aneuploidy, but either $x = 14$ or 13 may be basic for *Romulea* (de Vos 1972). Base number in *Crocus* is uncertain but the closely allied *Syringodea* clearly has $x = 6$. The primitive genus *Pillansia* has $x = 10$ but the genus is palaeotetraploid, thus $2n = 40$. *Micranthus* and *Thereianthus* also have $x = 10$. The related *Watsonia* has $x = 9$ and *Savannosiphon* $x = 8$. *Lapeirousia*, perhaps closely related to the two preceding genera, has an unusual bimodal karyotype with 1 large chromosome pair and 7–9 small pairs in most species. The genus may have $x = 10$ as the ancestral base. It exhibits a remarkable dysploid series in the tropical African section *Paniculata* with $x = 8$, 7, 6, 5 and 4, polyploid increase to $x = 6$, and subsequent stepwise dyploid descent to $x = 5$, 4 and 3 (Goldblatt and Takei 1993). Dysploidy has also been reported for several tropical African species of *Gladiolus* (Goldblatt et al. 1993). No ancestral base number has been postulated for Ixioideae. For the family as a whole, $x = 10$ is most likely ancestral, given the base numbers of $x = 10$ 19, 9 and 16 in the least specialised genera.

Pollination. Although the pollination ecology of Iridaceae is extremely diverse (e.g. Vogel 1954), it seems likely that pollination by bees is ancestral. In Nivenioideae, species of *Aristea* are mostly pollinated by bees foraging for pollen, the only reward. Some species of the southern African section *Pseudaristea* are, however, pollinated by monkey beetles (Scarabaeidae: Rutelinae: Hopliinae). The tubular flowers of *Nivenia* are pollinated either by long-tongued bees, mostly Anthophoridae, or long-tongued flies, *Prosoeca* spp. (Nemestrinidae), in the latter case while foraging for nectar. Among other Nivenioideae, *Klattia* is pollinated by sunbirds (*Nectarinia* spp.), as probably is *Witsenia*. Both genera produce large quantities of relatively dilute nectar that has fructose and glucose but lacks sucrose.

The radially symmetric flowers of most Iridoideae are pollinated by bees. The flowers of *Iris* and many species of *Moraea* function as meranthia, that is each flower consists of 3 separate and apparently zygomorphic units. Each unit consists of a platform provided by the limb of the outer tepal, a standard consisting of the style branch and its petaloid crests, and a gullet provided by the claw of the outer tepal and the closely opposed style branch. Bees, mostly foraging for nectar, probe the gullet for nectar located at the base of the tepal, in so doing brushing against the concealed anther. During visits to other flowers, pollen is transferred to a stigmatic lobe which lies above the anther. In *Bobartia* and several species of *Moraea* sect. *Homeria*, the anthers are prominent and these species are pollinated by a variety of bees foraging primarily for pollen. Several species of *Moraea*, notably those with strongly contrasting dark markings are pollinated by a combination of bees and monkey beetles, or the latter exclusively. Pollination by short-tongued muscid or carrion flies is characteristic of *Ferraria* and a few isolated species of *Moraea*. Associated characters are exposed nectar, dull coloration and foetid odor in *Ferraria* and most *Moraea* but in sect. *Homeria* these fly-pollinated flowers are often pale yellow or pink and have a rich, somewhat overripe, or fermenting or semenlike odors. Unspecialised generalist pollination is suspected for the small, radially symmetric flowers of genera such as *Olsynium*, *Orthrosanthus* and *Sisyrinchium*. In many New World Iridoideae oil is the primary reward for pollinators, a variety of bees.

Bee pollination is frequent in Ixioideae, both in species with small, actinomorphic flowers, and in many with short-tubed zyomorphic flowers. Important pollinators are *Apis mellifera*, species of *Rediviva* (Melittidae) and *Amegilla* and *Anthophora* (Anthophoridae).

Both pollen and nectar are available rewards, but in genera with zygomorphic flowers, e.g. *Gladiolus*, *Lapeirousia*, *Watsonia*, most often nectar. Pollen is brushed onto the insect's body passively as it probes the tube for nectar. Many species of *Crocus* are pollinated by female bumblebees. The tubular flowers of many species in several genera of the subfamily are pollinated by long-tongued flies, either *Moegistorhynchus* or *Prosoeca* (Nemestrinidae) or *Philoliche* (Tabanidae). Similarly coloured and marked species of several genera of Iridaceae (mostly Ixioideae) and *Pelargonium* (Geraniaceae) frequently form guilds and utilise the same pollinator(s). Long-tongued fly pollination is particularly common in southern Africa, especially in genera such as *Babiana*, *Lapeirousia* and *Gladiolus*, but it also occurs in species of *Geissorhiza*, *Hesperantha*, *Romulea*, *Tritonia* and *Watsonia* (Goldblatt et al. 1995). Floral coloration may be intense shades of red, blue or purple with pale markings or, more often, whitish to pale pink with red markings.

Moth pollination is assumed in the several species of *Hesperantha* that have pale flowers that open late in the day or in the evening and then produce a sweet scent, but some of these same species are also visited by bees. Pollination by hawk moths (Sphingidae) is recorded for *H. flava*. *Savannosiphon* and species of *Gladiolus* and *Lapeirousia* with tubes exceeding 6 cm and associated pale coloration and strong sweet odours are also presumed to be pollinated by hawk moths. Several more species of *Gladiolus* in South Africa with shorter tubes but strong sweet odours usually produced in the evenings are pollinated by a variety of moths, including Sphingidae and Noctuidae. The only other Lepidoptera important in pollination is *Aeropetes tulbaghia* (Saturnidae). This butterfly pollinates a guild of late summer-flowering species in southern Africa mostly with large, tubular, red flowers. In Iridaceae these include a few species of *Gladiolus* and *Tritoniopsis* and possibly *Hesperantha coccinea*.

Pollination by monkey beetles appears to be frequent in species of *Romulea* and *Ixia*, perhaps exclusively so in section *Ixia*, and also occurs in one species each of *Gladiolus* and *Hesperantha* and a few of *Sparaxis* and perhaps *Geissorhiza*. Such species usually have brightly coloured flowers with strongly contrasting markings, often shades of red, purple or orange with dark brown or black blotches, and usually lack nectar and fragrance. In section *Ixia* the slender tube is closed at the apex by united filaments also fused to the tepals bases.

Bird pollination, mostly by sunbirds (*Nectarinia* spp.), has evolved in several genera of Ixioideae and is associated with red to orange flowers, a long perianth tube, often slender below and wide and cylindric above, and the production of ample quantities of nectar. Often, these species have particularly sturdy stems and flowers with short lower tepals and well-exserted anthers. Sunbird pollination is best developed in *Gladiolus* and *Watsonia*, where is has evolved independently in several lineages in southern and, in the case of *Gladiolus*, tropical Africa as well (often bird-pollinated species were segregated in separate genera, but these are no longer recognised). Other genera in which bird pollination has evolved include *Babiana*, *Chasmanthe*, *Crocosmia*, *Tritoniopsis* and *Watsonia*, all predominantly or exclusively southern African.

Fruit and Seed Morphology. The fruit is a loculicidal capsule in nearly all genera. Capsules range from ± globose to oblong or cylindric, often 3-lobed and sometimes 3-angled in transverse section. The capsule walls range from firm and cartilaginous to coriaceous to hard and woody and are sometimes beaked (spp. of *Iris*, *Moraea* and *Dietes*). In *Moraea* sect. *Gynandriris* the sterile upper part of the ovary remains as a long rostrum. In species of *Dietes*, the woody capsules may be dehiscent only apically or are completely indehiscent. Capsules of *Aristea* are unusually variable and, together with the seeds, form the foundation of an infrageneric classification. Capsules of section *Pseudaristea* are cylindric, deeply 3-lobed, and tardily indehiscent while those of sections *Aristea* and *Racemosae* have wide, extremely thin radially oriented wings. In *Hesperantha* section *Radiata* capsule dehiscense is usually restricted to the upper part.

Seeds range from many to several to 2 (spp. of *Aristea*) or 1 (*Klattia*, *Witsenia*, most *Nivenia*) per locule. The seeds are typically hard, usually light to dark brown and have a smooth outline (without primary sculpturing). Cell outlines (secondary sculpturing) are often pronounced may be colliculate, foveate or occasionally tuberculate. In *Crocus* there is considerable variation in the seed surface, including papillate, digitiform and other epidermal cell types (Baytop et al. 1975; Mathew and Brighton 1977). A smooth glossy surface is characteristic of several genera of Ixioideae, including *Dierama*, *Ixia*, *Tritonia* and *Sparaxis*.

The seeds are relatively large, are ± globose to somewhat angled by pressure or are segmental often with a chalazal crest. In *Micranthus* and species of *Thereianthus* the seeds are ± fusiform. The seeds of many species of *Sisyrinchium* are unusual in being blackish and in having a depression (or sunken chalaza) between the micropyle and the funicle, sometimes including the micropyle. The depression may be as wide as $\frac{1}{2}$ the diameter of the seed. In many Ixioideae the chalazal end of the seed may be flattened or somewhat collapsed. Flat, disc-like, vertically compressed seeds are found in *Diplarrhena*, *Moraea* subgen. *Grandiflora* and *Monocephalae* and some species of *Iris*. In *Klattia*, *Nivenia* and *Witsenia* the seeds are shield-shaped and tangentially compressed. Seeds of *Aristea* spp. are especially variable, and include rounded, angular and radially compressed forms. In *Iris* series Hexagonae the seeds are large and corky and adapted for distribution by water. In *Iris pseudacorus* (series Laevigatae) the outer testa forms a flattened corky float while the seed body is globose. In *Gladiolus* and allies the seeds have a wide membranous circumferential wing that is an outgrowth of the raphe and the outer integument. The wing tissue is soft and consists of dead, inflated cells and large intercellular spaces. In many species of *Watsonia* the seeds have 2 rigid wings at opposite ends or a single wing either at the distal end or surrounding the whole seed. Winglike ridges on the angles of the seeds are also weakly developed in some species of *Hesperantha*. The large seeds of *Tritoniopsis* have an elaborately sculptured testa. A moderately developed spongy testa is characteristic of several species of *Moraea* subgen *Vieusseuxia* (e.g. *M. villosa*). A somewhat reticulate sculpturing characterises seeds of the South American *Solenomelus* and *Tapeinia*.

Arils are found in *Iris* sections *Hexapogon*, *Oncocyclus*, *Psammiris*, *Pseudoregelia* and species of section *Limniris* and subgenera *Scorpiris* (*Juno*) and *Hermodactyloides* (*Reticulata*). The aril, described as a white fleshy outgrowth of the funicle, is typically large and conspicuous. Species of *Crocus* also have arillate seeds. Arils are otherwise rare. A pale yellow or whitish aril or elaiosome containing copious lipid is present in several *Patersonia* spp. A whitish aril is present on the funicle of the seeds of *Aristea singularis* but in no other species of the genus.

A few species, mostly of forest habitats, have red, orange or yellow seeds, amongst which are *Chasmanthe* (Ixioideae) and *Iris foetidissima*, *Neomarica variegata* and *Libertia* spp. (Iridoideae). In *C. aethiopica* and *N. variegata* the seed coat is ± fleshy when first exposed but is always dry in *C. floribunda*. Species with such seeds are presumably dispersed by birds. In the monotypic east Asian *Belamcanda* the ripe seeds are smooth, spherical and glossy black. They remain attached to the axile placentas long after the capsule has dehisced and collectively resemble a blackberry. Seeds of *Babiana* and *Crocosmia* are also dark brown to blackish and stand out starkly against the light inner surface of the capsule, which is coloured orange in spp. of *Crocosmia*.

In several genera of Ixioideae, including *Crocosmia*, *Dierama*, *Ixia* and *Tritonia*, the seed vasculature is excluded during development, and often persists as a filiform appendage lying above the raphe.

Seed anatomy is known in the family mainly as a result of the studies of Huber (1969) and Manning and Goldblatt (1991). Both outer integument (testa) and inner integument (tegmen) are usually present. The outer integument typically consists of 4–6 cell layers, thus with 2–4 middle layers that remain intact. In Nivenioideae there are only 3 layers, thus 1 middle layer or in *Aristea* and *Geosiris* the middle layer is lacking. In most Iridaceae the outer epidermis is pigmented with brownish material (phlobaphene). In *Klattia*, *Nivenia* and *Witsenia* the outer epidermis lacks pigment and is transparent. In *Nivenia* and *Witsenia* the outer epidermis is usually partially exfoliated at maturity. In some species of *Syringodea* the outer integument is also crushed at maturity (de Vos 1974). Both outer and inner epidermis of the outer integument are usually present, the outer often collapsed, but in *Watsonia* it is palisadelike over the seed body. The outer epidermis wall is smooth, and sometimes hemispherically elevated. Papillae are occasionally developed in *Romulea* spp., where they are short and in *Crocus* spp., where they may be very long. Strongly thickened radial and inner walls are common but heavily thickened outer walls are found in some Ixioideae, where they form a hard smooth layer (e.g. *Ixia*, *Dierama*, *Sparaxis*). The tissue between the outer and inner epidermis usually consists of well-preserved cells often with brown-coloured phlobaphene. The inner layer regularly disintegrates in *Orthrosanthus*, *Sisyrinchium*, *Cipura* and many Ixioideae. In *Watsonia* the inner layer contains 1–3 cell layers with extraordinarily thick walls and a lumen packed with calcium oxalate crystals. The inner epidermis consists of cells with a fluid plasma of lipoid nature, either clear or with red-brown masses. This lipoid layer is frequent in Iridaceae. The lipoid layer tends to increase and is often 2-layered in Ixioideae and 2-several-layered in *Iris* and its allies. The cells of the inner epidermis are inflated and protrude inwards in *Libertia* and *Orthrosanthus*.

The inner integument or tegmen usually consists of a strongly flattened outer and a well-preserved inner epidermis that contains pigmented material. In *Aristea*, *Klattia*, *Nivenia* and *Witsenia*, where the outer integument is only 2- or 3-layered, the outer epidermis of the tegmen is well developed and sometimes thickened and unevenly enlarged, with the result that the seed coat becomes undulate at maturity. The cells of the inner epidermis of the tegmen are often enlarged and bladderlike and they impart a wavy sculpturing to the surface of the endosperm. In a few genera, notably *Orthrosanthus* and *Libertia*, and in some Ixioideae, the inner and outer layers form a ± homogeneous structure.

The cells of the endosperm have heavily thickened walls of hemicellulose and contain aleurone and oil bodies. Starch is occasionally present, e.g. in *Geosiris* and the African *Radinosiphon* (Ixioideae), in which a limited number of globose grains are present in the mature seed. The cell walls usually have large simple or weakly bordered pits (2–)4–10 mm wide. In many Ixioideae the walls are less heavily thickened and pits are often lacking except in the central part of the endosperm. The embryo is usually $\frac{1}{3}$–$\frac{2}{3}$ as long as the endosperm and is straight and fusiform. The cotyledon is terminal and the plumule lateral. Dwarf embryos less than $\frac{1}{3}$ as long as the endosperm are found in *Aristea*, *Nivenia* and *Patersonia*, all Nivenioideae.

Dispersal. Seeds are normally the most important dispersal units in Iridaceae, but vegetative reproduction by the branching or dividing of rhizomes, corms or bulbs is significant in many genera. In southern Africa dispersal of corms by mole rats is important, and some species are spread greater distances in this way than by seed. Mole rats store quantities of corms in parts of their burrows for later consumption, but some remain to sprout in the next growing season. Small basal or axillary cormlets, often found in many southern African species, are often inadvertently dispersed by mole rats while carrying larger corms to storage sites.

A few isolated species, sometimes polyploid and partly or entirely sterile, produce numbers of axillary cormlets on the flowering stems and inflorescences, notably *Hesperantha coccinea*, *Sparaxis bulbifera* and races of *Watsonia meriana*. In *Geissorhiza bolusii* bulbils are sometimes formed on the spikes instead of flowers or else they develop after the flowers fade.

In many genera of Ixioideae, the seeds are retained on the open valves of the dehisced capsules for a time. Such seeds are typically glossy brown and are presumably attractive to seed-eating birds and rodents. These animals may pass some of the seeds through their digestive tract, effectively dispersing the seeds some distance. Some forest species of *Libertia* and *Dietes* have indehiscent capsules and seeds are released gradually as the fruits rot on the stems or as they lie on the forest floor. Capsules of *Moraea* sect. *Gynandriris* are also indehiscent and are contained within the transparent inflorescence spathes. Seeds are released after deacy or abrasion. Specialisations for seed dispersal are limited. The globose to angular seeds of most genera apparently simply fall to the ground and may be gathered by insects (? ants) or small mammals and so dispersed over short distances. In *Moraea* sect. *Galaxia* and species of *Moraea* sect. *Gynandriris* and *Lapeirousia*, the whole plant may break loose and be dispersed for short or even long distances by wind, with the seeds being shed during this process. In some species of *Moraea* the stems become coiled during seed-ripening into a rounded structure that can roll short distances.

The discoid seeds of some *Moraea* spp. and the circumferentially winged seeds of *Gladiolus* are dispersed by wind. The seeds of many *Tritoniopsis* spp. have a light spongy seed coat and are probably also dispersed by wind. No observations of the efficency of this mode of dispersal are recorded.

The orange, occasionally fleshy seeds of species of *Libertia*, *Chasmanthe*, *Crocosmia* and *Neomarica* (often species of forest or woodland) are presumably dispersed by birds. Species in these genera are often more widespread than most other Iridaceae, perhaps due to their more efficient mode of dispersal. The seeds of *Melasphaerula* are probably dispersed by ants, but have no obvious modifications for this purpose. There are few published observations on the role in seed dispersal of the aril, characteristic of several species of *Crocus*, *Iris*, *Patersonia* and *Aristea singularis*. Their whitish or yellow colour suggests that they contain lipid and are thus elaiosomes and hence ant-dispersed. Ant dispersal has been confirmed in arillate seeds of *Crocus* (Mathew 1982).

Reproductive Systems. Outcrossing is the normal mode of reproduction. Most genera of Old World Iridoideae have a strong self-incompatibility system and anthers are, in addition, placed so that pollen does not come into contact with the stigmas. Stigmas are of the Dry type and the incompatibility system is sporophytic, where known. The few isolated self-compatible and autogamous species of *Moraea* are thus conspicuous. In Ixioideae a weaker self-compatibility system seems to be present and, in the absence of cross-pollination, autogamy often occurs with a lower seed production than for outcrossing. Morphological protandry is the rule in Ixioideae: anthers dehisce soon after flowers open but style branches remain appressed to one another for several hours or up to 1 day or longer, and only then unfold so that the stigmas are exposed and available for pollen deposition. Heterostyly is characteristic of *Nivenia*, six of the ten species of which have long- and short-styled morphs, 1 morph with short stamens and the other with long stamens. Herkogamy (spatial separation of anthers and receptive parts of the style branches) is the rule in the family. Prime examples are *Iris* and *Moraea*, where stigmatic lobes are held well above the anthers, but herkogamy is common elsewhere. In Ixioideae style branches may recurve and contact the anthers in old flowers that have not been pollinated and selfing may then occur. Three species of the southern African *Moraea* sect. *Homeria* are autogamous, ring-forming complex heterozygotes. Complex heterozygosity has also been reported in *Gelasine* (Kenton and Rudall 1987).

Phytochemistry. Iridaceae contain a wider range of phenolic compounds (Bate-Smith 1968; Williams et al. 1986) than do related families. A few species contain proanthocyanidins and the flavonol, myricetin, compounds which are otherwise characteristic constituents of woody dicotyledons. Another apparently primitive feature is the biflavone, amentoflavone, recorded in *Patersonia* and also present in *Isophysis*. The major flavonoids of the leaf and flower of Iridaceae are flavone C-glycosides, recorded in 65% of the species studied, with 2 flavonols, kaempferol and quercetin, occurring regularly. 6-Hydroxyflavones occur only in *Crocus*, a genus which is also characterised by the water-soluble yellow carotenoid crocein in floral tissues. Isoflavones, including irigenin, tectorigenin and irisolone, otherwise unknown in the monocotyledons, occur in *Belamcanda* and *Iris*. The unusual glucosylxanthone, mangiferin, occurs widely and sporadically in *Iris*, particularly in subgenus *Iris*, *Pardanthopsis* and the closely related *Belamcanda*, and *Moraea* sect. *Gynandriris* (Iridoideae-Irideae); in *Eleutherine*, *Gelasine* and *Tigridia* (incl. spp. treated elsewhere as *Rigidella*) (Iridoideae-Tigrideae) and in two species of *Crocus* (Ixioideae), but is lacking in several related genera that have been examined.

Meta-carboxy substituted aromatic amino and gamma-glutamyl peptides (Larsen et al. 1981) occur widely in Iridoideae, but not in *Sisyrinchium*. Gamma-glutamyl peptides have not been found in *Libertia*, *Orthrosanthus* and *Bobartia* of Iridoideae. Thus, gamma-glutamyl peptides have been confirmed in most genera of Iridoideae excepting members of Sisyrinchieae. These compounds are uniformly absent in the many Ixioideae and selected Nivenioideae studied. *Isophysis* has not been examined.

Chelidonic acid, a common compound in Asparagales and Haemodorales, but also characteristic of Liliales, is poorly represented in

Iridaceae. Cyanogenic compounds are rare but recorded in species of *Moraea*. The structure of the compounds is similar to the cyanogenic glycosides found in *Scilla* and *Urginea* (Hyacinthaceae). The leaves and corms of many *Moraea*, esp. sect. *Homeria*, are toxic to stock and to humans. The possibly toxic quinone, plumbagin, occurs in *Aristea* and in isolated species of *Sisyrinchium* and *Sparaxis*. Quinonoid pigments are also recorded in the bulb of *Eleutherine* and rhizome of *Libertia*. Other chemical classes known in the family and especially in *Iris* include steroidal saponins, fructan storage carbohydrates and hydroxycinnamic acid.

Subdivision and Relationships Within the Family. *Isophysis* is distinctive in Iridaceae in having a superior ovary and on this basis is accorded subfamilial status as Isophysidoideae. It accords in other features with the rest of the family (unifacial leaves; 3 stamens; styloid crystals present in some tissues). Some phylogenetic studies based both on morphological and molecular characters place the genus at the base of the family. Because of its taxonomic isolation, *Isophysis* has been placed in its own subfamily, Isophysidoideae.

The remaining genera fall into three groups based on a series of morphological specialisations in each. Iridoideae and Nivenioideae have radially symmetric flowers arranged in rhipidia and enclosed in large opposed spathelike bracts. The flowers are usually stalked and are raised sequentially out of the sheathing spathes. The pollen grains typically have a reticulate exine. The rootstock is basically a rhizome but bulbs and characteristic apically rooting corms are found in several Old world genera. In Iridoideae perigonal nectaries are located on the tepals and in many genera the style branches are elaborate and flattened to become petal-like (e.g. *Iris*). In Nivenioideae the rhipidia are typically paired in the same way, the flowers thus arranged in a double series and contained within the outer spathes of each of the constituent rhipidia. *Klattia*, *Nivenia*, *Witsenia* (southern African) and *Patersonia* (Australasian) have septal nectaries, but *Geosiris* and most species of *Aristea* lack nectaries. Only *A. spiralis* produces nectar in the subfamily from obscure perigonal nectaries at the base of the tepals.

Ixioideae are specialised in many features. The flowers have a distinct perianth tube, are actinomorphic or often bilaterally asymmetric, and are sessile, surrounded by a pair of opposed bracts. The flowers are arranged in a spike, this occasionally branched and paniclelike. The floral nectaries, when present, are always septal. The style branches are distinct and variously developed, although always slender. The pollen grains have a basically tectate-perforate structure, quite different from the rest of the family. The rootstock is a corm of the basal rooting type. The seeds of Ixioideae exhibit various specialisations including flattening at the chalazal end, and in one lineage, the vasculature is excluded during later development.

Ixioideae are relatively homogeneous and predominantly African. Three genera occur in Eurasia, *Crocus* restricted to the area and *Gladiolus* and *Romulea* shared with sub-Saharan Africa. Molecular data suggest that Ixioideae share a common ancestor with Nivenioideae (or may be nested within that subfamily) but there is no morphological evidence supporting this hypothesis. Unlike Ixioideae, Iridoideae are widespread and diverse. They fall into several large groupings for which tribal status seems appropriate. Amongst these are the largely New World and Pacific Sisyrinchoideae, two strictly New World tribes, Trimezieae and Tigridieae, and the predominantly Old World Irideae (with some spp. of *Iris* in N America). The circumscription and tribal allocation of genera seems reasonably clear, but the tribal interrelationships are uncertain and many details remain to be elucidated. The African *Bobartia*, usually aligned with Sisyrinchieae on the basis of floral similarities, has been shown by chloroplast gene sequence similarity to be allied to African Irideae. The affinities of the Australian *Diplarrhena* within Iridoideae are uncertain and await study.

In Ixioideae, which comprise nearly 2/3 of the species of Iridaceae, major generic groupings are unclear. The monotypic *Pillansia*, the only member of Ixioideae with a true panicle and lacking a leaf midrib, appears taxonomically isolated and is treated as comprising Pillansieae by Goldblatt (1990). Several genera with divided style branches and a corm that originates from a lateral rather than an apical bud seem to comprise a natural alliance, *Watsonieae*. This still leaves the majority of genera, some 20 genera, in Ixieae. Lewis (1954) and Goldblatt (1971) have recognised several small natural groups within Ixieae, as subtribes (e.g. *Crocus*, *Romulea*, and *Syringodea*, which have progressively reduced aerial stems). Although such immediate generic relationships are clear, broader affinities are not apparent. *Chasmanthe*, *Crocosmia*, *Devia*, *Duthieastrum*, *Sparaxis* and *Tritonia*, which have seeds with excluded vasculature and leaf marginal epidermis of colummnar, thickened cells, appear to contitute a large alliance. *Freesia*, which has similar leaf margins, has seeds without excluded vasculature. *Dierama* and *Ixia* also have seeds with excluded vasculature but the ancestral leaf margin anatomy.

AFFINITIES. In classifications of the Liliidae in which a series of families with septal nectaries and often seeds with blackish, phytomelan-encrusted seeds have been accorded ordinal status as Asparagales separate from Liliales, Iridaceae have been placed in either of these two large groups of families. The family is readily distinguished from related other liliod families by their 3 stamens, ancestrally unifacial leaves and, except for *Isophysis*, an inferior ovary. Iridaceae were included by Dahlgren et al. (1985) in Liliales, close to Colchicaceae, on the basis of their extrorse anthers, non-phytomelanous seeds, mottled tepals, presence of periogonal nectaries and Nuclear endosperm development. Perigonal nectaries are, however, now known to be derived in the family, and septal nectaries are the ancestral state. Mottled tepals are also derived in Iridaceae and Helobial endosperm development appears to be characteristic of the lower Iridaceae. Thus, the placing on Iridaceae in a narrowly circumscribed Liliales appears to have little support. Despite the absence of phytomelanous seeds, assignment of Iridaceae to Asparagales must be given serious consideration on morphological grounds. Molecular gene sequence similarities support the latter relationship. Sequences from the chloroplast gene, *rbc*L, place Iridaceae close to, but outside, the core taxa of Asparagales, which do have phytomelan-encrusted seeds. The molecular data place Iridaceae close to Ixioliriaceae, the Australasian Doryanthaceae, and somewhat more distant to Tecophilaeaceae, which, excepting Doryanthaceae and some Tecophilaeaceae, have phytomelanous seeds. None of these possible relationships has any particular merit on morphological grounds. The absence of a clearly identifiable ancestor for Iridaceae suggests that the family is extremely old, probably having had a late Cretaceous origin on the southern continents.

DISTRIBUTION AND HABITATS. Iridaceae have a nearly worldwide distribution but are especially well represented in S Africa and in temperate and highland S and Central America. Subfam. Ixioideae are largely African, and only *Crocus* exhibits radiation outside the continent, extending across Europe to the W Himalayas. *Gladiolus* and *Romulea* also occur in the Mediterranean and Middle East, but have their centre in the southern African winter-rainfall region. Several genera of Ixioideae occur in tropical Africa where *Zygotritonia* (4 spp.) and *Savannosiphon* (1) are endemic and *Gladiolus* has radiated significantly. Some 75 spp. of the genus are endemic to various centres across tropical sub-Saharan Africa. Most Ixioideae, however, are centred in the winter-rainfall area of South Africa, incl. the SW Cape, Namaqualand and the W Karoo (*Babiana*, *Geissorhiza*, *Ixia*, *Sparaxis*, *Watsonia*). *Dierama* is exceptional in being concentrated in eastern southern Africa, where there has been significant radiation in *Gladiolus* and *Hesperantha* as well. Notable disjunct geographical ranges include one sp. each of *Babiana* and *Romulea* on Socotra.

The monotypic *Isophysis* is restricted to Tasmania. In subfam. Nivenioideae *Aristea* is widespread in Africa and Madagascar, where *Geosiris* is endemic. The shrubby genera *Nivenia*, *Klattia* and *Witsenia* are restricted to the SW Cape and only the Australasian *Patersonia* occurs outside Africa, extending through Australia to the mountains of New Guinea, Borneo and Sumatra. Subfamily Iridoideae comprises Irideae, largely Old World (incl. Australasia) excepting *Iris*, which also occurs in N America. Tribes Trimezieae and Tigrideae are exclusively New World, and Sisyrinchieae largely so. *Sisyrinchium*, largest genus of the Sisyrinchieae, extends throughout S and N America including Greenland. Of the other genera of Sisyrinchieae, *Solenomelus* is restricted to the S Andes and *Tapeinia* to Patagonia, but *Orthrosanthus* extends north to Mexico as well as occurring throughout Australia. *Libertia* is found in Australia, New Guinea and New Zealand and in Andean S America. *Olsynium* has one sp. in western N America and the remainder occur in S America from Peru southward to Patagonia and one sp. is endemic on the Falkland Is.

Iris, largest genus of Irideae is exclusive to the northern hemisphere and is unusual in Iridaceae in its development in centres like California, China, Europe and the Middle East. Four of the six spp. of *Dietes* are restricted to southern Africa, one more extends from S Africa to Kenya and Uganda, and one more occurs on Lord Howe Island (Australia). *Moraea*, the other large genus of Irideae, is widespread in sub-Saharan Africa and centred in the southern African winter-rainfall region. *Moraea* sect. *Gynandriris* has an unusual disjunction with two of the nine spp. native to the Mediterranean and Middle East.

Tigrideae have radiated extensively in temperate S America where there are several small endemic genera (*Gelasine*, *Kelissa*, *Onira*) and in the Andes (*Cardenanthus*, *Mastigostyla*). A second New World centre is in northern Central America and Mexico, where *Ainea*, *Cobana*, *Fosteria* and *Sessilanthera* are endemic. *Tigridia*, the most speciose genus there, also occurs in Andean S America. *Nemastylis* is shared between the southern USA and Mexico and northern Mesoamerica. *Cypella*, one of the larger genera of Tigrideae, has a wide range in S and Central America including the West Indies. *Herbertia* has a notable disjunction with several spp. in Uruguay, S Brazil and Chile and one subsp. in southern USA.

Iridaceae are mostly small plants and thus often found in open scrub, desert or grassland. Members of the family also favour temperate climates and the distribution largely follows these conditions. Thus, species are concentrated in heathlands in S Africa and the Mediterranean (but poorly represented in similar vegetation types in Australia, Chile and western N America) and in the grasslands of cooler S America, the Andes or the higher elevation grasslands of Africa. Semi-deserts adjacent to these areas in temperate latitudes also have a fair representation of Iridaceae, notably the Middle East and Namaqualand and the Karoo in S Africa.

Few Iridaceae grow in forest habitats but *Neomarica* stands out in the New World and *Dietes* in the Old. Several species of *Iris* also grow in forest, blooming early in the spring before the canopy closes. In Africa several species of *Aristea*, *Crocosmia*, *Chasmanthe* favour forest margins.

Economic Importance. Iridaceae are important in horticulture, both for the cut flower industry and for garden display. *Gladiolus*, *Iris* (mainly *I. xiphium* cultivars for the florist trade) and *Freesia* are the most important and are grown commercially especially in Europe and the USA. Several species and cultivars of *Crocus*, *Freesia*, *Gladiolus*, *Iris* and *Watsonia* are grown in gardens throughout the world and the trade in the corms, bulbs and rhizomes is considerable. Less well known in gardens are *Belamcanda chinensis*, *Tigridia pavonia* and species of *Dietes*, *Ixia*, *Moraea*, *Sparaxis* and *Tritonia*, but they are available in the nursery trade.

The stigmas of *Crocus sativus* are the source of the spice saffron and the species is grown in parts of Europe, the Middle East and India for this spice. The corms of *Crocus* spp. are reported to be used as a food in Syria and corms of *Lapeirousia* are eaten today in parts of southern tropical Africa. Corms of many species were an important source of food for the early inhabitants of southern Africa. The corms of *Moraea fugax* are eaten today, but largely as a curiosity.

Key to the Subfamilies (with J.C. Manning)

1. Ovary superior; flowers without nectaries; inflorescence a solitary flower; perianth free to the base; style dividing into 3 short lobes apically **1. Isophysidoideae**: ***Isophysis*** (p. 318)
– Ovary inferior; either septal nectaries or perigonal nectaries or oil glands often present; inflorescence usually a rhipidium or spike 2
2. Nectaries septal; inflorescence a spike or by reduction flowers single on branches, always sessile and subtended by a pair of opposed bracts; tepals always united in a tube below; flowers lasting at least 2 days; style dividing into filiform branches, rarely simple; rootstock a corm producing roots from the base; pollen grains usually with perforate scabrate exine, rarely exine reticulate; pollen grain apertures usually with operculum unless zonosulculate **4. Ixioideae** (p. 327)
– Nectaries septal or perigonal or oil glands present; flowers in umbellate clusters (rhipidia) enclosed by a pair of opposed leafy bracts (spathes), rarely solitary on the peduncles or plants acaulescent but then style either dividing below the anthers into tangentially compressed, petal-like branches or dividing below or above the base of the anthers and obscurely 3-lobed apically, the lobes entire or fringed, individual flowers pedicellate, sometimes sessile; rootstock a woody caudex, a rhizome, a corm rooting from the apical bud, or a bulb; flowers with the tepals free to basally connate, or united in an extended tube; lasting a single day or 2 or more days 3
3. At least the terminal rhipidia united in pairs (binate); style notched or divided apically into 3 lobes, occasionally forming filiform branches; nectaries when present septal **2. Nivenioideae** (p. 318)
– Rhipidia never united in pairs; style dividing below the level of the anthers into three branches, these either filiform, extending between the stamens or thickened or flattened and lying opposite the stamens, usually each terminating in paired appendages (style crests); nectaries or oil glands when present perigonal **3. Iridoideae** (p. 319)

Key to the Genera of Nivenioideae

1. Plants perennial, either with aerial stems not woody or acaulescent, sometimes evergreen; rootstock a rhizome or corm; individual flowers stalked or sessile, the tepals free, basally connate, or united in a tube; lasting 1 day and deliquescing on fading 2
– Plants evergreen shrubs with woody aerial stems; rootstock a woody caudex; individual flowers sessile and the tepals always united in a well-developed tube and lasting at least 2 days 4
2. Plants achlorophylous, saprophytic and leaves reduced to small scales **3. *Geosiris***
– Plants green, autotrophic and leaves present and well developed 3
3. Tepals shortly connate basally; inner tepal whorl well developed; filaments free; style apically notched or lobed, the lobes sometimes fringed; flowers pedicellate, the pedicel sometimes very short **2. *Aristea***

- Tepals united in a tube included in the bracts; inner tepal whorl minute; filaments partly to completely united; style with 3 deeply fringed lobes; flowers sessile **7. *Patersonia***
4. Inflorescence compound, forming a compressed capitulum enclosed by enlarged green or coloured leaves; perianth tube shorter than the linear, spathulate tepals **5. *Klattia***
- Inflorescence either compound, forming branched panicles or corymbs, or flowers borne in isolated pairs, enclosed in green or brown spathes; perianth tube shorter or longer than the oblong to ovate tepals 5
5. Flowers shades of blue; tepals patent, not villous on the outside; stamens and/or style well exserted from the flower, species often heterostylous **4. *Nivenia***
- Flowers green to blackish with the tepals yellow on the reverse; tepals remaining closed during flowering, densely villous on the outside; stamens included in the flower **6. *Witsenia***

Key to the Genera of Iridoideae

1. Rootstock a creeping or erect rhizome or indistinct, plants evergreen or deciduous 2
- Rootstock a corm, bulb or tuber, plants deciduous 16
2. Stamens 2; anthers oblique on the filaments and unequal; flowers zygomorphic **36. *Diplarrhena***
- Stamens 3; anthers erect on the filaments and equal; flowers actinomorphic 3
3. Style usually dividing at apex of filament column (occasionally opposite or beyond the anthers) into 3 slender branches extending between or above the anthers, or style occasionally undivided; tepals not clawed and usually subequal, occasionally those of the outer or inner whorl smaller 4
- Style usually dividing into 3 thickened or radially or tangentially compressed more or less petaloid branches ramaining opposite appressed to the stamens; tepals subequal or those of the inner whorl smaller or lacking 12
4. Filaments free; pedicels sometimes hairy 5
- Filaments united below or entirely; pedicels glabrous; plants of Australasia and N and S America 7
5. Pedicels hairy below the apices; rhipidia usually several crowded apically and flowering stem consisting of 1 long internode and bearing a subapical leaf below the rhipidia, or if stem branched then sticky below the nodes; style branches extending between the bases of the anthers; plants of S Africa **37. *Bobartia***
- Pedicels glabrous; rhipidia not crowded apically on a single long internode and stems not sticky below the nodes; style branches extending between the anthers or above them; plants of S America and E Asia 6
6. Tepals red to orange, irregularly speckled with dark red; capsules with cartilaginous walls soon dehiscing but seeds remaining attached to axile placentas, globose and glossy black **41. *Belamcanda***
- Tepals not red to orange or irregularly speckled; capsules walls ± woody and seeds angular **15. *Pseudotrimezia***
7. Stems compressed, sometimes angled or winged; seeds globose or compressed globose, usually with the chalaza sunken; seed coat dark brown to blackish **12. *Sisyrinchium***
- Stems terete; seeds angular or oblong, usually with a chalazal crest; seed coat usually light to dark brown 8
8. Evergreen cushion plants to 5 cm high; flowering stems with a single rhipidium bearing 1 flower **10. *Tapeinia***
- Not evergreen, low-growing cushion plants, but occasionally acaulescent; flowering stems simple or branched, rhipidia with more than 1 (- several) flowers 9
9. Plants with leaves widely spaced, linear to terete, not forming a fan; stems occasionally with 1 or few slender branches; roots thick and fleshy; stem entirely to partly concealed by long sheathing leaf bases **13. *Olsynium***
- Plants with a basal fan of distichous, sword-shaped to linear leaves; stems usually branched, the lateral branches often fairly short; roots fibrous 10
10. Tepals united below in tube; flowers secund; filaments united entirely in a thick column **11. *Solenomelus***
- Tepals free to the base; flowers upright; filaments united in the lower half 11
11. Ovary sessile or nearly so, often pubescent; tepals subequal **9. *Orthrosanthus***
- Ovary usually stalked, never pubescent; outer tepals somewhat to much smaller than the inner **8. *Libertia***
12. Style branches thickened, often bearing acute paired terminal appendages; anthers lightly affixed to the abaxial side of the style branches on threadlike filaments not supporting the anther; inner tepals usually folded or geniculate near base of limb and bearing oil glands on short thichomes 13
- Style branches compressed tangentially and petaloid, each bearing flat paired terminal appendages; anthers appressed to the abaxial side of the style branches on sturdy filaments; outer tepals strongly clawed, the base of the limb usually held close to the apex of the style branch and with a prominent nectar guide, inner tepals not geniculately folded and not glandular 14
13. Stem compressed and broadly 2-winged; rhipidia crowded apically at the end of a long basal internode subtended by a subterminal leaf **16. *Neomarica***
- Stem terete; rhipidia single at the ends of branches, the stem usually branched and bearing leaves below **14. *Trimezia***
14. Tepals united below in a tube **38. *Iris***
- Tepals free 15
15. Plants deciduous; pedicels glabrous; (plants of Asia) **40. *Pardanthopsis***
- Plants evergreen; pedicels hairy above; (plants of Africa and Lord Howe Island) **35. *Dietes***
16. Rootstock a corm, bulb or tuber; leaves bifacial and channeled or unifacial or centric (square, oval or round in section) (plants mostly of Africa and Eurasia, except *Pseudotrimezia*) 17
- Rootstock a bulb; leaves unifacial (oriented edgewise to the stem) and plicate or sometimes subterete but at least once folded; (plants of S and N America) 21
17. Rootstock a bulb or tuber 18
- Rootstock a corm; tepals sometimes united below in a tube, more often free 20
18. Tepals free (S America) **15. *Pseudotrimezia***
- Tepals united below in a tube; leaves plane or centric and terete, or 4-many-sided. Eurasia and N Africa 19
19. Inner tepals present **38. *Iris***
- Inner tepals lacking **39. *Hermodactylis***
20. Leaves unifacial, oriented edgewise to the stem; old corms persisting for some years, with membranous tunics or tunics absent; tepals with crisped edges; style branches terminating in a feathery, plumose tuft **42. *Ferraria***
- Leaves bifacial and channelled to flat, oriented with the adaxial surface facing the stem, or terete; old corm usually resorbed annually, with persistent fibrous to woody tunics;

tepals with plane or undulate edges; style branches rarely feathery and plumose but then other characters not as above **43. *Moraea***

21. Style ± undivided or branches short and obscure, hardly distinct from the style, sometimes represented by more or less distinct lobes 22
- Style clearly 3-branched, usually dividing opposite or below the middle of the anthers, the branches variously simple and filiform or variously elaborated, sometimes divided almost to base and apparently 6 24

22. Style filiform throughout, minutely 3-fid apically; tepals with narrow claws **25. *Kelissa***
- Style thick or expanding above; tepals usually lacking well-defined claws 23

23. Style slender, widening gradually above, extending beyond the anthers **19. *Calydorea***
- Style thick throughout, not exceeding the anthers **18. *Cipura***

24. Filaments extremely short, less than 1 mm long; tepals not clearly differentiated into limb and claw, the surface plane and without glands 25
- Filaments usually well developed, at least 1 mm long; tepals usually distinctly clawed and often with secretory hairs on parts of the upper surface 26

25. Style branches undivided; anthers porose **34. *Cobana***
- Style branches each deeply divided almost to the base; anthers longitudinally dehiscent **33. *Sessilanthera***

26. Style branches simple, short or long, filiform or lobed, and extending between the anthers or dividing above them or remaining opposite the anthers 27
- Style branches deeply divided above, or flattened and truncate or with paired terminal crests, or if simple and filiform then branches opposite the anthers 30

27. Filaments at least partly united **22. *Gelasine***
- Filaments free entirely 28

28. Flowers white; style branches extending between the anthers; stem bearing a conspicuous subterminal cauline leaf **20. *Eleutherine***
- Flowers blue to purple, often with yellow markings; style dividing above the anthers; stem without a leaf inserted at the base of the rhipidia 29

29. Tepals with long conspicuous narrow claws; anthers clasping the style branches **21. *Onira***
- Tepals not obviously clawed; anthers not clasping the style branches **19. *Calydorea***

30. Style branches forked for at least $\frac{1}{2}$ their length into ± filiform arms 31
- Style branches flattened or thickened radially or tangentially, sometimes divided apically into lobes or crests and then stigmatic surfaces below the apices 35

31. Style reaching to about the apex of the filaments and arms of style branches extending outwards from the base of the anthers; anther connective narrow, not normally visible and anthers extorse 32
- Style dividing close to or beyond the anther apices; anther connective broader than the anther sacs and anthers latrorse 34

32. Anthers erect before dehiscence, collapsing spirally, and arms of the style branches horizontal; filaments occasionally free or partly to completely united **28. *Nemastylis***
- Anthers diverging and arms of the style branches arching outwards or suberect; filaments united 33

33. Style arms erect, papillose and apparently stigmatic throughout **32. *Fosteria***
- Style arms arching outwards, apically stigmatic **31. *Tigridia***

34. Tepals unequal, the inner much smaller than the outer; anthers linear; filaments free, supporting the anthers **29. *Ainea***
- Tepals subequal; anthers pandurate; filaments free or united; **30. *Alophia***

35. Filaments free entirely or in lower half 36
- Filaments united, widest below or cylindric throughout 37

36. Style branches truncate or with lobed appendages above the stigmas; inner tepals with the limbs suberect, concealing the anthers and style branches; rhipidia usually at least 2 per branch and subtended by well-developed leaf at their base **18. *Cipura***
- Style branches with paired terminal crests; inner tepal limbs spreading; branches terminating in solitary rhipidia distant from the leaves **17. *Cypella***

37. Tepals united below for at least 4 mm; plants usually more or less acaulescent **27. *Cardenanthus***
- Tepals free to base; plants with aerial stems 38

38. Anthers appressed to the style, the branches diverging above the anther apices **22. *Gelasine***
- Anthers appressed to the style branches 39

39. Filament column cylindric; anthers about as long or shorter than the style branches **26. *Mastigostyla***
- Filament column bottle-shaped or obconic 40

40. Column bottle-shaped; anthers exceeding the style branches; inner tepals less than $\frac{1}{2}$ as long as the outer **24. *Herbertia***
- Column obconic; anthers as long or somewhat shorter than the style branches; inner tepals about as long to slightly shorter than the outer **23. *Ennealophus***

Key to the Genera of Ixioideae

1. Corm developed entirely from an axillary bud at the base of the flowering stem 2
- Corm developed from the basal nodes of the flowering stem 8

2. Inflorescence a panicle, the individual flowers always stalked; flowers actinomorphic, bright orange; leaves without midribs; leaf margins without columnar epidermal cells; seeds glossy **44. *Pillansia***
- Inflorescence a simple or branched spike, or crowded in a basal tuft, the flowers sessile; leaf margins with subepidermal sclerenchyma associated with a marginal vein; seeds matte 3

3. Style branches undivided; floral bracts fairly short, coriaceous and the inner bracts always substantially longer than the outer; leaves sometimes with a long pseudopetiole, usually with more than 1 prominent vein, thus without a midrib; leaves often not contemporary with the flowers **50. *Tritoniopsis***
- Style branches each deeply divided; floral bracts short or long, coriaceous and the inner bracts shorter than the outer; a distinct midrib present unless centric 4

4. Stems compressed and angled or winged; seeds globose with the chalazal end sunken 5
- Stems round in section; seeds angular the angles sometimes extended into wings, the chalazal end a prominent crest 6

5. Corms bell-shaped, with a flat base; leaves plane or corrugate **48. *Lapeirousia***
- Corms globose, rounded at the base; leaf blades plane; pollen grains inaperturate **49. *Savannosiphon***

6. Stems with several basal leaves in a distichous fan; leaf blades plane, relatively broad, the margins moderately to

strongly thickened; flowers never blue or purple; basic chromosome number $x = 9$ **45. *Watsonia***
- Stems with a single basal leaf; leaf blades rounded in section or plane but then narrow and without thickened margins; flowers shades of blue to nearly white or purple; basic chromosome number $x = 10$ 7

7. Flowers small, less than 12 mm long, crowded in dense distichous spikes; bracts solid below with broad membranous margins; pollen grains zonosulculate, without an operculum; exine sculpturing reticulate **47. *Micranthus***
- Flowers medium, 20–40 mm long, in spiral spikes; bracts without membranous margins **46. *Thereianthus***

8. Plants acaulescent, with leaves usually entirely bifacial, usually channeled to adaxially grooved, sometimes terete; bracts tubular below; ovary subterranean at the base of an elongate tube; pollen inaperturate, spiraperturate or polyrugoidate 9
- Plants with aerial stems or rarely acaulescent but then ovary usually at ground level; the leaves unifacial or terete; bracts not tubular below; pollen monosulcate 10

9. Corm tunics woody; abaxial surface of the leaf rounded; adaxial surface of the leaf not translucent in the midline **60. *Syringodea***
- Corm tunics fibrous; abaxial surface of the leaf elaborately ridged and usually with a distinct keel; adaxial surface of the leaf translucent in the midline **61. *Crocus***

10. Seed vasculature excluded and the surface smooth and glossy 11
- Seed vasculature included 18

11. Outer and inner bracts coriaceous, green or dry and chaffy; leaf midribs consisting of more than 1 pair of vascular bundles 12
- Outer and inner bracts membranous to scarious, usually translucent to transparent with veins often dark-coloured, occasionally the outer solid below but then the margins lacerate; leaf midribs consisting of a single pair of veins 13

12. Flowers actinomorphic or zygomorphic but then the perianth tube not abruptly expanded at the top of the flower cylindric portion and the upper part forming a pouch at the base; seeds dark brown to blackish; basic chromosome number $x = 11$ **68. *Crocosmia***
- Flowers zygomorphic, the perianth tube narrow and cylindric below, abruptly expanded into a broad upper cylindrical part and pouched at the base of the upper part; seeds bright orange; basic chromosome number $x = 10$ **70. *Chasmanthe***

13. Leaf margins with subepidermal sclerenchyma associated with a marginal vein; flowers actinomorphic or the perianth regular and the stamens unilateral and the anthers pendent 14
- Leaf margins normally lacking submarginal sclerenchyma but the marginal epidermal cells often columnar and with thickened radial walls 15

14. Plants evergreen; leaves linear and fibrotic and often the midribs not evident in old plants; bracts subtending the branches dry and attenuate **63. *Dierama***
- Plants deciduous; leaves linear to lanceolate, not fibrotic, the midribs always evident; pollen grain operculum consisting of a single band **64. *Ixia***

15. Leaves centric, oval in transverse section with 4 longitudinal grooves; flowers dusty pink, actinomorphic with the stamens rotated anticlockwise; basic chromosome number $x = 10$ **69. *Devia***
- Leaves flat or rarely centric but then solid, without longitudinal grooves; flowers variously coloured, actinomorphic or zygomorphic, but stamens never rotated anticlockwise 16

16. Plants acaulescent with subterranean ovaries; basic chromosome number $x = 10$ **67. *Duthieastrum***
- Plants with aerial stems, these sometimes hidden by the leaf sheaths, the ovary never subterranean 17

17. Floral bracts pale or rust-coloured, membranous or dry but not papery and crinkled, sometimes streaked with dark flecks or veins; flowers usually zygomorphic and then with a tooth-like ridge in the midline of one or more of the three lower tepals; basic chromosome number $x = 11$ **66. *Tritonia***
- Floral bracts pale, dry, papery and crinkled, irregularly streaked with dark flecks or veins; flowers actinomorphic or zygomorphic but then never with toothlike ridges on the lower tepals; basic chromosome number $x = 10$ **65. *Sparaxis***

18. Inner floral bracts with broad dry scarious margins; leaves terete to oval in transverse section with narrow longitudinal grooves, or 4-winged; corm tunics composed of brittle woody concentric layers; flowers actinomorphic, solitary on the branches, not arranged in spikes or stems acaulescent **59. *Romulea***
- Inner floral bracts without scarious margins 19

19. Seeds smooth and glossy 20
- Seed surface sculptured and matte 21

20. Style branches undivided; leaf blades pleated, sometimes more or less linear and striate; stems, and or leaves or bracts pubescent to pilose; basic chromosome number $x = 8$ **51. *Babiana***
- Style branches divided for 1/2 their length; leaf blades plane; plants entirely glabrous; basic chromosome number $x = 11$ **62. *Freesia***

21. Flowers solitary on the branches; style branches each deeply divided; seeds angular; leaf margins lacking subepidermal sclerenchyma **55. *Xenoscapa***
- Flowers arranged in spikes, occasionally by reduction solitary; style branches undivided; seeds rounded, sometimes broadly winged; leaf margins with subepidermal sclerenchyma associated with a marginal vein 22

22. Corm asymmetric the roots produced from a lateral ridge near the base or corm obsolete and rootstock rhizomelike; basic chromosome number $x = 13$ 23
- Corm symmetric, the roots produced all around the lower half, rarely rootstock rhizomelike 24

23. Style branches dividing at the apex of the perianth tube or within the tube, the branches long and laxly spreading **57. *Hesperantha***
- Style branches usually dividing well above the mouth of the perianth tube, the branches relatively short and recurved, not laxly spreading **56. *Geissorhiza***

24. Style simple; leaves plicate; basic chromosome number $x = 8$ **52. *Zygotritonia***
- Style dividing into 3 slender branches; basic chromosome number $x = 15$ 25

25. Seeds usually with a broad circumferential wing, occasionally the wing obsolete and the seed irregularly angled to subglobose; leaves usually firm-textured, occasionally soft-textured **53. *Gladiolus***
- Seeds globose; leaves soft-textured 26

26. Stems lightly angled; corms rounded below and with fibrous tunics **54. *Radinosiphon***
- Stem terete; corms with a flat base and tunics firm and unbroken **58. *Melasphaerula***

Genera of the Iridaceae

I. Subfam. Isophysidoideae Takhtajan (1980).

Plants with a creeping rhizome; leaves unifacial; inflorescence a single flower enclosed in paired opposed leafy spathes (? a single-flowered rhipidium homologous to that in Iridoideae - see below); flower long-lasting, pedicellate, radially symmetric, nectaries reputedly absent; tepals free; stamens free; ovary superior, style slender, dividing into 3 short branches; capsule ± woody; seeds angular.

1. *Isophysis* T. Moore

Isophysis T. Moore, Proc. Linn. Soc. 2: 212 (1853); Cooke, Fl. Australia 46: 4–5 (1986).

Small evergreen perennial; leaves several in a distichous fan, coriaceous, the margins thickened and hyaline; flowering stems unbranched, with a terminal 1-flowered inflorescence enclosed in large leafy spathes; flower dark red or yellow, secund; tepals subequal, lanceolate, spreading; style short with recurved stigmatic branches; capsule narrowly ovoid. Only one sp., *I. tasmanica* (Hook.f.) T. Moore, mountains of Tasmania.

II. Subfam. Nivenioideae Schulze ex Goldblatt (1990).

Plants with a rhizome or woody caudex; rhizome, caudex and aerial stems sometimes woody, with secondary growth; leaves unifacial, occasionally terete, often persisting and plants evergreen; inflorescence units usually composed of paired and partly fused rhipidia (binate), each enclosed in opposed leafy spathes, the paired rhipidia varying in number and arrangement; flowers several to 2, occasionally 1 per paired rhipidium, pedicellate and exserted sequentially from the spathes or sessile, long-lasting or fugaceous, radially symmetric, heterostylous in *Nivenia*; tepals basally connate or united in a tube; nectaries, when present, septal, rarely perigonal; style slender, 3-lobed, minutely 3-forked apically or 3-branched, the lobes often fringed; capsules cartilaginous to woody; seeds angular to compressed laterally or tangentially, occasionally globose, rarely arillate.

2. *Aristea* Aiton

Aristea Aiton, Hort. Kew. 1: 67 (1789); Weimarck, Acta Univ. Lund II, 2, 36: 1–141 (1940); Goldblatt, Fl. Madagascar 45 (edition 2): 5 (1991); Goldblatt & LeThomas, Ann. Mo. Bot. Gard. 84: 236–284 (1997).

Small to large evergreen or seasonal perennials with a thick or slender rhizome; leaves lanceolate to linear or terete; flowering stem sometimes compressed or winged, variously branched or simple; inflorescences composed of 1-many binate rhipidia arranged in panicles or crowded in fascicles on short branches; flowers usually fugaceous, stellate, usually shades of blue, occasionally white or mauve, without nectar or rarely with perigonal nectaries; tepals occasionally with contrasting marks, basally connate, subequal or the inner or outer whorl smaller; style deflexed, barely divided apically or broadly lobed, the lobes fringed; capsules globose, oblong- to cylindric-trigonous, or broadly 3-winged; seeds several to many and angular to cylindric with oblique ends or 1–2 per locule, often compressed radially, rarely arillate; $x = 16$. About 50 spp., sub-Saharan Africa including Madagascar.

3. *Geosiris* Baill.

Geosiris Baill., Bull. Mens. Soc. Linn. Paris 2 (146): 1149–1150 (1884); H. Perrier, Fl. Madagascar 45: 2 (1946); Goldblatt et al., Bull. Mus. Natl. Hist. Nat. Paris IV, B, Adansonia 9: 239–248 (1987).

Achlorophyllous saprophyte with an underground rhizome; flowering stems bearing scalelike leaves, often branched; inflorescences composed of binate rhipidialike units; flowers sessile, numerous per binate rhipidium, mauve, without nectar; tepals spreading, basally connate; style 3-lobed; capsules ovoid; seeds many, minute. Only one sp., *G. aphylla* Baillon, forests of Central and E Madagascar.

4. *Nivenia* Vent. — Fig. 91A,B

Nivenia Vent., Dec. Gen. Nov. 5 (1808); Goldblatt, Woody Iridaceae: 51 (1993).

Woody evergreen shrubs with a woody caudex; aerial stem and caudex with extensive secondary growth; stems compressed, branched, often with short lateral shoots; leaves crowded in a distichous fan; inflorescences comprising binate rhipidia arranged in pseudoracemose or corymbose panicles, or solitary, rhipidial bracts green or dry and brown; flowers 2 or 1 per binate rhipidium, sessile, with nectar from septal nectaries, blue, often heterostylous; tube cylindric or funnel-shaped, often elongate; tepals spreading horizontally; filaments shortly or well exserted, unequal in one sp.; style 3-lobed to 3-branched, exceeding the stamens or in heterostylous spp. reaching below or beyond the anthers; capsules ovoid, woody; seeds 1(–4) per locule, tangentially compressed; $x = 16$. Ten spp., mountains of the SW Cape, S Africa.

5. *Klattia* Baker Fig. 91D,E

Klattia Baker, J. Linn. Soc., Bot. 16: 109 (1878); Goldblatt, Woody Iridaceae: 99 (1993).

Like *Nivenia* but inflorescence of sessile binate rhipidia in a terminal head enclosed by enlarged leaves, these sometimes brightly colored, rhipidial bracts membranous; flowers 2 per binate rhipidium, long-lasting, sessile, with ample nectar from septal nectaries, red, yellow or purple; tube short; tepals elongated and linear-spathulate; filaments long; style minutely 3-lobed; seeds 1 per locule. Three spp., mountains of the SW Cape, S Africa.

6. *Witsenia* Thunb. Fig. 91C

Witsenia Thunb., Diss. Nov. Gen. Pl. 2: 33–34 (1782); Goldblatt, Woody Iridaceae: 117 (1993).

Like *Nivenia* but stems slender and willowy; inflorescence of binate rhipidia arranged in a terminal, pseudoracemose panicle, rhipidial bracts membranous; flowers 2 per binate rhipidium, long-lasting, sessile, with ample nectar from septal nectaries, long-lasting, green and yellow; tube long, cylindric; tepals short, closed over the tube, densely pubescent externally; stamens included; style minutely 3-lobed; seeds 1 per locule. Only one sp., *W. maura* Thunb., coastal and montane marshes in the SW Cape, S Africa.

7. *Patersonia* R. Br. ex Ker Gawl., nom. cons.

Patersonia R. Br. ex Ker Gawl., Bot. Mag. 26: tab. 1041 (1807); Cooke, Fl. Australia 46: 13–26 (1986).

Small to medium seasonal perennials or somewhat shrubby with woody rhizomes sometimes with secondary growth; leaves sword-shaped to linear; flowering stems unbranched; inflorescences binate rhipidia, flowers sessile, shades of blue to purple, occasionally yellow; tube long, sometimes with nectar from septal nectaries; tepals unequal, the outer large, spreading, inner vestigial to absent; filaments partly to entirely united; style 3-lobed, the margins fringed; capsules cylindric; seeds angular to ellipsoid, arillate in one sp.; $x = ?\ 11$. About 20 spp., mainly Australia, also mountains of Borneo, New Guinea and Sumatra.

III. Subfam. Iridioideae Pax (1882).

Plants with rhizomes, bulbs or apically rooting corms, roots sometimes tuberous; leaves unifacial or bifacial and channelled to flat, occasionally terete or square in section; inflorescence units rhipidia enclosed in opposed leafy bracts or spathes, rhipidia varying in number and arrangement on the flowering stems, with several to few, occasionally 1, flowers; flowers frequently fugaceous, borne sequentially and exserted from the spathes on long pedicels, occasionally subsessile, radially symmetric (zygomorphic in *Diplarrhena*); tepals free or tubular below, often with nectaries at the base of the outer tepals or on the surface of the inner tepals; stamens only 2 in *Diplarrhena*, filaments free or united below or entirely, alternate with the style branches and or appressed or joined to the opposed style branch; style slender or more often short, divided into 3 variously developed branches, these sometimes filiform or flattenned and sometimes petaloid, then often divided above into paired appendages; capsules membranous to woody; seeds globose to angular or discoid, sometimes arillate.

1. Tribe Sisyrinchieae Baker (1878).

8. *Libertia* Spreng.

Libertia Spreng., Syst. Veg. 1: 127 (1824) non Dumort. (1822) (nom. cons.); Moore, N. Z. J. Bot. 5: 255–275 (1967); Cooke, Fl. Australia 46: 5–8 (1986).

Small to medium evergreen rhizomatous perennials; leaves several, lanceolate to linear; flowering stems several-branched, branches long or very short and sometimes clustered; inflorescence spathes short, pedicels exposed; flowers white or blue; tepals free, subequal or the outer whorl much smaller; filaments united in a short tube, anthers erect; style short, dividing above the filament tube into 3 slender branches extending between the stamens; capsules sometimes indehiscent; seeds brown and shed immediately or coloured and exposed in the open capsule; $x = 19$. About 10 spp., Australia, New Guinea, New Zealand and S America.

9. *Orthrosanthus* Sweet

Orthrosanthus Sweet, Fl. Austral.: t. 11 (1829); Cooke, Fl. Australia 46: 10–13 (1986); Henrich & Goldblatt, Ann. Mo. Bot. Gard. 74: 577-582 (1987).

Medium to large evergreen rhizomatous perennials; leaves ± linear, several; flowering stems few to several branched, the branches short to subsessile; inflorescence spathes relatively short; flowers actinomorphic, subsessile with ovaries usually included in the spathes, blue or white; tepals free or united in a short tube, usually subequal, spreading from the base; filaments united below; style short, dividing into 3 long, apically stigmatic branches extending between the stamens; capsules ellipsoid to cylindric, included or occasionally exserted; seeds angular to elongate, $x = 9$. Nine spp., Australia and Andean S America to Central America and Mexico.

10. *Tapeinia* Comm. ex Juss.

Tapeinia Comm. ex Juss., Gen. Pl.: 59 (1789); Moore, Bot. Not. 124: 81–86 (1971).

Small, evergreen cushionlike rhizomatous perennial; leaves crowded in a distichous fan, ± linear, flowering stems axillary, with a single terminal 1-flowered inflorescence; spathes herbaceous, smaller than the leaves; flower pale pink, greenish externally; tepals subequal, ± free; filaments erect, united in the lower half; anthers erect; style dividing near the mid- to upper part of the anthers into three slender branches extending above the anthers; capsules globose, ± woody; seeds obovoid, restricted to the lower part of the locules. Only one sp., *T. pumila* (Forst. f.) Baillon, Fuegia and S Chile.

11. *Solenomelus* Miers

Solenomelus Miers, Proc. Linn. Soc. Lond. 1: 122 (1842).

Seasonal perennials with short rhizomes; leaves lanceolate to linear, several; flowering stems erect, usually few-branched; spathes somewhat inflated; flowers secund, yellow or blue, campanulate with a short slender perianth tube and the ovary included in the spathes; tube cylindric, curving outwards; tepals subequal, cupped below; filaments united in a column, anthers contiguous; style slender, obscurely 3-lobed above; capsule and seeds unknown. Two spp., mountains of Chile and Argentina.

12. *Sisyrinchium* L. Fig. 93A

Sisyrinchium L., Sp. Pl.: 954 (1753); Bentham & Hooker, Gen. Pl. 3: 698–699 (1883); Goldblatt et al., Syst. Bot. 15: 507 (1990).

Small to medium seasonal perennials or occ. annuals with either rhizomes or rootstock indistinct, roots often thick and fleshy or tuberous, or fibrous; leaves lanceolate to linear or occ. terete; flowering stems simple or variously branched, often winged, comprising 1 or several internodes; flowers usually yellow or blue to purple with a yellow centre, tepals free, subequal, spreading from the base or cupped and then often including the stamens and style; filaments partly to entirely united, occ. free, often glandular-pubescent, anthers diverging or coherent; style short to long, dividing into 3 short to long branches, these usually extending between the stamens; capsules globose to cylindric, sometimes truncate; seeds globose, often with a deep depression on one side; $x = 9$. About 60 spp., throughout S, Central and N America, including Greenland.

13. *Olsynium* Raf.

Olsynium Raf., Fl. Tellur. 4: 29 (1837); Goldblatt et al., Syst. Bot. 15: 507 (1990).
Phaiophleps Raf. (1837).
Chamelum Phil. (1865).

Small to medium seasonal or evergreen perennials with a short rhizome or rootstock indistinct, and thickened roots; leaves linear or terete; flowering stems aerial and simple or branched or subterranean; flowers sometimes shortly pedicellate and borne at ground level, pink, red, orange, white or yellow, occasionally with dark stripes, often secund; tepals free or united below in a tube, subequal, often cupped below and including stamens and style; filaments united, anthers contiguous or diverging above; style dividing into 3 short to long branches, then extending between the stamens; capsules ± globose, sometimes truncate, often on elongate pedicels; seeds ± angular or subglobose; $x = 9$. About 12 spp., Andean and temperate S America and northwestern N America.

2. Tribe Mariceae Hutch. (1934).

14. *Trimezia* Salisb. ex Herbert Fig. 93B,C

Trimezia Salisb. ex Herbert, Edwards Bot. Reg. 30 (Misc.): 88 (1844); Ravenna, Wrightia 7: 90–95 (1982).
Anomalostylus R. Foster (1947).

Small to large evergreen or seasonal perennials with an erect rhizome with brownish, sometimes fleshy, or occasionally glutinous leaf bases; leaves lanceolate to linear or terete; stems erect, simple or branched, occasionally compressed; flowers usually shades of yellow, or purplish with contrasting dark or light patterning; tepals free, outer larger, inner with oil glands in the lower half, partly concealed by folds; filaments free, thickened below, slender and weak, anthers affixed to the style branches; style thickened above, dividing towards the anther apex into 3 branches, each terminating in paired acute crests and with an abaxial transverse stigma, or bilobed to truncate and apically stigmatic; capsules globose to cylindric, truncate; seeds angular; x = possibly 10. About 20 spp., S and Central America, Mexico and West Indies.

15. *Pseudotrimezia* R. Foster

Pseudotrimezia R. Forster, Contr. Gray Herb. 155: 8–9 (1945); Ravenna, Revista Jard. Bot. Carlos Thuys 2: 25–28 (1969); Onira 1: 48–52 (1988).

Small seasonal perennials with a bulb or bulb-like rootstock comprising a thick erect rhizome surrounded by somewhat thickened inner leaf bases and dry more or less fibrous outer tunics; leaves few, linear to terete or nearly square in section; flowering stems usually simple, terete; flowers ± stellate, yellow, sometimes with contrasting marks in the centre; tepals subequal, free, spreading from the base; filaments free or partly united, anthers erect, coherent or diverging, style slender, dividing towards the upper part of the anthers into 3 short to long and slender branches; capsules globose; seeds angular. Six to eight spp., Minas Gerais, Brazil.

16. *Neomarica* Sprague

Neomarica Sprague, Kew Bull. 1928: 28 (1928).

Medium to large, occ. small, evergreen perennials with a creeping rhizome; leaves lanceolate to ensiform, coriaceous; stems erect to inclined, compressed and broadly winged, with one long aerial internode, terminating in a leaf and 1-several, pedunculate or sessile rhipidia; flowers cream, yellow or blue, patterned with contrasting colours; tepals free, outer larger, spreading to reflexed distally, inner bearing oil glands in the lower half, partly concealed by folds; filaments free, slender and weak, thickened below, anthers affixed to the style branches; style thickened above, dividing towards the anther apex into 3 branches, each terminating in paired acute crests and with an abaxial transverse stigma lobe at the base of the crests; capsules obovoid to cylindric, truncate; seeds angular and brownish or globose, fleshy and reddish, then retained in the capsule; $x = 10$ or 9. About 12 spp., S and Central America and Mexico.

3. Tribe Tigrideae Baker (1878).

17. *Cypella* Herbert

Cypella Herbert, Bot. Mag. 53: sub tab. 2637 (1826); Baker, Handbk. Irideae 63 (1892).

Small to large seasonal perennials with bulbs usually with brown to blackish, papyraceous or glutinous tunics; leaves plicate, lanceolate to linear; flowering stems branched; flowers usually shades of yellow or blue to purple, patterned in contrasting colours; tepals free, unequal, the outer 3 larger, broadly clawed and ± widely cupped below, the inner 3 bearing oil glands in the lower half, these usually partly enclosed by folds; filaments weak and slender, thickened below, or sturdy, anthers attached at least apically to the opposed style branches; style branches well developed, often compressed, usually divided above into prominent acute crests and with a transverse stigmatic surface on the abaxial surface at the base of the crests, the stigma also occasionally with short paired crests in the centre; capsules obovoid to cylindric, truncate; seeds angular; $x = 7$. About 20 spp., S America to Mexico and Cuba.

18. *Cipura* Aubl. Fig. 93F

Cipura Aubl., Hist. Pl. Gui. 1: 38, t. 13 (1775).

Like *Cypella* but flowering stems with a conspicuous terminal cauline leaf extending above the apically clustered rhipidia; flowers yellow, blue to purple or white, sometimes marked with contrasting colours; tepals unequal, outer larger, ascending, inner erect, partly concealing the stamens and style, sometimes bearing oil glands below, these occasionally enclosed by folds; filaments free, weak and slender, occasionally united below, anthers attached to the style branches; style thick, either with sessile stigmatic lobes apically, or prominently 3-branched, compressed with an abaxial stigmatic lobe at the base of paired acute crests; $x = 7$. About six spp., S. America to Mexico and West Indies.

19. *Calydorea* Herbert Fig. 93E

Calydorea Herbert, Edwards Bot. Reg. 29 (Misc.): 85 (1843); Goldblatt & Henrich, Ann. Mo. Bot. Gard. 78: 504–511 (1991).
Catila Ravenna (1983).
Itysa Ravenna (1986).
?*Lethia* Ravenna (1986), not fully understood.

Small seasonal perennials with bulbs with dark brown, papyraceous tunics, occasionally leafless when in bloom; leaves few, lanceolate to linear, plicate; flowering stems simple or branched; flowers pale to dark blue to purple with yellow markings; tepals free, spreading almost from the base, often shortly clawed, subequal or the inner somewhat smaller, ? without nectaries or oil glands; filaments free, anthers ± erect; style shortly exceeding or shorter than the filaments, dividing into 3 short to long filiform branches sometimes extending between the stamens or broadly lobed; capsules ± globose to obovoid, truncate; seeds angular; $x = 7$. About eight spp., southern and northern S America, Central America and Florida.

20. *Eleutherine* Herbert

Eleutherine Herbert, Edwards Bot. Reg. 29: tab. 57 (1843); Goldblatt & Snow, Ann. Mo. Bot. Gard. 78: 945 (1991).

Small seasonal perennials with a bulb with dark reddish papery tunics; leaves few, plicate; flowering stem branched, often short, bearing a conspicuous subapical cauline leaf, the inflorescence spathes unusually short; flowers white; tepals subequal, spreading from the base, without nectaries or oil glands; filaments free, anthers diverging; style short, branches undivided, filiform, extending between the anthers; capsule obovoid, truncate; seeds angular; $x = 7$. Two spp., Mexico, West Indies, Central and S America.

21. *Onira* Ravenna

Onira Ravenna, Nord. J. Bot. 3: 197–205 (1983).

Like *Herbertia* but flowers violet with contrasting dark and white markings; tepals free, broadly clawed, unequal, the outer 3 larger, the inner 3 bearing oil glands in the middle part; filaments free, anthers strongly diverging, each slightly shorter than and appressed to an opposed style branch; style thickened below, deeply divided into 3 slender ascending branches, the apices slighly expanded and stigmatic. Only one sp., S America, S Brazil and Uruguay.

22. *Gelasine* Herbert

Gelasine Herbert, Bot. Mag. new ser. 13: tab. 3779 (1840); Ravenna, Nord. J. Bot. 4: 347–350.

Like *Cypella* but flowers blue to purple, often with white or darker markings; tepals subequal or the inner smaller; filaments united, anthers erect; style slender, reaching or well exceeding the anthers, the branches filiform or somewhat cuneate and laterally compressed below, terminating in relatively broad stigmatic lobes and sometimes with minute crestlike processes. Four spp., S America, Brazil, Uruguay.

23. *Ennealophus* N.E. Br.

Ennealophus N.E. Br., Kew Bull. 1909: 361 (1909); Ravenna, Not. Mensuel Mus. Nac. Hist. Nat. Santiago, Chile 249: 7–8 (1977); Wrightia 7: 232–234 (1983).
Eurynotia R.C. Forster (1945).
Tucma Ravenna (1973).

Like *Cypella*, but flowers only shades of blue to purple with contrasting markings; outer 3 tepals larger, usually broadly clawed, claws then forming a broad cup, inner smaller, spreading or erect and bearing oil glands; filaments united in a column, anthers appressed to the style branch, thecae lateral and divergent on a broad connective; style branches well developed, sometimes flattened, then with a transverse abaxial stigma and terminating in paired or multiple crests, or merely divided above and terminally stigmatic. Five spp., S America.

24. *Herbertia* Sweet Fig. 93D

Herbertia Sweet, Br. Fl. Gard. tab. 222 (1827).
Trifurcia Herbert (1840); Goldblatt, Brittonia 27: 382 (1975).

Small seasonal perennials with bulbs with dark brown papyraceous tunics; leaves few, ensiform to linear, plicate; flowering stems simple or few branched; flowers shades of blue to purple with contrasting dark and white markings; tepals free, unequal, spreading from the base, the outer 3 larger, inner very small, bearing oil glands in the lower part; filaments united, the column often bottle-shaped, anthers diverging, each appressed to an opposed style branch but exceeding it; style branches spreading, flattened, forked apically; capsules obovoid to cylindric, truncate; seeds angular; $x = 7$. About five spp., temperate S America and southern USA.

25. *Kelissa* Ravenna

Kelissa Ravenna, Bull. Mus. Nat. Hist. Nat., Paris IV, B, Adansonia 3: 105–110 (1981).

Like *Herbertia*, but tepals with long narrow claws, these ascending, ± cupped, yellow speckled with large purple spots; filaments united, anthers erect, contiguous; style enclosed by the stamens, sender, dividing at the anther apices into 3 minute branches. Only one sp., *K. brasilensis* (Bak.) Ravenna, from southern Brazil.

26. *Mastigostyla* I.M. Johnston

Mastigostyla I.M. Johnston, Contr. Gray Herb. 81: 85 (1925); R. Foster, Rhodora 64: 292–298 (1962).

Small to medium seasonal perennials with bulbs with brown to blackish, papyraceous tunics; leaves few, plicate, ensiform to linear; flowering stems simple or branched; flowers usually shades of blue to purple; tepals free, unequal, the outer 3 larger, clawed, claws ascending, concealing the anthers, inner often erect, small; filaments united in a column, anthers appressed to the style branches; style branches compressed, deeply divided, terminating in prominent slender crests, stigmas lateral on outer edges of the crests; capsules not known; seeds ? angular; $x = 14$. About 16 spp., Andean, Bolivia, Peru and Ecuador.

27. *Cardenanthus* R. Foster

Cardenanthus R. Forster, Contr. Gray Herb. 155: 3 (1945), Rhodora 64: 298–303 (1964).

Small subacaulescent perennials with bulbs with dark brown papery tunics; leaves few, linear, plicate; flowering stems very short, simple; flowers shades of blue to purple or white; tepals united in a short tube, unequal, the outer 3 larger, shortly clawed, the inner small, often with oil glands near the base; filaments united, anthers each appressed to an opposed style branch; style filiform, branches divided for half their length or almost to base, apically stigmatic; capsule ± ellipsoid, truncate; seeds angular. About eight spp., Andes of Bolivia, Peru and N Argentina.

28. *Nemastylis* Nutt.

Nemastylis Nutt., Trans. Am. Philos. Soc. 5: 157 (1835); Foster, Contr. Gray Herb. 155: 26–44 (1945); Goldblatt, Brittonia 27: 375–380 (1975).

Small to medium seasonal perennials with bulbs with brownish papyraceous tunics; leaves several to single, ensiform to linear, plicate; flowering stems simple to several branched; flowers shades of blue to purple, occasionally yellow, stellate, tepals free, spreading, subequal or inner much smaller; filaments free or united below to entirely, anthers longer than filaments, initially erect, collapsing spirally after dehiscence; style branches deeply divided almost to base into paired filiform arms extended horizontally between anther bases, apically stigmatic; capsules obovoid-truncate; seeds angular; $x = 7$. About five spp., Guatemala, Mexico and southern USA.

29. *Ainea* Ravenna

Ainea Ravenna, Bot. Not. 132: 467–469 (1979).

Small to medium seasonal perennials with bulbs with brownish papyraceous tunics; leaves few, ensiform to linear, plicate; flowering stems simple or few branched; flowers white or violet, secund; tepals free, spreading from base, the inner 3 much smaller, with oil glands in the lower part; filaments free, anthers longer than the filaments, latrorse, initially erect, collapsing circinately after dehiscence; style slender, much exceeding the anthers, branches short, divided for half their length into slender paired arms, apically stigmatic; capsules obovoid to cylindric, truncate; seeds angular. Only one sp., Mexico, Michoacan to Oaxaca.

30. *Alophia* Herbert

Alophia Herbert, Bot. Mag. new ser. 13: sub. tab. 3779 (1840); Goldblatt & Howard, Ann. Mo. Bot. Gard. 79: 901–905 (1992).

Small to medium seasonal perennials with bulbs with dark brown papyraceous tunics; leaves few, ensiform to linear, plicate; flowering stems simple or few branched; flowers shades of blue to purple with contrasting dark and white markings; tepals free, unequal, spreading from the base, outer larger, inner with oil glands scattered on the lower half, partly covered by a fold; filaments partly to entirely united, anthers with a broad pandurate connective; style branches ascending, short, deeply divided into filiform arms curving outwards above the anthers; capsules obovoid-truncate; seeds angular; $x = 14$. About five spp., S and Central America to Texas and Louisiana.

31. *Tigridia* Juss.

Tigridia Juss., Gen. Pl.: 57 (1789); Molseed, Univ. California Publ. Bot. 54 (1970).
Rigidella Lindley (1840); Cruden, Brittonia 23: 217–225 (1971).

Small to medium seasonal perennial with bulbs with dark papery tunics; leaves few, plicate, lanceolate to linear; flowering stems usually branched; flowers various colours, often yellow, purple or red with contrasting dark or pale markings, sometimes nodding; tepals free, unequal, often broadly clawed, claws either forming a wide cup or sometimes a narrow cylinder, limbs spreading to reflexed, the inner 3 usually smaller than outer, bearing oil glands and or nectaries in the lower part, often partly concealed by folds; filaments united, anthers diverging, pollen bisulcate; style dividing at the base of the anthers, the branches each divided to the base into 2 long diverging filiform arms extending either side of the opposed anther, usually with a distinct mucro in the sinus between the 2 arms, stigmas terminal on the arms; capsule obovoid to cylindric, truncate; seeds angular; $x = 14$ or 7. About 35 spp., Guatemala and Mexico, Andes of Peru, Bolivia and Ecuador.

32. *Fosteria* Molseed

Forsteria Molseed, Brittonia 20: 232–234 (1968).

Like *Tigridia* but flowers pale yellow with purple markings; anthers sterile below; style branches erect, each deeply divided into 2 suberect filiform arms, these papillate and apparently stigmatic throughout. Only one sp., *F. oaxacana* Molseed, mountains of Oaxaca, Mexico.

33. *Sessilanthera* Molseed & Cruden

Sessilanthera Molseed & Cruden, Brittonia 21: 191–193 (1969); Cruden, Brittonia 27: 108 (1975).

Like *Tigridia* but flowers yellow or white with yellow spots, usually with purple spots at the base of the tepals; tepals spreading from the base, without oil glands or nectaries, the inner 3 smaller than the outer; filaments vestigial; style with a distinct mucro in the sinus between each of the paired style arms. Three spp., southern Mexico and Guatemala.

34. *Cobana* Ravenna

Cobana Ravenna, Bot. Not. 127: 104–108 (1974).

Like *Tigridia* but flowers white, secund; tepals spreading from the base, without oil glands or nectaries, inner smaller than outer; filaments free, short (less than 1 mm), anthers dehiscing by apical pores; style branches undivided, filiform, extending between the anthers, stigmatic apically. Only one sp., *C. guatemalensis* (Standley) Ravenna, highlands of Guatemala and Honduras.

4. Tribe Irideae Klatt (1882).

35. *Dietes* Salisb. ex Klatt

Dietes Salisb. ex Klatt, Linnaea 34: 583 (1866) nom. cons.; Goldblatt, Ann. Mo. Bot. Gard. 68: 132–153 (1981).

Evergreen perennials with persistent flowering stems and woody rhizomes; flowering stems persistent, either paniculately branched or with few branches; leaves coriaceous, lanceolate, in a distichous fan; pedicels villous above; flowers fugaceous (long-lasting in one sp.), shades of white to yellow, often with contrasting dark nectar guides on the outer tepals and purple style branches; tepals free, clawed, outer larger than the inner; filaments usually free, anthers appressed to the opposed style branch; style branches broad and petaloid, with a transverse abaxial stigma lobe and divided above into paired erect crests; capsule large and woody, sometimes not or only partly dehiscent; seeds large, angular; $x = 10$. Six spp., S and E Africa and Lord Howe Island.

36. *Diplarrhena* Labill.

Diplarrhena Labill., Rel. Voy. Rech. La Peyrouse 1: 157 (1800) (as *Diplarrena*); Cooke, Fl. Australia 46: 26–27 (1986).

Medium sized ± evergreen perennials with persistent rhizomes; leaves lanceolate; flowering stems unbranched with a terminal rhipidium; flower bilaterally symmetric, secund, white with purple and yellow markings; tepals free, unequal, outer larger, upper 2 of these erect, the lower horizontal, inner small, upper of these hooded over the stamens, lower 2 longer, marked with yellow; fertile stamens 2, the anterior reduced to a cusp, filaments free, slightly unequal, anthers oblique, 1 larger than the other; style dividing into 3 unequal flat stigmatic lobes, upper of these largest and concealing the lower lobes; capsules ± cylindric; seeds discoid; $x = 16$. Two spp., SE Australia and Tasmania.

37. *Bobartia* L.

Bobartia L., Fl. Zeyl.: 17 (1747) et Sp. Pl. 1: 54 (1753); Strid, Opera Bot. 37: 1–44 (1974).

Evergreen rhizomatous perennials; leaves lanceolate in a distichous fan or terete and often long; flowering stems either several branched and sticky or simple with several apically crowded, sessile rhipidia; pedicels villous above, flowers usually yellow (blue in *B. lilacina*); tepals usually free (tubular below in *B. macrospatha*), subequal; stamens free, diverging; style branches long and slender, extending between the stamens, apically stigmatic; capsules woody, ovoid and truncate; seeds angular; $x = 10$. Fifteen spp., W Cape to Transkei, S Africa.

38. *Iris* L.

Iris L., Sp. Pl. 38 (1753); Dykes, The Genus *Iris* (1913); Rodionenko, Rod *Iris* (1961); Mathew, The *Iris* (1981).
Xiphium Miller (1768).
Juno Tratt ex Roem. & Schult. (1817).
Iridodictyum Rodion. (1961).

Small to large perennials with rhizomes, bulbs or swollen roots, the bulbs with papery or netted tunics; leaves several to single, unifacial, lanceolate to linear, or terete or square in section, or bifacial and channelled; flowering stems branched or simple, or entirely subterranean; flowers not articulated below the ovary, long-lived, many colours, mostly shades of yellow or violet to purple, occasionally white, pink or red; tube usually present and sometimes very long, as when stem underground; tepals unequal, outer larger, clawed, claw ascending, limb spreading or reflexed, often bearded or crested, inner sometimes erect, or reflexed completely, always smaller, occasionally reduced; filaments free, anthers appressed to the style branches; style branches flattenned and petaloid, terminating in flat paired appendages, the crests, stigma a transverse lobe on the abaxial surface below the crests; capsules globose to cylindric, sometimes triquetrous, often shortly rostrate; seeds usually angular, light to dark brown, rarely red, often arillate, occasionally discoid or testa corky; x = possibly 12 or 10. About 225 spp., northern hemisphere, especially Middle East to China.

The genus is diverse and several distinctive subgenera and sections are usually recognised, some of which have been treated as separate genera.

Subg. *Limniris* (Tausch) Spach: plants with rhizomes; leaves unifacial, plane; stem usually branched, limb of outer tepals without long hairs (beardless) but sometimes crested or finely papillate; seeds without arils [incl. sects. *Lophiris* (= *Evansia*) and *Limniris*, the latter with several subsections or series].

Subg. *Iris*: plants with rhizomes; leaves unifacial, plane; stem often unbranched, limb of outer tepals with long hairs (bearded); seeds without arils (sect. *Iris*) or arillate (sects. *Oncocyclus*, *Psammiris*, *Regelia*, *Hexapogon*, *Pseudoregelia*).

Subg. *Xiphium* (Miller) Spach: plants with bulbs with papery to leathery tunics; leaves bifacial and channelled; flowering stems aerial, unbranched; flowers with erect inner tepals; seeds without arils.

Subg. *Hermodactyloides* Spach (= *Iridodictyum* Rodion.; sect. *Reticulata* Dykes): plants with bulbs with netted tunics; leaves terete or square in section; flowering stem subterranean; flowers with a long tube; seeds arillate.

Subg. *Scorpiris* Spach (= *Juno* Tratt. ex Roem. & Schult.; subg. *Juno* (Tratt.) Baker): plants with bulbs and swollen tuberous roots; leaves bifacial and channelled; stems branched or simple, often subterranean; flowers with inner tepals usually completely reflexed, occasionally horizontal; seeds sometimes arillate.

Subg. *Nepalensis* (Dykes) Lawrence (= *Junopsis* Schulz; sect. *Nepalensis* Dykes): plants with a vestigial rhizome and tuberous persistent roots; leaves equitant, linear-ensiform; stems sometimes underground; flowers with inner tepals spreading; seeds arillate.

39. *Hermodactylis* Miller

Hermodactylis Miller, Gard. Dict. Abr. ed. 4: 28 (1754).

Small seasonal perennial with finger-like tubers; leaf single, linear, square in section; flowering stem unbranched; flower 1 per stem, *Iris*-like, long-lived, dull-coloured, usually green and blackish, or bronze to dull yellow; tepals unequal, outer large and clawed, claws concealing the anthers, limbs reflexed, inner small, erect; filaments free, anthers appressed to the style branches; ovary 1-locular with axile placentation, style branches flat and petal-like, terminating in flat paired appendages, the crests, stigma a transverse lobe on the abaxial surface below the crests; capsule cylindric, rostrate; seeds not arillate; $x = 10$. Only one sp., *H. tuberosus* (L.) Salisb., Mediterranean Basin, SE France and N Africa to Israel.

40. *Pardanthopsis* (Hance) Lenz

Pardanthopsis (Hance) Lenz, Aliso 7: 401–403 (1972).

Small seasonal perennial with a persistent rhizome; leaves several, ensiform, in a distichous fan; flowering stem repeatedly branched; flower fugaceous, pale bluish to cream with nectar guides on outer tepals; tepals free, unequal, outer larger, clawed, claw ascending, limb reflexed, inner smaller, erect; filaments free, anthers appressed to the style branches; style branches narrow slightly winged, terminating in flat paired crests, stigma a transverse lobe on the abaxial surface below the crests; capsule globose-truncate; seeds angular, ± winged at either end; $x = 16$. Only one sp., *P. dichotoma* (Pallas) Lenz, N China, Mongolia and Siberia.

41. *Belamcanda* Adans.

Belamcanda Adans., Fam. 2: 60, 524 (1763) (as *Belamkanda*), nom, cons.

Medium sized seasonal perennial with a persistent rhizome; leaves several, ensiform, in a distichous fan; flowering stem branched; flower fugaceous, orange or yellow spotted red to purple throughout; tepals free, unequal, not distinctly clawed, spreading, outer larger; filaments free, anthers erect, free; style slender, exceeding the anthers, tubular, terminating in short apically stigmatic branches bearing short crest-like appendages; capsule globose; seeds spherical, testa inflated, blackish, persisting on the placentas; $x = 16$. Only one sp., *B. chinensis* (L.) DC., E Asia, N India to Siberia and Japan.

42. *Ferraria* Burm. ex Miller

Ferraria Burm. ex Miller, Fig. Pl. 2: 187, tab. 280 (1759); de Vos, J.S. Afr. Bot. 45: 295–375 (1979).

Small to medium seasonal perennials with flattenned apically rooting persistent corms without tunics; leaves equitant, ensiform to linear, several, becoming bract-like above; flowering stems branched; flowers fugaceous, dull-coloured, usually cream to buff or brownish, occasionally bluish, spotted and blotched dark drown, green or purple; tepals free, clawed, claws forming a cup, limbs spreading to reflexed, margins crisped, inner somewhat smaller than outer; filaments united below, free and diverging towards apex; anthers appressed to style branches, thecae sometimes diverging; ovary often rostrate; style branches short, bilobed, densely fringed; capsules globose to ellipsoid, usually pointed or rostrate; seeds angular; $x = 10$. Ten spp., dry parts of central and SW Africa, mainly W Cape, extending to Zambia and Zaire.

43. *Moraea* Miller Fig. 89E–J

Moraea Miller, Fig. Pl. 2: tab. 238 (1758) (as *Morea*), nom. cons.; Goldblatt, Ann. Mo. Bot. Gard. 64: 243–295 (1977), Ann. Kirstenbosch Bot. Gard. 14 (1986), and Novon 8 (1998).
Galaxia Thunb. (1782); Goldblatt, J.S. Afr. Bot. 45: 385–423 (1974).
Homeria Vent. (1808); Goldblatt, Ann. Mo. Bot. Gard. 68: 413–503 (1981).
Hexaglottis Vent. (1808); Goldblatt, Ann. Mo. Bot. Gard. 74: 542–569 (1987).
Gynandriris Parl. (1854); Goldblatt, Bot. Not. 133: 239–260.
Barnadiella Goldblatt (1976).
Roggeveldia Goldblatt (1979).
Rheome Goldblatt (1981).
Sessilistigma Goldblatt (1984).

Large to small seasonal perennials with apically rooting corms with variously fibrous or woody and unbroken tunics; leaves bifacial, channelled to flat, or occasionally terete, several to only 1; flowering stems several to few branched, occasionally unbranched, sometimes with 1 long aerial internode and rhipidia and leaves crowded apically, entirely subterranean in several spp.; flowers fugaceous or long-lived, all colours but mainly yellow or blue to violet, usually with conspicuous yellow to white nectar guides on outer tepals, occasionally also on the inner; tepals free (united in several spp.), clawed, subequal to unequal, limbs spreading to reflexed or inner erect, the inner frequently smaller, sometimes tricuspidate to ciliate or lacking, claws occasionally short and clasping the base of the filament column; filaments occasionally free or united partly to entirely, anthers appressed to style branches; ovary occasionally beaked or forming a sterile tube above; style branches usually broad and flat, wider than the anther, terminating in large paired crests, stigma usually transverse and shorter than the width of the style branch or style branches occasionally reduced and narrow, 2-lobed, the stigma terminal and crests lacking or divided to base forming 2 filiform arms, or undivided, filiform and extending between the stamens; capsules obconic to globose or oblong, occasionally rostrate; seeds angular or discoid, testa sometimes spongy; $x = 10$. About 200 spp., sub-Saharan Africa, concentrated in the SW Cape and Drakensberg Mts, 2 ssp. in Eurasia.

IV. Subfam. Ixioideae Klatt (1866)

Plants with basally rooting corms; leaves lanceolate to linear, occ. ovate or terete; inflorescence usually a panicle or simple to variously branched spike, or flowers solitary on branches; flowers sessile, subtended by paired opposed bracts below the ovary, the inner bract usually smaller and bilobed, long-lived, radially symmetric or zygomorphic, perianth with a variously developed tube; tepals subequal or unequal, often bilabiate or the upper hooded, nectaries septal; stamens often unilateral, then arcuate or declinate; style slender, usually exserted or at least as long as tube, divided into 3 filiform branches, these sometimes each deeply divided; capsules membranous to woody, usually globose to ovoid or cylindric; seeds globose to angular, sometimes winged.

1. Tribe Pillansieae Goldblatt (1990).

44. *Pillansia* L. Bolus Fig. 89B

Pillansia L. Bolus, Ann. Bolus Herb. 1: 20 (1915); Oberm., Fl. Pl. Africa 25: tab. 1381 (1962).

Large, evergreen perennials with persistent flattened corms with coarse irregularly fibrous tunics; stem branched; leaves few, linear, coriaceous; inflorescence a panicle, the individual flowers sessile; bracts herbaceous below, dry above, inner often larger than outer; flowers orange-red, actinomorphic, stellate; tube short; tepals spreading; stamens erect, symmetrical; style branches short, apically bifurcate; capsules globose-truncate, woody; seeds few, large, somewhat weakly compressed and angular; $x = 20$. Only one sp., *P. templemannii* (Baker) L. Bolus, Caledon district, SW Cape, S Africa.

2. Tribe Watsonieae Klatt (1882).

45. *Watsonia* Miller Fig. 89A

Watsonia Miller, Figures of Plants 2: tab. 276 (1758), nom. cons.; Goldblatt, Ann. Kirstenbosch Bot. Gard. 17 (1989).

Large to medium seasonal or occ. evergreen perennials with ± globose corms with coarsely fibrous or irregularly broken tunics; leaves coriaceous and fibrotic, few to several, lanceolate to linear; stem branched or simple; inflorescence a distichous spike, usually many-flowered; bracts herbaceous or dry and brown throughout or above; flower usually shades of pink or orange-red, zygomorphic (*W. marginata* actinomorphic); tube dimorphic, slender below, flared and short or broadly cylindric, long and horizontal above; tepals spreading; stamens symmetrical or unilateral, then arcuate or declinate; style branches slender, deeply forked; capsules ± woody, globose to cylindric or fusiform; seeds angular to elongate and 1- or 2-winged; $x = 9$. Fifty two spp., southern Africa.

46. *Thereianthus* Lewis

Thereianthus Lewis, J.S. Afr. Bot. 7: 33 (1941).

Small seasonal, summer blooming perennials with small ± globose corms with irregularly broken or coarsely reticulate tunics; leaves few, only 1 basal, linear to terete; spike lax or crowded, with spirally arranged flowers; bracts coriaceous, dry and brown; flower shades of blue to purple or white, actinomorphic and salver-shaped to zygomorphic; tube cylindrical, usually straight; tepals subequal, spreading; stamens usually erect, occ. unilateral and arcuate; style exserted, branches deeply forked and recurved; capsules small, woody, globose to cylindrical; seeds small, angular to narrowly spindle-shaped; $x = 10$. Seven spp., mountains of the SW Cape, S Africa.

47. *Micranthus* (Pers.) Ecklon

Micranthus (Pers.) Ecklon, Top. Verz. Pflanzensamml.: 43 (1827), nom. cons.; Baker, Flora Cap. 6: 97 (1896).

Small seasonal perennials with small, globose, corms with tunics of medium to coarse fibres; leaves few, only 1 basal, ensiform to linear or terete and fistulose; scape erect, usually simple, slende; spike crowded, many-flowered, strongly distichous; bracts dry, membranous, small; flowers zygomorphic, very small, pale to deep blue; tube curved; tepals subequal, the upper hooded; stamens unilateral and arcuate; style branches deeply forked; capsules small, woody; seeds few per locule, narrowly spindle-shaped; $x = 10$. Three spp., Cape, S and SW Cape, S Africa.

48. *Lapeirousia* Pourret Figs. 89D, 90B,D

Lapeirousia Pourret, Mém. Acad. Sci. Toulouse 3: 79 (1788); Goldblatt, Contrib. Bolus Herb. 4: 1–74 (1972), Ann. Mo. Bot. Gard. 77: 430–484 (1990).

Large to small seasonal perennials with campanulate flat-based corms with woody tunics; leaves few to 1, often ribbed, or plane, sometimes falcate; stem simple to many-branched, sometimes winged, occ. subterranean; inflorescence a panicle, then often corymbose, or a spike, or occasionally contracted and fasciculate; bracts herbaceous, to membranous, occasionally outer large and keeled and crisped or ribbed; flower often blue, or white, yellow, reddish or violet, actinomorphic or zygomorphic; tube funnel-shaped or cylindric and then often very long; tepals equal and spreading, or upper largest and erect, lower 3 forming a lip; stamens symmetrical and erect or spreading, or unilateral and arcuate; style branches usually long, each deeply forked and recurved, occasionally entire; capsules membranous, globose to cylindrical; seeds many, globose or angled by pressure; chromosome numbers varied, $n = 4$–10. Forty two spp., sub-Saharan Africa, concentrated in Namibia and west coast of S Africa.

49. ***Savannosiphon*** Goldblatt & Marais

Savannosiphon Goldblatt & Marais, Ann. Mo. Bot. Gard. 66: 845–850 (1979).

Seasonal perennials with globose corms with fibrous tunics; leaves few, lanceolate, corrugate-plicate; stem winged above, unbranched; spike few flowered; bracts herbaceous; flowers large, white, opening in the evening, actinomorphic; tube long cylindric, widening apically; tepals subequal, the upper somewhat hooded; stamens apparently symmetrically arranged, included in the tube or the anthers partly exserted; style shortly exserted, branches deeply forked; capsules ovoid membranous; seeds globose, shiny; $x = 7$. Only one sp., *S. euryphylla* (Harms) Goldblatt & Marais, SE tropical Africa.

3. **Tribe Ixieae** Dumort (1822).

50. ***Tritoniopsis*** L. Bolus

Tritoniopsis L. Bolus, S. Afr. Gard. 19: 123 (1929); Lewis, J.S. Afr. Bot. 25: 319 (1959); Goldblatt, S. Afr. J. Bot. 56: 577–582 (1990).
Anapalina N.E. Br. (1932).

Seasonal perennials with deep-seated globose corms with coarse matted fibrous tunics; leaves few, sometimes hysteranthous, occasionally pseudopetiolate, ovate or lanceolate to linear, prominently (1–)3–5-veined; spike with flowers ± spirally arranged; bracts coriaceous short, often dry, the inner usually larger; flowers red, pink, white or yellow and brown, actinomorphic (*T. leslei*) or zygomorphic; tube short or long, cylindric, campanulate or tube slender below, curved and abruptly widening and broadly cylindric above; tepals subequal or unequal, the dorsal erect and hooded or horizontal and the lower 3 forming a lip; stamens symmetrical or unilateral and arcuate; style branches short, cuneate or apically bilobed; capsules globose, often inflated; seeds several, flat and 2-winged or prominently angular, testa often spongy; $x = 16$ (or 15). Twenty two spp., SW Cape to Transkei, S Africa.

51. ***Babiana*** Ker Gawl.

Babinia Ker-Gawl., Bot. Mag. 16: tab. 576 (1802); Lewis, J.S. Afr. Bot. Suppl. 3 (1959).
Anaclanthe N.E. Br. (1932).
Antholyza sensu auct. non L. (1753).

Small seasonal perennials with globose deep-seated corms, the tunics tough and fibrous, extending upwards in a neck; leaves several, usually plicate, occasionally plane or terete, sometimes twisted, usually pubescent; flowering stems erect, declinante, decumbent, sometimes subterranean, simple or branched, usually pubescent, the main axis sterile in one sp.; spike usually secund; bracts herbaceous, usually pubescent, often dry apically, occ. entirely brown, inner sometimes divided to the base; flowers often shades of blue to purple, occasionally red, pink or yellow, actinomorphic or zygomorphic, often bilabiate; tube short to long, in two spp. narrow and cylindric below, curving and abruptly wider and cylindrical above; tepals equal and spreading, or unequal with dorsal held apart and lower 3 forming a lip, the dorsal channeled and enclosing the stamens and style in two spp.; stamens symmetrical and erect or unilateral and arcuate; style usually exserted, branches short, slender, expanded apically; capsules ± globose, coriaceous; seeds brown to blackish, globose to angular; $x = 7$. About 63 spp., all but two in the SW Cape, Namaqualand and S Namibia, one extending to Zimbabwe and one in Socotra.

52. ***Zygotritonia*** Mildbr.

Zygotritonia Mildbr., Bot. Jahrb. Syst. 58: 230 (1923); Stapf, Hook. Ic. Pl. V, 2: tab. 3120 (1927); Goldblatt, Bull. Mus. Hist. Nat. Paris IV, B, Adansonia 11: 199–212 (1989).

Seasonal perennials with globose corms with fibrous netted tunics; leaves few, lanceolate to linear, prominently nerved to somewhat plicate; spike with spiralled flowers; bracts small, subequal; flowers yellow, orange or white, zygomorphic; tube short, cylindric; tepals unequal, upper larger, held apart and hooded, lower 3 geniculate, forming a lip; stamens unilateral and arcuate; style slender, undivided; capsules 3-lobed; seeds large, smooth, ± globose to weakly angled, 1–2 per loculus; $x = 7$. Four spp., tropical Africa, Malawi and Tanzania to Senegal.

53. *Gladiolus* L. Figs. 90C, 92

Gladiolus L., Gen. Pl. ed. 5: 23 (1754); Lewis et al., J.S. Afr. Bot. Suppl. 10 (1972); Goldblatt & de Vos, Bull. Mus. Natl. Hist. Nat. Paris IV, sect. B, Adansonia 11: 417–428. (1989); Goldblatt, *Gladiolus* in Fl. Trop. Africa (1996); Goldblatt & Manning, *Gladiolus* in Sthn Africa (1998).
Antholyza L. (1753).
Homoglossum Salisb. (1866).
Oenostachys Bullock (1930).

Small to large seasonal perennials, occasionally evergreen, with globose corms, the tunics variously papery and unbroken to fibrous or woody and split into claws; leaves occasionally not contemporary with the flowers, 1-several, lanceolate to linear or terete, blade plane or the margins and or midrib variously thickened; spike usually secund, occ. spiral or distichous; bracts herbaceous, usually large; flower zygomorphic and bilabiate, occasionally actinomorphic and stellate, variously coloured but often shades of purple, pink, yellow, white or red; tube short or long, cylindric or funnel-shaped, or slender below, abruptly widening into a wider cylindric upper part, occasionally with a short spur; tepals subequal and spreading, or unequal with the dorsal largest, often hoodlike or horizontal, the lower 3 narrow, often with contrasting marks, sometimes much reduced; stamens symmetrical or unilateral and arcuate; style exserted, branches short, spreading, apically expanded; capsules globose to obovoid or cylindric; seeds many, circumferentially winged, occasionally wingless; $x = 15$. About 255, spp., Africa, Europe and the Middle East, centred in S Africa.

54. *Radinosiphon* N.E. Br.

Radinosiphon N.E. Br., Trans. R. Soc. S. Afr. 20: 262 (1932); Carter, F. Pl. Africa 35: Pl. 1384 (1962).

Seasonal perennial with globose corms with soft fibrous tunics; leaves several, lanceolate, soft-textured; flowering stem compressed and angled, usually branched; spike secund; bracts herbaceous to membranous; flower pink to purple, zygomorphic; tube long, widening somewhat towards the throat; tepals unequal, upper largest, lower 3 smaller, forming a lip; stamens unilateral, arcuate; style exserted, branches short, slightly expanded apically; capsules membranous, globose; seeds small, ovoid to angular; $x = 15$. Only one sp., mountains of SE tropical Africa and eastern southern Africa.

55. *Xenoscapa* (Goldblatt) Goldblatt & Manning Fig. 90E

Xenoscapa (Goldblatt) Goldblatt & Manning, Syst. Bot. 20: 172 (1995).

Seasonal perennials with small globose corms with tunics of reticulate fibres; stem occasionally prostrate below, often branched; leaves 2, ovate and prostrate, soft-textured; inflorescence with flowers solitary on branches; bracts green; flowers white or mauve, zygomorphic; tube long, narrow and cylindric; tepals subequal or the dorsal larger and hoodlike; stamens unilateral and arcuate; style branches filiform, deeply forked and recurved, capsules cylindrical to oblong; seeds many, angular and matte; $x = 11$. Two spp., SW southern Africa.

56. *Geissorhiza* Ker Gawl. Fig. 89C

Geissorhiza Ker Gawl., Bot. Mag. 18: tab. 672 (1803); Foster, Contrib. Gray Herb. 135: 3 (1941); Goldblatt, Ann. Mo. Bot. Gard. 72: 277–447 (1985).

Seasonal usually small perennials with ± globose corms with concentric or imbricate, woody to papery, occasionally fibrous tunics; leaves 2–3 to several, occ. terete or coiled, sometimes pubescent, flat or strongly ribbed, often with margins and midrib prominent; flowering stems occasionally drooping, sometimes puberulous; spike usually drooping in bud, occasionally flowers solitary; bracts herbaceous to membranous, often dry above or entirely; flowers all colours except orange, often with contrasting colour in centre, actinomorphic and stellate to salver-shaped or zygomorphic and declinate; tube short and funnel-shaped to cylindric; tepals subequal, cupped to spreading; stamens erect or declinate, sometimes one filament shorter; style branches slender, recurved, occasionally included in the tube; capsules membranous; seeds angular; $x = 13$. Eighty four spp., S Africa, mainly SW Cape.

57. *Hesperantha* Ker Gawl.

Hesperantha Ker Gawl., König & Sims, Ann. Bot. 1: 225 (1805); Baker, Flora Cap. 6: 57 (1896); Goldblatt, J.S. Afr. Bot. 50: 50–141 (1984).

Seasonal perennials with ± globose or flat-based corms with concentric or imbricate ± woody tunics or a short rhizome; leaves 3 to several, occ. terete; flowering stems sometimes entirely subterranean; spike several to few-(1-)flowered; bracts herbaceous, occ. margins of the outer partly united around axis; flower often pale and opening in the evenings and then scented, or yellow, purple, pink or red, actinomorphic, salver-shaped, occ. zygomorphic and declinate; tube occasionally recurved, funnel-shaped to cylindric; tepals subequal, spreading; stamens occ. declinate, rarely included; style dividing at apex of tube, branches long, spreading, filiform, occasionally included; capsules membranous; seeds globose to angular; $x = 13$. About 65 spp., sub-Saharan Africa, mainly Drakensberg and SW Cape extending to Cameroun and Ethiopia.

58. *Melasphaerula* Ker Gawl.

Melasphaerula Ker Gawl., Bot. Mag. 17: tab. 615 (1803); Baker, Flora Cap. 6: 115 (1896).

Seasonal perennial with campanulate flat-based corms with woody, blackish tunics, usually cormiferous at the base; leaves several, soft textured; flowering stems wiry, much branched; spike horizontal to drooping; bracts small, herbaceous to membranous; flower white to cream, small, zygomorphic, bilabiate; tube short; segments subequal, attenuate, upper erect, lower streaked with purple; stamens unilateral, arcuate; style branches short, recurved; capsules membranous, deeply 3-angled; seeds 1–2 per locule, large, globose; $x = 10$. Only one sp., *M. ramosa* (Burm. f.) N.E. Brown, S Cape to S Namibia.

59. *Romulea* Maratti Fig. 90F

Romulea Maratti, Fl. Rom. Sat. 13 (1772); de Vos, J.S. Afr. Bot. Suppl. 9 (1972); Fl. S. Africa 7 (2, 2): 10–76 (1983).

Short-stemmed to acaulescent seasonal perennials with asymmetrical ± globose to compressed or symmetric campanulate corms with woody to coriaceous tunics; flowering stems short and solitary or branched or underground and emerging in fruit; leaves (1–)few to several, oblong to terete in section, nearly always with 4 longitudinal grooves, occasionally adaxial surface extending almost to apex; flowers solitary on the branches; bracts herbaceous, the inner with membranous margins or entirely membranous and dry; flower actinomorphic; campanulate to salver-shaped, variously coloured but usually shades of purple or yellow; tube short and funnel-shaped to long and cylindric; tepals subequal, cupped to outspread; stamens erect, filaments sometimes united; style branches each divided for about 1/2 their length; capsules membranous; seeds globose to ellipsoid; $x = 14$, 13 or 12. About 90 spp., S and tropical Africa, S Europe and Canary Is., concentrated in SW Cape and W Karoo, S Africa.

60. *Syringodea* Hook. f.

Syringodea Hook. f., Bot. Mag. 29: tab. 6072 (1873); de Vos, J.S. Afr. Bot. 40: 201–254 (1974); Fl. S. Africa 7 (2, 2): 1–9 (1983).

Acaulescent seasonal perennials with globose ± symmetrical corms with woody to coriaceous tunics, the outer becoming somewhat broken irregularly; flowering stems 1–several, subterranean, emerging in fruit; leaves either several, linear and ± bifacial, or solitary, terete and somewhat inflated; flowers solitary on each flowering stem; bracts membranous; flower actinomorphic, ± salver-shaped, usually shades of purple to blue; tube long and cylindric or somewhat expanded apically; tepals ± subequal, cupped to outspread; stamens erect; ovary subterranean, style branches 3, short and filiform; capsules clavate to obovoid with a narrow base, membranous; seeds ± globose to angular; $x = 6$. Eight spp., SW Cape, Karoo and Namaqualand to Vaal R., S Africa.

61. *Crocus* L.

Crocus L., Sp. Pl. 36 (1753); Maw, Monogr. Genus Crocus (1886); Mathew, The Crocus (1982).

Acaulescent seasonal perennials with globose us. symmetrical corms with fibrous, membranous or coriaceous tunics; flowering stems 1–several, subterranean, emerging in fruit; leaves several, bifacial, adaxially flat or channeled, usually with a white median stripe, abaxially strongly keeled; flowers solitary on each flowering stem; bracts membranous, inner bracts sometimes lacking; flowers actinomorphic, ± salver-shaped, usually shades of purple, blue or yellow; tube long and cylindric, somewhat expanded apically; tepals usually subequal, cupped to outspread; stamens erect; ovary underground, style well exserted, 3-

branched to multifid; capsules membranous; seeds globose to ellipsoid; chromosome number varied, $n = 4$–14. About 80 spp., Europe, the Mediterranean and Middle East to Central Asia.

62. *Freesia* Klatt

Freesia Klatt, Linnaea 34: 672 (1866); Goldblatt, J.S. Afr. Bot. 48: 39–91 (1982); Goldblatt & Manning, Syst. Bot. 20: 161–178 (1995).
Anomotheca Ker Gawl. (1804), in part.

Seasonal perennials with obconic corms with tunics of reticulate fibres; stem occ. prostrate below, often branched, flexed at base of spike, occ. winged; leaves several, ensiform to oblong, sometimes prostrate; inflorescence a ± horizontal secund spike; bracts coriaceous and green to membranous; flowers mainly yellow or white, pink, red or green, zygomorphic; tube long, narrow below, widening abruptly and funnel-shaped above or narrow and cylindric throughout, or funnel-shaped; tepals subequal or the lower smaller, sometimes forming a lip and upper largest and erect; stamens unilateral and arcuate; style branches filiform, deeply forked and recurved, capsules globose, surface rough-papillate; seeds several, round and shiny with an inflated raphe; $x = 11$. Fourteen spp., mainly S Africa, two in tropical Africa.

63. *Dierama* K. Koch

Dierama K. Koch, Ind. Sem. Hort. Berol. App.: 10 (1854); N.E. Br., J. R. Hort. Soc. 54: 193 (1929); Hilliard et al., *Dierama*: Hairbells of Afr. (1991).

Large evergreen perennials with flattened corms with fibrous tunics; leaves several, linear, tough and fibrous; flowering stems slender and wiry, drooping above; spike lax, drooping; bracts dry and papery, often lacerate, pale usually with brown streaks; flower shades of pink to purple, occ. white or red, pendent, actinomorphic, campanulate; tube short, funnel-shaped; tepals subequal; stamens straight, filaments coherent; style branches short and recurved; capsules firm-membranous; seeds hard, globose to angular, usually shiny; $x = 10$. Forty-four spp., E African mountains to E Cape, S Africa.

64. *Ixia* L.

Ixia L., Sp. Pl. ed. 2: 51 (1762), nom. cons.; Lewis, J.S. Afr. Bot. 28: 45 (1962).

Small seasonal perennials with ± globose corms with fibrous tunics; leaves few, ensiform, occ. margins crisped or margins and midrib prominent; flowering stems slender, often wiry; bracts short, membranous, translucent or occasionally dry and brown; flower various colours, even green, often with a contrasting dark colour in centre, actinomorphic, stellate to salver-shaped; tube narrow-cylindric or funnel-shaped; tepals subequal, spreading; stamens occasionally ± unilateral and pendent, filaments sometimes united; style included or exserted from tube, branches short and recurved; capsules membranous; seeds globose, usually smooth and shiny; $x = 10$. About fifty spp., Cape Province, S Africa, mainly SW Cape.

65. *Sparaxis* Ker Gawl.

Sparaxis Ker Gawl., König & Sims Ann. Bot. 1: 225 (1805); Goldblatt, J.S. Afr. Bot. 35: 219 (1969), Ann. Mo. Bot. Gard. 79: 143–159 (1992).

Small seasonal perennials with globose corms with tunics of pale, reticulate fibres; leaves several, soft-textured; flowering stems firm, erect, simple or branched; spike few- to several-flowered; bracts dry, scarious, irregularly streaked with brown markings, often lacerate or cuspidate; flower large, actinomorphic and stellate or zygomorphic and bilabiate; tube short and expanded above or long and slender; tepals subequal and spreading, or the uppermost larger and then often hooded and the lower 3 small; stamens erect and symmetrically disposed, or unilateral and spreading or arcuate; anthers straight or twisted; style exserted, branches either short with expanded apex, or long and filiform; capsules globose, firm-membranous; seeds several, globose, smooth and shiny; $x = 10$. Fourteen spp., SW and NW Cape, S Africa.

66. *Tritonia* Ker Gawl.

Tritonia Ker Gawl., Bot. Mag. 11: tab. 581 (1802); de Vos, J.S. Afr. Bot. 48: 105–163 (1982), 49: 347–422 (1983).

Small seasonal perennials with ± globose corms with fibrous tunics; leaves several, lanceolate to linear, sometimes terete, the margins occasionally expanded into broad wings; flowering stem occasionally ± subterranean; spike somewhat secund; bracts membranous to dry and scarious; flower occ. actinomorphic and stellate or zygomorphic, shades of yellow to orange, occasionally white, pink, purple or red; tube short to long, funnel-shaped or cylindric; tepals subequal or upper largest and hoodlike and lower 3 usually forming a lip and usually each bearing a toothlike callus on the base; stamens symmetrically disposed, erect to spreading or unilateral and arcuate; style dividing well above the mouth of the tube into 3 short, recurved occasionally shortly bifid branches; capsules obovoid; seeds many, globose to angular; $x = 11$. Twenty eight spp., SW Cape to SE tropical Africa.

67. *Duthieastrum* de Vos

Duthieastrum de Vos, J.S. Afr. Bot. 41: 91 (1975).

Small seasonal perennial with globose corm with tunics of fine reticulate fibres; leaves several, linear-falcate; flowering stem entirely subterranean; inflorescence extremely contracted, with 1 to several flowers borne sequentially; bracts almost entirely sheathed by leaf bases, membranous, green-tipped; flower actinomorphic, salver-shaped; tube slender, cylindric and straight, much longer than the tepals; tepals subequal, spreading horizontally; stamens symmetrically disposed, erect; style dividing above the anthers, branches short, filiform; capsules globose; seeds globose, smooth and shiny; $x = 10$. Only one sp., *D. linifolia* (Phill.) de Vos, interior S Africa.

68. *Crocosmia* Planch.

Crocosmia Planch., Flore des Serres 7: 161 & tab. 702 (1852); de Vos, J.S. Afr. Bot. 50: 463–502 (1984).

Large to medium-sized, seasonal perennials with ± globose corms with membranous to fibrous tunics; leaves ensiform, flat or plicate; flowering stems often inclined, usually with spreading branches; spike inclined to horizontal; bracts short, herbaceous with a dry apex or dry throughout; flower actinomorphic or zygomorphic, yellow to orange or reddish; tube straight or curved, funnel-shaped; tepals subequal and spreading or with upper larger and erect and lower 3 forming a lip; stamens symmetrical or unilateral and arcuate; style well exserted, branches short, recurved; capsules 3-lobed, depressed; seeds 2(–4) per loculus, large, globose, initially glossy; $x = 11$. Nine spp., tropical and eastern southern Africa and W Madagascar.

69. *Devia* Goldblatt & Manning

Devia Goldblatt & Manning, Ann. Mo. Bot. Gard. 77: 362 (1990).

Medium-sized, seasonal perennial with ± globose corms with coarse fibrous tunics; leaves linear, oval in transverse section and deeply 4-grooved on each surface; flowering stems sometimes branched; spike erect; bracts short, firm and dry throughout; flowers actinomorphic pale pink; tube straight or narrowly funnel-shaped; tepals subequal, cupped; stamens symmetrical and each twisted in clockwise direction; style well exserted; branches short, recurved; capsules 3-lobed, depressed; seeds 2 (–4) per loculus, large, globose, initially glossy; $x = 10$. Only one sp., *D. xeromorpha* Goldblatt & Manning, mountains of the western Karoo, S Africa.

70. *Chasmanthe* N.E. Br.

Chasmanthe N.E. Br., Trans. R. Soc. S. Afr. 20: 272 (1932); de Vos, S. Afr. J. Bot. 51: 253–261 (1985).

Medium to large, seasonal perennials with large depressed corms with papery to fibrous tunics; leaves several, lanceolate; spike secund or distichous; bracts herbaceous to dry-membranous; flower zygomorphic, shades of yellow, orange or red; tube slender below, widening abruptly to a wide cylindric upper part; tepals unequal, upper large, hoodlike, horizontal, others smaller and often reflexed; stamens unilateral arched under upper tepal; style dividing beyond the anthers, branches slender and recurved; capsules globose; seeds few, large, ± globose, orange, fleshy in one sp.; $x = 10$. Three spp., SW and S Cape, S Africa.

Selected Bibliography

Adamson, R.S. 1926. On the phylogeny of some shrubby Iridaceae. Trans. R. Soc. S. Afr. 13: 373–378.
Arber, A.R. 1921. The leaf structure of the Iridaceae, considered in relation to the phyllode theory. Ann. Bot. 35: 301–336.
Barthlott, W., Frölich, D. 1983. Mikromorphologie und Orientierungsmuster epicuticularer Wachs-Kristalloide: ein neues systematisches Merkmal bei Monokotylen. Plant Syst. Evol. 142: 171–185.
Bate-Smith, E.C. 1968. See general references.
Baytop, T., Mathew, B., Brighton, C. 1975. Four new taxa in Turkish *Crocus* (Iridaceae). Kew Bull. 30: 241–246.
Behnke, H.-D. 1981. See general references.
Cheadle, V.I. 1963. Vessels in Iridaceae. Phytomorphology 13: 245–248.
Dahlgren, R. et al. 1985. See general references.
Daumann, E. 1970. See general references.
Davis, G.L. 1966. See general references.
de Vos, M.P. 1970. Bydrae tot die morphologie en anatomie van Romulea II. Die blare. J. S. Afr. Bot. 36: 271–286.
de Vos, M.P. 1972. The genus *Romulea* in South Africa. J. S. Afr. Bot., Suppl. Vol. 9.

de Vos, M.P. 1974. Die Suid-Afrikaanse genus *Syringodea*. J. S. Afr. Bot. 40: 201–254.
de Vos, M.P. 1977. Knol ontwikkeling by sommige genera van die Iridaceae en die systematiese posisie. Tydskr. Natuurwet. 17: 9–19.
Goldblatt, P. 1971. Cytological and morphological studies in the southern African Iridaceae. J. S. Afr. Bot. 37: 317–460.
Goldblatt, P. 1982a. Corm morphology in *Hesperantha* (Iridaceae, Ixioideae) and a proposed infrageneric taxomomy. Ann. Mo. Bot. Gard. 69: 370–378.
Goldblatt, P. 1982b. Chromosome cytology in relation to supragereric systematics of neotropical Iridaceae. Syst. Bot. 7: 186–198.
Goldblatt, P. 1990. Phylogeny and classification of Iridaceae. Ann. Mo. Bot. Gard. 77: 607–627.
Goldblatt, P. 1993. The woody Iridaceae. Portland, OR: Timber Press.
Goldblatt, P. 1996. *Gladiolus* in tropical Africa. Portland, OR: Timber Press.
Goldblatt, P., Takei, M. 1993. Chromosome cytology of the African genus *Lapeirousia* (Iridaceae-Ixioideae). Ann. Mo. Bot. Gard. 80: 961–973.
Goldblatt, P., Henrich, J.E., Rudall, P. 1984. Occurrence of crystals in Iridaceae and allied families and their phylogenetic significance. Ann. Mo. Bot. Gard. 71: 1013–1020.
Goldblatt, P., Rudall, P., Cheadle, V.I., Dorr, L.J., Williams, C.A. 1987. Affinities of the Madagascan endemic *Geosiris*, Iridaceae or Geosiridaceae. Bull. Mus. Natl. Hist. Nat., B, IV, Adansonia 9: 239–248.
Goldblatt, P., Bari, A., Manning, J.C. 1991. Sulcus variability in the pollen grains of Iridaceae subfamily Ixioideae. Ann. Mo. Bot. Gard. 78: 950–961.
Goldblatt, P., Takei, M., Razzaq, Z.A. 1993. Chromosome cytology in tropical African *Gladiolus* (Iridaceae). Ann. Mo. Bot. Gard. 80: 461–470.
Goldblatt, P., Manning, J.C., Bernhardt, P. 1995. Pollination biology of *Lapeirousia* subgenus *Lapeirousia* (Iridaceae) in southern Africa: floral divergence and adaptation for long-tongued fly pollination. Ann. Mo. Bot. Gard. 82: 517–534.
Haeckel, I. 1930. Über Iridaceen. Flora 25: 1–82.
Hegnauer, R. 1963. See general references.
Heslop-Harrison, Y. 1977. The pollen-stigma interaction: pollen-tube penetration in *Crocus*. Ann. Bot. 41: 913–922.
Heslop-Harrison, Y., Shivanna, K.R. 1977. The receptive surface of the angiosperm stigma. Ann. Bot. 41: 1233–1258.
Huber, H. 1969. See general references.
Kenton, A., Heywood, C.A. 1984. Cytological studies in South American Iridaceae. Plant Syst. Evol. 146: 87–104.
Kenton, A., Rudall, P. 1987. An unusual case of complex heterozygosity in *Gelasine azurea* (Iridaceae), and its implications for reproductive biology. Evol. Trends Pl. 1: 95–103.
Larsen, P.O., Sorensen, F.T., Wieczorkowska, E., Goldblatt, P. 1981. *Meta*-carboxy-substituted aromatic amino acids and γ-glutamyl peptides: chemical characters for classification in the Iridaceae. Biochem. Syst. Ecol. 9: 313–323.
Lewis, G.J. 1954. Some aspects of the morphology, phylogeny and taxonomy of the South African Iridaceae. Ann. S. Afr. Mus. 40: 15–113.
Manning, J.C., Goldblatt, P. 1990. Endothecium in Iridaceae and its systematic implications. Am. J. Bot. 77:527–532.
Manning, J.C., Goldblatt, P. 1991. Seed-coat structure in the shrubby Cape Iridaceae, *Nivenia*, *Klattia* and *Witsenia*. Bot. J. Linn. Soc. 107: 387–404.
Mathew, B. 1982. The *Crocus*. London: Batsford.
Mathew, B., Brighton, C. 1977. *Crocus tournefortii* and its allies (Iridaceae). Kew Bull. 31: 775–786.
Molseed, E. 1970. The genus *Tigridia* (Iridaceae) of Mexico and Central America. Univ. Calif. Publ. Bot. 54: 1–128.
Rübsamen-Weustenfeld, T., Mukielka, V., Hamann, U. 1994. Zur Embryologie, Morphologie und systematischen Stellung von *Geosiris aphylla* Baillon (Monocotyledoneae-Geosiridaceae/Iridaceae). Bot. Jahrb. Syst. 115: 475–545.
Rudall, P. 1983. Leaf anatomy and relationships of *Dietes* (Iridaceae). Nord. J. Bot. 3: 471–478.
Rudall, P. 1984. Taxonomic and evolutionary implications of rhizome structure and secondary thickening in Iridaceae. Bot. Gaz. 145: 524–534.
Rudall, P. 1995. Iridaceae. In: Cutler, D.F., Gregory, M. (eds.) Anatomy of the Monocotyledons Oxford: Clarendon Press.
Rudall, P., Owens, S.J., Kenton, A.Y. 1984. Embryology and breeding systems in *Crocus* (Iridaceae) - a study in causes of chromosome variation. Plant Syst. Evol. 148: 119–134.
Rudall, P., Kenton, A.Y., Lawrence, J. 1986. An anatomical and chromosomal investigation of *Sisyrinchium* and allied genera. Bot. Gaz. (Crawfordsville) 147: 466–477.
Schulze, W. 1971. Beiträge zur Pollenmorphologie der Iridaceae und ihre Bedeutung für die Taxonomie. Feddes Repert. 82: 101–124.
Scott, D.H., Brebner, G. 1893. On the secondary tissues of certain monocotyledons. Ann. Bot. 7: 21–62.
Simpson, B.B., Neff, J.L. 1981. Floral rewards: alternatives to pollen and nectar. Ann. Mo. Bot. Gard. 68: 310–322.
Strid, A.K. 1974. A taxonomic revision of *Bobartia* L. (Iridaceae). Opera Bot. 37: 1–45.
Vogel, S. 1954. Blütenbiologische Typen als Elemente der Sippengliederung, dargestellt anhand der Flora Südafrikas. Bot. Studien 1: 1–338. Jena: G. Fischer.
Vogel, S. 1974. Ölblumen und ölsammelnde Bienen. Trop. Subtrop. Pflanzenwelt. (Akad. Wiss. Lit. Mainz) 7: 1–267.
Williams, C., Harborne, J.B., Goldblatt, P. 1986. Correlations between phenolic patterns and tribal classification in the family Iridaceae. Phytochemistry 25: 2135–2154.
Wunderlich, R. 1959. Zur Frage der Phylogenie der Endospermtypen bei den Angiospermen. Oesterr. Bot. Z. 106: 203–293.
Zavada, M.S. 1983. Comparative morphology of monocot pollen and evolutionary trends of apertures and wall structures. Bot. Rev. 49: 331–379.

Ixioliriaceae

K. Kubitzki

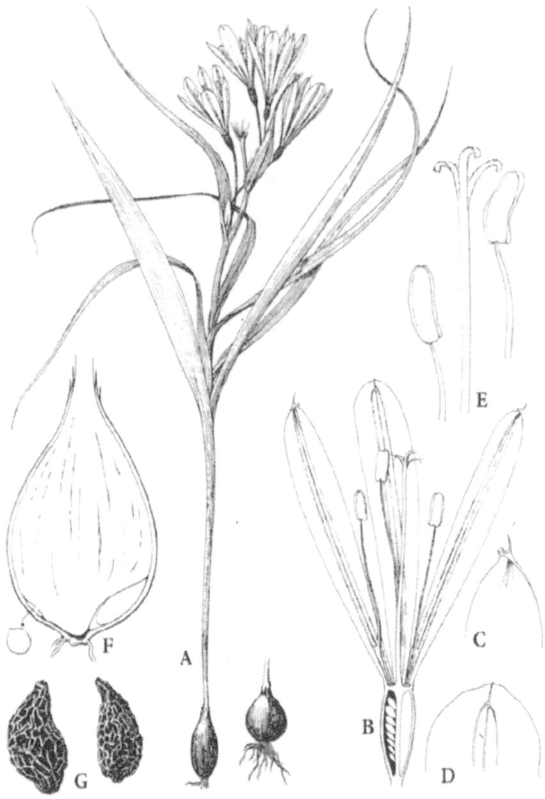

Fig. 94A–G. Ixioliriaceae. *Ixiolirion montanum*. **A** Flowering plant. **B** Flower, longitudinal section. **C** Apex of outer tepal. **D** Apex of inner tepal. **E** Style and 2 stamens, 1 of each whorl, in the relative position they have in the flower. **F** Corm, with lateral new corms formed at the base. **G** Seeds. (Dahlgren et al. 1985)

Ixioliriaceae (Pax) Nakai, Ord. Fam. Trib. Nov.: 234 (1943).

Perennial *herbs* from tunicated, bulb-shaped corms with basal innovation; aerial stem erect, leafy; *leaves* alternate, flat and linear, sheathing at the base, cylindrical-subulate at the apex. *Inflorescence* thyrses or pseudo-umbellas, many- to few-flowered. Pedicels often provided with a lateral prophyll. *Flowers* not articulated, hermaphrodite, actinomorphic, trimerous and epigynous; tepals 3 + 3, free nearly to the base, blue, violet or white, the outer ending in a prominent point emerging slightly below and outside the apex; stamens 3 + 3, inserted at the base of the tepals; anthers basifixed, tetrasporangiate, introrse-latrorse and longitudinally dehiscent; style erect, slender, apically tribrachiate, with dry stigmatic surfaces; ovary inferior, trilocular, with numerous ovules in axile placentation; ovules anatropous, crassinucellate; *capsule* loculicidal, dehiscent at the top; seeds numerous, ovate to pear-shaped, reticulate, testa black, phytomelanous, albuminous, the endosperm starch-free; embryo straight, fusiform, nearly as long as the endosperm.

A single genus of ca. three species of spring geophytes from semiarid regions from Egypt to lowland and mountains (up to 4000 m) of SW and inner Asia.

Morphology. The corm is massive, non-scaly and has basal innovations which were figured by Regel (1880) and Dahlgren et al. (1985). The leaves consist of a sheath and a dorsiventral blade, terminating in a cylindric apex. In contrast to Amaryllidaceae, *Ixiolirion* has no scapes, but leafy peduncles which, also in contrast to the Amaryllideace, are terminal. The ground plan of the inflorescence is a thyrse (*I. montanum*) with bostryces (helicoid cymes) as partial inflorescences; only in weaker individuals the inflorescences tend to be botryoids or corymbs (Huber 1969). The bracts of the partial inflorescences are ± leafy, in contrast to nearly all Amaryllidaceae, and the tepals are of a delicate texture, in contrast to Amaryllidaceae, but also Alliaceae and Hyacinthaceae (Fig. 94).

Anatomy. Uniseriate, 2-6-celled hairs are found on the lower leaf surface. The leaves are bifacial and have vascular bundles of varying size. The stomata are anemocytic. Vessels are restricted to the roots; they have scalariform perforations. The tissues are rich in raphides and mucilage cells (Oganezowa 1981; Arroyo 1982).

Pollen Morphology. The pollen is of a very generalised type, sulcate, the exine reticulate, homobrochate with simplibaculate muri and small (<1–2 μm) brochi, all three species studied by Schulze (1983).

Embryology. Microsporogenesis is simultaneous (Stenar 1925). A parietal cell is formed, and endosperm formation is Helobial (Davis 1966).

KARYOLOGY. Traub (1943) determined the chromosome number of *Ixiolirion tataricum* as $2n = 24$, the chromosomes varying considerably in size.

PHYTOCHEMISTRY. Chelidonic acid is present (Hegnauer 1963); saponins have apparently not reported; alkaloids are lacking (Dahlgren et al. 1985).

FRUIT AND SEED. In the seed coat the epidermis of the testa is phytomelan incrusted, while additional cell layers consist of flattened, empty cells. The cells of the tegmen are crushed and colourless. In the well-developed endosperm the radially elongated, pitted cells are free of starch, which is restricted to the cells surrounding the embryo (Huber 1969).

AFFINITIES. In his broadly construed Amaryllidaceae, Pax (1888) appended *Ixiolirion* as a special subtribe to his Amaryllidoideae (equivalent to the present-day Amaryllidaceae), but was fully aware that it is a strange element in that family. [In the second edition of the Natürl. Pflanzenfamilien, Pax and Hoffmann (1930) no longer commented on its problematic position within the family.] Traub (1943) placed it at the beginning of this family and considered it the most primitive element on base of its leaf-bearing peduncle and not fully (pseudo-)umbellate inflorescence. Nevertheless, characters of *Ixiolirion* such as the corm, the blue flowers (very rare in Amaryllidaceae), the precursory tips, the lack of alkaloids, and the simultaneous microsporogenesis cast doubt on such a placement. Huber (1969) stressed the great differences between *Ixiolirion* and Amaryllidaceae and commented upon the isolated position of the former, which, according to him, at best is comparable with *Aphyllanthes*, the Asphodelaceae, *Caesieae*, *Scilloideae* and *Simethis*. The molecular (*rbc*L) analysis, however, placed *Ixiolirion* with *Doryanthes* in the "Lower Asparagales" in close proximity to Iridaceae (Chase et al. 1995; Rudall et al. 1997). This placement supports the recognition of *Ixiolirion* as a family.
Only one genus:

Ixiolirion Fischer ex Herbert Fig. 94

Ixiolirion Fischer ex Herbert, App. Bot. Reg.: 37 (1821).
Kolpakowskia Regel (1878).

Characters as for family.

Selected Bibliography

Arroyo, S.C. 1982. Anatomia vegetativa de Ixiolirion Fisch. ex Herb. (Liliales) y su significado taxonomico. Parodiana 1: 271–286.
Chase, M.W. et al. 1995. See general references.
Dahlgren, R. et al. 1985. See general references.
Davis, G.L. 1966. See general references.
Hegnauer, R. 1963. See general references.
Huber, H. 1969. See general references.
Oganezova, G.G. 1981. Anatomical and morphological study in *Ixiolirion tataricum* ssp. *montanum*. Bot. Žh. (Moscow & Leningrad) 66: 702–713.
Pax, F. 1888. Amaryllidaceae. In: E.-P., Natürl. Pflanzenfamilien II, 5. Leipzig: W. Engelmann.
Pax, F., Hoffmann, K. 1930. Amaryllidaceae. In: E.-P., Natürl. Pflanzenfamilien, 2nd edn. 15a. Leipzig: W. Engelmann.
Regel, E. 1880. *Ixiolirion tataricum* (Amaryllis) Pall. γ Ledebouri. Gartenflora 1880: 193–195, t. 1014.
Rudall, P.J. et al. 1997. See general references.
Schulze, W. 1983. Beiträge zur Taxonomie der Liliifloren XIV. Der Umfang der Amaryllidaceae. Wiss. Z. Friedrich Schiller-Univ. Jena, Math.-Nat. R., 32, 6: 985–1003.
Stenar, H. 1925. Embryologische Studien I und II. II. Die Embryologie der Amaryllidaceen. Upsala: Dissertation.
Traub, H.P. 1943 ("1943"). The Ixiolirion tribe. Herbertia 9: 53–59.
Traub, H.P. 1963. The genera of Amaryllidaceae. La Jolla, California: The American Plant Life Society.

Johnsoniaceae

H.T. Clifford and J.G. Conran

Johnsoniaceae J.P. Lotsy, Vortr. Bot. Stammesgesch. 3(1): 731 (1911).

Perennial, tufted, shortly rhizomatous or stilt-rooted terrestrial *herbs* to 1 m tall, shrublike if the branched inflorescence is perennial and the leaves are caducous (*Corynotheca*). *Leaves* linear to 80 cm long, mostly basal and 2-ranked. *Inflorescence* a terminal spike or raceme, if branched with the flowers borne singly or in spicate (*Corynotheca*) or umbellate (*Tricoryne*) clusters on the branch apices; flowers and branches subtended by more or less conspicuous bracts. *Flowers* actinomorphic, sessile or pedicellate and then articulated or non-articulated on pedicels; perianth twisting or not twisting after anthesis, at length deciduous; tepals 6, more or less similar, at the base 3- to 7-nerved, free or united into a tube which may be $\frac{1}{2}$ their length; stamens free, 6 or 3 with or without staminodia, inserted on tepal bases, if 3, affixed to inner whorl of tepals; anther filaments unornamented; anthers dorsifixed, erect, dehiscing introrsely (and also extrorsely *Corynotheca*) by slits; ovary syncarpous, superior and 3-locular; style simple, filiform, rarely gynobasic (*Tricoryne*); stigma small, capitate; ovules anatropous or campylotropous, on axile placentas; ovules 1-few, usually 2 per placenta. *Fruit* a capsule, nut or schizocarp (*Tricoryne*); seeds black, shiny, arillate or lacking an aril (*Arnocrinum*), embryo linear-cylindrical, more than $\frac{1}{3}$ the length of the seed.

A family of 8 genera and about 50 species, largely restricted to Australia and New Guinea with one genus extending to Madagascar and S Africa.

Vegetative Morphology. The plants are shortly rhizomatous, with tuberous or fibrous adventitious roots. The leaves are radical, linear or terete and sometimes perennial (*Johnsonia*, *Hensmania*). Reduced or scalelike leaves may also form on the rhizome. When the plant is branched, the whole of its aerial portion may be interpreted as the inflorescence and the reduced foliage as bracts. According to Keighery (1984), the apparent leaves in *Hensmania*, *Hodgsoniola*, *Johnsonia* and *Stawellia* are better interpreted as phylloclades. The seedlings of *Caesia* and *Tricoryne* have a long, assimilating cotyledonary hyperphyll. In *Caesia*, there is no cotyledonary sheath, germination is epigeal, and the first leaf is similar to successive leaves, whereas in *Tricoryne*, there is a short cotyledonary sheath and the hyperphyll exhibits a hairpin-bend above the sheath, similar to the seedlings of *Dianella*. The subsequent leaves are linear to lanceolate, non-sheathing and spirally alternate in fascicles on short lateral branches (Thongpukdee 1989).

Vegetative Anatomy. The roots, if fleshy, have a parenchymatous cortex. The endodermis has conspicuous U-thickenings, except in *Arnocrinum*. The rhizodermal cells are alternately long and short. The stomata are anomocytic and occur on both leaf surfaces, except for *Hodgsoniola*, where the leaves are hypostomatic. In terete *Johnsonia* and *Stawellia* species, the stomata are restricted to grooves on the leaf surface and occur in paired rows within each groove.

Vessels with scalariform perforation plates occur in the stems of *Johnsonia*, *Caesia* and *Tricoryne*, although some *Caesia* species lack stem vessels, and there are also vessels with simple perforation plates in *Tricoryne* (Wagner 1977). The leaves of *Caesia* have undifferentiated mesophyll, whereas those of *Tricoryne* have a bifacial palisade and strongly lignified bundle sheath. Calcium oxalate raphides are present in all parts of the plants. Tannin cells are present, but mucilage canals or laticifers are absent (Thongpukdee 1989).

Inflorescence Structure. Within the family there is considerable diversity in inflorescence structure (Fig. 95). In five of the genera, the flowers are borne in single terminal spikes or racemes terminating a more or less leafy (*Hodgsoniola*) or leafless axis (*Hensmania*, *Johnsonia*, *Stawellia*). The bracts subtending the flowers may be small (*Hodgsoniola*) or large (*Stawellia*, *Hensmania*, *Johnsonia*). The remaining genera have branched inflorescences. That of *Arnocrinum* is much branched but only 1 or a few branches terminate in a dense spike of flowers, the remainder being sterile. All branches and flowers are subtended by small bracts. The inflorescences of *Tricoryne* and *Caesia* are similar, but here all or most branches terminate in few-flowered umbels (*Tricoryne*) or racemes (*Caesia*) and throughout, the bracts are conspicuous, being leaflike at the base but smaller distally. In *Tricoryne anceps* subsp. *pterocaulon*

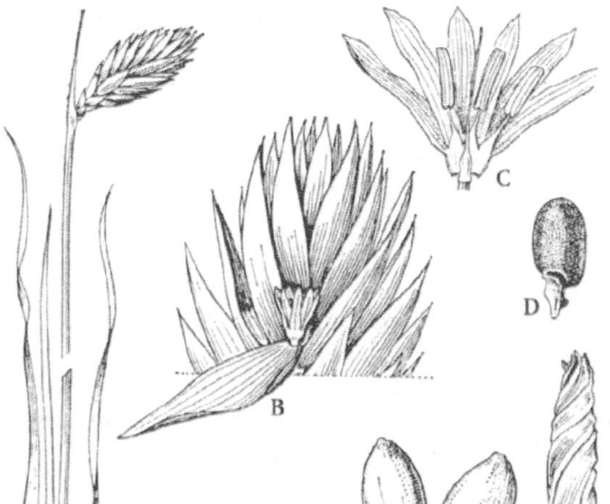

Fig. 95A–I. Johnsoniaceae. **A–D** *Johnsonia lupulina*. **A** Habit. **B** Bracts of terminal spike subtending flowers. **C** Opened flower. **D** Seed with aril. **E–I** *Tricoryne elatior*. **E** Habit. **F** Flower. **G** Stamen with tufted filament. **H** Twisted perianth of postanthetic flower. **I** Schizocarp. (Takhtajan 1982)

the inflorescence is flattened and cladodelike. The habit of *Corynotheca* is distinctly shrublike due to its perennial inflorescence and small, short-lived basal leaves. The flowers are sessile or shortly pedicellate and occur singly in bract axils along the inflorescence branches, each flower subtended by a pair of bracts. A short pericladium is developed in most genera, or the articulation is at the pedicel apex.

Floral Structure. The perianth consists of six, more or less similar tepals each of which has 3–7 veins at its base. These veins converge and fuse distally, except in *Johnsonia*, where they branch and neither fuse nor converge distally. The tepals are 5–12 mm long and either free and spreading radially, or are fused into a tube for up to ½ their length (*Arnocrinum*). The anther filaments are not fused; with respect to their relative lengths they are either longer or shorter than the anthers. If shorter, the anthers are connate about the style (e.g. *Arnocrinum*), if longer, then either well separated from the style (e.g. *Caesia*), or in most genera are erect and surround the style. The gynoecium is tricarpellary and 3-locular with 2 rows of ovules per placenta. The ovules may develop only towards the apex or base of the carpel thereby resulting in seeds being apparently pendulous or basal. The style is solitary and terminated by a small, Dry, capitate, stigma. In *Tricoryne* the style is gynobasic, similar to the Lamiaceae, and the 3 ovary chambers more or less free. Septal nectaries of the outer type (sensu Daumann 1970) are present at the very base of the gynoecium in at least *Johnsonia* and *Tricoryne*, but are poorly developed in both genera.

Embryology. Microsporogenesis is simultaneous (Rudall et al. 1997) and the 2-nucleate pollen grains are shed singly. The tapetum is binucleate and secretory. Anther wall thickenings are of the spiral type. The ovules are anatropous or campylotropous and tenuinucellate (at least in *Johnsonia* and *Tricoryne*). There is no obvious hypostase or central conducting passage of axially elongated cells (Zuleitungsbahn). Incomplete data for the megagametogenesis in *Johnsonia* and *Tricoryne* suggest Polygonum type embryo sac development, and a filiform apparatus is prominent (J.G. Conran, unpubl. observ.).

Pollen Morphology. The pollen grains are trichotomosulcate and rounded triangular or circular in polar view (Schulze 1982; Fig. 96) with a tectate-perforate or imperforate exine (*Caesia* etectate-columellate) on the polar face (Thongpukdee 1989).

Karyology. A chromosome number of $n = 12$ has been reported for species of *Johnsonia*, *Hensmania*, *Stawellia*, *Hodgsoniola* and *Arnocrinum* (Keighery 1984), which suggests a base number of $x = 6$ for the family.

Pollination. The flowers of all taxa are diurnal and last only 1 day. Keighery (1984) reported that nectar was absent in most genera, although rare in

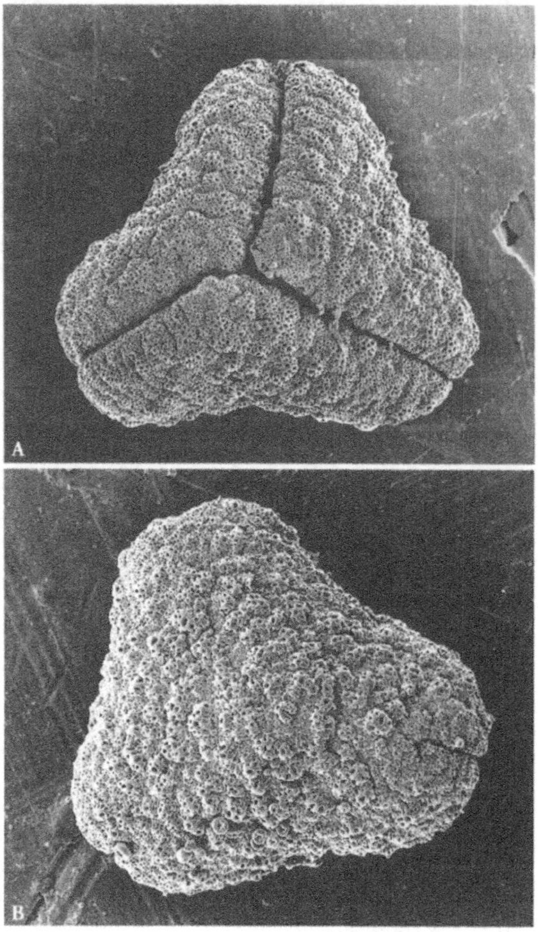

Fig. 96A,B. Johnsoniaceae. *Arnocrinum drummondii*. **A** Distal pole, **B** proximal pole of pollen. SEM ×1400. (Photo Palynological Laboratory Stockholm)

Tricoryne, and that pollination was through buzz-pollination by pollen-collecting bees. *Tricoryne* around Adelaide in S Australia is also visited by hoverflies (Syrphidae).

FRUIT AND SEED. The fruit is basically a loculicidal capsule which, as a result of only 1 ovule maturing and a failure to dehisce, gives rise to a nut in some taxa. If only a single seed forms in each of the locules and these separate, a schizocarp is formed. The inner integument collapses during development, and the seed is covered by a thick phytomelan deposit (Huber 1969). Strophioles are present in most genera, but absent in *Tricoryne* (where the seed is enclosed in an umbonate mericarp) and *Arnocrinum*. The testa is striately sculptured in *Caesia* species, *Corynotheca* and *Tricoryne*, smooth or papillate (*Caesia* and *Corynotheca* spp.), and reticulately ornamented in the remainder of the family (Thongpukdee 1989). The embryo is linear-cylindrical, curved in *Arnocrinum*, and more than ⅓ the length of the seed (Thongpukdee 1989). The endosperm is copious and non-starchy.

DISPERSAL. Nothing is known about dispersal, but the arillate seeds suggest animal, possibly ant dispersal for most genera.

PHYTOCHEMISTRY. Chelidonic acid was reported from *Johnsonia* by Ramstad (1953).

AFFINITIES. Bentham's (1878) tribe Johnsonieae (Liliaceae) has always been considered to be an odd and somewhat isolated group, although it has been variously related to taxa now included in the Aphyllanthaceae, Boryaceae and Lomandraceae (see Keighery 1984). Recent *rbc*L analyses (Chase et al. 1995) identified a Phormioid/Johnsonioid clade, and ongoing studies have added *Simethis*, *Johnsonia* and *Arnocrinum*, which also have trichotomosulcate pollen, to this clade (Rudall et al. 1996). The addition of new taxa also altered resolution within the clade, so that *Johnsonia*, *Stawellia*, *Caesia*, *Hensmania* and *Arnocrinum* are grouped together, with *Caesia* sister to the other four. Rudall et al. (1996) resurrected the family Johnsoniaceae (Keighery 1984) now expanded to include eight genera: *Arnocrinum*, *Hensmania*, *Hodgsoniola*, *Johnsonia* and *Stawellia* (i.e. Keighery's Anthericaceae-Johnsonieae), plus *Caesia*, *Corynotheca* and *Tricoryne*.

The *rbc*L resolution of the Johnsoniaceae agrees with morphological data. Keighery's Johnsonieae were grouped on the basis of several characters: a spicate or racemose inflorescence, anthers closely appressed to the style (forming a tube), base chromosome number of $x = 12$, and phyllodes (except in *Arnocrinum*). Keighery's tribal and subtribal groupings are still relevant. In addition, *Caesia*, *Corynotheca* and *Tricoryne* are closely related to each other, although *Corynotheca* has a nutlet, and *Tricoryne* a schizocarp (Henderson 1987). *Tricoryne* also has prominent hairs on the anther filaments, which might be taken to indicate a relationship with *Arthropodium*/*Dichopogon* (Lomandraceae), but the different origins of these hairs and trichotomosulcate pollen in *Tricoryne* (Schulze 1982) preclude this. Most Johnsoniaceae have arillate seeds, and this feature is rare elsewhere in the trichotomosulcate clade, although present in the sister clade, Asphodelaceae.

Distribution and Habitats. This is an Old World family restricted to the southern hemisphere. Its members grow in a wide diversity of habitats from sea level to about 1500 m between latitudes 6–43 °S, but preferentially in regions of low or markedly seasonal rainfall.

Economic Importance. The family are of limited horticultural interest.

Key to the Genera

1. Tepals twisting spirally after anthesis	2
– Tepals not twisting after anthesis	5
2. Fruit a schizocarp splitting into mericarps	1. *Tricoryne*
– Fruit a capsule or nut	3
3. Stamens 6	4
– Stamens alternating with 3 staminodes	2. *Hodgsoniola*
4. Stamens equally long	3. *Caesia*
– Stamens alternately long and short	4. *Corynotheca*
5. Stamens 3	6
– Stamens 6, alternately long and short	5. *Arnocrinum*
6. Peduncle clothed with pungent bracts	6. *Hensmania*
– Peduncle lacking bracts	7
7. Flowers hidden by floral bracts	7. *Johnsonia*
– Flowers projecting beyond floral bracts	8. *Stawellia*

1. *Tricoryne* R. Br. Fig. 95E–I

Tricoryne R. Br., Prod.: 278 (1810); Baker, J. Linn. Soc. Bot. 15: 361 (1876); Thongpukdee in Fl. Australia 45: 292–299 (1987).

Shortly rhizomatous perennial herbs to 1 m tall; stems branched, sometimes winged; leaves dispersed along stems or basal, bases equitant; inflorescence umbellate; tepals yellow to 8 m long; stamens 6, filaments with a tuft of clavate hairs immediately below the anthers; fruit a schizocarp, mericarps 3, seeds exarillate. About seven spp., Australia, except for *T. anceps*, which extends to Papua New Guinea.

2. *Hodgsoniola* F. Muell.

Hodgsoniola F. Muell., Fragm. 2: 176 (1861); Keighery, Fl. Australia 45: 306 (1987).

Shortly rhizomatous perennial herbs to 50 cm tall; stem unbranched; leaves terete, mostly basal, leaf-bases sheathing; inflorescence a raceme terminating in a bracteate scape; tepals united at base, purple, to 1 cm long; stamens 3, opposite petals and alternating with staminodes; fruit capsule. Only one sp., *H. junciformis* F. Muell., SW Australia.

3. *Caesia* R. Br.

Caesia R. Br., Prod.: 277 (1810); Jessop, Fl. Males. I, 9: 203–204 (1979); Henderson, Fl. Australia 45: 281–288 (1987).

Tufted perennial herbs to 70 cm tall; aerial stem weakly branched or unbranched; leaf blades linear, bases sheathing; inflorescence of cymes, sessile arranged along the branches or in simple or branched spikes; tepals green, white or blue, nerves often purple, to 10 mm long; stamens 6; fruit nut or more or less dehiscent capsule. About 11 spp., southern Africa, Papua New Guinea and Australia.

4. *Corynotheca* F. Muell. ex Benth.

Corynotheca F. Muell. ex Benth., Fl. Austral. 7: 49 (1878); Henderson, Fl. Australia 45: 299–306 (1987).

Tufted rhizomatous perennials; aerial stem short, leafy; leaf blades linear, reducing to bracts; inflorescence irregularly paniculate; tepals white, creamy pink or purplish, to 8 mm long; stamens 6, filaments straight or kinked, papillose; fruit a tardily dehiscing capsule or nut. About six spp., all Australian States (except Tasmania) and the Northern Territory.

5. *Arnocrinum* Endl. & Lehm.

Arnocrinum Endl. & Lehm., Plant. Preiss. 2: 41 (1846); Bentham, Fl. Austral. 7: 69–70 (1878); Keighery, Fl. Australia 45: 246–249 (1987).

Rhizomatous perennial to 60 cm tall; stems short leafy; inflorescence branched with scale leaves, flowers in condensed spikes; tepals to 12 mm, fused into a perianth tube, lobes purple, spirally twisted after anthesis; stamens 6, filaments alternately long and short, attached to apex of perianth tube; fruit capsule. Three spp., SW Australia.

6. *Hensmania* W. Fitzg.

Hensmania W. Fitzg., Proc. Linn. Soc. N.S.W. II, 28: 105 (1903); Keighery, Fl. Australia 45: 249–252 (1987).

Rhizomatous, tufted or stilt-rooted, perennial herbs; stems unbranched, to 20 cm tall; leaves terete with sheathing bases; inflorescence an umbel-like spike terminating a peduncle invested with pungent imbricate bracts; tepals united in the lower third, yellow or cream, to 6 mm long; fruit a capsule. Three spp., SW Australia.

7. *Johnsonia* R. Br. Fig. 95A–D

Johnsonia R. Br., Prodr.: 287 (1810); Keighery, Fl. Australia 45: 243–246 (1987).

Shortly rhizomatous perennial herbs to 70 cm tall; stem unbranched; leaves linear equitant; inflorescence a spike terminating a leafless scape; flowers concealed by conspicuous bracts; tepals connate at the base, green, white, pink or purple, to 8 mm long; stamens 3, attached to the inner tepals; fruit a capsule. Five spp., SW Australia.

8. *Stawellia* F. Muell.

Stawellia F. Muell., Fragm. 2: 176 (1861); Keighery, Fl. Australia 45: 252–254 (1987).

Tufted rhizomatous perennial herbs raised above the ground on stilt roots, to 40 cm tall; stems weakly branched; leaf blades terete, bases sheathing; inflorescence a gobular spike terminating a leafless scape; tepals shortly united at the base, purple or green, to 5 mm long; stamens 3; fruit a capsule. Two spp., SW Australia.

Selected Bibliography

Bentham, G. 1878. Flora Australiensis. London: L. Reeve.
Chase, M.W. et al. 1995. See general references.
Daumann, E. 1970. See general references.
Henderson, R.J.F. 1987. *Corynotheca*. In: George, A.S. (ed.) Flora of Australia. Canberra: AGPS, pp. 299–306.
Huber, H. 1969. See general references.
Keighery, G.J. 1984. The Johnsonieae (Liliaceae): biology and classification. Flora 175: 103–108.
Ramstad, E. 1953. Über das Vorkommen und die Verbreitung von Chelidonsäure in einigen Pflanzenfamilien. Pharm. Acta Helv. 28: 45–55.
Rudall, P., Chase, M.W., Conran, J.G. 1996. New circumscriptions and a new family of asparagoid lilies: genera formerly included in Anthericaceae. Kew Bull. 51: 667–680.
Rudall, P.J. et al. 1997. See general references.
Schulze, W. 1982. Beiträge zur Taxonomie der Liliifloren IX. Anthericaceae. Wiss. Z. Friedrich Schiller Univ. Jena 31: 291–307.
Takhtajan, A.L. 1982. See general references.
Thongpukdee, A. 1989. The taxonomic position of the genus *Tricoryne* R.Br. Ph.D. Thesis St. Lucia: University to Queensland.
Wagner, W.M. 1977. Vessel types of monocotyledons: a survey. Bot. Not. 130: 383–402.

Lanariaceae

P.J. Rudall

Lanariaceae H. Huber ex R. Dahlgren, Monogr. Syst. Bot. Mo. Bot. Gard. 25: 71 (1988).

Perennial *herbs* with short vertical rhizomes; roots emerging from lower part of rhizome; *leaves* distichous or spirally arranged, linear, basal, sheathing, without distinct petioles, dorsiventral, parallel-veined, glabrous; inflorescences and flowers densely woolly with branched hairs. *Inflorescence* a dense corymbose panicle with monochasial components; *flowers* bisexual, epigynous, small, covered with dense whitish hairs outside, pale mauve or purple inside; perianth with 6 equal tepals fused to a tube for their lower half; stamens 6, inner 3 slightly shorter than outer 3; filaments basally adnate to tepal tube; anthers dorsifixed, epipeltate, versatile, introrse; ovary inferior, syncarpous, trilocular, with septal nectaries and axile (marginal) placentation; style narrow, filiform; stigma small, capitate; ovules anatropous, 2 per locule, with micropyles facing downwards; *fruit* a 1-seeded capsule; seeds shining black, with phytomelan in seed coat.

A single monotypic genus from South Africa.

Vegetative Morphology. Lanariaceae are herbaceous geophytes with linear leaves with a distichous or spiral arrangement. The rhizomes are short and vertically oriented, enclosed by persistent fibrous leaf bases, and with persistent old fibrous roots (Dahlgren and Van Wyk 1988).

Vegetative Anatomy. (Schmidt 1891; Schnarf 1892; Schulze 1893; Geerinck 1968; Rudall et al. 1998). Stomata are paracytic and present on both surfaces of the leaf. Raphide crystals are absent, but small cuboidal styloid crystals are sometimes present, mainly in outer bundle-sheath cells surrounding sclerenchyma at larger vascular bundles. Hairs are normally absent from leaves, although thick ridges are present at leaf margins, and Dahlgren and Van Wyk (1988) reported occasional long simple (unbranched) hairs on the under surfaces of leaves (not seen here). Copious dense, highly branched (dendritic) hairs are present on the inflorescences and flowers. Leaves are thick and V-shaped in cross-section, with bulliform epidermal cells present over the midrib. Some groups of epidermal cells over vascular bundles and margins are lignified. The larger vascular bundles are linked to both epidermises by a sclerenchymatous girder. Larger vascular bundles are in a single row, but 2 rows of tiny randomly oriented bundles, lacking sclerenchyma, are present in spaces between them.

FLOWER STRUCTURE. Highly branched, dendritic hairs are present on the outside tepal surfaces; shorter unbranched papillae or unicellular hairs on the upper parts of the inside tepal surfaces (Fig. 97). Each tepal has 3 vascular traces. Each filament has a single vascular trace. The anther wall has endothecial thickenings in the subdermal layer. The style has a single central trilobed canal and 3 vascular traces, although Dahlgren and Van Wyk (1988) described the style as trilocular. The ovary is inferior, or slightly semiinferior, as the ovary top is conical (terminating in the style). Dahlgren and Van Wyk (1988) described the perianth as laterally fused to the ovary but free below, a feature not present in material examined here, indicating some variation. The ovary is trilocular, but with a central canal, and the carpel margins are barely fused in places. Septal nectaries are present throughout the length of the ovary, opening at the top. There are 2 ovules per carpel. There is an obturator present below each ovule, near the micropyle. Ovules are bitegmic and the inner integument forms the micropyle, which is long and curved towards the placenta. In the mature unfertilised ovule, the proximal (chalazal) region of the nucellus is relatively large.

Fig. 97A–D. Lanariaceae. *Lanaria plumosa*. **A** Stem and inflorescence. **B** Flower. **C** Flower, longitudinal section. **D** Ovary, longitudinal section. (Dahlgren et al. 1985)

EMBRYOLOGY. De Vos (1963) decribed the embryology of *Lanaria*. The tapetum is secretory. Microsporogenesis is simultaneous. Ovules are crassinucellate, with the archesporial cell forming a parietal cell. Embryo sac development is of the Polygonum type. In the mature megagametophyte the 3 small antipodals lie in a narrow constriction at the chalazal end. Endosperm development is Helobial.

POLLEN MORPHOLOGY. Pollen is sulcate. Non-apertural exine is tectate-columellate with a thick foot-layer. Apertural exine is absent (Erdtman 1952; Simpson 1983).

KARYOLOGY. Ornduff (1979) reported $2n = 36$ in *L. lanata* (as *L. plumosa*).

POLLINATION. No information known, although the presence of septal nectaries indicates insect pollination.

FRUIT AND SEED. Normally only 1 ovule develops into a seed, the others degenerating (De Vos 1963). Cell wall development in the endosperm is late, with over 100 free nuclei remaining in the full-sized seed. The endosperm tissue at first contains starch, later protein and oil. The mature embryo is slightly curved and about half the length of the endosperm tissue. The seed coat is formed mainly from the outer integument, and is black, hard and brittle, with phytomelan (De Vos 1963; Huber 1969). Germination takes place within 3 weeks of seed set.

PHYTOCHEMISTRY. Dora and Edwards (1991) established the absence of phenalones and the presence of a novel biflavonoid, lanaroflavone, in *L. lanata*.

Affinities. Dahlgren and Van Wyk (1988) and Rudall et al. (1998) discussed the systematic relationships of *Lanaria*. It was formerly included in Haemodoraceae (e.g. by Geerinck 1968), which it resembles in inflorescence morphology and tomentose flowers, but is now placed among the early-branching ("lower") Asparagales in Lilianae. There is now much evidence against a close relationship between *Lanaria* and Haemodoraceae, which belong in the commelinoid clade, characterised by the presence of cell-wall ferulates (Harris and Hartley 1980; Rudall and Caddick 1994). *Lanaria* lacks the phenalenone pigments that are characteristic of Haemodoraceae (Dora and Edwards 1991), and differs from them in its embryology (De Vos 1963) and pollen morphology (Simpson 1983). The presence of phytomelan in the seed coat indicates that it belongs in Asparagales, and from the simultaneous microsporogenesis (Rudall et al. 1997), paracytic stomata and *rbc*L (Rudall et al. 1998), amongst the "lower" Asparagales. A relationship with the "lower" asparagoid family Tecophilaeaceae has been considered; Dahlgren et al. (1985) included *Lanaria* in Tecophilaeaceae, but Dahlgren and Van Wyk (1988) and Dahlgren (1989) put it in its own family, Lanariaceae. Analysis of *rbc*L (Chase et al. 1995) placed it as sister to a clade including Asteliaceae, Hypoxidaceae and *Blandfordia*. By default, *Lanaria* has often been associated with the monotypic north American genus *Lophiola*, but differs from it in several respects, such as leaf morphology and anatomy (Rudall et al. 1997b). Analysis of *rbc*L confirms that *Lanaria* and *Lophiola* are not closely related (M.W. Chase, pers. comm.).

Distribution and Habitats. *L. lanata* grows in the south coast region of the Cape Province of South Africa, from Somerset West to Albany, in Fynbos vegetation at low altitudes.

Economic Importance. None known.

Only one genus:

Lanaria Aiton Fig. 97

Lanaria Aiton, Hort. Kew, ed. 1: 462 (1789); Benth. & Hook., Gen. Pl. 3: 675 (1883).

Only one sp., *L. lanata*(L.) Dur. & Schinz, S Africa.

Selected Bibliography

Chase M.W. et al. 1995. See general references.

Dahlgren, G. 1989. An updated angiosperm classification. Bot. J. Linn. Soc. 100: 197–203.

Dahlgren, R., Van Wyk, A.E. 1988. Structures and relationships of families endemic to or centered in southern Africa. Mon. Syst. Bot. Mo. Bot. Gard. 25: 1–94.

Dahlgren R. et al. 1985. See general references.

De Vos, M.P. 1963. Studies on the embryology and relationships of South African genera of the Haemodoraceae: *Lanaria* Ait. J. S. Afr. Bot. 29: 79–90.

Dora, G., Edwards, J.M. 1991. Taxonomic status of *Lanaria lanata* and isolation of a novel biflavone. J. Nat. Prod. 54: 796–801.

Erdtman, G. 1952. See general references.

Geerinck, D. 1968. Considerations taxonomiques au sujet des Haemodoraceae et des Hypoxidaceae (Monocotyledones). Bull. Soc. R. Bot. Belg. 101: 265–278.

Harris, P.J., Hartley, R.D. 1980. See general references.

Huber, H. 1969. See general references.

Ornduff, R. 1979. Chromosome numbers and relationships of certain African and American genera of Haemodoraceae. Ann. Mo. Bot. Gard. 66: 577–580.

Rudall, P.J., Caddick, L.R. 1994. Investigation of the presence of phenolic compounds in monocot cell walls, using UV fluorescence microscopy. Ann. Bot. 75: 483–491.

Rudall, P.J. et al. 1997. See general references.

Rudall, P.J., Chase, M.W., Cutler, D.F., Rusby, J., de Bruijn, A. 1998. Anatamical and molecular systematics of Asteliaceae and Hypoxidaceae. Bot. J. Linn. Soc.: in press.

Scharf, W. 1892. Beiträge zur Anatomie der Hypoxideen und einiger verwandter Pflanzen. Bot. Cbl. 52: 145–153, 177–184, 209–217, 241–249, 289–296, 321–327.

Schmidt, C. 1891. Ueber die Blätter einiger xerophiler Liliifloren. Bot. Cbl. 47: 1–6, 33–42, 97–107, 164–170.

Schulze, R. 1893. Beiträge zur vergleichenden Anatomie der Liliaceen, Haemodoraceen, Hypoxidoideen und Velloziaceen. Bot. Jahrb. Syst. 17: 295–394.

Simpson, M.G. 1983. Pollen ultrastructure of the Haemodoraceae and its taxonomic significance. Grana 22: 79–103.

Simpson, M.G. 1990. Phylogeny and classification of the Haemodoraceae. Ann. Mo. Bot. Gard. 77: 722–784.

Liliaceae

M.N. Tamura

Liliaceae Juss., Gen. Pl.: 48 (1789).
Tulipaceae Horan. (1834).
Fritillariaceae R.A. Salisbury (1866).

Bulbiferous or rhizomatous *perennials*. Aerial stem erect, not branching. Vernation convolutive to curved or flat. Foliage *leaves* alternate, or seemingly opposite or verticillate, sometimes crowded at basis of aerial stem, filiform to oval, apically acute to aristate, rarely cirrhose, sessile or spuriously petiolate, usually parallel-veined, rarely net-veined, sometimes sheathing at the base. *Inflorescence* often with bracts, usually a raceme, sometimes umbelliform, often 1-flowered. *Flowers* bisexual, hypogynous, usually actinomorphic. Perianth often infundibular, sometimes tubular, campanulate or cupular, sometimes horizontally patent or recurved. Tepals 3 + 3, free, the whorls often similar but rarely highly different, variable in color, often with patterns of striation, spotting, flecking or checkering, often nectariferous. Stamens 3 + 3. Filaments free, often filiform, sometimes flattened, rarely cylindrically swollen with an aciculus at apex, often glabrous, sometimes hairy. Anthers linear to globose, dorsifixed versatile or pseudobasifixed, extrorse or latrorse. Style long to rudimentary, apically trilobate to trisected. Ovary ellipsoid, more rarely cylindric or oval-globose, (1–2–)3-locular, lacking septal nectaries, with several to numerous anatropous bitegmic ovules. *Fruit* a loculicidal capsule or a berry. Seeds brown or pale, often flattened, with or without a marginal wing, placed like piles of coins on top of each other in the fruit, sometimes ellipsoidal or angled, and, in this case, often with an elaiosome. Endosperm lacking starch.

A family comprising 11 genera and about 550 species, distributed in the Northern Hemisphere, abundant in temperate Eurasia.

Vegetative Morphology and Anatomy. Lilioideae are bulbiferous, while Medeoloideae are rhizomatous. The bulbs of the Lilioideae are variable in construction: often tunicated (Fig. 98A,H), sometimes naked (*Fritillaria* spp., *Nomocharis*, *Lilium*, Fig. 99F, 100B), consisting of 1 (*Gagea* spp., *Fritillaria* spp.) to numerous fleshy nutrient scales (*Lilium*). The rhizomes are tuberous (*Medeola*) and decaying every year, or thickened (*Clintonia*) and surviving for some years. The bulbs of the Lilioideae are presumed to be derived from rhizomes. The roots have vessels with scalariform perforation plates with more than 10 bars (*Tulipa*, *Erythronium*, *Fritillaria*, *Lilium*; Cheadle and Kosakai 1971).

The aerial stem is erect, containing only tracheids and no vessels (*Tulipa*, *Erythronium*, *Fritillaria*, *Lilium*; Cheadle and Kosakai 1971). Foliage leaves range from 1 (*Gagea* spp., *Tulipa regelii*) to many. They are often alternate, sometimes seemingly opposite (*Tulipa* spp., *Erythronium*, *Fritillaria* spp.) or verticillate (*Medeola*, *Fritillaria* spp., *Nomocharis* spp., *Lilium* spp.), sometimes located at the basis of the aerial stem (*Clintonia*, *Gagea*, *Lloydia*, *Notholirion*). They are flat, dorsiventral, acute to aristate at apex, rarely cirrhose (*Fritillaria* spp.), sessile or spuriously petiolate, sometimes dark-spotted. Bulbils are sometimes produced in leaf axils (*Gagea* spp., *Lilium* spp.).

Inflorescence Structure. The aerial stem ends in an indeterminate or a determinate inflorescence. It is often simple, sometimes compound by lateral ramifications, and is often a raceme, which may sometimes be umbelliform (*Medeola*, *Clintonia umbellulata*, *Gagea* spp., *Lloydia* spp., *Fritillaria* spp.). Racemes may be reduced to the lowermost flower, which in this case is shifted into a terminal position (e.g., *Tulipa*, *Fritillaria* spp.). Bracts are sometimes lacking (*Gagea* spp., *Tulipa*, *Erythronium*, *Cardiocrinum giganteum*); often they are bracteose, sometimes (and then generally only the basal ones within an inflorescence) foliose.

Flower Structure. Flowers are trimerous, diplostemonous, usually actinomorphic, rarely weakly zygomorhic (*Fritillaria* spp.). The perianth is often infundibular, sometimes tubular (*Cardiocrinum*, *Lilium* spp.), campanulate (*Tulipa*, *Fritillaria*), or cupular (*Nomocharis* spp., *Lilium* spp.), sometimes horizontally patent (*Gagea*, *Nomocharis*), sometimes recurved (*Medeola*, *Erythronium*, *Lilium* spp.). The tepals are often similar in both whorls, sometimes the outer ones are wider (*Gagea* spp., *Erythronium*, *Fritillaria pallidiflora*), sometimes the inner (*Lloydia* spp., *Tulipa* spp., Lilieae spp.). Rarely, the difference extends to structure (*Nomocharis*). Perigonal nectaries are often present (Fig. 98–100), but sometimes only on the inner tepals (*Nomocharis*), and sometimes

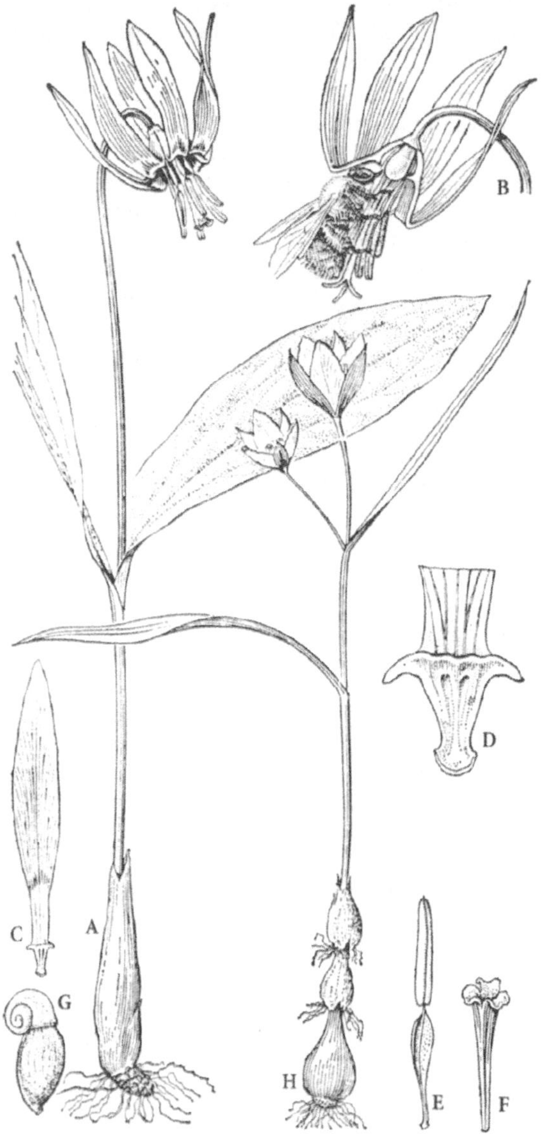

Fig. 98A–H. Liliaceae. A *Erythronium sibiricum*, flowering plant. B–G *E. japonicum*. B Flower, buzz-pollinated by *Xylocopa appendiculata*. C Tepal. D Base of tepal with perigonal nectary. E Stamen. F Style and stigma. G Seed with elaiosome. H *Tulipa biflora*, plant with depleted tubers of the 2 previous years. (Takhtajan 1982)

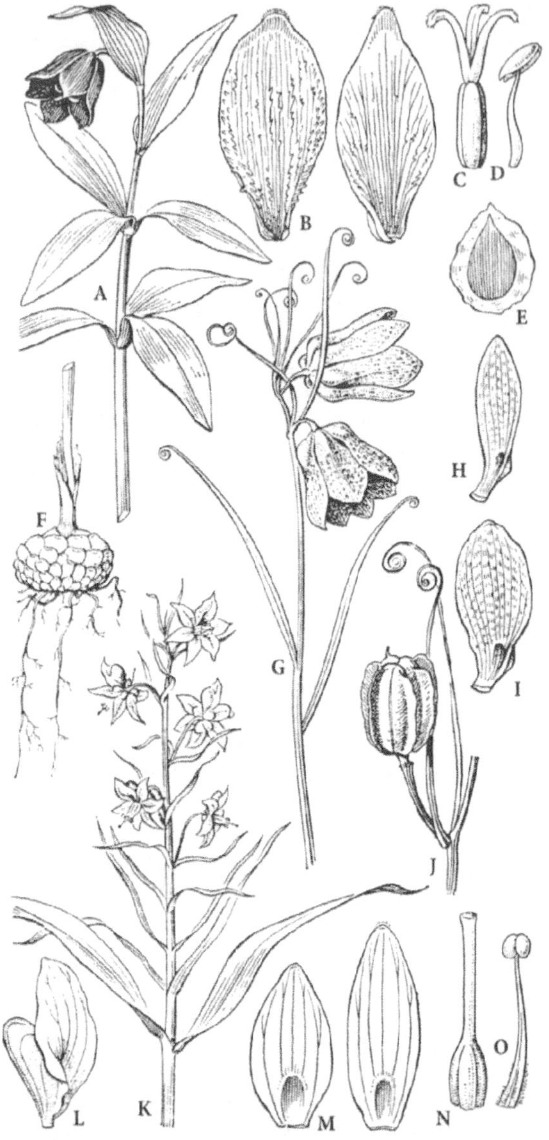

Fig. 99A–O. Liliaceae. A–F *Fritillaria kamtschatcensis*. A Flowering shoot. B Outer and inner tepals. C Pistil. D Stamen. E Winged seed. F Bulb. G–J *F. ruthenica*. G Inflorescence with tendrillar bracts. H, I Outer and inner tepals, nectariferous. J Capsule. K–O *F. karelinii*. K Flowering shoot. L Median upper spurred tepal, lateral view. M Inner and outer tepals, nectariferous, front view. N Pistil. O Stamen. (Takhtajan 1982)

completely absent (*Medeola*, *Clintonia*(?), *Gagea*, *Lloydia* spp., *Tulipa*).

The stamens are often similar in both whorls, sometimes different in length (*Gagea* spp., *Tulipa* spp., *Erythronium*). The filaments are often filiform, sometimes flattened, rarely cylindrically swollen or subulate with an aciculus at the apex (*Nomocharis*), often glabrous, sometimes with trichomes at the base (*Lloydia* spp., *Tulipa* spp., *Fritillaria karelinii*, *Lilium* spp.), free from each other, sometimes more or less fused to the tepals. The anthers are tetrasporangiate, extrorse or latrorse, but never introrse. They are dorsifixed versatile in *Nomocharis*, *Notholirion*, *Cardiocrinum*, and

Fig. 100A–H. Liliaceae. A–C *Lilium pennsylvanicum*. A Flowering shoot. B Bulb and shoot. C Inner and outer tepal. D, E. *L. longiflorum*. D Flower. E Basal part of tepal in transversal section, showing nectar grove. F–L *L. martagon*. F Flower. G Base of tepal. H Same, transversal section, with nectar grove. (Takhtajan 1982)

Lilium. In all other genera, including *Medeola* and *Clintonia*, the anthers are not capable of moving freely because of the tubular connective that more or less surrounds the filament tip (pseudobasifixed anthers).

The pistil is tricarpellary, generally trilocular with axile placentation. In *Medeola*, the carpels do not fuse completely with their flanks and the gynoecium is more or less unilocular with parietal placentation. Due to such incomplete fusions, a trimerous gynoecium may be even bilocular (several strains of *Clintonia*). The style is rudimentary (*Tulipa* spp.) or short to long, apically trilobate to trisected (*Medeola*). The stigma is developed along the free parts of the style; it is long, branchlike (*Medeola*) to short, crestlike (*Nomocharis*, *Cardiocrinum*, *Lilium*). Stigmatic surfaces are papillate, Dry (*Clintonia*, *Tulipa*, *Erythronium*, *Notholirion*) or Wet (*Fritillaria*, *Lilium*) (Heslop-Harrison and Shivanna 1977). The ovary is often narrowly ellipsoid or ellipsoid, sometimes ellipsoid to cylindric (*Nomocharis* spp., *Notholirion* spp., *Cardiocrinum*, *Lilium*) or oval-globose (Medeoloideae), comprising neither septal nectaries nor crystal raphides.

Floral Anatomy (based mostly on Utech and Kawano 1975b; Utech 1978). The pedicel base often contains a double ring of vascular bundles. The inner ring consists of 6 bundles, while in the outer sometimes 6 bundles are developed (*Medeola*), sometimes more numerous smaller ones (*Erythronium*).

The vascular supply of the tepals consists of 1 median and 2 lateral bundles. The lateral bundles sometimes branch secondarily in the tepal lamina. The median bundle of the outer tepal is derived from the pedicel without fusion and fission either from the inner ring (*Erythronium*) or from the outer one (*Medeola*). The lateral bundles of the outer tepal originate from the pedicel without fusion and fission. The median bundle of the inner tepal is sometimes derived from the inner ring of the pedicel after fission (*Erythronium*), sometimes from the outer ring without fusion and fission (*Medeola*). The lateral bundles of the inner tepal are derived from the inner or outer ring of the pedicel with or without fission. Each stamen contains 1 bundle, which is the product of a fusion of the bundles coming from the inner ring of the pedicel.

Each carpel has 1 dorsal bundle, which originates by fusion of the bundles from the inner ring. The ventral carpellary bundles, which supply the ovules, are often derivatives of the carpellary dorsal bundles (*Lilium*, *Tulipa*, *Erythronium japonicum*); in *Erythronium americanum* they are known to originate from additional lateral bundles extending from the outer ring of the pedicel.

Embryology. Apart from the epidermis and the fibrous endothecium, the anther contains 2 or 3 ephemeral middle layers and a glandular tapetum. The cells of the tapetum of *Lilium neilgherrense* were found to be binucleate (Subramanyam and Govindu 1949), those of *Tulipa clusiana* multinucleate (Wafai and Koul 1982). Cytokinesis in pollen mother cells is successive. The pollen grains are 2- or 3-celled at the dispersal stage.

The ovule is anatropous and bitegmic, with the micropyle formed by the inner integument. In the Lilioideae, the nucellus is usually small (rarely medium-sized: *Tulipa*), while it is large or medium-sized in the Medeoloideae (Björnstad

1970). There are no periclinal divisions in the apex of the nucellus, so that the 1-layered apex is retained on the growing embryo sac, which is a synapomorphic character of the Liliaceae. In *Lilium*, *Fritillaria*, *Erythronium*, *Tulipa*, *Gagea*, *Medeola*, and *Clintonia* the archesporial cell is known to function directly as the megaspore mother cell without cutting off a parietal cell (Björnstad 1970).

Although the common monosporic embryo sac development according to the Polygonum type is reported from close relatives of the Liliaceae such as the Calochortaceae and Colchicaceae, only tetrasporic embryo sacs are found in the Liliaceae. Among them, the Fritillaria type is the most frequent and common to the genera of Lilioideae and *Medeola* (Medeoloideae). *Clintonia*, the remaining member of the Medeoloideae, has the Clintonia type defined by Björnstad (1970), which resembles the Fritillaria type, and was once described as a reduced Fritillaria type by Maheshwari (1937). The Adoxa type and the Drusa type were also observed in addition to the Fritillaria type from the genera of Tulipeae (Lilioideae): the Adoxa type from *Erythronium* (eastern N American spp.) and *Tulipa* (Schaffner 1901; Bambacioni-Mezzetti 1931; Romanov 1938; Cooper 1939; Haque 1951), and the Drusa type from two *Tulipa* spp., *T. aitchisonii* and *T. clusiana* (Koul and Wafai 1981; Wafai and Koul 1982).

The synergids usually exhibit neither a filiform apparatus nor vacuoles, but rarely have a filiform apparatus in *Clintonia* and vacuoles in *Fritillaria*.

Double fertilization of the postmitotic type often occurs in the Fritillaria type embryo sacs of this family (Romanov 1961; Pechenitsyn 1972). The polar nuclei usually do not fuse before fertilization, but do so rarely in *Lilium* and *Lloydia*. Endosperm formation is of the Nuclear type.

Embryogeny of *Lilium* and *Erythronium* conforms to the Onagrad type. In *Erythronium americanum* (Jeffrey 1895) and *Tulipa clusiana* (Haccius and Hausner 1972), the embryo apex undergoes cleavage and produces 2 to 4 embryonic buds (cleavage polyembryony). In *Lilium* spp., an extra embryo (proembryo) develops from the synergid (Cooper 1943). However, these embryonic buds or proembryos do not reach maturity and collapse.

KARYOLOGY. $x = 12$ appears to be the basic chromosome number of the subfamily Lilioideae and $x = 7$ of the Medeoloideae, and numerical variation, apart from polyploidy, is brought about by ample aneuploid changes.

$x = 12$ is frequent in the Lilioideae, although $x = 9$, 11, and 13 are also reported from the genera *Cardiocrinum* and *Fritillaria*, where they are the result of Robertsonian rearrangements (Chauhan and Brandham 1985; Noda 1991). In the Medeoloideae, $x = 14$ of *Clintonia* appears to be derived from $x = 7$ through polyploidization.

The chromosomes are large (6.5–19 μm; Medeoloideae, Lilieae, *Erythronium*) or medium (2.2–8 μm; *Tulipa*, *Gagea*, *Lloydia*), although the Calochortaceae, the presumed closest relatives of Liliaceae, usually have rather small chromosomes (Tamura 1995).

Karyotypes are rather constant within the two subfamilies, with the exception of the tribe Tulipeae (Lilioideae). While the karyotypes of Medeoloideae show no peculiar differentiation, those of the tribe Lilieae (Lilioideae) are dimorphic. They have 2 long V-shaped (metacentric) chromosome pairs and 10 medium-sized I-shaped (telocentric or subtelocentric) pairs.

In the Tulipeae (always $x = 12$), at least 3 different karyotypes can be distinguished: (1) one having a single pair of long metacentric chromosomes (the majority of *Tulipa*, *Gagea*); (2) one consisting only of submetacentric or subtelocentric chromosomes (*Tulipa oculus-solis* group, *Erythronium*); (3) one having 3 metacentric chromosome pairs (*Tulipa clusiana* group), the remaining chromosomes varying from submetacentric to subtelocentric. The basic karyotypic structure of Tulipeae, as well as its chromosome number and size, is rather similar to that of *Tricyrtis* (Calochortaceae) (Tamura 1995).

The basic chromosome number of the Medeoloideae, $x = 7$, could be connected with $x = 12$ by stepwise Robertsonian rearrangements. Actually, 10 chromosomes out of the basic chromosome complement ($x = 12$) of the Lilieae are I-shaped, and their Robertsonian fusions could have led to the reduction from $x = 12$ to $x = 7$, although the Medeoloideae are hardly presumed to have a direct connection with the Lilieae.

POLLEN MORPHOLOGY. The pollen grains are spheroidal, 35–95 μm in longest axis, usually sulcate but rarely operculisulcate or provided with an irregular aperture (*Tulipa* spp.), usually reticulate but occasionally verrucate-areolate (*Tulipa* spp.) or gemmate (*Clintonia udensis*). When reticulate, the brochi of Lilieae (*Lilium* 4–15 μm; *Fritillaria* 4–5 μm; *Cardiocrinum* 3–4 μm) are often larger than those of the other liliaceous genera (*Lloydia* 1.5–2.5 μm; *Erythronium* 0.2–5 μm) (Shimakura 1973; Nakamura 1980; Kurosawa 1991).

In *Clintonia* and *Erythronium*, pollen grain characters of the Asian species seem to be advanced in comparison with the European and American species. Asian *Clintonia udensis* lacks an operculum, which is present in the remaining four American species (Takahashi and Sohma 1982). The reticulum of Asian *Erythronium japonicum* is composed of striate muri and a distinctive exine structure without columellae, while those of the European and American species have wavy muri and columellae (Takahashi 1987).

POLLINATION. All genera except *Medeola*, *Clintonia*(?), *Gagea*, *Tulipa*, and some species of *Lloydia* have perigonal nectaries, and all liliaceous genera are entomophilous; the flowers are often protandrous, which accounts for the outbreeding nature. Since the anthers in this family are usually extrorse, outbreeding is also promoted.

The flowers of *Erythronium* (Lilioideae-Tulipeae) are weakly protandrous; the stigma matures only after it has been elevated beyond the anthers, which may avoid autogamy. The flower is inverted with a reflexed perianth tip, and the stamens and pistil are downward-oriented and exserted. Auricles often exist near the base of the tepals (Fig. 98D) and function as a trap lid for the inverted nectary and a support for the stamens, which forces a bundle together with the style. In *E. japonicum*, the primary pollinator is *Xylocopa appendiculata* (Utech and Kawano 1975a; Fig. 98B), which holds fast to the exserted bundle of stamens and style and opens the auricle lid physically to seek nectar. Simultaneously, the anthers and stigma are touched by the thorax and abdomen of the bee and pollination is carried out.

Some species of *Fritillaria* (Lilioideae-Lilieae), which have large nectaries (4–12 × 1–4 mm) and show a distinct preponderance of fructose over glucose in nectar composition, are commonly pollinated by wasps (Vespidae). Other *Fritillaria* species, which have normal-sized nectaries (2–10 × 1–2 mm) and an overall balance between fructose and glucose in nectar composition, are pollinated by bumblebees (Apidae) (Rix and Rast 1975).

Medeola (Medeoloideae), based on its floral morphology, although nectar is lacking and scent is not perceptible, was presumed to be pollinated by insects such as flies and beetles (Berg 1962a).

FRUIT AND SEED. The capsule of the Lilioideae is loculicidal and narrowly ellipsoid to globose, sometimes 3-angled (*Gagea*, *Erythronium*) or 6-angled with more or less developed wings (*Fritillaria* spp., Fig. 99J). The fruit of the Medeoloideae is a globose, dark purplish blue berry, considered to be derived from a capsule.

The seeds of Lilieae and some genera of the Tulipeae (*Lloydia*, *Tulipa* sect. *Tulipa*, *T.* sect. *Eriostemones*, *Gagea* subgen. *Hornungia*) are flattened, orbicular to oblong, rounded deltoid or narrowly rhomboid in outline, placed like piles of coins on top of each other in the fruit, and often provided with a marginal wing (*Lilium*, *Nomocharis*, *Cardiocrinum*, *Fritillaria*). However, the seeds of the other genera of the Tulipeae (*Erythronium*, *Tulipa* sect. *Orithyia*, *T.* sect. *Amana*, *Gagea* subgen. *Gagea*) and Medeoloideae are often ellipsoidal or angled rather than flattened. They are not winged but sometimes have a lateral raphe crest or a raphal elaiosome (*Gagea* spp.), while a chalazal elaiosome occurs in *Erythronium* (except *E. grandiflorum*). The embryo of the ellipsoidal or angled seeds is always minute, while that of the flattened seeds is sometimes well developed (*Lloydia*, *Tulipa* sect. *Tulipa*, *T.* sect. *Eriostemones*, *Gagea* subgen. *Hornungia*, *Nomocharis*, *Lilium*).

The seed coat is formed by both integuments, although in general it lacks mechanical strength. The testa is thin, sometimes only 2-cell-layered. Its cells are often flattened or even collapsed, but sometimes palisadelike (exotesta of Medeoloideae). The cells of Liliaceae sometimes have lignified walls, especially on the outer epidermis. The tegmen is strongly compressed. The endosperm cells of mature seeds have thickened, pitted walls and contain aleurone and fatty oils but no starch (Berg 1962a; Huber 1969). The embryo is small and usually little differentiated, globular to ellipsoid, and only about 1/5–1/12 of the length of the seed.

DISPERSAL. The flatness of seeds in *Gagea* spp., *Lloydia*, *Tulipa* spp., *Fritillaria*, *Nomocharis*, *Notholirion*, *Cardiocrinum*, and *Lilium* may point to wind dispersal. The wings furnished along the seed margins in *Fritillaria*, *Nomocharis*, *Cardiocrinum* and *Lilium* seem to promote the dispersibility. The presence of elaiosomes in *Gagea* spp. and *Erythronium* points to dispersal by ants such as *Aphaenogaster* or *Formica*. This ant dispersal has been observed in *Erythronium japonicum* (Kawano et al. 1982). Berries in *Medeola* and *Clintonia* may be dispersed endozoochorically, although this has not been proved yet.

REPRODUCTIVE SYSTEMS. Both vegetative reproduction by rhizomes, bulbs, or bulbils and sexual reproduction by seeds occur, although the balance between the two is variable even among species. *Erythronium propullans* reproduces apparently only vegetatively, while *E. americanum* propagates not only vegetatively but also by seeds. In *E. japonicum* the recruitment of the individuals is exclusively dependent on sexual reproduction except for the accidental cases when old bulbs appended to a new main bulb become separated by mechanical strength (Kawano et al. 1982).

PHYTOCHEMISTRY (based mostly on Hegnauer 1963, 1986). The Liliaceae in the circumscription of this treatment appear as a homogeneous and distinct group by virtue of (1) the presence of flavonol glycosides based on quercetin and kaempferol (6 genera investigated); in the Colchicaceae, e.g., nearly exclusively flavone glycosides have been found to be present (Williams 1975); (2) the presence of di- and triferulic acid-sucrose esters, which have been demonstrated to accumulate in young anthers of *Tulipa*, *Erythronium*, *Fritillaria*, and *Lilium* species (Meurer et al. 1984); these compounds are rare in angiosperms and apart from the Liliaceae are known only from a few species of *Iris*; (3) the presence of the fungitoxic lactone α-methylene-γ-butyrolactone in *Gagea*, *Tulipa*, and *Erythronium* species, elsewhere known to occur in *Alstroemeria* and *Bomarea* species (Slob et al. 1975); (4) the absence of chelidonic acid.

Steroidal sapogenins are widespread in the bulbs and shoots of many species of the family, as in many members of the Lilianae. A great variety of steroidal alkaloids is known from species of *Fritillaria* and *Notholirion*, where these compounds are responsible for the toxicity of the bulbs.

Flower pigments (in *Tulipa* cultivars) are based on delphinidin, cyanidin, and pelargonidin.

DISTRIBUTION AND HABITATS. The Liliaceae are a holarctic element. The Medeoloideae and *Erythronium* are concentrated in N America, and especially *Medeola* is restricted to eastern N America. *Erythronium* is abundant in western N America.

The other eight genera are concentrated in Eurasia, and *Gagea*, *Tulipa*, *Nomocharis*, *Notholirion*, and *Cardiocrinum* occur exclusively there. *Gagea* and *Tulipa* extend mostly from the Mediterranean region to Central Asia; *Cardiocrinum* is distributed from the Himalayas to Japan, *Notholirion* from W Asia to western China. *Nomocharis* is restricted to Assam, Upper Burma, and western China. Among *Lloydia* species, only *L. serotina* reaches N America.

Most genera of the Liliaceae occur in temperate and arctic (alpine) zones, but *Lilium* extends into subtropical regions, while *Nomocharis* is restricted to the alpine belt (3700–4300 m alt.).

The members of the Lilioideae grow mostly on the forest floor, at forest margins, in grassland or alpine meadows. *Gagea*, *Tulipa*, and *Fritillaria* can be found also in steppe, *Tulipa* mainly there; *Gagea*, *Fritillaria*, and *Lilium* also on marshy ground; *Gagea* and *Lilium* also in sand dunes. *Lloydia* extends into alpine tundras. *Cardiocrinum* occurs only on the forest floor.

The genera of the Medeoloideae are restricted to the forest floor. *Gagea*, *Tulipa*, and several species of *Erythronium* are so-called spring geophytes.

SUBDIVISION AND RELATIONSHIPS WITHIN THE FAMILY. The Liliaceae can be subdivided into two subfamilies.

1. Subfam. Medeoloideae comprises *Medeola* and *Clintonia*. The two genera were sometimes considered as remote (Krause 1930; Huber 1969). However, the embryological (Berg 1962b; Björnstad 1970) and the karyological (Tamura 1995) similarities between them indicate their close relationship, as already proposed by Bentham and Hooker (1883).

On the other hand, Bentham and Hooker (1883) indicated an affinity between these two genera and the Trilliaceae. However, the embryological and karyological evidence points to a close relationship between *Medeola*, *Clintonia*, and Liliaceae, among which a relationship between *Medeola* and Liliaceae is also supported by *rbc*L sequencing data of cpDNA (Chase et al. 1993, 1995; Duvall et al. 1993).

2. Subfam. Lilioideae in the circumscription of this treatment is a lineage that has long been well recognized (Bentham and Hooker 1883; Huber 1969; Takhtajan 1980; Dahlgren et al. 1985; Thorne 1992), although also *Calochortus* was often included in it (Bentham and Hooker 1883; Thorne 1992); today *Calochortus* is considered as a member of the sister group of Liliaceae.

The Lilioideae can be further divided into two tribes. The Lilieae include *Fritillaria*, *Nomocharis*, *Notholirion*, *Cardiocrinum*, and *Lilium*, which share almost identical karyotypes and are considered as monophyletic. The Tulipeae include *Gagea*, *Lloydia*, *Tulipa*, and *Erythronium*. Within the Tulipeae, *Gagea* seems to be closely related to *Lloydia* and *Tulipa* to *Erythronium*, which is concordant with the suggestion of Melchior(1964).

Affinities. The closest realtives of the Liliaceae are undoubtedly the Calochortaceae, which comprise *Calochortus*, *Prosartes*, *Scoliopus*, *Streptopus*, and *Tricyrtis*. Lilioideae and *Calochortus* share a bulb; Medeoloideae, Tulipeae, *Fritillaria*, and *Calochortus* share pseudobasifixed anthers; some genera of the Liliaceae and *Tricyrtis* share dark-spotted foliage leaves and purple-spotted tepals. Lilieae, *Lloydia*, *Tulipa* spp., *Gagea* spp., and *Tricyrtis* share the flattened seeds, which are orbicular to oblong, rounded deltoid or narrowly rhombate in outline and placed like piles of coins on top of each other in the fruit; the seed morphological characters of *Medeola* and *Scoliopus* are similar to each other (Huber 1969). The karyological characters of *Medeola*, *Clintonia*, and *Scoliopus* resemble one another; those of Tulipeae and *Tricyrtis* resemble one another (Tamura 1995).

On the other hand, seed morphological (Huber 1969) and karyological (Tamura 1995) characters of the Calochortaceae (except *Scoliopus* and *Tricyrtis*) are different from those of the Liliaceae; embryological characters of the Calochortaceae differ from those of the Liliaceae (Berg 1962b; Björnstad 1970). *Scoliopus* and *Tricyrtis* (Calochortaceae) seem to occupy somewhat intermediate positions between the Liliaceae and the Calochortaceae.

In addition to the Calochortaceae, also the Colchicaceae are more or less related to the Liliaceae.

Curiously enough, the *rbc*L sequence data of cpDNA indicate a relationship between *Lilium*, *Medeola*, and *Smilax* (Chase et al. 1993, 1995; Duvall et al. 1993), and also *Tricyrtis* and *Smilax* share similar karyotypes and net-veined foliage leaves. It is possible that the Liliaceae and Calochortaceae are unexpectedly related to Smilacaceae.

Economic Importance. Bulbs of the Liliaceae are often used as drugs and food. Especially, bulbs of *Fritillaria* have commonly been used in China as drugs since ancient times. They are prescribed principally for cough and fever and sold under the name Bei-mu. Eighteen Chinese species are known to possess medicinal value (Chen and Hsia 1977). *F. thunbergii* contains several kinds of alkaloids.

Bulbs of *Lilium* spp., *Lloydia tibetica*, *Tulipa edulis*, and *Cardiocrinum giganteum* are also used as drugs for treating cough, fever, and injuries, and as diuretics and tonics. Those of *Lilium* spp., *Tulipa iliensis*, *T. edulis*, and *Fritillaria kamtschatcensis* are edible, sometimes much valued. In Iran, bulbs of *Tulipa* are roasted and eaten. Starch of good quality can be obtained from bulbs of *Erythronium japonicum* and *Cardiocrinum cordatum*.

Capsules of *Cardiocrinum cathayanum* and total plants of *Notholirion macrophyllum* may be used as drugs. Flowers of *Lilium dahuricum* are edible. Young foliage leaves of *Erythronium japonicum* are also edible when boiled. Foliage leaves of *Clintonia borealis* can be eaten as salad. Bulbs of *Lilium martagon* yield a black dye. Aromatics are often obtained from flowers of *Lilium*.

Tulipa, *Fritillaria*, and *Lilium* are frequently grown as ornamentals. *L. candidum*, one of the oldest ornamentals of the Western world, hybrids of *Lilium* spp., and some E Asian wild *Lilium* spp. such as *L. longiflorum* have recently become most important cut flowers. Tulips were introduced by Busbequius to the Netherlands in 1554 and have become there one of the most important articles of commerce.

Key to the Genera

1. Underground stem a rhizome; fruit a berry **(Medeoloideae)** 2
- Underground stem a bulb; fruit a capsule **(Lilioideae)** 3
2. Rhizome tuberous, decaying every year; foliage leaves cauline; flowers nodding; tepals pale greenish yellow, recurved; style trisected **1. *Medeola***
- Rhizome thickened, surviving for several years; foliage leaves basal; flowers erect to horizontal; tepals white, spreading; style trilobate **2. *Clintonia***
3. Anthers pseudobasifixed; seeds flattened, ellipsoidal or angled, not winged; if flattened, embryo well developed; axis of seed usually straight or slightly curved (Tulipeae) 4
- Anthers dorsifixed versatile or pseudobasifixed; if pseudobasifixed (*Fritillaria* spp.), seeds winged and embryo minute; seeds flattened; axis of seed moderately to strongly curved (Lilieae) 7
4. Foliage leaves 2, apparently opposite, more or less net-veined; flower nodding; tepals strongly recurved, nectariferous **6. *Erythronium***
- Foliage leaves 1-many, alternate, or apparently opposite or verticillate, parallel-veined; flower erect or horizontal; tepals spreading or more or less incurved, mostly without nectaries (except *Lloydia* spp.) 5
5. Bulb large, more than 1 cm in width; perianth campanulate to infundibular; tepals large, more than 2 cm in length, caducous; anthers linear to narrowly ellipsoid; style lacking or more or less developed **5. *Tulipa***
- Bulb small, less than 1 cm in width; perianth stellately expanded to infundibular; tepals small, less than 2 cm in length, persistent or marcescent; anthers ellipsoid to globose; style developed 6
6. Cauline foliage leaves present; perianth infundibular often with rounded segments at apex; tepals marcescent in fruit, often shorter than fruit; nectary present or absent; seeds flattened **4. *Lloydia***
- Cauline foliage leaves absent or present; perianth stellately expanded often with acute segments at apex; tepals persis-

tent, becoming green from yellow or white and thickening in fruit, often longer than fruit; nectary absent; seeds ellipsoidal, sharp-angled or flattened **3. *Gagea***

7. Flower nodding; perianth campanulate with segments erect in upper part; a nectarial pit, groove (furrow), or pouch (short spur) present at tepal base; anthers usually pseudobasifixed, rarely versatile; fruit sometimes winged; embryo minute **7. *Fritillaria***
- Flower erect, horizontal or nodding; perianth usually infundibular or more flat, rarely tubular; a nectarial pit and pouch (short spur) absent, but sometimes a nectarial groove or processes present at tepal base; anthers versatile; fruit not winged; embryo in ripe seed well developed, half the length of endosperm, or minute 8

8. Perianth usually horizontally patent, rarely cupular, not recurved; inner tepals entire, erose or erose-lacerate; nectarial processes present only at inner tepal base; filaments often swollen cylindrically, sometimes subulate, bearing an aciculus at apex; embryo well developed **8. *Nomocharis***
- Perianth usually infundibular, rarely tubular or cupular, sometimes recurved; tepals entire; a nectarial groove usually present at every tepal base, sometimes fringed with papillae or hairs; filaments usually filiform, rarely slightly flattened; embryo well developed or minute 9

9. Bulb with membranaceous tunica; basal foliage leaves present at flowering stage; style trifid, apically recurved; seeds not winged; embryo minute **9. *Notholirion***
- Bulb naked or with fibrous tunica; basal foliage leaf absent at flowering stage; style apically with 3 stigmatic crests; seeds winged; embryo well developed or minute 10

10. Bulb with fibrous tunica; foliage leaves ovate, cordate at base, net-veined, long petiolate, marcescent after anthesis; perianth tubular, not recurved; embryo minute **10. *Cardiocrinum***
- Bulb naked; foliage leaves linear to elliptic, attenuate at base, usually parallel-veined, sessile or shortly petiolate, persistent after anthesis; perianth usually infundibular, rarely tubular or cupular, sometimes recurved; embryo well developed **11. *Lilium***

I. Subfam. Medeoloideae (Benth.) M.N. Tamura, stat. nov.

Medeoleae Benth. in Benth. et Hook. f., Gen. Pl. 3: 762 (1883).

Underground stem a rhizome. Aerial stem erect. Foliage leaves rosettes, but in *Medeola* stalked rosettes. Bracts present. Tepals of both whorls similar, without a nectary(?), caducous. Stamens of both whorls similar, Filaments filiform, smooth. Anthers ellipsoid, pseudobasifixed, extrorse. Overy oval-globose. Ovules several. Fruit a globose, dark purple berry. Seeds ovoid, rounded or more or less angled. Embryo minute. Usually $x = 7$.

1. *Medeola* L.

Medeola L., Sp. Pl.: 339 (1753); R.Y. Berg, Skr. Norske Vidensk.-Akad. Oslo I. Mat.-Naturv. Kl. Ny Ser. 3: 1–55 (1962).

Rhizome tuberous, decaying every year. Aerial stem at vegetative stage producing a whorl of foliage leaves, followed by a second whorl subtending the flowers. Inflorescence a sessile sympodial umbel. Flowers 3–9, nodding. Perianth recurved. Tepals pale greenish yellow. Style trisected. Ovary 1-locular. Fruit erect after straightening of the pedicel. $x = 7$. Only one sp., *M. virginiana* L., forest floor of eastern N America.

2. *Clintonia* Rafin.

Clintonia Rafin., J. Phys.: 102 (1819).

Rhizome short, surviving for several years. Foliage leaves crowded at basis of orthotropic stem. Inflorescence often a raceme, sometimes an umbel [*C. umbellulata* (Michx.) Morong] or 1-flowered (*C. uniflora* (Menzies ex Schultes) Kunth). Flowers erect to horizontal. Perianth infundibular. Tepals white. Style trilobate. Ovary 2- or 3-locular. $x = 14$, rarely $x = 16$ [several strains of *C. borealis* (Ait.) Rafin.]. Five spp., forest floor of E Asia and N America, abundant in the latter area (four spp.) from temperate to subarctic zones.

II. Subfam. Lilioideae Engl. (1886).

Underground stem a bulb. Stem erect. Foliage leaves attached to middle part of aerial stem (cauline), sometimes also to its basal part (basal). Perianth often infundibular, sometimes tubular, campanulate or cupular, sometimes horizontally patent, sometimes recurved. Anthers extrorse or latrorse. Style trilobate or with three stigmatic crests to trifid. Ovary 3-locular. Ovules numerous. Fruit a loculicidal capsule. Usually $x = 12$.

1. Tribe Tulipeae Koch (1837).

Bulb tunicated. Inflorescence often restricted to a single terminal flower, sometimes a raceme or a corymb to an umbel. Bracts present or absent. Tepals persistent, marcescent or caducous, sometimes similar in width in both whorls, but sometimes outer wider (*Gagea*, *Erythronium*) or inner wider (*Lloydia*, *Tulipa*). Perigonal nectary often absent. Stamens of both whorls similar or different in length. Anthers pseudobasifixed. Ovary often narrowly ellipsoid, sometimes ellipsoid or obovoid (*Gagea*). Seeds flattened, ellipsoidal or sharp-angled, 2–8 mm long, not winged, sometimes with an elaiosome; axis of seeds usually straight or slightly curved. $x = 12$.

3. *Gagea* Salisb. Fig. 14K,M

Gagea Salisb., Ann. Bot. 2: 555 (1806).

Bulbils present inside or outside a bulbar tunica, sometimes also in cauline foliage leaf axils. Foliage leaves alternate; basal ones 1–4, cauline ones often absent, sometimes present (*G.* subgen. *Hornungia*). Inflorescence a raceme or corymose to umbelliform, sometimes 1-flowered. Bracts often present, sometimes absent (*G.* subgen. *Hornungia*). Perianth usually stellately spreading and flat. Tepals usually acute at apex, yellow or greenish, rarely white [*G. graeca* (L.) Dandy, *G. trinervia* (Viv.) Greuter], without a nectary, persistent, becoming green and thickening in fruit, often longer than a fruit. Filaments filiform or thickened in basal part. Anthers ellipsoid to globose. Style distinct, trilobate to trifid. Capsule ellipsoid to obovoid, usually 3-angled. Seeds often ellipsoidal or sharp-angled, sometimes flattened (*G.* subgen. *Hornungia*), sometimes with a lateral raphe crest or a raphal elaiosome. Embryo in *G.* subgen. *Hornungia* well developed, while minute in the myrmecochorous species (*G.* subgen. *Gagea*). Two subgen., about 90 spp., in temperate Eurasia.

4. *Lloydia* Salisb. ex Reichenb.

Lloydia Salisb. ex Reichenb., Fl. German. Excursoria: 102 (1830).

Tunica of bulb elongated upward, tubular. Foliage leaves alternate; basal ones 1–8 in addition to cauline ones. Inflorescence umbelliform, often 1-flowered. Bracts present. Perianth infundibular. Tepals rounded at apex, white, yellow or greenish, sometimes with a pit-shaped nectary, marcescent, often shorter than a fruit. Filaments filiform sometimes hairy. Anthers ellipsoid. Style distinct, trilobate. Capsule narrowly ellipsoid to broadly obovoid. Seeds flattened with a well-developed embryo. About 20 spp., temperate regions of the Northern Hemisphere, abundant in Eurasia, one sp., Europe [*L. serotina* (L.) Salisb. ex Reichenb.].

5. *Tulipa* L. Figs. 13 G–I, 98H

Tulipa L., Sp. Pl.: 305 (1753).
Amana Honda (1935).

Basal foliage leaves absent. Cauline foliage leaves 1 (*T. regelii* Krassn.)–12 (*T. schmidtii* Fomine), often 2–6, often alternate, sometimes apparently opposite. Flowers often solitary, sometimes 2 or more, rarely to 8 (*T. bifloriformis* Vved., *T. tarda* Stapf). Bracts usually absent, rarely present (*T.* sect. *Amana*). Perianth often campanulate, sometimes infundibular, white, yellow, or red, without a nectary, caducous. Stamens of both whorls similar (*T.* sect. *Tulipa*, sect. *Orithyia*) or different (*T.* sect. *Eriostemones*, sect. *Amana*) in length. Filaments thickened in middle or basal part, sometimes hairy (*T.* sect. *Eriostemones*, sect. *Lophophyllon*). Anthers linear to narrowly ellipsoid. Style often lacking, sometimes equalling the ovary in length (*T.* sect. *Orithyia*, sect. *Amana*), trifid. Capsule ellipsoid to subglobose. Seeds often flattened, sometimes ellipsoidal or sharp-angled (*T.* sect. *Orithyia*, sect. *Amana*). Embryo well developed in the flat-seeded species, not seen in *T.* sect. *Orithyia* and sect. *Amana*. Seven sect., about 150 spp., temperate Eurasia, abundant in W and Central Asia.

6. *Erythronium* L. Fig. 98A–G

Erythronium L., Sp. Pl.: 305 (1753); Applegate, Madroño 3: 58–113 (1935), rev. W N Amer. spp.

Bulb elongated, deep-seated. Foliage leaf at vegetative stage single; 2 at flowering stage, apparently opposite on the middle of aerial stem, more or less net-veined. Flowers nodding, usually solitary, rarely 2–5 in a raceme, usually ebracteate. Tepals spreading infundibuliform at anthesis, eventually reflexed, white, yellow, pink, or violet, caducous. Perigonal nectary present, often auricle-shaped on the tepal base. Stamens of both whorls different in length. Filaments often filiform, sometimes thickened in middle or basal part. Anthers linear to ellipsoid. Style distinct, more or less trifid. Capsule ellipsoid to subglobose, 3-angled. Seeds ellipsoidal or sharp-angled, with a chalazal elaiosome (except *E. grandiflorum* Pursh). Embryo minute. Twenty four spp., temperate regions of the Northern Hemisphere, abundant in western N America (15 spp.), two spp. Europe (*E. dens-canis* L., *E. caucasicum* Woronow).

2. Tribe Lilieae Ritgen (1831).

Bulb tunicated or naked. Basal foliage leaves usually absent. Inflorescence often a raceme, sometimes 1-flowered (spp. of *Fritillaria*, *Nomocharis*, *Lilium*). Bracts usually present [except *Cardiocrinum giganteum*) (Wall.) Makino]. Tepals usually caducous, rarely marcescent (*Fritillaria delavayi* Franch.), usually similar in width in both whorls or inner wider (except *Fritillaria pallidifora* Schrenk). Perigonal nectary present. Stamens of both whorls similar. Anthers often dorsifixed versatile, sometimes pseudobasifixed. Ovary often cylindric, sometimes narrowly ellipsoid (*Fritillaria* or *Notholirion* spp.), rarely turbinate (*Nomocharis* spp.). Style distinct. Seeds flattened, 5–15 mm long, usually winged; axis of seeds moderately to strongly curved. $x = 12$, rarely $x = 9, 11, 13$.

7. *Fritillaria* L. Fig. 99

Fritillaria L., Sp. Pl.: 303 (1753); D.E. Beetle, Madroño 7: 133–159 (1944), rev. N Am. spp.
Rhinopetalum Fisch. ex Alexand. (1829).
Korolkowia Regel (1873).

Bulb usually tunicated, rarely naked (*F. imperialis* L., *F. persica* L.), often of 2 or many fleshy scales, sometimes of a single scale (*F. persica* L.). Cauline foliage leaves often apparently opposite or verticillate, sometimes with cirrhose apex. Inflorescence a raceme or an umbel, often 1-flowered. Flowers nodding, usually actinomorphic, rarely weakly zygomorhic (*F.* sect. *Theresia*). Perianth campanulate with erect segments in upper part, white, yellow, greenish, purple or reddish, with nectarial pits, grooves (furrows), (*F. sewerzowii* Regel), or pouches (1 of them developed into a short spur) (*F.* sect. *Theresia*) at base. Filaments filiform or slightly flattened, sometimes papillose, rarely hairy [*F. karelinii* (Fisch.) Baker]. Anthers usually linear to ellipsoid, rarely subglobose (*F.* sect *Theresia*), usually pseudobasifixed, rarely dorsifixed versatile (*F. fusca* Turrill, *F.* sect. *Liliorhiza* spp.). Style trilobate to trifid. Capsule obovoid to globose, 6-angled, sometimes winged, erect after straightening of the pedicel. Seeds winged. Embryo minute. Four sect., about 130 spp., temperate regions of the Northern Hemisphere.

8. *Nomocharis* Franch.

Nomocharis Franch., J. Bot. (Morot) 3: 113, t. 3 (1889); Sealy, Bot. J. Linn. Soc. 87: 285–323 (1983), rev.

Bulb naked, consisting of several imbricate fleshy scales. Cauline foliage leaves often apparently verticillate, sometimes alternate [*N. saluenensis* Balf. f., *N. aperta* (Franch.) E. Wilson, *N. synaptica* Sealy]. Flowers often solitary, sometimes in a raceme. Perianth often horizontally patent, sometimes cupular (*N. saluenensis* Balf. fl., *N. basilissa* W.E. Evans). Tepals white, yellow, pink, or red; inner ones dark-colored and glistening at the base, entire, erose or erose-lacerate, more or less blotched inside in purplish red (except *N. basilissa* W.E. Evans). Nectarial processes present only at inner tepal base, often arranged flabellately but sometimes alongside a groove (channel) [*N. saluenensis* Balf. f., *N. aperta* (Franch.) E. Wilson]; nectaries usually absent at outer tepals. Filaments often swollen cylindrically, sometimes subulate [*N. saluenensis* Balf. f., *N. aperta* (Franch.) E. Wilson, *N. synaptica* Sealy], bearing an aciculus at apex. Anthers ellipsoid, dorsifixed versatile. Style apically with 3 stigmatic crests. Capsule ellipsoid. Seeds winged. Embryo well developed. Seven spp., alpine zone of Assam, Upper Burma, and western China.

9. *Notholirion* Wall. ex Boiss.

Notholirion Wall. ex Boiss., Fl. Orient. 5: 190 (1882).

Bulb with dark brown scarious tunica. Several to many bulbils outside the tunica present. Foliage leaves alternate; basal ones present at flowering stage in addition to cauline ones. Flowers 2–24 in a raceme. Perianth infundibular. Tepals pale purple to red. Filaments filiform. Anthers narrowly ellipsoid to ellipsoid, dorsifixed versatile. Style trifid, apically recurved. Capsule ellipsoid or obovoid. Seeds not winged. Embryo minute. Five spp., W Asia, Himalayas to western China.

10. *Cardiocrinum* Endl.

Cardiocrinum Endl., Gen. Pl.: 141 (1837).

Bulb with fibrous tunica. Cauline foliage leaves alternate, cordate at base, net-veined, long petiolated, convolute in bud, marcescent after anthesis. Flowers 3–20 in a raceme. Perianth tubular. Tepals white. Filaments more or less flattened. Anthers narrowly ellipsoid, dorsifixed versatile. Style apically with 3 stigmatic crests, elongate after anthesis. Capsule ellipsoid to subglobose. Seeds winged. Embryo minute. Three spp., Himalayas, China to Japan.

11. *Lilium* L. Figs. 12H, 13J–M, 100

Lilium L., Sp. Pl.: 302 (1753).

Bulb naked, of numerous imbricate fleshy scales. Rhizome sometimes connecting 2 bulbs (*L.* sect. *Pseudolirion*). Cauline foliage leaves often alternate, sometimes apparently verticillate. Bulbils sometimes present in the leaf axils. Flowers solitary or in a raceme. Perianth usually infundibular, rarely tubular or cupular, sometimes recurved. Tepals white, yellow, greenish, orange, pink, red, or purplish, nectariferous; nectaries usually narrowly grooved at tepal bases, sometimes fringed with papillae or hairs, rarely flat at outer tepals (*L. bolanderi* S. Wats., *L. grayi* S. Wats., *L. canadense* L., *L. parvum* Kellogg, *L. maritimum* Kellogg). Filaments filiform, sometimes hairy. Anthers linear to ellipsoid, dorsifixed versatile. Style apically with 3 stigmatic crests. Capsule narrowly ellipsoid to subglobose. Seeds winged. Embryo well developed. Six sect., about 110 spp., temperate regions of the Northern Hemisphere, abundant in E Asia.

Selected Bibliography

Bambacioni-Mezzetti, V. 1931. Sullo sviluppo dell' embrione in *Tulipa gesneriana*. Ann. Bot. (Rome) 19: 145–155.

Bentham, G., Hooker, J.D. 1883. Genera Plantarum, Vol. 3. London: Reeve; Williams & Norgate.

Berg, R.Y. 1962a. Morphology and taxonomic position of *Medeola*, Liliaceae. Skr. Nor. Vidensk.-Akad. Oslo Mat.-Naturvidensk. Kl. Ny Ser. 3: 1–55.

Berg, R.Y. 1962b. Contribution to the comparative embryology of the Liliaceae: *Scoliopus*, *Trillium*, *Paris*, and *Medeola*. Skr. Nor. Vidensk.-Akad. Oslo Mat.-Naturvidensk. Kl. Ny Ser. 4: 1–64.

Björnstad, I.N. 1970. Comparative embryology of Asparagoideae-Polygonateae, Liliaceae. Nytt Mag. Bot. 17: 169–207.

Chase, M.W. et al. 1993, 1995. See general references.

Chauhan, K.P.S., Brandham, P.E. 1985. The significance of Robertsonian fusion and monosomy in *Cardiocrinum* (Liliaceae). Kew Bull. 40: 567–571.

Cheadle, V.I., Kosakai, H. 1971. Vessels in Liliaceae. Phytomorphology 21: 320–333.

Chen, S.C., Hsia, K.C. 1977. A taxonomic study of the Chinese drug Bei-mu (*Fritillaria*). Acta Phytotaxon. Sin. 15: 31–46 (in Chinese with English summary).

Cooper, D.C. 1939. Development of megagametophyte in *Erythronium albidum*. Bot. Gaz. 100: 862–867.

Cooper, D.C. 1943. Haploid-diploid twin embryos in *Lilium* and *Nicotiana*. Am. J. Bot. 30: 408–413.

Dahlgren, R.M.T. et al. 1985. See general references.

Duvall, M.R. et al. 1993. See general references.

Haccius, B., Hausner, G. 1972. Frühembryogenese und Polyembryonie-Problem in der Gattung *Tulipa* L. Beitr. Biol. Pflanz. 48: 207–228.

Haque, A. 1951. The embryo sac of *Erythronium americanum*. Bot. Gaz. 112. 495–500.

Hegnauer, R. 1963, 1986. See general references.

Heslop-Harrison, Y., Shivanna, K.R. 1977. See general references.

Huber, H. 1969. See general references.

Jeffrey E.C. 1895. Polyembryony in *Erythronium americanum*. Ann. Bot. (Lond.) 9: 537–541.

Kawano, S., Hiratsuka, A., Hayashi, K. 1982. Life history characteristics and survivorship of *Erythronium japonicum*. Oikos 38: 129–149.

Koul, A.K., Wafai, B.A. 1981. Analysis of breeding system in the genus *Tulipa*. I. Sporogenesis, gametogenesis and embryogeny in diploid *T. aitchisonii*. Phytomorphology 31: 11–17.

Krause, K. 1930. See general references.

Kurosawa, K. 1991. SEM photomicrographs of pollen of angiosperms. Osaka: Osaka Museum of Natural History.

Maheshwari, P. 1937. A critical review of the types of embryo sacs in angiosperms. New Phytol. 36: 359–417.

Melchior, H. 1964. 3. Reihe Liliiflorae (Liliales). In: Melchior, H., Werdermann, E.(eds.) A. Engler's Syllabus der Pflanzenfamilien, 12th edn., vol. 2. Berlin: Borntraeger, pp. 512–543.

Meurer, B., Strack, D., Wiermann, R. 1984. The systematic distribution of ferulic acid-sucrose esters in anthers of the Liliaceae. Planta Med. 50: 376–380.

Nakamura, J. 1980. Diagnostic characters of pollen grains of Japan, parts I, II. Osaka: Osaka Museum of Natural History.

Noda, S. 1991. 27 Chromosomal variation and evolution in the genus *Lilium*. In: Tsuchiya, T., Gupta, P.K. (eds.) Chromosome engineering in plants: genetics, breeding, evolution, part B. Amsterdam: Elsevier, pp. 507–524.

Pechenitsyn, U.P. 1972. Double fertilization in species of *Tulipa* with *Fritillaria*-type embryo sac. Bot. Zh. 57: 465–469 (in Russian).

Rix, E.M., Rast, D. 1975. Nectar sugars and subgeneric classification in *Fritillaria*. Biochem. Syst. Ecol. 2: 207–209.

Romanov, I.D. 1938. Eine neue Form des Embryosackes von Adoxa-Typus bei *Tulipa tetraphylla* und *T. ostrovskiana*. Dokl. Akad. Nauk SSSR (Bot.) 19: 113–115.

Romanov, I.D. 1961. The origin of a particular structure of endosperm nuclei in *Gagea*. Dokl. Akad. Nauk SSSR 141: 984–986.

Schaffner, J.H. 1901. A contribution to the life history and cytology of *Erythronium*. Bot. Gaz. 31: 369–387.

Shimakura, M. 1973. Palynomorphs of Japanese plants. Osaka: Osaka Museum of Natural History.

Slob, A., Jekel, B., De Jong, B., Schlatmann, E. 1975. On the occurrence of tuliposides in the Liliiflorae. Phytochemistry 14: 1997–2005.

Subramanyam, K., Govindu, H.C. 1949. The development of the male and female gametophyte in *Lilium neilgherrense* Wight. J. Indian Bot. Soc. 28: 36–41.

Takahashi, M. 1987. Pollen morphology in the genus *Erythronium* (Liliaceae) and its systematic implications. Am. J. Bot. 74: 1254–1261.

Takahashi, M., Sohma, K. 1982. Pollen morphology of the genus *Clintonia*. Sci. Rep. Tohoku Univ. 38: 157–164.

Takhtajan, A.L. 1980. Outline of the classification of flowering plants (Magnoliophyta). Bot. Rev. 46: 225–359.

Tamura, M.N. 1995. See general references.

Thorne, R.T. 1992. See general references.

Utech, F.H. 1978. Floral vascular anatomy of *Medeola virginiana* L. (Liliaceae-Parideae = Trilliaceae) and tribal note. Ann. Carnegie Mus. 47: 13–28.

Utech, F.H., Kawano, S. 1975a. Biosystematic studies in *Erythronium* (Liliaceae-Tulipeae) I. Floral biology of *E. japonicum* Decne. Bot. Mag. (Tokyo) 88: 163–176.

Utech, F.H., Kawano, S. 1975b. Biosystematic studies in *Erythronium* (Liliaceae-Tulipeae) II. Floral anatomy of *E. japonicum* Decne. Bot. Mag. (Tokyo) 88: 177–185.

Wafai, B.A., Koul, A.K. 1982. Analysis of breeding system in *Tulipa*. II. Sporogenesis, gametogenesis and embryogeny in tetraploid *T. clusiana*. Phytomorphology 32: 289–301.

Williams, C.A. 1975. Biosystematics of the monocotyledoneae-flavonoid patterns in leaves of the Liliaceae. Biochem. Syst. Ecol. 3: 229–244.

Lomandraceae

J.G. Conran

Lomandraceae Lotsy, Vortr. Bot. Stammesgesch. 3: 761 (1911).

Erect, tufted, shrubby or arborescent *perennials* to about 10 m tall (*Cordyline*); roots fibrous or thickened and fleshy; vegetative axis elongate or condensed; *leaves* spiral or distichous, more or less sheathing, sometimes surrounded basally by a scarious sheath of former leaf bases; leaf blades dorsiventral, cylindrical or linear to linear-lanceolate, margins entire, scarious or prickly; ligules rarely present (*Sowerbaea, Laxmannia*). *Inflorescence* erect, simple or compound spikes, racemes, bracteate umbels or cymes; *flowers* hermaphrodite or dioecious, hypogynous, small to medium, white, blue, pink, mauve or purple; perianth of 2 similar trimerous whorls, persistent, marcescent, free or united, petaloid, scarious or fleshy, margins entire or fimbriate. Stamens 3 + 3 or 0 + 3, hypogynous, or all, or the inner 3 sometimes attached to the tepals, the outer 3 sometimes reduced to staminodes. Anthers basifixed, medifixed or dorsifixed, dehiscing introrsely or extrorsely, by slits (most genera) or pores (*Eustrephus, Thysanotus*), sometimes with hairy filaments or staminal appendages (*Arthropodium* and *Trichopetalum*). Ovary syncarpous, 3-locular with several axile ovules per locule. Style single, with 1–3 stigmatic lobes. Ovules anatropous or more or less campylotropous, single to numerous per locule. *Fruit* a berry, loculicidal capsule or nutlet. Seeds numerous to few, pale yellow to brown or black; testa thin, with or without phytomelan; endosperm copious, without starch; embryo linear, 1/2 to 1/3 the length of the seed, sometimes longer, straight, curved or more or less coiled.

A tropical to temperate family of 14 genera and ca. 180 species from Australasia, SE Asia, the Mascarènes, New Caledonia, New, Guinea, New Zealand, S and N America and the Pacific Islands.

Fig. 101A–N. Lomandraceae. **A–C** *Lomandra glauca*. **A** Habit. **B** Pistil. **C** Ovary, transversal section. **D–F** *L. micrantha*. **D** Habit. **E** Flower. **F** Stamen. **G, H** *Acanthocarpus preissii*. **G** Fruiting branch. **H** Dehisced fruit. **I, J** *Thysanotus tuberosus*. **I** Habit. **J** Flower. **K–N** *Sowerbaea juncea*. **K** Habit. **L** Leaf sheath. **M** Flower. **N** Stamen. (Takhtajan 1982)

Vegetative Morphology. The plants are generally rhizomatous or tufted-caespitose, with single or branched aerial stems (Fig. 101). In *Cordyline* the plants are tall woody shrubs or small trees. The rhizomes are subaerial with well-developed leaves and sometimes supported by short prop roots or are largely subterranean with scalelike leaves. The peculiar negatively geotropic branched rhizome of some *Cordyline* species resembles stilt roots, but its axial nature is well supported (Troll 1937: 722–724). In many taxa, the roots are fleshy or tuberous, the tubers often distant from the rhizome near the apex of otherwise fibrous or wiry roots. Stilt roots are present in most *Laxmannia* species (Pate et al. 1984), one *Romnalda* species, and an undescribed New Caledonian *Arthropodium* species.

The leaves are alternate, and in many of the *Lomandra* group are distichous or spirodistichously arranged. H. Huber (pers. comm.) observed that in the New Caledonian *Lomandra*, apart from the 2 major orthostichies, a 3rd orthostichy exists which bears smaller leaves. The leaves are linear or strap-shaped and range from a few mm to 1 m in length, although some *Cordyline* and *Arthropodium* species have broadly ovate-lanceolate leaves. The apex is sometimes erose and the margins sometimes lacerate. In cross-section the leaf blades are flat, circular, triangular or rhomboid in outline. Vernation in the Lomandra group of genera is mostly flat-curved, whereas the remainder of the family is predominantly supervolute. The lamina may contract towards the base or is pseudopetiolate. In *Sowerbaea* and *Laxmannia* there are short, ligulelike projections above the leaf sheath (Thongpukdee 1989), and the leaf bases are auriculate in *Acanthocarpus* (Stevens 1978).

Seedlings in the Lomandraceae show a range of the germination patterns described by Tillich (1995); *Arthropodium*, *Thysanotus* and *Trichopetalum* have a unifacial, elongated cotyledon and short ventrally closed sheath, germination is epigeal, and the first leaf is similar to successive leaves. In *Cordyline*, *Eustrephus* and *Laxmannia* the cotyledon is short and generally haustorial, although *Laxmannia* also shows a conspicuous hairpinlike bend where the middle part of the cotyledon joins the plumule. Taxa where the presence of a closed cotyledonary sheath and short, unifacial hyperphyll produce a tubular elongation of the cotyledonary sheath, include *Lomandra* and *Sowerbaea*, the latter with a leaflike region developed above the coleoptile sheath (Thongpukdee 1989; Tillich 1995). A short hypocotyl is developed in most genera examined. The first leaves in *Eustrephus* and some *Cordyline* species are cataphylls (Conran 1985; Tillich 1995). *Cordyline* is unusual in having some species with scalelike first leaves, and others with cotyledonary hyperphylls ranging from haustorial to long or more or less epigeal (Tomlinson and Fisher 1971; Tillich 1995).

VEGETATIVE ANATOMY. In *Lomandra*, *Acanthocarpus*, *Cordyline* and *Thysanotus spiniger*, the rhizomes, and sometimes aerial stems, exhibit anomalous secondary thickening (Stevens 1978; Waterhouse 1987 [1986]; Rudall 1995). The roots in at least *Acanthocarpus*, *Lomandra* and *Chamaexeros* have a sclerenchymatous cortical zone outside the root endodermis (Fahn 1954), and in *Romnalda*, the root decorticates to the innermost 2 layers of this lignified cortex (Stevens 1978).

The leaves are dorsiventral or unifacial, and in cross-section are flat to U- or V-shaped, terete or triquetrous. The *Lomandra* group of genera have characteristic sclerenchyma girders of inner bundle sheath cells and enlarged parenchymatous outer bundle-sheath cells, which extend to the epidermis on both surfaces, and are found even in the reduced leaves of *Xerolirion* (Fahn 1954; Rudall and Chase 1996), although similar girders also occur in *Cordyline*. Stomata are anomocytic and the epidermal cells are usually elongated, as is the generalised case in the Asparagales, although paracytic and tetracytic types are present in *Cordyline* and some *Lomandra* species, and more or less isodiametric epidermal cells occur in *Chamaexeros* and some *Cordyline* species. In *Lomandra* and *Acanthocarpus*, the stomata and tissues above the sclerenchyma girders are protected by unbranched papillose hairs (Fahn 1954). The leaves of *Cordyline* and most of the *Lomandra* group examined are amphistomatic. Many *Cordyline* species having complex sculpturing patterns on the outer surface of the costal, subsidiary and intercostal epidermal cells; these patterns appear to be species-specific (Conran in press). There is differentiated palisade mesophyll in those *Lomandra*-like taxa examined, except for *Romnalda*, and a hypodermis present in the latter.

With the exception of *Acanthocarpus* and *Arthropodium*, where vessels with scalariform perforations occur in the leaves, vessels are restricted to the roots; they have simple or scalariform perforations (Wagner 1977).

Raphide cells are known from all genera, and epidermal or hypodermal cells with rhomboidal or polyhedral crystals are found in some *Lomandra* and *Romnalda* species (Stevens 1978); silica bodies are absent (Fahn 1954). Tannin cells are not reported.

INFLORESCENCE MORPHOLOGY. The majority of the Lomandraceae have paniculate or contracted racemose or cymose inflorescences. In *Laxmannia* and *Sowerbaea*, there is a contracted umbel-like inflorescence which is covered by imbricate bracts, at least in bud (Keighery 1984b), and some *Thysanotus* species have contracted umbellate cymes (Brittan 1981) or the entire plant in some *Thysanotus* and *Murchisonia* species is a shrubby or twining, much branched, more or less leafless paniculate cyme. The inflorescence in *Chamaescilla* is a corymbose to conical panicle.

In *Romnalda* and *Acanthocarpus*, cymose flower clusters are borne on racemose side branches of a racemose axis. A single bract ensheathes each flower and the continuing part of the cyme. In many genera, each flower is surrounded by 2 or more bracts or prophylls, suggesting reduction from a compound inflorescence, but the precise structure of these flower clusters remains unclear (Stevens 1978). *Xerolirion* has strongly reduced inflorescences. In most genera, the flowers are articulate and have longer or shorter pericladia (Schlittler 1953).

Floral Morphology. The tepals whorls are either similar, or those of the inner whorl are broader, and have fimbriate margins in *Eustrephus*, *Thysanotus*, and *Trichopetalum*. There are also some *Arthropodium* species with finely fimbriate inner tepal margins. The anthers are poricidal in *Eustrephus* (Fig. 102C), some *Thysanotus* and some *Arthropodium*. *Arthropodium* and *Trichopetalum* also have hairy stamen filaments or staminal appendages. Studies in *Arthropodium* by Stringer and Conran (1991) have shown that at least for this genus, the hairy stamen appendages are fused to the filaments, creating the impression of hairy filaments. The stigmatic surface is Wet in at least *Lomandra*, *Arthropodium*, *Laxmannia* and *Sowerbaea* (Dahlgren et al. 1985; Thongpukdee 1989). Septal nectaries are present in most genera, including *Cordyline* (Daumann 1970) and *Lomandra* and related genera (P.J. Rudall, pers. comm.), despite reports to the contrary for the latter (e.g. Dahlgren and Clifford 1982).

Fig. 102A–E. Lomandraceae. *Eustrephus latifolius*. **A** Habit. **B** Partial inflorescence. **C** Stamen with poricidal opening. **D** Pistil. **E** Androecium. (Takhtajan 1982)

Embryology. Anther wall formation is of the monocotyledonous type. Microsporogenesis is successive, and the tapetum is binucleate and shows glandular-secretory degeneration. The endothecium has spiral or girdle (*Lomandra*) thickenings. Pollen grains are 2-nucleate at anthesis.

Ovules in the Lomandraceae are anatropous to campylotropous. Embryo sac development is of the Polygonum type and endosperm formation is Helobial. The proximal nucellar region is relatively long and multilayered. There is an enlarged dermal layer, little subdermal tissue and a central conducting passage of axially elongated cells (*Zuleitungsbahn*). The embryo-sac nuclei are enlarged, with the egg, synergids and antipodals almost filling the embryo sac in some taxa (Rudall 1994). Filiform apparatus are present in several genera. Lateral embryo-sac haustoria are present in *Arthropodium*, but haustoria are absent in

Lomandra (Schnarf and Wunderlich 1939). Although less marked in *Cordyline* and *Chamaescilla*, the enlarged dermal region was considered by Chase et al. (1996) to be the most reliable morphological synapomorphy for the family.

POLLEN MORPHOLOGY. The pollen grains are sulcate in most genera; zonasulcate (i.e. with a sulcus encircling the whole grain and dividing it into 2 halves) and negatively reticulate in *Acanthocarpus* and *Chamaexeros*; or spiraperturate and spinose, echinate or negatively reticulate in *Lomandra* (Chanda and Ghosh 1976).

KARYOLOGY. In the Lomandraceae, there are taxa with chromosome numbers based on $n = 4$ (*Sowerbaea* and *Laxmannia*); 6 (*Chamaescilla*); 7 (*Lomandra* and *Chamaexeros*); 8 (*Lomandra*, *Laxmannia* and *Acanthocarpus*); 10 (*Eustrephus*); 11 (*Chamaescilla*, *Arthropodium*, *Murchisonia* and *Thysanotus*); and 19 (*Cordyline*). Within many of these genera, there are diploid, and tetraploid, hexaploid and octoploid species, as well as taxa with infraspecific polyploids (e.g. Keighery 1984a). The karyotypes in the Lomandraceae vary considerably in size. In genera such as *Arthropodium*, the chromosomes are very small (Westphalen and Conran 1994), whereas those of *Laxmannia* and *Sowerbaea* are very large and acrocentric (Stewart and Barlow 1976; Keighery 1984b). The 38 somatic chromosomes in *Cordyline* represent 4 large chromosomes with submedian constrictions, and 34 small chromosomes which have subterminal or submedian constrictions (Satô 1942). In *Lomandra* there is a range of large to small chromsomes in different species, and submedian to subterminal constrictions, although their karyology has not been examined in detail (Briggs 1986).

POLLINATION. Buzz pollination occurs in *Thysanotus* and *Arthropodium* (including *Dichopogon*) by female bees of the genus *Lasioglossum* (Halictidae). These genera also act as models for the mimetic pseudantherous *Thelymitra nuda* (Orchidaceae), with pollination in the former through pollen reward via porate anthers (Bernhardt and Burns-Baloch 1986). A similar mechanism operates in the bee-pollinated nectarless flowers of *Sowerbaea* (Keighery 1984b). Insect pollination also occurs in *Cordyline*, where there is also evidence of self-incompatibility and widespread hybridisation, at least in the New Zealand taxa (Moore 1975; Beever 1981, Beever and Parkes 1996). The flowers in *Lomandra* were considered to have more or less fleshy tepals which are eaten by the pollinating beetles, although some species were thought to have wind pollination by Dahlgren et al. (1985). The flowers of some *Lomandra* and *Chamaexeros* species are fragrant and borne close to the ground (Macfarlane 1994; J.G. Conran, pers. observ.). Pollination in *Laxmannia* varies, with diurnal, nocturnal and autogamous species reported. The flowers are nectariferous and pollinated by flies or small native Australian bees (Keighery 1984b).

FRUIT AND SEED. Most taxa have loculicidal capsules. The fruit is a drupe in *Lomandra spicata* with a thin fleshy mesocarp. In the *Lomandra* and related taxa there are several ellipsoidal seeds with a cream or brown testa. The seeds are spheroidal to ellipsoid and usually smooth; the testa is foveolate in *Romnalda*. Seed coat anatomy in this group is known for *Acanthocarpus* and *Lomandra* (Huber 1969). The exotesta is colourless and the cells of the tegmen are collapsed but with cellular structure. The endosperm is thick-walled and pitted, storing aleurone, hemicellulose and fats. The embryo is usually vertical, centrally positioned and straight. The micropyle is near the base of the seed; in *Romnalda* the embryo is curved with the micropyle shifted to the centre of the outer part of the seed. Fruits in the majority of the family are capsular or berries with few to numerous seeds. The seeds have a thick phytomelan coat, and the outer epidermal cell walls of the outer integument are domed or papillate in many taxa. The remainder of the outer integument is sharply differentiated from the epidermis, and in many genera is flattened, but not collapsed, with thickened walls. However, in other genera, such as *Arthropodium*, the cells are more or less collapsed. The inner integument is strongly reduced or more or less collapsed (Huber 1969). Unlike the *Lomandra* group, the endosperm cell walls in the remaining genera examined (except for *Eustrephus*) are neither thick-walled nor pitted. In *Eustrephus*, the capsule is berry-like, the valves are fleshy and brightly coloured and the seeds bear a fleshy strophiole. Arillate seeds are present in *Thysanotus* and *Murchisonia*.

DISPERSAL AND REPRODUCTIVE BIOLOGY. The red or yellow drupes of some *Lomandra* species and red seeds of *L. banksii* suggest bird dispersal, and this is supported by the occurrence of several *Lomandra* species as occasional epiphytes in tree forks. The fruits in *Cordyline* are fleshy red, black, or blue-spotted white berries, and are dispersed by birds. The strophioles of *Eustrephus* are juicy, sweet-tasting and animal-dispersed. Some *Lomandra* species are ant-dispersed, as are probably the arillate genera. Burrows (1995) found that whereas most seeds germinated in autumn, some overwintered, germinating in spring. *Lomandra sonderi* similarly has inhibitors in the inner pericarp of the diaspore (Plummer et al. 1995). Germination is improved following pericarp removal or leaching, and the *Camptonotus* and *Iridomyrmex* ants which disperse the diaspores are thought to remove or consume the inner pericarp. The seeds were also found to have embryo dormancy, which could be overcome by the application of gibberellic acid (GA_3) or zeatin. Innate dormancy of 3 to 6 months occurs in *Arthropodium cirratum* in New Zealand, and secondary dormancy can occur in this species following exposure to low temperatures (ca. 2 °C) for at least 3 weeks, or after long-term storage of more than 9 months (Conner and Conner 1988).

PHYTOCHEMISTRY. Chelidonic acid and steroidal saponins are common. Phytosterol is reported from *Cordyline*, and napthoquinone from *Lomandra* roots. Leucanthocyanins are reported for *Eustrephus*. The shoots of *Laxmannia* were positive for HCN (Gibbs 1974). The seeds of *Arthropodium* and *Cordyline* store fatty acids; predominantly linoleic acid, with smaller quantities of oleic and palmitic acid (Hegnauer 1963; Moore and Edgar 1970). Fructans are present in the stems of *Cordyline* (Harris and Mann 1994), and root tubers of *Arthropodium* (incl. *Dichopogon*), *Chamaescilla* (Incoll et al. 1989) and *Thysanotus*, with the latter three genera also having amino-based storage compounds (Pate and Dixon 1982). Anthocyanins based on cyanidin were reported from *Arthropodium*, on pelargonodin and malvinidin from *Cordyline*, on delphinidin from *Sowerbaea* and *Thysanotus* (Gascoigne et al. 1948).

AFFINITIES. The affinities of the genera within the Lomandraceae have been controversial, the *Lomandra* group of genera generally being associated with taxa now placed in the Dasypogonaceae (Takhtajan 1987) and Xanthorrhoeaceae (Cronquist 1981) although their justifications for removal are summarised in Rudall and Chase (1996); the remainder of the family variously with taxa currently in the Anthericaceae, Asphodelaceae, Boryaceae and Hemerocallidaceae (Chase et al. 1997). The grouping together of these taxa into the Lomandraceae, and removal of other former allies stemmed initially from studies of their embryo sac morphology (Rudall 1994), vegetative anatomy (Rudall and Chase 1996), cell wall ferulates (Rudall and Caddick 1994) and *rbc*L molecular sequences (Chase et al. 1995, 1996). There are two informal grades recognised here, pending the results of ongoing study; the *Lomandra* group of genera (*Acanthocarpus, Chamaexeros, Lomandra, Romnalda* and *Xerolirion*); and the remainder (*Arthropodium, Chamaescilla, Cordyline, Eustrephus, Laxmannia, Murchisonia, Sowerbaea, Thysanotus, Trichopetalum*). The main differences between the grades relate to features such as the generally distichous phyllotaxy, xeromorphic anatomy, and the lack of phytomelan, storage roots or mucilage (fructans?) in the *Lomandra*-like taxa. Although the lomandroids have also been described as having unarticulated pedicels, chartaceous tepals and dioecious and possibly anemophilous flowers, and lacking septal nectaries (e.g. Dahlgren et al. 1985), none of these features is characteristic of the group. Reexamination of these features (J.G. Conran, pers. obs. P. Rudall, pers. comm.) shows that several are either inaccurate, such as the alleged lack of articulated pedicels or septal nectaries, or rare, such as chartaceous tepals (some *Lomandra* species, but also *Laxmannia*), possible anemophily (a few *Lomandra* species), or dioecy (only *Lomandra* and *Xerolirion*). The *Lomandra* group are morphologically a clearly circumscribed and readily recognisable group, in contrast to the remainder of the family, which although generally united by spiral phyllotaxy, mesomorphic anatomy, phytomelan, storage roots and relatively abundant mucilage are more difficult to define clearly, with no obvious synapomorphies. The relationships between most genera are uncertain, and the subject of ongoing research.

Eustrephus has generally been associated on its morphology with *Geitonoplesium* (Hemerocallidaceae) in the Smilacaceae (Cronquist 1981), Luzuriagaceae (Dahlgren et al. 1985) or Geitonoplesiaceae (Conran 1994), although it has also been included within *Luzuriaga* (Krause 1930). Recent molecular studies (Chase et al. 1995, 1997) have, however, placed it within the Lomandraceae. Within the latter, it shares elongate, porate anthers, tuberous roots, arillate seeds and fringed tepals with *Thysanotus* and, to a lesser extent, some *Arthropodium* species. It also has the family's distinctive embryo-sac morphology (Stenar 1952; Rudall 1994).

Sowerbaea is closely related to *Laxmannia* with which it shares possession of an unusual ligulelike structure above the leaf sheath, and similar inflorescence and chromosome characteristics. They were placed together as the tribe Sowerbaeae (Liliaceae sens. lat.) by Keighery (1984b).

PALAEOBOTANY. Fossil leaves assigned to the genus *Paracordyline* from the Eocene Golden Grove deposit of South Australia (Conran and Christophel in press) and the Oligocene subantarctic Îles Kerguélen (Conran 1997) are closely allied to *Cordyline*. In particular, there are strong cuticular similarities between the Australian Eocene fossil and those of the Australasian *Cordyline* species, whereas the cuticle of the Kerguélen fossil more closely resembles those of the extant New Zealand and Mascarene species. *Cordyline* pollen is known from the Lower Miocene onwards (Couper 1953).

DISTRIBUTION AND HABITATS. The family is centred in Australia (particularly the Lomandra group), with several genera confined to SW Western Australia. Several widespread genera such as *Arthropodium*, *Cordyline*, *Thysanotus* and *Eustrephus*, variously extend this range to New Zealand, New Caledonia, New Guinea, the Mascarènes and Madagascar, SE Asia and South America. The members of the family grow in a diversity of habitats from rainforest to open woodland, heathland, coastal dunes, arid *Triodia* (Poaceae) shrub steppe, and desert dunes. The stilt-rooted *Laxmannia* species in W Australia are thought to be protected from extreme diurnal summer soil surface temperatures (46–48 °C) both by raising the plants 2–4 cm above the soil into air temperatures some 6–10 °C lower than the substrate, and by providing extremely efficient water flow to the raised tissues (Pate et al. 1984). *Eustrephus* shows low levels of growth throughout the year (Conran 1988), and is both a food plant for adult *Aproida balyi* (Coleoptera: Chrysomelidae: Hispinae) and its obligate larval host (Samuelson 1989); the leaves are attacked by larvae of the leaf-rolling Tortricid moth *Cryptoptila immersana* (Common 1990). The arils are also attacked by *Drosophila* larvae (J.G. Conran, pers. observ.). *Acanthocarpus*, *Cordyline* and *Lomandra* are variously food plants for larvae of skippers (Lepidoptera: Hesperiidae) in the genera *Anisyntoides*, *Dispar*, *Sabera* and *Trapezites* (Common and Waterhouse 1981).

Mycorrhizae are reported for a number of genera in the Lomandraceae. There are both ectomycorrhizae and vesicular-arbuscular mycorrhizae in the roots of *Thysanotus* (McGee 1986b), although apparently only in the presence of companion plants. Vesicular-arbuscular mycorrhizae also occur in at least *Laxmannia* and *Arthropodium* species, but are apparently absent from *Lomandra* (McGee 1986a; F.A. Smith, pers. comm. Univ. Adelaide).

Lomandra longifolia in NSW appears to inhibit locally invasion of heathlands by bitou bush (*Chrysanthemoides monilifera*: Asteraceae) (Weiss and Noble 1984). Decreased frequency and height in *Lomandra filiformis* is indirectly associated with grazing by grey kangaroos in the Australian Capital Territory (Neave and Tanton 1989).

ECONOMIC IMPORTANCE. The roots, young shoots and fruits of *Cordyline* were eaten in New Zealand and Indonesia (Kunkel 1984). It is also being investigated for commercial fructan development in New Zealand (Harris and Mann 1994). In addition, the seeds of some New Zealand species have potentially commercial quantities of linoleic or oleic acid (Moore and Edgar 1970). *Cordyline* leaf fibre is used extensively in Polynesian and SE Asian societies, as is the fibre from the leaves of some *Lomandra* species by Australian aborigines. Most tuberous rooted taxa were used for food by Australian aborigines, as were the young shoots and strophioles of *Eustrephus* (Cribb and Cribb 1976; Gott and Conran 1991). Many species are cultivated as ornamentals, especially some *Cordyline*, *Thysanotus* species and *Arthropodium cirratum*. *Lomandra longifolia* is used for landscaping and on freeway embankments in E Australia, and *L. sonderi* is used for mine revegetation in SW Australia (Plummer et al. 1995).

KEY TO THE GENERA

1. Seeds with phytomelan; leaves spirally arranged; mucilaginous, drying slowly 2
– Seeds lacking phytomelan; leaves generally distichously arranged, scleromorphic, lacking obvious mucilage, drying quickly (the *Lomandra* group) 11
2. Inner tepals with fimbriate margins 3
– All tepals with more or less entire margins 5
3. Stamen filaments united **10. *Eustrephus***
– Stamen filaments free 4
4. Stigma trifid; anthers 2-tailed basally; seeds lacking arils **9. *Trichopetalum***
– Stigma capitate; anthers not tailed; seeds arillate **8. *Thysanotus***
5. Inflorescence a bracteate subumbel 6
– Inflorescence otherwise 7
6. Stamens 6; tepals scarious or papery **11. *Laxmannia***
– Stamens 3; tepals petaloid, membranous **12. *Sowerbaea***

7. Inflorescence cymose 8
- Inflorescence racemose or spicate 9
8. Inflorescence corymbose, erect; tepals twisting spirally after anthesis **13. *Chamaescilla***
- Inflorescence otherwise, prostrate to procumbent or twining, green and vinelike; tepals not twisting **7. *Murchisonia***
9. Plants woody, trees or shrubs **14. *Cordyline***
- Plants herbaceous **6. *Arthropodium***
10. Plants dioecious; flowers unisexual 11
- Plants hermaphrodite; flowers bisexual 12
11. Pedicels articulated or flowers sessile; plants not divaricately branched; leaves not scalelike and caducous **1. *Lomandra***
- Pedicels not articulated; plants divaricately branched; leaves reduced and scalelike, caducous **4. *Xerolirion***
12. Fruit a spinulose 3-lobed capsule; leaves borne on woody aerial stems **5. *Acanthocarpus***
- Fruit a berry or dry indehiscent; leaves basal, woody stems absent 13
13. Inflorescence bracteate and umbel-like **3. *Chamaexeros***
- Inflorescence racemose, not strongly bracteate **2. *Romnalda***

Genera of the Lomandraceae

1. *Lomandra* Labill. Fig. 101A–F

Lomandra Labill., Nov. Holl. Pl. 1: 92 (1805); Lee, Contrib. N. S. W. Natl. Herb. 3: 151–164 (1968); Lee, Contrib. N. S. W. Natl. Herb., Fl. Ser. 34: 16–42 (1966); Lee and Macfarlane, Fl. Australia 46: 100–141 (1986).
Xerotes R. Br. (1810), nom. illeg.

Dioecious herbs or shrubs with thick, woody rhizomes with monocotyledonous secondary thickening and fibrous roots. Stem leafy, leaves basal or cauline. Leaves perennial, sessile, sheathing, sheathes sometimes auricular, linear, triquetrous or terete, margins smooth, prickly or scarious. Male and female inflorescences similar or different; perigone short, pedicels articulated at the pedicel apex or pedicels absent; flowers single or clustered at each node, subtended by a single or several bracts; tepals free or united, petaloid or chartaceous, green, white, yellow or purple. Anthers oblong, outer whorl hypogynous or inserted on outer tepals, inner whorl inserted on inner tepals. Ovary small, ovoid; ovules single per locule. Style short, thick; stigma capitate or 3-lobed. Capsule globular, ovoid, or trigonous, seeds smooth, brown, reddish or yellow. $2n = 14, 16, 28, 32$. About 50 spp. largely restricted to Australia, but with two spp. extending to New Guinea and one of these to New Caledonia.

The genus is widespread throughout all habitats from rainforest to arid areas, with some moist forest taxa occasionally growing as epiphytes in branch axils. The genus is highly variable morphologically, and four sections were recognised on inflorescence and bract/prophyll structure by Lee and Macfarlane (1986).

2. *Romnalda* P. Stevens

Romnalda P. Stevens, J. Arnold Arbor. 59: 148 (1978); Henderson in Fl. Australia 46: 91–92 (1986).

Short herbs, with simple or few-branched, more or less procumbent stems and adventitious, fibrous, stilt roots. Leaves perennial, sessile, clustered, spirodistichously arranged, sheathing, linear. Flowers in scapose panicles; pedicels articulated; flowers several per node, in congested cymules, subtended by several bracts; tepals shortly united basally, membranous, white, persistent. Anthers oblong, introrse, filaments fused basally, inserted on the tepals, dorsifixed, dehiscing by longitudinal slits. Ovary small, ovoid; ovules 2 per locule. Style short, filiform, apically geniculate; stigma capitate. Capsule ovoid-asymmetrical, seeds ellipsoid to hemispherical, smooth or pitted, brown or yellow. $2n = 16$. Three spp.; two endemic to Australia, the other to New Guinea.

Distinguished from *Lomandra* by its hermaphrodite flowers and basally fused anthers by Stevens (1978). It is considered to be very closely related to *Lomandra*, and is included within it by Chase et al. (1996). Its relationships with the remainder of the family, and the *Lomandra* group in particular, are the subject of ongoing morphological and molecular study.

3. *Chamaexeros* Benth.

Chamaexeros Benth., Fl. Austral., 7: 110 (1878); George, Fl. Australia 46: 90–91; Macfarlane, Nuytsia 9: 375–382 (1994).

Tufted herbs with short to elongated and rarely vertical rhizomes, and wiry roots. Stems very short to elongated. Leaves perennial, sessile, clustered, linear, with scarious or serrate margins. Flowers in axillary, open, bracteate cymose panicles; perigone short, "pedicels" articulated at the base of the flower; flowers numerous, subtended by several bracts; tepals free, petaloid, cream or yellow, persistent. Anthers oblong, introrse, dorsifixed, dehiscing by longitudinal slits, filaments fused basally, outer 3 hypogynous, inner 3 inserted on the tepals. Ovary small, ovoid; ovules 2 per locule. Style filiform; stigma capitate. Capsule globular, seeds few, spherical, smooth, brown or yellow. $2n = 14, 28$. Four spp., endemic to SW Western Australia.

The biology and ecology of the 4 species was most recently studied by Macfarlane (1994). Several species have strongly scented flowers, and in all taxa, the flowers are borne at ground level and more or less hidden under the foliage. In *C. longicaulis*, there are elongated rhizomes which allow for vegetative reproduction, as well as elongate stems and vertical rhizome branches which are thought to act as food-storage areas.

4. *Xerolirion* A.S. George

Xerolirion A.S. George in Fl. Australia 46: 229 (1986).

Dioecious tufted herb, with short rhizomes and wiry roots. Stems few-branched, erect and divaricate. Leaves deciduous, sessile, alternate, basally stem-sheathing, linear. Inflorescence terminal; flowers in reduced cymes (males), or solitary (females); pedicels not articulated; tepals shortly united basally, petaloid, white, persistent. Anthers oblong, introrse, free, dorsifixed, inserted on the tepals, dehiscing by longitudinal slits; filaments free. Ovary small, ovoid, more or less 3-lobed; ovules 1 per locule. Style short, thick, tapering; stigma 3-lobed. Capsule more or less globular, seeds 1–3, obovoid, slightly rugose, with a prominent mamillate hilum, pale yellow. Only one sp., *X. divaricata* A.S. George (Basil's asparagus), endemic to arid S Western Australia.

Closely related to *Lomandra* and *Acanthocarpus*, but differing in the combination of dioecy, divaricate habit, terminal flowers and deciduous leaves.

5. *Acanthocarpus* Lehm. Fig. 101G,H

Acanthocarpus Lehm., Pl. Preiss. 2: 274 (1848); George in Fl. Australia 46: 92–98 (1986).

Tufted herb, with short rhizomes and wiry roots. Stems perennial, simple, becoming branched in later years. Leaves sessile, alternate, basally stem-sheathing, linear with prickly margins. Inflorescences terminal or in upper axils, reduced bracteate cymes or racemes; flower articulated on the pedicel; tepals shortly united basally, petaloid, pink or white with brown or purple midveins, persistent. Anthers oblong introrse, free, dorsifixed, dehiscing by longitudinal slits; filaments inserted on the tepals. Ovary small, ovoid; ovules 1 per locule. Style filiform; stigma simple. Capsule 3-lobed, rounded, papillate to muricate, rarely smooth, seeds 1–3, spherical to ellipsoid, rugose to reticulate, golden brown. $2n = 16, 32$. Seven spp., endemic to semiarid SW Australia.

6. *Arthropodium* R. Br.

Arthropodium R. Br., Prodr., 276 (1810); Payens, Nova Guinea II, 8: 389–390 (1957); Brittan in Fl. Australia 45: 341–348 (1987).
Dichopogon Kunth (1843).

Herbs with short rhizomes and fleshy roots, or with tubers at the apices of fibrous roots, or rarely with an elongate, aerial rhizome and stilt roots. Stem leafy at the base, scapose, sometimes enclosed by fibrous old leaf bases. Leaves annual or perennial, sessile, supervolute, sheathing, linear to lanceolate, sometimes pseudopetiolate. Flowers in loose racemes or panicles; perigone developed, articulated along the pedicel; flowers single to multiple per node, subtended by a single to numerous bracts; tepals free above the perigone, white, pink or purple. Anthers oblong or elongated, introrse, free, basifixed, dehiscing by longitudinal slits or apical pores, with 2 or 4 papillate or hairy appendages which are often partially fused to the filaments. Ovary small, ovoid; ovules numerous per locule. Style filiform, sometimes geniculate; stigma minutely capitate. Capsule ovoid, seeds angular, black, minutely papillate. $2n = 22, 44, 66$. About 20 spp., mostly from Australia, some taxa in New Zealand, New Guinea, New Caledonia and Madagascar.

The relationships between *Arthropodium* and the sometimes segregated *Dichopogon* have been the subject of a number of studies. Although traditionally separated on the presence of hairy stamens in the former and papillate anther appendages in the latter, Stringer and Conran (1991) found that these characters represent a continuum, and a phylogenetic study of the genera failed to separated them as discrete clades (J.G. Conran, in prep.).

7. *Murchisonia* Brittan

Murchisonia Brittan, J. R. Soc. W. Aust., 54: 95 (1971); Brittan, Fl. Australia, 45: 340–341 (1987).

Herbs with short rhizomes and somewhat fleshy roots. Stem leafy at the base and spicate-scapose, or leafless with the plant reduced to a twining, photosynthetic cymose panicle. Leaves annual, sessile, supervolute, sheathing, linear. Flowers in few-flowered, contracted umbels or solitary; pedicels not articulated, or absent; flowers several or single per node, subtended by 2 bracts; tepals free, pale purple, greenish white or white. Anthers elongate, introrse, free, basifixed, dehiscing by longitudinal slits, attached to the perianth bases. Ovary small, ovoid; ovules 2 per locule. Style shortly filiform; stigma minutely capitate. Capsule ovoid, seeds cylindrical or spheroid, black, arillate. Two spp., Western Australia, South Australia and the Northern Territory.

Murchisonia is very closely related to *Thysanotus*, differing mainly in the lack of tepal fimbriae, the possession of anthers adnate to the tepals and a persistent fragrance.

8. *Thysanotus* R. Br. Fig. 101I,J

Thysanotus R. Br., Prodr., 282 (1810); Brittan, Brunonia 4: 67–181 (1981); Brittan in Fl. Australia 45: 308–338 (1987).

Herbs with short rhizomes and fibrous or tuberous roots. Stem leafy at the base, enclosed by old leaf bases, or a leafless and branched erect or twining photosynthetic inflorescence. Leaves annual or perennial, sessile, supervolute, sheathing, linear. Inflorescence annual or perennial. Flowers in pedunculate racemes, terminal umbels, cymose panicles or flowers solitary; pedicels articulated; flowers single per node, subtended by a single bract; tepals free, pale purple; outer teplas linear lanceolate, entire; inner tepals broadly ovate, margins fimbriate. Anthers 3 or 6, introrse, free, basifixed, dehiscing by terminal pores or longitudinal slits; filaments hairless. Ovary small, ovoid; ovules 2 per locule. Style filiform, straight or curved; stigma minutely capitate. Capsule ovoid, seeds ellipsoid, black, arillate. $2n = 22, 44, 88$. About 50 spp., predominantly Western Australia, extending to SE Asia, Malesia and New Guinea.

There are a number of distinct forms within *Thysanotus*, with Thongpukdee (1989) finding two clearly separated clades between the 3-staminate and 6-staminate species.

9. *Trichopetalum* Lindley

Trichopetalum Lindley, Bot. Reg. t. 1535 (1832).
Bottinaea Colla (1834).

Herbs with erect rhizome and somewhat fleshy roots. Stem leafy at the base, scapose. Leaves annual, sessile, supervolute, sheathing, linear. Flowers in loose racemose panicles, articulated along the pedicel, several per node, subtended by several bracts; tepals free, white, outer whorl linearlanceolate, entire, inner whorl ovate, fimbriate. Anthers introrse, free, basifixed, dehiscing by longitudinal slits, filaments hairless. Ovary small, ovoid; ovules numerous per locule. Capsule ovoid, seeds compressed, round and papillate, black. Only one sp., *T. plumosum* (Ruíz & Pav.) MacBride, Chile.

Trichopetalum is morphologically close to *Thysanotus*, with which it shares (along with *Eustrephus*) fimbriate inner tepals.

10. *Eustrephus* R. Br. Fig. 102

Eustrephus R. Br. in Ker Gawl., Bot. Mag. 31: t. 1245 (1809); Schlittler, Ber. Schweiz. Bot. Ges. 61: 175–239 (1951); Conran and Clifford in Fl. Australia 46: 190–192 (1986); Conran, Muelleria 6: 363–369 (1987); Conran, Telopea 6: 39–41 (1994); Laferrière, Austrobaileya 4: 391–399 (1995); Laferrière, Fl. Malesiana 12: 731–736 (1996).

Shrub or twining from short rhizomes; roots fibrous with distal tubers. Stem leafy, branched, perennial. Leaves sessile, alternate, conduplicate, linear to broadly ovate. Flowers in axillary or terminal cymes; perigone developed, articulated on the pedicel; flowers numerous or single per node, subtended by a several bracts; tepals free, lanceolate, pale pink or white, margins of inner whorl fimbriate. Anthers elongate, introrse, basifixed, dehiscing by apical pores, filaments united, flattened. Ovary small, ovoid; ovules numerous per locule. Capsule berry-like, ovoid, more or less 3-lobed, fleshy, bright orange, seeds angular, black, strophiolate. $2n = 20$. Only one, very variable sp., *E. latifolius* R. Br. in Ker Gawl. from E Australia, extending to New Guinea, New Caledonia and the Pacific Islands; naturalised on Java.

Although there is a single species, because of its wide range and morphological variability, numerous infraspecific taxa were described by Schlittler (1951). Studies by Conran (1987) and Laferrière (1995) did not support the retention of any of Schlittler's taxa.

11. *Laxmannia* R. Br.

Laxmannia R. Br., Prodr., 285 (1810); Keighery in Fl. Australia 45: 254–264 (1987).
Bartlingia F. Muell. (1874).

Tufted herbs with short rhizomes and fibrous, often stilted roots. Stems leafy, scapose, perennial. Leaves perennial, sessile, sheathing, shortly ligulate above the sheath, linear, terete to triquetrous. Flowers in terminal scapose umbel-like bracteate heads; sessile or shortly pedicellate, not articulated; involucral bracts scarious; flowers single per node, subtended by a several hyaline, membranous bracts with jagged or divided margins; tepals free or shortly the inner whorl united basally, white, green, reddish or pink. Anthers introrse, free, dorsifixed, dehiscing by longitudinal slits; filaments hairless, the inner three adnate to the inner tepal whorl. Ovary small, ovoid; ovules 1–8 per locule. Capsule ovoid, seeds more or less triangular, mamillate, dull black. $2n = 8, 16, 24$. Thirteen spp., endemic to Australia, with a centre of diversity in SW Western Australia.

There are three informal groups of species in the genus, differentiated on their inflorescences and floral biology: the *L. sessiliflora* group, with sessile inflorescences and diurnal white flowers with similar tepal whorls; the *L. gracilis* group, with mostly pedunculate inflorescences and nocturnal or long-lasting coloured flowers with similar tepal whorls; and the *L. squarrosa* group, with large, generally pedunculate inflorescences and white, diurnal flowers, with the inner tepals smaller than the outer (Keighery 1987).

12. *Sowerbaea* Smith Fig. 101K–N

Sowerbaea Smith, Trans. Linn. Soc. Lond. 4: 218 (1798–1799); Henderson in Fl. Australia 45: 264–268 (1987).

Tufted herbs with fibrous, sometimes distantly tuberous roots. Stem condensed, leafy at the base. Leaves annual, sessile, sheathing, linear to filiform. Inflorescences terminal umbel-like clusters on simple or few-branched peduncles; pedicels not articulated; flower cluster subtended by an involucre of bracts, inner bracts fimbriate; tepals free, pink to purple. Anthers 3, opposite inner tepals, introrse, free, basifixed, dehiscing by slits, introrse, latrorse or extrorse, filaments hairless; staminodes 3, opposite outer tepals, or absent. Ovary small, ovoid; ovules 1–2 or 3–4 per locule. Style filiform; stigma minutely capitate. Capsule ovoid, seeds angular, black, minutely papillate. $2n = 8, 16, 32$. Five spp., Australia, predominantly on acidic or swampy soils.

13. *Chamaescilla* F. Muell. ex Benth.

Chamaescilla F. Muell. ex Benth. in Fl. Australia 7: 48 (1878); Henderson in Fl. Australia 45: 288–292 (1988).

Rosette herbs with fibrous or tuberous roots. Stem leafy at the base, scapose. Leaves annual, sessile, supervolute, sheathing, linear to filiform. Flowers in scapose, paniculate cymes; pedicels not articulated; flowers single per node, subtended by a single bract; tepals free, bright blue, twisting spirally after anthesis, persistent. Anthers introrse, free, basifixed, dehiscing by longitudinal slits, filaments hairless. Ovary small, oblong; ovules numerous per locule. Style filiform, slightly expanded apically; stigma papillose. Capsule membranous, strongly 3-lobed, seeds more or less discoid, biconvex, black, smooth. $2n = 6, 12, 24$. Two spp., southern Australia.

The two spp. are generally found in moist forest habitats on sandy soils. *C. corymbosa* (R. Br.) F. Muell. ex Benth. is highly variable, with three varieties recognised. Much of this variation is thought to be due to changes in ploidy levels (Henderson 1987).

14. *Cordyline* Comm. ex R. Br.

Cordyline Comm. ex R. Br., Prodr.: 280 (1810); Moore and Edgar, Fl. N. Z. 2: 46–51 (1970); Fosberg, Baileya 22: 180–181 (1985); Pedley in Fl. Australia 46: 81–86 (1986).
Cohnia Kunth (1850).

Woody shrubs or trees to 10 m with thick rhizomes and thick, somewhat fleshy roots. Stem leafy at the apices, woody with monocotyledonous secondary thickening. Leaves perennial, sessile or pseudopetiolate, conduplicate, sheathing, linear-lanceolate to broadly ovate. Flowers in terminal or lateral racemose or spicate panicles; pedicels not articulated; flowers single per node, subtended by several bracts; tepals free or united basally, white, pinkish, mauve-purple or dark purple-grey. Anthers introrse, free, medifixed, inserted on tepal bases, filaments hairless. Ovary small, ovoid; ovules 2-numerous per locule. Style shortly filiform; stigma capitate or trifid. Capsule ovoid, seeds few, rounded, black, smooth. $2n = 38$. About twenty spp., predominantly SE Asia, Australia and New Zealand, extending to India, the Pacific, South America and the Mascarènes.

The number of taxa varies, and in particular the relationships within the highly variable, widespread and ethnobotanically important *C. fruticosa* (L.) A. Chev. complex are difficult to determine (Fosberg 1985; Erlich 1989), including its variously segregated taxa from E. Australia (Pedley 1986). However, *Cordyline* taxa appear to be able to be distinguished on leaf cuticular features (Conran in press). A separate genus, *Cohnia* Kunth, is sometimes recognised for the free-tepalled taxa from New Caledonia and the Mascarènes.

Selected Bibliography

Beever, R.E. 1981. Self-incompatibility in *Cordyline kaspar* (Agavaceae). N. Z. J. Bot. 19: 13–16.

Beever, R.E., Parkes, S.L. 1996. Self-incompatibility in *Cordyline australis* (Asteliaceae). N. Z. J. Bot. 34: 135–137.
Bernhardt, P., Burns-Baloch, P. 1986. Floral mimesis in *Thelymitra nuda* (Orchidaceae). Plant Syst. Evol. 151: 187–202.
Boyd, L. 1932. Monocotyledonous seedlings. Trans. Proc. Edinb. Bot. Soc. 31: 1–24.
Briggs, B.G. 1986. Chromosome numbers in *Lomandra* (Dasypogonaceae). Telopea 2: 741–744.
Brittan, N.H. 1981. Revision of the genus *Thysanotus* R. Br. (Liliaceae). Brunonia 4: 67–181.
Burrows, C.J. 1995. Germination behaviour of the seeds of the New Zealand species *Aristotelia serrata, Coprosma robusta, Cordyline australis, Myrtus obcordata*, and *Schefflera digitata*. N. Z. J. Bot. 33: 257–264.
Chanda, S., Ghosh, K. 1976. Pollen morphology and its evolutionary significance in Xanthorrhoeaceae. In: Ferguson, I.K., Muller, J. (eds.) The evolutionary significance of the exine. London: Academic Press, pp. 527–559.
Chase, M.W. et al. 1995. See general references.
Chase, M.W. et al. 1996. See general references.
Common, I.F.B. 1990. Moths of Australia. Melbourne: Melbourne University.
Common, I.F.B., Waterhouse, D.F. 1981. Butterflies of Australia. Sydney: Angus and Robertson.
Conner, A.J., Conner, L.N. 1988. Germination and dormancy of *Arthropodium cirratum* seeds. N. Z. J. Bot. 15: 3–10.
Couper, R.A. 1953. Upper Mesozoic and Cainozoic spores and pollen grains from New Zealand. Pal. Bull. N.Z. Geol. Surv. Pal. 22: 1–77.
Conran, J.G. 1985. The taxonomic affinities of the genus *Drymophila* (Liliaceae s.l.). Ph.D. Thesis. St. Lucia: University of Queensland.
Conran, J.G. 1987. Variation in *Eustrephus* R. Br. ex Ker Gawler and *Geitonoplesium* Cunn. ex. R. Br. (Asparagales: Luzuriagaceae). Muelleria 6: 363–369.
Conran, J.G. 1988. The reproductive and vegetative phenology of some south east Qld rainforest monocotyledons. Proc. R. Soc. Queensl. 99: 35–43.
Conran, J.G. 1994. The Geitonoplesiaceae Dahlgren ex Conran (Liliiflorae: Asparagales) A new family of monocotyledons. Telopea 6: 39–41.
Conran, J.G. 1997. *Paracordyline kerguelensis*, an Oligocene monocotyledon macrofossil from the Kerguélen Islands. Alcheringa 21: 129–140.
Conran, J.G. in press. A preliminary investigation of leaf venation and cuticle features to characterise taxa within *Cordyline* Comm. ex R. Br. (Agavaceae s.l.). In: Coode, M. (ed.) Third International Flora Malesiana Conference Kew 1995. Royal Botanic Gardens, Kew.
Conran, J.G., Christophel, D.C. in press. *Paracordyline aureonemoralis*: an Eocene monocotyledon macrofossil from Adelaide, South Australia. Alcheringa.
Cribb, A.B., Cribb, J.W. 1976. Wild food in Australia. Sydney: Fontana Collins.
Cronquist, A. 1981. See general references.
Dahlgren, R.M.T., Clifford, H.T. 1982. See general references.
Dahlgren, R., Clifford, H.T. 1982. See general references.
Dahlgren, R.M.T. et al. 1985. See general references.
Daumann, E. 1970. See general references.
Erlich, C. 1989. Special problems in an ethnobotanical literature search: *Cordyline terminalis* (L.) Kunth, the "Hawaiian ti plant". J. Ethnobiol 9: 51–64.
Fahn, A. 1954. The anatomical structure of the Xanthorrhoeaceae Dumort. Bot. J. Linn. Soc. 55: 158–184.
Fosberg, F.R. 1985. *Cordyline fruticosa* (L.) Chevalier (Agavaceae). Baileya 22: 180–181.
Gascoigne, R.M., Ritchie, E., White, D.E. 1948. A survey of anthocyanins in the Australian flora. J. R. Soc N. S. W. 82: 44–70.
George, A. S. 1986. *Acanthocarpus*. In: George, A.S. (ed.) Flora of Australia, Vol. 46. Canberra: Australian Government Publishing Service, pp. 92–98.
Gibbs, D. 1974. Chemotaxonomy of flowering plants. Montreal: McGill Queens University Press.
Gott, B., Conran, J.G. 1991. Victorian koorie plants. Hamilton, Victoria: The Aboriginal Keeping Place.
Harris, W., Mann, J.D. 1994. Preliminary investigation of the suitability of *Cordyline australis* (Asphodelaceae) as a crop for fructose production. N. Z. J. Crop Hort. Sci. 22: 439–451.
Hegnauer, R. 1963. See general references.
Henderson, R.J.F. 1987. *Chamaescilla*. In: George, A.S. (ed.) Flora of Australia, Vol. 45. Canberra: Australian Government Publishing Service, pp. 288–292.
Huber, H. 1969. See general references.
Incoll, L.D., Bonnett, G.D., Gott, B. 1989. Fructans in the underground storage organs of some Australian plants used as food by aborigines. J. Plant Physiol. 134: 196–202.
Keighery, G.J. 1984a. Chromosome numbers of Australian Liliaceae. Feddes Repert. 95: 523–532.
Keighery, G.J. 1984b. The Johnsonieae (Liliaceae): biology and classification. Flora 175: 103–108.
Keighery, G.J. 1987. *Laxmannia*. In: George, A.S. (ed.) Flora of Australia, Vol. 45. Canberra: Australian Government Publishing Service, pp. 254–264.
Krause, K. 1930. See general references.
Kunkel, G. 1984. Plants for human consumption. Koenigstein: Kluwer.
Laferrière, J.E. 1995. Nomenclature and type specimens in *Eustrephus* and *Geitonoplesium* (Geitonoplesiaceae). Austrobaileya 4: 391–399.
Lee, A.T., Macfarlane, T.D. 1986. 5. *Lomandra*. In: George, A.S. (ed.) Flora of Australia, Vol. 46. Canberra: Australian Government Publishing Service, pp. 100–141.
Macfarlane, T.D. 1994. *Chamaexeros longicaulis* (Dasypogonaceae), a new species from Walpole, southwestern Australia. Nuytsia 9: 375–382.
McGee, P. 1986a. Growth response to and morphology of mycorrhizas of *Thysanotus* (Anthericaceae: Monocotyledonae). New Phytol. 109: 459–464.
McGee, P. 1986b. Mycorrhizal associations of plant species in a semiarid community. Aust. J. Bot. 34: 585–594.
Moore, L.B. 1975. Hybridism in *Cordyline* (Agavaceae). N. Z. J. Bot. 13: 305–316.
Moore, L.B., Edgar, E. 1970. Flora of New Zealand, Vol. 2. Wellington: Government Printer.
Neave, H.M., Tanton, M.T. 1989. The effects of grazing by kangaroos and rabbits on the vegetation and the habitat of other fauna in the Tidbinbilla Nature Reserve, Australian Capital Territory. Aust. Wildl. Res. 16: 337–352.
Pate, J.S., Dixon, K.W. 1982. Tuberous, cormous and bulbous plants. Nedlands, Western Australia: University of Western Australia Press.
Pate, J.S., Weber, G., Dixon, K.W. 1984. Stilt plants – extraordinary growth form of the Kwongan. In: Pate, J.S., Beard, J.S. (eds.) Kwongan: Plant life of the sandplain. Nedlands, Western Australia: University of Western Australia Press, pp. 101–125.
Pedley, L. 1986. Cordyline. In: George, A.S. (ed.) Flora of Australia, Vol. 46. Canberra: Australian Government Publishing Service, pp. 81–86.
Plummer, J.A., Crawford, A.D., Taylor, S.K. 1995. Germination of *Lomandra sonderi* (Dasypogonaceae) promoted by peri-

carp removal and chemical stimulation of the embryo. Aust. J. Bot. 43: 223–230.
Rudall, P. 1994. The ovule and embryo sac in Xanthorrhoeaceae sensu lato. Flora 189: 335–351.
Rudall, P. 1995. New records of secondary thickening in monocotyledons. IAWA J. 16: 261–268.
Rudall, P., Caddick, L.R. 1994. Investigation of the presence of phenolic compounds in monocotyledonous cell walls using UV fluorescence microscopy. Ann. Bot. 74: 483–491.
Rudall, P., Chase, M. 1996. Systematics of the Xanthorrhoeaceae sensu lato: evidence for polyphyly. Telopea 6: 185–203.
Rudall, P., Chase, M.W., Conran, J.G. 1996. New circumscriptions and a new family of asparagoid lilies: genera formerly included in Anthericaceae. Kew Bull. 51: 667–680.
Samuelson, G.A. 1989. A review of the Hispinae tribe Aproidini (Coleoptera: Chrysomelidae). Mem. Queensl. Mus. 27: 599–604.
Satô, D. 1942. Karyotype alternation and phylogeny in Liliaceae and allied families. Jpn. J. Bot. 12: 57–161.
Schlittler, J. 1951. Die Gattungen *Eustrephus* R. Br. ex Sims und *Geitonoplesium* (R. Br.) A. Cunn. Morphologische-anatomische Studie mit Berücksichtigung der systematischen, nomenklatorischen und arealgeographischen Verhältnisse. Ber. Schweiz. Bot. Ges. 61: 175–239.
Schlittler, J. 1953. Blütenartikulation und Phyllokladien der Liliaceae organphylogenetisch betrachtet. Feddes Repert. 55: 154–258.
Schnarf, K., Wunderlich, R. 1939. Zur vergleichenden Embryologie der Liliaceae-Asphodeloideae. Flora 133: 297–327.
Stenar, H. 1952. Notes on the embryology and anatomy of *Luzuriaga latifolia* Poir. Acta Horti Bergiani 16: 219–232.
Stevens, P.F. 1978. Generic limits in the Xeroteae (Liliaceae sensu lato). J. Arnold Arb. 59: 129–155.
Stewart, D.A., Barlow, B.A. 1976. Genomic differentiation and polyploidy in *Sowerbaea* (Liliaceae). Aust. J. Bot. 24: 349–367.
Stringer, S., Conran, J.G. 1991. Stamen and seed cuticle morphology in some *Arthropodium* and *Dichopogon* species (Anthericaceae). Aust. J. Bot. 39: 129–135.
Takhtajan, A.L. 1982. See general references.
Takhtajan, A.L. 1987. See general references.
Thongpukdee, A. 1989. The taxonomic position of the genus *Tricoryne* R. Br. Ph.D. Thesis St. Lucia: University of Queensland.
Tillich, H.-J. 1995. Seedlings and systematics in monocotyledons. In: Rudall, P.J., Cribb, P., Cutler, D.F., Humphries, C.J. (eds.) Monocotyledons: systematics and evolution. Royal Botanic Gardens, Kew.
Tomlinson, P.B., Fisher, J.B. 1971. Morphological studies in *Cordyline* (Agavaceae) I: Introduction and general morphology. J. Arnold Arbor. 52: 459–478.
Troll, W. 1937. Vergleichende Morphologie der höheren Pflanzen. Vol. 1, 1. Berlin: Borntraeger.
Wagner, W.M. 1977. See general references.
Waterhouse, J.T. 1987 [1986]. The phylogenetic significance of *Dracaena*-type growth. Proc. Linn. Soc. N.S.W. 109: 131–138.
Weiss, P.W., Noble, I.R. 1984. Status of coastal dune communities invaded by *Chrysanthemoides monilifera*. Aust. J. Ecol. 9: 93–98.
Westphalen, G., Conran, J.G. 1994. Chromosome numbers in the *Arthropodium-Dichopogon* complex (Asparagales: Anthericaceae). Taxon 43: 377–381.

Luzuriagaceae

J.G. Conran and H.T. Clifford

Luzuriagaceae Kunth, Enum. Pl.: 5, (1850) ("Luzuriageae").

Rhizomatous plants with perennial aerial shoots, terrestrial or epiphytic. Roots fibrous. Stems thin, branching or simple, erect or scrambling to 40 cm. *Leaves* alternate, distichous, sessile to shortly petiolate, without a sheathing base; blades ovate to lanceolate, resupinate; veins 3–5, parallel, with few cross-veins or higher venation reticulate, midrib prominent. *Inflorescence* an axillary, monobrachial cincinnus or flowers borne singly. *Flowers* small, hypogynous, actinomorphic, articulated on the pedicels (*Luzuriaga*) or at base of ovary (*Drymophila*); perianth marcescent (*Luzuriaga*) or caducous (*Drymophila*), not twisted after anthesis; tepals in 2 whorls of 3 (or rarely 4 in *Drymophila*), subequal, free, white or pale pink, nectariferous at the base. Stamens 3 + 3; filaments free; anthers 2-thecate, 4-sporangiate, basifixed, introrse or extrorse, dehiscing longitudinally by slits or apical pores. Gynoecium of 3 (or 4) united carpels, ovary 3-(or 4-) locular or rarely 1-locular, style simple, capitate or deeply trifid (*Drymophila*), erect; ovules several, hemianatropous, in 2 rows on axile or subparietal placentas. *Fruit* a berry. Seeds few to numerous, small, ovoid, pale brown/yellow; embryo 1/3 to 2/3 as long as the seed; endosperm copious, without starch.

A family of two genera and six species; southern South America, Australia and New Zealand.

Vegetative Morphology and Anatomy. The plants are long-rhizomatous and have erect, simple or branching stems (Fig. 103). The stems are wiry, tough and supple, up to 5 mm in diameter. The main stems and lateral branches appear to show determinate growth. The roots are fibrous.

The seedling in *Drymophila* has a short, little-differentiated cotyledon, a strong primary root, and there is no ligular sheath. The primary axis initially bears sheathing scale leaves, and the epicotyl and internodes are elongated as in *Smilax*. Later leaves are entire and 3–5-costate, ovate to lanceolate, bluntly acute. They have a prominent midrib, free vein endings (absent in *Luzuriaga*) and a well-developed blade and tapering base. The branches bear reduced scale leaves at their lower nodes. Venation is acrodromous with proximal convergence of the primary veins. There is slight (*Luzuriaga*) to pronounced acropetal weakening of the primary veins, and the secondary veins are parallel. Higher vein orders are parallel, and the areoles are isodiametric (*Drymophila*) or parallel (*Luzuriaga*) to the leaf axis. The leaves are resupinate by twisting of a more or less reduced petiole. Vernation is conduplicate (Conover 1983; Arroyo and Leuenberger 1988a).

The flowers are axillary and borne in reduced cincinni or singly. The peduncles are reduced and usually subtended by 1-several small bracts. In *Drymophila*, inflorescence production appears to be associated only with new growth, and inflorescences and new shoots are produced only during a short period in spring (Conran 1988b).

Vessels are present in the roots and stems. The perforation plates are simple or scalariform (Zweigelt 1913; Wagner 1977). The roots have a uniseriate exodermis which is unthickened. The cortex is mostly parenchymatous, but in mature roots of *Luzuriaga* a thickened inner cortex is present. The epidermal cells of the stem are small, with slightly thickened outer walls and a thick

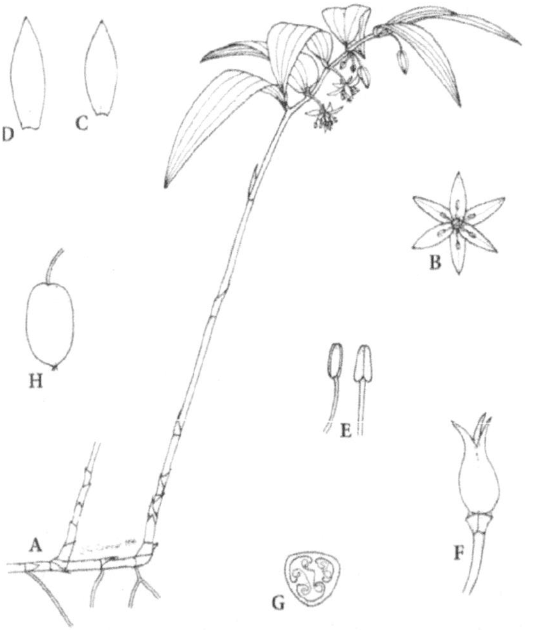

Fig. 103A–H. Luzuriagaceae. *Drymophila moorei.* **A** Flowering plantlet. **B** Flower. **C** Sepal. **D** Petal. **E** Stamens in side view and front view. **F** Pistil. **G** Ovary, transversal section. **H** Fruit. (Clifford and Conran 1987)

cuticle. A ring of fibrous bundles and small collateral bundles are found in the cortex, the outer part of which may be slightly collenchymatous. Scattered bundles occur in the centre of the stem. In all genera the cells are isodiametric to shortly elongate with sinous cell walls in *Drymophila* only. Stomata are present on the adaxial surface (where the leaves are resupinate). The stomata are anomocytic. The mesophyll is undifferentiated (Conran 1985b; Rodriguez and Marticorena 1988).

Tannin cells are present in at least *Drymophila*, but mucilage cells and laticifers are absent in all genera. Silica bodies are lacking in the family. Calcium oxalate raphides and druses are present in all tissues of *Drymophila* but are absent in *Luzuriaga* (Schlittler 1949).

Flower Structure. The perianth is usually 3-merous (occasionally 4-merous in *Drymophila*), regular, free and petaloid. Perianth venation is multinerved and reticulate in both genera. In *Luzuriga* the articulation of the pedicels corresponds to the prophyllar node (Schlittler 1953, see under Bibliography for Lomandraceae); in *Drymophila* the tepals articulate individually on the pedicel. The bases of the tepals have nectaries. The filaments are filiform and free. Anther dehiscence is mostly introrse, but extrorse in the single New Zealand species and one South American species of *Luzuriaga* (Ravenna 1969; Moore and Edgar 1970). The anthers are basifixed and dehisce by longitudinal slits or apical pores (some *Luzuriaga* spp.). The gynoecium is usually 3-carpellary and the ovary 3-locular with axile placentas or rarely, in *Drymophila*, 1-locular; occasionally, *Drymophila moorei* also produces 4-locular, 4-carpellary gynocia. There is a single stigma with capitate style in *Luzuriaga*, whereas in *Drymophila* the style is deeply trifid with 3 prominent stylodia. The stigma is smooth and Wet in *Drymophila* (Zweigelt 1913; Schlittler 1949; Wagner 1977; Conran 1985b; Rodriguez and Marticorena 1988).

Embryology. Anther-wall formation is of the monocotyledonous type in *Drymophila*. The well-developed endothecium has spiral thickenings in both genera, and the tapetum in *Drymophila* is binucleate and amoeboid (J.G. Conran, unpubl.). Microsporogenesis in *Drymophila* is successive (Conran 1985b). The ovule in both genera is bitegmic and tenuinucellate. In *Drymophila*, a linear tetrad is formed and the megaspore is chalazal, embryo sac formation is of the Polygonum type and an 8-nucleate embryo sac is produced. In both genera, the egg appartus, synergids and antipodal cells are small relative to the embryo sac and the polar nuclei fuse prior to fertilisation (J.G. Conran, unpubl.). [N.B. The report of the embryology in *Luzuriaga* by Stenar (1952) is based on *Eustrephus latifolius* (Lomandraceae)].

Pollen Morphology. The pollen grains are shed singly and are obovate and sulcate with a reticulate exine and 27–60 μm in length (Erdtman 1952; Schulze 1982a).

Karyology. The basic chromosome number for the Luzuriagaceae is apparently $x = 10$. The karyotype consists of 8 short and 2 long chromosome pairs. A diploid number of $2n = 20$ has been recorded in both spp. of *Drymophila* and one of *Luzuriaga* (Beuzenberg and Hair 1963; Fedorov 1969; Conran 1985a).

Pollination. Flowers of Drymophila moorei are visited and probably pollinated by bees (Halictidae) and hoverflies (Syrphidae). They are also visited by Chrysomelid beetles and a species of cricket (Gryllidae) (Conran 1988a), but the beetles and crickets also cause damage by feeding on the flowers, including the ovaries. Fruit set within individual plants of *D. moorei* is apparently very low (Conran 1988b). The flowers of *Luzuriaga marginata* were reported to be fragrant by Vallentin and Cotton (1921), and are self-compatible and possibly self-pollinated (Moore 1968).

Fruit and Seed. The fruits are berries. The perianth is deciduous in *Drymophila* and persistent in the other genera. Mature fruits are white in *Luzuriaga* and blue or orange-yellow in *Drymophila*. The pericarp is thick and fleshy and encloses 1–ca. 10 seeds which are ovoid or crescent-shaped. The seeds are pale yellow to brown and the seed surface is rugulose. The seed structure is uniform, with all cell layers thin-walled. The seeds sometimes shed the outer epidermis of the testa, which lacks phytomelan and collapses during its development. The endosperm is massive and consists of cells with thick, pitted walls containing aleurone and oil but not starch.

Dispersal. The berries are assumed to be dispersed by animals.

Phytochemistry. Saponins are present in all genera (Conran 1985b). Alkaloids are not reported.

Affinities of the Luzuriagaceae. The family has been related by features such as the baccate fruits, raphides, pedicel articulation, habit and anatomy to the Asparagalean families Ruscaceae, Herreriaceae, Hemerocallidaceae (incl. Geitonoplesiaceae), but by the lack of phytomelan on the seeds, and the presence of perigonal nectaries (Dahlgren et al. 1985; Conran 1989) also to the lilialean Uvulariaceae (=Colchicaceae) and Liliaceae. The traditional association of *Drymophila* with the Convallariaceae (Krause 1930; Dahlgren and Clifford 1982; Brummitt 1992) is based partly on its superficial resemblance to *Disporopsis*, and to the previous inclusion of members of the Colchicaceae within the Convallariaceae, and not supported by any recent study. The close association of *Drymophila* with *Geitonoplesium* in the molecular study of Chase et al. (1995) was the result of a sequence from misidentified material, and *Drymophila* was retained in the revised Luzuriagaceae of Chase et al. (1996). The family is now considered to be basal to a Colchicaceae-Alstroemeriaceae clade within the Liliales (Chase et al. 1995a), with which it shares features such as resupinate leaves (with the Alstroemeriaceae), perigonal nectaries and some taxa with reticulate venation or which lack calcium oxalate raphides (with the Colchicaceae).

Distribution and Habitats. The Luzuriagaceae are confined to the Southern Hemisphere. They extend from E Australia and New Zealand to S Chile, Argentina and the Falkland Islands (Malvinas); *Luzuriaga* to the Falkland Islands, upland forests of Chile and Argentina and New Zealand; *Drymophila* to SE Australia from SE Queensland and New South Wales to Victoria and Tasmania. The taxa grow in moist, shady temperate forests, and several *Luzuriaga* spp. grow on moss-covered rocks and tree trunks (Arroyo and Leuenberger 1988b; Rodriguez and Marticorena 1988), although on the Falkland (Malvinas) Islands, *L. marginata* grows in *Empetrum* heaths (Moore 1968). *Drymophila moorei* behaves as a cool-temperate understorey perennial with marked seasonality in growth and flowering responses (Conran 1988b).

Palaeobotany. Although there are no fossils definitely assignable to the Luzuriagaceae, similarities between the leaf venation of *Luzuriaga* and that of the early fossil monocot *Acaciaephyllum spatulatum* have been observed by Conover (1983) and Conran et al. (1994).

Economic Importance. *Luzuriaga radicans* "Quilineja" is used in Chile as a source of fibre for rope (Ravenna 1969; Rodriguez and Marticorena 1988). Several species are cultivated as ornamentals. Vallentin and Cotton (1921) list the fruits of *L. marginata* (cotton flower) in the Falkland Islands as poisonous.

Key to the Genera

1. Stigma deeply trifid; tepals caducous; pedicels not articulated **1. *Drymophila***
- Stigma capitate; tepals marcescent; pedicels articulated half-way along their length **2. *Luzuriaga***

1. *Drymophila* R. Br. Fig. 103

Drymophila R. Br., Prod. Fl. Nov. Holl.: 292 (1810).

Dwarf woody perennials with simple or branched, arching stems to 45 cm, with thin rhizomes; flowers single or sometimes 2–3, axillary, pendulous; tepals free 6(–8), white to pale pink; pedicels distally articulated; stamens free; anthers introrse; ovary 3(–4)-locular, rarely unilocular; ovules few to numerous; fruit an orange or dark blue, 1- to few-seeded berry; testa yellow-brown. $2n = 20$. Two spp., eastern Australia from SE Queensland to Tasmania.

2. *Luzuriaga* Ruíz et Pav.

Luzuriaga Ruíz et Pav., Fl. Per. Chil. 3, 65: t. 298 (1802), nom. cons.
Enargea Banks et Sol. ex Gaertn. (1788).

Shrub with a short rhizome; leaves oblong-ovate; inflorescence a few-flowered, axillary cincinnus, sometimes reduced to a single flower; flowers pendulous; tepals free, white; filaments free; anthers dehiscing by longitudinal slits or apical pores; style simple; ovary 3-locular; ovules numerous; fruit a white to dark purple, several seeded berry; seeds yellow to brown; $2n = 20$. Three spp., Peru to Tierra del Fuego, Falkland Islands and New Zealand.

Although the segregate genus *Enargea* was retained by Moore (1968) and Muñoz Pizarro (1973) based on poricidal anther dehiscence and extrorse anthers, the recent treatments by Arroyo and Leuenberger (1988a) and Rodriguez and Matricorena (1988) did not support its recognition.

Selected Bibliography

Arroyo, S.C., Leuenberger, B.E. 1988a. Leaf morphology and taxonomic history of Luzuriaga (Philesiaceae). Willdenowia 17: 159–172.
Arroyo, S.C., Leuenberger, B.E. 1988b. A note on *Luzuriaga marginata* (Philesiaceae) from Patagonia. Herbertia 44: 17–21.
Beuzenberg, E.J., Hair, J.B. 1963. Contributions to the chromosome atlas of the New Zealand flora. 5. Miscellaneous families. N.Z. J. Bot. 1: 53–67.
Brummitt, R.K. 1992. Vascular plant families and genera. London: Royal Botanic Gardens, Kew.
Chase, M.W. et al. 1995. See general references.
Chase, M.W. et al. 1995a. See general references.
Chase, M.W., Rudall, P.J., Conran, J.G. 1996. New circumscriptions and a new family of asparagoid lilies: genera formerly included in *Anthericaceae*. Kew Bull. 51: 667–680.
Clifford, H.T., Conran, J.G. 1987. Liliaceae (part). In: George, A.S. (ed.) Flora of Australia, Vol. 45. Canberra: AGPS, pp. 148–421.
Conover, M.H. 1983. The vegetative anatomy of the reticulate-veined Liliiflorae. Telopea 2: 401–412.
Conran, J.G. 1985a. Chromosome number reports on Colchicaceae, Petermanniaceae and Philesiaceae (IOPB Chromosome number reports 87, ed. Á. Löve). Taxon 34: 346–347.
Conran, J.G. 1985b. The taxonomic affinities of the genus *Drymophila* (Liliaceae s.l.). Ph.D. Thesis. St. Lucia: University of Queensland.
Conran, J.G. 1988a. Observations on the pollination ecology of *Drymophila moorei* Baker (Luzuriagaceae) in Southeast Queensland. Victorian Nat. 105: 43–47.
Conran, J.G. 1988b. The reproductive and vegetative phenology of some south east Qld rainforest monocotyledons. Proc. R. Soc. Queensl. 99: 35–43.
Conran, J.G. 1989. Cladistic analyses of some net-veined Liliiflorae. Plant Syst. Evol. 168: 123–141.
Conran, J.G., Christophel, D.C., Scriven, L.J. 1994. *Petermanniopsis angleseaënsis*: an Australian fossil net-veined monocotyledon from Eocene Victoria. Int. J. Plant Sci. 155: 816–827.
Dahlgren, R., Clifford, H.T. 1982. See general references.
Dahlgren, R.T.M. et al. 1985. See general references.
Erdtman, G. 1952. See general references.
Fedorov, A.A. (ed.) 1969. See general references.
Krause, K. 1930. See general references.
Moore, D.M. 1968. The vascular flora of the Falkland Islands. Br. Antarct. Surv. Bull. 60: 1–202.
Moore, L.B., Edgar, E. 1970. Flora of New Zealand, Vol. 2. Wellington: Government Printer.
Muñoz Pizarro, C. 1973. Chile: plantas en extinción. Santiago: Editorial Universitaria de Chile.
Ravenna, P.F. 1969. Liliaceae and Amaryllidaceae. In: Correa, M.N. (ed.) Flora Patagonica Parte II: Typhaceae a Orchidaceae (excepto Graminaea). Buenos Aires: Talleos Gráficos I.S.A.S, pp. 138–157.
Rodriguez, R., Marticorena, C. 1988. La especies del género *Luzuriaga* R. et P. Gayana Bot. 44: 3–15.
Schlittler, J. 1949. Die systematische Stellung der Gattung *Petermannia* F.v. Muell. und ihre phylogenetischen Beziehungen zu den Luzuriagoideae Engl. und den Dioscoreaceae Lindl. Vierteljahresschr. Naturforsch. Ges. Zuer 94, Beih. 1: 1–28.
Schulze, W. 1982a. Beiträge zur Taxonomie der Liliifloren VII. Philesiaceae. Wiss. Z. Friedrich Schiller Univ. Jena 31: 277–283.
Schulze, W. 1982b. Beiträge zur Taxonomie der Liliifloren VIII. Smilacaceae. Wiss. Z. Friedrich Schiller Univ. Jena 31: 285–289.
Stenar, H. 1952. Notes on the embryology and anatomy of *Luzuriaga latifolia* Poir. Acta Horti Bergiani 16: 219–232.

Takhtajan, A.L. 1982. See general references.
Vallentin, E.F., Cotton, E.M. 1921. Illustrations of the flowering plants and ferns of the Falkland Islands. London: Reeve.
Wagner, W.M. 1977. Vessel types of monocotyledons: a survey. Bot. Not. 130: 383–402.
Zweigelt, F. 1913. Vergleichende Anatomie der Asparagoideae, Ophiopogonoideae, Aletroideae, Luzuriagoideae und Smilacoideae nebst Bemerkungen über die Beziehungen der Ophiopogonoideae und Dracaenoideae. Denkschr. Akad. Wiss. Wien 88: 397–476.

Melanthiaceae

M.N. Tamura

Melanthiaceae Batsch, Tab. Affin. Regni Veg.: 133 (1802), nom. cons.
Heloniadaceae J. Agardh (1858).
Xerophyllaceae Takht. (1994).
Chionographidaceae Takht. (1994).

Perennial herbs, often with raphides. Underground stem usually a thick short rhizome or a bulb, rarely a corm. Roots fibrous. Aerial stem erect, simple, stout to slender, with foliage or scaly leaves, sometimes basally thickened, sometimes with fibrous leaf sheaths at the base. Foliage leaves all cauline (*Veratrum* spp.), or with large basal and small cauline leaves (*Xerophyllum*, *Zigadenus* spp.), or all leaves in a basal rosette (*Heloniopsis*), spirally arranged, bifacial, setaceous to broadly elliptical or spathulate, up to 50 cm long, deciduous or evergreen, sometimes sheathing. Inflorescence often a raceme, sometimes a panicle, a spike, or umbel-like, often glabrous, sometimes pubescent. Flowers hermaphroditic, andromonoecious, polygamous, gynodioecious, androdioecious, or dioecious, usually hypogynous, rarely half-epigynous (*Zigadenus* spp., *Stenanthium*) or epigynous (some *Veratrum*), funnel-shaped, cup-shaped to opening flat, trimerous, usually actinomorphic, rarely zygomorphic (*Chionographis*), bracteate or ebracteate, without bracteoles. Tepals 3 + 3, generally with similar whorls, petaloid, free, or connate in lower part, filiform to oval or spathulate, up to 20 mm long, white, green, yellow, pink, red, lilac, or dark purple-brown, persistent, marcescent or caducous, rarely clawed at the base (*Melanthium*); nectaries when present perigonal. Stamens 3 + 3, free or inserted at the tepal base; filaments filiform to subulate; anthers lanceolate to suborbicular, obcordate, cordate or hippocrepiform, small, up to 4.5 mm long (*Heloniopsis*), dorsifixed or basifixed, dehiscing with slits or valves, extrorse, with distinct or confluent thecae. Ovary ovoid to globose, 3-carpellate, with the carpels connate only in lower part (*Amianthium*, *Stenanthium*) to fully syncarpous (*Ypsilandra*, *Heloniopsis*), provided with 3 ventrally stigmatic, ± recurved styluli or a style ending in a capitate stigma; ovules bitegmic,

2-numerous in each carpel. Capsules narrowly ellipsoidal, conical to globular, obovoid, 5–25 mm long, often 3-lobed, septicidal, loculicidal or ventricidal; seeds linear to elliptical, 3–15 mm long, sometimes flat or angular, sometimes slightly curved, sometimes broadly winged, sometimes appendaged at both ends.

A family comprising 12 genera and ca. 97 species, distributed mostly in the temperate region of the Northern Hemisphere, five genera endemic in N America and three in E Asia, *Schoenocaulon* extending to S America.

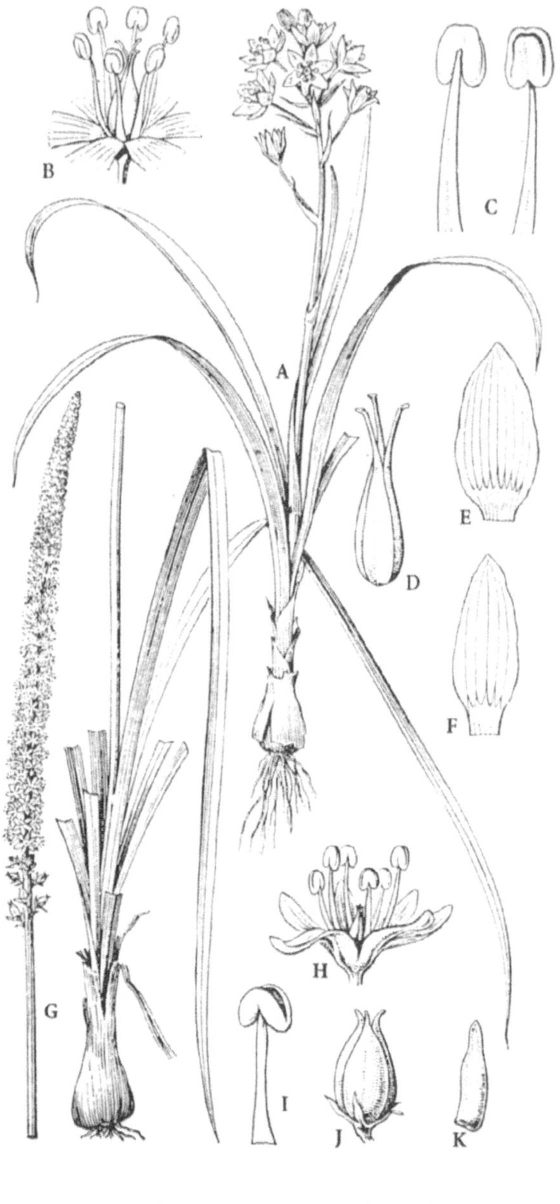

Fig. 104A–K. Melanthiaceae. A–F *Zigadenus fremontii*. A Flowering plant. B Flower, perianth partly removed. C Stamens in adaxial (*left*) and abaxial (*right*) view; note valvate anther dehiscence. D Pistil. E Outer and F inner tepal. G–K *Schoenocaulon officinarum*. G Habit. H Flower. I Stamen. J Fruit. K Seed. (Takhtajan 1982)

Vegetative Morphology and Anatomy. The underground stem is usually a thick short vertical or horizontal rhizome or a bulb, rarely a corm (*Schoenocaulon* spp.). The root cortex does not contain aerenchyma. The phloem does not contain scattered fibers (Ambrose 1980).

The vessels in the late metaxylem of roots of *Helonias* and *Ypsilandra* are small, 25–30 μm in diam. *Chionographis* has scalariform perforation plates with numerous perforations. In the Melanthieae and Heloniadeae the vessels are restricted to the roots. This may indicate that the vessel condition of this family is primitive among the Liliaceae s. lat. (Cheadle and Kosakai 1971).

The foliage leaves are all spirally arranged and bifacial but variable in shape: broadly elliptical (*Veratrum* spp.) to setaceous (*Xerophyllum*). They are primarily cauline (*Veratrum* spp., *Melanthium* spp., *Stenanthium* spp.), normal-sized basal and reduced cauline (*Xerophyllun*, *Veratrum* spp., *Melanthium* spp., *Stenanthium* spp., *Zigadenus*, *Amianthium*, *Helonias*, *Ypsilandra*, *Chamaelirium*, *Chionographis*) or all basal (*Schoenocaulon*, *Heloniopsis*, Fig. 105), although *Heloniopsis* has cauline scaly leaves.

The stomata lack subsidiary cells, a condition considered as advanced among the monocots (Goldblatt 1995). Epicuticular wax platelets in parallel orientation are known from *Zigadenus* (Frölich and Barthlott 1988). Raphides are generally present in Xerophylleae and Melanthieae, and occasionally in Heloniadeae.

The cotyledon has an assimilating elongated hyperphyll. *Xerophyllum* and *Zigadenus* always have a unifacial cotyledon, while *Veratrum* includes both unifacial and bifacial cotyledons. The haustorium is scarcely distinguishable from the free part of the cotyledon and, after shedding the seed, it represents only the green tip. The primary leaf is always bifacial. The primary root is weakly developed (Tillich 1995).

Inflorescence Structure. A raceme is found in *Xerophyllum*, *Zigadenus* spp., *Amianthium*, *Helonias*, *Ypsilandra*, and *Chamaelirium*. Among the Melanthieae, *Veratrum*, *Melanthium*, *Zigadenus* spp., and *Stenanthium* have panicles and *Schoenocaulon* a spike, and among the Heloniadeae, *Heloniopsis* has an umbel-like raceme and

Fig. 105A–E. Melanthiaceae. *Heloniopsis orientalis*. **A** Flowering plant. **B** Flower. **C** Stamen. **D** Pistil. **E** Ovary in cross-section. (Takhtajan 1982)

Chionographis a spike. Thus it appears that the racemes and spikes actually represent botryoids or stachyoids, respectively, and are derived from panicles or thyrses.

The racemes and spikes usually have perfect flowers, while the panicles have andromonoecious or polygamous flowers (*Veratrum*, *Melanthium*, *Zigadenus* spp., *Stenanthium*), which may be the result of constraints in nutrient allocation in the many-flowered inflorescences. Among the raceme- and spike-bearing taxa, sex distribution in the *Chamaelirium-Chionographis* lineage is exceptionally diversified. *Chamaelirium* is dioecious, while *Chionographis* is often hermaphroditic or andromonoecious, sometimes gynodioecious, and rarely androdioecious.

Flower Structure. Flowers are usually actinomorphic, but zygomorphic in *Chionographis*, in which 2–3(–4) lower tepals are reduced in length or sometimes completely suppressed. Floral bracts are present in *Xerophyllum* and the Melanthieae, but usually absent in the Heloniadeae. Bracteoles are always absent.

Xerophyllum has no perigonal nectaries. Among the Melanthieae, *Veratrum* spp., *Melanthium* spp., and *Zigadenus* have nectaries at the tepal bases. Presence or absence of perigonal nectaries has often been used to distinguish between *Veratrum* and *Melanthium* (Gleason 1952; Gleason and Cronquist 1963; Satake 1982), but many *Veratrum* spp. (Loesener 1926) and most *Melanthium* spp. possess perigonal nectaries.

Among the Heloniadeae, *Heloniopsis* has a nectarial gland in a deep pocket between each stamen and its subtending tepal. In *Helonias* and *Ypsilandra*, a nectarial gland is positioned basally on the adaxial side of each tepal but does not develop a pocket.

The anthers have some features appearing derived among the Liliales. The anther thecae are often confluent in this family: in the Melanthieae, the *Helonias-Ypsilandra-Heloniopsis* lineage, and in *Chionographis koidzumiana* and *C. chinensis*. The thecae of *Helonias* and *Heloniopsis* were often considered as distinct, but they are confluent at the top. The anthers dehisce by valves in the Melanthieae (Fig. 104C) and by slits in *Xerophyllum* and the Heloniadeae. They are always extrorse.

The carpels are often not fully connate and thus retain primitive features among the Liliales. They are connate only in the lower part in *Amianthium* and *Stenanthium*, but for most of their length in *Ypsilandra* and *Heloniopsis*, and intermediate in the remaining genera. In *Ypsilandra* and *Heloniopsis*, a single style is developed, while in the remaining genera the carpels end in distinct styluli. The ovaries lack septal nectaries in this family. The stigmatic surface is non-papillate and Dry in *Veratrum*.

In *Xerophyllum* and some Melanthieae, the ovaries are provided with deep septal clefts between the carpels, while ovaries of the Heloniadeae and some Melanthieae have only shallow or almost no furrows along the septa.

The carpellary tissue of *Amianthium* and *Schoenocaulon* contains numerous idioblasts with single crystals, druses, or occasionally bundles of raphides. Also *Melanthium*, *Chamaelirium*, and *Xerophyllum asphodeloides* have raphide idioblasts. In *Zigadenus*, idioblasts are usually numerous but vary in the contents of crystalline inclusions. Also in *Veratrum*, idioblasts are always present but are sometimes not crystalliferous. In *Stenanthium*, the quantity of raphide idioblasts varies from numerous to few. They are always rare in *Helonias*, *Heloniopsis*, and *Chionographis*, and absent in *Xerophyllum tenax*. In *Ypsilandra* spp., single prismatic crystals (probably of calcium oxalate) occur in the carpellary parenchyma.

In the ovaries, the carpellary margins are closed and fused to each other for most of their length but are open and free in the distal region of the ovary or only near the top of the locule in *Melanthium*, *Helonias*, *Ypsilandra*, *Heloniopsis*, and *Chamaelirium*; in *Amianthium*, *Chionographis*, and *Xerophyllum asphodeloides* the carpel margins are open and free from each other down beyond the lowermost ovular insertion level. In *Veratrum*, *Zigadenus*, *Stenanthium*, *Schoenocaulon*, and *Xerophyllum tenax* the carpel margins are completely closed and fused among each other. The Melanthieae sometimes have holes at the basal part of the center of the pistil (Sterling 1982).

In *Helonias*, *Ypsilandra*, and *Heloniopsis*, each carpel has 2 enlarged placentae, which bear numerous ovules in 2–6 vertical rows on each placental margin. In the Melanthieae except *Amianthium*, each carpel has 2–4 vertical placental rows. In *Xerophyllum*, *Amianthium*, *Chamaelirium* and *Chionographis*, only 2–4 (8–12 in *Chamaelirium*) ovules are borne on small placentae in each carpel.

The vascular supply of the flowers has been studied by El-Hamidi (1952), Utech (1978a,b), Utech and Kawano (1981), and Sterling (1980, 1982). The latter author recognized three major types of vascular supply, each typical of (1) the Melanthieae, (2) *Xerophyllum*, *Chamaelirium*, and *Chionographis*, and (3) *Helonias*, *Ypsilandra*, and *Heloniopsis*, respectively, thus showing some correlation with the tribal classification.

Embryology. Successive cytokinesis in the microspore mother cells accompanies meiosis in *Veratrum* (Sokolowska-Kulczycka 1973, 1976). The ovules of the Melanthiaceae are all bitegmic. In Xerophylleae and Melanthieae, they are campylotropous and basitropic, while in *Veratrum* (Sokolowska-Kulczycka 1973, 1976) and *Schoenocaulon* they are sometimes anatropous or plagiotropic. In *Chamaelirium* and *Chionographis*, they are campylotropous and epitropic. In *Helonias*, *Ypsilandra*, and *Heloniopsis*, they are anatropous and plagiotropic, but sometimes epitropic in *Helonias* and sometimes basitropic in *Heloniopsis*.

The archesporial cell cuts off a primary parietal cell in *Veratrum*, *Zigadenus*, *Amianthium*, *Heloniopsis*, and *Chionographis*, although in *Amianthium* it sometimes functions also directly as the megaspore mother cell. A nucellar cap is formed from periclinal divisions of the apical nucellar epidermal cells in *Amianthium* and *Chionographis* (Davis 1966).

Three types of embryo sacs are formed in *Veratrum*: monosporic modified Polygonum type, bisporic Allium type, and bisporic Veratrum lobelianum type. The modified Polygonum type and the Veratrum lobelianum type deviate slightly from the typical Polygonum type and the Allium type, respectively, in second meiotic division of the embryo sac mother cell: the mycropylar cell divides earlier than the chalazal cell, resulting in a temporal formation of a triad (Sokolowska-Kulczycka 1973, 1976).

The synergids of *Amianthium* exhibit the filiform apparatus. *Veratrum*, *Amianthium* and *Heloniopsis* have multinucleate antipodal cells (Davis 1966). Endosperm formation is Helobial (Dahlgren et al. 1985). Embryogeny conforms to the Asterad type in *Heloniopsis* (Davis 1966).

Karyology. The basic chromosome numbers of this family are highly variable: in Xerophylleae it is $x = 15$, in Melanthieae $x = 8, 10, 11, 16$, and in Heloniadeae $x = 12, 17, 21, 22$.

In Melanthieae, $x = 8$, which was observed in *Veratrum*, *Melanthium*, *Stenanthium*, and *Amianthium*, may be ancestral (Goldblatt 1995; Tamura 1995). $x = 16$ of *Zigadenus* and *Amianthium* may have originated from $x = 8$ by polyploidization. $x = 10$ and $x = 11$ were additionally observed in *Stenanthium* and *Zigadenus*, respectively. The origins of the $x = 10, 11$ are uncertain but may be derivatives from $x = 8$ (Tamura 1995).

In Heloniadeae, the counterparts of the paired Arcto-Tertiary elements share the same chromosome number: *Chamaelirium* in N America and *Chionographis* in E Asia $2n = 24$ ($x = 12$), and *Helonias* in N America and *Heloniopsis* in E Asia

2n = 34 (x = 17). 2n = 42 (x = 21) were counted for *Chionographis japonica* var. *hisauchiana* and var. *minoensis* and 2n = 44 (x = 22) for *Chionographis japonica* var. *kurohimensis*. From the viewpoints of cytology (Tanaka and Tanaka 1980) and allozyme variation (Maki 1992b), the numbers 2n = 42 and 44 are considered to have originated from 2n = 24 by allotetraploidization and successive reductions in chromosome number.

Although the origins of the basic chromosome numbers of some taxa of this family were affected by polyploidization, the present melanthiaceous genera are all diploid except *Veratrum*, in which ploidy levels vary from diploidy to dodecaploidy. Especially in *Veratrum album* subsp. *oxysepalum*, which extends from E Siberia to Japan, cytotypes based on 4x, 9x, 10x, and 12x exist.

Chromosome size of the Xerophylleae and the Melanthieae is medium, ranging 2.5–4.8 μm and 1.3–4 μm in length, respectively. Also the original chromosome size of this family may be presumed as medium (Tamura 1995). In the Heloniadeae, chromosome sizes are differently derived: *Helonias* and *Heloniopsis* have comparatively large chromosomes (1.8–6 μm) and *Chamaelirium* and *Chionographis* (southwest Japanese type) have small chromosomes (1–2 μm), although *Chionographis* (Normal type) retains the medium-sized chromosomes (1.7–3.2 μm).

Karyotypes of this family mostly consist only of metacentric and submetacemtric chromosomes, but in *Helonias* and *Heloniopsis* karyotypes are more or less asymmetric, containing 1-several subtelocentric chromosomes (M.N. Tamura and R. Ohfuji, unpubl.). Karyotypes of *Chionographis* are much specialized, consisting of holocentric chromosomes (Tanaka and Tanaka 1977, 1979, 1980), although *Chamaelirium*, its N American counterpart, has a localized kinetochore on each chromosome (Tanaka 1985; M.N. Tamura and R. Ohfuji, unpubl.).

Pollen Morphology. The pollen is mostly sulcate, but sulculate in *Heloniopsis* (K. Handa et al., unpubl.) and 4-porate in *Chionographis* and *Chamaelirium*. The grains are often spheroidal (21–32 μm in longest axis), sometimes spherical (*Chionographis* and *Chamaelirium*; 12–21 μm in diam.) (Takahashi and Kawano 1989).

Exine sculpture is often reticulate, although the muri are sometimes discontinuous; in *Chionographis* and *Chamaelirium* it is clavate and in *Veratrum stamineum* gemmate (K. Handa et al., unpubl.). In *Xerophyllum* the lumina are angular, 0.5–1.0 μm in diam., and in *Amianthium*, *Stenanthium*, and *Schoenocaulon* they are also angular and 0.3 μm in diam. In the latter three genera the muri are comparatively wide (0.3–0.6 μm thick), but narrow (0.1–0.2 μm thick) in *Xerophyllum* (Takahashi and Kawano 1989). *Helonias*, *Ypsilandra*, and *Heloniopsis* have the exine composed of an amorphous layer and spinules, which are ovate, 1 μm high, with acute tips.

The pollen grains of *Helonias*, *Ypsilandra*, and *Heloniopsis* and those of *Chionographis* and *Chamaelirium*, respectively, are similar in almost every detail, and both pollen types can be considered as advanced in this family.

Pollination. *Heloniopsis orientalis* is self-compatible but has protogynous flowers. The receptive stigma emerges from the perianth before it opens, and the anthers dehisce about 2 days after the opening of the perianth. Receptivity of the stigma is maintained for about 8 days. Many species of Diptera and Hymenoptera forage on the flowers, and almost all insects can be effective pollinators; self-pollination may be rather common when insect activity is low (Takahashi 1988).

Veratrum viride has protandrous flowers and is possibly cross-pollinated by insects (Mulligan and Munro 1987). *Veratrum album* and *V. nigrum* were also reported as protandrous (Loew and Kirchner 1911).

Fruit and Seed. The fruits of this family are always capsules but show various forms of dehiscence: septicidal in the *Chamaelirium-Chionographis* lineage (Heloniadeae), ventricidal in the Melanthieae, and loculicidal in Xerophylleae and the *Helonias-Ypsilandra-Heloniopsis* lineage (Heloniadeae).

The seeds are angular in *Xerophyllum* and part of the Melanthieae, while they are winged at one end in *Chionographis*, broadly winged around the body in *Veratrum*, *Melanthium*, and *Chamaelirium*, and longly caudate at both ends in the *Helonias-Ypsilandra-Heloniopsis* lineage. The seeds of this family lack phytomelan but sometimes possess phlobaphene deposits (*Amianthium*).

The embryo is often ovoid or globose, small, 1/5–1/9 of the endosperm in length, but sometimes linear, large, almost equalling the seed in length (*Melanthium* spp., *Zigadenus* spp.).

Dispersal. Winged seeds as in *Veratrum* (Mulligan and Munro 1987), *Melanthium*, and *Chamaelirium* and those with long caudations in *Heloniopsis* (Kawano 1989), *Ypsilandra*, and *Helonias* can be considered as adapted to wind dispersal.

In *Chionographis japonica* var. *japonica* and var. *kurohimensis*, individuals growing within 1 m distance share high genetic similarity, as evidenced by the spatial autocorrelation analysis of enzyme polymorphism (Maki and Masuda 1993, 1994). *Chionographis japonica* has no specialized mechanism of seed dispersal. Thus, gene flow in *C. japonica* via seed dispersal seems to be highly restricted.

Reproductive Systems. *Chionographis japonica* var. *japonica* is self-incompatible and outcrossing, while var. *kurohimensis* is predominantly selfing (>90%) (Maki 1992a, 1993a). *C. japonica* var. *kurohimensis* is andromonoecious and also gynodioecious. According to Maki (1993b), the ratio of male/bisexual flowers in an individual decreases when plant size increases. The ratio is not fixed genetically but varies reflecting environmental conditions. However, the ratio is not affected by the female individual frequency of the same population. This evidence suggests that *C. japonica* var. *kurohimensis* is not on the evolutionary pathway from bisexuality to dioecy, which was proposed by Lloyd (1976) and Charlesworth and Charlesworth (1978).

As for *Veratrum viride*, heavy seed-crop years are generally followed by 5 or more years of low seed production. The seeds germinate promptly with stratification at 3–5 °C for 4–4.5 months. The plants usually do not flower until 7–10 years after germination (Taylor 1956). They are estimated to persist occasionally for more than a century (Mulligan and Munro 1987).

Heloniopsis often produces propagules at tips of old leaves when the leaf tips touch the ground. Each leaf usually survives for 3 years.

Parasites. Two related species of the rust genus *Puccinia* (section *Caricinae*) attack only *Amianthium*, *Schoenocaulon*, *Veratrum*, *Zigadenus*, and *Xerophyllum* (Holm 1966; Savile 1979), which indicates that these genera form a genetically related group.

Phytochemistry. Alkaloids derived from steroidal precursors such as cholestanol as well as those of the C-nor-D-homo type (Seigler 1977) are present in most genera in the Veratreae (Hegnauer 1963, 1986; Dahlgren et al. 1985). The ceveratrum alkaloids are found in *Veratrum*, *Zigadenus*, and *Schoenocaulon*, and the jerveratrum alkaloids in *Veratrum* and *Amianthium* (Kupchan et al. 1961).

The alkaloids of *Veratrum* are esters of highly hydroxylated parent alkanolamine bases and include cevine, germine, protoverine, cevadine, veratridine, and protoveratrine (Kupchan and Gruenfeld 1959). As toxic alkaloids, *Veratrum* contains in the roots and rhizomes veratramin, jervin, and allied compounds (Campbell et al. 1985). *Zigadenus* contains zigadenin, especially in the bulbs. *Amianthium* contains jervin and amianthin in the bulbs and leaves. Steroidal alkaloids similar to those of Veratreae are also found in Liliaceae (*Fritillaria* and *Notholirion*).

According to Dahlgren and Clifford (1982), steroid saponins are found in *Veratrum*, *Zigadenus*, *Schoenocaulon*, *Heloniopsis*, *Chamaelirium*, and *Chionographis*, and chelidonic acid in *Veratrum*, *Melanthium*, *Zigadenus*, *Schoenocaulon*, and *Heloniopsis*. Both steroid saponins and chelidonic acid are widely distributed in the Lilianae (Hegnauer 1963).

Veratrum contains luteolin and sometimes also apigenin as leaf flavonoids but neither quercetin nor kaempferol (Williams 1975). *Xerophyllum tenax* contains a unique type of biflavonoid (Williams et al. 1988).

Distribution and Habitats. This family is distributed mostly in temperate regions of the Northern Hemisphere and is abundant in N America; a few *Schoenocaulon* spp. extend to S America.

The Xerophylleae are found in the USA. Among the Melanthieae, *Veratrum* extends over the temperate regions of the Northern Hemisphere. *Zigadenus* and *Stenanthium* are abundant in N America but have each one species in northern E Asia. The genera of the Heloniadeae comprise paired Arcto-Tertiary elements: *Chamaelirium* (eastern N America) and *Chionographis* (S China to Japan) forming a pair, and *Helonias* (eastern N America) and the *Ypsilandra-Heloniopsis* lineage (E Asia) another, in which *Ypsilandra* (W China to the Himalayas) and *Heloniopsis* (Taiwan, Korea to Japan) can be considered as having differentiated from each other in E Asia.

Veratrum viride Ait. is notable for tolerating winter temperatures to −40 °C in alpine areas of Canada (Mulligan and Munro 1987).

Subdivision and Relationships Within the Family. The Melanthiaceae can be subdivided into three tribes: (1) Xerophylleae, (2) Melanthieae, and (3) Heloniadeae.

1. The Xerophylleae are considered as monotypic in the present treatment, consisting only of *Xerophyllum*, which agrees with the concept of Dahlgren et al. (1985). *Xerophyllum* has sometimes been merged with the Heloniadeae (Bentham and Hooker 1883; Krause 1930), and a relationship between *Xerophyllum* and the Heloniadeae has recently been confirmed by the parsimony analysis of *rbc*L sequences (Chase et al. 1995). On the other hand, *Xerophyllum* and the Melanthieae share the possession of floral bracts and the susceptibility to 2 rust fungi: *Puccinia atropuncta* and *Uromyces veratri*. The latter characteristic has been regarded by Dahlgren et al. (1985) as a synapomorphy for the Melanthiaceae s.l., but Goldblatt (1995) limited this characteristic to *Xerophyllum* and the Melanthieae.

   Xerophyllum maintains some primitive characteristics among the Melanthiaceae, e.g., the possession of floral bracts, distinct anther thecae, anther dehiscence by slits, and styluli. Anderson (1940) indicated that there are no advanced anatomical characters present except for the reduction in ovule number in *Xerophyllum*. Sterling (1980) considered even the small number of the ovules (biovulate in *Xerophyllum asphodeloides*) as primitive.
2. The Melanthieae, which comprise six genera, are well defined morphologically (Bentham and Hooker 1883; Krause 1930; Hutchinson 1934; Melchior 1964; Schulze 1978). They share many characteristics also in respect to floral anatomy (Sterling 1982) and secondary chemistry (Kupchan et al. 1961; Hegnauer 1963). They share the basic chromosome number $x = 8$ (or $x = 16$) (Tamura 1995). According to the numerical analysis of anatomical and morphological characters by Ambrose (1980), the five genera of this tribe form a compact cluster (*Schoenocaulon* not having been analyzed). According to the *rbc*L sequence analysis by Chase et al. (1995), *Veratrum*, *Zigadenus*, and *Amianthium* form a strongly supported clade (decay value of 21).
3. The Heloniadeae consist of five genera, whose coherence was supported by Buxbaum (1925, 1927) and Schulze (1978). Among these, three (*Helonias*, *Ypsilandra*, *Heloniopsis*) resemble one another morphologically and anatomically (Gates 1918; Buxbaum 1927; Utech 1978b; Sterling 1980; Utech and Kawano 1981; Tanaka 1997), palynologically (Takahashi and Kawano 1989; K. Handa et al., unpubl.), and karyologically (Utech 1980; M.N. Tamura and R. Ohfuji, unpubl.).

   The remaining two genera, *Chamaelirium* and *Chionographis*, also resemble each other morphologically, palynologically (Takahashi and Kawano 1989), and karyologically (M.N. Tamura and R. Ohfuji, unpubl.). *Chamaelirium* has a raceme, actinomorphic flowers, and chromosomes with localized centromeres, while *Chionographis* has a spike, zygomorphic flowers, and chromosomes with dispersed centromeres, and seems to be advanced over *Chamaelirium*.

   This tribe has several characters that are derived within the family: the whole tribe usually lacks floral bracts; *Helonias*, *Ypsilandra* and *Heloniopsis* share sulculate (or sulcate) pollen grains with a spinulose exine and large chromosomes with a basic chromosome number $x = 17$; and *Chamaelirium* and *Chionographis* share 4-porate pollen grains with a clavate exine.

Affinities. Tamura (1995) suggested a separation of the Melanthiaceae s.l. into the *Narthecium* group and the *Heloniopsis* group. Primarily, the *Narthecium* group has bracteoles, introrse or latrorse anthers, exines with roundish (or nonangular) brochi, and very small chromosomes, while the *Heloniopsis* group lacks bracteoles and possesses extrorse or subextrorse anthers, angular brochi (provided that exine ornamentation is reticulate), and relatively large chromosomes. Goldblatt (1995) also suggested the same division, emphasizing the difference in nectaries, septal in the former and perigonal in the latter group. In the present circumscription, the Melanthiaceae correspond to the *Heloniopsis* group, and the *Narthecium* group is treated as the Nartheciaceae.

The Melanthiaceae s. l. have been considered as one of the most primitive plant groups in the monocotyledons (e.g., Hutchinson 1934; Walker 1986; Thorne 1992). In contrast to the Nartheciaceae, which have small chromosomes, the Melanthiaceae in the present circumscription usually have medium-sized chromosomes. The possession of extrorse or subextrorse anthers and perigonal nectaries strongly suggests the inclusion of the present Melanthiaceae in the Liliales (Goldblatt 1995; Tamura 1995). Also according to the *rbc*L sequence analysis by Chase et al. (1995), the Melanthiaceae in the present circumscription appear as part of a clade that further comprises the Colchicaceae, Alstroemeriaceae, Trilliaceae, and *Luzuriaga*. A close relationship between the families with small embryos (Melanthiaceae, Colchicaceae, and Trilliaceae) has already been suggested by Bentham and Hooker (1883), and that between the Melanthiaceae and Colchicaceae by Krause (1930). It is still uncertain which family is most closely related to the Melanthiaceae, but the Colchicaceae and the Trilliaceae seem to be the best candidates.

Dahlgren et al. (1985) suggested a relationship between the Melanthiaceae and the Campynemataceae. However, this relationship is not supported by embryology and karyology (Lowry II et al. 1987) and the *rbc*L topology (Chase et al. 1995).

Economic Importance. *Veratrum viride* was used by North American Indians in medicine and rituals. It was used by Cherokee Indians on skin abrasions, probably for the strong counter-irritant effect (Taylor 1940). Blackfoot Indians took powdered root for headaches and for itching. Juice from boiled roots was combed through children's hair to effectively kill lice. Pieces of root were also used as luck charms (Mulligan and Munro 1987).

Veratrum viride contains many alkaloids which have been useful in the treatment of hypertension disorders in humans (Osol and Farrar 1955). Active preparations caused a decrease in systolic and diastolic pressure, decreased heart rate, and increased peripheral blood flow (Kauntze and Trounce 1951). However, the modern formulation, veriloid, was removed from general use for hypotensive therapy in about 1961 because of side effects (Hightower 1979).

Veratrum viride is well known as a poisonous plant. Symptoms of poisoning in animals include salivation, muscular tremors, low temperature, and death resulting from heart paralysis (Kingsbury 1964). The plant is poisonous also for humans (cf. Mulligan and Munro 1987).

The alkaloids of the seeds of *Schoenocaulon* are medically useful. They are dissolved in acetic acid and frequently used as insecticides. The bulb of *Amianthium* is very poisonous.

Key to the Genera

1. Underground stem often a bulb, sometimes a rhizome; floral bracts present; styluli 3; $x = 8, 10, 11, 15, 16$ — 2
- Underground stem a rhizome; floral bracts usually absent or obsolete, rarely present; if floral bracts present, style 1, long, filiform; $x = 12, 17, 21, 22$ (**Heloniadeae**) — 8

2. Foliage leaves setaceous; bracts inserted at the basal to middle part of the pedicels; anther thecae distinct, dehiscing by slits; stigmas decurrent on inner side of styluli; capsules loculicidal; $x = 15$ (**Xerophylleae**) — 1. ***Xerophyllum***
- Foliage leaves linear to broadly elliptical; bracts inserted at the pedicel base; anther thecae confluent, dehiscing by valves; stigmas terminal on styluli; capsules ventricidal; $x = 8, 10, 11, 16$ (**Melanthieae**) — 3

3. Underground stem usually a rhizome, rarely a bulb (*Melanthium* spp.); aerial stem thickened at the base; upper part of aerial stem and inflorescence pubescent; flowers polygamomonoecious; seeds flat, broadly winged — 4
- Underground stem usually a bulb, rarely a rhizome (*Zigadenus* spp.); aerial stem without thickening; aerial stem and inflorescence glabrous; flowers usually hermaphrodite or polygamous, rarely staminate; seeds ovoid, ellipsoid or linear-cylindric, barely winged or wingless — 5

4. Foliage leaves plicate; tepals not clawed, often without nectar glands, sometimes with the glands at the base (*V. viride, V. californicum*) — 2. ***Veratrum***
- Foliage leaves not plicate; tepals clawed, often with nectar glands at the base, sometimes without the glands (*M. parviflorum*) — 3. ***Melanthium***

5. Tepals with 2 distinct or connate conspicuous nectar glands at the base — 4. ***Zigadenus***
- Tepals without glands or with inconspicuous nectar glands at the base (*Schoenocaulon*) — 6

6. Flowers hermaphrodite; capsules conical; seeds 1–2 in each carpel — 5. ***Amianthium***
- Flowers polygamous, hermaphrodite or staminate; capsules oblong-subcylindric to ovoid-urceolate; seeds several in each carpel — 7

7. Foliage leaves cauline and basal; inflorescence a panicle; stamens much shorter than tepals; ovary half-inferior to superior — 6. ***Stenanthium***
- Foliage leaves all basal; inflorescence a scapose spike or a raceme; stamens longer than tepals; ovary superior — 7. ***Schoenocaulon***

8. Inflorescence a raceme, sometimes umbel-like; flowers hermaphrodite; tepals usually spathulate, oblanceolate to oblong, rarely linear (*Ypsilandra alpinia*), 3–5-nerved; capsules 3-lobed to 3-parted, loculicidal; seeds caudate; $x = 17$ — 9
- Inflorescence a spike or a raceme; if a raceme, flowers dioecious; tepals usually filiform to linear, rarely spatulate (*Chamaelirium*), 1-nerved; capsules unlobed, septicidal; seeds winged; $x = 12, 21, 22$ — 11

9. Flowers 40–60; styluli 3, stigmatic on ventral side — 8. ***Helonias***
- Flowers 3–30; style 1, entire to 3-cleft, stigmatic terminally or on inner side — 10

10. Inflorescence a raceme; tepals without a pocket; stamens free from tepals; anthers hippocrepiform, ca. 1.8 mm long, basifixed — 9. ***Ypsilandra***
- Inflorescence umbel-like; tepals often with a pocket basally on the adaxial side; stamens often adnate to the tepals at the base, anthers lanceolate, 2.5–4.5 mm long, dorsifixed — 10. ***Heloniopsis***

11. Inflorescence a raceme; flowers actinomorphic, dioecious; ovules many — 11. ***Chamaelirium***
- Inflorescence a spike; flowers zygomorphic, often hermaphrodite, sometimes andromonoecious or dioecious; ovules 2 in each carpel — 12. ***Chionographis***

I. Tribe Xerophylleae Watson (1879).

1. *Xerophyllum* Michx.

Xerophyllum Michx., Fl. Bor. Am. I: 210 (1803); Utech, Ann. Carnegie Mus. 47: 147–167 (1978).

Underground stem a bulb. Basal foliage leaves densely tufted, setaceous, up to 100 cm long, dry, rigid. Aerial stem erect, simple, with many small filiform foliage leaves. Raceme terminal. Flowers cup-shaped to opening flat, hermaphroditic, actinomorphic. Bracts present. Tepals free, 5–7-nerved, marcescent, without nectar glands. Stamens as long as tepals; filaments subulate, inserted at tepal base; anthers dorsifixed versatile, dehiscent by slits, extrorse, with distinct thecae. Ovary superior, syncarpous with 3 recurved styluli; ovules 2 (*X. asphodeloides*) or 4 (*X. tenax*) in each carpel. Capsules loculicidal; seeds oblong, 3-angled, 3–4 mm long, without wings and appendages. $x = 15$. Two spp., USA.

II. **Tribe Melanthieae** Dahlgren et al. (1985).

2. ***Veratrum*** L.

Veratrum L., Sp. Pl.: 1044 (1753); Loesener, Feddes Rep. 24: 61–72, 25: 1–10 (1927/28), synopsis.

Rhizome thick, short. Foliage leaves spiral, only cauline or both basal and cauline, linear to broadly elliptical, 10–50 cm long, plicately nerved, sheathing. Aerial stem erect, stout, basally thickened. Panicle terminal, pubescent. Flowers funnel-shaped, cup-shaped to opening flat, white, yellowish, green, or dark purple-brown, andromonoecious or polygamous, actinomorphic. Bracts lanceolate to ovate. Tepals free or rarely connate at base, multinerved, marcescent, often without nectar glands, sometimes with glands at the base. Stamens inserted at tepal base; anthers cordate-orbicular, dehiscent by valves, extrorse, with confluent thecae. Ovary superior to inferior, syncarpous except for the shortly papillose styluli, containing 16–85 ovules in each carpel. Capsules ventricidal; seeds elliptical, flat, broadly winged. $x = 8$. About 50 spp., the temperate regions of the Northern Hemisphere.

3. ***Melanthium*** L.

Melanthium L., Sp. Pl.: 339 (1753).

Underground stem a rhizome or bulb. Foliage leaves spiral, linear to oblanceolate or elliptical, ca. 30 cm long, not plicate, sheathing. Aerial stem erect, leafy, basally thickened. Panicle terminal, pubsecent. Flowers funnel-shaped, cup-shaped to opening flat, cream, greenish to purplish, andromonoecious or polygamous, actinomorphic. Bracts present. Tepals clawed at base, marcescent, often with 2 nectar glands near base. Stamens inserted at tepal claws; anthers cordate to reniform, dehiscent by valves, extrorse, with confluent thecae. Ovary superior, syncarpous, with 3 styluli; ovules 20–22 in each carpel. Capsules ventricidal; seeds flat, elliptical to narrowly obovate, 5–10 mm long, yellowish brown, broadly winged. $x = 8$. About 5 spp., eastern USA.

4. ***Zigadenus*** Michx. Fig. 104A–F

Zigadenus Michx., Fl. Bor. Am. I: 213, t. 22 (1803).

From horizontal rhizome or bulb. Foliage leaves spiral, basal and cauline, linear to narrowly lanceolate, 10–50 cm long. Aerial stem simple, erect. Raceme or panicle terminal, glabrous. Flowers funnel-shaped, cup-shaped to opening flat, white, yellowish, greenish, brownish to purplish, hermaphroditic or polygamous, actinomorphic. Bracts narrowly lanceolate to oval. Tepals often more or less connate, persistent, with 2 distinct or connate nectar glands inside near base. Stamens inserted at tepal base, slightly shorter than tepals; anthers suborbicular, small, basifixed, dehiscent by valves, extrorse, with confluent thecae. Ovary superior to half-inferior, syncarpous, with 3 slender styluli. Capsule ventricidal; seeds ovoid, ellipsoid to linear-cylindric, 3–6 mm long, angular. $x = 11, 16$. About 10 spp., mostly N America, 1 sp. in Asia.

5. ***Amianthium*** A. Gray

Amianthium A. Gray, Ann. Lyc. New-York IV: 121 (1837).

Bulb tunicated, thick. Foliage leaves mostly basal, linear, up to 40 cm long. Aerial stem erect, with few much reduced foliage leaves. Raceme dense, terminal, glabrous. Flowers cup-shaped, white, yellowish, greenish, or purplish, hermaphroditic, actinomorphic. Bracts present. Tepals without nectar glands. Anthers cordate-ovate, dehiscent by valves, with confluent thecae. Ovary superior, carpels connate only in lower part; styluli 3, with minute stigmas; ovules 2–4 in each carpel. Capsule ventricidal; seeds 1–4 in each locule, ellipsoid, blackish brown. $x = 16$. Only one sp., *A. muscaetoxicum* (Walt.) Gray, eastern N America.

6. ***Stenanthium*** (A. Gray) Kunth

Stenanthium (A. Gray) Kunth, Enum. pl. IV: 189 (1843).

Bulbous. Foliage leaves spiral, linear, 20–40 cm long, conduplicate, keeled. Aerial stem erect, slender, leafy, with fibrous leaf sheath at base. Panicle terminal, glabrous. Flowers campanulate to funnel-shaped, white, greenish or bronze-purplish, andromonoecious, actinomorphic, sessile or shortly pedicellate. Bracts ca. 2 mm long. Tepals persistent, without nectar glands. Stamens much shorter than tepals; anthers obcordate, dehiscent by valves, extrorse, with confluent thecae. Ovary superior to half-inferior, carpels connate only in lower part; styluli short, stout, recurved; ovules numerous. Capsule ventricidal; seeds obliquely lanceolate, flat, 3–8 mm long, wingless. $x =$ 8, 10. About four spp., mostly N America, one sp. endemic to Sakhalin.

7. *Schoenocaulon* A. Gray Fig. 104G–K

Schoenocaulon A. Gray, Ann. Lyc. New-York IV: 127 (1837).
Sabadilla Brandt & Ratzebg. (1837).

Plant from bulb or corm. Foliage leaves basal, linear, long. Aerial stem simple, erect, without foliage leaves. Spike or raceme terminal on an elongate, glabrous scape. Bracts small. Tepals free, pale green to yellowish white, persistent, with serrulate margin, with inconspicuous nectar glands. Stamens longer than tepals; anthers suborbicular, small. Ovary superior; carpels connate except for short styluli; ovules 2–12 in each carpel. Capsules ventricidal; seeds linear, slightly curved, angular, blackish brown. $x = 8$. About ten spp., Florida and Texas to Venezuela and Peru.

III. Tribe Heloniadeae Reichb. (1828).

8. *Helonias* L.

Helonias L., Sp. Pl.: 342 (1753); Utech, Ann. Carnegie Mus. 47: 169–191 (1978).

Rhizome thick, short. Foliage leaves basal, rosulate, evergreen. Peduncle erect, simple, stout, hollow, with bract-like foliage leaves mostly in lower part. Raceme short, dense, terminal, with 40–60 flowers. Flowers funnel-shaped, hermaphrodite, actinomorphic, fragrant. Bracts obsolete. Tepals 3–5-nerved, persistent, with a nectarial gland basally on adaxial side. Inner stamens adnate to the ovary at base, outer free; anthers elliptical on abaxial side, hippocrepiform on adaxial side, blue, extrorse, dorsifixed, with confluent thecae. Ovary superior, syncarpous, except for the free or basally connate styluli; ovules 16 in each carpel. Capsule loculicidal; seeds linear, 3.5–4.8 mm long, longly caudate at both ends. $x = 17$. Only one sp., *H. bullata* L., eastern N America.

9. *Ypsilandra* Franch.

Ypsilandra Franch., Nouv. Arch. Mus. hist. nat. Paris II, 10: 93, t. 17 (1888).

Rhizome thick, short. Foliage leaves basal, rosulate, spathulate or oblanceolate to linear. Peduncle erect, simple, with several sheathing or bractlike, small foliage leaves. Raceme terminal, 4–30-flowered. Flowers hermaphroditic, actinomorphic, nodding in anthesis, erect in fruit. Bracts absent. Tepals persistent, with a nectarial gland basally on adaxial side. Inner stamens adnate to the ovary at base, outer free; anthers hippocrepiform, ca. 1.8 mm long, basifixed, with confluent thecae. Ovary superior, syncarpous; style 1–20 mm long, stigma 3-cleft to capitate; ovules numerous. Capsules loculicidal; seeds linear, longly caudate at both ends. Chromosome number unknown. Four spp., W China to the Himalayas.

10. *Heloniopsis* A. Gray Fig. 105

Heloniopsis A. Gray, Mem. Amer. Acad. II, 6: 416 (1859).

Rhizome thick, short. Foliage leaves spiral, basal, rosulate, narrowly oblong to oblanceolate or obovate, 5.5–20 cm long, evergreen, with smooth margin, sometimes minutely undulate. Peduncle arising from the center of leaf-rosette, hollow, with 2–8 scaly leaves. Umbel-like raceme terminal, 3–10-flowered. Flowers funnel-shaped, hermaphrodite, actinomorphic, nodding or horizontal in anthesis, erect in fruit. Bracts usually absent. Tepals often connate at base, often adnate to the ovary at base, sometimes free, spathulate to oblong, obtuse, 3–5-nerved, persistent, often with a deep pocket of a nectarial gland basally on adaxial side. Stamens often adnate to the tepals at base, sometimes free; anthers lanceolate, purple, dorsifixed, extrorse, with thecae confluent at the top. Ovary superior; style filiform, long, sunken; stigma capitate; ovules 60–180 in each carpel. Capsules loculicidal; seeds linear, 4.1–5.6 mm long, longly caudate at both ends. $x = 17$. Five spp., Taiwan, Korea, and Japan.

11. *Chamaelirium* Willd.

Chamaelirium Willd., Magaz. Ges. naturf. Fr. Berlin 2: 18 (1808).

Rhizome thick. Foliage leaves spiral, rosulate, spathulate to obovate, 10–20 cm long, evergreen, with entire or minutely undulate margin. Peduncle erect, with 11–17 small linear foliage leaves. Raceme terminal, slender. Flowers dioecious, funnel-shaped, actinomorphic. Bracts absent. Tepals free, linear to spatulate, 1-nerved, marcescent. Stamens rudimentary in pistillate flowers; anthers lanceolate to short-oblong, white, dorsifixed, extrorse, with distinct thecae. Ovary superior, syncarpous except for the linear-clavate to revolute styluli, rudimentary in staminate flowers; ovules 6–12 in each locule. Capsules septicidal; seeds linear-oblong, with white, coarsely cellular wing. $x = 12$. Only one sp., *C. luteum* (L.) A. Gray, eastern N America.

12. *Chionographis* Maxim.

Chionographis Maxim., Bull. Acad. St. Petersbourg 11: 435 (1867); Hara, J. Jap. Bot. 43: 257–267 (1968), rev.; N.Y. Tanaka, Shuseibutsugaku-kenkyu 8: 11–19 (1985).

Rhizome thick, short. Foliage leaves spiral, rosulate, spathulate to elliptical, 2.8–16 cm long, evergreen, with entire or minutely undulate margin. Peduncle erect, simple, with 3–30 small foliage leaves. Spike terminal. Flowers often hermaphroditic or andromonoecious, sometimes gynodioecious, rarely androdioecious, zygomorphic, sometimes scented. Bracts absent. Tepals free, 1-nerved, the upper 4-3(-2) filiform to linear, 3–20 mm long, the lower 2–3(–4) very short or suppressed, without nectar glands. Stamens inserted at tepal base; anthers suborbicular, extrorse, basifixed, with thecae confluent or distinct. Ovary superior to quarter-inferior, syncarpous except for revolute styluli; ovules 2 in each locule. Capsules septicidal; seeds fusiform, winged at one end. $x = 12, 21, 22$. Four spp., S China to Japan.

Selected Bibliography

Ambrose, J.D. 1980. A re-evaluation of the Melanthioideae (Liliaceae) using numerical analyses. In: Brickell, C.D., Cutler, D.F., Gregory, M. (eds.) Petaloid monocotyledons, 65–81. London: Academic Press. (Linnean Society Symposium Series, 8)

Anderson, C.E. 1940. Some studies on the floral anatomy of the Liliales. Ph.D. Thesis, Cornell University, Ithaca, New York.

Bentham, G., Hooker, J.D. 1883. Genera Plantarum, Vol. 3. London: Reeve; Williams & Norgate.

Buxbaum, F. 1925. Vergleichende Anatomie der Melanthioideae. Repert. Sp. Nov. Regni Veg., Beih. 29: 1–80.

Buxbaum, F. 1927. Nachträge zur vergleichenden Anatomie der Melanthioideae I. Beih. Bot. Centralbl. 44: 255–263.

Campbell, M.A., Brown, K.S., Hassell, J.R., Horigan, E.A., Keeler, R.F. 1985. Inhibition of limb chondrogenesis by a *Veratrum* alkaloid; temporal specificity in vivo and in vitro. Dev. Biol. 111: 464–470.

Charlesworth, B., Charlesworth, D. 1978. A model for the evolution of dioecy and gynodioecy. Am. Nat. 112: 975–997.

Chase, M.W. et al. 1995. See general references.

Cheadle, V.I., Kosakai, H. 1971. Vessels in Liliaceae. Phytomorphology 21: 320–333.

Dahlgren, R.M.T., Clifford, H.T. 1982. See general references.

Dahlgren, R.M.T. et al. 1985. See general references.

Davis, G.L. 1966. See general references.

El-Hamidi, A. 1952. Vergleichend-morphologische Untersuchungen am Gynoeceum der Unterfamilien Melanthioideae und Asphodeloideae der Liliaceae. Arb. Inst. Allg. Bot. Univ. Zürich A, 4: 1–50.

Frölich, D., Barthlott, W. 1988. See general references.

Gates, R.R. 1918. A systematic study of the North American Melanthiaceae from the genetic standpoint. J. Linn. Soc. Lond. (Bot.) 44: 131–172.

Gleason, H.A. 1952. Illustrated flora of the northeastern United States and adjacent Canada, Vol. 1. Lancaster: Lancaster Press.

Gleason, H.A., Cronquist, A. 1963. Manual of vascular plants of northeastern United States and adjacent Canada. Boston: Willard Grant Press.

Goldblatt, P. 1995. The status of R. Dahlgren's orders Liliales and Melanthiales. In: Rudall, P.J., Cribb, P.J., Cutler, D.F., Humphries, C.J. (eds.), Monocotyledons: systematics and evolution, pp. 181–200. Royal Botanic Gardens, Kew.

Hegnauer, R. 1963, 1986. See general references.

Hightower, C.E. 1979. Plants that kill and cure. Vet. Hum. Toxicol. 21: 360–362

Holm, L. 1966. Etudes uredinologiques, 4. Sur les *Puccinia caricicoles* et leurs alliés. Sven. Bot. Tidskr. 60: 23–32.

Hutchinson, J. 1934. The families of flowering plants, Vol. 2. London: Macmillan.

Kauntze, R., Trounce, J. 1951. The hypotensive action of veriloid (*Veratrum viride*). A clincial investigation. Lancet 260: 549–555.

Kawano, S. (ed.) 1989. Newton special issue, Shokubutu no Sekai, Vol. 4. Tokyo: Kyoikusha (in Japanese).

Kingsbury, J.M. 1964. Poisonous plants of the United States and Canada. Englewood Cliffs: Prentice-Hall.

Krause, K. 1930. See general references.

Kupchan, S.M., Gruenfeld, N. 1959. The hypotensive principles of cryptenamine, a *Veratrum viride* alkaloid preparation. J. Am. Pharm. Assoc. Sci. Ed. 48: 727–730.

Kupchan, S.M., Zimmerman, J.H., Afonso, A. 1961. The alkaloids and taxonomy of *Veratrum* and related genera. Lloydia 24: 1–22.

Lloyd, D.G. 1976. The transmission of genes via pollen and ovules in gynodioecious angiosperms. Theor. Popul. Biol. 9: 299–316.

Loesener, O. 1926. Studien über die Gattung Veratrum und ihre Verbreitung. Verh. Bot. Ver. Prov. Brandenb. 68: 106–166.

Loew, E., Kirchner, O. 1911. Melanthioideae, pp. 221–295. In: Kirchner, O., Loew, E., Schröter, C. (eds.) Lebensgeschichte der Blütenpflanzen Mitteleuropas, Vol. I, Abt. 3. Stuttgart: Ulmer.

Lowry II, P.P., Goldblatt, P., Tobe, H. 1987. Notes on the floral biology, cytology, and embryology of Campynemanthe (Liliales: Campynemataceae). Ann. Mo. Bot. Gard. 74: 573–576.

Maki, M. 1992a. Fixation indices and genetic diversity in hermaphroditic and gynodioecious populations of Japanese *Chionographis* (Liliaceae). Heredity 68: 329–336.

Maki, M. 1992b. Allozyme variation in Japanese species of *Chionographis* (Liliaceae). Bot. Mag. (Tokyo) 105: 149–160.

Maki, M. 1993a. Outcrossing and fecundity advantage of females in gynodioecious *Chionographis japonica* var. *kurohimensis* (Liliaceae). Am. J. Bot. 80: 629–634.
Maki, M. 1993b. Floral sex ratio variation in hermaphrodites of gynodioecious *Chionographis japonica* var. *kurohimensis* Ajima et Satomi (Liliaceae). J. Plant Res. 106: 181–186.
Maki, M., Masuda, M. 1993. Spatial autocorrelation of genotypes in a gynodioecious population of *Chionographis japonica* var. *kurohimensis* (Liliaceae). Int. J. Plant Sci. 154: 467–472.
Maki, M., Masuda, M. 1994. Spatial genetic structure within two populations of a self-incompatible perennial, *Chionographis japonica* var. *japonica* (Liliaceae). J. Plant Res. 107: 283–287.
Melchior, H. 1964. 3. Reihe Liliiflorae (Liliales). In: Melchior, H., Werdermann, E. (eds.), A. Engler's Syllabus der Pflanzenfamilien, 12th edn, Vol. 2, 512–543. Berlin: Gebrüder Borntraeger.
Mulligan, G.A., Munro, D.B. 1987. The biology of Canadian weeds. 77. *Veratrum viride* Ait. Can. J. Plant Sci. 67: 777–786.
Osol, A., Farrar, G.E. 1955. The dispensatory of the United States of America, 25th edn. Philadelphia: J.B. Lippincott.
Satake, Y. 1982. Liliaceae. In: Satake, Y., Ohwi, J., Kitamura, S., Watari, S., Tominari, T. (eds.), Wild flowers of Japan, herbaceous plants, Vol. 1, 21–51. Tokyo: Heibonsha (in Japanese).
Savile, D.B.O. 1979. Fungi as aids in higher plant classification. Bot. Rev. 45: 377–503.
Schulze, W. 1978. Beiträge zur Taxonomie der Liliifloren. IV. Melanthiaceae. Wiss. Z. Friedrich-Schiller-Univ. Jena, Math.-Naturwiss. Reihe 27: 87–95.
Seigler, D.S. 1977. Plant systematics and alkaloids. Alkaloids 16: 1–82.
Sokolowska-Kulczycka, A. 1973. Development of bisporic embryo sacs in *Veratrum lobelianum* Bernh. Acta Biol. Cracov. Ser. Bot. 16: 85–98.
Sokolowska-Kulczycka, A. 1976. Development of bisporic embryo sacs in *Veratrum nigrum*. L. Acta Biol. Cracov. Ser. Bot. 20: 11–24.
Sterling, C. 1980. Comparative morphology of the carpel in the Liliaceae: Helonieae. Bot. J. Linn. Soc. 80: 341–356.
Sterling, C. 1982. Comparative morphology of the carpel in the Liliaceae: Veratreae. Bot. J. Linn. Soc. 84: 57–77.
Takahashi, Hr. 1988. The pollination biology of *Heloniopsis orientalis* (Thunb.) C. Tanaka (Liliaceae). Plant Species Biol. 3: 117–123.
Takahashi, M., Kawano, S. 1989. Pollen morphology of the Melanthiaceae and its systematic implications. Ann. Mo. Bot. Gard. 76: 863–876.
Tamura, M.N. 1995. See general references.
Tanaka, N.Y. 1985. Variation of sexual system in *Chionographis* species. Shuseibutsugaku-kenkyu 8: 11–19 (in Japanese).
Tanaka, N.Y. 1997. Taxonomic significance of some floral characters in *Helonias* and *Ypsilandra*. J. Jpn. Bot. 72: 110–116.
Tanaka, N.Y., Tanaka, N. 1977. Chromosome studies in *Chionographis* (Liliaceae). I. On the holokinetic nature of chromosomes in *Chionographis japonica* Maxim. Cytologia 42: 753–763.
Tanaka, N.Y., Tanaka, N. 1979. Chromosome studies in *Chionographis* (Liliaceae). II. Morphological characteristics of the somatic chromosomes of four Japanese members. Cytologia 44: 935–949.
Tanaka, N.Y., Tanaka, N. 1980. Chromosome studies in *Chionographis* (Liliaceae). III. The mode of meiosis. Cytologia 45: 809–817.
Taylor, C.A. 1956. The culture of false hellebore. Econ. Bot. 10: 155–165.
Taylor, C.L. 1940. Plants used as curatives by certain southeastern tribes. Cambridge: Bot. Mus. Cambridge, Mass.
Thorne, R.T. 1992. See general references.
Tillich, H.-J. 1995. Seedlings and systematics in monocotyledons. In: Rudall, P.J., Cribb, P.J., Cutler, D.F., Humphries, C.J. (eds.), Monocotyledons: systematics and evolution, pp. 303–352. Royal Botanic Gardens, Kew.
Utech, F.H. 1978a. Comparison of the vascular floral anatomy of *Xerophyllum asphodeloides* (L.) Nutt. and *X. tenax* (Pursh) Nutt. (Liliaceae-Melanthioideae). Ann. Carnegie Mus. 47: 147–167.
Utech, F.H. 1978b. Vascular floral anatomy of *Helonias bullata* (Liliaceae-Helonieae), with a comparison to the Asian *Heloniopsis orientalis*. Ann. Carnegie Mus. 47: 169–191.
Utech, F.H. 1980. Somatic karyotype analysis of *Helonias bullata* L. (Liliaceae), with a comparison to the Asian *Heloniopsis orientalis* (Thunb.) C. Tanaka. Ann. Carnegie Mus. 49: 153–160.
Utech, F.H., Kawano, S. 1981. Vascular floral anatomy of the East Asian *Heloniopsis orientalis* (Thunb.) C. Tanaka (Liliaceae-Helonieae). Bot. Mag. (Tokyo) 94: 295–311.
Walker, J.W. 1986. Classification and evolution of the monocotyledons. Am. J. Bot. 73(6): 746 (Abstract).
Williams, C.A. 1975. Biosystematics of the Monocotyledoneae—flavonoid patterns in leaves of the Liliaceae. Biochem. Syst. Ecol. 3: 229–244.
Williams, C.A., Harborne, J.B., Mathew, B. 1988. A chemical appraisal via leaf flavonoids of Dahlgren's Liliiflorae. Phytochemistry 27: 2609–2629.

Nartheciaceae

M.N. Tamura

Nartheciaceae E.M. Fries ex J. Bjurzon, Skand. Växtfam.: 64 (1846).
Petrosaviaceae Hutch. (1934).
Tofieldiaceae Takht. (1994).
Japonoliriaceae Takht. (1994).

Perennial herbs, very rarely without chlorophyll and saprophytic (*Petrosavia*). Underground stem usually a rhizome, thick to thin, short to long, creeping to ascendant or erect, sometimes with scaly leaves, sometimes stoloniferous, rarely a corm (one sp. of *Aletris*). Roots fibrous. Foliage leaves mostly basal, often distichous, equitant, unifacial, sometimes spiral, tufted, bifacial, linear, ensiform, lanceolate or broadly oblanceolate, rarely absent. Peduncle erect, usually with small foliage leaves or scaly leaves, rarely leafless. Inflorescence a raceme or more rarely a spike, cyme or corymb, 1-flowered in *Harperocallis*. Flowers trimerous, bisexual, actinomorphic, often hypogynous, sometimes half-epigynous, often funnel-shaped, cup-shaped or opening flat, sometimes campanulate, urceolate, or tepals reflexed (*Aletris* spp.). Bracts present, rarely spathelike. Bracteoles usually present, 3-fid to 3-sected, calycular, or linear to lanceolate, rarely absent. Tepals 3 + 3 with generally similar whorls, petaloid, often free, sometimes connate in lower part, linear to ovate, obovate, spathulate or deltoid, 1–17 mm long, white, greenish, pale yellow, brown, pinkish, reddish to purple, rarely bright yellow (*Narthecium*), without spotted patterns, without perigonal nectaries and spurs. Stamens usually 3 + 3, rarely 9 or 10 (*Pleea*); filaments linear to subulate, sometimes dilated basally, usually glabrous, rarely pubescent (*Narthecium*); anthers linear-lanceolate to rounded, dorsifixed or basifixed, dehiscent longitudinally, introrse or latrorse. Ovary 3-carpellate, sessile or stipitate, superior to half- (rarely fully) inferior; ovules anatropous or campylotropous, bitegmic or more rarely unitegmic, usually numerous, rarely 4 in each locule (*Japonolirion*); the carpels connate only at the base or for most or all of their length, ending in recurved styluli or a single style; septal nectaries usually present, rarely absent. Capsules 2–15 mm long, septicidal or loculicidal; seeds linear to broadly elliptical, 0.3–11 mm long, brown, dark red, pale yellow, or whitish, sometimes with appendages at both ends, sometimes only at one end, sometimes with a wing, and sometimes with neither appendage nor wing.

A family comprising 10 genera and c 72 species, distributed mostly in the Northern Hemisphere, extending to northern S America (*Isidrogalvia*, *Nietneria*).

Vegetative Morphology. Tofieldieae, *Narthecium*, *Nietneria*, and *Lophiola* have distichous, equitant, unifacial foliage leaves, while *Japonolirion* and *Aletris* have spiral, bifacial leaves. In *Petrosavia*, which entirely lacks foliage leaves, the leaf primordia are formed in distichous succession (Stant 1970).

The cotyledon has an assimilating elongated hyperphyll. It is unifacial in *Tofieldia* and *Narthecium* and bifacial in *Aletris*. The haustorium is scarcely distinguishable from the free part of the cotyledon. The primary leaf is ensiform in *Tofieldia* and *Narthecium* and bifacial in *Aletris*. The primary root is weakly developed (Tillich 1995).

The primary leaf of *Petrosavia* is scaly, which may represent the primitive state of monocotyledons (Tillich 1995). The seedlings of the Nartheciaceae are similar to those of *Acorus* (Tillich 1995).

Vegetative Anatomy. The Narthecioideae have three unique root characters (Ambrose 1980). (1) The cortex is differentiated into 3 regions: the middle one with large air spaces and the inner one with thick-walled cells. (2) In the endodermis, cell walls are equally thickened. (3) Each phloem pole consists of 1 to 2 sieve cells surrounded by several contiguous parenchyma cells.

In *Petrosavia*, the root cortex consists of 4–6 layers of parenchyma cells containing mycorrhizal hyphae. Anticlinal and inner cell walls of the endodermis are greatly thickened and lignified with marked stratification (Stant 1970). In the Tofieldieae, the root cortex lacks air spaces. Both in the Narthecioideae and Tofieldieae, fibers and sieve cells are mixed in the phloem (Ambrose 1980), in contrast to Melanthiaceae.

Petrosavia is often vesselless throughout the body but sometimes develops vessels with scalariform perforations in the roots. *Aletris farinosa* constantly possesses vessels in the roots but lacks them in stems and leaves. The vessels have a kind of multiperforate plates that exhibit partially a netlike appearance. The vessel condition of this family can be considered as primitive among the Liliaceae s. lat. (Cheadle and Kosakai 1971).

The stomata lack subsidiary cells in this family, a condition considered as advanced among the monocots (Goldblatt 1995).

Raphide idioblasts are usually lacking in this family but occur in *Petrosavia* and *Aletris luteoviridis* (Utech 1978b; Sterling 1980; Takhtajan 1997). The Tofieldieae have unique crystalline inclusions, druses in parenchymatous tissues, and prismatic crystals in the bundle sheaths (Ambrose 1980). *Narthecium* and *Aletris* spp. have prismatic crystals but lack druses (Goldblatt 1995).

Epicuticular wax sculpture is parallel (Convallaria type) (*Narthecium asiaticum*) or unoriented (*N. ossifragum*), or unspecific or lacking (Frölich and Barthlott 1988).

Inflorescence Structure. The inflorescence is usually terminal and the peduncle stands inside the tuft of leaves, but that of *Japonolirion* is axillary and stands outside the tuft of leaves. The peduncles are usually furnished with small foliage leaves or scaly leaves but leafless in *Aletris luteoviridis*.

The ancestral form of inflorescences of this family may be a raceme, which is found in *Petrosavia* spp. *Pleea*, *Tofieldia* spp. (Fig. 106A), *Isidrogalvia*, *Japonolirion* (Fig. 107A), *Narthecium* (Fig. 108A), and *Aletris* spp. *Petrosavia stellaris* (Fig. 106G), *Nietneria*, and *Lophiola* have a cyme or a corymb, and *Tofieldia* spp. and *Aletris* spp. have a spike, which in both can be considered as a derived form of inflorescence. In *Harperocallis*, the inflorescence is reduced to a single flower. If, however, in future studies the racemes turn out to be botryoids, their derivation from a panicle or thyrse will be more likely.

The inflorescences are often glabrous, sometimes pubescent (*Isidrogalvia sessiliflora* and *Aletris spicata*), sometimes glandular (*Tofieldia* sect. *Triantha* and *Aletris foliata*), rarely woolly (*Lophiola*).

Flower Structure. Floral bracts are always present; rarely they are spathelike (*Pleea*). Bracteoles are usually present, 3-fid to 3-sected, calycular (Tofieldieae) or linear to lanceolate (Narthecioideae), but absent in *Petrosavia* sp., *Tofieldia* sp., and *Japonolirion*.

The tepals and stamens are often almost free but basally connate to a certain degree in *Petrosavia* spp., *Nietneria*, *Aletris* spp., and *Lophiola*. The carpels are connate at the base in *Petrosavia* spp., *Pleea*, and *Harperocallis*, for most of their length

Fig. 106A–J. Nartheciaceae. **A–F** *Tofieldia pusilla.* **A** Flowering plant. **B** Flower. **C** Stamen and tepal. **D** Pistil. **E** Fruit. **F** Seed. **G–J** *Petrosavia stellaris.* **G** Habit. **H** Flower. **I** Stamen. **J** Fruit. (Takhtajan 1982)

in Narthecioideae and *Isidrogalvia*, intermediate in the remaining genera, ending either in recurved styluli[1] or a single style.

[1] Because the fusion of carpels in this family and in Melanthiaceae is often incomplete, the stigmatic tips of the carpels do not meet and no style is formed, hence the term style branch or stylodium would be inappropriate. For such completely free stigmatic carpel apices Baumann-Bodenheim (1954) has coined the term stylulus, which is used here.

Fig. 107A–F. Nartheciaceae. *Japonolirion osense.* **A** Flowering plant. **B** Flower. **C** Stamen. **D** Fruit. **E** Ovary locule, longitudinally opened. **F** Seed. (Takhtajan 1982)

Fig. 108A–H. Nartheciaceae. *Narthecium ossifragum.* **A** Flowering plant. **B** Flower. **C** Tepal and stamen. **D** Pistil. **E** Infructescence. **F** Dehiscing capsule. **G** Capsule, longitudinal section. **H** Appendaged seed. (Takhtajan 1982)

In *Petrosavia*, septal glands, furnished with a well-marked epithelium, extend from just below the lowermost ovular insertion up to the level of carpellary separation, where they open on the lateral surfaces of the carpels (Sterling 1978) and produce nectar drops. In *Pleea* and *Tofieldia* spp., the carpel bases are separate for a short distance, where a septal nectary is differentiated. The glands extend from the center of the pistil base diagonally outwards, and to the carpel bases along the septum lines (Sterling 1979; van Heel 1988). Nectar secretion in *Tofieldia calyculata* is from the furrows of the ovary, i.e. from the flanks of the carpels, in *T. palustris* from the whole surface of the ovary (Loew and Kirchner 1911). Nectar secretion in *Narthecium* is often considered to be absent, but according to Loew and Kirchner (1911) may take place from the basis of the floral organs. In *Aletris*, septal pockets exist between the carpels as septal glands, which open in the floral tube zone via the septal indentations (Utech 1978b).

The carpellary sutures are open at anthesis and septal wing tips are loculicidally inrolled in *Pleea*, *Tofieldia*, and *Isidrogalvia*, while in *Petrosavia*, *Narthecium*, *Nietneria*, and *Aletris*, the sutures are closed from the base at least up to just above the uppermost ovular insertion and the septal wing tips are not inrolled.

The placenta is large in *Narthecium*, *Nietneria*, and *Aletris*, but rather small in *Pleea*. It varies from large to small in *Petrosavia* and *Tofieldia*. In *Pleea* and *Tofieldia* spp., the ovules are furnished with a filiform chalazal hook, which seems to be an extension of the funiculus (Sterling 1978, 1979; Utech 1978a).

The ovules are usually bitegmic but unitegmic in *Narthecium*, *Tofieldia iridacea*, and *Isidrogalvia duidae*. They are anatropous in *Pleea* and *Nietneria*, campylotropous in *Narthecium* and *Aletris*, and vary from anatropous to campylotropous in *Petrosavia*, *Tofieldia*, and *Isidrogalvia*. They are basitropic in *Narthecium*, plagiotropic in *Pleea* and *Aletris*, and vary from basitropic to plagiotropic in *Petrosavia*, *Tofieldia*, *Isidrogalvia*, and *Nietneria*. In the latter three genera, the ovules are oriented sometimes also epitropically (Sterling 1978, 1979, 1980).

Floral Anatomy. The vascular bundle supply of *Pleea*, *Aletris*, and *Japonolirion* was studied by Utech (1978a,b, 1984). It is noteworthy that in *Japonolirion* there is no interconnection between dorsal and ventral carpellary bundles once they branch off in the receptacle. Each dorsal bundle enters a stylulus and each ventral bundle supplies 2 ovules.

According to Sterling (1978, 1979, 1980), in *Petrosavia*, *Pleea*, *Tofieldia*, *Isidrogalvia*, *Narthecium*, *Nietneria*, and *Aletris*, an amphiphloic vascular bundle separates into an outer collateral dorsal bundle with normally oriented phloem and xylem and an inner amphiphloic (or collateral in *Nietneria*) central bundle, the latter of which further separates into 2 ventral bundles with invertedly oriented xylem and phloem.

In *Narthecium*, *Tofieldia* sect. *Triantha*, and *Isidrogalvia*, the two ventral bundles divide into 1–4 placental bundles and a septal bundle. The placental bundles give off ovular bundle branches. The septal bundle does not divide in *Narthecium* but further divides into 2 in *Tofieldia* sect. *Triantha* and *Isidrogalvia*. In these cases, the septal bundle(s), in addition to the dorsal bundle, enter(s) the style.

Embryology. The tapetum is secretory (Dahlgren et al. 1985). Simultaneous cytokinesis in the microspore mother cells accompanies meiosis at least in *Petrosavia* and *Tofieldia*.

The archesporial cell cuts off a primary parietal cell in *Tofieldia* and *Aletris* (Davis 1966). Three types of embryo sacs are formed in *Tofieldia*: monosporic Polygonum type, bisporic Allium type, and bisporic Veratrum lobelianum type (Sokolowska-Kulczycka 1980). The Veratrum lobelianum type deviates slightly from the Allium type in the second meiotic division of the embryo sac mother cell: the mycropylar cell divides earlier than the chalazal cell, resulting in a temporal formation of a triad. Embryo-sac formation of *Petrosavis* is according to the Polygonum type.

Endosperm formation is usually Helobial (Dahlgren et al. 1985), but in *Petrosavia* Nuclear (Takhtajan 1997). An integumentary obturator develops in *Petrosavia* and *Aletris aurea* but not in *Tofieldia* (Davis 1966).

Karyology. The basic chromosome numbers of Tofieldioideae-Tofieldieae are $x = 14$–16. Among them, $x = 15$ was observed in *Pleea*, *Tofieldia*, and *Harperocallis* and seems to be ancestral. Basic numbers such as $x = 14$ of *T. calyculata* (Sokolowska-Kulczycka 1980) and *T. pusilla* (Cave 1964) and $x = 16$ of *T. coccinea* (Zhukova 1967) and *T. racemosa* (M.N. Tamura and R. Ohfuji, unpubl.) seem to be derived.

The basic chromosome number of the monotypic Tofieldioideae-Petrosavieae was recently determined as $x = 15$ (Tamura and Takahashi 1998), which confirms the relationship between *Petrosavia* and the *Tofieldia* group.

The basic chromosome number of the monotypic Tofieldioideae-Japonolirieae was counted as $x = 12$ by Sato (1942) and as $x = 13$ by Kurosawa (1980).

The basic chromosome numbers of Narthecioideae are $x = 12, 13, 21, 22$. Among them, $x = 13$ was observed in *Narthecium*, *Nietneria* and *Aletris* and seems to be ancestral. $x = 12$ and $x = 22$ of *A. spicata* (M.N. Tamura and R. Ohfuji, unpubl.) seem to have been secondarily derived. The origin of $x = 21$ in *Lophiola* is uncertain, but it is hypothesized that this number arose in a manner similar to $x = 22$ in *A. spicata*. Another possibility is that the circumscription of the Narthecioideae is not correct.

Chromosome size in this family is mostly very small, ranging from 0.7 to 1.7 μm in length at mitotic metaphase, which can be considered a primitive characteristic in the monocots (Tamura 1995). Comparatively large chromosomes of the following taxa can be considered as derived: *Petrosavia*, 1.1–3.6 μm (Tamura and Takahashi 1998); *Pleea*, 1.5–3.4 μm (Utech 1979); *Tofieldia* sect. *Triantha* 1.1–2.9 μm; *Narthecium ossifragum* 1.0–2.6 μm; *Aletris spicata* with $x = 12$, 13 1.5–3.5 μm (M.N. Tamura and R. Ohfuji, unpubl.). The range of chromosome length in the karyotype of *Japonolirion* is much wider than that of other nartheciaceous genera: 0.4–3.1 μm (Sato 1942), the reason for which is unclear.

Usually Nartheciaceae are diploid, but rarely tetraploid forms exist, such as *Petrosavia sakuraii* ($2n = 60$), *Tofieldia japonica* ($2n = 60$), *Aletris luteoviridis* ($2n = 52$), *A. foliata* ($2n = 52$). Following Tanaka (1977), the interphase and mitotic prophase chromosomes can be categorized in the prochromosome type and the proximal type, respectively. The chromosomes at the mitotic metaphase are all metacentric or submetacentric in the sense of Levan et al. (1964) (M.N. Tamura and R. Ohfuji, unpubl.).

POLLEN MORPHOLOGY. Pollen grains are spheroidal, 14–36 μm in the longest axis, sulcate in *Petrosavia*, *Japonolirion*, *Narthecium*, *Aletris*, and *Lophiola*, and disulcate in *Pleea* and *Tofieldia* (Tofieldioideae-Tofieldieae).

The exine is usually reticulate, but gemmate in *Petrosavia* and *Japonolirion*, although the gemmae form a reticulum on the proximal face (K. Handa et al., unpubl.). In *Tofieldia* and *Aletris*, the exines vary widely in sculpture. For example, in *Tofieldia*, they are reticulate, rugulate, verrucate and gemmate (Ikuse 1956; K. Handa et al., unpubl.), and in *Aletris* they are perforate-foveolate (the pores 0.1–0.3 μm in diam.), reticulate (the lumina 0.5–1.5 μm in diam.), rugulate, and gemmate (Takahashi and Kawano 1989; K. Handa et al., unpubl.). *Aletris luteoviridis* has a reticulate exine with verrucate protrusions within the lumina (Takahashi and Kawano 1989). The lumina of reticulate exines are roundish rather than angulate, 0.2–0.7 μm in diam. in *Pleea* and *Tofieldia* (Tofieldioideae), 0.1–0.5 μm in diam. in *Narthecium* and *Lophiola*. The muri are often 0.2–0.3 μm thick, but narrow (0.1 μm) in *Lophiola* and wide (0.5 μm) in *Tofieldia* (Takahashi and Kawano 1989).

Pollen grains of *Petrosavia* and *Japonolirion* differ highly from those of the other members of the Tofieldioideae and are somewhat similar to those of *Aletris* spp., although the taxonomic significance of this similarity is still unclear.

POLLINATION. *Tofieldia calyculata* is protogynous, *T. palustris* homogamous. In *Petrosavia sakuraii* in Central Japan, anthers of the outer and inner stamens dehisce 1 and 2 days after the opening of the flowers, respectively. The stigmas on the nearly apocarpous pistil become receptive before flower opening and at that time are positioned in the center of the flower. At anthesis, the styluli move gradually outwards, and after about 5 days the stigmas touch the anthers of the outer stamens. The stigmas are receptive for more than 5 days after the opening of the flower, and the movement of the stigmas results in autonomous self-pollination. The selfing rate, including insect-mediated as well as autonomous self-pollination, is very high. Cross-pollination also may be performed by small *Lasioglossum* bees and some other insects, but these do not visit the flowers frequently (Takahashi et al. 1993).

Nectar secretion in *Tofieldia calyculata* as well as in *Petrosavia sakuraii* is from the furrows of the ovary, i.e., from the flanks of the carpels, in *T. palustris* from the whole surface of the ovary, and in *T. racemosa* as well as in *Pleea* from the basis of the floral organs. In *Aletris luteoviridis* nectariferous glands are formed in the lower perigynous zone. In *Narthecium*, Loew and Kirchner (1911) considered nectar secretion probably from the basis of the floral organs, while Utech (1978a) did not observe the nectariferous glands.

FRUIT AND SEED. The fruits of this family are always capsules. They dehisce septicidally in the Tofieldioideae and loculicidally in the Narthecioideae, although the capsule of *Harperocallis* (Tofieldioideae) was reported as loculicidal by McDaniel (1968). The capsules of Petrosavieae and Japonolirieae are small, 2.5–3 mm long, while that of *Narthecium* is large, 10–15 mm long. Those of Tofieldieae and *Aletris* are of medium or variable size, 2–10 mm long.

As in the Melanthiaceae, the seeds are always packed obliquely in the capsules, in contrast to the horizontal packing of the compressed seeds in Liliaceae and related families. The size of the seeds is also variable, but is more or less related to the size of the capsules and the shape of the seeds. The seeds of Petrosavieae and Japonolirieae are small, 0.3–0.8 mm long, broadly elliptical to elliptical, while those of *Narthecium* are large, 8–11 mm long, lanceolate to linear. Those of Tofieldieae and

Aletris are medium or variable in size, 0.6–4 mm long, elliptical to narrowly fusiform.

The seeds are often appendaged. Those of *Narthecium* and *Harperocallis* have filiform appendages at both ends. They are nearly equal in length and long in *Narthecium* and short in *Harperocallis*. In *Pleea* the chalazal appendage is much longer than the micropylar appendage. The seeds of *Isidrogalvia* have only an appendage at the chalazal end, which is decurrent along the seed body.

The seeds of *Tofieldia* are variable as to their appendages. In *T.* sect. *Triantha*, the chalazal appendage is always long and filiform, while the micropylar appendage is sometimes lacking. In *T.* sect. *Tofieldia*, seed appendages at both ends may be short and wide or not developed at all.

The seeds of *Petrosavia* are more or less winged. Those of *Japonolirion*, *Nietneria*, and *Aletris* have neither an appendage nor a wing and are rounded or acute at both ends.

The seeds of this family lack phytomelan but sometimes contain phlobaphene deposits (*Tofieldia*). The tegminal part of the seed coat is often thin and collapsed but sometimes retained although flattened (*Narthecium*).

Phytochemistry. Steroid saponins have been found in *Tofieldia*, *Narthecium*, and *Aletris*, and chelidonic acid in *Tofieldia* and *Narthecium*. The steroidal saponins of *Narthecium* are thought to cause liver disease (icterus) in nonpigmented sheep (Hegnauer 1963, 1986).

Quercetin and kaempferol glycosides as leaf flavonoids are known from *Tofieldia calyculata* but not from *Narthecium ossifragum*; flavones are lacking in both species (Williams 1975).

Distribution and Habitats. This family is centered in the Northern Hemisphere and extends to northern S America (*Isidrogalvia*, *Nietneria*).

Among the Tofieldioideae, *Tofieldia* sect. *Tofieldia* is distributed widely in temperate to subarctic regions of the Northern Hemisphere. *T.* sect. *Triantha* is abundant in N America, but *T. japonica* grows in swamps in Central Japan, *Japonolirion* is a vulnerable genus and restricted to serpentine swamps of central and N Japan. *Pleea* and *Harperocallis* are endemic to the southeastern USA. *Isidrogalvia* is distributed in Venezuela, Colombia, Ecuador, and Peru. *Petrosavia* is saprophytic and growing at the forest floor of tropical to temperate zones in Malaysia, Indochina and S China to Japan.

Among the Narthecioideae, *Narthecium* is a disjunct of temperate regions of the Northern Hemisphere, where it grows in swamps; it is lacking in China. *Nietneria* is restricted to savannahs and highlands in Venezuela and Guiana. *Aletris* is distributed in Asia and eastern N America, abundant in SW China. *Lophiola* is restricted to eastern N America.

Classication of the Family. The Nartheciaceae can be subdivided into two subfamilies: the Tofieldioideae and Narthecioideae.

Ambrose (1980) undertook a numerical analysis of morphological and anatomical characters of the Melanthiaceae s. l., which revealed three main clusters: (1) the *Tofieldia* group, (2) the *Narthecium* group, and (3) the *Melanthium* group, apart from the isolated genus *Xerophyllum*. Chase et al. (1995), based on a parsimony analysis of *rbc*L sequences, recognized similar groupings: (1) *Pleea* and *Tofieldia*, (2) *Aletris* (including *Metanarthecium*), and (3) *Heloniopsis*, *Chamaelirium*, *Xerophyllum*, *Amianthium*, *Zigadenus*, and *Veratrum* (the data referring to *Tofieldia* and *Metanarthecium* are quoted from Z.K. Shinwari and S. Kawano, unpubl.). Nartheciaceae in the present circumscription contain groups 1 and 2 of Ambrose (1980) and Chase et al. (1995) (see below under Affinities); group 1 corresponds to Subfamily Tofieldioideae and group 2 to Subfamily Narthecioideae.

Subfamily Tofieldioideae can be further subdivided into three tribes, the Petrosavieae, Tofieldieae, and Japonolirieae. The tribe Tofieldieae comprises four genera, *Pleea*, *Tofieldia*, *Isidrogalvia*, and *Harperocallis*, whose coherence is well supported, as was shown by Ambrose (1980) and Dahlgren et al. (1985). They share also disulcate pollen grains and the basic chromosome number $x = 15$.

Petrosavia, which was not analyzed by Ambrose (1980) and Chase et al. (1995), possesses the basic chromosome number, $x = 15$, along with almost free carpels and septicidal capsules. These characteristics indicate a relationship of *Petrosavia* to Tofieldieae rather than to Narthecioideae. *Petrosavia* has sulcate pollen with a gemmate exine, which is shared with *Japanolirion* (Tofieldioideae), although *Aletris pauciflora* and *A. khasiana* (Narthecioideae) have similar pollen.

In this treatment I include *Petrosavia* in the Tofieldioideae, although *Petrosavia* shares a single bracteole per flower not with the remaining genera but with the Narthecioideae. I distinguish *Petrosavia* at the tribal level from the other Tofieldioideae based on the characters derived from its saprophytic life style as well as its chromosome number (Japanolirieae: $x = 12, 13$) and pollen morphology (Tofieldieae: disulcate). Further studies, particularly in molecular systematics, will be needed in order to ascertain the precise position of *Petrosavia*.

Japonolirion, which was not analyzed by Ambrose (1980) and Chase et al. (1995), has the basic chromosome numbers $x = 12$, 13 (as *Aletris*), bifacial leaves (also as *Aletris*), sulcate pollen grains and gemmate or reticulate exine ornamentation with roundish brochi (also as *Aletris*), and lacks both septal and perigonal nectaries [as *Narthecium*(?), *Nietneria*, and *Lophiola*], which would favor an inclusion of *Japonolirion* in the Narthecioideae, from which it differs, however, in having 3 styluli and septicidal capsules. The relatively large chromosomes of *Japanolirion* and its lack of bracteoles and septal nectaries may suggest a relationship with the Melanthiaceae s. str. Nakai (1930); Kitamura (1964) and Shimizu (1983) emphasized the relationship between *Japonolirion* and Heloniadeae (Melanthiaceae s. str.). Nevertheless, unlike the Melanthiaceae, *Japonolirion* has introrse anthers and roundish brochi of the exine (see below under Affinities).

Based mainly on floral morphology and anatomy, Utech (1984) and T. Yahara (unpubl.) pointed to a relationship between *Japonolirion* and the *Tofieldia* group. In the present treatment, I primarily follow this suggestion and treat *Japonolirion* in the Tofieldioideae. *Japonolirion* differs, however, from the other members of the Tofieldioideae in characters such as leaf arrangement, presence or absence of bracteoles and septal nectaries, pollen and seed morphology, and chromosome number (see below under Key to the Genera). Accordingly, within the Tofieldioideae, *Japanolirion* is given the status of a monotypic tribe, Japonolirieae. Further molecular data on *Japonolirion* will help to determine the exact taxonomic position of this interesting genus.

Subfamily Narthecioideae consists of four genera: *Narthecium*, *Nietneria*, *Aletris*, and *Lophiola*, whose coherence is well supported (Ambrose 1980). The former three share the basic chromosome number $x = 13$; *Lophiola*, which previously had been included in Haemodoraceae, Amaryllidaceae, or Tecophilaeaceae, has $x = 21$. However, Huber (1969) indicated that *Lophiola* is anatomically close to *Aletris*, and M.N. Tamura and R. Ohfuji (unpubl.) recently proved that *Aletris* has also $x = 12$ and 22 in addition to $x = 13$, which demonstrates the instability of chromosome numbers in the *Aletris-Lophiola* lineage. Comprehensive karyological studies as well as molecular analyses of *Lophiola* are needed in order to confirm its taxonomic position.

AFFINITIES. The genera constituting the Nartheciaceae of the present treatment were previously included in the Melanthiaceae (Ambrose 1980; Dahlgren et al. 1985; Thorne 1992). Dahlgren et al. (1985) indicated as their only synapomorphy "plants attacked by the related fungal parasites *Puccinia atropuncta* and *Uromyces veratri*". This was recognized by Goldblatt (1995) as a characteristic only of the *Veratrum* group and *Xerophyllum*.

Tamura (1995) suggested a separation of the Melanthiaceae into the *Narthecium* group and the *Heloniopsis* group. According to him, primarily, the *Narthecium* group has bracteoles, introrse or latrorse anthers, roundish (or nonangular) exine brochi, and very small chromosomes, while the *Heloniopsis* group lacks bracteoles and possesses extrorse or subextrorse anthers, angular brochi (provided that pollen exine ornamentation is reticulate), and relatively large chromosomes. Goldblatt (1995) suggested the same division emphasizing that the former group has septal nectaries while the latter has perigonal ones.

On the other hand, Ambrose (1980) and Chase et al. (1995), based on morphological and anatomical characters and *rbc*L sequences, respectively, recognized three major groups in the Melanthiaceae s. l.: (1) the *Tofieldia* group, (2) the *Narthecium* group, and (3) the *Melanthium* group (see previous section).

In the present circumscription, the Nartheciaceae include the *Narthecium* group of Tamura (1995) and Goldblatt (1995) and the *Tofieldia* group and *Narthecium* group of Ambrose (1980) and Chase et al. (1995).

Petrosavia is also included in the present treatment of the Nartheciaceae, as proposed by Beccari (1871), Dahlgren et al. (1985), and Thorne (1992), although Deyl (1955), Stant (1970), and Takhtajan (1997) suggested a relationship between *Petrosavia* and the Triuridaceae, both of which are saprophytes. Stant (1970) revealed similarities between the anatomical structures of the stem, root, and leaf of *Petrosavia* and the Triuridaceae, but it should be verified whether this similarity reflects phylogenetic relationship or is a convergence due to saprophytism.

The Nartheciaceae occupy a basal position not only in a broadly construed family Liliaceae, as suggested by Hutchinson (1934), but also in the monoctyledons, a view first suggested by Thorne (1992) (for Melanthiaceae) and now also adopted by Takhtajan (1997). This is supported by the very small chromosomes (Tamura 1995), the often apocarpous or half-apocarpous gynoecia or at least free styluli, perhaps the lack of vessels in *Petrosavia*, and the variability of characters that in higher evolved lineages of the monocotyledons are more constant, such as the type of septal nectaries, type of embryo-sac formation, fruit dehiscence, chromosome basic number, etc. Also the molecular data of Chase et al. (1993, 1995) indicate, apart from *Acorus*, a basal position in the monocots for the genera of the Nartheciaceae, although *Pleea* and *Tofieldia* are sister to the Alismatanae and *Aletris* is isolated in the Lilianae. It may be significant that the Nartheciaceae share equitant leaves and also seedling characters with *Acorus* that are considered primitive in the monocots (Tillich 1995). Walker (1986) considered unifacial, equitant, distichous leaves, absence of raphides, and a basically (north) temperate distribution pattern as plesiomorphic for the monocots and therefore identified *Tofieldia* and *Narthecium* as two of the most primitive monocots. On the other hand, Huber (1969) considered the equitant leaves of *Tofieldia* as advanced, and a loss of raphides can be observed in many lineages of the monocotyledons. In my opinion, the unifacial, equitant, distichous leaves are advanced characteristics among the angiosperms, but might be plesiomorphic within the monocots. I consider that also the transference of nectar secretion from the base of the tepals to the furrows of the ovary might have once happened before the establishment of the monocots: the septal nectary found in the Nartheciaceae is possibly plesiomorphic within the monocots. Possibly advanced characters appearing among the Nartheciaceae include the fusion among ovary and tepals in *Aletris* and *Nietneria*, and the unitegmic ovules in *Narthecium* and elsewhere in the family.

I assume a scenario of evolution of the Nartheciaceae as follows. After the Acoraceae first branched off from the ancestor of the monocots, the Nartheciaceae became a core stock of the monocots. While several taxa including the Alismatales and the Asparagales-Liliales-Commelinales complex successively branched off from the core stock of the Nartheciaceae, the core stock of the Nartheciaceae has little modified its characteristics, possibly having adapted itself to swampy habitats. The evidence that the Acoraceae and the Nartheciaceae share the unifacial, equitant, distichous leaves might more or less support this scenario.

If my assumption is true, the coherence of the Nartheciaceae has not much changed from their origin until today. At the origin, i.e. before their having cut off several taxa, the Nartheciaceae must have been monophyletic and treated as a taxon; therefore also today I recognize the Nartheciaceae as a taxon, even though today they may be paraphyletic (see also Tamura 1996).

Due to their old age they appear somewhat isolated from other families.

Key to the Genera

1. Herbs saprophytic, without chlorophyll; leaves all scaly and cauline (**Tofieldioideae-Petrosavieae**) **1. *Petrosavia***
– Herbs with chlorophyll; leaves mostly foliage and basal 2
2. Bracteoles 3, free or united in lower part, or absent; styluli usually 3, rarely fused into a style (*Isidrogalvia*); capsule usually septicidal, rarely loculicidal (*Harperocallis*) (**Tofieldioideae**) 3
– Bracteole 1; style 1; capsule loculicidal (**Narthecioideae**) 7
3. Foliages leaves flat and spiral; scape axillary, i.e., standing outside a tuft of foliage leaves; bracteoles absent; ovules 4 in each carpel; septal nectary absent; x = 12, 13 (**Japonolirieae**) **6. *Japonolirion***
– Foliage leaves distichous and equitant; scape terminal, i.e., standing inside a tuft of foliage leaves; bracteoles usually present, rarely absent (*Tofieldia* spp.); ovules 5 to many in each carpel; septal nectary present (at least in *Pleea* and *Tofieldia*); x = usually 15, rarely 14, 16 (**Tofieldieae**) 4
4. Bracts large, spathelike, aristate at apex; bracteoles 3-parted, tubular, in middle part of pedicels; stamens usually 9, rarely 6, 10–12 **2. *Pleea***
– Bracts small to large, lanceolate to ellipticall or deltoid, obtuse or acute at apex; bracteoles 3-parted or 3-sected if present, cup-shaped, near apex of pedicels; stamens 6 5
5. Bract 0–1; bracteoles united in lower part; tepals 1.5–6 mm long, shorter than stamens and capsule, 1–5-nerved; anthers ovate to rounded, 0.4–0.6 mm long; seeds (except appendages) less than 1.5 mm long **3. *Tofieldia***
– Bracts 3–17; bracteoles free; tepals 6–15 mm long, longer than stamens and capsule, 5–12-nerved; anthers linear to oblong, 0.9–4 mm long; seeds (except appendages) 2–3 mm long 6
6. Inflorescence a raceme; anthers oblong, 0.9–2.2 mm long; ovary smooth, 2–3 mm long; style 1 in each flower; seeds dark red, with an appendage at chalazal end, decurrent along the seed body **4. *Isidrogalvia***
– One-flowered; anthers linear to lanceolate, ca. 4 mm long; ovary tuberculate, 7–8.5 mm long; styluli 3 in each flower; seeds pale yellow, with short appendages at both ends **5. *Harperocallis***
7. Inflorescence whitish-woolly **10. *Lophiola***
– Inflorescence glabrous or glandular-pubescent 8
8. Foliage leaves spiral, bifacial; seeds oblong to ovate **9. *Aletris***
– Foliage leaves distichous, unifacial; seeds linear to lanceolate 9
9. Inflorescence a cyme or a corymb; anthers ovate; seeds without appendages **8. *Nietneria***
– Inflorescence a raceme; anthers linear to lanceolate; seeds with long filiform appendages at both ends **7. *Narthecium***

I. Subfam. Tofieldioideae (Takht.) M.N. Tamura, stat. nov.

Tofieldiaceae Takhtaj., Bot. Zhurn. 79 (12): 65 (1994).

1. Tribe **Petrosavieae** Engler (1897).

1. ***Petrosavia*** Beccari Fig. 106G–J

Petrosavia Beccari, Nuov. Giorn. Bot. Ital. 3: 7 (1871); H. Ohba, J. Jap. Bot. 59: 106–110 (1984), rev.
Protolirion Ridley (1895).
Miyoshia Makino (1903).

Colorless or pale yellowish saprophyte. Underground stem thin, scaly. Aerial stem thin, erect. Leaves alternate, scaly, membranous, oval, oblong to narrowly deltoid, sometimes vaginate. Inflorescence a raceme, sometimes corymblike. Flowers erect, funnel-shaped, 2–4 mm diam. Bracteoles lanceolate to ovate, sometimes absent. Tepals persistent; outer 3 deltoid to ovate; inner 3 ovate. Stamens inserted at tepal base; filaments subulate; anthers ovate, dorsifixed, introrse. Ovary superior to half-inferior; ovules numerous; the carpels connate for 1/4 to 1/2 their length, sometimes only at the base; septal nectaries present; styluli short; stigmas capitate or slightly bifid. Capsules recurved, ventrally dehiscent in upper part; seeds elliptical, 0.3–0.5 mm long, brown, longitudinally striated, more or less winged. $x = 15$. Three spp., SE Asia to Japan.

2. Tribe **Tofieldieae** Kunth (1843).

2. ***Pleea*** Michx.

Pleea Michx., Fl. Bor. Am. I: 247, t. 25 (1803).
Tofieldia sensu Utech, Ann. Carnegie Mus. 47: 423–454 (1978), partly.

Rhizome thick, spreading. Foliage leaves mostly basal, equitant, linear, persistent, with smooth margin. Peduncle with 2–3 small foliage leaves. Raceme glabrous, 3–8-flowered. Flowers opening flat. Bracts spathelike. Bracteole 3-partite, tubular, in middle part of pedicel. Tepals free, linear to linear-lanceolate, persistent. Stamens usually 9 (outer 6 + inner 3), rarely 10, free, shorter than tepals; filaments glabrous, dilated basally; anthers linear-lanceolate, 2–3 mm long, dorsifixed, versatile, introrse. Ovary superior; ovules numerous; the carpels connate at base, stipitate; septal nectary present; styluli short. Capsules 3-ribbed, septicidal; seeds with a long filiform appendage at chalazal end and a short broad appendage at micropylar end. $x = 15$. Only one sp., *P. tenuifolia* Michx., coastal plains of SE USA.

3. ***Tofieldia*** Hudson Fig. 106A–F

Tofieldia Hudson, Fl. Angl. ed. 2: 157 (1778).
Triantha Nutt. (1818).

Rhizome short, ascendant, rarely stoloniferous. Foliage leaves distichous, mostly basal, equitant. Peduncle with 0–4 small foliage leaves in lower part, standing inside a tuft of foliage leaves. Inflorescence a raceme or a spike. Flowers erect, rarely nodding. Bracts small. Calycular bracteole often present. Tepals usually free, rarely connate at base, persistent. Stamens often free, sometimes connate at base, sometimes inserted at tepal base, often longer than tepals; filaments linear to subulate; anthers basifixed, introrse to latrorse. Ovary superior, stipitate or sessile; ovules 5–15 in each carpel; carpels connate for half to 4/5 their length; septal nectaries often present; styluli 3, slightly recurved. Capsules septicidal; seeds narrowly oblong to elliptical, often appendaged at both ends or at one end. Usually $x = 15$, rarely $x = 14, 16$. Two sections., about 20 spp., temperate to subarctic regions of the Northern Hemisphere.

4. ***Isidrogalvia*** Ruíz Lopez et Pavón

Isidrogalvia Ruíz Lopez et Pavón, Fl. peruv. et chil. 3: 69, t. 302 (1802); Cruden, Syst. Bot. 16: 270–282 (1991), rev.

Rhizomatous. Foliage leaves distichous, mostly basal, unifacial, sometimes ensiform, strongly ribbed, persistent. Peduncle with 3–17 small foliage leaves mostly in upper part. Inflorescence a raceme. Flowers shallowly cup-shaped. Bracts deltoid to ovate. Calycular bracteoles 3, free. Tepals narrowly elliptic to narrowly obovate, persistent, connate at base, longer than stamens, enclosing the capsules. Stamens free; filaments straight to narrowly subulate; anthers basifixed, latrorse. Ovary oblong; ovules numerous; the carpels connate for most of their length into a style with capitate stigma. Capsules 4.5–10 mm long, usually 9-ribbed, rarely 3-ribbed, septicidal; seeds fusiform, 2–3 mm long, dark red, with a white appendage at chalazal end, decurrent along the seed body. Five spp., northwestern S America.

5. ***Harperocallis*** McDaniel

Harperocallis McDaniel, J. Arnold Arbor. 49: 35–40 (1968).

Rhizomatous. Foliage leaves mostly basal, equitant. Peduncle erect, with 3–5 small foliage leaves. Flower solitary, opening flat. Calycular bracteoles 3(-4), free, persistent. Tepals oblanceolate, acute, persistent. Stamens free; anthers basifixed. Ovary elliptical, sessile, tuberculate, superior; ovules numerous; the carpels connate only at the base; styluli slightly recurved; stigmas capitate. Capsules elliptical, tuberculate, loculicidal; seeds narrowly fusiform, 2–3 mm long, straight to curved, often twisted, pale yellow, with short appendages at both ends. $x = ?$ Only one sp., *H. flava* McDaniel, W Florida, USA.

3. Tribe Japonolirieae (Takht.) M.N. Tamura, stat. nov.

Japonoliriaceae Takhtajan, Bot. Zhurn. 79(1): 97 (1994).

6. *Japonolirion* Nakai — Fig. 107

Japonolirion Nakai, Bot. Mag. (Tokyo) 44: 22–23 (1930); Utech, Ann. Carnegie Mus. 53: 447–461 (1984).

Rhizome creeping. Foliage leaves spiral, basal, tufted, bifacial, linear. Peduncle erect, with several scaly leaves, axillary in the basal leaves. Inflorescence an erect raceme. Flowers erect, shallowly cup-shaped to opening flat, pale yellow, sometimes greenish. Bracts oblong. Bracteoles absent. Tepals 6, free, membranous, persistent; outer 3 oblong; inner 3 obovate to spathulate. Stamens as long as inner tepals; filaments subulate, glabrous; anthers basifixed, introrse. Ovary superior; ovules 4 in each locule; the carpels loosely connate except for the recurved styluli; septal nectaries absent. Capsules elliptical, septicidal; seeds broadly elliptical, 0.8 mm long, without appendages. $x = 12, 13$. Only one sp., *J. osense* Nakai, serpentine swamps of central and N Japan.

II. Subfam. Narthecioideae (Benth.) M.N. Tamura, stat. nov.

Narthecieae Benth. in Benth. & Hook. f., Gen. Pl. 3: 760 (1883).

7. *Narthecium* Hudson — Fig. 108

Narthecium Hudson, Fl. Angl. ed. 1: 127 (1762).

Rhizome creeping. Foliage leaves distichous, mostly basal, equitant, unifacial, linear to ensiform. Peduncle with 1–10 small foliage leaves. Raceme simple, dense. Flowers erect, yellow. Bracts linear to lanceolate. Bracteole 1, linear to lanceolate, small, attached to middle part of a pedicel. Tepals connate at base, linear to oblong or oblanceolate, persistent. Stamens inserted at tepal base; filaments filiform, densely pubescent; anthers linear to lanceolate, dorsifixed or basifixed, introrse. Ovary superior to quarter-inferior, syncarpous, with ca. 35 ovules; style columnar; stigma capitate; septal nectaries probably absent. Capsules lanceolate, tapering at apex, loculicidal; seeds with long filiform appendages at both ends. $x = 13$. About 8 spp., highly disjunct in temperate regions of the Northern Hemisphere.

8. *Nietneria* Klotzsch & R. Schomb.

Nietneria Klotzsch & R. Schomb. in R. Schomb., Reise Brit. Guiana: 1068 (1848).

Rhizome short. Foliage leaves distichous, mostly basal, unifacial, linear to ensiform, stiff. Inflorescence a loose terminal corymb. Flowers campanulate. Bracteoles very small. Tepals persistent. Stamens inserted at tepal base; filaments filiform, shorter than tepals; anthers ovate. Ovary conical, superior to inferior, syncarpous; ovules numerous; septal nectaries lacking; style short; stigma capitate, small. Capsules loculicidal; seeds linear, small, without appendages. $x = 13$. Only one sp., *N. corymbosa* Klotzsch, Venezuela and Guiana.

9. *Aletris* L.

Aletris L., Sp. Pl. ed. 1: 319 (1753).
Metanarthecium Maxim. (1867); Hara, J. Jap. Bot. 42: 312–316 (1967); Utech, Ann. Carnegie Mus. 47: 455–477 (1978).
Meta-aletris Masamune (1938).

Underground stem usually a rhizome, rarely a corm. Foliage leaves spiral, mostly basal, bifacial, linear to lanceolate or broadly obolanceolate. Peduncle erect, with 0–11 small foliage leaves. Inflorescence a raceme or a spike, usually simple. Floral bracts linear to lanceolate. Bracteole 1. Perianth urceolate, campanulate to tubular in lower part and upright or ascendant in upper part, or opening flat or reflexed. Tepals connate at base, linear or connate up to for 3/4 their length. Stamens 6; filaments adnate to tepals at base or up to for most their length; anthers lanceolate to orbicular, dorsifixed or basifixed, introrse. Ovary superior to half-inferior, syncarpous, with 72–200 ovules; septal glands present; style short to long; stigma capitate to 3-lobed. Capsules conical to globose or obovoid, sometimes angular, loculicidal; seeds oblong to ovate, more or less acute at both ends. $x = 12, 13, 22$. About 30 spp., about 25 in E Asia and 5 in eastern N America.

Metanarthecium has often been separated from *Aletris*, but here I follow Franchet (1896), Hara

(1967), Ambrose (1980), and Kitagawa (1983) in considering *Metanarthecium* as congeneric with *Aletris*.

10. *Lophiola* Ker-Gawl.

Lophiola Ker-Gawl., Bot. Mag.: t. 1596 (1814).

Rhizome slender, stoloniferous. Aerial stem woolly in upper part, often solitary, sometimes subcaespitose. Foliage leaves basal and cauline, equitant, unifacial, linear. Inflorescence a terminal corymb, many-flowered, woolly. Bracts and bracteoles present. Tepals on the same whorl, more or less connate, pubescent outside except in upper part. Stamens inserted at tepal base; anthers basifixed, introrse. Ovary superior to half-inferior; syncarpous, without septal nectaries; ovules numerous; style 1. Capsules ovoid, green or fulvous, loculicidal; seeds oblong, whitish or drab, angular, rounded at both ends or caudate-tipped at base. $x = 21$. Two spp., eastern N America.

Selected Bibliography

Ambrose, J.D. 1980. A re-evaluation of the Melanthioideae (Liliaceae) using numerical analyses. In: Brickell, C.D., Cutler, D.F., Gregory, M. (eds.) Petaloid monocotyledons, pp. 65–81. London: Academic Press. (Linnean Society Symposium Series Vol. 8)

Baumann-bodenheim, M.G. 1954. Prinzipien eines Fruchtsystems der Angiospermen. Ber. Schweiz. Bot. Ges. 64: 94–112.

Beccari, O. 1871. *Petrosavia*: nuovo genere di piante parasite della famiglia delle Melanthaceae. Nuovo Giorn. Bot. Ital. 3: 7–11.

Cave, M. 1964. Cytological observations on some genera of the Agavaceae. Madroño 17: 163–169.

Chase, M.W. et al. 1993, 1995. See general references.

Cheadle, V.I., Kosakai, H. 1971. Vessels in Liliaceae. Phytomorphology 21: 320–333.

Dahlgren, R.M.T. et al. 1985. See general references.

Davis, G.L. 1966. See general references.

Deyl, M. 1955. The evolution of the plants and the taxonomy of the monocotyledons. Acta Mus. Nat. Pragae, Bot. no. 3, 11B: 1–143.

Frölich, D., Barthlott, W. 1988. See general references.

Goldblatt, P. 1995. The status of R. Dahlgren's orders Liliales and Melanthiales. In: Rudall, P.J., Cribb, P.J., Cutler, D.F., Humphries, C.J. (eds.), Monocotyledons: systematics and evolution, pp. 181–200. Royal Botanic Gardens, Kew.

Hara, H. 1967. The status of the genus *Metanarthecium* Maxim. J Jpn. Bot. 42: 312–316.

Heel, W.A. van 1988. On the development of some gynoecia with septal nectaries. Blumea 33: 477–504.

Hegnauer, R. 1963, 1986. See general references.

Huber, H. 1969. See general references.

Hutchinson, J. 1934. The families of flowering plants, Vol. 2. Monocotyledons. London: Macmillan.

Ikuse, M. 1956. Pollen grains of Japan. Tokyo: Hirokawa.

Kitagawa, M. 1983. *Aletris luteoviridis* (Maxim.) Franch. In: Ohwi, J., Kitagawa, M. (eds.) New flora of Japan: 436. Tokyo: Shibundo (in Japanese).

Kitamura, S. 1964. Liliaceae. In: Kitamura, S., Murata, G., Koyama, T. (eds.), Coloured illustrations of herbaceous plants of Japan, Vol. 3, pp. 90–157. Osaka: Hoikusha (in Japanese).

Kurosawa, S. 1980. Cytotaxonomic study on the spermatophytes in the Ose region. In: Hara, H. (ed.) Comprehensive research study of Osegahara and its allied area, pp. 46–48. Tokyo: Monbusho Scientific Research Program (in Japanese).

Levan, A., Fredga, K., Sandberg, A.A. 1964. Nomenclature for centromeric position on chromosomes. Hereditas 52: 201–220.

Loew, E., Kirchner, O. 1911. Melanthioideae, pp. 229–291. In: Kirchner, O., Loew, E., Schröter, C. (eds.) Lebensgeschichte der Blütenpflanzen Mitteleuropas, Vol. I, Abt. 3. Stuttgart: E. Ulmer.

McDaniel, S. 1968. *Harperocallis*. A new genus of the Liliaceae from Florida. J. Arnold Arbor. 49: 35–40.

Nakai, T. 1930. Plantae Japonicae and Koreanae. Bot. Mag. (Tokyo) 44: 22–23.

Sato, D. 1942. Karyotype alteration and phylogeny in Liliaceae and allied families. Jpn. J. Bot. 12: 57–161.

Shimizu, T. 1983. The new alpine flora of Japan in color. Osaka: Hoikusha (in Japanese).

Sokolowska-Kulczycka, A. 1980. Embryological studies of *Tofieldia calyculata* (L.) Whlb. Acta Biol. Cracov. Ser. Bot. 22: 113–128.

Stant, M.Y. 1970. Anatomy of *Petrosavia stellaris* Becc., a saprophytic monocotyledon. Bot. J. Linn. Soc. 63 (Suppl. 1): 147–161.

Sterling, C. 1978. Comparative morphology of the carpel in the Liliaceae: Hewardieae, Petrosavieae, and Tricyrteae. Bot. J. Linn. Soc. 77: 95–106.

Sterling, C. 1979. Comparative morphology of the carpel in the Liliaceae: Tofieldieae. Bot. J. Linn. Soc. 79: 321–332.

Sterling, C. 1980. Comparative morphology of the carpel in the Liliaceae: Helonieae. Bot. J. Linn. Soc. 80: 341–356.

Takahashi, Hr., Nishio, E., Hayashi, H. 1993. Pollination biology of the saprophytic species *Petrosavia sakuraii* (Makino) van Steenis in central Japan. J. Plant Res. 106: 213–217.

Takahashi, M., Kawano, S. 1989. Pollen morphology of the Melanthiaceae and its systematic implications. Ann. Mo. Bot. Gard. 76: 863–876.

Takhtajan, A.L. 1997. See general references.

Tamura, M.N. 1995. See general references.

Tamura, M.N. 1996. A recent history of classifications of Liliaceae with special reference to the influences of cladistics and molecular phylogeny. Shuseibutsugaku-kenkyu 20: 27–31 (in Japanese).

Tamura, M.N., Takahashi, Hr. 1998. Karyotype analysis of the saprophyte *Petrosavia sakuraii* (Makino) J.J. Smith ex van Steenis and its systematic implications. Acta Phytotaxon. Geobot. 49 (in press).

Tanaka, R. 1977. Recent karyotype studies. In: Ogawa, K., Kurozumi, K., Koike, J., Sato, S. (eds.), Plant cytology. 293–326. Tokyo: Asakura (in Japanese).

Thorne, R.T. 1992. See general references.

Tillich, H.-J. 1995. Seedlings and systematics in monocotyledons. In: Rudall, P.J., Cribb, P.J., Cutler, D.F., Humphries, C.J. (eds.) Monocotyledons: systematics and evolution, pp. 303–352. Royal Botanic Gardens, Kew.

Utech, F.H. 1978a. Floral vascular anatomy of *Pleea tenuifolia* Michx. (Liliaceae-Tofieldieae) and its reassignment to *Tofieldia*. Ann. Carnegie Mus. 47: 423–453.

Utech, F.H. 1978b. Floral vascular anatomy of the monotypic Japanese *Metanarthecium luteoviride* Maxim. (Liliaceae-Melanthioideae). Ann. Carnegie Mus. 47: 455–477.

Utech, F.H. 1979. Karyotype analysis, palynology, and external seed morphology of *Tofieldia tenuifolia* (Michx.) Utech (Liliaceae-Tofieldieae). Ann. Carnegie Mus. 48: 171–184.

Utech, F.H. 1984. Floral vascular anatomy of *Japonolirion osense* Nakai (Liliaceae) and its tribal relationship. Ann. Carnegie Mus. 53: 447–461.

Walker, J.W. 1986. Classification and evolution of the monocotyledons. Am. J. Bot. 73 (6): 746 (Abstract).

Williams, C.A. 1975. Biosystematics of the Monocotyledoneae-flavonoid patterns in leaves of the Liliaceae. Biochem. Syst. Ecol. 3: 229–244.

Zhukova, P.G. 1967. Chromosome numbers in some species of plants of the northeastern part of the USSR II. Bot. Zh. 52: 983–987 (in Russian).

Nolinaceae

D. Bogler

Nolinaceae Nakai, Ord. Fam. Trib. Nov.: 226 (1943).

Xerophytic, polycarpic, dioecious, or polygamodioecious *perennials*; shortly caulescent to large and treelike, stem fibrous, sometimes greatly swollen at base; trunks upright, reclining, or subterranean, often branched, covered by persistent leaves or scaly bark. *Leaves* in dense, terminal rosettes; blades linear with expanded spoonlike base, flat or concave, margins minutely serrulate or with enlarged prickles (*Dasylirion*), apices fibrous or brushy. *Inflorescences* in terminal, many-flowered, complex panicles, sometimes with dense axillary fascicles (*Dasylirion*); bracts large. *Flowers* mostly unisexual, trimerous, hypogynous, actinomorphic, articulated on jointed pedicels. Tepals 6, in 2 whorls, free, scarious, whitish yellow or tinged with purple. Stamens 6, ± exserted, rudimentary in pistillate flowers; filaments free, glabrous, inserted at base of tepals; anthers dorsifixed, introrse. Ovary with 1–3 locules, 3-sided, reduced and nonfunctional in staminate flowers; septal nectaries present; ovules 6, 2 per carpel; style short; stigma 3-lobed. *Fruits* dry, ± indehiscent, 3-winged samaras, 3-lobed ± inflated capsules, or 3-ribbed nuts; seeds 1–3 per fruit, phytomelan pigment not present; endosperm not starchy; embryo cylindrical, straight.

Four genera and about 50 spp., northern central America, Mexico, and southern USA.

Vegetative Morphology. Nolinaceae are true xerophytes and have many adaptations related to water conservation. Some features are probably adaptations to periodic burning (Dice 1988). Nolinaceae vary in habit from small and nearly acaulescent to large and treelike (Fig. 109). Several genera have trunks that are greatly enlarged at the base. The wood is fibrous and relatively soft. In *Dasylirion* the trunks frequently recline along the ground and are protected by a skirt of persistent prickly leaves. Subterranean lignotubers and underground branching stems are found in some taxa (Dice 1988; Bogler 1994). The thick and fleshy roots spread out from the thickened stem base. The roots may be expendable in time of severe drought, the plants sending out new ones when conditions improve. In *Beaucarnea recurvata* some epicotyledonary internodes start thickening at an early seedling stage and form a tuber (Troll 1937: 768), and the small plants are pulled underground by contractile roots. Some species of *Nolina* and *Dasylirion* remain shortly caulescent with a thickened stem base for their entire lives. In other species, and in *Beaucarnea* and *Calibanus*, the stem thickens and elongates into an aboveground trunk. Leaves are borne in dense, terminal rosettes. Leaf bases are spirally arranged, overlapping, and sometimes greatly thickened (*Dasylirion* and *Nolina*). The leaves are long-linear, flattened, concave, or occasionally quadrangulately thickened. They are borne upright, drooping, or recurved. The leaf margins are almost always lined with minute serrations. In most species of *Dasylirion* the margins have enlarged sharp prickles, and filiferous margins occur in *Nolina bigelovii* (Dice 1988).

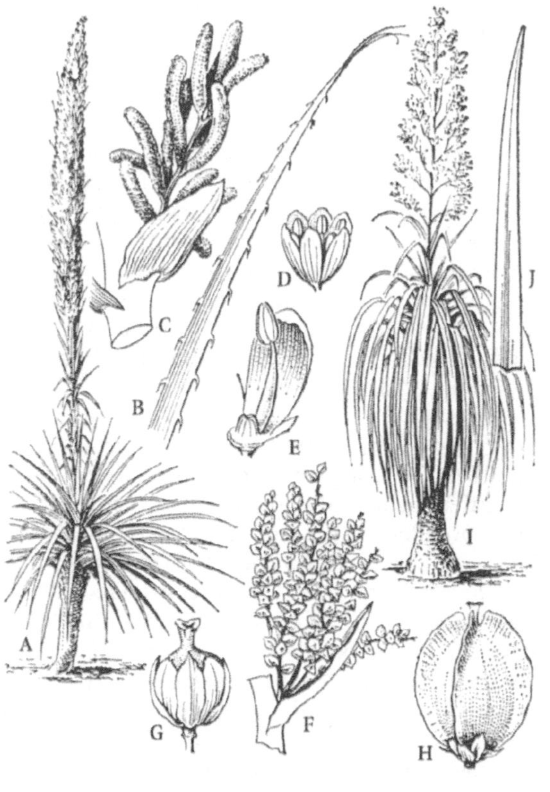

Fig. 109A–J. Nolinaceae. **A–H** *Dasylirion leiophyllum*. **A** Habit. **B** Leaf margin. **C** Staminate inflorescence. **D** Staminate flower. **E** Tepal and stamen. **F** Part of infructescence. **G** Young fruit. **H** Mature capsule. **I, J** *Nolina longifolia*. **I** Habit. **J** Leaf margin. (Takhtajan 1982)

Inflorescences. In all Nolinaceae the inflorescence is terminal, and further vegetative growth after flowering is accomplished by activation of an axillary bud which assumes apical dominance. In most genera the inflorescence is a woody, much-branched, open panicle bearing hundreds or thousands of flowers. There is a central primary branch or axis which bears smaller secondary and even smaller tertiary branches. The size of the branches generally decreases going up the inflorescence. The branches are subtended by papery bracts of varying size. In *Dasylirion* the secondary and tertiary branches of the panicle are highly condensed, resulting in dense, fingerlike, branched fascicles of flowers arrayed along the primary axis.

Vegetative Anatomy. The increase in girth and treelike form attained by many Nolinaceae is attributed to a secondary thickening meristem like that found in *Yucca* and other arborescent monocotyledons (Tomlinson and Zimmermann 1969). The leaves are isobilateral, with both the upper and lower surfaces having stomata and chlorenchyma. The leaves are tough and reinforced with sclerenchyma tissue. They have a thick cuticle and sometimes raised epidermal papillae or a layer of flaky wax which may help reflect sunlight and conserve moisture. Epicuticular wax scales of the Convallaria type have been demonstrated to occur in all four genera (Frölich and Barthlott 1988), in contrast to the wax sculpture in *Dracaena*. The stomata are located in parallel grooves that are often sunken and lined with protruding, interlocking papillae. The grooves appear to close when conditions become dry, thereby retaining moisture. The arrangement of subsidiary cells in the stomatal complex is variable, either paracytic or tetracytic. The guard cells have been reported to contain an oily substance (Ziegenspeck 1944). Alternating with the stomatal grooves are thick bands of sclerenchyma and vascular tissue. These are on both sides of the leaf or sometimes form girders extending from one side of the leaf to the other. There is sometimes an additional row of bundles in the center of the leaf. Chlorenchyma tissue occurs near the leaf surface in proximity to the stomatal grooves and vascular tissue. Crystals of calcium oxalate are found in some taxa. Vessel elements with scalariform or simple perforation plates have been found in the leaves and roots. The xylem in the inflorescence stalk apparently consists of tracheids only.

Flower Structure. The flowers are relatively small and mostly unisexual. All species of *Dasy-*

lirion, *Beaucarnea*, and *Calibanus* are dioecious. Some species of *Nolina* are dioecious but others are polygamo-dioecious with mixtures of pistillate, staminate, and perfect flowers (Thorne 1965). The flowers are borne on jointed pedicels, with the fruits articulating and eventually breaking away from the joint. In the pistillate flower there are reduced, nonfunctional stamens. Nectar is secreted from septal nectaries. In the staminate flowers the pistil is strongly reduced but still functions in nectar secretion. In fruits of *Nolina*, the 3 locules are well developed. In *Dasylirion*, *Beaucarnea*, and *Calibanus* the partitions between locules do not fully develop, resulting in a unilocular fruit. Initially, there are 2 ovules per carpel, but generally only 1 ovule matures per fruit, or up to 3 ovules in *Nolina*. The fruits of Nolinaceae are variously considered as capsules, samaras, or nutlets. They are dry, ± indehiscent, usually 3-winged or 3-lobed, and have 1–3 seeds at maturity. Like Dracaenaceae and Convallariaceae, the seeds of Nolinaceae lack the black phytomelan crust characteristic of other Asparagales (Huber 1969).

Embryology. Details of the embryology of Nolinaceae are unclear and confusing. Either Nolinaceae is quite variable in this respect, or some reports are in error. Microsporogenesis is successive, and pollen is shed in the 2- or 3-celled state (Thorne 1965). Ovules are anatropous with basal placentation. Embryo-sac development is reported to be of the Polygonum type (Thorne 1965). A well-developed hypostase is said to be present and the embryo sac develops a haustorium (Thorne 1965). The archesporial cell is said to divide and produce a parietal cell, which undergoes periclinal divisions in *Nolina* (Thorne 1965). However, Dahlgren et al. (1985, perhaps following Wunderlich 1950) say the opposite, that no parietal cell is cut off and there are no periclinal divisions. Likewise, endosperm formation is said to be Helobial (Thorne 1965) and Nuclear (Dahlgren et al. 1985). The mature embryo is cylindrical (Trelease 1911; Wunderlich 1950; Thorne 1965).

Pollen Morphology. Pollen is sulcate, ± elliptical, finely reticulate, and produced in monads (Johnson and Gale 1983; Ojeda et al. 1984; Bogler 1994).

Karyology. The most common chromosome numbers reported for Nolinaceae are $2n = 38$ and, less frequently, $2n = 36$ (Satô 1942; Hernández 1993). No multivalents are seen in meiosis, and polyploidy has not been reported in this family, although one case of aneuploidy, $2n = 41$, was reported for *Calibanus* (Johnson and Gale 1983). The chromosomes are submetacentric, ± evenly size-graded, and show little resemblance to the strongly bimodal karyotype found in Agavaceae. *Calibanus* and *Dasylirion* have a pair of satellites on the 10th chromosome pair, but these are lacking in *Nolina* (Johnson and Gale 1983). No difference in chromosome morphology between the sexes has been observed. A number of meiotic abnormalities were reported in three species of *Nolina* by Thorne (1965), including evidence of reciprocal translocations (anaphase bridges, dicentric chromatids, acentric fragments), and unusual microspore wall formation.

Pollination. A wide variety of insects visit Nolinaceae, but the primarily pollinators appear to be small bees in the families Andrenidae, Anthophoridae, Colletidae, Halictidae, Megachilidae, and Apidae (Dice 1988; Bogler 1994). Nolinaceae are also visited by a variety of beetles, butterflies, flies, and wasps. Insects are attracted by the nectar produced by both staminate and pistillate flowers, and the copious pollen, which is eagerly gathered by the bees.

Phytochemistry. Nolinaceae contain steroidal sapogenins, and a photosensitive compound in the leaves of *Nolina* is reported to be toxic to sheep and goats, apparently causing severe liver injury (Kingsbury 1965). Birds and most other animals do not seem to eat the fruits or seeds even though they are often abundant. The unusual flavonol 3-O-methyl-8-C-methylquercetin is found in Nolinaceae, and also occurs in Xanthorrhoeaceae (Williams et al. 1988).

Subdivision and Relationship Within the Family. Trelease (1911) considered *Nolina* to be older than or ancestral to the other genera because of its broader distribution, more variable growth habit, and fruit with 3 locules. *Beaucarnea* and *Dasylirion* have often been considered to be closely related to each other because they both have unilocular, single-seeded, winged fruits. In contrast to *Dasylirion*, *Beaucarnea* attains a much larger size, has a greatly swollen trunk, leaves that lack marginal prickles, and a large open panicle. In these traits there is no overlap between *Beaucarnea* and *Dasylirion*, and there are no intermediate taxa which could go in either genus. *Calibanus* has generally been considered related to *Nolina*. Recent chloroplast DNA and ITS rDNA sequence studies provide strong support for the basal position of *Nolina*, and indicate that *Beaucarnea* and *Calibanus* are closely related (Bogler and Simpson 1995, 1996).

Family Affinities. Nolinaceae were included in a broadly conceived family Liliaceae, usually positioned near *Dracaena* and *Yucca*. They have been placed in tribe Nolineae (Krause 1930), tribe Dracaeneae (Bentham and Hooker 1883), and tribe Yuccoideae (Baker 1881). Nolinaceae were included in Agavaceae by Hutchinson (1934) along with other arborescent or fibrous-leaved taxa such as *Yucca*, *Agave*, and *Dracaena*. This arrangement was popular and widely followed, but the placement of Nolinaceae in Agavaceae is not supported by karyotype, flower, or fruit and seed morphology. In recent systems the Agavaceae is treated more narrowly, with Nolinaceae and Dracaenaceae split off as one or two separate families (Dahlgren et al. 1985). Recent investigations using chloroplast DNA and ITS sequences support maintaining Nolinaceae separate from Agavaceae, and demonstrate a close relationship between Nolinaceae, Dracaenaceae, and Convallariaceae (Bogler and Simpson 1995, 1996).

Distribution and Habitats. Nolinaceae are generally restricted to dry regions of the southern US, Mexico, and northern Central America. They prefer mountainous terrain and are frequently found on steep well-drained slopes, rocky outcrops, and sandy ridges. Some of the habitats in which they occur are subject to periodic burning, and adaptation to fire may have been important in the evolution of the some taxa within the family. *Nolina* is the most widely distributed genus, occurring in several states in the southeastern US, throughout the southwestern US, Baja California, and northern Mexico. *Dasylirion* occurs in the interior regions of the southwestern US and Mexico south to Oaxaca, but does not occur in Baja California. It grows on dry rocky slopes and outcrops, hilly grasslands, and in open thorn forest. *Beaucarnea* occurs in northeastern Mexico to northern Central America. *Beaucarnea* generally prefers tropical deciduous forests and rarely overlaps with *Dasylirion*. *Calibanus* is restricted to arid hills in east-central Mexico.

Economic Uses. The leaves of Nolinaceae have been used since prehistoric times for weaving baskets, sandals, sleeping mats, hats, and other household items. The leaves are also used to make brooms, as roof thatching for houses, and as decorations. The enlarged bulbs and leaf bases of *Dasylirion* were formerly roasted in pits and eaten much like *Agave*. In addition, a strong, smoky-tasting liquor known as sotol is made from *Dasylirion* in a process similar to making mescal liquor and tequila from *Agave*. The roasted tubers are chopped up, mixed with water, allowed to ferment for a few days, and twice distilled. This beverage was formerly widely produced in Mexico and large quantities were smuggled into the US during prohibition, but now there are few sotol distilleries in operation (Tunnell and Madrid 1988).

Key to the Genera

1. Leaves armed with sharp enlarged marginal leaf prickles; inflorescence spikelike with dense axillary fascicles of flowers; tepals denticulate; fruits 3-winged samaras, unilocular with 1 seed **4. *Dasylirion***
- Leaves lacking enlarged marginal leaf prickles; inflorescence an open panicle; tepals not denticulate; fruits various 2
2. Trunk globose, low-growing, woody with scaly bark; leaf rosettes scattered over trunk; fruit leathery, subglobose or 3-sided, not 3-winged or lobed, 1-seeded **3. *Calibanus***
- Trunk not globose; leaf rosettes terminal; fruits dry, thin, 3 wings or lobes, 1–3 seeded 3
3. Arborescent with base of trunk greatly swollen above ground level; tepals apically entire; fruit 3-winged, unilocular samara with 1 seed **2. *Beaucarnea***
- Shortly caulescent to arborescent, base of trunk not greatly swollen above ground; tepals apically pilose; fruit deeply 3-lobed, 3-locular, often inflated, 1–3 seeds **1. *Nolina***

1. *Nolina* Michaux Fig. 109I,K

Nolina Michaux, Fl. Bor.-Amer. 1: 207 (1803); Trelease (1911); Thorne (1965); Dice (1988), rev.

Acaulescent to large and treelike, trunks sparingly branched, base of trunk sometimes enlarged at or below ground level; leaves linear, fibrous, with serrulate margins; inflorescences in large open panicles, unisexual or occasionally polygamodioecious with a mixture of staminate, pistillate, and perfect flowers (Thorne 1965); fruits 3-locular, with 3 ± inflated wings, 1–3 roundish seeds. About 23 spp., S Mexico to the SW and SE USA. Trelease (1911) divided *Nolina* into four sections: sect. *Graminifoliae* (=*Nolina*) which is acaulescent with small leaves; sect. *Erumpentes* with thick leaves, short inflorescences, small fruits that are not inflated and have protruding seeds; sect. *Microcarpae* with moderate-sized inflated fruits and small seeds; and sect. *Arborescentes* with treelike habit.

2. *Beaucarnea* Lemaire

Beaucarnea Lemaire, Illustr. Hortic. 8, Misc.: 57. (1861); Hernández (1993), rev.

Caudex greatly enlarged and swollen at base, trunks to 18 m tall, openly branched; inflorescences large, open panicles; flowers unisexual; ovary unilocular; fruits dry, 3-winged, 1-seeded samaras. About ten spp., northern Central America and Mexico. *Beaucarnea* was submerged in *Nolina* by Bentham and Hooker (1883), Hemsley (1884), Hutchinson (1934), and Johnson and Gale (1983). The *Nolina recurvata* that is often cited in European studies is really *Beaucarnea recurvata*. The leaves of *Beaucarnea* and *Nolina* are similar, but *Beaucarnea* is easily distinguished by its greatly swollen trunks and unilocular 3-winged fruits. Trelease (1911) proposed two sections: sect. *Beaucarnea* with recurved, glabrous leaves; and sect. *Papillatae* with erect, papillate leaves. Hernández (1993) suggests these divisions may not be clearly distinguished.

3. *Calibanus* Rose

Calibanus Rose, Contr. U. S. Nat. Herb. 10: 90 (1906); Trelease (1911), rev.

Caudex greatly swollen, subglobose, woody and often covered with scaly bark, upright trunk lacking; leaves in numerous scattered rosettes, narrow, blue-green; flowers in short open panicles, unisexual, purplish; fruit ovoid, 3-sided, 3-ribbed, leathery, indehiscent, 1-seeded. Only one sp., *C. caespitosus* Rose, east-central Mexico (Hidalgo and San Luis Potosi).

4. *Dasylirion* Zucc. Fig. 109A–H

Dasylirion Zucc., Allg. Gartenzeit. 6: 258 (1838); Bogler, Tax. phyl. *Dasylirion*. Ph.D diss., Austin, Texas (1994), rev.

Trunks short to elongate, sparingly branched, generally reclining and covered by old leaf bases; leaves in dense terminal rosettes, linear, usually with large and sharp marginal prickles (rarely absent), tips fibrous or brushy; inflorescence elongate and spike-like, with dense axillary fascicles of small unisexual flowers; tepals denticulate; fruit a 3-winged samara. About 16 spp., Mexico and southwestern USA. Trelease (1911) recognized two sections: sect. *Quadrangulatae* with 4-angled leaves and greatly reduced or absent leaf prickles; and sect. *Dasylirion* with flat leaves. Chloroplast DNA restriction site analysis indicates that the three species with brushy leaf tips in southern Mexico constitute a basal group in the genus (Bogler 1994) and prickleless leaves are derived.

Selected Bibliography

Baker, J.G. 1881. A synopsis of Aloineae and Yuccoideae. J. Linn. Soc., Bot. 18: 148–241.

Bentham, G., Hooker, J.D. 1883. See general references.

Bogler, D.J. 1994. Taxonomy and phylogeny of *Dasylirion* (Nolinaceae). Ph.D. Dissertation. Austin; Texas: University of Texas.

Bogler, D.J., Simpson, B.B. 1995. A chloroplast DNA study of the Agavaceae. Syst. Bot. 20: 191–205.

Bogler, D.J., Simpson, B.B. 1996. Phylogeny of Agavaceae based on ITS rDNA sequence variation. Am. J. Bot. 83: 1225–1235.

Dahlgren, R.M., Clifford, H.T., Yeo, P.F. 1985. See general references.

Dice, J.C. 1988. Systematic studies in the *Nolina bigelovii-N. parryi* (Nolinaceae) complex. M.A. Thesis. San Diego, California: San Diego State University.

Engler, A. 1888. Liliaceae. In: Engler, A., Prantl, K. (eds.) Die natürlichen Pflanzenfamilien. II, 5. Leipzig: W. Engelmann.

Frölich, D., Barthlott, W. 1988. Mikromorphologie der epicuticulären Wachse und das System der Monokotylen. Trop. Subtrop. Pflanzenwelt (Akad. Wiss. Lit. Mainz) 63: 135 p.

Hemsley, W.B. 1884. Biol. Centrali-Americana. Botany. 3: 371–374. London: Dulau & Co.

Hernández, L.G. 1993. Cladistic analysis of the American genera of Asparagales and the systematic study of *Beaucarnea* (Nolinaceae) and *Hemiphylacus* (Hyacinthaceae). Ph.D Dissertation. Austin, Texas: University of Texas.

Huber, H. 1969. See general references.

Hutchinson, J. 1934. The families of flowering plants, Vol. II, Monocotyledons. Oxford: Clarendon Press.

Johnson, M.A., Gale, R.M. 1983. Observations on the leaf anatomy, pollen, cytology and propagation of *Calibanus Hookeri* (Lem.) Trel. Bradleya 1: 25–32.

Kingsbury, J.M. 1965. Deadly harvest: a guide to common poisonous plants. New York, Chicago, and San Francisco: Holt, Rinehart and Winston.

Krause, K. 1930. See general references.

Ojeda, L., Ludlow-Wiechers, B., Orellana, R. 1984. Palinología de la familia Agavaceae para la península de Yucatán. Biotica 9: 379–398.

Satô, D. 1942. Karyotype alteration and phylogeny in Liliaceae and allied families. Jpn. J. Bot. 12: 58–161.

Takhtajan, A.L. 1982. See general references.

Thorne, F.M. 1965. The taxonomy of the genus *Nolina* Michx. (Agavaceae) in the southeastern United States. Ph.D Thesis. Athens, Georgia: University of Georgia.

Tomlinson, P.B., Zimmermann, M.H. 1969. Vascular anatomy of monocotyledons with secondary growth – an introduction. J. Arnold Arbor. 50: 159–179.

Trelease, W. 1911. The desert group Nolineae. Proc. Am. Philos. Soc. 50: 404–443.

Troll, W. 1937. Vergleichende Morphologie der höheren Pflanzen. Teil 1. Berlin: Borntraeger.

Tunnell, C., Madrid, E. 1988. Making and taking sotol in Chihuahua and Texas. In: Powell, A., Hollander, R., Barlow, J., McGillivray, W., Schmidly, D. (eds.) 3rd Symposium on Resources of the Chihuahua Desert Region, Nov. 10–12, 1988, Alpine, Texas.

Williams, C.A., Harborne, J.B., Mathew, B. 1988. A chemical appraisal via leaf flavonoids of Dahlgren's Liliiflorae. Phytochemistry 27: 2609–2629.
Wunderlich, R. 1950. Die Agavaceae Hutchinsons im Lichte ihrer Embryologie, ihres Gynözeums-, Staublatt- und Blattbaues. Oesterr. Bot. Z. 97: 437–502.
Ziegenspeck, H. 1944. Das Vorkommen von Öl in den Stomata der Monokotyledonen und die Bedeutung des konstitutionellen Vorkommens für die Systematik derselben. Feddes Repert. 53: 151–173.

Pandanaceae

B.C. Stone[1], K.-L. Huynh and H.-H. Poppendieck

Pandanaceae R.Br., Prodr.: 340 (1810), nom. cons.

Trees, shrubs or woody climbers; stems usually sympodially branched, annulate with leaf scars; adventitious roots sometimes branched and divaricate, functioning as prop roots or clasp roots, or, if early aborted, as spines or protuberances. *Leaves* spirally arranged, 3-ranked (*Freycinetia* and *Pandanus*) or 4-ranked (*Sararanga*), linear-ensiform, entire but armed on margins and the adaxial face of midrib with prickles, sometimes also in the adaxial face of lateral pleats, rarely unarmed. *Inflorescences* terminal, sometimes lateral or axillary, usually unisexual, at first erect but often soon nodding or pendent, in spikes or racemes of spikes in *Pandanus*, in pseudoumbels (rarely racemes) of spikes in *Freycinetia*, and in repeatedly branched panicles in *Sararanga*; lower bracts foliaceous, upper bracts variously coloured, sometimes fragrant and slightly sweet. *Flowers* strictly unisexual (most *Pandanus*), or with vestiges of the other sex, without perianth. Staminate inflorescence: flowers distinct in *Sararanga* and in most *Pandanus*, generally indistinct in *Freycinetia*; filaments smooth or papillate; anthers longitudinally dehiscent, retuse or acuminate. Pistillate inflorescence: flowers composed of one or several carpels; styles absent to distinct, sometimes spiniform (some *Pandanus*); stigmas reniform or hippocrepiform, sometimes elongated (some *Pandanus*); ovules anatropous, bitegmic. *Fruit* a drupe (*Pandanus*) or berry (*Freycinetia*, *Sararanga*). Seeds straight in *Pandanus* and *Sararanga*, straight or variously curved in *Freycinetia*; seed coat thin and membranous in *Pandanus*, thickened in *Sararanga* and *Freycinetia*; embryo small and straight; endosperm copious, oily, in *Freycinetia* starchy.

A palaeotropical family with 3 genera and 800–900 species.

Vegetative Morphology and Anatomy. Although very variable, the Pandanaceae can be

[1] Deceased.

readily recognised by their leaves (shape, anatomy, phyllotaxis). There are root climbers (*Freycinetia*), trees, shrubs, or epiphytes (*Pandanus*) (Fig. 110) or trees (*Sararanga*) (Fig. 111). In all genera, the phyllotaxis changes from tri-orthostichy in the juvenile stage to tri-spirostichy in the adult stage. This phyllotaxis is maintained in *Pandanus* and *Freycinetia* but becomes tetrastichous after the formation of the first inflorescence in *Sararanga*. In *Freycinetia*, the scandent habit is evident at the seedling stage already. In *Pandanus*, branching of the orthotropic stems may occur at the juvenile or only the adult stage. Phyllotaxis is dextrorse or sinistrorse,

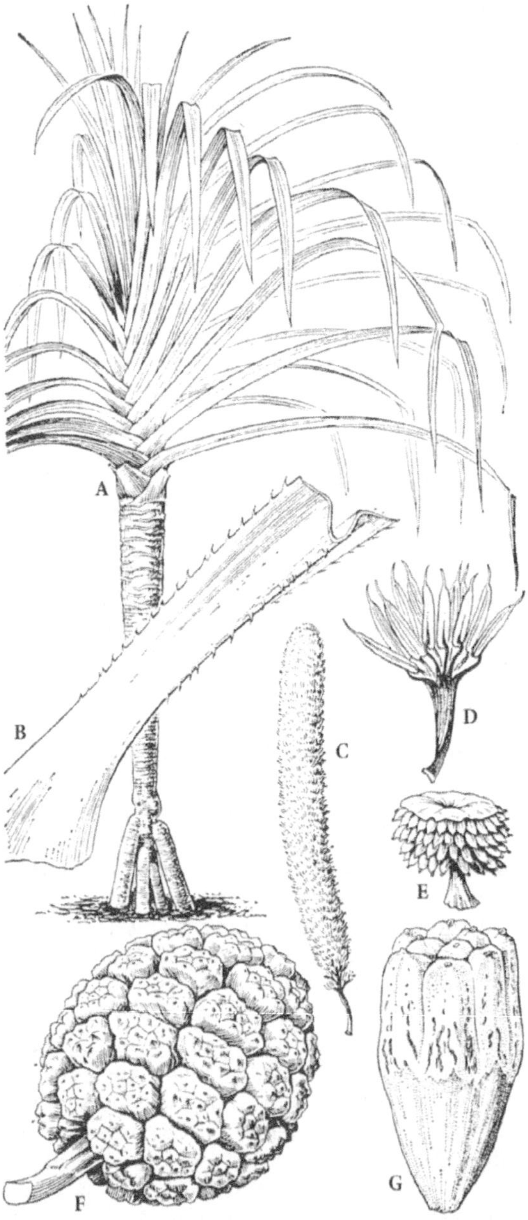

Fig. 110A–G. Pandanaceae. A–D *Pandanus furcatus*. **A** Habit. **B** Part of leaf. **C** Staminate inflorescence. **D** Staminate flower. **E** *P. macrocarpus*, staminate flower. **F, G** *P. tectorius*. **F** Infructescence. **G** Polydrupe. (Takhtajan 1982)

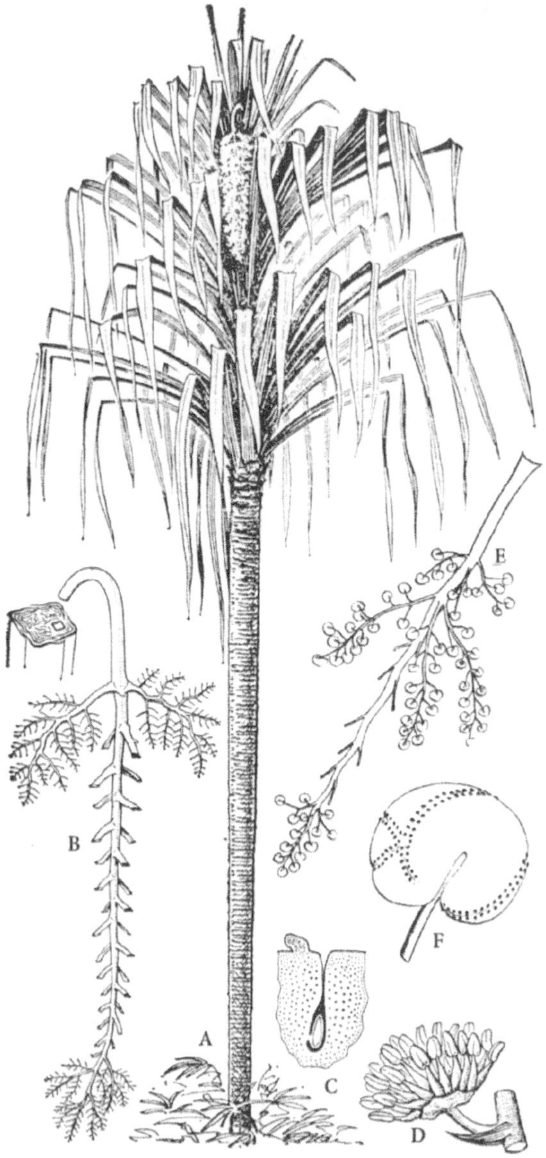

Fig. 111A–F. Pandanaceae. *Sararanga sinuosa*. **A** Habit. **B** Pistillate infructescence. **C** Sectional view of ovary and stigma. **D** Staminate flower. **E** Part of infructescence. **F** Mature fruit. (A–D Takhtajan 1982; E, F Stone 1961)

and both may occur on one and the same individual.

The leaves are generally M-shaped in transverse section, and may be amplexicaul or semiamplexicaul at the base, with or without distinct auricles.

Stomates are tetracytic throughout the family, each composed of 2 guard cells surrounded by 2 polar cells and 2 lateral cells (Tomlinson 1965; North and Willis 1970, 1971; Huynh 1974). They are smooth except for those of the leaves where the guard cells are smooth, but the lateral cells may be verrucate and the polar cells verrucate or papillose with simple or branched papillas, especially in *Pandanus* where these processes vary with sections and species. The lateral longitudinal veins in the leaves are connected by transverse veins, visible in species with tessellate leaves.

There are no true hairs except for stellate hairs on the axis of the inflorescence in *Sararanga philippinensis*. Both crystal cells, each with 1 prismatic crystal, and raphide cells are abundant. Oil cells may occur, especially in *Sararanga*. Fibre strands are frequent. In all genera, the leaf vascular bundles have a thick fibre strand at the adaxial and abaxial side (Warburg 1900; North and Willis 1970, 1971; Huynh 1974).

Vessels may be up to 2 cm long and have long and unspecialised perforation plates. Stem vascular bundles are either collateral, or bipolar with phloem tissues facing. The central cylinder of the stem is clearly delimited. In all taxa investigated, temporarily aggregated bi- or tripolar bundles have been found. Pandanaceae share these compound bundles with Cyclanthaceae (and some Araceae), but differ in having always discrete bundles not fusing basipetally (Zimmermann and Le Claire 1974; French and Tomlinson 1985). The aerial roots are polyarch with up to several hundred bundles in the central cylinder and additional strands of xylem and phloem in the pith (like some Musaceae, Palmae, Cyclanthaceae: v. Guttenberg 1968). Considering the morphogenetic plasticity and structural diversity of monocot stems and roots in general, however, these characters are not necessarily indicators of phyletic affinity.

Two types of sieve tube plastids occur in both *Pandanus* and *Freycinetia*, the PIIc type with wedge-shaped crystals, widespread in monocotyledons, and the PIIcf type with additional ring-shaped filaments (Behnke 1981). Epicuticular wax is of non-oriented platelets or amorphous; rod-like projections of the Strelitzia type were recorded once (Frölich and Barthlott 1988).

Floral Morphology. Inflorescences are generally terminal, sometimes lateral or axillary. True leaves gradually change upwards to foliaceous bracts and to floral bracts. The principal inflorescence axis may produce fertile branches branching further several times usually without bracts. Just below the base of the inflorescence, 1, 2, rarely 3 new vegetative shoots may occur. The highly reduced flowers in condensed inflorescences lack a macroscopic perianth and differ from other monocotyledons in many respect; primordia representing the non-functional sex organs seem to be usually present (Stone 1990).

In *Freycinetia*, the spikes are aggregated into a pseudoumbel (e.g. *F. whitmorei*) or raceme (*F. angustifolia*), of spikes, sometimes a solitary spike (e.g. *F. angustissima*). There is a correspondence between pistillate and staminate inflorescences as to the number and arrangement of spikes, and between pistillate and staminate flowers as to the number of components. This strict 1:1 relationship in the number of carpels and carpellodes, and of stamens and staminodes, and the fact that true hermaphrodite flowers and/or bisexual spikes are sometimes found in some species (Cox 1981; Poppendieck 1987; Huynh and Cox 1992), are considered as evidence of potential bisexuality and cryptical monoecy in a genus usually considered to be dioecius (Huynh 1991b, 1992; Huynh and Cox 1992). This was corroborated by the fact that both pistillate and staminate shoots might occur on one and the same individual (Poppendieck 1987) and that staminodes also normally have sporogenous tissue which aborts before meiosis (Huynh 1992). Staminate flowers are indistinct on the spike surface (except for some species e.g. *F. cumingiana*) but distinct beneath, which is revealed by the fact that the staminal vascular bundles of the flower have centripetal xylem (Huynh 1991b). The anthers are always retuse. In a very few species such as *F. samoensis*, the mature anthers have no endothecial thickenings. Such anthers have not been found in *Pandanus* and *Sararanga*. Neither a perianth nor a pedicel exist in either flower, but tepals may be traced at very early developmental stages (Cox 1990).

In *Pandanus*, the pistillate inflorescence is generally reduced in comparison with the staminate inflorescence, pistillate spikes being generally less numerous than staminate spikes. Both kinds of flowers are strictly unisexual, carpellodes and staminodes having been observed only occasionally except for the Madagascan sect. *Martellidendron*. Here, as in *Freycinetia*, vestiges of the other sex seem to be produced regularly (Huynh

1981, and unpubl.), and the staminodes of pistillate flowers have 4 pollen sacs (Huynh 1997). The flowers of *Pandanus* vary most in the family (Stone 1974). The pistillate flowers have one to several carpels, partially or entirely fused, with styles absent or distinct, sometimes spiniform, stigmas appearing centrifugal or centripetal in multicarpellate pistils, ventral or dorsal in unicarpellate pistils. The staminate flowers have 2–3 (*P. concinnus*) to several hundred stamens (*P. lamprocephalus*) or more (Huynh 1982). The anther is acuminate, rarely retuse, but in *P. concinnus* the curved apex of the anther renders the acumen invisible at first sight (Huynh 1982). The filaments are arranged at the apex or along the upper part of a variable stemonophore. Sometimes they are fused into a staminal tube without a stemonophore (Stone 1974; Huynh 1982). The pedicel is distinct in most staminate flowers but is lacking in pistillate flowers.

In *Sararanga*, the pistillate inflorescence is apparently not reduced in comparison with the staminate inflorescence. The pistillate flowers lack staminodes, but the staminate flowers have a pistillode. A perianthlike cupule is observed. Pedicels are distinct in *S. sinuosa*, but almost nil in *S. philippinensis.*

EMBRYOLOGY. The tapetum of the anther is glandular. The pollen grains are 2-celled when shed. The ovule is anatropous and bitegmic. In *Freycinetia*, the embryo sac conforms to the Polygonum type, in *Pandanus* to the Allium type. It has 3 antipodes in *Freycinetia*, but up to 64 or more in *Pandanus* because of migrating nuclei; a hypostase is differentiated; the endosperm is nuclear and becomes cellular at the final stage of development (Davis 1966).

POLLEN MORPHOLOGY. Pollen grains are subglobose or ovoid, uniporate monads. The exine is reticulate in *Sararanga*; spinulose, sometimes smooth, rarely reticulate in *Pandanus*; smooth in *Freycinetia*; spinulose in the fossil genus *Pandaniidites* (North and Willis 1971; Huynh 1980, 1981, 1982, 1991a; Hotton et al. 1994).

KARYOLOGY. Chromosome numbers have been scarcely studied. Most *Pandanus* species conform to the gametic number $n = 30$ although minor aneuploid variations may occur (Cheah and Stone 1971, 1973). Only *P. spiralis* is tetraploid ($n = 60$). In *Freycinetia* the gametic number is $n = 25$ in *F. funicularis* (Ganguly 1959), but $n = 28$ in *F. sumatrana* (Cheah and Stone 1973), and apparently $n = 30$ in *F. corneri* (Cheah and Stone 1971). Chromosome numbers are unknown in *Sararanga.*

REPRODUCTIVE BIOLOGY. *Freycinetia* is pollinated by vertebrates (e.g. bats, flying foxes, rats), *Pandanus* by the wind (probably also by small unspecialised insects), *Sararanga* by insects (Cox 1990). In all three genera the staminate plants collected are about one tenth or less of the pistillate ones, and most species of *Pandanus* and *Freycinetia* are known only by their pistillate plants. Apomixis is probably general in the family: in several populations in which staminate plants could not be found, fruits formed in pistillate plants, which is confirmed by bagging experiments and by observations in botanical gardens (Cox 1990). However, some *Pandanus* species never produced fruits (e.g. the well-known *P. amaryllifolius*) or never blossomed (e.g. *P. veitchii*). Most cultivars of *Pandanus* are reproduced only by cuttings. Fruit dispersal is by birds, mammals, sometimes by turtles and freshwater or ocean currents (Stone 1983; Cox 1990). However, in most species of the family, pollination and fruit dispersal have not been investigated.

FRUIT MORPHOLOGY. The fruit of *Freycinetia* is a fleshy or ligneous berry with 1 to some 12 carpels, a thin pericarp, being seminiferous in the lower part. The variable anatomy of the fruits provides reliable characters for distinguishing species (Huynh 1996). The fleshy berries have no lignified elements except for those in the vascular bundles. The ligneous berries have a central sclerenchyma in the upper part. Three types of ligneous berries have been found (Huynh 1995): (1) The berries may have fibre strands extending in a single stretch from apex to almost down to base through the ovary walls (e.g. *F. reineckii*); (2) isolated fibres and very characteristic fusiform/elliptic fibre bundles may be formed at successive levels along the berry as it develops (e.g. *F. impavida*); or (3) neither kind of fibre association may be observed. Raphide cells and crystal cells of varying density and distribution are frequently observed in ovary walls. The seeds can be freed from the thin ovary walls and germinate outside the fruit.

In *Pandanus*, the fruit is a drupe (monodrupe or polydrupe depending on the number of carpels and stigmas). The seeds are bound to the hard and bony endocarp within which they germinate, the seedlings making exit by the germination tubes at the base of the drupe. The seminiferous part, coinciding with the endocarp, divides the mesocarp into upper and lower mesocarp. In polydrupes, the endocarps are fused, as are the carpels either from base almost to apex or to about middle level of the upper mesocarp. The fruit always has fibre strands (generally called fibres) extending from apex to base, through the medullose (sometimes cavernous) mesocarp and the endocarp. In the upper mesocarp, the "axial fibres", one for each carpel, are generally broader than the lateral "fibres". Each of them contains a fertilisation canal extending from the corresponding stigma to seed locule. The endocarp is the most variable feature of the fruit (position, length, structure), and is therefore very useful in distinguishing species.

In *Sararanga*, the fruit is a polypyrenoid berry, with carpels in a characteristic biseriate sinuous row. It has no fibre strands. The pericarp is thin, the mesocarp thick but fleshy, the endocarp thin, the seminiferous part extending from base almost to apex of the fruit.

SEEDS. Seeds are small and straight in *Sararanga* and *Freycinetia*, sometimes lunate, hippocrepiform, fishhook-like or circinate in *Freycinetia*, larger, fusiform or obovoid in *Pandanus*. The endosperm is albuminous, white or coloured, fleshy, often edible. The seed coat is thin and membranous in *Pandanus*, thickened in *Freycinetia* and *Sararanga*. In *Freycinetia*, the funicle develops into a raphe; the part of the outer integument that is furthest away from the raphe is sometimes strongly developed showing lignified thickenings and is usually (but misleadingly) called strophiole (Huynh 1996). The seed of *Freycinetia* varies most in the family with regard to shape, size and ratio between raphe and the remainder of the seed, presence/absence of crystal cells in raphe, cell composition of raphe, presence/absence of a "strophiole", and is therefore very useful in taxonomy.

PHYTOCHEMISTRY. Despite their ethnobotanical importance, there is a severe lack of chemical information on Pandanaceae, and no taxonomic use can be made of the few data. In leaves, caffeic- and p-coumaric acid have been found, but flavonoids, proanthocyanidins, chinones, saponins and cyanogenic substances seem to be lacking (Hegnauer 1963, 1986). The aromatic substances in the leaves of *Pandanus amaryllifolius* were said to be derived from carotenoids by oxidation, while other authors have reported a mixture of 40 different substances including β-selinen and β-caryophyllene; more recently, piperidine alkaloids have been characterized from this species (Nonato et al. 1993). Staminate inflorescences of *P. tectorius* yield a perfume, while the pericarp and seed of *Pandanus* may contain a considerable amount of fatty oils.

DISTRIBUTION AND HABITATS. Although originally a west Gondwanian or west Laurasian group (see Palaeobotany, below), the present distribution of Pandanaceae is restricted to the Old World. *Pandanus* extends from W Africa far into the Pacific, while *Freycinetia* is wholly Austral-Asian, also reaching New Zealand. The two species of *Sararanga* are found from the Philippines to Melanesia.

Pandanaceae are found as separate individuals or in groups forming thickets, from sea level to 3000 m or higher altitudes, almost all in warm and moist regions. They grow on a variety of substrates such as granitic, siliceous, ultrabasic, muddy, peaty, lateritic or clayey soil, limestone or marine sediments, or as epiphytes.

RELATIONSHIPS. The genera of Pandanaceae are widely different in morphology, which does not facilitate establishing relationships between them. Generally speaking, *Sararanga* appears closer to *Pandanus* than to *Freycinetia*, considering its arborescent habit, leaves without auricles, staminate flowers with pedicels (also observed in *Pandanus*, not in *Freycinetia*), and reticulate pollen (also observed in *Pandanus eydouxia* and sect. *Martellidendron*: Huynh 1981, 1991a; not in *Freycinetia*). In pollen structure, *Sararanga* and *Pandanus* sect. *Martellidendron* consistently occupy a basal position in the family (Hotton et al. 1994), which is corroborated as to the former by its perianthlike cupule and the latter by its flowers having regularly vestiges of the other sex. *Freycinetia* and the other groups of *Pandanus* showed advanced features in various organs. *Pandanus* sect. *Martellidendron* appears to constitute the phylogenetic link between these two genera (Huynh 1997), considering its 3-layered exine, the rudimentary bisexuality, and its staminodes with four pollen-sacs, all three features being also observed in *Freycinetia*.

A close association between Pandanaceae and Cyclanthaceae is supported by morphological evidence (Erdtman 1952; Hotton et al. 1994) and morphological (Cox et al. 1995; Stevenson and Loconte 1995), molecular (Chase et al. 1995) and combined character analyses (Chase et al. 1995a), but an often suggested association with palms (e.g. Huber 1991) cannot be substantiated with molecular evidence available. Even the broader relationship of the family pair Pandanaceae/ Cyclanthaceae (either with Lilianae-Dioscoreales or the closely related "stemonoids": Chase 1995, 1995a; or with the "non-Liliiflorean" "Central Branch" of Huber 1991) is not well settled.

PALAEOBOTANY. *Pandaniidites*, a palynomorph referred to *Pandanus*, was found in the late Cretaceous and early Tertiary in N and S America (Elsik 1968; Leffingwell 1971; Jarzen 1983). However, the similarity of the fossil lemnaceous/araceous pollen of *Limnobiophyllum* (Stockey et al. 1997) with *Pandaniites* casts doubt on the identity of the latter. A fossilised fructification of *Pandanusocarpon*, attributed to Pandanaceae, was described from India and dated to the early Eocene (Bonde 1990). However, most of the numerous reports of Pandanaceous megafossils were later identified as *Nypa* (Jarzen 1983), and no fossil records from periods earlier than the late Cretaceous can confidently be attributed to *Freycinetia* or *Sararanga* (Cox et al. 1995).

ECONOMIC USES. Several Pandans are economically important depending on the regions where they are found: thatch for roof; weaving and plaiting for both home and commercial purposes; retted fibres of *Pandanus* leaves and *Freycinetia* roots make excellent cords. Fruit pulp (e.g. *P. conoideus*), endosperm (e.g. *F. dubius*), soft bracts (some *Freycinetia*) and even the soft inner tissues of the stem are edible. Fragrant leaves of *P. amaryllifolius* are used as a spice in cooking. *P. odoratissimus* and other species have fragrant pollen and bracts which are used for perfumes and to flavour popular drinks. No ethnobotanical research has been carried out so far, but medicinal (and magical) uses have been reported in ethnobotanical literature. Some *Pandanus* species are planted as household or garden trees, especially cultivars that have variegated leaves.

CONSERVATION. Pandanaceae are well known for their general endemism, sometimes very narrow. This is the primary threat for their survival, increased by their economic utility and the destruction of their natural habitats in several countries, and, in Africa, the present desertification. In Senegal, for example, *Pandanus heudelotianus* is already extinct, and *P. senegalensis*, recently described, may be the next to disappear if no suitable measures of conservation are taken. Also, several *Pandanus* species endemic to Mauritius do no longer exist. Conservation is necessary also because most Pandanaceae are known only partly. Staminate plants for example, are unknown for most species of *Freycinetia* and *Pandanus*.

KEY TO THE GENERA

1. Pistillate inflorescence (infructescence) a panicle; leaves in 4 spiral rows; flowers with a perianthlike cupule **3. *Sararange***
- Pistillate inflorescence (infructescence) a cephalium, or a raceme or pseudoumbel of cephalia; leaves in 3 spiral rows; flowers without a perianthlike cupule 2
2. Fruit a drupe with hard bony endocarp; carpels uniovulate, rarely 2-3-ovulate; mesocarp divided into upper and lower mesocarp; seeds bound to endocarp and always germinating within fruit **2. *Pandanus***
- Fruit a berry; carpels multiovulate (ovules much more numerous than 3); mesocarp not divided into upper and lower mesocarp; seeds capable of being free and germinating outside fruit **1. *Freycinetia***

1. *Freycinetia* Gaudich.

Freycinetia Gaudich., Ann. Sci. Nat. I, 3: 509 (1824); Warburg in Engl., Pflanzenreich IV.9: 1–100 (1900); Stone, Blumea 16: 361–372 (1968) (key to sections).

Woody climbers. Leaves in 3 spiral rows, remote or imbricate on stem, amplexicaul or semiamplexicaul, at base with distinct auricles, these persistent or caducous. Inflorescences in both sexes similar with regard to number and arrangement of spikes. Spikes pseudoumbellate, generally 3, sometimes 4–7, rarely 1, 2 or racemose. Flowers unisexual, the other sex vestigial or occ. missing. Staminate flowers indistinct. Pistillate flowers with multiovulate carpels. Fruit a berry. Seeds very variable in shape and structure, the funicle developed into a raphe. About 180 spp., tropical zones from Sri Lanka to the Marquesas, incl. Hawaii Is. and N Australia, also in New Zealand.

Since most species are known from pistillate plants only, the classification of *Freycinetia* has mostly been based on the fruits (Warburg 1900). Stone (1968) used additional evidence when dividing the genus into 17 sections; others were described subsequently. A partial grouping of sections into subgenera was tentatively proposed by Huynh (1995).

2. *Pandanus* Parkinson Fig. 110

Pandanus Parkinson, J. Voy. South Seas: 46 (1773); St. John, Pac. Sci. 14: 224–241 (1960); Stone, Bot. Jahrb. Syst. 94: 459–540 (1974) (key to sections).

Trees or shrubs, with or without prop roots, rarely epiphytes. Leaves in three spiral rows, amplexicaul, sometimes semiamplexicaul at base; auricles sometimes well differentiated (*P.* sect.

Dauphinensia). Inflorescences variable with sexes as to number of spikes, pistillate spikes being generally less numerous than staminate spikes. Flowers usually unisexual. Staminate flowers generally distinct, with stamens very variable in numbers, free on spikes or aggregated into phalanges of various forms. Pistillate flowers with 1-several, generally 1-ovulate carpels. Fruit a monodrupe or polydrupe (phalange). Seeds fusiform or obovoid; seed coat thin and membranous. About 700 spp., tropical zones from Africa to Asia and various archipelagos in the Pacific.

Classifications in *Pandanus* have been based essentially on fruits, only recently additional evidence has been utilised. Successive efforts of Warburg (1900), St. John (1960) and Stone (1974) led to the delimitation of some 60 sections, while others were subsequently described (e.g. by Huynh 1991a). At present, classification in *Pandanus* seems to be resolved at the sectional level but not at the subgeneric one. For example, the Asiatic sections (e.g. sect. *Rykia*) and Madagascan sections (e.g. sect. *Acanthostyla*) with spiniform stigmas are placed in two different subgenera (Stone 1974), apparently on the only basis of geographic distribution. Sect. *Martellidendron* is an isolated group and would be better treated as a separate genus (Hotton et al. 1994).

3. ***Sararanga*** Hemsley Fig. 111

Sararanga Hemsley, J. Linn. Soc. Bot. 30: 216, t. 11 (1894); Stone, Brittonia 13: 212–224 (1961), rev.

Large trees, branching at last; prop roots none; aerial roots forming a fine mesh or network at the base of the trunk. Leaves in four spiral rows on fertile shoots, not auriculate. Inflorescences paniculate, terminal, pendulous, richly branched to third or fourth order; axes quadrangular, glabrous (*S. sinuosa*) or with stellate hairs (*S. philippinensis*); flowers with a perianthlike cupule; pedicels distinct in *S. sinuosa*, almost nil in *S. philippinensis*. Staminate flowers with a pistillode; anthers retuse. Pistillate flowers without staminodes; carpels 1-ovulate, arranged in a biseriate sinuous row. Fruit a berry. Seeds small, straight; seed coat thickened. Two spp., from Philippines to Solomon Is.

Selected Bibliography

Behnke, H.-D. 1981. See general references.

Bonde, S.D. 1990. A new palm peduncle *Palmostroboxylon umariense* (Arecaceae) and a fruit Pandanusocarpon umariense (Pandanaceae) from the Deccan Intertrappean Beds of India. Proc. 3rd IOP Conference, Melbourne, pp. 59–65.

Chase, M.W. et al. 1995. See general references.

Chase, M.W. et al. 1995a. See general references.

Cheah, C.H., Stone, B.C. 1971. Pandanaceae. In: Löve, A. (ed.) IOPB Chromosome number reports 32. Taxon 20: 349–356.

Cheah, C.H., Stone, B.C. 1973. Chromosome studies of the genus *Pandanus* (Pandanaceae). Bot. Jahrb. Syst. 93: 498–529.

Cox, P.A. 1981. Bisexuality in the Pandanaceae: new findings in the genus *Freycinetia*. Biotropica 13: 195–198.

Cox, P.A. 1990. Pollination and the evolution of breeding systems in Pandanaceae. Ann. Mo. Bot. Gard. 77: 816–840.

Cox, P.A., Huynh, K.-L., Stone, B.C. 1995. Evolution and systematics of Pandanaceae. In: Rudall, P.J., Cribb, P.J., Cutler, D.F., Humphries, C.J. (eds.) Monocotyledons: systematics and evolution Royal Botanic Gardens, Kew, pp. 663–684.

Davis, G.L. 1966. See general references.

Elsik, W.C. 1968. Palynology of a Paleocene Rockdale lignite, Milam county, Texas. I. Morphology and taxonomy. Pollen Spores 10: 263–314.

Erdtman, G. 1952. See general references.

French, J.C., Tomlinson, P.B. 1985. Compound vascular bundles in monocotyledonous stems: construction and signification. Kew Bull. 41: 561–574.

Frölich, D., Barthlott, W. 1988. Mikromorphologie der epicuticulären Wachse und das System der Monokotylen. Trop. Subtrop. Pflanzenw. (Akad. Wiss. Lit. Mainz) 63: 1–135.

Ganguly, B. 1959. Chromosome number in Pandanales. Curr. Sci. 28: 82.

Guttenberg, H.v. 1968. Der primäre Bau der Angiospermenwurzel. 2. Aufl. Handb. Pflanzenanatomie. Band 8, Teil 5. Berlin: Bornträger.

Hegnauer, R. 1963, 1986. See general references.

Huber, H. 1991. See general references.

Hotton, C.L., Leffingwell, H.A., Skvarla, J. 1994. Pollen ultrastructure of Pandanaceae and the fossil genus *Pandaniidites*. In: Kurmann, M.H., Doyle, J.A. (eds.) Ultrastructure of fossil spores and pollen. Royal Botanic Gardens, Kew, pp. 173–191.

Huynh, K.-L. 1974. La morphologie microscopique de la feuille et la taxonomie du genre *Pandanus* I. Bot. Jahrb. Syst. 94: 190–256.

Huynh, K.-L. 1980. La morphologie du pollen de *Pandanus* subg. *Vinsonia* (Pandanaceae) et sa signification taxonomique. Pollen Spores 22: 173–189.

Huynh, K.-L. 1981. Pandanus kariangensis (sect. Martellidendron), une espèce nouvelle de Madagascar. Bull. Mus. Natl. Hist. Nat., B, Adansonia 1: 37–55.

Huynh, K.-L. 1982. La fleur mâle de quelques espèces de *Pandanus* subg. *Lophostigma* (Pandanaceae) et sa signification taxonomique, phylogénique et évolutive. Beitr. Biol. Pflanz. 57: 15–83.

Huynh, K.-L. 1991a. New data on the taxonomic position of *Pandanus eydouxia* (Pandanaceae), a species of the Mascarene Islands. Bot. Helv. 101: 29–37.

Huynh, K.-L. 1991b, 1992. The flower structure in the genus *Freycinetia*, Pandanaceae. Bot. Jahrb. Syst. 112: 295–328; 114: 417–441.

Huynh, K.-L. 1995. A new species of *Freycinetia* (Pandanaceae) from the Society Islands. New subgenus and new sections. Candollea 50: 231–245.

Huynh, K.-L. 1996. The genus *Freycinetia* (Pandanaceae) in New Guinea I. Bot. Jahrb. Syst. 118: 529–545.

Huynh, K.-L. 1997. The genus Pandanus (Pandanaceae) in

Madagascar (part 1). Bull. Soc. Neuchâteloise Sci. Nat. 120: 35–44.
Huynh, K.-L., Cox, P.A. 1992. Flower structure and potential bisexuality in *Freycinetia reineckei* (Pandanaceae), a species of the Samoa Islands. Bot. J. Linn. Soc. 110: 235–265.
Jarzen, D.M. 1978. Some Maestrichtian palynomorphs and their phytogeographical and palaeoecological implications. Palynology 2: 29–38.
Jarzen, D.M. 1983. The fossil pollen record of the Pandanaceae. Gard. Bull. Straits Settlem. 36: 163–175.
Leffingwell, H.A. 1971. Palynology of the Lance (Late Cretaceous) and Fort Union (Paleocene) Formations of the type Lance Area, Wyoming. Geol. Soc. Am. Spec. Paper 127: 1–21.
Martelli, U., Pichi-Sermolli, R. 1951. Les Pandanacées récoltées par Henri Perrier de la Bâthie à Madagascar. Mém. Inst. Sci. Madagascar, Sér. B, Biol. Vég. 3 (1): 1–174.
Muller, J. 1981. Fossil Pollen records of extant angiosperms. Bot. Rev. 47: 1–142.
Nonato, M.G., Garson, M.J., Truscott, J.W., Carver, J.A. 1993. Structural characterization of piperidine alkaloids from *Pandanus amaryllifolius* by inverse-detected 2-D NMR techniques. Phytochemistry 34: 1159–1163.
North, C.A., Willis, A.J. 1970. Contributions to the anatomy of *Freycinetia* species from the Solomon Islands. Bot. J. Linn. Soc. 63: 69–80.
North, C.A., Willis, A.J. 1971. Contributions to the anatomy of Sararanga (Pandanaceae). Bot. J. Linn. Soc. 64: 411–421.
Poppendieck, H.-H. 1987. Monoecy and sex changes in *Freycinetia* (Pandanaceae). Ann. Mo. Bot. Gard. 74: 314–320.
Stevenson, D., Loconte, H. 1995. See general references.
St. John, H. 1960. Revision of the genus *Pandanus* Stickman (part 1): Key to the sections. Pac. Sci. 14: 224–241.
Stockey, R.A., Hoffman, G.L., Rothwell, G.W. 1997. The fossil monocot *Limnobiophyllum scutatum*: resolving the phylogeny of Leumaceae. Am. J. Bot. 84: 355–368.
Stone, B.C. 1961. The genus *Sararanga* (Pandanaceae). Brittonia 13: 212–224.
Stone, B.C. 1967. Materials for a monograph of *Freycinetia* (Pandanaceae). I. Gard. Bull. Straits Settlem. 22: 129–152.
Stone, B.C. 1968. Materials for a monograph of *Freycinetia* Gaud. IV. Subdivision of the genus, with 15 new sections. Blumea 16: 361–372.
Stone, B.C. 1974. Towards an improved infrageneric classification in *Pandanus* (Pandanaceae). Bot. Jahrb. Syst. 94: 459–540.
Stone, B.C. 1983. Pandanaceae. Flore du Cambodge, du Laos et du Viêtnam 20: 3–48.
Stone, B.C. 1990. New evidence for the reconciliation of floral organisation in Pandanaceae with normal angiosperm patterns. In: Baas, P., Kalkman, K., Geesink, R. (eds.) The plant diversity of Malesia. Dordrecht: Kluwer, pp. 33–55.
Takhtajan, A.L. 1982. See general references.
Tomlinson, P.B. 1965. A study of stomatal structure in Pandanaceae. Pac. Sci. 19: 38–54.
Warburg, O. 1900. Pandanaceae. In: Engler, A. (ed.) Das Pflanzenreich IV.9. Leipzig: W. Engelmann.
Zimmermann, M.H., Le Claire, J. 1974. Vascular construction and development in the stems of certain Pandanaceae. Bot. J. Linn. Soc. 68: 21–41.

Pentastemonaceae

K. Kubitzki

Pentastemonaceae Duyfjes, Blumea 36: 552 (1992).

Juicy unbranched *herbs* of monopodial growth with scale leaves on rootstock; plants almost glabrous or with sparse uniseriate hairs. *Leaves* fleshy, but papyraceous when dry, dispersed, long-petiolate and sheathing at the base; blade ovate, with 6–8 pairs of arching basal and suprabasal veins connected by numerous transversal veinlets; margin entire. *Inflorescences* axillary, simple or compound racemes; flowers and bracts dotted with raphides; prophylls absent. *Flowers* functionally wholly or partly unisexual, pedicelled, 5-merous; tepals in 1 whorl, free or partly united, imbricate, outside papillose, persisting in fruit; stamens 5, in 1 whorl, ± sessile; filaments apparently absent and completely united into a conspicuous fleshy ring; connectives together with the top of the hypanthium and ovary grown into a complicated, swollen, disclike structure, leaving 5 pouches, in each of which 2 thecae of adjacent anthers are situated; anthers consisting of 2 short, broadly ovate thecae, each opening by a longitudinal lateral slit; ovary inferior, unilocular; ovules numerous, on 3 inward projecting parietal placentae; style short, inconspicuous; stigma broad, flattish, entire or 3- or 4-lobed. *Fruit* berrylike, sharply longitudinally 10-ridged. Seeds many, with a collarlike undulate arillode covering about one third of the seed; exostesta (sarcotesta) transparent; both testa and tegmen 2-layered; endosperm copious, embryo small.

The only genus comprises two species which are locally endemic to Sumatra.

Anatomy. The midrib is compound as in *Tacca*. The stomatal complex is tetracytic to cyclocytic; crystals are present as abundant styloids and raphides, and as minute, irregularly shaped, solitary crystals. Throughout the mesophyll minute druses have been observed (Baas in van Steenis 1982). *P. egregia* can be propagated with pieces of leaves.

Inflorescences and Flowers. The inflorescences are apparently simple or composed racemes, but Duyfjes (1991) has shown that some complications such as the "pseudoaxillary" position of lateral racemes and the presence of additional flowers in the axils of their supporting bracts point to some shifting of bracts or fusion of axes. The androecium shows a coronalike development of the basal staminal region like that in *Peliosanthes* (Convallariaceae). Stamens differentiate late into extensions that contact the stigma; in this and in the inferior position of the ovary *Pentastemona* resembles *Trichopus* (Trichopodaceae) and *Stenomeris* (Dioscoreaceae) (van Heel 1992).

Although all flowers are superficially similar and appear hermaphrodite, in the majority they are reduced to one sex; those which are truly hermaphrodite seem to be markedly protandrous; *P. egregia* may be dioecious (Duyfjes 1991).

POLLINATION. The excessive development of the stamens, by which access to the flowers is canalised, suggests the sapromyiophilous syndrome (Endress 1995).

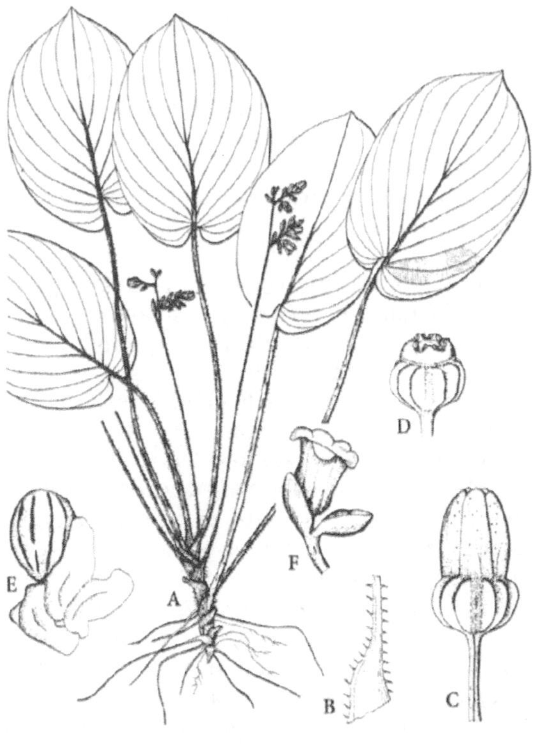

Fig. 112A–F. Pentastemonaceae. *Pentastemona sumatrana*. **A** Habit. **B** Leaf sheath. **C** Flower. **D** Flowers with tepals removed. **E** Ribbed seed, sarcotesta torn away. **F** *Pentastemona egregia*, flower. (Duyfjes 1993)

POLLEN MORPHOLOGY. The pollen is inaperturate, elliptic to circular in outline and relatively small (ca. 23 × 20 μm). The exine is intectate, scabrate and densely covered by Ubisch bodies that bear resemblance to exine elements. An massive intinal oncus marks the non-apertural portion of the pollen grains (van der Ham 1991). The grains in Stemonaceae are different.

KARYOLOGY. 2n = 14 (*P. egregia* (Schott) Steenis, Duyfjes 1991).

FRUIT AND SEED. The fruit of both species can best be characterised as a berry. The pericarp is thinly ribbed and juicy, as is the whole plant. After the opening of a mature fruit of *P. sumatrana*, the seeds lie in a jellylike mass of the arils and exotestas as one coherent mass (Duyfjes 1991); how they are dispersed is unknown.

The seeds of *P. egregia* are endotestal. The number of ridges corresponds to that of the rows of the U-shaped endotestal cells. The ridges are formed by elongation and thickening of the radial walls of the endotesta (Bouman and Devente 1992).

The exotesta is noticeable in the fresh condition but is only little visible when the seeds are dry.

AFFINITIES. Pentastemonaceae differ from Stemonaceae s. str. in the compound midrib, the tetracyclic or pentacyclic stomatal complex, the inferior ovary, the prominent stigmatic lobes, the extension of the stamens, which curve inwards and fuse with the stigmatic lobes, and the seeds ribbed by U-shaped endotestal cells but not by cell multiplication. The detailed studies of van der Ham (1991), Bouman and Devente (1992) and van Heel (1992) show that the relationship between them is not particularly close. There are also some indications for relationships to other taxa. Thus the histological structure of the ovule of *Pentastemona* strongly resembles that of *Tacca*. Also the compound midrib and the ridging of the seed coat show striking similarities between both genera. These genera share the marked tangential elongation of the endotestal cells with *Stenomeris*, but also some members of the Philydrales and Pontederiales. On the basis of floral structure and ontogeny, similarities have been found with *Trichopus* and *Stenomeris* and *Peliosanthes* (van Heel 1992). However, these similarities may be due to adaptation for sapromyiophily rather than to common descent, apart from those characters that are plesiomorphic. It may be noteworthy that compound midribs, cyclocytic stomata, epigyny and arillate seeds are a combination of characters

Pentastemona has in common with the Zingiberales. Molecular data relating to *Pentastemona* are still lacking.

Distribution and Habitats. The plants grow gregarious but very local as rock-dwellers in damp places in rain forest of the lowland and hill region of N and W Central Sumatra.

Only one genus:

Pentastemona van Steenis Fig. 112

Pentastemona van Steenis, Blumea 28: 160 (1982); Duyfjes, Fl. Males. I, 11: 393–398 (1992).

Characters as for family.

Selected Bibliography

Bouman, F. 1995. Seed structure and systematics in Dioscoreales. In: Rudall et al. 1995. See general references, pp. 139–156.

Bouman, F., Devente, H. 1992. Ovules and seeds in *Stemona* (Stemonaceae) and *Pentastemona* (Pentastemonaceae). Blumea 36: 501–514.

Duyfjes, B.E.E. 1991. Stemonaceae and Pentastemonaceae; with miscellaneous notes on members of both families. Blumea 36: 239–252.

Duyfjes, B.E.E. 1993. Pentastemonaceae. In: Flora Malesiana I, 11: 393–398. Leyden: Flora Malesiana Foundation.

Ham, R.W.J.M. van der 1991. Pollen morphology of the Stemonaceae. Blumea 36: 127–159.

Heel, W.A. van 1992. Floral morphology of Stemonaceae and Pentastemonaceae. Blumea 36: 481–499.

Steenis, C.G.G.J. van 1982. Pentastemona, a new 5-merous genus of monocotyledons from North Sumatra (Stemonaceae). Blumea 28: 151–163, 2 figs.

Petermanniaceae

J.G. Conran and H.T. Clifford

Petermanniaceae Hutch., Fam. Fl. Pl. 2: 113 (1934), nom. cons.

Perennial, rhizomatous, woody *climbers* with fibrous roots. Stems spinescent. *Leaves* alternate, petiolate, blades entire, lanceolate to ovate-lanceolate, with reticulate venation. *Inflorescence* terminal and leaf-opposed cymes, partly modified into tendrils. *Flowers* pedicellate, not articulated, ebracteate, hermaphroditic, epigynous, actinomorphic, small, numerous, erect to pendulous; tepals 3 + 3, 1-veined, basally nectariferous, caducous; stamens 3 + 3; filaments distinct, filiform; anthers basifixed, tetrasporangiate, extrorse, dehiscing longitudinally. Ovary unilocular, 3-carpellate with 3 parietal placentas, each bearing numerous anatropous ovules in 2 rows; ovary sometimes becoming incompletely 3-locular by fusion of the placentas; septal nectaries absent; style elongate, capitate; stigma 3-lobed, Wet. *Fruit* an epigynous berry. Seeds numerous, angular; testa dark brown; endosperm copious; embryo medium-sized, linear.

Only one species, restricted to northern temperate E Australia.

Vegetative Morphology. Rhizomatous plants with tall, branching, climbing or scrambling stems to 6 m. The stems are thick, tough and supple, densely spiny; they climb by leaf-opposed, branched, tendrillar, sterile inflorescences.

The seedlings have a compact cotyledon consisting of a short sheath and haustorium. There is a short hypocotyl and well-developed primary root. The first leaves are cataphylls, and the internodes of the primary shoot and the epicotyl are elongated (Conran 1985b). The following leaves are petiolate, reticulately veined, non-sheathing and alternate. The petiole is short relative to the lamina, and articulate against the blade. The blade is entire and tri- to pentacostate with 5 reticulate anastomoses. Vernation is conduplicate.

The pedunculate inflorescence is terminal, becoming leaf-opposed by overtopping sympodial growth (Fig. 113).

Vegetative Anatomy. The roots have a uniseriate velamen which is unthickened but suberised, and an extensive, unthickened cortex. The vascular strands contain vessels with scalariform perforations (Wagner 1977).

The epidermal cells of the stem are small, with heavily thickened outer walls and a thick cuticle. A ring of fibrous bundles and small collateral bundles are found in the cortex, the outer part of which may be collenchymatous. Scattered bundles occur in the centre of the stem (Schlittler 1949). Vessels with scalariform perforations and tracheids are present in the stem.

The epidermal cells of the leaves are isodiametric, irregularly polygonal, with sinuous walls. The stomata, present on the abaxial surface, are anomocytic. The mesophyll is uniform and densely packed. Vessels with scalariform perforations are present. The vascular bundles in the leaves are surrounded by a ring of fibres, and the bundle-sheath cells have chloroplasts (Conran 1985b).

Fig. 113A–G. Petermanniaceae. *Petermannia cirrosa*. **A** Flowering branch. **B** Flower and buds. **C** Sepal and petal. **D** Stamen in side view and front view. **E** Pistil. **F** Ovary, transversal section. **G** Fruit. (Conran and Clifford 1986a)

Calcium oxalate occurs as raphide bundles in all parts of the plant. Tannin occurs in the petiole and rhizome scale leaves (Tomlinson and Ayensu 1969).

Embryology. The tapetum is of the Amoeboid type. The well-developed endothecium has spiral thickenings. Microsporogenesis is successive.

The ovule is bitegmic, anatropous and crassinucellate. The archesporial cell cuts off a parietal cell and a nucellar cap is formed by periclinal divisions of the epidermis. Embryo sac formation is of the Polygonum type (Conran 1988a).

Pollen Morphology. The pollen grains are obovate, 43–47 μm long and sulcate with a thin and slightly spinose exine (Schulze 1982). The pollen contains starch grains at anthesis (Conran and Clifford 1986b).

Karyology. The chromosome number for *Petermannia* is 2n = 10 (Conran 1985a).

Pollination. The flowers are visited, and thereby possibly pollinated, by hoverflies (Diptera: Syrphidae).

Fruit and Seed. The fruits of *Petermannia* are berries. Mature fruits are bright red. The pericarp is thick, fleshy, and encloses numerous seeds. The seed surface is rugulose and dark brown at maturity. The seed has a thickened exotesta, a several-layered mesotesta and thick-walled endotesta. There is a cuticle between the testa and tegmen, and the tegmen consists of collapsed thin-walled parenchymatous cells, similar to those of several of the Colchicaceae (as Uvulariaceae) described by Fukuhara and Shinwari (1994). Starch granules are absent from the endosperm.

Phytochemistry. Saponins appear to be absent in *Petermannia* (Conran and Clifford 1986b).

Affinities. The Petermanniaceae have usually been associated with the Philesiaceae s.l. (Schlittler 1949) or Smilacaceae (Schulze 1982), and sometimes included within them. Cronquist (1981) included all three families within the Smilacaceae. Studies by Conran (1985b, 1987) and Conran and Clifford (1986b) suggested affinities with the Philesiaceae and Smilacaceae/Ripogonaceae, although the position of the Petermanniaceae was isolated from all of these. The recent study of the monocot relationships by Chase et al. (1995) and Stevenson and Loconte (1995) placed the Petermanniaceae on morphological grounds near the Luzuriagaceae, Philesiaceae and Smilacaceae. Incomplete *rbc*L sequence data for *Petermannia* place it with these families in the Liliales, possibly nearer the Colchicaceae, but the data are inconclusive at this stage (M.W. Chase, pers. comm.). The apparently similar seed anatomy between these two families is also supportive of this placement.

Distribution and Habitats. The family is confined exclusively to the central part of the east coast of Australia from far SE Queensland to NE New South Wales. *Petermannia* grows in rainforest and rainforest margins in temperate areas. The plants grow in either high or low light conditions, plants in bright light often having reddish foliage and more brightly coloured flowers. The species is most common in primitive *Nothofagus* Microphyll Mossy Forest, Notophyll Vine Forest (NVF) and adjacent *Eucalpytus acmenoides* wet sclerophyll forest, with or without NVF understorey, often at altitudes above ca. 800 m. This follows the Australian rainforest classification of Webb (1959). An account of the vegetative and reproductive phenology of *Petermannia* is given by Conran (1988b).

Palaeobotany. *Petermanniopsis angleseaënsis* Conran, Christophel & Scriven from the Eocene coal deposits in Anglesea, Victoria, Australia was found to be morphologically very close to the Petermanniaceae (Conran et al. 1994), in particular, being the only net-veined Lilianae with parallel primary and secondary venation, but reticulate higher-order veins. Nevertheless, the fossil's cuticle and dicraeoid (Y-shaped) marginal venation differed sufficiently from *Petermannia* for them to place it as *incertae sedis aff.* Petermanniaceae. This fossil has since been found in a number of different fossil rainforest types (Conran and Christophel submitted).

Only one genus:

Petermannia F. Muell. Fig. 113

Petermannia F. Muell., Fragm. Phytogr. Aust. 2: 92 (1860).

Characters as for family. Only one sp., *P. cirrosa* F. Muell.

Selected Bibliography

Chase, M.W. et al. 1995. See general references.

Conran, J.G. 1985a. Chromosome number reports on Colchicaceae, Petermanniaceae and Philesiaceae (IOPB Chromosome number reports 87, ed. Å. Löve). Taxon 34: 346–347.

Conran, J.G. 1985b. The taxonomic affinities of the genus *Drymophila* (Liliaceae s.l.). Ph.D. Thesis. St. Lucia: University of Queensland.

Conran, J.G. 1987. A phenetic study of the relationships of the genus *Drymophila* R. Br. within the reticulate-veined Liliiflorae. Aust. J. Bot. 35: 283–300.

Conran, J.G. 1988a. Embryology and possible relationships of *Petermannia cirrosa* (Petermanniaceae). Nord. J. Bot. 8: 13–17.

Conran, J.G. 1988b. The reproductive and vegetative phenology of some south east Qld rainforest monocotyledons. Proc. R. Soc. Queensl. 99: 35–43.

Conran, J.G., Christophel, D.C. Additional *Petermanniopsis* collections from Anglesea, Victoria, and an amended description. Trans. Roy. Soc. South. Aust. (submitted).

Conran, J.G., Christophel, D.C., Scriven, L.J. 1994. *Petermanniopsis angleseaënsis*: an Australian fossil net-veined monocotyledon from Eocene Victoria. Int. J. Pl. Sci. 155: 816–827.

Conran, J.G., Clifford, H.T. 1986a. Smilacaceae. In: George, A.S. (ed.) Flora of Australia, vol. 46. Canberra: Australian Government Publishing Service, pp. 180–196.

Conran, J.G., Clifford, H.T. 1986b. The taxonomic affinities of the genus *Ripogonum* J.R. et G. Forst. Nord. J. Bot. 5: 215–219.

Cronquist, A. 1981. See general references.

Fukuhara, T., Shinwari, Z.K. 1994. Seed-coat anatomy in Uvulariaceae (Liliales) of the northern hemisphere: systematic implications. Acta Phytotaxon. Geobot. 45: 1–14.

Schlittler, J. 1949. Die systematische Stellung der Gattung *Petermannia* F.v. Muell. und ihre phylogenetischen Beziehungen zu den Luzuriagoideae Engl. und den Dioscoreaceae Lindl. Vierteljahionchir. Naturforsch. Ges. Zuer. 94, Beih. 1: 1–28.

Schulze, W. 1982. Beiträge zur Taxonomie der Liliifloren VIII. Smilacaceae. Wiss. Z. Friedrich Schiller Univ. Jena 31: 285–289.

Stevenson, D.W., Loconte, H. 1995. Cladistic analysis of monocot families. In: Rudall, P.J., Cribb, P., Cutler, D.F., Humphries, C.J. (eds.) Monocotyledons: systematics and evolution. Royal Botanic Gardens, Kew.

Tomlinson, P.B., Ayensu, E.S. 1969. Notes on the vegetative morphology and anatomy of the Petermanniaceae (Monocotyledones). Bot. J. Linn. Soc. 62: 17–26.

Wagner, W.M. 1977. Vessel types of monocotyledons: a survey. Bot. Not. 130: 383–402.

Webb, L.J. 1959. A physiognomic classification of Australian rain forests. J. Ecol. 47: 551–570.

Philesiaceae

J.G. CONRAN and H.T. CLIFFORD

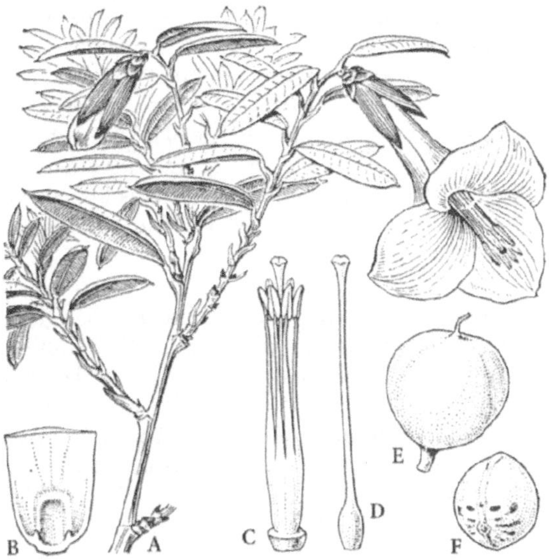

Fig. 114A–F. Philesiaceae. *Philesia buxifolia*. **A** Flowering shoot. **B** Petal with basal nectary. **C** Stamens and style. **D** Pistil. **E** Fruit. **F** Seed. (Takhtajan 1982)

Philesiaceae Dumort., Anal. Fam. Pl. 53, 54, 97 (1829) ("Phylesiaceae"), nom. cons.
Lapageriaceae Kunth (1850).

Much-branched erect or twining *shrub* from short woody rhizomes. Roots fibrous. *Leaves* alternate, spirally arranged or distichous, sessile or shortly petiolate, ± sheathing at the base; blades entire, with one or a few prominent parallel main veins and between them a distinctly reticulate or transverse venation. *Inflorescences* 1–3-flowered, terminal or axillary, with short bracteose peduncles. *Flowers* bisexual, actinomorphic, hypogynous, pendulous. Pedicels not articulated. Tepals in 2 whorls of 3, the outer whorl shorter and ± sepaloid (*Philesia*), or both whorls of similar length (*Lapageria*); inner tepals always large, petaloid and overlapping to form a campanulate perigone; tepals with basal nectaries, red or white, the inner sometimes spotted white. Stamens 3 + 3, filaments ± free from each other, or united in lower half; anthers 2-thecate, 4-sporangiate, subbasifixed, extrorse or introrse, dehiscing longitudinally. Gynoecium of 3 united carpels; ovary unilocular, with parietal intrusive placentae; style erect, filiform; stigma capitate; ovules anatropous or amphitropous, borne in 2 rows. *Fruit* a red berry; seeds few to many, ± globose, yellow-brown. Embryo $^1/_2$ as long as the seed, endosperm massive.

The family is restricted to southern Chile and contains two monotypic genera.

VEGETATIVE MORPHOLOGY AND ANATOMY. The plants are rhizomatous and have branching stems which may be erect (*Philesia*) or climbing to several metres (*Lapageria*). The stems are tough and supple and climb by twining dextrorsely. The main stems and lateral branches appear to show determinate growth. The roots arise from the rhizome; climbing roots are absent.

The seedlings of *Lapageria* have a compact cotyledon consisting of a short sheath and haustorium. There is a short hypocotyl and well-developed primary root. The first leaf alternates with the cotyledon, and is a cataphyll, as are several of the succeeding leaves, and the internodes of the primary shoot and the epicotyl are elongated (Conran 1985; Tillich 1995). The following leaves are ovate-oblong to oblanceolate, acute to mucronate, petiolate and reticulately veined. The leaves are non-sheathing and alternate. The petiole is short relative to the lamina, canaliculate and sometimes bent upwards. The blades are entire and 3-costate (*Philesia*) 5-costate (*Lapageria*); vernation is conduplicate. The primary venation is acrodromous and there is pronounced acropetal weakening of the primary veins. The secondary and higher vein orders are reticulate (Conover 1983).

The roots of *Lapageria* have a uniseriate exodermis which is unthickened, and an extensive, unthickened cortex. The endodermis is thickened. Vessels with scalariform end plates are present (Wagner 1977). The epidermal cells of the stem are small and rectangular with straight or slightly curved thickened, outer walls and a thick cuticle. A ring of fibrous bundles and small collateral bundles occurs in the cortex. Scattered vascular bundles occur in the centre of the stem. Vessels with scalariform perforation end plates are present. The epidermal cells of the leaves are isodiametric, irregularly polygonal with sinuous walls in *Lapageria* and straight walls in *Philesia* (Conran 1985; Conover 1991). The leaves of *Philesia* have palisade and spongy mesophyll while those of *Lapageria* have undifferentiated mesophyll (Schlittler 1949). The stomata are anomocytic, confined to the abaxial surface, and in *Lapageria* transversely oriented in relation to the leaf axis (Conover 1991). Calcium oxalate crystals are present in all parts of the plant as raphides. Tannin cells, mucilage canals and laticifers are absent. Starch is present in the leaves but absent from the seeds (Conran 1985). Silica bodies are absent.

FLOWER STRUCTURE. The flowers are pedicellate. The tepals are all multinerved with reticulate venation, and are shed singly. The perianth is deciduous. Basal tepal nectaries are present (Fig. 114B). The inner tepals are spotted in *Lapageria*. The filaments are free from each other in *Lapageria* and united below in *Philesia*. The anthers are 2-locular and 4-sporangiate. Anther dehiscence is extrorse in *Philesia* and introrse in *Lapageria*. The stigmatic surface is Dry in *Philesia* and Wet in *Lapageria* (Heslop-Harrison and Shivanna 1977).

EMBRYOLOGY. Microsporogenesis is successive. The ovules are anatropous or amphitropous and bitegmic. The nucellus is crassinucellate. A parietal cell is formed. Embryo-sac development in *Lapageria* is of the Polygonum type with 2 antipodal cells and synergids with a filiform apparatus. During meiosis, the nucellar epidermis divides anti- and periclinally (Cave 1966).

POLLEN MORPHOLOGY. The pollen grains are sphaerical, 60–76 μm in diameter, and inaperturate with a spinose exine (Schulze 1982; Fig. 115).

KARYOLOGY. Chromosome numbers are $n = 19$ in *Philesia* and $n = 15$ in *Lapageria*. In *Philesia* there are 1 extra large, 3 medium large and 15 small chromosomes and in *Lapageria*, 1 extra large, 3 medium large and 11 small chromosomes (Cave 1966).

POLLINATION. Pollination is not recorded for the family, but the large, pendulous red flowers are probably pollinated by hummingbirds.

FRUIT AND SEED. The fruits of the Philesiaceae are berries. The scales on the inflorescence axis are persistent but remain small. The pericarp is thick, fleshy and encloses the few to numerous, ovoid to globose seeds. In both genera, the testal portion of the seed coat disintegrates at maturity and the seeds assume the pale yellow colour of the well-

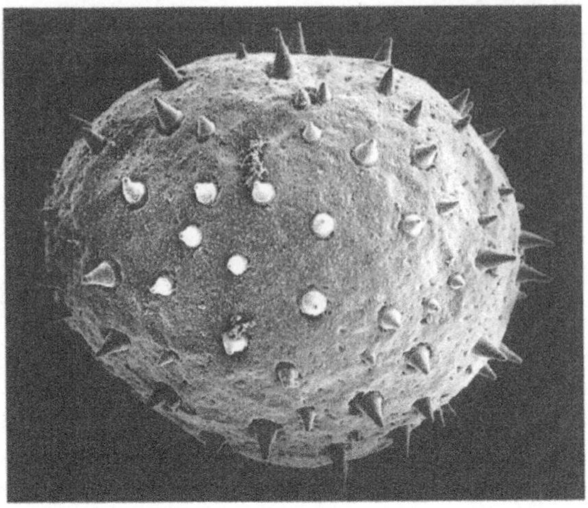

Fig. 115. Philesiaceae. *Lapageria rosea*, inaperturate spinulose pollen, SEM ×1500. (Palynological Laboratory, Stockholm)

developed tegmen. In *Lapageria* the tegmen is provided with a conspicuous, up to 10 μm thick inner and outer epidermis, in which the Philesiaceae agree with the Smilacaceae. The endosperm is massive and towards the inner portion, its thick walls are conspicuously pitted (Huber 1969). It contains aleurone and fatty oils, but not starch.

PHYTOCHEMISTRY. Saponins occur in both genera (Dahlgren and Clifford 1982) but alkaloids are not reported.

AFFINITIES OF THE PHILESIACEAE. The Philesiaceae have been variously placed near to, or in the case of Cronquist (1981), merged with, the Luzuriagaceae and Smilacaceae, and were considered to be a bridging group between the Asparagales and Dioscoreales by Dahlgren et al. (1985). Various genera now included within the Luzuriagaceae, Lomandraceae and Hemerocallidaceae were sometimes included within the Philesiaceae but differ in some aspects of leaf venation (Conover 1983), tepal structure, placentation, pollen morphology and in the case of *Geitonoplesium* (Hemerocallidaceae), the presence of phytomelan in the seed coat (Conran 1985). Studies summarised by Conran (1989), related the Philesiaceae s. str. to the Smilacaceae, and the recent molecular studies of the Lilianae by Chase et al. (1995, 1995a) supported a close relationship between the Philesiaceae and the Smilacaceae, as a sister clade to the Liliaceae s.str. and Calochortaceae in the Liliales. The spotted tepals with basal nectaries and the presence of raphides common to these families are also shared with the Alstromeriaceae, and the latter were part of a larger Liliales clade with these other families. There is no affinity of the Philesiceae to genera of the Behniaceae, Luzuriagaceae, Hemerocallidaceae or Lomandraceae such as *Behnia*, *Luzuriaga*, *Geitonoplesium* or *Eustrephus* with which they have been traditionally associated.

Distribution and Habitats. The family is confined to the southern parts of S America with *Philesia* in S Chile and the straits of Magellan, and *Lapageria* in S Chile only. The plants grow in cool-temperate *Nothofagus* and conifer forests.

Economic Importance. Both genera are cultivated for horticultural purposes. *Lapageria rosea* "Copihue" is the national flower of Chile. A horticultural hybrid between them, × *Philageria veitchii* is also cultivated. The fruits of *Lapageria* are edible.

Key to the Genera

1. Inner tepals much longer than the outer; plant erect, non-climbing 1. ***Philesia***
- Tepal whorls ± equal; plants climbing 2. ***Lapageria***

1. *Philesia* Comm. ex Juss. Fig. 114

Philesia Comm. ex Juss., Gen. Pl.: 41 (1789).

Erect, branching shrub; leaves oblong-obvate, 3-nerved, venation reticulate; outer tepals 3, short, acute, semi-sepaloid, inner tepals 3, longer, campanulate; stamens 6, filaments united in lower half; anthers extrorse; fruit a berry with an elongate stylar beak. Only one sp., *P. magellanica* J.F. Gmel., S Chile.

2. *Lapageria* Ruíz et Pav. Fig. 115

Lapageria Ruíz et Pav., Fl. Peruv. et Chil. 3: 64, t. 297 (1802).

Woody twining vine; leaves acute, ovate-lanceolate, 3–5-nerved, reticulate; tepals 6, subequal, campanulate, inner 3 spotted; stamens 6, filaments free; anthers introrse; fruit an elongate berry. Only one sp., *L. rosea* Ruíz et Pav., S Chile.

Selected Bibliography

Cave, M.S. 1966. The female gametophytes of *Lapageria rosea* and *Philesia magellanica*. Gayana Bot. 15: 25–31.

Chase, M.W. et al. (1995). See general references.

Chase, M.W. et al. (1995a). See general references.

Conover, M.H. 1983. The vegetative anatomy of the reticulate-veined Liliiflorae. Telopea 2: 401–412.

Conover, M.H. 1991. Epidermal patterns of the reticulate-veined Liliiflorae and their parallel-veined allies. Bot. J. Linn. Soc. 107: 295–312.

Conran, J.G. 1985. The taxonomic affinities of the genus *Drymophila* (Liliaceae s.l.). Ph.D. Thesis. St. Lucia: University of Queensland.

Conran, J.G. 1989. Cladistic analyses of some net-veined Liliiflorae. Plant Syst. Evol. 168: 123–141.

Cronquist, A. 1981. See general references.

Dahlgren, R., Clifford, H.T. 1982. See general references.

Dahlgren, R.T.M. et al. 1985. See general references.

Heslop-Harrison, Y., Shivanna, K.R. 1977. The receptive surface of the angiosperm stigma. Ann. Bot. 41: 1233–1258.

Huber, H. 1969. See general references.

Muñoz Pizarro, C. 1973. Chile: Plantas en extinción. Santiago: Editorial Universitaria de Chile.

Ravenna, P.F. 1969. Liliaceae and Amaryllidaceae. In: M.N. Correa (ed.) Flora Patagonica Parte II: Typhaceae a Orchidaceae (excepto Graminaea). Buenos Aires: Talleos Gráficos I.S.A.S., pp. 138–157.

Schlittler, J. 1949. Die systematische Stellung der Gattung *Petermannia* F.v. Muell. und ihre phylogenetischen Beziehungen zu den Luzuriagoideae Engl. und den Dioscoreaceae Lindl. Vierteljahionchir. Naturforsch. Ges. Zuer. 94, Beih. 1: 1–28.

Schulze, W. 1982. Beiträge zur Taxonomie der Liliifloren VII. Philesiaceae. Wiss. Z. Friedrich Schiller Univ. Jena 31: 277–283.

Takhtajan, A.L. 1982. See general references.

Tillich, H.-J. 1995. Seedlings and systematics in monocotyledons. In: Rudall, P.J., Cribb, P., Cutler, D.F., Humphries C.J. (eds.) Monocotyledons: systematics and evolution. Royal Botanic Gardens, Kew.

Wagner, W.M. 1977. Vessel types of monocotyledons: a survey. Bot. Not. 130: 383–402.

Ruscaceae

P.F. Yeo

Ruscaceae Spreng. ex Hutch., Fam. Fl. Plants 2: 109 (1934), nom. cons.

Monoecious, dioecious or subdioecious, shortly rhizomatous glabrous *perennials*, shrubs or lianes; *leaves* spiral, distichous or whorled, chartaceous or horny and either caducous or soon desiccated except those of seedlings of some species and inflorescence bracts of others; assimilatory leaves of seedling petiolate with an open sheathing base and a dilated blade, such leaves also occasionally present on the rhizome; green foliaceous inflorescence bracts sessile and similar to the phylloclades; stem simple or branched, green; upper stems and branches bearing green, leaflike phylloclades in axils of scale leaves; phylloclades subpetiolate, cordate-ovate to lanceolate; *inflorescence* a raceme terminating a lateral shoot (*Danae*) or a fascicle on the margin or surface of a phylloclade; bracts and bracteoles various; *flowers* perfect (*Danae*), articulated with the pedicel at or near its apex; tepals 6, free, in 2 unequal whorls, or partly united; filaments fused into a tube with 3 or 6 anthers at the apex; anthers 2-thecate, 4-sporangiate, with variable dehiscence; staminodia present in pistillate flowers; ovary 3-locular with 2 ovules in each locule, or 1-locular with 1–4 ovules; style evident or obsolete; stigma simple or ± 3-lobed; pistillodium present in staminate flowers; *fruit* a red or orange berry with thin flesh; seeds large, spherical (if solitary) or ± flattened, hard, whitish, with a thin testa, starchless endosperm and a small embryo.

A basically Mediterranean family, ranging from Macaronesia to the Hyrcanian Forests of N Iran and consisting of three genera (two of them monotypic) and 9–11 species.

Vegetative Morphology. In species with assimilatory leaves in the young plant some of the scales have incipient petioles or laminae (subulate or flattened, and green), and the first phylloclade-bearing shoot is axillary here. In species without assimilatory leaves this shoot is probably terminal. Succeeding vegetative shoots are produced from the lower axils of the seedling. The rhizome is sympodial and bears scale leaves. The distichous leaf arrangement occurs in seedlings of *Danae*, *Semele* and *Ruscus streptophyllum*, lateral branches of *Semele*, stems of *R. streptophyllum*, the rhizome of *R. aculeatus* (Hirsch 1977) and the

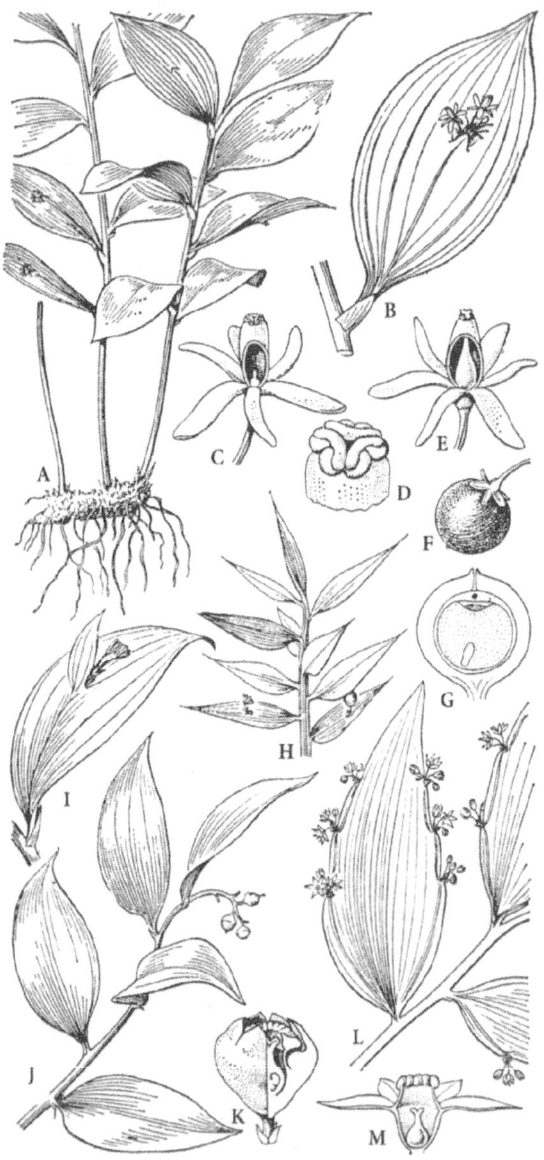

Fig. 116A–G. Ruscaceae. **A–G** *Ruscus colchicus*. **A** Plant with this year's and last year's aerial shoots. **B** Flowering phylloclade. **C** Staminate flower, staminal tube opened to show pistillode. **D** Apex of staminal tube. **E** Pistillate flower. **F** Fruit. **G** Fruit, longitudinal section. **H** *Ruscus aculeatus*, shoot. **I** *Ruscus hypoglossum*, phylloclade. **J, K** *Danae ramosa*. **J** Fruiting shoot. **K** Flower. **L, M** *Semele androgyna*. **L** Flowering shoot. **M** Flower, longitudinal section. (Takhtajan 1982)

primary stems and their branches in *Danae* (evident in young stages, see Arber 1924; Wenck 1935). The lowest scale leaves on the aerial part of the shoot in *Ruscus* are often whorled. Aerial shoots, if branched, produce only branches of the first order (phylloclades disregarded).

The phylloclades are resupinate in *Semele* and *Ruscus streptophyllum* and turned sideways to the axils in *Ruscus* series *Ramosae*. Some twisting also occurs in *Danae* (Arber 1924).

In *Ruscus* the ultimate phylloclade is terminal (i.e. not subtended by a scale leaf), whereas in *Danae* the terminal position is available for occupation by the raceme, should the branch be fertile. In *Semele* growth seems to stop at varying points in the meristematic cycle.

In *Semele* and *Ruscus* the development of phylloclades is affected by the presence and position of the inflorescence(s). In *Semele* 1 or more inflorescences are borne on 1 or both margins of the phylloclade; they are situated in notches, of which the proximal is the most prominent, or on the surface. Sterile phylloclade margins are entire. The proximal inflorescence on each margin is supplied by a conspicuous vein from the base; a branch near the apex of this vein supplies the next inflorescence, and so on. In *Ruscus* there is 1 (rarely 2) inflorescence on each phylloclade; it is situated centrally on the surface and is supplied by a vein which does not continue beyond it (but see Vegetative Anatomy, below). The inflorescence is constantly on one or other surface (characteristic for the species) except in *R. hypophyllum*, where its position varies. In *Ruscus* it is always subtended by a primary bract, the margins of which are fused at the base to the phylloclade (Arber 1924). In *R. hypophyllum* the inflorescence is sometimes "intermediate", being displaced towards one edge or actually on the margin (Motte 1939; Yeo 1968). In such displacements it is accompanied by a notch (as in *Semele*) and the part of the phylloclade distal to the marginal inflorescence may be organised into a pair of "wings" at right angles to the rest of the lamina. It thus appears that the inflorescence has polarity and requires a phylloclade-surface at its "back" and empty space "in front of" it, and that when it is displaced the phylloclade is organised accordingly. Arber (1924) has reported adaxial and abaxial inflorescences in *Semele*, the phylloclade above them usually being bifurcated, which again suggests that the position of the inflorescence influences the shape of the phylloclade.

For *Ruscus* three main interpretations have been presented for what are here regarded as phylloclades (see Arber 1924; Wenck 1935): one is that the part proximal to the inflorescence is an expanded portion of the shoot and the part distal to it is foliar; the second that it comprises a leaf (prophyll) and an adnate axillary shoot; the third is that it is simply a flattened shoot, i.e. a true phylloclade. The last interpretation is supported by Hirsch (1977) and Robert (1964), who find that neither anatomy nor ontogeny indicates a composite organ. The terminal phylloclade is produced by a tranformation of the shoot apex (Hirsch 1977).

Cooney-Sovetts and Sattler (1986) have called attention to the differential expression of stem and leaf characteristics in the phylloclades. The latter appear very leaflike in *Danae*, intermediate in *Ruscus*, and more shootlike in *Semele*.

Vegetative Anatomy. The roots may have a multiple velamen (*Semele*) and their vascular tissues have vessels with scalariform and simple perforation plates. Vessels are absent from the stems of *Ruscus* but present in the stems of the other genera and have scalariform perforation plates. The stomata are anomocytic. Raphides are present. The sieve-tube plastids have cuneate protein bodies and lack starch grains.

In the true leaf of the seedlings vascular bundles are orientated normally, with xylem towards the adaxial side; however in *Semele* the petiole tends to have a radial structure and there are inner bundles with xylem towards the abaxial side in addition to outer bundles normally orientated. The phylloclades have 1 or more radial or somewhat compressed steles and also vascular bundles outside the steles which are orientated opposite to those of the subtending leaf. In *Danae* the stele is present only in the basal part of the phylloclade. In *Ruscus* it is better developed in fertile phylloclades, but is much diminished and sometimes divided beyond the inflorescence, which it serves. In *Semele* the lamina of thin sterile phylloclades may have no stele, but those of more robust and fertile ones have several steles, those serving the inflorescences being wider.

Robert (1964) and Hirsch (1977) established that the phylloclade arises from a layer internal to the tunica, whereas the scale leaf arises from the tunica.

Inflorescence Structure. The inflorescence in *Danae* is lax and clearly racemose. In *Ruscus* the central axis is more condensed but here the inflorescence is probably also a raceme. In *R. hypophyllum* the prophylls encircle the inflorescence axis and have overlapping margins and a hooded apex; each bract envelops the flower it subtends as well as all later flowers. In *R. aculeatus* the inflorescence is fewer-flowered, in *R. colchicus* it may produce many flowers over a period of 2 or 3 years, and the rachis then becomes several mm long. In *Semele* the fascicles appear umbellate, but each pedicel is surrounded by several bracts, suggesting that the inflorescence is complex.

Flower Structure. The flowers are small and articulated with the pedicel at or just below the base of the perianth.

The perianth in *Danae* is urceolate, the syntepalous part thickened so as to occlude most of the internal space; in *Semele* and *Ruscus* it is stellate.

The androecium is partly adnate to the perianth in *Danae* and *Semele*, up to the point of separation of perianth segments in *Semele* but not so far as this in *Danae*. The filaments and connectives are ± united into a tube in all three genera. In *Danae* the thecae of adjacent anthers appear to be slightly coherent and each pair of thecae rests on a distinct swollen cushion (apparently the connective); the dehisced anthers face inwards. The uppermost part of the staminal tube in *Semele* also appears to be made up of connectives, but the structure is less clear. Here and in *Ruscus*, where the sporangia rest directly on the rim of the staminal tube, deshiscence appears to be upwards (neither introrse nor extrorse). However, Hutchinson (1959) and Huber (1969) state that anther dehiscence in the family is extrorse. In *Ruscus* the androecium is generally said to be triandrous but a reinvestigation appears to be desirable; the number of locules (sporangia) is 12 and at maturity they form 2 concentric rings round the apex of the staminal tube. In female flowers of *Semele* the androecium is like that of male flowers but slightly reduced in size, and with indehiscent anthers; in female flowers of *Ruscus* the filamental tube of the androecium is almost indistinguishable from that in the male but the thecae are papery and collapsed.

The ovary in *Danae* and *Semele* is 3-locular, each locule containing 2 collateral ovules; in *Ruscus* it is unilocular with 1–4 ovules. According to Pavillard (1919), the ovules of *Ruscus* are anatropous and, depending on their position on the placenta, may be apotropous or epitropous. This applies to the 2 superposed ovules of *R. hypophyllum* and *R. hypoglossum*, which have a parietal placenta. In *R. colchicus* the ovule is usually solitary and apical. In *R. streptophyllum* there are 2–4 collateral ovules; when there are 3 or 4 there is an incomplete median partition in the ovary. In *R. aculeatus* there are 1–4 ovules, collateral if more than 1. The style is elongate and divided apically into 3 papillate stigmatic branches in *Danae* and *Semele*. In *Ruscus* the swollen stigma is merely stipitate and is caplike or spherical, very obscurely lobed, epapillate and glutinous. A vestigial gynoecium is present in male flowers of *Semele*, complete with style and stigmatic branches. In *Ruscus* this vestige is present but usually less developed.

Floral Anatomy. The fusiform ovary of *R. hypoglossum* and *R. hypophyllum* is slightly curved and its stronger convexity corresponds with the median nerve of the carpel (Pavillard 1919).

The style in *R. streptophyllum* has a mucilage-filled central canal (P.F. Yeo, unpubl.), also observed in *R. hypoglossum* (Pavillard 1919).

Embryology. Microsporogenesis is successive and the tapetum glandular. The pollen grains are 2- or 3-celled when shed. The ovules are crassinucellate (P.J. Rudall unpubl.) and have a large funicle and 2 integuments, as in most asparagoids. The inner integument forms the micropyle. The embryo sac is long, and in *R. aculeatus* the lower end is contracted into a tubercle housing the 3 antipodal cells and enclosed in a sclerotised sheath of lignified cells (hypostase) (Pavillard 1919).

Pollen Morphology. The pollen grains are sulcate in *Ruscus*, 1-sulcoidate in *Danae* and inaperturate in *Semele*. Grains of *Danae* and *Semele* are tenuiexinate (Erdtman 1952; Schulze 1982).

Karyology. The basic chromosome number has been determined as $x = 20$ for all three genera (Satô 1942). The karyotypes of *Danae* and *Ruscus* consist of 6 large and 14 small chromosome, which are all submetacentric.

Pollination and Phaenology. Pollination in Ruscaceae has not been studied. The contrasting colours of the staminal column and tepals in *Ruscus* suggest insect pollination. Septal nectaries are absent and could not function anyway, for the staminal ring is tightly adpressed to the ovary. Perhaps the copious stigmatic exudate deceives possible pollinators such as Diptera.

Danae and *Semele* flower in early summer, *Danae* on new shoots with fruits ripening in the same autumn, *Semele* on year-old shoots with fruits ripening a year later. *Ruscus* is mainly winter-flowering, with a longer flowering season than in the other genera. Fruit development is rapid in *R. colchicus* and *R. hypoglossum* (ca. 6 months) but seems to take about 12 months in *R. streptophyllum* and *R. aculeatus*. In *R. hypophyllum* development is delayed throughout the summer but becomes rapid in autumn, with maturity around mid-winter.

Fruit and Seed. The berry is red in *Ruscus* and orange-red in *Danae* and *Semele*. In *Danae* the flesh is slightly sweet but in *Ruscus* tasteless though juicy. The skin is tough in the berries of at least *Semele* and *Ruscus*. In *Ruscus* all ovules usually develop into seeds; in the other genera 1 seed is usual (sometimes 2, collateral, in *Danae*). In the berry of *Semele* the style remnant is on the side, reflecting the necessarily asymmetric development.

The seeds in Ruscaceae are relatively large, globose if solitary, angular if 2 or more. The testa lacks phytomelan and disintegrates during development, whereas the 2-layered tegmen is well developed, and consists of colourless, thick-walled cells, the outer being longitudinally, the inner transversely elongated. The endosperm cells store aleurone, lipids and hemicellulose. They are radially elongated with pitted cell walls. The embryo is about $\frac{1}{2}$ or $\frac{1}{3}$ as long as the endosperm (Huber 1969).

Dispersal. The berries suggest bird dispersal but no observations are available. The berries of *R. streptophyllum* hang under the phylloclades and might only be found by ground-feeding birds.

Reproductive Systems. The family is marked by the presence of sex distributions other than hermaphroditism: monoecy in *Semele*, *R. streptophyllum* and *R. hypophyllum*, dioecy in *R. aculeatus*, *R. hyrcanus*, *R. hypoglossum* and *R. colchicus*. Dioecy tends not to be absolute, male plants occasionally producing a few fruits. In *R. aculeatus* freely fruiting plants are found in gardens; these produce both male and hermaphrodite flowers which self-fertilise automatically (the latter during only a small part of the total flowering season).

Nothing is yet known about sex determination in the dioecious species, but females predominate heavily in the progeny of crosses between monoecious and dioecious species.

Phytochemistry. Anthocyanins are present in the vegetative parts of at least *Semele* and *Ruscus* and in the flowers of *Ruscus*. Steroidal saponins and chelidonic acid are also recorded in the family. The roots of *Ruscus* contain chrysophanol (Hegnauer 1962, 1986).

Affinities. The relationship between Ruscaceae and Asparagaceae s. str. is a still unresolved problem. It is true that the similarity between the phylloclades of Ruscaceae and Asparagaceae is very suggestive. However, shoot architecture and seed coat anatomy are quite different in both groups. The clustering of up to 9 phylloclades in the axil of a single bract leaf, and the axillary position of inflorescences in Asparagaceae are not duplicated in Ruscaceae. The lack of phytomelan in the seed coats of the Ruscaceae is shared with Convallariaceae, Dracaenaceae, and Nolinaceae, but not with Asparagaceae. The *rbc*L analysis (Chase et al. 1995) supports this relationship, and shows *Ruscus* at some distance from *Asparagus*. The lack of septal nectaries in Ruscaceae can easily be reconciled with the structure of their flowers. This evidence indicates that it would be unwise to merge the two families; the matter needs to be further studied both from the morphological and molecular point of view.

Distribution and Habitats. The Ruscaceae extend from Macaronesia through the Mediterranean to N Iran. Only *R. aculeatus* is widespread, extending from the Azores to the W Caucasus, and northwards to S England and Belgium. *Danae* has a pronounced disjunction (N Syria and S Turkey and N Iran). *R. streptophyllum* is restricted to Madeira and *Semele androgyna* to the W Canary Islands.

The members of Ruscaceae grow essentially in woodland, often in heavy shade, under evergreens, but native populations of *R. aculeatus* is SW Britain grow mostly in maritime grassland.

Economic Importance. The family has minimal economic importance. Bunches of *R. aculeatus* were formerly used as brushes for cleaning butchers' chopping blocks and are now frequently cut, dried and painted for ornament. The young shoots may be eaten like asparagus. Seeds of *Ruscus* have been used for making coffee substitute. *R. hypophyllum* and *R. aculeatus* have been used for medicinal purposes. *Danae* and species of *Ruscus* are sometimes grown as ornamentals; *Semele* is an attractive liane but is grown mainly in botanic gardens. C. Shimony and E. Werker (in press) reported that *Ruscus* stems were formerly used for ropes and boat lashings in archaeological material.

Key to the Genera

1. Tall liane with shoots of unlimited growth; flowers in fascicles on the margin or surface of the phylloclades **2. *Semele***
- Perennials or low shrubs with shoots of limited growth; flowers in other positions 2
2. Branches alternate, each with 5–7 phylloclades; flowers in loose racemes terminating lateral branches **1. *Danae***
- Branches (the lowest) in 1 or more whorls, each with more than 7 phylloclades (unless very weak), or stem unbranched, inflorescences on surface of phylloclade **3. *Ruscus***

Genera of Ruscaceae

1. *Danae* Medicus Fig. 116J,K

Danae Medicus, Malvenfam.: 72 (1787).

Shrub to 1.2 m tall; scale leaves alternate, soon dry and often caducous; phylloclades ca. 5–7 per branch, ovate-lanceolate, asymmetric; racemes at apices of branches, 5-8-flowered; flowers hermaphrodite; tepals united for ⅔ of their length; perigone globose-urceolate; stamens 6, adnate basally to perianth tube; connectives conspicuous; ovary globose, trilocular, style distinct, stigma lobes papillate; ovules 2 per locule. Only *D. racemosa* (L.) Moench, disjunct between mountains in S Turkey/NW Syria and the Caspian region.

2. *Semele* Kunth Fig. 116L,M

Semele Kunth, Abh. K. Akad. Wiss. Berlin 1842: 49 (1844).

Monoecious climber with green stems 5–7 m, scale leaves soon dry, most of them subtending lateral branches each with up to 40 lanceolate to cordate-ovate phylloclades; flower fascicles in marginal indentations or on the surface of the phylloclades (var. *gayae*); flowers unisexual; tepals patent, united below; stamens 6, adnate basally to perianth tube; female flowers with indehiscent anthers; ovary globose, trilocular; stigma lobes (female flowers) papillate; ovules 2 per locule. Only one sp., *S. androgyna* (L.) Kunth, W Canary Isles (Tenerife, Gomera, Hierro, La Palma); also Madeira?

3. *Ruscus* L. Fig. 116A–I

Ruscus L., Sp. Pl.: 1041 (1753); Yeo, Notes R. Bot. Gard. Edinb. 28: 237–264 (1968).
Platyruscus Khokhrj. & V. Tikhom. (1993) (=*Ruscus* series *Simplices* Yeo).

Dioecious or monoecious perennials or shrubs; scale leaves of stems and branches subtending phylloclades, the lower sometimes whorled; inflorescences central on surface of phylloclades; flowers unisexual; tepals free, inner smaller than outer; filaments fused into a fleshy tube; anthers apparently 3, vestigial in female flowers; ovary unilocular, ellipsoidal to fusiform, invested by staminal tube; style short, stigma caplike or globose, smooth, glutinous; ovules 1–4. Six spp., in two series: Ser. *Ramosae* Yeo, aerial stems branched and phylloclades spine-tipped, with the widespread *R. aculeatus* L. and the eastern *R. hyrcanus* Woronov, and Ser. *Simplices* Yeo, aerial stems unbranched and phylloclades muticous, four vicariant species ranging from Madeira to W Caucasus.

Selected Bibliography

Arber, A. 1924. *Danae, Ruscus* and *Semele*: a morphological study. Ann. Bot. 38: 229–260.
Chase, M.W. et al. 1995. See general references.
Cooney-Sovetts, C., Sattler, R. 1986. Phylloclade development in the Asparagaceae: an example of homoeosis. Bot. J. Linn. Soc. 94: 327–371.
Erdtman, G. 1952. See general references.
Hegnauer, R. 1962, 1986. See general references.
Hillman, E.M., Warren, A. 1973. Survey of *Ruscus aculeatus* on Bookham Common: the first two years. Lond. Nat. 52: 93–103.
Hirsch, A. 1977. A developmental study of the phylloclades of *Ruscus aculeatus*. Bot. J. Linn. Soc. 74: 355–365.
Huber, H. 1969. See general references.
Hutchinson, J. 1959. The families of flowering plants, 2nd edn., vol. 2. Oxford: Oxford University Press.
Motte, J. 1939. Sur quelques aspects considérés comme tératologique du cladode de *Ruscus*. Bull. Soc. Bot. Fr. 86: 73–80.
Pavillard, G. 1919. Sur la structure femelle des *Ruscus*. C. R. Acad. Sci. Paris 168: 113–115.
Robert, P. 1964. Contribution à l'étude ontogénique de *Ruscus aculeatus* L. Développement du rhizome et formation des cladodes. Ann. Sci. Nat. Bot. Biol. Veg. 5: 739–752.
Satô, D. 1942. Karyotype alteration and phylogeny in Liliaceae and allied families. Jpn. J. Bot. 12: 57–161.
Schulze, W. 1982. Beiträge zur Taxonomie der Liliifloren. X. Asparagaceae. Wiss. Z. Friedrich-Schiller-Univ. Math.-Nat. R. 31: 309–330.
Shimony, C., Werker, E. 1997. Plant lashings. In: The Ma'agan Michael Ship (400 BC Merchantman) by E. Linder & Y. Kahanov. Israel Fiber Institute. (self-published – probably in 1997) (in press).
Takhtajan, A.L. 1982. See general references.
Wenck, S. 1935. Entwicklungsgeschichtliche Untersuchungen über die Assimilationsorgane von *Semele, Ruscus, Danae* und *Myrsiphyllum*. Beih. Bot. Centralbl. 53A: 1–25.
Yeo, P.F. 1968. A contribution to the taxonomy of the genus *Ruscus*. Notes R. Bot. Gard. Edinb. 28: 237–264.

Smilacaceae

J.G. Conran

Smilacaceae Vent., Tabl. Règne Vég. 2: 146 (1799), nom. cons.
Ripogonaceae Conran & Clifford (1985).

Tendriliferous climbers, or rarely twiners or shrubs or vines with perennial or rarely annual, sometimes spiny, aerial shoots up to many metres long from short, woody rhizomes with fibrous roots. *Leaves* alternate, opposite or verticillate, distichously or decussately arranged, petiolate, paired tendrils mostly present; blades entire, oblong-ovate to ovate or linear-lanceolate, main veins prominent, 3–7, venation distinctly reticulate. *Inflorescence* axillary or terminal panicles, simple or compound racemes, spikes or ebracteate umbels. *Flowers* dioecious or hermaphroditic, erect, actinomorphic, hypogynous; pedicels subtended by small scales; tepals in 2 petaloid trimerous whorls, free or fused to form a deep tube (*Heterosmilax*), with basal nectaries, white, cream, green or reddish. Stamens 3–12 (*Heterosmilax*) or 6, in 1–3 whorls, filaments free or connate below or for the whole length (*Heterosmilax* spp.); anthers basifixed, latrorse to introrse, dehiscing longitudinally, bisporangiate (Smilacoideae) or tetrasporangiate (Ripogonoideae). Gynoecium of 3 united carpels, rarely unicarpellate (*Smilax*); ovary 3-locular with basal or axile placentae; styluli short, spreading, rarely a rudimentary style; ovules atropous or anatropous, borne singly or paired in each loculus. *Fruit* a black, purple or red berry. Seeds single to several, globose to ovoid-angular, pale yellow to dark brown; embryo minute to $\frac{2}{3}$ the length of the seed, linear to capitate; endosperm copious, starchy (*Ripogonum*) or starchless.

The family comprises three genera and ca. 320 spp. and is pantropical, extending to temperate regions in both hemispheres.

Vegetative Morphology. The short-rhizomatous plants have tall, branching stems. The stems are perennial or annual, tough and supple or herbaceous, sometimes bearing prickles. The plants climb by means of paired tendrils (Fig. 117, 118), scrambling or by dextrorse twining (*Ripogonum*). The main stems and branches show determinate

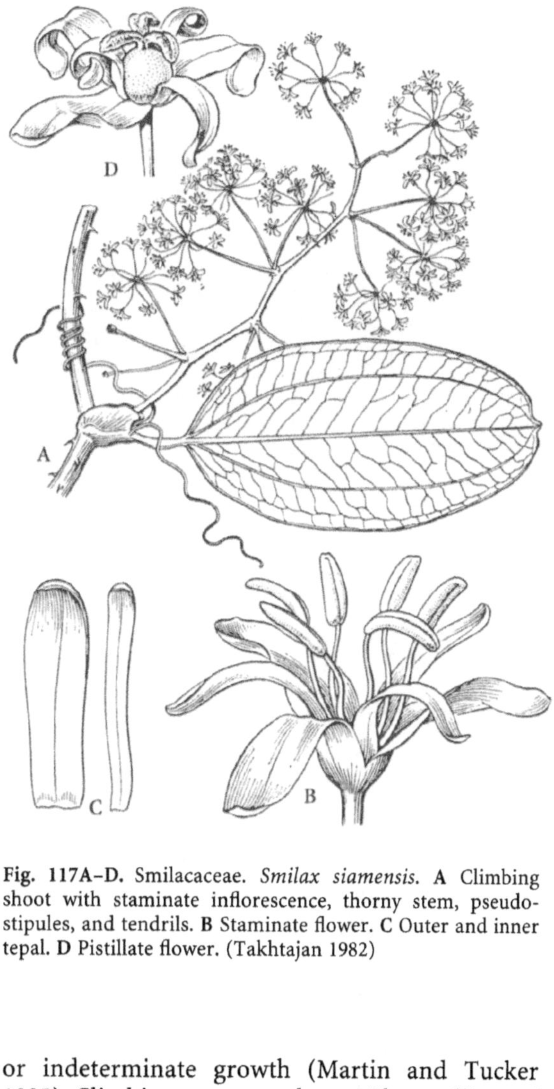

Fig. 117A–D. Smilacaceae. *Smilax siamensis.* **A** Climbing shoot with staminate inflorescence, thorny stem, pseudostipules, and tendrils. **B** Staminate flower. **C** Outer and inner tepal. **D** Pistillate flower. (Takhtajan 1982)

or indeterminate growth (Martin and Tucker 1985). Climbing roots are absent. The seedlings of *Smilax* have a compact cotyledon consisting of a short sheath and haustorium. There is a short hypocotyl and well-developed primary root. The first leaf alternates with the cotyledon, and is a cataphyll, as are several of the succeeding leaves, and the internodes of the primary shoot and the epicotyl are elongated (Conran 1985; Tillich 1995). The following leaves are ovate-oblong to linear-lanceolate, acute to acuminate with a well-developed drip tip, petiolate and reticulately veined.

The leaves are non-sheathing, alternate and distichously arranged, sometimes opposite, decussate or whorled (*Ripogonum*). The petiole is short relative to the lamina, canaliculate and sometimes bent upwards; the leaf bases of many species of the Smilacoideae bear a pair of narrow to inflated so-called stipules (not homologous

Fig. 118A–K. Smilacaceae. **A, B** *Smilax inversa*. **A** Staminate flowering shoot. **B** Leaf base with tendrils and a prophyll (*a*) at the base of the inflorescence axis. **C–G** *Heterosmilax polyandra*. **C** Twining shoot. **D** Pistillate flower. **E** Stamiante flower. **F** Androecium. **G** Pistil with 2 staminodes visible. **H–K** *Ripogonum scandens*. **H** Flowering shoot. **I** Flower. **J** Stamen. **K** Pistil. (Takhtajan 1982)

with the stipules of dicotyledous; see p. 2) and on the petiole, often a pair of tendrils. These tendrils have been variously interpreted as stipules, modified leaf segments or emergences (Troll 1939: 1966 et seq.). However, pseudostipules and tendrils co-occur in some species, and ternate leaf blades are unknown for any of the family's relatives. In species without tendrils, and on the flowering shoots of tendrillar species, 2 thorns are often present where the tendrils otherwise occur. This suggests that they are probably best interpreted as emergences equivalent to thorns (Velenovský 1907: 453).

The blades are entire and 3–7 costate with acrodromous to campylodromous primary venation, apical to proximal convergence and pronounced acropetal weakening, although Pole (1991, 1993) argues that because the primary veins fuse above the petiole, they are really secondary veins and the venation is brochidodromous. The secondary veins are pinnate; higher veins are reticulate with numerous free ends. The leaf areoles are irregular in both genera. Data mostly from Conover (1983). Vernation is conduplicate.

Vegetative Anatomy. Laticifers are absent. Vessels with scalariform perforation plates occur in the roots and stems and leaves of some species of *Heterosmilax* but are absent from the leaves of *Ripogonum* (Wagner 1977). The roots have a multiseriate exodermis with unthickened (*Smilax*; Mulay and Deshpande 1959) or thickened cells (*Ripogonum*), and an extensive cortex. The epidermal cells of the stem are small, with thickened outer walls and a thick cuticle. A ring of fibrous bundles and small collateral bundles are found in the cortex, the outer cells of which may be collenchymatous. The stems of *Ripogonum scandens* were found to have unusual vascular anatomy, relating to the decussate leaf arrangement (Simpson and Philipson 1969). The 2 leaf traces departing at each leaf node are intimately connected, and all the stem vascular tissues are derived from leaf traces. In the rhizome there is a specialised layer of vascular strands, the *réseau radicifère* of Mangin (1882), which is absent in the aerial stems. The vascular strands in the latter follow a more or less straight course, whereas those of the rhizome are helically arranged. The epidermal cells of the leaves are isodiametric, irregularly polygonal with sinuous or curved walls. The leaves of most taxa have arm cell mesophyll only, although some *Ripogonum* species tend towards palisade and spongy mesophyll (Conran 1985). The stomata are on the abaxial leaf surface, anomocytic and unoriented in relation to the leaf axis (Conover 1991). Raphides are present in all parts of the plant often associated with mucilage cells. Tannin cells are present.

Floral Morphology. The flowers are pedicellate and are subtended by small bracts which may be deciduous or persistent in fruit. The outer and inner tepal whorls are similar. The tepals are 1-nerved and free or fused into a deep, almost closed tube in *Heterosmilax*. The tepals are caducous. Nectaries are present at the base of the tepals. In *Smilax*, vestigial stamens are found in some female flowers and vestigial gynoecia in some male flowers. The anthers are bisporangiate in the Smilacoideae and tetrasporangiate in the Ripogonoideae. There are 1 (Smilacoideae) or 2 (Ripogonoideae) ovules per loculus. The stigmatic surface in *Smilax* is Dry (Heslop-Harrison and Shivanna 1977), but Wet and papillate in *Ripogonum* (Conran 1985).

Embryology. The tapetum is binucleate and amoeboid in *Ripogonum*. The endothecium in both subfamilies has spiral thickenings (Conran 1985). Microsporogenesis is successive (Humphrey 1914) and mature pollen grains are 2-celled. The ovules are bitegmic and atropous in *Smilax*, anatropous in *Ripogonum*. The nucellus is massive and crassinucellate. A parietal cell is formed. During meiosis, the nucellar epidermis divides anti- and periclinally. Embryo-sac development in *Ripogonum* is of the Polygonum type and 2–6 antipodal cells are formed. Endosperm formation is Nuclear (Schnarf 1931; Barker 1940).

Pollen Morphology. The pollen grains in the Smilacoideae are obovate, 13–25 μm long, inaperturate with a spinose exine. In the Ripogonoideae, the pollen grains are obovate, 29–34 μm long, sulcate with a reticulate and slightly foveolate exine (Schulze 1982).

Karyology. Chromosome numbers recorded for the family are $n = 14$, 15, 16, 30 and 32 in *Smilax*, $n = 31$ in *Heterosmilax* and $n = 15$ in *Ripogonum* (Beuzenberg and Hair 1963). Satô (1942) describes 28 long and 32 short chromosomes for *Smilax china* and 20 long and 42 short chromosomes for *Heterosmilax* in the diploid karyotype. Sen (1975), however, describes the Smilacoideae as having mostly medium length chromosomes. In *Ripogonum* there are several medium large and one very large chromosomes (Macmillan 1972).

Pollination. *Ripogonum scandens* has a sweet, spicy scent (Macmillan 1972), although the Australian species examined seem to be scentless (Conran 1985). Cranwell (1953) reported that *R. scandens* was insect- or wind-pollinated and Macmillan (1972) observed clouds of insects around its panicles. No insects have been observed visiting the flowers of *R. discolor* but the flowers showed damage caused by phytophagous beetles (Chrysomelidae) which also damaged the leaves of this species (Conran 1985).

Fruit and Seed. The fruits of the Smilacaceae are red, dark purple or black berries at maturity. The pericarp is thick, fleshy and encloses 1-several seeds. The seed surface is smooth to rugulose and pale yellow to dark brown at maturity , darkening on drying. In *Smilax* and *Ripogonum* the testal portion of the seed coat disintegrates during ripening. In *Smilax* and more pronounced in *Ripogonum*, the seed coat has a thick cuticle between testa an tegmen and on the inner epidermis of the tegmen. The endosperm is massive and consists of cells with thickened, pitted walls. It contains aleurone and fatty acids. Starch is present in *Ripogonum* seeds but absent in the Smilacoideae.

Dispersal. Several temperate *Smilax* species with annual aerial stems hold their fruit clusters after leaf fall, providing a food source for fruit-eating birds and mammals (Conran 1996). The seeds of *Smilax australis* have also been observed in bird droppings in SE Queensland by this author. The berries of *R. scandens* are eaten by a variety of New Zealand birds and rats. The seeds are deposited in the faeces (Macmillan 1953). Germination of fresh seed of *R. scandens*, *R. discolor* and *R. album* is good; in all species viability is short, and much lower after drying out.

Phytochemistry. Saponins and tannins are present in many *Smilax* and *Ripogonum* species (Farnsworth et al. 1971). Anthocyanin glycosides were reported in Australasian *Smilax* and *Ripogonum* species by Gascoigne et al. (1948) and Cain et al. (1962).

Subdivision and Relationships. The family contains three genera: *Smilax* with about 300 species, *Heterosmilax* with 12 species and *Ripogonum* with 6 species. The family is divided into two subfamilies: Smilacoideae with dioecious plants, tendrils, bisporangiate anthers and inaperturate spinulose pollen; and Ripogonoideae with hermaphroditic plants without tendrils, tetrasporangiate anthers and sulcate reticulate pollen (Conran and Clifford 1986a).

Affinities. The Smilacaceae, as defined by Cronquist (1981), included members of the Philesiaceae, Lomandraceae, Ripogonaceae,

Hemerocallidaceae and Petermanniaceae. Treatments by Dahlgren et al. (1985) and Conran (1985, 1989), however, did not supported this broad view. The inclusion of *Ripogonum* in the family was shown by Conran and Clifford (1986b) to be based, in part, on inaccurate information about the androecium (*Ripogonum* being 4-sporangiate), and the genus was removed to the Ripogonaceae, as first suggested by Huber (1977). Nevertheless, recent molecular and morphological studies of the monocotyledons summarised by Chase et al. (1995, 1995a) placed the Smilacaceae and Ripogonaceae as sister taxa, hence their recombination here as subfamilies. The Smilacaceae show affinities with the Philesiaceae, Liliaceae, Colchicaceae and Petermanniaceae in the Liliales.

Distribution and Habitats. The family, and especially *Smilax*, is pantropical, subtropical and temperate in both hemispheres with centres of diversity in S America, Africa and SE Asia. *Heterosmilax* occurs in India, China, Japan, Taiwan and Malesia. *Ripogonum* is centred in E Australia, extending to New Guinea and New Zealand.

The Smilacaceae are mostly understorey vines in moist forests, although many *Smilax* species extend into drier habitats as scrambling or thicket-forming creepers. Most species are woody and perennial, but some species in temperate regions have herbaceous, annual, aerial stems. The phenology of several *Smilax* and *Ripogonum* species was described by Conran (1988).

Palaeobotany. Numerous fossil *Smilax* species are reported for a number of Latest Cretaceous and Eocene and Miocene sites (e.g. Berry 1929; Morita 1931; Sun and Dilcher 1988) although Daghlian (1981) was reluctant to accept many of these records pending a more detailed study, and the difficulties of identifying fossil reticulate-veined monocot leaves are discussed by Conran et al. (1994). Fossil *Ripogonum* leaves (Pole 1993) and pollen (Couper 1953) are reported from Miocene New Zealand.

Economic Importance. *Smilax* species are used medicinally, with the leaves, roots and stems used to make infusions, and number of tropical species, especially from S America, are used medicinally or contain compounds which show antifungal activity (Caceres et al. 1991; Lavaud et al. 1993; Basualdo et al. 1995). The tuberous rhizomes of *S. china* are used to treat gout. *Smilax aristolochiifolia* roots are one of the natural sources of "radix sarsaparillae" which is a traditional antisyphylitic and today is used as a digestive tonic and as a flavouring for drinks and confectionery. The fruits of *S. megacarpa* are eaten in Java and used to make preserves (Dahlgren et al. 1985). The young shoots of numerous species are eaten as a vegetable (Kunkel 1984). Several species are cultivated horticulturally.

The stems of some *Ripogonum* species have been used to a limited extent for building and making baskets and fish traps. The shoots are reported as edible (Tanaka 1976).

Key to the Genera

1. Flowers bisexual; leaves without tendrils; anthers tetrasporangiate, longer than filaments — **3. *Rhipogonum***
– Flowers unisexual; leaves with tendrils; anthers bisporangiate, shorter than filaments — 2
2. Perianth segments free; stamens free — **2. *Smilax***
– Perianth fused into a deep tube; stamens usually fused basally — **1. *Heterosmilax***

Genera of the Smilacaceae

1. *Heterosmilax* Kunth — Figs. 118C–G

Heterosmilax Kunth, Enum. Pl.: 5, 270 (1850); Koyama, Brittonia 36: 184–205 (1984) rev.
Pseudosmilax Hayata (1920)
Oligosmilax Seeman (1868).

Smooth-stemmed, usually annual-stemmed climbers with tendrils; rhizome short; leaves distichous, petiolate, stipular, blade acute, ovate-lanceolate; inflorescence ebracteate umbels; tepals connate; stamens 3–12, connate basally, rarely free; berry black; seeds dark brown. Twelve spp., NE India, Indochina, Malesia, Japan and China. A separate genus *Pseudosmilax* has been widely recognised for the 9–12-staminate species (Krause 1930; Dahlgren et al. 1985) but was merged with *Heterosmilax* by Koyama (1984).

2. *Smilax* L. — Figs. 117, 118A,B

Smilax L., Spec. Pl., 1028 (1753); Koyama, Q. J. Taiwan Mus. 10: 1–61, rev.

Prickly or rarely smooth-stemmed, annual or perennial climbers with annual or perennial aerial stems; rhizome short; tendrils mostly present, leaves distichous, petiolate, stipular, blade acute, ovate to linear-lanceolate; inflorescence simple or compound racemes or racemose umbels; tepals free; stamens 6, rarely 3 or 12–18, free; fruit red to black; seeds dark brown. About 300 spp., worldwide in tropical, subtropical and temperate regions.

3. *Ripogonum* J.R. Forst. & G. Forst. Fig. 118H–K

Ripogonum J.R. Forst. & G. Forst., Char. gen.: 49, t. 25 (1776); Conran & Clifford, Fl. Australia 46: 188–190 (1986), rev.

Caulescent shrubs or woody twiners; rhizomes short; stems occasionally spinescent; tendrils absent; leaves distichously alternate, opposite or verticillate, shortly petiolate, blade acute to bluntly acuminate, often with a well-developed drip tip, 3-veined. Inflorescence axillary or terminal, a raceme, spike or panicle; flowers sessile or pedicellate, bracteate; tepals 3 + 3, free, petaloid; stamens 6, filaments free; style 1 or absent; stigma 3-lobed. Fruit a black or red berry, seeds yellow to brown. Six spp., five in eastern Australia with one extending to Papua new Guinea; one species endemic to New Zealand.

Selected Bibliography

Barker, A.J.D. 1940. A contribution to the life history of *Ripogonum scandens* J. R. et G. Forster. M.Sc. Wellington: Victoria University.

Basualdo, I., Zardini, E.M., Ortiz, M. 1995. Medicinal plants of Paraguay: underground organs, II. Econ. Bot. 49: 387–394.

Berry, E.W. 1929. A revision of the flora of the Latah Formation. Prof. Pap. U.S. Geol. Surv. 154: 225–266.

Beuzenberg, E.J., Hair, J.B. 1963. Contributions to the chromosome atlas of the New Zealand flora. 5. Miscellaneous families. N. Z. J. Bot. 1: 53–67.

Caceres, A., Lopez, B.R., Giron, M.A., Logemann, H. 1991. Plants used in Guatemala for the treatment of dermatophytic infections: 1. Screening for antimycotic activity of 44 plant extracts. J. Ethnopharm. 31: 263–276.

Cain, B.F., La Roche, S. Cambie, R.C. 1962. A New Zealand phytochemical survey. Part 5. The monocotyledons. N. Z. J. Sci. 5: 537–554.

Chase, M.W. et al. 1995. See general references.

Chase, M.W. et al. 1995a. See general references.

Conover, M.H. 1983. The vegetative anatomy of the reticulate-veined Liliiflorae. Telopea 2: 401–412.

Conover, M.H. 1991. Epidermal patterns of the reticulate-veined Liliiflorae and their parallel-veined allies. Bot. J. Linn. Soc. 107: 295–312.

Conran, J.G. 1985. The taxonomic affinities of the genus *Drymophila* (Liliaceae *s.l.*). Ph.D. Thesis. St. Lucia: University of Queensland.

Conran, J.G. 1988. The reproductive and vegetative phenology of some south east Qld rainforest monocotyledons. Proc. R. Soc. Queensl. 99: 35–43.

Conran, J.G. 1989. Cladistic analyses of some net-veined Liliiflorae. Plant Syst. Evol. 168: 123–141.

Conran, J.G. 1996. Corsiaceae, Smilacaceae and Xanthorrhoeaceae. The World of Plants (Japan) 105 (4/28): 258, 263–265, 268–271 (in Japanese).

Conran, J.G., Clifford, H.T. 1986a. Smilacaceae. In: George A.S. (ed.) Flora of Australia, vol. 46. Canberra: Australian Government Publishing Service, pp. 180–196.

Conran, J.G., Clifford, H.T. 1986b. The taxonomic affinities of the genus *Phipogonum* J.R. et G. Forst. Nord. J. Bot. 5: 215–219.

Conran, J.G., Christophel, D.C., Scriven, L.J. 1994. *Petermanniopsis angleseaënsis*: an Australian fossil net-veined monocotyledon from Eocene Victoria. Int. J. Plant Sci. 155: 816–827.

Couper, R.A. 1953. Upper Mesozoic and Cainozoic spores and pollen grains from New Zealand. Palyn. Bull. N. Z. Geol. Surv. Pal. 22: 1–77.

Cranwell, L.M. 1953. New Zealand pollen studies: The monocotyledons. Bull. Auckland Inst. Mus. 3: 1–91.

Cronquist, A. 1981. See general references.

Daghlian, C.P. 1981. A review of the fossil record of monocotyledons. Bot. Rev. 47: 517–555.

Dahlgren, R.M.T. et al. 1985. See general references.

Farnsworth, N.R., Blomster, R.N., Quimby, M.N., Schermerhorn, J.W. (eds.) 1971. The Lynn Index, vol. 7. New York: N. R. Farnsworth.

Gascoigne, R.M., Ritchie, E., White, D.E. 1948. A survey of anthocyanins in the Australian flora. J. R. Soc. N. S. W. 82: 44–70.

Heslop-Harrison, Y., Shivanna, K.R. 1977. The receptive surface of the angiosperm stigma. Ann. Bot. 41: 1233–1258.

Huber, H. 1977. See general references.

Humphrey, L.E. 1914. A cytological study of the stamens of *Smilax herbacea*. Ohio Nat. 15: 357–369.

Koyama, T. 1978. Smilacaceae. In: Li, H., Lui, T., Huang, T., Koyama, T., De Vol, C.E. (eds.) Flora of Taiwan, vol. 5. Taipei: Epoch Publ., pp. 110–137.

Koyama, T. 1984. A taxonomic revision of the genus *Heterosmilax* (Smilacaceae). Brittonia 36: 184–205.

Krause, K. 1930. See general references.

Kunkel, G. 1984. Plants for human consumption. Koenigstein: Kluwer.

Lavaud, C., Massiot, G., Le Men Olivier, L. 1993. Steroidal saponins of *Smilax krausiana*. Plant Med. Phytother. 26: 64–73.

Macmillan, B.W. 1972. The biological flora of New Zealand. 7. *Ripogonum scandens* J.R. et G. Forst. (Smilacaceae), Supplejack, Kareao. N. Z. J. Bot. 10: 641–672.

Mangin, L. 1882. Origine et insertion des racines adventives et modifications corrélations de la tige chez les monocotylédones. Ann. Sci. Nat. Bot. Biol. Veg. VI, 14: 216–363.

Martin, B.F., Tucker, S.C. 1985. Developmental studies in *Smilax* (Liliaceae). I. Organography and the shoot apex. Am. J. Bot. 72: 66–74.

Morita, H. 1931. On new species of the genera *Cinnamomum* and *Smilax* from the Miocene deposits of Oguni-Machi, Uzen Province, Japan. Jpn. J. Geol. Geogr. 9: 1–8.

Pole, M. 1991. A modified terminology for angiosperm leaf architecture. J. R. Soc. N. Z. 23: 297–312.

Pole, M. 1993. Early Miocene flora of the Manuherikia Group, New Zealand. 5. Smilacaceae, Polygonaceae, Elaeocarpaceae. J. R. Soc. N. Z. 23: 289–302.

Satô, D. 1942. Karyotype alternation and phylogeny in Liliaceae and allied families. Jpn. J. Bot. 12: 57–161.

Schnarf, K. 1931. Vergleichende Embryologie der Angiospermen. Berlin: Borntraeger.

Schulze, W. 1982. Beiträge zur Taxonomie der Liliifloren VIII. Smilacaceae. Wiss. Z. Friedrich Schiller Univ. Jena 31: 285–289.

Sen, S. 1975. Cytotaxonomy of Liliales. Feddes Rep. 86: 255–305.

Simpson, P.G., Philipson, W.R. 1969. Vascular anatomy in vegetative shoots of *Rhipogonum scandens* Forst. (Smilacaceae). N. Z. J. Bot. 7: 3–29.
Sun, Z., Dilcher, D.L. 1988. Fossil *Smilax* from Eocene sediments in western Tennessee. Am. J. Bot. 75: 118.
Takhtajan, A.L. 1982. See general references.
Tanaka, T. 1976. Tanaka's cyclopedia of edible plants of the world. Tokyo: Keigaku Publ. Co.
Tillich, H.-J. 1995. Seedlings and Systematics in Monocotyledons. In: Rudall, P.J., Cribb, P., Cutler, D.F., Humphries, C.J. (eds.) Monocotyledons: systematics and evolution. London: Royal Botanic Gardens, Kew, pp. 303–352.
Troll, W. 1939. Vergleichende Morphologie der höheren Pflanzen, vol. I, 2. Berlin: Borntraeger.
Velenovský, J. 1907. Vergleichende Morphologie der Pflanzen, vol. II. Prague: Fr. Řivnáč.
Wagner, W.M. 1977. Vessel types of monocotyledons: a survey. Bot. Not. 130: 383–402.

Stemonaceae

K. Kubitzki

Stemonaceae Engler in E.-P., Nat. Pflanzenfam. II, 5: 8 (1887), nom. cons.
Roxburghiaceae Wall., Pl. Asiat. Rar. 3: 50 (1832).

Twining, creeping, or erect perennials, with tuberous roots or rhizome; plants glabrous or with unicellular hairs. *Leaves* distichous, alternate, opposite, or verticillate, petiolate; blade ± ovate, with entire margin; longitudinal veins interconnected by fine transversal veinlets. *Inflorescences* axillary, uniflorous or in lax few-flowered cymose clusters, or several in subumbellate clusters, but never branched or compound, withering with the whole leafy stem; *flowers* articulated from pedicel with pericladium, dimerous, usually hermaphrodite, rarely (*Stichoneuron*) functionally unisexual with some dimorphism; tepals 2 + 2, free, or basally connate, valvate; stamens 2 + 2; filaments adnate to base of tepals; anthers bithecate, dehiscing introrsely by longitudinal slits; thecae sometimes protruding into sterile appendages the tips of which may be fused; connective sometimes apically with a tepaloid appendage and a median ridge separating the thecae; ovary superior or half-inferior, unilocular; ovules few to many, anatropous or hemi-anatropous; placenta basal (*Stemona*) or apical; style absent; stigma sessile. *Fruit* a 2-valved capsule, thin-walled; seed longitudinally ridged, without sarcotesta, but provided with elaiosome of juicy hairs originating from hilum, raphe, or micropyle; endosperm present; embryo minute.

A family of three genera and about 35 species, E and SE Asia and Australia, one species in the SE USA.

Vegetative Anatomy. The midrib is simple. The stomata are anomocytic; tannin cells and raphides are common; long prismatic crystals ("styloids") are known from *Stemona* and *Stichoneuron*. The stems of *Stemona* and *Stichoneuron* have the bundles in 2 closely interlaced peripheral rings, collateral in the outer ring, and amphivasal (i.e. the phloem completely encircled by the xylem) in the inner. *Croomia* has a single ring of amphivasal bundles. Vessels with scalariform perforations are present in the roots and stems of all genera (Lachner-Sandoval 1892; Ayensu 1968; Tomlinson and Ayensu 1968).

Flower Structure and Anatomy. In *Stemona* the stamens may have two kinds of apical extensions, which is rare in flowering plants. As described in detail by van Heel (1992), the inward extensions are due to the connection and elongation of apically converging sterile portions of the thecae, while the outward elongation represents an elongation of connective tissue. As to the gynoecium, the observations suggest a transition from trimery to monomery (the former manifest in some *Stemona*, the latter in all others).

Embryology. The pollen is binucleate when shed (van der Ham 1991). The ovule is anatropous, bitegmic and crassinucellar; the archespore cell cuts off a primary parietal cell which forms 2 parietal layers below the nucellar epidermis. The chalazal megaspore of a linear tetrade develops into a Polygonum-type embryo sac. Endosperm formation is Nuclear, but later becomes cellular throughout (Davis 1966).

Pollen Morphology. The pollen grains are sulcate or inaperturate (*Stichoneuron*), tectate-columellate with microreticulate or reticulate (*Croomia*) ornamentation, the intine is forming a thick oncus (thin in *Stichoneuron*), the shape is spheroidal to ellipsoid, the size relatively small (20–26 × 33–40 μm), the anther wall and in *Stemona* also the exine is sparsely covered with smooth Ubisch bodies. *Pentastemona* pollen is different (van der Ham 1991).

Karyology. The following chromosome numbers have been reported: *Stemona japonica* $2n = 14$; *Stichoneuron caudatum* $2n = 18$ and *Croomia pauciflora* $2n = 24$ (see Duyfjes 1993).

Pollination. No direct observations seem to have been made, but the detailed description of the flower of *Stemona tuberosa* (van Heel 1992) is pertinent. During development, the stamens broaden considerably and tangential and median wings develop on the stamens. These wings close up and touch in each flower, and lengthwise interstaminal spaces originate, in which the pollen of two neighbouring thecae is shed. By the terminal fusion of the inward apical extensions of the stamens, the flowers become a kind of window flowers. The pollen shed in the interstaminal spaces would only be accessible through these windows. The flowers of *Stemona tuberosa* and *S. javanica* are carrion-coloured (dirty green and purplish striped outside, and dark purplish inside) and produce a strong smell of carrion (Duyfjes 1991) or cheese (*S. tuberosa*, pers. comm. J. Bogner); they were also observed to be visited by flies.

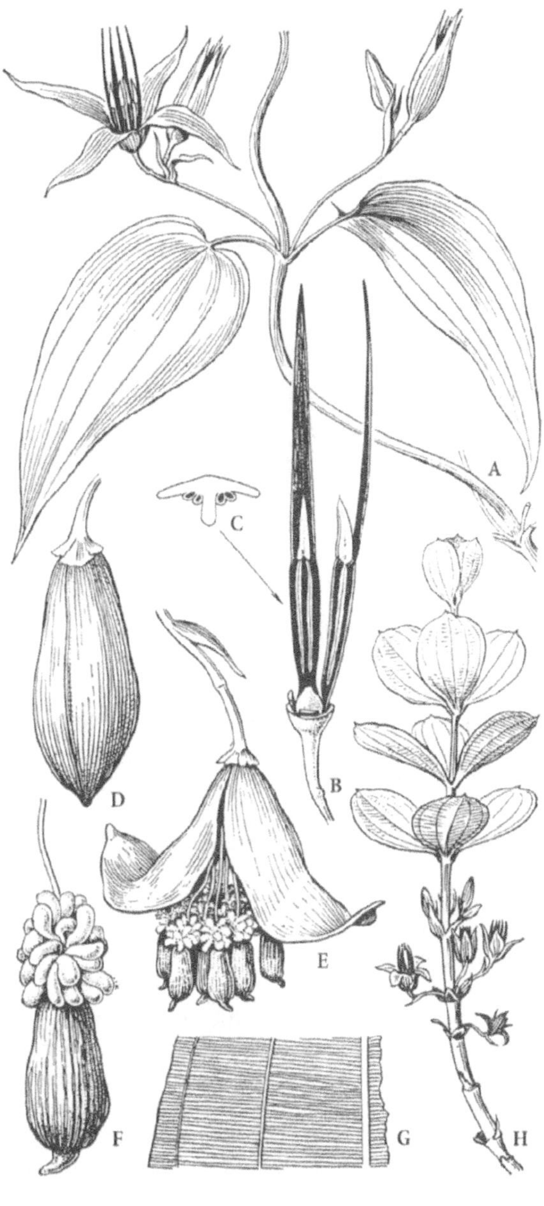

Fig. 119A–H. Stemonaceae. **A–G.** *Stemona tuberosa*. **A** Flowering shoot. **B** Two stamens. **C** Cross-section of stamen, showing median ridge separating the thecae. **D** Fruit. **E** Dehiscing fruit with seeds dangling on funicles, with coralloid elaiosomes. **F** Seed. **G** Leaf venation. **H** *Stemona sessiliflora*, flowering plant. (Takhtajan 1982)

Fig. 120A–E. Stemonaceae. *Stichoneurum membranaceum*. **A** Flowering plant. **B** Cymule. **C** Flower. **D** Arranged flower. **E** Stamen. (Takhtajan 1982)

FRUIT AND SEED. All members of the Stemonaceae share the possession of a ridged seed coat, which originates by periclinal divisions of testal layers and leads to a several-layered testa (Huber 1969; Bouman and Devente 1992), in contrast to *Tacca* and *Pentastemona*, where both integuments remain 2-layered. The endosperm stores aleuron, fat oil and varying quantities of starch (Huber 1969). The embryo is minute.

PHYTOCHEMISTRY. From the subterranean parts of *Stemona* and *Croomia* a large number of insecticidal alkaloids were isolated which represent a special type so far known only from these genera; tuberostemonin is the best-known representative. Saponins have been recorded for one *Stemona* (Hegnauer 1963, 1986).

AFFINITIES. Krause (1930) treated the Stemonaceae as close relatives of his Liliaceae, particularly Asparagoideae and Luzuriagoideae, as had been suggested by Lachner-Sandoval (1892). Hutchinson (1934) included them in his Dioscoreales. The latter placement receives support through characters such as the prolongation of the connectives; the distinct pith in the stem; tendencies towards an inferior position of the ovary, etc. Burkill (1960) thought that the family originated from "proto-Liliales" near the Dioscoreaceae. The position of the Stemonaceae in the Dioscoreales was also favoured by Dahlgren et al. (1985) and Huber (1969), while Huber (1991), rating high the articulated flowers, the entirely different testa (absence of a defined endostesta in *Stemona*), and the non-dioscoralean stem anatomy, included them in the Asparagales. In the recent *rbc*L study of Chase et al. (1995) the Stemonaceae form a weakly supported clade together with Pandanaceae, Cyclanthaceae and Velloziaceae. While the relationship between Pandanaceae and Cyclanthaceae is undisputed, it is hard to see what the dimerous Stemonaceae should have in common with the apparently tetramerous Cyclanthaceae.

DISTRIBUTION AND HABITATS. *Stemona* ranges from Japan and continental Asia through Malesia to N Australia; most of the species prefer a seasonal climate and occur in rather dry vegetation. *Stichoneuron* is distributed in continental SE Asia and the Malay Peninsula, and *Croomia* is disjunct between S Japan, E China and the SE USA (Rogers 1982). The two latter genera are low, delicate forest dwellers.

KEY TO THE GENERA

1. Usually scandent, some erect, or rarely creeping; roots spindle-shaped thickened; anthers with apical appendages; ovules numerous, basal; leaves only with basal nerves **1. *Stemona***
- Erect herbs; roots not spindle-shaped thickened; anthers without appendages; ovules few, apically inserted 2
2. Leaves with curved lateral veins; partial inflorescences racemelike, with densely set stiff pedicels; tepals valvate **2. *Stichoneuron***
- Leaves with basal veins only; partial inflorecences few-(mostly 1–2-) flowered; tepals imbricate **3. *Croomia***

1. *Stemona* Lour. Fig. 119

Stemona Lour., Fl. Cochinch. 1: 404 (1790); Schlechter, Notizbl. Bot. Gart. Berl.-Dahlem 9: 190–196 (1924), synopsis, classif.
Roxburghia W. Jones ex Roxb. (1795).

Climbers or creeping or erect herbs; leaves only with basal veins; petiole at base pulvinate; tepals up to 6 cm long; anthers with the thecae dorsally attached to the base of a long petaloid outgrowth of the connective, the thecae apically with or without a common subulate appendage; ovary superior; capsule 2-valved; seeds basally inserted, dangling on long funicles, with a coralloid or lobed elaiosome of vesicular hairs. About 25 spp., Japan through E and SE Asia to Australia.

2. *Stichoneuron* Hook. f. Fig. 120

Stichoneuron Hook. f. in Benth. & Hook., Gen. Pl. 3: 747 (1883).

Erect herb with short rhizome; leaves alternate, with pinnate lateral veins; petiole slightly sheathing; inflorescences axillary, with densely set stiff pedicels; flowers small; tepals patent, fused at base with each other and the ovary; stamens free, simple, without appendages; ovary half-inferior; ovules few, apically attached; fruit 1- or 2-seeded; elaiosome flimsy, of uniseriate hairs. Two spp., SE Asia.

3. *Croomia* Torr.

Croomia in Torr. & Gray, Fl. N Am. 1: 663 (1840).

Erect herb on creeping rhizome; leaves distichous, only with basal veins; inflorescences few-flowered; tepals imbricate, free at the base, patent; stamens free, with short filaments and unappendaged anthers; ovary half-inferior; ovules several, apically attached; capsule 2-valved, few-seeded. Elaiosome of uniseriate hairs. Three spp., two in E China, Ryu-kyu Isl., S Japan, one in SE USA.

Selected Bibliography

Ayensu, E.S. 1968. Comparative vegetative anatomy of the Stemonaceae (Roxburghiaceae). Bot. Gaz. 129: 160–165.

Bouman, F., Devente, N. 1992. A comparison of the structure of ovules and seeds in *Stemona* (Stemonaceae) and *Pentastemona* (Pentastemonaceae). Blumea 36: 501–514.

Burkill, I.H. 1960. The organography and evolution of Dioscoreaceae, the family of yams. J. Linn. Soc., Bot. 56: 319–412.

Chase, M.W. 1995. See general references.

Dahlgren, R.M.T. et al. 1985. See general references.

Davis, G.L. 1966. See general references.

Duyfjes, B.E.E. 1991. Stemonaceae and Pentastemonaceae: with miscellaneous notes on members of both families. Blumea 36: 239–252.

Duyfjes, B.E.E. 1993. Stemonaceae. In: Flora Melesiana I, 11: 399–411. Leyden: Flora Malesiana Foundation.

Ham, R.W.J.M. 1991. Pollen morphology of the Stemonaceae. Blumea 36: 127–159.

Hegnauer, R. 1963, 1986. See general references.

Huber, H. 1969, 1991. See general references.

Hutchinson, J. 1934. The families of flowering plants. Vol. 2. Monocotyledons. 2nd edn. London: Macmillan.

Krause, K. 1930. Stemonaceae. In: Engler, A. (ed.), Die natürlichen Pflanzenfamilien, 2nd ed., Vol. 15a. Leipzig: W. Engelmann, pp. 224–227.

Lachner-Sandoval, V. 1892. Beitrag zur Kenntnis der Gattung *Roxburghia*. Bot. Centralbl. 50: 65–70, 97–104, 129–135, 1 pl.

Rogers, G.K. 1982. The Stemonaceae in the southeastern United States. J. Arnold Arbor. 63: 327–336.

Steenis, C.G.G.J. van 1982. *Pentastemona*, a new 5-merous genus of monocotyledons from North Sumatra (Stemonaceae). Blumea 28: 151–163, 2 figs.

Takhtajan, A. 1982. See general references.

Tomlinson, P.B., Ayensu, E.S. 1968. Morphology and anatomy of *Croomia pauciflora* (Stemonaceae). J. Arnold Arbor. 49: 260–275.

Taccaceae

K. Kubitzki

Taccaceae Dumort., Anal. Fam.: 57, 58 (1829), nom. cons.

Perennial *herbs* from a solid, starchy, tuberous, globose, or elongated, vertical or horizontal rhizome; indumentum of minute pluricellular hairs present on young parts. *Leaves* all basal, large, long-petiolate with entire, pinnatifid, palmately partite, or palmatisect blade with pinnately divided segments, appearing together with the inflorescence; venation palmate or pinnate, with a secondary network. *Inflorescence* borne on a radical scape, involucrate, cymose, umbel-like; involucral bracts (2–)4(–12); floral bracts numerous, filiform, drooping, sometimes inconspicuous. *Flowers* bisexual, actinomorphic, epigynous; tepals 6, all dark-coloured, more or less petaloid; stamens 6 in 2 whorls, attached to perianth tube or to base of tepals; filaments short, flat and somewhat petaloid, forming, together with the broad connectives, a sort of hood over the inflexed anther; anther tetrasporangiate and dithecal, introrse, opening by longitudinal slits; ovary 6-ribbed, inferior, tricarpellate, unilocular, with more or less intruded parietal placentas, each with numerous ovules; septal nectaries present in several species; disc rarely (*T. leontopetaloides*) present but nectar secretion doubtful; style 1, provided with 3 incised wings; stigmatic branches 3, obcordate, often petaloid; ovules pendulous, anatropous, bitegmic, crassinucellar. *Fruit* a berry or a loculicidal capsule; seeds strongly ribbed, sometimes with a thin fleshy aril and fleshy raphe; embryo small; endosperm copious, more or less cartilaginous, storing protein and fat, but no starch and no or little hemicellulose. $n = 15$.

A pantropical monogeneric family, best developed in SE Asia and Polynesia, comprising ten species in the Old World and a single (Drenth 1972) or several (Limpricht 1928) in the New.

Vegetative Morphology and Anatomy (Limpricht 1928). The tubers are sympodially connected; in *T. leontopetaloides* ("*pinnatifida*"), for example, a fully developed plant comprises at the same time 3 bulbs of different age. Ptyxis is plicative (Huber 1991). Midrib compound. Elements interpreted as transitional between vessels and tracheids (Fahn 1954) but rather representing true vessels were found in the roots (3 spp. investigated) but not in the shoot; their perforation plates have up to 300 bars. Due to the presence of a meristematic ring, which surrounds the primary vascular bundles of the axis and produces additional vascular bundles, the rhizomes are capable of an anomalous secondary growth comparable to that of *Cordyline* and *Dracaena* (Limpricht 1928, in need of verification). Cells containing raphides are concentrated mainly along the veins, and calcium oxalate crystals are found preferably in the epidermal cells. Mucilaginous cells are known from the mesophyll. Peculiar trichomes consisting of a short row of cells bearing a multicellular body on which is another cell row are found on both sides of the young leaf. The stomata are anomocytic or, in Old World species, surrounded by 1 cell only and are randomly distributed over the leaf surface (Ling Ping-Ping 1981).

INFLORESCENCE AND FLORAL MORPHOLOGY. The peduncle terminates in an involucrum of leafy, mostly erect, and dull-coloured or snowy white bracts; sometimes these have different orientation (Goebel 1931). These bracts subtend groups of partial inflorescences which are umbellately arranged. In most species, there are further long, filiform and drooping bracts between the flowers (Fig. 121). Both the involucral and filiform bracts are bracts and prophylls of the cincinnate partial inflorescences. This interpretation was given by Eichler (1879), who also compared this inflorescence structure with that of various Amaryllidaceae.

By the thickening, broadening and incurving of the connectives the stamens become contiguous with the stigma between the stigmatic lobes. Thus, 6 separate interstaminal entrances are created that lead downwards to a circular hypanthial space around the style-stigma which is provided with a light-coloured bottom. For floral anatomy see Rao (1969).

EMBRYOLOGY. Anther wall formation follows the Dicotyledonous type (elsewhere in monocots known only from *Acorus*). The division of pollen mother cells is simultaneous (only shared with Dioscoreaceae) and the pollen 2-celled when shed. The micropyle is formed by the inner integument. The primary archesporial cell cuts off a primary parietal cell which forms a single parietal layer. Embryo-sac formation is according to the Polygonum type; somatic apospory has been observed in *T. plantaginea*. Endosperm development is Nuclear, but later the tissue becomes cellular (Davis 1966).

POLLEN MORPHOLOGY. The pollen grains are sulcate. The exine is tectate-columellate with a thick footlayer; an endexine is not present. The tectum is traversed by minute perforations and its sculpturing is verrucate to slightly striate (Zavada 1983).

POLLINATION. The dull colours of the flowers and structures such as the helmetlike stamens and broad stigmatic lobes very likely function as a trap, suggesting the syndrome of sapromyiophily (Endress 1995). Indeed, Saw (1993) was able to detect in the flowers of *Tacca integrifolia* a sweetish musky odour. The strikingly white, often ± striped bracts of *T. cristata* (a species not admitted by Drenth, but probably good) give rise to a spectacular pseudanthium that strongly resembles the flowers respectively pseudanthia (strictly speaking the Fahnenblumen) of *Paphiopedilum* and *Arisaema* occurring in the same area, not in the lowland (as *Tacca*) but in the hills.

FRUIT AND SEED (Huber 1969; Bouman 1995). The seeds are elongated and straight or slightly curved (see illustrations in Drenth 1972) with a noticeable raphe. According to Carter (1962), the seeds of *T. leontopetaloides* are surrounded by a thin fleshy aril, which is not evident when the seeds are dry; for this species the presence of an aril and a fleshy raphe could be verified; it is unknown, however, whether an aril is present in other species; some are lacking it. The seed coat is always longitudinally ribbed and remarkable because both integuments are 2-layered and well developed and take part in the mechanical function, which is restricted to the endotestal and exotegmic layers. Also the seed ridges originate from the localised stretching of exotegmic and/or endotestal cells. The minute embryo has a terminal growing point and the cotyledon is laterally inserted, as in *Dioscorea*.

DISPERSAL. *Tacca leontopetaloides* may have acquired its wide distribution from Africa through Malesia to Hawaii by dispersal through the sea, but man may also have interfered by widely disseminating this crop plant (Drenth 1976). Saw (1993), however, believes that it might rather be animal-dispersed because the seeds have "a spongy white testa with a sweetish taste", which very likely corresponds to what here is interpreted

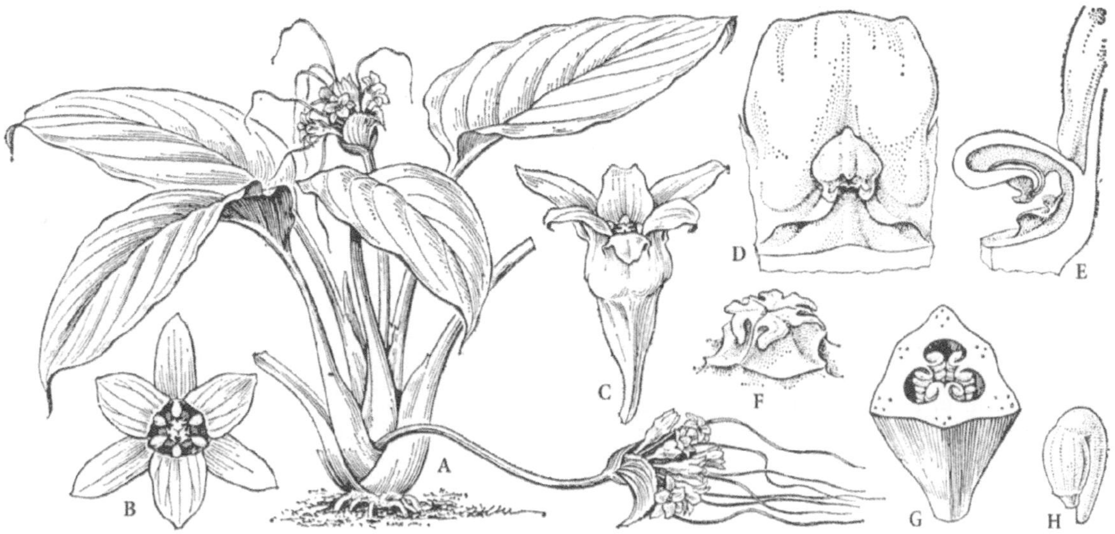

Fig. 121A–H. Taccaceae. *Tacca integrifolia.* **A** Flowering plant. **B** Flower, from above. **C** Same, side view. **D** Stamen, frontal view. **E** Stamen, median section. **F** Stigma. **G** Ovary, cross-section. **H** Developing seed. (Takhtajan 1982)

as the aril. Saw (1993) reported two kinds of fruit development in relation to dispersal. In the first form, after pollination the peduncle bends over and places the developing infructescence on the ground. *T. integrifolia*, *T. chantrieri*, and *T. palmitifida* belong to this group (Fig. 121A). Their fruits mature with a dull colour and have fleshy walls with a sweet taste and a jellylike texture; they are possibly dispersed by small mammals such as rodents. In the second form of develoment, the peduncle remains erect and the fruits mature in an erect position, remaining thin-walled and developing brightly coloured orange (*T. palmata*) or yellow (*T. leontopetaloides*) fruits.

Phytochemistry (Hegnauer 1963, 1986). The tubers contain alkaloids of unknown constitution and a bitter principle, taccalin, which is possibly a norditerpene of the diosbulbin type. The occurrence of saponins with diosgenin as the aglycon, of proanthocyanidins and possibly also of taccalin according to Hegnauer (1986) would not contradict the postulated close relationship between Taccaceae and Dioscoreaceae.

Subdivision. The previously distinguished genus *Schizocapsa* is not sufficiently distinct from *Tacca* and the sections that have been distinguished in *Tacca* are not clearly delineated (Drenth 1972) and cannot be upheld.

Affinities. In the past, many different suggestions as to the affinity of the Taccaceae have been made, but the one pointing to Dioscoreaceae, based entirely on plesiomorphies, received more credit than it deserves and considered the Taccaceae as closely related to Dioscoreaceae and Stemonaceae. In the classification by Dahlgren et al. (1985), the Taccaceae form part of a broadly construed order Dioscoreales. A position of *Tacca* near Dioscoreaceae is also supported by *rbc*L analysis (Chase et al. 1995). Huber (1969), however, has called attention to various characters which contradict a close relationship of *Tacca* with Dioscoreaceae, and according to him, the Taccaceae must constitute a branch rooted in the proto-monocotyledons and collateral to the Dioscoreaceae. For characters shared with *Pentastemona* see p. 407.

Distribution and Habitats. *Tacca* is pantropical and apart from being native to America, Africa and SE Asia, it radiates into China, Polynesia and Australia (distribution map in Drenth 1976), growing in the humid and seasonal (sub)tropical lowland in forests and open primary and secondary vegetation. *T. leontopetaloides* is a coastal species.

Uses. Highly productive strains of *T. leontopetaloides* with large (5–10-cm-long) tubers and with a starch content of up to 27% have spread from FiJi through Polynesia; occasionally, they are traded under the wrong name of East Indian Arrowroot.

Only one genus:

Tacca **J.R. & G. Forst.** Fig. 121

Tacca J.R. & G. Forst., Char. Gen. Pl.: 35 (1775), nom. cons.; Drenth, Blumea 20: 367–406 (1972), rev., and in Fl. Mal. I, 7:806–819 (1976).
Schizocapsa Hance (1881).

Description as for family. The distinction of *Schizocapsa* from *Tacca* had been based on its capsular instead of baccate fruit, and although this distinction holds, the difference is now considered as insufficient for generic separation.

Selected Bibliography

Bouman, F. 1995. Seed structure and systematics in Dioscoreales. In: Rudall, P.J., Cribb, P.J., Cutler, D.F., Humphries, C.J. (eds.) Monocotyledons: systematics and evolution. Royal Botanic Gardens, Kew, pp. 139–156.

Burkill, I.H. 1960. The organography and evolution of Dioscoreaceae, the family of yams. J. Linn. Soc. Lond. Bot. 56: 319–412.

Carter, S. 1962. Taccaceae. In: Flora of Tropical East Africa. 4 pp. London: Crown Agents.

Chase, M.W. et al. 1995. See general references.

Dahlgren, R., Clifford, H.T. 1982. See general references.

Daumann, E. 1970. Das Blütennektarium der Monocotyledonen unter besonderer Berücksichtigung seiner systematischen und phylogenetischen Bedeutung. Feddes Repert. 80: 463–590.

Davis, G.L. 1966. See general references.

Drenth, E. 1972. A revision of the family Taccaceae. Blumea 20: 367–406.

Drenth, E. 1976. Taccaceae. In: van Steenis, C.G.G.J. (ed.) Flora Males. I, 7: 806–819. Leyden: Noordhoff.

Eichler, A.W. 1879. In: LXVII. Sitzung vom 27. Juni 1879, pp. 106–108. Sitzungsber. Bot. Ver. Prov. Brandenburg 21: 106–108.

Endress, P.K. 1995. Major evolutionary traits of monocot flowers. pp. 43–79. In: Rudall et al. 1995. See general references.

Fahn, A. 1954. Metaxylem elements in some families of the monocotyledons. New Phytol. 53: 530–540.

Goebel, K. 1931. Blütenbildung und Sproßgestaltung. Zweiter Ergänzungsband zur Organographie der Pflanzen. Jena: G. Fischer.

Hegnauer, R. 1963, 1986. See general references.

Huber, H. 1969, 1991. See general references.

Limpricht, W. 1928. Taccaceae. In: Engler, A. (ed.) Das Pflanzenreich IV.42. Leipzig: W. Engelmann.

Ling Ping-Ping 1981. Stomatal studies in Chinese Taccaceae with a discussion on its taxonomical significance. Bull. Nanjing Bot. Gard. Mem. Sun Yat Sen 1981: 20–24.

Rao, V.S. 1969. The vascular anatomy of *Tacca pinnatifida*. J. Univ. Bombey 38. 65: 18–24.

Saw, L.G. 1993. Tacca: flowering and fruiting behaviour. Nat. Malaysiana 18: 3–6.

Takhtajan, A.L. 1982. See general references.

Zavada, M.S. 1983. Comparative morphology of monocot pollen and evolutionary trends of apertures and pollen wall structures. Bot. Rev. 49: 331–379.

Tecophilaeaceae

M.G. Simpson and P.J. Rudall

Tecophilaeaceae Leyb., Bonplandia 10: 370 (1862), nom. cons.
Cyanastraceae Engler (1900).

Erect, perennial, terrestrial *herbs*. Roots fibrous. Subterranean stem a globose to ellipsoid corm, 1–4 cm in diameter, in some genera with a membranous to fibrous tunic consisting of persistent sheathing leaves or fibrovascular bundles. *Leaves* basal to subbasal, or cauline in *Walleria*, spiral; base sheathing or non-sheathing, blades narrowly linear to lanceolate-ovate, or more or less petiolate in *Cyanastrum* and *Kabuyea*; entire, glabrous, flat, or marginally undulate; venation parallel with a major central vein. *Flowers* terminal and either solitary (or in small groups) and a panicle or (in *Walleria*) solitary in the axils of cauline leaves. Bracts and bracteoles (prophylls) often present on pedicel. Flowers 1–3 cm long, pedicellate, bisexual, trimerous. Perianth variable in color, zygomorphic or actinomorphic, homochlamydeous, basally syntepalous; perianth lobes 6, imbricate in 2 whorls, the outer median tepal positioned anteriorly; minute corona appendages present between adjacent stamens in some taxa. Androecium arising at mouth of perianth tube, opposite the tepals and either actinomorphic or zygomorphic, composed of either 6 fertile stamens or combinations of fertile stamens and staminodes; fertile stamens similar in size and shape or dimorphic, oriented equally in a circle or oriented in anterior and posterior groups. Fertile anthers 2-locular, poricidally dehiscent (or almost so). Gynoecium syncarpous; ovary semiinferior to almost superior; carpels and locules 3; placentation axile; ovules 2 to numerous per carpel. *Fruit* a loculicidal capsule. Seeds ranging from yellow to brown to black, and of varied size and shape.

Eight genera, ca. 23 species, Chile, southern and tropical Africa, Madagascar, and California.

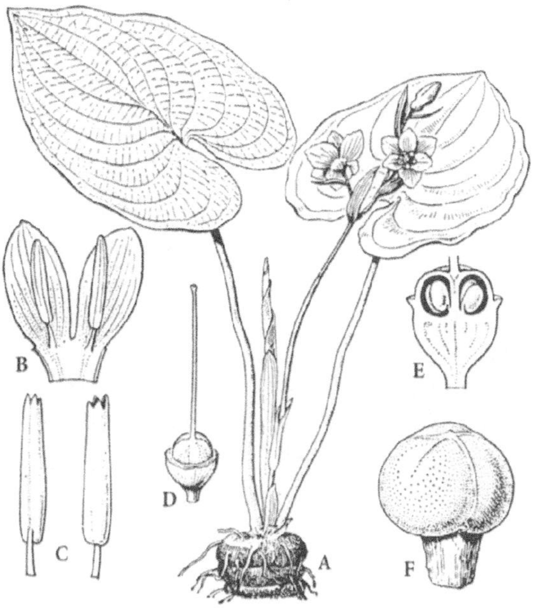

Fig. 122A–F. Tecophilaeaceae. *Cyanastrum cordifolium*. **A** Flowering plant. **B** Tepals with stamens. **C** Stamens. **D** Pistil. **E** Ovary, longitudinal section. **F** Capsule. (Takhtajan 1982)

Vegetative Morphology. All genera of Tecophilaeaceae have a subterranean, oblate to globose corm, 1–4 (generally 2) cm in diameter. In five genera (*Conanthera*, *Cyanella*, *Odontostomum*, *Tecophilaea*, and *Zephyra*) the corm is enveloped by a tunica, which consists of membranous to fibrous scale leaves or leaf bases or the reticulate fibrovascular remains of these scale leaves (Fig. 123). The tunic often continues above the corm, in some cases forming an apical tuft. Corms of *Walleria*, *Cyanastrum*, and *Kabuyea* lack a corm tunic (Fig. 122).

Leaves are bifacial and spirally arranged. Leaf venation is parallel, with 1 major central vein. Photosynthetic leaves of some genera are basal to subbasal with a tubular sheath surrounding the base of the aerial stem axis. One or 2 sheathing non-photosynthetic leaves are sometimes present at the base of the photosynthetic leaves. Leaf blades are glabrous, entire, flat or marginally undulate and are very narrowly linear, lanceolate, or lanceolate-ovate in shape, or more or less petiolate in *Cyanastrum* and *Kabuyea*. *Walleria* differs in having cauline photosynthetic leaves (basal leaves being absent) which are equally inserted along the aerial stem. Leaves of *Walleria* are sessile, lanceolate, entire, glabrous, and scabrous on the abaxial midrib.

Tillich (1995) described seedling structure in *Cyanastrum cordifolium*, *Cyanella*, and *Walleria*. In *Cyanella* and *Walleria* there is a haustorial hyperphyll, a short tubular sheath and a coleoptile, which becomes very long in *Cyanella*. In *Cyanastrum cordifolium* the hypocotyl and pri-

Fig. 123A–F. Tecophilaeaceae. *Cyanella capensis*. **A** Flowering shoot. **B** Tunicated corm. **C** Flower (the upper 5 stamens sterile. **D** Stamens in ventral and lateral view. **E** Dehiscing capsule. **F** Seed. (Takhtajan 1982)

mary root are missing (as in some other endospermless monocotyledons). In *Cyanastrum cordifolium* and *Walleria* the first few leaves are cataphylls.

Vegetative Anatomy. (Data from Schulze 1893; Cheadle 1969; Arroyo 1986; M.G. Simpson, unpubl.; P. Rudall, unpubl.) Vessels are absent from above-ground organs. Stomata are anomocytic (subsidiary cells absent), or occasionally indistinctly paracytic in cases where epidermal cells are irregularly shaped (e.g., in *Cyanastrum*). Trichomes are absent, except in *Cyanella* spp., which have tapering unicellular or short multiseriate trichomes, and *Walleria*, which has scabrate trichomes along the abaxial midrib and pedicel. Leaf mesophyll cells are relatively undifferentiated. Vascular bundles are in a single row, with sclerenchyma lacking or present in small amounts at larger bundles. Raphide crystals are present in the mesophyll of most species, but absent from *Cyanastrum* and *Kabuyea*. Secretory canals are present in the leaf mesophyll of *Cyanastrum* and *Kabuyea*, but not recorded in other genera. Occasional indigo blue anthocyanin cells are also present in *Cyanastrum* and *Kabuyea*.

Inflorescence Structure. In *Conanthera*, *Cyanella*, *Odontostomum*, and *Zephyra* the inflorescence is a panicle, either irregularly branched or consisting of a raceme of racemes (*Cyanella* rarely being 1-flowered). *Tecophilaea* has either a solitary flower or an irregular aggregate of 2–3 flowers. *Walleria* has solitary flowers in the axils of the cauline photosynthetic leaves. In *Cyanastrum* and *Kabuyea* the inflorescence is a simple raceme or the flowers are rarely paired (Brummitt et al. submitted).

Flower Morphology. Flowers are long-pedicellate with a single lanceolate to ovate bract subtending the pedicel and, in several taxa, a similar bracteole (prophyll) inserted opposite to the bract, approximately $^3/_4$ the length of the pedicel. Flowers are 1–3 cm in size, bisexual, and ascending to divergent in orientation. Perianth color is blue, purple, pink, white, or yellowish. The perianth is membranous, glabrous, actinomorphic or zygomorphic, syntepalous with a short to elongate, basal perianth tube and 6 perianth lobes, imbricate in 2 whorls (the outer median lobe anteriorly positioned). Outer and inner perianth lobes are glabrous, entire (to undulate), and variable in shape. Outer lobes have a distinctive mucronate to aristate (often involute-tubular) apical extension whereas the inner perianth lobes have an obtuse to rounded (rarely mucronulate) apex. Perianth lobes 3–7-veined (always from 3 major veins). Minute corona appendages are present between adjacent stamens in *Odontostomum* and *Zephyra*. The androecium is inserted at the mouth of the perianth tube, and consists of either: (1) 6 fertile stamens, (2)4 fertile stamens+2 staminodes (*Zephyra*), or (3) 3 fertile stamens + 3 staminodes

(*Tecophilaea*). In *Cyanella* the fertile stamens are dimorphic, either with (1) 3 small posterior + 3 large anterior stamens, or (2) 5 small posterior + 1 large anterior stamens (Fig. 123c), sometimes basally connate. Filaments of fertile stamens are terete, short to elongate. Fertile anthers are basifixed to subbasifixed, narrowly lanceolate, oblanceolate, or ovoid (with a basal caudate appendage in *Tecophilaea* and *Zephyra*, pubescent in *Zephyra*). Anthers are characteristically long and poricidally dehiscent by means of a small adaxial, apical pore (Fig. 122–124) or introrse slit. Thecae are parallel (or rarely adaxially swollen), and the connective has black spots in *Cyanella alba* and *C. lutea*. The gynoecium is syncarpous; the ovary is semiinferior (ranging from $^1/_4$ to $^2/_3$ inferior, or almost superior in *Walleria*), ellipsoid, ovoid, or obovoid, often 3-lobed, glabrous. There are 3 carpels and locules, the 3 carpels oriented opposite the outer perianth lobes, with the median carpel positioned anteriorly. In *Walleria* (Sterling 1974) and *Cyanastrum* and *Kabuyea* (P. Rudall, unpubl.) the style is sunken, i.e., lateral on each carpel. Placentation is axile, the placentae nonprotuberant, generally linear. The style is terminal, terete, straight, or curved with a rudimentary, minutely papillate stigma.

Fig. 124A–C. Tecophilaeaceae. *Walleria mackenzii*. **A** Flowering shoot. **B** Flower. **C** Same, longitudinal section. (Takhtajan 1982)

FLORAL ANATOMY. *Zephyra* is poorly known. *Walleria* was described by Sterling (1974). Other data are from M.G. Simpson (unpubl.) and P. Rudall (unpubl.). Septal nectaries are present throughout the ovary in *Conanthera*, *Odontostomum*, *Cyanastrum*, and *Kabuyea*, but absent or present as very short septal slits at the top of the ovary in *Cyanella*, *Tecophilaea*, and *Walleria*. Secretory canals are present scattered in the tepal mesophyll of *Cyanastrum*. Obturators are normally present. Ovules are bitegmic and campylotropous or anatropous and numerous, or 2 per carpel (in *Cyanastrum*, *Kabuyea* and *Odontostomum*). In mature ovules of many Tecophilaeaceae, the chalazal end of the nucellus is substantial, for example in *Cyanella* and *Tecophilaea* (where this region later persists as a postament: Rudall 1997), *Conanthera*, *Cyanastrum*, (P. Rudall, unpubl.) *Odontostomum* (Cave 1952) and *Walleria* (Sterling 1974). This substantial nucellus region is not the region which later becomes chalazosperm in *Cyanastrum*, which lies outside the raphal bundle, in the chalaza (see below).

EMBRYOLOGY. Data are largely from Cave (1952), De Vos (1950), Fries (1919), Nietsch (1941) Rudall (1997) and P. Rudall (unpubl.). Some genera, especially *Zephyra*, are poorly known. In all taxa examined, microsporogenesis is simultaneous (Rudall et al. 1997). The pollen grain generative cell is distal in position. The tapetum is secretory. Ovules are crassinucellate (the archesporial cell forms a parietal cell). Megasporogenesis is monosporic and embryo sac formation is of the Polygonum type. De Vos (1950) and Rudall (1997) reported a chalazal haustorium in *Cyanella*. Endosperm formation is variable, reported as Nuclear in *Cyanella* (De Vos 1950) and probably Helobial in *Odontostomum* (Cave 1952).

POLLEN MORPHOLOGY. (Data are from Erdtman 1952; Radulescu 1973; Schulze 1983; Simpson 1983, 1985; Zavada and Scott 1993). Tecophilaeaceae have sulcate, heteropolar, generally foveolate pollen grains (rugulose in *Zephyra*) with a tectate-columellate exine architecture, with a relatively thick, homogeneous foot-layer. Most genera (except *Cyanastrum* and *Kabuyea*) have an operculate aperture. The operculum consists of a band of tectate-columellate exine positioned median and parallel to the aperture. A characteristic electron-dense endexinous layer occurs inner to the ektexine both in the operculum and along the periphery of the aperture. *Walleria* differs in that the operculum consists of a band of granular ektexine atop the endexinous basal layer; otherwise the operculum of *Walleria* is ultrastructurally identical to that of other Tecophilaeaceae.

Karyology. (Data from Cave 1970; La Cour 1956; Nietsch 1941; Ornduff 1979; Satô 1942; Sen 1975). The basic chromosome number is mainly $x = 12$ or 11 ($x = 10$ in *Odontostomum*). Most species are diploid, with $2n = 24$, for example in *Cyanastrum cordifolium* (Nietsch 1941) (although $2n = 22$ was recorded by Sato 1941 for *Cyanastrum cordifolium*), *Cyanella alba*, *C. orchidiformis*, *Tecophiliaea cyanocrocus* and *Walleria*. *Cyanella* is the most variable genus, with $2n = 24$, 28, or 48 in *C. hyacinthoides*, $2n = 16$, 24, or 48 in *C. lutea*, indicating both aneuploidy and tetraploidy (Ornduff 1979).

Pollination. The poricidal anthers and flower structure of Tecophilaeaceae indicate buzz pollination, at least in the taxa with small, dry pollen (e.g., *Cyanella*) (Endress 1996). Only *Cyanella* has been studied with regard to pollination and reproductive biology (Dulberger and Ornduff 1978, 1980). No nectar is produced in observed species of *Cyanella*. Flowers are slightly scented. Insect guides are present in some species; in *C. alba* and *C. lutea* black spots occur on the connectives of the 5 posterior stamens; in *C. orchidiformis* tepals are light purple with dark purple bases. *Cyanella* spp. are heterantherous, having either 5 or 3 smaller posterior stamens and either 1 or 3 large anterior stamen(s). Pollen escapes only when the anthers are pulled down and released, the resultant snapping action producing a cloud of pollen. A springlike mechanism in the posterior stamen(s) may be enhanced by versatile anthers (in *C. lutea* and *C. orchidiformis*) and by the occurrence in the posterior anther(s) of an abaxial fleshy "cushion" (in *C. hyacinthoides* and *C. lutea*). Under natural conditions, pollen is thought to be discharged by rapid vibration from a heavy-bodied bee, as into other poricidally dehiscent flowers. The posterior stamens may function primarily in insect feeding and the anterior stamen(s) in pollination, although no significant differences in pollen size, morphology, or viability occur (Dulberger and Ornduff 1980; M.G. Simpson, pers. observ.).

Stylar enantiomorphy occurs in both *Cyanella alba* and *C. lutea* by the deflection of the style in a direction opposite to that of the anterior stamen. In these species pollen would tend to be deposited to one side of an insect's body in a location that would promote pollen flow only to a flower with a different stylar orientation. Thus, stylar enantiomorphy might tend to promote pollen transfer between different enantiomorphic flower types (Dulberger and Ornduff 1980). In *C. alba*, in which left-handed and right-handed flowers occur mostly on different individuals, stylar enantiomorphy would enhance outcrossing. However, moderate to strong self-incompatibility is reported in all investigated species of *Cyanella*.

Fruit and Seed. The fruit of Tecophilaeaceae is a globose, ovoid, or obovoid, generally 3-lobed loculicidal capsule (perhaps berry-like in *Walleria*). Laterally branched dorsal veins are often prominently visible in the fruit wall.

Seed structure is highly variable in the family (Huber 1969; Brummitt et al. submitted; P. Rudall, unpubl.). Seeds are ellipsoid, ovoid, or roughly spherical in *Kabuyea*. Seeds of *Conanthera*, *Cyanella*, *Odontostomum*, *Tecophilaea*, and *Zephyra* are small, whereas those of *Walleria*, *Kabuyea* and *Cyanastrum* are relatively large (up to 9 mm long in *Cyanastrum*). Seed colour ranges from phytomelan-black in *Cyanella* (Huber 1969) to brown in *Conanthera*, yellow to mid brown in *Kabuyea* and *Cyanastrum*, and dark brown in *Walleria*. In *Kabuyea* and *Cyanastrum* the testa is deeply pitted, either over the entire surface (*Kabuyea*) or at the chalazal end (*Cyanastrum*). In *Kabuyea* the pits contain rows of tiny microprickles. *Cyanastrum* seeds have an obvious swollen region at the chalazal end, which is lacking from the other taxa. In cross-section this region consists of loosely packed thin-walled cells in long strands, derived from the chalazal region outside the vascular bundle (i.e., not from the nucellus). Fries (1919) considered this tissue (which he called a chalazosperm, pointing out that it is not a perisperm) most likely to be a storage tissue, because mature seeds of *Cyanastrum* lack endosperm. However, he conceded that it has similarities with the elaiosomes found in seeds of other asparagoid taxa, and it could well have a role in seed dispersal. The role of this tissue is still unresolved. It is apparently lacking from most other Tecophilaeaceae, such as *Cyanella* (De Vos 1950), although Cave (1952) noted a similar, probably homologous, starch-rich chalazal tissue with loose strands of cells in the seed of *Odontostomum*. In *Tecophilaea* seeds there is no massive "chalazosperm", but the chalazal tissue proliferates and becomes loosely packed (Rudall 1997).

The presence of "chalazosperm" in *Cyanastrum*, which is a highly unusual tissue (if the storage function is confirmed), has been the main reason that many authors since Engler (1901) have separate family status accorded to *Cyanastrum* (see below), although Engler mistakenly called it perisperm.

Seeds are bitegmic, with the outer integument thick and multilayered. The epidermis of the outer integument is composed of thick-walled palisade cells. Endosperm is lacking in the mature seed of *Cyanastrum*, although present in the related genus *Kabuyea* and all other Tecophilaeaceae. The embryo is short to elongate, micropylar in position, and slightly swollen at the micropylar end.

PHYTOCHEMISTRY. Little is known about the chemistry of Tecophilaeaceae (see Gibbs 1974, and references therein). Saponins have been reported as absent in *Tecophilaea* and both absent and present in *Conanthera* spp. Leucoanthocyanins are absent from the leaves of *Cyanella orchidiformis*. Chelidonic acid is present in *Conanthera*. Slob (1973) reported small amounts of the potential allergens, tuliposides A and B, in *Tecophilaea cyanocrocus*.

RELATIONSHIPS WITHIN THE FAMILY. Tecophilaeaceae (including *Cyanastrum*, *Kabuyea*, and *Walleria*) are a well-defined family, characterized by the presence of a corm, mucronate to apically tubular outer perianth tepals, poricidal anthers (with minute apical introrse slits) and a semi-inferior ovary. All genera except *Cyanastrum* and *Kabuyea* have distinctive operculate pollen grains. The absence of operculate pollen grains in *Cyanastrum* and *Kabuyea*, and the presence of "chalazosperm" in *Cyanastrum* have been regarded as evidence against their inclusion in Tecophilaeaceae. However, the lack of "chalazosperm" in *Kabuyea* (Brummitt et al. submitted), which is clearly the sister genus to *Cyanastrum*, weakens this argument. Furthermore, *Odontostomum* probably shares the "chalazosperm" tissue (Cave 1952), and also resembles both *Cyanastrum* and *Kabuyea* in other respects, such as ovary and ovule structure. *Walleria* also resembles *Cyanastrum* and *Kabuyea* in some respects; for example, the corm is not tunicated and the style is sunken. Dahlgren and Van Wyk (1988) discussed the systematic relationships of *Walleria*, which was previously considered close to Uvulariaceae, and has sometimes been placed in its own family. Analysis of molecular data from *rbc*L (Chase et al. 1995) indicates that both *Cyanastrum* and *Walleria* are embedded in Tecophilaeaceae. Hutchinson (1934) also placed them in Tecophilaeaceae, and considered *Cyanastrum* close to *Cyanella*. Further morphological analysis is required to clarify these relationships. The Chilean genera *Tecophilaea* and *Zephyra* are probably sister taxa, by the common occurrence of staminodes. *Odontostomum* (California), *Cyanella* (South Africa), *Tecophilaea* and *Zephyra* are similar in the segregation of the androecium into a posterior and anterior group of stamens, *Odontostomum* and *Cyanella* having 1 anterior and 5 posterior stamens.

AFFINITIES. Analysis of data from *rbc*L (Chase et al. 1995), supported by the presence of simultaneous microsporogenesis (Rudall et al. 1997), indicated a placement for Tecophilaeaceae amongst the "lower" Asparagales, close to Iridaceae, Doryanthaceae, and the astelioid/orchid clade (including Lanariaceae). The presence of phytomelan in the seed coat of some taxa also indicates an asparagoid affinity. Serological analysis of seed proteins supports a relationship between Tecophilaeaceae and Iridaceae (Shneyer 1983).

Tecophilaeaceae were previously considered closely related to Haemodoraceae. However, there is now much evidence (including pollen structure: Simpson 1983) against this, since Haemodoraceae belong in the commelinoid clade, characterized by the presence of cell-wall ferulates (Harris and Hartley 1980; Rudall and Caddick 1994).

DISTRIBUTION AND HABITATS. Apart from *Odontostomum*, which is endemic to north-central California, USA, Teophilaeaceae occur either in Chile (*Conanthera*, *Tecophilaea*, and *Zephyra*) or Africa. *Cyanella* is restricted to South Africa; *Cyanastrum* and *Kabuyea* occur in tropical Africa, and *Walleria* is more widespread, with distributions in southern and tropical Africa and Madagascar.

Habitats vary only slightly. *Conanthera* grows in open hills and grasslands. *Tecophilaea* is found in rocky grasslands or river borders, and *Zephyra* in loose rocky soil of talus slopes. *Odontostomum* is reported in clay soil of grassy fields or forest openings. *Cyanella* is located in sandy or rocky soil of riverine slopes or grasslands. *Walleria* grows in generally sandy soil of open hill slopes or dense woodlands, and *Cyanastrum* and *Kabuyea* in woodland or stream banks.

ECONOMIC IMPORTANCE. *Conanthera* spp., *Cyanella* spp. and *Tecophilaea cyanocrocus* are utilized as ornamentals.

Key to the Genera

1. Corm lacking a tunic 2
- Corm with a membranous or fibrous tunic 4
2. Leaves cauline, ovary almost superior, seeds brown, warty **7. *Walleria***
- Leaves basal to subbasal, ovary semiinferior, seeds yellow, pitted 3
3. Leaves usually solitary, distinctly petiolate, perianth blue, seeds with chalazosperm **2. *Cyanastrum***
- Leaves 4 per corm, indistinctly petiolate, perianth white, seeds lacking chalazosperm **4. *Kabuyea***
4. Fertile stamens 6, staminodes absent 5
- Fertile stamens 3 or 4, staminodes present 7
5. Stamens dimorphic, of 2 different sizes and shapes **3. *Cyanella***
- Stamens all of similar size and shape 6
6. Stamens apically connivent, equal in a single ring; anthers narrowly lanceoloid **1. *Conanthera***
- Stamens distinct, positioned 5 posterior and 1 anterior; anthers obloid **5. *Odontostomum***
7. Fertile stamens 3, staminodes (posterior) 3 **6. *Tecophilaea***
- Fertile stamens 4, staminodes (latero-posterior) 2 **8. *Zephyra***

1. *Conanthera* Ruíz & Pav.

Conanthera Ruíz & Pav., Fl. Per. 3: 68, t. 301 (1802).

Herbs with tunicated corm; leaves basal, narrowly linear, glabrous, twisted and curled at anthesis; inflorescence an elongate panicle (raceme of racemes); bracts subtending only pedicels; flowers ca. 1.5 cm long; tepals blue to purple (sometimes blotched with black and white), zygomorphic; stamens 6; pollen operculate; ovary semiinferior; ovules numerous; seeds ellipsoid, laterally flattened. Five spp., Chile, open hills and grasslands.

2. *Cyanastrum* Oliv. Fig. 122

Cyanastrum Oliv., Hook. Icon. Pl. 10: t. 1965 (1891).

Herbs with nontunicated corm, bearing leaf and inflorescence separately, each surrounded by different cataphylls; leaves usually solitary, basal, petiolate, with ± cordate base; inflorescence a raceme; bracts and bracteoles sometimes absent; tepals blue; stamens 6, equal; pollen nonoperculate; ovary semiinferior, with sunken style; ovules 2 per carpel; seeds large, ovoid, yellow to brown, deeply pitted, with "chalazosperm". Three spp., tropical Africa.

3. *Cyanella* Royen ex L. Fig. 123

Cyanella Royen ex L., Gen. ed. 5: 149, et add. post index. (1754).

Herbs with tunicated corm; leaves basal, sheathing (outer leaf sheath large and enclosing other leaf bases in some spp.), lanceolate to acicular; inflorescence a panicle or flowers solitary; bracts subtending pedicels and bracteole inserted ca. 2/3 along length of pedicel; flowers ca. 2.5 cm wide; tepals white, yellowish, or violet (with dark purple bases in *C. orchidiformis*) zygomorphic (or slightly asymmetric by virtue of stylar enantiomorphy); stamens 6, dimorphic, either with 5 small posteriorly positioned stamens and 1 large anterior stamen or with 3 small posterior and 3 large anterior stamens; filaments stout, abrubtly incurved in posterior stamens; connectives of the 5 posterior stamens with black spots in *C. alba* and *C. lutea*; pollen operculate; ovary semiinferior; ovules numerous; seeds dark brown, ovoid, laterally flattened. Seven spp., S Africa.

4. *Kabuyea* Brummitt

Kabuyea Brummitt, Kew Bull. (in press) (*Cyanastrum hostifolium* Engl.)

Herbs with nontunicated corm, bearing leaf and inflorescence together, both surrounded by the same cataphyll; leaves basal, indistinctly petiolate; inflorescence a raceme; bracts sometimes absent; tepals white; stamens 6, equal; anthers dehiscing by a short clavate introrse slit near apex; pollen nonoperculate; ovary semiinferior, with sunken style; ovules 2 per carpel; seeds large, ± spherical, yellow, deeply pitted, lacking "chalazosperm". Only one sp., *K. hostifolia* (Engl.) Brummitt, tropical Africa.

5. *Odontostomum* Torr.

Odontostomum Torr., Pacif. Rail. Rep. 4: 150 (1856).

Herbs with tunicated corm; leaves basal, sheathing, very narrowly lanceolate; inflorescence a panicle (a raceme of 1–3 racemes); bracts very narrowly triangular; flowers ca. 1 cm long; tepals creamy white, slightly zygomorphic, with a relatively long (ca. 5 mm) ascending tube, minute coronal appendages arising at mouth of perianth tube between adjacent tepals; stamens 6, identical in size and shape, with 5 stamens positioned posteriorly in a semicircle and 1 stamen positioned anteriorly; pollen operculate; ovary mostly (ca. $^2/_3$) superior, slightly 3-lobed apically; ovules 2 per carpel; seeds brown, obovoid. Only one sp., California.

6. *Tecophilaea* Bertero ex Colla

Tecophilaea Bertero ex Colla, Mem. Acc. Torin. 39: 19, t. 55 (1836).

Herbs with tunicated corm; leaves two, basal, the outer a nonphotosynthetic tubular sheath, the inner photosynthetic leaf narrowly lanceolate; inflorescence with 1 or occasionally 2–3 flowers, racemose; peduncles of solitary flowers or inflorescences with a subtending bract and a subapical bracteole; flowers ca. 1.5 to 3 cm long; tepals purple, slightly zygomorphic, with constricted basal perianth tube; fertile stamens 3, anteriorly positioned, anthers with a basal caudate appendage arising from the 2 fused adaxial sporangia of adjacent thecae; staminodes 3, posteriorly positioned, with a minute terminal, adaxial cleft; pollen operculate; ovary semiinferior; ovules numerous; seeds ellipsoid. Two spp., Chile.

7. *Walleria* J. Kirk Fig. 124

Walleria J. Kirk, Trans. Linn. Soc. 24: 497. t. 52 (1864).

Herbs with nontunicated lobate corm (with a fibrous root arising from each lobe); aerial stem erect, scabrid (from inconspicuous hooks present on longitudinal ridges), foliose; cauline leaves progressively larger apically, spiral, flat, sessile (rarely ca. amplexicaul), linear to broadly lanceolate-ovate; flowers solitary in leaf axils; pedicels long, slender, scabrid; dorsal bracteole arising from pedicel near flower (at least in *W. nutans*); flowers ca. 2 cm long; perianth actinomorphic; tepals white or pink to light blue; stamens 6, equal, anthers yellow/blue or yellow/blue/yellow from base to apex; ovary almost superior (perianth tube fused to extreme base of ovary), with sunken style; pollen indistinctly operculate; ovules numerous; seeds dark brown, obloid, deeply lobed, each lobe bearing a tuft of minute hairs. Three spp., tropical and S Africa, Madagascar.

8. *Zephyra* D. Don

Zephyra D. Don, Edinb. N. Phil. J.: 236 (Oct. 1832).

Herbs with tunicated corm; leaves generally 2, basal, sheathing very narrowly lanceolate; inflorescence a narrow panicle (raceme of racemes) with bracts; flowers ca. 1.5 cm long; tepals white with blue on adaxial surfaces; perianth actinomorphic, with a short, basal perianth tube; fertile stamens 4, with short filaments, anthers with a basal pubescent, caudate appendage; staminodes 2, positioned opposite outer latero-posterior tepals, cylindrical, contorted, with a minute apical slit; pollen operculate; ovary ca. $^3/_4$ superior; seeds ellipsoid, laterally flattened. Only one sp., *Z. elegans* D. Don, Chile.

Selected Bibliography

Arroyo, S. 1986. Leaf anatomy in the Tecophilaeaceae. Bot. J. Linn. Soc. 93: 323–328.

Brummitt, R.K., Rudall, P.J., Banks, H., Johnson, M.A.T., Chase, M.W. Taxonomy of the former Cyanastraceae. Kew Bull. (submitted).

Cave, M.S. 1952. Sporogenesis and gametogenesis in *Odontostomum hartwegii*. Phytomorphology 2: 210–214.

Cave, M.S. 1970. Chromosomes of the Californian Liliaceae. Univ. Calif. Publ. Bot. 57: 1–48.

Chase et al. 1995. See general references.

Cheadle, V.I. 1969. Vessels in Amaryllidaceae and Tecophilaeaceae. Phytomorphology 19: 8–16.

Dahlgren, R., Van Wyk, A.E. 1988. Structures and relationships of families endemic to or centered in southern Africa. Mon. Syst. Bot. Mo. Bot. Gard. 25: 1–94.

De Vos, M.P. 1950. Die Ontwikkeling van die Saadknop en Saad by *Cyanella capensis*: 'n Gefal von Polyembryonie. S. Afr. J. Sci. 46: 220–226.

Dulberger, R., Ornduff, R. 1978. The breeding system in dimorphic *Cyanella alba* and *C. lutea* (Tecophilaeaceae). Bot. Soc. 27: 31–32.

Dulberger, R., Ornduff, R. 1980. Floral morphology and reproductive biology of four species of *Cyanella* (Tecophilaeaceae). New Phytol. 86: 45–56.

Endress, P.K. 1996. Diversity and evolutionary trends in angiosperm anthers. In: D'Arcy, W.G., Keating, R.C. (eds.) The anther. Cambridge: Cambridge University Press, pp. 92–110.

Engler, A. 1901. Beiträge zur Flora von Afrika XX: Berichte über die botanischen Ergebnisse der Nyassa-See- und Kinga-Gebirgs-Exped., Gyanastraceae. Bot. Jahrb. 28: 357.

Erdtman, G. 1952. See general references.

Fries, T.C.E. 1919. Der Samenbau bei *Cyanastrum*. Sven. Bot. Tidskr. 13: 295–304.

Gibbs, R.D. 1974. Chemotaxonomy of flowering plants. London: McGill-Queen's University Press.

Harris, P.J., Hartley, R.D. 1980. See general references.

Huber, H. 1969. See general references.

Hutchinson, J. 1934. The families of flowering plants. II. Monocotyledons. Oxford: Clarendon Press.

LaCour, L.F. 1956. In: Darlington, C.D., Wylie, A.P. (eds.) Chromosome atlas of flowering plants. London: Macmillan.

Nietsch, H. 1941. Zur systematischen Stellung von *Cyanastrum*. Oesterr. Bot. Z. 90: 31–52.

Ornduff, R. 1979. Chromosome numbers in *Cyanella* (Tecophilaeaceae). Ann. Mo. Bot. Gard. 66: 581–583.

Radulescu, D. 1973. La morphologie du pollen chez quelques Haemodoraceae. Lucr. Gradinii Bot. Bucur. 1972–73: 123–132.

Rudall, P. 1997. The nucellus and chalaza in monocotyledons: structure and systematics. Bot. Rev. 63: 140–181.

Rudall, P., Caddick, L.R. 1994. Investigation of the presence of phenolic compounds in monocot cell walls, using UV fluorescence microscopy. Ann. Bot. 75: 483–491.

Rudall, P. et al. 1997. See general references.
Satô, D. 1942. Karyotype alteration and phylogeny of Liliaceae and allied families. Jpn. J. Bot. 12: 57–161.
Schulze, R. 1893. Beiträge zur vergleichenden Anatomie Liliaceen, Haemodoraceen, Hypoxidaceen und Velloziaceen. Bot. Jahrb. Syst. 17: 295–394.
Schulze, W. 1983. Beiträge zur Taxonomie der Liliifloren. XI. Tecophilaeaceae und Cyanastraceae. Wiss. Z. Friedrich-Schiller Univ. Math.-Naturwiss. Reihe 32: 957–964.
Sen, S. 1975. Cytotaxonomy of Liliales. Feddes Repert. 86: 255–305.
Shneyer, V.S. 1983. The relationships between the Iridaceae and Liliaceae s.l. as revealed by the serological analysis of seed proteins. Bot. Zh. (Moskau) 68: 49–54 (in Russian).
Simpson, M.G. 1983. Pollen ultrastructure of the Haemodoraceae and its taxonomic significance. Grana 22: 79–103.
Simpson, M.G. 1985. Pollen ultrastructure of the Tecophilaeaceae. Grana 24: 77–92.
Slob, A. 1973. Tulip allergens in Alstroemeria and some other Liliiflorae. Phytochemistry 12: 811–815.
Sterling, C. 1974. Comparative morphology of the carpel in the Liliaceae: *Baeometra*, *Burchardia* and *Walleria*. Bot. J. Linn. Soc. 68: 115–125.
Takhtajan, A.L. 1982. See general references.
Tillich, H.-J. 1995. Früchte, Samen und Keimpflanzen bei den Cyanastraceae Engler 1900 und einigen vermuteten Verwandten. Feddes Repert. 106: 483–493.
Zavada, M.S., Scott, G. 1993. Pollen morphology of *Cyanella* species (Tecophilaeaceae). Grana 32: 189–192.

Themidaceae

K. Rahn

Themidaceae Salisb., Gen. Pl.: 85 (1866).
Alliaceae subfam. Brodiaeoideae Traub (1972).

Acaulescent or short-stemmed perennial *geophytes* without alliaceous odour; stem swollen, forming a starch-storing corm; shoot growth monopodial; each shoot forming a new corm every year on top of and from apex of the corm of the previous year; green leaves and scapes appearing from the upper part of the young corm; the dried leaf bases of previous year enveloping the young corm and forming a membranous, reticulate, or fibrous tunic. *Leaves* spirally arranged, linear, flat, angular, terete, or fistular, often fleshy, forming closed sheaths below; veins parallel. Scape usually solitary, from a leaf axil near the top of the new corm. *Inflorescence* an umbel, rarely reduced to a single flower. Pedicels often articulated at base and/or at apex, and subtended by an individual small bract (2 when the flower solitary). *Flowers* hermaphroditic, actinomorphic. Tepals 3 + 3, usually ± united below. Three or 6 stamens fertile, the missing ones often transformed into staminodes. Filaments inserted on the tepals, free from each other or united, often with lateral, dorsal, or apical appendices. A corona, scales or appendices often present between tepals and stamens. Anthers versatile except in *Brodiaea* and *Dichelostemma*, ± basifixed, introrse, opening with longitudinal slits. Ovary superior, sometimes on a gynophore, tricarpellate, trilocular. Style solitary, erect, at apex of ovary. Stigma capitate or trilobate with a Dry or sometimes Wet (*Bloomeria*) surface. Two or several ovules in each locule. Ovules anatropous. *Fruit* a loculicidal capsule with few to numerous seeds. Seeds 2–several in each loculus, angular or semiovoid (triangular in transection). Testa usually with a thick crust of phytomelan. Endosperm with fatty oils and aleuron, but no starch. Embryo short and straight.

Comprising 12 genera and about 60 species in western N America, a single species reaching Guatemala.

Vegetative Morphology. The leafy part of the stem is short in all species, and transformed to a

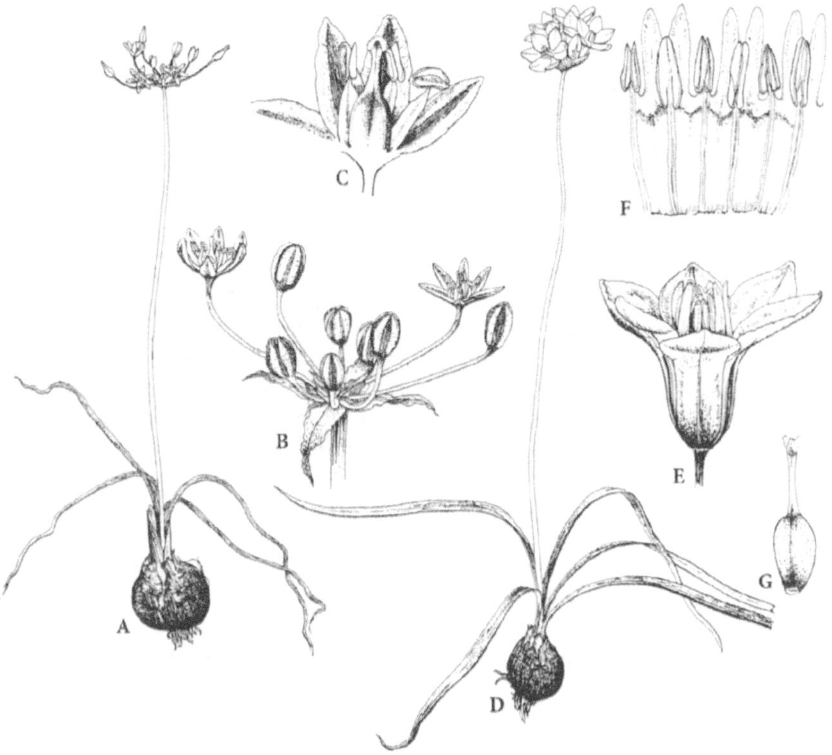

Fig. 125A–G. Themidaceae. A–C *Muilla transmontana*. A Flowering plant. B Pseudoumbel. C Flower. D–G *Dichelostemma pulchellum*. D Flowering plant. E Flower, side view. F Corona with stamens. G Pistil. (Cronquist et al. 1977)

usually depressed globose corm formed every year on top of that from the previous year (a monopodium). Green leaves appear from the upper half of the young corm, and scapes are axillary. Sheathing cataphylls envelope the young bud and are later transformed into a fibrous coat which covers the corm (Fig. 125). Beside the new apical, main corm other smaller corms are formed laterally from axils on top of the old corm (Smith 1930). The increase corm in *Triteleia* produces a single contractile root which pulls the increase corm away from the mother corm and deeper into the soil. Leaves are usually linear, flat, angular, terete or fistular (*Milla*, *Bessera*).

Vegetative Anatomy. Lacticifers are present in the two representatives of the family that have been tested, *Brodiaea* and *Milla* (Hegnauer 1963). Raphides of calcium oxalate are recorded as being present in *Brodiaea* and absent in *Milla*. The stomata are anomocytic.

Inflorescence Structure. The inflorescence is a condensed monochasium (umbel). The pedicels are subtended by scarious bracts, the lower or outermost being the largest, but when the umbel is reduced to a single flower, 2 bracts are present. The pedicels of many species are articulated basally or distally.

Flower Structure. A gynophore free from the tepal tube is developed in *Triteleia* and *Triteleiopsis*, and united with the tepal tube by 3 flanges in *Milla*, *Bessera* and *Petronymphe*.

Embryology. In *Muilla* and *Brodiaea* the embryo-sac development follows the Polygonum type; the endosperm is Nuclear in *Brodiaea* and Helobial in *Muilla* (Berg and Maze 1966; Berg 1978). Helobial endosperm was also recorded for *Triteleia* by Stenar (1949). The pollen grains are 2-celled when dispersed.

Pollen Morphology. According to Schulze (1980) all genera studied (*Muilla*, *Milla*, *Triteleia*,

Brodiaea and *Dichelostemma*) have sulcate pollen with a reticulate exine, their mean length ranging from 46–93 μm. In *Milla*, *Triteleia*, *Brodiaea* and *Dichelostemma* the sulcus reaches the proximal side of the pollen grain.

KARYOLOGY. Known chromosome numbers include for *Muilla* $2n = 14, 16$, and 20; *Bessera* $2n = 16$; *Bloomeria* $2n = 18$; *Dandya* $2n = 18$; *Milla* $2n = 14, 18, 20, 28, 42$; *Triteleia* $2n = 10, 14, 16, 18, 24, 28, 30. 32, 35, 42, 48, 50$ and 84; *Triteleiopsis* $2n = 33$; *Brodiaea* $2n = 12, 14, 24, 32, 36, 42$ and 48; *Dichelostemma* $2n = 18, 27, 30, 36, 45, 48, 54$ and 72; unknown in *Petronymphe*, *Androstephium* and *Jaimehintonia*. The wide range of aneuploidy makes speculations on a plausible base number impossible.

POLLINATION. All or most species are pollinated by insects. Septal nectaries are probably present in most or all species (seen in *Bessera* with the opening in the upper part of the ovary). In *Bloomeria* a small nectariferous cup is present at the base of each filament.

FRUIT AND SEED. The fruit is a loculicidal capsule, with 2 to several seeds in each of the 3 locules. Seeds are angular and more or less isodiametric, 1.7–4 mm long, and have a short and thick embryo. The epidermis of the testa has a crust of phytomelan 10–28 μm thick. The epidermis cells are isodiametric or slightly elongate. The inner layers of the testa are compressed or collapsed, as is the tegmen (Huber 1969). Endosperm is copious.

REPRODUCTIVE SYSTEMS. Vegetative propagation is restricted to the formation of increase corms.

PHYTOCHEMISTRY. (Hegnauer 1963, 1986). Saponins are recorded from *Brodiaea*, and not recorded as being absent in any species of Themidacae. Alliaceous odour is absent. Alkaloids are absent. Starch is present in the corms, but absent from the endosperm, where aleuron and fat are present.

SUBDIVISION AND RELATIONSHIPS WITHIN THEMIDACEAE. Traub (1963) divided Amaryllidaceae subfam. Allioideae into four tribes: Miluleae, Allieae, Gilliesieae and Agapantheae. The Allieae were divided into three subtribes: Alliinae, Brodiaeinae (with *Muilla*, *Bloomeria*, *Triteleiopsis*, *Triteleia*, *Dichelostemma* and *Brodiaea*), and Millinae (with *Dandya*, *Bessera*, *Petronymphe* and *Milla*).

AFFINITIES. After the rejection of *Gagea* (Schnarf 1948) the Alliaceae (including Themidaceae and *Agapanthus*) for long were considered an indisputably monophyletic taxon. The superior ovary placed it in Liliaceae, until Hutchinson (1934), mainly on account of the umbellate inflorescence, placed it in his Amaryllidaceae. Dahlgren et al. (1985) placed the Alliaceae in the Asparagales together with the Hyacinthaceae, Amaryllidaceae and 28 other families. The Alliaceae were separated from the Amaryllidaceae by having a superior ovary (like most other Asparagales), steroidal saponins, and lacking alkaloids. Based on a parsimony analysis of *rbc*L sequences Fay and Chase (1996) found that "Alliaceae sensu Dahlgren et al. were polyphyletic, with *Agapanthus* sister to Amaryllidaceae, and Brodiaeeae (with the exception of *Petronymphe* H.E. Moore) being more closely related to Hyacinthaceae . . . , than to the Amaryllidaceae-Alliaceae complex." Later, M.F. Fay (pers. comm.) announced that the "*Petronymphe*" proved to be a misidentified *Echeandia*.

DISTRIBUTION AND HABITATS. The Themidaceae are found in different habitats from sea level to 3000 m in southwestern Canada, the western United States and Mexico, and with a single species reaching Guatemala.

ECONOMIC IMPORTANCE. *Milla*, *Brodiaea*, *Triteleia* and *Dichelostemma* are sometimes grown as ornamentals.

KEY TO THE GENERA

1. Gynophore absent or short and free from the tepal tube; stamens 3 or 6 — 2
– Gynophore present; 6 fertile stamens — 7
2. Tepals free or shortly united below; stamens 6, free from each other; corona absent — 3
– Tepals united below, forming a distinct tube — 5
3. Filaments free — 4
– Filaments united at the base to form a shallow cup; perianth blue, lavender or white. Mexico — **3. *Dandya***
4. Filaments dilated at the base, without a cup or wing; leaves ± terete; pedicel not jointed at the summit; flowers white or greenish white. Western USA and Mexico — **1. *Muilla***
– Each filament winged or with a small nectariferous cup at base; leaves linear, carinate; pedicel jointed at the summit; flowers yellow. California and Baja California — **2. *Bloomeria***
5. Filaments free or slightly united below; appendices present in some species; fertile anthers usually 3, ± basifixed — 6
– Filaments united to form a tube with erect bifid appendices between the 6 versatile anthers. Western USA and Mexico — **12. *Androstephium***

6. Stigma with 3 recurving lobes; flowers in open umbels with stiff pedicels; leaves without keels or distinct vein impressions. Western N America **10. *Brodiaea***
- Stigma with 3 short lobes; flowers in dense umbels with flexuous pedicels; leaves with distinct keel and vein impressions. Western USA and Mexico **11. *Dichelostemma***
7. Gynophore united with the perianth tube by 3 flanges 8
- Gynophore free from perianth tube; leaves solid 10
8. Leaves solid; cross-section of leaf triangular; perianth yellowish green, ± 60 mm long. Mexico **6. *Petronymphe***
- Leaves fistulose 9
9. Stamens free from each other, inserted in throat, shorter than tepal lobes; perianth white, pink or blue, 25–220 mm long. Guatemala, Mexico, southern USA **4. *Milla***
- Stamens united below, forming a cup; stamens longer than tepal lobes; perianth red or violet, 18–35 mm long. Mexico **5. *Bessera***
10. Corona scales absent, anthers versatile 11
- Tepal segments alternating with coronar scales, anthers basifixed. Baja California **9. *Triteleiopsis***
11 One or 2 leaves, 2–25 mm wide. Filaments free from each other, often widened and with lateral appendices. Western N America **7. *Triteleia***
- More than 3 leaves, 1–1.5 mm wide. Filaments filiform or narrowly triangular, shortly united below. NE Mexico **8. *Jaimehintonia***

Genera of the Themidaceae

1. *Muilla* S. Watson ex Benth. Fig. 125A–C

Muilla S. Watson ex Benth. in Benth. & Hook., Gen. 3: 801 (1883); Ingram, Madroño 12: 19–27 (1953), rev.

Corm with a fibro-membranous coat. Leaves ± terete. Flowers not jointed. Perianth segments almost free, greenish or yellowish with a dark, 2- or 3-nerved midrib. Stamens 6, fertile, almost free. Filaments petaloid or filiform, thickened towards the base. Anthers versatile. Seeds 1–10 in each locule. $2n = 14$, 16 or 20. Four spp., Baja California, California and Nevada.

2. *Bloomeria* Kellogg

Bloomeria Kellogg, Proc. Calif. Acad. 2: 11 (1863); Ingram, Madroño 12: 19–27 (1953), rev.

Corm with a fibrous coat. Leaves linear, carinate. Pedicels subtended by membranous bracts. Flowers jointed, yellow. Perianth segments 6, free of each other, Stamens 6, fertile. Filaments free, at base winged or with a nectariferous cup. Anthers versatile. Each locule with 1-several seeds. $2n = 18$. Two spp., California and Baja California.

3. *Dandya* H.E. Moore

Dandya H.E. Moore, Gentes Herb. 8: 266–268 (1953); Lenz, Aliso 7: 313–320 (1971).

Corm with membranous tunics of parallel fibres. Leaves grasslike. Scape longer than leaves. Flowers articulate at base. Perianth tube 1 mm. Stamens 6, anthers versatile. Stigma 3-lobed. Ovary with 8–10 ovules. $2n = 18$. Four spp., Mexico.

4. *Milla* Cav.

Milla Cav., Icon. 2: 76 (1793 or 1794); Moore, Gentes Herb. 8: 278–294 (1953), rev.
? *Diphalangium* Schauer in Nees & Schauer, Linnaea 19: 702 (1847).

Corm with membranous tunics of parallel fibres. Leaves 2–7, linear, fistular. Pedicels not jointed. Perianth white, pink or blue, with tube longer than lobes. Stamens inserted in the throat, filaments distinct, filiform. Anthers basifixed. Stigma 3-lobed. $2n = 18, 20, 42$. Seven spp., southern United States, Mexico and Guatemala. The genus *Diphalangium* with one sp. in Mexico is dubious, and probably either related to or a synonym of *Milla*.

5. *Bessera* Schult. f.

Bessera Schult. f., Linnaea 4: 121 (1829), nom. cons.; Moore, Gentes Herb. 8: 268–277 (1953), rev.
Pharium Herbert (1832).
Behria E. Greene (1886).

Corm with membranous tunics of parallel fibres. Leaves 1 or 2, linear, planoconvex or terete, fistular. Scape hollow. Flowers articulate at base, red or violet. Filaments inserted on the perianth tube, united with each other below, forming a short tube. Anthers dorsifixed; 20–30 ovules in each locule. $2n = 16$. Two spp., Mexico.

6. *Petronymphe* H.E. Moore

Petronymphe H.E. Moore, Gentes Herb. 8: 258, fig. 104 (1951).

Corm with membranous tunics of parallel fibers. Leaves 3, narrow, cross section triangular, margins minutely denticulate. Scape slender. Flowers articulate with the pedicels. Tepal tube long, tubular, segments 6, short. Stamens 6, inserted at the throat. Anthers versatile. Capsule long, with many seeds in each locule. Only one sp., Mexico.

7. *Triteleia* Douglas ex Lindley

Triteleia Douglas ex Lindley, Bot. Reg. under t. 1293 (1830); Hoover, Am. Midl. Nat. 25: 73–100 (1941); Niehaus, The Four Seasons 6: 17–21 (1980), rev.
Calliprora Lindley (1833); Lenz, Aliso 8: 221–253, 353–377 (1975/76), rev.
Themis Salisb. (1866).

Corm with coarsely fibrous-reticulate coat. Scape slender, with 1 or 2 plane, linear leaves. Bracts scarious, acuminate. Pedicels sometimes jointed. Perianth tube of varying length. Stamens 6, all functional. Filaments variously inserted, variously dilated, and sometimes with 2 teeth. Anthers versatile. Ovary stipitate. $x = 7$ (and 8?). Eighteen spp. British Columbia, western USA and Mexico.

8. *Jaimehintonia* B. Turner

Jaimehintonia B. Turner, Novon 3: 86–88 (1993).

Corm with coarsely fibrous coat. Leaves filiform. Flowers 3–6. Perianth tube 16–20 mm, free part of tepals 10–12 mm long. Stamens 6, equal. Filaments united for 1 mm below. Anthers versatile. Ovary stipitate. Only one sp., NE Mexico.

9. *Triteleiopsis* Hoover

Triteleiopsis Hoover, Am. Midl. Nat. 25: 98 (1941).

Corm with a fibrous coat. Stem stout, with 3–8 leaves at base. Leaves keeled above the expanded base, with cormlets in the axiles. Bracts scarious, acuminate. Pedicels jointed. Perianth tube equalling the lobes in length. Perianth segments alternating with scales (corona). Stamens 6. Filaments papillate, inserted in the upper part of the perianth tube, free of each other. Anthers basifixed. Ovary stipitate. $2n = 33$. Only one sp., Baja California and Arizona.

10. *Brodiaea* Smith

Brodiaea Smith, Trans. Linn. Soc. Lond., 10: 2 (1811), nom. cons.; Niehaus, Univ. Calif. Publ. Bot. 60: 1–67 (1971), and in The Four Seasons 6: 11–21 (1980), rev.
Hookera Salisb. (1808).

Corm with coarsely fibrous-reticulate coat. Scape slender. Leaves linear, channelled. Pedicels subtended by membranous bracts. Pedicels sometimes articulate at base. Tepals 6, united below into a tube. Stamens 3, opposite the inner perianth segments, alternating with staminodia (these absent in one species), inserted in the perianth tube. Anthers basifixed. Stigma with 3 lobes. Ovary not stipitate. $x = 6$ and 8. Fifteen spp., southern British Columbia and western United States.

11. *Dichelostemma* Kunth Fig. 125D–G

Dichelostemma Kunth, Enum. Pl. 4: 469 (1843); Hoover, Am. Midl. Nat. 24: 463–476 (1940), rev., and in Plant Life (Herbertia ed.) 11: 13–24 (1955); Niehaus, The Four Seasons 6: 11–21 (1980); Keator, The Four Seasons 9: 24–39 (1991).

Corm fibrous-coated. Leaves linear, keeled below. Pedicels jointed to the perianth. Perianth with a ± inflated tube, 6 lobes of the same length. Outer stamens with reduced anthers or staminodial, the inner 3 anthers sessile on the throat, basifixed, with free lanceolate appendages, or (acc. to Lenz 1976) the appendages derived from the perianth, i.e. a corona. Style enlarged upwards. Seeds 3–4 in each locule. $x = 9$. Five spp., western United States and adjacent Mexico.

12. *Androstephium* Torr.

Androstephium Torr., Bot. Mex. Bound. 218 (1859); Uphof, Herbertia 11: 250–251 (1944).

Corm with fibrous coat. Leaves linear, channelled. Scape slender. Inflorescence with few flowers. Pedicels not jointed at the summit. Tepals 6, united below. Stamens 6, inserted in the throat, filaments united below and with bifid lobes between the versatile anthers. Two spp., southwestern United States.

Selected Bibliography

Berg, R.Y. 1978. Development of ovule, embryo sac, and endosperm in Brodiaea (Liliales). Norw. J. Bot. 25: 1–7.
Berg, R.Y., Maze, J.R. 1966. Contribution to the embryology of *Muilla* with a remark on the taxonomic position of the genus. Madroño 18: 143–151.
Cronquist, A., Holmgren, A.H., Holmgren, N.H., et al. 1977. Intermountain flora, vol. 6. New York: Columbia University Press.
Dahlgren, R.M.T. et al. 1985. See general references.
Fay, M.F., Chase, M.W. 1996. Resurrection of Themidaceae for the Brodiaea alliance, and recircumscription of Alliaceae, Amaryllidaceae and Agapanthoideae. Taxon 45: 441–451.
Hegnauer, R. 1963, 1986. See general references.
Huber, H. 1969. See general references.
Hutchinson, J. 1934. The families of flowering plants, Vol. 2 Monocotyledons. Oxford: Oxford University Press.
Lenz, L.M. 1976. The nature of the floral appendages in four species of *Dichelostemma* (Liliaceae). Aliso 8: 383–389.
Schnarf, K. 1948. Der Umfang der Lilioideae im natürlichen System. Österr. Bot. Z. 95: 257–269.
Schulze, W. 1980. Beiträge zur Taxonomie der Liliifloren. V. Alliaceae. Wiss. Zeitsch. Friedrich-Schiller-Univ. Jena

Math.-Naturwiss. Reihe 29: (Beiträge zur Phytotaxonomie 8:) 595–606.
Smith, F.H. 1930. The corm and contractile roots of *Brodiaea lactea*. Am. J. Bot. 17: 916–927.
Stenar, H. 1949. Zur Kenntnis der Embryologie und der Raphiden-Zellen bei *Bowiea volubilis* Harvey und anderen Liliazeen. Acta Horti Bergiani 15 (1951): 45–63.
Traub, H.P. 1963. The genera of Amaryllidaceae. La Jolla, CA: The American Plant Life Soc., pp. 15–43.

Trichopodaceae

H. Huber

Trichopodaceae Hutch., Fam. Fl. Pl. 2: 143 (1934), nom. cons.

Shortly rhizomatous *perennials*, unarmed throughout. Aerial shoots long-lived, elongate, twining to the right, or short and then foliage reduced to a single leaf. *Leaves* alternate, petiolate, the petiole with a proximal pulvinus only or lacking pulvini; blade lanceolate to cordate-ovate, principal veins palmate, acro- or camptodromous, secondary venation reticulate. *Flowers* perfect, epigynous, moderately small to medium-sized, long and slenderly pedicellate, solitary and axillary or arranged in axillary or terminal fascicles or short racemes; pedicel not articulated. Perianth regular, trimerous, 2-whorled, more or less campanulate; tepals similar; stamens 6, arranged in 2 cycles; filaments distinct, shorter than the anthers, these short, 2-thecate and tetrasporangiate, introrsely dehiscent, the thecae separated by a wide connective, the latter produced into a lanceolate appendage as long as or longer than the thecae and connivent over the stigma; thecae of adjacent stamens meeting with their edges. Ovary syncarpous, 3-locular at anthesis, style short, stigma 3-lobed, the lobes shortly bifid; placentation axile, ovules anatropous, bitegmic, tenuinucellate (at least in *Trichopus*), 2 and superposed in each loculus. *Fruit* 1-to 6-seeded, dry or slightly fleshy, indehiscent or opening by irregular ruptures (Bouman 1995), with three longitudinal keels or wings. Seeds ovoid or oblong, more or less compressed, unwinged, deeply and irregularely furrowed, the embryo minute, with a terminal cotyledon and embedded within abundant, hard, starch-free endosperm; both integuments 2-layered, and both contributing to the mechanical tissue.

A family of two monotypic genera, one in E Madagascar, the other on the S Deccan Peninsula, in Ceylon, and peninsular Malaya.

Vegetative Morphology. According to Burkill (1960), *Trichopus* preserves "through life the habit of the 1st-year seedling or *Dioscorea*, which lies in the arrest of the second leaf of the stem to the advantage of the first". In contrast to Dioscoreaceae (incl. *Stenomeris*), the petioles lack a distal pulvinus (*Avetra*) or pulvini at all (*Trichopus*).

VEGETATIVE ANATOMY. In that respect, Trichopodaceae agree largely with Dioscoreaceae but tannin cells are lacking and the number of phloem units at the abaxial side of the midrib and the petiole bundles tends to be larger (5–6 in *Avetra*, up to 9 in *Trichopus*) than in most Dioscoreaceae incl. *Stenomeris* (Ayensu 1972). *Avetra* differs from most Dioscoreaceae by the entire absence of foliar glands, whereas *Trichopus* bears glands with an 1-celled stalk and a crescent-shaped head consisting of several cells horizontally arranged in 1 row, but stretched transversely to the row which they form. The rhizome of *Trichopus* contains starch, but the starch type is not reported (Ayensu 1972).

ULTRASTRUCTURE. The epicuticular wax crystalloids of *Trichopus* agree with the Dioscoreaceae, but the sieve-tube plastids do not: in contrast to the latter family, they lack starch grains and correspond to those of most monocotyledons (Behnke 1981).

FLORAL STRUCTURE. The connective appendages connivent over and encaging the style (Fig. 126, 127) are a salient feature.

EMBRYOLOGY. This is only known from *Trichopus*. Unlike the Dioscoreaceae, microsporogenesis is successive; the ovules are bitegmic (both genera) and - in contrast to Dioscoreaceae - tenuinucellate; a parietal cell is not cut off. Embryo-sac formation follows the Polygonum type and the endosperm is initially Nuclear; as to embryogeny, Dahlgren and Clifford (1982) state that the Solanad type "is not recorded with certainty in the monocotyledons although suspected to occur in *Trichopus*, occurring also in Aristolochiaceae among the dicotyledons with several monocotyledonous features".

POLLEN MORPHOLOGY. The pollen of *Avetra* is 4(-5)-foraminate and spinulose (Erdtman 1952), while that of *Trichopus* is sulcate (Nagarjuna Rao 1955), as in several Dioscoreaceae.

KARYOLOGY. *Avetra* (G. Petersen, pers. comm.) and *Trichopus* (Ramachandran 1968) share the chromosome number $2n = 28$.

POLLINATION. *Trichopus* seems to be autogamous, at least in cultivation (H. Huber, pers. obsers.).

Fig. 126A–F. Trichopodaceae. *Avetra sempervirens*. **A** Twining shoot. **B** Androecium. **C** Stamens, ventral and side view, showing the ventral wing and terminal appendage. **D** Style with 3 bifid stigmatic lobes. **E** Fruit. **F** Cerebriform seed with adherent carpel bundles. (Takhtajan 1982)

SEED STRUCTURE. The seeds, deeply furrowed and cerebriform, are very distinctive. A crystalliferous endotesta was observed by H. Huber (1969) in *Trichopus* but is absent in the seeds studied by Bouman (1995), who found both layers of the testa collapsed in the mature seed; no crystals

Fig. 127A–H. Trichopodaceae. *Trichopus zeylanicus*. **A** Fruiting plant. **B** Condensed raceme with 2 pedicels. **C** Flower. **D** Flower, longitudinal section. **E** Stamen. **F** Style with stigmatic lobes. **G** Ovary, cross-section. **H** Seed in surface view and cross-section. (Takhtajan 1982)

have ever been seen in *Avetra*. In both genera the exotegmic cells are elongate, more strongly so in *Avetra* than in *Trichopus*, and are differentiated as a mechanical layer (Bouman 1995). Contrary to the Liliiflorae proper, well-defined pits are absent from the massive cell walls of the endosperm.

DISPERSAL. According to Burkill (1960), "the dispersal of the seeds (sc. of *Trichopus*) is undoubtedly by sudden floods which break the fruit from its slender peduncle and carry it away"; however, the fruiting pedicel, although slender, is tough and wiry and the fruit does not readily separate from it.

AFFINITIES. The rhizomes, perfect flowers and appendaged anthers (to which the non-pulvinate petioles could be added) led Burkill (1960) to view *Avetra*, *Stenomeris* and *Trichopus* as the most archaic representatives of Dioscoreaceae. At least *Trichopus*, however, which is the only of the three genera whose embryology is known, presents several progressions, such as the tenuinucellate ovules and an archesporial cell not cutting off a parietal cell. This shows that this genus is heterobathmic rather than conservative. Also the deeply furrowed seeds and the 2-layered testa can hardly be claimed to be plesiomorphic. Apart from their chromosome number, otherwise not known in Dioscoreales, *Trichopus* and *Avetra* share so many apomorphic flower and seed characters that a family Trichopodaceae can be satisfactorily defined.

USES. According to Pushpangadan et al. (1988), the unripe fruits of *Trichopus* are considered a highly rejuvenative tonic comparable to ginseng, used by the Kani tribe of Kerala to avoid fatigue.

KEY TO THE GENERA

1. Aerial stem twining, with numerous leaves. Perianth pure white, 1.5–3 cm long. Fruit 1-seeded. Seed glabrous **1. *Avetra***
– Aerial stems short, 1-leaved. Perianth dull, up to 1 cm long. Fruit up to 6-seeded. Seeds minutely puberulous **2. *Trichopus***

1. *Avetra* H. Perr. Fig. 126

Avetra H. Perr., Bull. Soc. Bot. Fr. 71: 25 (1925), and in Fl. Madag., Fam. 44^{e} bis: 1–3 (1950).

Wholly glabrous. Aerial stem twining to the right; leaves numerous, blade cordate. Flowers in short, few-flowered axillary racemes or more frequently the latter reduced to a solitary flower; racemes with numerous small chaffy bracts. Perianth segments linear-lanceolate, 1.5–3 cm long, shortly connate, pure white. Anther vertically winged on inner side between the thecae. Fruit 1-seeded, longitudinally 3-winged, 4–5 cm long., dry at maturity; seed glabrous. Only one sp., *A. sempervirens* H. Perr., E Madagascar.

2. *Trichopus* Gaertner Fig. 127

Trichopus Gaertner, Fruct. 1: 44 (1788); Sivarajan et al., Kew Bull. 45: 353–360 (1990), rev.

Minutely glandular-pubescent, at least on ovary and under surface of leaves. Rhizome short, simple or occasionally branched, densely covered with chaffy lanceolate scales. Aerial stems erect or ascending, 5–15 cm tall, axillary from rhizome scales, bearing 1 petioled leaf and terminating in a short raceme densely bracteate by scales resembling those of the rhizome but distichous and shorter. Perianth segments ovate-lanceolate, 3–10 mm long, shortly connate, dull purple or greenish. Anther not winged between thecae. Fruit up to 6-seeded, longitudinally 3-keeled, 1–1.5 cm long, slighty fleshy at maturity; seeds minutely puberulous. Only one sp., the variable *T. zeylanicus* Gaertner, SW Deccan Peninsula, Ceylon and peninsular Malaya.

Selected Bibliography

Ayensu, E.S. 1966. Taxonomic status of *Trichopus*: anatomical evidence. J. Linn. Soc. Bot. 59: 425–430.
Ayensu, E.S. 1972. Dioscoreales. In: Metcalfe C.R. (ed.) Anatomy of the monocotyledons, Vol. 6. Oxford: Clarendon Press.
Behnke, H.-D. 1981. Sieve-element characters. Nord. J. Bot. 1: 381–400.
Bouman, F. 1995. Seed structure and systematics in Dioscoreales. In: Rudall, P.J., Cribb, P.J., Cutler, D.F., Humphries, C.J. (eds.) Monocotyledons: systematics and evolution, Vol. 1. London: Royal Botanic Gardens, Kew, pp. 139–156.
Burkill, I.H. 1960. The organography and the evolution of Dioscoreaceae, the family of the yams. J. Linn. Soc. Bot. 56: 319–412.
Dahlgren, R., Clifford, H.T. 1982. See general references.
Erdtman, G. 1952. See general references.
Kale, N.N., Pai, R.M. 1979. The floral anatomy of *Trichopus zeylanicus* Gaertn. Proc. Indian Acad. Sci. B Plant Sci. 88: 63–67.
Nagarjuna Rao, A. 1955. Embryology of *Trichopus zeylanicus* Gaertn. J. Indian Bot. Soc. 34: 213–221.
Perrier de la Bâthie, H. 1950. Famille 44e bis: Trichopodacées. In: H. Humbert, Fl. de Madagascar. Paris: Muséum d'Histoire naturelle.
Pushpangadan, P., Rajasekharan, S., Ratheesh Kumar, P.K., Jawahar, C.R., Velayudhan Nair, V., Lakshmi, N., Sarada Amma, L. 1988. "Arogyapacha" (*Trichopus zeylonicus* Gaertn.). The Ginzeng of Kani tribes of Agasthyar Hills (Kerala) for evergreen health and vitality. Ancient Sci Life 8: 13–16.
Ramachandran, K. 1968. Cytological studies in the Dioscoreaceae. Cytologia 33: 401–410.
Sivarajan, V.V., Pushpangadan, P., Ratheesh Kumar, P.K. 1990. A revision of *Trichopus*. Kew Bull. 45: 353–360.
Takhtajan, A.L. 1982. See general references.

Trilliaceae

M.N. Tamura

Trilliaceae Lindl., Veg. Kingd.: 218 (1846), nom. cons.

Perennial *herbs*. Underground stem a rhizome, stout to slender, usually creeping, rarely erect, monopodial. Aerial stem simple, often glabrous, sometimes pubescent. Foliage *leaves* 3–22 in a pseudowhorl at stem apex, sessile to petiolate, narrowly to broadly ovate to obovate, at the base rounded, or sometimes cordate or tapering, sometimes variegated, glabrous or pubescent along main veins on abaxial surface. *Flowers* sessile to pedicellate, 3(–4–5)-merous, actinomorphic, perfect, solitary, terminal. Sepals 3 or sometimes 0 or 4–10, free, ovate to oblong, rarely clawed, green or sometimes maroon, yellow, yellowish green, or white, sometimes adaxially streaked or mottled with purple, maroon, or silver-white, ascending to reflexed, imbricate or contorted, persistent. Petals 3 or sometimes 0 or 4–8, free, imbricate or contorted, persistent, broadly ovate to filiform, rarely clawed, erect to spreading, maroon or sometimes pink, yellow, greenish yellow, green, or white. Stamens 6(7–24), in 2–6 whorls, persistent; anthers linear, basifixed, often with prolonged connective, dehiscing by longitudinal slits, extrorse, latrorse or introrse. Carpels 3(4–10), connate for more than 1/3 of their entire length; ovary superior; ovules numerous to several; styluli free or connate into a common style, linear to thickly subulate, erect, spreading or recurved, usually persistent, stigmatic on adaxial side, dry. *Fruit* baccate or capsular, usually green, maroon, dark purple or blackish, rarely white, yellow, or red. *Seeds* sometimes provided with an aril or a scarlet sarcotesta. $x = 5$.

A family comprising 4 genera and about 70 species, distributed from subarctic to subtropical regions of the Northern Hemisphere.

Vegetative Morphology and Anatomy. The foliage leaves are arranged in a pseudowhorl of 3 or more leaves at the apex of the aerial stem. *Trillium* sometimes lacks an aerial stem and possesses only a subterranean axis (*T. petiolatum*). Berg (1958) suggested that in this case the reduction of the aerial stem is correlated to the myrmecochorous seed dispersal. Kawano and Kato (1995), on the other hand, consider the reduction of the aerial stem of *T. petiolatum* as an adaptation to a dry and hot climate during the vegetation period and to volcanic activity.

According to Zomlefer (1996), the roots are contractile. Vessels are confined to the roots and have scalariform perforation plates. In the axes, the vascular strands are often arranged in 3 cylinders. Leaf venation is palmatireticulate with 3 or 5 (or 7) acrodromous primary veins, which are interconnected by reticulate venation. Conover (1991) found the epidermis structure of *Paris quadrifolia* and *Trillium erectum* very similar. The adaxial epidermal cells are very large, isodiametric, and have sinuous anticlinal cell walls. The stomata are restricted to the abaxial leaf surface and predominantly are oriented parallel to the midrib; they are surrounded by 4 stomatal contact cells. The guard cells are very large.

INFLORESCENCE STRUCTURE. The inflorescence is reduced to a single flower, which terminates the aerial stem (Fig. 128, 129).

FLOWER MORPHOLOGY AND ANATOMY. The perianth of the Trilliaceae is differentiated in sepals and tepals, which differ in size, shape, texture, and color. Septal nectaries and raphides are often present.

The ovary is 3(–10)-locular, 1-locular, or 3(–10)-locular below and 1-locular above, angular (*Daiswa*, *Trillium*) or nonangular (*Paris*, *Kinugasa*), sometimes topped by a disk or a ring of teeth (*Kinugasa*, *Daiswa*, *Trillium govanianum*). Placentation is axile in several-loculed ovaries or ovary portions of ovaries, and parietal in the 1-locular ovaries or ovary portions.

A prominent characteristic of the *Trillium* flower is the lack of an abscission tissue with the resulting persistence of all floral parts, so that usually the stamens, petals, and sepals remain attached to the receptacle until after fruit maturation (Berg 1958).

At anthesis *Trillium* has often an erect pedicel but sometimes this is declining (*T. cernuum*). Kawano and Kato (1995) interpret the change from the erect to the declining pedicel as correlated with a switch of pollinators from small flies, bees, and beetles to large bumblebees such as *Bombus*, *Xylocopa*, *Lassioglossum*, and also some butterflies. They contend that only the latter insects can visit and forage in hanging flowers, although such pollinator visitation has scarcely been observed (see below under Pollination). Species of *Trillium* subgen. *Phyllantherum* often lack a pedicel. Berg (1958) suggested that the reduction of the pedicel might correspond to the switch of seed dispersal from endozoochory to myrmecochory (see below under Dispersal).

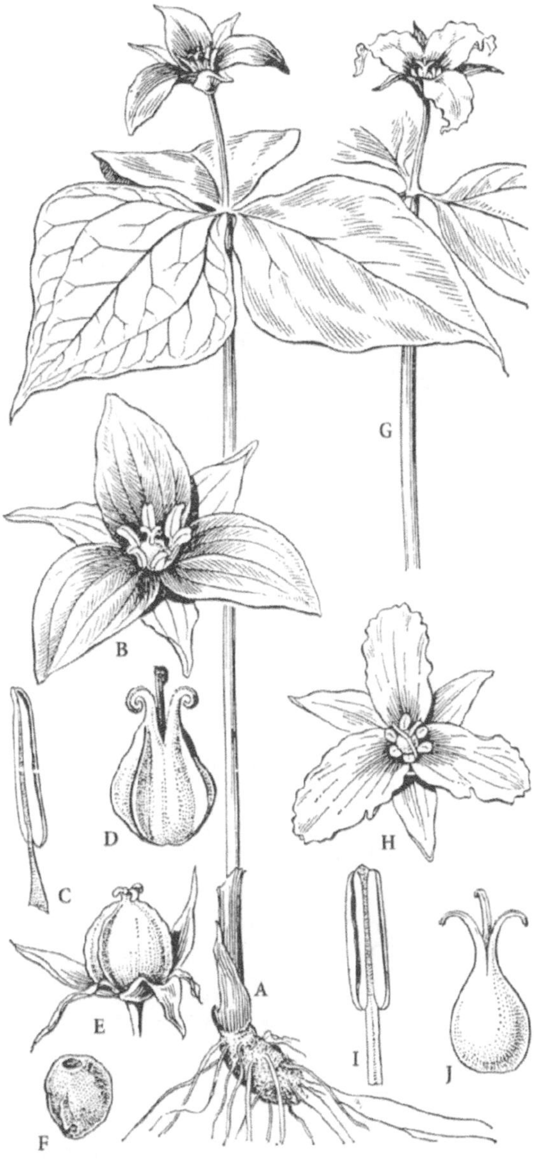

Fig. 128A–J. Trilliaceae. **A–F** *Trillium erectum*. **A** Flowering plant. **B** Flower. **C** Stamen. **D** Pistil. **E** Fruit. **F** Seed. **G–J** *T. undulatum*. **G** Flowering shoot. **H** Flower. **I** Stamen. **J** Pistil. (Takhtajan 1982)

EMBRYOLOGY (based mostly on Berg 1958, 1962). The embryologic characteristics of *Paris*,

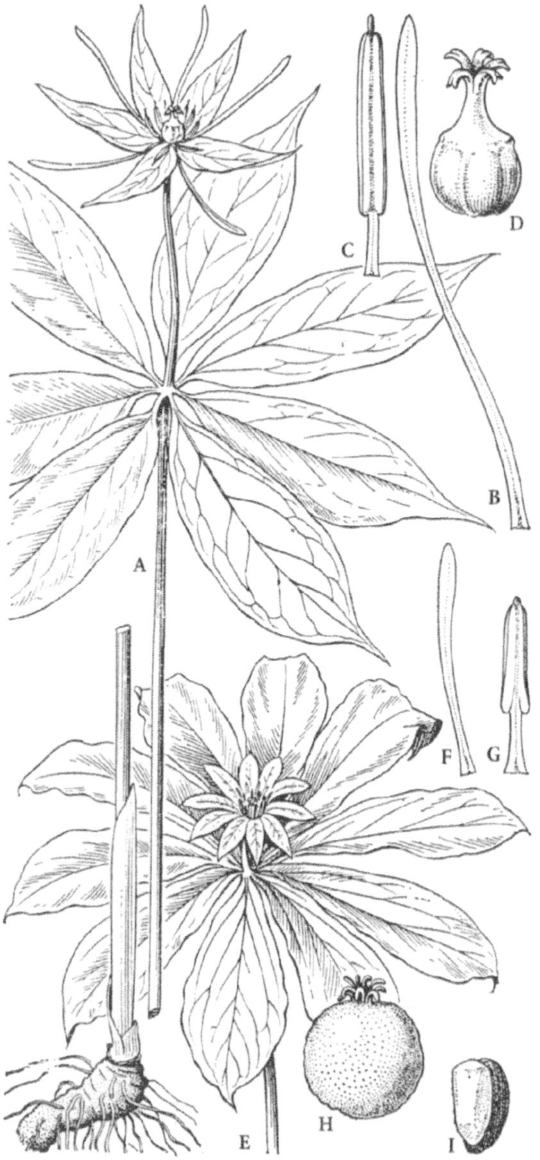

Fig. 129A–I. Trilliaceae. **A–D** *Daiswa polyphylla*. **A** Flowering plant. **B** Petal. **C** Stamen. **D** Pistil. **E–I** *Kinugasa japonica*. **E** Flowering shoot. **F** Petal. **G** Stamen. **H** Fruit. **I** Seed. (Takhtajan 1982)

Daiswa, and *Trillium* are very similar, although in *Daiswa* they have only partly been analyzed. The ovule is pleurotropous (before fertilization), anatropous, bitegmic, and crassinucellate. A distinct funiculus is absent. The cells of the placental epidermis are only slightly papillate, so that a typical stigmatoid tissue is lacking. Crystal cells containing raphide bundles are commonly present in the chalaza. The micropyle is formed only by the inner integument, and the outer integument does not overgrow the inner integument, even after fertilization.

The nucellus is well developed, although the basal part is short, consisting only of 1 or 2 cell layers. Periclinal cell divisions occur in the nucellar epidermis. Parietal cells are present. The embryo sac is bisporic (Scilla type). The synergids show no vacuoles. The antipodal cells are mononucleate and degenerate shortly after fertilization. The 2 polar nuclei do not fuse until the arrival of the sperm cell. *Paris* differs from *Trillium* in some embryologic characteristics, most notably in endosperm development, which in *Trillium* is Helobial, while in *Paris* it is Nuclear.

The embryo of *Trillium* is incomplete (Goebel 1922). When the seed becomes detached from the mother plant, the total length of the embryo amounts to less than 1/10 of the length of the endosperm. When the seeds are kept moist, their embryos slowly enlarge and differentiate, gradually acquiring the same length as the endosperm.

Karyology. The basic chromosome number is invariably $x = 5$. The chromosomes are all more or less large (8–36 μm in length). The karyotypes comprise 3 metacentric chromosome pairs in addition to 2 submetacentric to telocentric pairs.

The standard ploidy level of E to SE Asian *Paris* and *Daiswa* is $2x$ (Miyamoto et al. 1992), while that of European *P. quadrifolia* and E Himalayan *Daiswa* is $4x$ (Hara 1969) and that of *Kinugasa* $8x$. All N American *Trillium* spp. are $2x$, except for rare occasional $3x$ and supernumerary aneuploids. In contrast, Asian *Trillium* spp. comprise $2x$, $3x$, $4x$, and $6x$, including 5 hybrid derivatives (Kurabayashi 1958; Samejima and Samejima 1962).

Hong and Zhu (1987) and Li et al. (1988) recognized 2 basic karyotypes among *Paris* and *Daiswa*: the tropical type consisting of 6 metacentric and 4 acrocentric chromosomes, and the temperate type possessing 2 submetacentric chromosomes instead of the 2 acrocentric ones of the tropical type. *Kinugasa* and *Trillium* spp. correspond with the temperate type.

Tamura (1995), based on the data of Hara (1969), recognized 2 other basic karyotypes among *Paris*: the Tetraphylla type with large chromosomes (15–27 μm long), and the Quadrifolia type with relatively small chromosomes (8–17 μm). According to Hara (1969) and Tamura (1995), the chromosomes of *Daiswa* (22–36 μm) are generally larger than those of the other genera.

Haga (1934) and Hara (1969) distinguished the karyotypes of *Paris*, *Kinugasa*, *Daiswa*, and *Trillium* from one another by the positions of the satellites: *Paris* has a satellite only on an acrocentric chromosome, while *Kinugasa* one on a submetacentric and one on an acrocentric chromosome, *Daiswa* only on a metacentric chromosome, and *Trillium* lacks satellites altogether.

According to Hara (1969), the karyotype of *Daiswa* differs from those of *Paris* and *Kinugasa* also in other features: in *Daiswa* all 3 metacentric chromosome pairs are large, while in *Paris* and *Kinugasa* 1 of them is comparatively small, and in *Daiswa* supernumerary chromosomes are often observed, while in *Paris* and *Kinugasa* they are seldom found. According to Haga (1934), the karyotypes of *Paris*, *Kinugasa*, and *Trillium* are almost identical except for the possession of satellites.

The C-banding patterns as well as the cold-induced H-segment patterns suggest that *Kinugasa* may be of allo-octoploid origin with species of *Trillium* such as *T. erectum* and *Paris* as ancestors. The proximal C-bands on both arms of the metacentric chromosomes are shared by *Kinugasa*, *Paris tetraphylla*, and *P. verticillata*, all growing in Japan (Miyamoto et al. 1992).

The karyological evidence suggests that *Daiswa* is a descendent of the other trilliaceous genera.

Pollen Morphology. Pollen in *Paris*, *Kinugasa*, *Daiswa*, and *Trillium govanianum* is sulcate and ellipsoidal in shape; the sulcus has no marginal thickening. In contrast, pollen in *Trillium*, except *T. govanianum*, is omniaperturate and spherical in shape (Takahashi 1984).

The exine in *Paris* except *P. tetraphylla* and in *Daiswa* is psilate to foveolate or reticulate and tectate, that of *P. tetraphylla*, *Kinugasa*, and *T. govanianum* clavate to gemmate or rugulate and intectate (Takahashi 1983, 1984). The latter condition is considered as advanced in comparison with the former (Takahashi 1984). In *Daiswa*, pollen size correlates with the ploidy level (Hara 1969), and diploid *Daiswa* species have coarser reticulation than tetraploid ones.

Exine sculpture and structure of *Trillium* except *T. govanianum* is echinate, corrugate, granulate (the most common condition), verrucate, or spinulate and unstratified (Takahashi 1982), and can be considered as reduced and specialized in comparison with the other genera (Takahashi 1984). Thus the pollen morphology of *Paris* (except *P. tetraphylla*) and *Daiswa* is primitive, that of *Kinugasa*, *Paris tetraphylla*, and *Trillium govanianum* more or less advanced, and that of *Trillium* (except *T. govanianum*) strongly advanced.

Pollination. It is possible that the release of fragrant (*Trillium luteum*) or unpleasant odors (*T. foetidissimum*) might attract small bees and dung flies (Kawano et al. 1992). However, in both Japan and N America, effective pollinator visitations have scarcely been observed in *Trillium*, although this genus shares the capacity for outbreeding (Ohara 1989; see below under Reproductive Systems).

Fruit and Seed. The fruit is berrylike (baccate, pulpy or succulent) (*Paris*, *Kinugasa*, *Trillium* subgen. *Trillium* spp.) or capsulelike (nonfleshy, non pulpy, or mealy) (*Daiswa*, *Trillium* subgen. *Phyllantherum*, *T.* subgen. *Trillium* spp.). Anatomically, the berrylike fruits resemble the capsulelike ones except for the pulp (Berg 1958).

Baccate fruits are indehiscent. Capsulelike fruits of *Trillium* lack a mechanism for dehiscence and opens due to the pressure of the enlarging seeds and a degeneration of the wall tissue. Weak zones, along which the fruits can split, are found at the basis of the fruit and along the sutures of the carpels. The capsulelike fruit of *Daiswa* dehisces loculicidally.

The seed of *Trillium* is approximately 4 × 2.5 mm and is equalled or surpassed in volume by an elaiosome or strophiole (Sernander 1906). The seed is ellipsoidal but slightly curved and the appendage extends from its point of attachment at the concave or ventral side of the seed into all directions, often overarching and enclosing half of the seed. The seed of *Daiswa* has a sarcotesta. The seeds of *Paris* and *Kinugasa* lack both an appendage and sarcotesta.

Dispersal (based mostly on Berg 1958; Ohara 1989). Seeds of *Trillium* are dispersed by ants that are attracted to the elaiosome (Gates 1943; Berg 1958). Nesom and La Duke (1985), Ohara and Higashi (1987), and Higashi et al. (1989) reported that in Japan the seeds of *T. kamtschaticum* and *T. tschonoskii* are transported by the ant species *Aphaenogaster smythiesi japonica* and *Myrmica ruginodis*, while those of *T. nivale* in N America rely on *Aphaenogaster rudis* and *Formica neogagates*.

When seeds with an elaiosome are brought to an ant nest, the elaiosomes are cut off and the seeds are discarded near the nest. However, fresh elaiosomes attached to dropped seeds attract not only ants but also solitary ground beetles, which devour the elaiosomes during the night without transporting the seeds. Since ants are not interested in seeds without an elaiosome, a large proportion of dropped seeds without an elaiosome are left near the mother plants. Subsequently, clusters of seedlings are commonly discovered near the mother plants (Ohara and Kawano 1986; Ohara and Higashi 1987; Higashi et al. 1989).

It can often be seen in the field that mature berrylike fruits at the top of erect pedicels of species of *Trillium* such as *T. kamtschaticum* and *T. erectum* are eaten by small animals (Kawano and Kato 1995). This suggests a possibility of endozoochorous dispersal for species with a berrylike fruit on an erect pedicel. In contrast, the declination of the pedicel of several *Trillium* species with a berrylike fruit seems to weaken the exposure of the fruits towards flying birds. It is possible that endozoochory and berries are original attributes of *Trillium*, and myrmecochory and capsular fruits are specializations (Berg 1958).

Seedling Development (based on Ohara 1989). After the detachment from the mother plant, the seeds of *Trillium* take about $1^1/_2$ years to become established as seedlings. In the 1st year, the seedlings develop only roots and only in the 2nd year a cotyledon emerges as an aboveground organ (Samejima and Samejima 1962, 1987). The 1-leaf stage of *Trillium* lasts nearly for 10 years from germination. The seedlings have specific oblong-lanceolate leaves; in older individuals leaf shape gradually becomes ovate with a rounded base. After the 1-leaf stage, the *Trillium* plantlets usually suddenly become 3-leaved, skipping a 2-leaf stage.

Reproductive Systems (based mostly on Ohara 1989). In Japan, *Trillium* subgen. *Trillium* reproduces exclusively by means of seeds. In N America, *T.* subgen. *Trillium* and *T.* subgen. *Phyllantherum*, as far as growing in relatively stable habitats, depend mostly on seeds for their offspring recruitment, although vegetative offshoots are also occasionally formed. *Trillium* subgen. *Phyllantherum*, occurring in ecologically unstable flood plain habitats of the Coastal Plain of southeastern N America, to a large extent reproduces vegetatively, although sexual reproduction also plays a role in offspring recruitment.

In the course of vegetative reproduction, several small rhizome offshoots develop on the parental rhizome. They are not detached immediately and for several years increase in size. With the decay of the parental rhizome or their mechanical detachment, the small rhizome offshoots become independent.

Species of *Trillium* subgen. *Phyllantherum* which occur in unstable habitats and frequently propagate vegetatively have smaller numbers of ovules and lower seed-setting rates than the species of subgen. *Trillium* and those of subgen. *Phyllantherum* that occur in relatively stable habitats. Thus vegetative reproduction of subgen. *Phyllantherum* appears to compensate for the inefficiency of sexual reproduction under unstable environmental conditions.

According to a cold-induced banding analysis of the chromosomes (Fukuda 1990), *Trillium* includes both inbreeding and outbreeding populations. In N America *Trillium erectum* and *T. grandiflorum* are predominantly inbreeding. However, populations of *T. ovatum* in the Rocky Mountains are apparently outbreeding (see below under Distribution and Habitats). The Japanese species, *T. kamtschaticum*, has mixed populations of outbreeders and inbreeders. According to pollination experiments (Haga and Channell 1982), intraspecific seed-setting of *T. kamtschaticum* is indifferent upon self- and cross-pollination.

Population Structure (based on Ohara 1989). The *Trillium* species, depending mostly on seeds for their offspring recruitment (*T.* subgen. *Trillium* and a part of subgen. *Phyllantherum*), show a sigmoid population structure characterized by a conspicuous decrease in number of early juvenile stages, a subsequent leveling-off in intermediate stages, and a decrease again in older stages of the population.

On the other hand, most species of *Trillium* subgen. *Phyllantherum* show a nonsigmoid, convex-shaped population structure characterized by extremely small numbers of plants in the juvenile stage and a peak in numbers of plants in the intermediate class, which suggests that the offspring is not effectively recruited by seeds but is supplemented rather by the separation of vegetative offshoots from parental rhizomes.

Phytochemistry. According to Williams (1975), *Paris quadrifolia* contains kaempferol but lacks quercetin, luteolin, and apigenin. Zomlefer (1996) indicates for the Trilliaceae the presence of steroidal saponins (sapogenins), flavonols (kaempferol, quercetin), and chelidonic acid.

Distribution and Habitats. The Trilliaceae are generally woodland plants of temperate regions of the Northern Hemisphere. Most species of *Paris*, *Kinugasa*, and *Daiswa* are restricted to SE and E Asia, except for the European *P. quadrifolia* and the Caucasian *P. incompleta*. *Kinugasa* is endemic to Japan.

Trillium, which is often considered as an Arcto-Tertiary element, is disjunctly distributed in western N America (8 spp.), eastern N America (ca. 31 spp.), and from the W Himalayas to NE Asia (11 spp.). The Appalachian mountain region appears to be the center of diversity for *Trillium*.

Kawano et al. (1992) and Kawano and Kato (1995) recognized five types of habitats in *Trillium*, which are characterized as follows.

1. The mesic forest floor and gentle seepage slopes of summer-green broad-leaved hardwood forests of the Northern Hemisphere, dominated by various oak, maple, and beech species (Samejima and Samejima 1987), may be the original habitat for *Trillium*. Most species of *T.* subgen. *Trillium* (distributed widely in Asia and N America) are primarily associated with this type of habitat. *Trillium grandiflorum* shows postglacial range extension into once-glaciated territories in eastern N America (Samejima and Samejima 1987), while *T. sulcatum* is strictly confined to refugia in the southern Appalachian Mountains, and *T. pusillum* is limited to the unglaciated Ozarkian Mountains.
2. Coniferous or mixed forests with deep shade, as found, e.g., in eastern N America (with *T. undulatum*) and the Himalayas (with *T. govanianum*), seem to be derived conditions for *Trillium* species. Several species of *T.* subgen. *Phyllantherum* of western American coastal coniferous forests (*T. kurabayashii*) as well as those of the southeastern N American Coastal Plain, eastern lowland mixed forests, upland rocky woods, and floodplains of streams (*T. sessile* etc.) (Freeman 1975) occupy more or less similar habitats with rich light resources (with many scattered sunflecks) in spring. Considerable genetic diversity is known in the Rocky Mountain populations of *T. ovatum*, which occur under somewhat arid conditions beneath Ponderosa and Lodgepole pine forests, while those of the humid redwood and Douglas fir forests of the Coastal Range are contrastingly homogeneous in genetic structure, which may reflect the rather recent history of migration and population expansion of this species into the Rocky Mountains (Fukuda and Channell 1975).
3. Dark shady and damp river bank habitats covered by evergreen *Magnolia* and deciduous *Quercus* hardwood trees, and often exposed to flooding, notably in the Coastal Plain of eastern N America (Ohara and Utech 1986, 1988), may represent another more recently occupied habitat for *Trillium*. This is a typical habitat for most species of *T.* subgen. *Phyllantherum* (restricted to N America). A somewhat similar situation is found in the valley-bottom habitats of western N America, where *T. albidum* occurs.
4. Unstable dry rocky habitats with good drainage, as seen in *T. rivale*, developed on the marginal slopes of coniferous forests or on the floodplains under alders in western N America.
5. Xeric grassy rocky habitats under scattered conifers of *T. petiolatum* extending over the eastern side of the Cascades and the Palouse grassland region of the western N America represent two extremes of the habitat expansion.

Subdivision and Relationships Within the Family. *Kinugasa* and *Daiswa* are sometimes included in *Paris* (Krause 1930; Hara 1969; Li 1986). However, in the present treatment, following Takhtajan (1983), the three taxa are recognized at generic level.

Based on the *rbc*L sequence data of *Paris tetraphylla*, *Kinugasa japonica*, *Daiswa polyphylla*, *Trillium kamtschaticum*, and *T. sessile*, the Trilliaceae form a monophyletic group with 100% bootstrap support (Kato et al. 1995). However, a topology of the four genera within the Trilliaceae was not clarified.

Based on karyological evidence, *Daiswa* might be derived from the other three trilliaceous genera. Based on the ploidy level, *Kinugasa* (8*x*) might be derived. Based on pollen morphology, *Paris* (except *P. tetraphylla*) and *Daiswa* are the most primitive genera, and *Trillium* (except *T. govanianum*) is the most advanced. Putting all accounts at this stage together, *Paris* seems to be basal in Trilliaceae. However, more detailed molecular studies are needed in order to determine the exact relationship within this family.

Affinities. *Scoliopus* and *Medeola* are sometimes included in the *Trillium* group (Watson 1879; Bentham and Hooker 1883; Krause 1930; Hutchinson 1934) within a broadly construed family Liliaceae. Dahlgren et al. (1985) suggested a relationship between *Scoliopus*, *Paris*, and *Trillium*, which was supported by the study of leaf epidermal patterns (Conover 1991).

However, *Scoliopus* has a monosporic embryo sac and a basic chromosome number of $x = 7$ or 8,

Medeola a tetrasporic embryo sac (Fritillaria type) and $x = 7$, and the Trilliaceae a bisporic embryo sac (Scilla type) and $x = 5$. *Scoliopus* and *Medeola* differ from the Trilliaceae also in floral anatomy (Anderson 1940; Utech 1978, 1979), pollen morphology (Takahashi 1984) and plastid DNA sequences (Chase et al. 1995). In this treatment I put stress on the differences between *Scoliopus*, *Medeola*, and the Trilliaceae and propose a transfer of *Scoliopus* into the Calochortaceae and of *Medeola* into the Liliaceae s. str.

Based on analyses of chloroplast DNA restriction site variation (Davis 1995) and *rbc*L sequences (Chase et al. 1995; Kato et al. 1995), the Trilliaceae appear closely related to the Melanthiaceae. This is also reflected by morphological characters shared by the two families: (1) both always have parietal cells, and often (2) free styluli, (3) crystal raphides and (4) Helobial endosperm formation.

Nevertheless, the Trilliaceae differ from the Melanthiaceae in many respects: they have solitary flowers, berries in addition to capsules, latrorse or introrse anthers in addition to extrorse anthers, septal nectaries, very large chromosomes, and a basic chromosome number $x = 5$ (*Paris quadrifolia* has the smallest chromosomes in the Trilliaceae, 8–17 μm in length), while the Melanthiaceae have a 3-many-flowered panicle, raceme, spike, or umbel, always capsular fruits and extrorse anthers and lack septal nectaries; the chromosomes are small and based on $x = 8, 10, 11, 12, 15, 16, 17, 21, 22$ (*Helonias bullata* has the longest chromosomes in the Melanthiaceae, 2–6 μm in length). Possibly the Trilliaceae have recently been derived from the stock of the Melanthiaceae and have greatly changed their characteristics.

Based on leaf epidermal patterns (Conover 1991), the Trilliaceae seem to be related to *Tricyrtis* (Calochortaceae). Takhtajan (1997) suggested a relationship between the Trilliaceae and *Scoliopus* (Calochortaceae). It is therefore possible that the Trilliaceae are related to the Calochortaceae.

Key to the Genera

1. Foliage leaves usually 3, rarely 1, 2, 4–8; flowers usually trimerous; petals oval to broadly linear, often wider than sepals **4. *Trillium***
- Foliage leaves usually 4–14, rarely 3, 15–22; flowers usually 4-merous to polymerous; petals filiform to linear, much narrower than sepals 2

2. Rhizome slender, 2–5(–8) mm diam.; ovary without a disk at the apex; fruit a berry **1. *Paris***
- Rhizome stout, 8–30 mm diam.; ovary with a disk at the apex; fruit a berry or a capsule 3

3. Sepals totally white, petaloid; ovary nonangular; fruit a berry, indehiscent; seed without sarcotesta; 2n = 40 **2. *Kinugasa***
- Sepals green, rarely with whitish stripes; ovary angular; fruit a capsule, loculicidally dehiscent; seed with sarcotesta. 2n = 10, 20 **3. *Daiswa***

1. *Paris* L.

Paris L., Sp. Pl.: 367 (1753).
Paris subgen. *Paris* H. Li (1984); H. Li, Bull. Bot. Res. Kunming 6: 109–144 (1986), rev.

Rhizome slender, creeping. Aerial stem erect. Foliage leaves (3–)4–12, lanceolate or oblanceolate to elliptical or broadly obovate, often acuminate, sometimes attenuate, 4–15 cm long, sessile or shortly petiolate. Flowers usually pedicellate, rarely sessile. Sepals 3–5, narrowly lanceolate to oval, green, rarely purplish, patent to reflexed. Petals 0, (3–)4–5, filiform, linear to subulate. Stamens 8; anthers linear to oblong. Ovary roundish, 4–5-locular, usually lacking a disk at apex; ovules many to several; style slender, (3–)5–20 mm long, 4–5-cleft to dissected. Fruit subglobose, a berry, purple to black-purple, poisonous; seeds globose, ca. 2.5 mm long. 2n = 10, 15, 20. About five spp., Eurasia.

2. *Kinugasa* Tatewaki et Suto Fig. 129E–I

Kinugasa Tatewaki et Suto, Trans. Sapporo Nat. Hist. Soc. 14: 34 (1935).
Paris sect. *Kinugasa* (Tatewaki et Suto) Hara (1969).

Rhizome stout, 15–20 mm in diameter. Aerial stem erect, stout, terete, 30–80 cm high. Foliage leaves (6–)7–10(–12), broadly oblanceolate to obovate-oblong, abruptly acuminate, (10–)13–30 cm long, sessile. Flowers pedicellate. Sepals 7–10(–12), lanceolate-oblong, white, petaloid. Petals same as sepals in number, linear, white. Stamens 15–20. Ovary subglobose, non-angular, green, 6–10-locular, with a disk at apex; ovules many; style slender, 5–10(–12)-parted. Fruit a berry, globose, dark purple, edible; seeds elliptical, ca 2 mm long, brown. 2n = 40. Only one sp., *K. japonica* (Franch. et Savat.) Tatewaki et Suto, Central Japan.

3. *Daiswa* Rafin. Fig. 129A–D

Daiswa Rafin., Fl. Tellur. IV: 18 (1836); Takhtajan, Brittonia 35: 255–270 (1983), rev.
Paris sect. *Euthyra* (Salisb.) Franchet (1888); Hara, Jour. Fac. Sci. Univ. Tokyo, III, Vol. 10: 141–180 (1969).
Paris subgen. *Daiswa* (Rafin.) H. Li (1984); H. Li, Bull. Bot. Res. Kunming 6: 109–144 (1986), rev.

Rhizome stout. Aerial stem erect. Foliage leaves (3–)5–14(–22), narrowly lanceolate or broadly linear to oval, elliptical or broadly obovate, acuminate to aristate at apex, 3–27 cm long, usually petiolate, rarely sessile. Flowers pedicellate. Sepals 3–10, linear to lanceolate-oblong or oblanceolate, large, green, rarely with whitish stripes (*D. violacea* Levl.). Petals 0 or 3–12, long, filiform to linear. Stamens 3–21. Ovary subglobose to ovate pyramidal, angular, with a disk at apex; ovules numerous; style thick, cleft to dissected, branches 2–10, linear. Fruit a capsule, purple to deep purple, loculicidally dehiscent; seed coat scarlet, juicy. 2n = 10, 20. About 15 spp., SE Asia.

4. *Trillium* L. Fig. 128

Trillium L., Sp. Pl. 1: 339 (1753); Samejima & Samejima, Acta Horti Gotoburg. 25: 157–257 (1962), rev. E Asia spp.; Freeman, Brittonia 27: 1–62 (1975), rev. *T.* subgen. *Phyllantherum*; Samejima & Samejima, *Trillium* genus illustrated. 237. pp. (1987); Zomlefer, Harvard Papers in Botany 9: 91–120 (1996), rev. SE USA spp.
Trillidium Kunth (1850).

Rhizome stout to slender, short to long, horizontal to erect. Aerial stem 1(–2-more), usually erect, rarely decumbent. Foliage leaves usually 3, rarely 1, 2, 4–8, ovate, oblong to obovate, acute to obtuse at apex, rounded to cordate at base, sessile to petiolate, palmate-reticulate-veined. Flowers pedicellate (*T.* subgen. *Trillium*) or sessile (*T.* subgen. *Phyllantherum*), erect or nodding. Sepals and petals usually 3, rarely 0, 2, 4, or more, free, persistent; sepals ovate to oblong, green, often purplish or maroonish inside; petals oval to broadly linear, sometimes clawed or twisted, maroon, pink, yellow, green, or white, sometimes nectariferous at base. Stamens usually 6, persistent; filaments distinct; anthers linear. Ovary 3- or 1-locular, 3- or 6-angled or -winged, sometimes with septal nectaries; ovules numerous; styluli free or fused into a style with erect, spreading or recurved, persistent stylodia. Fruit a berry or capsule, indehiscent or irregularly dehiscent, usually green, maroon, blackish purple, rarely white, yellow or red; seeds ellipsoid to oblong, with white or yellow aril. 2n = 10, 15, 20, 30. Two subgenera, about 50 spp., Northern Hemisphere, abundant in N America.

Selected Bibliography

Anderson, C.E. 1940. Some studies on the floral anatomy of the Liliales. Unpublished Ph. D. dissertation. Ithaca, New York: Cornell Univ.

Bentham, G., Hooker, J.D. 1883. Genera Plantarum, Vol. 3. London: Reeve; Williams & Norgate.

Berg, R.Y. 1958. Seed dispersal, morphology, and phylogeny of *Trillium*. Skr. Nor. Vidensk.-Akad. Oslo. I. Mat.-Naturvidensk. Kl. no. 1: 1–36.

Berg, R.Y. 1962. Contribution to the comparative embryology of the Liliaceae: *Scoliopus*, *Trillium*, *Paris*, and *Medeola*. Skr. Nor. Vidensk.-Akad. Oslo I. Mat.-Naturvidensk. Kl. Ny Ser. no. 4: 1–64.

Chase, M.W. et al. 1995. See general references.

Conover, M.V. 1991. Epidermal patterns of the reticulate-veined Liliiflorae and their parallel-veined allies. Bot. J. Linn. Soc. 107: 295–312.

Dahlgren, R.M.T. et al. 1985. See general references.

Davis, J.I. 1995. A phylogenetic structure for the monocotyledons, as inferred from chloroplast DNA restriction site variation, and a comparison of measures of clade support. Syst. Bot. 20: 503–527.

Freeman, J.D. 1975. Revision of *Trillium* subgenus *Phyllantherum* (Liliaceae). Brittonia 27: 1–62.

Fukuda, I. 1990. Breeding systems in American and Asian *Trillium* species by means of chromosome analyses. Plant Species Biol. 5: 65–72.

Fukuda, I., Channell, R.B. 1975. Distribution and evolutionary significance of chromosome variation in *Trillium ovatum*. Evolution 29: 257–266.

Gates, B.N. 1943. Carunculate seed dissemination by ants. Rhodora 45: 438–445.

Goebel, K. 1922. Organographie der Pflanzen insbesondere der Archegoniaten und Samenpflanzen. 3. Teil: Spezielle Organographie der Samenpflanzen. 1. Heft: Vegetationsorgane. 2. Aufl., 1210–1492. Jena: G. Fischer.

Haga, T. 1934. The comparative morphology of the chromosome complement in the tribe Parideae. J. Fac. Sci. Hokkaido Univ. Bot. 3: 1–32.

Haga, T., Channell, R.B. 1982. Three species groups in American trilliums as revealed by compatibility with an Asian species. Bot. Mag. (Tokyo) 95: 77–80.

Hara, H. 1969. Variations in *Paris polyphylla* Smith, with reference to other Asiatic species. J. Fac. Sci. Univ. Tokyo, III 10: 141–180.

Higashi, S., Tsuyuzaki, S., Ohara, M., Ito, F. 1989. Adaptive advantages of ant-dispersed seeds in the myrmecochorous plant *Trillium tschonoskii* (Liliaceae). Oikos 54: 389–394.

Hong, D., Zhu, X. 1987. Cytotaxonomical studies on Liliaceae (s. l.) (1) Report on karyotypes of 10 species of 6 genera. Acta Phytotaxon. Sin. 25: 245–253.

Hutchinson, J. 1934. The families of flowering plants, Vol. 2. Monocotyledons. London: Macmillan.

Kato, H., Terauchi, R., Utech, F.H., Kawano, S. 1995. Molecular systematics of the Trilliaceae sensu lato as inferred from *rbc*L sequence data. Mol. Phylog. Evol. 4: 184–193.

Kawano, S., Kato, H. 1995. Evolutionary biology of *Trillium* and related genera (Trilliaceae) II. Cladistic analyses on gross morphological characters, and phylogeny and evolution of the genus *Trillium*. Plant Species Biol. 10: 169–183.

Kawano, S., Ohara, M., Utech F.H. 1992. Life history studies on the genus *Trillium* (Liliaceae) VI. Life history characteristics of three western North American species and their evolutionary-ecological implications. Plant Species Biol. 7: 21–36.

Krause, K. 1930. Liliaceae. In: Engler, A., Prantl, K. (eds.) Die natürlichen Pflanzenfamilien, Vol. 15a, 227–386. Leipzig: Engelmann.

Kurabayashi, M. 1958. Evolution and variation in Japanese species of *Trillium*. Evolution 12: 286–310.

Li, H. 1986. A study on the taxonomy of the genus *Paris* L. Bull. Bot. Res. Kunming 6: 109–144 (in Chinese).
Li, H., Gu, Z., Na, H. 1988. Cytogeographic study of the genus *Paris*. Acta Phytotaxon. Sin. 26: 1–10.
Miyamoto, J., Kurita, S., Gu, Z., Li, H. 1992. C-banding patterns in 18 taxa of the genus *Paris* sensu Li, Liliaceae. Cytologia 57: 181–194.
Nesom, G.L., La Duke, J.C. 1985. Biology of *Trillium nivale* (Liliaceae). Can. J. Bot. 63: 7–14.
Ohara, M. 1989. Life history evolution in the genus *Trillium*. Plant Species Biol. 4: 1–28.
Ohara, M., Higashi, S. 1987. Interference by ground beetles with the dispersal by ants of seeds of *Trillium* species (Liliaceae). J. Ecol. 75: 1091–1098.
Ohara, M., Kawano, S. 1986. Life history studies on the genus *Trillium* (Liliaceae) IV. Stage class structures and spatial distribution of four Japanese species. Plant Species Biol. 1: 147–161.
Ohara, M., Utech, F.H. 1986. Life history studies on the genus *Trillium* (Liliaceae) III. Reproductive biology of six sessile-flowered species occurring in the southeastern United States with special reference to vegetative reproduction. Plant Species Biol. 1: 135–145.
Ohara, M., Utech, F.H. 1988. Life history studies on the genus *Trillium* (Liliaceae) V. Reproductive biology and survivorship of three declinate-flowered species. Plant Species Biol. 3: 35–41.
Samejima, J., Samejima, K. 1962. Studies on the eastern Asiatic *Trillium* (Liliaceae). Acta Hortic. Gotob. 25: 157–257.
Samejima, K., Samejima, J. 1987. *Trillium* genus. Sapporo: Hokkaido Univ. Press.
Sernander, R. 1906. Entwurf einer Monographie der europäischen Myrmekochoren. K. Sven. Vetenskapsakad. Handl. 41, no. 7: 1–410.
Takahashi, M. 1982. Pollen morphology in North American species of *Trillium*. Am. J. Bot. 69: 1185–1195.
Takahashi, M. 1983. Pollen morphology in Asiatic species of *Trillium*. Bot. Mag. (Tokyo) 96: 377–384.
Takahashi, M. 1984. Pollen morphology in *Paris* and its related genera. Bot. Mag. (Tokyo) 97: 233–245.
Takhtajan, A. 1982, 1997. See general references.
Tamura, M.N. 1995. A karyological review of the orders Asparagales and Liliales (Monocotyledonae). Feddes Repert. 106: 83–111.
Utech, F.H. 1978. Floral vascular anatomy of *Medeola virginiana* L. (Liliaceae-Parideae = Trilliaceae) and tribal note. Ann. Carnegie Mus. 47: 13–28.
Utech, F.H. 1978. Floral vascular anatomy of *Scoliopus bigelovii* Torrey (Liliaceae-Parideae = Trilliaceae) and tribal note. Ann. Carnegie Mus. 48: 43–71.
Watson, S. 1879. Contributions to American botany. I. Revision of the North American Liliaceae. Proc. Am. Acad. Arts Sci. 14: 213–288.
Williams, C.A. 1975. Biosystematics of the Monocotyledoneae — flavonoid patterns in leaves of the Liliaceae. Biochem. Syst. Ecol. 3: 229–244.
Zomlefer, W.B. 1996. The Trilliaceae in the southeastern United States. Harv. Pap. Bot. 9: 91–120.

Triuridaceae

H. MAAS-VAN DE KAMER and T. WEUSTENFELD

Triuridaceae Gardner, Proc. Linn. Soc. Lond. 1: 177 (1843), Trans. Linn. Soc. Lond. 19: 160 (1843 '1845'), nom. cons.
Lacandoniaceae E. Martínez & Ramos (1989).

Achlorophyllous, mycotrophic, white, yellow or purple, usually perennial herbs. Stems erect, arising sympodially from a creeping or sometimes erect rhizome; rhizome and stem provided with small scalelike, alternate leaves. Roots filiform, mostly hairy, containing the mycelium of the mycorrhizal fungus. *Inflorescence* a terminal bracteate raceme. Monoecious or dioecious. *Flowers* unisexual or rarely bisexual, regular, pedicellate, white, yellow, purplish or red. Perianth of 3–6(–10) valvate tepals in a single cycle, the tepals persistent, mostly triangular to deltate, often with apical perianth appendages. Hermaphrodite flowers: stamens 2–6, free, carpels many, free. Staminate flowers: stamens 2–6(–8), sessile, provided with a filament or sessile on a small or large, central androphore. Anthers 2-, 3- or mostly 4-sporangiate, extrorse (to introrse), longitudinally to transversely dehiscing. Staminodes (*Seychellaria*) and connective appendages (*Andruris*, *Seychellaria madagascariensis*) rarely present. Pistillate flowers with 10-numerous free, ovoid carpels each containing 1 basal, anatropous ovule. Style (sub)basal to (sub)apical, linear, the stigmatic part either densely papillate to penicillate (brush-shaped style), or glabrous (awl-shaped style). *Fruit* a follicle or an achene. Seed with copious endosperm and small, undifferentiated embryo.

A family of 9 genera with ca. 45 species distributed throughout the Old and New World tropics.

VEGETATIVE STRUCTURES. (Review Rübsamen-Weustenfeld 1991: 11). In connection with the mycotrophic life style, the vegetative organs of these plants are very simple and reduced. Leake (1994), however, who had compiled the data on morphology and mycorrhizal infection, found the root systems of some Triuridaceae remarkable for their lack of adaptation to myco-heterotrophic nutrition. The subterranean, creeping or sometimes vertical (*Sciaphila purpurea*) rhizome gives rise to 1 or 2 cylindrical, well-developed roots from each node. The roots are filiform or rarely coral-shaped (*Sciaphila picta*) or form a "bird's nest" (*Sciaphila oligantha*, *Triuris* spp.), and are provided with long, white hairs; rarely (*Sciaphila purpurea*, *Peltophyllum*, *Triuris*), the roots are glabrous to sparsely hairy (Maas and Rübsamen 1986). The mycelium of the fungus is situated inside the cortex cells of the root as well as outside the plant and may probably be of the vesicular-arbuscular type (Leake 1994). An endodermis with wall thickenings has been demonstrated in the roots, surrounding the reduced and obscure central vascular tissue, and in some species it occurs also in the stems (Tomlinson 1982; Rübsamen-Weustenfeld 1991).

The shoots are short (2–30 cm, in *Sciaphila purpurea* rarely up to 1.4 m long). They are mostly unbranched but sometimes at the base of the old inflorescences provided with side branches. The stems of *Peltophyllum* and *Triuris* are hyaline and fragile (Maas and Rübsamen 1986; Maas-van de Kamer 1995). The vascular tissue of the stem is arranged in a single irregular cylinder of narrow vascular bundles either contiguous to the cortical sclerenchyma or enclosed in a continuous sclerotic ring (Tomlinson 1982).

The small scalelike leaves, 0–8(–15) in number, are spirally arranged, provided with one median vein, triangular to ovate, sessile, sometimes amplexicaul or basally slightly saccate; stomata and hairs are missing on them, as on the whole plant (Tomlinson 1982). True vessels are lacking in all parts of the plant (Wagner 1977).

Epicuticular wax structures of the Convallaria type have been reported for *Sciaphila purpurea* (Barthlott in Maas-van de Kamer 1995).

INFLORESCENCE. The inflorescence is a few- to many-flowered raceme (rarely reduced to a solitary flower) with bracts or rarely both bracts and bracteoles (*Seychellaria*), and not a cyme as has been suggested for *Triuris*. In *Triuris* the inflorescence is 1–4-flowered with a strongly flexuous main axis and amplexicaul, basally auriculate bracts. In Triurideae (except *Lacandonia*) the plants are usually dioecious. They are always monoecious in Sciaphileae and *Lacandonia*, where the pistillate flowers are produced first at the base of the raceme and plants are protogynous; the male flowers appear in the upper part of the inflorescence (Maas-van de Kamer 1995).

FLOWER STRUCTURE. Most flowers are unisexual but they are hermaphrodite in several species of *Sciaphila* and in *Lacandonia*. In the former the

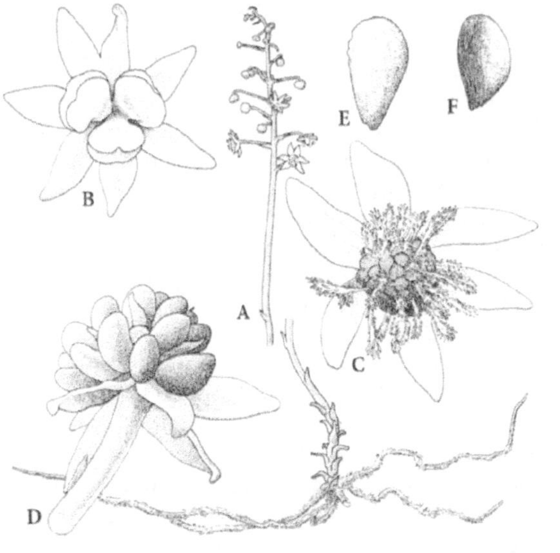

Fig. 130A–F. Triuridaceae. *Sciaphila albescens*. **A** Flowering plant. **B** Staminate flower. **C** Pistillate flower, note penicillate stigmas. **D** Infructescence. **E** Fruit. **F** Seed. (Maas et al. 1986)

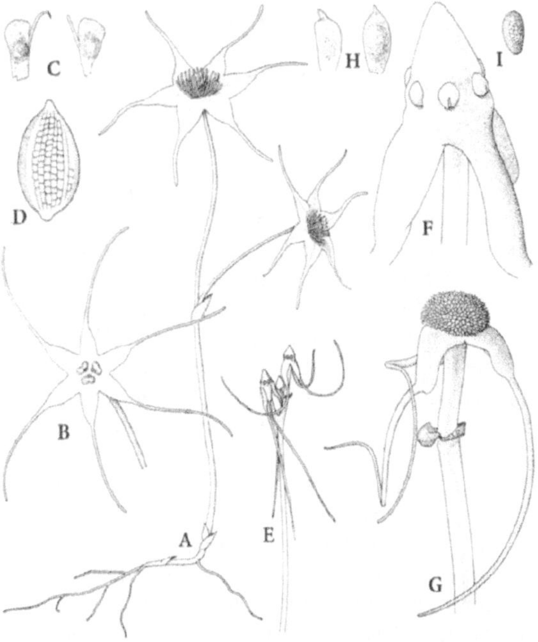

Fig. 131A–I. Triuridaceae. **A–D** *Peltophyllum luteum*. **A** Pistillate plant. **B** Staminate flower. **C** Two ovaries. **D** Seed. **E–G** *Triuris hexophthalma*. **E** Staminate flowers. **F** Staminate flower, enlarged. **G** Pistillate flower. **H** Two ovaries. **I** Seed. (Maas et al. 1986)

free stamens are sessile on the margin of the receptacle that bears the ovaries, while *Lacandonia* exhibits the inverse position: the gynoecium surrounds the 3 centrally arranged anthers (Márquez-Guzmán et al. 1989). In contrast to the uniformly reduced vegetative parts, flower structure shows a wide range of variation between species and even between members of one species. This concerns the distribution of male and female flowers within one inflorescence, the combination of male or female organs within one flower, as well as the position, number, size, shape and structure of floral organs (Rübsamen-Weustenfeld 1991; Maas-van de Kamer 1995).

The perianth consists of 3–10 (mostly 3 or 6) tepals; the tepal primordia are successively produced and may be arranged in a single or 2 whorls. The tepals are valvate in bud and stellately patent at anthesis, but in most species soon completely reflexed and appressed to the pedicel. The inner side of the tepals is often covered with papillae, hairs, or a raised glandular tissue along the midrib, while the apex sometimes bears tufts of hairs, globose knobs or long tails which are inflexed in bud.

The stamens range in number from (1–)2–6(8), with 3 as the most common number. In *Sciaphila* and *Soridium* the stamens are facing the outer tepals, while in *Peltophyllum* they are probably opposite the tepals of the inner whorl. In the first case they may represent the outer staminal whorl, in the second the inner one. The same may be true of *Triuris hyalina*, in which only 3 (outer) tepals alternate with the stamens. In *Triuris hexophthalma* and *Triuridopsis peruviana* 6 2-sporangiate stamens are evenly distributed around the androphore. They probably represent 6 half-stamens, 2 neighbouring ones of which originally belong to 1 stamen in front of a reduced inner tepal (Rübsamen-Weustenfeld 1991). The stamens are mostly 4-sporangiate, but 3-sporangiate in several species of *Sciaphila* and 2-sporangiate in *Soridium*, *Lacandonia*, *Triuris hexophthalma*, and *Triuridopsis*; they are sessile, filamented, or, as in *Triuris*, immersed in the conical androphore. The various ways of dehiscense of the anthers (longitudinally/transversally and introrse/extrorse) are discussed by Maas-van de Kamer (1995) and Rübsamen-Weustenfeld (1991), who concluded that these characters are of limited taxonomic value in the family. Three staminodes alternating with the 3 stamens are present in the centre of the male flowers of *Seychellaria*. Connective appendages overtopping the anthers are a characteristic feature of *Andruris* and *Seychellaria madagascariensis*. In *Triuridopsis* the centre of the staminate flowers is provided with a subulate projection (Maas-van de Kamer and Maas 1994).

The gynoecium consists of numerous free carpels, each containing 1 basal ovule. The development of the ovaries and ovules were described by Rübsamen-Weustenfeld (1991).

EMBRYOLOGY. (Summary: Rübsamen-Weustenfeld 1991, for *Lacandonia*: Márquez-Guzmán et al. 1989, 1993). The anther wall is 4-layered and formed according to the monocotyledonous type. The endothecial cells enlarge radially and develop U-shaped wall thickenings. The tapetal cells are 1-, 2- or 3-nucleate; the tapetum is secretory in the Triurideae, but predominantly periplasmodial or of an intermediate type in the Sciaphileae. The pollen mother cells develop successively forming square (isobilateral) or decussate tetrads. The mature pollen grains are 3-celled monads.

The ovules are anatropous, bitegmic (both integuments 2-layered), and tenuinucellate. Embryo-sac development is usually according to the Polygonum type, but in *Triuris* probably according to the Fritillaria type. Although apomixis frequently has been reported (Poulsen 1906; Wirz 1910; Giesen 1938), double fertilisation takes place at least in some cases (Rübsamen-Weustenfeld 1991: *Soridium*, Márques-Guzmán et al. 1993: probably within cleistogamous flowers of *Lacandonia*). Endosperm formation is Nuclear. In ripe seeds the endosperm is composed of 15 (Triurideae) to 150 cells (Sciaphileae). In addition to cell-wall thickenings of hemicellulose, the endosperm cells contain as food storage oil, protein and – only in immature Triurideae seeds – starch. The embryo is small and undifferentiated, and therefore has often been overlooked (Miers 1852) or confused with the endosperm. It is now known to be composed of a globose and multicellular embryo proper and a short suspensor, being disintegrated and resorbed, so that the older embryo appears to lie isolated in the endosperm of the ripe seed (Rübsamen-Weustenfeld 1991).

POLLEN MORPHOLOGY. The pollen grains are globose and 15–30 μm in diameter (Rübsamen-Weustenfeld 1991). They are inaperturate or probably sulcate (often only an indistinct aperture with thickened intine and interrupted or incomplete exine is recognisable). Exine sculpturing is finely to coarsely granulose to verrucose.

KARYOLOGY. Chromosome numbers are known for several species of the Sciaphileae. Reported counts of $n = 11, 14, 12$ (-16), 24 may point to $x = 12$, less frequently $x = 11$ and $x = 14$ as base numbers (for summary see Rübsamen-Weustenfeld 1991). The chromosome number of *Lacandonia* was determined as $n = 9$ (Davidse and Martínez 1990).

Pollination. Pollination of Triuridaceae has not yet been observed, and apomixis has often been suggested. Floral morphology, however, strongly suggests insect pollination. The family includes protogynous plants, and unisexual flowers, flowers emitting odour, or flowers with tepals provided with papils, glandular areas, hairs or appendages. Márquez-Guzmán et al. (1993) described autogamy and preanthesis cleistogamy in the hermaphrodite flowers of *Lacandonia*, which are proterandrous.

Fruit and Seed. Each fruit contains 1 seed that has approximately the same shape as the fruit: rounded at the chalazal pole and tapering towards the micropylar pole. The fruit is either a follicle dehiscing by a longitudinal slit at the abaxial side or an indehiscent achene. In the Sciaphileae the oldest fruits are the outermost on the receptacle whereas in *Triuris hexophthalma* the fruits placed in the centre of the flower mature first and are the first to be dispersed from the receptacle.

The epidermal cells of the 2- to 3-layered fruit wall are reticulately patterned by raised anticlinal and sunken periclinal cell walls, sometimes with papillate cells at the chalazal pole. Secondary wall thickenings found in the cells of the fruit wall are either radially oriented (Sciaphileae) or found to extend in longitudinal direction parallel to the surface (Triurideae) (Rübsamen-Weustenfeld 1991).

The seeds are compact and their surface sculpture (as seen by SEM) is characterised by differences in shape, size or cell pattern of the epidermal cells as well as by the structure of subepidermal cell layers; the surface of the outer periclinal cell wall is smooth, a micromorphological structure on the epidermis is lacking (Rübsamen-Weustenfeld 1991).

The testa is formed only by the outer integument; the nucellus and inner integument disappear during seed development. The inner cuticle of the testal layer is strongly thickened and directly adjacent to the endosperm tissue. In the Sciaphileae the testa is composed of both layers: the thin outer layer consists of longitudinally stretched cells and the inner layer of radially elongated cells with various types of wall thickenings or tannin contents in the different species. In the Triurideae, the testa consists only of one layer of flattened, isodiametric cells, which represent the inner layer of the outer integument (Rübsamen-Weustenfeld 1991).

Dispersal. Seed dispersal has never been observed but is supposed to be zoochorous (the diaspores becoming sticky by slime produced by the receptacle of flowers of Triurideae), anemochorous (yet the diaspores are larger than typical dust diaspores, the reticulate epidermal pattern is that of the fruit wall, the seeds are smooth, and ellipsoid to ovoid), or hydrochorous (the diaspores are water-resistant and able to float for some time) (Fleischer 1929; Rübsamen-Weustenfeld 1991).

Phytochemistry. The few data available on phytochemistry refer to the absence of UV fluorescence of the cell walls after treatment with NH_3 (Harris and Hartley 1980) and the rare occurrence of calcium oxalate raphides (Champion 1847) and tannin deposits (Rübsamen-Weustenfeld 1991).

Subdivision and Relationships Within the Family. Morphological and embryological differences confirm the subdivision of the Triuridaceae into the two tribes, Sciaphileae and Triurideae. They further support the distinction between *Sciaphila* and *Andruris* and a close relationship between *Andruris* and *Seychellaria*. *Lacandonia*, with 3 central stamens surrounded by many carpels, clearly falls within the range of variation of the Triuridaceae, where it belongs into the tribe Triurideae close to *Peltophyllum* (Rübsamen-Weustenfeld 1991; Maas-van de Kamer 1995).

Affinities. Since the publications of Miers (1842) and Gardner (1843a,b), the discussion about the relationships of Triuridaceae has not come to an end. No close relationship between Triuridaceae and the likewise heterotrophic Burmanniales could be established. However, Triuridaceae have several characters in common with both the Lilianae (*Petrosavia*) and the Alismatanae. With the latter they agree in some embryological characters (3-celled pollen, exine structure, 1-seeded carpels), but there are differences in the vegetative anatomy (Tomlinson 1982). There are several characters linking the Triuridaceae with the Lilianae: embryo sac development of the Fritillaria type, small embryos, the presence of endosperm, the Convallaria wax type, heterotrophy, and some anatomical traits

(Tomlinson 1982; Rübsamen-Weustenfeld 1991). As the Triuridaceae are a comparatively uniform family, which is clearly distinct from other systematic groups, they should be accommodated in a separate superorder, Triuridanae, or at least in an order, Triuridales, rather than being placed in the Lilianae.

Distribution and Habitats. The family is mainly distributed in tropical America, Africa and Asia, with most species in the Austral-Asian region. While *Sciaphila* is pantropical, the other genera have more restricted areas of distribution. Some genera occur in subtropical regions, such as *Peltophyllum* in Argentina and Paraguay and *Sciaphila* in Japan.

The plants prefer to live in dense and humid forests hidden under leaf mould at the base of large trees or along the bank of streams, occurring at an altitude of about 200 to 2200 m. Less often, they are found in temporarily inundated forests, forests on white sand, bamboo thickets or on termite nests (*Sciaphila purpurea*, *S. arfakiana*). They often grow in close association with other mycotrophic plants of various families.

Key to the Genera

1. Plants monoecious; flowers unisexual (some ssp. of *Sciaphila* hermaphrodite); tepals 4, 6 or 8; stamens 2, 3, 4 or 6; tepals not caudate; style basal; fruit dehiscent (*Soridium* excepted). Neotropics and Palaeotropics — Tribe Sciaphileae 2
- Plants dioecious; flowers unisexual (*Lacandonia* monoecious and bisexual); tepals 3 or 6; stamens 3 or 6; tepals caudate; style apical; fruit indehiscent. Neotropics — Tribe Triurideae 6
2. Staminate flowers provided with staminodes and/or connective appendages — 3
- Staminate flowers without sterile elements as staminodes or connective appendages — 4
3. Staminate flowers with 3 staminodes and sometimes also with appendaged connectives. Madagascar, Seychelles and E Africa — 1. ***Seychellaria***
- Staminate flowers without staminodes, but always with appendaged connectives. SE Asia — 2. ***Andruris***
4. Branches of inflorescence verticillate; staminate flowers with 4 stamens and 8 tepals. Sri Lanka — 3. ***Hyalisma***
- Branches of inflorescence alternate; staminate flowers with 2–6 stamens and 4–6 tepals — 5
5. Staminate flowers with 2 stamens and 4 tepals; fruit an achene. Tropical S America — 5. ***Soridium***
- Staminate flowers with 3 or 6 stamens and 4 or 6 tepals; fruit a follicle. Pantropical — 4. ***Sciaphila***
6. Flowers hermaphrodite; stamens 3, placed in the centre of flower within the carpels. Mexico — 7. ***Lacandonia***
- Flowers unisexual; plants dioecious — 7
7. Staminate flowers with anthers sessile on a conical androphore. Tropical America — 8. ***Triuris***
- Staminate flowers without androphore — 8
8. Staminate flowers with 3 tepals, 6 filamented monothecal anthers, and a sterile projection in the centre of the flower. Peru — 9. ***Triuridopsis***
- Staminate flowers with 6 tepals, 3 sessile dithecal anthers, and without sterile projection. S America — 6. ***Peltophyllum***

I. Tribe Sciaphileae Miers (1852).

Monoecious; flowers unisexual (some spp. of *Sciaphila* hermaphrodite); number of tepals 4, 6 or 8; number of stamens 2, 3, 4 or 6; tepals not caudate but bearded, papillate, or ending in a globose knob or short point; style (sub)basal; fruit dehiscent (*Soridium* excepted). Neotropics and paleotropics.

1. *Seychellaria* Hemsley

Seychellaria Hemsley, Ann. Bot. (Lond.) 21: 74 (1907); Giesen in Engler, Pflanzenreich IV. 18: 11 (1938), rev.; Vollesen, Fl. Trop. E. Africa 1 (1985).

Perennial rhizomatous herbs with scale-like leaves. Plants monoecious. Inflorescence a simple raceme or thyrse. Flowers unisexual. Tepals 6, unequal, glabrous with glandular tissue on the inner surface (only in female flowers), unappendaged. Male flowers with 3 stamens opposite the 3 larger tepals, alternating with 3 staminodes, sometimes connective provided with long appendage; anthers 4-sporangiate, with short filaments, basally connected in the centre of the flower. Female flowers with numerous free carpels; style (sub)basal, glabrous. Follicle dehiscent. Three spp., Seychelles, Madagascar and Tansania.

2. *Andruris* Schltr.

Andruris Schltr., Bot. Jahrb. Syst. 49: 71 (1912); Giesen in Engler, Pflanzenreich IV. 18: 15 (1938), rev.
Parexuris Nakai & F. Maek. (1936), partly.

Erect, simple or branched herbs with rhizomes and few scale-like leaves; monoecious. Inflorescence a raceme, sometimes branched. Flowers unisexual. Tepals (4–)6, equal (in most female flowers) or unequal, glabrous, in female flowers provided with a raised glandular tissue on the inner surface, in male flowers often ending in a globose knob. Male flowers with 3 stamens opposite the 3 larger tepals; anthers 4-sporangiate with short filaments, basally connected; connective with long, subulate appendage. Female flowers with numerous free carpels; style (sub)basal, glabrous. Follicle dehiscent. Five spp., Malesia, Polynesia, Micronesia, E India, S Japan, NE Australia.

3. *Hyalisma* Champ.

Hyalisma Champ., Calcutta J. Nat. Hist. 7: 466 (1847); Giesen in Engler, Pflanzenreich IV. 18: 29 (1938), rev.

Herbs, sometimes basally branched, with scalelike leaves. Plants monoecious or dioecious. Inflorescence a subumbellate raceme, at each node 2 to 4(5) flowers opposite each other or in verticils. Flowers unisexual. Tepals 8, equal, glabrous and unappendaged. Male flowers with 4 stamens, opposed the tepals; anthers 4-sporangiate, on short filaments, which are connected in the centre forming a central disk. Female flowers with numerous free carpels; style (sub)basal, glabrous. Fruit a dehiscent follicle. Only one sp., *H. janthina* Champ., S India and Sri Lanka.

4. *Sciaphila* Blume — Fig. 130

Sciaphila Blume, Bijdr. 10: 514 (1825); Giesen in Engler, Pflanzenreich IV. 18: 30 (1938), rev.; vau de Meerendonk in Fl. Males. I, 10: 110 (1984), reg. rev.; Maas & Rübsamen, Fl. Neotrop. 40: 21 (1986), reg. rev.

Simple or rarely branched erect herbs with rhizomes and scale-like leaves. Plants monoecious. Inflorescence a terminal raceme, sometimes branched from the base. Flowers unisexual or hermaphrodite. Tepals 4–10, usually 6, equal or unequal, inner side papillate, apex sometimes with tufts of hairs, or ending in a globose knob or short point. Male flowers with 2, 3 or 6 stamens opposed the tepals; anthers 3- or 4-sporangiate, sessile or on short filaments, often connate at the base forming a very small and flat circular androphore in the centre of the flower. Female flowers with numerous free carpels; style (sub)basal, apex papillate to penicillate or glabrous. Follicle dehiscent. About 30 spp., pantropical and subtropical, mainly in Indo-Malesia.

5. *Soridium* Miers

Soridium Miers, Proc. Linn. Soc. Lond. 2: 74 (1850); Giesen, in Engler, Pflanzenreich IV. 18: 71 (1938), rev.; Maas & Rübsamen, Fl. Neotrop. 40: 38 (1986), reg. rev.

Unbranched or basally branched herbs with rhizomes and scale-like leaves. Plants monoecious. Inflorescence a simple raceme. Flowers unisexual. Tepals 4, equal, inner side papillate, unappendaged. Stamens 2, epitepalous; anthers 2-sporangiate, sessile. Female flowers with numerous free papillate carpels; style (sub)basal, apex penicillate-papillate. Fruit an indehiscent achene. Only one sp., *S. spruceanum* Miers, Central America and northern S America.

II. Tribe Triurideae Miers (1852).

Dioecious (*Lacandonia* hermaphrodite); tepals 3 or 6, caudate, provided with a filiform appendage; stamens 3 or 6; style (sub)apical; fruit indehiscent. Neotropical.

6. *Peltophyllum* Gardner — Fig. 131A–D

Peltophyllum Gardner, Proc. Linn. Soc. Lond. 1: 176 (1843); Maas & Rübsamen, FL. Neotrop. 40: 39 (1986), reg. rev.
Hexuris Miers (1850); Giesen in Engler, Pflanzenreich IV. 18: 73 (1938), rev.

Unbranched herbs with filiform rhizomes, hyaline, flattened stems and scale-like leaves. Plants dioecious. Inflorescence a simple raceme. Flowers unisexual. Tepals usually 6, equal, glabrous, apex caudate. Male flowers with 3 stamens opposite the tepals; anthers 4-sporangiate, (sub)sessile on edge of concave centre of receptacle. Female flowers with numerous papillate carpels; style subapical, glabrous. Fruit an indehiscent achene. Two spp., S America (SE Brazil, adjacent Argentina, Paraguay, and Guyana). N.B.: *Peltophyllum* was based by Gardner on mixed material of a triuridaceous plant and a peltate leaf of a Menispermacea.

7. *Lacandonia* E. Martínez & Ramos

Lacandonia E. Martínez & Ramos, Ann. Mo. Bot. Gard. 76: 128 (1989); Martínez in Fl. Mesoamer. 6: 19 (1994), reg. rev.; Maas-van de Kamer & Maas, Plant Syst. Evol. 192: 258 (1994).

Hyaline herbs with rhizomes, simple stems, and scale-like leaves. Plants monoecious. Inflorescence a simple raceme. Flowers hermaphrodite. Tepals (4–)6, equal, densely papillate, apex caudate. Stamens (2–)3(–4), inserted around the edge of the concave centre of the receptacle; anthers 2(–3)-sporangiate, sessile. Gynoecium of numerous free, papillate carpels surrounding the stamens; style subapical, glabrous. Fruit an indehiscent achene. Only one sp., *L. schismatica* E. Martínez & Ramos, Mexico.

8. *Triuris* Miers — Fig. 131E–G

Triuris Miers, Proc. Linn. Soc. Lond. 1: 96 (1841); Giesen in Engler, Pflanzenreich IV. 18: 75 (1938), rev.; Maas & Rübsamen, Fl. Neotrop. 40: 47 (1986), reg. rev.

Unbranched herbs with short rhizomes, hyaline, flattened stems and 0–2 scale-like leaves; dioecious. Inflorescence a simple raceme. Tepals 3, equal, glabrous, apex caudate. Male flowers with 3 stamens alternating with the tepals or 6 (half-) stamens, 2 of them in front of each tepal; anthers 4-sporangiate (3 stamens) or 2-sporangiate (6 stamens), sessile, inserted in cavities at the base of the large, fleshy, conical to deltoid central androphore; female flowers with numerous glabrous carpels; style (sub) apical, glabrous. Fruit an indehiscent achene. Three spp., Central and S America.

9. ***Triuridopsis*** H. Maas & Maas

Triuridopsis H. Maas & Maas, Plant Syst. Evol. 192: 257 (1994).

Unbranched herbs with rhizomes and scalelike leaves; dioecious. Inflorescence a simple raceme. Tepals 3(4), equal, glabrous, with a reflexed subapical appendage. Male flowers with 6(8) (half-) stamens, 2 of them in front of each tepal; anthers 2-sporangiate; filaments distinct, basally connate; centre of the flowers provided with a sterile subulate projection. Female flowers with many free carpels; style apical, glabrous. Fruit an indehiscent achene, apically covered with large swollen cells. Only one sp., *T. peruviana* H. Maas & Maas, S America (Amazonian Peru).

Selected Bibliography

Champion, J.G. 1847. On two new Ceylon plants related to *Sciaphila* of Blume. Calcutta J. Nat. Hist. 7: 463–469.

Davidse, G., Martínez S., E. 1990. The chromosome number of *Lacandonia schismatica* (Lacandoniaceae). Syst. Bot. 15: 635–637.

Fleischer, E. 1929. Zur Biologie feilspanförmiger Samen. Bot. Arch. 26: 86–132.

Gardner, G. 1843a. Description of *Peltophyllum*, a new genus of plants allied to *Triuris* of Miers, with remarks on their affinities. Proc. Linn. Soc. Lond. 1: 176.

Gardner, G. 1843b ("1845") Description of *Peltophyllum*, a new genus of plants allied to *Triuris* of Miers, with remarks on their affinities. Trans. Linn. Soc. Lond. 19: 155–160.

Giesen, H. 1938. Triuridaceae. In: Engler, A. (ed.) Das Pflanzenreich. IV.18: 1–84. Leipzig: W. Engelmann.

Harris, P.J., Hartley, R.D. 1980. See general references.

Leake, J.R. 1994. The biology of myco-heterotrophic ("saprophytic") plants. New Phytol. 127: 171–216.

Maas, P.J.M., Rübsamen, T. 1986. Flora Neotropica. Monograph Number 40. Triuridaceae. New York: The New York Botanical Garden, pp. 1–55.

Maas-van de Kamer, H. 1995. Triuridiflorae – Gardener's delight? In: Rudall, P.J., Cribb, P.J., Cutler, D.F., Humphries, C.J. (eds.) Monocotyledons: systematics and evolution. Royal Botanic Gardens, Kew, pp. 287–301.

Maas-van de Kamer, H., Maas, P.J.M. 1994. *Triuridopsis*, a new monotypic genus in Triuridaceae. Plant Syst. Evol. 192: 257–262.

Márquez-Guzmán, J., Engleman, M., Martínez-Mena, A., Martínez, E., Ramos, C. 1989. Anatomia reproductiva de *Lacandonia schismatica* (Lacandoniaceae). Ann. Mo. Bot. Gard. 76: 124–127.

Márquez-Guzmán, J., Vásquez-Santana, S., Engleman, E.M., Martínez-Mena, A., Martínez, E. 1993. Pollen development and fertilization in *Lacandonia schismatica* (Lacandoniaceae). Ann. Mo. Bot. Gard. 80: 891–897.

Martínez, E., Ramos, C.H. 1989. Lacandoniaceae (Triuridales): Una nueva familia de México. Ann. Mo. Bot. Gard. 76: 128–135.

Miers, J. 1842 ("1845"). Description of a new genus of plants from Brazil. Trans. Linn. Soc. Lond. 19: 77–80.

Miers, J. 1852. On the family of Triuriaceae. Trans. Linn. Soc. Lond. 21: 43–59.

Poulsen, V.A. 1906: *Sciaphila nana* Bl. Et Bidrag til Stovvejens Udvikling hos Triuridaceerne. Vidensk. Medd. Dan. Naturhist. Foren. Kjobenhavn for 1906: 161–176.

Rübsamen-Weustenfeld, T. 1991. Morphologische, embryologische und systematische Untersuchungen an Triuridaceae. Bibl. Bot. 140: 1–113.

Tomlinson, P.B. 1982. See general references.

Wagner, P. 1977. See general references.

Wirz, H. 1910. Beiträge zur Entwicklungsgeschichte von *Sciaphila* spec. und von *Epirrhizanthes elongata*. Flora II, 1: 395–446.

Velloziaceae

K. Kubitzki[1]

Velloziaceae Endl., Ench. Bot.: 101 (1841), nom. cons.

Herbaceous or ± shrubby *perennials*; stem woody, fibrous, sometimes resinous, simple or dichasially branched, covered with persistent leaves or their imbricate sheaths. *Leaves* tristichous or spirotristichously arranged, clustered at end of the stem or its branches when new, the grasslike (rarely subterete) blades linear, dorsiventral, basally sheathing, entire or often dentate-spinulose along margins, reflexing with age, or caducous along a regular, usually straight transverse line. Adventitious roots alternating with the leaves, penetrating the lower, old, fibrous leaf sheaths in their downward growth. *Inflorescences* terminal, becoming lateral by elongation of axis, 1–many-flowered, their scapes caducous. *Flowers* perfect or rarely unisexual (*Barbacениopsis*), actinomorphic, usually brightly coloured, epigynous or rarely hemiepigynous (*Vellozia hemisphaerica* and allied spp.). Hypanthium equalling or often exceeding the ovary in length and adnate to it, often provided with a corona dorsal to the stamens and usually united to them. Tepals 6, in 2 whorls, generally all similar, blue, lilac, white, yellow, orange, red, purplish or green, erect to reflexed at anthesis. Stamens 6 to numerous, then in 6 bundles; filaments free, terete, simple and often bearing basal appendages ventrally (*Vellozia*), or adnate to the corona (*Barbacenia*), or fused with the tepals (*Barbaceniopsis*, *Xerophyta*); anthers elongated, 2-thecate, 4-sporangiate, exceptionally 10-sporangiate (*X. schnizleiniana*, Menezes 1980), basifixed to dorsifixed, introrse, or rarely extrorse; each theca (rarely each pollen sac) opening by a longitudinal slit. Ovary 3-locular, often depressed at apex or the walls prolonged upwards into a hypanthium; style slender but usually enlarged towards the stigmas; stigmas 3, apical or subapical, distinct to wholly confluent, linear to orbicular; ovules numerous; placentas stalked, expanded, 2 in each locule. Septal nectaries present in ovary walls. *Fruit* a capsule, loculicidally dehiscent by slits and/or pores; seeds numerous; embryo small; endosperm copious, starchy, non-farinaceous.

About 8 genera (number controversial) and over 200 species in S and Central America and Africa/Madagascar and the southern Arabian Peninsula.

Vegetative Morphology. The Velloziaceae are mostly erect, often profusely branched plants in which the leaves are aggregated in terminal rosettes; rarely, they have a creeping axis.

Alternating with the tristichous leaves, the adventitious roots sprout from the short apical, living portion of the stem and penetrate the dead matter of the leaf sheaths that surround the thin stem, growing downward towards the ground (Fig. 132). New roots are added every year from the growing stem apex in a constant rhythm; in *Vellozia incurvata* the number of roots averages 10 per year (Alves 1994). The roots add to the volume particularly of the basal stem portion, causing a peculiar form of secondary growth. When the stem base expands beyond a certain diameter, the cover of the leaf sheaths is lost and the stem base consists entirely of adventitious roots covering the old stem.

The leaves are arranged in three orthostichies or spirostichies; in various *Vellozia* species 2 or 3 parallel series of spirally ascending leaf sheaths can be observed that frequently change from sinistrorse to dextrorse and back again along the stem, usually after the formation of a flower. The growth rate was found to be low in *Vellozia* and Alves (1994), by extrapolating the annual growth, calculated an age of 105–551 years for *Vellozia* specimens 2.5–3 m tall. The stem of one of them was found to be about 12 mm in diameter. In reptant *Vellozia* species such as *V. crinita* the roots sprout from various points along the stem.

The function of this fibrous coat of roots is to gather moisture from fogs and rains and at the same time to protect the stem from savannah fires. The leaf sheaths in many *Vellozia* species are so densely packed and hermetically glued together by resin that, due to the lack of oxygen alone, they cannot burn. In contrast, the detached leaf sheaths of *Vellozia* species burn very well and stem segments about 50 cm long are sold at fairs in Bahia as fire starters; housewives claim that the sheaths will burn even in torrential rains (Alves 1994).

The leaves of most *Vellozia* have the stomata in abaxial furrows in the leaves, permitting contrac-

[1] Partly based on a draft manuscript prepared by the late L.B. Smith in about 1985.

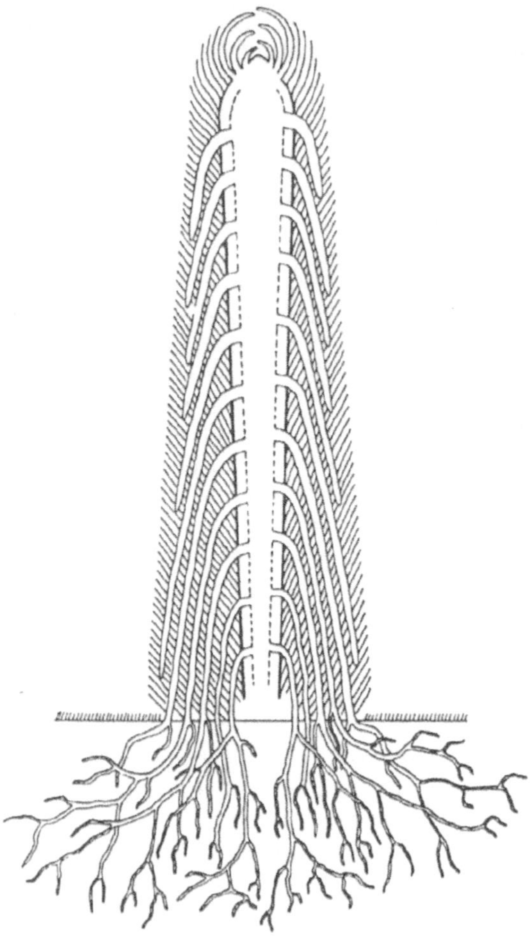

Fig. 132. Velloziaceae. Schematic longitudinal section of *Vellozia* plant, the stem producing adventitious roots which form a mantle around the stem, piercing the persistent leaf bases and spreading upon or in the substrate. (Weber 1954)

tions, inrolling, outrolling and lengthwise folding of the blades in response to water loss from the mesophyll. Such furrows occur also in most *Xerophyta* species and in *Barbaceniopsis* but are absent from *Nanuza*, *Talbotia* and nearly all *Barbacenia*. Epicuticular wax occurs in continuous layers or composite rodlets of the Strelitzia type (Barthlott and Frölich 1988).

Inflorescences. Many members of the family have solitary flowers which appear to be inserted laterally, but T. Stützel (pers. comm.) found the young stages to be initiated strictly terminally when examing shoot apices of a *Barbacenia*. Since many members of the family are reported to produce several-flowered racemes in a terminal position, which presumably have been reduced to a single flower in many species, the position of the latter, strictly spoken, is pseudoterminal. Later, when the scars of the peduncles of the flowers are found laterally on the axis, they have shifted there due to the sympodial structure of the shoots. This explains why flowers are found frequently in the axils of ramifications, which must be considered as dichasial.

Vegetative Anatomy. The anatomical structure of velloziaceous roots is identical with that of other monocotyledons. The vessels have simple perforations and are filled with a brown or yellow resin, especially abundantly so in *Vellozia*. The endodermis is thin-walled, sometimes starch-containing. The inner cortex has an open loose structure, whereas the outer is composed of a cylindric layer of stereomatic tissue. Outside these tissues is a hypodermis of a single layer of thin-walled cells and finally the epidermis, which sometimes develops root hairs. The hypodermis is stained blue by iodine and sulphuric acid.

Lipophilic (highly methoxylated) flavonoids are deposited on the leaf epidermis and can be washed off with methanol in *Vellozia* spp. (Harborne et al. 1994); it is unknown, however, in what form these substances are deposited there. Trichomes, when present, are branched and tend to be twisted or even coalesced with each other. In *Vellozia*, the stomata are mostly confined to the abaxial leaf surface and tend to be crowded in the furrows, often together with variously shaped papillae. The remaining genera have mostly amphistomatic leaves. Stomata are mostly paracytic, but often also tetracytic, i.e. provided with two parallel and two polar subsidiary cells. The leaves are either bifacial with distinct palisade and spongy tissue or aequifacial, mostly constantly so for genera or sections. The vascular bundles are collateral, with 1–several vessels in the xylem (absent from the genera *Xerophyta*, *Barbaceniopsis*, *Nanuza* and *Talbotia*: Menezes et al. 1994), and generally 2 phloem units lying laterally in flanges of the sclerenchymatous girder. Adaxially and abaxially the vascular bundles are surrounded by 2 sclerenchymatous girders, which are crescentiform, Y- or U-shaped. These caps extend to the lower and in some species also to the upper leaf epidermis (Fig. 133) and various combinations of these differently shaped caps are constant for taxonomic groupings in the family (Ayensu 1968, 1969, 1974). The extensions of the 2 caps do not meet in the middle of the bundles where the endodermis borders directly with the fibrovascular bundle and transfusion tracheids establish a contact between xylem and endodermis (Menezes 1975). In the genus *Barbacenia* s.l. an additional parenchymatous bundle sheath is added externally to the endodermis (Menezes 1980). Idioblasts containing raphide bundles seem to be rare, but tannin-containing cells are present in many species. The presence of vessels in the stems has been questioned (Wagner 1977).

Flower Structure. Although basically following the trimerous monocotyledonous pattern, flower structure shows considerable plasticity, obviously in adaptation to pollinators. This variation refers to the formation of shorter or longer hypanthial tubes in all genera, *Nanuza*, *Talbotia* and some species of *Vellozia* excepted; to the formation of a corona in *Barbacenia* s.l.; to elevated stamen numbers in *Vellozia* (9, mostly 18, occasionally up to 76; Fig. 134); to the variation from latrorse to introrse anthers (Menezes 1980); to the opening of anthers by a longitudinal slit common to the 2 pollen sacs of each theca vs. 1 slit for each pollen sac; to the structure of the stigma (trilobed-horizontal vs. elongated); and to the position of the stigma relative to the anthers (stigma above the stamens vs. below them). It is obvious that classifications based on these traits, which are highly integrated into the process of pollination, will reflect pollination syndromes rather than monophyletic groupings.

There has been much debate on the corona inserted between the tepals and the androecium. These extra appendages can be quite inconspicuous but sometimes, as in *Burlemarxia spiralis* (Menezes and Semir 1990), form a narrow petaloid tube densely enclosing the style and filaments, obviously excluding nectar consumers without elongated mouth parts. These appendages have often been interpreted as staminodial and more precisely filamentous (because anthers with a very short or without filament are often inserted on them). Menezes (1980, see also Menezes and Semir 1990) has studied their vascular bundle supply and has concluded that the appendages do not form part of the androecium but belong to the perianth, thus forming a corona such as in the Amaryllidaceae.

Embryology (mostly from Menezes 1976). The tapetum is secretory and microsporogenesis is successive. The pollen grains are dispersed in the 3-celled (Menezes 1976) or 2-celled (Johri et al. 1992) stage.

The ovules are anatropous, bitegmic and weakly crassinucellate to tenuinucellate. No primary parietal cell is cut off from the archesporial cell, but periclinal divisions in the nucellar epidermis usually give rise to a cellular cap above the embryo sac. Embryo-sac formation seems to follow the Polygonum type. Endosperm formation is Helobial, with the chalazal chamber small in size. The endosperm becomes rich in starch.

Pollen Morphology. The pollen grains of *Talbotia*, *Xerophyta*, *Nanuza* and *Barbacenia* s.l. are sulcate monads with a semitectate, reticulate to vermiform exine. Grains of *Vellozia* are connected in tetrahedral tetrads, apparently inaperturate and have a much thinner exine than the monads. The tectum in *Vellozia* varies from nearly compact to more or less disintegrated (Erdtman 1952; Ayensu and Skvarla 1974). *Barbaceniopsis boliviensis* has pollen in tetrads, dyads or monads (Menezes 1988).

Karyology. A considerable number of chromosome counts has now become available (Goldblatt and Poston 1988, Melo et al. 1997), which allow the following groups to be distinguished:

- one group with diploid species ($2n$ = 14 or 16) comprising the genera *Nanuza*, *Vellozia*, and the Brazilian representatives of *Xerophyta*;
- a second group of tetraploid species ($2n$ = 34), including all Brazilian species of *Pleurostima*, *Barbacenia*, and *Burlemarxia*;
- and a third group of hexaploid species ($2n$ = 48) including *Talbotia elegans* and the African members of the genus *Xerophyta*.

These results point to an American origin of the family with x = 8 or 9 as a base number.

An areticulate interphase nucleus structure is a constant feature within the material studied by Melo et al. (1997).

Pollination. Flowers of New World Velloziaceae are known to be pollinated by bees, hawkmoths, hummingbirds and bats (Vogel 1969; Ayensu 1973; Sazima 1977, 1978; Sazima and Sazima 1990). Most species of *Vellozia* are melittophilous and the main attractant of their dish- and bowl-shaped, violet or whitish and fragrant flowers is pollen, while a relatively concentrated nectar is produced in small amounts. Some hummingbird-pollinated Vellozias have tubular purplish flowers that produce much nectar; in these species, the filamental appendages (typical of the genus *Vellozia*) are particularly well developed and block the access to the nectar chamber. The white flowers of the former sect. *Radia* of

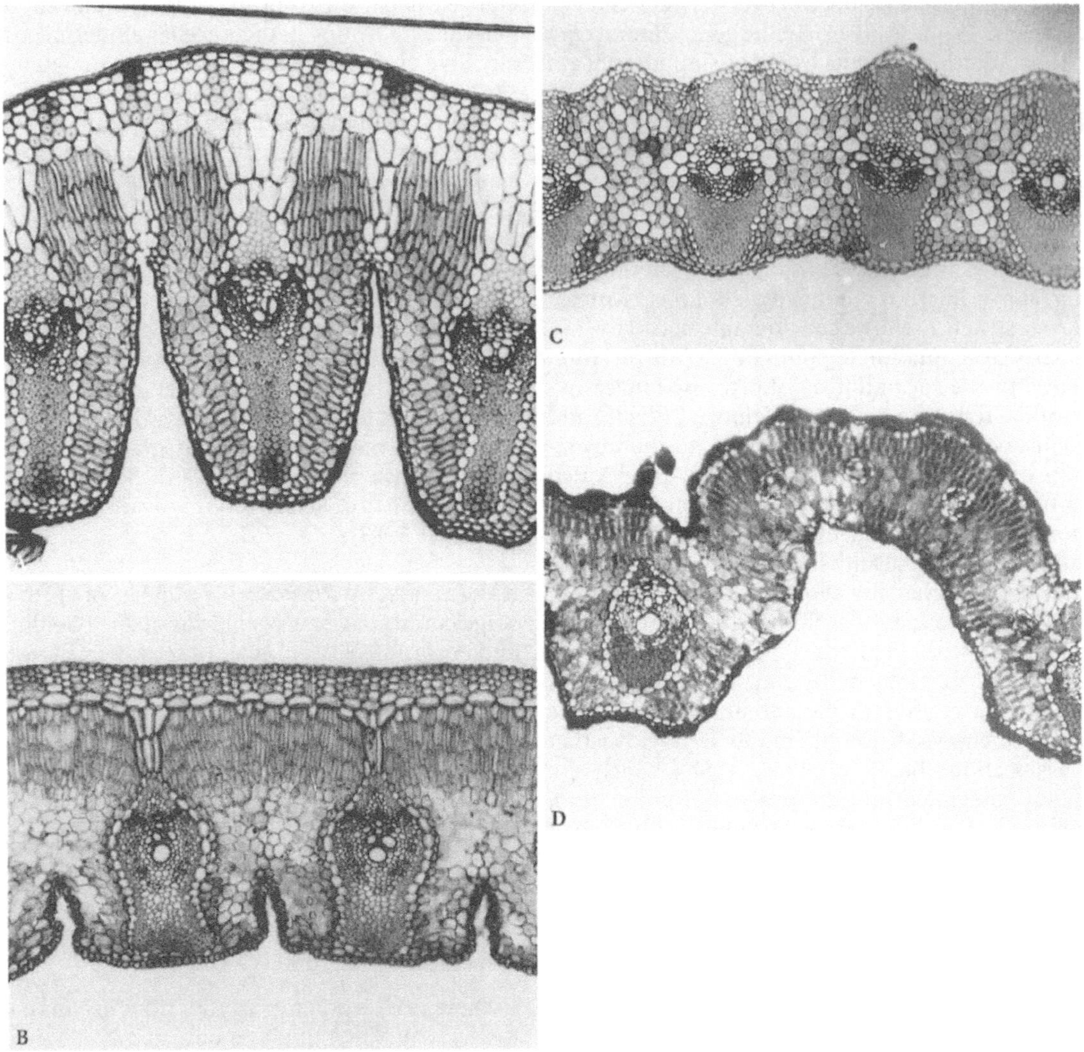

Fig. 133A–D. Velloziaceae. Transversal sections of leaves. **A** *Vellozia tubiflora*, with abaxial furrows and water-storing tissue above the fibrovascular strands and furrows. **B** *V. luteola*, less pronounced furrows and water-storing tissue only above fibrovascular strands. **C** *Barbacenia reflexa*, furrows lacking, water-storing parenchyma between the fibrovascular strands which extend from upper to lower epidermis. **D** *Nanuza plicata*, lamina plicate and with oil cells in the mesophyll; furrows, sclerenchymatic strands, and water-storing parenchyma lacking. (Photo R. Mello-Silva)

Vellozia and some *Pleurostima* are provided with very narrow floral tubes up to 105 mm in length and are supposed to be pollinated by hawkmoths. Pollination by hummingbirds, often trap lining, is known from many *Barbacenia*, whose flowers are yellow, orange, red or purplish and produce no scent but much dilute nectar. Some nocturnally flowering *Barbacenia* are known to be pollinated by long-tongued Glossophaginae. In the apparently ornithophilous *Barbacenia schwackei* bee pollination was found to be as effective as bird pollination in terms of reproductive success (Sazima 1978). It seems that in the Velloziaceae transitions between different modes of pollination can be studied in various instances.

Fruit and Seed. The variety in size and shape of the fruit capsule has been illustrated by Ayensu (1973). The surface of the capsules may be smooth or covered by bristles or glandular excrescences. The capsules are sticky, especially when atmo-

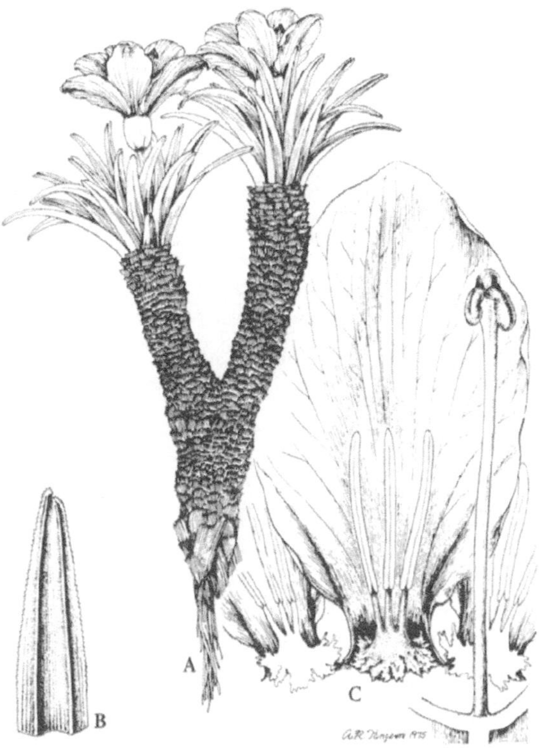

Fig. 134A–C. Velloziaceae. *Vellozia bicarinata*. A Habit. B Leaf. C Tepals with 3-staminate, basally appendaged phalanges, style, and lobed stigma. (Smith and Ayensu 1976)

spheric humidity mixes with the gummy substance that is present on the surface of all capsules. Opening of the capsule is either by 3 loculicidal fissures or by pores (for details see Menezes and Semir 1991).

The dried seeds of *Talbotia* are provided with distinctive hooks (Ayensu 1973); such hooks are not known from other genera. In the seed coat, the testa consists of 2 cell layers and does not produce phytomelan. The tegminal layer is strongly compressed. The endosperm contains copious starch, which together with the lack of phytomelan argues against a relationship with the Hypoxidaceae where the endosperm is starchless (Huber 1969).

PHYTOCHEMISTRY. Resins are most conspicuous in all parts of *Vellozia* and most *Barbacenia* s.l., even in the vessels of the roots and in excretion spaces in the leaves, but seemingly not in *Nanuza* and *Talbotia*, perhaps not in *Xerophyta*. The resins are complex mixtures of pentacyclic and tetracyclic triterpenes and diterpenes, all particularly highly oxidised, and many of them new for the family (Hegnauer 1963, 1986). Another characteristic feature are the highly O- and C-methylated and prenylated, lipophilic flavonoids that seem to occur on the cuticula and in the cytoplasmatic phase within the leaves; vacuolar flavonoids are mostly represented by C-glycosylflavones (Harborne et al. 1994). Leaf wax alkane distribution has been compared with taxonomic groupings in the family by Salatino et al. (1989). Cell wall-bound ferulate was found to be absent from the single species of *Xerophyta* tested by Harris and Hartley (1980).

SUBDIVISION AND RELATIONSHIP WITHIN THE FAMILY. Mello-Silva (1991) has given an account of the perplexing history of the family with a useful synopsis of the competing classifications proposed by Smith and Ayensu (mainly 1976) and Menezes (1980). These classifications are virtually intransigent in the choice of characters emphasised and the attempt to classify New World species in Old World genera and vice versa. While Smith and Ayensu (1976) based their classification on stigma shape, the shape of the sclerenchymatic caps accompanying the leaf vascular bundles, the structure of stamens etc., Menezes (1980) laid more stress on characters such as the presence of a corona, the simple vs. double nature of the bundle sheath surrounding the leaf vascular traces, and stamen number. In Smith and Ayensu's (1976) classification, in which preference is given to characters of the stigmas, *Xerophyta* is limited to the Old World. Menezes (1980), relying on the number of stamens, recognises *Xerophyta* as represented also in Brazil by a number of species, whereas Smith includes these species in *Vellozia* as sect. *Xerophytoides*.

In a cladistic study by Menezes et al. (1994), a wide array of characters and all taxa at generic and tribal rank that are in use in current classifications have been considered. Although the results are cautiously interpreted by the authors, their choice of the Bromeliaceae as an outgroup appears problematic and may bring distortions to the polarisation of characters and the resulting tree. I agree that *Talbotia* and *Nanuza* are early branches which may have originated prior to the extreme adaptation to xeric conditions and the origin of hypanthial tubes. *Vellozia* comes out as the sister group of the rest of the family and its polyandric condition and tetrad pollen (both previously considered as primitive for the family by Menezes 1980, 1988) may be apomorphic.

Talbotia in many ways differs from all other genera. Its floral morphology is unique in its unappendaged tepals with completely free filaments and extrorse anthers, while its leaf blades are broad and mesophytic in function.

Affinities. Although in the past the Velloziaceae and Hypoxidaceae have been considered as closely related, the characters they have in common, including the general habit, epigyny and paracytic stomata, are superficial similarities void of phylogenetic significance. Indeed, the Hypoxidaceae are typical members of the Asparagales with phytomelaniferous, starchless seeds. Dahlgren et al. (1985) included the Velloziaceae in the Bromeliiflorae, based on characters such as the para- and tetracytic stomata, Helobial endosperm formation, starchy endosperm, and the Strelitzia-type wax (Barthlott and Frölich 1988). However, the lack of UV fluorescence of the cell walls is at variance with this allocation. The placement of the family together with Cyclanthaceae, Pandanaceae, Stemonaceae and *Acanthochlamys* in the *rbc*L trees (Chase et al. 1995) does not seem to be accompanied by morphological characters; at best it indicates the strong isolation of the family.

Distribution and Habitats. In the Old World, the family is distributed from Madagascar and S Africa in the south to Angola in the NW and Ethiopia and further with a single species across to SW Arabia in the NE. In the New World, the distribution extends from Brazil and NW Argentina to NE Brazil, Venezuela (Fig. 135B) and Panama and includes also Andean countries; the Amazon Basin is largely excluded. Many species are highly endemic and grow in inaccessible localities. In the Brazilian central mountains the quartzitic Serra do Espinhaço (Fig. 135A) is the centre of diversity of the family. Most species prefer specific substrates such as bare granite or quartzite, some occur on sandstone, sand or humus; in Madagascar substrates such as granite, limestone or sandstone are populated.

With the exception of *Talbotia*, which grows in moist shade, the Velloziaceae are adapted to highly xeric conditions. When the dry plants are moistened, water is absorbed immediately from the leaves or leaf sheaths as if they were made of blotting paper, and the plants appear more or less explicitly poikilohydrous. Species growing in extreme aridity (*Xerophyta*) have been shown to have a drought tolerance of 0% relative humidity, i.e. they revive after losing all the water capable of diffusion at 28 °C (Gaff 1977).

Fig. 135A,B. Velloziaceae. **A** *Vellozia bicarinata*, Brazil, Serra do Espinhaço. **B** *Vellozia tubiflora*, Venezuela, Gran Sabana. Blackened leaf sheaths indicative of frequent fires. (Photo K. Kubitzki)

Key to the Genera

1. Stigmas vertical or nearly so, apical or sometimes subapical, separate to completely confluent, elliptic to linear; stamens always 6; appendages when present as a corona dorsal to the stamens 2
- Stigmas horizontal to somewhat reflexed, always apical and apically confluent, suborbicular; stamens 6–numerous, the filaments evident, slenderly cylindric; appendages when present always ventral to the stamens 7
2. Stamens adjacent to the style, the filaments free from the tepals and without any type of appendages, cylindric, the anthers latrorse to extrorse. S Africa **1. *Talbotia***
- Stamens adjacent or adnate to the tepals 3
3. Corona wanting; filaments flattened and almost wholly adnate to the tepals, or filaments missing; style shorter than stigma 4
- Corona present between tepals and stamens, usually the stamens adnate to the corona; anthers introrse; style longer than stigma 5
4. Flowers perfect; filaments triangular; leaf blades usually deciduous. Africa, Madagascar, Arabia **2. *Xerophyta***
- Flowers unisexual, plants dioecious; anthers sessile; leaf blades persistent. S America **3. *Barbaceniopsis***
5. Anthers dorsifixed. Brazil **5. *Barbacenia***
- Anthers basifixed 6

6. Filaments terete, free from corona; stigma apical. Brazil (only around Diamantina) **7. *Burlemarxia***
- Filaments fused to or adhering to corona; stigma subapical. S America **6. *Pleurostima***
7. Stem with leaf sheaths trigonous; leaf blades strongly plicate, without furrows; stamens 6; pollen in single grains. Brazil **4. *Nanuza***
- Stem with leaf sheaths terete; leaf blades not plicate, adaxially even or nearly so, abaxially nearly always with furrows; stamens usually more than 6; pollen in tetrads. S America, Panama **8. *Vellozia***

Genera of the Velloziaceae

1. ***Talbotia*** Balf.

Talbotia Balf., Trans. Bot. Soc. Edinb. 9: 190 (1868); Hook., Bot. Mag.: pl. 5803 (1869), as *Vellozia elegans*.

Stem short, erect; leaves tristichous; blades persistent, mesophytic; xylem in leaf composed mainly of tracheids; inflorescence a few-flowered raceme; tepals free, slightly unequal; stamens 6; filaments much shorter than anther, adjacent to the style; anthers erect, linear, basifixed, extrorse; stigma clavate, 6-furrowed; capsule obovoid, septa and walls breaking away from the 3 persistent angles. Only one sp., *T. elegans* (Hook.) Balf., S Africa.

2. ***Xerophyta*** Juss.

Xerophyta Juss., Gen.: 50 (1789); Perrier in Fl. Madag. 42: 1–15 (1950); L.B. Smith & Ayensu, Kew Bull. 29: 184–205 (1974), rev.
Vellozia sensu auct. quoad species africanae, non Vand.

Small to large highly xerophytic perennials; stem erect, simple to much-branched; leaves very narrowly triangular and filiform-attenuate or linear and obtuse, terete, ecarinate; blades caducous or rarely persistent; flowers solitary; perianth tube slightly exceeding the ovary; tepals nearly uniform; stamens 6; filaments flattened and almost wholly adnate to the tepals; anthers linear, latrorse or introrse, basifixed (sect. *Xerophyta*) or dorsifixed (sect. *Barbacenioides*); stigmas vertical, about the same length to much longer than the naked style base. Over 30 spp., Madagascar, tropical Africa, and southernmost Arabia, in three sections, these distinguishable by leaf anatomy. Sect. *Xerophyta*, leaves dorsiventral, two spp., Madagascar; sect. *Vellozioides*, leaves dorsiventral, about 12 spp., Madagascar and Africa; sect. *Barbacenioides*, leaves equifacial, about 16 spp., E Africa.

3. ***Barbaceniopsis*** L.B. Smith

Barbaceniopsis L.B. Smith, Contrib. U.S. Nat. Herb. 35: 270. (1962); L.B. Smith & Ayensu, Smithsonian Contrib. Bot. 30: 37–38 (1976), rev.

Xerophytic shrubs; stems simple or few-branched; blades long-persistent and gradually decomposing, linear; flowers solitary on long pedicels, unisexual, dioecious; perianth tube short; tepals subequal; stamens 6; filaments highly fused with tepals; anthers appearing sessile, basifixed, latrorse; stigmas linear, elongate. Three spp., Peru, Bolivia, and northeastern Argentina. Close to and perhaps to be included in *Xerophyta*.

4. ***Nanuza*** L.B. Smith & Ayensu Fig. 133D

Nanuza L.B. Smith & Ayensu, Smithsonian Contrib. Bot. 30: 38 (1976).

Xerophytic shrubs; stems sharply trigonous; blades broad, plicate, finally regularly caducous; flowers 1–3; perianth tube lacking; tepals subequal; stamens adjacent to the tepals; filaments very short; anthers linear, latrorse, auriculate at base, basifixed; style slender, exceeding the stamens; stigmas subhorizontal, orbicular, coherent; capsule loculicidally dehiscent. Only one sp., *N. plicata* (Mart.) L.B. Smith & Ayensu, E Brazil.

5. ***Barbacenia*** Vand. Fig. 133C

Barbacenia Vand., Fl. lusit. brasil. spec.: 21, pl. 1 (1788).
Aylthonia N. Menezes (1971).

Subxerophytic plants of small or medium stature; stem usually simple or few-branched; leaf blades mostly even; flowers 1–many; perianth tube more or less exceeding the ovary; coronoid appendages broad, flat, usually bilobed, free or fused to a tube, ± free from tepals; stamens 6; anthers dorsifixed, introrse, sessile on corona; stigmas globose to long and narrow, vertical or nearly so; capsules opening with loculicidal fissures and/or pores or many openings. About 90 spp., Brazil S and E of the Amazon Basin.

6. ***Pleurostima*** Raf.

Pleurostima Raf., Fl. tellur. 2: 97 (1836).

Stem simple or branched; leaves 3-ranked; perianth tube various, 2–45 mm long; corona lobes oblong or obcordate; anthers sessile, basifixed and auriculate, introrse, attached to corona lobes or hypanthium, provided with an apical appendage (always?); stigma clavate or capitate but stigmatic region subapical; capsule loculicidal with numerous intercostal openings. About 30 spp., Brazil, mainly Minas Gerais, and Venezuela. Sect. *Pleurostima*, with anthers attached to hypanthium or base of oblong corona lobes, and clavate stigma, about 20 spp.; sect. *Graziela*, with anthers attached to the obcordate corona lobes, and capitate stigma; some with long hypanthial tube, nine species.

7. *Burlemarxia* N.L. Menezes & Semir

Burlemarxia N.L. Menezes & Semir, Taxon 40: 418 (1991).

Stem simple or branching; leaves flat; perianth segments reflexed; hypanthium long or short; corona elongated, tubular, their lobes imbricate, largely free from filaments; stamens 6; filaments long; anthers basifixed, lacking auricles, apically with an appendage, 3 semiintrorse, 3 semiextrorse due to twisting of their filaments; style trigonous; stigma apical, trigonous and triangular; capsule opening with slits in the upper portion of the capsule or along its whole length. Three rupicolous spp., Minas Gerais, Brazil.

8. *Vellozia* Vand. Figs. 132, 133A,B, 134, 135

Vellozia Vand., Fl. lusit. brasil. spec.: 32, pl. 2 (1788); L.B. Smith & Ayensu, Smithsonian Contrib. Bot. 30 (1976), rev.

Xerophytic shrubs of low stature to over 6 m high; stem branched, terete or trigonous, including the old leaf-bases; leaf blades reflexing and persistent or caducous along a regular transverse line, not plicate; flowers perfect; perianth tube from equaling the ovary to several times as long; tepals subequal; stamens 6 or more, naked or basally appendaged; filaments terete; anthers basifixed, latrorse; pollen in tetrads; style elongate; stigmas horizontal or somewhat reflexed, suborbicular, confluent at apex. Sect. *Vellozia*, leaf blades flattened, stamens (6–)>9, tepals free, usually without claws, about 100 spp. centred in Serra do Espinhaço, but some in all Central, E and N Brazil and Bolivia; sect. *Xerophytoides*, mostly 6 stamens, rarely (*V. prolifera*) 12, leaf blades subterete, 12–25 mm long, six spp., Minas Gerais, Brazil. The few species of the widespread former sect. *Radia* with long, slender floral tubes and possibly sphingophilous flowers a polyphyletic assemblage (Mello-Silva 1996).

Selected Bibliography

Alves, R.J.V. 1994. Morphological age determination and longevity in some *Vellozia* populations in Brazil. Folia Geobot. Phytotaxon. 29: 55–59.

Ayensu, E.S. 1968. The anatomy of *Barbaceniopsis*, a new genus recently described in the Velloziaceae. Am. J. Bot. 55: 399–405.

Ayensu, E.S. 1969. Leaf anatomy and systematics of Old World Velloziaceae. Kew Bull. 23: 315–335.

Ayensu, E.S. 1973. Biological and morphological aspects of the Velloziaceae. Biotropica 5: 135–149.

Ayensu, E.S. 1974. Leaf anatomy and systematics of New World Velloziaceae. Smithsonian Contrib. Bot. 15: 1–125.

Ayensu, E.S., Skvarla, J.J. 1974. Fine structure of Velloziaceae pollen. Bull. Torrey Bot. Club 101: 250–266.

Barthlott, W., Frölich, D. 1988. See general references.

Chase, M.W. et al. 1995. See general references.

Dahlgren, R. et al. 1985. See general references.

Erdtman, G. 1952. See general references.

Gaff, D.F. 1977. Desiccation-tolerant vascular plants of southern Africa. Oecologie 31: 95–109.

Goldblatt, P., Poston, M.E. 1988. Observations on the cytology of Velloziaceae. Ann. Mo. Bot. Gard. 75: 192–195.

Harborne, J.B., Williams, C.A., Greenham, J., Eagles, J. 1994. Variation in the phenolic and vacuolar flavonoids of the genus *Vellozia*. Phytochemistry 35: 1475–1480.

Harris, P.J., Hartley, R.D. 1980. See general references.

Hegnauer, R. 1963, 1986. See general references.

Huber, H. 1969. See general references.

Johri, B.M. et al. 1992. See general references.

Melo, N.F. de, Guerra, M., Benko-Iseppon, A.M., Menezes, N.L. de 1997. Cytogenetics and cytotaxonomy of Velloziaceae. Plant Syst. Evol. 204: 257–273.

Mello-Silva, R. de 1991. The infra-familial taxonomic circumscription of the Velloziaceae: a historical and critical analysis. Taxon 40: 45–51.

Mello-Silva, R. de 1996. Revisão das *Vellozia* tubifloras (*Vellozia* sect. *Radia*) e caracteres para o aprimoramento da filogenia de Velloziaceae. Tese de Doutoramento. São Paulo: Universidade de São Paulo, Brazil.

Menezes, N.L. de 1971. Traqueídes de transfusão no gênero *Vellozia* Vand. Ciênc. Cult. 23: 389–409.

Menezes, N.L. de 1973. Natureza dos apêndices petalóides em Barbacenioideae (Velloziaceae). Bol. Zool. Biol. Mar. II, 30: 713–755.

Menezes, N.L. de 1975. Presença de traqueídes de transfusão e bainha mestomática em Barbacenioideae (Velloziaceae). Bol. Bot. Univ. S. Paulo 3: 29–60.

Menezes, N.L. de 1976. Megasporogenese, megagametogenese e embriogenese em Velloziaceae. Bol. Bot. Univ. S. Paulo 4: 41–60.

Menezes, N.L. de 1980. Evolution in Velloziaceae, with special reference to androecial characters. In: Brickell, C.D., Cutler, D., Gregory, M. (eds.) Petaloid monocotyledons: horticultural and botanical research. Linn. Soc. Symp. Ser. 8: 117–139.

Menezes, N.L. de 1988. Evolution of the anther in the family Velloziaceae. Bol. Bot. Univ. S. Paulo 10: 33–41.

Menezes, N.L. de, Semir, J. 1990. New considerations regarding the corona in Velloziaceae. Ann. Mo. Bot. Gard. 77: 539–544.

Menezes, N.L. de, Mello-Silva, R. de, Mayo, S.J. 1994. A cladistic analysis of the *Velloziaceae*. Kew Bull. 49: 71–92.

Salatino, M.L.F., Salatino, A., Menezes, N.L. de, Mello-Silva, R. de 1989. Alkanes of epicuticular waxes of *Velloziaceae*. Phytochemistry 28: 1105–1114.

Sazima, M. 1977. Hummingbird pollination of *Barbacenia flava* (*Velloziaceae*) in the Serra do Cipó, Minas Gerais, Brazil. Flora 166: 239–247.
Sazima, M. 1978. Biologia floral de espécies de Velloziaceae na Serra do Cipó, Minas Gerais. São Paulo: Tese de Doutoramento, Universidade de S. Paulo, Brazil.
Sazima, M., Sazima, I. 1990. Hummingbird pollination in two species of *Vellozia* (Liliiflorae: Velloziaceae) in southeastern Brazil. Bot. Acta 103: 83–86.
Smith, L.B., Ayensu, E.S. 1976. A revision of American Velloziaceae. Smithsonian Contrib. Bot. 30: 1–172.
Vogel, S. 1969. Chiropterophilie in der neotropischen Flora. Neue Mitteilungen III. Flora 158: 289–323.
Wagner, P. 1977. Vessel types in the monocotyledons: a survey. Bot. Not. 130: 383–402.
Weber, H. 1954. Wurzelstudien an tropischen Pflanzen 1. Abh. Math.-Nat. Kl. Akad. Wiss. Lit. Mainz 6: 211–249.

Xanthorrhoeaceae

H.T. Clifford

Xanthorrhoeaceae Dumort., Anal. Fam. Pl.: 60, 62, 103 (1829), nom. cons.

Perennial plants with thick woody stems varying from mainly subterranean, not appearing above the ground, to rather tall and arborescent, branched or unbranched. *Leaves* borne in dense tufts at the tips of the branches, spirally inserted, long and linear with thickened resiniferous bases persisting as a dense covering to the stem. *Inflorescence* massive, dense, spikelike, cylindrical, on a woody scape. *Flowers* bisexual, hypogynous, sessile, actinomorphic, in spirally arranged clusters surrounded by packed bracts. Perianth persistent and hardened at maturity; sepals 3, free, chartaceous or scarious; petals 3, free, membranous, white or yellow, with exserted apices. Stamens 3 + 3, free; anthers bithecate and tetrasporangiate, dorsifixed, dehiscing introrsely by slits. Gynoecium syncarpous, 3-carpellary; septal nectaries well developed; ovary 3-locular, each locule bearing 2 rows of ovules; style terminal, simple, subulate, tapering into an undivided stigma. *Fruit* a loculicidal capsule. Seeds 1–2 per locule, flattened; testa black; endosperm copious, lacking starch; embryo linear, transverse to long axis of seed.

A single genus, *Xanthorrhoea*, with about 30 species, restricted to tropical and subtropical Australia.

Vegetative Morphology. The stems are pachycaulous and either subterranean or arborescent, and then up to ca. 3 m in length. They are clothed with a persistent ring of resiniferous leaf bases to several cm wide. The roots arise adventitiously from the base of the stem; they are contractile. The seedlings are remotive, eligulate and germination is epigeal. The first leaves are linear and inserted distichously. Later leaves have a short broad base and then taper sharply into a narrow blade up to 1 m long, rarely only a few cm long. The blades are usually rhomboidal or triangular in section, especially towards the base. The leaves are at first erect but are later curved spreading or recurved. The inflorescence is a terminal leafless scape, i.e. a peduncle of a single internode, bearing the flowers in a dense "spike" in the upper part. The whole may be massive, up to 4m in length with the scape up to 10cm in diameter (Fig. 136).

VEGETATIVE ANATOMY. The vegetative anatomy has been described in some detail by Fahn (1954, 1961), Staff and Waterhouse (1981) and Rudall and Chase (1996). The roots are contractile and thereby drag the stem apex of the seedling below ground level. The cortex comprises a fleshy outer zone and a thickened inner zone. The outer cortex is short-lived and the surface of older roots is formed by the inner cortex. The endodermal cells have U-shaped thickening. The xylem includes vessels with simple perforation plates.

All *Xanthorrhoea* spp. exhibit secondary growth derived from a secondary thickening meristem. The rings of vascular bundles are grouped in alternate light and dark rings but their frequency of formation is not known. Vessels are lacking from the xylem of the stem.

The leaves are highly xeromorphic. The stomates are paracytic and restricted to grooves in the leaf surface. There is a well-defined sclerenchymous band below the epidermis. The chlorophyllous tissue assumes a palisade form 4–5 cells broad interrupted by bands of sclerenchyma. These tissues form a band below the hypodermis. The central portion of the leaf is made up of colourless parenchyma in which are embedded vascular bundles. Some of these are inverted relative to the remainder. The xylem includes vessels with scalariform perforation plates.

Silica bodies are lacking in the family but calcium oxalate raphides are widespread.

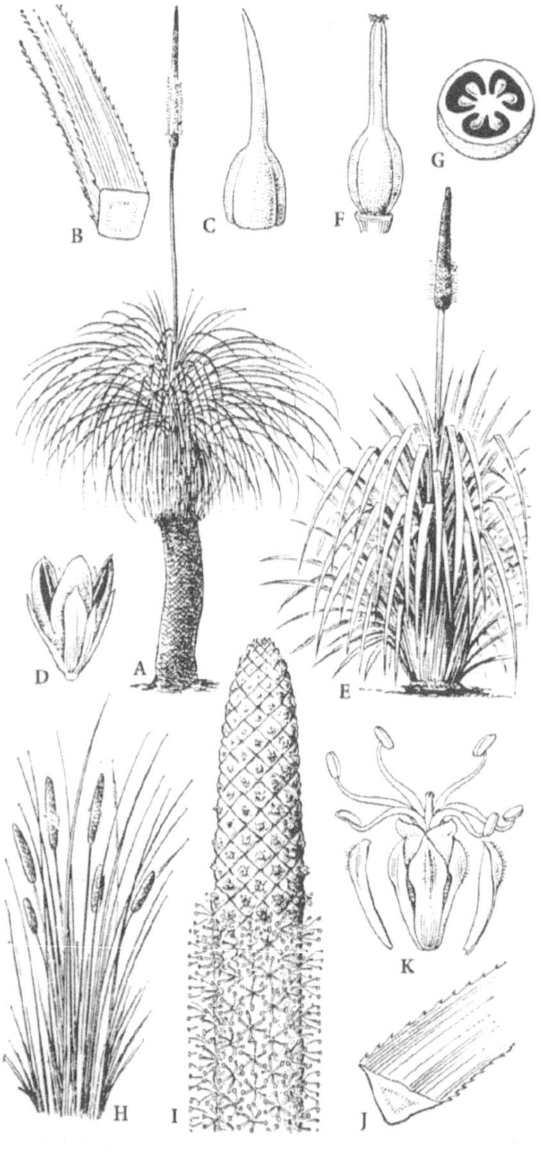

Fig. 136A–K. Xanthorrhoeaceae. **A–D** *Xanthorrhoea quadrangulata.* **A** Flowering plant. **B** Leaf. **C** Pistil. **D** Dehiscing fruit. **E–G** *X. resinosa.* **E** Flowering plant. **F** Pistil. **G** Fruit, cross-section. **H–L** *X. minor.* **H** Habit. **I** Upper portion of inflorescence. **J** Portion of leaf. **K** Flower with bracteoles. (Takhtajan 1982)

INFLORESCENCE STRUCTURE. The flowers are sessile with each surrounded by a series of floral bracts. The flowers occur in irregular clusters each of which is apparently a reduced cyme. The clusters are also subtended by bracts. The flower clusters are arranged in a spiral along the main axis. The whole inflorescence is a thyrse.

EMBRYOLOGY. The division of the pollen mother cell is successive (Rudall et al. 1997). The ovule of *Xanthorrhoea* is anatropous, bitegmic and markedly crassinucellate. The inner integument forms the micropyle and protrudes out through the outer integument. The chalazal region of the nucellus has a well-developed subdermal region and hypostase. The ovule corresponds closely to the Asparagalean type (Rudall 1994).

POLLEN MORPHOLOGY. The pollen morphology of several *Xanthorrhoea* has been described by Chanda and Ghosh (1976). The pollen grains are (extended-)sulcate monads. They are bean-shaped to ellipsoidal with a length of about 32 μm. The sulcus is variable in size extending from slightly to considerably more than the length of the grain. The exine is reticulate.

Karyology. Chromosome numbers are known for most species and in all those reported $2n = 22$ (Staff and Waterhouse 1981).

Pollination. Flowering inflorescences produce copious nectar and attract a wide range of possible pollen vectors. These include bees, butterflies, beetles, flies, moths, birds and possums (Clifford and Drake 1981; Staff and Waterhouse 1981). However, the relative effectiveness of these visitors as pollinators is not known.

Fruit and Seed. The fruit is always a capsule. The seeds are narrow elliptical and flattened with a black testa. The testal layer is more than 10 cells thick, and the outer cells are transformed in a thick (16–40 µm) phytomelan crust; the cells of the tegmic layer are collapsed and colourless; an inner cuticle is noticeable. The endosperm stores aleuron and oil; reserve cellulose is present in insignificant amounts (Huber 1969).

Dispersal. The seeds are shed by the swaying of the inflorescence shaking them from the capsule. It is possible that secondary dispersal is achieved by ground parrots, as these feed on the fallen seed.

Phytochemistry. All species produce copious resin as a yellow, red or brown secretion from the stem and leaf bases, the so-called acaroid resin. Included amongst this secretion is 2-hydroxy-4-methoxyacetophenone, 2-hydroxy-4, 6-dimethoxyacetophenone, xanthorrhein, chalcone, citronellol and derivatives of cinnamyl alcohol, p-coumaric acid and chrysophanic acid (Staff and Waterhouse 1981). Yellow resin contains a major part of polymerised derivatives of cinnamic acids. Anthraquinones form a possible link with Asphodelaceae (Hegnauer 1986). Cell wall-bound ferulate is absent.

Affinities. The phytomelan-incrusted seeds and septal nectaries indicate that Xanthorrhoeaceae are well placed in the Asparagales. In Cronquist's (1981) classification, *Xanthorrhoea* formed part of an expanded family concept, encompassing also *Acanthocarpus*, *Chamaexeros*, *Lomandra*, *Romanalda*, *Dasypogon*, *Calectasia*, *Kingia* and *Baxteria*. On the basis of anatomical, molecular and chemical data, the latter four genera have been recognised as forming a part of the commelinoid lineage, while the former four are well supported as a family, Lomandraceae (Rudall and Chase 1996). In the most detailed *rbc*L tree available (Rudall et al. 1997) *Xanthorrhoea* appears in an isolated position in proximity to Asphodelaceae. Here we follow the monotypic family concept for Xanthorrhoeaceae proposed by Dahlgren et al. (1985).

Distribution and Habitats. The genus is endemic in Australia (for distribution maps see Bedford in Fl. Australia 46, 1986). The genus *Xanthorrhoea* is almost exclusively restricted to open *Eucalyptus* forests and heathlands. With the exception of *X. thorntonii*, which grows in Central Australia, all species occur in regions receiving more than 250 mm annual rainfall, and most in regions receiving more than 500 mm.

Economic Importance. In bygone days the plants were harvested to extract resins, picric acid, alcohol and sugar. Likewise the resins were formerly used in the manufacture of varnishes, metal lacquers and wood stains. The resin was also used by Aborigines to fasten heads onto spears. The basal part of the stem of *X. preissii* and other species persists as a woody cone from which wood turners make bowls, vases and other ornaments. Some species have been implicated in cattle poisoning.

Only one genus:

Xanthorrhoea **Smith** Fig. 136

Xanthorrhoea Smith, Trans. Linn. Soc. 4: 219 (1798); Bedford in Fl. Australia 46: 148–169 (1986).

Description as for family. About 30 spp.

Selected Bibliography

Chanda, S., Ghosh, K. 1976. Pollen morphology and its evolutionary significance in Xanthorrhoeaceae. In: Ferguson, I.K., Muller, J. (eds.) The evolutionary significance of the exine. London: Academic Press, pp. 527–559.
Clifford, H.T., Drake, W.E. 1981. Pollination and dispersal in eastern Australian heathlands. In: Specht, R.L. (ed.) Ecosystems of the world, Vol. 9B. Heathlands and related shrublands, analytical studies. Amsterdam: Elsevier, pp. 39–49.
Cronquist, A. 1981. See general references.
Dahlgren, R. et al. 1985. See general references.
Fahn, A. 1954. The anatomical structure of Xanthorrhoeaceae Dumort. J. Linn. Soc. Bot. 55: 158–184.
Fahn, A. 1961. The anatomical structure of Xanthorrhoeaceae Dumort. and its taxonomic position. Recent Adv. Bot. 1: 155–160.
Hegnauer, R. 1986. See general references.
Huber, H. 1969. See general references.
Rudall, P. 1994. The ovule and embryo sac in *Xanthorrhoeaceae* sensu lato. Flora 189: 335–351.
Rudall, P., Chase, M.W. 1996. Systematics of Xanthorrhoea-

ceae sensu lato: new evidence for polyphyly. Telopea 6: 185–203.
Rudall, P.J. et al. 1997. See general references.
Staff, I.A., Waterhouse, J.T. 1981. The biology of arborescent monocotyledons, with special reference to Australian species. In: Pate, J.G., McComb, A.J. (eds.) Biology of Australian plants. Nedlands: University of West Australian Press, pp. 216–257.
Takhtajan, A.L. 1982. See general references.

Index to Scientific Names

References to main entries in **bold-faced** print, to illustrations in *italics*.

Printing and binding: Druckerei Triltsch, Würzburg

GPSR Compliance

The European Union's (EU) General Product Safety Regulation (GPSR) is a set of rules that requires consumer products to be safe and our obligations to ensure this.

If you have any concerns about our products, you can contact us on ProductSafety@springernature.com

In case Publisher is established outside the EU, the EU authorized representative is:

Springer Nature Customer Service Center GmbH
Europaplatz 3
69115 Heidelberg, Germany

Batch number: 10755805

Printed by Printforce, the Netherlands